MANAGEMENT, INFORMATION AND EDUCATIONAL ENGINEERING

PROCEEDINGS OF THE INTERNATIONAL CONFERENCE ON MANAGEMENT, INFORMATION AND EDUCATIONAL ENGINEERING, XIAMEN, NOVEMBER 22-23, 2014

Management, Information and Educational Engineering

Edited by

Prof. Hsiang-Chuan Liu
Asia University

Prof. Wen-Pei Sung
National Chin-Yi University of Technology

Dr. Wenli-Yao
Control Engineering and Information Science Research Association

VOLUME 2

CRC Press is an imprint of the
Taylor & Francis Group, an informa business
A BALKEMA BOOK

CRC Press/Balkema is an imprint of the Taylor & Francis Group, an informa business

Typeset by diacriTech, Chennai, India
Printed and bound in Great Britain by CPI Group (UK) Ltd, Croydon, CR0 4YY

Published by: CRC Press/Balkema
P.O. Box 11320, 2301 EH Leiden, The Netherlands
e-mail: Pub.NL@taylorandfrancis.com
www.crcpress.com – www.taylorandfrancis.com

ISBN: 978-1-138-02728-2 (set of 2 volumes)
ISBN: 978-1-138-02854-8 (Vol 1)
ISBN: 978-1-138-02855-5 (Vol 2)
ISBN: 978-1-315-73104-9 (eBook PDF)

Management, Information and Educational Engineering – Liu, Sung & Yao (Eds)
© 2015 Taylor & Francis Group, London, ISBN: 978-1-138-02728-2

Table of contents

VOLUME 2

Section 2: New technologies in education and sports

Management, Information and Educational Engineering – Liu, Sung & Yao (Eds)
© 2015 Taylor & Francis Group, London, ISBN: 978-1-138-02728-2

Preface

The 2014 International Conference on Management, Information and Educational Engineering (MIEE 2014) was held in Xiamen, China, on November 22-23, 2014. The aim was to provide a platform for researchers, engineers and academics as well as industry professionals from all over the world to present their research results and development activities in Management, Information and Educational Engineering.

For this conference, we received more than 700 submissions by email and through the electronic submission system, which were reviewed by international experts, and some 346 papers have been selected for presentation, representing 12 national and international organizations. The papers were grouped into three sections, as follows:

Computer science and information engineering: This section mainly deals with the application of computer science and communications technology to design and development of information systems.

New technologies in education and sports: This section mainly covers the topics of Education and Sports, research into new educational technologies at schools and other educational institutions and other issues.

Engineering management, production management, business and economics: This section mainly covers Engineering Management and Economics Management and other issues.

Finally, our sincere thanks to the conference chairs, organization staff, the authors and the members of the International Technological Committees for their hard work.

We look forward to seeing all of you next year at MIEE 2015.

October, 2014
Wen-Pei Sung

National Chin-Yi University of Technology

Management, Information and Educational Engineering – Liu, Sung & Yao (Eds)
© 2015 Taylor & Francis Group, London, ISBN: 978-1-138-02728-2

Committees

Conference Chairmen

Prof. Wen-Pei Sung, *National Chin-Yi University of Technology, Taiwan*
Prof. Jimmy C.M. Kao, *National Sun Yat-Sen University, Taiwan*
Prof. Gintaris Kaklauskas, *Vilnius Gediminas Technical University, Lithuania*
Dr. Chen Ran, *Control Engineering and Information Science Research Association (CEIS)*

Program Committee

Yan Wang, *University of Nottingham, U.K.*
Yu-Kuang Zhao, *National Chin-Yi University of Technology, Taiwan*
Yi-Ying Chang, *National Chin-Yi University of Technology, Taiwan*
Darius Bacinskas, *Vilnius Gediminas Technical University, Lithuania*
Viranjay M.Srivastava, *Jaypee University of Information Technology, Solan, H.P. India*
Ming-Ju Wu, *Taichung Veterans General Hospital, Taiwan*
Wang Liying, *Institute of Water Conservancy and Hydroelectric Power, China*
Chenggui Zhao, *Yunnan University of Finance and Economics, China*
Rahim Jamian, *Universiti Kuala Lumpur Malaysian Spanish Institute, Malaysia*
Li-Xin GUO, *Northeastern University, China*
Mostafa Shokshok, *National University of Malaysia, Malaysia*
Ramezan ali Mahdavinejad, *University of Tehran, Iran*
Anita Kovač Kralj, *University of Maribor, Slovenia*
Tjamme Wiegers, *Delft University of Technology, The Netherlands*
Gang Shi, *Inha University, South Korea*
Bhagavathi Tarigoppula, *Bradley University, USA*
Viranjay M.Srivastava, *Jaypee University of Information Technology, Solan, H.P. India*
Shyr-Shen Yu, *National Chung Hsing University, Taiwan*
Yen-Chieh Ouyang, *National Chung Hsing University, Taiwan*
Shen-Chuan Tai, *National Cheng Kung University, Taiwan*
Jzau-Sheng Lin, *National Chin-Yi University of Technology, Taiwan*
Chi-Jen Huang, *Kun Shan University, Taiwan*
Yean-Der Kuan, *National Chin-Yi University of Technology, Taiwan*
Qing He, *University of North China Electric Power, China*
JianHui Yang, *Henan University of Technology, China*
JiQing Tan, *Zhejiang University, China*
MeiYan Hang, *Inner Mongolia University of Science and Technology, China*
XingFang Jiang, *Nanjing University, China*
Yi Wang, *Guizhou Normal University, China*
ZhenYing Zhang, *Zhejiang Sci-Tech University, China*
LiXin Guo, *Northeastern University, China*
Zhong Li, *Zhejiang Sci-Tech University, China*
QingLong Zhan, *Tianjin Vocational Technology Normal University, China*

Co-sponsors

International Frontiers of Science and Technology Research Association
Hong Kong Control Engineering and Information Science Research Association

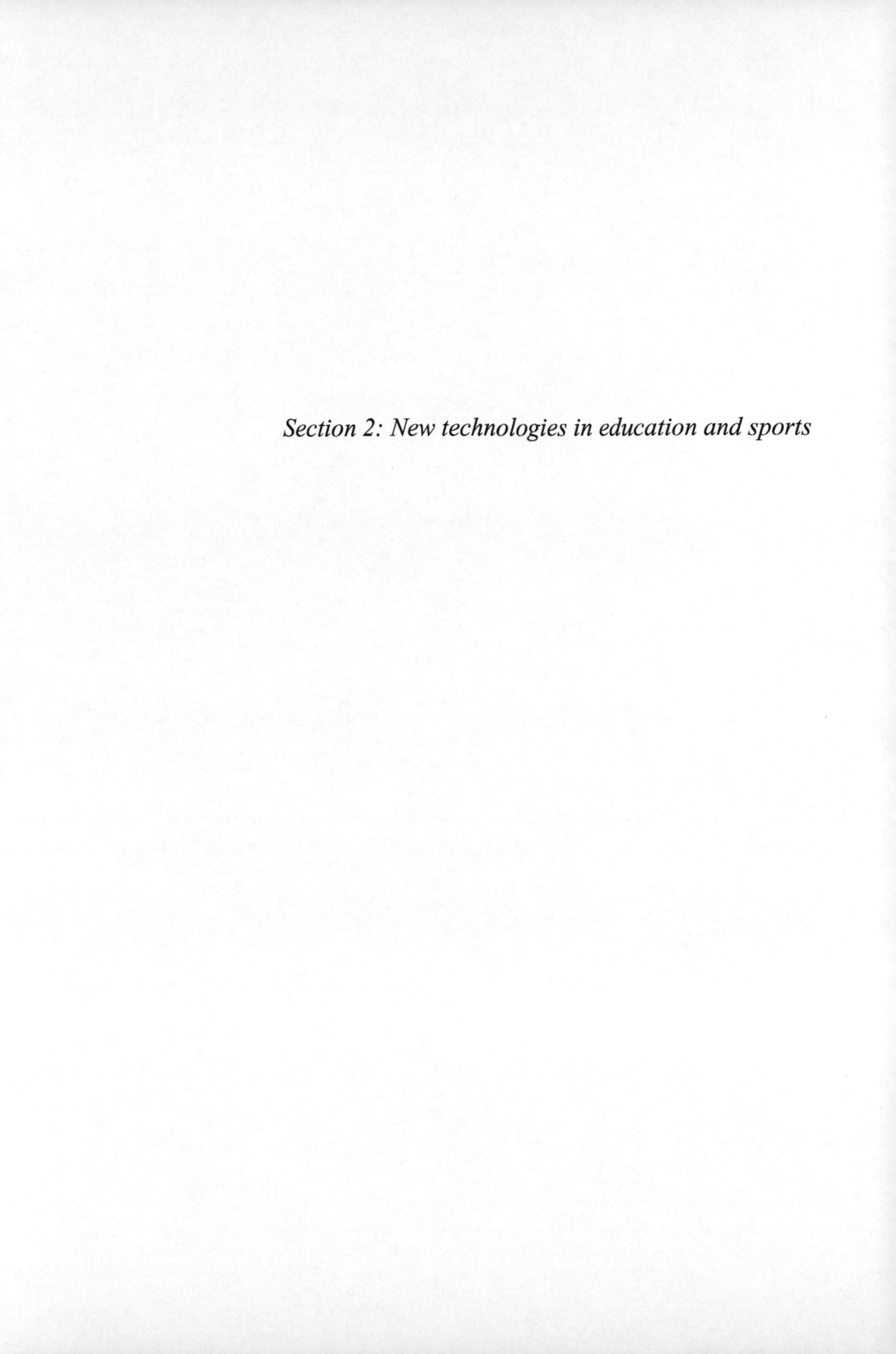

Section 2: New technologies in education and sports

Management, Information and Educational Engineering – Liu, Sung & Yao (Eds)
 ISBN: 978-1-138-02728-2

Research on the information literacy education of university embedded cognition

Ning Ning Sun & Li Hua Ma
Jilin Agricultural University library, Changchun, China.

ABSTRACT: Starting from the information literacy education of the university library, the teaching practice of students in the process of information demands an analysis, to construct a cognitive view to embed information literacy education service.

KEYWORDS: Information literacy; Service; Cognition.

Information literacy is a must in the library and information retrieval skills and computer operation skills to increase the ability of creating and solving ability. From the angle of cognition, cognitive activity learning behavior is in a variety of cognitive tools to support learners. To analyze the information literacy education of the University in the perspective of cognition, is in the process of study from the perspective of cognitive training initiative, enthusiasm, learning of students, and then to have the high mental ability, thinking and behavior of students' access to information, information literacy has become strong.

1 INTRODUCTION

In 1974 the president of Information Industry Association, American Paul based first put forward the concept of "information literacy", information literacy education has gradually replaced the library user education [1], global universities have started the research on the theory of information literacy and education model, and the research heat increase year by year. The information literacy of our whole education at the initial stage, the lack of a complete theoretical system as guidance, the theoretical research on the information literacy education emphasizes "action and effect", namely all sorts of software upgrades and hardware system construction. The statistics found that education, experience less relevant foreign study on information literacy, and few metrology data analysis and empirical investigation, the research contents are mostly concentrated in the basic theory, which relates to the systematic construction research literature deep rarely, the overall research level still needs to be strengthened and extended.

From the angle of view of the professional information on college, student information literacy education at present, the effect is not very ideal, college students in the major of information access and utilization compared to passive. The traditional information retrieval teaching belongs to the general teaching, although largely improving university students' information retrieval and analysis ability, but still unable to effectively improve the awareness of the need in the aspect of professional knowledge of students, unable to get information effectively integrated into personal knowledge base. Therefore, deepening the teaching process, following the students' information demand, the introduction of the idea of knowledge service, knowledge service model to construct the embedded in the teaching process, will greatly enhance the depth and level of the library service, to make a greater contribution to the teaching work.

2 THE RELATIONSHIP BETWEEN INFORMATION ILTERACY EDUCATION AND COGNITION

The students used the inherent knowledge and experience to new information for the selection and treatment during the study, to extract the essence characteristic of them, and then combined with the related knowledge, the choice of the main, useful information storing them up. Cognitive style, cognitive ability, knowledge structure, the emotion will have an impact on the information literacy. College Students' cognitive demand and information search behavior exist between the results of the study show, information search behavior scores, demand variability available cognitive interpretation 9%, explain the cognitive level of demand and there is a positive correlation between information search behavior. Individual high cognitive demand level, inclined to explore, reasonable selection of information, the judgment of its authority, reliability, novelty and accessibility, and actively strive to retrieve the desired information, adjust the retrieval strategy in the retrieval process, tended to have a comprehensive system for retrieval. Cognitive needs a relatively low level of individuals in seeking only pay attention to the problems to be solved related information, tend to rely on existing information and thus less effort, to evaluate the information and the judgment tends to be according to the external cues, enables its to search results of over generalization, cannot solve the current problems. They search for related information, does not care for its mode of action and mechanism to solve the problem, do not do to follow the thinking [2].

In the information literacy education for college students, there is little concern the individual life as well as the "human" existence. On the network today, our education should not only focus on cultivating students' information ability, we should strengthen the information ethics education, at the same time pay attention to the cultivation of students' individual characters [3] to better enable students to take the initiative to improve their information literacy and internalizes for own natural consciousness and natural demand. With the establishment of the research paradigm of cognitive view, information retrieval has appeared great change, gradually heavier technology and system paradigm to the cognitive paradigm shift. And the information literacy education is to emphasize information main body, namely the user's knowledge structure, cognitive context, user interaction with the system, the user task change. Our time for students in information literacy training should pay attention to the subjective initiative of its own, "on the premise of user cognitive oriented" as the cultivation of information literacy, pay attention to the study of the user's cognitive.

3 THE MAJOR POINTS OF THE INFORMATION LITERACY EDUCATION EMBEDDED IN COGNITIVE VIEW

The cognitive theory and the contents of the information literacy education to each other, to construct the embedded information literacy education based on cognitive theory, to the students learning guidance, to achieve the progressive training students' information literacy.

3.1 *The cultivation of information cognitive process of embedded teaching of middle school students*

Information literacy is a part of the overall quality of the people, to improve the students' information literacy is the effective way to the comprehensive implementation of quality education. Information emotional emphasis more on attitude and interest in the use of information technology, the cultivation of students' positive use of information technology to obtain information attitude is the first step in the information literacy education for embedded system. Students having a certain amount of information consciousness, he will automatically monitor their cognitive activities by this consciousness, make the cognitive behavior after smoothly[4].

3.2 *The development of students' cognitive structure of mutual inspiration*

The famous psychologist Piaget thinks, cognitive structure is the result of the structure of human internalization activity. He put the cognitive structure is understood as a dynamic transition system, embodies the essence of the development of cognitive structure, learning ability is a product of the environment and the innate potential interactions. The development of cognitive structure is in the students' original knowledge system on obtaining, understanding, processing and application of information. From a psychological perspective, cognitive structure through schema, assimilation, adaptation and balance through the interaction to gradually develop and grow. Assimilation and adaptation are the two important cognitive processes, when through the balance operation actively monitoring for them with the adjustment, metacognitive ability depends on the development of cognitive structure will gradually improve. Therefore, to improve the students' cognitive structure development of metacognitive ability plays a role in adding fuel to the flames [4]. And mutual inspiration can make the interaction between teachers and students or students and students discuss with each other, not only, but also to accept questioning and criticizing each other, showing a two-way flow of knowledge. In the process of the discussion content of obtaining recognition, gradually understand each other, not only let the students gain new meaning and reconstruction, improve their knowledge structure; teachers will continue to examine their own education idea and the behavior of reflection and reconstruction, improve teachers' ability of self, and promote their professional development [5].

3.3 *The information literacy curriculum embedded in other courses*

The integration of information technology course and other courses is the best mode to achieve the goal, and the reform of the teaching structure is the core task of integrating. Reforms in the teaching structure of teachers, students, teaching materials in the teaching media, the interaction of four factors, the cultivation of students' cognitive ability will become easier. Enhancing information literacy embedded in other courses, to students' self awareness and self monitoring feedback to create a favorable environment, provides the basic guarantee for the cultivation of students' cognitive ability, make the students' cognitive ability can obtain the full display of [3] under the environment of information technology.

4 SUMMARY

The main goal of the cognitive view of information literacy education is to improve the user use embedded on the literature information and absorption. It is the central focus point to "people", people used to integrate information and query, processing, and make the information literacy education towards the direction of the development of cognitive theory. At the same time, but also the information literacy education based on cognitive theory in the research field of information and other subjects combined, contribute to the development.

ACKNOWLEDGEMENT

The source of the project for the research school of Jilin Agricultural University special fund (No. 2014zx14).

REFERENCES

[1] Zhang Xiaojuan. Information Literacy: standard, model and implementation of [J]. knowledge of Library and information science, 2009, (1):17–22, 29.

[2] Guo Xihong, Zou Nannan, Cheng Wenying et al. Study on the correlation between demand and information search behavior. Journal of information 2014, 5:16–20.

[3] Zhang Yi Liu. The information literacy education for university students lack and Countermeasure of [J]. modern information, 2006, (5):197–199.

[4] the winter, Zhao Chengling. Application of information literacy to improve the means of training: Students' metacognitive ability. Distance education China, 2003, 3:57–58, 65.

[5] Ren Yingjie, Xu Xiaodong. The cognitive mechanism of mutual inspiration and its educational significance of E-education research. 2014, 6:29–33.

Management, Information and Educational Engineering – Liu, Sung & Yao (Eds)
© 2015 Taylor & Francis Group, London, ISBN: 978-1-138-02728-2

Problems and countermeasures of the cohesive models between secondary vocational education and undergraduate education system—a case of automotive major

Pei Tang, Rong Ying Zhu, Hong Ming Lv & Ji Sheng Xia
School of Automotive Engineering, Yancheng Institute of Technology, Yancheng, China

ABSTRACT: In this paper, the importance and necessity of the secondary vocational education link with the undergraduate education system were researched. The existing problems of the link were summarized, and the problems were analyzed respectively. The corresponding countermeasures were put forward to solve these problems. It could provide universal significance and the application value of the secondary vocational education link with the undergraduate education system.

KEYWORDS: Secondary Vocational Education; Undergraduate Education System; Automotive Specialty; Cohesive Models.

1 INTRODUCTION

At present, an important task in the reform and development of occupational education in China was the systematic construction of modern occupational education which strove to improve the ability of servicing national strategy. The occupation education had been made great development in China in recent years. It had been from quantity expansion to qualitative improvement. The connotation of occupational education and its extension are patulous. The occupation college was developed actively by the Chinese government.

To produce the integrated education system in secondary vocational education higher vocational education and undergraduate education. Jiangsu Province of China designed three kinds of cohesive models such as: 3+2" or "3+3" piecewise teach between secondary vocational education and higher vocational education. Namely, students first studied the secondary occupation education for three years, then through the registration mode in higher vocational education to study for 2 years or 3 years. At last they obtained higher occupational diploma in education; "3+4" piecewise teaching between secondary vocational education and undergraduate education. Namely, students first studied the secondary occupational education for three years, then through the registration mode in undergraduate education to study for 4 years. At last they obtained undergraduate diploma in education; 3+2" or "5+2" piecewise teach between higher vocational education and undergraduate education. Namely, students first studied the higher vocational education for 3 or 5 years, then through the registration mode in undergraduate education to study for 2 years. At last they obtained undergraduate diploma in education.

2 THE IMPORTANCE AND NECESSITY OF THE COHESIVE MODELS BETWEEN SECONDARY VOCATIONAL EDUCATION AND UNDERGRADUATE EDUCATION

At present the secondary occupation education and the higher occupation education were the main occupation education in China. The highly skilled talents mainly came from the relearn after general higher education and self-study and practice after vocational education. In this way not only the efficiency to develop high-end talents was low, but also the social cost was high and the success rate was low.

Occupation education had been reformed in China in recent years, It payed more attention to skill training. But subject to the educational system and the period was limited, the basic curricula such as physics, mathematics and some necessary professional basic curricula were compressed or canceled. The most visible consequence of this was that occupation education graduates were lack of knowledge renewal ability, innovation ability, technical transformation ability, occupation subsequent development ability and so on.

The modern manufacturing such as transportation, automotive design, machinery and electronics required highly skilled talents. The vocational education which lasted only three years must not cultivate high level technical talents. To extend the professional learning time for enhancing the training level, it was necessary to link the secondary vocational education with the undergraduate education. In order to establish and improve the occupation education system, the education department of Jiangsu province recently approved a number of modern vocational education system construction projects such as piecewise teach between secondary vocational education and higher vocational education, piecewise teaching between higher vocational education and undergraduate education. To improve the system of the modern occupation education and research the law of cohesive models between secondary vocational education and undergraduate education system, The talents training scheme and curriculum standard of automotive specialty should be formulated. The vocational colleges and universities should actively promote link of the education system between occupation education and undergraduate education.The cohesive contents included professional setting, curriculum, textbook, teacher, etc. The curriculum was the core and basis of cohesive models between secondary vocational education and undergraduate education system which laid the foundation for cultivating the high skilled talents.

3 PROBLEMS AND COUNTERMEASURES OF THE COHESIVE MODELS BETWEEN SECONDARY VOCATIONAL EDUCATION AND UNDERGRADUATE EDUCATION SYSTEM

3.1 *Problem in the connection of major*

At present, there were several major fields of automotive in secondary vocational education such as manufacture and repair of automotive, decorative of automotive, electronics of automotive, sheet metal of automotive, marketing of automotive, etc. While the automotive specialties of undergraduate education were automobile service engineering and automotive engineering. The training content and training methods of secondary vocational education and undergraduate education should be researched so that the automotive major would be docked with the automobile industry.

3.2 *Problem in the connection of cultivation targets*

According to the occupation post, the three parties which were composed of secondary vocational schools, undergraduate college and enterprise should be cooperated together combined with the training goal of occupational education. The occupation ability and the occupation quality requirements of graduate were determined.

The occupation ability was classified from low to high scientifically. According to national occupation qualification system properly, the graduates must get the certificates such as car mechanics, auto repair technician or engineer trainee of vehicle inspection and repair. In the meantime, the training program and curriculum arrangement came into being. The secondary vocational schools and undergraduate colleges should conduct special teaching in each level.

3.3 *Problem in the connection of curriculum structure*

The core of cohesive models was the curriculum. The vocational schools had attached great importance to the cultivation of practical ability. The proportion of practice teaching hours was very high, while the undergraduate colleges payed more attention to theoretical knowledge. It caused the following problems: The cultural basic curricula were disjointed between secondary vocational education and undergraduate education; The professional theory curricula were repeated sometimes. Therefore, we must revise training scheme. The principles and methods of curriculum design should be explored. Based on the investigation of talent demand, The target and standard of personal training between secondary vocational education and undergraduate education would be clarified respectively. According to the training laws of applied talents, the secondary vocational education graduates should be the talents who had operating skill, and the undergraduate education graduates should be the talents who had strong technique and independent study ability. Various levels of talents were cultured. It specified the direction for the development of teaching standards and curriculum standards.

3.4 *Problem in the connection of curriculum content*

The secondary vocational school emphasized the occupation training of basic knowledge and the training of practical skills. It was lack of sufficient attention to basic knowledge of culture so that the curriculum hours of culture were relatively sparse. If students didn't attach importance to the basic cultural curricula in the secondary stage, they would study difficultly in some curricula such as university English higher mathematics. They also would encounter some obstacles on the understanding of professional practice knowledge. Finally, because the students were lack of perfect knowledge structure and strong ability of self-study, the innovation ability and lifelong development of occupation were restricted.

The basic curricula played a basic role in quality education and lifelong learning.More wide basis curricula and professional basic curricula should be found in the connection of secondary occupational education and undergraduate education. The basic curricula such as physics and mathematics were essential in the study of secondary vocational education. It laid a good foundation for entering the stage of undergraduate education.This part of the basic curriculum, which provided the necessary knowledge for professional curricula were extended in the stage of undergraduate education.

3.5 *Problem in the connection of teaching mode*

The undergraduate emphasized the knowledge and academic, while the secondary occupation education emphasized the operability and practicality. The occupational ability was regarded as the center of teaching plan. The teaching materials were comprehensive and practical. The teaching methods were flexible. The teachers were called"the double teacher" which means that they were both teacher and engineer. All the above features had less common with undergraduate education. It would be a challenge for the cohesive models. The actual teaching was focused on typical work task as the main line. The typical work of the automobile 4S stores was regarded as the carrier. The learning activities were converted into work processes. The work process corresponded with the learning activities. The work process was refined. The theoretical knowledge and working skills were instilled in the process of all levels of teaching.

3.6 *Problem in the connection of teacher*

The cohesive models put forward a severe challenge to teachers, both secondary vocational schools and undergraduate colleges. The vocational school teachers would not only teach practical ability, but also they need to research automobile theory knowledge deeply. They would teach the preparatory knowledge completely to the students for implementation of undergraduate teaching. In the meantime, the university teachers would take more time and energy to a laboratory for strengthening the practical ability. To develop high level technicians, teachers of undergraduate colleges must possess the ability of engineering practice. The vocational school teachers and university teachers should often communicate each other. They would join on the teaching and research activities and explore a set of curriculum system for the cohesive models.

4 SUMMARY

According to the construction requirements of modern occupational education, the cohesive models between secondary vocational education and undergraduate education put forward that the principle, method and route of a curriculum system which followed three elements concluded job, task and capacity. These ideas and achievements had important significance for theory guiding. Through the combination of theory and practice, classroom knowledge and practical ability, school and enterprise, class and certificate, it could provide universal significance and the application value of the secondary vocational education link with the undergraduate education system.

ACKNOWLEDGEMENTS

The corresponding author of this paper is Tang Pei. This paper is supported by "Research and practice of the cohesive models between secondary vocational education and undergraduate education system system which were based on the occupation ability training".

REFERENCES

[1] Richard Arum, Yossi Shavitt. Secondary Vocational Education and the Transition from School to Work [J]. American Sociological Association, 1995(3):187–204.

[2] Lv, Hongming, Zhu, Longying. Construction of a practical teaching system for an automobile major[J]. American Sociological Association World Transactions on Engineering and Technology Education, 2013(3):282–292.

[3] Angelique Slaats, Hans G.L.C. Lodewijks. Learning styles in secondary vocational education: disciplinary differences[J]. Learning and Instruction, 1999(5):475–492.

[4] Jonathan Meer. Evidence on the returns to secondary vocational education[J]. Economics of Education Review, 2006(5):559–573.

[5] Huang Xingyun, Ou Qizhong. Network curriculum design for secondary vocational education. [J]. 2011 - 6th International Conference on Computer Science and Education, Final Program and Proceedings, 2011:592–594.

Management, Information and Educational Engineering – Liu, Sung & Yao (Eds)
© 2015 Taylor & Francis Group, London, ISBN: 978-1-138-02728-2

Study on innovation in ideological and political education based on the internet age

Hong Xu Guo
Student Work Department, Beihua University, Jilin City, China

ABSTRACT: Internet sweeps word with strong vitality, permeates all social areas and also has an increasingly broad and profound impact on the college students' idea. Facing the internet age, ideological and political education has to advance with time and hold reform chance to grasp the initiative of ideological and political education based on the internet to improve its effectiveness.

KEYWORDS: Internet age; Ideological and political education; Improve; Effectiveness.

1 INTRODUCTION

In recent years, the network as an important feature in IT-centric technology revolution has affected all areas of society. "Internet" at an alarming rate profound impact on the future of social processes and human, changing the way people live, learn, work and ways of thinking. At present, large-scale Internet has entered the lives of the Chinese people, and affects all aspects of society. Now, our domestic Internet users are to grow exponentially, according to the Internet Center, October 31, 1997 was 620,000 Internet users in China, 31 December 1998 reached 2.1 million, in June 1999 Internet users in China rose to 4 million. According to official forecasts, by the end of 1999, mainland China's Internet users will reach 7 million; to 2001 or 2002 domestic Internet users will reach 30 million to 50 million. With the expanding influence of the Internet, in the 21st century China will enter the "network society."

Universities are Chinese social "network" development frontier, with the popularity and development of network information education, Internet students will continue to increase. "Internet" for college students' behavior patterns, values, political attitudes, psychological development, moral values, etc. Will have a growing impact on how the "Internet age" to strengthen ideological and political education of college students has become an unavoidable a major issue and needs to be resolved.

2 THE POSITIVE ROLE OF THE INTERNET

With the rapid development of China's social networking today, we must clearly understand the far-reaching impact of the "Internet" to bring the current ideological and political work, which is both a positive side and negative side.

On the positive side is the "Internet" opened up a vast ideological and political education. Features networks with resource sharing, online information is shared by all mankind, everyone can get, can have. Meanwhile, online information sharing as well as another layer of meaning, that is, everyone can get information from the Internet, but also should make their own may provide information to others. Resource sharing network of ideological and political work in colleges and universities can occupy the market network, you can carry out ideological education of college students through the network, which overcomes the traditional ideological and political work of the smaller impact was weakness in a certain sense. Meanwhile, the network has to visualize, interesting features, network, graphics, animation, sound, image, fun and intuitive, so that it can be attractive. In addition, the network resource sharing can make the ideological and political workers learned from the Internet real idea of dynamic people, targeted at the right ideas and information posted online to educate and guide students, it sets up a good style of thinking, ideals, beliefs, and thus improve the ideological and political work of timeliness.

In the Internet age, due to new things emerging, increasingly competitive society, the ideological issues will be more complicated. Ideological and Political Education can take advantage of network-specific information highly integrated, two-way communication and selectively promote college education and targeted to achieve self-education. As the network has information can be copied, shared, real-time transmission of resistance, which makes the whole community college while receiving education possible, which is the traditional ideological and political education can not. In addition, the new network to adapt to ethics and code of conduct of the Internet age, the ideological and political quality of the proposed higher requirements, which prompted college students to be self-disciplined, realistic, unity, cooperation, adherence to morality and ethics.

Network can maximize the ideological and political work of socialization, ideological and political work network with government agencies, families, schools connected, which participate in the ideological and political work for the community to provide a convenient and can achieve the ideological and political education combine family and social forces at work, so that ideological and political work better results will help to further the formation of the great advantages of ideological and political work.

Workers learned from the Internet real idea of dynamic people, targeted at the right ideas and information posted online to educate and guide students, it set up a good style of thinking, ideals, beliefs, thereby improving the timeliness of ideological and political work.

3 NEGATIVE EFFECTS OF THE INTERNET

In the Internet age, due to new things emerging, increasingly competitive society, the ideological issues will be more complicated. Ideological and Political Education can take advantage of network-specific information highly integrated, two-way communication and selectively promote college education and targeted to achieve self-education. As the network has information can be copied, shared, real-time transmission of resistance, which makes the whole community college while receiving education possible, which is the traditional ideological and political education can not. In addition, the new network to adapt to ethics and code of conduct of the Internet age, the ideological and political quality of the proposed higher requirements, which prompted college students to be self-discipline, realistic, unity, cooperation, adherence to morality and ethics.

Network can maximize the ideological and political work of socialization, ideological and political work network with government agencies, families, schools connected, which participate in the ideological and political work for the community to provide a convenient and can achieve the ideological and political education combine family and social forces at work, so that ideological and political work better results will help to further the formation of the great advantages of ideological and political work.

But on the other hand, the network brings not only the ideological and political work opportunities, but also has had an impact and challenges; its main performance is as follows:

First, the Western culture and yellow unhealthy things are easy to produce adverse effects on young college students on the Internet. On the Internet between different countries and cultural traditions, moral values and way of life very different from their conflict is very intense. The important feature of the network is a shared and anonymity, freedom and openness, and now the Internet to get information about our efficient service must resort to a database developed Western countries, Western countries such values, lifestyle, awareness morphology, and yellow, it can be a lot of unhealthy content input. The young college students to form a view of the world and is a critical period of life, and therefore, online Western culture yellow, unhealthy things will be very easy to destroy young students inherent morality, values and cultural outlook, and thus poison the young students so its quagmire unable to extricate themselves, ideological and political workers painstakingly nurtured ideas and principles destroyed. College students have been accepted in Western culture unhealthy things. Light on the political apathy, weight may go reactionary road, which we must not be complacent, laissez-faire.

Secondly, because in the network, information dissemination speed, scale, scope, and the occult are far more than any previous media, it is easily the Western hostile forces and superstition, cults reactionary organizations to use to publicize our penetration. For example, in the "Falun Gong" is most prevalent from 1998 to the first half of 1999, Li Hongzhi and Falun Gong backbone had opened thousands of worldwide web site, use the Internet to quickly spread its fallacies and instructions, to lure the masses fooled. Another part of the remnant of Falun Gong, the Chinese New Year gathering in Tiananmen Square to engage the so-called "macro-Law" criminal activity, but also advance the use of Internet e-mail with domestic and foreign criminals in series and planning. Therefore, the ideological and political work must attach great importance to the ideological and political education, occupying new areas of the Internet, in order to curb illegal bad guys on the Internet for all college students, and incitement. Also, because the network of people who are mainly exchanged, machine dialogue or computer-mediated communication. People all day dealing with computer terminals, and a lack of interpersonal feelings that easy to make people tend to isolate, selfishness, indifference and non-socialized, easy to make people happy in real life and social development of others indifferent. College is an important period people interpersonal skills and interpersonal formed due to interaction with the traditional network, interpersonal rapport with a different, often difficult to form authentic and safe relationships, college students in the network exchanges once cheated it is easy to be fooled real doubt, pessimism and hostile attitude. Most of the time a lot of current college students and the Internet are playing online games, and online games are a lot of wars, violence, murder, etc. as the main content, which makes students addicted to online games could easily lead to cold, heartless and selfish personality, anxiety, depression and repressed emotions, and even serious will happen, "Internet syndrome", have an extremely adverse impact on their academic life.

4 HOW TO USE THE INTERNET TO CONDUCT EFFECTIVE IDEOLOGICAL AND POLITICAL EDUCATION WORK

How to make full use of the Internet has brought opportunities and favorable conditions for the ideological and political work to occupy positions in the network culture, to further strengthen the ideological and political education of contemporary college students to learn and to think it through, I think of the following aspects:

First, you must get rid of old ideas, take full advantage of the opportunities, make great efforts to ideological and political work networking work. The ideological and political work refers to the computer network at the core of the network party and government organs at all levels, political workers and transmission systems, communication systems composed. Ideological and political workers can use the network, switching, transmission of ideological and political education of information, including text, data, sound, graphics, animation and other forms of ideological education of college students, incentives, guidance and control. Ideological and Political is to improve the ideological and political education of information transmission rates, utilization. Spread through the network, enable advanced model of speech, counseling experts, the television network directly into the classroom education seminars that teach one person, the whole community college while receiving education possible. And because fun, image, intuitive advantage of online teaching, learning educated achieved so far superior to the ordinary classroom. Ideological and Political Work network can also be other mass media, such as newspapers, radio, television, books, video and audio information relocated outdoor advertising on the network, in order to achieve the ideological and political education and other mass media in conjunction with a complementary, thus greatly enhanced appeal, attractiveness and education role in guiding the ideological and political education of college students. Meanwhile, the network also allows learners and educators to achieve two-way communication, timely access to counseling information and timely feedback to improve the efficiency of the ideological and political education, to further enhance its educational function.

To improve the ideological and political work of the network must be established ideological and political education sites on the Internet? How to use the site to carry out the ideological and political education of it in addition to the previously mentioned web seminar lectures and counseling, but also can use the following methods of ideological and political education online: First, the use of the electronic letter delivery list. We can delivery server via an electronic letter on the website will be published in the electronic correspondence list of articles, magazines, newspapers, in the form of emails sent to the user's e-mail box, users can express their views on a particular issue, writing articles, the server will automatically forwarded to another user's electronic mailbox articles users go. Ideological and political work on the site, users can post on the issues college students ideological and political aspects of their views and discuss the delivery list through e-letter, and ideological and political workers are involved, make their own positive view, guide the discussion deepening the ultimate purpose of the ideological education of students. Second, the use is an electronic bulletin system (BBC). We can set up an electronic bulletin system on the ideological and political education websites, the contents of the ideological and political education to join them. We can build in the system forum; college Users can express their views on, views and discussions on certain issues. Ideological and political workers can show their stand, viewpoint and publicize the party's principles and policies to address the ideological issues to achieve the purpose of ideological and political education among. We can also create a chat room in the system communicate through chat this way and ideological understanding of their state of mind, published proper perspective, ideological education. With the use of the electronic bulletin system to establish forums and chat rooms should always play a guiding role in the ideological and political workers, in order to avoid off-topic, reach the purpose of ideological education. Third is the establishment of news services on ideological and political education website, namely the use of network news servers to provide customers with a variety of topics for discussion and mutual exchange service. Ideological and political workers can express their views in a newsgroup topic and play a guiding role, so that more students can get the correct ideological guidance. Taking into account that not every college users will visit the site of ideological and political education, participate in discussions, this part of the thinking person's online education should not give up, take the initiative to deal with its ideological problems, you can use e-mail to those who were there for the of ideological and political education, but also through the exchange and discussion of the involvement of other sites, its ideological education.

5 CONCLUSION

To improve the ideological and political work of the Internet age, we should continue to improve information literacy ideological and political workers. To meet the demands of the Internet age, ideological and political workers must have good information literacy, which has excellent information awareness, information and good information ethics. At present, the ideological and political quality of information workers needs to be improved, taking into account most of them have a high ideological and cultural quality, so long as the proper application of computer and network training them, it should be said that most people are capable online the ideological and political work. Ideological and political workers to improve information quality is directly related to the educational level of their thoughts, they only have a good sense of information and the ability to understand the network environment, and access to the ideological and political information from the network, to understand students' ideological dynamics, and to its for education and management to go; but only with good information ethics, in order to consciously safeguard the network information model laws and regulations, and resolutely resist all kinds of reactionary, superstitious, yellow garbage dissemination of information, but also in order to truly become qualified Internet age the ideological and political workers.

At the same time, we must also strengthen the college network ethics, education, the legal system of the network, it has the legal awareness of the network to establish a correct concept of network ethics. We want to guide students to consciously resist the erosion of network garbage, to do not throw garbage, not to engage in infringement, do not look at pornography, not for hackers, do law-abiding members of civilized network, and can consciously safeguard the network order.

In addition, the ideological and political workers must also strengthen the monitoring and management of network behavior. To establish a permanent agency network information management, and the development of network codes of conduct, through the review and monitoring to regulate network behavior of college students, on-line reactionary, yellow, unhealthy content to clean up, through analysis of monitoring to detect the presence of college students ideological problems and timely targeted education, so that "preventive measure", resulting in the formation of a healthy college campus, non-explicit, remove evil righting network environment. Only in this way can we take full advantage of the Internet age of ideological and political work, and training qualified personnel to complete the glorious task.

REFERENCES

Jiang Wenting. Students' Network Ideological and Political Education Research [D] Dalian: Liaoning Normal University Ideological and Political Department of Education, 2011: 16.

When cases Lin, Li Nan. Study ideological and political education [J]. SCIENCE & EDUCATION, 2011 (2).

Hu Yu. "Ideological and political work in new ways to explore the Internet Age" [J]. Tsinghua University, 2001 (1).

Pan Min, Chenzhong Run, at sunrise, "Summary of network ideological and political education of college studies". [J]. Ideological and political education, 2007 (3).

Zhang Xing again, Zhang Yu. "Strengthen the network Counselors occupied university ideological and political education of new positions" [J]. Theoretical Front in Higher Education, 2006 (5).

Chen Yan. "College counselors work research in the internet environment" [J]. Higher Education Research, 2007 (1).

Qiao Xiangping, Chen Yanfei. "Ideological and political education of college students New Exploration Network" [J] Hunan Social Sciences, 2007 (1).

Lin Lin. "On the Establishment of Ideological and Political Education of College Counselors Network Platform" [J]. Jilin Commercial College.

"Circular on Further Strengthening and Improving Ideological and Political Education" [N]. Wen Wei Po, 2004–10–15.

Huang Fafang. "Network Effects and Countermeasures on Ideological and Political Education" [D]. Huazhong Normal University master's degree thesis.

Huang Junguan. "Youth Internet Addiction Causes and Countermeasures" [J]. Education and occupation, 2006 (11).

Management, Information and Educational Engineering – Liu, Sung & Yao (Eds)
© 2015 Taylor & Francis Group, London, ISBN: 978-1-138-02728-2

Study on solfeggio teaching under MIDI environment

Yan An Wang
Music College, Beihua University, Jilin City, China

ABSTRACT: In accordance with current student condition under five-year-system pedagogical education, the author proposes the music educational model of taking students as the main body, and explores the solfeggio teaching way and method under the MIDI environment. The teaching actually affects and strengthens the innovation, and the education idea of the student multi-position whole development was enhanced by using the modernized teaching method.

KEYWORDS: MIDI environment; Solfeggio teaching; Auxiliary teaching.

1 INTRODUCTION

The current five-year teacher education is a hot "potato", the quality of students and student decline of traditional teaching mode "disdain" for higher teacher education have posed a severe test. How to use modern teaching methods, to change the traditional teaching model, build a student-centered learning environment, and effectively improve teaching effectiveness, and promote the overall development of students' multi-faceted, is the important issue of teacher education research. In this paper, high-division teaching ear training as an opportunity to focus on analysis, to explore ways of teaching ear training MIDI environment, methods.

Sight-singing, listening training, students of music hearing, memory, and the ability to accumulate musical vocabulary, etc., is to learn basic vocal music, instrumental music, music appreciation and other music professional skills, sight-singing and ear training to strengthen the two musical basic skills training enable students to better feel the music, expressive music and improve understanding of music.

Solfeggio Teaching MIDI environment, will make teachers lead the students to do lessons from the past that a lot of practice, a single boring teaching methods to get completely changed, and thus greatly liberating for piano teachers and students' dependence and bondage. Change this teaching model not only greatly enriches our teaching form and content, but also for our current advocacy of innovative education has a positive role in promoting.

2 USE OF PROFESSIONAL TEACHING EAR TRAINING SOFTWARE SYSTEM

Australian software vendor RISING of Auralia ear training software and Denmark Solfeggio multimedia music education software Ear Master Pro 4.0 is being recognized by excellent ear training software. For example Ear Master Pro 4.0 will have the following advantages:

Suitable for our operations habits Ear Master Pro 4.0 Chinese version only application 1.97M, lower demands on the hardware environment, it can run on almost all PC machines, friendly and user-friendly interface is very easy for beginners get started.

The software comes with a wide range of teaching, from debate to hear the tone interval, chords, tonality, rhythm and melody dictation, ear training encompasses virtually all teaching content; while, depending on the degree of learners, you can randomly choose schools for their own content, you can also sneak into the deep by progressively learning. There is a very important point is that in order to make better learners were listening to the debate, and at any time we can arbitrarily change the speed, which is almost all the music education software and music production software commonality.

It provides an interactive learning environment. Students learn basic ear training courses, software can give the students practice self-evaluation in the aid of the software, so that not only applies to the classroom, but also facilitate the students after-school learning. For example, listen to the rhythm of debate practice.

3 USE SONAR OR CUBASE ASSISTED INSTRUCTION SEQUENCER SOFTWARE

Known as "Soul Music", "Music Bones" reputation rhythm is a considerable proportion of teaching ear training in the training process, and also the most difficult training. Comes with almost is all the typical percussion mankind invention SONAR or CUBASE etc. sequencer software, which not only sounds realistic, but also rich and expressive. These can be any combination of percussion made backing (woven

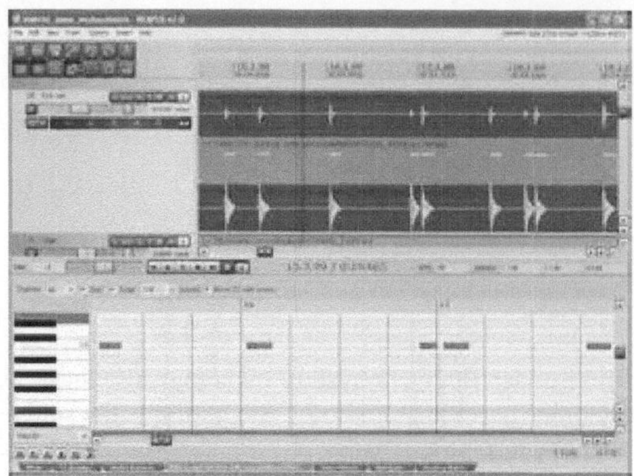

Figure 1. MIDI recording, editing, processing, mixing, and mastering environment.

Figure 2. Ear master pro 4.0 operating points.

rhythmic pattern), you can always change according to the teaching requirements of speed, strength, when traditional teaching and teacher stomp, clap their hands or mouth shouting "da da da" were compared its effects are worlds apart, so rich and vivid sound not only improve the efficiency of learning, but also easier for students to understand and accept. For example, Normal School music teaching "Music Integrated Course" in the second volume of a rhythm ear training exercises, the teacher playing the song "breathe", the students clapping, Paitui, stomping whom accompaniment.

First, to show the exact music in the classroom is our sight-singing, singing lessons and enjoy an important element in common with the traditional handwritten sheet music is neither beautiful nor accurate, but also waste a lot of energy. MIDI music system provides us with such a convenient, one can always render accurate music, and as needed, to emphasize the content of classroom amplification local music; Second, while listening to music while watching the play MIDI melody, which fully do static and dynamic binding, eyes, ears and mind open to students with great benefit; Third, these music files exported as JPG or GIF format, and then insert into your lesson plans or paper, that is what we should always do the job. Professional sheet music notation software such as Overture, Finale and other notation software such as Tony Tone, composers, TT composer and so can do this. As a sequencer software CUBASE3.0 also has a powerful music editing features, a variety of music symbols complete editing method is simple, those interested can search for relevant articles of learning online.

Secondly, multi-part solfege, by adjusting CUBASE3.0 acoustic phase (Pan), part volume, so that students in their own voices while sight-singing, to develop the habit of listening to other voices attention, which the bass auditory and intonation training is particularly important. For example, two part sight singing (children's song) "small pine tree." In this article, as the duet in the same rhythm down, both with degrees between the two voices, three degrees, there are six degrees of fauna carried out for accurate grasp of music image, singing quasi-pitch is the key, so the students singing and sound volume adjust timely between the upper and lower parts relative to solve the above problems have actively helped.

4 TT COMPOSER USING INTELLIGENT SOFTWARE TO EXPAND THE TEACHING

TT is a composer from the Central Conservatory of Music Alto Company independently developed computer music production software, installation and use are very simple, it is called "Everybody will learn" software. Related information, please consult our website Alto (http://www.centrmus.com). TT composer using intelligent software to expand our teaching can indeed play a very helpful role.

TT composer is colorful accompaniment figurations of traditional piano accompaniment a good supplement. TT composer can bring students to a variety of accompaniment styles full of imagination, to avoid monotony automatic accompaniment software brings problems, and we can be part of the accompaniment texture after deletion using either different Style, figurations more complete in combination in the same song.

The teacher let the students know, not just music composer things each of us can create their own works of creative fun give our lives to add a bright color, the use of TT were music composer Writing music makes us feel not so mysterious. At the same time, we believe that music, especially children's song writing into our music education, music education is not only the diversity of diverse needs, but also reflect the quality of education, innovation and education.

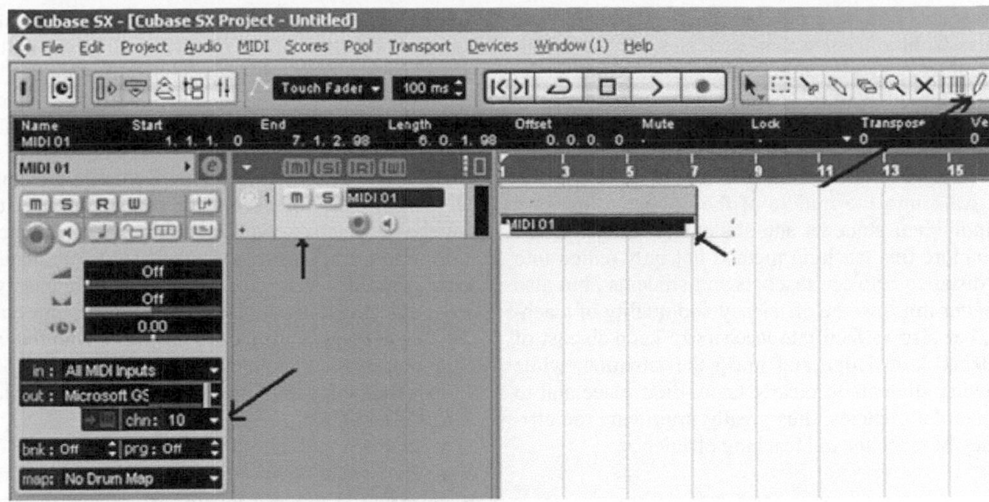

Figure 3. CUBASE3.0 in the process of establishing the rhythm track.

The teacher summarizes the students' work and gives correct me, then outstanding works entered into TT composer allocation chords and add accompaniment. Within a very short time, a big eye-opener for students to work with the birth, which makes traditional music teaching large Diego glasses, because this is no longer the "fancies", the "paper" becomes real "hearing art ", I think this is our match every teacher and eager to do. As another example, given a few key tones 1, 2, 3, 5, 6, please use their imagination to any combination of notes, to create an ethnic-style eight sections of the works, the end of the sound to the tonic "1" on. Also, students are under the guidance of a teacher, with a positive allocation harmony chord for the song and export audio files, and then burn a CD to prepare students to exchange and learning.

For sight-singing works, we feel the changes consequent on the style by changing its signature, key signature, speed, etc. This in-depth understanding of our students to learn by analogy works and other related disciplines singing are also very good.

The song's speed is moderate Allegro, emotions are warm and lively, and we change its time signature, key signature and the speed, and then add the appropriate waltz accompaniment style TT composer, the music, the mood here has undergone great changes.

5 AUXILIARY PRACTICING EAR TRAINING METHODS AND TECHNIQUES

Snoar has a rich timbre but also can load a variety of sources. It has all the typical percussion effects, sound very realistic, extreme efforts have feel, from electronic drums, bass drums, to the Japanese koto, and a hundred kinds of percussion instruments sanxian can be found. Teacher in the class is the actual audio sync to show students, instant playback, making rhythm training with different sounds and different sound. Difficulties encountered heavy rhythm, or difficult to grasp the rhythm of the students, teachers take varied sound to attract students of hearing and mobilize the enthusiasm of students, students can freely change their own particular type of rhythm instruments according to their preferences, and a time when speed is can be swapped. Change the past repeatedly listening fixed pattern. At the same time intervals, chords, melody and perception will not run because the low notes piano pop any deviation, its acoustics and teaching efficiency a big difference.

EarMaster ear training teaching abroad is many professional colleges in widespread use, referred to as an ear training guru. It is targeted; there is an intuitive precise statistical analysis, for different people to choose other advantages according to the learning focus. Figure 4 is a more comprehensive knowledge of EarMaster ear training software system, which the classification is easy to difficult, from simple to complex, progressive layers, has a highly targeted interaction and openness.

For example, teachers can follow the progress of teaching requirements and in accordance with the specific circumstances of the student's learning, individualized, targeted selection exercises. EarMaster can immediately provide the appropriate report card, showing the test of time, the number of examinations name and answer, the correct rate, draw rate, very intuitive and precise.

Teachers can base on these data, selects students in class solution assessment exercises, the more convenient for students learning purposes. As long as the user an overview of selected teachers, one can clearly know the lack of students. When teaching, the teacher in advance of the sight-singing melodies and listening content programming and playback, when you can walk into the middle of the collective exercise of individual students and student exchanges guide or explore this teaching method not only pulled into the distance between teachers and students , but also to better improve the efficiency and quality of teaching, but also to facilitate teachers to keep abreast of students' knowledge and grasp the situation, while allowing students to clearly know their place and to correct deficiencies, thus greatly improving the efficiency of teachers and teaching efficiency.

6 CONCLUSION

Solfeggio As music major institutions all normal basic courses, showing a particularly important role are particularly critical. However, with the development of the times, the problems of traditional teaching methods exist. Solfeggio paper spectrum of cases is difficult to produce the sound of hearing students' association, single and boring sound, intonation cannot be objective and quantitative analysis, no scientific accurate evaluation, etc., have been difficult to meet the needs of society. The technology used in computer music ear training is needed not only to the growing diversification of the times, or the transformation of classroom teaching and student needs. In this paper, several ear training software as the main object of study, research methods proposed computer music technology assisted teaching ear training. Traditional teaching carried out a detailed analysis of the problem, the use of computer music technology in electronic music can sound, audiovisual methods rich sound, digital audio analysis techniques, precise statistical analysis to solve the problem, for computer-aided teaching music technology feasibility and practicality were demonstrated.

However, in the domestic computer music technology in various music colleges teachers not common, most of the teachers and students do not understand this. How do the popularity of computer music technology knowledge is a question worth pondering. In this regard I have the following thoughts: for the provinces municipalities and school, the establishment of specialized lectures and presentations of academic research, and constantly update the relevant books and audio books and other books. Relatively simple to set up interesting computer music appreciation classes in schools, so that students have a general awareness of this. In the course of various types of normal or can be set to music colleges in elective courses on computer music technology to attract more non-professional students to learn, so that is very conducive to the popularity of computer music technology. Meanwhile, the era of the 21st century continues to progress, coupled with computer music technology, operational and professional, thus educators and educated to be life-long learning is a must.

With the rapid development of technology, the computer music technology to solfeggio in training is an essential modern means. It updates the traditional teaching model to optimize the teaching methods, expanding the horizons of students, but also to encourage students to take the initiative in various channels to seek ways of learning and acquiring new knowledge, provide teachers with modern technology teaching tools and teaching methods in solfeggio teaching do have a great prospect.

REFERENCES

Lisi xin. International Computer Music Research and Opinion of related professional disciplines [J]. Central Conservatory of Music, 2003 (2).

Zou LIANFENG. The use of multimedia technology in teaching ear training [J]. Yellow University of Science and Technology, 2007 (3): 107–108.

Jiang Lin, Li Meiping. Music software ear training Teaching [J]. Software Tribune, 2007 (3): 27–28.

Xiao Lei. On the practical application of Sibelius's music software [J]. Journal of Guangxi University (Natural Science), 2009 (3): 35–39.

YAN Jing-yu regulate its use singing resonance [J] Shandong: Qilu Art Gallery, 2002,4.

Zhang Xian sound technical and artistic resolve singing [J] Beijing: People's Music, 2005,8.

Cuiquan Xin Singing resonance cavity resonance and three practical exercises [J] Guangdong: Xinghai Conservatory of Music School Newspaper, 2001.

Xia Xianping vocal cords, glottis and vocal [J] Beijing: Biology Bulletin, 1989,10.

Zhu Jiming, Rick Chan, Lubbe real larynx anatomical study [J]. Shanxi: Anatomy, 1990,04.

Management, Information and Educational Engineering – Liu, Sung & Yao (Eds)
 ISBN: 978-1-138-02728-2

A study on English language acquisition of bilingual children with different bilingual proficiency

Feng Cun An, Yu Si Wu & Zhen Ai Zhang
Yanbian University, Yanji, Jilin, China

ABSTRACT: This thesis is aimed at Chinese-Korean bilingual children who are the beginners to study English in China. According to the results of English teaching experiments and statistical analysis of the achievement data of different types of bilingual students, this thesis analyzes English acquisition characteristics of these bilingual children with different bilingual proficiency. The result shows that different bilingual proficiency has significant impact on English language acquisition.

KEYWORDS: bilinguals; bilingual proficiency; foreign language acquisition.

1 INTRODUCTION

China is a multi-ethnic country. Most of the ethnic groups still maintain their own languages, culture and customs. At the same time, as a unified multi-ethnic country, the Chinese government requires the entire nation including the minorities to master a common language - Chinese Mandarin to strengthen the unity of the country, develop economy and culture and strengthen the ethnical communication. Thus, China's ethnic minorities also need to master a second language - Mandarin. So there is bound to be the phenomenon with two or more languages used simultaneously in the minority's areas of China. Those who can use two or more languages to learn, work and live are called "bilinguals" or "multilinguals". The "bilinguals" phenomenon is more common in minority nationality areas in China.

Chinese Korean is a typical "bilingual" group. In social life, work and study, they use both of Chinese and Korean. Due to the unique bilingual social environment, most of Korean ethnic people can use Korean Chinese and Chinese freely. Then do the bilingual proficiency have any effect on the ability to learn a foreign language or whether bilingual proficiency can promote the foreign language acquisition?

For this reason, the authors spent a semester doing a teaching experiment to find out whether different bilingual proficiency can affect the foreign language acquisition result.

2 EXPERIMENTATION

2.1 *Examinee*

As there were a few research data on bilingual foreign language acquisitions and a lack of previous experience, we chose bilingual children from Grade Three to be the examinees who had never learned a foreign language before. Thinking about the influence of the elements of students' intelligence, and language skills, etc., we selected middle-level students as the examinees. The ages, genders and some other factors were taken into account, too. And finally, we have three different types of classes with 12 students in each class. We also arrange a Han Chinese class as a contrast class. The followings are the information in details:

Class A: Han ethnic in Han primary school

Class B: Bilinguals in Han primary school

Class C: Bilinguals in Korean Primary school

Class D: Korean ethinc in Korean Primary school

2.2 *Experiment content*

The English teaching experiment lasted for a semester. There were 40 class hours totalled, at the same time for the four different classes. The main teaching contents were alphabets, pronunciation, vocabulary, simple sentences, the simple use of *do, have* and *be*, and some daily conversations. Through explanation and practice, the students were required to memorize 26 English letters, to master IPA and different combinations, to distinguish

Table 1. Descriptives Scores.

	N	Mean	Std. Deviati	Std. Error	95% Confidence Interval for Mean		Min	Max
					Lower Bound	Upper Bound		
A	19	85.54	3.76	.86	83.72	87.35	77.7	93.3
B	19	89.27	2.59	.59	88.02	90.52	84.9	94.0
C	19	80.37	7.01	1.60	76.98	83.75	69.4	91.5
D	19	74.44	7.60	1.74	70.77	78.10	61.8	90.7

different pronunciations, to master the simple use of *do*, *be* and *have*, to make phrases and simple sentences with the words, to use the conversation learned flexibly and to gain the ability of using simple English.

The teaching content, teaching time, progress, assessment content and assessment methods of each experimental class were the same. The only difference was the teaching method. For Class A and D, teachers just used one language, that is Chinese and Korean, respectively, but for Class B and C, teachers adopted the method of bilingual teaching through the contrast of Chinese and Korean. But they had a slight emphasis on the teaching language. For example, the teacher of Class B mainly wrote Chinese on the blackboard while the teacher of Class C wrote Korean.

2.3 *Data collection*

There were assessments on the teaching contents after each lesson. Each time at the end of class, the four teachers would have a discussion about the teaching situation and the problems appeared in class. Then they came up with the way to deal with the problems and made preparations for the specific content for the next lesson. After each assessment, they had a serious statistic on the results, kept detailed records and analyzed how these difficulties and errors appeared then compared among different class types. After the end of the teaching experiment, they collected all data, carried on the statistics and analyzed the results.

3 DATA ANALYSIS

We used SPSS12.0 for the 19 assessments of different classes. Table 1 is the result table of descriptive statistics. The table can show the average score sample content N, Mean, standard deviation, standard error, 95% confidence interval, minimum and maximum of the four classes.

As it could be seen from Table 1, the average scores of Class B were higher than Class A; Class C was higher than Class D. In that way, could we infer that there were differences between monolingual classes and bilingual classes and questioned whether these differences were significant? Was there a significance on foreign language acquisition caused by different bilingual proficiencies? We could not draw a conclusion just according to Table 1.

If we wanted to analyze the statistics further, we had to make sure that whether the variance of all data of each class had homogeneity, see Table 2.

Table 2. Test of homogeneity of variances Scores.

Levene Statistic	df1	df2	Sig.
9.707	3	72	.000

Of significance, $p < 0.05$, we can find that there were significant differences at the level (• = 0.05) in the variance of each class, so the variance of each class had no homogeneity. This conclusion was an important condition for selecting multiple comparison method.

Table 3 was a single-factor analysis of variance. It mainly used the variance of select items to analyze the results. The output showed squared deviations within the groups and between groups, the Sum of Squares, degrees of freedom (df), Mean Square, F value and probability p values. For the results $p < 0.05$, there was a significant difference in the mean at • = 0.05

Table 3. Anova Scores.

			Sum of Squares	Df	Mean Square	F	Sig.
Between Groups	(Combined)		2366.031	3	788.677	24.641	.000
	Linear Term	Contrast	1691.882	1	1691.882	52.861	.000
		Deviation	674.148	2	337.074	10.531	.000
Within Groups			2304.469	72	32.007		
Total			4670.500	75			

*The mean difference is significant at the .05 level.

level among the groups. So there was significance to carry on the statistics of each team's mean.

Since there were four different types of classes in this teaching experiment, we needed to carry out multiple comparisons about the mean with the LSD method and Games-Howell method, see Table 4.

Table 4 is the results of multiple comparisons. According to the results of the test of homogeneity of variances in Table 2, $p < 0.05$ which illustrated that there were significant differences in the variance at• = 0.05 level of each class, so each class had no homogeneity of variances. So we only adopt Games-Howell method to analyze the results and made conclusions on all analytical data. To see from the results shown on the table, there were significant differences at the • = 0.05 level among A, B; A, C; A, D; B, C; B, D; but there was no significant difference between the C and D. In the table the one which was marked with "*" showed that the mean difference was significant at the .05 level.

Table 4. Multiple comparisons dependent variable: Scores.

	(I) CLASS	(J) CLASS	Mean Difference (I-J)	Std. Error	Sig.	95% Confidence Interval Lower Bound	95% Confidence Interval Upper Bound
LSD	A	B	−3.7284*	1.8355	.046	7.3874	6.9393E-0
		C	5.1726*	1.8355	.006	1.5136	8.8317
		D	11.1000*	1.8355	.000	7.4410	14.7590
	B	A	3.7284*	1.8355	.046	6.939E-0	7.3874
		C	8.9011*	1.8355	.000	5.2420	12.5601
		D	14.8284*	1.8355	.000	11.1694	18.4874
	C	A	−5.1726*	1.8355	.006	8.8317	1.5136
		B	−8.9011*	1.8355	.000	12.5601	5.2420
		D	5.9274*	1.8355	.002	2.2683	9.5864
	D	A	11.1000*	1.8355	.000	14.7590	7.4410
		B	14.8284*	1.8355	.000	18.4874	11.1694
		C	−5.9274*	1.8355	.002	9.5864	2.2683
Games-Howell	A	B	−3.7284*	1.8355	.006	6.5712	−.8856
		C	5.1726*	1.8355	.040	.1794	10.1659
		D	11.1000*	1.8355	.000	5.7617	16.4383
	B	A	3.7284*	1.8355	.006	.8856	6.5712
		C	8.9011*	1.8355	.000	4.1489	13.6532
		D	14.8284*	1.8355	.000	9.7105	19.9464
	C	A	−5.1726*	1.8355	.040	−10.1659	−.1794
		B	−8.9011*	1.8355	.000	−13.6532	−4.1489
		D	5.9274	1.8355	.078	−.4693	12.3241
	D	A	−11.1000*	1.8355	.000	−16.4383	−5.7617
		B	−14.8284*	1.8355	.000	−19.9464	−9.7105
		C	−5. 9274	1.8355	.078	−12.3241	.4693

* The mean difference is significant at the .05 level.

To demonstrate the mean comparison among the classes visually, the following Mean Scatter figure was designed with common factor variance Class as the horizontal axis and independent variable Scores as the vertical axis. The distribution of the mean of each class could be seen from the figure.

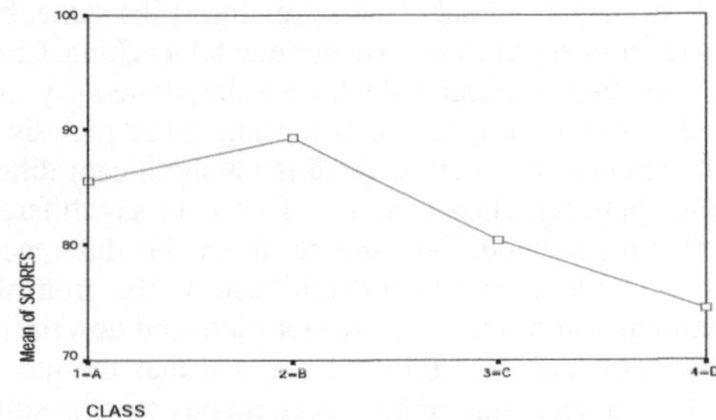

Then, based on the analysis above, bilingual comparison education had no significant difference in the scores of Class C and D. However, the difference of bilingual proficiency had significant differences in foreign language acquisition. This difference also appeared in the process of bilinguals' and monolinguals' foreign language acquisition. Causes of the differences would be analyzed below.

4 CONCLUSION DESCRIPTION

First, look at Class C and D. The natural circumstances of two experimental classes are almost the same, but in the process of English experimental teaching, Class C uses both Korean and Chinese simultaneously, and only Korean for Class D. According to the analysis of the results of statistics, there is no significant difference between class C and D. That is to say different teaching methods will not result in the differences of students' achievement significantly. But from the comparison of the mean scatter plots and descriptive statistics analysis table, we can see that the use of bilingual teaching in the classroom has more positive impact on students' language acquisition than monolingual teaching, but it is not very significant.

Next, let's see the differences among Class A, B, C and D. From result, it can be seen that there are significant differences between classes A, B and classes C, D. As it has already been mentioned above that different types of languages have influences on foreign language acquisition. Although Chinese and English belong to different types, there are many similarities between them, such as SVO word order, but Korean is SOV structure. This is the impact of foreign language acquisition caused by different types of languages. For this reason, many Korean ethnic students learn Japanese well. And for these children, their cognitive capacity for language is still not strong, so they do not have a full understanding of the differences between languages. In that way, to support the transfer function of language in the process of foreign language acquisition will bring great effect.

The following focuses on the differences between Class B and other classes:

Class B and Class C, D: The students in Class B, C, and D are all Korean ethnic, they are all in the same grade with ten years' experience, all master two kinds of languages with Korean as their native language and Chinese ad their second language. The problem is that they have different bilingual proficiencies and different frequencies of using the two languages.

The students in a Class B study in Han ethnic school, during school they can use Chinese to study. At school, the active language code in the brain is Chinese. But when these students go back home, they communicate with their parents in Korean. Because of the needs of daily life, they have to learn Korean, then they can have no barriers in their lives. At home, Korean language code is active in the brain. Moreover, the Korean language and cultural knowledge are also acquired with native language. Although most of these students can't write or read Korean, it does not matter because the original form of the language is oral expression. Therefore, it does not affect their bilingual proficiency. School - family these two different locations to promote them to convert language codes frequently and have different language thinking. Over the years, they have been accustomed to this frequent code switching. Accordingly, the two languages almost exist and develop in the brain simultaneously. In daily communication, they can start communication in different languages at any time with different language. The students who formed the habit to think and express in bilingual have strong bilingual proficiency.

The students in Class C and Class D are Korean ethnic learning in Korean Schools.

Although they also learn and use Chinese, the feeling and use of Chinese are different from the students in Class B. The students in the Class C study scientific and cultural knowledge in Korean, and the frequency of using Korean in the family and social interaction is also high. Generally, Chinese is just a course or a branch of knowledge for them. They have to use Chinese for some environmental needs, although they can use Chinese to communicate with others, Chinese is just a "second language" for them. For most students, the Chinese have not become necessary language for survival yet, so the level of their Chinese is not high, and they can not use Chinese knowledge flexibly. Thus, they can't use code-switching as freely as Class B. Although these students can use Chinese in their life, the consciousness by using Chinese is less than the students in Class B, and so are their bilingual proficiencies. In foreign language acquisition, the role of Chinese can not be fully mobilized. However, they have great potential, with the improvement of their Chinese and the increasing frequency of using Chinese, their bilingual proficiencies are enhancing gradually.

It is clear that the differences of bilingual proficiencies have significant effects on foreign language acquisition. Thus, we must strengthen the Chinese-Korean bilinguals' bilingual proficiencies which can not only develop their bilingual advantage, but also play a potential role in foreign language acquisition.

5 CONCLUSION

It was found through the teaching experiment, there is a significant relationship between foreign language acquisition and bilingual proficiency. For a bilingual, the higher his bilingual proficiency is, the more advantages it has in foreign language acquisition. Therefore, for Chinese Korean ethnic students, they should start from childhood to strengthen their bilingual proficiencies. Language is an important tool for thinking. Language skills and thinking skills are mutually reinforcing, and the development of language skills can develop children's intelligence. In the process of foreign language acquisition, since bilingual have demonstrated their advantages, they should play to their strengths in foreign language acquisition.

Making good use of bilingual foreign language acquisition will eventually lead to multi-lingual people.

The research on foreign language acquisition based on bilingual is a systematic project. The foreign language acquisition research in bilingual is still in its infancy, both theory and practice are inadequate. There is still much to be researched in this area, but it also requires more manpower, material resources. The study did not find out how bilingual learn a foreign language. To get enough evidence and theoretical support, there is still much work to do. The function of this article is just throwing a stone to clear the road.

ACKNOWLEDGEMENT

Thanks to the national social science fund project (10XMZ042) from which the article obtained the support. And the author Zhen'ai Zhang is the corresponding author of this article.

REFERENCES

Zhang, Zhen'ai. 1998. Korean-Chinese Bilinguals and English Eduaction. *The Journal of Yanbian University*, (1):

Gui, Shichun. 1997. *Linguistic Methodology*. Beijing: Foreign Language Teaching and Research Press.

Li, Shaoshan. 2001. *Statistics in Linguistic Research*. Xi'an: Xi'an Jiaotong University Press.

Lu, Wendai. 2000. *SPSS for Windows*. Beijign: Electronic Industry Press.

Ha, Jingxiong. & Teng Xing. 2001. On the Education for Minorities. Beijing: Science of Education Press.

David, W. Carroll. 2000. *Psychology of Language*. Beijing: Foreign Language Teaching and Research Press.

Herbert, H. Clark & Eve V. Clark. 1977. *Psychology and Language*. Harcourt Btace Jovanovich, Inc.

Management, Information and Educational Engineering – Liu, Sung & Yao
© 2015 Taylor & Francis Group, London, ISBN: 978-1-138-02728-2

Exploitation on school-based traditional ethnic sports course in primary and middle schools of ethnic region in northwest Guangxi

Li Chun Wei
Institute of Physical Education, Hechi University, Yizhou, Guangxi, China

ABSTRACT: Study on ethnic sports resources in Hechi, northwest of Guangxi, and exploration of school-based P.E. course in primary and middle schools show that exploitation on school-based course can be of assistance to schools to better establish their own special running characteristics. Also, it may upgrade the teacher's proficiency, enhance the development of the student's personality, realize individualized P.E. education and satisfy the plural needs of student; moreover, it will serve to popularize the inheritance and development of ethnic sports. The exploitation of text and course should be discussed in aspects of content selection, course exploitation, course implementation, and teacher's training.

KEYWORDS: Northwest Guangxi of China; Primary and middle schools in ethnic region; Traditional ethnic sports; Exploitation on school-based course.

1 INTRODUCTION

The new round of course reform has infused new vitality to P.E. education in primary and middle schools in ethnic region. It has altered the over-conformed situation of the course management by practicing a three-level system of state management, regional management and school management, which has promoted the course's adaptive capacity to certain region, teacher and student as well. It goes along with the international trend of course reform and highlights the school-based course in the project. The exploitation on school-based course can fully explore student's potential individual advantages, and enhance the all-rounded development of their personality; it conforms to the development of modern education, and gives expression to the superiority of traditional ethnic sports, in favor of giving play to teacher's initiatives and improvement in teaching and researching level; northwest of Guangxi is the gregarious dwelling of ethnic peoples, it is rich in traditional ethnic sports resources, and every project there possesses strong ethnic culture characteristics, which boasts infinite potential of exploitation. Utilize the traditional ethnic sports as school-based course can not only solve the problems of insufficient fund, venues and facilities, but also enrich the teaching content in class. It will also render systematic inheritance and further development of the traditional ethnic sports in northwest Guangxi.

2 SUMMARY OF TRADITIONAL ETHNIC SPORTS IN NORTHWEST GUANGXI

Northwest Guangxi in China is the gregarious dwelling of several ethnic groups with the Zhuang people as the body. It consists of 16 cities and counties and populates seven minority peoples: Zhuang, Yao, Miao, Molao, Dong, Monan, and Shui people. Among them, Molao and Maonan Autonomous County of Guangxi is the only ones for the people in China. The distribution of minority peoples feature in that" large area inhabited by several nationalities and small inhabited by their own". The people in northwest Guangxi is industrial, courageous, and earthy, and they have created and developed various traditional ethnic sports with diverse forms, rich contents, strong ethnic characteristics and profound appreciation of the people. The traditional ethnic sports is a gorgeous pearl in our treasury vault of our national culture, and our precious ethnic cultural heritage.

There are multiple traditional ethnic sports events in northwest Guangxi. According to incomplete statistics, there are totally 128 traditional ethnic sports in Guangxi, of which over 60 has been listed in the sports section of the General Annals of Guangxi, and 28 remains to be listed.

For example, in Zhuang people's culture, there are embroidered ball throwing, shoulder pole beating, turtledove jumping, Zhuang Kung Fu, frog Kung Fu, knife circle penetrating, fire circle penetrating, chair dragon line, wooden wheel rolling, headstand walking race, three-person board shoes race, hunting, promotion picture, pack basket ball, Liao ball, bee drum beating, etc. In Yao's, there are bronze drum beating, dustpan beating, hasten laborer dance, wrestling, Spring buffalo dance, broze bell dance, Hongmen performance, strap beating, bamboo bar pulling, archery race, spinning top, knife mountain climbing, trumpet ball, pulling out the Huluxiao wine, Bamboo lyre drum beating,etc. In Molao's, there are elephant step and dragon dance, grass dragon dance, pearl grabbing, phoenix's egg protection, zongzi grabbing, fenghuo ball, etc. In Maonan's, there are "with the top"," with the filler", "with the patchwork","with the back", stone lock carrying, and wrestling with tied waist, etc. In Shui people's tradition, there are horse race, archery race on horseback, and Huamei scraping, etc. Performance events are also quite abundant in Hechi, there has been 91 performance events exploited and collected since 1984.[1] In terms of their characteristics, they are rich in deep cultural implication, distinguishing in styles, and create diverse phenomena. In terms of their contents, some take their roots in agricultural production activities, some originated from ethnic cultural specialties, some evolved from religions and sacrificial ceremonies, and some derived from wars. They can serve to put forward local ethnic cultures, facilitate national coherence, and they boast commercial exploitation values.

According to the manifestation and the characteristics of the traditional ethnic sports, we divide them into four categories: the competitive events, the performing events, the fitting and entertaining events, and the events of the game.

3 DEFINITION OF THE SCHOOL-BASED COURSE AND ITS EXPLOITATION

School-based course is a scheme with teachers as its center. In the practical implementation of the national *Spots and Health Standard* and regional *Implementation Scheme of Sports and Health Course*, it will assess the students' demand in the school;on the basis of making the bast of sports resources in schools and communities, it aims at promoting student's health.[2] School-based course exploitation refers to a series of reform activities under the guidance of the national *Spots and Health Standard* and regional *Implementation Scheme of Sports and Health Course,* which is intended to satisfy students' need of exercise and enhance their healthy growth according to the specific characteristics, conditions and educational resources of certain schools.

4 SIGNIFICANCE OF LAUNCHING SCHOOL-BASED COURSE EXPLOITATION IN PRIMARY AND MIDDLE SCHOOLS IN NORTHWEST GUANGXI

4.1 *Be assistant to build up sports teaching characteristics and establish special running styles of schools*

The characteristics of a school refers to its own distinguishable traits in comparison to those schools in the equivalent categories. In other words, schools should take full advantage of its own characteristics and merits in its running process, put the education scheme in effect creatively, and establish itself special and stable characteristics and style.[3] At the same time, its emphasis on the uniqueness and differences of teachers and students has involved in the schools particular running philosophy, which contributes to the creation and development of the tradition and running style of a school. The economy, culture, and education of Northwest Guangxi is relatively undeveloped, which cast an influence on the disadvantageous condition of the education facilities in local middle and primary schools compared with their urban counterparts. The schools should thus enhance their advantages and avoid theie weaknesses, actively initiate sports teaching characteristics, and establish their special running styles, which lies in the utilization of the blessed ethnic cultural resources in Northwest Guangxi. In addition, the exploitation of school-based course will help to improve the sports teaching facilities and the establishment of special sports curriculum system.

4.2 *Be assistant to improve the professional developing level of P.E. Teacher*

The exploitation of the school-based P.E. course focus on the teachers, which established the position of professional autonomy, entitles them to participate in the exploitation and make decisions in it, and confer on them the right and responsibility in the process, reinforcing their sense of responsibility and obligation. However, this exploitation is one of creativeness with difficulties of multiple aspects; therefore it requires a higher standard of professional consciousness and quality. Teachers are supposed to go deeper into the exploration of the course, of the students and of the society, and be bald to innovate with breakthroughs of the out-of-date teaching pattern. The processing and integrating of traditional ethnic sports resources is a procedure of studying and innovating, which will elevate the individual thinking and innovative ability of teachers and promote their professional spirit, knowledge and competence.

4.3 *Be assistant to develop students' individuality, realizing individualized education and satisfying the diversified sports needs of students.*

One prominent trait of modern society is the principle of" people first", which highlights the development of people's individuality and creativeness. Sports activities lies the premise on people's physical activities. People will have various needs due to their differences in physical qualities, capacities, health conditions, interests, and living environment. As a result, on the basis of general unified requirements, the setting of P.E. course should consider the specific condition of individuals, make the most of the circumstances to accomplish individualized education and satisfy the diversified sports needs of students. There are plentiful ethnic groups settling in Northwest Guangxi, whose interests and customs are different from each other more or less. Consequently, the sports activities are abundant and diverse: some are competitive, some fitness-based, some entertaining, others feature in customs, or blend various characteristics together. These diverse sports activities provide schools with numerous available choices and expand the space of development, which will absolutely benefit the accomplishment of individualized education and satisfaction of students' diversified needs.

4.4 *Be assistance to inheritance and development of traditional ethnic sports in Northwest Guangxi*

The popularization and mass recognition of Chinese traditional sports is bound up with the rise and fall of traditional ethnic culture and inheritance and development of Chinese traditional culture. In the light of educational anthropology, education is the major way of generating culture. Also, inheritance of culture is an important part in education. For a long time, schools have always been the cradle of culture, the carrier of culture, the manufacturing location of culture, and an essential front of advanced culture. To cast our eyes on history, many sports activities started their origins, popularization in schools. So to carry out the exploitation of school-based P.E. course in primary and middle schools in Northwest Guangxi, and involve traditional ethnic sports in modern society genuinely, and complement the modern sports mutually is of great significance. And then, those graduating from schools with abundant and excellent knowledge of traditional ethnic sports can serve as the inheritor and transmitter of ethnic culture, which will be conducive to the inheritance and development of ethnic culture.

5 THE ADVANTAGES OF LAUNCHING SCHOOL-BASED COURSE IN PRIMARY AND MIDDLE SCHOOLS IN NORTHWEST GUANGXI

5.1 *Geographical resources advantages*

Hechi, city of Northwest Guangxi, populates 8 minority groups: Zhuang, Yao, Molao, Maonan, Shui, Dong, Miao, etc., which adds up to 3 million and 400,000, representing over 85% of all in Guangxi. The area is one of those dwells most minority groups.[4] In the history spreading thousands of years, the minority groups integrate and live with each other, together they creating and inheriting the traditional ethnic sports of distinctive styles, leaving precious cultural heritages for human civilization. Events like spinning top beating, board shoes race, board shoes dance, high-heel horse, embroidered ball throwing, bamboo ball, shoulder pole beating, all these iconic events of Northwest Guangxi, not only constitute the daily exercise activities of local people, but belong to the local original ethnic culture dated back to thousands of years ago. As a result, infusing these events in the school-based exploitation process in primary and middle schools in Northwest Guangxi can serve to radiate to a full extent the contents, forms, and cultural inheritance to teachers and students in that area, and further the integration of traditional ethnic sports into primary and middle school classes and establish its own characteristics.

5.2 *National policies lay a foundation of school-based course exploitation*

In 2001, it was clearly announced in the *"Basic Edu cation Curriculum Reform(trial implementation) "* in the National Education Work Meeting that" (schools are supposed to) alter the over-centralization of the curriculum management, and carry forward a three-level management system of state management, regional management, and school management, and strengthen the course's adaptive capacity to different locations, schools, and students." The new outline has proposed a curriculum management system that puts functions as its core, according to which the division of curriculum is conducted. On this background, the school-based course exploitation has broken through the old centralized management system, and provide schools, teachers, and students wider space to choose and make decisions. *School Management of Curriculum Guidelines* declaimed that schools should" explore students diversified sports developing needs, exploit and select text resources that match with their own characteristics and optional for students." It has also asserted concrete requirements of the exploitation process and modes of execution, which consolidates a firm guarantee of school-based course exploitation of traditional ethnic sports.[5]

6 THE IDEA OF SCHOOL-BASED EXPLOITATION IN PRIMARY AND MIDDLE SCHOOLS OF NORTHWEST GUANGXI

6.1 *Selection of contents*

The selection of course contents should reflect the educational purpose and reality. The traditional ethnic sports enjoy a time-honored history, and it mainly consists of various sports events and physical exercises. Its materials are abundant and extensive, which lead to the broad openness of its unfolding modes, and possess unique ethnic characteristics. The birth and development of traditional ethnic sports are tightly connected with the mass productive labor, custom, songs and dances of local folks, embedded with strong regional and ethnic flavor, which offers extensive space of the school-based course exploitation. Consequently, the selection of course contents should emphasize its ethnic features: lay a key consideration to the beloved traditional ethnic sports, correspond to the discipline of students' mental and physical health development as well as school's characteristics, and appropriately select suitable course contents of traditional ethnic sports. And then schools should take into account the local circumstances when selecting the course contents, comprehensively regarding the characteristics of different schools and students, available sites, faculties, and current situation of P.E. teaching to establish course contents with their regional characteristics. For example, firework grabbing, embroifered ball throwing, bronze drum beating, bar climbing, dancing with lusheng, elephant tug-of-war, these are all sports events demonstrating local customs and practices varied in styles. They mingle with the local customs, cultures, folklore, and historical evolution, constituting the ethnic characteristics of course contents in local schools.

6.2 *Course exploitation*

In the *Course Exploitation and Research Guidelines*, Mr. Steinhaus has come up with the famous "procedure schema", which states the teacher's important role in course exploitation. Accordingly, we should pay heed to the following aspects in the process of exploitation: firstly, we should highlight the principle of "health first". The various forms of traditional ethnic sports in Northwest Guangxi can involve in multi-sensory experiences, which will boost the students' capacity of reacting, flexibility, power and endurance, stimulate them to think actively, and promote their healthy development physically and mentally. Secondly, we should stick to the philosophy of "people first", adequately concern about the needs of students and achieve ideal teaching effects. Thirdly, we should respect the law of P.E. teaching. To incorporate traditional ethnic sports in the P. E. course system should always commit to the precondition of students' capacity, recognition, a strong willingness to participate, and the convenience to practice and popularize of the events.[6]

6.3 *Course implementation*

Course implementation is the process of carrying out a course plan in reality by teachers, and it is the key to realizing the predetermined course plan. It includes: inverting the traditional ethnic sports in textbooks, making teaching plans, carrying out P.E. Teaching, and recording the teaching results. The first one requires P.E. teachers to systematically arrange the theoretical, skills, and practice parts of traditional ethnic sports, providing text documents for teaching. The theoretical part includes traditional ethnic sports knowledge, the principle, traits, structure, exercising values and significance of the events. The part of skills include the essentials of skills and teaching methods of traditional ethnic sports. The practice part includes the practical purpose, significance, methods, steps, and requirements of traditional ethnic sports. Teaching plan is the direct documents to carry out teaching process, which consists of annual plan, term plan, unit plan and teaching period plan. It is also the embodiment of teaching contents and teaching methods. Teachers should choose a feasible teaching organizing method and make a scientific distribution of the teaching period and the contents, according to course standard and the principles of compiling teaching plans. P.E. teaching is the practice section in course implementation, which should strictly follow the requests of course exploitation. We ought to have the foresight of the possible problems emerging in the specific teaching process, and solve them on time once confronted. The teaching process should be innovative, which can properly utilize new teaching and organizing methods and accomplish predetermined results.

6.4 *Training of the teachers*

In the school-based course exploitation, teachers are not only the executors of the curriculum, but also the master of it. The exploitation requires of theoretical supports and professional training. The teachers' level of the relative proficiency will directly exert an influence on the quality of school-based course exploitation.[7]

The training of P.E. Teachers in primary and middle schools should base on the autonomous reading and mutual communication of the teachers by flexible, plural, lively and vivid measures. The teachers should get handle of the relevant skills, practicing methods, and general knowledge about the sports. We encourage the teachers to go deeper into the local folk's life, and experience to a full degree the geographic and cultural environments, and observe the sports activities carried out by local people. Only in that way will they be able to understand the profound values of local culture and the genuine spiritual substance of traditional ethnic sports, which will advance the inheritance and development of the sports. Schools can also employ some exports out of the campus to be the guide of the exploitation. The training can solve multiple problems: strengthening the course consciousness of the body part in the exploitation, establishing specific education and course concepts, distinguishing some incorrect perceptions and behaviors in the implementation; and then understanding the true connotations and traits of traditional ethnic sports school-based course exploitation, mastering the basic theory, basic method and basic characteristics of teaching and learning, developing lively and vivid, excellent and effective teaching activities.

7 SUMMARY

Traditional ethnic sports in Northwest Guangxi of China is one of the excellent national cultural treasures, tinted with deep-rooted cultural gene, and shoulder the historic responsibility of cultural inheritance. They not only bring together the traits of interests, competitiveness, fitness, entertaining, and performing, but also possess the economic values for the local materials can be drawn on and the requests of site are quite humble. They provide more optional materials for the school-based course exploitation in primary and middle schools of Northwest Guangxi, and have tremendous values and advantages of exploitation. To infuse traditional ethnic sports in the P.E. course will ease the shortage of fund, facilities, sites, and moreover, enrich the contents and connotation of P.E. teaching, activate the sports and cultural life on campus, and expand the practicability and operability of the physical facilities in schools. Meanwhile, the implementation and popularization of traditional ethnic sports through schools may dig out and collect more valuable and distinctive projects, which in turn will accelerate the advancement and standardized development of traditional ethnic sports.

ABOUT THE AUTHOR

Wei Lichun,(1965), female, professor,Zhuang people, citizen of Yizhou, Guangxi,dean of Physical Culture Institute of Hechi University.
Discipline of research: Traditional ethnic sports, P.E. teaching and training methods.

ACKNOWLEDGEMENTS

[1] Study of Establishment and Practice of Characteristics Educational System in Colleges and Universities of Ethnic Regions,key funded project of Guangxi higher education curriculum reform project(Project code:2013JGZ154).
[2] Predicaments of Molao's Traditional Sports and New Developing Path in Modern Society, project of research on philosophy, sociology and science in Guangxi(Project code:2013FTY004).

REFERENCES

[1] Zhu Lantao, Chen Wei, Study and Research on the Traditional Ethnic Sports Resources in Guangxi, [J], Study of Ethnics in Guangxi, 2012, (3):146–153.
[2] Dong Cuixiang, Zhou Dengchong, Definitions of School-based Course and Relevant Concepts, [J], Jounal TianJin Physical Culture College, 2005, 20(1):51–53.
[3] Zhou Jianping, Li Yong, A Simple Analysis of School-based course Exploitation of Primary and Elementary Schools in Miao and Dong Ethnic Group Autonomous County in Southeast Yunnan [J], Journal of Kaili College, 2012, 30(6),184–186.
[4] Wang Hongying, Yang Zaizhun, Theoretical Discussion and Practice of School-based Course Exploitation in College and University, [J], China Physical Science and Technology, 2007, 6(30):132–135.
[5] Lou Lanping, Rational Reflection on School-based Course Exploitation [J], Journal of Beijing Sports University, 2004, 27(10):1389–1390.
[6] Dong Cuixiang, Li Xingyan, Wang Shan, The Ideological Foundation of School-based Course Exploitation[J], Journal of Shanghai Sports University, 2007, 31(5):91–94.
[7] Hong Yan, Jin Yule, The Ideological Foundation of School-based Course, [J], Journal of Southwest China Normal University, 2003, 3, (86).

Management, Information and Educational Engineering – Liu, Sung & Yao (Eds)
© 2015 Taylor & Francis Group, London, ISBN: 978-1-138-02728-2

A discourse-based English passive voice teaching

Xue Feng Zhai
Faculty of Foreign Language & Culture, Kunming University of Science and Technology, Kunming, China

Guo Feng Ding
Law School, Kunming University of Science and Technology, Kunming, China

ABSTRACT: Different from the traditional grammar teaching, in the discourse-based approach, teachers can teach target forms together with authentic or simplified discourse, which can supply learners with abundant examples of contextualized usages of the target structure to promote the establishment of form-meaning relationships, and accelerate students' master of the items, especially in their use. Thus, this paper, taking functional grammar as its framework and the teaching of English passive voice as an example, proposes grammar teaching of discourse-oriented approach.

KEYWORDS: discourse-based approach; English passive voice; grammar teaching.

1 INTRODUCTION

Grammatical competence is now considered as an integral component of communicative competence, and is extremely important for learners to improve their communicative competence and to achieve the desirable proficiency in their language learning. However, research shows that Chinese college students' grammatical competence is not satisfactory.

Li Qi's investigation of English major students in the North-eastern Normal University revealed that although the participants possessed solid grammatical knowledge, they lacked high level of accuracy and native-like appropriateness of actual use in given contexts (23). Kong Yan's error analysis of Chinese students' acquisition of English passive voice showed that for English learners of intermediate and advanced levels, despite lower error rates in passive structure, the error rates in use, especially in discourse, are still high (43).

To help develop college students' grammatical competence, a new and effective approach to college grammar teaching is in urgent need, one which should be able to overcome the shortcomings of the traditional approach, and should teach the rules of the language along with grammatical rules. Under this circumstance, this paper, taking functional grammar as its framework and the teaching of English passive voice as an example, proposes grammar teaching of discourse-based approach.

2 LITERATURE REVIEW

2.1 *Different approaches to grammar teaching*

1 Traditional sentence-based grammar teaching
The traditional approach to grammar teaching is form-oriented and is restricted at the sentence level. When presenting a grammatical item, it usually takes no account of its meaning and functional aspects, and isolates the structure from both its social and linguistic contexts. Thus, it can leave learners with the impression that they know something, but in fact they have often learned a structural pattern without understanding its context, register, and general appropriateness.

2 Discourse-based grammar teaching
From the functional perspective, grammar is a general description of how language operates, a study of how syntax (form), semantics (meaning), and pragmatics (use) work together to enable individuals to communicate through language. Grammatical competence, as an integral component of communicative competence, should also be the aggregation of three dimensions, namely, grammatical form, semantic meaning and pragmatic use.

In contrast with the sentence-based traditional grammar teaching, a discourse-oriented approach acknowledges the indissoluble link between structures and functions in contexts and takes appropriateness and use as the heart of the explanation. In the discourse-based grammar teaching, teachers can teach target forms together with authentic or simplified discourse, which can supply learners with abundant examples of contextualized usages of the target structure to promote the establishment of form-meaning relationships (Li Qi 36).

2.2 *Different approaches to the passive voice*

The passive voice has attracted a great deal of attention in the linguistic literature. In the analysis of issues concerning the passive, different approaches have been adopted and various views about its form, meaning, and function has been offered.

1 Structural view of the English passive
Structuralism takes a purely formal approach to the passive. It treats passive structures in the framework of subject analysis, which views the passive as a structure in which the subject has the meaning of the undergoer of an action and occurs in the pattern of auxiliary be plus the past participle of the verb. This rigid approach to the passive fails in many respects and has received a lot of criticism.

2 Transformational study of the English passive
In dealing with the passive, Chomsky, the founder of the well-known transformational-generative grammar adopts the traditional idea of an active-passive correlation in terms of passive transformation. He insists that grammar rules can produce grammatical sentences and the passive is derived directly from the active and is thus analyzed in light of the passive sentences themselves with no consideration of the contexts in which they occur.

3 Functional study of the English passive voice
Functional study of the English passive voice is based on Halliday's functional grammar. Functional grammarians think that the active and the passive have the same ideational and interpersonal functions, but they are different in textual function since the focus of the speaker varies with the voice he selects and that, though the passive is not directly generated from the active, there exists an indirect systemic relation between them. What is more, both linguistic and extralinguistic, contextual elements are taken into account in their study of the passive with a view to discovering the motivation behind each passive that is used in a given text or utterance.

In the following part, English passive voice will be introduced from the perspective of functional grammar in terms of its use, to show how to carry out a discourse-based teaching of grammar.

3 TEACHING OF THE PASSIVE VOICE UNDER FUNCTIONAL GRAMMAR

In functional grammar theory, English voices are divided into two categories: the middle and the non-middle. "A clause with no feature of 'agency' is neither active nor passive but middle. One with an agency is non-middle or effective in voice" (Halliday 168). If the clause is non-middle, since either participant of the two (Medium and Agent) can become Subject, there is a choice between active and passive. The reasons for choosing passive are as follows: (i) to get the Medium as subject, and therefore as unmarked Theme; (ii) to make the Agent either a) latest news, by putting it last, thus achieving the end-focus and prominence; or b) implicit, by leaving it out, by which the process itself becomes focus information, thus being strengthened.

The above-mentioned reasons are the general functions of PV analyzed on its syntactical level. In the real discourse, the use of PV may be affected by many factors like the discourse producer's intention, the textual factors, and pragmatic factors as well. Its functions are briefly summarized as follows:

1 Syntactic function: the stress of the important part of a sentence
   a. To stress the Agent by using the passive structure with an explicit Agent. See examples (4.4) and (4.5)
   b. To stress the process by using the passive structure with an implicit Agent. See example (4.6).

(3.1) A: How was the dam damaged?
B: It // was damaged *by the flood.*

(3.2) A: The flowers are all gone. What happened to them?
B: They // were destroyed / *by the rain.*
Given New Focus

(3.3) And until recently this hostile attitude towards daydreaming was the most common one. Daydreaming // <u>was viewed as a waste of time</u>. Or it // <u>was considered an unhealthy escape from real life and its duties.</u> But now some people are taking a fresh look at daydreaming. Some think it may be a very healthy thing to do.

– taken from *Daydream a little,* Para. 2, Unit8 (text A, Band 3)

2 Textual functions: Keeping the textual topics; introducing a new topic or changing topics; textual cohesion and coherence.

Firstly, by application of the passive voice, comprehension of the semantic entailment of the textual topic could be achieved. In developing a text, to achieve thematic prominence in the textual layer, the writer often, by thematization, reorganizes the elements in a clause, and put what is wanted as the textual topic at the theme position. The passive voice is just such a grammatical device for realizing such a function.

(3.4) ...*Mussolini, Hitler's fellow fascist dictator and partner in aggregation,* had met his end, *and* it had been shared by *his mistress, Clara Petacci.* (topic: Mussolini and his mistress) ...*They* (Theme/topic) // had been caught by Italian guerrillas on April 27 while trying to escape to Switzerland and executed after a brief trial. On the Saturday night of April 28 *the bodies* // were brought to Milan in a truck and dumped on the town square. The next day *they* // were strung up by the heels from lampposts and later cut down so that throughout the rest of Sunday, *they* lay in the gutter. On May Day *Benito Mussolini* // was buried beside his mistress in the pauper's plot of a Milan cemetry. In such a horrible climax of degradation *Mussolini and Fascism* (Theme / topic) passed into history.— taken from *The Death of Hitler*, Para. 2 & 3, Unit9 (text B Band 3)

Secondly, by the alteration of the information structure and that of the information focus, the introduction or the transformation of the topics is achieved.

(3.5) The Three Gorges Dam, which is the biggest construction project in China since the building of the GreatWall and the Grand Canal, has been built to control flooding and provide hydro-electric power for the central region of China. The dam is nearly 200 meters high and 1.5 kilometers wide. *It* is the largest hydroelectric power station and dam in the world and has cost more than any other construction project in history.

Thirdly the passive voice could play the role of combining and linking the text, that is, promoting the continuous development of the topic.

(3.6) *Concrete* is produced by mixing together cement, water, and mineral aggregates. *This mixture* is placed into suitable mold, compacted, and allowed to harden.

3 Pragmatic functions

a. The euphemistic tone of the speaker or writer

The passive may be used as a way of deliberately obscuring who is doing what to whom. The passive had better be employed when the speaker does not wish to mention the actor and when it is inconvenient to mention the actor in consideration of certain communicative strategies, interpersonal relationship as well as politeness. See the following three examples. Compare and make choice under the given situations (the situations are omitted here).

(3.7) a. The dishes must be washed up./ b. You must wash up the dishes.

(3.8) a. You are requested to give a performance./ b. We request you to give a performance.

b. Formal and objective statement of the fact

The passive is resorted to when one wishes to use an impersonal style, for instance, in science writing and news report, in which the agent is left implicit and the process is stressed. See the following examples:

(3.9) Concrete *is produced* by mixing together cement, water and mineral aggregates. This mixture *is placed* into a suitable mold, *compacted* (impacted) and allowed to harden. It is similar to building stone, but has the advantage that it *can be* easily *molded* into any suitable shape, and also that it *can be* conveniently *reinforced* with steel rods to improve its structural properties.

(3.10) News Item

Japan (1) Police have arrested the man reported to be the second highest leader in the Aum Shinri Kyo (Japanese AUM doomsday organization) religious group. (2) The man is the fifth leader of the group to be arrested on a number of charges (accused of). (3) The group's leader is still missing. The group was suspected of the nerve gas attack in Tokyo's underground train system last month. (4) However, the charges against those arrested are not directly linked to the attack. (5) The group had denied any link. (6) Also in Japan, as many as 300 people became sick after smelling a poisonous gas in the main railway station in Yokohama. (7) No serious injuries were reported. (8) Police say they do not believe the incident is connected to the attack in Tokyo

—VOA Special English

4 CONCLUSION

Discourse-based grammar teaching is an important component of recent approaches to grammar teaching (Nassaji & Fotos 128). Different from the traditional teaching method, in the discourse-based approach, teachers can teach target forms together with authentic or simplified discourse, which can supply learners with abundant examples of contextualized usages of the target structure to promote the establishment of form-meaning relationships. As is demonstrated in the teaching of passive voice in the paper, only presented in discourse, can the various functions of PV be fully explained by the teacher and be deeply and really appreciated and understood by the students. And only when grammar is used in discourse, the appropriateness of the use of PV can be achieved.

REFERENCES

[1] M.A.K. Halliday. *An Introduction to Functional Grammar* (2nd ed.)[M]. Beijing: Foreign Language Teaching and Research Press / Edward Arnold (Publishers) Limited, 2000.

[2] Li, Qi. A Survey of Grammatical Competence of English Majors[M]. North-eastern Normal University, 2006.

[3] Kong Yan. Error Analysis of Chinese Students' Acquisition of English Passive Voice[M]. Central University of Nationalities, 2006.

[4] Nassaji, H.& S. Fotos. Current developments in research on the teaching of grammar[J]. *Annual Review of Applied Linguistics*. 2004 (24):126–145.

[5] Zhai Xiangjun, etc. *College English Intensive Reading Book3*[M]. Shanghai: Shanghai Foreign Language Education Press, 2006.

Management, Information and Educational Engineering – Liu, Sung & Yao (Eds)
© 2015 Taylor & Francis Group, London, ISBN: 978-1-138-02728-2

Study on the talent training model of safety engineering specialty based on career orientation

Xiao Yun Liu, Guan Hua Liu, Yang Jie Xiong, He Sheng Wang & Zhang Qi Xia
College of Resources and Environmental Engineering, Wuhan University of Science and Technology, Wuhan, China

ABSTRACT: This paper aims to strengthen the pertinence of safety engineering professional training, and ease the contradictions of industry security talent shortage, through methods such as literature research, data access and data statistics, based on clarifying the status of professional development and training of safety engineering in China, from the two dimensions of employment and employment number of students majoring in safety engineering employment market demand. We built in the talents training goal of "guide", "interactive" theory and the practice teaching system, "collaborative" talent training methods as the main content of the employment guidance of safety engineering personnel training mode.

KEYWORDS: Career orientation; Safety engineering; Talented person cultivation; Market demand.

1 INTRODUCTION

The upgrading of safety engineering disciplines effectively promoted the reform of the safety engineering education system and the innovative of the personnel training mode which is of great importance to broaden the field of safety education and directions of student employment [1]. With society and economy developing, the need of safety engineering professionals has extended from traditional mining, metallurgy industry to oil, chemicals, machinery, construction, transportation, information and other fields [2]. Data show that in 2014 there are up to 7.27 million of university graduates, the difficulty of graduate employment has shown normalization. Whether safety engineering can withstand the impact of the employment pressure will be key to its survival and development [3]. Based on this, the analysis of safety engineering discipline development, personnel training, market demand, graduates employment situation, building safety engineering professional training mode, which has employment oriented industries to alleviate the shortage of safety professional conflicts, to promote the realization of the national safety production situation fundamental improvement to provide personal support and intellectual protection is of very great significance.

2 THE DEVELOPMENT OF SAFETY ENGINEERING DISCIPLINE AND TALENT TRAINING STATUS IN CHINA

2.1 *The development of the discipline of safety engineering*

1 Slow development period. The development of safety engineering major in China started in the 1950s, from 57 to 90s development has been slow. Xi 'an Institute of Mining in 1957 in the country opened the "mine ventilation and safety". In 1958, Beijing Institute of Labor opened the "industrial safety technology" and "industrial hygiene technology" undergraduate majors. In 1982, at the university of Hengyang-the affiliated school of Hunan University, setting safety engineering three-year specialist. In 1983, Shenyang Institute of Aviation Industry setting safety engineering four-year undergraduate programs, China university of Mining set up domestic mine ventilation and safety undergraduate majors for the first time. During this period, safety engineering, while development is slow, but still has trained a large number of professional and technical personnel, laid a solid foundation for the development of safety engineering.

2 Rapid development period. In 1993, the national standard "discipline classification and code" began to implement, the implementation name the safety science and technology as the standard of class discipline, breaking the natural science and social science in subject classification, set up "management, safety and environmental" comprehensive disciplines [4]. In 1998, the ministry of education promulgated the history of our country's higher education for the fourth time to modify the catalog of major of undergraduate course of common colleges and universities, and further integration of undergraduate specialties, cancelled the mine ventilation and safety professional, by safety engineering to cover all the safety problems of the industry, as a primary secondary under safety and environmental science disciplines, a subject of science and technology. After ten years fast developments in the 1990s, the scale of safety engineering began to develop fast.

3 Golden development period. In the 21st century, security engineering was in the golden period of development. In 2002, with the implementation of the Production Safety Law of the People's Republic of China, the enterprise demand for safety production management has increased dramatically, universities set up the number of safety engineering and related professionals also began to increase significantly. At the same time, on a safety engineering specialty in colleges and university curriculum reform, in order to adapt to the needs of society, more conducive to the cultivation of applied talents. In 2011, the state council degree committee, the ministry of education take "the safety science and engineering" as a graduate student discipline education level, the development of academic disciplines, which is of great significance to further advance security optimization security personnel knowledge structure, to speed up cultivating high-level security talents [5].

2.2 *Safety engineering talents cultivation conditions*

Because safety discipline is a very wide range of comprehensive interdisciplinary sciences, a lot of colleges and universities set up different types of safety engineering, including military industry, chemical industry, petroleum, mining, metallurgy, aviation, civil engineering, transportation, energy, environment, economy, etc. [6]. The Table 1 shows that since 2000, the numbers of new start safety engineering specialty in colleges and universities has been rising trend, the numbers has reached 153 now.

At present the national recruit safety engineering doctoral about 280 per year, about 1000 master's degree students. Safety engineering talents cultivation had already formed the bachelor, master and doctor multi-level, three-dimensional talent cultivation system. Set up safety engineering colleges and universities of safety engineering of the nation's provinces (municipalities) distribution from the point of geography, in addition to Hainan, Qinghai, Tibet, open safety engineering college of provinces and municipalities and autonomous regions of 28; From the point of quantity, Shandong, Jiangsu, Hubei, and Liaoning colleges and universities set up safety engineering quantity is more, it is 10, 11, 10, and 12.

3 THE MARKET DEMAND ANALYSIS OF SAFETY ENGINEERING STUDENTS' EMPLOYMENT

3.1 *The classification of job market demand*

1 Government department. At the national level, our country has the Production Safety Supervision Administration and the State Coal Mine Safety Supervision Bureau. Meantime, every province has Administration of Work Safety which bears the supervision and management of safety production in different industries. With the growing government's emphasis on safety production and gradual standardization of safety supervision and regulatory system, the government management of security has become a relatively stable industry. Nowadays, those people who want to work in security management departments mainly go through the public recruitment of the national civil service examination. This examination requires that candidates should have high comprehensive qualities and fully grasp the laws, regulations and industry standards, technical specifications in the security field.

2 Enterprise. The State Safety Production Law clearly indicates that if production units employ more than 300 people, they should set up a security management agency or employ professional security managers. If under 300 people, they should employ full-time or part-time security managers. On the one hand, enterprise internal security management requires relatively stable managers. On the other hand, hidden risks prevention in production processes within the enterprise also requires safety production technicians. In order to accelerate enterprises' development and enhance the competitiveness of enterprises, companies should attach importance to safety work and employ more and more security managers[7]. In other words, safety engineering students mainly work in enterprises.

3 Intermediary agency. Government departments often rely on intermediary agencies (third party) to evaluate safety production condition in the enterprise. Meantime, the government has strict requirements about the qualification and operation management of safety assessment intermediary agencies. These agencies need employees majoring different profession as well as conferred of National

Table 1. The numbers of safety engineering university in China.

Year	1957	1985	1990	2000	2002	2004	2006	2008	2010	2012	2014
number	1	16	17	30	45	68	85	114	127	139	153

Safety Appraiser. We can select a number of safety appraisers in the engineering field to organize a team to appraise enterprises' safety production condition, thus a relatively stable security assessment industry forms.

4 Universities, research institutes and etc. These institutions undertake the analysis and study of key technologies and theories, including the existing state of hidden risks, analysis and research the mechanism of the accident, the accident prediction and control, and accident damage assessment. Select a project from the practical production, after in-depth study, the obtained results are applied in practical production, to solve various problems in safety production. With the party and government attaching great importance to safety production work, constant evolving of safe disciplinary and economic society, universities, research institutes and other organizations will need more specialized superior safety professionals.

3.2 *The number of the market demand*

1 Statistical analysis of working field of graduates. The field of safety engineering Professionals is very broad. With the level of social development and the needs of different levels of safety professionals varying, the main areas of employment are also subject to change. As a result of the different colleges possess different safety engineering features, its field of employment is also different. Take 118 graduates who graduate from Wuhan University of Science and Technology from 2009~2011 employment data for example, from the employment industry distribution,72 graduates went to construction enterprises (China State Construction Engineering Corporation, China Railway Engineering Corporation, China Communication Construction Corporation, China National Nuclear Corporation, etc.), accounting for 61%; 15 chose Corporation Mining companies (China National Petroleum Corporation, Wuhan Iron and Steel Mining, Henan Coal, Jiangxi Jin shan gold deposit, etc.), accounting for 13%;19 went to other companies (China Construction Bank, Gree, Kai bang Motor, China Resources Gas, etc.), accounting for 16%;12 to Administrative institutions (Ningbo Municipal Group, the Second Artillery Corps, Hubei tobacco, Three Gorges Navigation Authority), accounting for 10%.Properties of employment units are mainly state-owned enterprises, accounting for a higher proportion. The number of graduates into the mining business and Administrative institutions shows a declining trend, the number of graduates employed in other business units increase yearly. From the geographical point of view of employment, graduates mainly work in Hubei, Guangdong and Jiangsu and Zhejiang. Among them, 40 graduates in Hubei (Wuhan, Yi chang, Shi yan, Xiang yang, etc.), accounting for 34%;20 in Guang dong (Guang zhou, Shen zhen, Zhu hai, etc.), accounting for 17%;14 in Jiangsu (Nanjing, Lianyungang, Zhen jiang, etc.), accounting for 12 percent, Zhejiang (Hang zhou, Ning bo, Zhou shan, etc.) 9, accounting for 8%.Visible,enrollment base and economically developed areas and urban centers are still the priorities of graduates.

2 Statistical Analysis about employing unit's position for demand .According to China's largest search engine's statistical information, alumni set in August 2014, the position of safety engineering requirements of the top 10 cities, nine cities position requirements are more than 100, Beijing, Shanghai, Shenzhen ranking top three. Beijing has the biggest demand for 516, as shown in Figure 1.

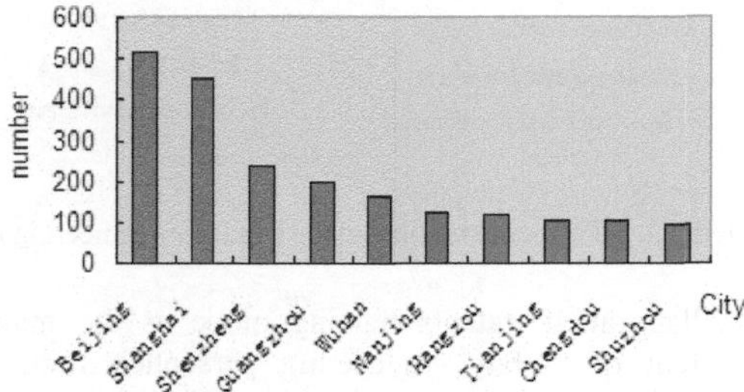

Figure 1. Safety engineering requirements of professional position chart.

Using Wuhan city as an example, according to a 2013 yearbook data in Wuhan city, according to the regulations of the state, there are a total of 1001 enterprises which are staffed with full-time security personnel. In addition there are all kinds of big companies a total of more than 40000 which don't need to has to be stuffed with full-time security personnel. According to the regulations of the state, enterprises above designated size must equipped with full-time safety-tech-management personnel, the safety engineering professional talent gap of up to 3000 people. The output of graduates of the safety engineering specialty which has been set up in Hubei's colleges and universities is less than 1000 people a year. Apparently, the gap of demand for the safety engineering professionals will not change in short term.

4 THE EMPLOYMENT GUIDANCE OF SAFETY ENGINEERING PERSONNEL TRAINING MODE'S CONSTRUCTION

Based on the above analysis, this paper tries to build the "Guide" to the talent training goal, the theory of the "interactive" and the system of practice teaching,

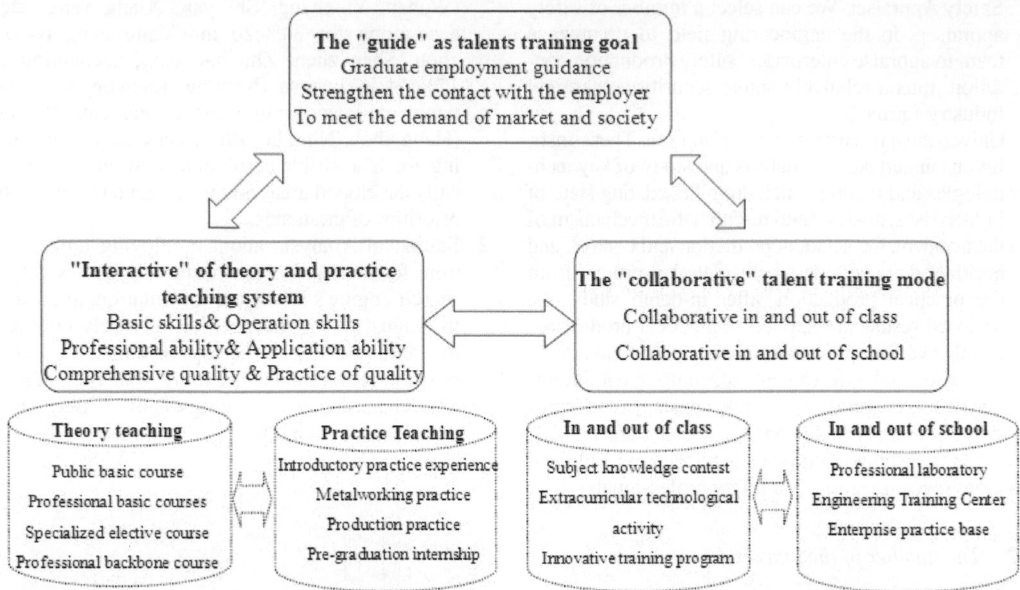

Figure 2. Personnel training mode of safety engineering major based on employment orientation.

"collaborative" talents training mode as the main content of "safety engineering personnel training mode based on employment guidance", as shown in Figure 2.

4.1 *Set up the "guide" as talents training goal*

Talent training objective is the aim and key of personnel training, which plays an affecting role in talent cultivation system, belonging to the first link Multidisciplinary cross and blend of industries are the features of the safety engineering. For different industry, different social and economic development stage, different environment's conditions, the requirements of safety and conditions of guarantee are very different. Safety engineering professionals must have a unique professional quality and engineering ability, in order to meet the needs of social and economic development [8].

Oriented talents training goal is to regard graduate's employment as guidance, realize the docking with unit, to meet the needs of market and society, improve safety engineering students' employment ability, to strengthen the pertinence of talent training. In training scheme, curriculum setting, discipline direction, focus, platform construction, etc., with the employment as the guidance, training the persons who can satisfy the needs of the development of regional economy and senior engineering and management. In particular, some schools which open safety engineering professional recently determine the scale of personnel training on the basis of characteristics and teaching conditions and the location factors, not blindly following the trend, going with the flow.

4.2 *Build the theory of "interactive" and the practice teaching system*

Theory and practice teaching system construction, which should be on the basis of the labor market demand for safety professional ability, unit of choose and employ persons (basic abilities and operation skills, professional skills, comprehensive ability and practice ability and application) to highlight the characteristics of "interactive". By making feasible theory and practice of teaching evaluation system, rely on the industry to identify the core competence of professional requirements and various abilities and skills of the inspection standards, being strict on appraisal system, to ensure the quality of teaching.

"Interactive" teaching theory and practice, is to build the "public basic curriculum & professional foundation courses and professional courses& professional curriculum" to be the main course of theory teaching and "metalworking practice & cognition practice & production practice & graduation practice" as the main content of practice teaching system of interactive teaching, promoting the interaction between the students' knowledge and skills of ascension, the interaction between intelligence and ability. The theory of "Interactive" and the practice teaching system will be the focus of teaching reform to improve the current theory and the practice teaching link, improving students systematically using knowledge, the skills to solve problems and the ability to adapt, to avoid cultivating the students who are unfit for a higher post but unwilling to take a lower one and have difficulty living well in society. Through interactive teaching, making the effects to cultivate the students' core competitiveness of employment.

4.3 *Use "collaborative" talent training mode*

Collaborative talent training mode, is the strength of talent training, method and platform effectively integrate and full of the resource elements such as polymerization, advantageous to realize the superposition effect of talent training, to improve the quality of personnel training. To implement security engineering students "collaborative" talent training mode, explore the in-class and after-class collaborative, schools and enterprises. Classroom is given priority to with theoretical teaching, extracurricular is given priority to with practical teaching, the school focuses on basic ability training, enterprise focusing on development and improve the comprehensive ability.

Formed by the subject knowledge contests, extracurricular activities of science and technology, the innovation training program of innovation ability cultivation system together. Fully grasp the "earthquake disaster reduction day", "national safety production month", "11, 9" on fire, and other important time node, through academic lectures, social practices, safety classes of the community, such as form, let the students take an active part in these activities, guides the student to walk into society, understanding of production safety, cultivate their consciousness of security responsibility. Through professional characteristic laboratories, engineering training center, the teaching practice base and so on the synergy of subject platform and make the students in the process of practice gradually form a good professional habit, cultivate the students' professional quality, asked the students to the standard requirement of enterprise employees themselves, train the ability of effective engineering.

5 CONCLUSION

1 The development of the discipline of safety engineering major has entered into the golden stage of development, professional set up in colleges and universities showed a trend of increasing year by year, the number of personnel training and quality continuously strengthen.
2 The government departments, enterprises, intermediary organizations, universities and research institutes is the main safety engineering graduates employment direction, the enterprise is the main body of safety engineering professionals demand. Safety engineering talents cultivation quantity cannot meet the demand of market supply and demand contradiction ease and still I need to ensure the quality of talent cultivation under the condition of further safety engineering personnel training scale.
3 This article constructed in the talents training goal of "guide", "interactive" theory and the practice teaching system, "collaborative" talent training mode as the main content of the training mode, to set up safety engineering specialty in colleges and universities education teaching reform has provided the beneficial reference and exploration, to relieve the contradictions of industry security talent shortage, in order to promote the national production safety situation to achieve a fundamental improvement in the talent support and intellectual protection is of great significance.

REFERENCES

[1] Hua Li, Jiang-ping Zhao, Xiao-hong Cui. Study on the telent training model of the safety engineering specialty [J].Journal of Safety Science and Technology, 2012, 8(8):143–146.
[2] Hong-jie Zhang, Xiao-dong Xiang, Wang-sheng Chen. Safety engineering teaching methods reform and innovation talents training[J]. Journal of Safety Science and Technology,2009,19(12):111–114.
[3] Chao Wu. Initial Study of the Science of Safety Science[J]. China Safety Science Journal, 2007,17(11):1–11.
[4] [EB/OL].http://www.gov.cn/gzdt/2011–06/01/content_1874759.htm,2011–06–01/2014–07–15.
[5] Jian-chun Yang, Jing Wu, Qiong Wu, et al. Market demand and talents training mode of safety specialty[J].Journal of Safety and Environment, 2006,6(z1): 172–173.
[6] Kai Wang, Shan Li, Xia Pan, et al. Analysis on current situations of the development for higher education of safety engineering discipline in China [J]. Journal of Safety Science and Technology,2012,08(5):172–173.
[7] Xiao-wei Zhai, Jun Deng, Zhen-min Luo, et al. Discussion on Monder Enterprises' Requirements for Safety Engineering Personnel and Methods of Students Cultivation[J].China Safety Science Journal,2007,17(4):111–114.
[8] Hui Shao, Dong Zhang, Xiu-kun Ge, et al. Study and Practice on High-quality Professionals'Training Mode of Safety Engineering Specialty[J].China Safety Science Journal,2009,19(12):111–114.

Foundation Project: Teaching research project in Colleges and universities of Hubei Province (2013227);The students' science and Technology Innovation Fund Project of Wuhan University of Science and Technology (13SHA166).

Management, Information and Educational Engineering – Liu, Sung & Yao (Eds)
© 2015 Taylor & Francis Group, London, ISBN: 978-1-138-02728-2

The compared research of modern sports teaching mode in China and the USA

Lin Jie Wei
Xi'an University of Architecture & Technology, China

ABSTRACT: This article lists the Chinese modern physical education modes: "Three basic" teaching mode; enlightenment teaching mode and comprehend teaching mode, compared with the health-optimizing physical education mode and sports education teaching mode, which have been popular in the United States in recent years in order to find the gap and promote the development of modern physical education mode in China.

KEYWORDS: China; America; physical education mode; compare research.

For a long time, China followed the traditional sports teaching mode that was dominated by teachers: the teaching method is single, the teaching effect is poor. It damages the enthusiasm and initiative of students learning, and causes students to "prefer sports to physical education classes".

1 ANALYSIS OF THE AMERICAN MODERN MODEL OF TEACHING

In 1910, Clark Hetherington earned the title of "father if modern physical education" with his landmark paper "Fundamental Education." Hetherington described both the scope and the categories of the new physical education. Hetherington's four phases became the four primary objectives of the new physical education: physical development objective; motor development objective; mental development; social development objective. Lesson plans and the model of teaching organized around the four objectives quickly became the standard in the physical education curriculum in schools. Typically, the Health-Optimizing Physical Education Mode and the Sports Education Mode.

1.1 *The health-optimizing physical education mode*

Health-Optimizing Physical Education (HOPE) has as its primary goal that children and youths develop and value a physically active lifestyle. In HOPE mode, physical educators typically have an expanded role that goes beyond their responsibilities to plan and teach the physical education classes. HOPE mode typically aims to help students learn ways of managing their physical-activity behavior through planning, goal setting, self-monitoring, self-reinforcement, and resisting negative influences that prevent engagement in physical activity. It makes students not only in physical education classes, but also at the rest time, which student in the school will engage in physical activity.

1.2 *The sport education mode*

Recently, the Sports Education mode has already entry into the United physical, educational curriculum and instruction literature (Siedentop, 1994; Siedentop et al., 2011. The sport Education mode is based on the assumption that good competition is both fun and educationally useful. It has five defining characteristics that distinguish it from more traditional forms of physical education: Sport Education is divided into seasons that are longer than typical education units in a multi-activity program; Students are organized immediately into teams, and they retain that affiliation throughout the season; Seasons are built around a series of competitions that grow increasingly complex as students master the techniques and tactics involved in the activity for that season; Accommodations are experiences by employing graded competition and contests pit teams of similar skill levels; The season ends with a culminating event that not only determines the seasonal champion but also provides a festive way to conclude the experience; Records are kept throughout the season so that students and teams can mark their progress.

In Sport Education, teams are organized so that each team has a mixture of more and less-skilled students. Each team member has a role to play in ensuring that the team performs well and that the season is a success. In addition to employing a smaller team sized, the mode also encourages teachers to modify the conditions of play to help students acquire techniques and tactics.

2 ANALYSIS OF THE CHINESE "THREE BASICS" TEACHING MODE

2.1 *"Three basics" teaching mode*

"Three Basics" teaching mode is the most common mode that used in Chinese sports teaching. "Three Basics" teaching mode is central in learning basic sports knowledge, technology and skills, and follow the law of students' cognition and skill forming, divided the teaching course into perception, understand, consolidate, application and so on. Teaching process follows the educational thought of the former Soviet Union educator Kelof. The Chinese sports education aim is pass on the Basic knowledge, basic technology and basic skill, that is "Three Basics".

After the teacher raising the teaching content, object and task, the teacher using some teaching methods to promote students' perceptual knowledge of learning content, form the visual imagery. Students will establish action kinesthetic representation, the correct muscles' feeling and skill formation through the imitation and representation practice. Then, teacher summarizes the content of the learning, evaluation and points out the problems and teaching feedback effect. "Three basic" teaching mode emphasizes the teacher's dominant and control function, teacher is the soul of the teaching and the main power source. Teacher arranging the teaching aim, task, requirements, practice time, speed and rate, and evaluation, students will successfully complete the study task as long as take the teacher's command and requirement.

2.2 *Comprehend teaching mode*

Comprehend teaching mode by trying to start learning (understand) movement technology, changing the defect of pursues skills, but ignore students' awareness of the whole sports and grasp of the characteristics of the movement to improve the teaching quality. Comprehend teaching mode arises the idea of that the students try to understand the importance of learning sports skills, using integrity teaching before use decompose teaching mode, then organize the teaching activities in the form of competition in order to improve the students' learning enthusiasm. The operating procedures are: introduce project—tentative competition—find the problem—technology teaching focus on the problem one by one—integrity practice—practice again. Comprehend teaching mode is suitable for all kinds of ball games.

3 ANALYSIS OF THE MODERN TEACHING MODE IN THE USA AND CHINA

3.1 *The analysis of the teaching aim and guiding ideology*

Chinese ministry of education promulgated by the national general school sports working guiding points out that: School physical education curriculum is to promote the healthy development of students, using the physical exercises as the main means, enhancing the students' constitution, improve their health, promote the harmonious development of students' body and mind through the reasonable physical education and scientific physical training process.

The guiding ideology of the United States school physical education emphasis on the education—through—the physical mode started with optimal individual development within a democratic social framework as its primary value orientation. Chinese guiding ideology focuses on the students' individual wellness while the United States see the students' development as the center of physical education.

3.2 *The analysis of the operation process*

Wu Jian analyzed nearly 100 cases of physical lessons over Chinese country. He comes to the conclusion that the order of Chinese physical teaching course is that: Preparation part —the main teach and exercise — relax. We can deduce that the teaching process is designed as the main line of the forming the movement and the exercise load. Preparation part encloses the introduce, warming activity, game and so on. Basic part is the core of the teaching, teacher arranging different teaching content according to the task, choosing the teaching method, and guide students get the movement skill. Ending part usually have physical fitness exercise, relax and conclusion.

The United States physical class is starting from the teacher introducing the class's content. From the warming up, the teacher will arrange warm up exercise around the teaching content or directly into the main content without a warm up. They think that some teaching content doesn't warm up because the tension of the activity won't injure the student and some activities have the effect of warm up itself. We hardly find the preparation, basic and ending structure in the United State physical class, but the teaching purpose and content consistently throughout the whole class.

3.3 *The analysis of teacher and student relationship*

Chinese physical teaching mode emphasizes the class is the basic organizational form of teaching, it fully affirmed teacher play a leading role in the teaching course. They decided the teaching content, teaching method and evaluate students. The student will get the exercise knowledge and skill as quickly as possible and little detours. However, student in the teaching process is subordinate position, they passively accepting knowledge is badly to cultivate their personality, ability and interest. There are bigger difference comparing with the American student participate in the teaching design and planning. The students have higher enthusiasm and creativity into the course of learning.

In the United State physical teaching, teachers pay more attention to training students' potential, self-learning and self-design ability, they also give priority to the learning and emphasize " taking students as the center" and "interesting center". Teacher arranging suitable activity for students who made the learning aim and program by themselves and freely seek and create the activity apply to their style, and self-assessment of learning effects. Students are both the learner and the policymaker, they are completing the teaching task through their autonomous learning activities such as exploration, discovery, and communicate with each other. Teachers are guiding the process, but mainly inspire and encourage the student learning.

4 EPILOGUE

In conclusion, the difference between China and the United States is dramatic. Although Chinese sports, education has some significant changes after reform and development in recent years, but we still need to learn some advanced experience in teaching philosophy, teaching content setting, teaching subject, teaching methods and operating procedures to make the physical education apply to the rule of students' body and mental development. We should take the student as the teaching body, fully arouse the students' enthusiasm, creativity and social intercourse ability, at the same time of trains the student to have a healthy body, give full play to the sports teaching in shaping the students form a complete personality of an irreplaceable role in the process, make sports become an indispensable part of the students' life.

REFERENCES

[1] Daryl Siedentop, Hans van der Mars Introduction to Physical Education, Fitness & Sport eighth edition 2011.p57.

[2] Jianyu Wang, Wei Bian, Aifeng Huang, The comparison of the school sports teaching materials content, [j] Sports Journal, 2010.9p62.

[3] Neil J. Dougherty. Modern Sports Teaching Theory. 1987.14.

[4] Shuzhi Wang, The comparison of sports teaching mode between China and Foreign country.[j] Scientific and Technological information.2007.35.p236.

Management, Information and Educational Engineering – Liu, Sung & Yao (Eds)
© 2015 Taylor & Francis Group, London, ISBN: 978-1-138-02728-2

Network information foraging behavior strategy of virtual scientific research team members

Qi Wang & Wen Yong Chen
Library, Jilin Agricultural University, Changchun, China

ABSTRACT: By investment-revenue analysis, the article discussed network information foraging strategy of a virtual scientific research team, and expounded the behavior mechanism of the most favorable information source, the optimal information spectrum, the most favorable information patchy and so on, and the significance of information behavior.

KEYWORDS: Virtual scientific research team; Information foraging behavior; Information foraging strategy.

1 INTRODUCTION

Information is an important resource element which the virtual scientific research team requires. To a great extent, the acquisition time, cost, and obstacles of useful information affect the users' willingness and satisfaction. When mankind entered the network-centric world, they had been drowned by the huge amount of information they created. "Information explosion" or even "information overload", which make users encounter more difficulties and challenges when they seeking and utilizing information. Although information can be transformed into knowledge, and then it can be into behavior, decision-making ability, whereas in the chain of information to knowledge to capabilities, information is the starting point, our primary task is to obtain information on the behavioral decision. Because of the complexity of the network information environment, we must take appropriate information foraging strategies to obtain information. One purpose of researching information foraging is to explain and predict how people taking the most optimal information foraging strategies to change their behavior to adapt the changing of information environment.

2 INFORMATION FORAGING THEORY

The discussed model is given as follows:

In the 1970s optimal foraging theory was first developed by anthropologists and ecologists to explain how animals hunting for food. It suggested that the eating habits of animals revolve around maximizing energy intake over a given amount of time. For every predator, certain prey is worth pursuing, while others would result in a net loss of energy. In the foraging process, animals need to constantly evaluate the predation food contained energy and their consumption energy, by selecting different environment and different food to optimize profit, the assessment results determine the animals remain to pray there or to find another one.

In the early 1990s, Peter Pirolli and Stuart Card from PARC noticed the similarities between users' information searching patterns and animal food foraging strategies. Working together with psychologists they analyzed users' actions and information landscape that they navigated (links, descriptions, and other data), they showed that information seekers use the same strategies as food foragers. People also need to achieve optimal balance between time, money, energy and the required information during searching for information. Therefore, they first proposed the information foraging theory [1], and pointed out that in the process of searching and absorbing information, people often need to constantly adjust their information foraging strategy according to their information environment to maximize information earnings.

Information foraging theory consists of three major components: information patches, information scent, and diet information.

Information foraging is the process of people seeking, acquiring, and absorbing information. Information foraging strategy refers to various methods and measures which people adapt to obtain maximum information foraging efficiency. Information patches are an aggregate with a rich information resource as a physical metaphor in the information environment, it can be a website, a paper, a book, a web and a collection of documents people seeking information through them. Information scent is a conceptual extension of search scent; it is a subjective evaluation between the information and the information correlation during the information seeking process, with a navigation action determining the information seeking methods. In short, the role of information scent helps people step closer to the information they require. In the process of searching for Internet information, people use information related marker to find the best information patches. Information diet refers to the kind of foraging information. Because of the regularity of network information resource distribution, and time, money and energy, etc. scarce resources people spending in foraging information, people face how to select information in seeking information. If they seek scope is too narrow, it will make people spend more time to build a seek mode and increase the seeking results one-sidedness; if the selected range is too wide, people could be flooded with the retrieved information again. Thus, it's important that people select the appropriate information diet to search information under the networked information environment.

So in seeking information, we evaluate the strength of the scent (determined by relevancy), the index of patch richness (how much relevant information are we likely to find) and the distance of the patch (how difficult is it to get).

Information foraging theory is a theory which applied the animal optimal foraging theory to how people searching for information under the network environment. Its assumptions are: (1) Before seeking information, there was a "built-in" information foraging mechanism, which people judged the information (resources) by information classification based on the existing contents in their mind, then make different information foraging strategies on the basis of the specific tasks required information. By revealing, mastering and using this mechanism, it can be better to understand people's searching behavior in the network, and guide technical personnel to improve design quality and usability of a website and searching engines for any other users. (2) People's browsing behavior is guided by the information scent on the Webpage. Since Peter Pirolli and Stuart Card put forward the information foraging theory, PARC researchers developed detailed and systematic on information foraging theory [2–4], and researched the SNIF-ACT information foraging model (Models of information foraging). At present, the research on information foraging countermeasures are developing rapidly and deeply, it has become one of the hotspots of information behavior and information retrieval theory.

3 INFORMATION FORAGING STRATEGIES

3.1 *Increasing the net value of the information*

According to the information foraging theory, there are a variety of information resources in a network environment, and different information resources' quality, enrichment and distribution pattern are not identical. Because of the time, money and energy limitation, people choose information resources will play an important role in the gain information efficiency. Then, how people search for and digest the information in the network environment?

During the process of seeking and absorbing information, people need to consume time, money and energy obtain the necessary information, but also get some net value of information, the formula is:

Net value of information = total energy information-seeking energy dissipation-processing and absorbing energy dissipation

The ratio between the net value of information and information processing time is a measure of information resources advantage, in which the information processing refers to the time from seeking to absorbing [4]. The information behavior study found, people always choose the more advantageous information resources in the information foraging activities. According to the research of Gursoy and Umbreit [5], people try to choose the maximum net revenue of information resources in the network information foraging. For information seekers, they will spend more money on collecting information from the higher cost information resources, so the information net income is higher in unit search time; though low cost information resources will spend less money, the useful information resources are too little, so the information net income provided will be smaller. Therefore, people should choose middle cost information resources in the information foraging, its net income in unit searching time is almost the highest, and the advantage is also great.

3.2 *Enhancing the information density*

In people's information foraging activities, if the most advantageous information density in the information resources is small, information acquisition efficiency is low; on the contrary, the advantageous information density is large, the acquisition efficiency is higher, so people would choose some smaller advantageous information resources. It is because that if the most favorable information density and acquisition efficiency is low, the recognition time of people choose the most favorable information will be longer, acquired information net income of unit total processing time (total processing time, including recognition time and processing time) will be small; if the people have no strict selection on the information types, the recognition time will shorten and the information net income will increase. Under the network environment, people's information spectrum should include how much smaller advantageous types of information resources? Sandatrom, Agata and Spink have conducted in-depth investigation and discussion about this [6–8], they put forward the optimal information spectrum elements, respectively, the main points are as follows: ① if the advantageous information availability increases, the information category in the optimal information spectrum will decrease; ②during people's information foraging, recognition time and information average benefit will decrease with the expansion of information spectrum range; ③according to the information advantage and recognition time, the types in the optimal spectrum information could be presumed. Due to the need of information efficiency, people's unit time in taking information energy should be maintained at a certain level [9]. It can be speculated that: ① if there is enough favorable information opportunity to meet to ignore the poor information, then no matter the number of poor information will not affect the information foragers' specific selection of favorable information; ② if the favorable information number increases, information foragers will immediately transform various types information foraging into a single, favorable information; ③ the assumption that people have been foraging optimal information spectrum range of types of information resources, there are two kinds of new information resources of X and Y, when forage X, the intake information energy in unit total processing time more than meet the information value people demand; when forage Y, the intake information energy in unit total processing time below information numerical which people's information needs, people always forage information resource X, and refuse to Y.

Under the networked information environment, the optimal information spectrum should also include specific information or information elements (such as some data information, factual information and picture information etc.), which fulfills people's necessary information needs. Foraging this kind of information, cannot target information net income energy, but fulfill to increase people's knowledge or change the knowledge structure. In addition, the optimal information spectrum is usually with the change of information requirements and the networked information environment.

3.3 *Reducing the information foraging investment behavior adaptation*

Information patches is an important theory in information foraging. Information foraging theory considers [10], the network information environment presents patches structure and information patches is a spatial and temporal characteristics of cluster information. Like selecting the most favorable information resources, under the network environment, people will seek the most favorable information patches to forage, namely people prefer to foraging in big amount and good quality information.

People's information foraging activities usually cause the "reduction" of information patches' valuable information, or need more cost to obtain valuable information which make information patches availability "lower" by time [3]. Therefore, people should choose the appropriate time to stop information foraging behavior, or transfer another information patches to search for information in the condition of less expense or overcoming the low technical. What mechanism makes people transfer one information patch to another at the right time? In 2005 Ingwerson and Jarvrvo [11] from the cognitive view, pointed out: "like animal foraging behavior in nature, patches structure in network information environment has great influence on people's information foraging behavior, even a decisive role on. In the uncertain network information environment, people need to constantly assess the information seeking expected cost and information expected value to determine their information foraging behavior." Therefore, people stay or leave the information patches is determined by giving up time (Which is the longest time between obtaining useful information first time to the next time?). In the information patches with large enough information density, two information foraging time interval is short, no more than the giving up time, so people will not transfer; but with a reducing value of information density, the interval time gets useful information is gradually extended, once more than the giving up time, people will give up this information patches and transfer to another. Giving up time is related to the seeking time, information transfer costs and the difficulty degree of information foraging, it equals the time from the last obtaining the useful information to leave the information patches. In the "built-in" information foraging mechanisms, people may have a relatively fixed giving up time, as if there is an alarm clock after each information foraging, people will leave when they cannot forage information.

3.4 *Information sampling foraging behavior*

From the economics view, when people choice the most favorable information patches, they will reduce investment and improve efficiency of information foraging with the help of information behavior. So, how people learn the relative advantages of information patches? When people enter a new information environment for information foraging, they usually obtain information patches status in information environment by information sampling foraging (test retrieval) behavior, and make the most favorable information foraging decisions according to these information. Generally speaking, people focus on the maximum information density of information patches, when the most favorable information patches quality decline, people will transfer to the second favorable information patches to continue foraging information. In addition, people usually depend on the previous web or one time information seek behavior accumulating all kinds of information resource abundance on the network, and store relative advantageous of every information patches for use.

4 CONCLUSIONS

The information foraging theory development is very fast although it appeared not long. At present, China has begun to study the theory [12–13].After some general overview of the information foraging theory, this paper discussed the network information foraging strategy of virtual scientific research team by investment-benefit analysis method to cause the attention of people.

ACKNOWLEDGMENTS

The corresponding author of this paper is Wenyong Chen. This paper is supported by the Youth Foundation of Jilin Agricultural University (Grant NO. 201337), CALIS Programs of the National Agronomy Literature Information Center (Grant NO. 2014026).

REFERENCES

[1] Pirolli P, Card S K. Information foraging in information access environments[C/OL]. In Proceedings of the Conference on Human Factors in Computing Systems, CHI '95. New York: Association for Computing Machinery. 1995, 51–58 [2010-07-08]. http://www.sigchi.org/chi95/proceedings/papers/ppp_bdy.htm.

[2] Pirolli P, Card S K. Information Foraging Models of Browsers for Very Large Document Spaces[C/OL]. In the Proceedings of the Working Conference on Advanced Visual Interfaces, AVI '98, L'Aquila, Italy. New York: ACM Press. 1998, 83–93 [2010-07-08].http://citeseerx.ist.psu.edu/viewdoc/download?doi=10.1.1.25.2997&rep=rep1&type=pdf.

[3] Pirolli P, Card S K. Information foraging. Psychology Review, 1999, 106(4):643–675.

[4] Pirolli, P. Information Foraging Theory: Adaptive Interaction with Information. New York: Oxford University Press, 2007,30–46.

[5] Gursoy D, Umbreit W T. Tourist Information Search Behaviour: Cross-cultural Comparison of European Union Member States. International Journal of Hospitality Management, 2004, 23(1):55–70.

[6] Sandatrom P E. An optimal foraging approach to information seeking and use. The Library Quarterly, 1994, 64(4):414–449.

[7] Agata I, Kim S-Y. The application of optimal foraging to information seeking behaviour: Suggestion for measuring information gain. Library and Information Science, 1996, 35(1):51–57.

[8] Spink A, Cole C. Human information behaviour: integrating diverse approaches and information use.Journal of the American Society for Information Science and Technology, 2006, 57(1):25–35.

[9] Pirolli P. An Elementary Social Information Foaging Model. M A Boston (Eds.), In Proceedings of the 27th international conference on Human factors in computing systems. New York: ACM Press, 2009, 605–614.

[10] Chen C, Cribbin T, Kuljis J, Macredie R. Footprints of informationforagers: behaviour semantics of visual exploration. International Journal of Human-Computer Studies, 2002, 57(2):139–163.

[11] Ingwerson P, Jarvrvo k. Turn: Integration of Information Seeking and Retrieval in Context (The Information Retrieval Series. New York : Springer-Verlag, Inc. Secaucus, NJ, USA, 2005, 27–28.

[12] Yang Yang, Zhang Xinmin. Advance in Information Foraging Theory. New Technology of Library and Information Service, 2009, (1):73–79.

[13] Yang Yang, Zhang Xinmin. Empirical Research of Network Information Environment Based on Foraging Theory.Journal of the China Society for Scientific and Technical Information, 2010, 29(1):169–176.

Management, Information and Educational Engineering – Liu, Sung & Yao (Eds)
© 2015 Taylor & Francis Group, London, ISBN: 978-1-138-02728-2

Exploration on adaptability education for college students

F. G. Meng, K.Wang* & M.X. Zhu
College of Electrical & Information Engineering, Beihua University, Jilin City, Jilin Province, China

ABSTRACT: University life is a new world for high school students. Teachers should fully understand the characteristics of freshmen, and based on these, develop effective adaptability education for freshmen from the psychology, learning, environment and economy, and get students into roles as soon as possible to shorten the period of adaptation to the new learning environment and start to study and live in the university in good mind.

KEYWORDS: University students; Adaptability education.

1 INTRODUCTION

Entrance education for freshman is the foundation of the ideological and political work in the whole work of institutions of higher learning. Whether this work is successful or not, it directly determines whether the students adapt to university life and future development as soon as possible and the ideological and political work is smoothly carried out in institutions of higher learning.

Therefore, it has practical significance to fully understand the characteristics of freshmen, develop adaptability education on this fruitful basis, shorten the adaptation period to the new learning environment, promote the healthy growth of college students, guide students to establish a right outlook on life and the right outlook on the world, make a right self-evaluation and develop all aspects of their ability positively.

Li, Hui 2010. analyzed situation change at the beginning of the college life, the freshmen who have just passed the college entrance examination have to face a new world, bidding farewell to the familiar environment and being away from their parents, they start the collective campus life and tackle all to affairs in life independently in an unfamiliar environment, if lacking of care and warm, they may have a sense of loneliness naturally. Therefore, adaptability education on Freshman should pay attention to lead the students to adjust their mentality, adapt to the new environment, have a further understanding of themselves, adopt into the new collective as soon as possible, find their own position and finish the "four adaptations" as soon as possible.

2 ADAPTABILITY EDUCATION METHOD

2.1 *Adaptation to the psychology*

One is to accurate self positioning, and complete role transformation successfully. The students who can enter university are the outstanding ones in the high school stage usually win the attention from the parents, teachers and students, and usually are the central figure in life. After entering the university, all kinds of talents get together, almost everyone here has a glorious past, and everyone is talent in the study and the master wizards, if rescheduled for seating, only a few people can keep the original center position and role.

In contrast, most of the students will change from a central role to the general role, superiority of many students suddenly will disappear, as well as the confidence in their own ability, even they will have a strong sense of inferiority and lose power and enterprising, which will make many students fail in finding their positions. Therefore, it is necessary to teach students to adjust their own state of mind, learn to find their own strengths, accept the "not perfect" yourself, relax the ropes binding their spirit and adopt into university life in a cheerful mood.

2.2 *Adjust expectations*

Establish and maintain a good attitude. Most freshmen have higher expectations of the university that have admitted them, and make a tentative plan and outline about the future of alma mater in their mind preschool. While after entering university, they find there is a large gap between the reality and expectation, including school hardware facilities, as well as the school's various rules and regulations, some may be disappointed, frustrated and even rebellious.

In Lin, Lixia & Ru, Zhengkai, it is necessary to guide the students to face the reality and their own, lower their expectations, accept the "not perfect" environment, learn to self build good, optimistic, healthy, upward attitude, for university life, actively adapt into college life, and actively seek their own development under the existing conditions rather than blindly complain. Otherwise, the pessimistic disappointed mood will only contain the normal development and growth of themselves.

2.3 *Adaptation to the study*

Zou, Changhua & Han, Jiantao (2011) pointed out that college students' learning is a professional learning in a domain, adapting to a professional is one of the objectives of learning adaptation for college students. College students' professional adaptation levels directly affect the achievements of professional learning and the future employment for college students.

On the one hand, the university class site is "hit a shot for a place" which is different from the "a carrot a pit" in high school. And the library, Internet bar, dormitory, study room also has a big difference in the status of students in the study. Therefore, the Freshman adaptation education must help the students understand how to use the library, how to make use of the network, how to through various channels to obtain more information and how to use modern means of science and technology to master the use of their knowledge and improve their ability.

On the other hand, after entering the University, the teaching mode has changed from the teacher-centered mode to the student-centered self-study mode. The knowledge the teacher taught in the classroom teaching are just outline and framework, students not only have to digest classroom learning content, but also read a lot of related books and documents related, achieve mastery through a comprehensive study of the subject finally.

Effects of learning method for the learning results are self-evident, while the learning method in the University is different from that in primary and middle schools, self-learning ability has become an important factor affecting academic achievement, many students find it difficult to adapt to it. Therefore, it is necessary to carry on a series of learning methods introduction and learning experience exchange. With the help of the planned introduction and explanation of The characteristics of the University and the general method; the constantly exploring and learning in the actual study, active observation and thinking; learning experience introduction and exchange between the new and the old, these can enable students to gradually adapt to the learning form and characteristics in university to learn the form and characteristics of learning and pass learning adaptation period smoothly.

2.4 *Adaptation to the environment*

Tian, Jiaqi (2013) analyzed the economic interaction between the university and the surrounding area is strong, which has a very strong correlation function. Because of the "school economic effect", the entertainment places around the colleges and universities attract students' consumption and hit on their likes. The students, who have been depressed by the college entrance examination students for a long time, once excessively relax themselves, various problems will occur. Some students indulge in surfing the internet and playing computer games all night, chatting and watching movies, which abandon their studies. Therefore, we need to carry out "self control, self education and self management" education among freshmen, teach students to deal with the relationship between learning and entertainment, learn how to be webmasters and carry out a number of preventive education.

2.5 *Adaptation to the economy*

In Zhang Yuanhong & Kong Qingna (2013) for the vast majority of students, entering the University is the first step for them to begin independent living in the true sense. After the University, the living environment has changed a lot, without the daily care of the parents and the elders, many things need to handle alone. From the family life that cannot live without parents to university life that everything needed to be handled independently, everything should be learned from the very beginning. Therefore, learning how to manage money is the first lessons to students learn to live independently for students.

With the national college recruitment of students, paying for university has been accepted by the students and parents. But as the increase of tuition fees and living consumption level, the economic capacity can bear by more and more families is very limited.

Therefore, the adaptability education in freshmen should involve in teaching students to consider, which costs in your life is a must, which is not. At the same time, they should learn to make their own consumption plan according to family economic ability, develop the habit of thrift rather than following the fashion or love ostentation.

In order to make the students adapt to the university life as soon as possible, it should require every freshmen to adjust attitude actively, and make comprehensive understanding and relocation of themselves; do not have the psychological fear because of not adoption; do not have a sense of loss because of the disappear of the learning advantage; do not have a sense of disappointment because of the not adaption to the living environment; do not have a sense of inferiority because of temporary economic difficulties. To do this work well will play an effective role in the whole educational process.

ACKNOWLEDGEMENT

Thanks are due to corresponding author Kai Wang for his assistance with the surveys and valuable discussion.

REFERENCES

Li, Hui. 2010. Optimization Principle and Method for Adaptability Environment to the University Students. Beijing, China: People's Publishing House.

Lin, Lixia & Ru, Zhengkai. 2010. A Discussion on preparation for university freshmen education with counselors, Chinese Power Education Press 6(1):13–16.

Tian, Jiaqi. 2013. Related research on interaction between University and the surrounding economic. Beijing, China: Chinese Business.

Zhang Yuanhong & Kong Qingna. 2013. Discussion on the Psychological Educational Course Reformation to College Students under New Situation. Research on Ideological and Political EducationNo.02:24–26.

Zou, Changhua & Han, Jiantao. 2011. Analysis on the current situation of professional adaptability of college students. Chaohu College Journal 4(2):33–35.

Management, Information and Educational Engineering – Liu, Sung & Yao (Eds)
© 2015 Taylor & Francis Group, London, ISBN: 978-1-138-02728-2

An empirical analysis of competitiveness of foreign trade in Guizhou province

Wei Liu & Yan Shang
Guizhou University of Finance & Economics, Guiyang, Guizhou, China

ABSTRACT: In recent years the economy grows rapidly in Guizhou, and foreign trade has developed fast. This paper reveals an index of comparative advantage and trade competitiveness through the share index of the international market, and evaluates the status of the competitiveness of foreign trade from many angles, which will promote the reasonable and healthy development of the foreign trade in the province.

KEYWORDS: Foreign trade; Guizhou; Competitiveness; Index analysis.

1 THE STATUS OF FOREIGN TRADE IN GUIZHOU PROVINCE

Guizhou province is located in China's southwest, it has a pleasant climate, nationalities, resources, tremendous development potential. However, due to historical reasons, the province's lags of economic development is behind and low degree of opening to the outside world, foreign trade and the utilization of foreign capital started too late. In 30 years of reform and opening up, foreign trade enters the stage of rapid growth, and it has made a positive contribution to the healthy and stable development of the economy. Especially since it joining in WTO, foreign trade in Guizhou presents the following features:

1.1 *The total of foreign trade has grown steadily*

Since 2002, the total volume of foreign trade has increased by 9.6 times, the absolute surplus rose by $5.94 billion. Along with exports rising from 440 million in 2002 to 4.95 billion in 2012, 12 times, the absolute surplus rose by $4.51 billion. Imports rose from 250 million in 2002 to 1.68 billion in 2012. This figure was 6.7 times bigger than before, while the absolute surplus rose by $1.43 billion. Surplus of foreign trade increased from 190 million in 2002 to 3.28 billion in 2012. Increasing the scale of foreign trade has given a strong push to foreign economy, and improved the upgrading of industrial structure and export-oriented economy.

1.2 *The growth fast and huge potential*

From 2002 to 2012, the average growth rate of the total amount of foreign trade in Guizhou annually has reached 29%. Export has grown by an average of 30.5% in a year, and import has grown by an average of 27% in a year. In 2003–2011, the average annual growth rate of import and export in China was 21.7% and export was an average of 21.6%, and import was an average of 21.8%. It is obvious that the growth speed of foreign trade in Guizhou is higher than the national average. Especially in the severe financial crisis of 2008, the export of Guizhou province remained 30% growth, higher than the 24.8% growth of the whole country. Rapid development of foreign trade gives a drive to GDP, and it is an important source of leapfrog development in the province.

1.3 *Structure optimization of export commodity*

Structure of export commodity is the important basis to measure a foreign trade's situation of the structure, and the change of structure reflects a large extent the change of industrial structure. Export continued to maintain growth in 2011, of which the export of primary products stood for $331 million, accounting for 11.07% of the total amount of export in the province, decreased by 5.2% than last year. Export of manufactured goods was $2.65 billion, accounting for 88.9% of the total amount of export in the province, increased by 4.8% than last year. The decline of the primary products and the proportion of manufactured goods suggest that structure has been optimized, and Guizhou is in the rising stage of industrialization.

1.4 *Foreign trade gives priority to Asia, more diversified import and export market distribution*

According to the statistics in 2000–2010, the proportion of trade that Guizhou with Asian partners is 57% or more every year. The nations within the association of Southeast Asia is the largest trading partner and the largest export market for Guizhou. In 2011 the export trade in Guizhou was $507 million, accounting for 26.7% of the whole exports. Export trade with Hong Kong, Japan and South Korea are respectively $228 million, $107 million and $067 million. This suggests that these countries have done an international trade with Guizhou very often because of the Asian geographical advantages. By 2010, export of Guizhou has spread to 139 countries and regions, and the top 10 are the United States, Indonesia, Australia, Japan, Vietnam, India, Taiwan, Korea, Hong Kong and Thailand, which suggest that Guizhou has gradually formed a key and full range of diversified pattern of foreign trade market.

2 AN INDEX ANALYSIS ON COMPETITIVENESS OF FOREIGN TRADE IN GUIZHOU PROVINCE

Foreign trade competitiveness refers to the ability that gets marketed and profit from a country or region. The country with the competitiveness can trade their products, industry. We generally selected indicators visually that can indicate the result of the international competitiveness when analysis and evaluating the international competitiveness of foreign trade. Here we choose RCA and TC to make analysis on the situation of competitiveness of foreign trade of Guizhou province in recent years.

2.1 *The share analysis of the international market*

The share index of the international market is the most simple and important index to reflect the competitiveness of a country's foreign trade. It refers to the products export of a country or region accounted for the proportion of the total amount of the world. The greater the proportion is that the stronger the international competitiveness of the export commodities is; the smaller the proportion is that the weaker the international competitiveness of the export commodities is. In 2006–2011, the scale of export trade expands unceasingly in Guizhou province, accounting for the proportion of China's export rising from 0.009% to 0.016%, while the proportion comparing with other provinces is in a large gap. But this figure presents a rising trend overall, which suggests that the trade competitiveness of Guizhou export commodities is enhanced.

2.2 *Analysis of revealed comparative advantage index*

The revealed comparative advantage index refers to the size that the proportion of a certain type of the products in export of one country or region relative to the product in total world. In general, the RCA value is greater than 1, and it is said this product has a comparative advantage in the international market and a certain international competitiveness; the RCA values are less than 1, and it is said this product didn't have a comparative advantage in the international market, and the international competitive power is relatively weak. Due to the limit statistical data, this article divides the products into two major categories of primary products and manufactured goods, from 2006 to 2011. The RCA index analyzes the general situation of export trade in Guizhou. We know that the RCA index of primary products and manufactured goods in Guizhou less than 1 from table 1, but the RCA index is slightly higher than agricultural manufactured goods, which shows Guizhou is in a comparative disadvantage in foreign trade. It should be done to improve the added value of manufactured goods, while increasing the export volume at the same time and the international competitiveness of primary products.

Table 1. 2006–2011 RCA index of the export product in Guizhou.

year	primary	manufactured
2006	0.01063	0.02287
2007	0.01081	0.02855
2008	0.02067	0.02569
2009	0.01401	0.02404
2010	0.01437	0.02681
2011	0.01318	0.03519

2.3 *Analysis of trade competitiveness index*

TC refers to the proportion of a region's difference between the import and export trade for which accounts the total amount of import and export trade, namely the TC index = (exports – imports)/(exports plus imports). The indexes as a relative value are always between −1~1, eliminating the economic expansion and fluctuation. Its value is close to −1 indicates the competitiveness is weak, −1 is said the industry is only import without any export; the more close to 1 indicates the greater competitiveness, 1 is said the industry is only exported without any import. In 2006–2010, the TC indexes of Guizhou in the aspect of food, chemical industry and related industrial products, machinery and electronic products are greater than zero, and in which the competitiveness of food export is very strong with the TC index remained steady at 0.98 above. This figure shows that the food industry is at specialization level. Chemical industry and related industrial products, during 2006–2010, the TC index were 0.925, 0.954, 0.929, 0.936, 0.929; Mechanical and electrical products had kept rising, grew steadily from 0.104 in 2006 to 0.327 in 2010. The TC index of minerals had been negative, which showed this industry was at imported professional level.

3 CONCLUSION

We can get the following conclusions from the multi-angle analysis of foreign trade competitiveness in Guizhou province that the foreign trade has been growing at a speed, while the share of international market increasing. But because of a lower base, smaller scale, the overall competitiveness of foreign trade is still relatively backward. In view of the situation of foreign trade competitiveness, in order to realize the development of foreign trade faster and better, it is urgent to need us to adjust the optimization structure of export commodity and develop the advantage industry; to encourage and guide all relative enterprises to participate actively in international competition and achieve the scale of economies; to increase the intensity of investment and capital introduction and expand the international market share of foreign trade in Guizhou; to strive to improve the soft environment for investment in order to improve the export competitiveness of foreign trade.

REFERENCES

[1] Cao Xiaolei. The empirical research of foreign trade competitiveness in Jiangsu province[J]. The price issue, 2011(2).

[2] He Weifu. The countermeasure research of improving international competitiveness of foreign trade in guizhou[J]. Economic problems, 2008(7).

[3] Li Mingqiao. The analysis of openness and present situation of international trade in Guizhou province[J]. The regional economic, 2008(8).

Management, Information and Educational Engineering – Liu, Sung & Yao (Eds)
© 2015 Taylor & Francis Group, London, ISBN: 978-1-138-02728-2

Evaluating the function and effect of an educational game

J.J. Tang & W.Q. Qu
Institute of Educational Technology, Beijing International Study University, Beijing

ABSTRACT: In order to survey its utility, we evaluated an educational game which was popular in China mainland. The research method was questionnaire survey. According to the statistical analysis of the survey, the effect of education is far below than expected. The users are not satisfied with the educational function of the game. The game designer is supposed to improve game to help users developing positive emotions and attitudes and correct living values.

KEYWORDS: gamification, game-based learning, educational game.

1 INTRODUCTION

Game-based learning is a process which makes students acquiring knowledge, skill and attitude with pleasant studying experience through using educational games in classroom or self-study[1]. In a broadest sense, educational game includes all games owing educational values. In a narrow sense, educational game is a computer game designed for certain teaching objects. As a new teaching and learning tool, educational game is getting so popular that the US government holds a National STEM Game Design Challenge every year[2]. In china, more and more games companies occupy educational game markets and invest more and more in developing educational games.

2 THE NECESSITY OF EVALUATING THE EDUCATIONAL FUNCTION OF EDUCATIONAL GAME

In recent years, many educational games have been put published, such as Mole manor, Roco Kingdom, Disney Club Penguin, webkinz, and poptropica etc.

By means of attracting many users, these games became popular quickly. However, the weak educational function leaves a negative impression on users' mind. Meanwhile, a lot of people worried about the violent images arising from the game. Even some people doubted the aim of the designer's objective on education. They also have questions on whether this product suits, children's cognitive ability.[3]

So it is necessary to evaluate the function of these games. We chose a popular educational game called Umfun in china to evaluate. This game declares that its users are the students at the age of 12 to 16. The game aims at creating a happy learning environment, making knowledge learning through games.

3 THE METHOD OF EVALUATING

In order to evaluate the function of the Umfun game on education, A scale was designed. Based on the scale, a survey was made by the online system. After a statistical analysis of the survey's result, a conclusion has been drawn. In the survey, six hundred and forty-six students from 4 schools were being invited to complete this survey online. Among them, 534 are valid. Among the 534 students in case, 48.7% are boys, 51.3% are girls; 58.9% are 13 years old, 41.1% are 14 years old.

4 DESIGN OF THE SCALE

There are 35 items in this scale, divided into 3 dimensions: accessible UI, entertainment, education respectively.

Dimension one, the accessible UI can be seen from three aspects: rational layout, beautiful interface, convenience. Dimension two, the entertainment elements can be measured from the following five points: easy and comfortable hand control, authentic scenario experience, happy role players experience, wonderful success experience, pleasant communicating experience.[4] The third Dimension, the educational function is indicated in three parts: skills and knowledge, emotion , attitude and values.[5]

Answers and Score of each item:

For each item, there are five different answers. For each answer, different score is given.

Table 1. Answer and score of each answer.

answer	A. excellent	B. good	C. fair	D. poor	E. disappointing
score	5	4	3	2	1

5 THE STATISTIC ANALYSIS OF THE SCALE

5.1 *The average score and standard deviation of each dimension*

Table 2. The average score and standard deviation of each dimension.

Dimension	Average	Standard deviation
Accessible UI	3.562	.78359
entertainment	3.361	.45736
education	3.038	.28154

Table 2 shows the average score of each dimension and its standard deviation of each dimension. In this table, the average score of all dimensions is above "fair" and less than "good". However, the average score of educational dimension is the lowest among the three dimensions while the entertainment dimension is in the middle and the UI is highest, which indicates the design of the game's interface is best and the players like its accessibility and convenience.

Table 3. The average score and standard deviation of UI's three factors.

Dimension	Average	Standard deviation
rational layout	3.394	.35434
beautiful interface	4.221	.58764
convenience	3.073	.36879

The above table shows the average score and standard deviation of the three aspects of UI dimension. All of these average scores are above "fair". Especially, the average score of beautiful interface is high than 4, which indicates users thought the game's UI is beautiful.

Table 4 shows the average score and standard deviation of the five points of entertainment. Most of these average scores are higher than three, which indicates the entertainment of this game is in the middle level. However, the average score of authentic scenario experience is "poor". It is probably due to the effect of the 2-D image of the game, which the users are not satisfied.

Table 4. The average score and Standard deviation of entertainment's five points.

Dimension	Average	Standard deviation
Easy and comfortable hand control	3.116	.82334
Authentic scenario experience	2.814	.38728
Happy role players experience	3.535	.56322
Wonderful success experience	3.976	.67323
Pleasant communicating experience	3.669	.48729

Table 5. The average score of and Standard deviation from the three parts of educational function.

Dimension	Average	Standard deviation
emotion and attitude	2.911	.37657
skills and knowledge	3.383	.42325
value	2.821	.47845

Table 5 shows the average score and Standard deviation of the three parts of educational function. Among them the average score of skill and knowledge is higher than 3 while the other two are lower than 3. It indicates that the users can gain some knowledge and skills, while they don't think the game is helpful for developing proper attitude and value.

5.2 *Influence of age and gender on the result*

Table 6. The influencing factors of age.

Dimension	age	N	average	Standard deviation	T value
Accessible UI	13	289	3.573	0.3755	0.478
	14	245	3.549	0.4128	
entertainment	13	289	3.401	0.7233	1.373
	14	245	3.313	0.8761	
education	13	289	3.031	0.3564	0.528
	14	245	3.046	0.3126	

Table 6 shows the influence of age of students in case on the survey result. Through T value, we know there is not distinctive difference in accessible UI and education, while there is a distinctive difference in entertainment, which indicates different age users have different evaluations. The younger users prefer to play this game.

Table 7. The influencing factors of gender.

Dimension	Gender	N	Average	Standard deviation	T value
Accessible UI	Boy	253	3.573	0.4768	0.981
	Girl	281	3.552	0.3656	
Entertainment	Boy	253	3.454	0.8433	2.35
	Girl	281	3.251	0.7876	
Education	Boy	253	3.051	0.2541	0.967
	Girl	281	3.026	0.2178	

Table 7 shows the influence of gender of students in case on the survey result. Through T value, there is distinctive difference in entertainment, which indicates that the boys prefer to play this game, while the girls are not satisfied with the game's entertainment.

6 CONCLUSION

Through the analysis of the above data, one conclusion can be drawn that the overall level of the game is good because the most average scores of the dimensions are higher than the average level. However, taking the game's original object into consideration (creating a happy learning environment, making knowledge learning through games), the effect is far below. In the meanwhile, the average score of "emotion and attitude" and "value" is much lower than expected. This shows that the users are not satisfied with the educational function of the game. The game designer is supposed to improve the effect of education as well as eliciting users' positive "emotion and attitude" and "value".

The purpose of the educational game is to promote learning. However, the educational function of deficient. Thus, if the educational game only attracts users playing games without helping users to achieve the educational object, it is a failure. So the finding of this article is to remind the game designer putting more energy into developing new process and method to improve the educational efficiency in the future. New games are expected to produce for users to form a good habit of learning, to help users developing positive emotions and attitudes, to build correct living values.

REFERENCES

[1] Si Zhiguo. The developing and application of gamification learning community [D].Beijing: captical normal university, 2005.

[2] National STEM Video Game Challenge [EB/OL]. http://www.stemchallenge.org, 2013-05-29.

[3] Alan Gershenfeld. Will computer game change education? Guang Ming Daily, 2014–5–3 (003).

[4] Tao Kan. From gaming to learning: the game-based learning under the view of game [J]. China Educational Technology, 2013(9):22–27.

[5] Zhang Honggao. The integration of Knowledge and skill, process and method, attitude and values [J]. research of basic education, 2004(5):16–18.

Management, Information and Educational Engineering – Liu, Sung & Yao (Eds)
 ISBN: 978-1-138-02728-2

Research into validity of implementation of humanistic quality-based education in integrated English teaching of English major

Zhi Hong Wang
Anhui Xinhua University, Hefei, China

ABSTRACT: The current situation of the English major in China shows that humanistic qualities are indispensable factors in the cultivation of English majors. An experimental study is adopted to compare academic achievement and humanistic qualities of the experimental class with those of the control class both before and after the teaching experiment, aiming to explore the validity of the implementation of humanistic quality-based education in Integrated English teaching. The results indicate that humanistic quality-based education is helpful to enhance the students' overall qualities, improve academic achievement and reform the teaching mode.

KEYWORDS: Integrated English; Humanistic Qualities; Validity; Academic Achievement.

1 INTRODUCTION

1.1 *Background*

Since 1970s, the concept of humanism, as well as humanistic approach, has gained attention from people at home and abroad. Hamachek (1977: 149) believes that "humanistic education starts with the idea that students are different, and it strives to help students become more like themselves and less like each other." Williams and Burden (1991: 38) summarize a number of humanistic messages for the language teacher, like creating a sense of belonging, involving the whole person, developing personal identity, encouraging creativity, developing knowledge of the process of learning, allowing for choice, etc. In China, some scholars emphasize the importance of humanistic quality in the English major. He (2004) holds that the features of English major lie in the fact that students majoring in English should not only have a higher proficiency in language than non-English major students, but also have stronger humanistic qualities than common English learners. Hu and Sun (2006) analyze the disadvantages in foreign education: just because humanistic education only makes up a lower percentage in the whole education, foreign language major students are weak in humanistic foundation and critical thinking, and have a superficial understanding of western culture and Chinese culture. Therefore, English education should revert to humanistic education, focusing on the training of talents by general humanistic quality-based education.

1.2 *Objective*

The present research aims to research the validity of the implementation of humanistic quality-based education in Integrated English teaching, focusing on three issues: whether humanistic education is helpful to the enhancement of students' overall qualities, whether humanistic education is conducive to the improvement of academic achievement, and whether humanistic education is advantageous to the reform of teaching mode.

2 CONNOTATION OF HUMANISTIC QUALITIES

Many scholars believe that humanistic qualities are comparatively stable internal qualities, but they differ in the understanding of the components of humanistic qualities (Shi 2008:12). Hu (1998: 29) argues that humanistic qualities are composed of knowledge, ability, concept, emotion, willpower, etc. Qian (2001:184) asserts that humanistic qualities cover humanistic scientific knowledge, social psychology, cultural cultivation, and humanistic spirit, and so on.

According to the practical teaching situation, research purpose, and the definition of humanistic qualities given by Baidupedia, human qualities can be defined as the integration of three aspects, including humanistic knowledge, humanistic competence and humanistic spirit. Humanistic knowledge is the basic knowledge about humanistic fields, like history, literature, politics, law, arts, philosophy, morality, language, etc. Humanistic competence refers to the ability to apply humanistic knowledge to practice, including practical ability, interpersonal ability, creation ability, organizational ability, communication ability, and aesthetic ability. Humanistic spirit constitutes the base of world view and value stemming from humanistic thoughts and humanistic methods, represented in the concern for current affairs, creative spirit, morality, psychological qualities and legal awareness, etc.

3 RESEARCH PROJECT

3.1 *Research plan*

For English major in China, the course of integrated English lasts two academic years. Approaching the end of the first academic year, a survey is conducted to examine the situation of humanistic qualities of the English major. Then two classes are selected as the experimental class and the control class respectively. During the second academic year, the experiment on implementation of humanistic quality-based education is carried out at the teaching of Integrated English. In order to control the influence of other factors on teaching effects, the same teacher adopts two different teaching methods to teach the two classes.

In the control class, the teacher plays the role of a lecturer, spoon-feeding students with language points and humanistic knowledge in each unit; students follow the rote-learning pattern; interaction between teacher and students is lacking in the classroom.

On the contrary, the teacher informs the students in the experimental class of the teaching method, and classifies the whole class into 6 small learning groups according to sex, academic achievement, learning style, etc. Each group leader organizes group members to learn the assigned task. The teachers' role transfers from a sole lecturer to a guider, advisor and supporter.

At the end of the fourth semester, the experiment is over. Then another questionnaire is used to compare the two classes from the perspectives of academic achievement, humanistic quality, and teaching effect.

3.2 *Research subjects*

The research takes the students in the parallel two classes as subjects, with 23 students in each class.

3.3 *Research tools*

Before and after the experiment, a questionnaire with the same structure but different contents is used to examine students' present humanistic qualities. The first part includes 12 questions, covering politics, economy, culture, literature, etc., aiming to investigate students' humanistic knowledge. The second part and the third part are presented in the form of Likert Scale, with five options, i.e. "Completely Agree."(5 points), "Agree"(4 points), "Neither Agree Nor Disagree"(3 points), "Disagree"(2 points), and "Completely Disagree"(1 point). In the second part, there are 12 questions of multiple choices, helping students reflect on their humanistic competence, including practical ability, interpersonal ability, creation ability, organizational ability, communication ability and aesthetic ability. The third part contains 10 questions of multiple choices, exploring students' humanistic spirit, like the concern for current affairs, creative spirit, morality, psychological qualities and legal awareness.

4 RESULTS

4.1 *Pre-test result*

Before the experiment, the first questionnaire is used to examine the students in the two classes, aiming to collect their humanistic quality data. The academic achievement of the second semester is taken as the pre-test score. SPSS 13.0 is adopted to do Independent-Samples T Test, thus obtaining the results of academic achievement and humanistic qualities.

Table 1. Pre-test result of academic achievement and humanistic qualities of the students.

Items	Class	Mean	Std. deviation	Std. error mean
Pre-test	1	81.3043	5.68439	1.18528
Score	2	80.9565	6.63653	1.38381
Humanistic	1	29.13	5.354	1.116
Quality	2	29.61	6.900	1.439
Humanistic	1	3.09	1.676	0.350
Knowledge	2	3.17	1.267	0.264
Humanistic	1	14.26	3.583	0.747
Competence	2	14.22	4.348	0.907
Humanistic	1	11.78	2.522	0.526
Spirit	2	12.22	2.999	0.625

In the tables through the article, Class 1 is the control class while Class 2 is the experimental class. A conclusion can be drawn from Table 1 that the two classes do not differ very much in academic achievement and humanistic qualities, for the difference of the mean is less than 1.

Table 2. Pre-test result of independent test samples.

		Levene's Test for Equality of Variances		t-test for Equality of Means						
		F	Sig.	t	df	Sig. (2-tailed)	Mean Difference	Std. Error Difference	95% Confidence Interval of the Difference	
									Lower	Upper
Pre-test Score	Equal Variances assumed	.596	.444	.191	44	.849	.34783	1.82204	-3.32425	4.01990
	Equal variances not assumed			.191	42.985	.850	.34783	1.82204	-3.32670	4.02235
Humanistic Qualities	Equal variances assumed	.659	.421	-.263	44	.794	-.478	1.821	-4.148	3.192
	Equal variances not assumed			-.263	41.442	.794	-.478	1.821	-4.155	3.198
Humanistic Knowledge	Equal Variances assumed	.821	.372	-.198	44	.844	-.087	.438	-.970	.796
	Equal variances not assumed			-1.98	40.947	.844	-.087	.438	-.972	.798
Humanistic Competence	Equal variances assumed	.701	.407	.037	44	.971	.043	1.175	-2.324	2.411
	Equal variances not assumed			.037	42.449	.971	.043	1.175	-2.327	2.414
Humanistic Spirit	Equal variances assumed	.216	.645	-.532	44	.597	-.435	.817	-2.082	1.212
	Equal variances not assumed			-.532	42.740	.597	-.435	.817	-2.083	1.213

From Table 2, it can be seen that significance level of equal variances > 0.05, indicating that equal variances is assumed. Besides, a significant level of pre-test score is 0.849 > 0.05, and significance level of humanistic qualities is 0.794 > 0.05. Therefore, under the condition of 95% confidence interval of the difference, there is no significant difference in students' academic achievement and humanistic qualities. This fact proves that the precondition of the teaching experiment is valid.

4.2 *Post-test result*

After a school year's teaching experiment, the second questionnaire is used to examine students' humanistic qualities. In addition, the final examination result is taken as post-test score. Independent-Samples T Test is conducted again.

Table 3 shows that the students in the two classes differ in academic achievement and humanistic qualities. Compared with the control class, the students in the experimental class have higher numerical value in post-test scores (>3.26), humanistic quality (> 3.65), the three components of humanistic quality (> 0.92, 1.35, and 1.44 respectively).

Table 4 shows that significance level of the post-test score, humanistic quality and its three components of the two classes < 0.05, so under the condition of 95% confidence interval of the difference, there is a significant difference in the two teaching methods. The result proves that the students in the experimental class have improved their academic achievement and humanistic quality after one school year's experiment.

Table 3. Post-test result of academic achievement and humanistic qualities of the students.

Items	Class	Mean	Std. deviation	Std. error mean
Post-test Score	1	80.7826	5.96929	1.24468
	2	84.0435	4.65616	0.97088
Humanistic Quality	1	83.35	2.534	0.528
	2	87.00	3.943	0.822
Humanistic Knowledge	1	4.43	0.992	0.207
	2	5.35	1.774	0.370
Humanistic Competence	1	40.52	2.150	0.448
	2	41.87	2.160	0.450
Humanistic Spirit	1	38.39	2.426	0.506
	2	39.83	2.348	0.490

4.3 *Interview*

Interview with the students from the two classes reveals the following information. The students in the experimental class are satisfied with the teaching method, for they have improved their learning habit and take initiative to gain knowledge and practice skills. By contrast, the students are accustomed to indoctrination by means of spoon-feeding method. They feel they lack practical ability, but they do not know how to improve themselves.

Table 4. Post-test result of independent samples test.

		Levene's Test for Equality of Variances		t-test for Equality of Means						
		F	Sig.	t	df	Sig. (2-tailed)	Mean Difference	Std. Error Difference	95% Confidence Interval of the Difference	
									Lower	Upper
Post-test Score	Equal Variances assumed	1.868	.179	-2.066	44	.045	-3.26087	1.57856	-6.44224	.07950
	Equal variances not assumed			-2.066	41.538	.045	-3.26087	1.57856	-6.44757	.07417
Humanistic Qualities	Equal variances assumed	5.903	.019	-3.737	44	.001	-3.652	.977	-5.622	-1.683
	Equal variances not assumed			-3.737	37.522	.001	-3.652	.977	-5.631	-1.673
Humanistic Knowledge	Equal Variances assumed	5.578	.023	-2.155	44	.037	-.913	.424	-1.767	-.059
	Equal variances not assumed			-2.155	34.537	.038	-.913	.424	-1.774	-.052
Humanistic Competence	Equal variances assumed	.052	.821	-2.121	44	.040	-1.348	.635	-2.629	.067
	Equal variances not assumed			-2.121	43.999	.040	-1.348	.635	-2.629	-.067
Humanistic Spirit	Equal variances assumed	.001	.970	-2.038	44	.048	-1.435	.704	-2.854	-.016
	Equal variances not assumed			-2.038	43.953	.048	-1.435	.704	-2.854	-.016

5 DISCUSSION

5.1 *Influence of humanistic quality-based education on individual overall quality*

After one-year experiment, the students in the experimental class have improved themselves in the three aspects of humanistic qualities. The reasons for the students' progress can be analyzed in the following points.

First, proper learning tasks guide students to take the initiative in gaining humanistic knowledge. The teaching course book contains all kinds of humanistic knowledge points. The teacher designs some learning tasks centering around the knowledge points related to the theme of the text, and assigns them to smaller learning groups who are responsible for searching relevant information and sharing the knowledge with the whole class. The group work enables students to play an active role in the learning process and grasp the knowledge through practice.

Second, effective group work offers students the opportunity to improve humanistic competence. Each small group consists of 4 or 5 students with diversified learning abilities and learning styles. The group leader organizes each member to finish the assignment. Someone is responsible for searching useful materials from websites or library, others for writing down a coherent article, and still others for presentation. In the cooperative activity, students improve interpersonal ability and an organization's ability from cooperation, cultivate practical ability, creation ability and aesthetic ability from preparing the presentation material, and enhance communication abilities from presentation and oral defense. After experiencing various learning tasks, naturally, students heighten their humanistic competence.

Third, the fair encouragement mechanism promotes humanistic spirits. The teacher helps students make the fair and equal regulations concerning the encouragement and punishment. So every student is motivated to make progress and develop the honor of collectivism in the atmosphere of cooperation and competition. While doing learning tasks, students heighten morality, temper psychological quality, and cultivate the awareness of observing regulations.

Fourth, extracurricular practice guarantees the internalization of humanistic knowledge and humanistic competence into humanistic quality. Humanistic knowledge can be gained from teaching and learning, while humanistic quality cannot be acquired from teaching. Because the cultivation of humanistic quality follows a certain rule: it is only during practice that a person can improve his humanistic quality by drawing on human knowledge and humanistic competence to solve the problems or meet the challenges. As a result, humanistic quality-based education requires the integration of classroom teaching and extracurricular practice. The teacher should encourage students to participate in all kinds of activities, like English competition, English speech contest, translation competition, social practice, volunteer service, etc. Students may enhance the humanistic quality of these useful activities.

5.2 *Influence of humanistic quality-based education on academic achievement*

Implementation of humanistic quality-based education in teaching helps students make progress in human knowlcdgc, humanistic competence and humanistic spirit. Humanistic knowledge and humanistic competence guide students to develop overall quality in all-round, while humanistic spirit imbues students with healthy personality. All these qualities stimulate the students' learning ability from both IQ and EQ, and further exert a positive influence on academic achievement.

5.3 *Influence of humanistic quality-based education on teaching mode*

During the process of implementation of humanistic quality-based education in teaching, the teacher does not need spoon-feed students with language knowledge, but guides students to accomplish learning tasks by means of task-based teaching or cooperative learning. While students are taking initiative in gaining language knowledge and humanistic knowledge, they practice humanistic competence and acquire humanistic spirit.

6 CONCLUSION

Implementation of humanistic quality-based education in integrated English teaching reveals the following three conclusions.

First, implementation of humanistic quality-based education is conducive to the heightening of students' overall quality. It is the whole-person education that plays a subtle impact on students' healthy growth. Accordingly, students perfect their overall quality while accomplishing learning tasks.

Second, implementation of humanistic quality-based education is advantageous to the improvement of academic achievement. Students may experience humanistic love in the harmonious yet competitive atmosphere, and take full advantage of subject awareness to form sustained inner driven power, thus helping individual self-learning, self-development and self-actualization.

Third, humanistic quality-based education is beneficial to the reform of teaching mode. It is different from the spoon-feeding teaching mode, for it can realize the teacher's guiding role and students' initiative.

ACKNOWLEDGEMENT

The research is supported by Humanities and Social Sciences Program of Anhui Province Office of Education (2010sk528).

REFERENCES

Baidupedia. (n. d.) *Humanistic Qualities*. Retrieved August 21, 2011, from http://baike.baidu.com/view/14506.

Hamachek, D. E. 1977. Humanistic Psychology: theoretical-philosophical framework and implications for teaching. In D. J. Treffinger, J. Davis and R. E. Ripple (Eds.) *Handbook on Teaching Educational Psychology*. New York: Academic Press.

He Zhaoxiong. 2004. Thoughts on the States Quo and the Future of EFLT for College English Majors in China. *Shan Dong Foreign Language Journal* 04(6): 3–4.

Hu Wenzhong & Sun Youzhong. 2006. On Strengthening Humanistic Education in the English Language Curriculum. *Foreign Language Teaching and Research* 06(5): 243–247.

Hu Xianzhang. 1998. Enhancing our Understanding, Transform our Concept, Improving College Students' Humanistic Quality-based Education. In *Humanistic Inspiration of University in China* (2nd Volume). Wuhan: Huazhong University of Science and Technology Press.

Qian Yuanwei. 2001. *Quality-oriented Education in Society*. Guangzhou: Guangdong Education Press.

Shi Yajun. 2008. *On Humanistic Qualities*. Beijing: China Renmin University Press.

Williams, M. & R. L. Burden. 1997. *Psychology for Language Teachers*. Cambridge: Cambridge University Press.

Management, Information and Educational Engineering – Liu, Sung & Yao (Eds)
© 2015 Taylor & Francis Group, London, ISBN: 978-1-138-02728-2

A brief talk about harmonious ideological-education

Yu Ting Dong
College of Software, Jiangxi University of Science and Technology, Nan Chang, Jiang Xi, China

ABSTRACT: The ancient great history recorder Sima Guang said "the one who is intelligent is the one who has morality, but the one who already has morality may not be intelligent". College- and university students are our nation's precious resources, and they are the hope of our people and the future of our country. At the same time, they are the treasures in our people's eyes. Moreover, history makes them take the responsibilities to revive our nation. So, it appears that it is very necessary to give our students an education on morality. In order to achieve this point, many colleges and universities have already put efforts into it, but it appears to have little effect. This text will talk about the method of building a harmonious mind by teaching them "forgiveness" "benevolence" "bravery" and "independence".

KEYWORDS: Characters; View of value; Harmonious mind.

As far as I'm concerned, many colleges and universities have already put efforts into it. And to that end, the colleges start a new course named <<Ideology with Morality & Basic Law Items>>. What's more, colleges also equipped with specialists and psychological counseling rooms. That indicates colleges take ideological education very seriously. But it appears to have little effect. So I think Whether we can educate our students with a harmonious mind, we can start with teaching them "forgiveness" "benevolence" "bravery" "independence", and that means, if you have the ability to forgive others, you will find peace; if you are kind enough, you will have nothing to worry about; if you are brave, you will be fearless; if you are independent, you live on your own.

"Forgiveness", put it in a simply way, means forgive people, don't hurt anyone, and don't force anyone to do anything, even if someone had already hurt you, you should try your best to forgive him, and here I want to quote Confucious' well-known saying "If you don't want something, then you should not throw that to others". This is thc basic attitude towards the things and the surrounding people.

Well, when facing the person who has hurt you before, the person who cheated on your friends, you should use your big heart to forgive him, to let it go. Should we forgive the people who show no respect for you? The answer is yes, we professors ought to make students be aware of that forgiveness is a kind of let it go, shake off the pain which you don't have to take, and don't over thinking about the pain. If you still have the pain inside, you will lose in great depression, I once met a student, abused by another one for no reason, as time goes by, he cannot let it go, always thinking about the dark side, and wondering why should a man insulted him like that,

He just buried that bitter story in his deepest heart, which makes him act abnormal in campus. From this example, what I tasted is the pain, that the student can never forget about it, he just let the pain growing inside, take the heart completely, and make it be numb, What I suggest for this kid is to look forward, let it go, he has a bright life waiting for him, give the student who abused him a chance and give him a chance.

Sometimes, we should not only teach students how to deal with people, but also teach them how to help people, and we professors should set a sample of "benevolence", give them a hint that we human beings always need each other, when somebody needs help, we must try our best to help them out. As the Holy Bible says "It is more blessed to give than to receive", to be a man with a kind heart, like I said before, you will have nothing to worry about.

We professors, are the people to give directions to students, to teach them that " If you have a mind of kindheartedness, you have nothing to worry about". In brief, don't let the small things take the place of your heart, with a positive attitude towards everything, for example, you will have disagreements with your friends, you will have an argument with your lover, inevitably, you will have problems with strangers. All the things are have ways to deal with, we don't have a standard to judge it right or wrong. Just have a big heart, and accept what it is, don't haggle over trifles. Our professors also should let them be aware of the true meaning of benevolence and what will make a person have a mind of kindheartedness. In fact, it is actually teaching them how to deal with people how to make a stand in society.

Furthermore, there still have a question about how to help with our students have a kind heart. Confucius once said "If you want to make a stand in society, then you should help others first; If you want your dreams to come true, then you should help others dreams to come true first". "Do something within your ability to help others, feel the feelings in other mind, that is the way to be a kind man" <<The Anatects of Confucious*what is harmony>>

Judged by these rebellion cases, we professors must send a right message to our students, tell them what is the true brave. Here I quote Confucious words "There is nothing wrong for a person who is brave, but bravery should be limited by righteousness, if your bravery carried by your righteousness, then you have the true bravery"<< The Anatects of Confucious>>

First of all, to envisage the regrets of your life, you should use your bravery to dissolve it. The king of Yue has suffered a lot, then have the power to retake his kingdom back, Han xin was insulted greatly, but he is one of the great men to build the empire of Han. Everybody knows that not every day is a sunny day, we just should be energetic and have a brave heart to the obstacles. Only brave people can find a light in the dark tunnel, finally get to the destination.

Secondly, to enhance the cultivation, self-examine yourself and have the courage to correct the bad. If you can do this every day, you can get everything under control, even if you made big mistakes. Briefly speaking, shake off your bad habits. There is a story about Confucious and his son. Confucius told his son to do something, his son just listened with an open mind, and correct his behaviors immediately. What about our students? Whether our students could take the good instructions and correct their bad acts or not? Did they just eager to find excuses for their mistakes?

Thirdly, college students should be brave enough to correct others bad behaviors. And have the courage to build a harmonious environment.

The emperor Taizong of Tang was known as the great man. The man who is not only good at warfare, but also good at culture. More importantly, he has the ability to feel others feelings and he always self-examines himself. As an emperor, it is not easy to act like that. It is also tough for his servant WeiZheng to survive around him. The old saying goes "serve the emperor is just like serve the tiger". What I respect him is that he has the courage to say something that others can't say, and he can do something that others can't do. This courage just like an arrow, straight and sharp, not afraid of nobilities. And from that example, we can draw a conclusion that there exist two kinds of courage. One is that you have the courage to admit your faults and correct them spite of your status. The other one is you have the courage to point our others mistakes with a brave heart.

We often hear that a man should be independent when he is thirty. To be independent is that you can make a living on your own. Last but not the least, I want to say something about what is independent. And how can our students to be a man with self-reliance.

From the beginning, I want to talk the faith. Roman Roland once said "the most terrible enemy is the lack of strong faith" And I was always wondering how the faith plays a significant role in building a person's character? What makes me have an open mind is because I read an article named<<faith>>, I thought it was really enlightening. And now I want to share my own views with you. The faith is just like your backbone, once you have faith in your mind, you're one step closer to your success, once you have faith in your mind, you have the courage to pursuit your dreams, once you have the courage, you will have the strength to build your own life. SiMagang was sentenced to death by cruel way, but he didn't suicide, because he had the faith to finish his <<historical records>>. Zuchongzhi was insulted by his enemy, but he didn't panic, because he had the faith to carry out his policy. Luxun suffered a lot, but his faith makes him stand up and be a critic, to use his words to wake up the whole nation.

If you want to cultivate an attic faith in yourself, you need work with perseverance and have the spirit to strive to be stronger, be patient and do everything you have to do and trained yourself with basic skills.

Then I want to talk about the honesty. Almost every college has the problems about the students cheating in exams. As for another group of people, they didn't pay their tuition on purpose. These situations reflect a very serious issue. Now in this era, students at colleges are lack of attic faith. To improve their abilities to have good faith, we should grab the root, and teach them to be honest with people. The honesty is that if you say you would do something them you do it. Of course, our actions cannot break the law. For that, I propose several points to follow.

No.1: To strengthen the basic function of your league. And carry out more meaningful activities to help students build an attic faith.

No.2: To put more efforts to broadcast our will, And make our will be popular by recommending students with great movies, published books, radio broadcasting, fancy posters, press crops, and newspapers. And let the students grow a mind in their deep heart that being honest with people is how the society asking for Let the college press to report the dishonest behaviors to make students be aware of them

No.3: Establish and improve the students credit archives, and strengthen the supervision, broaden the students' feedback channels and take every feedback seriously.

I think it is better to put "forgiveness" "benevolence" "bravery" and "Independence" together to use it as a direction, to point out the way for our schools in ideological education. In order to build a harmonious campus, and educate our students ready to build a harmonious society. I always hold a belief, that with all the help of our school faculties, the experiences accumulated before, and the meaningful experiments, our professors would make more effective methods to educate our students, to send a fresh flood to build our harmonious to society.

CONCLUSIONS

Above all, in my view, it is better to put "forgiveness" "benevolence" "bravery" and "Independence" together to use it as a direction, to point out the way for our schools in ideological education. In order to build a harmonious campus, and educate our students ready to build a harmonious society. I always hold a belief, that with all the help of our school faculties, the experiences accumulated before, and the meaningful experiments, our professors would make more effective methods to educate our students, to send a fresh flood to build our harmonious to society.

REFERENCES

Lu,X.H.2007.the exploration of harmonious ideological and political education in colleges and universities. *School party construction and ideological education magazine*:66.

Zang,Jing.2008. The theory of harmonious culture of The Times. *Shanxi University Journals*31(1): 23–26.

Wang,Y.T.2008. Under the background of harmonious culture construction in colleges and universities education administration theory and practice. *Modern education science magazine*:46–50.

Wei,M.C.2008.Cultivating college students' harmonious concept.*Jiangxi university of science and technology journals*29(2): 43–45.

Guo, Q.J. 1982. *A History of Chinese Educational Thought*. Beijing: Foreign Language Press.

Yu, Dan. 2009. *Confucius from the Heart*. Shanghai: Zhong Hua Book Company.

Zhang, S.M. 1965. *Theory and practice of harmonious campus in colleges and universities*. Shandong:Shandong University Press.

Wu, J.Q.2007.*Higher education research in harmonious society*.Beijing: China social publishing house.

Tang, Y.J.1996. *Confucianism and China in the 21st Century*. Shanghai:Xuelin Press.

Fu, Z.P.2005. *An introduction to a harmonious society*. Beijing: people's publishing house.

Management, Information and Educational Engineering – Liu, Sung & Yao (Eds)
© 2015 Taylor & Francis Group, London, ISBN: 978-1-138-02728-2

Discussion on innovative talent cultivating mode under the ideology of CDIO engineering educational concept

Z.R. Bai, X.M. Liu, J.J. Liu & Y.L.Wang
Hebei University of Architecture, Zhan, Jiakou, Hebei, China

ABSTRACT: This thesis analyzes that in our country, presently, hysteresis of engineering educational concept, the curriculum is mainly theory-centered while weak in practice; the teaching staff structure is unreasonable and low in practice level and so forth. It proposes the establishment of an innovative cultivation mode under the CDIO engineering educational concept which includes building engineering educational concept of "major projects", building the cultivation mode of coalition of college and enterprise, 3.3 establish curriculum system with integration, and improve students' knowledge structure, establishing new-type teaching staff to draw lessons from for engineering educational reform in our country.

KEYWORDS: CDIO; engineering education; innovative talents; cultivating mode.

The development of society and economy and the innovation of technology require that higher education should provide talents that adapt to social and economic development. In Occident, engineering education has attracted much attention, and the key point in reform is under the prerequisite of continuing to keep the scientific basis to strengthen engineering practice training, and enhance to cultivate the comprehensive ability of students. How to qualify students with favorable moral integrity and with certain innovation ability teamwork ability and management ability within a limited educational system is the key point to reform and explore the higher engineering education.

1 THE DEFINITIONS OF CDIO ENGINEERING EDUCATIONAL CONCEPT AND INNOVATIVE TALENTS

CDIO engineering educational mode is the latest achievement of internal engineering educational reform in recent years. In 2000, MIT and RSIT and two other universities from the transnational research organization, and probe into and carry out a new mode of engineering education. After four years' intensive research, it puts forward the CDIO engineering educational concept. CDIO applies engineering projects (including products, production process and system), product development and life cycle of a product running as a carrier, and stresses that learning the theory and practice of engineering through the process of conceive-design-implement-operate the system and product of the real world. The educational mode of CDIO requires that the engineering talents from colleges and universities be proficient in these four links. CDIO requires the engineering educational graduates have four knowledge capabilities: engineering basic knowledge, personal ability, interpersonal teamwork ability and engineering system ability. CDIO educational concept advocates "learning through practice", encourages students to obtain engineering ability actively and practically and arouses students' learning interests. The engineering educational concept of CDIO is the inheritance and development of the long-term engineering educational reform of Occident and it covers the certain shortages of engineering professionals through many teaching methods and aims at fostering innovative engineering technological talents of with full development.

Innovative talents refer to the creative talents with innovative spirits and innovative consciousness and innovative passion and with questing spirits of critical and inquisitive, i.e, top talents with innovative consciousness, innovative thoughts and innovative ability who can have new inventions, new discoveries, new ideas and new explorations constantly in economy, scientific research, military and culture and can have innovative achievements.

2 THE PROBLEMS THAT EXIST IN THE PRESENT COLLEGES AND UNIVERSITIES ENGINEERING EDUCATION

There exists the problem of engineering education disjointing engineering practice in the traditional engineering educational cultivating mode in our country.

That is, the cultivation of engineering technology talents is mainly completed by colleges and universities, and the enterprises and employers seldom participate in the cultivating activities of students and the alleged practical teaching, many colleges and universities are simply the practice designed according to a certain course and lack of the specific requirements of the talents actively, making practical teaching a mere formality. Therefore, the apparent shortcoming of the traditional engineering education is to foster the uniqueness of the subject and to foster the unicity of the fostering approach. This not only leads to the waste of social resources, but also restricts the development of engineering education and it is mainly reflected in three aspects:

2.1 *Hysteresis of the engineering educational concept*

Engineering educational concept lags behind the demand of talent cultivation of colleges and universities. Teachers working on engineering education do well in theoretical teaching, but lacking of comprehensive knowledge of practical ability, they can only teach directing at a certain subject and the comprehensive quality cannot be enhanced. Under the engineering educational concept of CDIO, to become prominent professional ability, innovative spirits, strong practical ability, Comprehensive applied talents mastering basic laws and regulations and industry regulations, colleges and universities and teachers, as educational subjects, should transform the educational concepts and explore innovative talent cultivating modes appropriate to social demands.

2.2 *The curriculum focuses on theory teaching, and practical teaching is weak*

Developed countries in the Occident, the fostering of engineers centers on the fostering of practical ability and the present engineering educational mode has still stayed in the educational mode of rich theoretical knowledge but scanty practical ability. Weak practical ability is the bottleneck restricting the employment of graduates of colleges and universities. The fundamental problem lies in that on one hand, the teaching staff of engineering education lacking structural diversity, and most teachers conform to the college-college schema, and master and doctor graduates can get teaching posts with little practical experience. Lack of systematic, practical experience in engineering scheme, project approval, design, implementation and operation, resulting in teachers' repeating what the book says and lack of perceptual cognizance of practice. On the other hand, short of communication between colleges and universities and enterprises, enterprises scarcely take part in the talents cultivation of colleges and universities, and do not attach due importance to the engineering practical joints of students.

2.3 *The structure of teaching staff is unreasonable, and the practical level is low*

At present, there exist the phenomena of seeking high academic qualifications of Undergraduate colleges and universities, making the teachers recruited mostly fresh graduates, lack of related academic, practical experience and a few of them have job qualification certificates whereas those with ample practical experience and professional skills but without high academic qualifications are rejected from colleges and universities. The lack of practical experience of professional teachers has directly led to the teaching mode of theory-based, more than practice-based during the teaching process. Furthermore, many colleges and universities focus their attention on the academic achievements of teachers while paying no attention to the practical ability during assessment and evaluation of professional titles. Thus, teachers put their energy mainly into scientific research, publishing papers and apply for projects, but do not attach due importance to the practical ability of students. The students many colleges and universities fostered have poor practical ability, have weak innovative ability, and have low comprehensive quality, so they do not gain recognition from enterprises and society. It is the lack of engineering practical background of the teaching staff of colleges and universities seriously that causes such a situation.

3 ESTABLISH THE INNOVATIVE TALENT CULTIVATION MODE UNDER CDIO ENGINEERING EDUCATIONAL CONCEPT

3.1 *Setting up the engineering educational concept of "major project"*

Innovating engineering educational concept is the nuclear and soul of the innovative engineering educational system and is the forerunner of building innovative engineering talent cultivation mode of colleges and universities. Only by the guidance of advanced and scientific engineering educational concept for the educational practice can realize the prospective goal, and cultivate large numbers of prominent engineering technology talents with innovative ability. Innovating engineering educational concept should first stick to the educational system of student-centered with many innovative talents showing up. To change the traditional educational concept of imparting knowledge, based on the new concept of knowledge-transference and cultivating innovative ability as the core; to change the traditional educational concept of teacher-centered and establish the new concept of teacher-directed, and highlight the dominant position and function of students.

3.2 *Establish the cultivation mode of "coalition of college and enterprise" and strengthen students' practical ability*

The cultivation mode of coalition of college and enterprise is to divide students' learning time into two stages of which three years is to receive systematic professional knowledge education, and one year is to practice in enterprises. During the stage of learning on campus, colleges reestablish the curriculum and teaching contents and strengthen the cultivation of qualified versatile talents according to the practical need of enterprises and the accurate orientation of talent cultivation of colleges centering on engineering practical ability, engineering design ability and engineering innovative ability. During the year, practice in enterprises can be divided into different periods and also practice in enterprises throughout the year. If practicing in different periods, students are distributed to different enterprises to practice according to the degree of teaching contents. If practicing in enterprises for a whole year, students go to enterprises to practice after nearly finishing the teaching task. The cultivation mode of coalition of college and enterprise puts double-tutor system into effect. That is, colleges designate intramural teachers while enterprises appoint tutors with professional engineers with rich experience to instruct students learn and practice in enterprises. The students receive preliminary training of engineers through learning and practice to enhance engineering consciousness, engineering quality and engineering innovative ability to cultivate the ability to analyze and solve problems applying professional knowledge and participate in the design, research and development, debugging and operation management to promote their own engineering technological ability.

3.3 *Establish curriculum system with integration, and improve students' knowledge structure*

Based on the CDIO curriculum system and its ability system, the engineering knowledge structure includes four graduations of knowledge: one is the base layer, including science of mathematics needed to establish engineering science, natural science reflecting the laws of cognition of man to the objective world, humanities and social sciences needed by cultivating humanistic quality of engineers and scientific engineering outlook; the second is tool-approach layer, including the foreign language ability needed for engineers to refer to document literature and engineering softwares on the platform of internet and philosophy methodology instructing engineers to contemplate; the third is theory-experiment layer, including the professional theory knowledge needed for engineers to work, basic technological skills gained from experiment and disciplinary history containing historical engineering case experience and engineering scientific thinking essence; the fourth is experience-scientific research layer, including the empirical special knowledge for engineers accumulated from practice and exploration in different engineering circumstances, the frontier knowledge obtained from scientific thinking about the new engineering questions, and that is the nuclear knowledge for engineers to innovate.

Establishing the engineering education system with integration is to establish the interdisciplinary curriculum and break through the restrictions of traditional unitary discipline and specialty. According to the need of students' professional development and the practical need of enterprises, integrating the latest achievements of modern engineering science into it and the curriculum content reflects the intersectionality, comprehensiveness and forwardness of subjects, and combine the unitary and repetitive curriculum content, increases the proportion of comprehensive curriculum on the premise of mastering basic knowledge of the subject

3.4 *Enhance teachers' practical ability, and build new teaching staff*

To cultivate excellent engineers, it is needed to have innovative teaching staff with the high engineering practical ability. Colleges should perfect the training system of teachers, and let teachers take part in the teaching experiment and practice. Creating such chances for teachers to improve their teaching ability in key universities, to have titular positions in enterprises, participate in technological reform, technological innovation and technological research and development promote their theoretical competence.

In addition, having more ways to recruit teachers and implementing the appointment system of full-time and part-time teachers. On one hand, inviting famous professors domestic and overseas and experts in first-tier affair departments to give lectures to widen information range of teachers and students. On the other hand, it is feasible to invite professionals with ample practical experience and high technological operation ability and with a certain education background to work as part-time teachers. These part-time teachers bring about the latest technology of the industry, narrowing the distance between colleges and society, and it integrates theoretical courses closely with practical teaching which not only intensifies students' theoretical knowledge but also enhances students' practical operational capacity.

REFERENCES

[1] Z.R.Bai & G.Lv,2013.The application type undergraduate colleges and universities "double teacher" the teacher troop construction research.

[2] Z.R.Bai et al.2014.The school enterprise Cooperation Training Mode Research Excellence Engineer carrier.

[3] S.Wang. "Ways of cultivating engineering education model innovation ability based on". JA: Economic Research Guide,vol:2,No:220,2014:105–107.

Management, Information and Educational Engineering – Liu, Sung & Yao (Eds)
© 2015 Taylor & Francis Group, London, ISBN: 978-1-138-02728-2

The predicaments of specialized English listening and speaking bilingual teaching for international trade majors

Yong Cai Han
College of Economics and Management, Wuzhou University, Wuzhou, Guangxi, China

ABSTRACT: The significance of learning and application of specialized English is presented by characteristics of the international trade specialty. It is undoubted that the listening and speaking course is the critical step to the application of professional knowledge. Currently, some predicaments such as scarcity of faculty, old-fashioned instruction and lack of textbook make the course difficult both to teach and to acquire. All teaching staff need to explore the innovative and scientific teaching and assessment methods, to prepare in-depth teaching materials and pre-courses by specialty, to create a good professional English training environment in listening and speaking.

KEYWORDS: international trade major; specialized English; listening and speaking course; predicament; bilingual instruction.

1 INTRODUCTION

The demand for professional communication in English by enterprise and society sees increasing, according to the characteristics of the international trade specialty and the inclination of graduate career-taken. All the teachers and the tutors face an arduous task that how to encourage students to read and comprehend English data related to specialty and speak it out fluently without any tongue slip. Because of the very reason, all colleges, universities and institutes energetically advocate instruction, specialty English bilingually and attaching great importance to cultivate the application skill to learn professional English. They also spare no efforts to reform and make some nucleus courses bilingual-based exquisite and high-end courses at all levels. There is no exception in our university, that is, we have also made a stride in syllabus design of the international business specialty courses in an all-round, from big class of over 100 students to about 40, from college teachers to professional teachers, from unilateral listening in class to various activities such as regular foreign trade English corner, from trial bilingual instruction to the success of applying for national project for special high-quality course, international trade special English bilingual teaching has gradually evolved into a teaching system with distinctive characteristics and played a crucial role in the curriculum construction and development.

Currently, we have set many courses in the way of bilingual instruction of all types and levels, such as Foreign Trade Practice English Reading Textbook of professional-driven optional course, Excerpts from English Newspaper of professional main course, and International Trade Documentation Practice, Foreign Trade Talk in English, Foreign Trade Contract and International Business Correspondence of both the main professional courses and the practice round. We still have a long and uphill journey to finish in enhancing student's market tapping and business negotiation skills although we have made stride to integrate the in-class learning with extracurricular activities and synchronize the teaching practice with the theory study.

2 EXISTING PREDICAMENTS

2.1 *Scarcity of eligible teachers*

The subject, professional English listening and speaking, is designed for senior students majored in international business and trade, who has laid a good foundation for English learning some knowledge in the field and grasped the terminology in the whole course of business transactions. The majors are generally showing interest in listening and speaking and strong desire to overcome the barriers of communication with business partners. So the demand for teaching staff keeps upward increasing because of the high expectation placed by his teaching objects. On the one hand, it is requested that teachers should master knowledge in all dimensions, including language listening and speaking instruction method and academic theoretic knowledge. Spoken English is expected as fluently and correctly as the natives. Teachers who are eligible should learn the knowledge in the field of foreign trade systematically. On the other, technical handling is among others a must for teachers, who should be familiar with various advanced technologies with the popularity of multimedia. Teaching staff face more challenges when they change from the traditional grammar-translation method to situational one as they should study methodology. All these make teaching staff spend much more time and energy in preparing courseware. The hard job cools teachers' passion for they feel exhausted to be up to expectation. In a word, the resource for the eligible faculty sees scarce and can not meet the demand from society, university and knowledge recipients in the near future.

2.2 *Lack of innovative instruction*

It is not advisable to copycat traditional audio-lingual method or oral approach or both for their shortcoming could be brought out in teaching. Teachers have the inclination to use all-English teaching language in order to build up student's vocabulary and exercise their English practice ability. Students fail to practically use English deal with the problems arising from business negotiation or other difficulties in a transaction if the field terminology is unintelligible to them. Further, students who are used to normal English courses tend to be tired and get a sense of tediousness if the teacher pays most part of his attention to academic impartation or explanation. As a result, students by no means can get something from economic lectures, let alone sitting over the table as the trading specialist. What is more, multimedia-aided instruction may lead to a pure intensifying listening class because the technology which becomes a means of learning has almost replaced the teacher-student communication. Currently, language lab cannot facilitate students well for they find difficult to access to useful academic data, which make autonomous exploitative learning unrealistic.

2.3 *Lack of in-depth and systematic textbook*

At present, In addition to the international trade course teaching, China has published many versions of textbooks dealing with English listening and speaking in international trade, such as Foreign Trade Talk in English (by Chuan Qin), Listening and Speaking for International Business (by Jizhi Ruan), A Speaking Course of Foreign Trade English (by GeLin Zhu, Chun Jiang), A New Spoken English Course for International Business (by Ying Liao), English for Business Communication (by Baoguo Zhou) and International Business English Conversation (by MeiRong Teng), etc. These function-or-subject-centered textbooks have their own limitations although these textbooks do good to students communicative skills to some extent. Some focus on building up vocabulary, some put emphasize on syntax practicing, others just list several dialogues and fall shortage of in-depth content, still others quite substantial in content, but as old as Adam, which means that they need to update or follow up so that can keep pace with the times. Just as mentioned above, our department has come across the predicament in continuity of textbook or teaching material selection. We had planned to offer the course of professional listening and speaking, as long as 3 terms but later on we had to reluctantly cut it down to only one semester. The very deep-rooted reason lies in the lack of international trade professional English textbook written in a systematic and continuous way.

2.4 *Difficult to integrate professional knowledge with language acquisition*

The teaching objects have acquired an English learning ability, especially the listening and speaking ability significantly in the year of freshman and sophomore. They will be coming into a plateau and experience temporarily fossilized state. Senior students begin to learn professional English listening course and actually they have just begun to put their book learning into academic practice. Here comes a contradiction, whether language acquisition and professional knowledge gaining can develop side by side. As such, for foreign trade, negotiations in English, business development and formal occasions for negotiation must follow the rule of Language appropriateness. Teachers guide students in combination with accuracy and fluency. Students are supposed to master the cultural background and the natives' texture of life. Students should not only master professional knowledge in a foreign language, but also develop cross-cultural communication skills. The fact that students must face pressures from a backlog of subject learning tasks and knowledge digesting makes difficult for them to focus on language acquisition. They may not distinguish the simplest terminology with a certain tone of voice, let alone to dialogue and exchange ideas. Many learners can't listen and practice speaking activities in spite of their better understanding in the language grammar and vocabulary. Learners interviewed by author said that they have no idea about the listening materials and retell or paraphrase it out, although the vocabulary is among the least difficulty and the speed the least quickly. That is to say, words appear individually could be intelligible before they being put together with some proper words or special terms. This phenomenon highlights that students have not yet built a framework of language acquisition during the activities and neither internalized received data as the language cognitive structure. It's hard for them to enhance ability to intake language learning in the real sense. Some basic courses and professional backbone courses have not been taught in bilingual way, so it's common for students that they accumulate less special words and lower reading intelligibility. Therefore, students are likely to be frustrated by the psychological barriers. This tends to lead to a vicious circle that, according to human psychology, disadvantage of receipt of information may in turn hinder professional knowledge learning and transformation.

3 PROBE THE WAY OUT

3.1 *To meet the demand for faculty*

In the first place, school authorities should attach importance to the training of bilingual teachers, stay focused on the communication-oriented bilingual talents and build a platform for English teachers to learn international trade professional knowledge. Only in doing so, can we improve academic level and teaching method of newcomers and to improve professional competence by the way of on-job training, self-taught, applying for further study, participating scientific research project, going abroad to equip teachers with strong ability of specialized subject (professional knowledge and their own literacy) and the technical ability (design and organization of classroom activity ability), at the same time avoiding too professional but little general knowledge staff. Teachers should not only probe into language knowledge and advanced teaching theory, but aptly explore teaching content involving the other knowledge. Secondly, university can import some talents abroad who are really aspired in teaching if the conditions permit and invite foreign trade experts, scholars and specialists to make professional knowledge of the full English lecture. Thirdly, it's highly recommended that universities employ some staff engaged with practical work experience or foreign international business personnel to present and participate in professional English teaching. The most practical way is to choose some from the economic teacher team who have got undergraduate courses in English major and then gets a master's or doctor's degrees in the economy, or the quality international English teacher who has got the further education experience abroad because they have solid English speaking skills with a wide range of professional knowledge. Their English, economic and trade experience in the field together will make the professional English teaching, especially in the listening and speaking teaching a new breakthrough.

3.2 *To innovate teaching and assessing methods*

The faculties are supposed to comprehensively use various teaching methods and absorb the essence of situational teaching method and audio teaching while putting aside limitations. Direct teaching method should be cast away and neither should the grammar translation teaching method be copycatted. Teachers cannot completely apply American structuralism theory, which emphasizes on grammar and neglects the context. Grammar is instructed in an inductive or deducing way. Teacher, in his initiatives to innovate teaching method such as dictation training method, interactive listening and discussion methods with academic data to boldly challenge old-fashioned and stereotype teaching approaches. To put the students in the central place can stimulate their creativity for knowledge exploration, founding probing, and work cooperation to realize the meaningful construction of language acquisition. To adopt different listening strategies according to learners' language level can make the teaching efficient and develop students' ability to cope with all kinds of difficulties in listening and speaking. The challenging questions designed during teaching can exercise students' thinking over some complicated problems and professional knowledge outside the class. Teachers as a guide should not only excel in organizing activities in class, but be equipped with comprehensive knowledge in addition to teaching materials, especially cultural background knowledge to deal with the problem of information surfacing and deep-processing. Teachers guide learning tasks timely and effectively and organize the class activities closely based on the topic so that giving full play to teacher's professional knowledge and teaching skills. To make full use of modern multimedia in teaching is a good way to arouse student's passion for independent learning, so the early fossilization in listening could be avoided. In practice, teachers should be well aware of the auxiliary role multimedia facility played and write suitable and applicable courseware for students of this specialty. In addition, teachers are encouraged to come up with some new methods of assessments such as setting up an archive for each student's performance and to adopt the combination of process evaluation and goal assessment.

3.3 *To select teaching materials carefully*

institutes can introduce original edition of teaching materials related to economy and trade from Hong Kong or America to solve the problem of continuity and gradation of the teaching materials. The selection should pay attention to professional characteristics and highlight international trade rules and practice in transaction. The contents should be arranged intensively and extensively and helpful for learners further exploration. For example, the books may have bearing on how to solve the foreign trade disputes, especially resulted from the terms of the contracts and documents, how to do well preparing for business negotiations, how to maximize the profit while concluding a deal pleasantly as well as how to identify the authenticity of L/C, how to use the red clause of L/C and so forth. All these require learners' oral expression. Therefore, to select practical professional materials is one of the ways leading the predicaments out. It can be supplemented by articles or excerpts from foreign newspapers and network, which collect the latest international trade practice and application. There may not be perfectly customized materials for such kind of teaching system because of the complexity of international trade. So teachers can write teaching materials according to the actual situation and have learners help to errata every edition of the materials, list the work of follow-up in the long-term research planning, and download search the relevant international video conference as auxiliary materials.

3.4 *To offer fundamental pre courses*

Use Talents of international trade should grasp modern economics and the basic theories of management, master international business operation theory, practice, and professional techniques, be familiar with different cultures and economic and social status in the world. Their knowledge can meet requirements of economic globalization and regional development and can independently conduct multinational management and business planning. Therefore, major in international trade need to improve the language acquisition and application ability on the basis of accurate understanding and proficiency of language step-by-step. It's important to open some basic professional courses in bilingual teaching, international economics, International trade theory, International trade practice and international payment are among the list. The curricula can facilitate bilingual teachers with updating knowledge on the one hand and, on the other, learners can naturally realize perceptible internalization of professional knowledge by uniting theory digestion and language acquisition into case study. This job can get done better when teachers keep pace with the trend of world economy and have a keen eye in the world's latest economic dynamics and share the field frontier research with students and reading a piece of economic news every day and study economists' achievements. Students can learn more from a prepared speech and improvisational economic reports.

To boil down, if teachers want to create an ecological learning environment in professional English listening and speaking through both inside classroom and extracurricular activities, it is desirable to adopt the small group of classroom like the earliest army teaching to maintain students' concentration inside classroom or grade the classes according to the level of listening ability. Meaningful classroom activities will turn superficial learning to underskin learning and improve the proficiency of the listening. Constructivism holds that the learner is not a passive knowledge receiver, but the knowledge structure builder. The best way is to make them experienced in the real surroundings with the view to achieving the goal of cognitive meaning building. Students can improve the application skills in actual or quasi-professional simulation. Organizing students to pay a visit to enterprises or to take in professional post to get situational experience can compensate for the shortcoming of the in-class practice and develop three-dimensional practicing environment.

REFERENCES

Baimei Shu & Youlin Chen 1999. *Foreign Language Teaching Methods*. Higher Education Press, Beijing, China.

Liming Yu et al. 2009. *Bilingual Education— The Implications of the Canadian Immersion Education to the Bilingual Instruction in Chinese Universities*. Foreign Language Teaching and Research Press, Beijing, China.

Binhua Wong. *Bilingual Education and Bilingual Teaching*. Shanghai Education Publishing House. Shanghai, China.

Mimi Wong. To Explore Legal English Listening and Speaking Teaching. *Journal of Hebei Law Science*, 2011.06.

Qiang Niu. The Fossilization in Transitional Language and Revelation for Teaching. *Journal of Foreign Languages and Their Teaching*.2000 (5): 29–30.

Xiaoli Ra. *On Ecological Teaching and Its Construction for English Listening and Speaking Course in Vocational Colleges*. The Proceeding of Sichuan Education College: 2009.10.

Management, Information and Educational Engineering – Liu, Sung & Yao (Eds)
 ISBN: 978-1-138-02728-2

Comparative study of open source e-learning systems

V. Maněna, M. Maněnová & K. Rybenská
Faculty of Education, University of Hradec Králové, Czech Republic

ABSTRACT: The paper is based on studies that compared the individual types of tools in open source Learning Management Systems Moodle and Claroline. Selected LMS systems are used in secondary schools in the Czech Republic. These systems are used primarily for their availability. In comparison system, we proceeded from the description of each tool, their functions and options. Individual instruments have been validated for both system in teaching practice, to their mutual comparison was valid. Results of the survey summarizes the comparison table.

KEYWORDS: LMS; E-learning; open source, comparative study.

1 INTRODUCTION

E-learning is one of the ways to effectively use information and communication technologies in education. Elements of e-learning began to be used initially in high schools and aimed to improve the quality of the educational process. Information and communication technologies are used to support the cognitive processes and social and psychological aspects of education. The teacher training is an indispensable factor, but his role is changing now. At the same time, the design of learning materials is changing too. As we already know, suitability of technology can not be assessed separately from the basic strategy approach to their use in the educational process. E-learning technology, we can use as a component of the educational process itself, as an administrative tool and also as a development tool for creating multimedia learning materials. On the market today there are several LMS systems in various price categories that provide these features. Many universities in the Czech Republic and the world have developed their own tools and systems. Some of these systems are also available for free, under some kind of open source license. E-learning has gradually expanded to high school and begin to use it as primary school many years ago. (Hansen, 2000)

2 THEORETICAL BACKGROUND

2.1 *LMS moodle*

The system Moodle is designed to support teaching and his name was originally an acronym for Modular Object-Oriented Dynamic Learning Enviroment. The first creator software was Martin Dougiamas and the first version was published in 2002. Moodle is developed as a tool to support social constructivist approach to learning. Moodle can be used to support full-time, combined forms of education and distance education. It is provided free of charge as a free and open source software licensed under the GNU General Public License. To work with this software, user just need ordinary computer, tablet or smartphone with a web browser connected to the internet. (Moodle, 2014)

Moodle was designed to promote constructivist pedagogical approach to education. The advantage is that the courses can be developed within the system, resulting in considerable savings. The system can be used both for distance study, as a supplement to traditional teaching.

User interface is intuitive and easy for every user, which is at least minimally computer literate. Moodle is now extended with so-called repositories, which are used for data storage. All users can download or upload information into the repository and are not limited only to their own computer. Appearance of the system can be modified to match the original web site of school or organization.

The course is built from components called modules. Each module is characterized by its specific properties that can be set differently, according to user's needs. The standard Moodle package contains several modules, which can be used in any course, but it is also possible to install external modules. One type of modules represents study materials and the second type represents the activities of students. These two types of modules differ in level of interactivity of student. Study material is a unidirectional flow of information and creates a kind of basis. Moodle is based on the fact that the most important in the learning process is the active involvement of students in various educational activities, and based on this, students acquire new knowledge. On the one hand, it is possible to use Moodle only as a system that enables the presentation of the study material on the Internet, on the other hand, it is possible and advantageous to use all his potential, which offers us, and to emphasize the student activity modules.

Standard modules of activities can be included in various sections of the course or in a block on side of the course. Part of the course can be arranged thematically or by week. Each week or topic is graphically separated. Modules can represent activities below:

- Poll. Teachers create a poll with multiple choice. The aim is to inspire students to arouse interest in the topic, think about it, etc.
- Chat. It enables each participant to send text messages to other in real-time discussion.
- Database. Allow students and teachers to create and view files of records that relate to any topic.
- Forum. This is the most common form of discussions among all participants of the courses. Participants have the opportunity to comment at any time on the topic. It is also possible to adjust the assessment of individual contributions by others. Users can also take advantage subscribe posts by email.
- Survey. A module that contains several questionnaires that are used to collect information about students, about teaching, etc.
- Lecture. Programming-based learning. It consists of several pages. On each page there is a question that has several choices. Student answers the question on the basis of the answers get farther or returns. The program may be either linear or branched.
- Dictionary. A module that provides students and teachers to create and edit a list of definitions.
- Test. Very often used module that allows teachers to build different kinds of tests. It is the choice of several types of tasks. For example matching, fill in blank, multiple choice, short text answer, numerical answer, true/false.
- Test Hot Potatoes. Tests are created in software Hot Potatoes and then are imported into Moodle.
- Assignment. The module, in which the teacher arranges tasks (may be an essay, papers, projects, etc.) and the student is responsible for its contents saved in the system. The teacher can enter the date of submission, if that expires, students no longer have the option to submit.
- Wiki. With this module, students can work together on web pages.

In addition to these basic modules that are included in each Moodle package, there are also external modules that are constantly growing and are regularly updated.

Study materials form one of the cornerstones of the course. Moodle gives you the opportunity to choose from several different types of learning materials:

- Poll. Teachers create a poll with multiple choice. The aim is to inspire students to arouse interest in the topic, think about it, etc.
- Label. Text, which can be supplemented by pictures, which is located directly on the course page the links between different activities.
- Page. Content that is written as a text, klterý may contain images and other multimedia elements.
- Link to file or web. The teacher inserts a link to a file or web page.
- Folder. This is the option of opening a specific folder. A teacher can insert into a folder multiple files at once.

The philosophy of the system is based on a social constructivist theory of learning. Instructional tasks, activities or projects can be designed to allow for cooperation between teacher and students or between students. Students can be divided into groups to communicate with each other via chat synchronously or asynchronously in the forum. The wiki tool (module) allows students to write text together. Discussion forums can be set so that only the teacher or a student can start a new discussion topic (thread).

2.2 *LMS claroline*

It is a case of representative of opensource of LMS systém, which started to develop at the Université catholique de Louvain in 2000. Followingly Claroline Consortium originated in 2007, which is formed by 15 universities at present time (e.g. from Belgium, France, Canada, Spain, Chile) and Belgian and French universities are predominating. This is the reason why the starting language of system is French, English is in the second place. Basic feature of LMS Claroline is simplicity of user's dividing line and at the same well though-out implementation of the separate functions. (Šín, 2013)

Administration of the users is easy. Claroline distinguishes three types of users – student, teacher, administrator. Courses could be incorporated in different categories (defined by user).

In Caroline learning curriculum consists of separate topics, which affiliate educational materials (HML, pages, sets, references, tests, etc.). This thematical arrangement plays here an important role and to each of the topics is added its statistics with he survey of students' progress. Problematic seems to be possibility to record of one set only. In case of recording more sets it is necessary to zipp them.

Test is formed in the offer of the same name Training. Each test must have certain title and description. It is possible to select, if all tasks should be depicted together, or one question on one page. Self-evident are even "extended choices" concerning the possibility to set data and time of starting and closing of test, time limit and test behviour after its completing (imagination or not imagination of correct answers, etc.) Tests could be used both for evaluation of students e.g. by setting of "one permitted attempt" and for training, by means of unlimited number of atempts. We can summarize generally that to the positives of the system could be placed easy installation and administration, administration of classes, limiting of students number in course, side menu and its understandability, statistics and their clear arrangement. As a problematic fact is considered insufficient localization in Czech language, less topics and extending modules than in Moodle, less extending setting.

3 THE PROJECT

3.1 *Project aim*

The goal of our research project was to compare the options open source learning management systems. At the same time, we aimed to determine which LMS are used in secondary schools in selected locations.

3.2 *Methodology*

Based on the goals we set as a basic research method, we chose comparative analysis. We used descriptive statistics to determine the type of LMS in secondary schools in selected locations. In this investigation, we analyzed websites of secondary schools, or contacted the school management. In total, our survey contained data from 89 high schools in region and 202 secondary schools in the capital city of Prague. For mutual comparison of LMS, we identified the following criteria:

- Tools intended for generating contents.
- Communication tools.
- Tools for collecting and evaluating activities.
- Tools for cooperation.
- Price.

Each of the selected LMS system was evaluated by four independent assessors - teachers who work with the system in the classroom.

3.3 *Research results*

From a total of 291 schools, 64 were using e-learning (31.7%). Moodle was used in the 69 % of schools and 31% used a different system. There were versions of Claroline 1.10.7, Moodle 2.3–2.5.

First we compared basic functions of both e-learning systems. Moodle has more features, but Claroline's features are enough for needs of primary and secondary schools.

Moodle has more communication tools than Claroline, but most of users prefer chat and discussion forum. Both of these tools are in Claroline and Moodle.

Both systems have equal tools for evaluation and self-evaluation.

Table 1. Basic tools of selected LMS – generating content of study.

	Claroline	Moodle
URL	+	+
File	+	+
Folder	+	+
Legend	–	+
Book	–	+
Lecture	–	+
Syllabus	–	–
Lesson plan	+	–
Integration with study contents of other LMS	–	+

Table 2. Basic tools of selected LMS – communication tools.

	Claroline	Moodle
Discussion forum	+	+
Chat	+	+
Reports	+	+
Inquiry	–	+
Comments	+	+
Blogs	–	+
Survey (question-form)	–	+
Quickmail	+	+

Table 3. Basic tools of selected LMS – collection and evaluation of activities.

	Claroline	Moodle
Assignment (on-line text, set, off-line activity)	+	+
Workshop (Self and Peer Assessment)	+	+

Moodle has more tools for cooperative learning. Both systems have wiki, which is in many cases the most preferable tool for collaborative learning.

Table 4. Basic tools of selected LMS – tools for cooperation.

	Claroline	Moodle
Dictionary (index)	–	+
Database	–	+
Wiki	+	+

Table 5. Basic tools of selected LMS – additional features.

	Claroline	Moodle
Tests	+	+
Tracking	+	+
Statistics	+	+
Group mode	+	+
Language adjustment	+	+
Gradual loosening	–	+
Calendar	+	+
Internal mail	–	+
Certificates	–	+
Virtual classroom	+	–
Price	–	–

Generaly speaking, Moodle has more and advanced features compared to Claroline. Buth both systems can be used very well on primary and secondary school. Moodle's advanced features are suitable for higher education, but can be used on primary school as well.

4 CONCLUSION

We can say that the LMS Moodle is a sophisticated system for creating e-learning courses. It offers a fairly wide range of applications. We would like to emphasize in particular tools for collaborative teaching. The Claroline is simpler than Moodle, but may well serve the needs of schools.

REFERENCES

Hansen, R. E. 2000. *The Role and Experience in Learning: Giving Meaning and Authenticity to the Learning Process in Schools.* Journal of Technology Education. Vol. 11 No. 2, Spring 2000. Available from http://scholar.lib.vt.edu/ejournals/JTE/v11n2/pdf/hansen.

Moodle. 2014. *Moodle philosophy.* Online. Available from https://docs.moodle.org/27/en/Philosophy.

Šín, M. 2013. *LMS Claroline – e-learningový systém očima uživatele Moodle.* Online. Linuxexpres, 2013-05-06. Available from: http://www.linuxexpres.cz/software/lms-claroline-e-learningovy-system-ocima-uzivatele-moodlu.

Management, Information and Educational Engineering – Liu, Sung & Yao (Eds)
© 2015 Taylor & Francis Group, London, ISBN: 978-1-138-02728-2

Transformation of higher vocational education in the context of economic transition

Hui Qing Li & Ping Li
Shenzhen Polytechnic, Shenzhen, China

ABSTRACT: The economic transition drives and forces the transformation of higher vocational education, which in turn promotes the economic development. The transformation of school-running system and mechanism for higher vocational education involves the diversification of school-running subjects, the marketization of resource allocation and the socialization of quality assessment; the adjustment of talent training mode is represented by synchronization of vocational education programs and economic planning, correspondence of major setting to economic restructuring, course development jointly completed by schools, industries and enterprises and multi-level vocational education; the transformation of higher vocational education philosophy should start with education philosophy, learning philosophy, scientific research philosophy and success philosophy.

KEYWORDS: economic transition; transformation of higher vocational education; mode adjustment; philosophy transformation.

China's economy has made remarkable achievements through 30 years of fast development, but at the same time it is confronted with a series of problems like growing environmental pollution, resource depletion and aging of population. The development mode characterized by low labor growth and high resource and energy consumption that has long been observed by China has been difficult to continue and entails economic transition imperatively. Any transformation ultimately involves the transformation of people. As higher vocational education features simultaneous development with the economy, it is particularly important to explore the transformation of higher vocational education in the context of economic transition.

1 TRANSFORMATION OF SCHOOL-RUNNING SYSTEM AND MECHANISM FOR HIGHER VOCATIONAL EDUCATION

Like the economic transition of China, the transformation of higher vocational education also has its intrinsic motivation and external pressure. First of all, the higher vocational education is separated from higher education system. It achieves rapid development by innovating major setting, curriculum system and talent training mode, but the operation system continues to follow that of higher education which is featured by low management efficiency and simple school-running mode, thus restricting the development of higher vocational development. Therefore, the system and mechanism must be transformed if the higher vocational education wants to get thoroughly reborn. Next, the economic development is the foundation for higher vocational education development. The economic structure restricts the structure of higher vocational education, and the economic transition includes system and mechanism and economic restructuring. Thus, the effect of economic transition in driving and forcing the transformation of higher vocational education is quite apparent. Thirdly, the economic transition strives for the transformation from manufacturing to innovation as one of its goals, so it poses higher requirements for the overall higher education. In this context, the transformation of system and mechanism for higher vocational education plays the role of a pilot.

Most of the current higher vocational colleges are transformed from previous secondary technical schools and institutions of higher education. They contribute to the formation of vocational education system together with the training institutions from the society and enterprises. In fact, the system remains quite imperfect, such as unreasonable structure, poor school-running conditions, etc. In short, the system is far from powerful, so it fails to adapt to the requirements of economic and social development. The transformation of higher vocational education should divide universities into research-oriented type and applied technology type at first, of which the latter are only allowed to engage in higher vocational education. In this way, the higher vocational education can be developed into a large-scale and multi-level system that adapts to or even leads the economic and industrial development.

At present, the higher vocational education colleges are mostly public institutions, characterized by unbalanced development between the east and the west and between public colleges and private colleges, simple subjects of school-running and monotonous school-running mode. The current economic transition promotes the development of mixed ownership economy, open to the domestic market in order to stimulate domestic demand and endow Chinese economy with endogenous vitality. The transformation of higher vocational education also applies the form of mixed ownership. That is to say, governments, industries, enterprises and individuals all can become the subjects of school-running by resorting to resources or technology, thus forming a diversified school-running pattern. In this way, it will help to expand the participation of the whole society in higher vocational education and improve school-running conditions.

The resource allocation for higher vocational education is also unreasonable because governments often allocate physical or monetary resources directly or indirectly as per the number of enrollment. The resource allocation varies greatly depending on school-running conditions and regions, so the practice is contrary to educational fairness. The process of resources is not targeted and fails to highlight the benefit-based principle, and resources often become school's "private property" once allocated, with low utilization rate. Therefore, the transformation of resource allocation must be made for higher vocational education as the economic transition progresses in order to realize market-oriented allocation of resources. The marketization of resource allocation is reflected in two aspects: marketization of resource source and sharing of resource utilization. As the subjects of school-running are diversifying, the marketization of resource allocation is the inevitable result of higher vocational education development. The governments may set up education funds and participate in the resource allocation for higher vocational education through such funds. As a result, the governments change to the subject of supervising resource allocation and utilization from the subject of educational resource allocation.

With the diversification of school-running subjects and modes, the role the governments play in higher vocational education is also changing accordingly. Specifically, the governments should perfect the education legislation and the school-running and access system for higher vocational education, establish the supervision system for higher vocational education quality, improve the assessment mechanism on the operation of higher vocational education system, and identify the elimination mechanism for unqualified colleges or majors and the withdrawal system for school-running subjects, so as to provide institutional guarantee for the transformation of higher vocational education and its positive operation after transformation.

The diversified school-running pattern of higher vocational education has actually introduced the market mechanism, so the assessment on education quality is bound to socialization. The teaching assessment features inspection function and guiding function. The latter often exerts a greater social influence, for example, the university rankings arising from social assessment directly influence the enrollment, employment and recognition of universities. The assessment on teaching quality in higher vocational education should be carried out by third parties, and the results should be reported timely to governments, schools, industries and enterprises. The index system of quality assessment should seek for public opinions and feature openness and foresight, so as establish unique higher vocational education in combination with economic transition.

2 ADJUSTMENT OF TALENT TRAINING MODE DRIVEN BY ECONOMIC RESTRUCTURING

For the past many years, the higher vocational education in China has been exploring a higher vocational education mode suitable for China all the time according to the psychological and mental factors of students by drawing on advanced foreign experience, such as "dual system" from Germany and "teaching factory" from Singapore. The project teaching, case teaching and other teaching methods based on working process have enjoyed great popularity through teaching practice. However, the economic transition denotes changing national conditions, so the talent training mode for higher vocational education should also be adjusted accordingly.

In the past, the fast economic development of China was achieved at the cost of huge consumption of material resources, energy and labor forces, while the economic transition exactly aims to improve the quality of economic operation and enhance the comprehensive competitiveness by relying on technological progress and independent innovation. Therefore, higher vocational education must serve as the distribution and processing center of new technologies. The governments should consider vocational education planning while making economic planning, or even incorporate it into economic planning. After that, schools, industries and enterprises should refine their own vocational education planning according to government planning and in combination with school characteristics, so that the higher vocational education can develop simultaneously with economy. The higher vocational education should not only adapt to economic transition, but also participate in technological innovation and play a leading role in new technology, new processes and new products. An assessment and adjustment mechanism should be established for major setting and curriculum system of schools. The majors should be classified by categories as far as possible, the specialized courses should be determined based on market demands, and the curriculum development should be completed by schools, industries and enterprises and should not be put into implementation before approval by relevant governmental authorities.

The higher vocational education should strengthen practical teaching and insist on teaching activities which combine learning with working, manual laboring with thinking and study with practice. According to constructivism, if learners want to complete meaning construction for the knowledge learned, that is, to have a profound understanding of the nature and rules of the things reflected by such knowledge and the connection between these things and other things, the best way is to feel and experience in real world instead of merely listening to other's description and explanation of such experience. Here the best way to feel and experience is practice and training. To improve the process of practical training, the first is to strengthen the cooperation between schools and enterprises. The governments can transfer partial stock rights of state-owned enterprises to schools and the stock rights of higher vocational education colleges to enterprises, thus laying a solid foundation for school-enterprise cooperation. Next, in addition to strengthening the construction of training rooms, schools may also operate factories to serve as the experimental base of technological innovation.

The faculty is a key point in the transformation of higher vocational education. To strengthen the construction of double-qualified faculty, teachers must be familiar with industry developments and have international vision, and must be experts in both teaching and subject matters. The faculty training should be provided on a regular basis and integrate production and teaching.

The aging of population which influences labor supply is an important factor in economic transition from labor-intensive to technology-intensive. In addition to serving as the distribution center of technologies, the higher vocational education should also be adjusted structurally and provided at multiple levels for different objects. The first level is provided for students graduating from junior high schools, adhering to 3 + 2 educational system. The students graduate from secondary technical schools in the first three years and then graduate from junior colleges during employment if they can pass the examination in the last two years. In this way, the students have the freedom to decide whether to graduate from secondary technical schools or junior colleges. They can choose to work for some time after graduating from secondary technical schools and then continue to finish the courses of junior colleges in the last two years. The second level is provided for students graduating from senior high schools, also adhering to 3 + 2 educational system. The students graduate from junior colleges in the first three years and then graduate from undergraduate colleges during employment if they pass the examination in the last two years, with equal freedom to choose. The third level is undergraduate and graduate education, and the fourth level is just-in-time training for re-employment and pre-job training for other people.

3 TRANSFORMATION OF HIGHER VOCATIONAL EDUCATION PHILOSOPHY DRIVEN BY ECONOMIC TRANSITION

The economic transition is bound to drive cultural transformation, but the fact is, new culture has often waken up when the original economy remains rushing forward with inertia. From the time perspective, economic transition and cultural transformation are like twin brothers. During pregnancy, the economic transition is the elder brother and drives the younger one, but during delivery, the roles are changed and the cultural transformation awakens the economic transition. From the relationship between the two, the economic transition decides the cultural transformation which then drives the economic transition again. The sensitivity of culture to economy is the characteristic of higher education, and the cultural transformation determines the transformation of higher vocational education philosophy.

Firstly, transformation of higher vocational education philosophy. The vocational education is often inevitably misunderstood, because people tend to think that it aims to teach students professional skills and make them useful for the society, and that the vocational education should base teaching on working process and combine learning with working as it is anyway for creating employment for students. In fact, this understanding is not accurate. The nature of education is to make learners perfect and transcend themselves, but not to serve as the tool of training for employment. The vocational education is a path, a grip and an environment for learners to constructed by themselves. Making the learners useful is indeed the result of vocational education, but not the overall objective. The teaching method that bases learning on working process and combines learning with working should focus on learning rather than working. The vocational education is based on vocation, but also goes beyond it. It aims to foster a mind of seeking perfection and nourish humanistic reflection and critical spirit, which, if absent, would make use hard to imagine the transformation of China's economy from manufacturing to creation.

Secondly, transformation of learning philosophy. Currently, the new enrollments for higher vocational colleges remain junior or senior high school students, and the social training institutions continue to set up such positions to the society as cook and barber. Our vocational education remains in its primary stage. The economic transition leads to changes in employment, and re-employment training will strengthen the concept of life-long learning. The economic transition promotes social transformation to a learning-oriented society, so the education should not be limited to children or the undereducated nor for getting a job. The higher vocational education should play a leading role in the formation of a learning-oriented society and step into advanced stage.

Thirdly, transformation of scientific research philosophy. The pyramid structure of scientific research in China has its historical and practical reasons. The top-down research habits make us doubt how higher vocational college students engage in scientific research. People with such doubt often focus more on the results of scientific research, but ignore its target, viz. innovation. For higher vocational education, innovation can be cultivated as a kind of consciousness. Through vocational education, vocational scientific research can be made and innovation and entrepreneurship can be deeply rooted in the mind of learners and get popular in the society. For example, when Li Ka-shing asked his son of several years old to attend Board meetings, it is just like the students of vocational institutions engaging in scientific research. Therefore, vocational scientific research projects should be established and become the compulsory course of vocational colleges.

Fourthly, transformation of success philosophy. The success philosophy is actually the concrete representation of values, and the core of cultural transformation is the adjustment of values. The economic transition of human society is mostly established due to economic crisis which is usually caused by violation of economic ethics due to wrong economic values. This is the indirect result arising from the humanity philosophy that success is understood as behaving more excellent and getting wealthier than counterparts. Therefore, higher vocational education should incorporate social sense of accomplishment, sense of social responsibilities and the spirit of unity, cooperation and mutual benefit. In this way, the pursuit of materials by human can be restricted by humanity.

4 CONCLUSION

China's economy which is restricted by labor, resources and environment starts to transform. The subsequent transformation of higher vocational education should embark on systems and mechanisms. The transformation of systems and mechanisms will naturally lead to the change of talent training mode and the update of education philosophy. The transformation of higher vocational education will embrace the great development of vocational education and make due contribution to the high-quality comprehensive transformation of China's economy.

REFERENCES

Li Yuanchao. June 28, 2013. Eliminate Crisis Fundamentally through Win-win Innovation and Development. http://www.china.org.cn/chinese.

Zhang Yingqiang. Spirit and Mission of University Culture[M],Hefei: Anhui Education Press,2008:35–50.

Xiao Fengxiang & Fu Weidong. Crossing the "Economic Transition Inflection Point": China's Vocational Education Facing Missions and Challenges[J], Journal of Tianjin University (Social Sciences),2014 (3): 142–147.

Cheng Wei. On the Characteristics of Higher Education Reform in the Economic Transition of China[J],Journal of Liaoning University (Philosophy and Social Science Edition),2006 (9):1–6.

Wang Xiaohua. The Study of Higher Vocational Education Based on Industrial Upgrade Professional Setting - To Hangzhou for example[J], China Higher Education Research,2013 (2):107–110.

Management, Information and Educational Engineering – Liu, Sung & Yao (Eds)
© 2015 Taylor & Francis Group, London, ISBN: 978-1-138-02728-2

Analysis on the Japanese language teaching reform in the institutions of higher learning

Fan Wu
College of Foreign Language, Beihua University, Jilin, China

Zhen Chen
College of Information Technology and Media, Beihua University, Jilin, China

ABSTRACT: As the economic exchanges between China and Japan become more frequent, the number of Japan's companies shows an increasing trend in China in recent years. Under such a background, there are increasingly more schools in China to offer a Japanese major and simultaneously there are increasingly more Japanese learners. As a result, the traditional education model becomes unable to meet the need of the modern society for Japanese language talents. How the teachers train the students who will be the most popular to be employed by enterprises and possess an excellent comprehensive quality after graduation is an issue necessary to be studied in this paper.

KEYWORDS: institutions of higher learning; Japanese language teaching; teaching reform; teaching methods.

1 THE CURRENT SITUATION OF JAPANESE LANGUAGE TEACHING IN INSTITUTIONS OF HIGHER LEARNING

In recent years, the vigorous development of Japanese major based on meeting the market demand for Japanese talents is closely inseparable with the high importance attached by the Japanese Language Subcommittee of the Ministry of Education. At present, the contents of the courses offered for Japanese language students are diversified; the tests for verifying the ability of students in Japanese language include not only JLPT (the Japanese-Language Proficiency Test), J. TEST, but also College Japanese Test Brand 4 and College Japanese Test Brand 8. Therefore, the learning of Japanese language students is tested through all sorts of Japanese language proficiency tests in terms of listening comprehension, grammar, reading comprehension, and composition, etc. However, a great number of students learn Japanese language only for passing the tests or obtaining certificates because they have no enthusiasm for Japanese language learning and also think lightly of the practice, so that they only acquire all kinds of certificates during the period of the upcoming graduation, but know little about Japan's economy, politics, and culture. Worse, the communication with Japanese people is beset with difficulties. Finally, they are incapable of making authentic Japanese daily communication with Japanese people although they can apply the learned grammars and sentence structures to answering sheets. Considering this existing phenomenon, the problems in Japanese language teaching are analyzed as follows according to the actual teaching conditions.

1.1 *The divorced closeness between the japanese language teaching and the practical application*

According to student's employment directions, Japanese language courses such as Business Japanese, Japanese Economy, Tourism Japanese, and Science and Technology Japanese are offered in the curriculum provision; in order to let students gain a better understanding of Japan's history and culture, the courses such as an Introduction to Japan, Japanese Literature, and Japanese Linguistics are offered for students; in order to let students practice oral Japanese better, Japanese conversation courses taught mainly by Japanese language teachers are offered; in order to improve the students' listening ability, listening comprehension courses are offered for students. In the setting of these courses, the requirements in the accumulation of students' listening, speaking, reading, writing, and translating are considered. In the primary Japanese courses, top priority is given to comprehensive Japanese, Japanese listening, and

Japanese conversation courses. When students begin to learn Japanese language in institutions of higher learning, their problems are slow adaptation and poor pronunciation; the time for reciting pseudonym and practicing pronunciation will occupy students' a lot of time, making the time of the students too longer to learn basic Japanese language. In the teaching process, teachers often provide students with explanations according to the teaching materials. In this way, the designed teaching contents in each semester can be completed, but the teaching time is limited and there is no enough time for the teachers to let students practice in the classroom when many text contents are completed. The students in non-key institutions of higher learning are poor in self-study, memory, and ability in induction, comprehension, and practical application, so that they cannot make a connection between what they have learned.

1.2 *Fuzzy training objective*

At present, the institutions of higher learning usually set up Japanese major orientations such as business Japanese orientation, Japanese translation orientation, and IT Japanese when Japanese major is offered, in order to meet the needs of social development and market demand for the talented personnel. In the learning process, students are required to learn other professional knowledge in addition to Japanese professional knowledge. In terms of curriculum provision, the teaching materials marked with business and IT are basically chosen; the teachers only repeat what the book says, but do not get a good understanding of the essence in the major orientation, so that the interest of the students in learning is not stimulated or most students cannot understand the teaching materials once the contents are a little bit difficult. For the students in the ordinary institutions of higher learning, their initiative to self-study is very poor, and also the sheer Japanese knowledge has made them overwhelmed, making them feel unable to do what they hope to do plus the study of other professional knowledge; they can only scrape through the final exam of each semester by relying on the answers from teachers. During the period of the upcoming graduation, they cannot speak native Japanese at all or cannot basically speak Japanese without grammatical mistakes. This is a problem to emerge as schools only pursue the sources of students and make an expansion of enrollment, because enrollment expansion makes the major segmented and the originally unsound subject system is further into chaos, and the students do not obtain a solid foundation for the sheer language learning, and also their overall proficiency in Japanese is not improved plus the incomplete understanding of the professional knowledge. Thus, they only know a little but are not proficient in the related fields.

1.3 *Necessary to strengthen the faculty*

The development of Japanese major is restricted by the teaching methods. At present, an expository teaching method is commonly applied to Japanese education in China's ordinary colleges and universities, in which teachers are responsible for providing explanations and students only need to learn. However, it is not suitable for students to learn a foreign language, because students only receive the knowledge mechanically and passively, and the training in the language application ability of students is ignored by the teachers in the teaching process. In addition, the education will only be a mechanical copy and no top students will be trained if the teachers are not regularly engaged in advanced studies, keep divorced from the social environment, or do not possess practical experience in the business, along with the upgrading of the knowledge and the frequent change of the cultures. Therefore, it is necessary to improve the teaching methods in schools.

2 THE MEASURE FOR IMPROVING THE JAPANESE LANGUAGE TEACHING IN INSTITUTIONS OF HIGHER LEARNING

2.1 *Improving the construction of the japanese language subject system*

In the construction of the Japanese language subject system, the training objective must be placed on how to train the students' ability to meet Japanese comprehensive practice, and also it can be geared to the different levels of college students for further highlighting the combination of college students' universality and individuality. The teaching guiding ideology and teaching objective of Japanese education decide that Japanese language teaching should closely keep pace with the advanced development of the times, reform and innovation must be constantly implemented in the teaching process, and then a teaching system to integrate theory with practice and the teaching with scientific research can be established. On the subject design, the basic education centered on the subject must be tightly implemented, and attention can be paid to offering both Japanese public courses and elective courses on the basis of the well-done basic education. In the public courses, attention must be paid to improving the comprehensive quality and knowledge structure of students. In the elective courses, the characteristic of Japanese society and culture should be stressed.

Besides, the special characteristics of Japanese major must be specifically introduced in the construction of the Japanese language subject system, and then the Japanese personnel training objective and a Japanese personal training plan must be further defined. Also, a subject system must be constructed according to the difference of the major.

2.2 *Updating the teaching contents and strengthening the characteristics of the practicality and the times*

The teaching contents must be updated from the following points.

1. Teaching contents about Chinese and Japanese cultures can meet the requirements of the times change and the students
2. The spread of Chinese and Japanese cultures must be focused and also helps must be produced to train the ability of the students in comprehending and appreciating the Japanese culture
3. Teaching contents must be diversified and highlight the practicability and targets
4. Teaching contents can fit with the modern teaching means such as computer, multimedia, and network and facilitate the application of the modern technologies to teaching
5. Teaching contents must be interesting and ideological, which are conductive to improving the students' interest in learning Chinese and Japanese cultures and their ability in intercultural communication.

2.3 *Paying attention to the input of Japanese culture in teaching*

To infiltrate Japanese social and cultural knowledge in Japanese language teaching, the teachers must first get an understanding the social and cultural differences between China and Japan to let students feel aware of cross-culture as soon as possible. Through the following channels, Japanese social and cultural knowledge can be input into Japanese language teaching.

First, the selected teaching materials must be originally edited. In the primary stage of teaching, a percentage of the original Japanese textbooks must be chosen, in which many Japanese cultural backgrounds and customs are involved. In the easy-to-understand dialogues, all aspects of the Japanese culture may be involved.

Second, the cultural knowledge-related materials in the Japanese cultural background, social customs, and social relations must be intentionally accumulated by reading literary works, newspapers, and magazines, or surfing the Internet. The students must be encouraged to read many materials at the extracurricular time and especially in the reading days.

Reading the famous Japanese literary works is also an important method to learn Japanese culture, because the literary works of any nation should be the essential part of its national culture, the accumulation of the traditional culture, and the most vivid and abundant material to know well its national character, inner world, cultural background, customs and habits, and social communication and so on.

In addition, Japan's latest information can be accessed through online reading. Also, online reading is the most direct, the most efficient way to know well, Japan's current social dynamics, social problems, and social relations. The information is not involved in the teaching materials.

Third, electronic and audio-visual teaching methods can be applied. All aspects of Japanese culture can be reflected by letting the students viewing pictures, videos, or films, so that Japanese society, interpersonal relationship, way of thinking, and other social culture are visually known well by the students. At the same time, the ability of the students to speak and listen to Japanese and accurately communicate in Japanese can be improved.

Fourth, the role of native Japanese teachers must be fully played. Native Japanese teachers can vividly introduce Japan's social situation, cultural life, and local conditions and customs to the students according to their personal experience, so the knowledge in social culture unavailable in the teaching materials are available for students. Moreover, the students, who learn Japanese at home, are in shortage of native Japanese social culture and language environment and have a few chances to make face-to-face communication with Japanese, so the first Japanese to communicate with them are the native Japanese teachers. Thus, the students must be encouraged to boldly communicate and frequently talk with the native Japanese teachers, aiming at speaking a fluent, native Japanese language.

ACKNOWLEDGMENT

This work is supported by the Social Science Research Project by Department of Education of Jilin Province (Grant 2014178).

REFERENCES

[1] J. Green. Proceedings of ISTP Conference on JSCIT. 2009.2344–2348P.

[2] M. Petter. Proceedings of ISTP Conference on JSCIT. 2012.578–582P.

[3] K.T. Se. Proceedings of SCCI Conference on CVPR. 2010.211–218P.

[4] L. Muller. Proceedings of MMUS Conference on SSRE. 2012. 156–165P.

[5] M. Teferel. Proceedings of ISTP Conference on JSCIT 2013. 2158–2162P.

[6] H.T. Heteres. Proceedings of ISTP Conference on JSCIT.2013. 588–594P.

Management, Information and Educational Engineering – Liu, Sung & Yao (Eds)
© 2015 Taylor & Francis Group, London, ISBN: 978-1-138-02728-2

Application and research of multimedia teaching technique

Lei Xia
Wuhan Business School, China

ABSTRACT: Multimedia technology is rising rapidly, by its effective forms of pan, graphic, combining movement and extraordinary performance beyond time and space. Furthermore, it enhances people's understanding and feeling of abstract things and processes, and gradually gets into the classroom, introducing a new realm of teaching, and also makes modern multimedia technology become an important part of education.

KEYWORDS: multimedia; teaching technology; teaching application; advantages.

1 INTRODUCTION

With the development of society, things people pursue are constantly changing, especially today's next-generation successor of the new century naturally have a lot of curiosity to new things and courage to face them. Therefore, the use of multimedia teaching has become extremely popular. But at the same time, the widely use of multimedia teaching gives us a bigger challenge from which aspects of the matter. From multimedia teaching to traditional teaching and from traditional teaching to multimedia teaching, through analysis of their similarities and differences, profound observations are obtained about multimedia applications in language teaching.

2 MULTIMEDIA SOFTWARE TECHNOLOGY INTRODUCTION

2.1 *PPT software*

Currently PPT software is the most common software technology in multimedia teaching application. People can use it to design and create any slides in accordance with their own ideas and slide whatever they want to lay the foundation for the further development of multimedia teaching of a foundation to make a teaching model richer and more interesting.

2.2 *D studio MAX dimensional animation software*

In order to make multimedia software more dynamic can give people a lifelike effect, 3D Studio MAX dimensional animation software came into being, which gives the appearance of vitality and realistic effects we want. It is composed of windows9x or NT operating system three-dimensional animation software.

2.3 *Director multimedia authoring software*

This is a different technique from the first two ones, which specializes in mapping. If you need to map, it is the best choice, a multimedia authoring software based on the timeline. Through it, the producer can make various graphical effects through different production processes and procedures, if necessary, can produce different effects picture.

2.4 *Flash web animation software*

Flash is suitable for making small capacity animation files, in this regard it has a great advantage over the Director. Secondly. Since Flash is small Director, naturally it is animation is relatively easy, so it is very popular in current animation software.

3 ADVANTAGES OF MULTIMEDIA TEACHING

With the development of information society, especially today's fast development of science and technology provides a convenient condition for multimedia teaching to step ahead. Now people's lives are becoming more beautiful for the assistance of multimedia. Advanced multimedia technology has far exceeded the traditional teaching and demonstrated its unique advantages.

3.1 *Visual novel multimedia teaching methods to improve the reproducibility of the scene*

The practice has proved that various components of multimedia teaching methods make most students concentrate on learning something of interest, resulting in higher learning efficiency. Multimedia teaching has the features of a large amount of information and strong inclusion and its efficiency is quite impressive. Learn more teaching knowledge through computer platform and network channels. During the process of teaching, the teacher only shows course contents to the students by sending the program, in this way completely change the teaching methods and teachers' resources are constantly expanding. Now there is a big difference between knowledge acquisition channels and traditional way of knowledge sources, multimedia teaching methods for teachers reduce handwritten costs, enhance teaching capacity and more effectively improve teaching quality and enhance the rhythm of classroom teaching and also students' awareness of hard work to promote students to have more information in a short time. Teachers can more easily achieve their educational objectives.

3.2 *The script show of new generation of calligraphic*

In modern multimedia teaching, ppt software production can effectively help us to understand the process about the course and the overall summary of the whole class. Thus, we can make all the class knowledge by different slides, and then through the appropriate adjustments and decor, complete the full integration of all information to demonstrate the students a simple analog calligraphy style of teaching. I believe this has a great necessity and enforceability.

3.3 *A combination of static and dynamic beauty show*

Multimedia teaching combines static and dynamic enough to make students feel the beauty, of course, and this unique animated teaching method must be consistent with the teacher's teaching requirements in order to obtain extraordinary teaching achievements. In other words, in multimedia teaching, we can purposely have different information member organically combined together to achieve this unique mode of teaching and make students feel the animated beauty anywhere.

3.4 *The beauty of making a courseware background picture*

The use of multimedia teaching needs different steps to produce a good effect slide to highlight the uniqueness of multi-media teaching, from which making the screen will have a great impact on the multimedia courseware teaching, because no matter where the class is, which the teacher is in the class will not present knowledge to the students for a whole class. Sometimes for students to better absorb and understand knowledge what they have learned, it is necessary for everyone to leave enough time alone to digest and deepen knowledge. In this process, the courseware help students to integrate knowledge atmosphere through different background images. Adding in some pictures to courseware will make the effect change dramatically.

4 COUNTERMEASURES AND SUGGESTIONS OF MULTIMEDIA TEACHING IN MULTIMEDIA APPLICATIONS

4.1 *Application multimedia teaching should grasp the degree*

As the saying goes: everything should have a degree. That for multimedia teaching is no exception. Although it has a lot of unique advantages, it will never allow us to use anytime, anywhere with impunity, or it would only be a departure from its advantage and make it had a negative side. Maybe some people think that any problem in teaching will be solved by the use of multimedia teaching. If only so I will warn you that you will go into the abyss. At any time, we do not forget what the purposes are we using multimedia education, if you ascribed to the degree that the negative impact is likely to harm you, for example, so that we gradually lose their ability to deal with things and passion.

4.2 *Misunderstandings of multimedia teaching*

The development of multimedia technology really alleviates a lot of the burden for contemporary teachers and save a lot of time. But technology advancement does not mean that the teacher in class using multimedia software alone is sufficient but not necessary to practice the behavior of some operations. Whether, to what extent technologically developed, the most fundamental principle that we still can not lose is the so-called teacher behavior. If so, the teacher is a dereliction of duty; emergence of multimedia technologies and applications is a mistake. For the better applications of multimedia education. Multimedia technology must be closely combined together with the practice acts of teachers, and then give students a better education.

REFERENCES

[1] Tan Haoqiang. *C Programming* (3rd edition) [M], Tsinghua University Press, 2005.

[2] Tang Yan'er. *Chinese Modern Distance Education Calls for New Policies and Regulations* [M]. Academic Forum, 2007 edition.

[3] Huo Liting. *Herbart Pedagogy Rebellion and Restoration* [J]. Inner Mongolia Agricultural University (Social Science Edition), 2008 (4).

Management, Information and Educational Engineering – Liu, Sung & Yao (Eds)
© 2015 Taylor & Francis Group, London, ISBN: 978-1-138-02728-2

Study on the needs of distance education in the development of the transportation industry

Pei Pei Tang
Transport Management Institute, Ministry of Transport, Beijing, China

Meng Bo Gao
Qijiankaixuan Residential Quarter, Tongzhou District, Beijing, China

ABSTRACT: This paper analyzes the work characteristics and the basic situation of transportation industry employees. Based on the requirements for quality it also researches the employee's needs in job training, law enforcement training, vocational skills training, continuing education of professional personnel and academic education of in-service personnel. The analysis shows developing modern distance education by the application of modern information technology tools in the transportation industry can integrate training resources. It is the most effective way to realize large-scale training of practitioners and building high-quality team of employees.

KEYWORDS: Distance education; Transportation industry; Continuing education; Training; Modern information technology.

1 INTRODUCTION

Transportation industry changing patterns of development have brought new challenges and opportunities to all levels of education and training of cadres' transportation. Distance education is a new form of education using the Internet and multimedia. Teachers and students are separated and teaching and learning through a varicty of actions to achieve contact with educational technology and media resources. Distance education can make interaction and integration of various educational and training resources. This open and convenient feature is an effective way of educating. Situation industry professionals will be analyzed in this paper. It also explores industry practitioners to the demand for distance education in various fields.

2 SITUATION ANALYSIS OF THE TRANSPORT SECTOR EMPLOYEES

2.1 *Transportation industry employees work characteristics*

1 **Service:** Transportation industries are a modern service industry in providing services for social and public needs, and provide society with products mainly transportation. In the process of management and service it reflects its social value through social recognition and reputation. The most basic responsibility of the transport sector is to improve the management level and easy line to protect people and achieve a smooth flow of goods. Transport sector employees should have a good sense of service and high level of service. It is necessary to receive regular continuing education and training.
2 **Security:** With the improvement of the level of China's economic development, road transport safety is increasingly being everyone's attention. The primary purpose is to ensure transport safety during construction and operation of construction safety. As the transportation industry employees should have a good safety awareness.
3 **Differences in the working environment:** Due to the wide distribution of traffic, road maintenance workers working and living environment is not the same. Workers at the bustling traffic on the working and living environment is relatively better. Workers in remote mountainous areas, working and living environment conditions are difficult. Regional differences in the transport sector have also led to uneven distribution of human resources.

Due to the transportation industry employees' work in the field and remote areas they cannot accept the long-term academic education. Open and flexible distance education become the most economical and convenient way of education. Features of distance education are a two-way, real-time, interactive and controllable. Its significant advantage is: any person at any time and any place to start learning from any section of any course. Online Education convenient and flexible learning model directly reflects the learning and active learning features. It fully reflects the development of the basic requirements of modern education and lifelong education.

2.2 *Analysis on the basic situation of employees in the transport sector*

Source transportation industry employees are very extensive. Including national fixed staff, national contract workers; there is a considerable part of the temporary workers. This group of people is generally low cultural quality and greater disparity. Type the number of high-tech talents is small. This situation cannot meet the needs of China's rapidly growing transportation industry.

Transportation employees are infrastructure-based conservation and transportation and production personnel. "Continuing Education transportation industry comprehensive study report" for industry practitioners' qualifications, titles, and skill levels were analyzed.

Table 1. Transportation industry practitioners' qualifications constitute.

Education (%)	Graduate and above	Undergraduate	College
Road infrastructure	0.2	6.1	12.2
Transportation and production areas	0.2	3.8	11.8
Education (%)	High School	Junior high school and below	
Road infrastructure	27.9	53.5	
Transportation and production areas	39.4	44.7	

Table 2. Transportation industry practitioners and technical titles constitute.

Quality Structure (%)	Advanced	Intermediate	Primary
Road infrastructure	1.5	7.0	12.2
Transportation and production areas	0.8	4.4	7.6

Data show that the transportation industry employees degree or above accounted for only a small part. High school education accounted for about 80%. Primary education accounts for about 50%. The number has the highest proportion of junior titles, the lowest proportion of senior titles. Most of which are junior titles below that constitute a technical level ratio cannot meet the needs of the industry.

3 THE DEVELOPMENT OF THE TRANSPORTATION INDUSTRY REQUIREMENTS FOR QUALITY OF EMPLOYEES

National Cadre Education and Training conference emphasized the need for reform and innovation to do a large-scale training of new cadres. "Implementation Opinions 2008–2012 large-scale training of cadres" a new round of large-scale training of cadres for the Organization Department of the CPC Central Committee General Office issued the direction. It puts forward new requirements under the new situation of cadres training. This new requirement is to enhance the ability to focus on scientific development, improve the quality and efficiency of education and training of cadres and cadres to actively promote education and training reform and innovation. Learn what is necessary to insist on doing. Through the implementation of full coverage, multi-tools, high-quality training, promote learning party and learning society. Cadre education and training is in order to make work better for the scientific development of services and services for the healthy growth of cadres.

China will strive to achieve transportation "three changes." First, by relying on infrastructure investment and construction changes to the conservation, management and transportation services; Second, by relying heavily on increased consumption of material resources to science and technology, industry, innovation, resource-saving and environment-friendly changes; third change is by a single mode of transport to the development of an integrated transport system. This requires integrated management of the transport sector departments and leading cadres at all levels. They should be familiar with the highway and waterway transportation, but also master the aviation postal aspects of the business. They should also understand the knowledge of rail transport and integrated transport. In recent years a number of new situations often encountered snow disasters, earthquakes, and events such as the taxi industry groups. These events have on traffic cadres to improve crisis management capabilities and solve practical problems raised new demands. This also means that the industry needs a new management cadre system to support its knowledge management capabilities.

4 DEMAND FOR DISTANCE EDUCATION OF TRANSPORTATION INDUSTRY PRACTITIONERS

4.1 *Vocational qualification training needs analysis of the transport sector*

With the economic and social development and technological progress the industry demands on the quality of employees increases. Employees need to be trained in all aspects including basic vocational knowledge, skills and professional ethics, rules and regulations, knowledge and technology updates and posts adaptability. First is job training such as registered civil engineer and highway engineer supervising practitioners In the report of "Vocational Education Development Strategy" number of vocational qualification training needs of the transport sector were predicted.

Table 3. Future industry professionals predict the overall size (Unit: million persons).

Category	Year: 2010	Year: 2020
Trading of road passenger and freight transport and ancillary services	3256	5228
Motor vehicle maintenance and inspection	288	368
Transportation infrastructure construction and maintenance	1218	1345
Waterway transport	230	253
Port	106	123
Industry Administration	36	36
Total	5134	7354

Table 4. Future industry professional qualification training needs (Unit: million persons).

Strategy session	Year: 2011~2020
Total vocational qualification training needs	2200
Annual training requirement	220

Data shows that from 2020 the industry is expected to reach more than 73 million employees. 2010–2020 total transport sector vocational qualification training needs will increase by nearly 22 million people. Annual training needs of about 200 million people. Thus, the demand for vocational qualification training for industry professionals is very large. Training task is very heavy we can remotely implement a variety of training to meet the demand.

4.2 *Analysis of transport sector law enforcement training needs*

Transportation Management Institute in the "About our hospital professional and technical personnel to update their knowledge engineering training situation report" is mentioned as of October 2012, the National Transportation law enforcement officers a total of 41 million people. Where the number of secondary school education or below accounted for about 21% of the total number of tertiary education accounted for 53 percent; bachelor's degree accounted for 23%; master's degree or above accounted for 3%. In all law enforcement officers, the legal profession and the professional for 19 million people the number of traffic, accounting for 46%; other professional persons for 22 million people, accounting for 54%. Transportation employees in administrative law enforcement professionals are seldom especially legal type and technical talents. In some remote areas, units of graduate law enforcement officer are less. Most law enforcement officers had not received the knowledge of the education system. Overall quality of transport law enforcement officers needs to be improved.

In 2012 the National Transportation legal work meeting, Vice Minister Gao Hongfeng stressed that carrying out the work of transportation administrative law enforcement training to is necessary. We strive for three to five years for all law enforcement personnel rotations once. And this massive training relies on traditional forms of face to face training is very difficult to popularity. Distance education training methods can be more grassroots law enforcement team and expand the scope of training.

4.3 *Analysis of transportation industry management cadre training needs*

Ministry of Transport proposed that in the next five years to further increase the size of the industry in the education and training of cadres. Training targeted has effective growing. Cadres according to law and public service capacity have been raised. That is to carry out large-scale industry management cadre education and training for one of the main tasks of this goal. The size of the whole industry education and training of cadres reached 2.2 million people. Faced with such a large-scale training mission, organize training not only can remotely be fully utilized and also is one of the most convenient and effective channels of high-quality education and training resources.

5 SUMMARY

Analysis shows that the transportation industry employees in education and training and personnel training in all aspects put forward higher requirements. Industry practitioners' vocational qualification training demand is great. Professional and technical personnel are very scarce. Increasing the size of the industry management cadre training education and training leading to the traditional methods cannot meet the need. Transportation industry employees in urgent need of large-scale training of open and distance education and training methods.

REFERENCES

[1] People's Republic of China Ministry of Transport. Highway and waterway transportation and long-term talent development program(2011–2020)[Z].Beijing: People's Republic of China Ministry of Transport, 2011.

[2] National Bureau of Population and Employment Security Department. China Labor Statistical Yearbook [Z]. Beijing: China Statistics Press, 2011.

[3] Beijing Jiao tong University. Transportation industry comprehensive study of Continuing Education [R]. Beijing: Beijing Jiao tong University, 2009.

[4] L. Z.ping. Vocational Education Development Strategy Research [M]. Beijing: People's Communications Press, 2005.

[5] People's Republic of China Ministry of Transport. Highway and waterway transportation education and training development plan [Z]. Beijing: People's Republic of China Ministry of Transport, 2010.

Management, Information and Educational Engineering – Liu, Sung & Yao (Eds)
© 2015 Taylor & Francis Group, London, ISBN: 978-1-138-02728-2

Research on the application of task-based language teaching for the English pronunciation course in high schools in China

Yi Lu Nie*
School of Foreign Languages, East China Normal University, Shanghai, P.R. China

ABSTRACT: Though the importance of English education has long been recognized in China, the teaching of pronunciation has not received adequate attention. In this paper, we research the application of task-based language teaching method for the English pronunciation class in high schools in China. A 10-week teaching program was designed, involving three different groups consisting of 25 students. The two experimental groups, respectively received instructions by using the Task-Based Language Teaching method and the conventional teaching method. The third group, namely the control group, received no particular pronunciation instruction. Through analyzing the scores with two-way ANOVA and post-hoc method in SPSS, we find that the Task-based Language Teaching group showed more significant progress both on segmental and suprasegmental aspects in pronunciation than the other two groups, and the conventional group is better than the control group, which confirms the effectiveness of Task-Based Language Teaching method in high school English pronunciation course.

KEYWORDS: Pronunciation teaching in China's high school, Task-based Language Teaching Method.

1 INTRODUCTION

Cliché as it seems, Rajagopalan (2004) pointed that English is recognized as a world language. And China has set English as an important subject in elementary education since China's Opening and Reform in 1978. Despite the importance of English in China, "silent English" is still a label for Chinese students. One of the reasons is that they put most of their efforts in the training of grammar and reading to get a high score in the national college entrance examinations. This phenomenon poses a great challenge to English acquisition because the core language use lies in the communicative function of a language. And communication in English starts with the correct production of sounds and the combination of them, as Gimson & Ramsaran (1970) noted that "to speak any language, a person must know nearly 100% of its phonetics, while only 50%–90% of its grammar and 1% of the vocabulary may be sufficient". So, the current situation of pronunciation teaching calls for adjustment both applicable in English class under the Chinese education system and practicable in improving the pronunciation of secondary school students.

The teaching of pronunciation has long been influenced by two principles that are contradictory to each other—the nativeness principle and the intelligibility principle, pointed out by Levis (2005). The first principle states that English as Second Language (ESL) learners are likely to achieve native-like pronunciation while the second principle holds that understandability is the key to one's pronunciation acquisition. The latter one has now been the guiding principle in the teaching of pronunciation as researchers agree that foreign language learners' ultimate goal is to achieve intelligibility. Rivers & Temperley (1978) have examined the effect of segmental instructions through phonetic transcriptions and imitation/repetition practice, as well as the effect of suprasegmental instructions which was pointed out by Levis & Pickering (2004). Moreover, Macdonald et al. (1994) probed into pronunciation instruction in authentic classrooms. Derwing et al. (1998) carried out a study where a 12-week course of instruction was conducted. They tried to evaluate the effects of 3 types of instruction (segmental accuracy, general speaking habits and prosodic factors, and no specific pronunciation instruction) on ESL learners. Students were tested in terms of accentedness, comprehensibility, and fluency before and after the course. In this research, they concluded that the choice of pronunciation instruction depended on the target of it, the way of measurement, and the aspects being measured. Also, they stressed that attention to both suprasegmental and segmental instructions benefited ESL learners. Moreover, Levis used two matrixes to analyze the importance of the two factors on pronunciation teaching, of which the first was a two-by-two Speak-Listener Intelligibility Matrix, and the second was a three-by-three World Englishes

*Corresponding author: yilunie@gmail.com

Speaker-Listener Intelligibility Matrix. Results showed that the two factors weighed differently if different contexts were provided.

Task-Based Language Teaching (TBLT) was put forward in the 70s of the 20th century in the west, but it was not until the middle 80s that the systemic TBLT theory came into being. Several definitions of "task" appeared afterwards. The widely adopted definition of TBLT is the one put forward by Skehan (1998). He summarized the features of TBLT proposed by the former scholars, and concluded that a task is an activity in which meaning is primary, and learners are not given other peoples meaning to regurgitate, and there is some sort of relationship to compare real world activities, and task completion has some priority.

In this paper, we research on the application of task-based language teaching method to English pronunciation class in high school in China. Section 2 is the research design. Section 3 contains date collection and analysis. Section 4 is the conclusions and limitations.

2 RESEARCH DESIGN

2.1 *Participants*

This research has been undertaken in a Chinese high school. Altogether 75 students take part in the program, and they are divided into three groups of the same starting level according to their performance in the pre-test. Two groups, as the experimental groups, will receive pronunciation instruction, respectively by using TBLT method or conventional PPP (presentation, practice, and production) approach. While the third group, as the control group, will not have any particular pronunciation instruction session. For the two experimental groups, English pronunciation sessions will be given to students once a week. Each pronunciation session on the 10-week program will consist of two 40-minute parts with an interval of 10 minutes in between.

2.2 *The operation of the TBLT class*

The design of TBLT sessions is based on Willis' (1996) framework, which includes pre-task, the task cycle, and language focus.

The arrangement of the TBLT sessions is as follows: the teacher introduces the topic of the specific session, such as the distinction of the monophthongs: /e/, /Q/ and /E/ and the correct pronunciation of the monophthong /E/ and the diphthong /ei/. A quick elaboration on the topic will be given to help with students' primitive acquisition. Then, the task of the session is supposed to be described clearly until all students understand the steps. A model of task conduction may be presented if necessary. During the task phase, students will be split into pairs or groups accordingly, and do a meaning-focused activity. In the next procedure, students complete the task and give reports, in which process the teacher serves as a monitor and lead students to focus on the language topic. Eventually, the teacher gives feedback and helps students consolidate their pronunciation acquisition.

2.3 *Case study*

The topic of the session is the correct pronunciation of the pair of short vowel /i/ and the long vowel /i:/ and sound link. The topic is set as two major problems often occur in students' pronunciation of these two vowels. The first one is that students tend to replace some vowels with Chinese pinyin. The other is that many students consider the difference between the short vowel and the long vowel lying only in their duration. As for sound link, students are rarely conscious of this concept even though they have heard it in the recordings. Sound link happens when a word ends with a consonant sound and the next word starts with a vowel sound and only when two words are in the same sense group, can they be linked.

In the pre-task phase, the teacher begins the session with some lead-in introduction to the pronunciation of /i/ and /i:/, including the correct pronunciation and the distinction. The teacher asks students to pronounce /i/, /i:/ and Chinese "yi". If students find themselves mixing them, they must have done wrong. Then, the teacher demonstrates the correct pronunciation of /i/ and /i:/, and tells students the right positions of the lips and tongue when pronouncing the sounds. What is important is that the teacher should emphasize the difference of degrees of mouth opening when pronouncing the two sounds, as students may only see the difference in the lengths. Also, sound link will be explained and exemplified. The teacher reads the following phrase "get a move on" in two ways, with or without sound link, and asks students to tell the difference. If it is pronounced with sound link, the phrase may sound like two words that have the following phonetic symbols: /getE/ and /mu:vOn/.

The major part of the session is the task cycle and the main task of the session is a story making-up competition. The teacher introduces the task and the rules. After that, students get into groups of 7 or 8 to make up a story with given phrases. And each group has a group leader to make sure all the processes are undergoing according to the rules. Each student is required to make one or two sentences with a given phrase in the given order. When a group of phrases is used up, all the sentences made by the students should make up a coherent story. Different groups will be given a different phrase list, thus different stories will be made and assessed. The stories will be assessed in the following areas: completeness, coherence and the correct pronunciation of the given phrases and the use of sound link.

The topic of the session covers both segmental instructions and suprasegmental instructions. In the TBLT class, meaning is primary. And students know that task completion is somewhat prior to other requirements. Students in the class are not required to do drills and much repetition. They focus on group work and they have to do close-to-life communication with each other. In this process, there is no escape for them to employ all kinds of skills in English pronunciation. The close-to-life communications enhance students' communication ability, which is the ultimate goal of English speaking. And different skills are practiced and reviewed time and time again in every session. Comparatively, conventional class, mostly offers rigid exercises and patterns. Students often do repetitive practice which helps with the improvement of one or more particular skill. They probably know the skills, but it is not certain that they know how to put them into use in real-life communication.

3 DATA COLLECTION AND ANALYSIS

3.1 *Data collection*

Research data come from the pre-test and the post-test results.

Three English teachers are the listener-raters responsible for both pre-test and post-test, which are conducted on all participants before and after the program with no time for preparation beforehand. And both tests consist of two parts—reading phonetic symbols and given words, and reading given sentences, both of which account for 100 points. The first part will be assessed in terms of accurate pronunciation and the second part of overall pronunciation, including accuracy, stress, fluency and intonation. To be specific, the result of the first part will provide data concerning segmental aspect and the latter can more or less reflect the level of student' suprasegmental pronunciation skills.

The first part includes 48 phonetic symbols and 26 words covering all phonetic symbols, each phonetic symbol accounting for 1 point and each word accounting for 2 points. The second part includes 10 sentences, each one accounting for 10 points. Accuracy, stress, fluency and the appropriateness of intonation and overall impression respectively account for 2 points in each sentence.

3.2 *Data analysis*

For students' pre-test and post-test results, two-way ANOVA is used for data analysis for the two parts respectively, in which two factors can be measured at the same time. For each part, after submitting the data to the two-way ANOVA with pre-test and post-test as dependent variables and the type of instruction as the fixed factor, one can get three sets of F-value/p-value, two of which are supposed to check if the factors cause any difference in students' pronunciation respectively, and the third one is to prove the significance of the two factors' interaction. A post-hoc test is also carried out in order to compare the effect of different types of instruction, which can prove if TBLT instruction is more effective than conventional instruction.

Table 1 and 2 are multiple comparisons of data from the pre-test concerning segmental and suprasegmantal aspects respectively, which show that the significance values among the three groups are larger than 0.05. Apparently, there is no statistical difference in students' performance in the pre-test, which indicates that they have the similar starting line in terms of both aspects.

Table 1. Multiple comparisons of segmental aspects.

(I) Group	(J) Group	Mean difference (I-J)	Sig.
Control Group	Conventional Group	−0.0560	0.981
Control Group	TBLT Group	0.2280	0.924
Conventional Group	TBLT Group	0.2840	0.906

Table 2. Multiple comparisons of suprasegmantal aspects.

(I) Group	(J) Group	Mean difference (I-J)	Sig.
Control Group	Conventional Group	−0.0973	0.966
Control Group	TBLT Group	0.8013	0.723
Conventional Group	TBLT Group	0.8987	0.691

Tables 3 and 4 are pairwise comparisons of between-subjects segmental and suprasegmantal aspects. From table 3, the different effects of three types of instruction on the segmental aspects can be seen clearly. Both experimental groups show more significant changes than the control group, which has received no particular pronunciation instruction, as the mean difference between the conventional group and the control group is 4.213 and the sig is $.026 < .05$ and the mean difference between the TBLT group and the control group is 7.974 and the sig is $.000 < .05$. In terms of the difference between the TBLT group and the conventional group, their mean difference is 3.761 and the sig is $.046 < .05$, which implies the existence of statistical difference between the two groups and the effect of the TBLT group is better than the conventional group with regard to the segments. In terms of table 4, which is concerned with the suprasegmental aspects, though the mean value of the conventional group is slightly higher than the control group, the sig value is .656 > .05. Compared to the control group, the conventional group does not show noticeable changes, either. The mean difference between the TBLT group and the control group is 5.644 and the sig value is .013 < .05, and the figures between the TBLT group and the conventional group is 4.560 and .040 < .05. The effect of the TBLT instruction is apparent compared with the other two groups. Only the TBLT group improves significantly. This table indicates that the TBLT method has a positive effect on the improvement of pronunciation regarding suprasegmental aspects while the conventional method does not show a significant effect. The TBLT method focuses more on meaning than form, and thus students have more opportunities to do meaningful activities rather than repetitive drills. Students are encouraged to communicate with each other, which leads to students being exposed more often to suprasegmental practices despite part of the focus on segmental training in pronunciation class.

Table 3. Pairwise comparisons of between-subjects at segmental aspects.

(I) Group	Group (J)	Mean difference (I-J)	Sig.
Control Group	Conventional Group	−4.213	0.026
Control Group	TBLT Group	−7.974	0.000
Conventional Group	TBLT Group	−3.761	0.046

Table 4. Pairwise comparisons of between-subjects at suprasegmantal aspects.

(I) Group	(J) Group	Mean difference (I-J)	Sig.
Control Group	Conventional Group	−.994	.656
Control Group	TBLT Group	−5.644	.013
Conventional Group	TBLT Group	−4.650	.040

4 CONCLUSIONS AND LIMITATIONS

In this paper, a 10-week program is designed to test the applicability of the TBLT method in a Chinese high school course.

Based on the students' test scores, SPSS is used to analyze the result. Two-way ANOVA and post-hoc are applied to prove the different effect of different kinds of instructions. According to the result of the analysis, the TBLT method proves to be more effective than no particular instruction and the conventional method both in terms of segmental aspects and of suprasegmental aspects.

However, in terms of the application of the TBLT method to EFL pronunciation class, there are several concerns. Just like Carless summarized in 2003, the language teacher may not fully understand the tasks at hand in The TBLT class. And even when the teacher does, he/she may not be able to apply TBLT method in pronunciation class successfully because it is not always easy to relate the pronunciation topics to tasks in real-life or close to real-life activities. On the other hand, not all the students welcome TBLT method as some of them may not have a correct understanding of it and they think it better not to shift the teaching method easily.

REFERENCES

Derwing, T. M., Munro, M. J., & Wiebe, G. 1998. Evidence in favor of a broad framework for pronunciation instruction. *Language Learning* 48(3): 393–410.

Gimson, A. C., & Ramsaran, S. 1970. *An Introduction to the Pronunciation of English*. London: Edward Arnold.

Levis, J. M. 2005. Changing contexts and shifting paradigms in pronunciation teaching. *TESOL Quarterly* 39(3): 369–377.

Levis, J., & Pickering, L. 2004. Teaching intonation in discourse using speech visualization technology. *System* 32(4): 505–524.

Macdonald, D., Yule, G., & Powers, M. 1994. Attempts to improve English L2 pronunciation: The variable effects of different types of instruction. *Language Learning* 44(1): 75–100.

Rajagopalan, K. 2004. The concept of 'World English' and its implications for ELT. *ELT Journal* 58(2): 111–117.

Rivers, W. M., & Temperley, M. S. 1978. *A Practical Guide to the Teaching of English as a Second or Foreign Language*. Oxford: Oxford University Press.

Skehan, P. 1998. *A Cognitive Approach to Language Learning*. Oxford: Oxford University Press.

Willis, J. 1996. *A Framework for Task-Based Learning*. Harlow, Essex: Longman.

Management, Information and Educational Engineering – Liu, Sung & Yao (Eds)
 ISBN: 978-1-138-02728-2

Strategies on improving listening ability

Hong Mei Xing
Public Foreign Language Educational Institute, Beihua University, Jilin, Jilin, China

Lei Sun
Computer Institute, Changchun University, Changchun, Jilin, China

Yan Zeng
Beijing Webrate Technology Co., Ltd., Beijing, China

ABSTRACT: At present, a college English teaching reform aimed at fully improving the English comprehensive application ability has been carried out. The main purpose of the reform is to improve the students' language application skills. The ability of listening is one of the most important skills in language application. However, many students can do nothing to help improve their listening ability. Although the students spend a lot of time practicing listening, the listening ability does not increase significantly. In order to improve the listening ability and quality, the listener has to get to know the related factors that have an effect on the listening ability, and take the corresponding measures to practice. The factors influencing the listening ability and quality can be divided into two aspects: language obstacles and non-language obstacles.

KEYWORDS: language obstacles; non-language obstacles; listening comprehension.

1 LANGUAGE OBSTACLES

For learning a foreign language, the main obstacle affecting the listening ability is from the language. It includes four aspects: pronunciation, vocabulary, grammar and comprehensive understanding.

1 Pronunciation obstacles
Phonetics is the basic element of language. The first thing needed to improve listening ability is the correct pronunciation. If the listener's pronunciation is not accurate, when the correct pronunciation of the same word comes into the ears, it will not be responded in the mind, or get an incorrect response, mistaken for another word by the mind. Sometimes, the words heard are familiar. But due to listener's different pronunciation from correct one, they will not get to know this until reading the tape script.

2 Vocabulary obstacles
Vocabulary is the foundation of the language. If the students want to quickly improve listening ability, they have to grasp enough vocabulary. Most of the students from colleges and universities in mainland of China have grasped much greater vocabulary in reading than in listening. This has much to do with our emphasis on reading ability, ignoring listening, speaking and writing ability for many years.

3 Grammar obstacles
Many students have a much more solid foundation of grammar learned in high school. But the problem is that they can only apply the grammar knowledge into reading. Few students can correctly apply them into listening.

4 Comprehensive understanding obstacles
For many students, the most difficult task of listening comprehension tests is a comprehensive understanding of sentence and passage. Pronunciation, vocabulary and grammar are necessary bases in order to construct comprehensive understanding ability. But these are not enough, it is still hard to improve listening ability. Because this knowledge is embedded in the comprehensive understanding of the listening material. In other words, there is none of the questions is to examine how to distinguish pronunciation, vocabulary and grammar in listening tests. Some students focus their attention on the word, not the whole sentence or the passage when doing listening practice. On the surface, they may understand every word, but do not know what the dialogue or passage is saying when put all the words together.

2 NON-LANGUAGE OBSTACLES

Non-language obstacles refer to barriers that have nothing to do with language, yet they have much effect on listening ability. There are two aspects: psychological obstacles and cultural obstacles

1 Psychological obstacles
When doing listening practice, some students can't focus their attention, and are very nervous. This is a type of psychological obstacles. Inattention, mainly refers to the student body is sitting with the mind absent in listening practice. Even if this situation only lasts for a few seconds, the input of the listening material can be interrupted. It will naturally affect listening ability. Nervousness mainly refers the blank in the mind due to the tension in the tests. Students can't digest the input content, and worse is, listening material cannot be input into the brain.

2 Cultural obstacles
Language is the carrier of culture. From macro scope, no matter what language is learnt, one has to learn the culture in which the language takes root. At the same time, cultural differences between the native language and target language have to be understood. From microscope, the individual's width of knowledge also affects listening ability.

3 CORRESPONDING MEASURES

From the above analysis, we get a picture about the factors influencing listening ability. In order to improve the listening comprehension ability, university teachers can take measures from the following aspects:

1 Strengthen the input of culture
First, teachers should guide the student to read Chinese and foreign language books that are related to foreign culture, getting students to understand human geography and customs of the target language. Second, with the aid of multimedia, teachers help students get visual information through films, television, etc. In the end, foreign novels are strongly recommended for students. Foreign novels contain a large amount of authentic foreign language, and are also the best way to understand the culture of the target language.

2 Improve the students' interest in listening course
The listening ability is improved through keeping practice and practice. According to the content of the listening material, the differences between Chinese and foreign culture, teachers should make listening course full of fun, with the aid of multimedia. It can fully mobilize students' subjective motivation and as much as possible to make every student participate in the classroom teaching.

3 Combine intensive listening and extensive listening together
Intensive listening is the basis of listening ability. When choose to listen material for students, teachers should choose those suitable for the students' present listening ability. The listening materials have to be accompanied with the tape scripts.
Extensive listening is an add-up to intensive listening. Extensive listening emphasizes that the amount of listening materials should be large. It only requires students to understand the main idea of each listening material.

No matter intensive listening and extensive listening, the listeners must grasp the key points when doing listening practice. They can take notes to summarize the main ideas of general dialogues or the passages. Taking notes can help listeners memorize the information that has been input. Listeners can adopt abbreviations, image code, etc. according to their own habits.

4 Emphasize the importance of vocabulary and grammar
Vocabulary is the basis of listening, speaking, reading and writing translation. It is very important to learn a foreign language. If one wants to improve listening ability, he must grasp a certain amount of vocabulary. This requires teachers to guide students to memorize words by using the pronunciation rules. There are no new grammar rules to be learnt in universities. The teacher needs to help students review the grammars in the process of teaching. Some important and difficult grammars need to be practiced. These lay a solid foundation to improve listening comprehension ability.

REFERENCES

[1] Carter, David. Interpreting Anaphors in Natural Language Texts [D]. Ellis Horwood Limited, Chichester, 1987. In English.

[2] Brian Seaton. A Handbook of English Language Teaching [D]. London: Terms and Practice. The Macmillan Press 1982. In English.

[3] Levinson, Stephen C Pragmatics. Cambridge University Press [D]., 2002. In English.

[4] Ungerer. F & Schmid, H. J An Introduction to Cognitive Linguistics. [D]Foreign Language Teaching &Research Press, 2001. In English.

[5] Jeremy Harmer. The Practice of English Language Teaching [D]. London: Longman Press, 1983. In English.

Management, Information and Educational Engineering – Liu, Sung & Yao (Eds)
© 2015 Taylor & Francis Group, London, ISBN: 978-1-138-02728-2

Study on college students' learning motivation

Hong Mei Xing
Public Foreign Language Educational Institute, Beihua University, Jilin, Jilin, China

Lei Sun
Computer Institute, Changchun University, Changchun, Jilin, China

Yan Zeng
Beijing Webrate Technology Co., Ltd., Beijing, China

ABSTRACT: Many ancient and modern, Chinese and foreign educators and psychologists have paid special attention to the role of learning motivation in the process of learning. For example: Gardner and Lambert, the most influential foreign language learning motivation researchers stated there were two types foreign language motivation: mating motivation and instrumental motivation.

KEYWORDS: learning motivation; learning requirement; learning expectation.

Motivation is the individual's internal psychological activity process that triggers and maintains individual activities, and directs efforts towards a goal. There is another definition of motivation that state as internal motive which govern people's action to achieve the purpose. Motivation can be divided into surface motivation and deep motivation. Surface motivation has a direction correlation to the future of the individual and motive is driven from the external; Deep motivation is generally not related to learners' future and economic interests, with learning motive coming from interest in learning knowledge or culture itself. Motivation has the following three functions: (1) the activation function, namely, motivation can make people act; (2) the directional function, that is, with the function of motivation, individual's action is directed to an aim; (3) the intensifying functions, namely after the activities, motivation can maintain and adjust activities.

Accordingly, learning motivation is an internal process or psychological state to inspire the individual to learn, maintain the triggered learning activities, directing the behavior towards certain learning objectives. It can trigger and reinforce learning activities. The reverse holds true. Once learning motivation is formed, it will stay in the whole process of learning activities. Two basic elements of learning motivation are learning requirements and learning expectations.

1 LEARNING REQUIREMENTS

Learning requirements refer to the individual's state of mind in which he feels the lack of the learning activities and strives to gain satisfaction. It includes learning interests, hobbies and learning belief, etc.

1 Learning interest

Interest (the deep motivation) is crucial to the success of a career. Without interest in specialty, there is no inspiration to learn. A minority of students have born interest in their chosen specialty. They enjoy learning specific subject as a kind of enjoyment, with learning initiative, relaxing, and the learning effect is obvious. Most people get interested in specialty out of work or life need. This kind of interest generated from the need is passive, weak at the beginning. When learning or advantage gets progress, the interest will become positive and intense. Great interest in learning will make people fully concentrate attention, think positively.

As the dynamic mechanism of generating learning activities, learning motivation is important qualification to launch, sustain and complete learning activity, and affects learning effect. The reason why learning motivation can influence the learning effect is because it directly constraints learning enthusiasm. Learning enthusiasm refers to the students' serious, tense, initiative and strong state showed in the learning activities. These mental states are mainly embodied in three aspects which separately are the students' attention level towards study, emotion tendency and perseverance and determination. They are external manifestations of learning motivation, and are consistent with the qualities and level of the learning motivation.

Emotion, is also a mood. Emotion is both spiritual, and at the same time, material. Science shows different emotions directly affect people's behavior. The joyful mood can generate an upward force on people, leading to active behavior and strengthened self-confidence. In the teaching practice, to create a relaxing and joyful learning atmosphere, is the guarantee and prerequisites to make students in high spirits.

Perseverance and determination is the highest form of self-confidence. Unshakable perseverance comes from self-confidence. It is very normal that in the process of learning there is ups and downs, fast and slow progress sometimes even backwards and stops. In the climax and fast forward, learner's self-confidence will be enhanced. They feel proud, even smug about themselves. If there is a slow progress or stop in learning, learners prone to be anxious. Some of them question their own ability of learning, losing confidence, and has never recovered. Generally speaking, the students with stronger anxiety have flaws in such aspects as grasping key points of examination, coding information, resulting in poor academic performance. On the other hand, poor performance is not only related to the situation of the exam itself, but also has much to do with the students' disorder in information processing, especially the unreasonable material organization.

2 LEARNING EXPECTATIONS

Learning expectation is the individual's subjective estimates on the goal to be reached in learning activities. It is another basic component in learning motivation structure. Learning expectation is closely related to learning goals, but these two can't be equally treated. Learning goal is the expected results that individual wants to achieve. The expected results stay in the mind in the form of ideas. And learning expectation is reflection of learning objectives in the individual's mind. The experiments showed that the better grades that college students want to get into college, the stronger the learning motivation is.Without doubts, the expectations should be practical, and in accordance with the actual situation of their own. A reasonable learning goal would make them continue to work hard, in order to achieve more. Even if the temporary score is not outstanding, but if the learner continues to work hard, and with confidence to get good grades in a certain period of time, these learners will also be very hardworking.

Learning requirements and learning expectations are two basic components of learning motivation and are closely related. Learning requirements are the most fundamental driven force for individuals to engage in learning activities. Without this self-produced motivation, individual's learning could not occur. Thus, learning requirements hold the dominating place in the structure of learning motivation.

The role of motivation in the learning activities is complicated. For college teachers, to understand and grasp the types and features of the students learning motivation is of much benefit to effective teaching. Since there are no significant differences in most people's intelligence, it is particularly important to fully mobilize students' non-intelligence factors and stimulate their learning motivation. The approach and method to mobilize enthusiasm is varied. Out of teaching practice, in view of different specialties, different classes and different students, college teachers should adopt different methods to improve the students' motivation level, to encourage students to achieve learning goals, and produce more positive and effective learning motivation.

REFERENCES

[1] Carter, David. Interpreting Anaphors in Natural Language Texts [D]. Ellis Horwood Limited, Chichester, 1987. In English.

[2] Levinson, Stephen C Pragmatics. Cambridge University Press [D]., 2002. In English.

[3] Ungerer. F & Schmid, H. J An Introduction to Cognitive Linguistics. [D]Foreign Language Teaching & Research Press, 2001. In English.

Management, Information and Educational Engineering – Liu, Sung & Yao (Eds)
© 2015 Taylor & Francis Group, London, ISBN: 978-1-138-02728-2

Analysis and solutions to the private college students' mental health education

Yan Deng
Sichuan TOP IT Vocational Institute, Chengdu, Sichuan, China

ABSTRACT: Purpose: Recently, most of the colleges and universities have paid attention to this issue, but there still are so many deficits in specific strategies and policies. This thesis is based on the analysis of the college students' mental health education. Through the perspectives of enhancing the environmental building of college students' mental health education, gathering high quality professional teachers, enriching the methods of teaching in class and constructing the benchmark system of mental health, the thesis wants to investigate and conclude sufficient solutions to the college students' mental health education. Methods: Experiential summary. Conclusion: 1. Integration of resources to carry out college students' mental health education. 2. To form high quality teachers. 3. Rich private colleges college students' mental health education class teaching means and methods.

KEYWORDS: mental health education, background analysis, explore and think.

1 AN INTRODUCTION TO THE RECENT SITUATION OF COLLEGE STUDENTS' MENTAL HEALTH EDUCATION IN PRIVATE-OWNED COLLEGES

1.1 *Ideological and political education as the mental health education*

The ideological and political education tends to make judgments according to students' behaviors and problems and lead them to the mainstream. On the contrary, mental health education does not help the students to make any decisions, or lead them to make any decisions by their own value system. The golden idea of mental education is that, educators or consultants must keep themselves neutral, not to offer any instructions for students, meanwhile, many mental health education theories emphasizes the function of men's selves, and believe that everyone has the ambition and capability to become better.

Some privately-owned colleges think that mental problems are thinking problems, so they can easily help the students realize their faults and step to the right direction via sentimental and rational convincing. It mixes the mental education with character building, so it is very difficult for the mental education's function to work.

1.2 *Low qualities and rare opportunities of the privately-owned mental education teachers*

1.2.1 *Education staff's lacking of occupation belongings leads to scarce career planning and stagnant education development.*

Mental health education organizations vary in different privately-owned colleges, some attached to student's union, some attached to Youth League Committee, and some attached to related teaching branches. These layouts make the mental education jobs unorganized and inefficient for lacking of overall planning, communication and ambiguity of liability.

1.2.2 *Serious shortage of mental educational teachers*

Some privately-owned colleges seriously lack of psychological teachers, especially the professional teachers and researchers. The main sources of mental education staff are psychology, education and political theory teachers and supervisors, college hospital staff and part timers. A large number of them have no training on mental help and consulting.

Taking a Sichuan-based privately-owned college as an example, there are only three full-time professionals in mental center, but for education burdens, they are also taking great portions of time in teaching tasks. So, it is very difficult for them to implement other mental educations. Because of cost issues, it is hard for privately-owned colleges to implement both professional and part-time mental educational teams. The part timers always work extra hours with little even without payment, in the long term; they will lose enthusiasm and quit the job.

1.2.3 *Shortage in mental education's training system of privately-owned colleges*

Compared with other developed countries and regions, we lack a high quality teaching team in mental education; the main reasons are that our mental education is still in its infant stage with low efficiency, and the training organizations are rare. It is worse in college mental education.

The mental education staff in many privately-owned colleges adopted their profession rather late in lives. Limited by their systematic study and training, they cannot handle the problems of students.

Meanwhile, survival issue and practice of operation offers almost no opportunities for the teachers to get leveled up. Sometimes, they have to pay the tuition by themselves to take training courses, as the charge kept rising, a lot of teachers have to drop the chance to take any training courses. So, the efforts of the mental educationists are not sufficient, and they need colleges to support them either in charge or in time, by doing so, it can help the educationists polish their skills and enrich their knowledge.

1.3 *Stiffness in privately-owned colleges' mental education*

College students' mental health has aroused more and more attentions, but there are still some colleges, which do not understand it well, and this makes the education worse.

Part of the privately-owned colleges included mental education into their curriculum, but for the practical part, they are always professional courses based on printed books. Their education mainly focuses on knowledge, so many students can not acquire any benefits from this course. It is totally not going to work with this pattern, on the contrary, it has a negative effect on students' minds and cannot achieve the goal of its original purposes.

From the perspective of mental health courses, different colleges have different ways. Some colleges set mental health courses as compulsory course, some set them as optional or limited optional course, and others set no such courses.

Taking a Sichuan-based privately-owned college as an example, there are only 16 class hours in freshman period, including mental adaptation, communication, characteristics, personality, studying psychology, internet psychology, love and sex psychology and career psychology. However, all its courses are easy and hard to further explore, courses are always theories and knowledge that the students are always confused with. Limited by the curriculum, students cannot use this knowledge to practice.

For students, the mental courses have high popularity, and they want to relax themselves and study some skills from mental education. The survey supports my point, for there are over 78% questioned students thinking the mental education courses are insufficient and 85% of them needing variable instructions on mental health according to stage characteristics and missions.

1.4 *Wrong understanding of college students' mental health education*

Many teachers have always targeted the minority who has mental diseases as their focus. Since the job of mental health education is to popularize its importance and to shape one's personality, recently mental health education has not only limited the contents, but has no merit to solve the mental problems of college students. The worse is, by some degree, that it brings more burdens to the students, because they are always afraid of being treated as a psycho or being labeled as sick. That makes it very difficult to pave out the daily operation of mental education, not mention the solving of problems.

1.5 *Lack of mental education's measuring system of private-owned colleges*

First, privately-owned colleges pay much attention on survey to the new enrollments, but they neglect the overall understanding of whole schooling years. Recently, most colleges will take some valuation jobs over freshmen, but these jobs vary in time and focus. The SCL-90 chart is the most popular one to test the students, but after the first test, there is always no further test. Some problems such as communication, anxiety and body can be picked out from the test, so some colleges are taking some measurements to follow and probe these students. The lack of further investigation will lead to some problems, such as the issues of study, communication and emotion in sophomore, mental burdens of job hunting in junior and so on. The college cannot direct and release this burden for lacking of further investigations.

Second, some colleges will revalue the process and have talks with certain students who have such problems, and report them to related departments. But the process just ends here, they will not compare these data with previous ones and analyze why it happens. So, departments, which have direction relationship with the students' living, cannot realize the students from mental characteristics, and hard to think and explore the way to educate them sufficiently.

Last, archival management in some colleges is improper. In previous years, most colleges kept their archives by paper, and after the survey, all archives are isolated without proper care. Recently, after the popularity of computer software's, most colleges kept their archives on the computer, but still, they lack of data filing and handling.

2 THE EXPLORATION AND THINKING ON HOW TO IMPROVE THE MENTAL HEALTH EDUCATION OF PRIVATE-OWNED COLLEGES

2.1 *Improve the environmental construction of privately-owned mental health education*

2.1.1 *Constructing and completing mental health evaluation systems, checking and tracking students' mental problems*

Colleges should adopt some methods like mental consulting or mental measuring to operate mental health survey and construct college student mental health archive, so as to find out problems and solve them immediately.

On the other hand, freshmen always run a higher degree in survey, but their mental conflicts will release or vanish when they gradually adapt themselves to the new environment. Meanwhile, new conflicts will come after their lives stepped in colleges. So, privately-owned colleges should take survey in every stage to follow and track students' mental health.

Except measurements, private-owned colleges should also try different channels to grasp the students' mental status. For example, a Sichuan college grasps their students' mental status monthly by filling monthly mental health chart. It is useful to find out the students' mental status so as to direct and intervene them immediately.

2.1.1.1 Setting mental health courses as compulsory course to popular mental health knowledge

The mental health course has always been the common channel to popularize mental health knowledge. It not only makes the students realize themselves but erect the right health idea. Mental department has the liability to introduce the channel to improve mental health, and make the students grasp a scientific and sufficient way to study. It can develop students' potentials and cultivate their innovation and practical abilities. Besides, these courses should also convey the way of mental adaptation to make the students learn how to make mental adaptation and release mental problems sufficiently. Finally, these courses let the students learn the origins of mental problems and the main phenomena of mental problems to treat them with scientific attitude.

Furthermore, according to the situation, mental health teachers should improve the interactive communication between teachers and students in conveying mental health knowledge. Through the combination of theory and case, based on interactive analysis, teachers should let the students pay attention to self-adaptation. Meanwhile, the teachers must positively stimulate their students' enthusiasm and initiation to create good chances by scenario experience and discussion. Finally, teachers must research and learn from experience, emphasize the interactive process and probe the development rules with students. All these will improve students and complete their personality in the study and living.

2.1.1.2 Combination of resources, developing mental health activities and improving students' mental qualities

The mental health education in private-owned colleges should cultivate the atmosphere, which care students' mental health and improve students' health quality. Colleges should take lectures, mental salon, mental knowledge competition, mental community and mental magazine as models to propagate and popularize mental health knowledge. Colleges should also support the mental health association, and help them to operate some special activities such as mental training, movie watching, mental salon and discussion to improve their abilities by helping themselves and others. Colleges should focus on the content and methods of mental health education, especially the Internet platform, because it can greatly improve the sufficiency of mental health education.

2.1.1.3 Campus atmosphere of mental health education

First, colleges should move forward the development of infrastructure. Colleges need make the campus beautiful and green, and make the campus buildings different from others. Meanwhile, colleges should let the students feel that campus is the right place to learn knowledge; another way to improve the mental health education is to complete the teaching facilities such as enough books and good supporting services.

Second, colleges should pay attention to the development of software. The impact of good college and class atmosphere is profound to the students. Also, colleges should emphasize the importance of cultural propaganda, college ratio, college magazines and other media to fully utilize their functions.

2.1.2 *Professional teams with high quality*

2.1.2.1 Professional mental education teams

The job's orientation of private-owned colleges requires the professional training background and a certificate of its teachers. Besides this, the professional teachers should be respective and sincere to step into the students' inner heart.

Meanwhile, professional teams should improve their quality through basic experience and opportunity to get promoted. That will guarantee a sufficient and professional team to handle daily problems.

2.1.2.2 Supervisor's quality of mental health education

The supervisor is the most familiar one to most college students. The supervisor knows their students better and they are the perfect one for students to confide. So it is very important to improve the quality of supervisors in colleges.

Although mental consulting avoids the double relationship between consultant and visitor, supervisors' daily job is not fully consulting, so improving the training of supervisors and letting supervisors handle daily problems with their mental health knowledge can guarantee the operation of mental health education efficiently.

2.1.2.3 Introduction of professionals outside of the college

It is critical to introduce the professionals from outside. It can not only tighten the line between college and society, but also improve the relationship between full time and part time teachers. Colleges can hire some social, mental professionals to take some courses in colleges, and they can also precede some lectures to stimulate students' interests. By this way, the improvement of students' mental health has great guarantee.

3 THE MEASUREMENTS TO ENRICH THE MENTAL HEALTH COURSES IN PRIVATE-OWNED COLLEGES

Recently, some college mental courses are simple and unitary; basically, they utilize the traditional way of teaching, so this kind of teaching lacks the interaction between teachers and students. Lacking of positivity made the education inefficient or even ended in failure, and in class, they have no innovation, so the course lost its original target as popularity and became boring to most of the students. So, it is vital for colleges to change the original design of courses. Increasing interactive content to class makes students experience different feelings in rehearsals, it not only improve the curiosity, but also make the students use the way into their daily lives.

Colleges should enrich the mental courses, and set a systematic, targeting and hierarchical course for their students. For example, they can set a system of studying the method, career training, living instruction, and emotional control based on different stages and grades.

As a conclusion, college students' mental health has a great impact on their growth and the success of total higher education in China. So, colleges should take great emphasis on students' mental health education to provide basic hardware or software support, and to contribute every possible resource to let the students learn knowledge and skill.

REFERENCES

Wei Tongru, The development trend of college students' mental health education in our country, Beijing, 2008.

Guan min, Our country college students' mental health education present situation analysis and countermeasure thought, Hubei, 2005.

Qu Zhengliang, Current situation analysis and countermeasures of college students' psychological health education work, Hu nan, 2006.

Management, Information and Educational Engineering – Liu, Sung & Yao (Eds)
© 2015 Taylor & Francis Group, London, ISBN: 978-1-138-02728-2

A survey of the reform of college physical education teaching

Xiao Mei Zhao & Qiang Wei
Physical Education Department, Tangshan College, Hebei, China

ABSTRACT: This article analyzes the process of the reform of college physical education and the present situation, illustrates the principles that we stick to during the reform of physical education in colleges and some reform achievements, points out the existing mistakes and problems, and puts forward the corresponding strategy and the concrete operation scheme, in the hope of providing additional insights into the development of the reform of physical education in colleges and universities.

KEYWORDS: College physical education teaching; Reform; Teaching mode.

In recent years, the reform of college physical education teaching has brought rapid changes. Teaching methods and teaching thoughts have developed constantly. The single and rigid teaching modes have become various and flexible. Many new things appear. The new ideas, new teaching terms, and teaching methods have sprung up. The reforms of college PE teaching have been taken at any time. At present, the reform of college physical education teaching is stepping into a new era.

In the study of the reform of college physical education teaching, we should teach students in accordance with their aptitude, taking the physical differences of students into consideration, to make every student gets a proper physical education and proper physical exercises. At the same time, we should tap the potential of students in the process of the reform and cultivate their personalities, trying to create one kind of physical education class which can enhance the students' physical quality comprehensively. The reform should explore new teaching methods and teaching modes courageously, and break through the traditional shackles to change the traditional teacher-centered mode. Although the reform of college physical education teaching in recent years has made great achievements, some problems still exist The author analyzed the reform of college physical education teaching and carried on a scrupulous research, pointed out some existing problems, and put forward some measures for improvement.

1 THE ANALYSIS OF THE PRESENT SITUATION

1.1 *The unchangeable principles of the reform of college physical education teaching*

After studying the reform process of the university physical education teaching and exploring the meaning and the essence of the reform, we can see that it is not hard to find, the principle is always the same although education reform is constantly changing. All of the reform measures are to make full use of the physical education time, to reflect the value of physical education, to tap the potential of physical education curriculum. Tapping the potential of physical education and embodying the value of physical education are the unchangeable principles of the reform of college PE teaching. They are the constant theme of sports reform, also the goal of the reform of college physical education teaching. A new era of reform of physical education is still to adhere to the constant principle, aiming at embodying the value of physical education and tapping the potential of physical education to push the reform through.

2 THE ACHIEVEMENTS OF THE REFORM OF COLLEGE PHYSICAL EDUCATION TEACHING

2.1 *The deepening perception of the reform*

Knowledge is the precondition of all and it is the foundation of practice. Without the understanding of the things early, everything is hard to move on. For many years, the reform of college physical education teaching has been a process during which the perception is deepening and the thought is constantly improved. (1) The understanding of sports is constantly deepening. The understanding of sports has changed from the original thoughts of simple exercise into the increasing awareness of the theory of sports teaching now. The understanding of sports is becoming more and more perfect, and more and more rich. It is something from nothing. (2) The deepening understanding of the sports teaching purpose. Teaching purpose and the value of physical education changed from one dimensional biological concept of sports in the multidimensional concept of sports in the aspects of biology, psychology and society.

The change of the value of physical education promotes understanding of sports. A lot of people realize that sports is not only a purely physical fitness, but also the guarantee of healthy life. A great number of people begin to attach importance to sports training and fitness. At school, for students, physical education curriculum changed from the original single recreation and exercise with the implementation of quality education, which put emphasis on cultivating comprehensive talents. More and more physical education curriculum is taken seriously by the school. The teaching contents are becoming more and more rich which can stimulate the interest of students and involve more students into the physical education. The understanding of sports is changing, and the value of sports is changing, as well as individual's inner perspective. Only in this way, can we pay attention to the sports in daily life, and take exercise intuitively and let people find new sports to enrich the sports world.

2.2 *The broader vision*

The changes of vision, to a certain extent, promote the reform. After the establishment of new China, the reform of college physical education has been based on the Soviet model, and then the view gradually expanded to the world, especially the western developed countries. After the study of other's teaching modes, and realizing the shortage of our own teaching mode, we began to take teaching reform in our country, and gradually got rid of the Soviet model, starting to absorb outstanding achievements in the reform of physical education all over the world. Second, the reform of college physical education was taken under the background of reform and opening up. It was comprehensive and powerful. The reform was taken from a single sports science development in a combination with a variety of other disciplines, including natural science, social studies, cultural and economic, social and so on.

2.3 *Diverse teaching modes and human-centered contents*

Compared with the previous physical education curriculum, the biggest change in the reform of college physical education is that the teaching modes have been increasingly diversified. The teaching contents are more and more humanized. The courses are becoming more and more attractive. The change of teaching modes and teaching contents is the practice of the deepening reform of physical education, and it is the greatest achievement of the reform of college physical education. The change of physical education teaching modes is from top to bottom. The national education ministry issued the corresponding documents to guide the schools to carry on the reform the physical education mode. The various universities took deeper and broader reform in modes and contents in accordance with the guidance of the state, taking the actual situation of staff, venues and students into consideration and breaking the shackles. The single and monotonous education mode becomes rich and scientific. The teaching contents are well-designed and more attractive. In terms of teaching contents, the reform is directed at the philosophy of people-oriented; embodying the humanization of the reform of teaching content and well connecting the newly-emerged sports with traditional sports, providing students with a variety of choices, and teaching students according to their aptitude has been possible. The diverse and humanized teaching contents will be able to attract students and stimulate students' interests. In the diversification of teaching content, we should set limitation; the contents shouldn't be too many, adapting to the students' interest. Humanization should be given priority to.

3 THE EXISTING PROBLEMS

3.1 *Indistinct aim of the reform*

When making their own teaching tasks, different colleges cannot reflect their own personalization and lack innovation. Some of the colleges cannot embody their identity and fail to emphasize the importance of the lifelong physical exercise, the physical quality development and the cultivation of their own personalities. It is difficult to make students form a kind of lifetime sports faith to fulfill the mission of the college sports education. What's more, a lot of theoretical stuff stays on the books. And students learned a lot, but there is no practice, those things are just theoretical. Although some knowledge is comprehensive, but empty, lack of focus and goals, and without enthusiasm.

3.2 *The physical education classes are often normalized by the teachers*

The physical education classes are often normalized and simplified by the teachers. The teachers often organize the student to practice queue formation to enhance the coordination and mobilization, and put these as one of teaching evaluation. This form is too simple, monotonous and boring, and it can't let students play their own special skill and look for their favorite sports. This kind of sports classes is difficult to meet students' needs which are diverse and human-centered. And it is unfavorable for students to improve the interest in sports, and it cannot cultivate the students' faith of lifelong exercise.

3.3 *Too much autonomy in the reform of college PE teaching*

The reforms of college physical education are probably of two trends. The first trend is that the schools give students too much autonomy. They advocate the open policy overly, ignoring the supervision status of teachers. The students are neglected, inactive and at a loss, let alone physical exercise in advance. The second trend is that the schools are against passive teaching process. To stimulate student interests, meet the needs of the students, and give full play to the students' creativity and enthusiasm, the schools organize students to choose their own learning contents. Therefore, some of the teachers think that if we want to stimulate students' interests and meet the requirement of students, the teachers should just provide guidance to the students without teaching organization. But students in the class are lack of organization, discipline and regulation. In the long run, the college physical education teaching is difficult to become scientific and standardized, and the students are unable to master the spirit of sports and to tap the potential. This kind of teaching which is just to satisfy the students' a temporary need, denies the role of the teachers and it is very undesirable. The students' cooperation and active learning should be combined with the supervision under the guidance of teachers in order to create a positive learning atmosphere.

3.4 *Misunderstanding of competitive sports*

In quite a Long time, competitive sports have been strongly rejected and thought the cause of fights and disputes between students. Some people even think that competitive games should be expelled out of school. Are competitive sports not really suitable to exist in the campus? What's the matter with competitive sports exactly? The changes over the years have proved that there is nothing wrong with competitive sports. In recent years, the contents of the competitive sports have been weakened, while students are becoming increasingly weak. This shows that competitive sports are suitable for the campus. And competitive sports can arouse students' interest. The human nature had made it. But some colleges are lack of competitive sports; many students can only choose other items which are not appealing enough instead of funny competitive sports.

3.5 *Backward teaching materials*

Nowadays, the college physical education teaching materials are single and dull, which can't be made full use of by the students. The extracurricular practice teaching materials are less than the theoretical physical education materials. It is difficult for students to get more useful knowledge. Now, composing sports teaching materials has been in the traditional and stable style. There is no innovation, and less new science of sports knowledge, even less new ideas in accordance to the needs of social development, let alone new contents. Sports teaching material are out of date, lack of local features and corny. The contents of teaching material are too single, lack of entertainment, and ignoring the cultivation of students' skills and interests, and not in favor of the students to take the initiative to find suitable ways to exercise.

4 COUNTERMEASURES FOR THE REFORM

4.1 *Regard the reform in the philosophical perspective*

The reform of college sports education should pay attention to a combination of teaching and learning. The reform should be well-organized, carefully planned and with clear purpose. As for the teaching, the teachers should set good examples, improve the sense of responsibility and take each class seriously. The simplification and formalization of class should be removed. As for the learning, it is from the perspective of students' obligations. The students should cooperate with the teacher for each class, give full play to their own expertise, and find their potential. The teachers and students are equal in the terms of value and status. The teachers should do their own things, and take responsibility, to create a learning atmosphere for students to learn actively with enthusiasm. The students should learn critically, explore new knowledge, and construct their own knowledge structure.

4.2 *Keep competitive sports in perspective*

All things exist for reasons. In the sports world, the existence of competitive sports has its reason. Competitive sports embody the technology and skills of sports, and the technology and skills of sports are the essence of sports. In addition to that, competitive sports and sports are also closely linked. The athletics sports are the most important way to improve students' physical fitness and to execute the objective of physical education. It is the essential means to promote students' development. Competitive sports play an indispensable role in education, socialization, and the development of their own personality.

4.3 *Improving the teaching material system*

Teaching materials which can appeal to more students should be designed. The relationship between the traditional and the new should be handled. We should be groundbreaking and rational. A complete system of teaching materials, which can highlight the key points and be well-proportioned, should be established. The teaching materials should be modernized and localized to show the features. The curricular and extracurricular teaching materials should accord with these each other, coping with the relationship between the comprehensive development and individual personality.

5 CONCLUSIONS

To sum up, the reform of college physical education should be seen as a whole. The present reform should be combined with previous data, and be perfect gradually. Education reform is a process of rational thinking and constant practice. Only a moment of passion and impulse cannot work. In addition, going in an extremely way in the reform of college physical education should be avoided. Therefore, in the face of the problems in the reform of college physical education, proper judgment and measures must be taken.

REFERENCES

[1] Ma Linxin. The reflection on the reform of college physical education teaching. [J] China Management Magazine,2013(7);100–101.

[2] Wang Sen. A brief analysis of the reform of college physical education teaching [J] City Tutor,2012(6);110–111.

[3] Yan Lisha, The mistakes and measures for the development of the reform of college physical education teaching [J] ,Journal of Chifeng College, 2013(8):122–124.

Management, Information and Educational Engineering – Liu, Sung & Yao (Eds)
© 2015 Taylor & Francis Group, London, ISBN: 978-1-138-02728-2

Study on the application of art education based on multimedia teaching

Bao Quan Pan & Mei Ni Xiong
College of Fine Arts, Beihua University, Jilin, China

ABSTRACT: The significance and function of the multimedia teaching in art education were analyzed according to the special feature of multimedia technique, and the application of multimedia teaching in art education from three aspects was discussed. Prospects for the multimedia teaching in art education are given at last.

KEYWORDS: Multimedia skill; Teaching course software; Art teaching.

1 INTRODUCTION

With the rapid development of the information age, computer multimedia technology is widely used in the teaching of subjects, promoting education and educational technology change and progress, so that modern education in the information age. In recent years, more and more people have realized that IT is playing an increasingly important role in modern education teaching. In this environment, the art of teaching, great changes have taken place in the traditional fine arts teaching gradually integrated multimedia technology and theory, in the art of teaching has injected new vigor and vitality. However, most of the existing school art teaching professional application of multimedia technology is still in the primary stage, has not yet formed a complete teaching system. As a complement and supplement traditional teaching, the popularity of multimedia features and the computer so that the computer multimedia teaching has become an indispensable part of the art of teaching. Multimedia systems usually consist of multimedia hardware systems, multimedia operating systems, multimedia authoring tools, and multimedia applications in four parts. The multimedia hardware system includes a computer, CD- ROM, audio input/output, video input/output and other equipment, is the foundation of a multimedia system.

Real-time operating system includes multimedia task scheduling, and synchronization of multimedia data conversion control, drive and control of multimedia devices and a user interface with a graphical and pan functions and the like. Multimedia authoring tool is a tool to create multimedia applications software, collectively, it can be text, graphics, images, animation, video and audio and other multimedia information control and management, and put them into a complete connection required for multimedia applications. Multimedia application system developers to use computer language or multimedia authoring tools to create multimedia applications software products, is directly facing the user. In the art of teaching are the multi-use multimedia applications. So, computer art multimedia technology compared with the traditional teaching of art teaching, advantages and practical significance were discussed.

2 THE ROLE OF COMPUTER MULTIMEDIA TECHNOLOGY IN TEACHING ART

The role of computer multimedia technology in teaching art has the following main aspects:

1 multimedia art teaching can improve the quality of teaching, improve teaching efficiency. Full use of modern teaching media for teaching knowledge and skills is promote the development of students' knowledge and skills to improve their mastery of knowledge, ability quality. The use of modern teaching media for teaching and vivid is infection and strong, easy to stimulate students' interest in learning and internal motivation. Students develop knowledge and skills in all aspects of perception, understanding, memory, applications can play a beneficial impact. Student learning and mastery of knowledge is a variety of senses (eyes, ears, nose, and body) to pass information to the outside world and the formation of the brain centers. These different functions of the senses, in the study, eyes, ears, brain function, coordinate better the play, the higher the efficiency of learning, which is a rule. The use of modern teaching media can fully mobilize the student's eyes, ears and brain function; improve the efficiency of student learning.

Multimedia art teaching offers a personalized teaching environment, so individualized and personalized learning a reality. Modern art education is education for the future of innovation. Individualized teaching is that we adhere to the objectives pursued, because there is no personality, there is no innovation. Modern educational technology, especially the development of information technology, multimedia computers and networks are increasingly expanding range of applications in art teaching, to personalize teaching created unprecedented conditions. Better able to adapt to the individual differences of students to achieve individualized, so that students get a real subject of this innovative liberation.

Multimedia courseware easy to upgrade, you can make the art of teaching always walk in the forefront of teaching. Multimedia courseware upgrades can continue to improve rapidly add new content to accept the most advanced ideas and design philosophy of art, which is the traditional way of education as the main textbook is difficult to match.

4 multimedia art teaching art educations can expand and promote the popularity of art in a larger context. With the development of science, has been an increase in home computers, computer technology for multimedia teaching methods used for art education to create a good physical condition.

3 COMPUTER MULTIMEDIA TECHNOLOGY IN TEACHING ART

Computer multimedia technology in an art foundation course in computer applications in art teaching universities is generally placed in specialized courses designed art lesson, rarely involves basic course, most teachers believe that the traditional art of teaching is the key to solving the shape of basic skills students. But the study, we found that the traditional teaching of basic courses in computer multimedia teaching should be assisted. In this regard, we can learn on the computer and briefly studied art analysis. We know that the traditional art of teaching is mainly on the basis of sketches and color teaching, teaching in the sketch is mainly to solve our modeling capabilities, modeling training, perspective, anatomy, sense of space, texture, composition and other content; color teaching in our main solution hue contrast, color contrast, complementary color contrast and color space and color contrast and well-being and so on. We discuss these issues now can be done in a number of software, that is, for color sketch studies can be carried out entirely in the computer. Such teaching and research can save a lot of time, avoiding the production of hand-painted dull duplication of effort. Take a color mosaic effect work constitutes an example, if you want to complete a form, you need to repeatedly create and modify, wasting a lot of time and effort. The application of computers to complete simply enters an image in Photoshop software, and then performs Filter- Pix elate- Mosaic command, a mosaic work done in an instant.

We also found that there are some modern software can simulate the effect of a variety of drawing tools, has now developed software such as Fractal Design Painter is good painting software. It provides a lot of painting tools, such as pens, pencils, brushes, watercolor pen, oil paints, airbrush, crayons, etc., but also offers a wide variety of strokes, to a certain extent, truly mimics the effect of freehand drawing. Such as when using painting tools, mimic Van Gogh, Cezanne and other masters of the brush strokes; when the brush tool, combined with wet drawing paper, the effect of Chinese painting on rice paper for the imitable. Although it cannot fully replace the painting, the painting can be used as a research method and means. In this way, students can save a lot of freehand drawing practice time, students can also develop visual thinking and analytical skills, help cultivate innovation ability.

Computer multimedia technology in art appreciation class application is an important part of art class art teaching content. The purpose of art appreciation, mainly in fine arts teaching in two ways: first, through the appreciation of works of art, to understand the history of art development, grasp the law of development of fine arts; second, through appreciation, inspire students' creative thinking, and thus create outstanding works. Through appreciation, teaching courses on pictures or slides are operations due to the complexity of the fuzzy, often not very efficient. Using computer multimedia teaching, art appreciation class effect is greatly changed. As long as teachers have a multimedia computer, there are art appreciation information and data to a disc loaded into the optical drive (CDROM), students in the appreciation of classical works, a little mouse, a painter's masterpiece and biographies will appear, not only can get a glimpse of picture works, you can also zoom in partial observation; not only a comprehensive view of the works, you can also choose to enjoy or have a work print.

Students in the classroom can break time, space, geographical restrictions, and foreign gallop across the ancient and modern, art gallery stroll. Appreciate the art of architecture, when not only the usual way to appreciate the buildings form of visual perception in the stationary state, but also can do dynamic display, allowing the building to spin up, enjoy the multiple sides of the building. You can also incomplete, or damaged historic buildings were restored. Enjoy a graphic design or cover design time, but also can be altered according to their preferences, but it is also indispensable for students to creative material. These are the traditional appreciation class unmatched.

Computer multimedia technology in the computer is professional courses in art, as a modern tool, first used in art class in design teaching. It is widely used in the design, liberated the previous complex design work, improve work efficiency, the designer's right-hand man.

The accuracy is concerned, with the computer to produce an object the size of an absolute right, almost no errors. Such as the use of CAD software or 3DMAX produced graphics, with an accuracy of less than 0. 1mm, it can show the most delicate part of the design, which is the traditional hand-painted difficult to achieve, but its camera angles, pleasant to the eye point of view, depending on from fully simulate the real, not the existence of handmade randomness and imprecision. Therefore, the authenticity of computer design by the majority of designers agrees. In recent years, the development of software used in graphic design software is gradually increasing, a common two-dimensional CorelDraw, Photoshop, AutoCAD, and three-dimensional 3DMAX other software.

In painting courses teaching, there must be a lot of painting techniques require students to master, every teacher of painting styles, techniques have their own characteristics. The instructor is necessary to allow students to expand their horizons, to absorb. Application of multimedia technology, breaking the teachers' knowledge of this one-way transmission of information sources, expanding the scope of information exchange teaching, breaks the teaching space, so that unrestricted access to information and delivery time and space. For example the Central Academy of Fine Arts masters editing techniques textbooks, covering drawing, painting, oil painting techniques, such as content, teachers can combine curriculum content for teaching demonstration, in the classroom so that students can see the famous painter's lecture demonstrations, but the key place to repeat Play, learn to understand the true meaning of painting masters. Similarly, professional instructor's techniques can also be made from their own multimedia courseware demonstration teaching, improve teaching effectiveness.

4 IMPACT OF CONTEMPORARY CHINESE PAINTING ART EDUCATION DEVELOPMENT

Chinese painting as a symbol of Chinese art, has developed a number of years, has experienced a long history. Its development and development of traditional art education is basically synchronized. A traditional art education model for studying the development of self-contained Chinese painting provides a strong guarantee, but the crisis has brought to the development of Chinese painting, Chinese painting may lead to the development of stagnation and rigid.

With the introduction of Western culture, contemporary art education has been given a new content and features, traditional art education has been seriously challenged, and Chinese painting is no exception. Chinese and Western cultures fierce collision, given the contemporary art education conform to the trend of the times, and the Chinese have a tendency to weaken the function of painting, we can see the impact on Chinese painting contemporary art education development is enormous. Therefore, Chinese painting to have sustainable development, we must look for a path.

Given the cultural heritage and cultural traditions there are significant differences among different ethnic, fusion of contemporary Chinese painting art education and the presence of significant challenges, mainly as follows: Western contemporary art education curriculum is based on the theory, although trying to achieve Western reconcile, western system in order to mark the traditional Chinese painting art education is based on the poetry of learning-based, it is trying to achieve a comprehensive innovation in the original basis. Since the starting point of contemporary art education and traditional art education, the focus is different. Should contrary to objective laws, not according to the actual situation of contemporary art education and the development of Chinese painting, blindly impose two very different systems together, will generate a lot of negative impact, but we should make an objective assessment, can not ignore the contemporary the positive role of art education for the development of Chinese painting.

Contradictions are opposites, we study the impact of Chinese contemporary art education in the process of painting effects, specifically requires an objective analysis of contemporary Chinese painting art education teaching, writing produced. This article from the aspect of two dimensions, the use of comparative approach, in teaching, through the teaching objectives, teaching methods, teaching materials, teaching content, teaching evaluation analysis of differences in traditional and contemporary art education art education; in the creative aspects of, will be an inspiration on Chinese painting, technique, emotion comparative study, summed up the pros and cons of contemporary art education for the development of Chinese painting, but overall, pros and cons, more harm than good.

Because of our long-term in a feudal society, and the ancient social, political, cultural and systems are designed to meet the needs of the ruling class. The prevalence of the traditional hierarchy is art education, just to satisfy the interests of the ruling class and the service. In this case, the role of art education is major with a strong "into enlightenment, helping Fallon," the. In this historical context, the traditional art education does not have a complete and standardized education system. In particular the teaching process, the main idea is to master personal education based, then painting learners main objective things through observation and copying works of the classical way to achieve the purpose of drawing creation. Contemporary Art education is committed to the community to cultivate creative talents, and the fine arts education in the country gaining in popularity, becoming a national art education. Up to now, we have been gradually achieving the transformation of quality education by exam-oriented education to the "moral, intellectual, physical, aesthetic, labor" comprehensive development.

Visible, education has become the most basic rights of every person, and not a few aristocratic privileges, which will benefit the construction of a socialist harmonious society. Contemporary art education teaching objectives conducive to the development of our comprehensive quality education, improve the overall level of China's national art knowledge. For the construction of a beautiful home that has everything to gain but no harm. However, we also see the shortcomings cannot be ignored him an objective reality. Universality corresponds to the peculiarities of contemporary art education is universal knowledge for all students in the art of Chinese painting teaching goal setting, and not as a traditional art education that, according to each student's specific conditions, can amount tailor the most appropriate teaching painting learning objectives and teaching goals at any time to adjust according to the needs of learning. Contemporary Art by differences in education and Chinese painting teaching objectives analysis, summed up contemporary art education exists on the pros and cons of teaching objectives. We can see that they are closely related, the latter is dependent on the former, while drawing on Western art education.

5 CONCLUSION

Computer art multimedia technology in teaching, not only to speed up the progress of the art of teaching, improve learning outcomes, more important is that it can give students a more novel stimuli, resulting in the best area of the cerebral cortex related development, so as to stimulate their creative evolving thinking. To innovative works are art majors. The information age, computer multimedia teaching the art of teaching is the direction of development, however, stressed the role of modern teaching media is not to deny the traditional teaching, but rather calls attention to a combination of both in teaching practice, flexible use. Contemporary art teaching focus on the integrated is use of multimedia in order to continuously improve the quality and efficiency of the art of teaching. The main trends in the art of teaching is the traditional teaching media development and integration of modern teaching media, and gradually form a complete, current information society to adapt to the development of art education system.

ACKNOWLEDGEMENT

Jilin Province Department of Education Projects [2012–344]

REFERENCES

Li Si Hui. On Myth and Art Education of College Students Comprehensive [J]. "Heilongjiang Science and Technology Information." 2008.07.

Liangtai Sheng. Reform of university teaching of Chinese painting Rethinking [J]. "Zong Tai'an College of Education Science" .2003 01.

Han Jing. Confront contemporary art education principal institutions of higher art China Forum on –2010 [J]. "Art Watch." 2010.05.

ZHANG Yao-guang. Lilley. Chinese modern art education in the "Western painting" Complex [J]. "Grand Art" . 2011 09.

Ge Xintong. Reflections on Contemporary Art Education [J]. .2010. O5 "business culture."

Zhang Jun to stay. On Chinese painting teaching traditional culture [J]. "Arts education research" .2011.10.

Lee slip down modern art education model and the concept of cultural and ecological harmony - of Our universities Normal Art Education in the 21st Century [J]. "Inner Mongolia Normal University (Educational Science Edition)." 2006.11.

Reflections by Lisa Shuai. Contemporary art trends in China [J]. "Industry and Technology Forum." 2012.

Zhang Bing. Explore the aesthetic perspective of Art Education [J]. "China-school education" .2010.05.

Management, Information and Educational Engineering – Liu, Sung & Yao (Eds)
© 2015 Taylor & Francis Group, London, ISBN: 978-1-138-02728-2

Research on higher art education based on the aesthetic intuition theory

Hui Jie Ji
School of Chinese Language and Culture, Beihua University, Jilin City, China

ABSTRACT: The problem of higher art education has been collated and analyzed by combing aesthetic intuition theory, investigating the actual educational situation and education experimental. The theory of aesthetic intuition solutions is proposed based on the basic law of higher art education.

KEYWORDS: Intuition; Aesthetic intuition; Art education; Comprehensive education.

1 INTRODUCTION

Art is a human and often by this mental activity, but also constitutes a history of human civilization is an important part, as early as in ancient times, arts education has existed. It is a unique way to penetrate into the depths of people's lives. Art was once thought to be in this science, as opposed to the discipline and rationality was shelved, art education and education are also divorced from the overall goal. Until the 1960s, some art-depth study of psychologists and psychological mechanisms of aesthetic theorist of art, perception and cross-cultural, just to make people aware of the value of art itself, an emerging trend of arts education grew and developed.

Arts education throughout the education foundation stage should play what role, what kind of arts education is the most effective? United States is the world's first art courses included in the core subjects of basic education; it would mean the so-called core subjects will stand on the same level with math, science, language these disciplines, equitable division of teaching time. As we all know, in today's education, discipline increasingly fine division. Such an increase implies something the students want to learn more and more, but their times are fixed. As a result, the constant increase in core subjects or other subjects only make students more overwhelmed. For this reason, the development trend of the moment the best education is infinitely split from discipline to a new integration. Arts education in this, should implement the existing curriculum standard, advocated artistic ability is to train students and humanity's integration development. It is an integrated aesthetic intuition all sense and sensibility to artistic ability into aesthetic activity. Therefore, how people perceive the aesthetic intuition, how to understand the process of aesthetic intuition in art education's role, and what kind of approach to training is worth exploring issues such as students' aesthetic intuition.

2 AESTHETIC INTUITION THEORY

Intuition aesthetics is one of the modern Western aesthetic genres, produced in Europe in the early 20th century's. The main representatives are the French and British Whitehead Bergson. Is a philosophy of life as the philosophical foundation of an aesthetic genre, life philosophy that life is the essence of all things, the basis of existence, reason and experience can only grasp stationary things, and only intuition can experience the presence of life? Intuition is above life philosophy Aesthetic Theory in the aesthetic field. Bergson believed the ordinary way of things intuition and expression, not the expression of a direct perception of the phenomenon of life, resulting unique feelings, it's missing a number of specific characteristics of life. The artist has created a life can show stretches of perception; intuition is the beauty of this unique intuition, that is, artistic motives. Theoretically knowledgeable people also believe that intuition is less than the young and innocent little children, the artist must not lose the innocence of a child. Intuition aesthetics and performance aesthetics of the 20th century laid the aesthetic thought iconoclastic, anti-rationalism in the direction of the symbols on the aesthetics and aesthetics phenomenology had some impact. Intuition has become an important concept in modern aesthetic theory.

In the field of aesthetics, it is also known as the aesthetic beauty of intuition, the two words mean the same thing. In the 19th century, the Western emphasis on logic and experimental cultural background, once the aesthetic intuition irrational and mystical tendencies to the extreme. But it is precisely because of this phenomenon, the West intuition theoretical psychology research shifted from philosophical speculation. People are with psychology, physiology and brain science to study the aesthetic intuition. Although each school on a range of issues there are differences, but in the presence of rational precipitation aesthetic intuition on this point is more consistent. On an aesthetic intuition emphasized the rational side, but it is not wholly logical. Intuition is not a simple and intuitive aesthetic is not in accordance with the usual "syllogism" deductive logic or inductive logic reasoning. This means that it has a strong insight and creativity. Aesthetic intuition can be used for art appreciation, but also can be used for artistic creation. Since the research on intuition and aesthetic intuition is derived from a branch of science refinement-visual psychology. Visual psychology is a psychological phenomenon in visual perception and visual perception caused for the study subjects.

3 REALITY ART EDUCATION CONFUSION

The overall goal of art education, new curriculum standards referred to "the learning process of students in art, rich visual, tactile and aesthetic experience, access to art of enduring interest in learning, form the basic art literacy", "improve the aesthetic ability to understand art the unique role of cultural life and social development." Here is more clearly explained the ultimate goal of art courses. In the "targets" in the four fields of study made specific teaching requirements, these instructional purposes basically not much difference with older versions of curriculum standards. It is worth mentioning that in the "comprehensive exploration of the area," the teaching objectives, calling for "understanding the relationship between art and nature, art and life, art and culture, between art and technology, to explore and comprehensive art activities, and to published in various forms of learning outcomes", where clearly pointed out a comprehensive exploration of objects and methods, while the new curriculum standards also require our art curriculum to make students feel a sense of pleasure and success.

As we all know, in today's education, discipline increasingly fine division. Such an increase implies something the students want to learn more and more, but their time is fixed. As a result, the constant increase in core subjects or other subjects only make students more overwhelmed. For this reason, the development trend of the moment the best education is infinitely split from discipline to a new integration. Arts education in this, should implement the existing curriculum standard, advocated artistic ability is to train students and humanities integration development. It is an integrated aesthetic intuition all sense and sensibility to artistic ability into aesthetic activity. Therefore, how people perceive the aesthetic intuition, how to understand the process of aesthetic intuition in art education's role, and what kind of approach to training is worth exploring issues such as students' aesthetic intuition.

See some natural or man-made works of art, we tend to happen on an emotional excitement, perhaps pleasant excitement, perhaps sad passionate, whether it is the former or the latter, in short, we were moved, so emotional the excitement, called appreciation, that is, we are looking at things from the beauty. Education is when a person learned in school has forgotten the rest of the stuff. Into their own behavior within that part of the way of thinking is not visible knowledge is permanent. No matter what kind of art education means education methods, we want to achieve the ultimate goal is to train a person's right or even aesthetic intuition is a national beauty bright positive psychological intuition and aesthetic.

Because aesthetic intuition without hesitation and a look that is beautiful aesthetic phenomenon, people tend to go for the emotional behavior aesthetic intuition, put it on the shelf, and scientific and rational opposition. As Freud and psychoanalysis, led by Western modernist art is the artistic expression of this anti-rational. After careful thought is not difficult to find, aesthetic intuition itself is permeated with a rational component, bears deep and rich social and cultural connotations, this intuition has all the ability to think. It has a rational component, needs to be guided art education and promotion. Aesthetic intuition is an important measure of the strength of the aesthetic capacity through artistic training of artists generally has relatively strong aesthetic intuition.

4 LIGHT ART EDUCATION THEORY

Aesthetic intuition is the most common aesthetic procedure in a state of mind. In ancient times, people's living environment is difficult to see what kind of rules, and art to create simple, graphical rules, and created a sense of order in the midst of the confusion. Therefore, the original artist was regarded as a shaman; human ethnic groups have a higher cultural knowledge. In the original sense of the human simple, the artist represents the gods and prophets. Because of the lack of art and art-depth understanding of artistic creation is often attributed to divine inspiration, is a sacred and inviolable power. "Divinity" art naturally ruled mankind for a long time. "Aesthetic intuition" of the prototype, from Plato and Aristotle to Plotinus, has gradually formed. It is worth mentioning that in the "comprehensive exploration of the area," the teaching objectives, calling for "understanding the relationship between art and nature, art and life, art and culture, between art and technology, to explore and comprehensive art activities, and published in various forms of learning outcomes", where clearly pointed out a comprehensive exploration of objects and methods, while the new curriculum standards also require our art curriculum to make students feel the sense of pleasure and success.

Plato believed that only God's creation is the ideal aesthetic. Just imitating what the artist craftsman made, like a mirror copy out the same and it is an imitation of God's creation artisans creates useful objects. Plato's "ecstasy," said the same divinity with color, referring to nothing more than a gifted or possessed by the god's subconscious creation. Among these, already contains intuitive grasp the whole meaning of the United States. Aesthetic point of view Plato's divinity with a strong color, his art holding a pessimistic attitude that mankind is impossible to achieve, "God," the realm of abstract painting cannot grasp the nature of the object.

To Aristotle, he turned to the topic of the aesthetic theory of the art technical aspects of the creative process, to avoid the continued discussion of the divine. Make aesthetic break free from the rich colors of theology, divinity Aristotle optimism after Chinese scholars with the views coincide. The impact of this transition is far-reaching, is the source of the integration of Chinese and Western aesthetic concepts, but also from the theological aesthetic research laid the foundation for the later. Pontius is to distinguish it from the general aesthetic feeling activities, and the creation of the visual aesthetic concepts. Research can feel the gradual transition from theological aesthetics of color visual description to describe the concept of rationality.

5 PRACTICE OF ART EDUCATION OF THE ROAD

Observation is an acquired skill acquisition, called the skill is learned through training. While some people are born with the advantages of focus, eye-hand coordination advantage, but most people need to be trained on its sensory organs. Here I am more willing to put into intuition and observation records, the most honest observation is that we see something intuitive, able to record directly to the most essential thing is intuition courses on other subjects very helpful thing. Just about the ability to observe these courses are often submerged in the culture keen on realism in art education, preferences technical than artistic.

About appreciation activities, no doubt, are to be developed through art education. A key material from the late finishing aesthetic experiment, you can find and enjoy their psychological types and performance capabilities there is a certain relationship. From a narrow perspective, to appreciate a work of art is the expression of others to make one kind of feedback. A picture is good or bad, is neat piecemeal, everyone really is there so a standard, but the standard spots carefully grind down, you will find a subconscious standard. Factors for each type of art society in general have a similar psychological basis, whether children or adults in painting performance will be with a certain idea, but rejected the direct manifestation of an adult psychological tendency increases the complexity of the painting expression sex.

Few students lack the test standard of beauty; there are also various types of painting equivalent of each individual's psychological tendencies. This is the most original aesthetic intuition, give a very simple example, let an intense phobia of people to enjoy a dense line of painting is not very realistic. Observation and appreciation of the culture are to comply with the students' aesthetic level and preferences. Different preferences for each student, but always from the many types of art found in the subjects they are interested in, integrated teaching art classes cut from a different perspective, not only enrich the students' knowledge, but also for aesthetic intuition made a very good bedding.

6 MATERIALS AND METHODS

Selection of Sprague-Daley rats, pregnant 9–10d. Pregnant rats were sacrificed by cervical dislocation, cut the skin, muscle and peritoneum along the midline to expose the uterus. Separated out the rat embryos were placed in vials containing penicillin Bouin's fixative fixed 24 h. Under the rat tail with a scalpel cut the number of rat tail collagen fibers extracted root, first into penicillin vials containing Bouin's fixative fixed 24 h. 2~3 min and then stained with eosin, and make it red. According to requirements of the paraffin-embedded tissue dehydration, transparent, dipping wax, pour the melted paraffin-embedded slot, and then head up the vertical placement of rat embryo groove surrounding the embryo is inserted vertically slightly longer than three embryos the longitudinal axis of rat tail collagen fiber bundles as a positioning line, so that rat tail collagen fiber bundles in the embryo outside evenly distributed and parallel to the longitudinal axis of the embryo.

Selection of 5 mm is serial sections of a sheep embryo (HE staining, 500). Sliced shot taken using a digital microscope image, due to the larger slices, each slice will be divided into several parts were taken and saved as JPEG format. Then run Motic images assembly1.0, the number of rows and columns load pictures selected, and sequentially load images, select the appropriate consolidation method, the optimal scanning step and scan mode, mosaic image, and save it as JPEG format.

Using Adobe photoshop7.01, and the use of hand-positioning and computer positioning method combines two-dimensional image acquisition to be corrected positioning. Get image method in Java then the captured image from the DICOM format into JPEG format, select 256 colors (16-bit depth) image, to obtain black and white DICOM images while the image capture, compression processing. Finally, all converted to DICOM format two-dimensional image into a specific directory, run the three-dimensional reconstruction of medical software, complete three-dimensional reconstruction of a two-dimensional image.

A picture is good or bad, is neat piecemeal, everyone really is there so a standard, but the standard spots carefully grind down, you will find a subconscious standard. Factors for each type of art society in general have a similar psychological basis, whether children or adults in painting performance will be with a certain idea, but rejected the direct manifestation of an adult psychological tendency increases the complexity of the painting expression sex.

7 CONCLUSION

In this study, the specific impact statements and interviews statistical data analysis, from all areas of the Jilin-based art education art teacher and student perspective, a comprehensive understanding of the current status quo in particular the basis of the art of teaching and the problems Jilin area, and from the aesthetic intuition the paper analyzes the causes.

Outstanding cases related to the foreign culture intuitive analysis of finishing and modern educational environment, designed two different levels of student learning, teaching cases, and adjust the implementation of the specific circumstances. Finally, analyzes and summarizes the information obtained teaching experiment results showed that significantly improve the students' interest in fine arts disciplines, changed their concept of aesthetic appreciation. Art Education for the training of basic aesthetic ability to provide certain significance. Meanwhile, the study of this subject in statistics and interviews enrich the research results of this project.

REFERENCES

Xiao Ying. Aesthetic strategies of Contemporary Aesthetic Culture [J]. "Academic Monthly", 1995, 2.

Chen Meimin. Aesthetic under Contemporary Aesthetic Culture [J]. Hainan Radio and Television University, 2005,3.

Lu Chang. On the relationship between arts education and aesthetic education [J]. Theory and practice of education, 2002,2.

Chen Chi-yu. Chinese modern aesthetics and art theory [J]. Central China Normal University, 2000,2.

Tan Hui before parsing school children aesthetic intuition and creativity collaborative development [J] Art Education Research, 2011.

Lijiang Jing since Reform and Opening adolescent aesthetic changes - Taking an example of aesthetic art and culture [J] China Youth Research, 2009,6.

Liu new. Aesthetic Intuition evolution in China in the 20th century literary theory [J]. Literary theory, 2002.03.

Li Feng reality. Guangqian "aesthetic intuition" theory building process [J]. Shenyang College of Education, 2006.6.

Management, Information and Educational Engineering – Liu, Sung & Yao (Eds)
 ISBN: 978-1-138-02728-2

Application of new multimedia teaching in music education

Ning Sun
Teacher's College, Beihua University, Jilin, China

ABSTRACT: According to the special feature of multimedia techniques, the significance and function of the multimedia teaching are analyzed in music education, then the application of multimedia teaching in music education is discussed from three aspects. Prospects for the multimedia teaching in art education is given at last.

KEYWORDS: New Multimedia Techniques (NMT); Music education; Application.

1 INTRODUCTION

With the continuous deepening of quality education in the regular teaching, with the "curriculum reform" in the conventional teaching development, multimedia technology as a new means of education has appeared in the music teaching in universities. It breaks through the limit of traditional teaching in time, space, greatly broaden the students' vision of the music, to arouse students' interest in music learning, enrich the teaching content, widen teaching view, stimulating students' thinking. Therefore, paying attention to the organic integration of multimedia and music teaching is very important. But we also have to admit that some teachers have in this respect gone to another extreme, too much emphasis on multimedia materials show, reflect and ignore the teacher's own technical quality and teaching ability, this is a new problem worthy of our attention.

"Compulsory education music course standard" explicitly pointed out: "modern education technology represented by the information technology greatly expands the capacity of music teaching, enrich teaching methods and teaching resources, has broad application prospects in music education, teachers should strive to master the modern information technology, the use of the combination of audio-visual, audio one, strong image, a large amount of information, resources and broad service for teaching." So, under the new curriculum standard university music classroom should be how to apply multimedia technology, make its full service for classroom teaching. Here, the author of this problem from the following aspects to talk about their own views:

Multimedia music class", refers to a variety of multimedia of combinatorial optimization music class", also refers to "multimedia computer-assisted teaching of music class". Especially the latter, integration and control of multimedia computers, diversification, diversification of information media, has broken through the limitations of time and space, the characteristics of the use of a variety of art forms to expand capacity. It gives the description language of music and text content into shape, sound combination of the picture, let the relative aesthetic object single static active again become the aesthetic object dynamic, and accelerate the speed of information about beauty, increase the information capacity of beauty, so that students are directly influenced by the United States, in order to obtain the optimal effect of education.

The teaching of music teaching with other subjects has the same Blackboard Design and appropriate amount of exercise. If the content is made in the courseware, it can reduce the classroom blackboard writing time; increase the capacity of teaching, the completion of more teaching task. With the multimedia music teaching can compensate for the lack of status of traditional teaching. Teaching multimedia before or after class to make a good spectral sequence, when the class with large screen projection is very clear and novel step-by-step or synchronous display and analysis of music and sound, some courses can also be using multimedia (audio and video) with demonstration teaching. And select the best effect through stereo feeling.

2 USE OF PROFESSIONAL TEACHING EAR TRAINING SOFTWARE SYSTEM

Music is a thaw in ideology and art as one of the disciplines, to the students to carry out ideological and moral education and aesthetic education is an important task of music teaching. Therefore, teachers in the teaching of music knowledge to the students, the cultivation of students' ability in music expression and the ability to create music at the same time, must organically into the ideological and moral education and aesthetic education. As in singing teaching, through the combination, multimedia material into the expression of the song, vividly depicted the beautiful natural scenery, colorful scene...... Let the students get the experience of beauty while watching and listening.

Visible, the use of multimedia technology not only breaks through the limitation of traditional music teaching in time, space and region, is conducive to the creation of music aesthetic situation, provide the conditions and environment for the teachers and students of music emotional experience, but also has interactive ability is very strong, so conducive to change the kind of Teacher centered traditional music the teaching mode, the initiative of teachers and students to fully mobilize. Especially for students, multimedia technology provides a convenient tool for the realization of new music learning goals, for students to understand the rich and colorful music world opened the door, help student organization, construction and completion of a number of music learning tasks, can effectively develop students' thinking ability in music. The application of the way students learns from a single classroom learning to the multi way, multi way of development, students not only in the classroom through multimedia-aided teaching mode to get the knowledge of music, but also by using the multimedia courseware or information, network is facing in the computer room or secondary field of individualized learning.

Although the multimedia technology in the teaching practice has been widely used, but it is not a panacea, it also has its inherent defects, of which the most prominent is the machinery of its. Music education is an education of the emotions. It pays attention to love moving, so it needs the music teachers' emotional investment, and asked the teacher to the students to show in activity in the emotional analysis to determine the specific, and take corresponding measures. These multimedia technologies are not competent. Some music teachers walk into a misunderstanding, think the culture of a class containing the culvert is more wide, the use of teaching methods, teaching aids more, more can show the excellent teachers thought the novel and the teaching level, and whether the open class, or a survey course, or rating class, class almost all have the audio-visual media, and frequent replacement, see things in a blur, too busy to attend to all students.

Memory is a profound one time to participate in the teaching activity, listen to the music open class section, the topic is "small crow love my mom", because the content of the song with the plot of a certain, teachers in order to stimulate the students' interest, the whole teaching process into a story, with the form of animation, with music playing, students eat with appetite after watching cartoon, enjoy the pictures, while still under the guidance of the teacher role play and so on, this course, student interest is aroused, but students' auditory association has been ignored, so that a class down, leaving the students impressive just animation image and story, but the main melody the song is not familiar with, so that music lessons can be said to be successful? I'm not by confused: why must use multimedia? A lesson in the most time to appreciate the picture, there is no time for students to carefully to experience music, feel the music, the students then what passion and inspiration? Ask the teacher after class, she said: "this is a public class, innovation should have the teaching form, can manifest the new curriculum idea!" See light suddenly, originally she put the new curriculum reform just understood as adding multimedia in the traditional teaching means, just understand innovation for teaching form!

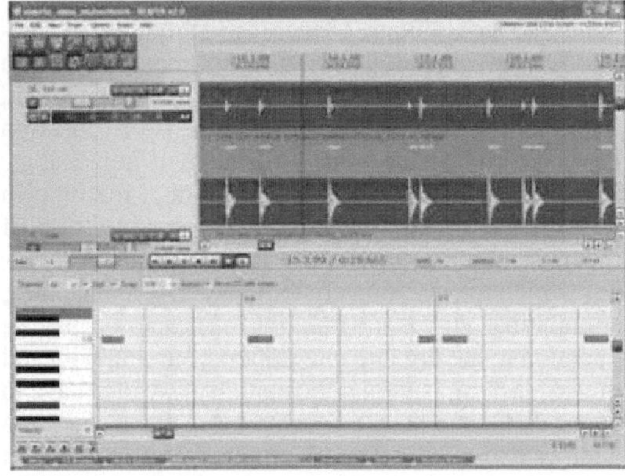

Figure 1. NMT recording, editing, processing, mixing, and mastering environment.

A music teacher should work hard to improve the level and ability of using modern teaching media teaching, into the realm of art, on the other hand, attention should also be paid to abandon the wrong tendency to exaggerate the role of multimedia. The modern teaching media, is the teaching activity tools, teaching media must depend on the teachers carefully design, operation, in order to play its effectiveness, it is not possible to substitute teachers' work.

Famous music educator in our country teacher Liu Dechang he stressed in lesson comment: "to the appropriate and reasonable use of multimedia, teachers don't forget to reflect their own teaching ability, not completely by means of sound; teachers' teaching language gradually scattered color, from the process of emotional communication with students. It is necessary to the dialogue between teachers and students, man–machine dialogue replace human dialogue. As the music education of emotional education is even more so. For example: in the fifth grade ninth volumes of Professor Songs "Grandma's Penghu bay", the students may have a smattering of the rhythm of the song, but for the song, and the scene not too understands.

In order to let the students a deeper understanding of the expression of artistic songs, so I will be Taiwan's geography, local customs and practices and the beautiful scenery of Penghu Bay and other video data into the courseware, let the student through the most intuitive picture to understand people and landscape of Taiwan and Taiwan, and the combination of language into the teacher carefully organized the class project. In time to enable students to appreciate the Penghu Bay scenery spot, I will "Grandma's Penghu bay" with the song of the fan to sing, not allowing them to listen to the tape's fan to sing. At that time, I found that the students seemed to be the teacher emotional fan sing infection, also be overcome by one's feelings sang! Based on this, a song from the songs of Taiwan, in the process of teaching my timely guide students to understand the Taiwan is an inalienable part of China, want to love the motherland, the motherland blessings for unification cause! I let the students talk about their own ideas, some say: "from now on Study hard scientific and cultural knowledge, grow up to contribute to the reunification of motherland."...... The children watched heartfelt passion Zhuang, I was moved. Finally, all teachers and students together to sing "my heart" Chinese end of classroom, classroom was immediately surrounded by thick patriotism; everyone's heart a long time can not be calm.

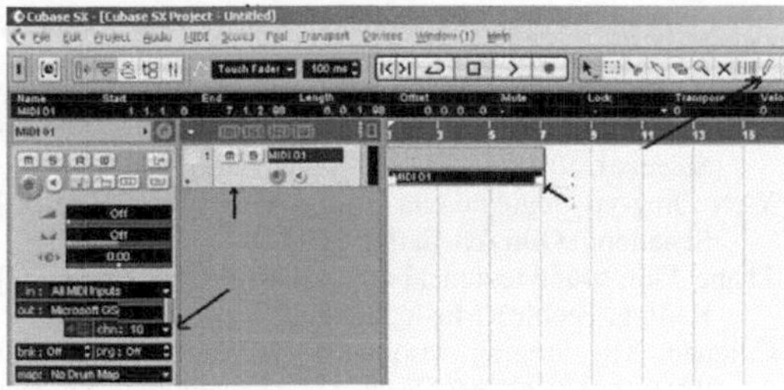

Figure 2. Ear Master Pro 4.0 operating points.

3 USE SONAR OR CUBASE ASSISTED INSTRUCTION SEQUENCER SOFTWARE

Multimedia technology in classroom teaching can only play a supporting role, but can not replace the leading role of teachers in teaching. Our famous educator Liu Dechang teacher also told: "teachers should grasp and effect of its dominant position in teaching, multimedia only as an auxiliary tool in music classroom, don't rely too much on." Multimedia for music teaching is like a double-edged sword, if used properly, it can improve the teaching efficiency, received a very good teaching effect and on the contrary, if used improperly, distracting, also will dilute the music teaching itself, and even influence and reduce the quality of teaching. For example, blindly rely on the multimedia technology, the courseware design into the structure of sequence, the teacher have a class does not stop button edge, into a "announcer" which not only takes the computer thinking limitations and replaced the teachers' and students' ideas, people become slaves of machines, but also changed the music teaching as the nature of the subject of the humanities courses, harmonious human relationship between teachers and students is replaced by cold man-machine relationship.

Teaching is not as intuitive and visually, as teaching media and the teaching media, but to highlight the characteristics of music course, outstanding music, not to engage in "showy", more can not proceed. For example: in the learning of the first grade seventh class enjoy the song "the first volume of small frog", I designed a small frog, lotus leaf, the moon, the wind, rain, different image cock. With the continuous change of music melody, rhythm, dynamics, the beautiful moonlight respectively appear on screen; the little frog, insect, playing on the lotus jump scene; the little frog bravely face the storm; after the wind and rain, the sun rises at cockcrow sound, small frog to situational labor on the lake the. Then let the students to imitate, and according to the change of music strength, speed, the tone of the complete show the story start, after and results, the student deepen feeling for music and understanding. This requires teachers to pre analysis of teaching targets and contents do a good job of teaching design, determine the use of multimedia in teaching objective, teaching methods, to achieve the final purpose of auxiliary, obtain actual effect.

The choice of making multimedia to consider from a variety of factors, choose those who accord with the practical, the content is correct, form beautiful, making the economy, multimedia technique on the innovation requirements, in order to achieve the overall optimization.

For example, the third volumes of "the level of sound" lesson, students most likely to put the size of the sound and the sound level confuse. In teaching, I put the "pitch" concept to explain over to the "music hall" to complete the. Have a lively and interesting cartoon elf software, which he led the children swim music palace. Open the "knowledge palace", appeared in front of us is a decorated more messy cartoon room, the elf tells us: this room is put the music related items. According to the need of the course content, click a doll in the radiator, this is the doll side to jump, as he explained what kind of sound is a soprano, and a female voice screaming for example, lively image: click on a doll in the stove, as he jumped down, side explain to the cello, and vigorous bass effect demonstration. Like to watch cartoons students in learning, interest is very high. After the demo, students can quickly and accurately answer the concept of "Treble" and "bass". Good media teaching works not only need to design excellent teaching, need more two-dimensional animated, excellent 3D image art production personnel, need high configuration of computer and large production software. It can be said: the people are need of various in very professional participation. So if we can have from the investigation on the market of educational software to recognize the advantages and limitations, relationship, will be the most valuable part of the learning activities into the teaching activities, each in his element, overall optimization.

4 CONCLUSION

As music major institutions all normal basic courses, showing a particularly important role are particularly critical. However, with the development of the times, the problems of traditional teaching methods exist. Paper spectrum of cases is difficult to produce the sound of hearing students association, single and boring sound, intonation can not be objective and quantitative analysis, no science accurate evaluation, etc., have been difficult to meet the needs of society. The technology used in computer music ear training is needed not only to the growing diversification of the times, or the transformation of classroom teaching and student needs. In this paper, several ear training software as the main object of study, research methods proposed computer music technology assisted teaching ear training. Traditional teaching carried out a detailed analysis of the problem, the use of computer music technology in electronic music can sound audiovisual methods rich sound, digital audio analysis techniques, precise statistical analysis to solve the problem, for computer-aided teaching music technology feasibility and practicality were demonstrated.

However, in the domestic computer music technology in various music colleges Teachers not common, most of the teachers and students do not understand this. How the popularity does is computer music technology knowledge is a question worth pondering. In this regard I have the following thoughts: for the provinces municipalities and school, the establishment of specialized lectures and presentations of academic research, and constantly update the relevant books and audio books and other books. Relatively simple to set up interesting computer music appreciation classes in schools, so that students have a general awareness of this. In the course of various types of normal or can be set to music colleges in elective courses on computer music technology to attract more non-professional students to learn, so that is very conducive to the popularity of computer music technology. Meanwhile, the era of the 21st century continue to progress, coupled with computer music technology, operational and professional, thus educators and educated to be life-long learning is a must.

With the rapid development of technology, the computer music technology in training is an essential modern means. It updates the traditional teaching model to optimize the teaching methods, expanding the horizons of students, but also to encourage students to take the initiative in various channels to seek ways of learning and acquiring new knowledge, provide teachers with modern technology teaching tools and teaching methods in teaching do have a great prospect.

REFERENCES

Lisi xin. International Computer Music Research and Opinion of related professional disciplines [J]. Central Conservatory of Music, 2003 (2).

Zou LIANFENG. The use of multimedia technology in teaching ear training [J]. Yellow University of Science and Technology, 2007 (3): 107–108.

Jiang Lin, Li Meiping. Music software ear training Teaching [J]. Software Tribune, 2007 (3): 27–28.

Xiao Lei. On the practical application of Sibelius's music software [J]. Journal of Guangxi University (Natural Science), 2009 (3): 35–39.

YAN Jing-yu regulate its use singing resonance [J] Shandong: Qilu Art Gallery, 2002,4.

Zhang Xian sound technical and artistic resolve singing [J] Beijing: People's Music, 2005,8.

Cuiquan Xin Singing resonance cavity resonance and three practical exercises [J] Guangdong: Xinghai Conservatory of Music School

Newspaper, 2001.

Xia Xianping vocal cords, glottis and vocal [J] Beijing: Biology Bulletin, 1989,10.

Zhu Jiming, Rick Chan, Lubbe real larynx anatomical study [J]. Shanxi: Anatomy, 1990,04.

Management, Information and Educational Engineering – Liu, Sung & Yao (Eds)
© 2015 Taylor & Francis Group, London, ISBN: 978-1-138-02728-2

Study on ideological and political education under new media technology background

Xiao Wei Wang
Jilin Railway Technology College, Jilin, China

ABSTRACT: The blog, instant messaging tools, and streaming media as the main indicator of new media technology has a tremendous impact on the current ideological and political education, with the rapid development of information technology. Under the new media technology in the background of the ideological and political education of college students, the study can solve the current college students' education of ideological and political difficulties and problems, it can further assure ideological and political education of the law, help to enrich students' ideological and political education research, and pioneering vision of ideological and political education, in order to better guide the practice.

KEYWORDS: Ideological and political education; College students; New media technology.

1 INTRODUCTION

Since the beginning of the last century, new media technology with its unique mode of transmission has developed rapidly and fundamentally changed human ways of working, learning, lifestyle and way of thinking, For college students, especially. Currently, the new media have become college students receive information, express their feelings, an important platform to show themselves.

New media technologies from the form of expression can usually be divided into the chat category, display type and forums category. Chat category of new media technology is based on individual chat as the main form, emphasizing the interactive communication between individuals, such as SMS, QQ, e-mail, MSN and so on. Display category is the use of new media technology network platform, in some specific form of the carrier, through dynamic or static display, to achieve the purpose of dissemination and exchange, such as the blog, digital TV, Internet TV, portals and so on. Forum category of new media technology is based on students' individual as the basic unit, with the help of the network setup process, and gradually extended from the individual to a different unit classes, schools, and gradually form a mass communication platform. Such as the school network, QQ group, alumni, paste it, and other campus BBS, a virtual community forum.

2 FEATURES NEW MEDIA TECHNOLOGIES

New media technologies rapidly gaining popularity with its unique advantages, changing the way college students learn the lifestyle and way of thinking radically. Varieties of new media technologies generally have the following characteristics:

Email, BBS, personal websites, etc. is a relatively early emergence of new media, information exchange, so that students can demonstrate good self, self-expression. Then there is a blog also can more easily and quickly publish personalized information, with creativity, flexibility and constructive, college students can make use of new technology, the freedom to distribute personalized information through text, pictures, video and other forms to reflect the individual original ecology life, thought and so on. In addition, college students face to face can be directly accessible to the network, personal, emotional language is more prone to proximity and identity, to narrow the distance between people. The features of the new media are relatively easy to meet the internal needs of college students to express themselves, by the group of college students of all ages and this knowledge blitz.

Media and traditional ideological and political education is different, most of the new media venue is not determined in the real environment, but in a virtual environment. This makes the new media technologies operator who has occult, such as college students in the network can use a nickname instead of a virtual ID, concealing the true identity. Uncertainty and conceal the identity of the performance of the main scenarios to ease the college education of psychological readiness, enhance the educational effect. In this situation, the exchange between the two sides to minimize the traditional face to face communication process a variety of possible confounding factors objective. Meanwhile, with the protective layer, college students can freely express their speech through the new media blog, forum, logs, and other forms of participation opinions, freedom of communication which meets the college's behavior considerably.

New media technology in the expression of the essence to achieve a spread equal interaction, mainly for the equality of information dissemination and views expressed. For example, regardless of status, anyone can easily apply for the adoption of new media technology's own place in the virtual space, equal to express their views and opinions. Each person is educated in an interview with the information, the information in the release time for the trainer, two perspectives on the timely exchange of interest, communication, initiative and enhance two-way interaction on the basis of equality of the. This duality of both the main qualities of knowledge, ability, and quality of the main psychological and raised higher requirements, but also for effective communication between educators and educational communication object provides psychological basis, to enhance the ideological and political education, equality laid a good foundation to improve educational efficiency. The features of the new media technology, the weekdays are generally acceptable for others to education of students, will no doubt get a great sense of respect and satisfaction, and thus further stimulate the enthusiasm of their participation.

Through new media technology to build up in the virtual world space can be stored for a long, easy for people to feel free to browse their record bit by bit, all-round understanding of their mature the whole process. In addition, the adoption of new media technology, information can be delivered instantly; the Earth has become a village narrow space shortens the distance, psychological distance between people closer. This unbounded makes us to keep abreast of the situation of college students receive information, thus greatly improving the ideological and political education targeted, effectiveness, timeliness, and coverage.

Compared with the traditional means of education, the application of new media technologies is to transmit information with more timeliness, fast. Currently, more and more students mainly get the latest information through the network, so that the advantages of traditional advertising media gradually lost; more and more people to chat through the network, communication, make calls, letters, and other traditional contact unprecedented impact. The efficiency of information transmission, the new media technology can mobilize university students to participate in the activities of the enthusiasm, initiative. Therefore, educators should make full use of the Internet and other new media technologies; expand the ideological and political education means and space, rich educational means.

3 NECESSITY OF NEW MEDIA TECHNOLOGY BACKGROUND IDEOLOGICAL AND POLITICAL EDUCATION OF COLLEGE STUDENTS

New media technology is a new networking tool, application mode, and dissemination of information carrier network environment, since the end of the century, with its unique mode of transmission of rapid development. "Currently, the" new media have become college students receive information, express feelings, an important platform to show themselves, and with unprecedented breadth and depth of involvement to the ideological and spiritual world, fundamentally changing the way college students learn, lifestyle and way of thinking, to the ideological and political education of college students brought increasingly profound impact.

In the new media age, information dissemination has barriers', 'space without barriers', 'information without barriers' characteristics, information dissemination, and use of more freedom, and difficult to monitor, some irresponsible, negative and backward ideological and cultural even anti-socialist rhetoric widespread use of new media technologies, the impact of our culture theme, to the ideological and political work has brought great impact and difficulties. The so-called ideological and political education environment, the ideological and political education are facing the object that surrounds education and its impact on the objective reality. With the development of computer science and technology, new media, college students should get a great deal of popularity, becoming an indispensable tool for learning and life, the campus has become a new media culture and surround college education have a significant impact on its environment. However, the new media information pervasive, campus new media culture may also bring some ideology has obvious tendency to disseminate bourgeois mainstream culture for the purpose of colonial culture, as well as immoral, unhealthy thoughts, forming some of the ideological and Political Education unfavorable educational environment.

There are a lot on the information superhighway beneficial health information, but there are not objective, scientific, decadent wrong things. On the one hand, new media, especially English is the main language of the Internet, the US-led Western countries by means of language advantage, relying on its economic and technological advantages brazenly to other countries, especially developing countries, with a clear tendency to spread ideological content, giving young college students to instill the wrong political views and ideas; on the other hand, some domestic anti-socialist anti-evil people in countries using new media to disseminate reactionary information, confuse abetting youth astray. Such as "Falun Gong" cult had a virus on the network wanton dissemination of ideas, and if some of the separatists oppose reunification of the motherland through new media to promote "Taiwan independence ideology," "Tibetan separatist ideology"; addition, "may spread pornography through the campus of new media, violence , murder and other information, bring some immoral, unhealthy trend, which stained the campus new media environment in which the ideological and political education environment complicated. Every one of us are more or less suffering in our main affect the environment in which the activities of the spirit. "College students are in the formative period of life, worldview, values, and ideas extreme, full of curiosity about new things, it is easily affected by the environment, how to deal with the environment in the process of ideological and political education change, while avoiding disadvantages, become a serious problem.

For the purposes of educators can choose a lot of information in a targeted, persuasive, the latest information, as educational materials. From the point of view of traditional ideological and political education, educators due to limitations of subjective and objective conditions, information and knowledge reserves less involved in the area of small, narrow coverage, the impact of the effect of the ideological and political education. Today, the Internet as a representative of the new media makes the global sharing of information resources possible. Ideological and political education to college can be collected through a network of ideological and political education resources in different contexts from different regions, and with the network and convenient interactive features to achieve strong interaction and thus maximize the sharing of educational resources, so since the original narrow, enclosed space of ideological and political education colleges into a whole society, open educational space, making channels ideological and political education becomes unblocked and for ideological and political education to provide a very vivid and rich educational resources.

First, the amount of information inherent in the nature of new media and information technology oversized, so that the content of education to become full and rich, both selectivity and objectivity; Second, the new media technology to make educational content form three-dimensional, variable dynamic, tend Macross. This way, you can make collectivism, patriotism, etc. These original abstract and difficult to grasp the ideological and political education, through the collection sound, color, light, painting and other new media technology as one of the deduction, so that the image becomes abstract, dull change was lively, attractive and practical effect of greatly enhancing the ideological and political education. Currently, the new media information fast alternation, with the new media, ideological and political education colleges can complete collection of ideological and political education in a short time, screening, selecting those times stronger, strong educational ideological and political education, which greatly improve the ideological and political education of timeliness, reflect the ideological and political education work requirements.

4 IDEOLOGICAL AND POLITICAL EDUCATION AND THE REASONS FOR THE PROBLEMS IN THE NEW MEDIA TECHNOLOGY BACKGROUND

Although college students showed overall new media's rational attitude, but active, broad interests, the pursuit of novelty and lack of self-control college students are also likely to be all sorts of temptations, so that the new media for a variety of ideological and emotional catharsis way difficult to manage. University educators in terms of ideology, both for understanding new media ideological dynamics provide an effective way, but also in the ideological and political work to bring an unprecedented problem.

For now, the ideological and political education major problem in the communication process relies on new media should be attributed to the presence of the main part of the new media, new media literacy main difference, thus affecting the new media of communication effects. New media literacy is poor, mainly in the following aspects of performance, first of all, the main advantage of the new media applications of new media communication ability and quality difference.

Ideological and Political Education and the reasons for the problems in the new media technology background

Although college students showed overall new media's rational attitude, but active, broad interests, the pursuit of novelty and lack of self-control college students are also likely to be all sorts of temptations, so that the new media for a variety of ideological and emotional catharsis way difficult to manage. University educators in terms of ideology, both for understanding new media ideological dynamics provide an effective way, but also in the ideological and political work to bring an unprecedented problem.

For now, the ideological and political education major problem in the communication process relies on new media should be attributed to the presence of the main part of the new media, new media literacy main difference, thus affecting the new media of communication effects. New media literacy is poor, mainly in the following aspects of performance, first of all, the main advantage of the new media applications of new media communication ability and quality difference.

Competency here mainly refers to the subject of new media communication networks and other new media technologies to use to communicate the basic qualifications and qualities. The most important thing is flexibility in the use of modern technology and new media feature the ability to communicate effectively. Obviously, if the user's ability is too poor, the quality of new media certainly affects the results of communication, generating communication barriers. Secondly, the main ideological training part of the new media communication is not high. Outstanding performance for not applying the Marxist stand, viewpoint and method to analyze and solve problems, and do not pay attention to the actual conditions, cannot grasp the core and key issues. Issues such as new media information for intricate scientific use Marxism cannot stand not to be discriminating sieve, losing the ability to identify useful information, not thinking of the students to meet the growing idea of finding valuable resources.

The main difference is the ability to participate in the core values that have not mastered the new media. In the modern information society, new media have become an important channel and means for people to obtain information, but due to some students' more one-sided understanding of the new media, the use of new media technology has mainly stayed in some of the surface of things, far from excavations new media value and fun. This affects their understanding and acceptance of the new media to a certain extent, necessarily preclude an effective new media communication.

5 CONCLUSION

Firstly, a large number of references by reading, to define the concept of new media at home and abroad on the basis of research status of new media technology concept to define, classify, and then the new media technologies originality, occult, equality, long-term, timeliness and other characteristics of presentation, in-depth analysis of the important reasons for the students of new media technologies are pleased to accept. The impact of new media technology in learning and life at the same college, for ideological and political education of college students also provided a rare opportunity. New media technologies are to expand the content and space ideological and political education, enriching the means and methods of ideological and political education, and enhance the relevance and effectiveness of the ideological and political education.

In order to ensure the effectiveness of ideological and political education from strengthening supervision, improve the system to start, regulate the content of education, processes, and to form from the source, to establish a relatively complete defense, control, and guide system, pay attention to the online information collection and analysis, strengthening the network monitoring to ensure network security; We shall be open to the characteristics of the culture, and guide students to establish a correct "view of the Internet," insisted ethical construction of the network, to implement a virtue net. Ideological and political work, some do not because college students do not have the ability and level of human-induced, but due to the mechanism. As a large and a wide range of systems engineering, and only build coordination, balance, and efficient operation mechanism, ideological and political work in order to have planned and arranged, organized, focused unfolded to obtain tangible initiative. To do this, try the following: the establishment of the various departments of the Propaganda Department, Student Affairs, Network Information Center and other joint working mechanism; establish the effect of new media ideological and political education evaluation mechanisms. Strengthen ideological and political education team building, ideological and political education is an important measure in the new media conditions. For example: the establishment of a hierarchical network of ideological and political education work force; through training to enhance the overall quality of the team; guide the ideological and political education of workers in continuous learning and innovation.

REFERENCES

Wen Lin, Li Nan. Study ideological and political education [J]. SCIENCE & EDUCATION, 2011 (2).

Hu Yu. "Ideological and political work in new ways to explore the Internet Age" [J]. Tsinghua University, 2001 (1).

Pan Min, Chenzhong Run, at sunrise, "Summary of network ideological and political education of college studies". [J]. Ideological and political education, 2007 (3).

Zhang Xing again, Zhang Yu. "Strengthen the network Counselors occupied university ideological and political education of new positions" [J]. Theoretical Front in Higher Education, 2006 (5).

Chen Yan. "College counselors work research in the internet environment" [J]. Higher Education Research, 2007 (1).

Qiao Xiangping, Chen Yanfei. "Ideological and political education of college students New Exploration Network" [J] Hunan Social Sciences, 2007 (1).

Management, Information and Educational Engineering – Liu, Sung & Yao (Eds)
© 2015 Taylor & Francis Group, London, ISBN: 978-1-138-02728-2

Training model for young teachers based on competency for China's application-oriented universities

Yong Fa Li, Fu Cheng Liu & Ya Ying Feng
Business Institute, Anhui University of Finance and Economics, Bengbu, Anhui, China

ABSTRACT: The construction of a high level application-oriented university has become the goal of China's local undergraduate colleges and universities. This requires building an excellent team of application- oriented teachers. This paper starts with the analysis of the young teacher training model based on competency, which includes five components: the concept of training, competency assessment, training goals, training resources and training processes. Next, it will identify the problems in each of these components, and finally state some suitable solutions.

1 INTRODUCTION

A revolution has been undertaken nowadays. Higher education in China has taken a great-leap-forward development in the last decade, and has formed three types of schools. These are research-oriented university, application-oriented university and teaching-oriented university. "The Resolution to Accelerate the Modern Vocational Education Development" issued by China's education ministry in 2014 will drive a large number of undergraduate colleges and universities transitioning to application –oriented teaching.

The comprehensive quality of teachers is the key factor influencing the quality of higher education, and young teachers in colleges and universities are the main group of teaching staff. Competency is the main component in the professional quality of teachers and the solid base of the teachers' professional development. The young teacher is highly educated, trained, active, energetic and receptive. Nevertheless, they all may suffer from difficulties in the initial stage of their professional careers, such as lack of experience, and the situation being unclear. They will then be eager for direction and help. China's Ministry of Education issued two documents in 2007, which are "Opinions on further improving the undergraduate teaching quality and teaching reform project" and "Several opinions on further improving the undergraduate teaching quality and teaching reform project". These documents highlighted the importance of cultivating young teachers. Thus, a young teacher training model based on competence will directly determine the core competitiveness of colleges and universities. However, most of them are all facing the following dilemma: how to design and perform an effective personnel training model to enhance young teacher's occupation competence.

2 CORE CONCEPT AND DEFINITION

2.1 *Teacher competency*

Generally speaking, competency refers to a primary characteristic of an individual to produce a superior performance on the job. In 1973, the famous psychologist David C. McClelland published an article "Measurement of Competency Instead of Intelligence", which defined the term of competency as talent, knowledge, skills, capabilities, traits and motivation to do one's jobs well. The Iceberg Model of competency built by McClelland, has two different types: (a) explicit competency, which includes knowledge and skills, and (b) implicit competency, which includes self-conception, personality traits, motivation and need. Guasch et al. (2010) supports the view that competency means a system of complex actions, including knowledge, abilities and attitudes that are required for the successful completion of tasks.

Therefore, teachers' competency refers to the specialties and characteristics of excellent teachers who have a successful teaching career, and help to improve the studying efficiency and motivation of their students. This distinguishes them from ordinary teachers. Liu (2008) contends that competency of teachers in colleges and universities, mainly contains four aspects: (a) concept competency, which refers to the capacity to understand and expound the professional theory, (b) technology competency, which means skills and expertise in professional teaching, (c) integration competency, which refers to the capacity to combine theory and technology with practice, and (d) professional attitude or values.

2.2 *Training model of young teachers*

The training model of young teacher refers to the overall process of training teachers in relatively stable teaching contents, teaching methods, and evaluation systems. This is done in line with a particular training target and talented person specification, which guided by certain modern educational theory and ideology (Li, 2014). Liu (2011) has identified five models of young teachers' training in universities or colleges, which are (a) pre-service training, (b) tutorial systems, (c) school-based training, (d) in-service training and (e) continuing education. Wang (2009) puts forward the pluralism of the teaching models, such as the international training model, the integration training model of industry, education, research, and the teacher morality training model.

2.3 *Application-oriented university*

Application-oriented university focuses on cultivating application-oriented undergraduates. This type of university is something between a research university and a technical university. It is aimed primarily at fostering local economic development and trains applied talents for this purpose. (Shanghai Liu 2013). The application-oriented university has four distinctive features: (a) focusing on training application-oriented talents, (b) focusing on undergraduate education, (c) focusing on the quality of teaching, and (d) focusing on serving local economic and social development (Pan 2010). However, whether or not application-oriented universities can be considered as an independent way of running schools is controversial (Hu 2013). Regardless, there are special requirements for young teachers in applied universities, such as applied teaching, applied research and applied talent training.

3 TEACHER TRAINING MODEL BASED ON COMPETENCY

In our view a teacher training model based on competency is, in essence, a transformative tool that can make an incompetent or an ordinary teacher into a better one. Furthermore, different teacher training models have different conversion efficiency. A teacher training model based on competency includes 5 elements: training concept, competency assessment, training goal, training resources and training process, as shown in Figure 1. The teacher training concept affects other components. A competency assessment, training goal, training resources and training process should be considered in order when designing a good teacher training model.

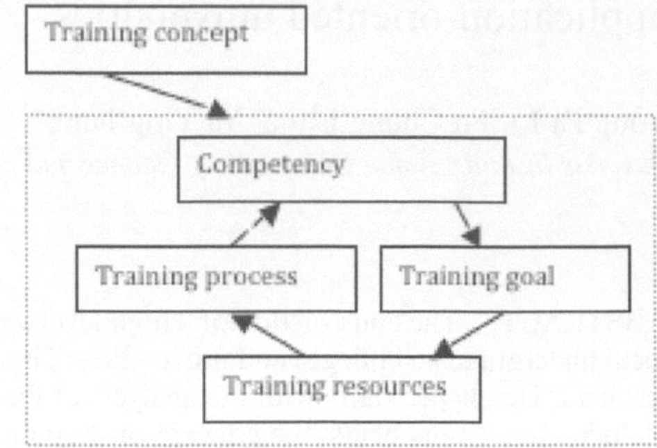

Figure 1. Teacher training model based on competency

3.1 *Training concept*

The teacher training concept refers to the knowledge of the supervisor or organizer of teacher training about the teachers' idiosyncrasies, training objectives, training missions, needs, intrinsic and extrinsic motivation. Also, the knowledge should cover training principles, which includes quality, teaching and learning, teachers and students, etc. The teacher training concept is exactly about what kind of person a teacher should become, and how to become. At present, China's mainstream training concept highlights two points: moral education and competency. Moral education comes first, which refers to the requirement of moral cultivation, while competency is key, which is a basic requirement for working as a teacher.

3.2 *Competency assessment*

Teacher competency is affiliated with the individual characteristics of teachers, including the specialized knowledge, skills and qualities of successful teaching. Therefore, competency assessment requires a series of activities to determine whether or not he is suited to teaching and has the capacity for education and teaching work.

3.3 *Training goal*

Different colleges and universities should have diverse goals. The teacher training goal is the basis of the teacher training model. So, establishing the accurate goal is a prerequisite for teacher training. Guasch et al (2010) highlights the need to observe in an integrative manner the diverse teacher roles or functions, whilst designing training proposals for the development of competencies.

3.4 *Training resources*

At present, there is an urgent problem in how to obtain higher-quality training resources which seriously affect training effects. Training sources contain trainers, training venues, training facilities, online training and training materials, etc. Training resources should be extensively studied and developed in and out of colleges and universities.

3.5 *Training process*

The training process refers to each activity necessary to achieve the goal of teacher training, and to allocate resources efficiently. The key activities include the planning, organizing, directing, and controlling of teacher training. Experiments and practical teaching are a necessary part in the process of teacher training.

4 CURRENT PROBLEMS IN APPLICATION-ORIENTED UNIVERSITIES

4.1 *Teacher training concepts are outdated*

Obviously, teacher training concepts for colleges and universities lag far behind the current situation. There are three common misunderstandings in our mind. (a) Because young teachers qualify for the teaching by self-study and exploring the truth without outside approval, teacher training is an irrelevance to some extent. (b) Because of some difficulties for teacher training, such as lacking of funds, absence of good trainers and other training equipment, some colleges and universities try to avoid the systematic teacher training. (c) Because the traditional standard for a qualified teacher is to pass all relevant exams. The goal of teacher training is simply the examinations nowadays. Therefore, teacher training concepts for application-oriented universities are short-sighted, neglect humanities, and are insufficient sectors for practice. They also lack systematic design and organization. Furthermore, the evaluators have paid little attention to the young teacher's professional development, social responsibility and improvement of the teaching process.

4.2 *Teacher competency assessment is not professional*

Teacher competency assessment should be based on reality, and should grasp the requirements of practice in order to continue into a career. Currently, there are no clear dividing lines among different positions or development paths, so competency appraisals adopt a similar pattern. Moreover, there are other obvious problems with teacher competency assessment, such as the lack of standardized testing, out-of-date methods of assessment, and not enough importance placed on assessment. For these reasons current teacher competency assessment is unreliable and ineffective.

4.3 *Teacher training goals are unclear*

Application-oriented universities are now struggling to develop and construct application-oriented subjects, a practice teaching system and resources and facilities. But, it is absolutely inappropriate that application-oriented universities and the other types of universities have nearly identical training goals.

Current training goals of application-oriented universities are unclear, which is illustrated by these three facts: (a) no clear-cut distinction in professional direction and the nature of the work, (b) overly focusing on scientific research, and (c) giving too little care to moral education and scientific methods.

4.4 *Teacher training resources are inadequate*

The main issues affecting teacher training are lack of funds for teacher education, lack of spare time for teachers, inadequate training materials and guidelines, and poor training facilities.

4.5 *Teacher training process is irrational*

In the process of teacher training, traditional teaching styles, which couldn't satisfy the needs of training talents, are still being adopted by many organizations. For example, some applied universities equate specialized knowledge with teaching skills, and treat research as the standard for teaching and learning, instead of the acquisition of educational theory, concept and idea. Some teacher training focuses on form and neglect the content and real effect.

5 COUNTERMEASURES

As mentioned above, the teacher training model based on competency consists of five parts, and there are problems with each part of application-oriented universities. Solving these problems is complicated and full of challenge. Following are the countermeasures displayed by key words for each component.

5.1 *Innovation*

The innovation of the teacher training concept has the function of guiding and leading teacher training in application-oriented universities. Applied undergraduate education is developed in China because it is difficult for university graduates to obtain employment. One reason for this is a shortage of properly trained applied graduates. A key issue is how to optimize the faculty team for developing applied talents. Therefore, the innovation of the teacher training concept is necessary for promoting the quality of teacher training.

5.2 *Effectiveness*

There are no reliable competency-assessment-methods for teachers from application-oriented universities. Because teacher competency consists of several aspects, such as knowledge, skills, attitudes, and values, no single assessment method can be used to review the full range of an individual's characteristics. If it is not clear how to assess teacher competency, it is impossible to develop and implement an effective education plan, or to improve young teachers' motivation. Wang (2014) set up an index system for teacher competency in applied universities which contains five dimensions: self-development, openness and innovation, motivation and values, communication and coordination, and social orientation. Because outstanding young teachers are diverse in personal behavior, teacher competency assessment has to take these idiosyncrasies of young teachers into account. This is an effective assessment tool.

5.3 *System*

Teacher training goals should be developed into a coherent system based on clear structure and the ability to be reused. More importantly, the planning system should combine the career plans of the teachers themselves with the development of application-oriented universities. The applied university needs a large number of staff consisting of double qualified teachers who are not only masters of theoretical knowledge, but also guiders in practical skills. Therefore, good teacher training goals include theoretical knowledge, specialized practical skills, emotional element, teaching capability, and applied research ability.

5.4 *Trans-boundary*

When it comes to teacher training resources, application-oriented universities should do their best to achieve a balance between the allocation of what already has and outside resources, even looking for foreign resources. There is no doubt a kind of trans-boundary activities should involve governments, firms, universities and research institutes from both China and abroad, sharing their knowledge. Each application-oriented university should also take advantage of the strategic value of information technology in teacher training.

5.5 *Experience*

It is very important to enhance teachers' practical experience in order to fully inspire the initiative of young teachers. Young teachers at application-oriented universities should obtain practical experience and the ability to combine teaching with problem solving skills. Thus, they can train higher technology applied talents for working on the first line of production, construction, management and service. To improve teacher training level, application-oriented universities should support young teachers to collaborate with government offices, enterprises, and other organizations, and encourage them to attend advanced studies and academic conferences at home and abroad. The aim of this is to exchange experiences with other teachers in relevant colleges and universities. Chongqing University of Science and Technology in China has implemented "three types of experiences" as a training model for young teachers, which require that young teachers have tree experiences consisting of production practices, refresher courses and student affairs.

ACKNOWLEDGEMENTS

Financially supported by the provincial teaching research projects in Anhui, China (Grant No.: 2013jyxm542 and No.: 2012jyxm861).

REFERENCES

David C. McClelland. 1973.Testing for competency rather than intelligence, *American Psychologist*, 28: 1–14.

Li Li. 2012. Analysis on "three types of experiences" training model for young teachers in colleges and universities[J]. *Journal of Chongqing University of Science and Technology (Social Sciences Edition)*, 11: 173–175.

Maoyuan Pan & Qi Wang. 2010.The development of China's characteristic university from the view of higher education classification [J]. *China Higher Education*, 5: 19–21.

Qian Wang. 2009. Research on teacher training model and system in new undergraduate institutions[J]. *Educator (Higher Education Forum)*, 5: 38–39.

Shanghai Liu. 2013. Research on applied university's development strategy [J]. *Journal of Chongqing University of Science and Technology (Social Sciences Edition)*, 4: 161–163.

Teresa Guasch, Ibis Alvarez &Anna Espasa. 2010. University teacher competencies in a virtual teaching/learning environment: Analysis of a teacher training experience[J].*Teaching and Teacher Education*, 26: 199–206.

Tianyou Hu. 2013. Logic and Problems of the Construction of Application-oriented University[J]. *China Higher Education Research*, 5:26–31.

Xiao Liu. 2013.Analysis on the suitability of teaching staff construction in applied university [J]. *Education Review*. 1: 36–38.

Xianfeng Liu. 2008. A preliminary study on teacher competency and its development strategy in colleges and universities [J].*China Adult Education*,11(4) 418–427.

Yiyu Wang. 2014. Research on setting up an index system for teacher competency in applied universities[J]. *Education Review*, 6: 50–52.

Zhongguo Li. 2014. From strategy to business models and onto tactics[J]. *Journal of Linyi University*, 36(1): 15–20.

Management, Information and Educational Engineering – Liu, Sung & Yao (Eds)

The problems and countermeasures of the practical application of case teaching

Xiu Lai Gu
Bengbu Automobile NCO Academy, Bengbu, China

ABSTRACT: Case teaching is a teaching method which, under the guidance of teachers, according to the requirements of the syllabus, uses cases to organize cadets to learn, research, and train their ability. The method can stimulate cadets' study activities, make them understand the obtained knowledge deeply, and cultivate their ability to analyze and solve problems and make theory more close to practice. The essay discusses the current situation and problems in practical application of case teaching and puts forward corresponding countermeasures.

KEYWORDS: case; teaching method; practice; application

1 THE CONNOTATION OF THE CASE METHOD

The case method, according to the requirements of teaching aims, takes cases as basic material, and puts the cadets into specific situations, and then identifies problems, analyzes problems and solves problems, to improve the theoretical level and practical ability of cadets. Firstly, the case method takes cadets as subject and changes passive study into active participation. In this way, Cadets can actively master knowledge, ability, emotion, attitude, and sense of worth in virtue of selected cases. Secondly, case method combines theory to practice, and focuses on the creative spirit and skills training of future job for cadets. Thirdly, its teaching purpose is to make cadets know how to think, explore, develop, consistently consummates their knowledge system, enrich student experience, cultivate a scientific attitude to seek truth from facts and really reaches the aims to strengthen knowledge, cultivate ability and comprehensively improve the quality, so that teaching quality makes progress.

2 THE CURRENT SITUATION AND PROBLEM ANALYSIS OF CASE TEACHING IN PRACTICAL APPLICATION

2.1 *The current situation*

At present, case teaching has been applied greatly, and exerts positive effect to mobilize the student enthusiasm of the cadets, develop ability, and promote the teaching benefits teachers as well as cadets. Firstly, the lessons of case teaching are insufficient. Although teachers in general realize the importance of case teaching, they often hesitate when designing a semester teaching plan. Generally, the case discussion needs to be arranged from 1 to 2 times in each course, several teachers do not arrange that. Secondly, the exertion of case teaching is very random. In case teaching, teachers need not teach hardly on the platform and take case teaching as easy teaching, so that they think they can deal with teaching easily, and made a perfunctory effort. Thirdly, case teaching does not reach its teaching effect. According to the writers' investigation, in the case teaching of Automobile Technology Use course, only 20% cadets are satisfied to case teaching effect and 80% not.

2.2 *Existing Problems*

2.2.1 *The deficiency of proper cases.*

The typical case is true typical events containing difficult problems or situation. The aim of case teaching is to make cadets learn how to think and explore, and really reaches the aims to strengthen knowledge, cultivate ability and comprehensively improve quality, so that teaching quality makes progress. In concrete teaching, some teachers want to do case teaching, but the trouble is that there is no case. On the other hand, the writing of case is not standard, and is lack of pertinence, which influences the correct analysis and judgment of cadets. Therefore, the lack of teaching case resources with high quality has become the bottleneck of case teaching.

2.2.2 *Insufficient preparation before class.*

A lot of work about case teaching need be fulfilled before class. If the preparation is insufficient, the exertion of case teaching will be unbending. Such insufficiency embodies two aspects. Firstly, the teaching design and teaching plan preparation are inadequate, including the careful selection of cases, appropriate dispensing timeliness of case material, confirmation of discussion outline, concrete organization design of discussion and the arrangement of discussion place, etc. Secondly, cadets have no adequate preparation. For example, they have not carefully read the given cases or they are lazy to think and communicate with classmates after reading. They just wait the ready-make resolution form the lesson or teachers.

2.2.3 *Insufficiency of effective encouragement.*

Case teaching needs active participation and interaction of teachers and cadets. The positivity from teachers naturally influences case teaching. But due to the defects in the appraisal system of teaching performance, the academy has not taken to the corresponding content of case teaching into the appraisal system of teaching performance. Especially, the work of case composition not only has not been included in evaluation, but also has no any financial aid, so that teachers, in general, have no motive power to compose cases with high quality. On the other hand, in case teaching, seldom do teachers appropriately praise and encourage cadets for their active performance to stimulate their interest and enthusiasm. In particular, the performance in case discussion of cadets has not been connected with corresponding score and evaluation, as result in deficiency of encouragement.

3 THE COUNTERMEASURES TO IMPROVE CASE TEACHING

3.1 *Scientifically select and write cases*

Good case material can play a multiplier effect to case teaching. In practice, what cost much time and make teachers feel hard is the writing and collection of the teaching case with higher applied value.

3.1.1 *Choose cases from available material.*

When selecting cases, teachers should pay attention to the following problems. The first is necessary powerful pertinence. That is, the selection of cases must be combined with cadets' specialty and be operated with pertinence in accordance with the teaching purpose to ensure cadets obtain the corresponding knowledge form the cases. The second is the authenticity of the cases. What cases display and reflect on is a real event and is not fictional and divorced from reality, and one or more difficult problems get involved in it. The third is typicality. The given event can make cadets think deeply, and, when the cadets meet the same or similar events, the case has the reference significance and value. The fourth is timeliness. The selected cases must embody times characteristics and adapt to the problems and requirements the grassroots units meet at present. The fifth is the integrality. Each case must have a complete plot from the beginning to the end, because fragmented and broken story cannot give cadets overall sense.

3.1.2 *Write case by teachers themselves.*

In order to improve teaching effect, teachers must go to the grassroots units to conduct research or function in an acting capacity, so that they can transfer the problems and living examples existing in troops training, or investigate and collect corresponding material through troops, and especially collect and reorganize those cases reflecting on the fundamental, exploration and prospectiveness of the specialty, an then, after analyzing, draw forth problems and compile cases to solve the problems that teachers have no smooth channel to get cases, the supply of cases material is insufficient and lack of quality. However, the construction of case teaching resource library is complicated system engineering and a long-term hard work, which need the participation of plenty of teachers coming from the first line of teaching and the common effort from the majority of teachers.

3.2 *Careful preparation before class*

3.2.1 *Teachers' preparation.*

Teachers, before class in each semester, should constitute semester case teaching overall plan, considering the teaching purpose and content, confirm the quantity of concrete teaching cases, and, according to teaching process, confirm the time for each case. At the same time, teachers should carefully prepare and study cases, connecting teaching content, and concentrate on the pertinence, authenticity, typicality, timeliness, and integrity of cases to move cadets' study initiative and discussion enthusiasm. Two weeks or more before case teaching class, teachers should provide case teaching material and discussion outline to ensure that cadets have enough time reading cases, looking up corresponding material and thinking and communicating. After Understanding cadets' situation, teachers provide necessary support and help cadets solving the difficulty and problems met in preparation period. If the case itself have some problems, teachers must replenish, revise and perfect it in time to make sure that case discussion can be fulfilled on schedule. The concrete procedure of case discussion must be designed according to the implement requirement, number of cadets, available teaching conditions and the difficulty of the cases, considering any problem that may come along, corresponding countermeasures must be prepared.

3.2.2 *Cadets' preparation.*
Case teaching puts forward higher requirement to cadets, which need cadets to support and coordinate. As for cadets, firstly, they must have necessary theoretic knowledge for the discussion. Secondly, they should be familiar with the materials according to teachers' requirements, and think and communicate around the difficult problems in the discussion outline, freely explore, audaciously query, carefully look up material, actively investigate, and positively think and obtain knowledge to experience the pleasure of study.

3.3 *Construction of effective incentive mechanism*

3.3.1 *As for teachers.*
The first is to increase input. The teachers, devoted to the research of case teaching and case composition, must obtain financial support. The second is to perfect performance appraisal system. The third is to encourage teachers to participate training and created conditions so that they have the opportunity to investigate in grassroots units or function in an acting capacity to enrich the practical experience and case teaching content.

3.3.2 *As for cadets.*
The first is to erect role models and establish typical case. Teachers must openly affirm and [raise those have excellent performance in case teaching. The second is to launch a competition. Whether it is the preparation before class or a case discussion, teachers can take team as a unit to compete. The third is to combine the performance of each cadet in case teaching with their examination score. The fourth is to give suitable praise and award, under the correct circumstances, to the outstanding individual and company.

In a word, the advantages of the case teaching method are obvious. It extends the teacher' view, enriches the teaching content, burst into the vitality and creativity of classroom teaching, and has played a positive role in cultivating talents. But at present the application and development of the case teaching method are not maturing, and many problems in the actual teaching are worthy pondering and discuss.

Management, Information and Educational Engineering – Liu, Sung & Yao (Eds)
© 2015 Taylor & Francis Group, London, ISBN: 978-1-138-02728-2

Three mistakes of the teaching methods and means reform

Xiu Lai Gu
Bengbu Automobile NCO Academy, Bengbu, China

ABSTRACT: Teaching methods and means are the important factors to realize the classroom teaching purpose and promote teaching quality in Academies. To reform teaching methods and means on the teaching researches in academies is an eternal subject. Combined with the reform of teaching methods and means in the academies, the writer thinks that it is necessary to discuss some practicalities, clear some vague cognitions, as soon as possible, out of the three mistakes, and realize the new spanning of teaching methods and means.

KEYWORDS: teaching method, means, reform.

1 CLASSROOM QUESTIONING IS NOT EQUAL TO THE HEURISTIC TEACHING METHOD

Asking questions at classroom is one of common teaching methods that teachers in academies adopt question - and - answer method of teaching in the classroom teaching process according to teaching aims and study situation, which is an important form of bilateral activities between teachers and students in teaching. The functions of classroom asking include the promotion of the new knowledge study, benefit of review and reinforcement, cultivation of psychological quality and the improvement of the ability of expression. However, the heuristic teaching method takes cadets as the subject, and stimulates cadets to think with the illumination from teachers. That is, the educated, illuminated by teachers, actively obtain knowledge, develop intelligence, and cultivate and form perfect personality. The core of the heuristic teaching method is to guide cadets to think, and make them infer other things from one fact. Therefore, there are relations and evident discrimination between the two models. Firstly, Questioning with high quality can guide cadets to think. In this way, it has the heuristic function. Secondly, the questions, which can be answered just by memory and repeat, almost have no heuristic function, so that they do not belong to the heuristic teaching method. Thirdly, some questions with low effectiveness, even no effect, not only have no any help to developing cadets' thought and ability, but also consume plenty of time and reduce the efficiency of classroom teaching. Fourthly, although some teachers take questioning as the breakthrough point of teaching, the whole teaching process has been introduced one by one according to the question pre-designed. Such questioning become a personal show, which cannot mobilize the enthusiasm of students thinking and is harmful to cultivate cadets' innovation ability and favorable personality and quality. So that, questioning is the one method of heuristic teaching method rather than the heuristic teaching method itself. Focusing on heuristic teaching method, the writer has three advices. The first of all, the heuristic teaching method is not only a teaching method but also a guidance of ideology or a teaching idea. Its basic essence is to entirely stimulate cadets' inner motivation of study and activity to promote them to actively think and obtain knowledge. In the second, the forms of the heuristic teaching method are different. In teaching practice, we should have a strong idea of the heuristic teaching method, and pay attention to mix them with concrete teaching activities at any time. In the third, "illumination" and "development" in teaching is the dialectical relation and has the reciprocal causal relationship. "Illumination" is the prerequisite of "development", and "development" is the development and result of "illumination". In order to make cadets develop with illumination, teachers should have good methods of illumination. No good illumination, no great development. Undoubtedly, it is a very good heuristic teaching method that knowledge teaching combines with a learning method comprehensively.

2 ABUNDANT TEACHING MEANS IS NOT EQUAL TO THE GOOD TEACHING EFFECT

Teaching means are all kinds of material carriers, which directly get involved in teaching and transfer, deal with teaching information in certain forms and methods. It is the indispensable important part of teaching activity. The modern education technology is widely applied in the college classroom teaching, and it has changed the traditional way of teaching and learning and improved teaching quality. But at present in the process of widely application of modern educational technology means, some phenomena of the excessive application have aroused our attention. Otherwise, it will get the opposite of what we want and will reduce the teaching effect.

2.1 *No audio-visual means, no lesson*

Several teachers think that audio-visual means is indispensable at present, and they must use teaching media in each class, which seems to they cannot teach without the help of teaching media. Teaching situations are so different that not all the teaching content are necessary for multimedia aided teaching. To jump on the bandwagon would run counter to the desire. Under the condition of modern educational technology, teachers should endeavor to the combination of teaching content and form, and even the combination of human beings and computer, and focus on improving their own teaching ability combining to multimedia means.

2.2 *Multimedia abuse*

When teaching, teachers not only teach knowledge, but also communicate with cadets in heart. The rigorous teaching attitude, responsibility consciousnesses will potentially impact cadets, and even affect their whole life. In the traditional teaching process, teachers can spread knowledge information, thought information and personal information, and exchange emotion with cadets through body movement, language and eyes, etc. Teaching with multimedia in the whole process, however, become to see films and teachers become announcers. Instillation from teachers turns into from computers, which is very unfavorable to students' active learning and positive thinking, Only when the traditional teaching means combine with teachers' personal characteristics and modern teaching means, can modern teaching means exert their effect in classroom.

2.3 *Fancy multi-media courseware manufacture*

The computer multi-media technology show its strong ability to content expression, at the same time, it also exposes problems that format is more important than content, which is a reversal of the order of host and guest. At present, in courseware manufacture, several teachers seek pattern face-lifting, and make the courseware fancy and bewildering. The pictures, animation, video and background attract too much the attention of the cadets, so that some important concepts and theoretical principle are ignored and the results run counter to the teaching desire.

Therefore, the use of modern educational technology means must be timely, appropriate, proper and refined, which does not mean that more teaching means more effectual. If the modern educational technology means are used improperly or the excessively, the negative influence will be brought to the teaching, so that the expected teaching effect cannot be achieved easily. According to the characteristics of cadets and the course and the needs teaching content, teachers should reasonably dispatch and appropriately apply to the need teaching means to exert the best teaching efficiency, and avoid taking teaching means as teacher-student props.

3 TRADITIONAL TEACHING METHODS ARE NOT EQUAL TO BEING OBSOLETE

In recent years, there is a phenomenon should arouse the attention of people. That is how to treat the traditional teaching method of "problem" in the exploration of teaching method reform. The current is all traditional teaching methods will be abandoned. Such phenomena, the one-sided and with tainted glasses look to traditional teaching methods, must be attached importance to. The writer thinks that such phenomena is unfair to traditional teaching methods, and p advocate traditional teaching methods must be treated dialectically, because the traditional teaching method is not equal to obsolete.

Firstly, the advanced teaching method depends not upon its tense. Every evolution is a kind of innovation on the basis of inheritance of history and civilization. In fact, our traditional teaching method is also in style. Tracing to Confucius, do not his discussion teaching method and the latter method in official school, college, home school with a private tutor, such as, communication, travel, reading and chant, etc. we now advocate? Now, some class with improper methods, especially the fashionable mistake which weighs form light effect, is created and invented because we ourselves lack enough study. It is said that predict the future by reviewing the past. If we just devide quality depending upon its tense, our views must be prejudiced. Each coin has two sides. There is the essence of the traditional teaching method. And the modern teaching methods do not reach the acme of perfection.

Secondly, that the teaching method is modern or not depends mainly on its connotation. Education is the fundamental guarantee for the continuation of human society. The teaching method should keep pace with the times and the needs of the development, and change in time from local conditions. Especially, teachers should enrich the connotation and performance of teaching methods. Such enrichment is the result of consistent development of thought, thinking and philosophy. So that, in order to grasp teaching methods, teachers must distinguish them from connotation instead of purely evaluate them from the angle of form.

Thirdly, the advances of the teaching methods are mainly embodied in its advanced ideas. If the traditional teaching methods are not clear, teachers will impossibly know what the modern teaching method is. The advances of the teaching methods are firstly embodied in its advanced ideas rather than its form and even tense. As long as we erect advanced thought and idea, we can find colorful form and various forms.

4 CONCLUSION

In brief, the reform of teaching method should not abandon past advantages. It is a process which develops in succession and perfects in development. No outdated method, only the thought out of date. There are a lot of modern ideas in traditional methods and not all the traditional teaching methods are antiquated. We must get a corrcct attitude that we should not negate the past entirely when the modern is mentioned.

Management, Information and Educational Engineering – Liu, Sung & Yao (Eds)
© 2015 Taylor & Francis Group, London, ISBN: 978-1-138-02728-2

The utilization of question-guided teaching model in job-oriented education

Xiu Lai Gu
Bengbu Automobile NCO Academy, Bengbu, China

ABSTRACT: Question-guided teaching is an opening teaching model, including constructing the scene, asking questions, exploring, concluding and putting forward new questions, according teaching contents. The teaching objects in job-oriented education in general have a certain theoretic and practical basis. They have clear study requirements and know what needs to be resolved and what they hope to learn to solve. Therefore, they are a good practical basis of Question-guided teaching in job-oriented education.

KEYWORDS: job-oriented education, teaching, question-guided model.

1 THE TEACHING PURPOSE OF THE QUESTION-GUIDED TEACHING MODEL

Question-guided teaching model can change cadets from passive study to active study and entirely explore their personality and creativity. With Emotion, it can stimulate cadets' interest and guide their thought, communicate sentiment and guide their study, and attach importance to EQ and guide their understanding. Firstly, the model can make cadets be good at observation, and understand the whole process to develop cadets' thinking ability to migration, extension and linkage. Therefore, cadets can understand clearly the growth points of knowledge and the corresponding posts on the whole work net, and extract, purify, digest and absorb what they had learned. Secondly, it makes cadets imagine, and understand methods to develop a cadets' divergent thinking ability to multi-angle and multi-level. Thirdly, the model can be given extended application and make cadets understand the generality to develop their dialectical thinking ability. Fourthly, it is good at seeking common ground while reserving differences and makes cadets understand characteristics to develop their ability of creative thinking. Fifthly, it can judge error easily and make cadets understand the essence to develop the ability of critical thinking.

2 THE TEACHING PRINCIPLES OF THE QUESTION-GUIDED MODEL

2.1 *Simplicity*

Simplicity is the thinking resource to create the best ways of thinking. It can create not only the best ways of thinking, but also the beauty in practice and study. The model, to realize the classroom teaching effect with high effectiveness and quality, strives to simple and reasonable approach, concise and accurate language expression, a simple and lively way of thinking and simplified and refined summary and induction to improve the effectiveness and the speed to remember and master knowledge.

2.2 *Practicality*

Practice consciousness is the thought of the function and application of post-practice, which means the cognition of practice effect and the application extent. Teachers should pay attention to the cultivation of cadets' consciousness to apply the practice, including that cadets can observe with practice eye in practical work, explain the problems in job with practical knowledge and analyze and resolve problems with practical ways.

2.3 *Individualization*

Each person has specific personality. Only those, demonstrating the unique individuality, may become the real subject, the excellent talents in the job. To develop personality not only develops their physical strength but also their psychological energy. Teachers should develop their intelligence to improve IQ and cultivate their non-intelligence factors to improve EQ. Only the persons with harmonious physical and psychology and the integration of non-intelligence into intelligence can be harmonious persons, with whole function and all-round development.

2.4 *Integration*

In classroom teaching, teachers should pay attention to the Whole optimization and balance development. They must concentrate on teaching with layers, teaching in accordance with cadets' aptitude, and the concordance between credit management and classroom teaching reform. Furthermore, teachers should attach importance to psychological education to develop cadets' individuality and the concordance between EQ promotion and the opening practical classroom teaching.

3 THE APPLICATION TO THE QUESTION-ORIENTED MODEL

The question-oriented teaching must adhere to certain steps, and the program must accord with cognitive law. To be specific, the process must be controlled in accordance with the sequence: acquisition of perceptual knowledge, questioning and explanation, and acquisition of rational thought.

3.1 *To guide cadets to ask questions*

With teachers' encourage, help and education, cadets have courage to ask classmates and teachers about the difficulty of knowledge and the exercises they cannot solve individually, rather than allow them to continue or just copy their homework.

3.2 *To guide cadets to simply simulate questioning*

With the teachers' guidance, cadets learn preliminarily query according to teachers' methods to ask questions. For example, when teachers teach the course of Tactic Logistics, they can ask cadets to simulate the questioning methods they often use, such as, the relations between logistics and tactic, the position of logistics. At the stage, the questions given by cadets may be direct.

3.3 *To guide cadets to ask questions after initial thinking*

On the basis of simulation questioning, cadets can think with consciousness and put forward some questions with stimulation. For example, after reading and thinking, they may further ask,"Is the consumption of logistical materials increase or decrease under information condition?", and "Is the link of tactical logistics and strategic logistics tighter or looser?" On the stage, the questions from cadets have certain maturity.

3.4 *To guide cadets to study with their questions and ask questions after study*

After cadets asked the questions mentioned in the last paragraph, teachers need not answer immediately rather than return the questions to cadets, and guide the cadets to have a discussion focusing on the thread, logistic support ability. After cadets master the characteristics of logistics and understand entirely its concrete representations, some questions which come from the teaching book and high above the book will be asked. On the stage, questions from cadets have to be depth to some extent and difficult and embody the level to some degree, and even make teachers feel difficult.

3.5 *To guide cadets to ask questions after mastery and deep thinking*

When cadets ask the mentioned questions, teachers must get involved in time, and act cadet temporarily and guide cadets deepen the questions and attract them ask questions at a higher level. For example, are quantized logistics and quantization in logistics the future direction of logistic development under information condition? At this stage, questions asked by cadets often can hit the crux, which have the characteristics of discovering the rules, original and finding, and then they can "solve problems", and even can be a good essay. Those cadets have the primary base to study in the field of the course and specialty. With query, explanation and level control, teachers can obtain undreamed-of result in teaching practice and reach the prospective teaching purpose.

4 THE STRATEGY OF QUESTION-ORIENTED TEACHING

4.1 *Stimulation*

Simulation is an effective method that teachers succeed in fulfill cadets' non-intelligence factors and make them be willing to ask questions. The first is to stimulate interest. That means teachers should create an interesting classroom atmosphere, focusing on cadets' learning desire and taste need, to attract them to ask questions. The second is to stimulate emotion. Emotion is a catalytic agent to stimulate cadets' subjectivity, teachers must create a harmonious and relaxed teaching environment with friendship to open the emotional channel of communication between teaching and study, and make their personal emotion integrate with the emotional factors of the teaching book, So that resonance will appear among the heart of the book and the hearts of the cadets, and among the hearts of teachers and cadets.

4.2 *Method guidance.*

From the existing cognition structure and thinking level of cadets, teachers ask questions skillfully and make cadets obtain questioning methods, by asking questions, and master questioning skills, and make them know how to ask questions and like asking questions.

Stimulate questions with revolt. Using the argument points in teaching books, teachers can stimulate cadets to argue and ask questions, and guide students to find out the problems in personal work practice.

Scene causes suspicions. Teachers deliberately appear mistakes in teaching, so that cadets can find out and put forward them. In this way, cadets' able to find have been guided and developed.

Leave behind doubt with blankness. Teachers with experience in teaching seem to just open a window to science world stretching as far as eye can see, and deliberately leave behind something unexplained, which make temporary knowledge blankness. When cadets find that they cannot get the lost knowledge from teachers and the teaching books, they will be too impatient to wait and look for them in the vast knowledge ocean. Such teaching does not abandon the knowledge points, but play cat and mouse. The purpose is to give cadets chances to digest and think, and, on this base, make them put forward their own questions, which is helpful to further teaching and study.

Teaching program is in the heart rather than in hand. Question-guided teaching is a kind of random guided teaching model, which need teachers to guide cadets according their characteristics and random situations in the classroom, so that the unified teaching programs do not exist. However, teachers must have a divergent guidancc map in mind and can give multi-dimensions divergent guidance in the classroom. Such teaching method has higher requirements for teachers with the ability to act in an emergency and control the whole situation. They should have prepared and have ready plans to meet any situation.

Management, Information and Educational Engineering – Liu, Sung & Yao (Eds)
© 2015 Taylor & Francis Group, London, ISBN: 978-1-138-02728-2

Listen-to-Write: A computer-assisted approach to improving college English writing in mainland China

Qing Song Gu & Jin Long Liu
The Institute of Translation and Intercultural Studies, Shanghai University of Engineering Science, Shanghai, P.R. China

ABSTRACT: For a long time, College English Test has not delivered the desired results, especially in terms of writing. A primary reason may be a lack of linguistic information in learners' long-term memory. Listen-to-Write is proposed as a computer-assisted approach to improving college English writing in mainland China.

KEYWORDS: Long-term Memory; Listening; Writing.

1 INTRODUCTION

1.1 *College English writing in China*

Since the end of the 1980s, College English Test (CET) has remained the most important national English test in mainland Chinese universities. Carried out twice a year, CET has helped create a storm of learning English in China, but meanwhile, brought about some problems. For example, students in mainland China have made very little progress in English writing over these years, although both students and teachers have made great efforts to improve English writing either for CET or for practical use. There are many factors involved, but we believe that the writing consciousness developed by the controlled writing format in CET and lack of linguistic information in students' long-term memory are two primary factors because the former has refrained teachers from trying new strategies and the latter highly limits students' output in writing.

1.2 *Computer-assisted language learning*

Computer-assisted language learning is abbreviated as CALL, which is not a method but rather a tool that teachers can use to facilitate language learning processes. Since the emergence of computers, a new information age has started and language learning has benefited a lot of computer science. There are varieties of CALL tools, such as software programs for computer use, the Internet, multi-media facilities centered by computers, and the like.

1.3 *Listen-to-Write*

Listen-to-Write is a computer-assisted approach we first proposed in 2009 at the Fifth International Conference on Chinese and East-Asian Learners. Listen-to-Write is designed for second language acquisition (SLA) in the classroom, which can effectively and efficiently improve college English writing in mainland China through intensifying the influence of listening on writing ability.

2 THEORETICAL STUDIES

2.1 *Long-term memory*

Memory, which is thought of as an organism's ability to store, retain, and recall information, comprises sensory memory, short-term memory, and long-term memory in terms of its span. Sensory memory is the ability to retain impressions of sensory information, both visual and auditory. Short term memory is the capacity for holding a small amount of information actively in the mind, in a readily available state for a short period. Long-term memory can last as little as a few days or as long as decades. It differs structurally and functionally from working memory or short-term memory, which ostensibly stores items for only around 18 seconds. Long-term memory also encodes information semantically for storage. It can store information for as long as a lifetime.

2.2 *Memory and listening*

The important role of auditory memory takes shape in infancy, and so it plays a long-term role in language development over the course of a lifetime. Moreover, auditory information may be retained more efficiently in long-term memory than visual information. We do not mean to suggest that visual activities are less important in language input than auditory ones, but rather we suggest that it is practical to give preference to auditory input in language learning while attaching necessary importance to visual input. Thus, in writing education, listening may be a more suitable supporting skill (at least for some learners) than reading.

2.3 *Listening and writing*

Language skills are often categorized as receptive or productive. Speaking and writing are productive skills. Listening and reading are receptive skills. However, in terms of comprehension, listening is an active, purposeful process of making sense of what is heard. Traditionally, reading is accepted as the main supporting skill for writing, and listening as that of speaking. But actually, it is a fact that listening reinforces memory and, therefore, it does indeed support writing. Listening contributes not only to comprehension but also to memory, or rather auditory sensory memory. With regard to writing, it is both a physical and a mental act. Its purpose is both to express and impress. It is both a process and a product. While writing, information is being actively taken out and purposefully processed in the form of memory. To be exact, writing depends much more on long-term memory than on short-term memory.

3 MAIN CLASSROOM ACTIVITIES

3.1 *Listen-and-write*

Listen-and-write is a bottom-up activity, according to CALL, which focuses on linguistic details needed in writing. Students are asked to read aloud after the speaker and write down exactly what is heard. This activity is quite similar to dictation and can be extended to intensive listening. All the materials shall be selected to meet the needs of writing, usually at three linguistic levels: word, phrase, and sentence, and shall be recorded as auditory files by human native speakers or by non-human software programs (even better than human voice). By the way, gap-filling can be used as an effective listen-and-write activity.

3.2 *Listen-and-guess*

Listen-and-guess is a top-down activity according to CALL. This activity focuses on how to write and targets those who have already had a good understanding of sentence patterns, of the ways to develop a paragraph (exemplification, cause-effect, etc.), and of the often-used structures or organizations applied in writing (e.g., what-why-how). For example, the teacher may play the mp3 of the first sentence of a paragraph (beginning, body, or ending) (e.g. *In the early days of nuclear power, the USA made money on it.*) and ask the listeners to guess what the paragraph is like – including the beginning, body, or ending – and give reasons orally. Then, the teacher plays the whole paragraph (e.g. *In the early days of nuclear power, the USA made money on it. But today's opponents have so complicated its development that no nuclear plants have been ordered or built here in 12 years.*), asks the listeners to guess again and announces the answer (e.g. *This paragraph is most likely to be a beginning paragraph to present a problem or phenomenon.*). Finally, the teacher plays the whole paragraph and asks the listeners to read aloud after the speaker. This activity focuses very much on the process of listening-oriented teaching in the classroom.

3.3 *Listen-and-translate*

In first language acquisition, translating seems unconscious in writing, but in second language acquisition, it plays an obvious role in writing. In some sense, writing in SLA is a kind of translating from the source language to the target language. Practice in writing, when translating is employed together with listening, the effect shall be reinforced.

4 RECOMMENDED ASSESSMENT

At the beginning of the Listen-to-Write course, students are asked to write a 200-word composition on the topic to be practiced. Then a database shall be built for each student, in which the students' mistakes made in the composition are recorded and classified at word, phrase, clause, sentence, paragraph levels. During the implementation of the approach, students are asked to write a composition every two weeks. Add their mistakes to their databases and, more importantly, the listening materials should be selected and designed to help students correct those mistakes consciously and unconsciously. At the beginning, most mistakes are probably similar to what have been made previously. Certainly, improvements will emerge week after week. Meanwhile, students' databases should include features of their meritorious performance in using the language, which mirror their progress in writing. It is important to design a syllabus for each stage, attached with some quizzes and a final test, and all mistakes in the quizzes and the test can be added into students' databases, as new references to seeking and designing new listening materials. At the end of the whole course, students are asked to write a 300-word composition. The work is evaluated in detail according to both discourse organization and linguistic and editorial details. Both merits and mistakes are recorded in the database. The students' progress is evaluated according to what is recorded in their database. Most results are intended to be encouraging.

5 CONCLUSION

Listen-to-Write is a proposal for a language learning approach aimed at improving college English writing in mainland China. It is convincing that listening is an effective way to expand and sustain memory, and it is through persistence and repetition that information is converted from auditory sensory memory to short-term memory and in the end to long-term memory. While it is true that persistent, repeated and targeted listening does improve writing by converting linguistic information to long-term memory, it cannot be said that listening is versatile in improving writing. No matter how advanced computer technology is used, listening needs patience; persistent and repeated listening needs far more patience; and therefore, for every mainland Chinese university student who wants to improve writing in English, Listen-to-Write is not only a recommended strategy, but a substantial challenge as well.

ACKNOWLEDGEMENT

This paper was financially supported by the Program of Educational and Scientific Research of Shanghai University of Engineering Science, titled "Research about the system construction of college English expanding courses" (ID: y201421003).

REFERENCES

[1] Baddeley, A.D. (1966). "*The influence of acoustic and semantic similarity on long-term memory for word sequences*". The Quarterly Journal of Experimental Psychology 18 (4): 302–309. PMID 5956072.

[2] David Nunan. (2003). *Practical English Language Teaching*. New York, NY: The McGraw-Hill Companies.

[3] Peterson, L.R.; Peterson, M.J. (1959). "*Short-term retention of individual verbal items*." Journal of Experimental Psychology 58: 193–198. PMID 14432252.

[4] Qingsong Gu. (2014). *A Theoretical and Practical Study of the Listen-to-write Approach*. Shanghai: Shanghai Jiao Tong University Press.

[5] Qingsong(Pine) Gu. et al. (2011). *The Listen-to-write Approach: Using Auditory Memory and CALL in Chinese EFL College Writing Education.* http://newsmanager.commpartners.com/tesolslwis/issues/2011–03–03/4.html.

[6] Shichun Gui. (2000). *A New Psycholinguistics*. Shanghai: Shanghai Foreign Language Education Press.

Management, Information and Educational Engineering – Liu, Sung & Yao (Eds)
© 2015 Taylor & Francis Group, London, ISBN: 978-1-138-02728-2

Problems existing in entrepreneurship education in colleges and universities under the new situation of employment and discussion on countermeasures

X. Yao & L.C. He
Sichuan Agricultural University, Chengdu, Sichuan, China

ABSTRACT: Under the circumstance where the employment situation is austere at present, the job of entrepreneurship education in colleges and universities has a direct effect on self-employment of university students and also on the issue of employment among university students. For the time being, the entrepreneurship education in colleges and universities is still at its primary stage and it is inevitable that problems of this kind or that kind may arise in the process of development. On the basis of analyzing the current situation of entrepreneurship by university students, this paper proposes a lot of problems existing in the entrepreneurship education in colleges and universities. Then, it makes an analysis with regard to these problems and puts forward corresponding countermeasures.

KEYWORDS: Colleges and universities; Entrepreneurship education; Problems and countermeasures.

1 INTRODUCTION

With constant deepening of the policy of reform and opening up in China and the rapid development of the economic society, new development opportunity has also been witnessed in the higher education career. In the past 15 years of expansion in recruitment in colleges and universities, the average gross enrollment rate across the whole country increased to approximately 30% in 2013 from 8% in 1999. In 2013, there was a number of nearly 10 million candidates enrolled for the college entrance examination and the average acceptance rate was up to 75%. On one hand, the source of students for colleges and universities has been on a constant increase. On the other hand, the large scale of graduates gives rise to the huge pressure on employment and increasingly austere employment situation. According to statistics by the Ministry of Human Resources and Social Security (HRSS), the number of graduates in general institutes of higher education across the whole country reached 7 million in 2013, which was the highest over the years, approximately 0.2 million more than in 2012. Nonetheless, it was shown in 2013 in "The Report on Employment among Chinese University Students", for the past few years, there was approximately a proportion of 10%-15% students each year who were unable to take up an occupation. Thus, accumulated in this way, the pressure on employment for the years to come will become larger.

At present, the issue of difficulty in employment by university students has escalated from "quantitative change" to "qualitative change", and has become the topic of general focus in the society and an issue to be urgently resolved by the government at all levels. The Party Central Committee and the State Council attach great importance to the employment and entrepreneurship among university students and have promulgated, in succession, multiple relevant policies to stimulate employment among university students and to encourage self-employment among them. The government at all levels also takes the initiative in formulating supportive plans, especially supporting self-employment of university students, providing a good entrepreneurship environment, fully mobilizing the initiative and enthusiasm of self-employment of university students, offering support in terms of policy and funds and alleviating the pressure on employment of university students. Therefore, under the circumstance when the employment situation is unoptimistic, quite a large number of university students choose entrepreneurship one after another and quite a few successful models spring up. However, generally speaking, the entrepreneurship of university students is still at its primary stage and the entrepreneurship education in colleges and universities is still deficient. Thus, a lot of problems are worthy of reflecting on.

2 THE CURRENT SITUATION OF ENTREPRENEURSHIP AMONG UNIVERSITY STUDENTS

2.1 *Lack of the belief in entrepreneurship among university graduates*

Currently, the large majority of university students are the generation after 90s. Although the generation after 90s goes after personality, they are reluctant to obey too much their parents. What's more, edified by the traditional Chinese family education and school education, most university students have no systematic or scientific idea of the concept of entrepreneurship. Therefore, after graduation, a large number of university students choose to seek a suitable job by means of registering for examination of civil servant and public institution, online recruitment and on-site recruitment and even get a job through pull by their parents. There are rarely any students who choose self-employment. It is indicated by relevant statistical data that, the entrepreneurship rate in the same year of graduation is less than 1%. Not only is this rate far below the level of 20%-30% in developed countries, but also the number of university students who choose self-employment is a rarity of the rarities compared with the number of graduates each year.

2.2 *Lack of vigorous support for entrepreneurship among university graduates*

On one hand, at present, the entrepreneurship atmosphere of the entire society is not strong and a great many university students fail to get understanding and support from their parents, the school and the society in their entrepreneurship. In some areas, the strength of policy support on self-employment of university students is inadequate, where the approval process is complicated and the promise is not realized in due course. This causes university students to lack the initiative and enthusiasm in entrepreneurship. On the other hand, there areas are short of stable capital chain support. It is true that, at present, it is possible to obtain petty loan or loan with low interest in virtue of some policies, but the amount of such loan is quite limited, as it is "only on a small scale". Due to lack of other investing and financing channels, it is quitc difficult for the entrepreneurship projects to obtain bank loan or venture investment. And due to lack of stable capital support, quite a lot of entrepreneurship projects have to be terminated.

2.3 *Lack of practical experience in entrepreneurship among university graduates*

An online survey finds that, approximately half of the university students hold the view that they are short of entrepreneurship experience, and that they have an idea of neither corporate management and project operation nor usage and flow of capital. Lots of students report that they have no precise idea of the market quotation and have no ability to rapidly analyze and deal with complicated information. Even quite a few students have no idea of the basic flow in undertaking a company. All this is closely connected with the fact that they do not receive systematic entrepreneurship education while at school.

3 PROBLEMS EXISTING IN ENTREPRENEURSHIP EDUCATION IN COLLEGES AND UNIVERSITIES

3.1 *Lack of adequate emphasis on entrepreneurship education in colleges and universities*

In the eyes of many commercialists and investors, there is no country all over the world that has so many entrepreneurship opportunities as in China. Nevertheless, these opportunities have been ignored by most people, especially by the Ministry of Education. For quite a long period of time, cramming-based knowledge imparting and guidance-based scientific research is still the main teaching method in higher education in China. It is true that the domestic entrepreneurship education has, in the past few years, presented a trend of gradual warming up with the change in the employment situation, and especially after 2002 when the Ministry of Education approved such nine universities as Renmin University of China and Tsinghua University, etc. to take the initiative in the experimental work of entrepreneurship education, quite a large number of colleges and universities at home have made tentative exploration in entrepreneurship education. Afterwards, the three typical types of entrepreneurship education modes are formed, namely, "practical entrepreneurship education", "classroom-based entrepreneurship education" and "synthesized entrepreneurship education". However, entrepreneurship education still has not been brought into the discipline construction of the nation. Neither is it a first-level discipline, nor it is a second-level discipline. Being not a discipline as the precondition means that entrepreneurship education is unlikely to be brought into the teaching system of the country. As a result, it can only be taught in the form of a selective course or a practice course, which seriously confines its development.

Even in colleges and universities which have carried out entrepreneurship education, the concept of entrepreneurship education among the executive level leaders and teachers is also weak and obsolescent. On one hand, they fail to escalate the importance of entrepreneurship education to a height in relation to the employment rate of graduates. Neither do they realize that entrepreneurship is also one of the employment means nor they are aware that entrepreneurship can also stimulate employment. Although some colleges and universities have carried out entrepreneurship education, they merely provide several vocational counsel training for graduates and encourage students to achieve the desired by means of undertaking an enterprise. They fail to give a scientific and systematic introduction, not to mention enabling students to acquire knowledge and improvement of knowledge and quality. Consequently, quite a large number of students have no idea of how to undertake an enterprise because they have never received systematic entrepreneurship education while at school although they have the enthusiasm in entrepreneurship. On the other hand, currently, the primary method of assessment on students is still examination and school performance is the main determinant of students' study effect. All this is not to the advantage of cultivating students' capacity in entrepreneurship. The idea of entrepreneurship education and the concept of carrying out entrepreneurship training among university students have still not been established in all institutions of higher learning.

3.2 *Lack of teams with professional teachers in entrepreneurship education*

Due to different degrees of emphasis on entrepreneurship education and limited size of personnel force and expenditure, there is only a few number of full-time entrepreneurship teachers who are mainly teachers teaching the "two courses" of Marxism theory and ideological cultivation and part-time teachers of student administrative staff. Besides, the number of these teachers is limited, which is unlikely to form a scientific and research teaching team. Most of them are busy with their own work at ordinary times, so they have no much time for theoretical innovation and course research of entrepreneurship education. They can only manage to finish the daily classroom teaching requirements, which leads to quite slow construction of the curriculum system of entrepreneurship education. The Report to the Eighteenth National Congress of the Communist Party of China proposes new orientation and objective of higher education in the next one decade. In the same year, the Ministry of Education also issued notice of "Basic Requirements for Teaching of Entrepreneurship Education in Regular Undergraduate Universities (For Trial Implementation)". In the notice, colleges and universities are required to rationally assess and determine full-time teachers' staffing level and assign adequate full-time teachers with high quality in accordance with the number of students and the actual teaching tasks", so as to assure the number of full-time teachers of entrepreneurship education.

3.3 *Low quality of teachers*

Teachers of entrepreneurship in a lot of colleges and universities haven't received systematic learning in entrepreneurship, so their knowledge structure and practical experience are unlikely to satisfy thc nced of entrepreneurship education and teaching. Entrepreneurship education is a multidisciplinary course and its greatest distinction from other general courses is that teachers of entrepreneurship have to integrate the entrepreneurship idea and entrepreneurship cases into the professional teaching and guide students to form entrepreneurship thought and entrepreneurship thinking in the classroom to exercise their practical experience. Therefore, teachers of entrepreneurship education are not only required to possess professional knowledge reserve, but also have practical experiences in entrepreneurship, namely, kind of inter-disciplinary talents with high quality. However, the fact is that, currently, most of teachers of entrepreneurship education are deficient of practical experiences in entrepreneurship, and haven't undertaken an enterprise personally. Thus, they are unable to combine the theory with practice and their teaching in the class is nothing more than one-way cramming, lacking in case analysis and open-ended research discussion or communication. And the entrepreneurship education they offer to the students can only be "an armchair strategist", incapable of achieving a duc tcaching effect.

3.4 *Incomplete supporting policies*

On one hand, there is not a perfect assessment and performance rewarding mechanism in relation to teachers of entrepreneurship education and the traditional assessment and rewarding method is no longer suitable for promotion of teaching of entrepreneurship education. Furthermore, the high requirements of entrepreneurship on teachers, to a large extent, add to difficulty of teachers engaged in entrepreneurship education. This cuts down on their enthusiasm in actively exploring teaching methods and innovating teaching means and also gives rise to the low sense of identity among other teachers and students and frustration in promotion, which directly impacts the working enthusiasm of teachers of entrepreneurship education. On the other hand, popularity of entrepreneurship education in colleges and universities also needs support of the whole society. Currently, university students at school have no way to get powerful support from the employment department, corporate management department, industry and commerce department or social media in their practical activities of entrepreneurship. The coordination effect of colleges and universities is almost one in a million. As a consequence, the university students who have the intention to undertake an enterprise are unlikely to rely on any relevant policy, or they have no idea of the procedures of extant policies, so they lose the motive in entrepreneurship.

4 PROBLEMS EXISTING IN ENTREPRENEURSHIP EDUCATION IN COLLEGES AND UNIVERSITIES AND COUNTERMEASURES

4.1 *To have correct understanding in the orientation of entrepreneurship education in higher education*

Development of the entrepreneurship education of a country is closely bounded up with improvement of its overall educational level. As further extension of daily teaching and scientific research, entrepreneurship education is updating of the educational idea instead of re-construction that breaks away from the traditional educational mode. Therefore, colleges and universities ought to have correct understanding in the orientation of entrepreneurship education and take advantage of the opportunity in teaching reform to endow the traditional education with new connotation and vitality. The curriculum system, teaching syllabus, teaching content and teaching method should be combined with professional education and practical education. The idea of entrepreneurship education should run through classroom teaching and extracurricular activity to constantly reinforce the entrepreneurship consciousness of university students, enhance their comprehensive quality and satisfy the need of the contemporary society on inter-disciplinary talents.

4.2 *To accelerate recruitment, introduction and cultivation of teachers for entrepreneurship education*

In the face of the huge pressure on employment, quite a lot of colleges and universities have made trial implementation of entrepreneurship education by regarding entrepreneurship education as emphasis in the work in the future for some time and proposing high requirements on entrepreneurship teachers. Therefore, on one hand, colleges and universities should attach great importance to recruitment and introduction of teachers of entrepreneurship and determine job qualifications for teachers of entrepreneurship that fit with professional features of their schools. They can organize an interview group consisting of executive leaders, distinguished university teachers, middle and senior corporate managers and relevant governmental personnel, and organize an interview and written examination to make preferential selection of entrepreneurship teachers with high starting point and high level in an open, fair and impartial way. In the meantime, the colleges and universities may also resort to all kinds of talent introduction plans formulated by the Organization Department of the Central Committee of the CPC, the Ministry of Education, the Ministry of Science and Technology and the provinces and cities where they are located to vigorously bring in excellent working teachers of entrepreneurship both at home and abroad and recruit by invitation the middle and top-level technical and management personnel in cooperative enterprises to hold the post as part-time teachers to make up for deficiency of front line teachers and optimize constitution of the teaching staff. On the other hand, at the time of vigorously "external bringing in teachers", colleges and universities should also pay more attention to "internal training". Before they take up the position, the teachers should, first of all, receive pre-employment training which involves such professional courses as management and marketing that are related with entrepreneurship education and apply flexible and interactive teaching means to motive students' interest in learning. The colleges and universities should also fortify the strength of training on teachers of entrepreneurship education who are on the post. For instance, they can employ such patterns as panel discussion, enterprise training, post practice and putting on field practice to realize combination of classroom and factory and unification of theory and practice. It is feasible to enhance the capacity of teachers in entrepreneurship education by means of continuing education, bring it into the management system of teachers as one of the requirements for the occupational ability of teachers, and realize "recharging" of teachers on the post. In addition, the colleges and universities should collect and sort out relevant cases regarding entrepreneurship of teachers according to the features of the schools and form systematic internal educational teaching materials. After trial lecture for some period of time, the teaching materials can be taught to the entire society and can be learned by all teachers related in the form of textbook.

4.3 *To further strengthen matching and completeness of relevant mechanism, policy and system*

First of all, the governmental section and the colleges and universities should intensify mechanism establishment and offer good learning and entrepreneurial environment for university students' entrepreneurship.

The former should simplify the approval procedures of entrepreneurship and reduce the threshold conditions for entrepreneurship. The administrative sections of examination and approval at all levels should relax restrictions on registration capital and occasions of entrepreneurship and cut down on taxes and dues appropriately to offer vigorous support for university students' entrepreneurship. At present, quite a lot of local governments have set up small amount soft loan or offered support for university students to undertake an enterprise by the means of financing. Then, for those small and medium-sized technology-based enterprises initiated by university students, the government section should draw up more favorable incentive policies of entrepreneurship to form an atmosphere of supporting innovation and cultivating originality. In the meantime, it is necessary to set up a quantity of rewards and offer further policy inclination and consideration to those university students who achieve excellent performance in entrepreneurship. It is also a must to further reinforce entrepreneurial credit aid, entrepreneurial subsidy and reduction on taxes and dues.

Then, the government section ought to establish information disclosure mechanism and provide such all-round and three-dimensional services as project information, entrepreneurial starting guidance, project extension, capital integration and achievement transformation by means of constructing entrepreneurial information sharing platform, entrepreneurship personnel training platform, entrepreneurial project exchange platform, entrepreneurial achievement incubation center or incubator. Furthermore, it is necessary to provide all kinds of entrepreneurial training, information and market service system to serve for entrepreneurship at all levels in an all-round way, standardize such factor markets as capital, property right, technology and labor force, construct a standard market system and let entrepreneurs compete in a fair way on the precondition of information equivalence.

Finally, the colleges and universities are supposed to set up scientific entrepreneurship education evaluation system and incentive system. To this end, it is necessary to define the assessment criteria consistent with the teaching content of entrepreneurial teachers, consummate teaching guidance information feedback system and bring all work in relation to entrepreneurship education within the scope of assessment to have rules to follow. At the same time, the colleges and universities need to build up the rewarding system that suits entrepreneurial teachers and ensure that their recruitment through invitation, promotion and increment of salary will proceed on schedule. It is also feasible to offer effective guidance for broadening the developmental space of teachers, encourage and support teachers in the aspects of teaching material compilation, thesis publication, further education and training and experience exchange, make a thorough plan for their occupational path and mobilize their enthusiasm in the work. Only if the evaluation system and the incentive system are combined effectively, can construction of the teaching staff of entrepreneurship step on a track of benign development.

4.4 *To strengthen entrepreneurship practice and to cultivate the innovation spirit of university students*

Entrepreneurship practice is the most important constituent part in entrepreneurship education among university students which can motivate university students' entrepreneurship consciousness and entrepreneurship potential and cultivate their willpower and spirit of hard working and plain living, courage of innovation and fearlessness of difficulty. Therefore, at the time of carrying out entrepreneurship education, the colleges and universities should further reinforce the strength of entrepreneurship practice. In the first place, it is necessary to enlarge investment in manpower, material resources and financial resources and cooperate with the cooperative enterprises to construct entrepreneurship practice bases or entrepreneurship parks for university students. The colleges and universities can also combine classroom teaching with practice by inviting relevant enterprise personnel and successful entrepreneurial university students to give a lecture in the practice base or graft a project and organize university students to take part in entrepreneurial activities under the guidance of teachers, which can exercise university students' capacity in entrepreneurship. In the second place, on the basis of the existing university students occupational guidance center, the colleges and universities should set up entrepreneurial guidance center and furnish counseling and support to the university students who intend to undertake an enterprise in terms of policy, regulation and service procedure, etc. If the condition allows, the colleges and universities can also establish entrepreneurial supporting fund and offer necessary capital assurance to the entrepreneurial projects with developmental prospect. In the third place, the colleges and universities ought to take an initiative in launching all kinds of entrepreneurial competitions or encourage students to go out of the campus for a competition and select excellent teachers to lead the students in constituting a competition team. The competition helps to strengthen university students' understanding in entrepreneurship and enables them to give full play to their intelligence and wisdom. Furthermore, it also publicizes positively the entrepreneurial achievements of the schools and attracts funds.

Entrepreneurship education in colleges and universities is a systematic project that involves multiple disciplines and needs common efforts of colleges and universities, the society, the parents and the students themselves. Thus, the administrative service section at all levels and the colleges and universities should take the initiative in carrying out the requirements proposed by the Party Central Committee and the State Council and continuously fish out a set of entrepreneurship education system that suits local economic and social development and complies with professional features of the colleges and universities to open an achievable path to facilitate employment of graduates.

REFERENCES

Guo, X.Q. & Zeng, C. 2009. Discussion on Cultivation of University Students' Entrepreneurship Capacity and Entrepreneurship Consciousness. *Entrepreneur World* (2): 104–105.

Yao, L.N. 2009. Reflections and Exploration on Entrepreneurship among University Students. *Inner Mongolia Science Technology and Economy* (21): 55–56.

Zhang, M.C. 2010. An Explorative Analysis of Construction of Teachers in Entrepreneurship Education in Colleges and Universities. *Henan Education* (8): 75–76.

Zhang, Y., Lv, Z.H., Luo, Y.J. & Zhang, W. 2010. An Analysis of the Current Situation of Employment among University Students in the Current Economic Situation and Discussion on Countermeasures. *School Party Construction and Ideological Education* (2): 78–79.

Zhu, C.F., Li, W. & Ran, J.C. 2006. Discussion on the Operational Mechanism of Colleges and Universities in Implementation of Entrepreneurship Education. *Jiangsu Higher Education* (4): 114–116.

Management, Information and Educational Engineering – Liu, Sung & Yao (Eds)
© 2015 Taylor & Francis Group, London, ISBN: 978-1-138-02728-2

Study on middle school students' overweight lessons-burden in Gansu province

Ruo Fan Zhao & Jian Min Han
College of Humanities, Gansu Agricultural University, Lanzhou, Gansu, China

ABSTRACT: Middle school students' overweight lessons-burden in Gansu province is becoming increasingly fierce, which seriously affects the physical and mental health of adolescents, and also affects the healthy development of basic education. The author mainly applies the method of questionnaire investigation and interview; the survey shows that scores on education evaluation and examination tend to be serious, the uneven allocation of educational resources and the poor implementation of alleviating burdens policy are the main causes of middle school students' overweight lessons-burden. Gansu province should actively promote the reform of the way of education evaluation and examination, optimize the allocation of educational resources and increase the regulatory enforcement on implementation of alleviating burdens policy, to relieve middle school students' lessons-burden thoroughly.

KEYWORDS: Gansu province; Middle school students; Lessons-burden.

The problem of too heavy lessons-burden exists in the fundamental education commonly. In poor education Gansu province, the problem exists as well. The Gansu Education Department has made the policy of reducing lessons-burden for many times in order to reduce the middle school students' lessons-burden completely. As a result, the middle school students' lessons-burden is heavier and heavier. From 1999 to 2013, the rate of NMET enrollment increased from 27.8% to 77%. The senior education enrollment rate increased to 82%. The enrollment rate increased while the enrollment reduced, but the lessons-burden didn't reduce, on the contrary, it went on increasing. In the end, the was a problem that reality was not equal in theory. The problem above has exposed many disadvantages in the present education system, on the other hand it has been proved that the education problem needs reforms in many aspects.

In the process of researching the middle school students' too heavy lessons-burden in Gansu province, collected data mainly used questionnaires and talks. Above all, the sampling survey was used in Lanzhou city, Xifeng district and Heshui country. The middle school students' lessons-burden questionnaires were handed out to 555 students, but 518 questionnaires were handed inefficiently. Next, there was a talk between parents and teachers, including 40parents and 20 teachers.

1 THE SITUATION OF GANSU MIDDLE SCHOOL STUDENTS' LESSONS-BURDEN

There are junior and senior school students, although they are taught in different education levels, they and also have too much lessons-burden in common. The lessons-burden refers to their own responsibility and duty which the objective lesson contents give and the subjective feelings of pressure. But all come from the learners who are studying according to their previous experiences. Through the survey analysis, we find that the lessons-burden of the middle school students in Gansu is very serious beyond their own bearable abilities. Lessons-burden mainly comes from two aspects: first, the school puts lessons-burden pressures on them; second, the outclass coaching organizations bring homework loads and the coaching lessons occupied the students weekends and holidays. Middle school students' overweight lessons-burden can cause students' physical qualities and PE grade reduce obviously, which can cause the nearsighted rate, reduce obviously, which can cause great damage to the students' physically and mentally.[1]

1.1 *The lessons-burden situation of middle school students in and after school*

1.1.1 *The time situation at school of Gansu province middle school students every day on average*

The survey shows that 72.01% junior school students stay in school more than 7 hours; 77.99% senior school students stay in school more than 8 hours every day on average. But Gansu Education Department requires that the junior school students should stay in school less than 7 hours, the senior school students less than 8 hours.As for this reason, 80% teachers think there has not been reform in the exam form since the new curriculum was carried out, the teaching contents have increased by one third and the teaching is more difficult than before, it is hard to finish the teaching contents in the scheduled time. Studying too long in school makes the students not fix their attention and makes the studying efficiency low.

1.1.2 *The amount of Gansu province middle school students' homework*

Gansu Education Department requires that the time of finishing homework should be limited in 1.5 hours in junior school, that the time of finishing homework should be limited in 2 hours in senior school.But the survey shows that 12.93% junior school students spend 2 hours doing their homework every day. 41.8% junior school students spend 3 hours finishing all the homework every day, 32.05% junior school students spend 4 hours doing homework every day, only 12.16% junior school student can finish their homework in 1.5 hours. 24.8% senior school students spend 3 hours doing their homework, 67.18% senior school students spend 4 hours doing their homework, only 8.01% senior school students can finish their homework in 2 hours.

1.1.3 *The sleeping situation of Gansu province middle school students every day on average*

The survey shows that 75.0% junior school students sleep less than 8 hours, that 80.86% senior school students sleep less than 7 hours every day on average, which obey the national rules that the middle school students should sleep 9 hours at least. The middle school period is the key period that the teenagers body and brain grow, not enough sleep for long stops the middle school students' body and brain from growing and influence the growth of the teenagers.

1.2 *The coaching situation of the middle school students*

1.2.1 *The situation that middle school students attend the coaching classes after school*

The outclass coaching of all subjects is the important source for the students' lessons-burden.85.93% junior school students attend the outclass coaching at the weekends or in summer and winter holidays.Among the students,13.14% junior school students attend one subject,55.87% junior school students attend two subjects,30.99% of them attend three subjects at least. The senior three students attend the lessons organized by school,80.15% senior school students attend math and English coaching lessons during their free time.

1.2.2 *The situation that middle school students have outclass coaching datas*

The students who have been looked into reflect that Gansu middle school students have bought the outclass coaching datas willingly or unwillingly. 56.25% junior students and 61.07% senior students think that the outclass coaching datas are too many. Besides, 54.3% junior students that each subject has two outclass coaching data. 33.2% junior students think each subject has three outclass coaching datas. But 49.24% senior students think each subject has three outclass coaching datas, 38.17% senior students think each subject has four outclass coaching datas.A number of outclass coaching datas take up the students' free time, which makes the students trapped into all kinds of exercises.

2 THE REASON OF GANSU PROVINCE MIDDLE SCHOOL STUDENTS' OVERWEIGHT LESSONS-BURDEN

2.1 *Education evaluation and examined grades tend to seriousness*

Exam rule is not only a simple education rule in China but also a social resource and redistributed rule, especially it is a way that the social classes flow.[2] In the modern education rules, exam has become a main method with the students changing the society levels. High grade is the aim for the middle school students to chase each other, became of grades, the middle school students extend the studying time, studying pressure is larger, lessons-burden becomes the important problem that the students have to bear. Grade is the only standard of education evaluation, it also becomes a social evaluation standard gradually. The school fame, the government achievement also depends on the grades.School and society play an important role in increasing the students' lessons-burden.[3]In the process of survey, 92% parents and teachers think NMET rules is unreasonable, which is the direct reason leading to the students' overweight lessons-burden.

First of all, as far as middle school students are concerned, education evaluation and checked grades tend to stopping the steps of reducing load directly. The education practice has proved that the middle school students lessons-burden is more and more serious when the grade is an only standard to evaluation education. With the direction of NMET, the upper students' lessons-burden can't be reduced and their lessons-burden is more and more.As a result, the upper students lessons and homework become the attractive reason of the lower students, because it can force the lower education to deal with the appearance of the lower lessons and requirement in advance, which can lead to the too heavy lessons-burden for the middle school students.[4]

Secondly, as far as Gansu middle schools are concerned, grades can evaluate the teaching quality, it is an only standard to get fame for school, whether the school teaching work goes well can be controled by the grades.Modern school management in Gansu province has strong planned economy time management.The school function is to carry out the higher authorities' orders and finish the aim of going to a higher school, which is made by the upper Management Department, and reach the fundamental requirements of the upper Management Department. Under the limit of administration and economy, grade becomes an only standard to evaluate schools, school is the important resource for the students to have overweight lessons-burden.

Finally, as for outclass coaching organization, it is influenced by the grade, the parents and middle school students need to raise their grades, the coaching organization can meet their needs.There are all kinds of coaching organization in Gansu province for their benefits, it is another important resource for overweight lessons-burden of middle school students.

2.2 *Education resource equipments are not equal*

2.2.1 *Education resource equipments are not equal among the middle school*

In 2010, the world education funds reached 4.9% of GDP on average, but in China education funds reached 4% of GDP in 2012, the education resource was not much enough.[5] Under this condition, Gansu province Fundamental Education meets a lack of education resource for a long time. The local government is the first responsible person of Gansu province education resource equipments. But in the process of education resource equipments, the government thinks of its achievements, and regards efficiency as its first aim, meanwhile it neglects fairness, so weak schools come into being while the key middle school are developed. For example, the key middle schools

Table 1. These four middle school educational resources.

	Key university rate	Teachers numbers: Amounts	Teachers numbers: Masters numbers	School area (Mu)	Stored books(ten thousand copies)	Students numbers (person)
The middle school attache to Northwest Wormal university	94.62%	190	51	156	20	3200
LanZhou first middle school	50%	165	22	77	-	2800
QingYang first middle school	36.4%	132	5	80	13	2500
HeShui first middle school	3.5%	167	2	91	8	3207

attached to the Northwest Wormal University. The key middle schools depend on the policy, and attracts a great deal of investment, best students and best teachers. The weak schools are not concerned, if so, the strong is the strongest, the weak is the weakest. Such as: the middle school attacked to Northwest Wormal University, Lanzhou first, middle school, QingYang first, middle school and HeShui first, middle school, we can find that Gansu province education resource equipments are not equal seriously.

The table above shows that the enrolled rate of the middle school attached to the Northwest Wormal University is the highest—up to 94.62%, but the enrolled rate of HeShui first, middle school is 3.5%, the rate of going to a higher school in key middle school is higher than other schools. Every teacher in the middle school attached to the Northwest Wormal University can teach 16.84 students, every teacher in HeShui first middle school can teach 19.2 students on average. The middle school attached to the Northwest Wormal University can cover an area of 156 Mu, it is twice as large as other schools. The stored books can reach two hundred thousand copies in key middle school, only eighty thousand copies in HeShui first middle school. Though the enrolled rate of the middle school attached to the Northwest Wormal University is higher, a large number of thus the education resource is waste seriously. The key middle school enrolled rate is found above the best educational resource monopolization and unequal competitions.

The key schools can reflect the unequal education resources equipments in Gansu province and bring the lack of high quality education resources.The key schools get benefits instead of non-key schools in the process of developing and encroach the legal benefits of the non-key school students. The non-key school students have no chances to get an equal education. Gansu province's educational resources are wasted, education systems are destroyed. First of all, the high quality education resources in the schools can lead to competitions among the students, it can increase the middle school students' lessons-burden. Secondly, there is fierce competition among the key schools, it can increase the middle school students' lessons-burden. Finally, because of these unequal education resources, equipment, choosing schools strongly not only cause too heavy lessons-burden for the students, but also bring an unburdened economy load for the parents.[6]

2.2.2 *There are unequal chances to get education resources among all classes*

Different classes have unequal education resources, the competitions are unfair for the students in the process of learning. Raftery and Hout put forward to the maximally maintained inequality, it is called for, short of MMI theory, the MMI theory thinks. Expanded education can't lead to equal education chance distribution. On the contrary, only advantageous groups have the education chance to promote them, unfair education chances can go on. Its main reason lies in that the social resources owed by the upper classes have advantages to create education chances for their children. Only the upper class education is saturated, the unfair education among different classes can drop.[7]

Through the survey from the champion of NMET from 2004-2013, we find that MMI theory is right in Gansu province. The parents who are from the champion 23 students in NMET, only six of their parents were not educated, four of their parents are farmers, one of their parents is a worker, one of their parents is an individual economy. Other students' parents have advanced education diplomas, among these 34 parents, 19 people are teachers, 10 people are clerks, 2 people are engineers, one is an army, one is an accountant, one has no work. The increase of education chance didn't change this unfair situation, instead it led to unfair education in the wide range, this unfair competition has caused the students' overweight lesions-burden in the Gansu middle schools.

2.3 *The policy of reducing burden is not carried out strongly*

After many measures of reducing burden failed, Gansu province raised the strictest reducing the burden in history in 2013, and limited the students' sleeping time strictly, the amount of homework and forbade any grades having missed lessons except senior 3 at the weekends or during the summer and winter holidays in order to reduce the students' lessons-burden completely. However, though the survey, we find that the carried results are not ideal become of not being strict management and beings carried out strongly.

First, reducing burden refers to the school's benefit without being carried out efficiently.The survey shows that the policy has been carried out in part schools in LanZhou, but in XiFeng ang HeShui the policy doesn't work.The teachers attending the talk think the schools in LanZhou have rich education resources and enough ability to deal with the reducing burden orders and face the danger of lowering the rate of going to a higher school, but in other places education resources are very short and are hard to deal with the influence of the policy. Because of the rate of entering higher school, banned making up for missed lessons order only becomes a great wish, it is hard to exist here.

Second, although the reducing burden policy had developed, but because the policy is hard to supervise, all the schools ignore it, 86.87% junior and 91.99% senior middle school students can't finish their homework in scheduled time, the policy is hard to get the aim to reducing students burden like unreal things.

Third, the policy is replaced to perform.The efficiency of reducing burden policy limited to schools, extracurricular coaching has increased dramatically. The students are busy with the lessons in school and the coaching classes.The parents are busy looking for best coaching classes and teachers for their children. The students lessons-burden is still too heavy, the wages of coaching lessons become the economy burden for the parents to bear.

3 THE ANALYSED METHODS TO REDUCE THE STUDENTS LESSONS-BURDEN

The students lessons-burden caused without enough protection for its existing rightly and developing rightly, at the same time it led to various education aims and various teaching functions in school. The government, school and society should carry out constant reform in all levels and all aspects, explore the practical measures for reducing lessons-burden, promote the education develop healthily.

3.1 *Improve education evaluation methods*

NMET decides on a person's whole life, the examination is regarded as the only testing tool, the grade is the only goal of evaluation, it is not scientific. The education evaluation ways must be diversified.

Individual differences theories think that every individual is different. There are differences among individuals and individuals, the ability has early and late differences, sex differences and character differences. According to this theory, Gansu province should take measures to reform in education evaluation methods. First, the reform should break the usual rule to have an NMET each year, and offer the students to have many chances to exam. Second, the education evaluation standards must be diversified, they are based on the students morals, intelligence, physical training, beauty and labor. Gansu province must adjust the requirement of education developing rules, and get rid of the only education evaluation way by the grade.

3.2 *Promote Gansu province education resource equipment by making up for weak machine-made*

First, the education department must invest much money for Gansu province's education resources, and make sure to have enough education resources to meet the student's requirements. In the situation of enlarging needs, keeping growth and adjusting structures. Saving money is enlarging quickly, the flowing funds in society is increase quickly, China has stepped into the period of rich funds, and raise the possibility to enlarge Gansu province education resource invest.[8]

Second, promote Gansu province education resource equipment by making up for weak machine-made, raise the education quality and realize the leaping development. The realized ways to make up for weak machine-made can be diversified. For example, gansu province government should tend to the weak groups properly in the process of making education policy, this tendency should be based on the education developing ability of raising the weak groups and competition abilities. In addition, depend on the third department's power, the third department is based on promoting public benefits, but it has flexible ways in the process of handling, and help the weak groups to ensure education benefits, which will promote fairness in education and reduce the students' lessons-burden.

3.3 *To strengthen the supervision of implementation of the reducing burden policy*

Realizing reducing burden not only depends on the government to make out policy, and it is useful to carry out the policy. To realize the Gansu province middle school students' reducing burden goal must depend on the policy to carry out. On the one hand, the government should strengthen scientific and manageable policies while make policy, and should be able to make the policy goal planned in amounts so as to be checked evaluated and supervised. On the other hand, the government should make rules to punish the schools seriously, which violate the reducing burden policy. The government also should make the inspired policy, to praise the schools which observe the rules, direct the schools to observe the rules actively, and raise the policy's public authorities. Finally, the government should strengthen the supervision of public opinion, and offer many ways to complain, make a good environment for supervision of public opinion.

ACKNOWLEDGMENTS

I very gratefully acknowledge the assistance of professor Han Jian Min in guiding me, and providing data from his project of the influence of urbanization on local education in Gansu province. Project number is 1105ZCRA235.

REFERENCES

[1] Song Naiqing&YangXin. A quantitative analysis of the heavy course load of Chinese primary and secondary school students.Educational research2014(3):25.

[2] ZhangLing&Huang Xuejun.Introduction to lighten students' lessons-burden:based on the perspective of difference hypothesis.Journal of the Chinese society of education2012(2):14–15.

[3] Cheng Pingyuan.China's education problem investigation.Peking.Tsinghua university press.

[4] Shanzi.Overweight lesson burden:an analysis of export congestion and swollen lesson.Journal of schooling studies2013(3):6–7.

[5] Yang Dongping.China's education development report in 2013.Peking.Social sciences academic press.

[6] DongHui&YangLan.Research summary on school-level variables of academic burden.Global education2012 (12):45.

[7] Li Chunling.Higher educational expansion and unequal educational opportunities——investigation on the equalization effect of expansion of higher education. Sociological studies2010(3):84–85.

[8] Liu Daoxing.The revolution of education investment. Peking.Social sciences academic press.

[9] Chen Chuanfeng&Chen Wenhui.On excessive workload of secondary school students:degree of severity, causes and counter measures.Journal of the Chinese society of education2011(7):14.

Management, Information and Educational Engineering – Liu, Sung & Yao (Eds)
© 2015 Taylor & Francis Group, London, ISBN: 978-1-138-02728-2

Teaching strategy on English major in China

Rui Dai
Anhui Sanlian University, Hefei, China

ABSTRACT: English, as a language for international communication, is essential in the world today. With the rapid change and development of the society as well as thc change of the qualified people needed, university students have changed their learning needs and desires of English learning. Therefore, most universities own English-majors. The English teaching methods directly affect the achievement of the aims of English teaching and the satisfaction of the learners' needs. We should analyze English major learners in Chinese universities and find the precondition for implementing the teaching plan and improving teaching methods. The university English teaching has emphasized on improving their practical application abilities of listening, speaking, reading and writing instead of the specific explanations of words, sentences and other basic language points. In this thesis, the writer will mainly research the present teaching and learning situation, explore learners' needs, pros and cons of current teaching methods, and aim to find effective teaching methods to better suit the needs of university English Major Learners, supported by a survey designed by the author. Also, this thesis intends to find out what kind of teaching methods are most frequently adopted in the process of teaching, what kind of possible effective teaching approach might be greatly encouraged and used in the future, and accordingly give some suggestions for English educators and learners to choose effective ways of teaching and learning.

1 INTRODUCTION

What kind of needs should be analyzed among university students in China? We should think over that analysis involves objective needs and subjective needs. A large number of students majoring in English in university in China learn English as their second foreign language. Somc of them are successful learners, while most of them have encountered various kinds of difficulties. It is vital and necessary for language teachers to help those language learners who have problems during their learning process. And the first and foremost step is to find the reasons and identify the needs of the students.

Objective needs include all the objective conditions of the language learners, such as their original language level, learning conditions and environments, learning methods, problems and difficulties in their learning process while subjective needs refer to the cognitive and affective needs of the language learners such as learning motivation, requirements for teachers, expected learning outcome, self-esteem and attitude towards learning.

It is common that in the purpose of passing the university entrance examination, teachers lay much emphasis on the practice of English reading and writing. The ignorance of the importance of listening and speaking has lead to a worrying and embarrassing situation: a large number of students majoring in English in university do well in exams but perform terribly bad in English communication. As far as English major students are concerned, what they pursue is not to grasp some easy English for daily communication, while English major students deem that they need to perform very excellently both in oral and written English. It is noticing that Chinese students majoring in English in university who are negatively influenced by the university entrance examination find it much more difficult to do a good job in listening and speaking than reading and writing. This, to a large extent, is mainly caused by the exam-oriented teaching mode and the unreasonably designed content of the exam itself. The majority of students majoring in English in university consider studying English a means for their careers.

2 OBSTACLES AND REASONS ON LISTENING, SPEAKING, READING, AND WRITING

The four main components of English are: listening, speaking, reading and writing. Teachers have to know clearly about the present teaching situations and existing problems of each of the above four aspects aiming at solving them, implement teaching plan and improve teaching methods according to needs analysis.

The following chart illustrates the percentage of students who felt they most urgently need to improve:

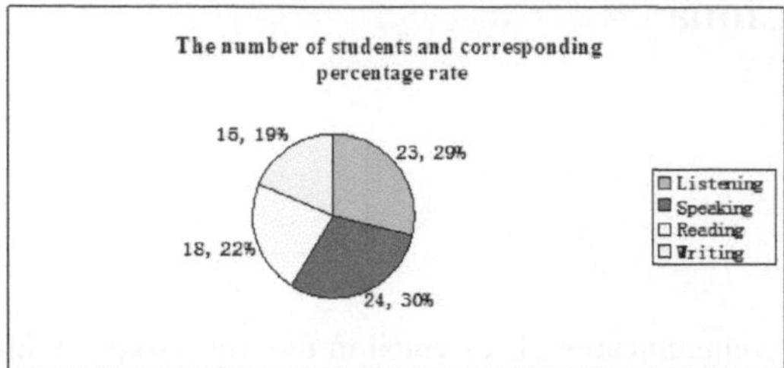

Listening is the combination of listening and understanding. The process of listening comprehension is the process of using all kinds of knowledge and skills. In the listening process, how much a student master the basic knowledge of language , to a large extend, determines his comprehension ability and his reaction rate. Pronunciation barriers and speed barriers. A large number of university students are so poor at pronunciation that they can not understand the meaning of a passage, even a word of a native speaker. Knowing what others are saying is the basis of making a conversation. Besides, much emphasis has to be put on some English words with similar pronunciation and correctly identifying some confusing phonemes, such as sheep - ship, house - horse. Some English teachers speak English very slowly in order to make sense.

As time passes, students form this habit of adapting to the slow or normal speed while can not follow the speed of the listening materials. If they meet some words with pronunciation of weak, stress, loss of detonation, stress transfer and the change of pronunciation and intonation, they will find it extremely difficult to understand.

There are a variety of factors that make students be afraid of speaking out. Firstly, they feel nervous and embarrassed when they can not continue speaking. Besides, most students fear to be laughed at and criticized by teachers when making mistakes. Secondly, a great number of students find it difficult to pick up appropriate topics and communicate with foreigners due to the lack of intercultural competence. A perennial issue in the Chinese oral English class is that teachers will always keep saying while students keep listening. Traditionally, students are assumed as passive receivers of knowledge but not active executors of English learning. Therefore, it is not surprising at all Chinese students are still deaf and dumb in English even if they have learned English for almost 12 years. This is probably the result of lacking of practice.

Reading can be divided into intensive reading and extensive reading. Teachers cannot use the same teaching method for both in that these two kinds of reading require differently from each other. The aims of intensive reading are to understand the context; to know clearly what the main idea was, why did the author tell the story, how it went on, when did it happen. However, it is enough for readers to get the main idea without knowing every detail. Thus, it is ineffective and a waste of time for teachers to analyze every detail of a text.

Many English teachers I interviewed all agreed that most university students were poor at English writing with so many errors that the teachers could not tell what they wanted to say sometimes. The students usually find in difficult to use right words or phrases to express appropriately in their writings. There are a lot of linguistic errors, such as, lexical, syntactical, and sentences, influenced by L1 direct transfer, seem confusing.

In China, English learning is quite a boring and daunting task while English teaching is also a tedious and arduous work. And they are still "deaf and dumb in English", since the English courses are test-oriented and grammar-based Chinese university students find it difficult to communicate with foreigners and they are discontent with their learning outcome for years of hard-working in spite of hardships.

3 THE METHODS AND STRATEGIES OF TEACHING FOR STUDENTS MAJORING IN ENGLISH

3.1 *Listening teaching*

The first action for most students is to translate English to Chinese while listening English. This is indeed not a good way to improve listening for that people have waste time for note-taking which is very important for getting the main idea and answering questions. It is better for students to think in English. It is an effective way to get a great deal of knowledge of western culture and literature as well as living habits, local conditions and customs and ways of life.

First, learning phonetic symbol is the premise of learning English. The phonetic symbol determines students' pronunciation, intonation. Therefore, the first and essential task for teachers is to make sure that students pronounce correctly. Second, on the condition of making sense, teachers should work on speeding up and try to speak faster to train students' listening. Third, different people show different preference for British English or American English. However, it is better for teachers to attach importance to both. No one can predict the speaker's accent in the exams or in a conversation. Fourth, the dominating reason why students find it difficult to express themselves is that their vocabulary is too small. There's no doubt that teachers should help students with enlarging their vocabulary efficiently, such as, assigning reading homework regularly, reciting some good articles.

3.2 *Speaking teaching*

In the student-centered situation, students are actively performed in the learning process deciding what to learn, how to learn, while a teacher is defined as a co-communicator, an instructor, an organizer of resources, a needs analyst. Teachers should give clear instruction to students and allow them to develop their own speaking habits and create their own goals. Besides, it is teacher's duty to choose the appropriate learning content based on students' interests, needs, expected goals, for example writing E-mail, interpreting, etc. Another essential part for teachers is to encourage students to speak English anytime and anywhere.

Another effective way of oral English teaching I highly recommend is cooperative learning. In general, English of students from economically developed costal provinces and urban area is better than those from inland rural areas (Hu, 2003). Therefore, teachers have to make sure that members of each group have the similar English proficiency both in spoken English and in listening. So they can share their ideas freely without fear and cooperate harmoniously.

Teachers should make efforts to help students to recall their own culture while introducing the target culture, to make a comparison of the two cultures and to find out and analyze the similarities and differences. So students are capable of connecting the target culture with their source culture and making good performance in the real communication situation.

3.3 *Reading teaching*

Reading can be divided into intensive reading and extensive reading. The intensive reading requires students to completely understand the context as possible as they can and try their best to grasp every part of it. So it is better for teachers to guide students to acquire intensive reading skills and know every detail well. Besides, students themselves should think over in person and make effort to get the necessary information and knowledge with their teachers' guidance. As for the approaches of extensive reading, scanning and skimming (to select text that are worth spending time on and read carefully) ,scanning and skipping (to glance through the context rapidly to get some specific information with some part skipped) are highly recommended.

From a large number of English students, the reason they cannot understand the text is due to small vocabulary and lack of previous knowledge. An effective method to solve the problem is to enlarge students' vocabulary and enrich their knowledge by reading more and reciting more.

3.4 *Writing teaching*

In such an interpersonal communicative society, English for communicative purposes (ECP) is much more vital than English for academic purposes (EAP). Much more emphasis on people' communication ability has been put, not only in speech but also in writing.

Writing is closely linked to Reading, which provides methods for acquiring linguistic knowledge and writing style. Reading and writing are considered as a complementary process. Reading can help students develop the techniques of reading and guide them to write. So the author lists two ways of the integration of reading and writing. Firstly, summarizing provides students with valuable practice in getting the main idea and restating them in their own words. It is an effective way for students to write a summary based on the articles they have read to improve their reading and writing competences. Besides, imitation is also an important means of acquiring the skills of English writing. Generally speaking, the teachers should choose some good materials concerning with western cultures and social phenomena which can serve as the models for the students to imitate not only in writing styles but also in the construction of sentences.

In addition, it is convenient for mutual talks and giving timely feedbacks through blog. So students can reflect themselves and revise their writings through teachers' and classmates' evaluation. In foreign countries, numerous linguists and language educators have made enormous studies considering the blog as a good method of teaching and communicating. Meanwhile, students make full use of the functions of computer, such as, spelling, revision and grammar examination to improve their own writing and help teachers save lots of time in correcting unimportant mistakes to a large extent.

4 CONCLUSION

Language is a kind of cultural form. Students, who consider English as a foreign language, must know the history and knowledge of the western culture and literature .They need to know some British and American national people's living habits, cultural background, local conditions and customs and ways of life. With a good command of such information, listening comprehension will be much easier.

Besides, the students' dissatisfaction of Chinese traditional teaching methodology shows that effective reforms on English education in China are necessary and expected. Some western education concepts, such as, Communicative Language Teaching approach (CLT), learner-centered teaching method, English for

Communicative Purpose (ECP), task-based teaching strategy are greatly encouraged to be adopted based on China's teaching and learning condition.

ACKNOWLEDGEMENT

The paper is one of the results of the provincial quality project teaching and researching topic. Project name: trial point of major comprehensive reform, Code: 2013zyzm086.

About the Author: Rui Dai, graduate student degree, associate professor in the English Department of Anhui Sanlian University, visiting scholar of the Ministry of Education in Shanghai International Studies University.

REFERENCES

[1] Brindley, G. (1989). *The Role of Needs Analysis in Adult ESL Program Design.* In R. Johnson (Ed.), *The Second Language Curriculum* (pp. 14-18), Cambridge: Cambridge University Press.

[2] Brindley, G. (2000). *Needs Analysis*. In M. Byram (Ed.), *Routledge Encyclopedia of Language Teaching and Learning* (pp. 439-445). London: Routledge.

[3] Brown, J. D. (1995). *The Elements of Language Curriculum: A Systematic Approach to Program Development.* Boston: Heinle and Heinle.

[4] Chang, C-S. (2008). *Listening Strategies of L2 Learners With Varied Test Tasks.* TESL CANADA Journal, 25(2), 1-22

[5] Haiyan Zhang. *An Empirical Study of University English Writing Teaching Based on Blog [J].* Journal of Changchun University of Science and Technology, 2011, 5.

[6] Keene, E. O. (2008). *To understand: New horizons in reading comprehension.* Portsmouth, NH: Heinemann.

[7] Kemp, E. (1998). *Design Effective Instruction. Englewood Cliffs*, New Jersey: Prentice-Hall, Inc.

[8] Lefrancois, Guy R. (2004). *Theories of Human Learning*, Beijing: Foreign Language Teaching and Research Press.

[9] Liao, X. (2004). *The need of Communicative Language Teaching in China.* ELT Journal, 58(3), 270-273.

[10] Nunan, D. (1988). *The Learner-Centered Curriculum: A Study in Second Language Teaching.* Cambridge: Cambridge University Press.

[11] Ou, Yangjing & Guo, Xiaohua. (2010). *A New Exploration of College English Writing Based on Error Analysis Theory*. Foreign Language and Literature, (2), 137-140.

[12] Othman, J., & Vanathas, C. (2006). *Topic familiarity and its influence on listening comprehension.* The English Teacher Journal, 34, 19-32.

[13] Poole, A. (2009). *The reading strategies used by male and female Colombian university students.* Profile, 11(1), 29-40.

[14] Richards, J. C. (2008). *Teaching Listening and Speaking*. Cambridge, England: Cambridge University Press.

Management, Information and Educational Engineering – Liu, Sung & Yao (Eds)
© 2015 Taylor & Francis Group, London, ISBN: 978-1-138-02728-2

Research on innovation of teaching model of office advanced training

Xia Cao & Fa Hai Li
School of Electrical and Information Engineering, Hubei University of Automotive Technology, Shiyan, China

ABSTRACT: In order to improve the senior Office training effect, we proposed an innovation of the teaching model which consists of "training + level exam + contest". First, we did some revision on the teaching content and optimized it to ensure the teaching content is advanced and scientific. Secondly, we improved the teaching methods and quality teaching: we did innovation on practical teaching again and enhanced students' application capabilities. Then we implemented the innovative teaching mode and enhanced the ability to develop practical applications. Finally we made an evaluation of innovative reforms to promote further comprehensive evaluation. Practice has proved that the new teaching model ensured the realization of teaching objectives, providing an important reference for other education reforms.

KEYWORDS: Office Advanced Training; Teaching Model; Innovation.

1 INTRODUCTION

Office software family is the most frequently used office software in practical work. Now Colleges and universities, and even high school set up a basic computer application course, this course focuses on the basic operation of the main part of the Office suite of office software, word, excel, PowerPoint, outlook and other software. Students only know fewer of these functions in the computer software Application Foundation and most of the students did not learn the advanced features of Office software. So these functions are not used. With the continuous development of information technology, Office software is more and more often used. In practical work, the advanced features of Office software are always needed. Most of the jobs (marketing finance financial, administrative management, enterprises English in the non-IT Recruitment Management, Business English), requires the employer candidates familiar with the Office functions and to be quite skilled and able to take advantage of the flexibility Office related affairs. According to the graduate feedback information, our department adjusted the teaching plan, opened the "Office Advanced Training" course in management and liberal arts majors, and majority of the students highly recognized and praised this.

2 AN IDEA ABOUT INNOVATIVE TEACHING MODE

2.1 *Revision of teaching curriculum syllabus to improve teaching content*

In the revised syllabus, we should start from the application of Office advanced training courses and take advanced applications and integrated application as the main point. We establish a "student-centered, teacher-led" philosophy of education, and actively carry out the reform of teaching content. On the one hand, we teach students to master Office advanced training projects, on the other hand, we should focus on students' ability in solving daily problems though the knowledge they have learnt. Meanwhile, we take the update and integration of the teaching contents and curriculum as one, and follow office software cutting-edge technology, to introduce the new technology into the Office software Practical Teaching in time to ensure advanced scientific teaching content.

In optimizing the teaching content, we should combine professional training programs, to develop training content in line with the actual needs. First when we organize the training content we should have a global system view, to distinguish between levels, both have concrete and abstract ways to form a "three-dimensional" knowledge system. Secondly, we must tap of training in vertical and horizontal sequence organic links between knowledge. We should organize knowledge content from whole to part, from the general to the individual, progressively, from easy to difficult, from the known to the unknown. Finally, to determine the appropriate depth and breadth, to control the depth, breadth, difficulty, speed to make student receive mental potential range.

2.2 *Improve teaching methods to improve teaching quality*

Teaching methods should fully reflect the student-oriented, respect the students, develop students' ability to do a facilitator. First, we should translate the "teaching-centered" education model to a "student-centered" education model. The traditional mode of education is always put student passive and subordinate position, but just suppress the students' creativity and motivation to learn, we stressed the need to respect the student's status and rights, and fully consider the student's personality, personality dignity. Chinese students too passive operation, innovation and problem-solving skills of students with respect to advanced Western countries is relatively weak, the prevalence of high scores, low-energy phenomena [1]. Therefore, we must encourage students, using heuristic teaching, the teaching of the transition from demonstration, validation of the comprehensive training, design and innovation training, develop an interest in their own learning, active, positive, creative participation in learning as to establish a good way of thinking is more important than knowledge itself. Furthermore, it is necessary to traditional single training model into focus the development of students' personality and creative diversity training model, students' creative spirit and the desire to create, develop students' ability to adapt skills, social skills, innovation capacity.

Teaching methods' reform and innovation of "Office Advanced Training" reflect disciplined management and the characteristics of liberal arts majors, in practical teaching activities by heuristic teaching to come to succinctly and training, thus creating a harmonious, dynamic, open classroom atmosphere, stimulating students' curiosity and potential ability, guiding students in active learning. Through teacher-student interaction, extracurricular school, the network will complement the classroom teaching. Combine with modern information technology to enrich the network teaching, teaching in the form of a more comprehensive, open for students to create a self-learning environment.

2.3 *Innovative teaching, enhance application capabilities*

In people's traditional impression, for the cultivation of professional talents, mostly abstract theoretical knowledge, boring lectures [2]. Training "practical, personalized" application-oriented talents, to implement the "strong, heavy application of" the principles of personnel training, change the abstract to the concrete, becomes boring as vivid. Throughout the practical teaching, students practice the concept, skills training, internship project participants, promoters, completely out of the role of passive recipients under examination-oriented education. The new training model has been in use since the reform so far.

To manage discipline and practice of project as an example of liberal arts majors, based on professional training objectives, teaching has always been throughout the project, a top-down design teaching programs at all levels. The most advanced project is the professional-level projects are derived from the practice of project outcomes or enterprises in the domestic game. Based on the level of project analysis and dismantling, teachers designed a series of projects to support the secondary level capability required for the project, these two projects is the project contains a set of related industry background proficiency requirements, which is equivalent to a comprehensive training program, the curriculum knowledge associated organically combine to make students understand the curriculum between the organic and the associated knowledge base rather than isolated knowledge points, while a strong support level projects teaching and learning. For example, "Batch Processing Report Card" project, the students report card design, send bulk mail merge with technology to master the requirements put forward a comprehensive, integrated learning for these capabilities, and ultimately makes students proficient in computers. Requirements set closely around the entire course of social development of talent, and college "Excellence" and "industry personal training programs", education reform idea of combining theory and effectively melt in practice. After a good training for students to put their theoretical knowledge into practice to improve their ability to cultivate a sense of team spirit and cooperation for the college application delivery to the enterprise-level expertise and laid a solid foundation. This comprehensive, full of, full of learning, effectively increasing the joy of learning, improve learning efficiency, greatly enhance the ability of students to practical application.

2.4 *Combined with the experimental teaching demonstration center system, promote diversified learning modes*

Before the training course, students are supposed to download training guide book and material demonstration center teaching system through the internet, they are also allowed to preview the training target and content. Let the students enter the laboratory with questions, thus students can avoid wasting time during the training; teachers can improve teaching efficiency, improve the students' interests in advanced training course of Office, cultivate students' autonomous learning ability and innovation ability. By means of the teaching system, changing teaching method and the interaction between teachers and students, the exchange between teachers and students can be achieved a wide range, democracy and targeted stage, the relationship between the teachers and students can be equality and harmony. During teaching, teachers design teaching situation and teaching plan according to students' actual situation. This can let students immersed in the ocean of knowledge, thus out of interest, students can keep themselves in the problematical situation, began to take the initiative to solve the problem answer. Thorough training on the analysis, teachers can directly display the whole operation process and operational details, and repeated reproduction in whole process which are abstract and difficult to express in the traditional teaching, even can't express the problem vividly demonstrated. In the comprehensive training, students learn through observation, seek the solution with a problem, all this improved the students' learning enthusiasm and the aspiration to explore, so that a considerable part of the students change the way they used to just like take the medicine according to prescription into voluntarily do the research. The way of learning from the academic society, produced a qualitative leap, improve the students' ability to use knowledge and solve practical problems [3].

2.5 *With "training + grade exam + competition" mode of teaching, strengthen the practical application of ability*

"Training + Level Test + Contest" teaching mode that allows students to learn through training to apply knowledge in practice, through the National Computer Rank Examination check students' practical ability, through competition to stimulate students' creative potential through the three effective integration of students' learning ability and creative thinking skills. In the "Training Level Test + + Contest" for students to learn the process of the integrated use of knowledge, learn to think independently and solve problems, learn to take the initiative to learn.

Practical teaching as an important means of application-oriented personal training, as to cultivate students' ability to apply technical expertise, ability to analyze problems and solve problems plays an irreplaceable important role. Through the machine training, students are able to appreciate what they have learned how it should be applied to generate interest in learning, but also accumulated some basic experience to solve problems. With the basic problem-solving experience, encounter problems in the future when we can solve the problem through self-study, in order to cultivate students' ability to think and solve problems.

NCRE examination of society, which is closely coordinated with the needs of the community, welcomed by employers. Depending on its assessment of the extent and content of the different sectors of society need to use a computer designed, reflecting the high degree of integration of knowledge and practical application of the university and the community to enhance their employability and competitiveness of an important reference. In teaching, on the one hand to encourage students to take the exam ideologically courage, on the other hand, we are closely linked to the National Computer Rank Examination points, repeated training to students to improve their exam pass rate, forming a "grade exam + training"Teaching model to improve the students' ability to apply the true sense.

Subject to the game way contest to inspire students to integrate theory with practice and the ability to work independently, pay attention to the cultivation of students' awareness of innovation, science and spirit, sense of collaboration, organizational ability and competitive strength. The "ITAT contest" on the application and expansion of innovative applications have higher requirements, the contest's events "Office Automation Advanced Applications" single subject not only the game, as well as integrated application game, students can enrich the cultural life entertaining, students sense of competition in the workplace and improve professional competence. We propaganda "ITAT contest" will teach students a comprehensive understanding of the contest. Held annually in advance campus "ITAT contest", from the selection of the best players to participate in a national "ITAT contest" to form a pattern, "Training + Contest" teaching, students' interest in learning from the perspective of academic competition, thereby improving students' ability to apply and innovation.

2.6 *Innovative pricing model*

Evaluation methods no longer a simple test on the machine, but to pay attention to assess students' practical ability, so as to put the culture of teaching focus on the ability to effectively train students to analyze problems, problem-solving skills, to develop students' innovative spirit, improve their overall quality, faster and better integration with social needs [4].

Due to the practical aspects of project design based on the particularity implementation of teaching, assessment and summative evaluation process to take a combination of at the end of the student's ability to learn a comprehensive evaluation also changed the traditional model, adopt a project design based, write, on the plane, experimental, oral, contests and other forms of comprehensive evaluation of the respondent way [5]. Self-assessment and peer assessment, group self-assessment and peer assessment aspects of the teacher evaluation scores given by students during the final evaluation. Meanwhile, the Teaching Quality Management and Assurance college teaching process and results of the Ministry of Training faculties to assess aspects of the whole, to improve the practice of the next school year to prepare.

3 THE EFFECT OF TEACHING PRACTICE

Since the implementation of the reform, Office advanced training courses pilot innovative teaching mode after several years, not only fundamentally change the traditional teaching methods, but also improve the management disciplines and undergraduate liberal arts majors practical aspects, to achieve the traditional computer applications and the combination of new technologies and methods to implement a student-centered enlightening discussion, research and comprehensive teaching methods, training of the students' basic skills, to develop observation, analyze and solve problems in basic ability to develop a way of thinking particularly comprehensive analysis of the problem [6], to stimulate the sense of innovation, training and achieved good results, so Office advanced training to become a welcomed and loved by the students of the course, in the "training + rating exam + contest" teaching mode excitation and driven, self-improvement and enrich the design and made many close to the actual creative new good works, effectively ensuring the realization of the new teaching objectives.

4 CONCLUSIONS

Teaching through practice and training, basic quality of students get a comprehensive analysis of the use of knowledge and ability to get training and enhanced, they have a lot of people get the computer certificate. According to statistics, since the practice of teaching reform in recent years, our school management science and liberal arts majors, 80% of students received a National Computer Rank Examination Certificate, in addition to a number of students received the National Information Technology Application Contest certificate. After graduation, most students entering the enterprises, institutions, and has been universally praised by the employer.

REFERENCES

[1] Wang, C. & Liang, M. 2011. An innovative model of practice teaching. *Research and Exploration in Laboratory* 30(7): 152–154.

[2] Yao, X.S. & Jiao, Y.Z. 2011. Study and practice on mode innovation in experiment teaching of agricultural engineering. *Journal of Henan Institute of Education (Natural Science Edition)* 20(1): 60–64.

[3] Liu, C.H. & Zhang, H.Q. 2010. Innovation and experimental of teaching mode of laboratory opening. *Experimental Technology and Management* 27(10): 188–190.

[4] Zhu, W.Y. 2012. Innovation and exploration of experimental teaching of economics and management and its operation mechanism. *Research and Exploration in Laboratory* 31(8): 435–438.

[5] He, C.L. & Li, M.D. 2012. Teaching model of innovation and personnel training practices for IT professionals in normal university. *Journal of China West Normal University (Natural Sciences)* 33(2): 218–220.

[6] Wang, L.F. & Dong, D.L. 2011. Exploration and practice of the multi-mode multi-level electronic and electrical practice teaching model. *Research and Exploration in Laboratory* 30(3): 94–96.

Management, Information and Educational Engineering – Liu, Sung & Yao (Eds)

A study on dimensions of university teachers' research performance

F. Ren
School of Management, Tianjin Normal University, Tianjin, China

ABSTRACT: We cannot make an accurate judgment about university teachers' scientific research situation if we explain research performance from the perspective of scientific research results, and it is necessary to extend the connotation of the research performed. The paper redefines the connotation of research performance, and puts forward that scientific research performed should include not only the teacher's scientific research output, but also the actions which are beneficial to scientific research achievements. Through the questionnaire survey of 215 teachers in colleges and universities as well as the exploratory factor analysis and confirmatory factor analysis, we develop a four-factor model to reflect the dimensions of university teachers' research performance, the four dimensions are research achievements, direct research behavior, altruistic behavior and indirect research behavior. Finally, the results are discussed and further research directions are pointed out.

KEYWORDS: University teachers; Research performance; Dimensions; Behavior performance.

1 INTRODUCTION

The study of research performance management has always been the focus of academia, the evaluation of research performance and the antecedents of research performance are always hot issues of theoretical research. However, few scholars have discussed the meaning of the research performance, the vast majority of scholars believe that the research performance is a series of scientific research achievements, including research projects, published papers, academic monograph, rewards, economic value of scientific research, and so on (Giovanni et al. 2011). In fact, this is a performance point of view based on the perspective of scientific research results (Giovanni et al. 2013). With the development of the theory of performance management, more and more scholars began to realize the deficiency of performance point of view based on results. They think that the result of work will be affected by many factors, the effect of individual has been just one of them, so it will be unfair to evaluate the employee's job performance completely from the perspective of the work results, in many cases, the cause of the unsatisfactory work result is the deterioration of the external environment (Neal & Griffin 1999). At the same time, too much emphasis on the result of the work can cause employees to pursue short-term interests, and ignore the concrete manifestation of employees in the process of work, it will be unconductive to performance improvement. Because of the deficiency of result performance view, the researchers started to explain the connotation of performance from another perspective, in these studies, the concept of performance based on behavior is one of the most influential point of view, and the performance is defined as a series of employees' behavior which is beneficial to the organization. This study believes that if we understand the research performance as a series of scientific research achievements, such as projects, papers, monograph, it will be unconductive to our objective and accurate evaluation of research performance, and the teachers' subjective effort and cooperation behavior should also be incorporated into the concept and category of research performance, it is necessary to extend the connotation of the research performance. This paper uses the related theories of behavior performance and redefines the connotation of research performance from two aspects of behavior and results, and on this basis, the dimensions of research performance are analyzed.

2 CONNOTATION OF RESEARCH PERFORMANCE

The concept of behavior performance has enriched our understanding of the performance, taking the performance as the employees' behavior is more advantageous to determine the adaptability of employees to work. However, the existing strucure model of behavior performance may not be suitable for university teachers, at the same time, evaluating the university teachers' research performance completely from the the category of behavior performance and taking the research performance as teachers' behavior are inappropriate, which will be unable to fully embody the unity betwween teachers' personal and university on the development target, and ignoring the results completely could harm efficiency improvement.

Therefore, this study puts forward that research performance is result first, that is, university teachers' research output, including undertaking and participating in the research projects, published papers, science and technology works, prizewinning achievements, patents, economic value of scientific achievements, etc. On the other hand, research performance should also include a series of good behavior which will promote research achievements of themselves, others and the research team, results and behavior are equally important for the evaluation of university teachers, and they are the indispensable parts of research performance.

3 RESEARCH METHOD

Through the analysis of existing literature and discussing with the relevant experts, the initial questionnaire reflecting the content structure of university teachers' research performance was developed. Furthermore, we used this questionnaire as tool and select a certain number of university teachers for survey research. After obtaining research data, we used SPSS statistical software to do exploratory factor analysis, and through repeated screening, the dimensions of university teachers' research performance were determined. Then we used LISREL software to do a confirmatory factor analysis of the structure model of university teachers' research performance, which was determined by exploratory factor analysis, the validity of the model was tested, and finally the dimensions of university teachers' research performance were confirmed.

From the point of the present studies, most scholars thought that the research achievements should include project, paper, monograph, reward, and so on. Therefore, in this paper, on the basis of existing research, at the same time considering the opinions of the related experts, research performance should contain research projects, published papers, science and technology works, prizewinning achievements, and patents from the perspective of the result. Around the above several aspects, the questionnaire designed seven items.

Few research had payed enough attention to research performance from the perspective of behavior. In this paper, through in-depth interviews of experts, combined with the existing research, we explained research performance in three aspects under the perspective of behavior performance, including the behavior of directly beneficial to the research achievements of their own, behavior of beneficial to the research achievements of others, and behavior of beneficial to the improvement of scientific research levels of their own. Around the good behavior of the above three aspects, the questionnaire designed twenty-one items.

The questionnaire included measurements of basic respondent information, result performance, and behavior performance. Except for basic respondent information, all the items are measured by five-point Likert scales (1 = "very unimportant" and 5 = "very important").

With the help of favorable conditions of an author working in the University, we actively took advantage of colleagues, classmates, friend and other social relations, and directly asked respondents to complete the paper questionnaire or fill the questionnaire by E-mail. In this study, a total of 250 questionnaires was distributed, after a week, 233 questionnaires were collected, deducting 18 incomplete questionnaires, 215 questionnaires were used for analysis. Among the respondents, 30 years old and under, 35 people, accounted for 16.3%, 31–40 years old, 109 people, accounted for 50.7%, 41–50 years old, 33 people, accounted for 15.3%, 51–60 years old, 29 people, accounted for 13.5%, 61 years old and above, 9 people, accounted for 4.2%; bachelor, 17 people, accounted for 7.9%, master, 94 people, accounted for 43.7%, doctor, 104 people, accounted for 48.3%; teaching assistant, 5 people, accounted for 2.3%, lecturer, 96 people, accounted for 44.7%, associate professor, 80 people, accounted for 37.2%, professor, 34 people, accounted for 15.8%. From the point of distribution, the sample covers different age, educational background, professional title, subject of university teachers, and the respondents have certain representativeness.

4 DATA ANALYSIS AND RESULTS

This study adopted the method of factor analysis to explore the dimensions of university teachers' research performance. Before factor analysis, we must confirm whether the sample is suitable for factor analysis. We used SPSS statistical software to do data analysis, the results showed that the KMO value of the sample was 0.917, in Bartlett's test of sphericity chi-square value was 2729.198, significance level was 0.000, and the sample data was normally distributed. From what has been discussed above, the sample data was suitable for factor analysis. We used principal component analysis with varimax rotation to extract common factors, and selected the eigenvalue greater than 1 factors. Four common factors were extracted at the first time, and the four common factors can explain 61.700% of the variation. However, as the first using questionnaire, there existed a serious cross loadings phenomenon, the items of the questionnaire needed to be adjusted. We deleted the items of which the loadings on four factors were less than

Table 1. Remaining 15 items.

No.	Items
1	Undertaking and participating in the research projects
2	Publishing academic papers
3	Publishing academic monographs
4	Prizewinning achievements
5	Making full use of time to engage in scientific research activities
6	Insisting on overcoming the difficulties in the process of scientific research
7	Doing scientific research with honesty and preciseness
8	Always keeping passion for research work
9	Helping others to solve the difficulties in the scientific research
10	Sharing research information with others
11	Giving positive affirmation of the research achievements of others
12	Encouraging others to overcome the difficulties with scientific research
13	Constantly tracking the academic frontier areas
14	Learning new theories and research methods
15	Taking an active part in academic meetings

0.4 and the ones of which the loadings on two factors were greater than 0.4. After several adjustments, this study removed 13 items from the first 28 items. The remained 15 items are shown in Table 1.

The study continued to use principal component analysis method to extract the common factors from the remained 15 items, and the principle of extraction was still characterized by the eigenvalue greater than 1. Three common factors were extracted, however the eigenvalue of the fourth common factor was 0.961, very close to 1. Therefore, this study changed the extraction principle of the common factors and fixedly extracted four common factors. At this time, the four common factor models can explain 71.539% of the variation, and the model validity had improved significantly than previously. At the same time, the KMO value of the sample was 0.899, the result of Bartlett's test of sphericity was significant, the sample data was suitable for factor analysis. After deleting inapposite items, the Cronbach's alpha coefficient was 0.919, and the questionnaire had good reliability. In conclusion, four factors model is more ideal, and the results of factor analysis are shown in table 2.

According to the items that the four factors contained, this paper named the four factors as follows. Factor 1 is scientific research achievements, mainly including Undertaking and participating in the research projects, academic papers, academic monographs, and prizewinning achievements, and the variance contribution ratio of factor 1 is 20.327%. Factor 2

Table 2. Factor loadings of items.

Item No.	Factor 1	Factor 2	Factor 3	Factor 4
1	**0.745**	0.314	0.207	0.225
2	**0.776**	0.214	0.247	0.221
3	**0.749**	0.304	0.147	0.175
4	**0.784**	0.212	0.215	0.136
5	0.355	**0.736**	0.178	0.244
6	0.261	**0.722**	0.236	0.130
7	0.289	**0.736**	0.369	0.021
8	0.212	**0.809**	0.016	0.273
9	0.138	0.003	**0.833**	0.162
10	0.186	0.152	**0.820**	0.038
11	0.189	0.320	**0.760**	0.048
12	0.273	0.324	**0.531**	0.208
13	0.064	0.357	0.124	**0.691**
14	0.295	0.114	0.037	**0.818**
15	0.371	0.129	0.380	**0.606**

is the direct research behavior, mainly including making full use of time to engage in scientific research activities, insisting on overcoming the difficulties in the process of scientific research, doing scientific research with honesty and preciseness, and always keeping passion for research work, and the variance contribution ratio of factor 2 is 20.071%. Factor 3 is altruistic behavior, mainly including helping others to solve the difficulties in the scientific research, sharing research information with others, giving positive affirmation of the research achievements of others, and encouraging others to overcome the difficulties with scientific research, and the variance contribution ratio of factor 3 is 18.557%. Factor 4 is an indirect research behavior, mainly including constantly tracking the academic frontier areas, learning new theories and research methods, and taking an active part in academic meetings, indirect research behavior actually refers to the behavior which is conductive to the scientific research level, and the variance contribution ratio of factor 4 is 12.583%.

The structural model of university teachers' research performance, which was determined by exploratory factor analysis still needed to be further verified. We used confirmatory factor analysis method to test the fitting degree of data and model. Using the related function of LISREL 8.70, the main fitting indexes are shown as follows: the value of χ^2 was 263.71, the value of χ^2/df was 3.02, the value of RMSEA was 0.097, the value of NFI was 0.94, the value of NNFI was 0.95, the value of CFI was 0.95, the value of IFI was 0.96, the value of GFI was 0.86. In general, the main fitting indexes were satisfactory, the result of exploratory factor analysis was verified, and the four factor model of university teachers' research performance was well supported.

5 DISCUSSION AND LIMITATIONS

This study integrated behavior performance point of view, extended the connotation of the research performance, explained the university teachers' research performance from two aspects of behavior and results, and deeply analyzed the dimensions of university teachers' research performance. Using survey research method, through the exploratory factor analysis and confirmatory factor analysis, four factors model was built to reflect the dimensions of university teachers' research performance. According to the eigenvalue and variance contribution, in the order the four factors are scientific research achievements, direct research behavior, altruistic behavior, and indirect research behavior, it means that the university teachers' research performance is composed of the above four main dimensions.

In the four dimensions of research performance, the variance contribution of scientific research achievements is the highest, scientific research achievements still are the most important factor reflecting the university teachers' research performance level, research performance content includes teachers' direct results in the first place, such as research projects, academic papers, academic monographs, and prizewinning achievements. At present, the theory researchers and practitioners understood the research performance primarily based on the point of view of performance results. This study suggested that scientific research achievements were important for the evaluation of teachers' research performance, but they were not all, the behaviors which were conducive to scientific research achievements should also be part of research performance. At the same time the results of data analysis in this paper also confirmed the opinion, factor 1, factor 2, and factor 3 extracted from factor analysis reflected the behaviors which were conducive to scientific research achievements, specifically including direct research behavior, altruistic behavior, and indirect research behavior. Direct research behavior refers to the behavior of directly beneficial to research achievements of teachers, including making full use of time to engage in scientific research activities, insisting on overcoming the difficulties in the process of scientific research, doing scientific research with honesty and preciseness, and always keeping passion for research work, variance contribution rate of direct research behavior is similar to one of the scientific research achievements, and direct research behavior is an important content of research performance. In the two-dimensional model of task performance and contextual performance, contextual performance includes a series of behavior which has a positive influence on others' work (Motowidlo & Van Scotter 1994). Altruistic behavior Proposed in this paper has the similar connotation with contextual performance, and refers to the behavior of beneficial to the research achievements of others, including helping others to solve the difficulties in the scientific research, sharing research information with others, giving positive affirmation of the research achievements of others, and encouraging others to overcome the difficulties in scientific research. Altruistic behavior will not affect research achievements of their own, but it will have a positive impact on others, most behavior performance model proposed by previous researchers took the behavior of beneficial to the work of others as an important part of the performance. Finally, this study found that research performance should also include indirect research behavior. Although indirect research behavior will not directly promote the generation of scientific research achievements, it will affect personnel scientific research level and have indirect influence on scientific research achievements.

In this paper the questionnaire was used for the first time, the reliability and validity of the questionnaire were the result of a one-time test, therefore the questionnaire still needs more empirical research to verify. Secondly, the questionnaire used in this study was designed on the basis of previous studies of related issues and the results of in-depth interviews of experts, the beneficial behaviors which were conducive to scientific research achievements were divided into three main aspects, all the items of the questionnaire were all around the scientific research achievements and the above three aspects. The division way of research behavior has universal meaning, however, whether contains all the beneficial behaviors still needs further research.

ACKNOWLEDGMENT

The research work was supported by Tianjin "Twelfth Five-Year" Plan of Education Science under Grant No. HEYP5009.

REFERENCES

Giovanni, A., Ciriacl, A.D. & Francesco, R. 2013. The importance of accounting for the number of co-authors and their order when assessing research performance at the individual level in the life sciences. *Journal of Informetrics* 7(1): 198–208.

Giovanni, A., Ciriacl, A.D. & Marco, S. 2011. The relationship between scientists' research performance and the degree of internationalization of their research. *Scientometrics* 86(3): 629–643.

Motowidlo, S.J. & Van Scotter, J.R. 1994. Evidence that contextual performance should be distinguished from task performance. *Journal of Applied Psychology* 79(4): 475–480.

Neal, A. & Griffin, M.A. 1999. Developing a model of individual performance for human resource management. *Asian Pacific Journal of Human Resource* 37(2): 44–59.

Management, Information and Educational Engineering – Liu, Sung & Yao (Eds)
© 2015 Taylor & Francis Group, London, ISBN: 978-1-138-02728-2

The impact of network embeddedness to the enterprise's technological innovation performance in the environment uncertainty: Considering the mediation and adjustment

J.J. Duan
School of Economics and Management, Tianjin Vocational Institute, Tianjin, China

ABSTRACT: This paper takes innovation network theory as research basis, to explore the differentiated impact on network embeddedness to enterprise technological innovation performance in different network embeddness relationship. Analysis of network embeddedness how to influence "enterprises social capital" and "across organizational knowledge management", thus effecting the performance of enterprise technological innovation, and studies the moderating effect of external environment variables in the process of network embeddedness and network competence influencing on enterprise's technological innovation performance. It could help enterprise to understand the technological innovation performance enhancing path in a network environment, and could make enterprise to grasp the effect mechanism of network embeddedness, network competence and technological innovation performance, and could establish the theoretical foundation for enterprises to strengthen innovation network construction, to improve network competence and enhance competitive advantages.

KEYWORDS: Network embeddedness; Environment Uncertainty; Technological Innovation Performance.

1 INTRODUCTION

In the fast changing market environment, the complexity in the process of enterprise innovation increases so much, the innovation has become a complex network constructed by the interaction of many factors. The technology innovation network formed by different innovation body has become the important organization form of enterprise technology innovation activates. Some enterprises join the innovation network under the formal institutional framework, and some enterprises embedded in the innovation network by informal contact. Through formal or informal ties established contact and obtain and transfer knowledge and information. Whatever enterprises embedded into the innovation network in any way, the relationship between organizations will impact on the performance of technological innovation, and the embedded feature of enterprises is especially significant impact on the performance of technology innovation. In order to improve the performance of enterprise technology innovation, urgently needs to research the mechanism of network embedded features to enterprise technology innovation performance under uncertainty environment.

2 THEORETICAL BACKGROUND

2.1 *Embedded network and enterprise technology innovation performance*

Granovetter thought that embeddness refers to the economic action and its consequence will be impacted by agent relations and the whole network. The earliest and the most traditional classification of network embeddedness is divided into relationship embeddedness and structural embeddedness. Relationship embeddedness measured mainly from content, direction and degree of mutual benefit. Structural embeddedness measured by network size, density and the position in the network. Some scholars have carried out some research of relationship of network embeddedness and enterprise technology innovation performance. Laursen (2006) through to the survey of British manufacturing enterprises found that the enterprises who accessed resources from the network have achieved better performance than who innovate independently. Password (2007) found that relationship embeddedness could impact on knowledge obtained from network in turn impact on enterprise innovation performance. Wu&Wei based on the research on the pharmaceutical companies found that the strong coupling of enterprise network can promote knowledge transfer and improve technological innovation capability. Also the longer coupling, the more positive impacted on enterprise technology innovation by relationship embeddedness. Liu through the empirical research found that the dimensions of network embeddedness can impact on enterprise technology innovation by two types of learning. Existing research shows that network embeddedness has an important impact on the performance of enterprise technology innovation, but most of the researches are based on relationship embeddedness, the relationship of structural embeddedness and performance is relatively small, and the research, which considers the interaction of relationship embeddedness,, and structural embeddedness impact on technological innovation performance is more rare. At the same time in the research of relationship embeddedness, few scholars classify the network type which the enterprise embedded, lack of consideration about the difference between the impact of different embedded network relationship to enterprise technology innovation performance, and the research on the role of intermediate variables in the relationship of network embeddedness and enterprise technology innovation performance is still less.

2.2 *Enterprise social capital and enterprise technology innovation performance*

Enterprise capital is regarded as a kind of network resources, which rooted in the internal of network relationship and utilized through the network. The research on social capital is divided into three categories, one is considering social capital as characteristics and contact between enterprise and external entity. The second category considers the contact between enterprise internal departments as social capital. The third category integrated internal and external perspective to research the efficacy of enterprise social capital. Most scholars focus on the research into the efficacy of social capital, has made some achievements, confirmed the enterprise internal social capital has positive impact on enterprise performance, considered social capital has promoting effect to enterprise resource acquisition, it is the key to the success of the enterprise. As the importance of enterprise technology innovation increasing and deepening research on social capital, in recent years, some scholars began to study the effect of social capital to the enterprise technology innovation. Landry research the impact of social capital of enterprise innovation from innovation process. Wu&Wei research the relationship of social capital and enterprise technology innovation from the angle of coordination. McFadyen&Cannella analyzed the relationship between social capital and achievements of enterprise knowledge creation. Liu (2012) argued that social capital can help knowledge transfer in the cooperation of enterprise industry-university-institute. Hou discussed the impact of social capital to technology innovation performance based on the moderating effect of absorbing ability. Most of existing research are based on enterprise's transaction cost, innovation process and knowledge learning from which enterprise internal perspective, the perspective of external of enterprise slightly. At the same time, the existing research has confirmed that social capital has a positive impact on enterprise innovation, but did not clear the mechanism of every characteristic dimension of social capital how to impact on technological innovation performance. Most of existing research take enterprise social capital as independent variables to study the relationship with innovation performance, the research of antecedent of social capital is less.

2.3 *Across organization knowledge management and enterprise technology innovation performance*

The rapid development of economy and technology makes the knowledge and becomes the important weapon of enterprise competition, so knowledge management has been a concern topic in academia. The current research on knowledge management is divided into three schools, the first is a technical school, to analysis the importance of knowledge management from the perspective of information. The second is behaviorism school. It thinks that people's behavior can be changed and promoted by knowledge management. The third is strategic school. It thinks that thorough knowledge management can satisfy the future needs of the organization. The process of knowledge management is the process of the promotion of competitiveness. Although three schools have different definitions of knowledge management, they all consider that it is a very complicated process, the process includes recognition, collection, creation, organization, sharing, application and creation of knowledge. Some researchers described the process of knowledge acquisition, knowledge creation and knowledge storage. Then the scholars studied the impact of every segment of the process of knowledge management to innovation activities. Most of these researches based on the enterprise internal, mainly based on an atomic assumed model of an enterprise, namely regarded enterprise as individuals rather than network perspective. Because of the innovation network, make the knowledge management of an enterprise from the individual level up to the network level, from a pure internal knowledge management become across organizational knowledge management. The representative research on the relationship of across organization knowledge management and enterprise innovation such as: Wiklund think that knowledge resources in the network can improve the innovation ability of the enterprise; Smith thinks that the knowledge which acquire from enterprise external can increase the wealth of knowledge and improve the ability to create new knowledge. These researches study the direct impact of across organization knowledge management as which independent variable to enterprise innovation, considering across organizational knowledge management to be the outcome variable to investigate the impact of network elements to innovation is relatively less. Most of researches aimed at the impact of network elements to one segment of across organization knowledge management, with the one- sidedness. The research of considering the across organization knowledge management as an intermediate variable investigate the effect of network elements to innovation performance is more rare.

2.4 *The regulatory role of environmental uncertainty in innovation network*

Dynamics and evolution are an important essential feature of innovation network. Currently, most scholars research focus on the impact of the change of the network relationship of each behavior body and interact with each other in an innovation network to the other, the research on the impact of constantly updated with technology knowledge and information of external network to innovation network concerned less. In the existing references, Khoja thought that technology and market factors have an important influence on innovation network formation; Balaji thought that changing of environment can impact on the types of network relations between enterprises; Wen-Cong Ma found that environment uncertainty has regulated the relationship between product innovation and market performance; Li Wang (2012) thought that environment dynamics has positively related to the intensity of enterprise innovation. These results confirmed that the uncertainty of external environment can impact on the evolution path of innovation network, but the research concerned with the impact of its two network relationship and network structure is relatively less. The research which discussed the effect of environmental uncertainty in the relationship of network elements to the technology innovation performance is more rare. Both network relationship and network structure can directly impact on enterprise technology innovation performance, but technology innovation is greatly influenced by the external environment, put aside the effect of environment to study the relationship between network elements and innovation performance apparently is not comprehensive, scientific and practical.

3 MECHANISM OF IMPACT

The proposed theoretical model is shown as Figure 1.

3.1 *The impact of embedded network relationship type of enterprise technology innovation performance*

Enterprises to achieve technology innovation need factors such as ideas, knowledge and capital, these elements gained by network, the network between organization can be divided into technical network and commercial network which is based on connection objects (concloude suppliers, competitors, customers and research institutions), Considering the reality of the period of Chinese economic transition, the resource allocated and power approved most of the projects controlled by government departments and industry associations, so the government network must be attention. In this paper, enterprise embedded

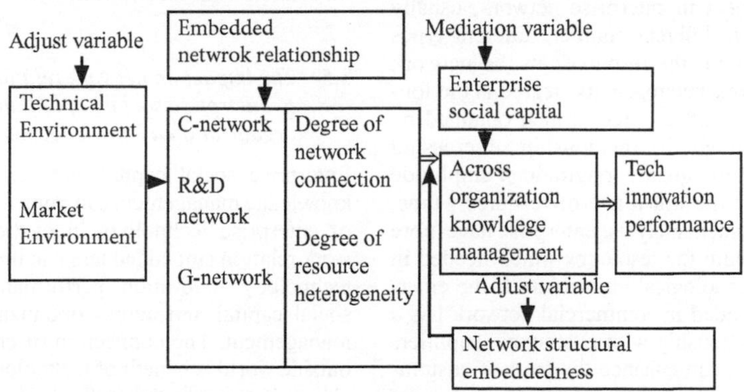

Figure 1. The proposed theoretical model of impact of embedded network relationship to enterprise technology innovation performance.

network relationship types divided into three kinds, they are commercial network, government network and R&D network. Embedded commercial network can make enterprise identify market demand better, through the communication with customers can more accurately determine the value of new technological transformation; embedded R&D network can make enterprise to realize the technology transfer faster, and to realize the innovation function by use new technology more effectively; Embedded government network can make enterprise obtain the "non-market" and special resources, and improve the reputation and legitimacy of enterprise, and to promote the innovation activities for enterprises.

3.2 *The impact of network connection degree to enterprise technology innovation performance*

Granovetter mearsured the strength of network by interaction frequency, emotional foundation, intimacy and mutual benefit, the mutual trust between the two sides is the basis of dual relationship. The resources of which enterprises innovation need obtain from the social relation network, the stronger of the connection degree of enterprise network means the more social relations it has, thorough positive communication with different subjects in the network can share information and opinions. However, different external partners provided different resources for enterprise innovation, namely in different embedded network relationship enterprise can obtain different resources. In the process of enterprise innovation, because of degree of network connection in different types network has strong or weak, will also impact on final technology innovation performance.

3.3 *The impact of resource heterogeneity degree of embedded network relation to enterprise technology innovation performance*

The different subject in enterprise network, usually has resources with different number, different types and different content, the resources in the network are so rich. Network heterogeneity degree is the longitudinal amplitude of resources which obtained by an enterprise through the relationship of crossing the level position, the greater longitudinal amplitude means the greater the difference of resource's type, quantity and content, namely the enterprise have more opportunity to obtain the resources which needed in the process of technological innovation. If the enterprise which embedded in commercial network has a stronger joint relationship with customers, suppliers and competitors, it can enhance the trust of customers, and obtain more information which customers need, and make technological innovation more guiding, at the same time suppliers can provide sufficient support when it develop the market of technology innovation products. For competitors, can avoid the possibility of malicious competition, enterprises can rapidly respond to innovation activities through obtaining imitation capability of competitors, than to improve the chance of successful innovation.

3.4 *The impact on network structure embeddedness of the relationship between network relation embeddedness and enterprise technology innovation performance*

The position of enterprise in the network determines the opportunities and constraints for actors, network structure embeddedness can describe by network scale, network density and mediation Centricity. The larger the network scale of enterprise, the more joint relationship with other subjects in the network, then build more connection, obtain more resources and easier to satisfy the demand of technology innovation. Network density refers to the ratio of real contact number to capable contact number in the network. The higher network density, the more relationships between organization, the resources and information in network flow faster. High density is helpful to develop the trust relationship, sharing information and common behavior will be increased, thereby to improve the transfer of tacit knowledge. The height of network density can impact on the number of resources and information which obtained by the enterprise from the network. Mediation Centricity means the position of middlemen which occupy on the shortest path of actors connected with other actors and connecting the two sides, show the potential control power to others. If an enterprise has higher mediation Centricity in the network, so that it is in the position of span structure hole and has information advantage and control advantage, is helpful to obtain the resources and information needed by technology innovation from the network.

3.5 *The impact on network relation embeddedness to enterprise social capital and across organizational knowledge management*

Enterprise social capital and across organizational knowledge management can improve the development of enterprise technology innovation activities, network relation embeddedness can impact on enterprise technology innovation performance by enterprise social capital and across organizational knowledge management. The connection of enterprise from the outside world is benefit of technology innovation, and also can promote the technology innovation performance. Social capital is a kind of network resources, it is based on trust, norms and relation network, it is a union of various kinds of actual or potential resources which can contribute to target. The enterprise social capital consists of structure dimension, cognitive dimension and relationship dimension. Structure dimension describes the connection structure between organizations, cognitive dimension include common language and identification culture between organizations, relationship dimension refers to relationship property of organization members, especially trust and responsibility. Enterprise obtains more social capital can promote technology innovation performance. The higher degree of enterprise embeddedness network relationship, the more contact with its partners, it is helpful to understand each other, to deepen the mutual trust, and to promote the effectiveness of communication. The higher resource heterogeneity degree of enterprise embeddedness network relationship, the stronger connection with partners in the network, the degree of trust between partners is higher, and the more communication channels, the knowledge transmission in the network more easily.

Across organizational knowledge management activities are divided into four processes, they are knowledge acquisition, knowledge transfer, knowledge sharing and knowledge application. This paper regards organizational knowledge acquisition and across organizational knowledge integration as the two dimensions of cross organizational knowledge management. Across organizational knowledge acquisition refers to enterprise obtained valuable knowledge from other organizations by itself network relationship, this is the foundation of enterprise to carry out technical innovation activities. Across organizational knowledge integration is the dynamic process of combining; integrating and refining the external knowledge resources according to enterprise own network position and network relations. The knowledge obtained by enterprise only used through integrating new knowledge system, knowledge through integrating make the enterprise have competition ability which can not imitate. Across organizational knowledge management can improve the performance of enterprise technology innovation. The higher degree of embedded network relationship, the closer the relationship between enterprise and other subjects in the network, the higher the frequency of knowledge exchange, and the greater chance to make up for own knowledge gap. The higher degree of resource heterogeneity of embedded network relationship, enterprise is easier to establish the mechanism of information sharing and trust with other organizations in the network. Thus realize the sharing and transferring of visualization knowledge to obtain the core competitiveness, at the same time realize the transferring tacit knowledge by cooperation with other organizations.

3.6 *The impact on relationship between environment uncertainty and relationship of embedded network to enterprise technology innovation performance*

Environmental uncertainty refers to enterprise's external environment dynamics in the process of technology innovation activities, it can be divided into technical environment and market environment, including the degree of disorder market, the degree of market competition and the degree of technical progress. The innovation network of an enterprise is dynamic; it is closely related to the change of technical environment and market environment. Under the impact of rapid change of technical and market environment, the innovation network of enterprise needs continuously update the network behavior, relationships and members guarantee the injected fresh blood to the enterprise innovation. As the change of technical and market environment, enterprise needs more external members to establish relationship in order to promote the degree of network connection, the degree of resource heterogeneity of embedded network relationship will also increase, which impact on performance of enterprise technology innovation.

4 CONCLUSIONS

In this paper, research on the mechanism of network embeddedness for enterprise technology innovation performance, according to different subjects in the network classifying embedded network relationship. Emphasize on in the different embedded network relationship, the degree of the network connection and degree of resources, heterogeneous has a different impact on enterprise technology innovation performance. At the same time introducing embedded network structure into the model analyses the adjustment effect of this variable. Considering enterprise social capital and across organizational knowledge management as intermediate variable, pointed out its intermediary role in the relationship between embedded network and enterprise technology innovation performance. Considering the impact of the external environment to technology innovation, introducing the environment variable into model, discussed the impact of embedded network to technology innovation performance under external environment. It could help enterprise understanding the technological innovation performance enhancing path in a network environment, and could make enterprise grasp the effect mechanism of network embeddedness, network competence and technological innovation performance, and could establish the theoretical foundation for enterprises to strengthen innovation network construction, to improve network competence and enhance competitive advantages.

ACKNOWLEDGEMENTS

The research work was supported by Humanity and Social Science Youth foundation of the Ministry of Education of China NO.13YJC630028, Humanity and Social Science of Tianjin, China NO.TJGL12-125 and cultivation project of the Tianjin Vocational Institute.

REFERENCES

Fang. L 2012. The emprical research on effect of social captial to performance of I-C-U knowledge transfer. Management of research and development, 24(1):103–111

Laursen K & Salter A. 2006. Open for innovation: the role of openness in explaining innovation performance among U.K. manufacturing firms. *Strategic Management Journal*, 27:131–150

Li. W 2012. Manager decision preference, environment uncertainty and innovation strength: based on the emprical research on Chinese enterprises. *Journal of Science of Science and Management of S.&T.*, 30(7):1101–1109

Password F 2007. Learning to reduce interorganizational learning:an analysis of architectural product innvoation in strategic alliances. *Journal of Product Innovation Management*, 24(4):369–391

Management, Information and Educational Engineering – Liu, Sung & Yao (Eds)
© 2015 Taylor & Francis Group, London, ISBN: 978-1-138-02728-2

Strategic analysis of the entrance of traditional ethnic sports into classrooms of higher educational institutions—taking Guangxi of China as an example

Z.F. Huang & Y.Q. Wei
Institute of Physical Education, Hechi University, Yizhou, Guangxi, China

ABSTRACT: Traditional ethnic sports possess distinguishing characteristics, excellent functions, and extensive popularity, at the same time its performing, competitive, and entertaining features are outstanding. Its entrance into the classrooms of colleges and universities will highlight the regional and ethnic characteristics in schools, allowing the students to grab a deeper understanding of ethnic cultures and accelerating the development of traditional ethnic culture. Its entrance can be carried out in the following aspects: first, assert the specific goal of exploiting traditional ethnic sports course; secondly, analyze the contents of traditional ethnic groups; thirdly, study the methods of exploiting and utilizing the traditional ethnic sports resources in Guangxi; finally, realize the rapid and favorable development of traditional ethnic sports course in higher education in Guangxi through real case analyses.

KEYWORDS: Traditional ethnic sports; classrooms of higher educational institutions; strategies.

1 INTRODUCTION

Currently, the entertaining and social functions of traditional ethnic sports tend to be bestowed with new educational and exercising values which conforms to modern education and these two develop beside each other, infusing robust vitality into entertainment and exercise of modern society. The values of traditional ethnic sports are attracting attention of experts and educators, transplanting to schools and entering classrooms. It breaks out of a sub-cultural ideology and integrate into an optimizing education system on a daily base. So far, there have been traditional ethnic sports course set up in some colleges and universities in Guangxi, which demonstrates the P.E. philosophy of certain institutions, and goes along with the running traits and goals of regional higher educational institutions. The establishment of traditional sports course has broken through the limitations of P.E. teaching in higher educational institutions, and shows the substance of physical education—healthy and lifelong sports. It has optimized the teaching contents, provoked students' learning interests, put forward ethnic cultures, and cultivated the students' ability to practice lifelong sports. In addition, it has cultivated student's practice skills in multiple aspects, and explored a new way of implementing education reform in higher educational institutions concerning regional and ethnic situations, and it provides precious referential and practicing values for education reforms in local colleges and universities.

2 DEVELOPING STRATEGIES OF CARRYING OUT REFORMS IN TRADITIONAL ETHNIC SPORTS COURSE IN HIGHER EDUCATIONAL INSTITUTIONS IN GUANGXI

2.1 *Asserting the goal of exploiting traditional ethnic sports*

Educational theories hold that cultivating goal is the core and essence of a curriculum. It is not only the embodiment of education proposals, but also the starting point of the whole curriculum; it is at the same time the fundamental standards of reforms in curriculum structures, teaching contents, and assessments of certain courses. The setting of teaching goal is a dynamic process, which should conform to the needs of developments of modern times, correspond to the special law of physical education in colleges and universities and the law of developments of college students mentally and physically, uphold the principle of "health first", and enhance the healthy growth and development of college students. It should stimulate students' interests in exercising, and inculcate in them the sense of lifelong sports; it should put students' health at the center, and attach importance to the major role of students; it should pay attention to individuality and different demands of different students, assuring that every student is benefited through the course. We should set up our teaching goals, according to the above perceptions.

2.1.1 *Goal of knowledge*
Students should handle the knowledge of scientific exercise and the fitting knowledge of traditional ethnic sports. As a curriculum of regional and ethnic characteristic, traditional ethnic sports should exert its functions on enriching students' experiences, rendering them abundant practice and everyday skills, and cultivating their sense of lifelong sports. Also, through this curriculum, students ought to further their understanding of ethnic cultures.

2.1.2 *Goal of exercising skills*
Students should obtain knowledge of traditional ethnic sports; learn and practice the basic modes and skills of the sports. Moreover, they are supposed to handle at least two traditional ethnic sports events and develop their individual advantages.

2.1.3 *Goal of mental and physical health*
Students should understand the positive influences of traditional ethnic sports on mental and physical health, understand correctly the connections between the ethnic sports and dignity and self-confidence. They are supposed to use appropriate methods to adjust their temperaments and optimize their emotional conditions, get over their psychological obstacles, and polish their persistence.

2.1.4 *Goal of adaptive capacity of society*
Students should establish good relationships with others, demonstrate excellent sports ethics and cooperative spirits, and properly balance competition and cooperation.

2.1.5 *Goal of Comprehensive ability development*
They should be able to correct assessment of health and physiques conditions; manipulation and establishing self-exercising plans; cultivation of autonomous exercise and lifelong exercise; acknowledgement of competing modes of traditional ethnic sports; the capacity of appreciating and commenting on the traditional ethnic spots competition.

2.2 *Analysis of traditional ethnic sports events in Guangxi, China*

In the chronicle practice and life experience, the various minority groups in Guangxi have created and developed colorful traditional ethnic sports, which adequately demonstrate the people's spice of life, featured in regional and ethnic characteristics. These events display the differences and rich customs, lifestyles, faiths of different ethnic groups, exhibit the relics of different social formations, and show the distinctive geographical and folk features. In the light of cultural anthropology, these various kinds of sports are closely related to reproduction, labor, and military activities, which are rich in cultural connotations. Their entrance into schools will infuse a fresh atmosphere into schools. The exploitation of traditional ethnic sports will contribute to the establishment of curriculum with regional characteristics. In the process of exploitation, some courses should be transformed instead of directly taken into effect, some need not. The transformation aims at adapting to the physical and mental characteristics of the students. It is mainly about the simplification of rules and skills, the reduction of difficulties, and transformation of facilities. The traditional ethnic sports events in Guangxi are divided into two major categories (as shown in Chart I), the first group consists of primarily entertaining events, including performing, dancing or playing events, which are intended for leisure, entertainment, and fitness. Their rules, yet exist, are quite loosen for their gist remain to be recreated. The second group are competitive events, which focus on the competition of physical power, skills, and techniques.

Chart I. Classification and percentage of different categories of traditional ethnic sports in Guangxi (n=283).

Categories	Competitive	Performing	Dancing	Playing
Number	47	123	49	64
Percentage	16.6%	43.5%	17.3%	22.6%

There are rich resources of traditional ethnic sports in Guangxi, which boast distinctive features and excellent exercising values. Schools can perfectly select these events in their courses according to local situations and students realistic needs. The course selection should embody the conformity of subject characteristics of traditional sports and student's practical life; the inheritance of ethnic physical cultures, and the conformity of the effectiveness of physical education and interests of students.

2.3 *The methods of exploiting and utilizing traditional ethnic sports*

Guangxi populates multiple minority groups, whose traditional sports are quite different from each other in their forms and contents. These events not only comprise the gorgeous ethnic cultures, but also become a significant way of exercising. Traditional ethnic sports events are important resources of P.E. course, whose exploitation can not only enrich the teaching contents of higher educational institutions, but also promote the inheritance of ethnic cultures, so they have infinite exploiting values.The methods of exploiting are as follows:

2.3.1 *Filter*

Filter refers to selecting suitable courses according to certain standards among numerous events. It is a preferential process of analyzing and judging traditional ethnic sports depending on the teaching goal. Firstly, we should assess the values of certain traditional ethnic sports events, concerning its contribution to students' physical development, acquisition of physical knowledge and skills, sufficient experience in sports,cultivation of good morality, and its effects on the inheritance of ethnic cultures; then, we should analyze their contribution to the accomplishment of teaching goals; finally, we should estimate their feasibility from real study conditions,i.e., analyze the possibility of students' internalization of the knowledge according to their mental and physical conditions; guarantee the course is moderate in its level of difficulty; evaluate the standards of required hardware facilities such as sites; and weigh the probability of their development into a whole security system of teaching, managing and organizing.

2.3.2 *Transformation*

Transformation means processing, altering and modifying some elements of existent traditional ethnic sports events in the view of specific objects and conditions in its tangible implementation. It is actually one that innovates and restructure the traditional ethnic sports resources. After the transformation, those events, though retaining some of the old characteristics and elements, have been transformed substantially. There are abundant traditional ethnic sports resources in Guangxi waiting to be intensively exploited. Institutions can creatively alter them into courses welcomed by students in order to satisfy and adapt to the needs of students. Specific methods of transformation are varied in forms, primarily including simplification of rules(reduction of competition time, minimization of competition terrain, and decrease of body contact), simplification of techniques and tactics(lowering the requirements of techniques and tactics, and reserve simple and basic ones), reduction of the level of difficulty(reducing that of sports, motion, and letting go of the details of motion), transformation of facilities(reducing their weight and changing their functions). The contents of transformation should be provided with pluralism, may be functional or structural; it can aim at single or several or even the entire elements of traditional ethnic sports; it can be general and systemic transformation, or partial change. There are four fundamental principles to take into account of the transformation: the principles of interestingness and playability, education and culture, adaptability and feasibility, practicability and closeness to life. The events transformed should serve to stimulate students' interests in studying, cultivate students' interests and habits of doing sports, improve the general health of "mentality, physiques, and social adaptability", and inculcate in them the sense of lifelong sports.

2.3.3 *Integration*

Integration is to organically combine the elements of traditional ethnic sports in line with certain means and bring along a new P.E. Course system. The levels and methods of integration are of various kinds, which may lead to integration in space, in functions, in structure, or in the elements. It can carry on among sports resources of the same categories, for example, among knowledge resources; it can also carry on among those of different categories, for example among knowledge and physical exercising resources. Moreover, integration can be interdisciplinary, such as integrating P.E. with military, dancing, music,medicine,etc.The methods of integration are diversified: we can integrate single events together, or comprehensively integrate multiple events together. The key of integration is to extract the essences.

3 REAL CASE ANALYSIS: THE EXPLOITATION AND UTILIZATION OF TRADITIONAL ETHNIC SPORTS IN GUANGXI–FIREWORK GRABBING

Guangxi University for Nationalities is one of the largest comprehensive university in Guangxi, with over 20,000 students, of whom students from minority groups take up a considerable proportion. As a university in ethnic region, the institution always pays close attention to the practice of traditional ethnic sports with regional characteristics, especially the development of students' consciousness of exercising and their all-round improvement through the practice. We rely on the key education department project in the 11th Five Plan–To Give Full Play to the Advantages of Traditional Ethnic Sports and Promote Reform of P.E in Higher Educational Institutions of Ethnic Region, discover breakthrough in the exploitation and utilization of traditional ethnic sports in view of the real situation of our school, and establish it to be the special cause. Owing to the joint efforts of teachers and students of the whole university, we have entered the initial stage of a course system characterized with the ethnic sports event–firework grabbing.

3.1 *Goals of exploiting traditional ethnic sports*

3.1.1 To stimulate students' interests in traditional ethnic sports, encourage them to take an active part in the sports, strengthen their physiques, cultivate their good habits of exercising, and facilitate the balanced development of their mental and physical health, and the social adaptability.

3.1.2 To make students get hold of basic knowledge and practice skills of traditional sports, and cultivate an active habit of doing sports, and enrich their practice and life experience.

3.1.3 To cultivate good psychological qualities, active and optimistic attitudes, good moral characters and healthy lifestyles of students.

3.1.4 To give full play to regional geographic advantages, highlight the institutional characteristics and individuality, and put forward the culture of traditional ethnic sports.

3.2 *The practical steps of traditional ethnic sports exploitation*

The first stage: preparation for exploitation. We should learn documents and theories related to course reform in physical education, establish an organization and confirm the number of staffs to participate and carry out training of them. We are supposed to analyze the traditional ethnic sports resources, make clear the goals of exploiting goals, display the characteristics of the university, draw up an exploiting plan, and collect relevant information.

The second stage: practice of exploitation. The university should determine the specific division of labor and responsibility of our staffs, making sure that the practice is carried out promptly, effectively, and orderly. The staffs should analysis and sort the data in correspondence to their own obligations, and carry out the practice according to the practice plan.

The third stage: Summary and utilization. The data connected to the research should be analyzed, sorted, integrated and refined, and finally experience should be summed up. Firework grabbing ought to be popularized in the scale of the entire school. P.E. courses should guide the students to acquire the basic knowledge and skills of traditional ethnic sports and cultivate them to be fans of the sports.

3.3 *Analysis of the contents of firework grabbing course*

Firework grabbing is a traditional ethnic sports events, which is initially popular with Zhuang' Dong, and Molao people and enjoys an enduring prosperity of several hundreds of years. There is a strong, distinctive ethnic ring to it, along with its competitiveness, entertainment and antagonism, which brings along with it the reputation of "the oriental football". It is prevalent among folks, particularly teenagers in Sanjiang Region. Traditional firework grabbing doesn't confine the number of participants or specific sites; everyone can participate in every round, and three rounds usher in the termination. The sites of this event are arbitrary: usually river basins or hillsides, and there is no boundary of the sites. To reinforce the development and inheritance of this event, the Athlete Committee of Guangxi has organized exploitation and sort of it, and altered the rules and practice methods in order to promote its implementation among all the people.

Firework grabbing simply has a moderate requirement of the sites and facilities: basketball or soccer courts can do. And there is no strict limitation on the number of participates. It is fit to carry out in higher educational institutions in Guangxi. The event not only build up the students' bodies and develop their physiques, but also exert positive effects on the cultivation of their morality and capacity of observation, as well as the sense of collectivism. The practice of firework in schools should focus on its functions of exercise and entertainment. It can be adapted according to the particular local conditions and individual differences. Training conditions or facilities permitting, schools can organize representative teams and take part in competitions, in order to strengthen the physiques of the students and enrich the cultural life on the campus.

3.4 *Transformation and integration of firework grabbing*

3.4.1 *Transformation*

We substitute a ruby ring 5cm in diameter for the original "firework", an iron ring 50cm in diameter wrapped with a piece of red cloth or silk.

As for competition terrain, it can be flexible in size according to original and newly-set rules to adapt to the teaching of the schools and competitions.

3.4.2 *Integration*

We integrate the ways of playing firework grabbing with those of football, handball and soccer, so as to provoke students' interests in it.

4 SUMMARY

The exploitation and utilization of traditional ethnic sports resources should be in line with physical educational goals of higher educational institutions, the characteristics of students' psychological and physical conditions, and objective situations. The principles of fundamentality, practicability, ethnic traits and scientificity should be followed. The courses should embody the conformity of the subject characteristics and students' real conditions, the transmission of traditional ethnic sports knowledge and practical skills and its contribution to cultivating skills, temperament, values and virtues, thus realize the unification of the effectiveness of physical education and students' interests in learning it.

ACKNOWLEDGEMENT

Study of Establishment and Practice of Characteristics Educational System in Colleges and Universities of Ethnic Regions, key funded project of Guangxi higher education curriculum reform project (Project code:2013JGZ154).

REFERENCES

[1] Zhao Ming, Hu Xiaowen , Reflections on Developing Ethnic Sports[J], Sports and Science, 2000(3):18–20.

[2] Liu Jingnan, Reflection on the Establishing Traditional Ethnic Sports as Required Course[J], Journal of Guangxi College for Nationalities, 2002,6(2):55–58.

[3] Wu Yong, The Study of the Current Situations of the implementation of Traditional Ethnic Sports Courses in Colleges and Universities of Hubei Province[J], China University of Geosciences,2013,05,01.

Management, Information and Educational Engineering – Liu, Sung & Yao (Eds)
© 2015 Taylor & Francis Group, London, ISBN: 978-1-138-02728-2

The current situation in curriculum settings of physical education major and its future conceptions in new undergraduate colleges and universities—Taking new undergraduate colleges and universities in Guangxi province of China as an example

Z.F. Huang & Y.Q. Wei
Institute of Physical Education, Hechi University ,Yizhou , Guangxi, China

ABSTRACT: With the development of society, the monotonous training goal in new undergraduate colleges and universities in China cannot meet the need for social development, the lag of curriculum setting divorces from the market requirement, the imbalanced curriculum system structures, the repetitive and complicated teaching contents, the equivalences of optional and compulsory courses and the unformed patterns of discretionary diverse course selecting, which causes the students' unprofessionality. The practical teaching system is incomplete and in the meantime, the students' abilities of practicing, innovating and pioneering are weak. The conception of curriculum setting which is oriented to social needs is put forward in new undergraduate colleges and universities through diagnoses. It is proposed that the training programs are supposed to be reformed according to social demands for those physical education majors in new undergraduate colleges and universities and the curriculum settings to be foresighted and flexible. Meanwhile, more emphasis should be laid on guiding career planning for students majoring in physical education, in order to combine the training goal, the social demand and the expectation of students into one.

KEYWORDS: Physical education major; curriculum setting; current situation;conception.

1 INTRODUCTION

With the continuous enlargement of higher education in China, it becomes increasingly difficult for employment of college students, especially for those students from new undergraduate colleges and universities. 'Unemployment upon graduation' has now become a severe problem for college graduates and it is more prominent for those physical majors. According to the statistics from MyCOS, physical education ranks to the sixth as the most lightly regarded career prospects in 2011. In 2012, physical education was warned by a red card because of its low employment rate and wages. For a long time, China's physical education talents mainly serve for physical education career and competitive sports career, whose training patterns are monotonous and the feature of specialization and education is marked. However, they are lack of socialization and popularization. The maladjustment between teaching contents and social needs, the untight connection between applied talents training and the sports industry and the disjoint of talent cultivation and local economic development, which result in the monotonous employment of physical education students. With the saturation of various kinds of physical faculties, this kind of contradiction is more intense. However, those fitness industries which are relevant to physical industries are now springing up in a large scale. The number and scale of health clubs are sharply enhancing and those quality development and field survival training programs are highly regarded. In contrast with the keep-fit fever, the development level and speed of trainers comparatively lag. On the one hand, students major in physical education are confronted with the difficulty of employment. On the other hand, there is a great demand for talents of fitness industries. The national medium and long-term plan for education reform and development (2010-2020) pointed out that new undergraduate colleges and universities are supposed to firmly establish the conception of serving of social on their own initiative, combine the production, study, research and use together in order to create a new mechanism of talents cultivating between colleges, research institutes and enterprises. Therefore, reforming talent training and broadening the caliber of physical education to adapt social needs are the most urgent issue at present. The caliber of physical education to adapt social needs are the most urgent issue at present.

2 THE EXISTING PROBLEMS OF CURRICULUM SETTINGS ON PHYSICAL EDUCATION SPECIALTY IN NEW UNDERGRADUATE COLLEGES AND UNIVERSITIES

Training goal is the starting point and home to all educational activities, which concentrated reflects some social need and the basis of curriculum setting. In order to reform the curriculum setting, the first step is to make a scientific and clear position of the talent training goal. At present, Chinese physical education specialty in colleges and universities continue to use the same training goal as the ministry for national education issued in 2003 called 'The Physical Education Curriculum Program in National Colleges', which has implemented for 10 years up to now. During the ten years, social needs have a sea change, which causes Chinese higher education changing from the monotonous elite education to the diversified one and elite education is gradually replaced by mass education. But the training goal, curriculum setting and the teaching period of physical education specialty have no changes at all. The widening enrollment, the general decline of students' physical fitness and the fast-changing social needs are all conflicted with curriculum settings of physical education.

2.1 *The monotonous training goal is difficult to adapt the demand for social development*

At present, the training goal of physical education specialty in new undergraduate colleges and universities is to cultivate talents who are qualified to school physical education and competitions and also those inter-disciplinary talents who can work on physical research, management and social guidance. The training goal is mainly aimed at training gym teachers. However, there is a trend of saturation of gym teachers in middle school and high school in the developed area in China. Owing to the success implement of single-child policy and the improvement of economic conditions in the countryside, the number of primary and secondary students is gradually declining, which result in the merger of many schools. Though physical teachers in primary and secondary schools in the countryside are in shortage, the phenomenon of part-time job is quite common. Along with the negligence of physical courses and the authorized strength of schools, it forced that the area of employment of physical education talents cannot limit in teachers. In the survey of the employment situation of physical education graduates from Hechi College in Guangxi province in 2008–2015 (as shown in Table 1), there is a diversified trend of employment form and channel. A large part of talents from physical education engage in administrative institutions and enterprises or some would like to be a freelance, which make the employment scale much larger as the demand for employment expands. As Hu Jintao talked on the 100th anniversary of Tsinghua University, the focus of comprehensively improving quality of higher education is to serve for economic and social development. Therefore, new undergraduate colleges and universities have to change their former training goal of merely training gym teachers. Only in this way can they provide the strong backing to the employment of physical education graduates, can they serve for economic and social development. Otherwise, it will only cause the waste of resources and talents.

2.2 *The detachment between curriculum setting and market demands*

The main improving project of physical education specialty in 2003 was track and field, basketball, volleyball, football, gymnastics and martial arts, which was the basis of all physical projects. However, with the improving of people's living standard and the continuous progressing of socialization of physics, the mass physical projects aimed at body building, relaxation and entertainment become more popular among people. The Taekwondo coach, tennis coach, training partner, sports dancing coach, health club coach, fitness instructor and quality development trainer are in great demand. But the physical education specialty in the six new undergraduate colleges and universities in Guangxi province, these courses are not set. The traditional sports such as track and field, volleyball

Table 1. Statistics of employment area of physical education graduates in Hechi College in the past five years.

Session	Gym teachers or other relevant industries	Administrative institutions and enterprises	Non-public enterprises	Freelances	Host graduate	Recruitment
2008	31.25%	6.25%	52.65%	6.25%	4.10%	0
2009	57.33%	2.66%	22.86%	6.66%	4.00%	6.49%
2010	51.94%	10.38%	23.41%	3.89%	3.89%	6.49%
2011	60.02%	7.95%	10.22%	9.09%	1.16%	11.36%
2012	63%	4%	12%	0	0	21%
2013	38.4%	3.4%	16.2%	1.2%	1.2%	39%

and gymnastics are so professional that students from physical education specialties can only engage in physical education career. But as one of the main contents of curriculum setting, all students are required to study these courses, which results in the disunion of curriculum setting, students' expectation and social needs.

2.3 *The unreasonable curriculum system structures and the repetitive and complex teaching contents cause students as unprofessional*

First, curriculum arrangement for primary and secondary is unclear. The curriculum scheme of physical education specialty in 2003 stipulated the total course period was 2600-2800 hours, including 720 hours of public elementary courses such as Marxist philosophy principle, college English, basic of computer engineering and basis of law and 2000 hours for specialized courses. Students have to study around 20 main courses and general compulsory courses, which makes the students' lack of energy and be worn out. Over the years, the physical education specialty has advocated mastering many skills while specializing in one, in order to cultivate students to universal geniuses. However, the unclearness for primary and secondary causes the insufficient period of main improving projects. Many colleges and universities do not arrange main improving projects until the fifth semester. But in the seventh semester, the education internship is arranged for students. Therefore, it only costs one year for a main improving project studying, which results in the fact of students as unprofessional.

Secondly, there are many repetitive contents of theory courses, causing unnecessary waste of course period. For instance, repetitive teaching contents appear in School Physical Education, Sports Teaching Theory and Middle School Sports Teaching. And students are confused with many complex course names such as Sociology of Sports and Sports Social Science.

Thirdly, with the increasing enrollment scales, teachers are in shortage in colleges. There is a little connection between physical class teaching in primary and middle school and the teaching contents of physical education specialty. Students are almost blind to the health curriculum in primary and middle school, which makes the employment of graduates much harder. The transplant of competition events is obvious in the contents of compulsory courses. For example, Aerobics is one of the most popular courses in fitness clubs and meanwhile, it is one of the main courses in physical education specialty. However, as teachers are busy with teaching, which causes their unknown about the trends of Aerobics in fitness clubs. So the employment of graduates cannot perfectly joint with enterprises.

2.4 *The pattern of selecting course arbitrarily has not been formed*

The survey of the six new undergraduate colleges and universities in Guangxi province shows that owing to the limitation of site, equipment and faculty, the specialized optional courses are replaced by restricting optional courses. Roller-skating, swimming and rock-climbing these leisure sports fail to open in time. Those popular arbitrary optional courses are not completely opened such as Marketing, Public Speaking, Dealing Skills, Modern Social Etiquette, Job Hutting Skills and Communication Skills etc.

2.5 *The practice teaching system has not well developed and students' ability of practicing, innovating and pioneering are weak*

The high practicalness is one of the features of physical education. In order to form its own characteristics, the step of professional practice teaching must be strengthened. And this could be achieved through the combination with teaching and social needs. However, the teaching system at present in new undergraduate colleges and universities in Guangxi province is not well developed, which is mainly reflected in the aspects as follows:

First, the step of curriculum practicing is imperfect. For instance, courses like Sports Anatomy, Sports Health Care, Sports Training are supposed to combine theory tightly with practice. But owing to the limitation of faculties and equipment, students have few chances to operate by themselves during the teaching process, which cannot motivate students to learn and apply their knowledge to all.

Secondly, the step of education internship is also not well developed. With the increasing enrollment scale, much internship of physical education students is a mere formality. The survey of 280 physical education graduates of 2013 from the six new undergraduate colleges and universities in Guangxi province shows that there are 40% students do not attend the education internship. Although 4 of them arranged the internship, a large number of students found an internship by themselves instead of attending the unified internship.

Thirdly, the training step of innovating and pioneering is weak. Because the employment fields of physical education students are not limited in physical teachers, which means a large number of them will engage in some other administrative institutions or enterprises or be a freelance. Therefore, it is required that physical education courses should be tightly connected with social needs. However, owing to the poor supervision, there is little impact on practical activities, though colleges and universities arranged them every summer vacation. Again, the cooperation between colleges and enterprises is not enough, which causes the undeveloped mechanism of the evaluation for practical education.

3 THE CONCEPTION OF CURRICULUM SETTING WHICH IS ORIENTED TO SOCIAL NEEDS IN PHYSICAL EDUCATION SPECIALTY.

According to the training goal and the current situation of physical education, conceptions in the curriculum setting of physical education are as follows:

3.1 Change the teaching pattern of basic theory courses and make theory tightly combined with practice. Simplify and compress those repetitive, complex and theoretic courses such as The Introduction to Sports, Physical Education of School, Sports Teaching Theory, Sports Teaching Method etc.

3.2 Simplify the optional course to let students quit those unsuitable courses according to their own cases.

3.3 Set up the main improving projects when students are in grade one. After four years studying, every student will have a good command of skills and this will polish their career development.

3.4 Make the career planning when as a freshman and make adaption and revise during the study. Confirm the employment direction in grade three and adapt to the development of society. Set different curriculum according to different employment directions, for instance, the direction of advanced study, civil servants, physical teachers, health clubs, quality trainers, cross major employment, self-employed.

3.5 According to the school's characteristics and social needs, set up some innovative optional courses and the pattern of training talents both for social and educational. And the students' ability of practicing, innovating, social adaption can be improved in this way.

3.6 Set up the optional courses on cross-major, widen students' knowledge and improve their comprehension abilities. For example, Economics, Marketing, Management, The Common Sense of Administration, Dealing Skills, Public Speaking, Document Writing Skills, Modern Social Etiquette, Job Hunting Skills and Communication Skills.

4 ATTENTION SHOULD BE PAID DURING THE IMPLEMENTATION OF CURRICULUM SCHEME

4.1 Strengthen the step of practicing. Combine social with classes to improve students' ability of practicing and job hunting.

4.2 Schools are supposed to cooperate with enterprises such as inviting gold medal trainers to give lessons to students, cooperating with health club coaches or letting students intern in some health clubs. In the meantime, increase periods of students' internship and teaching observing.

4.3 Attach importance to optional courses of cross-major. The university education is not only for vocational education, but also can educate people. And one's quality can be improved through these courses. His spirit can be cultivated and moral quality will also be perfected in order to be a better man.

5 CONCLUSION

The main purpose of education in colleges and universities is to cultivate graduates who adapt to social needs, which must do research in deep and reform training schemes according to social needs. The training goal is supposed to be foresighted and the curriculum setting to be flexible. Meanwhile, lead physical education students treating their career plan rationally so that they can understand the matching between career requirements and their own qualities and also, their striving directions. Only in that way can they adapt to the rapid development of society and promote the employment rate.

ACKNOWLEDGEMENT

Phased objectives of the higher education reform project in Guangxi, 2014–Build a New Personnel-cultivation Mechanism of the 'Integration of Society and Class' in the P.E. Education Faculty of Newly-founded university in Guangxi (Project code:2014JGA210).

REFERENCES

[1] Top of College students' ability of employment in 2011 [EL/ OL] http: / /ca-reer.eol.cn.2011.

[2] Chang Guochao: The Study of Current Training Pattern of Physical Education Students in Henan Province [D]. Henan University.

[3] Fang Qianhua, Huang Hansheng, The Evolution of Curriculum Setting of Physical Education Since Reform and Opening up[J]. The Journal of Physical Institute in Xi An 2006(1).

[4] Tian Ying, Zheng Yuxia, Li Fengjuan The Establishment of Practical Teaching System of Physical Specialty in Colleges and Universities [J]. The Journal of Shenyang Normal University 2011(1).

Management, Information and Educational Engineering – Liu, Sung & Yao (Eds)
© 2015 Taylor & Francis Group, London, ISBN: 978-1-138-02728-2

The past meets the future: The inheritance of ethnic opera and reform of opera teaching in local college—taking the "Mulam Opera" in Luocheng autonomous county as an example

Hai Yan Wei
Institute of Art Education, Hechi University, Yizhou, Guangxi, China

ABSTRACT: The opera teaching in local colleges shoulders the significant task of inheritance and preservation of ethnic opera, responsible for 3 major tasks: the cultivation of inheriting personnel, the study of ethnic opera, and the training of new blood. Confronted with the almost extinction of ethnic opera such as Molao Opera, local colleges should do better inheritance and preservation by carrying forward education reform, establishing educational opera teaching workshops, intensifying the construction of bipolar teaching groups, building training center and practice base, etc.

KEYWORDS: Ethnic opera; college education reform; cultural inheritance.

Located in northwest mountain region of Guangxi Autonomous Region, LuoCheng is the major settlement of Molao people, who have long been keen on songs and opera. In the long-term process of production development and practice, the people have created a bunch of operas that demonstrate the unique characteristics and charm of their culture. However, at present, when global economy and cultural diversity dominate, the opera has been confronted with external strikes and is close to extinction. How can we maintain the inheritance and vitality of the non-renewable ethnic opera? It is a problem worth studying that how can we take advantage of education reform in local college to realize the practical and effective inheritance and preservation of the ethnic opera, and to bestow it with new connotation and function.

1 INTRODUCTION AND CURRENT SITUATION OF MOLAO OPERA

1.1 *The artistic characteristics of molao opera*

Molao Opera is an emerging ethnic opera, with the folk religious sacrificial ceremony and Nuo opera as its basic artistic form, blended with other forms such as Cai Diao. It constitutes a more comprehensive artistic form, rich in local and ethnic features.

1.1.1 *Comprehensiveness*

Opera is a comprehensive art form. The development of the Molao Opera through generation is based on other artistic forms, organically integrating Cai Diao, Molao folk song , Nuo opera in the traditional Yifan Festival and the music and dance of the singer. It is a genuine "blended art". For example, the arrangement and performance of *A Brief Biography of Pan Man*, the first Molao opera, used the methods that bond symbolism realism. It does not go with the trend nor pursue gorgeous effects, just stresses a kind of beauty —simple but elegant. The dance section infuse the "Hand tactic" and "Gang step" in the traditional dancing in Yifan Festival; the singing use folk music and professional Molao singers' song; as for instrument, it use relatively popping electronic keyboard, popular drum set, violin, flute, pipa and dulcimer, incorporating western and national instruments to make the opera possess ethnic characteristics as well as freshness of western electronic music. Molao Opera absorb the essences of other ethnic traditional arts, and move forward toward musical opera, creating a special singing and dancing opera with the distinctive artistic character and psychological feature of Molao people.[1]

1.1.2 *Jocosity*

As a kind of ideology, the emergence and development opera are certain to be influenced by ethnic regional culture. Molao Opera has been set up and developed on the foundation of the singing and dancing in the traditional Yifan Festival and Molao folk songs. As a result, the opera has always reflected the psychological features of Molao people through its developing process. For example, A Brief Biography of Pan Man is adapted from the folk tale: The Story of Pan Man. Its the "HA-HE-HA" way of singing in Yifan Festival as the basic style, which is very funny and hilarious. Many folktales in Molao tradition demonstrate this kind of ethnic psychology in a range of degrees, such as *Corps, Mountain Kenwang, Pusheng Col*, etc. Wisdom and humor of Molao people permeate the plots and details of the opera, showing profound allegory and intensive love and hatred. [2]

In the founding process of Molao Opera, it adapted itself to the appreciation customs of different people and maintain its own ethnic flavor at the same time. Molao Opera has demonstrated its aesthetic perceptions and appreciation of the ethinic aesthetic formulary; therefore, it has won appreciation of its own people, and long been alive and developed on the ground of the mountain region.

1.2 *The inheritance of and difficulties facing Molao Opera*

1.2.1 *The current situation of Molao Opera*

It was in less than 30 years that Molao Opera witnessed its origin after the first exploitation. In the chronic practice, the old artists have continuously explored and studied the opera, they were expert at innovating the performing arts. After the birth of A Brief Biography of Pan Man, Molao Opera kept blooming brilliantly on stage.

Table 1. The awards list of Molao Opera.

Date	Name	Awarding details
December, 1983	A Brief Biography of Pan Man	Poineering work of Molao Opera, Excellence Award in the second Guangxi opera exposition, Excellen Director Award, Excellent Music Award, Excellent Theatrical Design Awards, The main character won Excellent Performing Award
December, 1991	Raw Chicken and Cooked Duck	Third Prize in "Osmanthus Awards" of Guangxi Autonomous Region
December, 2001	Red Straps	Silver Prize in the National Stars Award
December, 2012	The Romance of Jade Flute	Took part in the 8th Opera Exposition in Guangxi, the script won Golden Prize in the National Opera Culture Award
January, 2013	Returning Home	The script won Silver Prize in the 8th National Opera Culture Award

1.2.2 *The predicament of Molao Opera*

With the change in society and people's values and appreciation perceptions, the inheritance of Molao Opera are confronted with predicament. By analyzing related documents ,and visiting Molao art troupes and villages, the author discovered that problems existing in the inheritance of Molao Opera are as follows:

1.2.2.1 The drain of professional personnel

Due to the reform of the institutional system of art-troupes ,professional personnel tend to change their profession or quit the profession.

1.2.2.2 Lack of scripts

There is severe lack of scripts which going along with the main-stream social values and people's appreciation demands. The art troupes are in short of talented scriptwriter. And classic scripts are absolutely insufficient.

1.2.2.3 Aging of the actors

The inheritors of Molao Opera are generally aged, while the present young actors in art troupes are often lack enthusiasm for and study of Molao Opera. Some young ones even don't know how to speak the local dialect or sing local folk songs, and they are relatively indifference to the ethnic culture.

1.2.2.4 The marginalization of ecological culture

As an artistic phenomena in mass society, Molao Opera has ecological contents and forms, and unique family identity. But it isn't difficult to find out it is up against the predicament of marginalization under the strikes of contemporary main-stream culture.

2 IT IS THE MISSION OF COLLEGE OPERA EDUCATION TO INHERIT AND PRESERVE ETHNIC OPERAS

At present, many ethnic operas in China are at the edge of extinction. It is the responsibility that falls on the whole society to rescue them. Colleges will play an important role in the preservation and inheritance of non-material cultural heritage such as ethnic opera.

Firstly, cultural inheritance is one of the major function of colleges. As is known to all, there are four major functions of colleges and universities: personnel cultivation, scientific research, social service, and cultural inheritance and innovation. The last is an important one that has been newly listed. The State Council of China has pointed out on *Notice About Strengthening Culture Heritage Protection* that " Education departments should include the knowledge of cultural heritage and its protection in the teaching plan and textbook, and organize viewing and emulating activities, stimulating the passion for national excellent traditional cultures in the youth."

Inheritance is the foundation, and innovation is the impetus. The inheritance of local operas bestows colleges with three obligations: cultivation of professional personnel, study and research of ethnic operas, and development of new driving force. The core lies in cultivation of professional personnel; the problems are mainly caused by the severe insufficiency of inheritors. To cultivate personnel who are capable speakers, singers, and dances is the urgent mission of the inheritance of ethnic opera; the study and research of ethnic opera is the foundation, it is through research that we can explore more excellent ethnic operas and analyze the regulations in their inheritance and preservation, in order to cultivate personnel in a specific manner and protect the culture; the development of new driving force refers to cultivating a batch of opera enthusiasts who concerns about and support the course of inheriting the culture, especially the mass group of students, who are the protectors and attendants of the future of ethnic opera. [3]

Secondly, colleges have the special advantages of cultural inheritance and protection. For one thing, there are many professional teachers in colleges. Most of them come from the local ethnic group. Having grown up along with the ethnic operas, they are familiar with the local culture, they are likely to integrate the opera in their teaching and pass them on to the students. For another, there are research groups in colleges. Among them are experts and scholars in folklore, theater art, sociology, and linguistics. They can approach the change, connotation, artistic forms and values of ethnic opera in different lights, and provide theoretical support and rational guidance for the inheritance and preservation of ethnic opera.[4] The greatest advantage of colleges is that they boast the vast and vigorous groups of students, of whom the art majors are undoubtedly the best inheritor and transmitter of ethnic opera; students who study other disciplines are also the potential audience, future managers and supporter of ethnic opera. A great many undergraduates are fond of opera: they join opera clubs and take active part in related activities. Given a stage, they will light up the passion for ethnic operas.

Thirdly, the inheritance of ethnic opera will be of help to enhance the development of opera teaching in colleges. In the process, the curricular setting, contents of education, teaching methods, practice education, and education research in colleges will all be improved. For example, in curricular setting, colleges can break through the present performing system capitulated on conformity, and infuse the essence of ethnic opera in teaching, which will result in the diversity and distinction of various colleges. The inheritance of ethnic opera can also further the teachers' and students' understanding of the folk life—the understanding of the rich connotations in the performing contents, the profound meanings of body languages, and the features of the accessories such as garments. It will immerse them into the charisma of ethnic opera, so that they can better their understanding of the opera as well as the culture and the religious belief. [5]

3 THE INHERITANCE OF MOLAO OPERA AND REFORM OF OPERA TEACHING IN COLLEGES

As a theatrical art form, Molao Opera has absorbed the art characteristics of the cultures of Molao people, Han people, and other ethnic groups. It has inherited the folk songs, fair tales, prayers, tributes and dances in sacrificial ceremonies, costumes ,and architectural techniques of Molao people, and has become a excellent cultural accumulation loved by the Molao masses.

3.1 *Bring in the resources of ethnic opera, and reform the classroom teaching*

Classroom is an important spot for knowledge transmission, personnel cultivation, and culture inheritance in colleges. Ethnic opera must be involved in the classroom as that it can get better inherited and preserved. Firstly, colleges should bring in the resources of ethnic opera and reform the contents of classroom teaching. Colleges should organize relative experts, government agents in culture departments, and opera inheritors to edit the textbook and implement it as one of the main textbook in opera teaching; secondly, colleges should innovate the teaching methods, combing traditional teaching with viewing and emulation in the field, videos and music appreciation, practice and blog writing, etc. They should arrange the students to go into the midst of the villages, festivals, dwellings of inheritors, communities in ethnic regions, so as to let them appreciate and emulate the opera, practice and create their own pieces. Consequently, colleges can expand the scope of their classes, and improve the proficiency and enrich the experience of their students in the connotation and skills of ethnic opera; Thirdly, colleges should alter the testing modes in the past. The modes based on memorizing, general writing, even graduation examination should give way to new ones such as opera creation including ethnic opera, stage performance, and property making, etc. Moreover, the examiners can expand to schools, art troupes, inheritors and teachers, instead of refraining to schools and teachers, allowing the students to walk out from schools and classroom, and enter the ethnic group and the society. [6]

3.2 *Establishing special educational opera teaching workshop*

What is educational opera? It refers to opera that uses theatrical methods and add theatrical elements in teaching or social activities and makes the students learn through practice. Colleges located in ethnic regions can consider establishing a teaching system incorporating multiple workshops, and invite experts to teach and transmit the inheritance mission in schools by means of workshops and seminars. The teaching section can be designed as: introduce the art of Nuo Opera—let students make special Nuo masks on their own—let students conduct scenario shows with the masks. This workshop pattern will be an innovation in teaching pattern, which break through the traditional ones that focus on instruction and lecture by the teacher. It manages to combine theory with practice, and education with artistic practice, by which the students will be able to learn through practice and deepen their understanding of national non-material cultural heritage. Therefore, this pattern will strengthen the cultural consciousness and faith of the young generation, and realize the educational perception of performing faculty in local colleges to cultivate practical personnel.

3.3 *Intensifying the construction of bipolar teaching groups*

For a long time, under the influence of traditional art education, ethnic opera has been excluded from the orthodox opera performing education, it is not considered as a refined art form, let alone being projected as a particular major lesson. College teachers mostly graduated from modern opera performing major, who lack due knowledge of ethnic opera. For example, few teachers in the college the author working in concern about ethnic operas, such as Molao Opera. In fact, there are many students interested in ethnic opera. But they cannot carry on it for lack of guidance from professional teachers. Therefore, colleges can appoint groups of teachers to study in the cradles of ethnic opera and get the handle of related skills; on the other hand, colleges can employ professional inheritors to teach the students in schools.[7] This bipolar teaching pattern will be an important guarantee to realize the modernization and sustainable development of ethnic opera, to build up special teaching methods in colleges, and to cultivate practical opera personnel.

3.4 *Building training center and practice base*

The training center and practice base of ethnic opera can provide a favorable platform for students to improve their practical skills. The college the author working in is located on the northwest of Guangxi Autonomous Region, where there are abundant resources of ethnic opera. In the celebrations of Festival of Molao people, Feitao Festival of Maonan people, Frog Festival of Zhuang people, Zhuzhu Festival of Yao people, there are affluent contents for performing. To utilize the opera and dancing resources to a full extent, the college has established a dancing training center in Northwest Guangxi, the ethnic joint performance training center in Northwest Guangxi, researching and practice teaching center of ethnic art resources, Liu San Jie College Student Art Troupe,etc, and set up a dozen off-campus training centers in ethnic villages. These centers and bases have organized students to participate in the Bronze Drum Folk Song Art Festival, regional folk celebrations, regional ethnic performances, and to discover more about the ethnic culture through field investigation and learning, which has fully take advantage of social and cultural space, created a second classroom, and obtained satisfactory achievements. For example, the musical drama Sweet Honey, The White Jeans and the Tanned Brother, the Caidiao opera, The Suriculturist Girl Choosing Her Husband, etc. have won prizes at various contests ranging from provincial to national rank.

4 CONCLUSION

It is the inevitable destination of the development of ethnic opera to deepen the exploration and analysis of the ethnic opera resources, which is also the obligation and source of the opera education in local colleges. Local colleges should set foot in ethnic opera art, commit to the special curricular training, integrate ethnic opera resources with college opera education contents, and enrich the performing techniques and elements, so as to maximize the function and value of the art of ethnic opera in college teaching. At the same time, colleges should carry forward reform in teaching methods, intensify the construction of bipolar teaching groups, deepen the study of ethnic opera. Only in this way can ethnic opera education reinforce the cultural connotation of the opera, as well as get it better accepted by the outside world; only in this way can ethnic opera satisfy the aesthetic demand of modern audience and achieve its inheritance and innovation.

ACKNOWLEDGEMENTS

1 The National Social Science Fund Project in 2012 "Characteristic Culture Resource Industrialization of Mulam Ethnic Minority and Its Preservation and Research" (project number: 12BMZ034).

2 The Planning Project of Ministry of Education Subsidized by the Unit in 2011 "Study on the Transmission Mechanism of National Education and National Culture of Mulam Ethnic Minority in Guangxi" (project number: FMB110045).

REFERENCES

[1] Lai Ruimin, The Origin of Molao Opera, [M], Nanning, Guangxi People Press,1994.

[2] Lai Ruimin, The Origin of Molao Opera, [M], Nanning, Guangxi People Press,1994.

[3] Fu Jin, The Protection of Non-material Cultural Heritage and Development of Ethnic Opera[J], Sichuan Opera, 2010(04).

[4] Liu Wenfeng, The Values and Protection of Traditional Opera in Non-material Cultural Heritage[J], Art Review, 2012(07).

[5] Zheng Xuesong, The Study of the Inheritance of Non-material Cultural Heritage in the Light of Anthropology—Taking that in Henan Province as an Example[J], Journal of Henan University (Social Sciences),2013(05).

[6] Wan Ping, Lin Lin,Ma Li, The Inheritance and Preservatio of Sichuan Opera in the Light of 'Non-material Cultural Heritage Protection Law'—Taking Sichuan Opera Research Institute as a Example[J], Sichuan Opera,2013(09).

[7] Qian Yongping, The Study of Qunqu Opera Protection under the Circumstances of Becoming Heritage[J], Cultural Heritage,2011(02).

Management, Information and Educational Engineering – Liu, Sung & Yao (Eds)
© 2015 Taylor & Francis Group, London, ISBN: 978-1-138-02728-2

Research on the value of establishing traditional ethnic sports course in colleges and universities of ethnic regions—taking Guangxi as an example

Qing Song Zhu
Institute of Physical Education, Hechi University, Yizhou, Guangxi, China

ABSTRACT: Ethnic regions are the cradle of traditional ethnic sports, and establishing related courses in those regions has become a trend. This article carried out research on the value of establishing traditional ethnic sports in colleges and universities of ethnic regions through documents and field research, discovering that the establishment has high values of constructing top quality curriculum in higher educational institutions, inheriting and protecting cultures, impressing moral principles and spiritual beliefs on students, prompting bodybuilding and creating entertainment.

KEYWORDS: Ethnic regions; colleges and universities; traditional ethnic sports; values.

1 INTRODUCTION

During its chronicle development, traditional ethnic sports have formed a distinctive cultural system on its own. There are various sports events featured in local cultures, for example, embroidered ball throwing, firework gabbing, archery race, pearl ball, and spinning top beating, adding some 150 events under exploitation and sorted. The events are rich in contents, varied in forms. However, in a time when western sports dominate, our traditional ethnic sports are confronted with unprecedented challenges: either cater to western culture and become competitive, or struggle on the edge of extinction due to insufficient participants. It is obliged that a nation should protect and inherit the unique national characteristics attached to its own traditional cultures; therefore, researching on establishing traditional ethnic sports courses in colleges and universities is of great importance to the optimization and reform in higher educational institutions courses and inheritance and transmission of traditional ethnic sports cultures.

Table 1. The distribution of traditional ethnic sports in Guangxi.

Minority groups	Major distributing regions	Representative events
Zhuang	Nanning, Liuzhou, Chongzuo, Laibin, Baise, Hechi	embroidered ball throwing, board-shoe race, firework grabbing
Yao	Liuzhou, Guilin, Hezhou Baise, Hechi, Laibin	bronze archery race, drum dance, monkey drum dance, chairs dragon, trumpet bell
Miao	Rongshui, Longlin, singing, Ziyuan Xilin, Longsheng, Nandan, etc.	Bar is climbing, grass ball bearing
Dong	singing, Rongshiu, Longsheng.	Firework grabbing
Malaya	Hechi	Tiger Zhang, grass dragon dance
Maonan	Hechi	"bumping together", "with the top", "struggling together"
Hui	urban areas of Nanning, Liuzhou, Guilin, affiliated counties: Lingchuan, Lingui, Yongfu, Luzhai	wrestling, zha quan
Jing	Wanwei, Wutou, Shanxin islands of Pingjiang county	bamboo bar jumping, "dog beating"
Yi	Longlin, Napo, Xilin	Damoqiu, tug-of-war, Swing
Shui	Rongshui, Yizhou,Huanjiang, Nandan	Horse race, seed of tung tree darting
Qilao	Longlin	egg peeling, Hualongbeating, spinning top beating sp

2 AN OVERVIEW ON TRADITIONAL ETHNIC SPORTS EVENTS IN GUANGXI AUTONOMOUS REGION

There are 11 minority groups, besides Han people, living in Guangxi, including Zhuang, Yao, Miao, Dong, Yi, Hui, Molao, Maonan, Shui, Jing, Qilao peoples. Detailed information about the settlement distribution of these peoples and their representative traditional ethnic sports events can be seen in Chart1.

From Table 1, it is known that these 11 ethnic groups have spread in an extensive scope involving many cities and counties; their colorful traditional ethnic sports are rich in variation and local characteristics, yet sufficient in interestingness.

There are primarily 19 colleges and universities in Guangxi, which locate in 9 cities including Nanning, Guilin, and Liuzhou. Among them, Nanning boasts the most institutions, following it is Guilin and Liuzhou, respectively. There are also ethnic groups settling in these 9 cities. Detailed information is listed in Table 2.

Table 2. Distribution of colleges and universities in Guangxi and ethnic groups attaching to the cities.

Location	Colleges and Universities	Ethnic Groups
Nanning	Guangxi University, Guangxi Medical University, Guangxi Traditional Chinese Medicine University, Guangxi Teachers College, Guangxi Arts College, Guangxi University for Nationalities, Guangxi Finance and Economics College	Zhuang, Hui
Guilin	Guilin University of Electronic Technology, Guilin Engineering College, Guangxi Normal University, Guilin Medical College	Yao, Hui,Dong
Liuzhou	Guangxi Technical College	Zhuang, Yao, Hui
Hechi	Hechi University	Shui, Molao, Maonan
Baise	Youjiang Medical College, Baise University	Zhuang, Yao, Qilao, Yi
Wuzhou	Wuzhou University	Zhuang, Yao
Hezhou	Hezhou University	Yao
Yulin	Yulin University	Zhuang, Yao, Miao

According to Chart 2, there are also ethnic groups living in the 9 cities that locate major colleges and universities in Guangxi, rendering on the establishment of traditional ethnic sports course abundant resources. Local colleges and universities can fully utilize these resources on the spot. On the other hand, Liuzhou, Guilin, Yizhou, Hezhou, Wuzhou are close neighbors to each other, which allows the universities in these regions to share resources with each other. Institutions can join hands in exploiting local traditional ethnic sports events, or they can establish courses of their own characteristics, and learn from each other.

3 VALUE ANALYSIS

3.1 *The value of establishing top quality courses in colleges and universities*

Compared to ordinary ones, top quality courses are courses of advanced level, excellent quality, and distinctive characteristics. Course establishment is one of the fundamental construction of colleges and universities. Its level and qualities are direct reflections of the personnel cultivating goal, quality, degree and characteristics of an institution. As one important form that embodies the effectiveness and high standards of course construction, the syllabus, teaching objectives, teaching contents, teaching methods, textbooks, and the first-rate, effective, scientific teacher groups of top quality courses can be a radiating influence of course construction in colleges and universities.

Guangxi Autonomous Region, as an ethnic region, has an advantageous condition of numerous cultural heritages of traditional ethnic cultures, which can be significant elements and contents of course construction in colleges and universities. The regional characteristics will facilitate the effective integration between course contents and local ethnic cultures.

Local folks have their own lifestyles, customs, and spiritual cultural beliefs, and therefore schools should draw on local resources to set up the courses, endowing the courses with not only local characteristics, but also modernism and creativeness.

Consequently, schools should combine the course contents with local cultures, and realize the shared goal of course construction of" famous teacher, famous teaching techniques, and famous effects.", more importantly, inherit and carry forward the traits of top quality curriculum.

3.2 *The value of cultural inheritance and protection*

Colleges and universities are important fields of protecting, transmitting, and putting forward traditional ethnic sports. Because they are rich in resources regarding various aspects as compared to primary schools and middle schools, for example, they are in a favorable position with high-qualified teachers, teaching facilities, and holdings of traditional ethnic cultures, besides, they have relative platforms for studying and researching, and they have more frequent communication with the outside world. In addition, university students are relatively keen on thinking and understanding, have civilized qualities, and are more active-minded. They are likely to experience and try new things. Moreover, colleges and universities gather together students from all over the country, as a result, the establishment and implementation of traditional ethnic sports course will not only pass down to them the cultural heritages, but also penetrate the culture in every part of the country. Because those students from external provinces will carry with them the more or less impressions of traditional ethnic groups with them, no matter they'll stay in Guangxi or not, so that in a sense, they put the traditional ethnic sports culture forward.

The establishment of traditional ethnic sports courses should begin with "school-based establishment", which in fact inherits and protects traditional ethnic sports cultures. To put the sports resources into textbooks is the "school-based establishment.", during which schools are supposed to carry out all-round collection and exploitation of related resources, and adapt them into textbooks that are knowledgeable for students. The final school-based textbooks may represent some differences compared with the original sports events, but this process of adaptation in fact inherits and protects traditional ethnic sports cultures, which is of great value.

3.3 *The value of the education of moral principles towards university students*

Traditional ethnic sports are embedded with profound cultural connotations and time-honored history. Moral principles are significant in carrying the inheritance of traditional cultures, which is often potential and intangible. These moral principles act on traditional ethnic sports as heteronomy, as time passes, they will edify and civilize the participants, and have become a common code. For example, kindness, righteousness, courtesy, wisdom, and trustworthiness have always been widely popularized and become the fundamental principles followed by Chinese people for thousands of years.

In the globalized modern society, university students are under cultural shocks of fast-food characteristics, and cultural impacts of the western world. They are gradually losing the sense of commitment and belongings toward their native culture. The moral principles carried by traditional ethnic sports will serve to guide, restrain, and regulate students' behaviors.

Students of local colleges and universities can in the first place absorb the local cultures to improve themselves morally. Secondly, they can take part in traditional ethnic sports events and learn through practice. Thirdly, they can follow the prominent examples of their fellows coming from ethnic cultures. As they constrain themselves, acquire moral principles in practice and learn from their minority fellows, things going on like this, they will establish themselves moral principles, demonstrating that traditional ethnic sports indeed have the value of educating moral principles towards university students.

3.4 *The value of establishing spiritual beliefs of university students*

As a great and enduring country, China has generated abundant and content-rich traditional cultures, including national spiritual beliefs. Traditional ethnic sports is in itself a system which various expressions and rich contents, including etiquette, beliefs and philosophy, the typical yin-yang theory, and the syncretism between heaven and men developed on the basis of the former.

Some traditional ethnic sports originated from the ancient fairy tales in Chinese history; therefore there exist events in sacrifice to the gods, and in honor of ancestors, such as Maogusi Dance. Some are in memory of certain ancient celebrities: the Dragon Boat race is in memory of Quyuan, and the miracles of "climbing the mountain of swords, dipping into the ocean of fire" are for the ancient hero Long Jiulang; Some derived from particular ethnic traditions, for example, there are four categories of Man people: the Red, the Yellow, the Blue, and the White, who, in Spring Festival, put up flags with their corresponding colors on their doors. These beautiful and vivid flags are manifestations of auspicious prospects of the year to come.

Traditional ethnic sports are endowed with values of spiritual beliefs besides being competitive and entertaining events. University students will be impressed by the traditional ethnic sports cultures, better understand the Chinese philosophy, and establish their own spiritual beliefs through taking part in the events. More importantly, they will learn to respect the beliefs of the people with different cultural backgrounds, which will enhance the solidarity among people from different groups and strengthen the national coherence.

3.5 *The value of bodybuilding and entertaining*

Traditional ethnic sports possess distinct characteristics of bodybuilding and entertaining. They are created by ancient people in labor, which is intended for catering their own spiritual needs. They felt free to have fun on the condition of their time and space. Particularly, the integration between traditional ethnic sports with music and dance display the entertainment of the events. Apart from simple, entertaining events, there are some complicated and demanding events with strict rules, which adds to its function on body-building. Traditional ethnic sports are always rich in recreation, as a result, they are always popular with the people.

University students constitute an important part of the future of our nation, whose physical and psychological health not only influence their own development but that of the country. Accordingly, colleges and universities in their education should pay attention to students' all-round health. P.E. courses are indispensable in this cause. Modern sports events, football, basketball, volleyball, and various small ones, to just name a few, can strengthen the students' physiques, but traditional ones will temper their mood and cultivate their personalities as well, which makes them stand out of modern events. The implementation of traditional ethnic sports courses in university can build up students' health and educate them morally. It allows them to experience the entertainment of traditional ethnic sports, and exercise in a relaxing and delightful atmosphere. Qigong can cultivate their mind. Martial arts will build up their physiques. The iconic one combining both is Taiji. All the above can be competitive events as well as performing events. Traditional sports events are truly diverse and colorful.

4 SUMMARY

4.1 Traditional ethnic sports are a bright wonderful flower in the garden of traditional Chinese cultures, the inheritance and protection of which is our historical responsibility and mission. Particularly, ethnic regions should shoulder this task and serve as major battlefields. The establishment of traditional ethnic sports will also benefit the inheritance of traditional ethnic cultures, therefore colleges and universities can function as the major battlefield of cultural inheritance.

4.2 The establishment of traditional ethnic sports course will promote the construction of top quality courses in colleges and universities, displaying the characteristics of the institutions, inculcating in students moral principles, spiritual beliefs and sense of belongings towards traditional Chinese culture.

4.3 Guangxi is a typical settling place for minority peoples with abundant ethnic cultural resources. Local colleges and universities should make the best of these advantageous resources, integrating traditional ethnic sports with higher educational courses and promoting its development.

ACKNOWLEDGEMENT

Study of Establishment and Practice of Characteristics Educational System in Colleges and Universities of Ethnic Regions, key funded project of Guangxi higher education curriculum reform project (Project code:2013JGZ154).

REFERENCES

[1] Bai Jinxiang, Tian Zuguo, The Recognition and Modern Development of Chinese Traditional Ethnic Sports on the Background of Globalization[J], Culture Exposition(Theory), 2011(1):53–55.

[2] Chen Bo, Feng Hongjing, Research on the Cultural Values of Traditional Ethnic Sports[J], Culture Sports Guide, 2008, 10:27–28.

[3] Hu Xiaoming, The Dimensional Values of Traditional Ethnic Sports[J], Journal of Physical Education, 2007,14(8):5–9.

[4] Liu Yamei, Liu Guozhao, Guo Qiang, The Study of Possibility and Feasibility of Inheriting Traditional Ethnic Sports in Higher Educational Institutions[J], Chinese Martial Art·Study,2012,1(1):94–96.

[5] Wang Jianli, Chen Yabin, The Historical Mission of Colleges and Universities in the Inheritance of Traditional Ethnic Sports[J], Internet Wealth Sports Study, 2010,5:114–115.

[6] Tao Zhichao, Lu Qing, The Inheritance of Traditional Ethnic Sports in the Light of Higher Educational Institutions[J],Boxing and Martial Art Science, 2011, 8(5):93–94.

Management, Information and Educational Engineering – Liu, Sung & Yao (Eds)
© 2015 Taylor & Francis Group, London, ISBN: 978-1-138-02728-2

An empirical study of the application of functional grammar theories to the teaching of the English passive voice

Xue Feng Zhai
Faculty of Foreign Language & Culture, Kunming University of Science and Technology, China

Guo Feng Ding
Law School, Kunming University of Science and Technology, China

ABSTRACT: The paper puts forward functional grammar teaching based on Larsen-Freeman's framework of grammatical competence. An empirical study is done on the application of functional grammar theories to the teaching of the English passive voice. The results show that functional grammar is useful and can be applied to college grammar class and that compared with the traditional approach of grammar teaching, functional approach is effective in improving college students' grammatical competence.

KEYWORDS: grammatical competence, traditional grammar, functional grammar.

1 INTRODUCTION

Grammatical competence is essential for communication. Certain level of grammatical competence serves as the base for the development of communicative competence. Without it, learners cannot achieve high target-level proficiency and their language learning may become fossilized at an early stage.

Then, how about the grammatical competence of college students? Researches and college English teachers' teaching experience tell us that the situation is not optimistic. Li Qi's investigation of English major students in North-eastern Normal University revealed that although the participants possessed solid grammatical knowledge, they lacked high level of accuracy and native-like appropriateness in actual use in given contexts (23). Considering the actual conditions of Chinese college students' grammar teaching and learning, a new approach of grammar teaching is called for, an approach that can suit college students' need and can teach the use of the grammatical rules.

Functional grammar, as a meaning-centered grammar, which focuses on the notion of language functions and tells us what language does and how it works, is believed to be able to serve the demand of college grammar teaching reform. Thus, to teach college English grammar in functional approach has become the agreement of many linguistic experts and researchers. However, most of the researches in this field stop at the theoretical level, and few of them have initiated an experimental study from the profile of college students' grammatical competence.

Initiated by the reasons mentioned above, the present study, selecting English passive voice as the example grammar item and non-English major students from two classes as experimental subjects, attempts to make an empirical study of English grammar teaching under Halliday's functional grammar theories.

2 LITERATURE REVIEW

This part provides a detailed review of the related literature on the definitions of grammatical competence and different approaches to grammar teaching.

2.1 *Definitions of grammatical competence*

For several decades, the definition of grammatical competence, along with that of communicative competence, has been a question of controversy.

The notion of grammatical competence was first proposed by Chomsky to refer to the grammatical knowledge of the ideal language user, disregarding the actual use of language in concrete situations. To react against it, Hymes and other linguists proposed the conception of communicative competence, among which grammatical competence is an essential component. Canale and Swain refer to grammatical competence as the mastery of language code such as features and rules, including vocabulary knowledge, word formation, syntax, pronunciation, spelling and linguistic semantics. Different as the above various definitions of grammatical competence are, they are similar in the fact that all of them deal solely with the formal and structural aspects of the language, while excluding the aspect of pragmatic use.

Holding a different opinion with the previous linguists, Larsen-Freeman worked out a framework of 'grammatical competence', which is composed of three dimensions, namely form, meaning (semantics), and use (pragmatics). The dimension of form consists of the visible or audible units: the sounds, written symbols, inflectional morphemes, function words, and syntactic structures. The dimension of meaning refers to the semantic meaning encoded in language, which is the essential denotation of a decontextualized form. The third dimension is the pragmatic use, which is not the meaning encoded in language, but what people mean by the language they use and which consists of two main units: one is the unit of social functions, which is concerned with the use of grammatical knowledge in social context, and the other the unit of discourse patterns, which deals with the use of grammatical knowledge in linguistic context. Here, Larsen-Freeman's definition of grammatical competence is adopted as one of the theoretical bases for the present research.

2.2 *Different approaches to grammar teaching*

This section offers an overview of the historical development of grammar teaching.

2.2.1 *Traditional grammar-based syllabus*

During the period from the 1950s to the 1960s, much of foreign language teaching was influenced by both structural linguistics and behaviorism. Language teachers at that time were mainly concerned with the description of patterns of structures(sounds, words, sentences) of a language and teaching a foreign language was seen as a process of gradual accumulation of discrete linguistic items segmented into independent parts for learners during instruction, and that the role of learners was to synthesize these parts into 'whole chunks' during communication.

The traditional grammar-based syllabus has been severely criticized for the lack of the authenticity of the input which may result in the learners' inability to use structures in real contexts (Shu Dingfang 102). These interactions will not necessarily transfer to actual language use in real-life situations.

2.2.2 *Communication-based syllabus*

In 1970s, some researchers, influenced by the concept of communicative competence, attempted to abandon the grammar-accuracy orthodoxy in favor of more communication-oriented syllabus that focused on language use. However, rules of language use would be useless without rules of grammar. In other words, a pure communication-based language approach may lead to fossilization of acquisition in that it can lead to the development of a broken, ungrammatical, pidginized form of the language beyond which students can never really progress.

It is believed that appropriate grammar teaching instruction could serve as one of the effective methods to redress problems existing in learners' grammatical competence. And grammar should never be taught as an end in itself but always in light of meaning, social factors and discourse or a combination of these factors.

3 METHODOLOGY

3.1 *Research questions*

This study is intended to answer the following research questions:

1 Is functional grammar applicable to grammar teaching of college students?
2 If applicable, is grammar teaching of functional approach more effective than the traditional approach?

3.2 *A general description of the experiment*

To answer the above two research questions and to explore an workable approach to college functional grammar teaching, first, the writer develops a set of teaching contents and teaching procedure, mainly based on *Functional Grammar* by Halliday and *A General Introduction to Systemic-Functional Linguistics* by Hu Zhuanglin, and Zhang Zhenbang's *Essentials of A New English Grammar*. Then, specific teaching materials is organized according to this set of teaching contents and teaching procedure.

The 61 subjects are sophomore students from two intact classes (class 2 and class 3) of School of Engineering, Anhui Science and Technology University. The grammatical item to be taught is English passive voice. Class 3 (30 students) is chosen as the Experimental Group, receiving functional grammar passive voice teaching; whereas class 2 (31 students) as the Controlled Group, is taught in traditional approach. Before the experiment, a pre-test of English passive voice is held for the purpose of ensuring that there is no significant difference in the mastery of the grammar item between the two groups. Another purpose of the pre-test is to record and collect experimental data for future comparison with the post-test. After the teaching experiment, a post-test is done. Meanwhile, interviews are made with the Experimental Group to check and improve the teaching effects during the functional grammar teaching phases and after the experiment, and to collect data for future qualitative analysis as well. Besides, statistical tools Independent Samples T-test and Paired

Samples T-test by SPSS12.0 are employed for the quantitative data analysis.

4 RESULTS AND DISCUSSION

In this section, we attempt to represent the results, discuss and try to account for the findings, thus answering the two research questions proposed in the experiment.

4.1 *Pre-test and post-test*

Table 1. A comparison between CG and EG in the pre-test.

Group	Number	Mean	Std. Deviation	P
CG	31	33.3145	6.5287	.990
EG	30	33.2917	7.0506	

From Table 1, one can see that the two groups are homogeneous before the treatment; therefore the two groups are comparable. Consequently, these two groups can be used as subjects of the experiment.

Table 2. A comparison between the two groups in the post-test.

Group	N	Mean	Std. Deviation	P
CG	31	35.3871	5.9129	015
EG	30	39.0333	5.4058	

From Table 2, one can easily find that there is a great gap in the performance between EG and CG. The results of T-test show that the Experimental Group is obviously better than the Controlled Group in the use of English passive voice after the seven-week experiment.

The significant difference between the two tests with EG is shown in **Table 3**. Generally speaking, a statistically significant difference exists between the pre-test and post-test of EG and the grammatical competence in passive voice of EG is gradually increasing.

Table 3. Paired samples test: A comparison between pre-test and post-test with EG.

		Mean	N	Std. Deviation	P
Pair1	Pretest	33.2917	30	7.05064	.000
	Posttest	39.0333	30	5.40583	

4.2 *Interview after the post-test*

After the post-test, an interview is done to find out the effects of the English passive voice teaching done under functional approach and to testify and learn the strengths and weaknesses of the present teaching process. The results of the data analysis are presented as follows:

Good points: (1) As for the teaching materials, since most of the examples and practice exercises given in the classes are from the band 3 textbook, the authentic Band 4 test papers and writings done by the students themselves, they are less difficult to be accepted and understood.

(2) As for the teaching approach, the presentation of grammatical rules are combined and integrated into the teaching and training of reading and writing skills. This is, to them, a quite effective method of grammar teaching.

Complaints and Suggestions:

1 The consolidation practice and exercises in and after the class are inadequate.

2 In the presentation section, sometimes there are many difficult words in the examples cited, thus increasing the difficulty level of the points being taught.

In the following part, we will discuss the results obtained from the pre-test and post-test and the interview after the post-test.

4.3 *Discussion*

4.3.1 *Positive results*

The results showed in Table2 and Table 3 indicate that the EG receiving functional grammar teaching obviously performs better than the CG receiving traditional grammar teaching. This could be attributed to the advantages of functional grammar teaching over the traditional grammar teaching method.

The traditional approach is restricted at sentence level, and when presenting a grammatical item it usually takes no account of the meaning and function of the structure, and isolates the structure from both its social and linguistic contexts. This kind of practice often results in learners' inability to apply what they have learned about the grammatical forms into use communicatively. However, in contrast with the traditional way, functional grammar, taking discourse as its starting point, combines the grammatical form with its semantic meaning and context. Therefore, the teaching of functional grammar can, to the largest extent, integrate grammar teaching with the development of students' communicative competence, and facilitate the transfer of their grammatical knowledge into their corresponding grammatical competence(Zhang Delu 76). Besides, the positive feedbacks from students' interview about the teaching process and practice material also contributes to explaining the success of the experiment.

4.3.2 *Complaints from the interviewers*

It is widely believed that practice is of vital importance in the teaching and learning of grammar. Grammatical competence cannot be attained solely through exposure to the target language or meaningful input, unless enough output practice is done. And as for the difficulty of the input materials, according to Krashen's Input Hypothesis, only comprehensible input, i.e. the kind of input that is simplified and with the help of contextualized and extralinguistic clues can cause L2 acquisition. According to the above teaching and learning theories, the above complaints and suggestions seem quite reasonable, though they are only a few, not most of the interviewees' opinions.

5 CONCLUSION

In this part, we will conclude the major findings of the research, and put forward the implications of the study to College English grammar teaching.

5.1 *New findings*

The major findings of the research are as follows:

1 Functional grammar is useful and can be applied to college students grammar class;
2 Compared with traditional approaches of grammar teaching, functional grammar teaching is effective in improving college students' grammatical competence.

5.2 *Pedagogical implications of the study*

To teach functional grammar to college students, generally speaking, two sub-questions should be considered. First, what should be taught in a functional grammar class; then, how to teach functional grammar. Next we shall dwell on them one by one.

5.2.1 *What to teach*

Larsen-Freeman sees grammar as a higher-order concept within linguistics, and argues that it has three interrelated dimensions: form, meaning and use. She also asserts that grammar should be seen as a skill like reading and writing rather than an area of knowledge (14). Linguists' opinions on grammar can provide some inspirations for our language teachers. Grammar teaching should not merely focus on the form, but also on meaning and use as well. Since the learners are college students who have learned systematically certain grammatical knowledge in middle schools, it is suggested that the grammar teaching at college should focus on the use of grammatical knowledge.

5.2.2 *How to teach*

1 Teaching functional grammar in discourse

In discourse-based grammar teaching, teachers can teach target forms together with authentic or simplified discourse, which can supply learners with abundant examples of contextualized usages of the target structure to promote the establishment of form-meaning relationships. It also suggests the combination of grammar and writing, grammar and speaking, grammar and reading and so on.

2 Teaching functional grammar with enough comprehensible input and output practice

Input and output of enough volume is vital for grammar teaching. Functional grammar is grammar in use, the theories of it are somehow abstract and difficult. Therefore, enough input and output practice in and after class are necessary and can help learners understand the complex grammar rules, consolidate what they have learned in class, and then put them into communicative use. Besides, abstract theories or complex grammar rules in functional grammar should be simplified or adjusted to meet students' needs and understanding capability.

Due to the comparatively short period of research and experimental time and relatively small sample size, this research surely has its limitations. It is hoped that this study can shed some new lights on the further research in this field and more researches are to be induced to make up for the deficiency of the present study to explore the application of functional grammar to college English teaching, thus letting more college students benefit from functional grammar.

REFERENCES

[1] M.A.K. Halliday. *An Introduction to Functional Grammar* (2nd ed.)[M]. Beijing: Foreign Language Teaching and Research Press / Edward Arnold (Publishers) Limited, 2000.

[2] Larsen-Freeman, D. *Teaching Language: From Grammar to Grammaring*[M]. Boston: Heinle&Heinle, 2003.

[3] Li Qi. A Survey of Grammatical Competence of English Majors. North-eastern Normal University, 2006.5.

[4] Shu Dingfang, Zhuang Zhixiang, *Modern Foreign Languages Teaching—Theory, Practice and Methodology* [M]. Shanghai: Shanghai Foreign Languages Teaching Press, 2006.

[5] Zhang Delu, etc. Functional Grammar and Foreign Language Teaching[M]. Beijing: Foreign Language Teaching and Research Press, 2005.

Management, Information and Educational Engineering – Liu, Sung & Yao (Eds)
© 2015 Taylor & Francis Group, London, ISBN: 978-1-138-02728-2

The effect of teaching methodology on students' interest in volleyball class

Hui Wang
School of Physical Education, Yan'an University, China

ABSTRACT: In the process of volleyball teaching, interest is one of the important factors to influence students' consciousness and initiative. And because volleyball needs high skillfulness and tight collectivity, it is difficult for the new learners to master the volleyball skills within a short period. Additionally, that the present teaching methods are inflexible, and the pursuit of correct movement is excessive has made this course boring and influenced the students' interest greatly. So it is necessary to innovate teaching methods so as to improve students' interest and initiative. To improve the quality of volleyball teaching and to promote the popularization and the development of volleyball in Yan'an University, on the basis of the idea "Happy Sports", this essay attempts to explore the effect of the application of happy teaching on volleyball teaching in colleges.

KEYWORDS: volleyball teaching; teaching methodology and means; interest.

Nowadays, the development of volleyball is not optimistic. Few students play volleyball as a physical exercise except in volleyball lessons and volleyball team training. To improve the teaching quality of volleyball, enrich teaching connotation and enhance volleyball popularization and development is the urgent problem to be solved. And this essay applies literature review, questionnaire, mathematical statistics to investigate the present interest of some students in Yan'an University. It aims to change the present volleyball teaching methods by advocating happy volleyball so as to stimulate students' interest in learning volleyball and puts forward some relevant measures and suggestions.

1 SUREVY OF THE CURRENT SITUATION OF STUDENTS' INTEREST IN VOLLEYBALL

According to the survey, students still do not get rid of the original examination- oriented education concept a are not highly interested in volleyball. There are many factors causing this situation, such as limits on sports facilities, insufficient attention of school authority, and decrease in inter-college volleyball exchanges. Students are forced to learn volleyball, and they learn only for grades. Therefore, it is difficult to mobilize the students' enthusiasm and initiative in volleyball lessons. Compared with football and basketball, volleyball is less popular. Due to the characteristics of volleyball itself, difficult to grasp and skillful, it is difficult to establish the confidence for students. Meanwhile, the conventional teaching pattern, which embodies in teachers' improper and boring teaching, undermines the students' interest. Accordingly, students' interest in and duration of volleyball is far from that of other sports, e.g. football, basketball.

1.1 *The degree of students' interest in volleyball*

According to the current survey, the situation of the student interest in volleyball is not optimistic. Volleyball has been put to the edge in the school sports. Since in all groups, the proportion of boy choosing basketball and football is higher than that of volleyball, while the opposite in girls, gender differences inevitably exist. From the current situation at colleges, physical education has not been treated properly, as students lack formal P.E. teaching system, are physically weak, lack athletic ability with faint sports consciousness. And the data show that only 19% of students enjoy volleyball class. It is safe to say that students are not interested in it, and this is a commonly existing problem in college PE teaching.

1.2 *Reasons for the absent treatment of volleyball at colleges*

1.2.1 *Over-emphasis on techniques*

According to the survey, students think volleyball needs more techniques and is difficult to master compared to basketball and football. And the traditional teaching mode hinders the student interest in learning. According to the survey of personal interest, boys choose such competitive sports as basketball and football, while girls choose volleyball, for it is less competitive.

1.2.2 *Lack of overall volleyball teaching*

Sports teaching is the central part of the school sports work and an important way for students to gain sports knowledge, to grasp sports techniques and to improve sports skills. Volleyball was taken into college physical education syllabus as an elective in the early 1990s in China. One of the major reasons why volleyball is less popular is that volleyball teaching is not fully carried out at school. In terms of the actual situation, volleyball teaching is offered in few schools: even if it is offered, teaching is not complete, with digging, passing and no other contents; and the teaching proportion of this subject is rather small in physical education. This teaching method which lacks systematic, sequential, fun and is contrary to personal development education that is advocated by quality education will inevitably lead to the decline in students' interest in volleyball and the decay of teaching skills with teachers.

1.2.3 *Lack of curricular- extracurricular combined teaching methods*

To the students in Yan'an university, spare time is rich that should be made full use of. In accordance to the survey, there are few volleyball matches or no at all, but basketball and football activities are carried out very well, For example, in the annual school sports meeting, students can actively participate in many matches and games, but volleyball competition and extracurricular activities are rather rare.

Effect of teaching methods and means on students' interest

2 EFFECT OF TEACHING METHODS AND MEANS FOR STUDENTS' INTEREST

2.1 *Game teaching method*

Sports games are considered lively, vivid and diverse with great fun, amusement and broad adaptability. Now, since the idea of Happy PE teaching is widely advocated, putting sports games into volleyball skill teaching is a new teaching perspective.

2.1.1 *Digging relay*

Method: The students are divided into several groups (not necessarily equal number of members), with a better player selected by the team who stands face to face about 3 meters away right in front of the team members standing in a column line. When the game begins, the best player passes the ball to the opposite team player who is expected to dig the ball back to the best player and then run back to the rear of the tea; meanwhile, the best player digs the ball to the next player in the line who does the same. And so on and so forth. When the teacher stops the game, the number of continuous repeated digging will be calculated and the winner goes to the beam that digs the most.

Rules: Students stand in a column line and each takes its turn to dig the ball. Each team member is allowed to dig the ball with their hands only, otherwise the team plays foul and have to restart the cycle. Double hit is a fault, and then restart. It is expected that such games help students in enhancing heir competitive consciousness and improving their learning interest.

2.1.2 *Shuttle relay*

Methods: Students are divided into teams with equal members, and each team is further divided into group A and group B that stand face to face in a column line about 3 meters away. The first student of each group holds a ball. When the game begins, the player with the ball in group A passes the ball to the first student of group B, and then runs quickly to the rear of group A. The first student in group B digs the ball back to the second member of group A, and then runs quickly back to the rear of a group B. And so on. The team that dig the most will be the winner. Rules: Students are not allowed to pass the ball, drop the ball and hold the ball with two hands, otherwise the team plays foul; the distance between the two groups would not be decreased; in the case of "double hit" or the drop of the ball of one group, restart.

2.2 *Mini-match teaching method*

Competitions are one important goal of sports training and learning, therefore, from the point of view of the interests and needs of students, to help students master competition skills and methods, to change of competitive volleyball ideas bravely, to regard the student as the core and to stimulate the interest of students will help students understand the techniques as soon as possible to improve the teaching effect. Mini-competition teaching method is a kind of training and of diverse forms that looks competition at the core of teaching to foster the students' sense of competitiveness and cooperation. This teaching form, firstly, will arouse students' learning enthusiasm, promote students' extracurricular exercises and improve the quality of teaching; secondly, it helps students grasp the basic skills of digging, passing, spiking and blocking, grasp and understand offensive formation, the tactical coordination and volleyball competition rules and the law of judges.

2.3 *The teaching methods of changing the external environment*

Volleyball teaching and training is conducted to improve the students' interest, and we may need to change the external environment so as to achieve the purpose of teaching. For example, in a volleyball match, we may obtain better results with a reasonable changing rule and a proper training method which is based on the students' training level and emphasis. In addition, if necessary, we may complement an offense and a defense to the opponent team so that could get extraordinary effects when opponent attacks and defenses. By doing so, students' enthusiasm of learning will be greatly inspired. In daily training and teaching, teachers' conscious changing external environment could not only improve students' adaptive capacity, but also greatly inspire students' interests of learning.

3 CONCLUSION

With the teaching methods given above, students could learn the kinds of volleyball skills and tactics in a leisurely way in the course of volleyball teaching. What's more, their volleyball skills will be improved and their interests for volleyball will be greatly inspired. In this way, students will long for the chance of playing volleyball and the improvement of volleyball skills as well. In all, the methods will effectively improve today's volleyball teaching.

Copy the template file B2ProcA4.dot (if you print on A4 size paper) or B2ProcLe.dot (for Letter size paper) to the template directory. This directory can be found by selecting the Tools menu, Options and then by tabbing the File Locations. When the Word program has been started opening the File menu and choose New. Now select the template B2ProcA4.dot or B2ProcLe.dot (see above). Start by renaming the document by clicking Save As in the menu Files. Name your file as follows: First three letters of the file name should be the first three letters of the last name of the first author, the second three letters should be the first letter of the first three words of the title of the paper (e.g. this paper: balpcc.doc). Now you can type your paper, or copy the old version of your paper onto this new formatted file.

REFERENCES

[1] Qi Xiaohong, Investigation and Analysis on the Motive System of Volleyball General Course for P.E Students [J] Journal of Nanjing Sports Institute (Social science edition), 2007.(03).

[2] Zhang Xiaokun, Investigation and Analysis on College Volleyball Elective Course about Student's Emotional State[J] Journal of Harbin Institute of P.E., 2007.(02).

[3] Huo Hangqi, The present Situation of Volleyball Teaching in Universities in Hebei Province and the Corresponding Reform Tactics[J] Journal of Hebei Normal University(Natural science edition), 2007.(02).

[4] Zhu Jianyu, Application of the learning guided teaching in Volleyball Course[J] Journal of Changsha University, 2007.(02).

[5] Dai Ke, Li Ding, Program Teaching Experiment of Volleyball Specific Elective Course in colleges[J] Journal of P.E, 2007.(01).

Management, Information and Educational Engineering – Liu, Sung & Yao (Eds)
© 2015 Taylor & Francis Group, London, ISBN: 978-1-138-02728-2

The study of campus sports culture and its construction in higher education institutions

Hui Wang
School of Physical Education, Yan'an University, China

ABSTRACT: Campus culture is an important ingredient of culture. Campus sports culture embodies the cultural charming as an aggregate culture, and highlights in the university campus culture with its unique characteristics. At the same time, the construction of campus sports culture in college is one of the key points in the current higher education sports reform and development. Based on documentation and investigation, analyzing the current situation of sports construction, illustrating its connotation, essentiality and existing problem, the study put forward how to improve the construction of the campus sports culture, aiming to provide valuable reference and advice for the construction of campus sports culture in higher education institutions.

KEYWORDS: Higher education institutions, Campus sports culture, Construction.

1 INTRODUCTION

The rich practical experience in the cultural reform and development was summed up in China Seventeenth CPC Central Committee Sixth Plenary Session, pointing out that extensive cultural activity must be developed among common people, complete with community culture, village culture, corporate culture, campus culture and so on. With the development of the national economy, the ordinary people are thirsty for improving their cultural level. The government pays more and more attention to the construction of cultural activities in order to meet people's demands. As an important part of the modern campus culture, the construction of campus sports cultures was concerned with the Communist Party and the country. However, in contemporary China, there existing many problems in the college campus sports culture which has apparent gap and unbalanced development compared with the other developed countries. Therefore, it is necessary to study college campus sports culture and its construction. This paper analyzes the shortcomings of the current college campus sports culture and discussing how to construct it, aims to provide valuable reference and advice for the construction of campus sports culture in higher education institutions.

2 THE CONNOTATION OF COLLEGE CAMPUS SPORTS CULTURE

As an important carrier of culture, Colleges and universities play an important role in the aspect of cultural construction, transmission and protection. As a kind of social culture, the sports culture can affect the human values, ways of thinking, the idea of management, the wisdom of survival and social atmosphere. Especially, it plays an integration function in establishing and maintaining social order. The sports culture contains material, system and spirit culture on human sports, which can be divided into three aspects: sports material culture, sports system culture and sports concept culture. Furthermore, the sports culture is a special kind of sports culture phenomenon in human sports life. The campus sports culture, as the most vigorous and important part of campus culture, inherits the internal material and spirit of sports. The campus sports culture is also a kind of unique cultural phenomenon of profound connotation and plentiful extension. In this specific university environment, based on students and teachers, the campus sports culture is a kind of group culture with the main content of extra-curricular sports activities and the characteristic of school spirits. As one of social cultures, the campus sports culture is the sum of sports material and spiritual civilization, which was created by the school staff and students in the sports practice process under certain social conditions. Campus culture has rich connotation and wide extension, firstly it forms a campus culture group together with moral education, intellectual education, aesthetic education, music education and so on; secondly, it consists of sports cultural groups with social sports culture, military sports culture and so on.

3 THE IMPORTANCE OF COLLEGE CAMPUS SPORTS CULTURE

If national culture is the soul of a nation, then the campus sports culture is the soul of campus sports activities.

The campus sports culture is the sum of material civilization and spiritual civilization, created by the school staff and students, its influence on students is comprehensive and profound. A good campus sports culture can not only cultivate student sentiment, strong physique, moral education, but also can divert students' psychology and make students integrate into society smoothly. In addition, the construction level of campus sports culture affects the quality of education indirectly. So it is significant for students to create a healthy and harmonious cultural atmosphere on campus through rich content and various forms of sports, which can provide a good learning environment and psychological development space, and promote students' comprehensive development physically and mentally.

4 THE SITUATION OF COLLEGE CAMPUS SPORTS CULTURE CONSTRUCTION

The college campus sports culture is a multi-level cultural form, which includes four kinds of basic culture morphology of material culture, system culture, spiritual culture, behavior culture. On the whole the construction of campus sports culture in our country has achieved great progress in recent years, but there are still some problems. With the rapid development of society, the research and construction of college campus sports culture have fallen behind. The survey indicates the university has many problems in the construction of campus sports culture. Such as, the lack of teaching material about humanism, insufficient humanities landscape and cultural venues, nonstandard of community management, and without characteristics of campus sports culture.

5 THOUGHTS ON THE CONSTRUCTION OF CAMPUS SPORTS CULTURE

5.1 *To strengthen the characteristics of campus sports culture construction*

The construction of campus sports culture in higher education institutions should adopt suitable measures to local conditions and highlight the characteristics. School type, school conditions and geographical location are different, therefore higher education institutions should construct campus sports culture according to their own specific circumstances, thus form their unique traditions and characteristics. Such as, the ethnic university can be built campus sports culture with ethnic characteristics, and attract more students do physical exercise. In China, the characteristics of campus sports culture in Tsinghua University and other universities is significant. The campus sports culture in Tsinghua University is to work hard, exercise more, build up physique, and healthily work for our country fifty years, which embodies a unique feature of Tsinghua University campus sports culture construction. As the American scholars believe that "Harvard University has its training mode, and Yale University has its own. If all teachers at Harvard are asked to teach at Yale University, they cannot create students who bear Harvard style completely." [3]. This explains that the unique campus culture affects the students intrinsic, the charm lies in this. Therefore, the key of the construction of campus sports culture in universities should emphasis on the characteristic construction.

5.2 *To strengthen publicity and education of humanity's knowledge with the network and other media*

With the popularization of network in the campus, surfing on the internet for everyone has become a reality. A survey shows that "rate of surfing on the Internet" of college students is very high. Nearly 80% students use the network at the average of half an hour to 5 hours a day. The network has become an important tool of their lives and learning. Therefore, as a campus important medium, the campus networks play an important role in the propaganda and education on sports humanity's knowledge. College students can learn the sports humanities knowledge and physical techniques theory via internet, and then has the power of participating in sports activities. As a result, the campus sports culture atmosphere are formatted favorably. Not only the campus network can play an important role in the publicity education of sports humanities knowledge, but also the newspaper, campus radio and cable TV play a corresponding role. This has the positive significance for enriching the theoretical knowledge for sports college students, enhancing the enthusiasm for sports, improving the sports culture of college students, and promoting the construction of campus sports culture in colleges and universities.

5.3 *Increasing the sports facilities and extending free time open to students*

Due to the continuous expansion of universities, the amount of students on campus is increasing. The campus sports venue has been unable to meet the needs of the students. According to the relevant regulations of Chinese "sports law", the universities are confronted obsolete and lack situation of the sports venues and facilities. The leaders of education departments and universities should attach great importance to the problem of lacking sports venues, equipment and venue, and carry out strictly relevant national regulations about the construction of school sports venue. The universities should add sports facilities to ensure the normal development of physical education and provide the necessary material conditions for the all-round development of college students. In addition, in order to pursuit economic benefits, the free time open to students is too little in most universities. It seriously hinders the participation enthusiasm of the students in sports, and is not conducive to the construction of college campus sports culture. Therefore, the universities should increase the sports facilities and extend the free time open to students, which can stimulate the participate enthusiasm of students in sports activities.

5.4 *Standardization of the construction of campus sports clubs and associations*

The campus sports clubs and associations organize and participate in various forms with rich content and colorful sports activities based on common sports interest. It is also a kind of campus sports culture phenomenon and an important part of the college campus sports culture. The survey indicates that college sports club in our country lacks certain management criterion at present. The problems of loosing organizational discipline and low efficiency seriously affect the construction of campus sports culture. Therefore, the universities should bring out a series of practical assessment system based on reality, which give guidance and supervision for campus sports clubs and associations. Particularly, the relevant department should control strictly in approval of sports clubs and associations and ensure its necessity.

5.5 *Learning and innovating*

The construction of campus sports culture cannot explore the road in a single and closed environment, cannot follow the conservation track, but should actively absorb the mode of developed countries in the construction of campus sports culture. On the one hand, it should discard its dross and use the essence to construct the campus sports culture. On the other hand, in reference foreign experience, it should combine local characteristics and innovate, thus create a new perspective of the construction of campus sports culture.

REFERENCES

[1] CCP decisions on deepening the reform of the cultural system and promoting the socialist cultural development and prosperity.

[2] Yu Kehong. Xie Xiang, Xia Siyong. Sports culture [M] Guilin: Guangxi Normal University Press. 2003, 5–6.

[3] Jiang Zhiming, Qu Xinyi. Problems and thoughts on contemporary sports culture construction in University [J]. Journal of Harbin Institute of Physical Education, 2010, 28(5): 16–17.

[4] Chu Yonghe. Study On the history of campus sports culture [J]. Journal of Dalian Education University, 2003(1): 30–31.

[5] Zhang Yuansheng. Internet use preference and presentation[N]. Journalism Review.

[6] Yu Xiuytao. Research on extracurricular sports culture of College Students[J]. Journal of Xinxiang University, 2011(8).

[7] Cao Linlin, Zhu Zhaoyong. Discussion on the function and construction of university campus sports culture [J]. Science and technology aspect, 2011.

Management, Information and Educational Engineering – Liu, Sung & Yao (Eds)
 ISBN: 978-1-138-02728-2

Transformation and development strategies for EGP teachers in higher vocational colleges

Xiao Mei Ping & Shan Hu Ma
Shi Jiazhuang Vocational College of Finance & Economics, Shi Jiazhuang, Hebei, China

Zhi Gang Liu & Ai Min Zhang
Shi Jiazhuang Jingying Future School, Shi Jiazhuang, Hebei, China

ABSTRACT: With the rapid development of higher vocational college, as well as the purpose to cultivate graduates with high competence and vocational adaptability, EGP teachers are facing challenges to complete their transformation and redesign their professional development. The thesis analyzes the necessities for the reason EGP teachers have to transform their professional development direction and it also indicates that EOP has been a mainstream in vocational English teaching. The thesis repositions English teachers' role in terms of EOP concept and lists requirements for EGP teachers to apply EOP teaching model. Besides, based on practical exploring, the author gives some developing strategies and requirements for EGP teachers' transformation and professional development.

KEYWORDS: EGP; higher vocational college; EOP; transformation; professional development.

1 INTRODUCTION

With the fast development of higher vocational education in our country, the number of the EGP teachers in higher vocational colleges is increasingly large. However, the composition and the quality of such a large group may have a big difference. These English teachers hold a large sum of different kinds of educational theories and teaching models. Thus, the above factors may cause a great divergence to future development of the vocational education. According to *strengthen the work on talent education and training in the higher vocational college* issued by Ministry of Education, as for the public English teachers, they ought to desalt the limitation between the basic course teachers and the specialized course teachers and make themselves become an expert in one field through possessing all-round knowledge and ability in a gradual way. However, how to make a successful transformation and what measures should the higher vocational colleges take to promote the EGP teacher's transformation and professional development have becoming an urgent issue. Based on the practical experience, the author has made a deep discussion on the necessities of transformation and development strategies and training measures for the EGP teachers in higher vocational colleges.

2 THE NECESSITIES OF TRANSFORMATION

2.1 *To meet the requirements of the national program on vocational education*

General Secretary Xi Jinping made an important instruction on accelerating the development of vocational education at national conference of the vocational education work on June 23, 2014. He emphasized that the nation endeavors to construct the vocational education system with Chinese characteristics. During the meeting with the representatives before the formal conference, Premier of the State Council Li Keqiang also emphasized that the nation ought to cultivate the coach-like teaching staff with distinguished characteristics of vocational education. That means all the teachers in vocational colleges should set their professional development goal to meet these requirements. To achieve this goal, the urgent thing for teachers is to shift their traditional ideas and thoughts about education and throw themselves into learning about professional knowledge and try to obtain enterprise practical training.

2.2 *To comply with the trend of the English teaching reform in the higher vocational colleges*

Under the guidance of vocational education policy, English teaching reform is blending with vocational education features. That is to say, The traditional teaching methods are out of date, especially focusing on the knowledge and imparting and ignoring the cultivation of the language skills, which make students lack interest and motivation in English learning. Although some English teachers strive to use different methods and simplify the teaching contents, the teaching effect is still not satisfying. The key point is that the teachers don't grasp the direction of the English teaching reform. At present, the main direction of the English teaching reform in the higher vocational colleges is about teaching English for occupational purposes (EOP) and emphasizes the combination of the language knowledge and professional quality to promote the language skills and occupational skills at the same time. Therefore, the EGP teachers in higher vocational colleges have to confront the challenges, complete the self-transformation and catch up with the trend of English teaching reform. It is certain that there will be huge hardship and test.

2.3 *To meet the requirements of talent training aims of higher vocational colleges*

The task of higher College education is to cultivate comprehensive quality talents. Moreover, with the high rate of the increasingly opening up and the great needs of international communication, the talents who possess a higher English level and technical skills must be more popular in the future. The English teaching aim of the higher vocational colleges should be to cultivate the specialized talents with both professional and language skills. *The fundamental requirements of English course teaching in higher vocational colleges (on trial)* in 2006 published by Higher Education Press pointed clearly that the education in the higher vocational college is to develop the higher application-oriented talents in the fields of technique production management and service. In this way, the English teaching ought to not only lay the language foundation, but also concentrates on fostering the language skills for practical use, especially the ability to use English in handling the daily and business activities related to the foreign affairs. Therefore, it also becomes an exploring research field for the EGP teachers in the higher vocational colleges.

3 THE DIRECTION OF TRANSFORMATION

From the view of both the needs of talent training and personal occupational development of the EGP teachers, the transformation trend of the EGP teachers in the higher vocational colleges is imperative. Cai Jigang, a professor at Fudan University, once indicated that the English teaching ought to be developed and carried out following the direction of EAP (English for Academic Purposes) or EOP (English for Occupational Purposes) Hence, the English teachers in higher vocational colleges ought to adapt to the new situation and transform to the EOP teacher positively.

3.1 *Conception and features of EOP*

EOP (English for Occupational Purposes) is an important component of ESP (English for Special Purposes). Hutchinson & Waters (1987) made an authorized definition on ESP: ESP is an approach to language teaching in which all decisions to content and method are based on the learner's reason for learning (ESP is related to certain occupational subjects and objectives. The teaching content and method of ESP are determined by the special purposes and the needs of learners). ESP has a larger connotation and what is comparatively accepted is that it centers on the ultimate language use and language environment of the learners.

The ultimate purpose of EOP is to foster the students to exercise the communication ability in certain practical occasions, not simply emphasizes on the student's ability of listening, speaking, reading, writing and translating, but attaches more importance on fostering the occupational abilities which the students will need and use in their future jobs.

3.2 *Requirements for EOP teaching model*

The EOP teaching model aims to teach English language points and skills related to certain professions at the superficial and practical level. The features of the EOP teaching model require that the English teachers don't only have a solid foundation in English language, but also learn about some related occupational knowledge and gain practical experience. What's more, the EOP teachers should be very familiar with EOP theory, be good at grasp features of the target language and meanwhile have the ability to analyze learner's needs.

4 SUGGESTIONS ON DEVELOPMENT STRATEGIES FOR EGP TEACHERS

4.1 *Set up a reasonable course system*

Curriculum design is the core of the training of talents. Scientists and reasonable course system are the key to the realization of the goal of talents' cultivation. In Shi Jiazhuang Vocational College of Finance and Economics, English course is set up as EGP+EOP. EGP, English for General PurposeS, is set up at the first school year, which focuses on the basic language points and culture tips to help students go along with their freshmen life. While EOP, English for Occupational PurposeS, is set up at the third term, aiming to cultivate students professional accomplishment and communication ability. The course model of EGP+EOP conforms to the positioning of vocational English teaching as well as the social needs and talent training needs. At the same time, it gives EGP teachers time and space to complete transformation.

4.2 *Establish the effective training system*

Staff construction is the important task of educational reform and development, even one of the core tasks in Institutions of Higher vocational Education. An effective training system is a guarantee and support for English teaching staff. In our college, we put forward a training plan named as two kinds of practices, one combination. We have established the Foreign Language Bridge Translating Studio as the in-campus practice base and have made Shi Jiazhuang Shenzhou Hotel as the out-campus practice base. Meanwhile, we ask English teachers to involve in professional research projects, joint together with professional teachers and help each other to make improvement. This training system establishes a new platform and effectively helps EGP teachers design their professional development.

4.3 *Obtain related certificates*

Through learning to obtain some professional certificates is a fast and effective way to reach to a double-qualified teacher. But what kind of certificates are suitable for public English teachers is still a disturbing problem because there is no specific document to define it. Through investigating and considering every factor affecting teaching and teacher professional development, we advise the EGP teacher to prepare for the certificates like TOEIC (Test of English for International Communication, CATTI (China Accreditation Test for Translators and Interpreters) and BEC (Business English Certificate). These examinations are related to English and meanwhile aim to practice learners' practical competence in real business situations. If English teachers get to the intermediate level of these exams, they will be competent as an English teacher in vocational colleges.

5 REQUIREMENTS FOR EGP TEACHERS

EGP teachers should not forget to strengthen basic skills in terms of the language ability, education concept, knowledge structure and teaching methods.

5.1 *To raise language skills*

English language is the tool for English teachers to teach knowledge and organize teaching activities. Regardless of language skills, it is unlikely for an English teacher to talk about education career. To teach is to learn. EGP teachers must keep a good learning habit to absorb new knowledge and consolidate the language basis continuously. EGP teachers still have to learn from each other to make great effort to explain the profound knowledge in a simple way and guarantee the teaching effect.

5.2 *To update teaching concept*

Teaching idea and concept are the soul of the teachers, which constrain every teaching activity. EGP teachers should firstly learn EOP teaching theories, have a clear awareness of the importance of the EOP teaching model, and secondly pay close attention to recent news about education at home and abroad.

5.3 *To optimize the structure of knowledge*

The new turn of vocational education reform urgently asks teachers to the optimization of their knowledge structure. Through investigation, we made a statistic. It shows that 70 percent students hope the teachers give much more practical illustrations or examples to expand knowledge, and 55 percent students demand that their teachers had better know their major knowledge. From this view, students do not remain on the level of getting pure language knowledge, but have a more clear need. EGP teachers should make themselves get ready to enrich professional knowledge through various channels and approaches.

5.4 *To upgrade the standards of teaching and researching*

At present, most students lack interest in English learning due to the boring teaching materials and methods. The CBI teaching method is widely adopted in the EOP teaching model. It is a kind of experience teaching method. It must follow certain principles, such as to put together a variety of language skills; to make students active to each phase of the learning activities. Through the 160 papers (from CNKI) published between 2009 and 2013, we have found that only 12 papers make research on CBI teaching. Among which, Shang Weixia and Zhu Jinlan have made the empirical research and their results can be referenced for the English teachers. Therefore, there is a large research space for the English teachers to make the attempt and innovation. Only when a teacher has initiated in teaching and researching, would he be progressive and his career success.

6 CONCLUSIONS

The transformation development for EGP teachers has become a trend and necessity. A successful transformation may not optimize the teacher's knowledge structure, but raise the teacher's professional quality, help them to reposition the traditional role in teaching.

REFERENCES

Cai, J.G. 2007. The *English Teaching Features and Countermeasure Study of Chinese Universities during the Transition Period [J]*. Foreign Language Teaching and Research.

Chen, X. 2012.*A Study on the Career-oriented Public English Teaching Reform-based on the Theory CBI.* Journal of Tongling Vocational and Technical College.

Department of Higher Education, Ministry of Education of P. R. China,2000. *The Fundamental Requirements of English Course Teaching for Higher Vocational Colleges(on trial) [m]*. Beijing: Higher Education Press.

http://baike.baidu.com/view/3452792.htm?fr=aladdin

http://news.xinhuanet.com/politics/2014–06/24/c_12.

Hutchinson T, Waters A. 1987.*English for Specific Purpose: A Learning- centered Approach [m]. Cambridge:CUP.*

Ministry of Education,2006.The fundamental requirements of English course teaching for higher vocational colleges(on trial). Higher Education Press.

Shang,W.X. 2010.A *Case Study of CBI Teaching method on the English Teaching in the Higher Vocational College*. Journal of Harbin Institute of Vocational Technology.

Zhu, J.L.2012. *A Teaching Case Study on effect of ESP on EGP teaching based on CBI concept*. Journal of Nanjing Institute of Industry Technology.

Management, Information and Educational Engineering – Liu, Sung & Yao (Eds)
© 2015 Taylor & Francis Group, London, ISBN: 978-1-138-02728-2

Enlightenment of research themes on the Poyang Lake ecological economic zone on academic tourism research

Ci Pin Jin
The Center for Tourism Planning and Research, Nanchang University, Nanchang, Jiangxi, China
Tourism Department of Economics & Management School, Nanchang University, Nanchang, Jiangxi, China

ABSTRACT: In December 2009, China's State Council approved the Poyang Lake Ecological Economic Zone Project, which has made Jiangxi's Poyang Lake ecological economic zone to become one of the development strategies of the state construction. Related academic research is rapidly increasing, but the research focuses mainly on regional comparisons, development countermeasures, economic development, and ecological industry. However, there is less fundamental research on regions, systematic economic theory and practice. This paper reveals that the tourism academic studies should be based on relevant research to strengthen the research on regional development stage, regional growth, and regional development space in order to provide research results for the relevant decisions of the government.

KEYWORDS: The Poyang Lake ecological economic zone; Research themes; Tourism research.

In 2009, China's State Council approved the implementation of the Poyang Lake Ecological Economic Zone Project. In order to explore the new horizons of the academic studies of tourism, this paper takes the "Poyang Lake ecological economic zone" as title search terms to retrieve the related research results from the China Journal NET and the Chinese Social Science Citation Index (CSSCI), and the retrieval period is from 1958 to 2013. This research differs from the article of *The Advances of the Research on Eco-economic Zone in China in the perspectives*. And the time span of the research is much greater.

1 TEMPORAL DISTRIBUTION CHARACTERISTICS

Table 1. Quantity distribution of research articles on Poyang ecological economic zone, 2008–2013.

Year	Quantity/%	Year	Quantity/%	Year	Quantity/%
2013	104/13	2012	204/26	2011	238/30
2010	162/21	2009	47/6	2008	33/4

According to the retrieval, a trace number of research topics in relation to the Poyang Lake ecological economic zone can be searched from 1958 to 2007, so they are not listed in Table 1. Along with the continuous improvement of economy in China and Jiangxi province, and in particular the deepening of the reform and opening up, there is a rapidly increasing number of the research literature on the Poyang Lake ecological economic zone, as shown in Table 1 (Literature: 766). This indicates that the studies of the Poyang Lake ecological economic zone are rapidly increasing in abundance, which provides the rich theoretical and practical research results for the suggestion and recognition of the concept of the Poyang Lake ecological economic zone as well as the construction and development of the ecological economic zone.

Table 1 has shown that the proportion of the research on the Poyang Lake ecological economic zone was still lower in 2008 and 2009, but it reached a peak in 2011. This indicates that the Poyang Lake ecological economic zone is increasingly showing its importance in the ecological strategic position along with the deepening of the reform and opening and the improvement of people's living standards. Especially, since the State Council approved the Poyang Lake Ecological Economic Zone Project in December 2009, Poyang Lake, as the core ecological economic zone, has risen up to a national strategy in China construction, causing immediately more extensive attention of scholars. In 2010, the amount of published academic articles grew up rapidly. This also illustrates that there is a significant correlation between regional academic research and government policy. However, as time goes on, regional research tends to be the normal.

Table 2. Distribution of research themes of the Poyang Lake ecological economic zone, 2008–2013.

Year and article number	2013	2012	2011	2010	2009	2008	Total
Comparative Study	11	36	46	36	19	15	163
Education, Talent	15	25	22	13	2		77
Low-carbon, Resources, Energy	12	21	20	17	5		75
Culture, Promotion, Information	9	29	18	5	2	3	66
Industries, Property Rights	14	12	18	13	6	1	64
Tourism, Sports	6	14	19	16	4	1	60
Agriculture, Crop	7	13	18	12	2		52
Environmental Protection	9	12	14	10	3		48
legality, Auditing	3	10	16	9			38
Finance, Tax, Insurance	1	13	10	4	3	7	38
Farmland, Land, Forestry	1	11	9	1	1	2	25
Economy, Trade, Logistics	4	2	7	9		1	23
Infrastructure, Town	3	3	3	8	1	2	20
Rural area, Farmer	4	6	3	5		1	19
Industry, Quality supervision			5	6	1		12
Hydrology, Meteorological phenomena	5		1	2			8
Total	104	207	229	166	49	33	788

2 RESEARCH THEME DISTRIBUTION

The themes of the research literature shown in Table 2 focus mainly on single, other topics coming from the same article are not included in this table. Because there are only several published articles about the Poyang Lake ecological economic zone before 2008, Table 2 lists the distribution of research themes mainly from 2008 to 2013. The distribution is sorted by the volume of literature.

2.1 *Regions, countermeasures and comparative study*

Major research themes in Table 2 account for 27% of the total articles. Among them, the proportion of comparative study is 9.1%, the proportions of countermeasure and regional researches are 7.7% and 10.2% respectively. These types of research are as follows.

2.1.1 *Regions*

Regional research mentioned in this paper refers to geographical research which is based on the Poyang Lake ecological economic zone and counties and cities included. The main research contents are as follows: region-wide development policies (environment, economic level, brand-building, space, zoning, planning, the economic circle, trade, etc.) and county economy (development strategies, economy and diversification, competitiveness, integration into the economic circle, difference comparison, resource transformation, public finance, etc.).

2.1.2 *Countermeasures*

The main contents include ecological compensation (civilization, industry, protection, and economy), industry development (system structure, economic growth pole, optimization, industry-study-research combination, and industrial park), sustainable development (low-carbon economy, environment coordination, science development, technology innovation, science and technology support, town construction, education, integrated evaluation, etc.), and regulatory mechanism (legal construction, conflict mediation, village council, village governance, resource appraisal and protection, social security, credit rating, engineering advisory, etc.).

2.1.3 *Comparative study*

Comparative study was conducted mainly from the following areas: the development of lake districts at home and abroad, wetland resource conservation of the Xining River, the pattern of the national regional development, the Tennessee Valley development, eco-industrial park development at home and broad, the ecological economic development of Nanchang ancient town lakes, the development strategies of other regions, Pudong District and New Binhai District, ecological city construction in foreign countries, and the cluster analysis of science and technology innovation ability.

2.2 *Other closely-related economy research*

In addition to the above, other studies which are closely related to economy include nine areas as follows.

2.2.1 *Education*

Education research accounts for about 7.1% of the total articles. The main contents are higher education (competitiveness, leveling, eco-development of services, ecological view, and art), public education (ecological view, the development of the regional society, and self-taught examination), and vocational education (industry-study-research combination, characteristic specialty, educational technology, grouping, professional, educational faculty, and ecological view).

Talent is 1.9%, which includes: training mode (professional and high-end talent), resource allocation (resource development and team-building), and information sharing and rural talent.

2.2.2 *Industry*

Industry accounts for 7.7% of the total articles. The contents focus mainly on characteristics, creativity, industry transformation (undertaking), planning, layout, cluster, leading industry and effect expansion, and high and new technology.

2.2.3 *Low-carbon*

Study on low-carbon economy accounts for 6.3%. The main contents are as follows: low-carbon economy industry, tourism economy, industry, energy-saving and emission-reduction, regional policy, circular economy, comparative study, ecological ethics, legal system, and evaluation), low carbon agriculture (agricultural and sideline products, Eco-Park).

Resources research is only 1.6%. The main contents are resource protection (trade, mining, forestry, agricultural productivity, coupling, and evaluation).

2.2.4 *Region tourism*

Tourism takes 7% of the total articles. The research contents mainly include the integration of resources and the development of eco-brand characteristics.

2.2.5 *Agriculture*

Agricultural study accounts for 5.6%. The main contents are the modernization of agriculture, special features, brands, leading enterprises, industrial clusters, and farmer cooperation.

2.2.6 *Environmental protection*

Environmental protection is 4.2%. The research contents are mainly on factors affecting the environmental protection, forestry, urban ecology, water ecology, management mechanism, corporate responsibility, environmental protection industry, rural environmental protection and so on.

2.2.7 *Finance*

Finance accounts for 4%. The main contents focus on eco-fund investment, financial size (structure, system, institution, and coordination), program marketing, as well as investment and financing.

2.2.8 *Land*

Land accounts only for 1.7%. The research contents involve use efficiency, eco-security, sustainable use, cultivated land pressures (loss), and management.

2.2.9 *Economy*

Economy takes less than 1% of the total articles. The main contents include eco-economic mode, circular economy, empirical analysis, ecological forestry, and regional collaboration.

2.3 *Overview of the state funded research projects*

Research projects supported by the National Natural Science Foundation and the National Social Science Fund are the high level representatives of academic research in China. The research articles about the Poyang Lake ecological economic zone relying on the two funds are shown in Table 3.

Table 3. Research literature overview of the Poyang Lake ecological economic zone relying on the state funded research projects.

Type	Content summaries
Industry	Low carbon, symbiotic networks, industrial transfer, industrial manufacturing, photovoltaic efficiency, competitiveness, eco-tourism.
Urbanization	Cities, towns, urban spatial structure.
Land	Intensive Utilization of Resources, sustainable use of resources, fragility of soil erosion.
Economy	Hours communication area, Economic Contacts Variability, comparative research on economic differences.
Agriculture	Rural public cultural service, circular economy, disaster Emergency Information resources Planning, agricultural Productivity, pig green supply chain.

The projects supported by the National Natural Science Foundation and the National Social Science Fund, and CSSCI indexed literature is a hallmark of acknowledged high scientific standards of domestic scientific research. CSSCI has collected the research articles about the Poyang Lake ecological economic zone in the same abundance as China Journal NET, but in much less (absolute) number. There were totally 68 articles within the retrieval period. The articles were largely published between the year 2009 and 3013. Except 10 pieces of articles published in 2009, there was the similar number published in the other years.

Since 2009, only a small amount of the projects on the National Science Foundation have focused the research on the Poyang Lake ecological economic zone. The studies include hydrology, meteorology and environmental monitoring. In the last four years, there was an average of 1~2 national social sciences Fund projects emphasizing the research on the Poyang Lake ecological economic zone each year. These studies include the new models of river basin comprehensive development, low-carbon tourism, and industry spatial layout.

Between 2001 and 2011, a total of four projects on the national social sciences foundation were approved about the Poyang Lake ecological economic zone, which involved in century ecological evolution of the river basins , the loss value accounting of wetland ecological environment, the cultural creative industry development, and the main body function partition (zoning) the counties respectively.

3 ENLIGHTENMENT TO ACADEMIC RESEARCH ON TOURISM

3.1 *Strengthen the research on regional development stage*

According to 2.2.1~2.2.9, there are rich comparative research on the Poyang Lake ecological economic zone and countermeasure study on locality, but inadequate on regional development stage, which indicates that no in-depth articles can be found for the development model represented by the theories of both the core-periphery suggested by Khufu-Fischer, Rostow and Friedman, and the new international division of labor. Therefore, when making research on regional tourism investment and the competitiveness of tourism service trade (reflecting the economic level of the region), it is vital to strengthen the research on intra-regional development stages of the level of rural economy, the structures of agricultural production, the level of industrialization, economic growth stage in order to provide decision-making basis for intra-regional tourism investment and tourism service trade.

3.2 *Strengthen the research on regional development space*

In terms of geographical space, the Poyang Lake ecological economic zone occupies 1/3 of the total land area, 1/2 of the total population, and 60% of GDP of Jiangxi province. In terms of ecological function protection, it is one of the most important ecological areas in the world, performing a variety of ecological functions of regulating and storing large basin flood, regulating the climate, and degrading the pollution. It can be said that the Poyang Lake ecological economic zone is not only the leading economic development in Jiangxi province, but also one of the leading influential factors on China environments. There are some regional research topics on "leading" features, as listed in 2.2.1~2.2.9. The research contents involve leading industry, industry groups, growth pole effect, industry transfer (gradient development theory), and so on, but there are still less tourism-related research. Therefore, it is important to strengthen the research on leading tourism enterprises and industry groups (including the expansion effect), external economic (interregional economic linkages), the growth poles of the tourism industry, the gradient of the tourism industry, leaping development to provide research results for the tourism development space of the Poyang Lake ecological economic zone.

3.3 *Strengthen the research on regional growth*

Theoretical study on regional growth focuses mainly on economic take-off, innovation and dissemination, the dual economic structure and production complexes within some certain region. By analyzing comprehensively the research results in Table 2 and the second point above (RESEARCH THEME DISTRIBUTION), this paper has found that there is still inadequate basic research on the regional growth of the Poyang Lake ecological economic zone. It is time to strengthen more intra-regional academic tourism research on capital transfer, regional economic innovation (including enterprise innovation), the development stage of city economy within the region, and the effects of the tourism industry to improve the dual economic structure in order to better promote the tourism economic effects on the national economy.

4 CONCLUSION

Early research on the Poyang Lake ecological economic zone is relatively scattered. However, with the economic development of China and Jiangxi province, especially after the reform and opening up, the economic growth has been promoting the development of the related studies, what's more, the Poyang Lake Ecological Economic Zone Project approved by China's State Council in 2009 has brought the research on the Poyang Lake ecological economic zone to a climax, but the basic research on the Poyang Lake Ecological Economic Zone and systematic economic theory and practice still need further. Regional academic tourism research should be based on related studies to emphasize on regional development stage, regional growth, and regional development space in order to provide research results for the relevant decisions of the government.

REFERENCE

WANG Shu-ming et al, Advances of the Research on Eco-economic Zone in China[J], Journal of Anhui Agri. Sci. 2011,39(29) : 18080–18082, 18157.

Management, Information and Educational Engineering – Liu, Sung & Yao (Eds)
© 2015 Taylor & Francis Group, London, ISBN: 978-1-138-02728-2

The integration of learning management system and other related information systems in open learning

Ming Jie Tan & Pei Ji Shao
University of Electronic Science and Technology of China, Chengdu, P.R. China

Yi Gong
Sichuan Radio and TV University, Chengdu, P.R. China

ABSTRACT: As open learning is developing into a significant scale, information isolated island, caused by the unsatisfactory integration of information systems, has attracted much attention. This paper describes other common information systems related to learning management system in open learning, puts forward a framework for system integration based on an analysis of the demand for system integration and key technologies. A data center framework is then proposed based on an analysis of data integration.

KEYWORDS: Open Learning; Information System; System Integration.

1 INTRODUCTION

Open learning is a form of education without the barriers that allows learners to receive education at their convenience without the constraints of time, space and other objective conditions [1]. Open University of the UK, the earliest organization to offer open learning, has now nearly 300,000 registered students, about one fifth of whom come from areas outside Britain. Countries like Canada, South Africa and India have also followed suit. The Open University of China has now over 3 million active students, making the university No.1 in student number in the world.

Recent years have seen the emergence and prevalence of MOOC (Massive Open Online Courses) and its future development seems imaginable [2]. In an age of fast progressing information technology, the popularization of communication devices has been the foundation for the booming of open learning. Learning management system, the core information system in open learning, has been used by educational institutions to keep and release learning resources and by teachers and students to interact in the education process. However, to conduct teaching and learning activities cannot solely rely on learning management system [3]. In order to provide a better information technology support to open learning, there is a need to set up peripheral information systems, including virtual lab, online examination system, online video teaching system and online customer service system.

Information isolated island, a problem faced by many educational institutions in the process of information construction refers to the situation where the databases of various information systems (which do not have connectors) are isolated from each other and the systems have bad organic integration [4]. It is essential for open learning, where teaching and learning activities are conducted solely through information technology, to integrate different information systems effectively. This paper systemizes integration demand, integration technology and data integration to explore into the integration of learning management system and other peripheral information systems in open learning.

2 PERIPHERAL INFORMATION SYSTEMS

Many systems provide support for open education: virtual lab which allows students to experiment virtually online, online video teaching system which supports online video teaching, online examination system that allows students to take tests and examinations online, and online customer service system that provides support to students in learning. These systems are briefly introduced below.

2.1 *Virtual lab platform*

Using multimedia and computer analog technology, virtual lab enables experimenters to obtain similar to real experiment results through simulation and expansion of real experimentation environment [5]. The virtual lab platform is constituted by experiment bases facing different course experiment demands and the corresponding function modules for experiment management.

As online learners are mainly learning online based on the Internet, it is the reality of this form of education that there is the separation in time and space between students and teachers, and between students and the physical teaching environment. Students are thus unable to make real experiments on the spot. Virtual lab provides an online open virtual platform for students based on virtual reality technology. The system framework of the virtual lab platform is shown below.

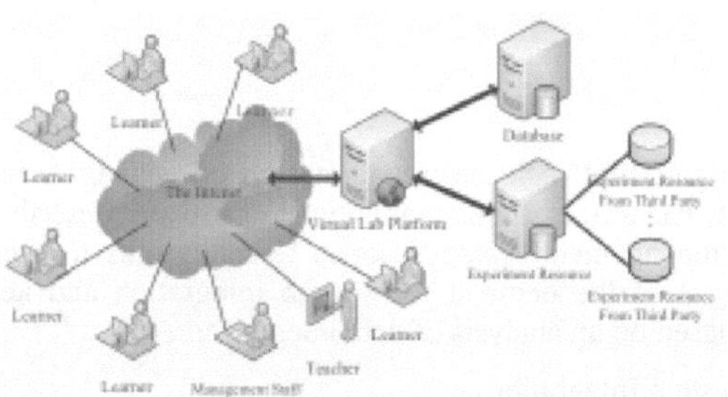

Figure 1. The system framework of virtual lab platform.

Teachers can make plans for the content of experiments in advance as necessary to meet teaching requirements. Learners can carry out experiments in the virtual platform through the Internet anytime and anywhere, and teachers can later instruct students on their experiments and evaluate experiment results either in real time or not.

2.2 *Video teaching system*

Video teaching system is the video conference system for online teaching, which realizes real-time interactive communication through the spread of audio, video and document materials based on transmission lines and multimedia devices. It can reduce learners' loneliness caused by the separation with teachers in time and space and enhance the learning experience.

Meanwhile, teaching videos can be kept on learning management system as teaching resources for students to view anytime. Students can also be timely reminded by the learning management system of teaching activities and enter the video teaching system via a single sign-on to interact with teachers in real time.

2.3 *Online examination system*

Tests and exams are key steps in the whole teaching process. Online examination system allows open learners to take formative tests in the learning process and take online terminal exams at the end of the course [6]. Online examination system is comprised of question base management, exam module and evaluation module.

2.4 *Online customer service system*

A real-time communication software program based on web-pages, online customer service system allows web-owners to communicate in real time with visitors without the need for visitors to download any client program. Through this system, learners can make real-time consultations online if they need, which can improve their user experience, and the management and service of the education provider.

The online customer service system consists of three parts—client server, customer service provider server and management server. Client server is the interface where visitor asks questions to service provider; service provider server is the interface where the provider answers questions and it includes basic management functions such as dialogue dispatch, phrase management, etc.; management server provides functions including the overall setup of the system, user management, log query and data analysis. The basic framework of the system is shown below.

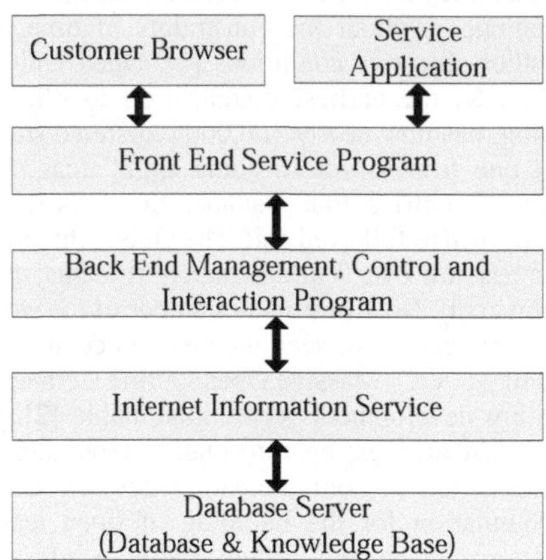

Figure 2. The system framework for online customer service.

3 SYSTEM INTEGRATION

Along with the teaching management system, learning management system is included in two of the core information systems of open learning [7]. The integration of learning management system with other systems includes the synchronization with data in teaching management system and the integration with peripheral learning assistant systems.

3.1 *System integration requirement*

3.1.1 *Learning management system*

The learning management system can obtain information such as major, curriculum, teaching plan, school system structure, student basic information and selected courses from the teaching management system, and can also return the grades of formative tests to the teaching management system. The learning management system allows students to check information such as academic credentials, selected courses and grades.

3.1.2 *Online video teaching system*

The online video teaching system can obtain information such as student basic information and selected courses from the teaching management system. The application and approval for the online video classroom can be made in learning management system, and the online video teaching system will arrange for the classrooms after obtaining related information. Courses available online are also provided to students in learning management system, and students can click on the corresponding links into the online video teaching system.

3.1.3 *Online examination system*

Online examination system obtains information such as major, curriculum, teaching plan, school system structure, student basic information and signed exams from the teaching management system. Exam systems targeting different contexts can interact with exam data through connectors. Student exam information is reported to students in learning management system, where students click on related links into the online examination system.

3.1.4 *Virtual lab*

The virtual lab system obtains information such as major, curriculum, teaching plan, school system structure, student basic information and selected courses from the teaching management system. Available courses online are provided to students in the learning management system, where students click on the related links to study in virtual lab courses. Evaluation from the platform is then returned to the teaching management system. The virtual lab platform is well expandable and compatible with new virtual lab teaching resources.

3.1.5 *Online customer service system*

Once students log in the learning management system, they can carry their basic information to the online customer service system by clicking on the link to it. Service providers can view student basic information such as name and major at the provider server.

The integration framework of learning management system and other peripheral systems is shown below.

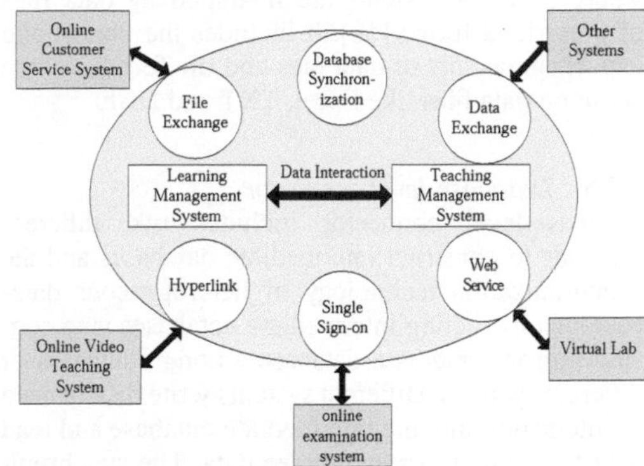

Figure 3. The integration framework of learning management system and other peripheral systems.

3.2 *Key techniques*

To meet practical requirements, the integration of learning management systems and other peripheral systems will involve these techniques: hyperlink, single sign-on, Web Service, file exchange connector and database-level connector.

3.2.1 *Hyperlink*

Hyperlink generally refers to the directing link from one web page to another, and hyperlink here is one from one system to another. In the redirection process, user information of the original system and other information needed should be carried to the target system, and the information carried should be encrypted to ensure system security if necessary.

3.2.2 *Single sign-on*

Single sign-on means user needs only one sign-on to have the access to various systems that are mutually trusted [8]. The certification system will certify user's register information, and if the information is passed, a certification ticket will be returned to the user, which will be certified for assessing permission by other systems that the user might be visiting.

3.2.3 *Web service*

Web Service is a service-oriented architecture technology that provides services through standard Web protocols in order to ensure the application services of different platforms can interoperate [9]. With this technology, different systems can interact and share data in real time.

3.2.4 *File exchange connector*

File exchange connector is a kind of data connector designed by software system to meet specific demands facing specific data interaction purposes, where different systems are mediated by data files of prescribed forms [10]. It includes the import and export connectors of data files and file forms include common data files like Excel, TXT and DBF.

3.2.5 *Database-level connector*

Database-level connector includes two different methods to construct intermediate databases and the synchronization technology of heterogeneous databases. Constructing intermediate databases is to construct an intermediate database among databases of different systems. Different systems write data in need of interaction into the intermediate database and read data from there to synchronize data. The synchronization technology of heterogeneous databases mainly includes: trigger technology, log analysis, middleware technology, etc.

4 DATA INTEGRATION AND DATA CENTER

The data center in open learning uses data integration technology to integrate heterogeneous data sources from individual information systems into the center database so as to share data and information precisely and timely. Based on normalized data standard, data center can store various data, information and resources together so that heterogeneous database systems can integrate, exchange and share data, and the data in each system will be updated timely and highly consistently.

The framework of data integration and data center in open learning is shown below. The system framework of data center is comprised of data source layer, data integration layer, data sharing layer and application service layer.

4.1 *Data source layer*

Data source layer, the source of data in data center, includes heterogeneous operation system databases, such as teaching management system, learning management system, system of personnel and payroll and databases from other information systems, as well as peripheral data sources.

4.2 *Data integration layer*

Data integration layer contains data extraction and application transfer as well as data synchronization and application. Data extraction and application transfer are to extract, transfer, integrate and clean data from heterogeneous data sources based on prescribed data standard and pass them on to data center for storage. Data synchronization and application is to obtain data from data center based on practical requirements to ensure the data synchronization in each operation system either in real time or not.

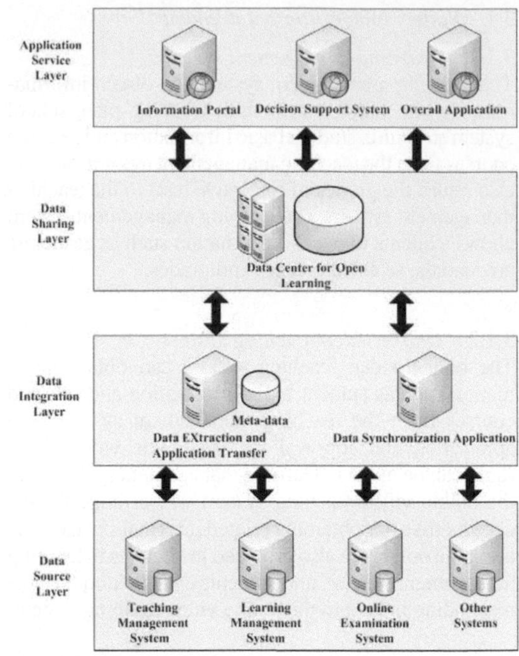

Figure 4. The system framework of data center.

4.3 *Data sharing layer*

Data center in data sharing layer keeps integrated shared data from different databases. Data sharing layer summarizes information and provide accurate shared data for application service layer in real time.

4.4 *Application service layer*

Using data from the data center, application service layer makes statistical analysis to produce information and support for decision-making. Targeted at the overall application of cross-application systems, an overall application database is created using data center, from where an overall application program can be built.

5 PREFERENCES, SYMBOLS AND UNITS

Open education, a form of education with open learning resources for students who can study anytime and anywhere only with a terminal device.

While providing conveniences for learners, open learners often feel lonely because of lack of communication caused by the separation with teachers in time and space. Learning management system and other peripheral supporting information systems, if well integrated, can provide students and teachers with an interactive platform that is more timely and effective and better improve the utilization efficiency of teaching resources. This paper provides an effective way to integrate learning management system and other related information systems in open education so as to enhance the learners' learning experience.

REFERENCES

[1] Mulligan B, Passmore D, Baker R, et al. Open Learning Badges as a Currency for Higher Education Qualifications[J]. Organising Committee, 2012, 10(5): 50.

[2] Mackness J, Mak S, Williams R. The ideals and reality of participating in a MOOC[J]. 2010.

[3] Kumar S, Gankotiya A K, Dutta K. A comparative study of moodle with other e-learning systems[C]// Electronics Computer Technology (ICECT), 2011 3rd International Conference on. IEEE, 2011, 5: 414–418.

[4] Xuan P, Ferguson K, Marshall C, et al. An Infrastructure to Support Data Integration and Curation for Higher Educational Research[C]//Proc. 8th IEEE International Conference on e-Science, Chicago. 2012: 8–12.

[5] Ogilvie R, Sawyer R, Greenwold M, et al. Evolution of a cross-institutional asynchronous online 500 level college histology course with interactive lectures and virtual lab component (530.1)[J]. The FASEB Journal, 2014, 28(1 Supplement): 530.1.

[6] Geyuan D, Yang S, Zhongchen Y, et al. Design of general computer online exam system based on C/S mode multi-layer structure[J]. Microcomputer & Its Applications, 2011, 14: 003.

[7] Cassidy T. Education Management Information System (EMIS) Development in Latin America and the Caribbean: Lessons and Challenges[J]. Tersedia: http:// www. iadb. org/IDBDocs. cfm, 2005.

[8] Dare T S, Ek E B, Luckenbaugh G L. Method and system for authenticating users to multiple computer servers via a single sign-on: U.S. Patent 5,684,950[P]. 1997–11–4.

[9] Ankolekar A, Burstein M, Hobbs J R, et al. DAML-S: Web service description for the semantic web[M]// The Semantic Web—ISWC 2002. Springer Berlin Heidelberg, 2002: 348–363.

[10] Hass Y, Vishlitzky N, Raz Y. File transfer utility which employs an intermediate data storage system: U.S. Patent 6,438,586[P]. 2002–8-20.

Management, Information and Educational Engineering – Liu, Sung & Yao (Eds)
© 2015 Taylor & Francis Group, London, ISBN: 978-1-138-02728-2

Research and practice on teaching and research team construction of a newly-built undergraduate academy

Wei Wei & Jun Dai
Xi'an Unversity, Xi'an, China

ABSTRACT: For teaching and research of a newly-built undergraduate academy, on the basis of elaborating the importance and substance of teaching and research team, the relevant factors that affect team building are analyzed and researched, the teaching and research team building belongs to the mechanical engineering discipline of Xi'an University, and, for example, describes the relevant practical works as well as the preliminary results achieved. It also lays a good foundation for further in-depth study on teaching and research team building of a newly-built undergraduate college, and plays a reference for team building of other newly-built colleges.

KEYWORDS: Newly-built undergraduate academy; teaching; research; team.

1 INTRODUCTION

With the development of higher education and science, teaching and research issues are increasingly complex, personal research model in the traditional sense can not meet the needs of development. For new-built undergraduate academy, how to highlight teaching and strengthen scientific research are real problems placed in front of every educator. Building efficient teaching and research team is an effective way to solve these problems. Therefore, the study and practice of teaching and research team building has an extremely important significance.

In teaching, teaching model is no longer scattered, closed form, to emphasize collaboration, open form, and establish a curriculum centered teaching team, main courses need to have a number of teachers who have similar professional backgrounds to participate, according to their own features to complete interrelated and complementary teaching plans and tasks. In research, use of different teachers' professional characteristics to complete both integrated and complex research tasks, emphasizing the collective team operation to adapt to changes and demands of scientific research.

Therefore, teaching and research team composed of teaching and research staffs who have complementary skills and willing to bear responsibility for co-teaching and research purposes and methods of work. The age structure of the population, the educational structure, title structure, study relationship background should be scientific and rational, it not only bears the task of teaching and research, but also bears the responsibility to cultivate teaching and research personnel.

2 THE INFLUENCING FACTORS OF TEAM BUILDING

2.1 *Location of teaching and research*

Features are the key of universities and profession development, it is crucial to maintain innovation capability of teaching and research team. Therefore, team building must be based on their actual situation, the existing foundation and conditions, combined with the historical tradition to find their own characteristics, and then locate the direction of teaching and research. Looking teaching and research direction that is distinctive and in accordance with the schools and profession development is key to achieve breakthroughs in teaching and research and the core of discipline cultivation.

2.2 *Personnel and composition*

The factor of a person plays a decisive role in teaching and research team. At this level, the choice involves two types of people. The first is the choice of the team leader (academic leaders), the good and bad of leader decides the academic level of the entire team, research style and culture. Its responsibilities include: developing team goals, coordinating the relationship between team members and team, creating a good team atmosphere, managing them effectively through the system. According to its mandate, selection of team leader needs to consider the academic level, management talent, communication capabilities and charisma. Followed by the selection of team members, team members are cells of teaching and research team, each cell has its own professional background, educational background and knowledge background. In order to ensure echelon structures and sustainable development, selection of members should pay attention to the professional complementary, age structure and relevant background, try to choose a high-quality staff, has unique innovation, perseverance, rigorous scholarship and meticulous study spirit.

The composition of the teaching and research team are mainly two types: The first is the temporary team set up in accordance with the course or research characteristics. The second is the conventional team run-in naturally with practice. Temporary team is set up according to the specific objectives, because of its temporary, also caused running of team members, team rules, team culture needing a long process. Conventional team has an advantage in team course characteristics, research orientation, team rules and team culture, but the knowledge structure of the team members has certain limitations. Therefore, teaching and research team of university should mainly rely on a conventional-type team, in accordance with the purpose of teaching and research and research goals, carry on dynamic management for team, set up temporary team on the basis of conventional-type team.

2.3 *Cultivation environment*

Teaching and research team of university is based on teaching and research, and its environment of survival and development must be able to stimulate innovation and implement innovation achievements rapidly. In this sense, there are external and internal environment. The external environment is the ability school and secondary college can provide hardware and software support to team building outside the team. In terms of hardware, it's necessary to consider whether the teaching and research equipment, facilities and venues can be met. On the software side, the main consideration is whether there are relevant policies. The internal environment is the innovation sense among staff within the team, collaboration, cooperation, the management incentive mechanism. If these factors can be combined, the internal environment will be aggressive.

2.4 *Mechanism construction*

To build a united and efficient, innovative teaching and research team, establishing and improving the management mechanism is essential, only sound management mechanism to ensure the normal operation and sustainable development of the teaching and research team. Through the study of teaching and research teams found that they have mature management systems in line with its own characteristics, summed up the team management mechanisms have a lot in common, such as: sound and reasonable evaluation mechanism, advanced and mature incentives and stable and efficient communication mechanisms.

For the new-built teaching and research team, there cannot be a sound management mechanism, but with the development and practice of the team, through the gradual accumulation, a feature of management mechanism will be worked out, the team members can maximize their initiative and innovation, contribute their own strength to the development of the team.

3 THE PRACTICE OF MECHANICAL ENGINEERING TEAM BUILDING

3.1 *Location, composition of the staff, cultivation environment and management mechanism*

On the location and direction of teaching and research side, under the guidance of local, applied, open orientation, according to the professional features and the direction and extent of the local economy, the specialty in mechanical engineering combined with the actual situation to determine teaching and research position, namely: production-oriented line, focusing on the practical engineering application, servicing in the equipment manufacturing industry. Research directions of undergraduate are mechanical design, manufacturing, electronics and automation, the research direction of professional master is mechanical remanufacturing.

In terms of personnel and composition, the conventional team was set up based on the specialty in mechanical engineering. The specialty has ten teachers, including two professors, three associate professors, three lecturers and two teaching assistants. Conventional teaching based on the majority staff, as well as focusing on curriculum development and teaching reform. Teaching echelon centralized on curriculum was developed, and teaching methods were innovated. The staffs of part of teaching reform projects are shown in Table 1. In scientific research, the major established the temporary team centralized on a research project on the basis of conventional team. The staffs of part of research projects are shown in Table 3. The conventional team was headed by the academic leader of the mechanical engineering discipline, the temporary team was headed by the host of research project.

In the cultivation environment of team, on the external environment, the school nurtured teaching and research team, which belong to key disciplines and ascendant major priority, and gave greater support to mechanical engineering discipline in terms of both hardware and software. In 2013, the mechanics specialty was approved as key major of university, the school applied for provincial key laboratories in mechanical remanufacturing and professional master's degree in mechanical remanufacturing based on mechanics major in the same year, as an opportunity, the professional laboratory construction and teacher team building had a great progress. On the internal environment, team members worked together to create a good aggressive internal environment under the coordination of the team leader.

In terms of the management mechanisms, through investigation, evaluation mechanisms, incentives and communication mechanisms were established initially within the team, on this basis, relevant management mechanisms will be improved gradually through continuous accumulation and practices, and mature management system will be formed eventually.

3.2 *The achievements of teaching and research team building*

3.2.1 *On teaching side*

With the practice of mechanical engineering conventional teaching team building, main courses in professional realized one person-more courses and one course-more personality types teaching echelon gradually, while teaching method and teaching model had been innovated with elevated laboratory conditions, more attention to practical ability, for example: in version 14 personnel training programs, the experiments of key course and centralized practical aspects had further been increased. The teaching model of "Engineering Drawing", "mechanical principles" and "mechanical design "had been changed to Scene teaching mode. In addition, a series of teaching reform had been implemented. Teaching reform projects of mechanical engineering major in 2012–2014 are shown in Table 1.

Because mechanical engineering teaching team explored constantly in innovative teaching model, mechanical engineering major had made considerable progress in improving the quality of teaching, provincial and national university student training projects list in 2012–2014 were shown in Table 2.

3.2.2 *On scientific research side*

With the practice of mechanical engineering conventional research team building, the development of temporary research team centralized on research project was smooth. The past two years, the team had made significant progress in terms of longitudinal projects. Part of the longitudinal research projects lists in 2012–2014 was shown in Table 3. With the application of provincial key laboratories in mechanical remanufacturing and professional master's degree in mechanical remanufacturing deepening, the research team of mechanical engineering major will be further expanded, while the level of scientific research will be further enhanced.

Table 1. Teaching reform projects list in 2012–2014.

Year	Project name	Host	Members	Level
2012	Study on Engineering Quality Culture and Science Innovation of Mechanical and Electrical Major Students	Ling Liu	Hao Tian, Zhuqing Zhao	School level
2013	Research and Practice on Teaching and Research Team Construction of Mechanical Engineering	Wei Wei	Jun Dai, Yuming Zhou, Yanli Zhang, Ling Liu, Binfeng He, Peiying Bian	School level
2013	The Modular Teaching Reform in Mechanics of Materials	Yanli Zhang	Jun Dai, Yuming Zhou, Wei Wei	School level

Table 2. Provincial and national university student training projects list in 2012–2014.

Year	Project name	Level
2012	Development of Intelligent Household Wastewater Recycling Device	national level
2012	Automatic Glass Cleaning Device for Building	national level
2013	An Intelligent Dustproof Blackboard	provincial level
2013	Intelligent Garbage Can Used Solar and Infrared	provincial level
2014	Energy Saving and Environmental Protection Type Third Generation Automatic Washing System	provincial level
2014	Auto Tracking System about Solar Energy	provincial level
2014	Logic Control System Design of Intelligent Railway Crossing Signal	provincial level

4 CONCLUSION

Research and teaching are two important functions of universities, and they are important support to achieve personnel training, discipline construction and service social. Teaching is the foundation of scientific research, and research is the development of teaching.

Table 3. Part of the longitudinal research projects list in 2012–2014.

Year	Project name	Host	Members	Level
2012	Development of Intelligent AC Parameter of Capacitor and Inductor Measurement Instrument	Wei Wei	Yuming Zhou, Hao Tian, Jianbo Lv	Municipal level
2012	Research on Automatic Casting Process Design and Simulation Software	Binfeng He	Jun Dai, Zhuqing Zhao, Guofa Mi, Zhian Xu	Municipal level
2013	Study on Remanufacturing Process of Key Parts of Motor Rotor	Yanli Zhang	Jun Dai, Binfeng He, Fuxing Fu, Wei Wei, Jiayin Gu	Municipal level
2014	Research on The Key Process of Size Recovery and Strengthen for Parts of Machine Tool Spindle	Yanli Zhang	Chaohui Liu, Binfeng He, Jun Dai, Fuxing Fu, Gengrong Chang, Wei Wei	Provincial level
2014	Research on Intelligent Vehicle Collision Avoidance System	Ling Liu	Yuanhua Zhou, Jun Dai, Hao Tian, Yuming Zhou, Yanli Zhang, Yanmei Jiao, Peiying Bian	Provincial level

For the New-built undergraduate academy, in the period of changing type to technology type university, external environment of teaching and research team building had a good momentum, but the university is still in its infancy in teaching and research team building, there are a lot of major have not teaching and research team, and some majors have teaching and research team are not very clear in the positioning and orientation, personnel and composition are not very reasonable, nurturing environment and mechanism construction are not mature.

History and actual situation have fully explained the university can not horizontal comparison with veteran universities in the teaching and research team building, education level can improve will not happen overnight, and it requires constant accumulation and solid work practices, only in this way teaching and research team will truly be set up, education level can really be improved.

ACKNOWLEDGEMENTS

This work was supported by education and teaching reform Project of Xi'an university in 2012–2014: 12C115, and the key specialty construction Project of Xi'an university in 2013.

REFERENCES

Xiaoli, Bai. & Jinglan, Ran. 2009. Create theory and practice of Innovative teaching and research team. China electric power education(18).22–23.

Yanchao, Sheng. 2010. Structure of university teaching and research team and its optimization strategies. Research in Teaching33(2):30–34.

Yanchao, Sheng. 2008. Setting up and running protection of higher education teaching and research team. Research in Teaching31(3):215–218.

Guwu, Xu. 2011. Scientific research innovation team incentive policies. Research in Teaching34(6):32–34.

Xuexi, Chen. & Fumei, Song. 2012. Discussion on the problems of university teaching and scientific research team building. Journal of north china institute of science and technology9(1):100–102.

Wentao, Wan. & Lihua, Zhou. 2006. Exploration for the formation conditions of scientific research team. China Higher Education Research(11):42–45.

Yang, Zeng. 2007. scientific research team and university innovation. Higher Education Exploration(2): 134–135.

Management, Information and Educational Engineering – Liu, Sung & Yao (Eds)
© 2015 Taylor & Francis Group, London, ISBN: 978-1-138-02728-2

Study on ideological and political education of college students in the background of new media

An Bo Zheng
College of Fine Arts, Beihua University, Jilin, China

ABSTRACT: The significance and function in the background of new media in the ideological and political education of college students was analyzed according to the special feature of multimedia technique, and the application in the background of new media in the ideological and political education of college students from three aspects was discussed. Prospects for the background of new media to ideological and political education of college students are given at last.

KEYWORDS: New media; Teaching course software; Ideological and political education; College studends.

1 INTRODUCTION

New media is defined relative to a letter, telephone, newspaper, broadcasting, film, television and other traditional media using digital technology, Internet technology and mobile communication technology and other new technologies to provide information services to the audience of the new media. New media appeared, with its digital, multimedia, real-time and interactive advantage, which greatly affects people's work, study and life, especially the influence of ways of thinking and behavior of contemporary college students. Therefore, ideological and political educators should conform to the trend of the times, change the way of thinking and working mode, the innovation of Ideological and political education work for college students.

2 THE INFLUENCE OF NEW MEDIA ON COLLEGE STUDENTS

In the new media environment, students to obtain information more quickly more diversified. At present, the way for students to obtain information from the traditional reading, read newspaper, watch TV and listen to the radio waiting to extend the network, the way of expressing information from the traditional single to multi view transformation, information forms are also from static to multiple and dynamic change. In this case, the present college students in obtaining information through the new media technology and the role of the subconscious mind has changed, which has passive acceptance into in order to actively search. However, much also useless information is bad information faster into groups of students, thus affecting the healthy growth of University students.

In the new media environment, people's communication is anonymous. This reduces the interference from other individuals or social factors, is conducive to the protection of personal privacy, is conducive to the freedom of speech, and is conducive to better communicate each other thoughts. Therefore, the current college students are more willing to express their ideas by means of BBS, e-mail, QQ, blogs, mobile phone messages, find themselves unable to get in real life satisfaction and confidence in even. However, reliance on new media reduce their contacts with people in real life the opportunity, easily lead to interpersonal relationship in reality indifferent, causes the interpersonal obstacles in reality.

New media with its fast and mass content and other advantages is gradually changing the way students learn, makes their learning more flexible. In the new media environment, college students are no longer adhering to the traditional book learning and classroom learning, but also to use the search engine, the forum post and electronic bulletin board, etc. However, due to the limitation of the knowledge, experience and other factors by way of thinking of the restrictions, so in the face of massive information, they often exhibit poor choice, showing criticism rather than discrimination weaknesses. At the same time, the new media's reliance will reduce their learning ability; also can make their disregard for the accumulation of knowledge.

New media for students is to provide a different from the real life of self display arena. In the open and virtual space, students can be freely and vividly to self display, publicity of their own unique personality, MSN, QQ space, blog and microblog can become students express their emotions, to show his unique individuality and the real carrier of personality. However, due to weak self-control, college students in such an environment, but also easy to indulge themselves, is very easy to forget to yourself and to others and to the social responsibility.

3 FACING THE IDEOLOGICAL AND POLITICAL EDUCATION OF COLLEGE STUDENTS UNDER THE NEW MEDIA ENVIRONMENT OPPORTUNITY

From the previous analysis can be seen, the new media technology is a double-edged sword, it brings convenience and positive influence, but for their study and life, will bring the negative influence to them. The characteristics of new media technology to strengthen and improve college ideological and political education work to bring both opportunities and also challenges, but generally speaking, the opportunities outweigh the challenges.

The new media have the big bearing capacity, high speed, multimedia, three-dimensional, wide coverage and strong interaction advantages. In this way, we on the ideological and political education can not only with the help of the rich educational resources in new media, and can be a greater range, active, quick to the students to publicize and spread the correct ideas, theory and policy, can overcome the traditional education way limit, provided a new and unprecedented a broad platform for ideological and political education of College students.

In the age of new media, mobile phone short message, blog, forum network become the new carrier of a new ideological and political education work and new means with the characteristic of flexible, fast, and shows its unique advantage. Through the new media, in some cases, students don't have traditionally within the specified time to the specified location to receive education, and can obtain their required knowledge and information through mobile phone text messages and the Internet and other ways, which greatly facilitates the students, but also greatly enriched the College Ideological and political education work methods.

In the new media environment, students can change from passive learning to active learning, is the "indoctrination" for independent reading. Through the Internet and other new media, students can freely choose what to learn the content or access to the information they want, and can take different ways to respond to the information source, feedback and recreation can be conveniently and timely participation information. Thus, the interaction of information communication and exchange mode makes the educated to accept passively by transformation in order to actively participate in, we can make the ideological and political education work effect greatly upgrade.

New media technology has broken the boundaries between the real world and the virtual world, fundamentally changed people's way of communication. Between people in the conditions of new media communication, everyone can put their real situation and hidden, can freely speak one's mind freely, to express their own opinions. In this way, will be conducive to the educators understand college students real ideas from, so that their ideological and political education work, do have a definite object in view; is conducive to a more in-depth discussion on the related problems, so that the work of Ideological and political education should produce some actual effect.

In Ideological and political education work, the degree of trust between teachers and students how to have been important factors influencing quality of education and education and the effect of control. In the traditional teacher-student relationship, both teachers and students are always in a state of inequality, which makes the students are often reluctant to teacher to tell the truth, it hinders the effectiveness of Ideological and political education promotion. In the mobile phone short message, blog, forum and other new media's help, between education and communication with a certain degree of concealment, and also brought the two sides in the personality, rights and equality of feeling, is conducive to the formation of a relaxed and harmonious atmosphere, from which to remove the barriers between teachers and students, thereby enhancing the degree of trust of both teachers and students, the ideological and political education to have the good teaching effect.

4 THE INNOVATION OF THE IDEOLOGICAL AND POLITICAL EDUCATION WORK IN THE NEW MEDIA ENVIRONMENT STRATEGY

The application of new media means for college students' ideological and political education work to develop a new space, so the ideological and political education workers should grasp this opportunity, improve the ways of thinking and working way, innovation of the ideological and political education of College Students under the new media environment, the implementation of.

In the ideological and political education work for college students, we should be based on the use of traditional media, pay attention to the use of new media and new technology, so that the old and new media can complementary advantages, in order to give full play to the role of the new and old media respectively, all-round strengthening and improving college ideological and political education work. For example, network media, mobile media and aging in the dissemination of information on strong, interactive modes, we can open the "mobile phone forum" and "Network Interview" column, discuss the hot and difficult problems the students care about in this kind of forum or interview; newspapers and magazines such as a clear direction, strong credibility. We can use it to report the advanced deeds of advanced characters, use it to guide the correct guidance of public opinion on campus.

In the new media era, network forum, blog, mobile phone text messages, Internet chat and other emerging media, in new ways to spread information at the same time, also gradually but also naturally formed a new campus network culture forms. So, in the new media environment, the initiative to grasp the ideological and political education work, we must pay attention to this kind of change in the new media, under the new technology conditions for college campus culture construction, we must further strengthen the construction of campus network culture. Strengthen the construction of campus network culture, must pay attention to its advanced nature, and must depend on the advanced thought and theory to guide the construction of campus network culture. To promote the socialist ideology, must use the advanced culture of scientific, popular, nation and socialism to occupy the university campus media position. Therefore, should produce a lot of vivid and intuitive new media material, theme website theme of Ideological and political education should be the choice of ideological, knowledge and humanity in one, to try to make the site operation can achieve fast, mass and interaction, as far as possible so that it can close to campus, close to the students' study, life and reality thought. Ideological and political education workers should make full use of new media communication fast, Union maw and audio-visual features, trying to improve the campus network position attraction and appeal, to attract more students into the construction of college campus network culture to so as to hold the initiative of Ideological and political education work.

In the age of new media, mobile phone short message, blog and QQ has been infiltrated into all aspects of students' learning and life. Ideological and political education workers must face up to the situation, and try to play the new carrier of network media, mobile media and other functions, with interactive, experiential, guided and infiltration and other ways to carry out ideological and political education work, realize the harmony and unification of virtual space and realistic space. Ideological and political education worker should conform to the requirements, make full use of the college students are widely used in communication and communication means, will carry forward the main melody education into the college students' study and daily life, so as to enhance the ideological and political education work perspective.

With the rapid development of the information age, computer multimedia technology is widely used in the teaching of subjects, promoting education and educational technology change and progress, so that modern education in the information age. In recent years, more and more people have realized that IT is playing an increasingly important role in modern education teaching. In this environment, the ideological and political education of teaching, great changes have taken place in the traditional fine arts teaching gradually integrated multimedia technology and theory, to the ideological and political education of teaching has injected new vigor and vitality. However, most of the existing school ideological and political education teaching professional application of multimedia technology is still in the primary stage, has not yet formed a complete teaching system. As a complement and supplement traditional teaching, the popularity of multimedia features and the computer so that the computer multimedia teaching has become an indispensable ideological and political education of the ideological and political education of teaching. Multimedia systems usually consist of multimedia hardware systems, multimedia operating systems, multimedia authoring tools, and multimedia applications in four parts. The multimedia hardware system includes a computer, CD- ROM, audio input/output, video input/output and other equipment, is the foundation of a multimedia system.

Real-time operating system includes multimedia task scheduling, and synchronization of multimedia data conversion control, drive and control of multimedia devices and a user interface with a graphical and pan functions and the like. Multimedia authoring tool is a tool to create multimedia applications software, collectively, it can be text, graphics, images, animation, video and audio and other multimedia information control and management, and put them into a complete connection required for multimedia applications. Multimedia application system developers to use computer language or multimedia authoring tools to create multimedia applications software products, is directly facing the user. In the ideological and political education of teaching are the multi-use multimedia applications. So, computer ideological and political education multimedia technology compared with the traditional teaching of ideological and political education teaching, advantages and practical significance were discussed.

Ideological and political education workers are the backbone to effectively carry out the ideological and political education of college. In the new media environment, ideological and political education workers must constantly improve their understanding and use of new medium level, must try to network technology and mobile communication technology means applied to daily on the students' Ideological and political education work to. Should pay attention to the construction of their on Ideological and political education network platform, students use the Internet to study guide; should pay attention to the online information and comments, timely to comment as the guidance of students; should pay attention to ideological dynamic networks, in a timely manner to carry out related research of public opinion; should use the new media. In addition, universities thought political education worker should also closely follow the trend, pay attention to understand new media especially loved by students on the network language and communication, timely fill in Ideological and political education content, improving education, in order to make their work of Ideological and political education of college students can be more closer to the life, it is easy for college students to understand and accept.

5 CONCLUSION

Computer ideological and political education multimedia technology in teaching, not only to speed up the progress of the ideological and political education of teaching, improve learning outcomes, more important is that it can give students a more novel stimulus, resulting in the best area of the cerebral cortex related development, so as to stimulate their creative evolving thinking. The innovative works are ideological and political education majors. The information age, computer multimedia teaching the ideological and political education of teaching is the direction of development, however, stressed the role of modern teaching media is not to deny the traditional teaching, but rather calls attention to a combination of both in teaching practice, flexible use. Contemporary ideological and political education teaching focus on the integrated is use of multimedia in order to continuously improve the quality and efficiency of the ideological and political education of teaching. The main trends are in the ideological and political education of teaching is the traditional teaching media development and integration of modern teaching media, and gradually form a complete, current information society to adapt to the development of ideological and political education system.

REFERENCES

Li Si Hui. On Myth and Ideological and political education Education of College Students Comprehensive [J]. "Heilongjiang Science and Technology Information." 2008.07.

Xiao Xue bin, Julie. Influence of new media on College Students' Ideological and political education and its countermeasures (J). The study of ideological education, 2009 (7).

Wang Huan cheng (J). The new trend of development of Ideological and political education in the new media environment. The forum of contemporary education (Management Research), 2010 (8).

Lai Yong. Exploration of Ideological and political education of College Students under the new media environment J. The wealth of networks, 2010 (7).

Li Yan, Zeng Waylon, He Haitao. Under the new media environment, the new carrier of Ideological and political education of college students on J. Journal of Chong qing University of Posts and Telecommunications: Social Science Edition, 2010 (5).

Exposing the red blue, Wang Qisi. The college counselors use of new media platforms to carry out ideological and political education J. Science and Technology Plaza, 2009 (12).

Liangtai Sheng. Reform of university teaching of Chinese painting Rethinking [J]. "Zong Tai'an College of Education Science" .2003 01

Han Jing. Confront contemporary ideological and political education education principal institutions of higher ideological and political education China Forum on –2010 [J]. "Ideological and political education Watch." 2010.05.

ZHANG Yao-guang. Lilley. Chinese modern ideological and political education education in the "Western painting" Complex [J]. "Grand Art" .2011 09

Ge Xintong. Reflections on Contemporary Ideological and political education [J]. 2010. O5 "business culture."

Zhang Jun to stay. On Chinese painting teaching traditional culture [J]. "Arts education research" .2011.10

Lee slip down modern ideological and political education model and the concept of cultural and ecological harmony - of Our universities Normal Ideological and political education in the 21st Century [J]. "Inner Mongolia Normal University (Educational Science Edition)." 2006.11

Reflections by Lisa Shuai. Contemporary ideological and political education trends in China [J]. "Industry and Technology Forum." 2012.

Zhang Bing. Explore the aesthetic perspective of Ideological and political education [J]. "China-school education". 2010.05.

Management, Information and Educational Engineering – Liu, Sung & Yao (Eds)
© 2015 Taylor & Francis Group, London, ISBN: 978-1-138-02728-2

Research on theory and practice of teaching mode and teaching design based on network environment

Chun Hua Liu
College of Education Science, School of Economic Management, Beihua University, Jilin, China

Hai Long Tang
School of Chinese Language and Culture, Beihua University, Jilin, China

ABSTRACT: With the development of new network technology, the possibilities are provided for the reform of the theory and practice of the teaching mode and teaching design, which made traditional education pattern face a new challenge and chance. In this paper, the main application of the network during the teaching process in different aspects was discussed, how to guide students to get resources on the internet was summarized, and a reference for education in network environment was provided.

KEYWORDS: Teaching mode and teaching design; New network technology; Teaching process; Reform.

1 INTRODUCTION

Open education pilot time has passed, it is also a process of the teaching mode reform continuously explores, constantly sum up. In many of the modern distance education workers are thinking about such a question: in the open education, how to build a suitable model to organize the teaching work, to ensure the quality of personnel training, to promote the construction of the lifelong education system, is conducive to higher education and better for the local economic construction and make a contribution. As the practice of open education and the specific design and application discussion, we are engaged in the work is to carry out the teaching mode, and in practice to test its theoretical rationality and practical feasibility. To this end, we design a T DSE open education teaching mode, and applied in practice.

The system design of the T based on DSE teaching mode: "organization of autonomous, collaborative learning" is the core connotation of the teaching organization form of open education, which consists of a plurality of teaching information processing, processing, transmission links constitutions, namely: the design of Teaching—the teaching support service—teaching quality assurance (monitoring and assessment). It has a strong practical teaching mode system we call TDSE mode. The so-called T refers to the teaching and the D Teaching, refers to the teaching design and the design of the Design, S refers to the teaching support service in the service and Service, E refers to the guarantee of teaching quality monitoring and Evaluation in the. The implementation of the guiding ideology of this model is: to change the traditional education idea, establish the open and distance education concept as the guide, to carry out teaching activities centered autonomous learning, cooperative learning students and students, taking the teaching mode reform as the core, take the teaching design, the teaching support service system construction, the supervision of teaching quality as the key point. In modern distance education conditions, and gradually formed a "take the student as the center, teaching mode and operation mechanism of organizational autonomy, collaborative learning".

2 THE THEORETICAL THINKING OF THE DESIGN OF TEACHING MODE

Advanced education to advanced idea as the instruction, advanced technical support. Remote knowledge transmission, in the technology of computer network, certainly better than letters, radio and television, telephone and other means there are more advantages. Distance education, including knowledge transformation and then using distance of imparting knowledge and distance learning, and on this basis, rather than just text information transmission and inflexible, also requested to transmit audio, image (static, dynamic), even require the use of computer network tools, between teachers and students can timely or non timely communication, achieve interactive teaching on the web, emotional interaction purposes. At present and in the foreseeable future, only the computer network can complete these requirements. So the design of teaching mode, we should primarily be based on computer network.

However, as a technical means, the network is only a kind of knowledge transmission channel; it does not of itself produce knowledge. Knowledge comes from? The network is a professional web site source of teaching. "Technical attributes, link education website should become the study population, different teaching units for the teaching of information exchange and communication." These sites provide teaching resources, the choice of students learning through the teaching resources to acquire knowledge. Therefore, the construction of a rich in content, practical teaching platform has become the primary premise of computer remote teaching development. All the teaching activities should be based on such a platform to carry out. Or that of Distance Higher Education in the university campus is in fact the website content, form a virtual reality.

To carry out teaching activities naturally involves the application of teaching media. Both of TVU open education, or the Ordinary University School of network or other types of school, computer multimedia courseware used in although does not occupy a decisive position, but in the future development of the modern distance education, computer multimedia courseware will become the mainstream of teaching media. Especially with the popularity of home computer use, broadband into people's homes also makes this trend more and more obvious, more and more into reality. But need to note is that the computer multimedia courseware is a kind of important electronic textbooks, but does not exclude other teaching media and the existing equipment, such as paper textbooks, recording video, broadcast satellite equipment, they also have some of its advantages, such as: easy to carry, technical support is simple and related technology matures, popular and easy to accept, can carry out learning anytime and anywhere. Especially in the number of existing computer multimedia courseware is insufficient, quality is not high enough cases, the more obvious effect. To take into account other factors, such as: fettered by the traditional teaching mode, requires a process of students' learning conception renewal, learning methods, learning adaptation competence. Therefore, the open education teaching should be based on computer multimedia courseware based, multi media teaching integration of other courseware.

Adult higher education stage of the students already has the independent learning ability, and can perform some of the individual educational investment in the economy (such as the purchase of a personal computer), have certain experience of social practice, the development with modern information transmission, processing technology, which makes the way of learning to autonomous learning based possible. This is in fact also can urge students to effectively use and improve their ability of self-study.

3 OPEN EDUCATION IS OPEN THE WHOLE PROCESS OF EDUCATION OBJECT

On the other hand, the old teaching mode is all the students in a class in the same teaching environment, through similar ways of learning, with nearly identical to the "spoon feeding" method of learning to accept the same teaching content. Students cultivated in this way, just mechanical receivers of knowledge, social adaptation ability, professional competence, learning ability, innovation ability will undoubtedly inhibit the traditional teaching mode. In fact, every student has their own individual characteristics in learning, is expressed in many aspects. For example, the basic knowledge, the knowledge structure, learning ability, for adult education should pay attention to the learning environment difference, the expenditure level of education consumption, even the working industry difference. Therefore, in the design of teaching mode, the need to fully consider how this model can be adapted to the individual characteristics of students, truly "teaching students in accordance with their aptitude", from "teaching".

Open education is open the whole process of education object, teaching resources, learning methods, teaching process. To ensure that students can truly individualize autonomous learning, we should change the role of teachers. In the double subject teacher — student theory, really play a decisive role is the teacher, the teacher assigned materials, unified arrangement of the teaching progress, at the same time state, to impart knowledge by natural means, the individual characteristics of student learning is neither neglected, nor independent essence. While in the open education teaching mode, the relationship between teachers and students has become the leading role of the teacher, the student as the main body.

First, open the change of a teacher's role to reflect the opening teaching mode. In guiding the students to complete their studies in the process, we should fully consider the individual characteristics of students. For example, the course studying, not unified requirements choose the same course, but gives a flexible, have greater choice space selection suggestions, let the student self selection; in the choice of the use of teaching resources, students can choose the suitable for my use of materials, and even can choose teachers; the ways of learning, students can take the online learning, group discussion, completing a course of study to accept face to face a variety of ways. In this process, fade the subject status of teachers. Only if we take the student as the center, respect the student's personality can reflect the openness of open education and to guarantee the students accept the fair educational rights.

4 FUNCTION OF THE TARGET TEACHING MODE

The teaching resources are the production elements of the industrialization of education. To guarantee the teaching activities carried out smoothly, especially to meet the open education students independent distance on the selection of resources need to learn, the number of essential resources, resource quality will not be lower. The two points are the current restrictions, we can really carry out, where the bottleneck of remote teaching. There is a high quality, rich teaching resources, how to guide students to choose, how to guide students to make good use of resources in the self-study is the internal requirement of students' Autonomous learning. Therefore, the construction of teaching mode is designed to promote resources more efficiently and provide students.

Whether the traditional university campus or the new network university, any kind of educational organizations exist in practice that students are not able to fully autonomous learning, he needs counseling. Here said the teaching guidance is not only correct answers, the traditional sense of the operation. The specific design of open education teaching patterns needs to consider how to provide students with the more profound connotation of teaching, counselling, the whole learning process throughout the students. For example: learning the knowledge of doubt, learning methods of training, learning ability and learning and psychological adjustment, to study the effect of self detection, collaborative learning and the improvement of the quality of the. This is another function requirement of teaching mode.

The practice teaching mode of open education, even in the pilot phase should also take into account should be based on reality, but also focus on the future development. As work theory, theory of right and wrong also need to practice to identify. After the designed ideal teaching mode, it is ultimately extending and applying in teaching organization activities. Therefore, in the model design, we must consider the local education present situation, including the pilot unit condition of the current education foundation. This is the mode of open education and teaching of third functional requirements.

Reform the old teaching model, is to make our higher education is not only cultivate students. Opening education should be able to use the opportunity of various academic and non academic education in the lifelong education system, train a large number of understand the theory and practice of higher applied talents with strong ability. Therefore, the students' learning process if there are opportunities to practice and practice, whether the effect is very important in the teaching mode design. Practice can be considered: the experimental course, social investigation, production practice, graduation thesis, practice base activities, etc. In the specific model design, the teaching management is regulations on these aspects. This is the fourth functional requirements.

Adult higher education of students on a stage, are generally social work experience, the specialty is also more than your current is engaged in industry. So that students in learning do not just accept the "what", "why should the pursuit of". Most of the time, they will learn the knowledge and practical work combined to make some thinking, or with some doubts in the work to learn, want to answer. Therefore, the better teaching mode should be able to provide the opportunity for students to study learning. For example, our goal is to cultivate higher applied talents, but modern people need to accept is the lifelong education. Only have a certain amount of autonomous learning and research learning ability, can better accept continuing education.

Quality is the eternal theme of. As mentioned earlier, we test the standard of talent is the comprehensive quality of the concept of quality, and the inspection of open education model is effective or not is the important criterion to cultivate high-quality personnel with all-round development. The operation mode and the premise is to achieve this is according to a predetermined path, whether the program into practice, whether the standard mode of operation. Therefore, quality control is very important. There are two aspects of meaning: one of, in the mode of teaching designs, whether the design quality control system complete; second, system has really played the role of control. This is the mode of open education and teaching of sixth functional requirements.

5 TEACHING DESIGN AND APPLICATION OF TDSE MODE

The teaching design is the basic link, the teaching process of open education pilot first step, teaching design; rules are instructive to the teaching process. Design of open education teaching process mainly includes: the professional teaching implementation details of the course design, teaching implementation details of the design, course guidance design, practical aspects of the design, independent of the course design etc.

The teachers' professional responsibility according to the Open University of China open education of the professional teaching implementation plan, combined with local social development of the national economy, the need for qualified personnel, design and develop an "open education pilot * * professional this (the) science to carry out detailed rules" (hereinafter referred to as the "professional teaching implementation rules").

"Rules for the implementation of the design principle of" professional teaching is: to meet the needs of on personnel training mode reform and open education pilot project, give full play to the functions of modern distance education technology means, achieved by the enclosed education to open education, change the teacher centered to student-centered. The teachers' professional responsibility in the implementation of professional teaching design of "rules", should be related to the Central Radio and TV university professional teaching implementation plan as the basis.

6 THE PRACTICAL APPLICATION TDSE MODE—TO ENSURE AND MONITOR THE QUALITY

Constructing a complete and reasonable, high running efficiency of the system of teaching service is not only the general requirements of teaching rules, is under the mode of open education, how to students' individual autonomous learning better provides the teaching service performance. Open education teachers is not only the traditional teaching in the classroom, including the entire teaching activity design and guidance, provides the teaching service products and teaching support personnel. A teaching system of open education can be formed in the following framework:

a. Subject experts. Experts from the ordinary university have the professional senior title personnel as. Mainly in charge of the work of experts in the following aspects: the auspices to formulate the professional teaching plans; the detailed rules for the implementation of the detailed rules for the implementation of examination of professional teaching and curriculum; hosted and participated in the resources construction since the open course hosted the seminar; professional teaching; professional development experts put forward reference to the pilot units, in all other important teaching link and on major teaching affairs proposed expert advice etc..
b. Teachers' professional responsibility. Teachers' professional responsibility is by the open education pilot units as a full-time teacher. The design, management and service in professional liability shall be fully responsible for the professional teachers in the teaching process of open education, at the same time as the course coordinators, professional counselors; course tutor provides instruction and management.
c. Pilot project teachers' professional responsibility. The pilot units should be the Central Radio and TV University and the provincial RTVU the relevant provisions, the establishment of pilot project, teachers' professional responsibility, professional liability shall be assumed by the professional teachers with subtropical high title or have postgraduate qualifications of part-time teachers, professional guidance and accept the provincial radio and TV University Teachers' Professional responsibility.

7 CONCLUSION

This paper has made the analysis, the research in theory and practice of teaching mode reform and open education, the general view is: the transverse structure of the model lies in the teachers' leading — teaching platform (resources) — the student main body; the longitudinal links in the teaching process is the teaching design — supporting service — quality control. Thus, we propose reflected this view TDSE mode.

We think: the pilot project of open education in the reform of teaching mode bring us two kinds of situation: one, since it is a pilot project, there is no generally accepted and fixed mode can be used (so is TDSE mode), requires each pilot unit comprehensive innovation, practice, nature has the certain difficulty; second, as a new pioneering work, fewer rules, are not subject to the old mode of binding, which is conducive to the advance and application in practice of the new things. As long as the premise to follow the laws of education, focuses on the educational development of the situation, accurate positioning, bold reform, after all, can open up space for development in the construction of lifelong education system.

Corresponding author: Hai-long Tang

REFERENCES

Sun Luyi. The teaching process of distance education and awareness of J. Chinese distance education 2002.5.

Liu Xiping. Modern Distance Open Education Teaching and learning model to investigate the J. Chinese distance education 2002.6.

Zhang Weiyuan, Ouyang Lihong. The adult learning style theory, classification and measurement of the review of J. open education research, 2002.2.

Sun Guolin. The present situation of radio and TV university teaching organization form analysis and thinking of J. Journal of Chengdu University (NATURAL SCIENCE EDITION) total 61 2001.

Wang Qi, Yang Zheng. Study on Open Education Learning on the J. 2002.3 study based on Network Environment.

Yao Limin, Chen Jianguang. The background of multimedia network technology development tendency of higher education J. China Distance Education 2002.6.

Management, Information and Educational Engineering – Liu, Sung & Yao (Eds)
 ISBN: 978-1-138-02728-2

The application of ecological educational ideas to the English teaching activities

Xu Tian
English Department, Foreign Languages School, Harbin University of Science and Technology, Harbin, China

ABSTRACT: Ecology is the science that studies the relationship among creatures and that of biotic and abiotic environments. Educational ecology studies the discipline and mechanism of education and its ecological environment. Presently, there exists the educational situation which is inappropriate to the ecological, educational ideas in the college English class, therefore, the application of ecological educational ideas to English teaching is highly significant.

KEYWORDS: Ecological education; Ecological system; English teaching.

1 THE INTERPRETATION OF EDUCATIONAL ECOLOGY

There has been the ecological trend in the modern scientific development in the recent years, so numerous experts and researchers apply ecology to many fields, thereby appearing cultural ecology, social ecology and ecological anthropology, etc. Since these fields have some relationship with education and teaching development, they provide the theoretical foundation for the research of educational ecology (Yu 2006). In 1976, Lawrence Cremin, who was the dean of Teachers College in Columbia University and education critic, tried to apply the ecological methods to the education research activities. He was the first man to put forward the theory of educational ecology and tried to solve the teaching problems by ecological theory and methods in his book *Public Education*.

Ecology is the science which studies the relationship among creatures and that of biotic and abotic environments. Educational ecology is to study the discipline and mechanism of education and its ecological environment (Wu & Zhu 2000). With the increasing communication between China and foreign countries, the need for the students with a good command of English is huge. Therefore, English teaching is more and more important, at the same time, ecological theories are applied into English teaching activities, among which exist the interaction between teachers and students, among students themselves and between students and learning environment (Wang 2010). Hence, ecological theories are beneficial for the improvement of English teaching activities.

2 STATUS QUO OF COLLEGE ENGLISH TEACHING

2.1 *The unbalance among ecological factors*

According to the point of view of ecology, the ecological system is composed of several ecological factors. In this system, each factor performs its own function and they are in a balanced position. As for the college English teaching, the ecological education system is composed of three factors, i.e. teachers, students and class environment. However, in the English classroom, teachers and students are not equally balanced. To be specific, present college English class is teachers-oriented. That is to say, teachers control the whole teaching process, from drawing up and carrying out the teaching plan to the choice of the teaching materials. It is teachers who give lectures, and it is teachers who raise questions and students who answer the questions. Many students may think the teaching process is plain and dull. Even worse, some teachers impose their own opinion on the students in their class. All the above phenomena result in students' disinterest to the English study and lack of initiative. Or they consider English as a burden for their study. Therefore, it is hard for the students to improve and develop their practical English ability.

2.2 *The unbalance among students' cognitive level*

In the natural and harmonious ecological system, every ecological group should be balanced. However, in real English teaching activities, due to the different levels of students' cognitive and the limitation of objective conditions, teachers have to carry out the teaching activity based on their own designed teaching plan. There is no chance and energy for the teachers to measure students' English level in a short time. Consequently, some students may consider the teaching content too easy to pay attention to the lecture, while some students who are with the weak foundation could not keep up with the teachers, resulting in the frustrated mood facing English. Thus, in the teaching activities, students cannot accept the same amount of English knowledge of the teachers.

2.3 *The limitation of teaching environment*

The ecological environment in nature is dynamic, which changes with the interaction between the ecological factors. The minor change of one factor in the ecological system must bring the changes in the whole system. However, the present English teaching system is not ecological, in which the English teaching plan in most colleges does not change for several years. What is more, English teachers have a large amount of preparation, so the teaching materials may not be replaced for several years, which results in the consequence that the teaching content cannot keep up with the progress and requirements of the modern age. Additionally, due to the enrollment expansion of the colleges, the large class size (more than 30 students in average class) makes teachers be not able to pay attention to every student in the class. Due to the limitation of class time, it is hard for the teacher to take care of every student in the teaching activities. Most students only listen to the lectures from the teacher without interacting with the teacher or among students themselves. Therefore, the teaching environment is not dynamic.

3 THE IMPLICATION OF ECOLOGICAL EDUCATION IDEAS FOR ENGLISH TEACHING

Based on the status quo of the English class, it is necessary to instruct English teaching by the ecological education ideas in the teaching system. The ecological education ideas tell us we should emphasize equality, openness and development in English class teaching.

3.1 *The equality between teachers and students*

Ecological system is the focus and key points in the ecology. In terms of the function in the nature, each ecological factor in the ecological system is equal. In the teaching system, teachers and students, as the main ecological factors, interact and affect with each other. Therefore, they should be equal. In the process of English teaching, knowledge and information are the connection between teachers and students. More specifically, it should be clear to the teachers that knowledge and information is passed on to the students only through the medium—class. Teachers are not the creators of knowledge or information, at the same time, students are not the mere recipients of the knowledge. The knowledge or information they receive from the teachers is dealt with and processed in their brain, and students build their own knowledge hierarchy based on the acquired knowledge. That is to say, though teachers and students are connected together through knowledge and information in the teaching system, the two ecological factors (teachers and students) have hierarchical relationship and interact with each other. But in terms of function in the ecological system, they are equal. Therefore, in the teaching ecological system, teachers and students should have the necessary communication in order to achieve the equal, harmonious and sharing environment to learn the language.

Furthermore, the ecological teaching environment should, as well, be reflected in the equal attention to various subjects. English teachers should help students establish the consciousness of equal attention to the different subjects, for example, the integrated development of listening, speaking, reading and writing; the equal emphasis on the subjects in other fields, such as sociology, psychology and economics. Only when students give equal attention to every subject, can their knowledge system be comprehensive and can they become competitive in the future society.

3.2 *The openness of the teaching environment*

In the nature, the ecological system is always having the material, energetic and informational communication with the outside world, even the relatively independent ecological system has the same connection with the outside, and the continuous energy and materials enter or come out. The openness principle points out that people, when researching the ecological system, should adopt open and dynamic thought. Only when we put the research objects and ecological system together into the surrounding environment, can we reveal the essence comprehensively and profoundly. (Zhu 2010)

The English teaching environment should be open as well. Only in the open ecological environment, can input and output happen. To be specific, output is the effect of input, and input is the cause. The variation of input may touch off the change of output. Therefore, in the real English teaching environment, teachers should not always confine the teaching content into the ordinary class activities and teaching materials. What they should do is to find out some fresh materials and interesting methods to carry out the English teaching. Colleges and teachers, being the provider and creator of the teaching environment, should provide a real language environment, and make use of the modern technological appliances, such as internet, multi-media or language laboratory, to create an English learning friendly circumstances. For example, in the listening class, teachers can require students to retell the story after listening to some exciting materials. Or in the reading class, teachers can spare several minutes before class begins, during which students can share what they read or hear recently with the classmates. They can even discuss the bestsellers or the new movie. All these activities can motivate students' interests towards English. Moreover, the teaching plan should be open to the students at the very beginning of the semester, so that students (the main part of the teaching ecological system) can preview and review the knowledge before and after the class. What is more important is that students can have an overall concept towards the knowledge acquired. If the students consider something inappropriate, teachers can immediately adjust the teaching plan which is beneficial to the students.

According to the ecological theory, openness enhances the communication among the factors. In other words, openness makes every factor in the ecological system exchange continuously and always be in a dynamic state. For instance, the individuals in the ecological system can improve themselves in the open environment. Applying this principle into the teaching activities, openness can promote the communication between teachers and students and that among students. The continuous exchange of the information between teachers and students and that among students can build up a harmonious relationship between teachers and students, in the meanwhile, can build students' confidence to learn English and can add the opportunities for students to cooperate with each other. Only in the open English teaching circumstances, can students be provided the real and natural language materials. Teachers and students can obtain the information and knowledge through books, reference materials, internet and through other channels. The knowledge can be updated through the flow of information and resources, thus students can improve their independent learning ability and adaptability, which is the basis for the students' sustainable development.

3.3 *The development of the curriculum*

In the natural ecological environment, with the advancement of the industrial civilization, the conflict between human and nature is increasingly acute. For the existence and development of human beings, the idea of sustainable development was accepted by all the nations in the world in the year of 1992. In the same way, for the benefits of students, education should also be sustained. In the teaching circumstances, the needs of students are always changing, especially the needs of English, therefore, the English curriculum should be developed continuously.

In the ecological teaching environment, English curriculum should be developed. In the first place, as for the students, apart from the basic English courses, they should, as well, take some courses which are related with language and culture, such as, poem appreciation, literature criticism, etc. These subjects can add cultural deposits and do good to the English study. They can additionally choose some selective courses, such as legal English, tourism English, and economic English, etc., which helps students master some fundamental interdisciplinary knowledge.

Secondly, setting up the selective courses in other majors, on one hand, can have some knowledge about other majors by means of English in order to broaden the horizon, so the students can apply what they learn in other subjects. On the other hand, certain knowledge about other majors can promote students' confidence and enhance their core competitiveness in the society. The college or university should set up the structure of curriculum systematically, integrate knowledge from various subjects and cultivate comprehensive talents (Zhou & Liu 2011).

The core value of ecological education system is to promote the development, that is, the common and comprehensive development of both teachers and students. To ensure the harmonious development of education, universities, experts and teaching staff must make sure it develops sustainably. More exactly, setting up certain selective courses can promote the sustainable development of the education ecological system. For instance, integrating certain selective courses into college English teaching system can let students broaden their minds and build up a comprehensive knowledge system, so students can have better competitiveness when entering the society.

4 CONCLUSION

The human being should conform to the natural ecological rules, and only when they conduct correctly under the guidance of ecological views, can they live in a harmonious way with the nature. As for the university English education, teachers should construct the open, equal and a developing English studying system, and pay attention to the comprehensive development of the students. At the same time, in the English class, students should be treated equally as the teachers in order to raise their learning interests, inspire their potentials and cultivate excellent English talents.

REFERENCES

Wu, D.F. & Zhu, W.W. 2000. *Educational Ecology*. Nanjing: Jiangsu Education Publishing House.

Wang, X.H. 2010. The construction of college ecological English class environment. *Heilongjiang Higher Education Research* 10(9): 173–175.

Yu, J.Y. 2006. *The Theory and Research of Ecological Teaching*. Nanjing: Nanjing Normal University.

Zhou, G.X. & Liu, Y.B. 2011. On the comprehensive education in college from the perspective of ecological civilization construction. *Zhejiang Social Sciences* 11(8): 129–134.

Zhu, Z.W. 2010. *Steady State and Environment*. Beijing: People's Education Press.

Management, Information and Educational Engineering – Liu, Sung & Yao (Eds)
© 2015 Taylor & Francis Group, London, ISBN: 978-1-138-02728-2

The effect and practice research on the order-oriented personnel training mode

Jun Ling Zhou & Ling Feng Xie
Baoli Garden, Haizhu District, Guangzhou, Guangdong Province, China

ABSTRACT: "Order-oriented Personnel Training Mode" is a mode that directly meets the requirement of the enterprises who want to cultivate practical talents, in which way, they avoid blindness and strengthen pertinence. As for the students, their aim of study turns much clearer, their initiative and enthusiasm is much higher, in which case, this mode is efficient in formation of the students' career competence, and shortens the graduates adjustment period into the enterprise. The training mode has transformed from the traditional knowledge imparting to improvement of employment ability and the cultivation of comprehensive professional quality. This mode reconstructs the competence-based talent cultivation so as to improve the quality of vocational education personnel training effectively.

KEYWORDS: Order-oriented training, practical teaching, comprehensive professional quality, practical talents.

1 INTRODUCTION

The essence of "Order-oriented Personnel Training Mode" is combining vocational education with social and economic development, to realize effective interaction and "win-win" for everyone's situation. The core of "Order-oriented Personnel Training Mode" is collaboration between industry and school, the combination of working with learning, and two-way participation. The aim of "Order-oriented Personnel Training Mode" is strengthening the efficiency in school management and the enterprises' talent competitive advantage

The final purpose is to promote the benign development of discipline construction.

From the personnel training mode architecture, personnel training mode stage and the feasibility of three parts-depth analysis and study of the specific implementation.

More and more enterprises try to cooperate with colleges and universities, to sign the cooperation agreement in order-oriented personnel training. The enterprises participate in the graduates training program formulation and the training process management; allocate the professors who have abundant theory knowledge and strong practical ability to teach and instruct the students to do field work and finish the graduation design; provide training equipments, venues, and internship opportunities for students, after graduation some students are employed directly by the enterprise.

In the sight of international situation, "Order-oriented Personnel Training Mode" originates from Herman Schneider, academic dean of College of Engineering University of Cincinnati, America. He initiates the cooperative education in 1996. The students are divided into two groups, one group of students studied in school and the other group worked in factories; after one week, two groups reversed. This mode was called the earliest classic mode for college-enterprise cooperative education; it strengthens the education on practice and on-the-spot teaching, to provide the diversified practice experience. Through this kind of training mode, the graduates have internship experience in the university, thus they can be confident to work immediately after graduation. The training mode has transformed from the traditional knowledge imparting to improvement of employment ability and the cultivation of comprehensive professional quality. This mode reconstructs the competence-based talent cultivation so as to improve the quality of vocational education personnel training effectiveness. Through a variety of platforms of university-enterprise cooperation, with the horizontal subject intervention, and promoting teaching by scientific research, the training mode is trying to get more tremendous achievements. This training mode provides reference for the reforming of the teaching content and methods according to the actual requirements of the enterprises. This mode directly meets the requirement of the enterprises who want to cultivate practical talents, in which way, they avoid the blindness and strengthening the pertinence. And for the students, their aim of study turns much clearer, their initiative and enthusiasm is much higher, in such a case, this mode is efficient in formation of the students' career competence, shortens the graduates adjustment period into the enterprise. It manifests the talent training characteristics of "school-enterprise cooperation, the combination of working with learning", and realizes the effect of "triple-win" for school, enterprise and students. Graduates who are trained through this mode, have sturdy professional knowledge and practical ability, and own good professional accomplishment, so they are employed by enterprises and put into an important position. This mode breaks the limitation of employing graduates by simple recruitment. By getting along with the students and project training for a long time, the enterprises have deep and comprehensive understanding of the students, so that they can choose high-quality talents. This mode is different from school-base, enterprise-base and society-base education modes; it is a way to cultivate talents by the vocational colleges and enterprises, and is based on market and social requirements. This mode, imitating true enterprise operation, is a kind of exploration for the vocational education to meet the needs of the enterprise personnel quality; it's a kind of cooperation between vocational colleges and extramural organizations such as enterprises, industries, and service departments, with the target that social education meets the needs of enterprises. The most essential factor is the emphasis on the cooperation between colleges and enterprises, and taking full use of each other's advantage to guarantee the personnel training specification and quality. The mode combines the theory study with the practical operations and training tightly for students; it takes training the students' comprehensive profession, ability, and employment competitiveness as the key point, to deliver high quality applied talents for the society.

2 TALENT CULTIVATION MODEL FRAMEWORK

Vocational colleges, enterprises, government and market organize the core of university-industry cooperation. Vocational colleges and enterprises are the subjects of this activity. Considering the need for their respective interests, through the university-enterprise cooperation, both the colleges and enterprises want to achieve resource sharing, complementary advantages and win-win interaction, mutual support, and based on their respective social demand target; to restrict each other, and create a community of interests.

The model framework as follows:

a. The government has driven training mode
This model is mainly driven by national plans and promoted by the government, so that some spontaneous cooperation is only a supplement. This model is a kind of high-end project source, with funds guarantee.

b. The market driven training mode
Under the condition of a market economy, market mechanism as the basis and means of resource allocation, both of the parties (the vocational colleges and enterprises) must take the initiative to adapt to the market competition environment, or they will be far out of the market demand, which will finally threaten their survival at any time, let alone the development. Therefore, vocational colleges must actively seek the cooperation with enterprises on the basis of national talent training policy, to cultivate talents for the future, develop training plan, to ensure that the cultivation conforms to the needs of enterprises.

3 SPECIFIC IMPLEMENTATION PHASES OF PERSONNEL TRAINING MODE

Phase I : When the students begin their graduation design (paper) and field work course in the fifth semester, the enterprises come to visit the college to give explanatory sessions, lead the students to visit the enterprise, organize skill contests, etc. to make the graduates have a thorough understanding of the features and characteristics of the cooperative enterprises. The students can get to know what they are really interested in, and through the initial contact with enterprise face to face, and will try to get further cooperation.

Phase II : Make a decision on cooperative enterprises and the design direction according to the mutual choice in the first phase. Begin to designing by integrating with the graduation design curriculum content. The enterprise will send leading, designing personnel to do early brainstorming intellectual exploration; make a decision on the emphasis on the graduation design and the central part which needs training, to lay a solid foundation for the students, in order to adapt to the job requirements.

Phase III : In the midterm defense of the dissertation, the enterprise will send leading, designing personnel to our college to participate in the question-and-answer session, and give guidance according to the actual condition adjust requirements and goals.

Phase IV: Before terminal defense, the designing personnel from the enterprise communicate with graduates 2 days a week (face-to-face or network real-time communication), to modify the design scheme, and to control the quality of design requirements and specifications strictly.

Phase V: Hold a graduation design exhibition and participate in the competitions at home and abroad, through various types of skills contest, to test the result of effect of personnel training mode.

The implementation of "Order-oriented Personnel Training Mode" made the curriculum content of vocational education break through the limitation of the campus walls, converts the college education into combination between college education and practice in the enterprise, which construct the ability-based combination of learning with working personnel training model. At the same time, it requires the imitation of real market mode in the teaching process, to promote and complete a full set of design conforms to the market demand with the help and support from the leading personnel in the enterprise. In this process, the confidence, the sense of proud and achievement is cultivated, and the social and personal ability, the comprehensive quality and language ability has been improved comprehensively. The enterprise can also cultivate their own designers flexibly, according to the job requirement, and emphasizes in some respects to take advantage of each graduate's personal characteristic. Both of the enterprise and the students can find the best breakthrough point in the adaptation and break-in in this period. After graduating, enter into the enterprise, they can make development plan with the enterprise. In the terms of personnel development, this is conducive to the stability of the team.

4 FEASIBILITY ANALYSIS

The implementation of "Order-oriented Personnel Training Mode" made the curriculum content of vocational education break through the limitation of the campus walls, converts the college education into combination between college education and practice in the enterprise, which reconstruct the ability-based combination of learning with working personnel training model. Many vocational colleges benefit much on the terms of discipline construction, curriculum system, talent, training etc., and have made real progress.

1 "Order-oriented Personnel Training Mode" makes the talent, training scheme based on market demand, according to professional ability requirements to establish the corresponding training goal, lay emphasis on the combination with social talents requirements especially for enterprise and industries. Through this kind of training, the students have a precise and clear sense of employment after finish their graduation design.
2 The implementation of the "Order-oriented Personnel Training Mode", involves the curriculum formulation, scientific evaluation, with the participation and the guidance from the experts in industries and enterprises; and it adjusted constantly according to the labor market situation. In the course design, it takes professional activities as the core; in teaching content, it lays emphasis on the cultivation of professional quality, besides improving the ability to solve problems by using the knowledge and skills they have learned; and in the way of training, it ignores the boundary between theory and practice, and put what they have learned into practice.
3 Strengthen the practical and on-the-spot teaching, to provide more job opportunities for students. After the students finished their graduate design, they have got some working experience and can be confident when they get involved in their jobs at once.

"Order-oriented Personnel Training Mode" combines the theory and practice, transforms the personnel training mode of traditional knowledge imparted to the improvement of employment ability and professional ability, and reconstructs the ability-based combination of learning with working personnel training mode. It improves the quality of vocational education personnel training effectiveness. Through a variety of platforms of university-enterprise cooperation, with the intervention of graduate designing, and promoting teaching by scientific research, the training mode is trying to get more tremendous achievements. The essence of "Order-oriented Personnel Training Mode" is combining vocational education with social and economic development, to realize effective interaction and "win-win" for everyone's situation. The core of "Order-oriented Personnel Training Mode" is collaboration between industry and school, the combination of working with learning, and two-way participation. The aim of "Order-oriented Personnel Training Mode" is strengthening the efficiency in school management and the enterprises' talent competitive advantage

The final purpose is to promote the benign development of discipline construction.

REFERENCES

[1] *The Conditions, Countermeasures and Suggestions of the Construction of "Order-oriented" Training Mode for Vocational Education*, Academic Journal of Nanjing Institute of Industry Technology, Wu Hongyu, Guo Libin.

[2] *To Build* a *Innovative Teaching Model* of *Practice Education* on *the Higher Vocational Art Design*, *Researches on Higher Education*, Shen Jianfei. 2009.7.

[3] *The Exploration of Higher Vocational Education Mode [J]*. Vocational & Technical Education Forum, Huang Yuanshan, Zhang Jie. 2000, (12).

[4] *Industry-University-Research Cooperation is the Only Way for the Development of Higher Vocational Education [N]. Guangming Daily*, Zhou Ji. 2002–11–28.

Management, Information and Educational Engineering – Liu, Sung & Yao (Eds)
© 2015 Taylor & Francis Group, London, ISBN: 978-1-138-02728-2

Research of college student employability under the perspective of regional economic differences

Su Xia Cui, Guang Hong Li & Wei Hua Peng
University of Jinan, Jinan, China

ABSTRACT: Taking ordinary university graduates as the research object and based on the matter-element analysis method, the paper analyzes current college student employability under the perspective of regional economic differences. The study found that graduates employability in the east is 3.13% higher than the midland and 8.34% more than the west. Among dominant indexes, "social practice experience" is the biggest difference. However, the variations of "professional category", "personal temperament" and "professional certification" between different areas are almost zero. Based on the government's macro-control, correct guidance of universities and students' employment consciousness, the paper raised "Trinity" strategy to curb the area vicious cycle.

KEYWORDS: Regional Economic Differences; College Student; Employability.

1 INTRODUCTION

With the development of economic globalization, China's economic development and the pattern of regional produce changes greatly. And employability is becoming increasingly apparent between different areas. The college students' employment market has changed significantly. On the one hand, the employers claim that it is hard to find the excellent human resources in recruitment. On the other hand, the graduates complain that it is not easy to find the suitable job. Students focus on employment can not only achieve their value, but will also generate agglomeration effect, which will make opportunities for regional development, and promote development of economic and social continuously and efficiently, Li & Sun (2013). On the contrary, the flow of college student imbalance will affect and cause an imbalance of regional economic development. To curb the vicious cycle, reducing the gap between economically developed areas' "talent high consumption" and underdeveloped areas' "talent shortage" becomes the fundamental problems to be solved.

2 MODEL CONSTRUCTION

2.1 *Sample selection*

To confirm the College Student Employability in regions of economic difference, the research group selected 2000 college graduates from 11 colleges and universities in 20 provinces as the study sample, considering the school level, regional factors, universities and school characteristics and other aspects of nature. There are four "985" colleges, five "211" colleges, eleven provincial and provincial general focus on undergraduate colleges, from the view of colleges' region. There are nine colleges in the east, five colleges in the midland and six colleges in the west. The research group took a multi-step random sampling method to select the wanted sample. The sampling units of the first step are colleges and universities. The sampling unit of the second step are classes of senior. And the sampling unit of the third step is individual student. Then the paper summarized the college student employability of regional economic differences by questionnaires.

2.2 *The college student employability model in the different regions based on the matter-element analysis*

Matter-element analysis is a theory to solve the incompatibility problem, which is under the existing conditions with ideas to achieve certain goals, Guo & Liu (2013). The proper research the College Students Employability under the different regions based on the Matter-element Analysis.

2.2.1 *Determine the classical field*

The key factors that affect the College Student Employability contain 23 evaluation indexes which are professional certification, awards situation, social practice experience, practical ability, problem-solving ability, innovation ability, and so on. These evaluation indexes are divided into five levels, they are ineffective, largely ineffective, more effective, efficient and very effective and are represented by N_1, N_2, N_3, N_4, N_5 respectively, whose scores were 0 to 60 points, 61 to 70 points, 71 ~ 80 points, 81 to 90 points, 91 to 100 points. According to this five rating scale, the paper established the appropriate classical matter- element R_0:

$$R_0 = \begin{bmatrix} N & N_1 & N_2 & N_3 & N_4 & N_5 \\ c_1 & \langle 0,60 \rangle & \langle 61,70 \rangle & \langle 71,80 \rangle & \langle 81,90 \rangle & \langle 91,100 \rangle \\ c_2 & \langle 0,60 \rangle & \langle 61,70 \rangle & \langle 71,80 \rangle & \langle 81,90 \rangle & \langle 91,100 \rangle \\ \vdots & \vdots & \vdots & \vdots & \vdots & \vdots \\ c_n & \langle 0,60 \rangle & \langle 61,70 \rangle & \langle 71,80 \rangle & \langle 81,90 \rangle & \langle 91,100 \rangle \end{bmatrix} \quad (1)$$

2.2.2 *Determine the segment field*
Segment Field matter-element R_p:

$$R_p = (p, c, v_p) = \begin{bmatrix} p & c_1 & \langle 0,100 \rangle \\ & c_2 & \langle 0,100 \rangle \\ & \vdots & \vdots \\ & c_n & \langle 0,100 \rangle \end{bmatrix} \quad (2)$$

$c_1, c_2, \cdots, c_n$ in the formula are three grade indexes., which are corresponded with the second grade indexes in the College Student Employability evaluation index system. According to the evaluation index system, n of 7 two grade indexes are 6, 3, 1, 3, 4, 3, 3.

Other two grade indexes of classical domain matter-element and section domain matter-element is decided by n values. Finally, the research group concluded 7 two grade indexes, including the base case, the advantages of job, career ideas, professionalism, expertise, skills of job search, ability to work.

2.2.3 *Determine matter-element to be evaluated*
Apply the actual assessment of three grade indexes, which are corresponded with 7 two grade indexes, to determine the matter-element of being evaluated. There are total 7 two grade indexes, which are corresponded with 7 matter-elements to be evaluated in the College Student Employability evaluation system. Specifically as follows:

$$R_{0p}^{11} = \begin{bmatrix} X_{11} & c_1 & X_{111} \\ & c_2 & X_{112} \\ & c_3 & X_{113} \\ & c_4 & X_{114} \\ & c_5 & X_{115} \\ & c_6 & X_{116} \end{bmatrix} \quad (3)$$

$$R_{0p}^{12} = \begin{bmatrix} X_{12} & c_1 & X_{121} \\ & c_2 & X_{122} \\ & c_3 & X_{123} \end{bmatrix} \quad (4)$$

$$R_{0p}^{21} = [X_{21} \quad c_1 \quad X_{211}] \quad (5)$$

$$R_{0p}^{22} = \begin{bmatrix} X_{22} & c_1 & X_{221} \\ & c_2 & X_{222} \\ & c_3 & X_{223} \end{bmatrix} \quad (6)$$

$$R_{0p}^{23} = \begin{bmatrix} X_{23} & c_1 & X_{231} \\ & c_2 & X_{232} \\ & c_3 & X_{233} \\ & c_4 & X_{234} \end{bmatrix} \quad (7)$$

$$R_{0p}^{24} = \begin{bmatrix} X_{24} & c_1 & X_{241} \\ & c_2 & X_{242} \\ & c_3 & X_{243} \end{bmatrix} \quad (8)$$

$$R_{0p}^{25} = \begin{bmatrix} X_{25} & c_1 & X_{251} \\ & c_2 & X_{252} \\ & c_3 & X_{253} \end{bmatrix} \quad (9)$$

The third column of matter-element expressions are three grade indexes' measured values.

2.2.4 *Determine the evaluation indexes weights*
The research group uses the combination of the Analytic Network Process (ANP) and expert scoring to determine the weight of each evaluation index. Get the importance of each index by expert scoring method. Given the situation that College Students' Career Guidance persons are more familiar with graduate, the research group organizes relevant managers to score, which can compare the real reactions. According to the basic principle of ANP and Super Decisions (SD) software, the research group made weight of each index scientifically.

2.2.5 *Determine the correlation of each evaluation rating*
Input classical matter-element, segment field matter-element, matter-element to be evaluated and evaluation of three grade evaluation indexes, which are corresponded with two grade evaluation into the prepared computer program, then the research group

obtained correlations of each evaluation level by running the program.

Determine correlations of 7 two grade indexes related to each evaluation rating. Get $K_{N_k}(X_{ij})$, the correlation of 7 two grade indexes related to each evaluation rating by running the program. In the correlation of $K_{N_k}(X_{ij})$, k = 1,2,3,4,5; The values of "i" and "j" are shown as follows, i = 1, j = 1, 2;i = 2, j = 1, 2, 3,4;i = 3, j = 1, 2,3. Take the base case X_{11} of two grade indexes as an example. First, make classical matter-element R_0:

$$R_0 = \begin{bmatrix} N & N_1 & N_2 & N_3 & N_4 & N_5 \\ c_1 & \langle 0,60\rangle & \langle 61,70\rangle & \langle 71,80\rangle & \langle 81,90\rangle & \langle 91,100\rangle \\ c_2 & \langle 0,60\rangle & \langle 61,70\rangle & \langle 71,80\rangle & \langle 81,90\rangle & \langle 91,100\rangle \\ c_3 & \langle 0,60\rangle & \langle 61,70\rangle & \langle 71,80\rangle & \langle 81,90\rangle & \langle 91,100\rangle \\ c_4 & \langle 0,60\rangle & \langle 61,70\rangle & \langle 71,80\rangle & \langle 81,90\rangle & \langle 91,100\rangle \\ c_5 & \langle 0,60\rangle & \langle 61,70\rangle & \langle 71,80\rangle & \langle 81,90\rangle & \langle 91,100\rangle \\ c_6 & \langle 0,60\rangle & \langle 61,70\rangle & \langle 71,80\rangle & \langle 81,90\rangle & \langle 91,100\rangle \end{bmatrix} \tag{10}$$

Segment Field matter-element:

$$R_p = (p, c, v_p) = \begin{bmatrix} X_{11} & c_1 & \langle 0,100\rangle \\ & c_2 & \langle 0,100\rangle \\ & c_3 & \langle 0,100\rangle \\ & c_4 & \langle 0,100\rangle \\ & c_5 & \langle 0,100\rangle \\ & c_6 & \langle 0,100\rangle \end{bmatrix} \tag{11}$$

Matter-element to be evaluated:

$$R_{0p}^{11} = \begin{bmatrix} X_{11} & c_1 & X_{111} \\ & c_2 & X_{112} \\ & c_3 & X_{113} \\ & c_4 & X_{114} \\ & c_5 & X_{115} \\ & c_6 & X_{116} \end{bmatrix} \tag{12}$$

The weights related three grade indexes;

$$w_{111} = 0.26546, w_{112} = 0.07679, w_{113} = 0.26350,$$
$$w_{114} = 0.18373, w_{115} = 0.10476, w_{116} = 0.10577,$$

Put the above calculation into the associated program, get correlations of X_{11} on five evaluation ratings by calculating. Other correlations of 6 two grade indexes are calculated as above.

Determine correlations of one grade index related to each evaluation rating. In determining the correlations of 7 two rating indexes, then determine the correlations of 2 one grade indexes related to 5 evaluation ratings. The formula is as follows:

$$K_{N_k}(X_i) = \sum_{j=1}^{n} W_{X_{ij}} K_{N_k}(X_{ij}) \tag{13}$$

Where k=1, 2, 3, 4, 5; In the formula, X_i is one grade index, which is corresponded with X_{ij}, two grade index. $W_{X_{ij}}$ is the weight of two grade index. Put the weights of 7 two grade indexes and correlation data into the above equation, then calculate correlations of 2 one grade indexes.

Determine the target level –correlations of college student employability on each rating. In determining the correlation of one grade, then determine the correlations of target level related to 5 evaluation ratings. The formula is as follows:

$$K_{N_k}(X) = \sum_{i=1}^{3} W_{X_i} K_{N_k}(X_i) \tag{14}$$

Where i = 1,2,3; k = 1,2,3,4,5. In the formula, X is target level related to X_i of one grade index, W_{X_i} is weight of X_i related to one grade index. Put the correlation data of X_i related to one grade index and weight of W_{X_i} into the above equation, calculate the correlation of target level related to college employability corresponded with each evaluation level.

3 THE EMPIRICAL RESULTS

3.1 *The comparison of evaluation indexes related to college student employability*

Table 1 mainly compared college student employability in the different regions of the east, midland and west. The study found that, graduates employment ratio in eastern and coastal areas is highest, 3.13% higher than the central region and 8.34% more than the west. Among dominant indexes, "social practice experience" has biggest difference. The index of the eastern area is higher than central and western areas about 8% ~ 10%. The variations of "professional category", "personal qualities" and "professional certification" between different areas are almost zero.

From single index factor, in addition to social practice experience, college student employability in the west is 19.8% lower than the east, 16.8% lower than the midland, whose differences are evident. However the only difference between the central and eastern is about 3%. In terms of working loyalty and entrepreneurship, the differences of regions are similar, the

Table 1. Evaluation indexes' comparison of the college student employability in different regions.

Evaluation Index	East	Midland	West
Professional Category	73.9	72.8	74.5
Personal Qualities	76.6	76.6	74.5
Professional Certification	67.3	67.2	65.5
Social Practice Experience	78.8	70.6	60.1
Professionalism	81.1	78.1	61.3
Working Loyalty	82.2	79.9	71.2
Entrepreneurship	72.3	69.1	61.4
Innovation Capacity	74.8	70.1	62.2
Academic	76.1	77.2	74.4
External Image	76.2	77.1	74.3
Awards	68.1	68.5	61.9
Practical Ability	76.9	70.1	64.8
Problem-solving	78.1	70.2	65.9
Learning Ability	79.9	71.1	66.3
Social Adaptability	78.1	72	72.1
Resilience	76.5	70	63.9
Job Target	77.2	79.8	77.1

east is 2%~3% higher than the west and 11% higher than the west.

3.2 *The college students employability of eastern and midland was significantly better than the west*

Using one-way ANOVA to analyze each of evaluation ratings of college student employability, it was found that the college student employability in the east and central is significantly higher than the west.

ANOVA results indicated that 13 evaluation indexes of different regions are significant differences in the evaluation index of social practice experience, professionalism, job loyalty, and so on. Other evaluation indexes results are no longer list display. Overall, the College Student Employability in the east and midland are significantly better than the west.

Table 2. ANOVA results.

Evaluation Index	F	P<
Social Practice Experience	27.202	0.000
Professionalism	8.288	0.003
Working Loyalty	20.551	0.000
Entrepreneurship	80.376	0.000
Practical Ability	11.33	0.001
Problem-solving Skills	26.03	0.000
Learning Ability	39.179	0.000
Innovation Capacity	34.404	0.000
Career Planning Capacity	42.776	0.000
Presentation Skill	20.518	0.000
Resilience	27.013	0.000
Teamwork	11.116	0.000
Organizational Communication Skills	18.723	0.000

4 CONCLUSIONS

Through the above empirical research, the college student employability is discriminate in different areas. Because of the imbalance of regional economic development, the college student employability in east and midland is superior to the west obviously. Employment areas will impact on graduates' employment. Though universities are distributed throughout Finland, most jobs are located in the core area around Helsinki, Kivinen (2000). Based on this, regional unbalanced development, which causes differences of college employability, is mainly for the huge gapbetweeng the east and the central and the west in our country.

With the transformation of economic growth mode and economic structure, quantity type expansion shifted to quality growth and labor-intensive industries transformed into to knowledge-intensive industries gradually. These adjustments and changes will increase the demand for intellectual talent, which will provide space for college students to display their talent, Jiang & Zhang & Geng (2013). So it is important for governments to increase reform and give full play to the role of macro-control. The policy will promote balanced flow of college graduation and enhance the college student employability of each region. When colleges start the conduct of the employment situation and employment guidance education, they should help graduates to analyze the situation of economic and social development correctly, make rational goals of development industry and scientific choice of employment. At the same time, students should strive to overcome the reluctance to go the grass roots and the midland, transform the concept of employment. Identify their positions in the labor market and realize the value of their lives in more challenging places, so as to lay a solid foundation for their employment.

The above conclusions, as well as "Trinity" strategy based on the empirical results about government, universities and students, will solve unbalanced situation relevant regional economic development and college student employability. To figure out the influences of regional economic development on the college student employability, it is important to balance regional economic development and protect college students full employment by analyzing the level of economic development of the provinces within the same region of the county and school level as well as other universities.

ACKNOWLEDGEMENTS

This paper is supported by the China National Social Science Fund Project 2011 (Grant No. 11BGL065): The Talent Screening Criteria Evolution and Promotion Path of College Student Employability; The Shandong Social Science Planning Project 2012 (Grant No. 12DGLJ11): Coupling Mechanism of Industrial Structure Adjustment and Human Capital Transfer in Shandong Province; the Shandong Social Science Planning Project 2014 (Grant No.14DGLJ03): Research of Industrial Structure' Optimization Mechanism under the Perspective of Employment Difficulties of Shandong Province. The paper is also supported by professor Guang Hong Li, who is master tutor of University of Jinan and the corresponding author.

REFERENCES

Li, G. H.& Sun, L.L.& Li, W.X. 2013.Research of Key Factors and Development Path of Talent Gathering Under the Perspective of Evolutionary Game. *J. Donyue Tribune*(11):52–58.

Guo, W.& Liu, C. J.& Pan,F.2013. Comprehensive Evaluation of Regional Development Difference Based on Extension Engineering Method. *J. Statistics and Decision*(15).

Osmo Kivinen .2000.Higher Education and Graduate Employment in Finland. *European Journal of Education* (2);165–177.

Jiang, Y.& Zhang, T. Q.& Geng X.2013. Employability of College Students: Concepts, Dimensions and Measurement. *J. Journal of Shandong University: 45–54.*

Management, Information and Educational Engineering – Liu, Sung & Yao (Eds)
© 2015 Taylor & Francis Group, London, ISBN: 978-1-138-02728-2

The research on the present state of college students' physical constitution and improvement measures

Yong Zhou
Institute of Information Technology of Guilin University of Electronic Technology, Guilin City, People's Republic of China

ABSTRACT: Doing research on the current physical health of college students from IIT of GUET in the last three years of documentation, mathematical statistics and an interview with experts. It turns out that there is variation of the present state of college students' physical constitution in the index of body shape, cardiopulmonary function and the power of the human limb and so on, which displays a decreasing trend in total. Cornering these changes, improved countermeasures are put forward: Take control of sports intensity and density in gym classes; put physical examination in the prior place; build a network platform of physical health; provide query and exercise prescription; strengthen the facilities construction of the gym building; standardize the campus morning exercises.

KEYWORDS: College students; Physical health; Present state; Countermeasures.

Quality education takes a leading place in our current national education. The key objective of quality education is to cultivate professionals with full development of morality, wisdom, physique and aesthetics. Nowadays, there is no denying that our students are facing the crisis of the continuous physical decline, especially for those college students, and this also becomes a challenge faced with the administrative departments of education and educational institution.[1] As we all know, China will be strong if the young is strong. How to enhance the youth health in this material prosperous and highly developed society has been not only a task of training talents to be fulfilled by colleges, but also a trend towards talents' cultivation in the new era. Considering the date of students' physical health from relative physical fitness tests in our school, students' physical quality witnesses a decline, so enhancement of physical strengthens of younger generations is a practical problem seriously examined and weighed by higher education.[2]–[3]

1 RESEARCH OBJECTS AND METHODS

1.1 *Objects*

Mainly take the last three years, students in IIT of GUET for example, doing a tracking investigation of their physical fitness in the school.

1.2 *Methods*

1.2.1 *Documents*

Has a basic understanding of research trends towards current physical health of college students through library database, network resource and some research results related to physical fitness and improved measures, which provides the rationale and the basis for research.

1.2.2 *Mathematical statistics*

Excel does some help for the statistics, classification and computational analysis to the physical test marks of students in IIT of GUET for the last three years, which offers a reference for this subject.

2 RESEARCH AND ANALYSIS IN CURRENT PHYSICAL HEALTH OF OUR SCHOOL STUDENTS

2.1 *Analyze the indexes of body build*

According to the Table 1, all the participants (including male and female), from the indexes of body build, the percentage of overweight students rise from 4.7% to 5.5%, meanwhile, the obesity rate is up to 12.8% from 10.3%. While, the number of normal weight students is gradually declining, and the percentage of people who suffer from malnutrition increases

Table 1. Indexes of body build.

	2011~2012	2012~2013	2013~2014
participants	6531	7347	8124
Overweight (%)	310 (4.7)	396 (5.4)	445 (5.5)
fat (%)	673 (10.3)	823 (11.2)	1040 (12.8)
malnutrition (%)	647 (9.9)	724 (9.9)	930 (11.4)
Normal weight (%)	2197 (33.6)	2477 (33.7)	2553 (31.5)
Low weight (%)	2704 (41.4)	2927 (39.8)	3156 (38.8)

by 1.5%. The indexes of low weight students have achieved positive growth, from 41.4% in 2011 to 38.8% in 2013. From this point of view, the students in our institution turn to be obesity. More specifically, it can be seen in table 1 and line charts the trend.

2.2 *Cardio-pulmonary function*

Table 2. The indexes of vital capacity in recent three years.

	2011~2012	2012~2013	2013~2014
participants	6531	7347	8124
fail (%)	2094 (32.1)	2496 (33.9)	2061 (25.4)
excellent (%)	396 (6)	539 (7.3)	1046 (12.9)
pass (%)	2867 (43.8)	3034 (41.3)	2812 (34.6)
well (%)	1174 (17.9)	1278 (17.4)	2205 (27.1)
average Male	3673.9ml	3702.6 ml	3972.7 ml
Female	2521.4 ml	2527.8 ml	2728.7 ml

Cardio-pulmonary function is a key index in current college students physical health assessment. According to the demand of the trial program of the standards for Students' Constitution and Health, the test items of cardio-pulmonary function are vital capacity and endurance running(male:1000m, female: 800m). The trend of average males' and females' vital capacity is reflected in chart 2 and the statistics is in table 2. It can be seen from them, in recent three years, the indexes of students' vital capacity saw an increase, the average of boys' and girls' rise from 3673.9 and 2521.4 in 2011 to 3972.7 and 2728.7 in 2013.The rise in statistics is not only due to expertly mastering vital capacity test skills, but also related to high-precision test instrument newly purchased.

1000m and 800m are always listed in the endurance test. If the indexes of vital capacity are influenced by instruments and test skills to some extent, students' indexes of Cardio-pulmonary function can be directly reflected by middle-distance race, for this is slightly affected by external factors. So test scores are the best explanations to the power of Cardio-pulmonary function. In table 3, the percentage of students who fail in the test in our institution rises year by year, from 8.4% in 2011 to 16.9% in 2013. Excellent rates fall from 5.5% to 2.8%. In chart 3, the total time students spent on middle-distance race gradually increase, so the test scores are getting lower and lower. The time of boys' and girls' scores decrease to 266 seconds and 255 seconds from 253 seconds and 248 seconds. This feature is remarkable in males' performance in the test. Meanwhile, in terms of the number of students who fail in the test, it was 548 in 2011, which increased to 1373 in 2013. While the number of excellent students fell to 228 from 357. Therefore, the student's physique in the indexes of cardio-pulmonary function witnesses a decrease from data.

Table 3. The indexes of endurance in recent three years.

	2011~2012	2012~2013	2013~2014
participants	6531	7347	8124
Fail (%)	548 (8.4)	927 (12.6)	1373 (16.9)
Excellent (%)	357 (5.5)	393 (5.3)	228 (2.8)
Pass (%)	4261 (65.2)	4693 (63.9)	5394 (66.4)
Well (%)	1365 (20.9)	1334 (18.2)	1129 (13.9)
Average Male	253 s	257 s	266 s
Female	248 s	251 s	255 s

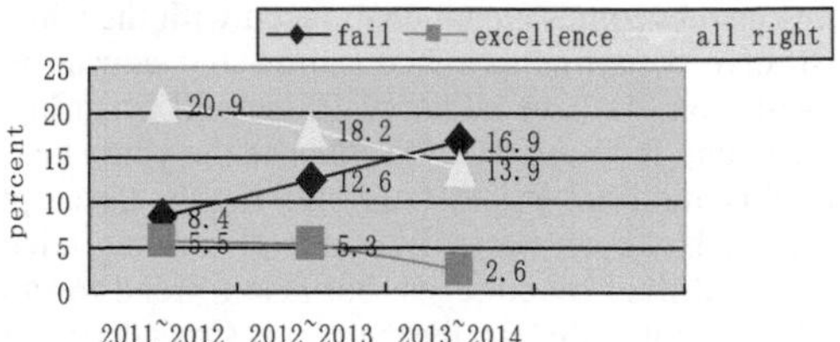

Figure 1. The trend of indexes of endurance (%).

2.3 *Analysis of limb strength function*

Physical strength is mainly inspects the student of upper limb, lower limbs and waist strength quality, the index can distinguish the body quality condition of individual specimens to a great extent. My courtyard limb strength skills mainly measured by standing long jump, in the past three years, measuring detection through uninterrupted on the college all boys and girls, the statistical results in table 4. In Table 4, do not pass the number increased year by year, increased from 346 in 2011 to 1379 in 2013, an increase of nearly 5 times, and the excellent rate is a linear decline. In Figure 5, the excellent and good, failed three indicators, from 2011 to 2012 is the fastest decline in first gear, the indicators are in this two-year detection decreases dramatically.

Table 4. The standing long jump index in recent 3 years statistics.

	2011~2012	2011~2012	2011~2012
The number of test	6531	7347	8124
Fail (%)	346 (5.2)	1000 (13.6)	1379 (16.9)
Excellent (%)	469 (7.2)	357 (4.9)	289 (3.6)
Pass (%)	4005 (61.3)	4540 (61.8)	4915 (60.5)
Good (%)	1711 (26.2)	1450 (19.7)	1541 (18.9)
Average Male	2.31 m	2.26 m	2.25 m
Female	1.71 m	1.66 m	1.64 m

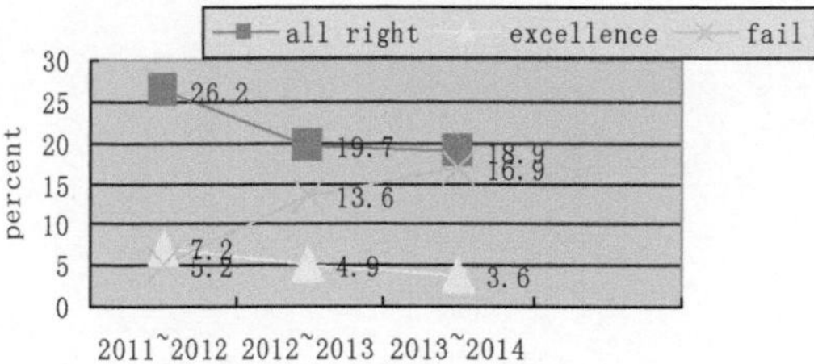

Figure 2. Nearly 3 years of standing long jump (%) overall trend.

In table 4, our college in recent 3 years standing long jump boys and girls of average value trend. In the figure, our college boys standing long jump from an average of 2.31 meters down to 2.25 meters in 2013, dropped by an average of 6 cm; the girls from 2011 1.71 meters down to 1.64 meters in 2013, dropped by an average of 7 cm. An important index of standing long jump as a test students thigh and abdominal strength, the strength of the quality, will directly affect the heart and lung function indexes of students is good or bad, because the middle and long distance operation cannot do without the movement of limb movement, the two have a direct correlation.

3 IMPROVEMENT STRATEGY

3.1 *The sports classroom should control the movement density and intensity of exercise*

According to the current status of physical health of students, exercise density of University PE classroom should be controlled between 40~50% and exercise intensity should be controlled between 1.5~1.8, especially the basic part of physical education and Sport density should be controlled at about 30%. In the whole class, the duration of the average heart rate in 130times/min should account for more than half of all class, only reach a certain density, can students get effective physical exercise and enhanced in PE class.[4]

3.2 *Sports achievement weight increases physical examinations, and focuses on an assessment index of cardiopulmonary function*

To further increase the weight of physical ability examination sports scores, the structure of sports achievement should be 50% skills +30% physical +20% usually attendance for frame, wherein the 30% physical examination content mainly for heart and lung function index assessment, will be 800 meters (female) and 1000 m (male) as a compulsory part of physical examination, body strength training as auxiliary content. At thc same time, for the performance of difficulty and beauty sports, such as Tai Chi, yoga, fitness evaluation weight suggestion will increase to around 35%.

3.3 *The construction of the physical health network platform, to provide query and "exercise prescription"*

To further increase the funding for sports scientific research support, constructing the healthy network platform will be the physical fitness test, physical test data on the Internet, students can query the download in a certain authority, timely understanding of the dynamic of their own physical health, and according to the network platform for the "exercise prescription" for fitness guidance.[5] In addition, the test data shall be informed and in a certain range to inform, to avoid the current most universities just measured less publicity notice of the defects.

3.4 *Strengthening the construction of sports facilities, improve student sports land*

Sports facilities are the most fundamental guarantee of students physical fitness, according to the Ministry of education "ordinary college sports facilities equipped with 'Directory' " standard, strengthen the construction of sports facilities, increase the student sports land, according to the college student sports present situation, suggestions for outdoor is 5.6 square meters, 0.4 square meters of indoor.

ACKNOWLEDGEMENTS

This research was supported by the University Scientific Research Fund in 2014 from the Guangxi Department of Education: Study on Students' Physical health Diagnosis Warning System (LX2014660);

It is also supported by the Higher Education Teaching Reformation Project from Institute of Information Technology of Guilin University of Electronic Technology: Study on College Students' Physical Health Situation and Improvement Strategies (2013JGY25).

REFERENCES

[1] Yang Huanan. "National student physical health standard" implementation research [J]. Journal of physical education, 2014 (4): 127–130.

[2] Chai Jiao. Survey and the present situation of the network of College Students' physical fitness test attitude evaluation system research [J]. Journal of Xi'an Physical Education University, 2007 (1): 114–117.

[3] Fangzhi. Southeast University students physique present situation and countermeasure research of [J]. Sports scientific literature bulletin, 2013 (5): 98–99+112.

[4] Kong Peihong. PE course in the amount of exercise volume and density [J]. Journal of Beijing Sport University, 1981 (3): 29–31.

[5] Wang Ruiyuan. Exercise physiology [M]. Beijing: People's sports press, 2012.

Management, Information and Educational Engineering – Liu, Sung & Yao (Eds)
 ISBN: 978-1-138-02728-2

The research of college English learning strategies based on networks

Shi Fang Wen
School of Foreign Languages, Shandong University of Traditional Chinese Medicine, Shandong, China

Wen Shuang Bao
School of Accountancy, Shandong Management University, Shandong, China

ABSTRACT: College English learning is not the same as it was in junior or high school, learners are made to learn autonomously and interact with the students or the instructors in a collaborative way which is based on networks, but a lack of learning strategy use leads to some students' English anxiety even boring. The present study presents a strategy of English-learning about different kinds of aspects, mainly at a view of the instructors, focusing on memory and affective strategies, aiming to help the instructors make a better teaching-program and at the same time help the learners enjoy the process of learning.

KEYWORDS: English Learning; Memory and Affective Strategies.

1 INTRODUCTION

Since the 1990s, modern information technology centering on computer internet has developed swiftly. College English Teaching Curriculum Requirement issued by Higher Education Bureau of the Ministry of Education of China In January, 2004 made the personalized, autonomous internet teaching mode become goal and trend of college English teaching reform, which leads to a significant educational shift in English teaching from teacher-centered to student-centered, and it's now no longer the pattern of learning English that you do what the teacher ask you to without thinking much which is the so-called exam-oriented education.

The development of internet technology enables learners to learn autonomously and interact with the students or the instructors in a collaborative way and then conduct distance learning, which testifies the superiority and feasibility of computer-based teaching mode compared with traditional teaching mode (Alison & Stephan, 2000).

However, some problems which limit internet language teaching and learning are found out in reality. That is, lack of learning strategy use leads to the students' language anxiety. It displays in the following aspects: firstly, learners are confused with considerable information. Secondly, sometimes they cannot obtain effective and timely feedback. Thirdly, they are so deeply indulged in online entertainment activities that they can't control themselves (mainly to male). Finally, they are unfamiliar with computer and internet operation (mainly to female) and they are not used to screen reading comparable to paper reading. All these impede the learning efficiency. Therefore, it is a key issue for the instructors to promote the teaching reform how to help learners to employ reasonable and effective learning strategies and reduce language anxiety in the network-based environment.

2 THE DEFINITION AND CLASSFICATION

2.1 *The definition of learning strategies*

Many scholars abroad have given the definition on language learning strategies, and most of them have something in common in a way. Rubin (1975) defined it as "the techniques or devices which a learner may use to acquire knowledge". Stern (1983) defined it as "particular forms of observable learning behavior, more or less consciously employed by the learner", but the mostly used is the onc defined by Oxford in 1989, "specific actions taken by the learner to make learning easier, faster, more enjoyable, more self-directed, more effective, and more transferable to new situations".

According to the definitions, we confirm that language learning strategies are the conscious thoughts and actions that learners take in order to achieve a learning goal, mainly focused on two points: 1) The purpose is to improve learning efficiency. 2) It is an action more than a way to the learners in English learning.

2.2 *The classification of English learning strategies*

The same as to the definition of strategies, there are various kinds of classification of strategies from different aspects and angles. O'Malley and Chamot classify the strategies into metacognitive, cognitive, and affective/social strategies. According to the relationship between strategies and language materials, Oxford classified English learning strategies into two categories including six sorts, shown as Fig.1.

Direct strategies are to deal with new language material, and make a direct relationship with the language itself in the case of a certain specific task. Indirect strategies are to deal with the management of learning things in macro point of view.

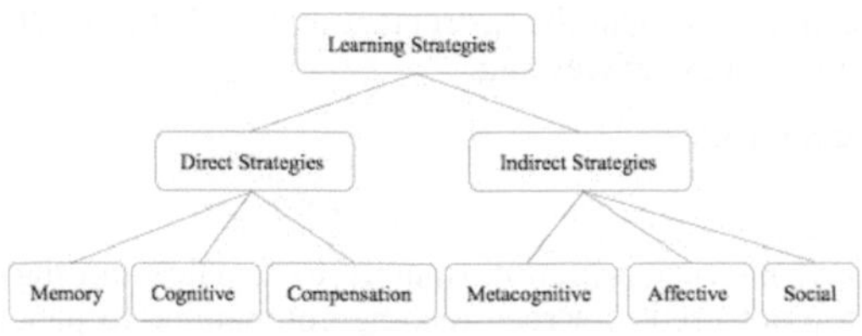

Figure 1. Oxford classification of English learning strategies.

3 CERTAIN STRATEGIES

A considerable amount of students in college gets some troubles in English learning based on networks because of lack of learning strategy, Therefore, the instructors are required to try their best to help students to get rid of that problem. The objective in doing so is to create a more effective language learning model and to instill in students increased interest and motivation to learn another language. Here are some strategies to solve the problems at a point view as an instructor.

3.1 *Being a popular instructor*

There is an old saying, "Trust your master, follow his way", When the students love their instructors, they will be naturally willing to get close to them, believe in them, and perform actively in the class.

To be a popular English teacher, he/she should not only be an expert in linguistics, psychology, and language teaching methodology but also qualified with sympathetic and noble qualities. Based on learner-centered model of education, teachers need to be sensitized to their new role. They should be a facilitator rather than a lecturer. They are not on superior and dominant position any longer, the current communicative language teaching approach allows the teacher to be seen as more of a facilitator whose responsibility is to provide students with opportunities to communicate in English in situations as authentic as possible with authentic materials in a network-based environment.

3.2 *Change learner's beliefs*

It is noted that many students don't volunteer to answer the teacher's questions because they don't believe opinions ought to be expressed in English until they can be said with a standard accent and without any mistake. The freshmen expect more help from teachers and other people, and still believe that the teacher should tell them how to learn as they did during their high school. Therefore, the instructors should help them change wrong beliefs by discussion or other efficient methods. They can consider making more use of these media to tutor their students.

However, some teachers themselves have to get rid of such beliefs that they spend a lot of time making PPT on teaching materials, and read them words by words during the classes, which often makes students frustrated. In fact, instructors should realize that only computer and network are integrated into the English teaching curriculum and regarded as incentive tools in cognitive and affective aspect, will the learners make their own efforts to be more self-directed and increase motivation and confidence, thus reducing language anxiety.

3.3 *Integrate culture into English teaching*

Cultural values together with the beliefs of the learners play an important role in English learning strategy and anxiety. H. Nostrand (1996) recommended the injection of careful 'doses' of culture shock in the foreign language classroom. The instructor should guide the students to pay attention to some clues in the process of cultural learning. The western students, in our mind, are personalized, open-minded, aggressive and energetic, which involves in their beliefs, emphasizing logicality, rationality and individuality. Therefore, the students must respect cultural differences between the motherland and other countries and manage to cross the cultural gap.

Teachers can play a positive role in helping learners to move through stages of acculturation and increase the learners' chances to succeed in both foreign language learning and language culture learning. Teachers can help learners to change that experience into one of the increased cultural awareness and self-awareness, although some cross-cultural experiences have a negative impact on foreign language learning, they do have positive values to foreign language and culture learning.

3.4 *Do strategy training*

Since the strategy in learning is of great significance, researchers have put forward different training models. Generally, the models are stepped by the following steps:1) to raise students' awareness; 2) to help them brainstorm the strategies used; 3) to model the strategies; 4) to have them practice the strategies; 5) to guide them in selecting the strategies that address their particular needs; 6) to evaluate their progress and strategy use.

According to the investigation, the students use memory strategies the least frequently and effective strategies are adopted the second least frequently in English learning. Although a lot of students spend a lot of time in memorizing words, they have no good efficiency. So it's better for the teachers to focus attention on the memory and affective strategies.

As a matter of fact, in the computer-based environment some good memory strategies can be provided to the students. Above all, the new vocabularies will stand out by underlining, emboldening or changing character and colors in specially-made file of vocabularies, as is helpful to improve memory. Furthermore, vivid relationship is established through pictures, voices and cartoon between new words and them. Besides these, new words are connected with old ones and try to create a situation to have a deep impression on the new vocabularies. Anyway, only through different ways of repetition can the brains be stimulated again and again and form a long memory.

As for the effective strategies, here are some suggestions that the instructors can do to help the learners to enhance their emotion. Firstly, to reduce learners' anxiety, the instructors can add some light music or funny talk shows when they feel tired in the learning process. Secondly, ask them just to speak out or write down some positive statements to feel more confident. In addition, help them to keep an English learning diary and discuss their feelings about English learning with persons.

3.5 *Stimulating students' motivation*

The results of this study indicate that there is a positive correlation between motivation, especially intrinsic motivation, and autonomy in English learning. Therefore, teachers should attach more importance to triggering students' intrinsic motivation. Deci and Ryan (1985) claim that intrinsic motivation leads to more effective learning. English teachers are entitled with new roles as facilitators and helpers for students' English learning nowadays, hence they should take some measures to stimulate the students' motivation of intrinsic interest, for example, providing the knowledge concerning the target culture when having classes as much as possible, which can absorb the students' attention to learning English because they can learn much more about the western world.

3.6 *Encouraging students to take responsibility for English learning*

Learner autonomy comes into play as learners begin to take responsibility for their own learning. The teacher should make it clear to the students that they have to take responsibility for their own learning and should always bear in mind that the focus of teaching English in the classroom is on developing the students' ability to take on more responsibility for their own learning so as to become effective learners in the classroom, and more importantly, out of the classroom without the help of their teacher. English teachers are supposed to make students aware of their duties in English learning and understand their learning process by providing students with enough knowledge of learner autonomy, including the necessity of autonomous learning and the essence of it. They should also give the students more freedom in the teaching process by encouraging them to set learning goals for themselves, to design classroom activities and evaluate the textbook that they are using and other materials, to monitor their own learning behavior, to assess their own learning and performance or their peers'. After that, they can help students to reflect on their learning and solve some problems cooperatively.

3.7 *Preparing teachers for autonomy*

Preparing teachers for autonomy is as important as preparing learners for autonomous learning. Only when teachers identify autonomy as a goal, and identify the teacher behaviors that promote autonomous learning, can they be aware of their changes in their roles when working with their learners in the autonomous learning preparation. In order to foster autonomy among learners, teachers should be aware of the importance of their own role in the process of helping learners take greater control over their learning. It might be assumed that teachers themselves need to receive the training or the learning experiences of autonomous learning.

4 SUMMARY

As the old saying goes,'Give a man a fish and you feed him for a day. Teach a man to fish and you feed him for a lifetime.'Learner autonomy has been a very important topic in the field of foreign language learning. It's of great honor for a teacher to teach his students how to learn, but not what to and it's of great significance for the students to get the skill of how to learn. In the present study, a simple strategy of English-learning was presented to both instructors and learners, we truly hope that the instructors can make a better teaching-program and at the same time the learners can enjoy the process of learning.

REFERENCES

Alison, L. & Stephan, 2000. A. Dealing with computer-Related anxiety in the Project-Oriented CALL classroom Computer Assisted Language Learning, Vol. 13, No. 4–5, 377–395.

Amuzie, G. L. & Winke, P. 2009. Changes in language learning beliefs as a result of study abroad. *System*.

Andrex D. C. & Ernesco M. 2010. Language learner strategies: thirty years of research and practice. *Oxford University Press*.

Deci, E. L. & Ryan R. M. 1985. Intrinsic Motivation and Self-determination in Human Behavior. *New York: Plenum*.

Larsson W, Aspelin P. 2013. Lundberg N. Learning strategies in the planning and evaluation phase of image production [J]. *Radiography*, (4):347–352.

Oxford, R. 1989. Use of language learning strategies: A synthesis of studies with implications for strategy training [J]. *System*, 17(2), 235–247.

Pappamihiel, N.E. 2002. 36. English as a Second Language Students and English Language Anxiety: Issues in the Mainstream Classroom. *Research in the Teaching of English*, 327–355.

Rubin, J. 1975. What the "Good Language Learner" Can Teach Us [J]. *TESOL Quarterly*, 9(1):41–49.

Stern, H. H. 1983. Fundamental Concepts of Language Teaching [M]. *Oxford: Oxford University Press*.

Management, Information and Educational Engineering – Liu, Sung & Yao (Eds)
© 2015 Taylor & Francis Group, London, ISBN: 978-1-138-02728-2

Improve the quality of applied talents based on strengthening mathematical culture teaching

H.S. Liu & S.F. Yan
Basic Department, North China Institute of Science and Technology, China

ABSTRACT: Mathematics is a kind of advanced culture and an important part of human civilization. Starting from the actual situation of mathematics teaching, expound the necessity to cultivate the mathematical culture quality, give the detailed measures, point out some problems needing attention. Practice shows that strengthening the students' mathematical culture quality can fully mobilize the enthusiasm of students, promote the forming of mathematical good habits, improve the quality of training applied talents.

KEYWORDS: Mathematical culture; Quality; Applied talent; Teaching quality.

1 INTRODUCTION

Einstein said that the creative principle resided in mathematics. Not only as the basis for other related learning courses, mathematics is also the foundation of the whole high level applied talents, even is the foundation of lifelong education. However, at present, quite a number of university mathematics curriculum teaching only pays attention to the mathematical knowledge and skills, theorem proving, derivation and examples exercises, but ignores the cultivation of students' understanding about the mathematical cultural connotation. [1] Mathematics is just number and formula in many college students mind, it is abstract, abstruse, mysterious, and even boring, then students lost interest and enthusiasm of learning mathematics. These cause the mathematics quality education can not fully, correctly implement, teaching quality education cannot be promoted. In addition, quite a few students on the mathematics understanding is very superficial, macroscopic understanding and the overall grasp of mathematics is poor; and this is the precisely essence benefit of a person. Therefore, it should pay attention to the cultivation of students' mathematics culture quality in mathematics teaching, improve teaching quality, and enhance the quality of applied talents.

2 NECESSITY

Mathematics is an important part of human culture, it belongs to the culture of science and is a kind of rational culture. Mathematical culture as a hot spot of mathematics education research, more and more causes attention of educators.

In foreign countries, America, Germany, Australia and some other developed countries have been focused some mathematical curriculum goals on "understanding and appreciation", cleared about the humanities education function, highlighted the education of mathematical culture. Its main features are: pay attention to the interest in content, emphasize on the history and the actual contact. Use heuristic teaching methods and emphasize the role of students' subjects. Highlight the cultivation of students' innovation ability in the whole process. In China, scholars have discussed that mathematics is culture. For example, the Ma Zunting thought that mathematics was a kind of culture, put forward views of "culture and mathematics were the mutual function " in 1933. [2] Li Daqian academician puts forward: "mathematics was a kind of advanced culture, was the important foundation of human civilization. Its emergence and development in the process of human civilization played an important role in promoting" in 2005. [3] Mr. Gu Pei has given a precise definition: "the connotation of mathematical culture was defined as mathematics thought, spirit, method, and their formation and development. Broadly said, also contained a number of mathematicians, history, mathematics beauty, mathematics education, mathematics development and cultural relations, etc." [4] Researches of Nankai University and some key domestic university have also clearly pointed out that mathematics should be humane educational goal, from philosophy, aesthetics, culture and so on various levels of understanding and appreciation of mathematics, aimed at through implementation of mathematics curriculum, improved the rational spirit and humanistic spirit of students.

College students are important absorbers of cultural inheritance and the major transmitter of culture. Improving the college students' mathematics culture quality, not only is the time need, but also the students' needs. Therefore, the training of mathematical culture quality has important significance of education.

3 MEASURES

3.1 *Update traditional concept*

In Chinese mathematics education history, it is not difficult to find that traditional mathematics education has always attached importance to its value as a tool, especially focused on the cultivation of mathematical knowledge and skills. Although after many years of study, Students' experience and the feeling are still very superficial about mathematical thought, spirit, but these mathematical literacy are accompanied by the growth of life resources. With the social progress, more and more people discover that learning mathematics not only may master the key opened the door to science and technology, but also realize the cultural value of mathematics from the deep understanding. As a culture, mathematics should become everyone's lifelong learning and good knowledge and enhance the cultural quality. As educators, we should look at the mathematics culture into the broader field, not let mathematics free from culture and improve the traditional mathematics education into the education of mathematics culture level. The final purpose of mathematical culture education is to improve students' mathematics accomplishment, lay a good mathematical foundation for their lifelong sustainable development. Ignoring the value of mathematical culture, education will inevitably lead to that the students' innovative spirit and practical ability cannot be effective and the mathematics quality education will also not comprehensively and correctly implement. Not only that, mathematics accomplishment enhancement also improves the cultural quality, thinking quality. With the increasing of the scale of higher education expands, social needs, and strive to create a consistent with the requirements of the development of the times educational mode should university mathematics education, follow the laws of education, innovation training mechanism, to explore the educational scheme is effective in practice, so as to achieve the purpose of training talents with all-round development. North China Institute of Science and Technology has carried out the idea of education activities in the great debate three times, advocated the school staff to actively carry out educational research and reflection and guided them to further emancipate the mind, renew the concept, in the implementation of the fundamental task of strengthening and improving, training application-oriented talents, and strive to improve the quality of education and teaching. I think that strengthening mathematical culture teaching must enhance the quality of math education for college students. That is the best idea practice of an educated mind and big discussion activity.

3.2 *Use efficient teaching methods*

Because mathematics courses are abstract, esoteric, boring, so we should fully consider the major features of the students and use the combination of different teaching methods in the classroom. Teaching methods should be flexible. You can use "discussion" or "two-way" teaching form, You also can carry out from applying by the professional fields. It is worth mentioning that paying attention to "sample" teaching process. From the comparison of Chinese and foreign education history can be seen in the traditional mathematics teaching China, mainly to teach "examples and pithy formula", while western teaching lays particular stress on "theorem and its proof". At present in our country, a considerable part of mathematics teaching in Colleges and universities tend to ignore the former, tend to have too much emphasis on the "formal" effect. Practice shows that: with a good example (preferably with students' professional related or similar) to explain instead of boring, tedious example proof of the theorem of calculus, which helps the students to learn the content understanding and grasp, and can make the students feel mathematics activity, fully mobilize the enthusiasm of them, its main role play teaching. If can seize the opportune moment in the teaching process of mathematics culture introduction increases, more can get twice the result with half the effort, training is more helpful to application talents.

3.3 *Use advanced teaching means*

With the rapid development of modern information technology, exchange of mankind has become more efficient, the multimedia teaching is popular in universities; and due to the popularity of a group of powerful mathematical software system, to make the teaching and learning of mathematics has undergone profound changes. Therefore, in the course of university mathematics teaching, the use of advanced teaching methods, the introduction of mathematical software system represents the general trend. Using modern teaching methods in mathematics teaching, can not only image, vividly some abstract concept concrete, enables the students to have a perceptual awareness, to deepen the understanding of knowledge, improve students' learning efficiency; and can increase the classroom information, greatly improving the teaching efficiency, improve the quality of education and teaching.

3.4 *Enrich activity*

Strengthen mathematical culture teaching must have the carrier, in addition to outside the classroom teaching, also should be based on expert lectures, mathematical modeling, mathematical experiment and other forms, to carry out various types of mathematics cultural education. The second classroom activities, as a complement to the student's mathematics, cultural knowledge, stimulate students' interest in learning mathematics, improve the students' mathematical culture quality, promote the quality of applied talents culture.

4 ATTENTIONS

4.1 *Clear the importance*

When emphasis on the main mathematical courses, we should not adopt the mode of thinking of metaphysics, but should be clear: mathematical culture is not decorated in mathematics, but overall; not attached, but organic; not overwhelming, but apt; not draw a forced analogy, but natural; not a long and minute statement, but the finishing touch [5]. Therefore, in the outstanding mathematical culture teaching, we must fully consider the overall coordination, not overwhelming, destruction of the original teaching system.

4.2 *Take mathematical knowledge as the carrier*

As Engels said that like all other disciplines, mathematics was generated from the actual needs of people's lives, mathematics was scientific research about the space form and the relationship. Therefore, the teaching of mathematics culture must take the mathematics knowledge as the carrier, otherwise it is meaningless. That is, only teach students mathematical knowledge at the same time, make their thinking method to learn mathematics, grasp the spiritual essence of mathematics, know the mathematics sequence of events, master the theoretical knowledge in mathematics culture. In this way, they would no longer feel mathematics concept is boring, mathematical theorems and formulas are the wood, passive water, and then contribute to the understanding to abstract knowledge, improve the ability of using mathematical knowledge to solve practical problems.

4.3 *Step by step*

Mathematics culture is a kind of advanced culture, broad and profound, rich, is the important foundation of human civilization. Therefore, when teaching of mathematical culture, we should not be too much, be just perfect, draw a forced analogy, meaningless. As the application background of mathematics should be concise and to the point introduction, don't do things sloppily, must carry on the organic connection with the existing content. The penetration of mathematical culture is a gradual and long process, to follow its rules.

5 CONCLUSION

Mathematical literacy is not innate, is training in learning and practice. Mathematical knowledge is the most basic carrier of mathematical diathesis training, teachers in the mathematics teaching process, not only to impart basic knowledge to students, but also to mathematics knowledge as the carrier, to let students understand the broad and profound mathematical thinking, ingenious mathematical method. Teachers should cultivate students' mathematical way of rational thinking, grasp the spirit of Science in mathematical research, to let the students felt the value of learning mathematics, and thereby achieve enlightenment thinking, edify sentiment, enhance innovation capability, enhance the ultimate objective of mathematical literacy. These initiatives will stimulate students' interest in learning mathematics, improve math, promote the school quality education practice of North China Institute of Science and Technology, the philosophy of education and talent cultivation target view.

ACKNOWLEDGEMENTS

In this paper, the research was sponsored by the Central Universities Science Foundation (Project No.3142014127), The educational fund and Key disciplines Fund Project of North China Institute of Science and Technology (Project No. HKXJZD201402, HKJY201436, HKJY201439).

REFERENCES

[1] Shuqin BING.2013.Mathematics Culture and College Mathematics Teaching. Modern Computer 20(7):29–32.

[2] Zunting MA.1933.Mathematics and Culture. Mainland Magazine2(3): 59–61.

[3] Daqian LI.2006. The idea of mathematical modeling into mathematics course. Chinese University Teaching5(1):9–11.

[4] Pei GU.2008. Mathematical Culture. BeiJing: Higher Education Press:1–2.

[5] Changyi GOU, Pei GU.2008. Improve Higher Mathematics Teaching in Liberal Art by Integrating Mathematical Culture. Journal of Mathematics Education,17(6) : 5–7.

Management, Information and Educational Engineering – Liu, Sung & Yao (Eds)
© 2015 Taylor & Francis Group, London, ISBN: 978-1-138-02728-2

Spoken English teaching strategies in colleges based on grammatical competence

Lin Wang
Anhui Xinhua University, Hefei, China

ABSTRACT: This paper explicates the importance of spoken English and its connotation. Based on the implication of grammatical competence and investigation of the status quo of college English majors' spoken English, this paper proposes a set of strategies for use in oral English teaching practice, including input strategy and output strategy, interactive strategy, competition strategy, affective strategies, and evaluation strategies. Teachers play a leading role in the teaching process, by making flexible use of various strategies, with a view to carrying out oral activities, and guiding students' development of oral language. Finally, a viable model for teaching strategies is proposed.

KEYWORDS: Spoken language; Grammatical competence; Teaching strategies; Teaching model.

1 INTRODUCTION

Among four language skills, namely listening, speaking, reading and writing skills, speaking is a skill involving both listening and speaking, which means, in the speaking process, a speaker and a listener often need to interact with each other, so the listening accuracy will directly affect the relatedness and appropriateness of the response. Oral communication reflects multiple skills, such as pronunciation, grammar, vocabulary, intonation, fluency, accuracy, appropriateness and flexibility. It is its multidimensional nature that makes spoken language a quick means to examine a person's linguistic competence.

2 THE CONNOTATION OF SPOKEN LANGUAGE

Declarative knowledge and procedural knowledge are two basic types of knowledge. Declarative knowledge is the type of knowledge that is, by its very nature, expressed in declarative sentences or indicative propositions, while procedural knowledge is the knowledge exercised in the performance of some task (Wikipedia). For example, knowing the location of the keys on the piano is declarative knowledge; knowing how to move one's fingers across the key board is procedural knowledge. Therefore, speaking is a skill, which can be improved and consolidated through training and constant training.

3 GRAMMATICAL COMPETENCE

Grammatical competence refers to knowing how to use the language correctly by forming well-formed utterances. It is an integral part of the overall linguistic competence. However, for most oral activities, the focus is on the delivery of fluent speech to achieve the purpose of communication. This does not mean grammatical competence is not important. Rather, it contributes to the clarity of the message conveyed. Therefore, teachers should balance both fluency and accuracy when it comes to both designing oral activities and evaluating students' oral performance.

4 STATUS QUO OF ENGLISH MAJORS' ORAL ENGLISH IN CHINA

In college, English majors face a strenuous task of developing their oral competence to achieve fluent, natural, accurate and decent use of the spoken language. With further study, students' awareness of the importance of spoken language is greatly enhanced, and they show more enthusiasm in practicing speaking English. Yet sometimes, they may also experience no significant progress in oral competence, and lose confidence. Some students even experience frustrations and give up the practice of oral language. Therefore, it is an urgent task for teachers, especially those teaching integrated courses, to adopt various teaching strategies in their teaching practice to facilitate the development of students' spoken language. Students' enhanced awareness and their enthusiasm are favorable factors to be utilized rationally. Therefore, it is find proper means to motivate students' initiative to participate in the oral activities.

5 ORAL ENGLISH TEACHING STRATEGIES

To train students' oral skills, teachers should establish scientific goals, develop effective plans, and then use reasonable teaching strategies to effectively promote the development of students' speaking ability. Then what are learning strategies? Learning strategies to promote learning and recall information on language and content areas, skills students have taken, methods and intentional action (Ernesto Macaro, 2008). Oral teaching strategy refers to teachers in the classroom teaching process, aimed at improving the efficiency of spoken language training, to improve students' oral proficiency and the use of teaching methods, measures collectively. Oral teaching strategies include the following sub-strategies: input strategy, output strategy, interactive strategy, competitive strategy, affective strategies, evaluation strategies, teachers in the teaching process should be based on the psychological characteristics of students, teaching content, class size, actual topic tasks , conscious and flexible use of various strategies, targeted to carry out oral activities, and guide students' oral development.

5.1 *Input strategy*

Input is used to refer to the language that is addressed to the L2 learner either by a native speaker or by another L2 learner.(Rod Ellis) According to Krashen's input hypothesis, for foreign language learners, the most important thing is to provide them with comprehensible input. Krashen holds that by understanding the input which is slightly higher than the learner's current level of linguistic competence and comprehensible to him, the learner acquires the language. Comprehensible input can be formulated as $i + 1$, and i indicates the current level of learner language skills, and $i + 1$ is the next stage of language acquisition. The input hypothesis advocates that learners understand the language input that contains $i + 1$ and make progress. It is clear that in the learning process, teachers also play a role of providing language input; therefore teachers' language should be smooth, accurate, tailored to the current levels of students. Input can take the form of the aural and visual input. Language Acquisition Device is a prerequisite for language acquisition. Limited verbal input can trigger off an infinite output. But the amount of minimum input has been a mystery (Shan Xingyuan, 2004). Since we can not determine the amount of the minimum input suitable for students, to be on the safe side, when providing language materials as input to students, teachers should expose students to the way native speakers use the English language in different situations, encouraging them to use VOA, BBC, CNN, and the White House and CCTV-9, etc as supplementary input channels, and participate in the English corner, and Christmas parties organized by foreign teachers and other activities in order to broaden the language input channels; watch English movies and learn English songs. In a word, the importance of written language input should not be overlooked.

5.2 *Output strategy*

Swain(1985) proposed the output hypothesis, whose theoretical basis is that language output, different from language input, can compel learners to develop their output language through the form of speaking and writing. Allwright(1984) claims that it is through interaction that acquisition takes place. Due to the inadequacy of the learner's linguistic knowledge, he may make mistakes in interaction with others, which leads to misunderstanding. When misunderstandings arise in the course of interaction, both parties need to make compensations through meaning negotiation. Applied linguists think that the process of meaning negotiation is conducive to language acquisition. Output can take different forms such as prepared speech, spontaneous speech, etc.

5.3 *Interaction strategy*

Students can adjust their language learning according to the teacher's and peer students' feedback. By interacting with the teacher and peer students, students become a part of the teaching activities. They can use interaction to clarify their misunderstanding, voice their opinions and so on. Interaction between teacher and student can take the form of question, error correction, feedback, brainstorming and free discussion, etc. To better interact with students, the teacher should have a repertoire of interaction strategies.

5.4 *Competition strategy*

In competition, students can fully demonstrate their language ability and critical thinking ability. In competition, students can also demonstrate their good psychological quality of being able to carry out purposeful interaction and communication, using language skill and strategies, under a certain external pressure. Students should be encouraged to constantly take part in competition at ordinary times in order to fit in with the challenge of competitions in the future. Similarly, competition can take a variety of forms like free talk, simulated debate, and speech contest.

5.5 *Affective strategy*

Affective factors have important influence over the students' foreign language learning. Affect is a filter through which the input goes before it is acquired. Under positive affect, it is easy for the input to pass through the filter; while it is difficult for the input to pass through the filter if learners hold negative feelings concerning the input.

5.6 *Evaluation strategy*

The criteria of assessment can be approached from several aspects, such as quantity, quality and freedom of oral speech. For evaluation mode, there are summative assessment and formative assessment. For way of rating, there can be students' self-assessment, assessment by peer students, assessments by teacher. Different means of evaluation are used to enhance the objectivity of the result.

6 CONCLUSIONS

To sum up, teachers are designers of the oral activities in the process of oral English teaching, playing a vital role, while students are the main participants of teaching activities. Students' oral development is a result of factors interacting with each other, such as subject and object, subjectivity and objectivity, internal and external factors. In order to develop students' oral language, teachers need to plan globally, providing students with extensive opportunities to practice their oral language. Teachers can flexibly adopt teaching strategies according to the real characteristics of teaching contents, cognitive level of students, and their thinking patterns. Oral English teaching strategies are a dynamic system. This paper proposes an advisable model for oral English teaching strategies. However, there is no such thing as good strategies or bad ones. One strategy might be workable in one situation, yet is problematic in another. Therefore teachers should have a balanced understanding of these strategies so that they can have them at disposal.

REFERENCES

David W. Carroll. 2003. *Psychology of Language* 3rd Edition Foreign Language Teaching and Research Press 2003.

Ernesto Macaro. 2008. *Learning Strategies in Foreign and Second Language Classrooms* World Publishing Corporation.

Keith Johnson. 2002. *An Introduction to Foreign Language Learning and Teaching* Foreign Language Teaching and Research Press 2002.

Rod Ellis. 2004. *Understanding Second Language Acquisition* Shanghai Foreign Language Education Press 127-.

Shan Xingyuan. 2004. *Non-macro linguistic Acquisition and measurement of Speech Input Amount* Foreign Language Research, third issue, 2004.

Tricia Hedge. 2008. *Teaching and Learning in the Language Classroom* Shanghai Foreign Language Education Press.

Wang Qiang. 2006. *A Course in English Language Teaching 2nd edition* Higher Education Press 2006 http://en.wikipedia.org/wiki/Declarative_knowledge.

Management, Information and Educational Engineering – Liu, Sung & Yao (Eds)
© 2015 Taylor & Francis Group, London, ISBN: 978-1-138-02728-2

A study on function and application of culture context in translation teaching

Xiang Li
Anhui Xinhua University, Hefei, China

ABSTRACT: With the advent of the 21st century, within today's globalization era, a nation or virtually the whole world is striding for a higher human civilization level. During the cross-linguistic and intercultural communication between various nations, preeminent translation work from an intercultural perspective serves as vital links to bridge those gaps and discrepancies due to the impact of culture context. Like an art, translation teaching, as well as training, demands ceaseless in-depth exploration into the more interdisciplinary domains, its tactics and techniques go far beyond the linguistic category. While proceeding with this work, teachers focus on the cultural connotations of both original language and target language in the given culture context. In the undergraduate syllabus of the English major, *Translation Studies*, a compulsory course, is conducted to enhance student's translating competence and cultural appreciation capability, assist them to solve some corresponding issues induced by cultural discrepancies during transformation between original language and target language, and advance cross-cultural communicative awareness and quality, by introducing general theories and principles of translation, with languages as the cultural carrier, which involves studies such as the English language, British and American cultures as well as different customs, cultural elements of inter-lingual activity and its strategic option, etc.

KEYWORDS: Translation teaching; Culture context; Function; Application; Intercultural communication.

1 INTRODUCTION

Culture context, also termed as cultural background, is the totality of material and spiritual civilization of a nation or country, which does not only affect the meaning of vocabulary and discourse, but also endow them more newly-developed derivative connotations. Culture context comprises lots of elements:

In a narrow sense, there is an inter-relatedness between cultural context and personal language variation, involving age, psychology, gender, profession, education background. To take gender variation for instance, the female social status, owing to their own gender feature relating to social, historical, physiological factors, is subject to the male in the social hierarchies, thus their linguistic features unconsciously tend to employ a more indirect, concessional and euphemistic pattern reflecting their inferior, subordinate status, while their counterpart, in a way, take advantage of the direct, aggressive or even arbitrary sexist language to defend the male-dominated position intentionally or unintentionally. Providing a qualified "go-between" facilitates a good marriage, it is the precondition to be acquainted with both his and her multi-dimensional background. During translation teaching, teachers do implement such approach by introducing their profound culture connotation behind the surface level of language. Based on well-constructed universal grammatical teaching, all successful teaching of translation depends on the pragmatic repertoire of what they can do via the intercultural permeation ranging from age to gender to psychology to profession to educational background.

In a broad sense, culture context includes geographical features, religious beliefs, historical tradition, national customs, socioeconomic stratification and sociopolitical status, etc. It can also be paraphrased with over-all cultural circumstances where certain a group of language users do mostly inhabit, that come under influence of material condition and spiritual accumulation, and they combine with specific space-time continuum to form the linguistic performance settings, just like their geographical territory, historical era, social environment, cultural tradition and heritage, and the like. Both generality and individuality bears every piece of land across boundaries. The comprehension to generality benefits the exotic cultural ingredients immigration which will pave the way for grasping the individuality. However, the cognition to individuality does deepen the digestion of its specific cultural information, and distill some essences from the generality. When instructing translation, if a translator fails to transcend the limitations of collective knowledge and detect the individualistic features by means of contrastive study, it is no possible to reflect the cultural charm of the source language, which makes the translating works tasteless and pulseless. In other words, if translators sometimes violate the general features, to put the common expressions into target language may also bring about cross-linguistic communicative barrier and even failure.

In fact, as far as the connotation of cultural context is concerned, only combining the narrow sense with broad sense, refraining from the isolation of two aspects, the translation can offer insight into abstract concept and its profound meaning embedded in alien cultural soil.

2 THE FUNCTION OF CULTURE CONTEXT IN TRANSLATION TEACHING

According to survey of translation activities, the mistake is mainly attributed to inaccurate understanding of the source language within a cultural context. Actually speaking, in the process of translating, the culture context produces positive effect. The communicative function of the language is at the center of human life; the connotation is rooted in context; context defines its meaning. There is no way to convey the original thoughts or implication in isolation from cultural context. Concretely speaking, the significance of culture context in translation teaching is embodied in three aspects as follows:

2.1 *Elimination of ambiguity*

It is prerequisite to introduce the precise and intellectual translation techniques into translation teaching activities. And as for instructors and their teaching target, cultural differences lead to errors or defaults arising due to cultural dissimilarity. Lack of culture appreciation can trigger immediate preconception and misinterpretation. When a word with multiple meanings is applied to inadequate text, it creates ambiguity. Malinosky points out: "they regard context as the sole determiner of meaning without which meaning doesn't exist." Culture context not only exerts a great influence on the determination of meanings of words, but also changes the essential meanings of words sometimes. "Y*ouyi*" in Chinese used here for exemplification refers to friendships in general. In the feudal days, a man should not touch the hand of a woman in giving or accepting thing, let along monk and nun, in the excerpt from a Chinese classic literature, *Hartong Anecdote*: "In a fit of bad temper, he was disillusioned with the mortal world, retreated into remote mountains, simply shaved his head and became a monk in a small Buddhist monastery. Shortly after some people rumored that he had established 'youyi'–friendship with nun, provoking discussion and gossip, growing more violent and hotter." Among this specific cultural background, the so-called "*youyi*" implies a sexual relationship or flirtatious behavior. The close relation between cultural context and determination is revealed obviously. But how to make an accurate choice of equivalent in translation activity is the task that needs further discussion. The "landlord" and "trade union" provide outstanding examples with instructors to illustrate the cultural context function in translation teaching. In Chinese culture, "landlord, or *dizhu* in Chinese" usually refers to someone who owned a lot of farmland and got money by renting land without working. It used to be associated with exploration, naturally carrying a negative overtone. On the contrary, it is not equal to "landlord" with effectively neutral color in western countries. Moreover, in western countries, in order to shorten working hours, win higher wages, better working conditions, deal with employers as an organization, "Trade Union" is formed by employees, to employees, for employees, which aims at the maintenance of benign employment relationship and reduces the constant struggle especially in a particular trade and profession, formed to represent their interests and deal as a group with the employer. Against this special cultural background, the term has strong political overtones. The organization, which is established purposefully and expected to stage constant struggle. In China, the term bearing no political overtones only refers to a kind of mass organization in each working unit under the leadership of the party.

Therefore, what strategy instructors adopt in translation teaching is to place concrete expressions in the respective cultural background to eliminate ambiguity and a present equivalent accordingly in intercultural communication. Translation teaching tries to develop the student's culture awareness and provide specific training for conscious attention to the similarities and differences between learning intercultural translation and learning other subjects.

2.2 *Indication of referents*

Because the meanings of words are conditioned and influenced to a great extent by the cultural context, the meanings of Lexis enjoying flexibility and the reference may be of uncertainty. So in the different contexts, a new or novel meaning is attached to them. All successful translation teaching depends on the teacher's and student's well-formed personality, well-organized mind and solid expertise in this field. Thus, a more precise referent can be achieved chiefly through expanding their scope of knowledge and enforcing the intensive translation practice. Kinship terminologies illustrate important distinctions in the different cultures. The feudal society lasting for several thousand years in China, the patriarchal clan system is forged. The semantic field of the same concept does not contain the same components. In Chinese, the kinship terms "Bofu", "Shushu", "Jiufu", "Gufu", "Yifu", "Biaoshu" absolutely differ in blood relationship. English words, however, are loaded with larger quantity of information than Chinese terms, the term "Uncle" is always used as the corresponding word in all the Chinese terms mentioned above. Lexical gaps are of common occurrence in English-Chinese translation, even for Brother and Sister, respectively, including elder brother or sister, and younger brother or sister, which generates the lexical gap that characterize the difference between two descriptions of an object by different linguistic representations and usually results in much-puzzled referent. In translation teaching, it is a vital procedure to explore and formulate contextual knowledge in a natural language. During cross-cultural translating activities, the patriarchal clan system should be taken into careful consideration and cultural clashes caused by lexical lacuna should be carefully fathomed before you can find a method to adapt the target language to the source language. There is an example, extracting from *A Dream of Red Mansions*: "Her whole air is so distinguished! She does not like 'Waisunnv', but look like a direct descendant of Jia." In feudal China, a married lass like spilled water breaks away from the direct relation of the original male-dominated family, the lass and her offspring could be viewed as relatives of indirect blood relations. Lin Daiyu, the heroine of the historical novel, is Jiamin's daughter, and Jiamin is the Old Ancestress's. The discourse of Wang Xifeng emphasizes subconsciously the core of the male blood relationship instead of the female. But either "Waisunnv" or "Sunnv" in English is expressed with "granddaughter", that doesn't distinguish direct blood from indirect blood relationship. Comparatively speaking, Mr. Yang, Hsien-Yi fulfills an accurate translation related to faithfulness, expressiveness and elegance: "Her whole air is so distinguished! She doesn't take after her father, son-in-law of our Old Ancestress, but looks more like a chia." Under the teacher's supervision, the process of implementing successful, smooth translation teaching is to indicate the choice of referents on the basis of the deep-seated cultural elements rather than meeting the surface level of a powerful language.

2.3 *Provision of clue for inferring word meaning*

Cultural context may prove extremely valuable in deduction of what the meaning refers. In many cases, consulting the dictionary is an effective way to secure accurate equivalent between source language and target language. Unfortunately, it fails to solve the problem sometimes. For instance, "And if, with the decay of vitality, weariness increases, the thought of rest will be not unwelcome." In the above sentence, the word "rest" can't be interpreted as inactivity as a way of regaining one's strength, but it reveals death with its extended meaning. The contextual details in the sentence entertain sufficient hints and clues for the word "rest". In Chinese culture, the indirect expression, "Changmian", hedges against the verbal taboo and does indicate the euphemistic term for "death", which embodies the historical tradition, cultural psychology and national values. So the cultural context facilitates transfers epithet during further cross-linguistic exchange with the help of valuable clue between lines to confirm the connotation as well as denotation. Here is another thing cited from *A Dream of Red Mansions*. Lin Daiyu is late to visit Baoyu, but not single could she enter. "Never do such things again. . ." she sobbed at the last. Under the special cultural circumstance, the utterance is although short enough, it is of profound and complicated meaning, since it decodes not only the deep sorrow in her heart and sympathetic emotion for her beloved, but also the extremely frustrating distress after failure in the revolt against feudalism. The connotative meaning can be hardly translated until the interlaced factors, such as the exact situation in the Jia's mansion as the context, the special relation between Lin Daiyu and Jia Baoyu, are taken into a full consideration.When conducting translation teaching, the stating point is to take some translating techniques which teachers have observed from authentic linguistic materials, and then employ the strategies to spur student's highly integrative motivation for any alien cultural components with particular topics and exercises.

3 APPLICATION OF CULTURE CONTEXT IN TRANSLATION

3.1 *Application of color's implication*

Duo to the dissimilarities between the linguistic form and cultural context which consists of cognitive process, thinking pattern, ethnic tradition, national psychology, and so forth, all sorts of colors are different in associative meaning and implied meaning caused by the above-mentioned factors. In Chinese, "Hongyanbing" is originated from jealous psychology. Supposing translators put literally into "being red-eyed", the target reader in western countries will get confused abruptly. To westerners, they express the jealousy, envious feelings with the color, "Green" instead. Similarly, "Lvmaozi" denotes that a married woman cuckolds with other man, if students translate it into "to wear a green hat", the western reader must neglect its implied meaning in Chinese seeing from the literal meaning. According to adaptation tactics, translation teaching must fully detect the reader's cultural context so as to achieve the pragmatic equivalence. Therefore, "to be cuckolded" is the more acceptable version for target readers to some degree.

Table 1. Mark these color connotations between Chinese and English on a scale from 1–7.

Translation teaching must focus on the study of dissimilarities in concrete cultural context

Items	RED in Chinese	RED in English
warmth	●	●
wrath	○	●
hostile	○	●
vitality	●	○
felicity	●	○
evil	○	●
anxiety	●	●
Items	**GREEN in Chinese**	**GREEN in English**
cool	●	●
peace	●	●
safety	●	●
calm	●	●
control	○	○
envy	○	●
health	●	●

3.2 *Application of pun and euphemism*

Pun is endowed with a strong cultural color. Many renowned writers are inclined to use puns to add literature charm into their works, such as homograph, homophone, homonym and polysemant. But because of the disparities between languages, it is not very easy to render its two-fold meaning explicitly. The following conversation between customer and waitress takes place in a restaurant.

> Waitress: "Sir, why do not you eat the fish?
> There is so much suffering in your stomach?"
> Customer: "Oh! Long time no see (sea)."

In this quote above, "see" and "sea" belong to homophone in which the humor sense is conveyed. If the core word cannot be handled well from the pragmatic perspective, the translation is doomed to failure. In the light of homophone characteristics, homonymic appliance on interchange between "sea" and "see" demonstrates a radiant look in the original context and provokes a pragmatic effect vividly and humorously.

Euphemism stems form the Greek word, "euphemismos", and indicates "to speak with good words." A euphemism is a mild, indirect or less offensive expression. When the speaker or writer fears that the more direct language might be harsh, unpleasant or aggressive, the euphemism is employed. In general, the existence of taboo words as well as taboo ideas stimulates the emergence of euphemisms. In many counties, people avoid employing the direct words that relate to death. This, therefore, arouses a large numbers of euphemisms to death subject. In English-speaking world, people don't use "die", but "pass away" or use an idiomatic expression "kick the bucket". In Chinese, there exist lots of words to convey "death" concept, including "Shengtian", "Zhoule", "Shuizhaole". . . . With regard to pragmatic factor during translating, translators must abide by one certain cultural text and contribute a more exact equivalent in view of different styles.

3.3 *Application of national values*

Under the influence of historical culture, the gradually-developed national values and cultural psychology determine the linguistic behavior modes. Polite language is of pervasiveness, but in English and Chinese, there are various approaches to actualize and standards to appraise. Traditional semi-colony and semi-feudal society of China forged an introspection-oriented cultural mentality in consequence of isolation mechanism and self-sufficient agricultural economy model in the feudal society. Personal value is based on the collective consciousness, which reflects it too. When Chinese are exposed to praise and compliment from others on social occasion, they used to adopt self-deprecation ways to show his or her courtesy involving "Bu", "Nali", "Guojiang", "Yiban", etc. Nevertheless, to embody personal value, to pursue freedom and to exploit territory are the general tendency to follow in western culture. The response to the compliment, "You are so beautiful today!" is "Thank you!" So in the cross-linguistic and intercultural translation teaching, in order to attain translation equivalence, conform to the translation principles: faithfulness, expressiveness and elegance. The objective of translating practice should be concentrated on comparison of the cultural components from multi-perspectives and multi-levels in the long run.

4 CONCLUSIONS

As Huston Smith (1991) said: "when historians look back on our century, they may remember it most, not for space travel or the release of nuclear energy, but as the time when the people of the world first came to take one another seriously." In short, it is a real mistake not to attach great importance to culture context in translation teaching. By means of contrastive and comparative studies of the cultural background of source language and target language, to spotlight the prominent and dominant cultural individuality based on cultural overlap and to strengthen immigration of the target language culture are the important approaches to enhance the competence in translation teaching and practice. Nowadays, society is thirsty for cross-linguistic, intercultural professionals with great creativity and responsibility to make connections between Chinese and western culture.

REFERENCES

He Shanfen. 2002. *Contrastive Studies of English and Chinese Languages*. Shanghai:Shanghai Foreign Language Education Press:361–364; 390–396.

Larry A. Samovar & Richard E. Porter & Lisa A. Stefani. 2000. *Communication between Cultures*. Beijing:Foreign Language Teaching and Research Press.

Ouyang Lifeng & Xu Huijuan. 2003. Of Pragmatic Translation from Cultural Context. *Journal of Anhui University(Philosophy and Social Sciences)*.02(3): 18–19.

Steve J. Kulich & Michael H. Prosser. 2007. *Intercultural Perspectives on Chinese communication*. Shanghai: Shanghai Foreign Language Education Press.

Vivian Cook. 2000. *Second Language Learning and Language Teaching*. Beijing:Foreign Language Teaching and Research Press: 95–99.

YANG HSIEN-YI & GLADYS YANG.1978. *A Dream of Red Mansions*. Beijing:Foreign Languages Press.

Zhang Weiyou. 2000. *English Lexicography*. Beijing:Foreign Language Teaching and Research Press.

Zhou Fangzhu. 2008. *Principles of Translation between English and Chinese*. Hefei:Anhui University Press.

Management, Information and Educational Engineering – Liu, Sung & Yao (Eds)
© 2015 Taylor & Francis Group, London, ISBN: 978-1-138-02728-2

On the curriculum setting, teaching mode and teacher allocation of ACCA in universities

Sha Lv
Sichuan Normal University, Chengdu, China

Yuan Feng
Jinan University, Zhuhai, China

Hui Su
Sichuan Normal University, Chengdu, China

ABSTRACT: As the demand for high quality internationalized accounting talents is strong in China, bringing ACCA teaching to cultivate applied accounting undergraduates with international competitiveness has generated considerable excitement in the undergraduate education. This paper addresses the issues of curriculum setting, teaching mode and teacher allocation in ACCA courses in universities. The above-mentioned areas are highlighted to optimize the combination of academic education and vocational certification education.

KEYWORDS: ACCA; Curriculum setting; Teaching mode.

1 INTRODUCTION

As economic globalization deepens and domestic capital market grows, demand for accounting talents with international vision has increased. Accordingly, accounting profession requires interdisciplinary talents equipped with management skills and strategic thinking, specialized in business and familiar with market rules and international practices. To respond to this trend, in recent years, the Association of Chartered Certified Accountants (ACCA), one of the largest and fastest-growing global accountancy bodies, expands cooperation with universities in cultivating still more innovative accounting talents with international vision. Under this joint running mode, university students, having completed general undergraduate courses and acquired advanced financial knowledge in ACCA courses, will be afforded double certificates (qualification certificate and graduation certification) upon graduation. Till now, there are 70 odd universities and colleges having signed cooperation agreement with ACCA, joining the cultivation of talents with international vision. However, considering the different social context, accounting standards and teaching methods between China and Britain or other western countries, it is crucial for Chinese universities to achieve the optimization of curriculum setting and teacher allocation; find out the effective ACCA teaching mode.

2 ACCA CURRICULUM SETTING

The ACCA curriculum system is divided into two phases or four modules. Phase I: Foundation Level consists of two main modules: Knowledge and Skills—with 3 and 6 lower divisions respectively—involving core knowledge, such as accounting, tax law, auditing and financial management. Phase II: Profession Level is subdivided into two main modules: 3 core courses and 4 optional courses, among which each student selects two. Professional Level, the higher level of Foundation Level, focuses on occupational skills and knowledge needed for senior accountants. Those 14 courses in the curricula are carried out from the easy to the difficulty in order to enable students to gain the basic principles and concrete operation of accounting, develop comprehensive ability, and foster critical thinking. Details of ACCA course programs are as follows.

Table 1. ACCA Course Structure.

Phase I: Foundation Level		Phase II: Profession Level
Knowledge	Skills	Core courses
F1 Accountants in Business	F4 Corporate and Business Law	P1 Professional Accountant
F2 Management Accounting	F5 Performance Management	P2 Corporate Reporting
F3 Financial Accounting	F6 Taxation	P3 Business Analysis
	F7 Financial Reporting	Optional courses (select 2 out of 4)
	F8 Audit and Assurance	P4 Advanced Financial Management
	F9 Financial Management	P5 Advanced Performance Management
		P6 Advanced Taxation
		P7 Advanced Audit and Assurance

To increase the pass rate of ACCA, universities need to carry out all the courses required in ACCA exams step after step and set curriculums and teaching mode in light with ACCA examination syllabus. As teaching ACCA is time consuming owing to its large and complex content, many universities and colleges have replaced the accounting course which is supposed to open with some ACCA courses so as to make use of the limited credits. However, given the ACCA curriculum has its particularity in content and required capability, the replacement of accounting courses with ACCA courses fails to present students with a complete framework of accounting, especially the principles and practices of China's accounting. As a result, graduates applying for accounting positions in China are put in a disadvantaged position. It also means students need to pay extra time and effort in specialized courses since some basic knowledge are missed out in teaching procedure, breaking down the linkage between knowledge units, which goes against the primary purpose.

Besides, comparing with other accounting courses, ACCA class provided in higher education is exam-oriented. Given the limited total teaching time, it is almost impossible for teachers to touch all the content and fully analyze them. Also, students only care about content relative to the exam. All these make it more difficult to equip students with comprehensive and systematic knowledge and skills. More specifically, the study material of ACCA is a lack of fundamental knowledge, such as background information, principle and accounting theories. Instead, it stresses questions and keys targeting on the exam, which is more suitable for employees with theoretical basis. But for students, who are lack of practical experience, this kind of teaching mode is demanding. Hence, to make teaching more productive and improve students' ability, it is crucial to establish a long-term effective ACCA teaching mode.

2.1 *Add chinese foundation courses and bridging courses*

Comparing with Chinese traditional teaching textbook, ACCA's is informative and contains a lot of points without full explanation. Besides, there are gaps between points and lack of logical linkage. All these call for opening additional foundation courses and bridging courses (in Chinese) in curriculum setting. Take Sichuan Normal University, the university the author serving as an example, there are corresponding Chinese courses prior to each ACCA course. Specifically, we have Primary Accounting and Financial Accounting before F3 (Financial Accounting) and Financial Management Principles before F9 (Financial Management). Under the curriculum structure, curriculum of general education, fundamental subjects and practice account for 70% of the total while core professional courses contain 30%. At the same time, we developed more optional courses so as to provide more choices for students. In this way, students could learn systematically and gradually become a master in ACCA. Besides, having a good command of domestic accounting affairs plays an important role in students' career development since most ACCA graduates work in China's enterprises and public institutions. Thus, this paper proposes to add preliminary courses and Chinese elements (e.g. Chinese Tax Law) into the curriculum as long as it does not cost extra class hour.

2.2 *Establish gradient teaching mode*

In terms of degree of difficulty, the F1–F9 courses at Foundation Level are similar to undergraduate courses and the P1–P7 courses in Professional Level are similar to postgraduate courses. Universities, therefore, could imitate the teaching model of Combined Bachelor-Master's Degree Program, i.e. move some content taught in the Professional Level to postgraduate period to free students from the overloaded exam burden and help lay a sound professional foundation. This gradient teaching mode is more reasonable since it complies with the logical law of digesting and absorbing knowledge solves the problems in ACCA curricula setting to some content and soothes the contradiction between exam-oriented education and cultivating highly competent people. Furthermore, the content of some courses canceled because of the limitation of credits could be integrated into the relative courses. In addition, universities could directly give more credits to ACCA. After all, this major aims to cultivate accounting talents with international competitiveness.

2.3 *Enhance practice curriculum*

Now that accounting features practice and operation, the examination of each class requires students have the ability to analyze and solve problems with learning knowledge. Hence, accounting teaching must focus on improving students' ability to take down practical problems by implying knowledge into practice. And to meet the teaching goal of cultivating high-caliber personnel, practically, this paper propose to cooperate with overseas universities and colleges or professional bodies, mimic global situation and collaborate with domestic and foreign agencies to establish practice bases.

2.4 *Highlight academic education*

ACCA courses in higher education are based on academic education and supplement by exam-oriented education. Generally, academic education enables students to acquire comprehensive, systematic and in-depth knowledge, laying a solid foundation for applying knowledge. However, if exam-oriented education plays the dominate role in teaching, students could barely master comprehensive professional basis and wide knowledge, which hinder their future career development.

To achieve better integration between exam-oriented education and academic education, universities may encourage students to participate in global ACCA exams and improve the assessment and management of specialized courses. To be specific, schools should make scores on global ACCA exams transferable to that of correlated courses, and then convert them to credits based on the difficulty level of the exams. Thus, students will be fully motivated.

3 TEACHING MODE OF ACCA

ACCA courses have universal textbooks and papers (English version). Currently, most ACCA courses in higher education are picking up textbooks provided by the BPP teaching material, a detailed and informative version. With limited class teaching hours, it is hard for students to grasp so much information in such a short term. Worse yet, students, whose native language is Chinese, have difficulty in understanding and analyzing problems in English and would easily fall into learning fatigue if the teaching language is English.

Most ACCA courses in higher education require students to take corresponding ACCA exams at the end of the term. That is to say, students have to take two to three exams at the end of each semester, which is demanding. More productive and rewarding teaching mode is required in ACCA training courses.

3.1 *Emphasize the "independent learning" mode*

The teaching effect largely depends on the subjects' initiative and participation since knowledge acquirement is an active process. Especially in ACCA teaching, we should attach great importance to students, the learning subjects. Only by enhancing students' learning motivation and participation and arranging course content and pace properly, can we make the complex professional knowledge in English digestible and comprehensible. Students must be informed that ACCA courses are time-limited and preview and review are required. As for teachers, before class, they need to tell students the main idea and difficulties of the next class and ask them to preview; during class, teachers should adopt various methods to explain the requirements and content of the subject, especially the important and difficult points of the course and strengthen interaction with students such as asking questions and case discussions so as to deepen students' understanding of the knowledge through teacher-student interaction such as questions and case discussions; after class ,teachers proclaim assignment and teaching requirements, for example, summarizing the learned knowledge in English and handing in case analysis, aiming at improving students' English level and enhancing their master of learned knowledge.

3.2 *Adopt flexible and different teaching methods*

Teachers should adopt different teaching methods according to different training goals, subjects and students. Multimedia teaching is recommended, in which the main content of the course is presented in English, making it easier for students to catch the main points and deepen their understanding and absorbing of the knowledge. Meanwhile, in order to improve students' overall ability of English—listening, speaking, reading and writing and provide students with opportunities to master professional knowledge through English, we might encourage group discussion and create a learning environment featured of interaction and cooperation. In this way, we also develop students' sense of cooperative learning and creative thinking. Practices show that students can gain a large quantity of knowledge through communicating with comrades. Cooperation with partners is especially important in the learning process of ACCA advanced courses. We can also make use of web-based teaching platform to enhance interaction between teachers and students, such as opening study forums. On the teaching platform, teachers can pose questions and assignment while students can put up and answer questions. Whatever teaching methods adopt, the ultimate goal is to improve the effectiveness teaching professional knowledge.

3.3 *Implement progressive bilingual teaching method*

ACCA courses in higher education usually use Chinese and English in teaching, so mastering English is the prerequisite to learn ACCA and pass ACCA exams. Take the author's university for an example, in the first semester, we do not offer professional courses of ACCA but intensive English courses to prepare students with language ability to receive bilingual education and improve their comprehensive abilities step by step. In the second semester, students start to learn the professional knowledge of ACCA and would feel difficult when expose to a large number of specialized vocabularies. At this time, teachers should guide them to adapt to the curriculum. Moreover, teachers should keep checking whether students can keep up with the class and make corresponding adjustments of the teaching methods and difficulty coefficient. And teachers also need to constantly adjust the time using English in class according to the English level of students. Early in ACCA courses, teachers may as well explain jargons and main principles in Chinese; in the middle of courses when students have got familiar with the jargons, teachers can cover simple part through English, but still using Chinese to explain complex matters; in the later period of courses, teachers should interact with students in English and deliberately improve students' ability in using English. Nevertheless, professional knowledge remains the focal point in ACCA courses.

4 TEACHER ALLOCATION

In ACCA courses, teachers serve as a guide. It means teachers should adopt varied teaching models and methods at different teaching stages and particular situations. At present, teachers of ACCA courses can be classified into three types: external foreign teachers, external teachers from domestic training institutions and the faculty members of own schools. Those three types of teachers distinct from each other in teaching experiences, mobility, cultural background, language proficiency, master of international and Chinese accounting principles and the like. However, ACCA courses require teachers be experts of the entire ACCA teaching system, have solid domestic and international accounting professional knowledge, and have the ability to understand, write, and express in English. Therefore, we must establish a long-term mechanism of optimizing teachers' allocation of ACCA training courses and build a high-quality ACCA teaching faculty.

4.1 *Introduce talents to teach ACCA*

Universities that offering ACCA courses can introduce teachers who have passed ACCA exams and have an English teaching background from other universities both home and abroad as full-time ACCA teachers. These teachers are familiar with the content and proficient in English, ensuring the high quality of bilingual teaching of ACCA. And universities should evaluate the performance of teachers based on their teaching ability and teaching results instead of their achievement in scientific research to make sure they have enough time and energy to dig into ACCA study and teaching.

4.2 *Train selected teachers*

Universities send prominent teachers who have a solid foundation of professional knowledge, rich teaching experience and are proficient in English to famous colleges and universities with a long history of accounting disciplines and strong scientific research and teaching ability in the UK, to have two or three years' study and training. Consequently, on the one hand, teachers can get chances to know the social, economic and cultural backgrounds of UK, exercise and improve their English and get familiar with the thought pattern of English-speaking countries; on the other hand, they can get to know more about the knowledge system, the teaching model and the training methods of ACCA courses through the professional learning and training in the UK.

4.3 *Encourage teachers to attend domestic ACCA teaching workshops*

Universities should encourage ACCA teachers to attend domestic ACCA teaching workshops, build teaching teams and hold teaching symposiums within the team. Through those workshops and experience exchange conferences, ACCA course teachers would be motivated to keep learning from the prominent teachers' experience and improve the level of expertise and quality of teaching.

REFERENCES

Ming Li and Huijuan Zhao. ACCA brings opportunities and challenges to universities. *Heilongjiang Researches on Higher Education*. 2013(8):201–202.

Jianhua Guo. Discussion on the Orientation of ACCA in universities. *Communication of Finance and Accounting*. 2014(2):41–43.

Jianping Zhang. On ACCA teaching mode—based on the theory of multiple intelligences. *Science and Technology Information*. 2013(13):1–3.

Gaoliang Tian, Junrui Zhang and Fangjun Wang. The innovation and practice on ACCA teaching mode based on international talent cultivation. *University Education*. 2014(2): 50–52.

Management, Information and Educational Engineering – Liu, Sung & Yao (Eds)
© 2015 Taylor & Francis Group, London, ISBN: 978-1-138-02728-2

Investigation of the present situation of Nanchang middle school football development

Q. Xiong, L.H. Chen, F. Chen, Y.H. Wang & J.H. Li
Key Laboratory of the State Sports General Administration, Nanchang, China
Institute of Physical Education, Jiangxi Normal University, Nanchang, China

ABSTRACT: This paper analyses the current status of Nanchang middle school football development from development situation of sports class, teacher strength, grand equipment and students' interests by the method of questionnaire, literature, mathematical statistics, interviews. And put forward some reasonable suggestions to promote the professional quality of the coaches, encourage more participation in sports. It offers reference value for the development of students' football sport in Nanchang city.

KEYWORDS: Nanchang City; middle school students; football development; present situation.

1 INTRODUCTION

For correct settings of margins in the Page Setup dialog box (File menu) see Table 1. It is known as football not only is the first biggest activities that swept the world, but also is an important task of school physical education. It is important to improve the students' physical quality, strengthen the fighting spirit, the spirit of unity and cooperation. Developing football well not only is good for the training of the reserve players to play a very important role, but also can strengthen the teaching content of physical education, the cultivation of students' interest in sports, enrich the cultural life in school. Football is a part of the daily life of the people's spiritual and cultural life. A country's football level was affected by many factors, but the cultivation of young's training is the most basic and most important factors (Zhuang 2011). Large and small scale schools as the cradle of China's football talent occur the phenomenon that football activity development situation is wilting in different degree and the outlook is not optimistic (Liu 2011). The aim of this paper is to find the cause of football reserve wilting phenomenon in Nanchang city through the investigation and analysis of the present situation of middle school students.

2 METHODS

2.1 *Objects*

Extracting students from Nanchang City fifteenth Middle School, Nanchang City seventeenth Middle School, Nanchang City twenty-third Middle School, Hongdu middle school and Nanchang University attached middle school.

2.2 *The method of literature*

Through the Chinese journal full text database; China How Net; Wan-fang database this paper describes the present situation of the development of football of middle school students as well as through the internet to relevant domestic and foreign documents. This paper summarizes the literature material and provides the reference for the research.

2.3 *Questionnaire survey*

Granting questionnaires to students from Nanchang City fifteenth Middle School, Nanchang City seventeenth Middle School, Nanchang City twenty-third Middle School, Hongdu middle school and Nanchang University attached middle school. Granting 200 copies, and 180 copies of questionnaires were recovered. (Table 1) The effective recovery rate is 90%.

Table 1. The recovery of the questionnaire.

School	Granted	Recovered
fifteenth Middle School	40	40
seventeenth Middle School	40	33
twenty-third Middle School	40	35
Nanchang University attached middle school	40	37
Hongdu middle school	40	35

2.4 *Statistics*

The questionnaire obtained data for mathematical operation statistics

3 RESULTS AND DISCUSSION

3.1 *Investigation and analysis of students' interest in football*

Interest is the best teacher as well as is the important driving power of students to obtain important dynamic knowledge, broaden one's horizon, and enrich psychological activities. Interest can promote human make progress. The excellent football players all showed great interest since they are young. The interests that they adhere make them successful (Li 2011).

Table 2. Survey of student's interest to football.

	Interested	General	Non-interested
Number	76	94	30
percent (%)	38	47	15

It can conclude from the questionnaires filled with 200 students of table 2 that 47% students interested generally in football, 15% students are not interested in football at all, the reason which many men and women interested in differences. The reasons include the differences between male and female as well as recognize the deviation of football. But many students still interested in football, there are 38% students usually like to join the football activities or pay attention to football. From the survey results, cultivating the interest of football does not only depend on the students themselves' understanding of foot, but also need to lead students reverse cognitive deviation of football, which can prompt them to cultivate their interests of football as well as participate in it.

3.2 *Investigation and analysis of teachers that Nanchang city middle school physical education teachers*

Physical education teachers are the important prerequisite of football development. The number of spacial football teachers and academic structure are important guarantees for high school football teaching and amateur training (Zeng 2011).

According to the relevant documents of China's teacher with the request, the middle school physical education teachers and students is the ratio of 300:1, which means every three hundred students is in need of a physical education teacher, although in recent years the schools are vigorously improve teachers, but the current situation still cannot meet the requirements of the development of physical education, physical education is also in the reform, every week 3 hours, so the lack of teachers directly affect the physical education (Shen 2011).

Table 3. The teacher resource questionnaire.

School	Number of non-specialized teachers	Number of specialized teachers
fifteenth Middle School	4	0
seventeenth Middle School	5	1
twenty-third Middle School	4	0
Nanchang University attached middle school	6	1
Hongdu middle school	4	1

Table 4. Teacher education survey.

The average education	Undergraduate course (%)	Junior college (%)	Senior college (%)
Number of non-specialized teachers	40	40	20
Number of specialized teachers	0	100	0

Table 5. Teacher's age survey.

Age (year)	20-30	30-40	40-55
Number of non-specialized teachers	8	9	6
Number of specialized teachers	2	0	0

As Table 3 shows, the 5 schools in the survey, professional PE teachers in 4–6 people, professional PE teachers can basically meet the daily work of teaching; football special teachers, have special football sports teachers only 2 schools, three other schools are made by professional PE teachers are responsible for training, the game and other football related work. Thus, it can be seen that the lack of special football teachers is an important factor to restrict the current high school football sports in our city development (Ma 2010). In addition, the investigation found that through our physical education teacher and football special teachers' age and education structure (Table 4, 5), full-time teachers of our province sports are mostly in the 30–40 years old, undergraduate college which accounted for 40%, accounted for 40%, secondary school accounted for 20%, mostly in the junior college level; 2 football PE teachers are in the 20–30 years old, or college level, educational level is relatively low. It could be seen that low teachers in our city young and structure of educational background, lack of teaching experience, the impact will be different degrees of quality in our province sports teaching and the teaching of soccer.

3.3 *Investigation and analysis of present situation of Nanchang city high school football sports venues*

As everyone knows, each school is equipped with a football field condition of students are not the same, there are many schools due to various reasons, no space or smaller venues, which restricts the development of students interested in different extent, the choice of extracurricular activities.

Table 6. The school site conditions.

School	Large football field	Small football field
Fifteenth Middle School	0	0
seventeenth Middle School	1	0
twenty-third Middle School	0	1
Nanchang University attached middle school	0	1
Hongdu middle school	1	0
Percentage (%)	40	40

Investigation shows, only fifteen is neither large nor small football field, soccer field, they are influenced by the interest of site factors and the student movement of the transfer will be converted into a football field original basketball more. There are a large football field seventeen and South high school, a small football field twenty-three and Hongdu middle school. From the survey results, only 20% of the school is not equipped with a football field, 40% of the school has a large football field, also has a small football field 40%. Schools generally have enough trouble to land resources and construction of a standard football field, basketball so determined to take football second construction of small or used for building covers an area of small, which makes the soccer movement development lack the most basic guarantee of material, field problems hindered the development of football from the front.

3.4 *Football teaching present situation investigation was carried out in Nanchang city middle school P.E*

Football teaching can make the students more direct contact with the football, which affects the students' understanding of football. Football teaching can be divided into the basic rules of teaching, teaching simple technology and tactics teaching. Only involves the rule of teaching and simple technique teaching in general high school football course. Only have a certain understanding of football, the interest can be generated, so the teacher of football teaching is particularly important (Huang 2004).

Table 7. Soccer School PE Teaching.

School	Football teaching courses
Fifteenth Middle School	N
Seventeenth Middle School	N
Twenty-third Middle School	N
Nanchang University attached middle school	Y
Hongdu middle school	Y
Setting percentage (%)	40%

But through the survey (Table 7), five schools have football teaching only two, the situation is not optimistic. It is understood that the present those not developing football teaching of school physical education mainly in the "sheep" type, the so-called "sheep" is a free, what sports do you like to play anything, do not want to exercise in the playground for a walk, a lot of students in football teaching formal did not come into contact with, it checks a lot of students' interest, like to play but I didn't know how to play, not to mention no interest in football teaching, carrying out the teaching of physical education and even is the normal development of worrying.

3.5 *Investigation and analysis of the parents to attend the football sport attitude of middle school students*

Students usually involved in what kind of exercise is largely determined by the parents of the sport attitude. No matter what the movement, if the child love, parental support, so it will be easy to become the child's lifelong hobby If the child love, parents objected to, then it is very likely to have been strangled in the bud (Ji 2003). Half of the parents of their children to participate in football to keep support attitude, 37.5% of them are fully supported (Table 8). They think that football can strengthen the body, also has a 12.5% to utilitarian attitude, think children like football at least in vivo test a good result. While the other half of parents opposed to 30% are also afraid of children Wanwusangzhi, will delay their studies, 20% of the parents with their fierce fight to see on television, subjectively think football is a vulnerable movement, which opposed. Soccer movement development is inseparable from the social attention and strong support, parents treat their children play football attitude largely determines whether children will be football as his lifelong interest in sports, but we must from the conceptual change some parents' prejudice of football (Pelletier & Vallerand 2010).

Table 8. Parents' attitudes toward the investigation situation children to participate in football.

Attitude	Enhanced physique	Handle sport Examination	Delay the studies	Easily hurt
Members	75	25	60	40
percentage (%)	37.5	12.5	30	20

Function and motor task internal motivation and external motivation is related to the complexity of. Task complexity is greater, the greater the role of internal motivation: task complexity is small; the effect of external motivation is greater (Zhu 1992). According to the characteristics of football, football is the task of complex projects, then the internal motivation more. Most early research and practitioners believe that external motivation and internal motivation is complementary to each other, that there are two kinds of motivation than only one kind of motive is good. However, many modern research shows that external motivation could weaken the internal motivation, may also strengthen the internal motivation, the key lies in how to use the appropriate external reward (Zhu 1992). In Nanchang young football players in external motivation, the motivation of external constraints the highest intensity, external constraint refers to the exogenous control and participates in the activities, such as by the material rewards to control or the control of another (Ji 2001).

4 CONCLUSION

To improve the students amateur league system, strengthen the training and management of school work, increase the sports education funds, enhance the quality of the teachers' team, reform the football match system, may be helpful for the development of students' football sport in Nanchang city.

ACKNOWLEDGEMENTS

Corresponding author: Jiang-Hua Li. This research work was supported by National Natural Science Foundation of China under Grant No. 21365013 and Teaching Reform Project of Colleges and Universities in Jiangxi Province under Grant No. JXJG-13-2-6.

REFERENCES

Ding, J. 2004. How to stimulate students' interest in sports training. *The middle school sports network*. 85(20):96–104.

Huang, W.X. 2004. How to stimulate students' interest in learning. *Primary school sports*21(23):32–43.

Ji, L. 2001. Physical education curriculum standard. *Beijing Normal University press*. 52(7):84–92.

Ji, L.& Wang, X.Z. 2003. New methods of teaching primary school sports curriculum.*Higher Education Press*.61(7):63–71.

Li, W.D. & He, Z.L. 2011. The national youth campus football sustainable development issue. *Journal of Physical Education* 45(03):64–72.

Liu, Y.N. 2011. Research on the development of football on campus in Shanghai. *Journal of Physical Education* 84(36):86–94.

Ma, Y.X. 2010. Research and practice of football teaching characteristics and regional promotion. *Zhejiang* 87(22):36–58.

Pelletier, L.G. & Vallerand, R.J. 2010. Toward a New Measure of Intrinsic Motivation ,Extrinsic Motivation,and Amolivation in Sports:The Sport Motivation Scale. *JOURIIAL OF SPORT&EXERCISE PSYCHOLOGY*. (17).35–53.

Shen, S.Y. 2011. Carry out the campus football China football will develop. *Sports Expo 74*(16):54–62.

Zeng, G.M. & Yu, J.J. 2011. Model analysis and development suggestion of Shanghai City Campus football status. *Sports scientific research* 58(01):45–49.

Zhuang, Q. 2011. The analysis of development condition of youth campus football-Taking Ji'nan city as an example . *Sports* 65(10):78–83.

Management, Information and Educational Engineering – Liu, Sung & Yao (Eds)

Construction of teaching faculty in the stratified teaching

Hai Yan Huang, Ai Min Gong & Chao Ying Hu
College of Water Conservancy and Hydroelectric Engineering, Yunnan Agricultural University, Kunming, China

ABSTRACT: The methods and application of teaching faculty in the stratified teaching are discussed in this paper for water conservancy talents training. The teaching faculty includes the teaching team, the expert team and academic team. The target of construction of the teaching faculty is improving the teacher's teaching ability and research ability. The core of construction is taking part in the teaching and scientific research by the way of expert guidance, the course leads, project lead, resource guide through the special plan and special events. The construction achievements show that the abilities of mutual integration, solidarity and cooperation, interaction of the faculty are improved and the scientific and academic level is enhanced.

KEYWORDS: Construction; Teaching faculty; Stratified teaching; Talents training.

1 INTRODUCTION

Yunnan province has not only large mountains, but also many rivers and lakes. The region of Yunnan province spans seven climate types, so it obviously has three-dimensional climate. And there are significant changes in water resources during the year. The developed region is the central region of Yunnan province, which has better water infrastructure. The poorest region is the northern region of Yunnan province, which has weak water infrastructure. There are many ethnic minorities in the southern edge of Yunnan province, and the water resources development and utilization relationships are more complex. Therefore, there are the unique regional characteristics and the characteristics of the typical topography needs in Yunnan province to train hydraulics and technical personnel of adaptable, personalized, and diversified, to work independently under specific environmental.

Thinking from the frontier facts, diversity and specificity of the regional economic development process in Yunnan province for water conservancy talents, a new training plan of "Professional recruitment, Training categories, Shunt in the middle period, Stratified teaching" is advised to culture the students and train the teaching faculty. The practice of the teaching reform plan starts at 2008; and the shunt in the middle period is completed in the February 2011; and then the stratified teaching is in the way. The construction achievements are given in this paper that how to train students to the technology type, excellent engineer type, academic research type personnel, and how to build a teaching system for different teach method.

2 STRATIFIED TEACHING

Stratified teaching is for the teaching strategies of all students, focusing on individual differences, teaching students in accordance with their aptitude as a starting point, combining each student's own objective reality, implementing hierarchical lesson planning, teaching, practice and evaluation, coordinating the relationship between the teaching objectives and teaching requirements, targeting students at different levels to choose the ways and means of education, promoting each level students' learning ability and teaching requirements to better adapt to each other, so that each level of the students get a good education.

According to the characteristics of the stratified teaching, training objectives for water resource management, taking personnel training quality as the core, the students are divided into three levels: technology talents, outstanding engineer talents and academic research talents. There are three kinds of training methods: basic quality training, engineering quality training and study quality training. There are three kinds of abilities: engineering capabilities, engineering design capability and technological innovation capability. The training model is shown in Figure 1.

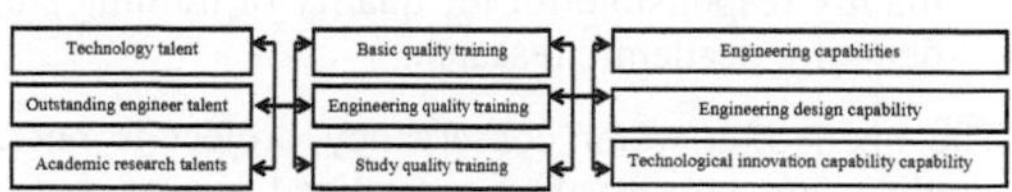

Figure 1. Stratified teaching model.

3 CONSTRUCTION OF TEACHING FACULTY

Three course groups are advised that they are a group of technical and applied courses, a group of excellent engineers' courses, and group of academic research courses. On the base of the training model and according to the market demand and local industry expertise required, all the courses are integrated as a course system shown as Figure 1.

All the teachers, including the teaching team, team of experts, and academic team service for the students around these three course groups. They are trained in the methods of teaching research, discipline construction, scientific research, production projects, social services and so on.

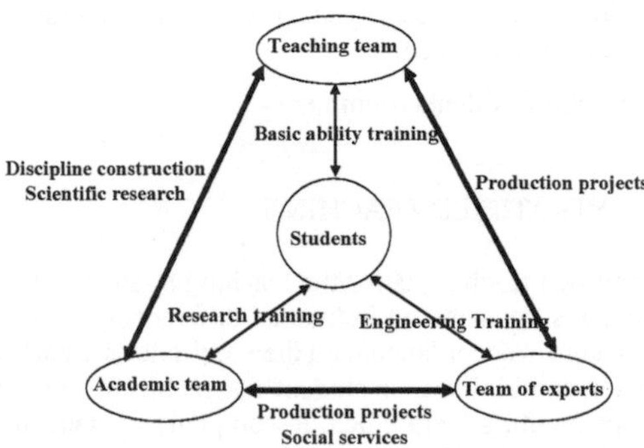

Figure 2. Training model.

1 Teaching team: It is the teacher composition to improve the quality and effectiveness of teaching and take promoting teaching reform as the main task. Relying on the each course group's foundation course, it is mainly responsible for the basic professional quality of training students;
2 A team of experts: It is the teacher composition to enhance the influence to carry out production projects and social services as the main task. Relying on technology applied course group, outstanding engineers course group of core courses, they are mainly responsible for the engineering quality of training students;
3 Academic team: It is the teacher composition to raise the level of scientific research, academic research as the main task. Relying on academic research course group's core courses, they are mainly responsible for the quality of training students for academic research.

The need of a water conservancy project is not a single water conservancy specialized talents, but a combination of multi-disciplinary talents. It includes the planning, design, flood control, drought relief, rural water conservancy, resettlement, construction management, engineering management, watershed management, reservoir operation, hydraulic power, scheduling of water resources, hydrology, weather, etc. Therefore, the teaching faculty is the group who are mutual integration, mutual support, solidarity and cooperation and include the teaching team, the team of experts and the academic team. It is a shared responsibility, high-quality teaching faculty. At present, the teaching team has been formed a team that composed of Yunnan Province of provincial teaching team, Yunnan Agricultural University of school teaching team, it is a stable teaching team. The team of experts has formed that include school teachers and external experts who have a certain influence in their field. The academic team has formed a consisting of professors and associate professors, which has rational structure and stable research direction.

4 THE PRACTICE OF THE CONSTRUCTION OF TEACHING FACULTY

The construction of teaching faculty adheres to the idea that is "morality is important, educating is the base, teaching and research are simultaneously developed, and inheritance and innovation are the goal". The target of construction of the teaching faculty is improving the teacher's teaching ability and research ability. The core of construction is taking part in the teaching and scientific research by the way of expert guidance, the course lead, project lead, resource guide through the special plan and special events.

4.1 *The construction way of teaching faculty*

1 The experts guide

Based on the water conservancy industry in Yunnan province, we act the province of the construction of water conservancy development, technical problems as the research direction and goal, give full play to the lead role of the school and province of inside and outside teaching experts, integrating the rich expert resource and promoting the harmonious development of the teachers' time.

We open "expert rostrum", according to the overall arrangement of the education system, organizing regular teaching experts, engineering and research experts, who parse and comment on the teaching, production and scientific research of the "frontier problem", "hot spots", "difficulty" and "blind spots" and "pathological problems". It will promote the teaching team, a team of experts and academic team's mutual fusion, support and solidarity.

2 The course guide

With the construction of technology applied course group, excellence engineers group and the academic research, the curriculum group of as the platform, through the organization as the course teaching center of a series of activities, which improves teachers "practical" ability, seeks breakthroughs in classroom teaching, improves teachers' education teaching ability and academic level of scientific research.

It will need open exhibition of the altar and invite famous teachers, host excellent courses, student's favorite teachers, and combines classroom teaching to show class quality. At last, we also employ relevant professional experts to review.

3 The project guide

This is in the form of subject to support the teachers' teaching theory research, teaching reform and innovation, production, scientific and technological research projects. Research is in the form of diversity, which provides a platform for the school of provincial and national teaching masters and the scientific research project.

4 The resource guide

Build a platform is used to teach, product and research scientific information resources, which for teachers meets the need of modern teaching, modern water conservancy construction, and scientific research teaching and scientific research information resources. Using the Internet to carry out series of online training and communication, such as network questionnaire way to understand a line of teachers' teaching, production and scientific research needs; It also can open online teacher training curriculum (including information technology training, teaching theory and method of training, etc.), provide text materials and video tutorials, and online learning for teachers; Recommending and introducing the open course group of world-famous universities and related video download service, which closes to teachers and lets them feel the world famous universities teachers' teaching style.

4.2 *The teachers' faculty construction work*

In order to promote the teaching team, team of experts and academic team between mutual fusion and mutual support, solidarity and collaboration and interaction with, and form teacher autonomy and the development of consciousness, which will through four special plan and four thematic activities to carry out the teachers construction work.

1 The special plan

Through teacher incentive plan to encourage the leader to play a leading role, and establish a series of teachers' incentive system and reward programs, the key to the plan is that supporting discipline leaders, academic leaders, teaching leaders, and inspiring a batch has extensive influence field, provincial and national masters.

By the young and middle-aged teachers support program, this will support a number of outstanding young teachers, and grow up to be at an early age in both have certain popularity of expert teachers and scholars.

Through the youth academic talent cultivation plan, which achieves a batch of with academic potential, the key to this plan is that development guidance and training of young teachers. For example, we support the collocation of the new teachers and the old teachers, support the long-term personal development plan, and continuing education plan, take participate in the school subject construction, teaching reform, social production and scientific research.

Through the international teachers complete program, which speeds up the pace of internationalization of teachers, promotes the perfect development of the bilingual teaching of specialized teachers.

2 The project activity

Academic cycle activities: the third primary school period the first week of every school year for academic titles, swap project teacher training, seminars.

Teachers' salon activities: through theme exchange, workshop, production workshop, the famous teacher teaching seminars, research forum, teachers communicate seminars, to promote the teaching team, a team of experts and academic team communication between, experience sharing and mutual cooperation.

Teacher development database: The database includes teachers' information database and expert repository. By the teachers' information convenient database implementation of the declaration, management and consulting services, we provide professional guidance, advice and estimates suggestion for the construction of a teaching team, a team of experts and academic team. The databases write the "teachers handbook", "manual of experts", "the scientific research guide", "the teacher development", guide to launch demonstration materials, and announce teacher development related promotional materials and information.

Improve teachers' incentive system: the system improves the teaching team, a team of experts and academic team's selection, evaluation and assessment system. It will need the selection and recognition of outstanding individual and team on a regular basis.

4.3 *The construction achievements*

After four years of stratified teaching reform practice, the following results are obtained:

1 The team has been developed as the provincial "Comprehensive reform of water resources and hydropower engineering project", the provincial "Water conservancy and hydropower projects the professional excellence engineers education training plan construction project", the provincial

"Agricultural soil and water conservation engineering teaching team" in 2012.

2 The teaching team, a team of experts and academic team have been obtained 12 teaching achievement prizes and 4 awards of progress of science and technology the progress prize in science and technology of teaching achievement prizes. They published 17 textbooks, won 16 national patents, computer software copyright, and published 161 academic theses.

3 In the national construction engineering cost member training center, the rate of students who get a certificate rate is more than 70%. In Yunnan province water conservancy engineering training center of our school, the rate of students getting the certificate more than 95% who take part in the training of water conservancy project budget, the quality inspectors, and the inspector' student. Authorized by provincial ministries of water resources, Yunnan provincial water resources bureau, our school will train county (city) bureau chief in Yunnan province for on-the-job training four periods. The students' number is more than 1780 people.

5 SUMMARY

For the regional economic development in Yunnan province which the demand of the water conservancy talents has the characters of diversity and particularity, the College of water resources and hydropower and architecture of Yunnan agricultural University advise and practice the new methods of teaching faculty in the stratified teaching. The teaching faculty includes the teaching team, the expert team and academic team. The target of construction of teaching faculty is improving the teacher's teaching ability and research ability. The core of construction is taking part in the teaching and science research by the way of expert guidance, course lead, project lead, resource guide through the special plan and special events. The construction achievements show that the abilities of mutual integration, solidarity and cooperation, interaction of the faculty are improved and the scientific and academic level are enhanced.

REFERENCES

[1] The China University of Geosciences Development Eesearch Center. Modern higher education information [R]. October 15, 2011.

[2] Huang Haiyan. practice and exploration of basic mechanics experimental teaching reform[J]. Experimental Technology and Management. 2008,25 (1): 119–122.

[3] Sun Haiyan. Discussion of specialized course teaching employment under the guidance [J]. Journal of Jilin College of education. 2012,8 (6): 17–18.

[4] Sun Haiyan, Gong Aimin, Zhang Ling. A discussion on the teaching method of Building Material [J].Science Journal. 2012,18:155–156.

Management, Information and Educational Engineering – Liu, Sung & Yao (Eds)
© 2015 Taylor & Francis Group, London, ISBN: 978-1-138-02728-2

The research of digital language feature extraction and recognition

Qi Ping Zou
Physical and Electrical Engineering School, HeChi University, GuangXi, Yizhou, China

ABSTRACT: The isolated speech recognition system based on the training method, based on vector quantization and neural networks is trained on speech feature, the recognition phase using the recognition method of BP neural network and the support vector machine, and the isolated word speech recognition system. Before the training, the orderly clustering network time warping of speech signal feature, and the feature vector dimension transient redundant information of the speech signal is reduced, improves the speech recognition rate; put forward a kind of [MFCC] based parametric speech recognition algorithm. Simulation results show the effectiveness of speech recognition algorithm based on [MFCC] parameters.

KEYWORDS: Voice recognition; feature extraction; support vector machine; (SVM) neural network.

The nature of the speech recognition system is a kind of multidimensional pattern recognition system, the basic structure is shown in figure 1.

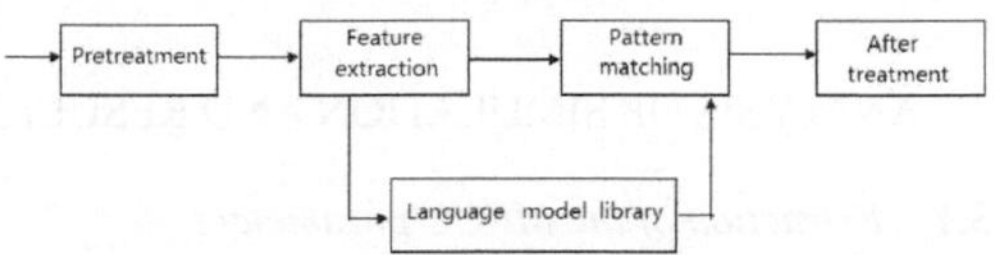

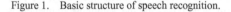

Figure 1. Basic structure of speech recognition.

1 FEATURE EXTRACTION

A speech signal DFT is:

$$X_a(k)=\sum_{n=0}^{N-1}x(n)e^{-j2\pi nk/N} \qquad 0\le k<N \tag{1-1}$$

The X (n) for speech signal input, N says the Fu Liyetransform points.

The definition of an M filter, the filter for the triangular filter, the center frequency is f (m), m = 1,2, M,..., is the triangle the frequency response of the filter press type (3-2) definition:

$$H_m^{'}(k)=\begin{cases}0 & k<f(m-1)\\ \dfrac{2(k-f(m-1))}{(f(m+1)-f(m-1))(f(m)-f(m-1))} & f(m-1)\le k\le f(m)\\ \dfrac{2(f(m+1)-k)}{(f(m+1)-f(m-1))(f(m+1)-f(m))} & f(m)\le k\le f(m+1)\\ 0 & k>f(m+1)\end{cases} \tag{1-2}$$

Mel filter of the center frequency is defined:

$$f(m)=\frac{N}{Fs}B^{-1}(B(f_l)+m\frac{B(f_h)-B(f_l)}{M+1}) \tag{1-3}$$

The highest frequency respectively and filter group and the lowest frequency, Fs is the sampling frequency, the unit Hz. M is the number of filters, N to FFT transformation points, in the formula $B^{-1}(b)=700(e^{\frac{b}{1125}}-1)$. Logarithmic energy output from each filter bank for:

$$S(m)=\ln(\sum_{k=0}^{N-1}|X_a(k)|^2H_m(k)) \quad ,0\le m<M \tag{1-4}$$

The MFCC coefficients are obtained by cosine transform:

$$C(n)=\sum_{m=0}^{M-1}S(m)\cos(\pi n(m+0.5/M)), \quad 0\le n<M \tag{1-5}$$

The extraction of MFCC parameters, the available type (2-7) difference feature parameter extraction ΔMFCC, ΔΔMFCC parameter.

$$d_t=\begin{cases}c_{t+1}-c_t & t<\Theta\\ c_t-c_{t+1} & t\ge T-\Theta\\ \dfrac{\sum_{\theta-1}^{\Theta}\theta(c_{t+\theta}-c_{t-\theta})}{2\sum_{\theta-1}^{\Theta}\theta^2}\end{cases} \tag{1-6}$$

In the formula, d_t In the T representation of a first-order differential cepstrum coefficient, T is the dimension cepstrum coefficients, a derivative of the time difference, the value is 1 or 2, $1 \leq \theta \leq \Theta$, c_t said the T cepstral coefficients .

2 IDENTIFICATION

2.1 *BP neural network identification*

The output error definition of network $E = \frac{1}{2}(d - O)^2 = \frac{1}{2}\sum_{k=1}^{l}(d_k - o_k)^2$, To expand to the hidden layer and the input layer, the error decreases under the principle, should be proportional to the negative gradient adjustment and error weights to make, i.e.

$$\Delta w_{jk} \propto -\frac{\partial E}{\partial w_{jk}}, j = 0,1,2, \cdots m; k = 1,2, \cdots l$$

$$\Delta v_{ij} \propto -\frac{\partial E}{\partial v_{ij}}, i = 0,1,2, \cdots n; j = 1,2, \cdots m$$

The formula can be given to adjust the weights of all layers, written in vector form as:

$$\Delta W = \eta(\delta^o Y^T)^T, \quad \Delta V = \eta(\delta^o X^T)^T$$

In the formula, $X = (x_1, x_2, \cdots, x_n)^T$ is input vector, $Y = (y_1, y_2, \cdots, y_m)^T$ is the hidden layer output vector, $O = (o_1, o_2, \cdots, o_l)^T$ is ouput vector, $d = (d_1, d_2, \cdots, d_l)^T$ is the desired output, while $W = [w_{jk}]_{m \times l}$ and $V = [v_{ij}]_{n \times m}$ are the hidden layer to the output layer to the input layer and the hidden layer weights matrix.

2.2 *Support vector machine (Supportvectormachine, SVM)identification*

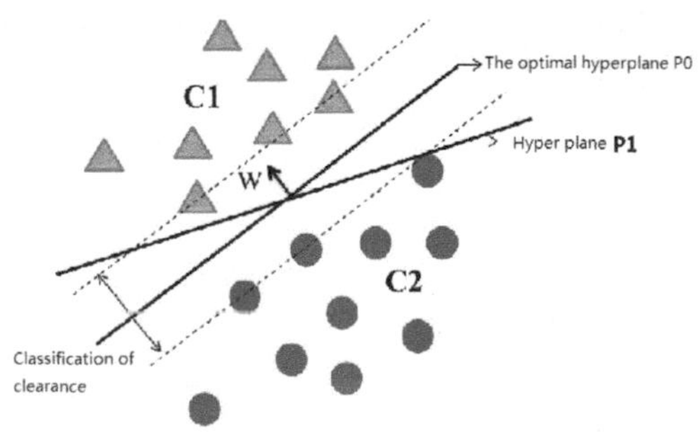

Figure 2. Schematic diagram of the optimal hyperplane.

C1 and C2 represent two types of data sample, the sample shown in Figure 2.1 in a two-dimensional, linear in P0, P1is a classification function. If a linear function can be completely separated the two class of all samples, so that the data is linearly separable; otherwise known nonlinear separable. The training data samples of two kinds of linear separability assumption, $\{(x_1, y_1), (x_2, y_2), \ldots (x_N, y_N)\}$, $x_i \in R^d$ (d is sample length of x_i), $y_i \in \{+1, -1\}$, $i = 1,2, \ldots , N$. The general expression of linear discriminant function is $f(x) = w * x + b$, classification of surface equations of the function corresponds is:

$$w * x + b = 0$$

Linear discriminant function value is generally continuous real number, while the classification output is discrete values. Such as the use of numerical representation of –1class C1, and numerical +1 category C2. all the samples are only represented by the values –1 and +1. Then wecan set a threshold, discriminant function by judging thevalue is greater than or less than the threshold to judgebelongs to a kind of. If we take this threshold is 0, when $f(x) \leq 0$, the samples for category C1 (–1); when $f(x) \geq 0$,the samples for category C2 (+1).

3 ANALYSIS OF SIMULATION AND RESULTS

3.1 *Extraction of the MFCC parameters*

MATLAB, the Mel filter order number is 24, the FFT transform length is 256, the sampling frequency is 8000Hz.The pre-emphasis, the speech signal into frames (each 256 points into a frame), MFCC parameters of each frame,calculating the differential coefficient. Combined with the MFCC parameter and the first-order differential MFCCparameters can be obtained, the results are as shown in Figure 3 and figure 4.

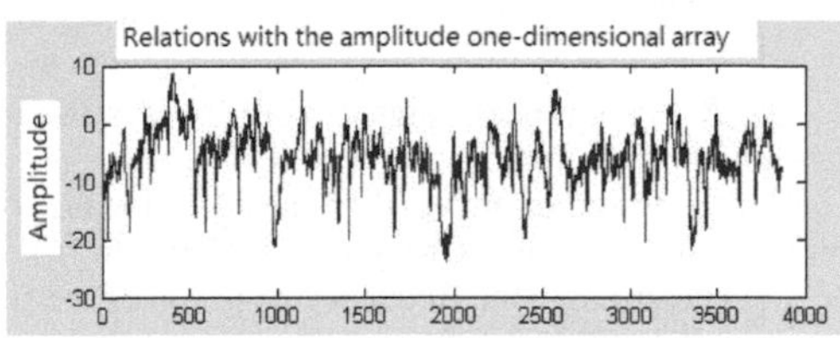

Figure 3. Relationship of one-dimensional array and its amplitude.

3.2 *Analysis and design of the simulation system of digital voice*

The whole experimental platform based on Matlab software, the simulation experiment of speech

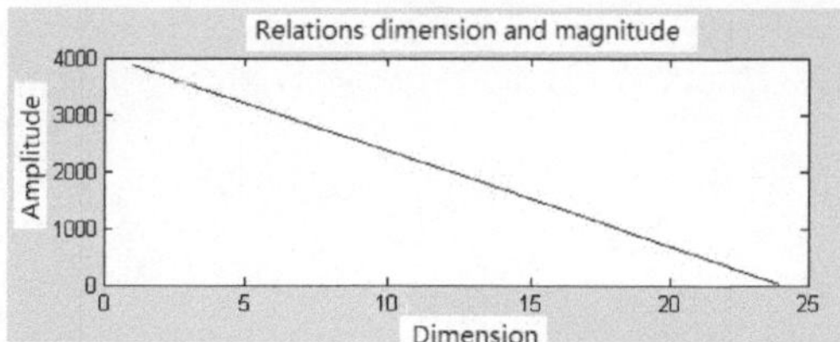

Figure 4. Relationship dimension and amplitude.

database recorded by5 people, each digital sound 8 times (0–9), a total of 400samples. The sampling frequency of 22050Hz, single channel, 16 bit sampling accuracy of the speech signal recording. In this experiment, taking the first 320 samples are used for training, 80 samples for testing, the use of BPneural network and support vector machine two recognition methods, the results are as shown in Figure 5 and Figure 6.

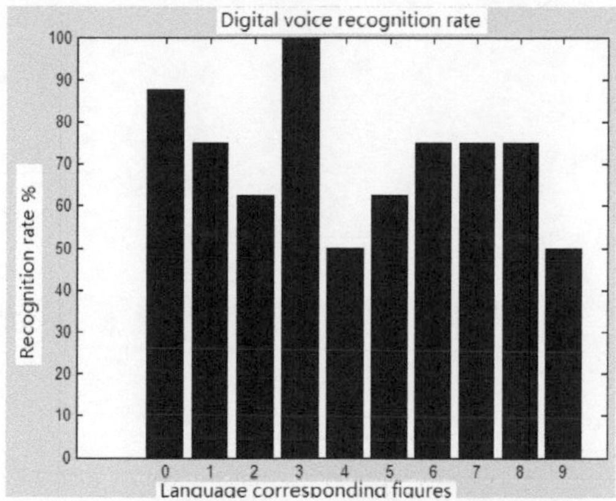

Figure 5. BP neural netwok recognition results.

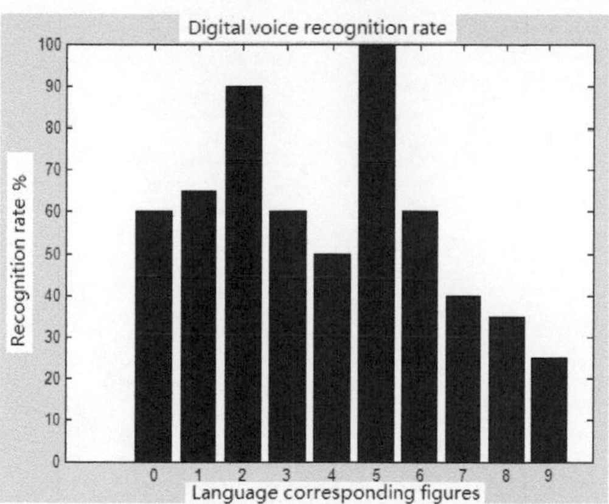

Figure 6. Machine recognition results support vector diagram.

The Figure 5 and Figure 6 show the recognition, support vector machine for speech rate than the BP neural networkbased on speech recognition based on high.

4 CONCLUSION

Support vector machine classifier is a two class problem, it can only answer belong to the positive class and negative class problem, but encountered many problems in the actual application process will be. By the SVM to multi class SVMthere are two main methods: (1) in an optimal formula of all data and global optimization (2) to multi-class problem is decomposed into a plurality of two classification problems. When data are the same, the more the former than the latter complex calculation. So in actual use, the multi class SVM problem is decomposed into two classification problems. The multi class classifier commonly used the value of two classifiers having one to many, one by one, as DAGSSVM three kind.

REFERENCES

[1] Wang Bingxi, Qu Dan, Peng Xuan. Practical fundamentals of speech recognition [M]. Beijing: National Defence Industry Press, 2005.

[2] Zhang Xiongwei, Chen Liang, Yang Jibin. Modern-speech processing technology and application [M]. Beijing:Mechanical Industry Press, 2003.

[3] Hu hang. Speech signal processing [M]. Harbin Institute of Technology press, 2005.

[4] Wang Xiuli. Research on [4] and the method of feature extraction and endpoint detection in speaker recognition system DSP realization [D]. Jilin: Jilin University, 2006.

[5] Wang Xiaoya. Application of [J]. radio engineering-Cepstrum in pitch and formant extraction process, 2004,34(1): 57–58.

[6] [D]. of Jilin University Wang towers. Study on Extraction of speech feature in speaker recognition, 2009.

Management, Information and Educational Engineering – Liu, Sung & Yao (Eds)

The research and practice of project teaching method in the teaching of DSP applied technology

Guang Yang & Bin Guo
Changchun University of Science and Technology, Changchun,China

ABSTRACT: The paper introduces the teaching content, teaching method and teaching practice and other aspects of project teaching method in the DSP applied technology curriculum in details, and provides the implementation scheme. The project teaching method can make students understand the knowledge deeply, combining the theory with practice, stimulating their learning interests and improving the teaching quality and students' engineering practical ability.

KEYWORDS: DSP Applied Technology; Teaching Reform; Project Teaching Method.

1 INTRODUCTION

DSP applied technology includes the hardware technology and software technology of DSP[1]. In recent years, some emerging subjects are all closely related to the DSP technology, such as digital communications, speech processing, artificial intelligence, neural networks, digital image processing and consumer electronic. Due to its advantages of power control function, fast processing speed and strong real-time, the DSP technology has been widely used in various industrial control fields, and has played an important and indispensable role in the field of industrial control. At present, the DSP technology has become a cutting-edge technology field with wide application and rapid development.

With the wide development of DSP applied technology in various industrial control fields, its urgent needs a large number of high-quality professionals to carry out the development, design and algorithm description of DSP software and hardware. The DSP applied technology has become one of the important cutting-edge high-tech that the students major in electronic information must master. And DSP applied technology is the course with a close integration of theory and practice[1]. In the DSP applied technology, the theory and practice are closely linked and complementary to each other by using the theory to guide the practice and strengthening the theory with the practice. Therefore, the research on the teaching method of DSP application technology has extremely important practical significance. It has become a great event in the current teaching reform that how to better perform DSP applied technology in teaching and how to make students have good capabilities of DSP application development.

2 THE TEACHING SITUATION OF DSP APPLIED TECHNOLOGY CURRICULUM

DSP applied technology has been widely welcomed by college students for its engineering, technical and practical features set in one[2]. The course refers to a wide range of knowledge and rich teaching content. Through studying the theoretical and practical course of DSP applied technology, students are required to master the hardware structure of the DSP system, command systems, link command file and documentation, the interrupt vector table complication as well as the development platform, etc., to master the description of common algorithms and the basic theory of DSP and the skilled application of the assembly language, C language programming and code generation and debugging of DSP. The students are also required to verify the theoretical knowledge and methods they have learned in the experiments, to independently and skillfully apply the DSP development platform for the hardware design, software programming and system debugging, etc.. DSP applied technology courses in the undergraduate teaching focuses on training students to independently complete the digital signal processing system, hardware system design, software programming and the system layout design by regarding the general programmable digital signal processor as the core controller in the electronic system design.

3 THE COMPARISON OF THE TRADITIONAL TEACHING METHOD AND THE PROJECT TEACHING METHOD

Project teaching method is a kind of teaching method that students learned through the completion of a complete project under the teacher's guidance. It transfers the knowledge and content of the traditional subject system into several teaching projects. Around the project, students are organized to study and directly take part in the whole process of teaching, and actively experience, comprehend and explore. Specially, it aims to create the actual develop situation of engineering project for the students by using the project as the oriented and choosing the actual operational project as the platform. A student can complete the project through the understanding of the actual project so that they can master the knowledge, the use of the skill and improve the relative theoretical knowledge and actual practical skills on the basis of the project platform.

Traditional teaching method is aiming for imparting knowledge and skills, dominating by the teacher lectures. The students are passive to study and obey the arrangement of the teacher. They are obviously affected by the external dynamic factor. Finding out the insufficiency of students, teachers strengthen the teaching so as to improve the students' knowledge level.

4 THE IMPLEMENTATION SCHEME OF PROJECT TEACHING METHOD IN THE COURSE OF TEACHING

4.1 *The selection of project*

In the project teaching of undergraduate courses, the project should take the teaching task of courses and the skills should be reached as the core, and implement the modularity management of the core knowledge of specialized courses. It not only highlights the important and difficult points to make students understand problems deeply, but also emphasizes the cultivation and application of basic skills. Meanwhile, the project also should have some rich information and problems that are interdisciplinary, cutting-edged and fun, thus improving students' abilities to analyze and solve problems.

4.2 *The project implementation plan*

The content of the project implementation plan includes the time of the teacher's guidance, the time of student's self-study, the time of t finishing the project, the schedule arrangement of the project and the effect of the completion of the project, etc..

Project implementation plan project is to make the plans and overall arrangements for the development activities involved in project teaching process, such as how to allocate the personnel, where to get the information, the making of the project implementation plan. The students in the project group should focus on the learning project to form the study groups and make the detailed division of labor[3]. The members of the project team should actively discuss, consult books through various channels, collect data, discuss research methods, jointly formulate detailed implementation plans and make preparation for researching the project work. The project group should consider the cooperation and complementarity. For example, some students are good at hardware design, so they should be mixed to form the members of the project group.

Through the division and cooperation of labor among the members of the project team, the digestion in the subject knowledge, and the complementary advantages in the aspects of intelligence, it could innovative thinking, build their own new system of knowledge and enhance their own comprehensive practical abilities. In this process, the teaching interaction between the teacher and student is very important. It not only emphasizes the students' subjectivity that students independently publish their own views and opinions through their autonomous learning, but also requires teachers' dominant that the teachers should timely understand the students' learning dynamic, provide the timely and appropriate guidance.

4.3 *The project implement*

Project implementation is the core part of project teaching. Project teaching method emphasizes the principal role of students, encouraging students to be self-study, self-education, self -initiative for acquiring knowledge. Let the students comprehend the knowledge, understand the knowledge and improve themselves in the process of participating in the project. Through the summary of the teachers, the students could transform the knowledge into their own knowledge and increase their own practical experience. Because the students have the firsthand experience and practice, they will never be forgotten once mastering the knowledge and skills. In the whole process of project implementation, as the principle, the students should actively participate in every link of the project. When you meet the problem, you should self-query data, analyzing the question and solving problems independently.

Under the conditions set by the teacher, according to project requirements, completing the project and learning on purpose will be the autonomous behavior of students. The students involved in the project independently or in small groups will independently complete a typical project task. And in the process of completing the task, the students will continually discover, analyze and solve the problems by improving the ability. The teacher's role is changed into the learning mentor and counselor for the students. The students' learning is changing from the passive acceptance education to creative educational transformation[4].

4.4 *The exchange and assessment of project achievements*

The key point of the project teaching method is the detailed implementation process of the project, not the final achievements[4]. Hence, the assessment of the implementation process of the project is very important.

In the assessment, the teacher should summarize and evaluate the problems that the student meets and the complete situation of the project in the process of project implementation. The teachers must point out the causes of the problem and the solutions, summarize the advantages and disadvantages[3] of the project implementation plan as well as the improvement measures. Meanwhile, the teachers also should guide the students to learn other advantages, and improve and enhance their own ability. Therefore, the students can enhance their comprehensive ability in the process of completing the project. For the outstanding aspect of the project, the teachers should give the adequate encouragement and affirmation to the students in order to simulate their learning interest and the enthusiasm to participate in the project, enhance their confidence and the sense of pride and accomplishment to make the students keep the good learning state. At the same time, the students participating in the project are required to do a self-assessment. Each project team sent a representative to introduce the project achievements, the project implementation situation, problems, solutions, advantages and disadvantages of achievements, the room for improvement, harvest and feelings, and finally make a summary evaluation. After each project team introduces the situation, the project team should evaluate each other, and encourage the students to make the discussion and communication[2], adequately express their own opinions, questions or improvement measures.

5 THE PRACTICE OF THE PROJECT TEACHING METHOD IN THE "DSP APPLIED TECHNOLOGY" TEACHING COURSE

5.1 *The selection and conditions of the project*

1 Select ICETEK-2407 DSP teaching experimental box as the supporting conditions ICETEK-2407 DSP DSP is teaching experiment box provides a complete set of solution for DSP teaching and research. The biggest feature of this system is the modular design which can meet the basic needs of the current teaching and research, such as: DSP experimental teaching, video, voice, network graduation design based on DSP, the development of embedded system based on DSP.

2 Choose the students major in electronic information engineering as experimental subjects, setting the project title.

According to the project and requirements set by the project teaching method are shown in Table 1.

Table 1. Project and Requirements.

No.	Name of Project	Specific requirements
1	Date collection, storage and transmission experiment	Grasp the configuration and program of the TMS320LF2407 DSP serial communication. Learn the design of asynchronous serial communication program. Grasp the data collecting method of the initialization and starting of the A / D in the TMS320LF2407 DSP. Grasp the expansion method of the external data storage RAM and the method of accessing the external RAM on the experiment box.
2	DC motor control experiment	Control the TMS320LF2407 DSP general I / O pin to produce the PWM signal with the different duty cycle. Learn the control principle and control method of a DC motor. Realize the speed regulation and applications of a DC motor.
3	Stepper motor control experiment	By using the expansion I / O port of TMS320LF2407 DSP to control the peripheral devices information. Grasp the control principle of the general timer of the LF2407 DSP and the programming method of the interrupt service routine. Understand the controlling method of a stepper motor, and flexibly apply for the position control, angle control and other industrial fields.
4	Sampling theorem experiment	Grasp the basic process and program processing process of the conversion of the A / D chip in the TMS320LF2407 DSP; Familiar with the application of the FFT and the achieving method of the DSP assembly language and C language. Grasp the contents, the principle and its practical application of the sampling theorem.

5.2 *The implementation plan and effect*

The teaching methods have a teaching try in DSP Applied Technology course design of the electronic information engineering. Time is a week (24 periods). The equipment is ICETEK-2407 DSP teaching experiment box. Four is in one group, totaling 10 groups. The project topic is data collection, storage and transmission test. The guiding time is 4 periods. The practice time is 16 periods. The time for the comprehensive design report is 4 periods. The comprehensive design report requires the students to design their own hardware system and compile software program according to the basic module circuits. Through the trial of the eight classes in the second session, the teaching effect is remarkable. The students' learning initiative and enthusiasm are obviously improved as well as the ability to hardware design and software debugging.

6 CONCLUSION

Through practice, it proves that using project teaching method in the DSP Applied Technology course, the students' learning purposes are clearer, and the learning, enthusiasm has been greatly improved, thus improving the efficiency of learning. In the process of making project, students not only have learned the teaching content of the course, but also have expanded the thinking, mastering the self-study approach of learning other hardware system design and software debugging. What is more important is that the students should combine the theoretical knowledge in the textbooks with the practical application, and improve their abilities to the comprehensive design and application. At the same time, the use of project teaching method demands the requirements for the teachers. It requires that the teachers should have strong experience of hardware system design and software debugging, good teaching organizational capability, being good at playing the principle status of students and dominant position of teachers.

The use of the project teaching method in DSP Applied Technology does not mean to completely abandon the traditional teaching methods. It not only learns the advantages of traditional teaching methods to further improve the project teaching method and make students have solid theoretical knowledge, but also trains the students have skilled skills of hardware design and software debugging.

REFERENCES

[1] Qiu Xun, Yan Youjun. The Exploration and Practice of the Project Teaching Method in the DSP Course [J]. Journal of Su Zhou Vocational University, 2008,19(4),99–101.

[2] Gao Guowang, Dang Ruirong, Ren Zhiping. The Exploration and Practice of DSP Teaching Course Reform and Innovation. [J]. Journal of Technology College Education, 2010.1.

[3] Wang Ping. The Practice of the Project Teaching Method in the Professional Course. [J]. Education and Vocation 2009,634(30):129–130.

[4] Liu Yongnian. The Investigation of the Computer Education Reform of the Non-Computer Specialty Bases on the Project Teaching Method. [J].Education and Vocation. 2011,700(24):105–106.

Management, Information and Educational Engineering – Liu, Sung & Yao (Eds)
© 2015 Taylor & Francis Group, London, ISBN: 978-1-138-02728-2

How corporate growth influences earnings management

Lin Jiang
Sichuan University, Chengdu, China

ABSTRACT: This study examines whether the extent of earnings management of China's listed companies is related to their corporate growth levels. In doing so, I compared 2290 observations' actual growth rate and their sustainable growth rate for the period of 2004-2013 of Chinese listed companies to divide them into two groups: fast growing firms and slow growing firms. The study finds a significant positive association between corporate growth and earnings management in the fast growing group and a significant negative association between them in the slow growing group. The finding reveals that in the fast growing group, firms with higher growth rate tend to be more likely to manage earnings while in the slow growing group, firms with lower growth rate are more likely to do so.

KEYWORDS: Corporate growth; sustainable growth; earnings management.

1 INTRODUCTION

Earnings management behavior has been existed in companies for a long time as a worldwide phenomenon. As Jackson once put that Earnings management is the manipulation of accounting numbers within the scope of the Generally Accepted Accounting Principles (GAAP) (Jackson and Pitman, 2001). According to Katherine Schipper, 1989, earnings management is the act of intentionally influencing the process of financial reporting to obtain some private gain. And Healy and Wahlen, 1999 says that earnings management involves the alteration of financial reports to mislead stakeholders about the organization's underlying performance, or to influence contractual outcomes that depend on reported accounting numbers.

From the above definition, it's easy to find out that the purpose of earnings management is to cover a bad financial status. So in this case, lots of financial index which indicates a financial condition have an impact on earnings management. Gunny, 2010 suggests a positive association between earnings management (EM) and return on assets (ROA), it means the larger ROA is, the higher the extent of EM. Asset-liability ratio (LEV) is a financial index, which uses to measure a firm's debt paying ability. According to the research of Fung and Goodwin, 2013, higher LEV indicates higher possibility of earnings management.

Although there are many studies about factors that influence the extent of EM, rare of them studies if and how corporate growth influence EM. So this paper will find out whether this influence exist and how it works thus to provide a new way of recognizing earnings management for information users such as investors and regulators.

In order to estimate firms' growth levels so to see if they are growing too fast or too slow, the corporate sustainable growth rate as the reference line is needed. In a financial context, the sustainable growth rate is defined as the maximum pace at which a company can grow revenue without depleting its financial resources (Higgins, 1977). In this sense, companies should try to sell as much as goods or service to reach this rate but never to exceed it. Because if a firm is growing too fast that its actual growth rate (AGR) is greater than its sustainable growth rate (SGR), the firm cannot sustain a rapidly growing activity without financing it (Yu-Chun Chang, 2012); But if it grows too slow that it AGR is less than its SGR, there will be a drop in the efficiency of resource utilization. The two status both have a negative impact on financial situation, so may lead to earnings management, which aims for covering this bad situation.

The remainder of this paper is organized as follows. Section 2 provides the theory, analysis and the research hypothesis. Section 3 introduces the sample selection, variables and the model. Section 4 presents empirical test on whether corporate growth levels have an influence on earnings management and analyzes the results of the test. Section 5 concludes.

2 THEORY ANALYSIS AND HYPOTHESIS

As analyzed above, fast growing firms have to finance themselves because of lack of capital (Yu-Chun Chang, 2012). And financing is always a motivation for companies to involve in earnings management. Obviously, the faster it grows, the worse need for financing it is in. So the first hypothesis is as follows:

H_1: In fast growing firms, the ones with higher growth rate are more likely to involve in EM.

On the other hand, a slow growing firms' sale is inadequate and they still have room for growing. Companies of this kind may artificially increase their earnings in the financial statements to avoid an awkward situation of low sales. Hence, the second hypothesis is as follows:

H_2: In slow growing firms, the ones with lower growth rate are more likely to involve in EM.

3 SAMPLE SELECTION, VARIABLES AND MODEL

3.1 *Sample selection*

I use ten years' data from the year 2004 to 2013 of Chinese listed companies from both the Shanghai and Shenzhen Stock Exchanges as the primary sample. Then the screening process is as follows: (1) Eliminate those firms in the finance and insurance industry, for the particularity in their financial index that differ from other industries. (2) Eliminate those firms in ST status because when a firm is marked as ST(special trade) and faced with a very urgent situation that it may be delisted by the stock exchange, then its financial strategy and activities will be quite different from those of other firms. (3) Eliminate firms whose financial data of current year or last year is not complete.

At last I get 2290 observations of 229 companies. The information of corporate SGR and AGR and other financial data all come from the GTA database.

3.2 *Variable definition*

3.2.1 *Earnings management*

Scholars around the world usually use the Jones model (Jones, 1991) or Cross-sectional modified Jones model (Dechow et al., 1995) to measure the extent of firms' earnings management. This paper is based on the modified Jones model and include intangible assets in the model under the consideration that the land use right of Chinese listed companies is contained in intangible assets rather than fixed assets. So the final model is as follows:

$$\frac{TA_{i,t}}{A_{i,t-1}} = \beta_1 \frac{1}{A_{i,t-1}} + \beta_2 \left(\frac{\Delta REV_{i,t}}{A_{i,t-1}} - \frac{\Delta REC_{i,t}}{A_{i,t-1}}\right) + \beta_3 \frac{\mathrm{PPE}_{i,t}}{A_{i,t-1}} + \beta_4 \frac{\mathrm{IA}_{i,t}}{A_{i,t-1}} + \varepsilon_{i,t}$$

Where $\mathrm{TA}_{i,t}$ is measured by the difference between net profit and net cash flow from operations and means the total accruals of firm i for period t; $\mathrm{A}_{i,t-1}$ is total assets of firm i at the end of period t-1; $\Delta REV_{i,t}$ is the change in sales of firm i for period t; $\Delta REC_{i,t}$ is the change in receivables of firm i for period t; $PPE_{i,t}$ is net value of fixed assets of firm i at period t; $IA_{i,t}$ is net value of intangible assets of firm i at period t; $\varepsilon_{i,t}$ is the residual of this regression model and its absolute value is the extent of each sample's earnings management(EM).

It is important to note that this paper uses net value of fixed assets and intangible assets rather than original value of these assets because the depreciation and amortization of these assets are also factors that influence a firm's earnings management so that should be taken into account.

3.2.2 *Corporate growth*

In my study, companies' sales growth rate is used as a proxy of the AGR, which can be obtained from the following formula:

$$AGR = \frac{\text{(Current period's sales revenue-Previous period's sales revenue)}}{\text{(Previous period's sales revenue)}}$$

The index GAP is defined as the difference between AGR and SGR. If GAP is above zero, it means the firm's AGR is larger than its SGR and is a fast growing firm, and these samples will be divided into group 1. And group 2 contains the other samples whose GAP is below zero, which means a slow growing status.

3.2.3 *Control variables*

In order to better study the relationship between earnings management and corporate growth, this paper also includes some of the control variables: Firm size (SIZE); Return on Assets (ROA); Asset-liability Ratio (LEV); Cash Recovery for all assets (CR).

Firm size affects discretionary accruals (Gu et al., 2005), hence this study controls for total assets as a proxy for firm size. Then, ROA is also included since Gunny, 2010 suggests a positive association between EM and ROA. Asset-liability Ratio is also described as leverage, it is measured based on the ratio of total liabilities to total assets. According to Fung and Goodwin. (2013), leverage increases the potential for EM. So this paper also controls for Asset-liability Ratio. At last, Cash Recovery for all assets is also included in the model. Detail definitions are in table 1.

Table 1. Variable definitions.

Variables	Implication	Definition
EM	Earnings management	The absolute value of abnormal accruals
GAP	The level of corporate growth	Actual growth rate(AGR) - sustainable growth rate(SGR)
SIZE	Firm size	Log of total assets
ROA	Return on assets	Net income divided by average total assets
LEV	Asset-liability ratio	Total debt divided by total assets
CR	Cash recovery for all assets	Operating activities cash flows divided by final total assets

3.3 *Estimation model*

The hypotheses above predicted that higher growth rate firms are more likely to manipulate earnings in fast growing group and less likely to do it in a slow growing group. Thus, the coefficient of GAP, β_1, is expected to be positive in group 1 and negative in group 2. The estimation model is presented as follows:

$$EM = \beta_0 + \beta_1 GAP + \beta_2 A + \beta_3 ROA + \beta_4 LEV + \beta_5 LR + \beta_6 CR$$

All the variables are as previously defined and the model is tested separately in two groups.

4 EMPIRICAL TEST AND RESULTS

4.1 *Descriptive statistics*

Table 2 provides descriptive statistics of the variables used in this study. The observations are much more in group 1 than in group 2, which means most companies are in a fast growing status. And the lager span between Min and Max in group 1 shows a great extent of this fast growing situation.

4.2 *Regression results*

Table 3 below gives the regression statistics of the estimated model. The coefficient of GAP is positive in group 1 and negative in group 2. This is in line with previous hypotheses. In fast growing firms, the faster a firm grows, the more likely it will be involved in earnings management. This is because fast growing firms are short of capital so they have to financing themselves. For this purpose, companies may manipulate their earnings in order to attract funds from all kinds of investors; In slow growing firms, the slower a firm grows, the more likely it will manipulate earnings, this is mainly because a slow growing firms' sale is inadequate and they still have room for growing, thus they may artificially increase their earnings in the financial statements to avoid an embarrassing situation of low sales.

4.3 *Robustness test*

In this paper, companies' sales growth rate is used as a proxy of the AGR, and now to test the reliability of the above results, the robustness check is conducted. That is, replacing the sales growth rate with total asset growth rate. And the results are shown in table 4.

Table 2. Descriptive statistics.

Group	var.	N	Mean	Median	Min	Max	Std.
1	EM	1484	0.058774	0.040054	0.000041	0.716405	0.065237
	GAP	1484	0.292136	0.161764	0.000184	7.419738	0.575841
	SIZE	1484	22.375864	22.289035	19.703010	26.895395	1.159366
	ROA	1484	0.057988	0.047539	0.001043	0.311311	0.040532
	LEV	1484	0.489847	0.497836	0.037253	0.866991	0.167401
	CR	1484	0.073961	0.072456	0.249224	0.616577	0.077872
2	EM	806	0.061618	0.038509	0.000166	1.423702	0.086029
	GAP	806	0.139920	0.092505	0.938814	0.000094	0.146626
	SIZE	806	22.383604	22.263659	19.942590	26.146754	1.116745
	ROA	806	0.062481	0.049797	0.000274	0.399900	0.048756
	LEV	806	0.458823	0.468112	0.044432	0.838230	0.176650
	CR	806	0.065207	0.063885	0.565469	0.423061	0.092937

Table 3. Regression results.

Variables	Group 1	Group 2
β_0	0.0936	0.2100
GAP	0.0187***	−0.0463**
	(0.000)	(0.014)
SIZE	−0.0045***	−0.0098***
	(0.004)	(0.000)
ROA	0.5028***	0.6203***
	(0.000)	(0.000)
LEV	0.0893***	0.1168***
	(0.000)	(0.000)
CR	−0.1725***	−0.4352***
	(0.000)	(0.000)
Adjusted R^2	0.1069	0.2183
F-statistic (Sig)	36.5015	45.9594
	(0.000)	(0.000)
N	1484	806

*** Indicates that estimated coefficient is significant at two-tailed 1% level.

** Indicates that estimated coefficient is significant at two-tailed 5% level.

Table 4. Robustness test.

Variables	Group 1	Group 2
β_0	0.1391	0.2100
GAP	0.0859***	−0.0419*
	(0.000)	(0.067)
SIZE	−0.0065***	−0.0096***
	(0.000)	(0.000)
ROA	0.7147***	0.2826***
	(0.000)	(0.000)
LEV	0.0844***	0.0763***
	(0.000)	(0.000)
CR	−0.4097***	−0.0313*
	(0.000)	(0.0881)
Adjusted R^2	0.2823	0.0836
F-statistic (Sig)	117.8967	15.8033
	(0.000)	(0.000)
N	1487	803

*** Indicates that estimated coefficient is significant at two-tailed 1% level.

* Indicates that estimated coefficient is significant at two-tailed 10% level.

5 CONCLUSION

This paper studies the relationship between earnings management and corporate growth. The effect of the regression model is good with the adjusted R^2 to be 0.126 and 0.218 respectively after controlling for variables as Firm Size(SIZE), Return on Assets(ROA), Asset-liability Ratio(LEV), and Cash Recovery for all assets(CR).

In light of the regression results, GAP is positively related with EM in group 1 and negatively related to it in group 2. That means in fast growing firms, the faster a firm grows, the more likely it will be involved in earnings management. But in slow growing firms, the slower a firm grows, the more likely it will manipulate earnings. The finding is consistent with previous hypotheses.

REFERENCES

Higgins, R.C.1997,How Much Can a Firm Afford? [J]. Financial Management,8:186–198.

Yu-Chun Chang.2012.Strategy formulation implications from using a sustainable growth model.[J].Journal of Air Transport Management,20:1–3.

Katherine Schipper.1989.Commentary on Earnings management[J].Accounting Horizons. 3(4):67–69.

Jackson,S., Pitman,M.2001.Auditors and earnings management.[J].The CPA Journal.7:39–44.

Fuxiu Jiang,Bing Zhu,Jicheng Huang.2013.CEO's financial experience and earnings management.[J].Journal of Multinational Financial Management, 23(3):134–145.

Fengyi Lin, Shengfu Wu. 2014.Comparison of cosmetic earnings management for the developed markets and emerging markets: Some empirical evidence from the United States and Taiwan.[J]. Economic Modelling,36:466–473.

Jones, J. 1991.Earnings management during import relief investigations.[J]. Journal of Accounting Research, 29(2):193–228.

Dechow, P.M., Sloan, R.G., Sweeney, A.P., 1995.Detecting earnings management.[J]. The Accounting Review,70:193–225.

Jaggi, B. and Lee, P.2002. Earnings Management Response to Debt Covenant Violations and Debt Restructuring.[J]. Journal of Accounting, Auditing & Finance,17(4):295–324.

Simon Y.K. Fung and John Goodwin.2013.Short-term Debt Maturity, Monitoring and Accruals-based Earnings Management.[J]. Journal of Contemporary Accounting & Economics,17(4):295–324.

Gu, Z., Lee, C.W.J. and Rosett, J.G. 2005.What Determines the Variability of Accounting Accruals?[J]. Review of Quantitative Finance and Accounting,24:313–334.

Gunny, K. 2010.The Relation between Earnings Management Using Real Activities Manipulation and Future Performance: Evidence from Meeting Earnings Benchmarks.[J]. Contemporary Accounting Research, 27(3):855–888.

Paul M Healy, James M Wahlen.1999. A Review of the Earnings Management Literature[J]. Accounting Horizons.13(4):365–383.

Management, Information and Educational Engineering – Liu, Sung & Yao (Eds)
© 2015 Taylor & Francis Group, London, ISBN: 978-1-138-02728-2

Application of action-oriented teaching method in courses of art history and theory

Hai Ying Liu
Jingdezhen Ceramic Institute, Jingdezhen, China

ABSTRACT: Action-oriented teaching method in the cources of art history and theory is helpful to improve the arts students' theoretical knowledge learning, to activate the creative potential of students, and to improve the students' autonomous learning awareness. This teaching method can change the traditional teaching mode and improve the students' interest in learning, can take the initiative to "learn" drive "teaching" strategic, can also apply the behavior orientation method in art professional course. It make art students' unity between theory and practise , and put knowledge to use.

KEYWORDS: Art History and Theory; College Students; Action-Oriented Teaching Method; Teaching Reform.

Aimed at the increasingly detailed profession division and the complexity and abstraction of occupation, the field of vocational education in Germany put forward, in the 1970s, the action-oriented teaching method, the gist of which is that teachers, during the whole process of teaching, should put students on top priority on stages of information collection, independent work planning, decision making, implementing, examining, and evaluating, and that the teaching quality should be reflected in a student's comprehensive quality. The method is a series of strategy and idea which would let students study by themselves, explore with others, and complete the whole process of their study. In China, the method has already been known by college teachers and gradually used in their daily teaching. This article makes an effort to apply the method to the teaching of art history and art theory courses, hoping to give a strong support for future teaching.

1 DILEMMA OF ART HISTORY AND ART THEORY TEACHING

In the teaching programs for college art students, there are also some courses concerning art history and theory except major skill practicing courses. Art history and theory are courses of art theory for college art students, including history and theory. The historical means art history, such as Chinese and Western art history, history of sculpture, history of porcelain art, and history of design. The theory includes art theory, design theory, design psychology, and art planning. The goals of these courses are to make students know what and how they do in their art creating and become "perfect" people with the abilities of art recognition, art appreciation, and art creation. On the one hand, teachers of professional skills put a large amount of time on students' independent practice and exercise, leaving limited time for the elaboration of art history. In the practicing of art, on the other hand, students may have some good questions and doubts, which need to be addressed with reason by teachers of art history and theory. Therefore, it is a necessity as well as a must to set up such courses.

In today's classroom, however, courses in art history and theory have become a nuisance of students. First of all, in students' point of view, college art students put great emphasis on the professional skill learning while ignore the learning of basic theories. At the same time, they may spend much of their time focusing on their professional learning and therefore fail to acquire a solid general knowledge, even with some students attending classes only for checking. Utilitarianism and hedonism are spreading among current college students, and the idea of the uselessness of study has been accepted by a few people. New technology gadgets are ubiquitous in classrooms and watching movies on mobile phones, playing games, and chatting online are commonly seen. If teachers still follow traditional teaching methods, with a cramming style and boring class, it will be difficult to attract students' attention.

Secondly, in teachers' point of view, teachers are prominent in the traditional teaching, which is boring and repetitive. Students are lack of interest in the full-text teaching materials, causing widespread early leaving and even class skipping. And because teachers are teaching without thinking about students' professions, theory is alienated from the practice, causing the prevalence of the idea of uselessness of theory. Therefore, students are unwilling to attend the class and the efficiency of learning is low. In some cases, college courses in art history and theory are even performed by technique teachers who possess no experience of skill learning, or by art history and theory teachers who majored in liberal arts and have no experience of art creation. Various reasons have affected the teaching of art history and theory. As a teacher, he or she needs to improve their teaching methods and linguistic attraction rather than pass the buck to students all the time. Ye Shengtao, a famous Chinese educator, once said: "What teachers do is to guide and inspire while students to work hard and learn. It is wrong that teachers lecture while students listen and accept." Thirdly, in textbook's point of view, art students are good at perceptual cognition and satisfied with visualized images and videos. Textbooks of art history and theory, however, are full of characters with little pictures. Totally different from the colorful pictures in the textbooks of major courses, some pictures in textbooks of art history and theory are printed in black and white, greatly reducing their effects. Moreover, the content of these textbooks is lengthy and jumbled, with some ideas related to interdisciplinary courses like philosophy and psychology, of which many art students have a little accumulation of knowledge. As a result, it is difficult for students to understand and absorb these ideas, thus they gradually lose interest in these courses.

2 APPLICATION OF ACTION-ORIENTED METHOD IN COURSES OF ART HISTORY AND THEORY

According to the above-mentioned, it is imperative for the teaching of art history and theory to change. If traditional teaching methods and spoon-feeding courses are taken, the effects of teaching are worrying, not to mention art talents training. This is contrary to China's education guidelines. With abundant pictures and vivid videos, modern multimedia teaching alone is no stranger to students and does not guarantee students' attention. Tao Xingzhi, another famous Chinese educator, said: "It is better to give students a couple of keys to vaults of culture and the universe, then stuff them like Tianjin ducks some trifles of knowledge."

2.1 *Orientation of the teaching process*

In order to make a change in art history and theory teachers' monologue and prompt students to play a more active role in the teaching process, teachers may hand over the teaching task to students to accomplish by themselves. The way of cases may be taken. Students collect questions confronted in the creating process and deal with them by teaching themselves. Then practice and theory can be combined by using the textbooks of art history and theory. In this way, the theory will be better understood and better works of art will be created. At last, teachers make comments on and summarize the students' work. The whole teaching process is performed in "chorus" instead of teachers' boring "monologue".

2.2 *The transformation of class arrangement*

In the new interchangeable and interactive classroom, students will play the leading role and be guided by teachers in their learning process. They find questions and seek answers by themselves or from their teachers. The closed-book exams will become more flexible, and forms like a summary of feelings and experience, investigation report, or open-book exam can be adopted. Thus the students' ability of self-study can be combined with their self-understanding of the knowledge learned.

2.3 *Role transition of teachers and students*

The role of teachers as the lecturer and organizer of teaching will be transformed into an assistant and director of students' learning. Take the ceramic history, for example. When they are taking courses like porcelain painting or porcelain molding, students may fail to perform molding and decoration in their minds if they do not well master Chinese and Western porcelain development history. In the classroom, teachers need students to grasp the meaning and changing of traditional patterns, as well as characteristics of the times. Students may, in the first place, do some market research, and then make explanation in the classroom. After that, teachers make some supplement and give their summary. In this way, the content of porcelain history and theory courses will be remembered well and quality teaching will be achieved.

Action-oriented teaching method can change art students' ignorance of art history and theory, promote their activeness and enthusiasm for learning, employ art theory as the guide of art creation, and uplift students' artistic quality and accomplishment. As people and students oriented, the ultimate goal of this method is to improve students' creative mind. The following should be carefully noted in carrying out this teaching method. First, teachers should possess the ability of mastering the overall situation, keeping the classroom in good order and prompting students to actively complete their assignment. Second, teachers should improve their learning and cultivation, as well as artistic talent, abandoning the step-by-step style of lesson preparation. Teachers need to address students' unexpected questions in the classroom. Third, the method may make students think that they are freer in the classroom, leading to their indolence. How to evaluate students' performance in a more reasonable way has become particularly important.

ACKNOWLEDGMENT

The research leading to these results has received the support of Colleges and universities teaching reform research subject of Jiangxi Province "Application of Action-Oriented Teaching Method in Courses of Art History and Theory" (NO. JXJG-13-11-17).

REFERENCES

[1] Y. Xu. Improve college art design history teaching thinking[J]. Decoration, 8(2006).
[2] X.F. Du. High and new technology of art history and the trend of development[J]. Art technology, 11(2012).
[3] W.J. Li. Main problems of art theory class teaching and cracking[J].Modern education science, 9(2013).
[4] M.Li. Research on The university art education development present situation and countermeasure[J]. Sichuan opera,10(2013).

Management, Information and Educational Engineering – Liu, Sung & Yao (Eds)
© 2015 Taylor & Francis Group, London, ISBN: 978-1-138-02728-2

Research on evaluation mechanism of university students' autonomous learning

Lin Ying Jiang, Li Ping Huang, Hong Juan Liu & Shuo Ren
Software College, Northeastern University, Shenyang, Liaoning, China

ABSTRACT: The characteristic of higher education determines the fact that university students need a much more autonomous learning than before. The main body, methods and its individuation pattern of autonomous learning are all really different from the traditional acceptance study mode. Thus, the traditional evaluation mechanism of study cannot satisfy the need of university students' autonomous learning. So, based on the results and existing problems of analysis of traditional evaluation mechanism, a new appropriate evaluation mechanism of university students' autonomous learning is proposed in this article. It is illustrated in aspects of assessment principles, assessment contents, assessment forms and concrete operations. After the practice in the teaching process, this new evaluation mechanism can stimulate the enthusiasm of students and improve the learning effect.

KEYWORDS: Autonomous learning; evaluation mechanism; arbitration management.

1 INTRODUCTION

Higher education is very different from elementary education where the knowledge in university is not just restricted to books and it takes orders of magnitude deeper and broader than elementary education's. For the most part of the time, the university teachers only lecture their students about key points of certain knowledge, and some teachers even introduce the advanced technology of the subjects. The amount of information is huge and the process of classes is fast. The only way that university students want to learn those subjects well is that they need to study on their own.

The development of information technology provides students a better ubiquitous learning environment than before. But, meanwhile, there are problems in the ubiquitous learning environment (Yang et al. 2013). For instance, students may be confused about which thing to learn in front of massive amounts of information; may lack ability of thinking with gaining others' research accomplishment easily by convenient communicating methods; and those various kinds of online learning conditions may distract students. As a matter of fact, teachers should provide necessary support and guidance for those students who learn by themselves in a ubiquitous learning environment to make it more efficient.

According to psychological research, the feedback of study result influences the efficiency of study. On the one hand, learners can adjust their learning activities and improve their learning strategy on the basis of feedback information; on the other hand, learners can be initiative and positive to study by the desire of getting a better grade and avoiding making the same mistakes again (Hua. 2014). It is a good way to strong the learner's faith in study by getting the timing and specific feedback. Learners can not only diagnose and evaluate the effect of their study, but also modify the mistakes in time.

Therefore, to establish a correct assessment concept becomes a hot topic. It is also the key to stimulating the students' enthusiasm of study. This article is about research on how to evaluate the autonomous learning of university students, and presents a new evaluation mechanism to detect the outcome of autonomous learning, and improve the study strategy and efficiency. This evaluation mechanism is not about the complete informal autonomous learning; it is about autonomous learning of a certain specialized course, focusing on students' ability to find problems, analyze problems and solve problems when studying on their own.

2 RESEARCH OF TRADITIONAL LEARNING EVALUATION MECHANISM

As feedback and adjustment mechanism, learning assessment plays an important role in the process of teaching and learning. It is the measurement and assessment of students learning condition (Yi. 2013).

The essence of study evaluation is to assess the extent which study result equals to education objective. The key point to evaluate autonomous learning is about setting the clear and operable objective which is also means, by observing, measuring and recording the learning content, learning process and learning result, to make identification and value judgments of the learning effect as well as reflection and revision of learning objectives.

There are three kinds of evaluation in the traditional classroom learning process, diagnostic assessment, formative assessment, and summative assessment. Diagnostic assessment aims to ensure students meet the requirement of current teaching goal, then teaching students in accordance of their aptitude. Formative assessment is a kind of development assessment based on observation, recordation and reflection. Its purpose is to motivate students to learn, to help students effectively regulate their own learning process. Summative assessment, also known as "ex-post evaluation", means to evaluate the final results of the teaching activities after teaching. Examinations or assessments, all belong to this kind of evaluation.

In the current teaching process, most of the universities use summative assessment methods, which the most common manifestations are tested and examinations.

3 THE PROBLEMS IN THE TRADITIONAL LEARNING EVALUATION MECHANISM

Due to the constraints of the current teaching conditions, student-staff ratio is large, and individual evaluation methods of students are difficult to implement. Therefore, during the teaching process, it is common to emphasize the final outcomes of students but neglect formative assessment. This is presented as the large proportion of the final exams. Formative assessment just becomes a mere formality, which wants to replace the assessment of the whole students with the check of homework or asking a few students in class.

I made researches about the evaluation methods of the curriculum in my university. Research results are shown in Table 1.

According to the research, evaluation methods mainly base on teacher-initiated evaluation and students tend to be passive to accept the evaluations. The methods are simple and lacking of pertinence. The main types of evaluation are formative assessment and summative assessment. But summative assessment is larger proportion which is generally greater than or equal to 70%. This assessment is of single feedback and is mostly given in the form of the final transcript.

Table 1. Research Results of Traditional Learning Evaluation Methods.

Evaluation types	Evaluation methods	Evaluation scope	Unique feedback	Feedback form
Formative assessment	Attendance / Participation	Sampling	No	Explanation in class
	Answering in class	Sampling	Yes	Explanation in class
	Homework / Exercises / Tests	Entirety / Sampling	No	Grade
Summative assessment	Mid-term exam	Entirety	No	Grade
	Final exam	Entirety	No	Grade
	Course paper	Entirety	No	Grade

From table 1, we can see that traditional learning evaluation methods emphasize a unified exam and are lack of individualized and targeted evaluation. The role of this evaluation mechanism in formal classroom learning process is far from satisfactory. In autonomous learning with relatively strong personalized process, this mechanism will not achieve the desired effect (Xie et al. 2013).

4 EVALUATION MECHANISM OF UNIVERSITY STUDENTS' AUTONOMOUS LEARNING

The principles, content, form and operations are discussed in this evaluation mechanism according to the patterns and features of autonomous learning of university students.

4.1 *Assessment principles*

According to the new mechanism, students are the body and active participants of the evaluation rather than passive recipients. Summative assessment is replaced by the new formative assessment.

The traditional method, which involves single tests and exams, is discarded. The new mechanism should be designed to be friendly, which means students get less pressure and be less offensive. It should also be objective and accurate. In a word, it follows the principles like initiative, procedural, diversity and humanity.

The procedure of evaluation can promote the communication between students and help them learn from each other. Thus, students can analyze and think in different ways, which helps them discover more questions.

By encouraging moderate competition, which are fair and open, students can get progress together.

4.2 *Assessment contents*

Learning pattern is the method that can enable individual achieve the best learning state. An inquiry-based learning pattern is generally applied to teaching process of higher education.

The advantages of Inquiry-based learning pattern are providing creativity room for students, developing innovative thinking and operational ability of students and urging learners on thinking. It mainly organizes students' learning activities and learning contents around coming up with and solving problems. Thus, the contents of the evaluation mechanism in this article involve the ability to study independently, including collecting, analyzing, and using information, communicating and cooperating with other group members, levels of hardworking, learning effect, etc.

4.3 *Assessment forms*

The assessment forms consist of the evaluation of students by their study groups, the self-evaluation of students themselves and mutual evaluation and mutual test between each student as well as evaluated by their teachers. The detailed process is as Figure 1 shows.

Firstly, the teacher initiates a study process with certain ranged topics. All the students should look up the references about the topic. And then try to find and come up with some meaningful questions in the ubiquitous learning environment. Next, analyze the problems and give answers. Finally, students evaluate themselves on their own.

Secondly, teachers will collect all the problems which the students give. After a sketchy evaluation of those problems and solutions, teachers send those problems to the students who are not the presenters of the problems in a random order. All the students will evaluate their received problems considered how difficult the problem is and how the problem inspires them. Then they must give their own solution from a various view.

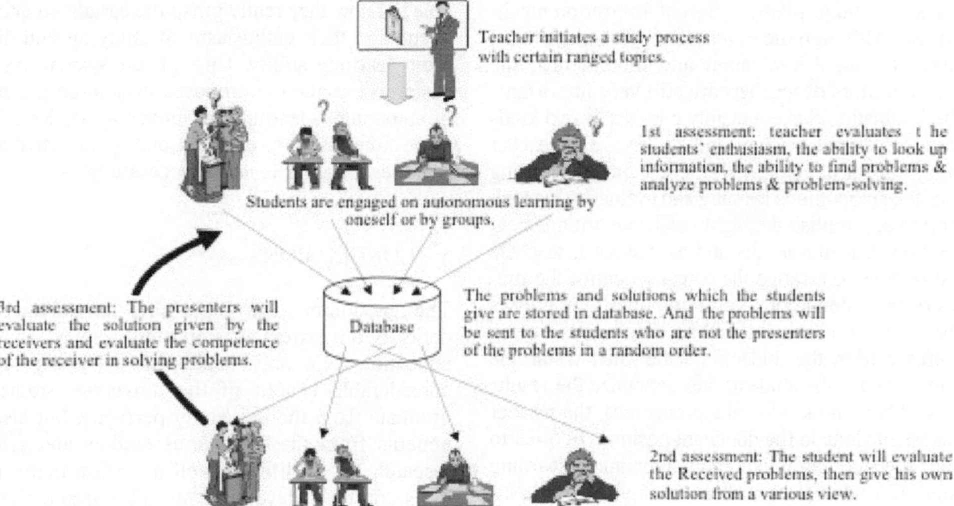

Figure 1. Evaluation process of university students' autonomous learning.

Eventually, the solutions of the problems from the receiver students will be sent back to the problem presenters. The presenters will evaluate the solution given by the receivers and evaluate the competence of the receiver in solving problems. The above process gives the students a chance of mutual analyzing and mutual evaluating. If any students are unpleasant with the evaluation, they can put forward their objections to teachers. And teachers can guide them and join the evaluation to coordinate it, which highly reflects the self-regulation of students.

4.4 *Concrete operations*

For evaluation mechanism of autonomous learning, its results should be quantitative, and its methods should be operational, which fully uses modern information technology to achieve electronic evaluation.

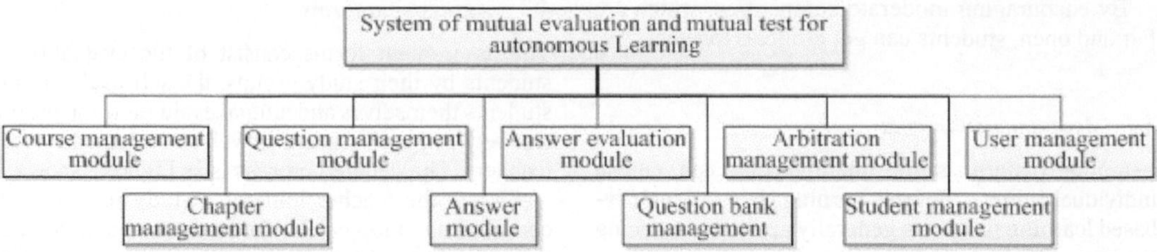

Figure 2. The function module diagram of the system.

Thus we need to develop an autonomous learning platform of mutual evaluation and mutual test system for university students to effectively implement learning evaluation mechanism.

In addition, it needs to have an appropriate security support to ensure the quality of the evaluation results. A full-time teacher is an important guarantee of autonomous learning evaluation quality. Teachers should also support the designing of the evaluation standard, the designing and monitoring of the evaluation process, and implementation of arbitration mechanism, etc. Although the evaluation mainly adopts the method of mutual evaluation and mutual test, the roles as monitors of teachers are still very important.

The Evaluation System mainly consists of two kinds of users, which are teachers and students. The teacher plays a role in the director, while the student is the leading actor of the autonomous learning and evaluation system. Aiming to accomplish the highly efficient arrangement of the daily teaching affairs and assessments, teachers should be able to arrange the courses, control the processes of the autonomous learning, publish the notices for the students, collect as well as distribute the questions presented by the students, decide when to start the examinations for the students and announce the results of them. Throughout all of the procedures, the teacher should be all alone in the dominant position. In order to enhance the studying interest and autonomous learning abilities of students, as well as strengthen the communication among students, students should take the dominant positions and be able to come up with questions, upload questions, answer and evaluate questions presented by others and evaluate the answers from other students. To guarantee the fairness of the assessments, the students are able to apply for the arbitrations to the teachers if they have the contentiousness of the results of the assessments. According to the analysis above, autonomous learning platform of mutual evaluation and mutual test system for university students is designed and divided into 9 functional modules, which are question management module, course management module, answer module, chapter management module, answer evaluation module, arbitration management module, question bank management, student management module, and user management module. The structure of the system function module is shown in figure 2.

Via the system, the teacher should make an evaluation of the questions and answers from the students in multiple terms through the teaching activities, so that they are able to know the extent of mastery of students and adjust the plans of teaching activities. Students can also get their assessment results, so that they are able to know they really grasp the certain chapters and encourage their enthusiasm of studying and autonomous learning ability. Through the system, it's much easier to execute the proposed evaluation mechanism of autonomous learning for university students. At the same time, in-time, efficient and quantitative assessment results are able to do successfully.

5 CONCLUSIONS

The autonomous learning ability for university students is an extremely important subject of higher education. Not only could the university students considerable benefit of the university studies and graduate from the university perfectly, but also gain benefits from the continuous studies and scientific research for all life, as well as create better conditions for their developments after graduation. The establishment of evaluation mechanisms of students' autonomous learning will improve the cultivation of the ability of students' autonomous learning.

An appropriate evaluation mechanism of university student autonomous learning is proposed in this article. The mechanism abandons the traditional assessment method which is dominated by the teachers, but enable the students to participate in the evaluation process. This pattern aims to be open, fair, in-time, efficient and operable. Through the practice of the teaching procedures, the mechanism will stimulate the enthusiasm of students and improve the learning effect, which is worth generalizing.

REFERENCES

Hua Qing. 2014. Cultivation of autonomous learning ability in ubiquitous learning environment for university students specialized in Japanese. *University education.* (1): 29–30.

Xie Yuntian (ed.). 2013. Difference Evaluation Application in Student Assessment. *Journal of Teaching and Management*. (6):72–73.

Yang Xianmin (ed.). 2013 The Design of Ubiquitous Learning Environment from the Perspective of Ecology. *Educational Research.* (3):98–105.

Yi Jin. 2014. Construction of Classroom Learning Assessment to Improve Teaching and Learning. *Journal of Educational Studies.* 9(5):61–67.

Management, Information and Educational Engineering – Liu, Sung & Yao (Eds)
© 2015 Taylor & Francis Group, London, ISBN: 978-1-138-02728-2

An evaluation system for autonomous learning in the ubiquitous learning environment

Hong Juan Liu, Li Ping Huang & Lin Ying Jiang
Software College, Northeastern University, Shenyang, Liaoning, China

ABSTRACT: With the arrival of the education information age, the autonomous learning ability of college students obtains more and more attention. In order to evaluate the effect of the students' autonomous learning effectively, based on J2EE platform, a mutual evaluation and mutual test system for autonomous learning of university students is designed and developed. Through this system, students can put forward questions under the guidance of teachers, answer questions from other students, evaluate the answers of the others, while teachers can evaluate the behavior of students in the whole process comprehensively. Hence, the ability to ask questions, analyze problems and solve problems of students can be improved to a certain extent.

KEYWORDS: Mutual evaluation and mutual test system; Autonomous learning; Ubiquitous learning.

1 INTRODUCTION

With the rapid development of computer, network and communication technology, great changes have taken place in the traditional of learning theory or learning mode. In the light of the continuous development of ubiquitous computing (Li et al. 2006), pervasive computing and cloud computing technology, ubiquitous learning is proposed and has gotten a broad attention. The ubiquitous learning is a kind of learning form that can break the limitation of time and space and integrate more advanced concepts and technology. That is, anyone can obtain any desired information in any place at any time. The learning form provides for students with a convenient learning with some portable terminal device in 5A (Anyone, Anywhere, Anytime, Any device, Anything) (Yang. 2011). Fundamentally, the ubiquitous learning is a learner-centered, task-focused, on-needed, moderate and immediate learning from. In that sense, ubiquitous learning has great role in promoting autonomous learning.

Autonomous learning is a new kind of education idea, also known as self-regulated learning, and refers to a practical activity that can develop the initiative and creativity of students in the learning process. USA famous expert, Professor Zimmerman of the City University of New York gives a definition of autonomous learning. Namely, based on the feedback to the learning efficiency and learning skills, autonomous students select and apply autonomous learning strategies to obtain the desired learning outcomes (Yang. 2012).

Psychology research shows, variety of feedback information from the study results has obvious influence of learning effect. On the one hand, the learners can adjust learning activities according to the feedback information and improve learning strategies. On the other hand, the learners will enhance motivation to get better grades or avoid making mistakes again, to keep the initiative and enthusiasm of learning (Hua. 2014). If learners can get timely and explicit feedback in certain ways, they can diagnose and evaluate their own learning effect, realize the mistakes in the learning process and correct them in time. Meanwhile, they can see heir learning achievements and their confidence to continue studying is strengthened.

Therefore, it is necessary to put forward an evaluation mechanism that can check the learning effect of autonomous learning of college students. The evaluation mechanism can effectively evaluate students' learning to professional courses, thereby improving autonomous learning strategies, enhancing learning effect. In this paper, mutual evaluation and mutual test system for autonomous learning is designed and implemented, which can significantly improve the ability to ask questions, analyze questions and solve questions of students. Application results show that it cannot only effectively detect the effectiveness of student learning, but also can improve the enthusiasm and initiative of students in a certain extent.

2 THE MUTUAL EVALUATION AND MUTUAL TEST SYSTEM

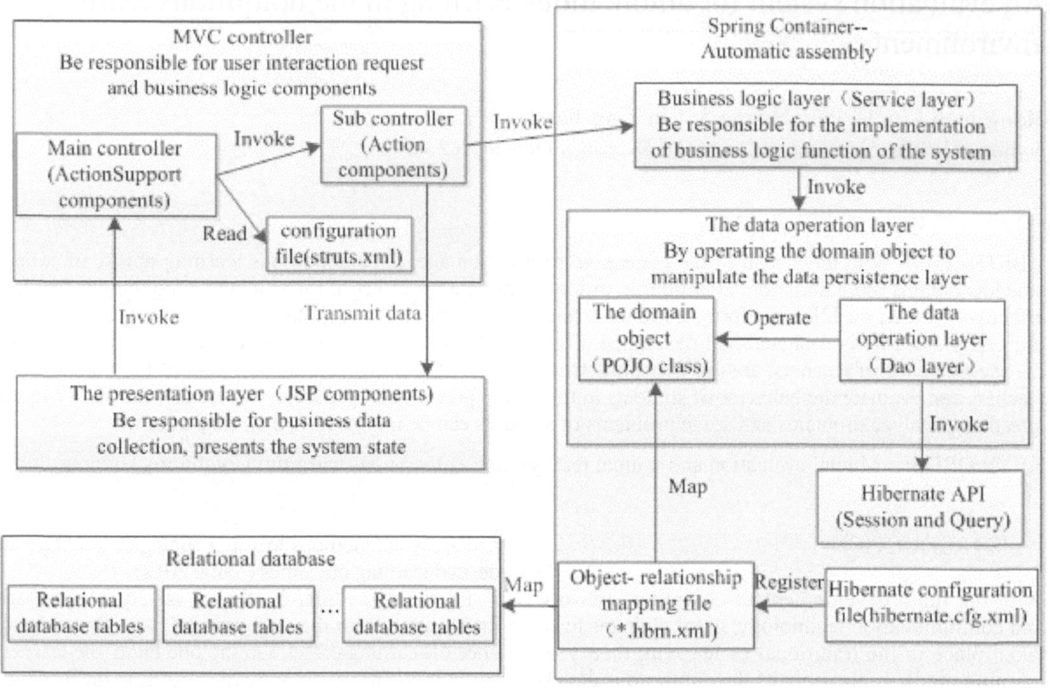

Figure 1. The SSH framework of system.

2.1 *The design of framework*

The mutual evaluation and mutual testing system for autonomous learning of the university is based on three layer structure of J2EE, namely, the model layer, control layer and the view layer. Using the development model, we cannot only realize the complete separation of view, controller and model, but also realize the separation of business logic layer and persistence layer. In this way, the system reusability can be greatly enhanced. Meanwhile, because the coupling between different layers is small, it is convenient to work in parallel for team members, which will greatly improve the efficiency of development. Specifically, SSH framework is used to achieve the separation and coordination of the three layers. The model layer uses the Hibernate framework, uses JavaBean to generate the tables and associations in the database and operates the database through JavaBean. The control layer uses Struts framework to connect the data layer and the view layer, which receives, processes and transmits data. The view layer using JSP pages technology to show the user and interact with users. Spring framework agglutinates Hibernate and Struts, which manage the whole framework transparently. The SSH framework of the system is shown in Figure 1.

2.2 *The analysis and design of function*

The mutual evaluation and mutual test system for autonomous learning consists of two users, students and full-time teachers. In order to realize the efficient management of teachers to daily teaching and student evaluation, the teacher should be able to carry on the management of the course, control question chapters, issue a notice to inform students, check the questions students upload, collect and distribute the questions, monitor the answer process for students and decide when to perform the evaluation between students and the publish the evaluation results. After the comprehensive evaluation of the questions and answers of the students, teachers can grasp students' mastering situation to the knowledge points of a certain chapter. In order to improve the learning interest and autonomous learning ability of students and strengthen the communication between students, students can put forward questions, upload questions, answer the questions of others and evaluate the answers by other students. In order to ensure the evaluation is fair, if a student has the objection for his answers, he may apply in arbitration. For his own questions, if he thinks his score is lower for his questions, he may also apply for arbitration.

Through analyzing the function demand of mutual evaluation and mutual test system for autonomous learning, the function of the system is divided into nine modules: question management module, course management module, answer module, chapter management module, answer evaluation module, arbitration management module, question bank management, student management module, and user management module. The structure of the system function module is shown in figure 2.

For the use of students, the related function modules include question management, course management, answer management, answer evaluation and arbitration management. Question management module includes the function to add, view, modify, and delete questions. The question mainly includes choice question, judging question and short answer questions. According to the specified questions chapters, question type and question difficulty by teacher, students can add a new question with its answer, reference source and other information. In answer management module, students can examine the questions that will be answered and the questions that have been answered. The interface to examine the questions is shown in Figure 3. If a student does not complete the task to add questions, he is not allowed to answer any question. Only when the student completes the task, the system will search a question by another student as his question to be answered. Students can also query, evaluate and modify the answers by another student. Finally, if a student has objections to his question score or answer score, he can apply for arbitration.

For the teachers in the system, they can manage questions, manage a question bank of students, manage the information of students, manage the information of users, manage the courses, manage the answer evaluation and manage the arbitration.

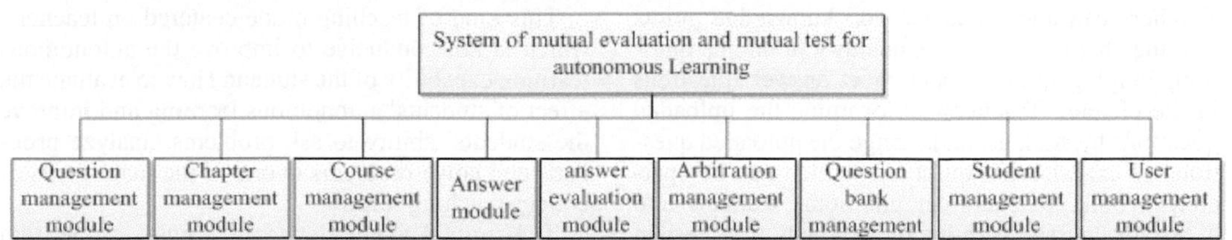

Figure 2. The function module diagram of the system.

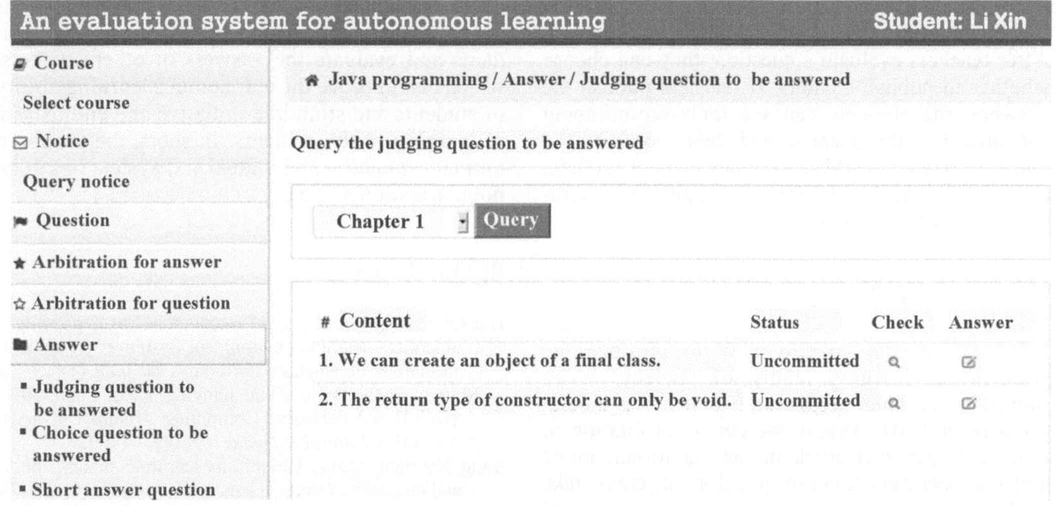

Figure 3. The interface to examine the questions by students.

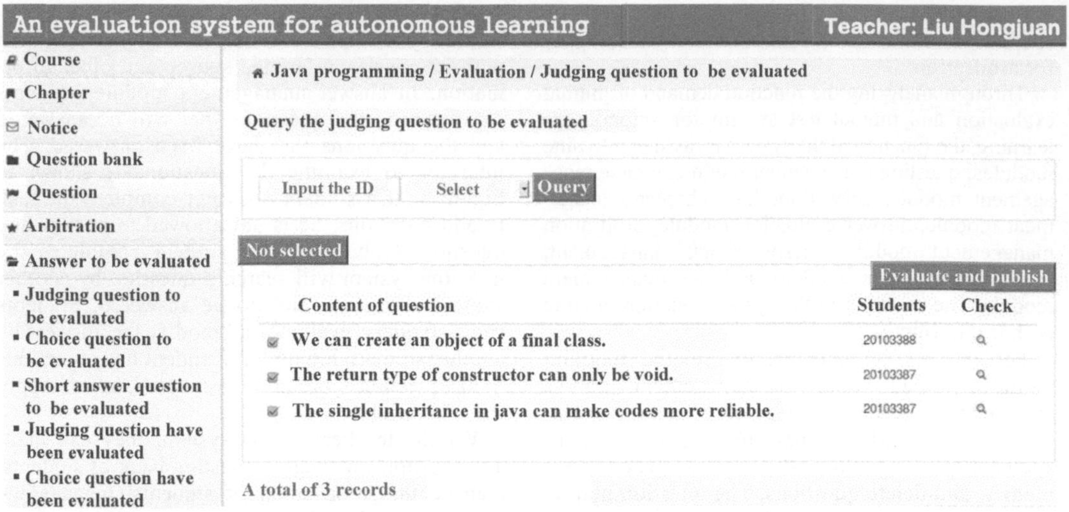

An evaluation system for autonomous learning | Teacher: Liu Hongjuan

- Course
- Chapter
- Notice
- Question bank
- Question
- Arbitration
- Answer to be evaluated
  - Judging question to be evaluated
  - Choice question to be evaluated
  - Short answer question to be evaluated
  - Judging question have been evaluated
  - Choice question have been evaluated

Java programming / Evaluation / Judging question to be evaluated

Query the judging question to be evaluated

Input the ID | Select | Query

Not selected

Evaluate and publish

Content of question	Students	Check
We can create an object of a final class.	20103388	
The return type of constructor can only be void.	20103387	
The single inheritance in java can make codes more reliable.	20103387	

A total of 3 records

Figure 4. The interface to query the answers that need to be evaluated.

Teachers can add a chapter, add knowledge points into the chapter and set the quantity of choice question, judging question and short answer questions in the chapter. Teachers can examine the uploaded questions by students and change the uploaded questions as examination questions, which can be operated in bulk. Teachers can add some students into a certain course, examine the students in a certain course and delete a student in a certain course. If a student has been added in a course, a repeat adding tip will be given. In answer evaluation module, teachers can query the answers that need to be evaluated, evaluate the answers and publish the results of the evaluation. The interface to query the answers that need to be evaluated is shown in Figure 4. After teachers evaluate a question, they can choose whether to publish answers. If teachers publish the answers, the students can see their scores about the answer to the question and their scores on the question itself. In arbitration management module, teachers may deal with, query or modify the arbitration proposed by students.

3 CONCLUSIONS

For most of the college, how to improve the autonomous learning ability of students is an important problem to research and solve curriculum reform. At present, the course evaluation of most colleges still stuck in the traditional mode and teachers puts forward questions in class, take back the answers of students and record the score of every student.

This kind of teaching mode centered on teachers, which is not conducive to improve the autonomous learning capability of the student. How to evaluate the effect of students' autonomous learning and improve the students' ability to ask problems, analyze problems and solve problems is one of the main goals of college teaching reform.

The mutual evaluation and mutual test system for autonomous learning presented in this paper utilizes a new concept of teaching and examination methods. Using this system, teachers can examine the learning situation of the students from many aspects and teaching quality and efficiency can be improved. The students put forward questions independently, answer the questions of other students and evaluate the answers of other students, which can improve the autonomous learning ability of students and stimulate initiative and enthusiasm of learning of the students. In short, the proposed mutual evaluation and mutual test system has good popularization value.

REFERENCES

Hua Qing. 2014. Cultivation of autonomous learning ability in ubiquitous learning environment for university students specialized in Japanese. *University education*. (1): 29–30.

Li Shiqun, Shane Balfe, Zhou Jianying, Kefei Chen. 2006. Trust Based Pervasive Computing. *Wuhan University Journal of Natural Sciences* (6): 1477–1480.

Yang Xiaotang. 2011. Ubiquitous learning: theory, model and resources. *Distance education of China* (6): 69–73.

Yang Yanan. 2012. The strategy of graduate independent study. *Inner Mongol: National university of the inner Mongol.*

Management, Information and Educational Engineering – Liu, Sung & Yao (Eds)
© 2015 Taylor & Francis Group, London, ISBN: 978-1-138-02728-2

Research on aesthetic teaching modes of basketball under new curriculum standards

Shou Yu Yang
Langfang Radio and TV University, China

ABSTRACT: According to aesthetic teaching theories and health curriculum concepts, the whole teaching process of basketball is examined and handled in the view of aesthetics which is based on the unique aesthetic factors in basketball and combined with the characteristics of the students' aesthetic psychology, in order to explore aesthetic teaching modes of basketball. Thus, by promoting aesthetic teaching and learning, the students' quality can have an integrated and harmonious development and the rethinking of the aesthetic teaching process and the objective evaluation of the basketball can also be deepened.

KEYWORDS: Basketball teaching; Aesthetic teaching; Aesthetic experience.

Under the requirements for a new round of reform of the basic education curriculum, physical education is not only to impart sports skills and knowledge, to develop intelligence, to develop ability and to make the students have moral education, but also lays stress on cultivating the students' aesthetic ability. The cultivation of quality should be emphasized throughout the entire process of teaching physical education. The aesthetic factor of sports presents everywhere during the process. If teachers use these aesthetic factors to teach, they can arouse the students' aesthetic perception, rich their emotional experience, improve their ability of aesthetic evaluation and aesthetic creation, and make the students naturally obtain sports skills and knowledge which will be developed actively in the process of experiencing athletic beauty. But as for teaching physical education in the classroom, due to different teaching materials, there will also be variations in terms of aesthetic characteristics and forms. The design for aesthetic teaching should be targeted in the teaching process. This paper focuses on the discussion of the aesthetic mode of basketball in teaching materials.

1 AESTHETIC TEACHING OF BASKETBALL

1.1 *Aesthetic characteristics of basketball*

Basketball is one of the world's most widely developed and most influential sports items. Its characteristics include rich content, exquisite skills and fierce competition, etc. and also contain rich aesthetic resources which include almost all the elements of beauty. It has formed a relatively complete aesthetic system. The superb, rapid and orderly rhythm of a game and the unpredictable, exciting results of it include many aesthetic factors such as posture beauty, beauty of mind, spectacular beauty, harmonious beauty, motion beauty, agile beauty, flexible beauty, beauty of strength, competitive beauty, breathtaking beauty, beauty of clothing and equipment, etc. which make people produce a feeling of gracefulness and a sense of lofty in the appreciation process and promote a harmonious unification of individual personality to achieve a high degree of integration of the "humanization of nature" and "naturalization of human".

1.2 *Aesthetic teaching of basketball*

Aesthetic teaching refers to by turning all teaching factors (such as teaching objectives, content, methods, evaluation, context, etc.) into aesthetic objects the whole teaching process is transformed into activities of appreciation of beauty, expression of beauty and creation of beauty, so that the entire teaching process can become a harmonious unity of static and dynamic state and also become a highly harmonious integration of internal logical beauty and external formal beauty during which the teaching quality can be greatly improved, the burden of students can also be reduced. Aesthetic teaching is a kind of teaching ideas, operating modes and methods which enables teachers and students to get a full physical and mental pleasure. The theories of aesthetic teaching should have relevant teaching modes, so as to promote the aesthetic teaching of physical education and health curriculum and also to promote the implementation of the concepts of new curriculum standards.

In the traditional basketball teaching process, the knowledge of basketball skills is the focus of basketball teaching, but emotional education and aesthetic education are always neglected during the teaching process. Such teaching activities which only care about the instruction of basketball skills, but pay less attention to the students' emotional life and experience will cause great damage to the students' learning, enthusiasm for basketball, which will form a strange phenomenon among students who like basketball but worry about basketball classes. There is no doubt that it is indispensable to teach skills in basketball teaching, but to learn basketball skills doesn't mean it can replace the students' emotional experience and aesthetic psychology of basketball. Only learning skills are difficult to achieve the expected teaching objectives due to the lack of students' active participation and such way of teaching cares less about the existence and development of human life which will lead to the loss of function of sports to carry forward humanity and appreciation of beauty.

Compared with the traditional basketball teaching, aesthetic basketball teaching is the process of combining aesthetic viewpoints with handling basketball teaching which will help make the basketball teaching achieve the best condition to promote aesthetic education. It puts the teaching focus on the emotional experience and on the cultivation of the students' aesthetic ability. Emotional education should be paid more attention. To cultivate the students' perceptive ability of basketball skills and their performance ability should also be strengthened in order to achieve the purposes of making the students consciously perceive beauty, appreciate beauty, get nurtured and infected by beauty to help them cultivate their spirit, rich their emotion, acquire basketball skills and improve their aesthetic and creative thinking ability in the teaching process, which will help form a healthy behavioral beauty and a healthy spiritual beauty.

1.3 *The definition of basketball aesthetic education*

Based on the above analysis, basketball aesthetic education refers to an organic integration of basketball skills in the teaching materials and the theories of aesthetic education by using a variety of aesthetic teaching means to fully explore aesthetic connotation of basketball according to the aesthetic characteristics of basketball. Aesthetic infiltration needs to be carried out consciously, regularly and effectively to develop the students' healthy physical beauty, healthy behavioral beauty and healthy spiritual beauty.

2 BASIC MODES OF AESTHETIC TEACHING OF BASKETBALL

2.1 *Basic ideas of mode*

The basic mode of basketball aesthetic teaching is constructed by combining the students' characteristics of aesthetic psychology during the process of practical exploration, which is based on aesthetic theories that have already been well studied, theories of physical education and health curriculum and special aesthetic factors of basketball in the teaching materials.

It is a new, scientific and the feasible teaching mode, which can guide students into the beautiful world of basketball from a certain aesthetic standpoint. In the vivid practical activity of appreciating beauty, students can experience beauty in a free, happy, lively atmosphere which can fully motivate the students' initiative and enthusiasm to learn basketball and can stimulate the students' existing aesthetic psychology to help them better grasp aesthetic ideas of basketball skills in the aesthetic experience and aesthetic enjoyment and also help them develop healthy physical beauty, healthy behavioral beauty and healthy spiritual beauty which fully reflects basic concepts of the new curriculum standards.

2.2 *Operational programs of mode*

Aesthetic experience is the principal line running through basketball teaching process, any teaching step can't proceed without the students' aesthetic experience. The first step is for both teachers and students to finish leading-in, which can achieve the purpose of arousing aesthetic interest and warming up. It is the primary step to realize the aesthetic teaching of basketball. A good leading-in is like the "curtain" in a performance to create an effect of revealing feelings. The teacher can utilize music, excellent basketball postcards and classic basketball game pictures to make the teaching atmosphere lively and interesting, so that the students can stay in an exciting, stimulating and passionate learning state. The next step is for both teachers and students to study and explore the viewpoint, namely, to reveal and analyze knowledge points of skills and aesthetic viewpoints. This is the core part of aesthetic mode of teaching basketball. To determine the "point" requires achieving "accuracy" and "concision". "Accuracy" means to determine the "point" according to the characteristics of the specific teaching content; "concision" means that knowledge points and aesthetic viewpoints for a class shouldn't contain too many, one or two is acceptable. The knowledge points of skills and aesthetic viewpoints for a class should be reflected accurately and concisely to differentiate key points and difficult points. In order to fully play the main role of the students, students are allowed to have self-directed practice, discussion and exchanges; teachers guide the students' actions to help them master and solve difficult points. This is followed by strengthening viewpoints, which means that based on revealing viewpoints before knowledge points of skills and aesthetic viewpoints can be developed through the students' experience activities. The next is the extension of the viewpoints which means that the extension of skills and aesthetic viewpoints. The teaching material is only one example and a platform which provides the students with a clue of learning. It needs extensive contacts to enhance and consolidate their basketball skills and application ability. Then it is the detection of the viewpoints which aims to feedback the teaching effect by detecting the masters degree of skills and aesthetic viewpoints and the achievement of the teaching objective. It is the most important link of knowledge transfer in the teaching process and it needs to be flexible to fully play a main role of the students who can detect the skills and aesthetic viewpoints and to guarantee the multidirectional feedback. Finally, it is the regression of viewpoints which means that relaxation and stable aesthetic emotion. Teachers and students can use various means to relax and experience the nourishment of the soul.

But in practical teaching, it can't be carried out in accordance with the basic operational mode because of the differences in teaching content, such as different techniques and tactics. Hence certain targeted operational mode needs to be designed according to specific teaching content. If we call the above mentioned operational mode as "a routine one" having a general and universal meaning, then those specific and diversified operational modes can be called "variant ones". Only by creating more targeted operational modes in practice, teaching activities can truly achieve aesthetic expression and aesthetic creation.

3 COGNITION AND EVALUATION OF AESTHETIC TEACHING OF BASKETBALL

The aesthetic teaching of basketball requires us to break through traditional understanding and a single curriculum objective. Cognitive objectives and aesthetic objectives should take an equally important position. The implementation process of aesthetic teaching of basketball puts emphasis on the psychological process of "experiencing", and especially pays attention to the student's emotional experience and accumulation. Its purpose is to cultivate the students' beauty of health and at the same time to cultivate their beauty of mental health and beauty of behavior through practical experience. Based on the above understanding, the evaluation of the aesthetic teaching of basketball should include not only the evaluation of cognitive objectives, but also the evaluation of aesthetic objectives; it should not only evaluate the results, but also conduct a comprehensive examination demonstrating the students' wisdom, ability, attitude, belief, etc. during the process of experiencing beauty. Aesthetic experience, aesthetic appreciation and aesthetic creation should be evaluated thoroughly and comprehensively based on the overall learning level of the students. That is to say, aesthetic evaluation means there is a conversion from the evaluation of a single sports skill in the evaluation of a comprehensive physical quality.

REFERENCES

[1] Guo Cheng, Zhang Jinglan. Aesthetic teaching principles and practice [M]. Chongqing: Southwestern Normal University press, 2001.9.

[2] Hu Xiaoming. Sports aesthetics. [M]. Chengdu: Sichuan Education Press, 1988.7.

[3] Fan Wei. Aestheticism in classroom teaching under the new curriculum standard [J]. Journal of Southwestern Normal University, 2003 (5).

Management, Information and Educational Engineering – Liu, Sung & Yao (Eds)
© 2015 Taylor & Francis Group, London, ISBN: 978-1-138-02728-2

Reform of basic chemistry experiment course for non-chemistry specialties in independent college

Ting Liu & Bo Quan Jiang
Department of Biology and Chemistry, Nanchang University College of Science and Technology, Nanchang, Jiangxi, P.R.China

ABSTRACT: Basic Chemistry Experiment Course (BCEC) is one of the most important basic courses in independent college. Reform of the basic chemistry experiment course has a great significance for cultivating creative talents. Aimed at the present problems of this course teaching, several measures, including cultivating the right attitude of the students' learning BCEC, practicing the open experiment teaching mode of the BCEC, implementing the online teaching of the BCEC and establishing scientific and objective experiment assessment mechanism, were proposed and practiced. The results showed that these measures stimulated the students' activities and interests of learning BCEC, improved the students' experimental abilities of standard operation, increased the students' abilities of independently analyzing and solving practical problems and cultivated the students' innovation consciousnesses and abilities.

KEYWORDS: Basic chemistry experiment, Independent college, Reform, Non-chemistry specialty.

1 INTRODUCTION

Independent college is made up of ordinary undergraduate course colleges and universities and social forces (as partners) including enterprises, institutions, social organizations or individuals and other agencies having cooperation ability. Independent college is held for undergraduate level education of institutions of higher education and is a secondary college of undergraduate level in accordance with the new mechanism and model [S H Zai, 2013 and C X Zhang, 2014]. In recent years, some leadership of the ministry of education has pointed out that independent college is to ensure sustained and healthy development of higher education and is an important part of higher education development in China. Independent colleges were produced in the late 90 s of the last century in China. With its rapid development, there have been more than 300 independent colleges with over 200 millions of students at present in China. However, independent college is a new thing because it is a new school-running mode produced based on the undergraduate education, thus reform is needed in the aspects such as teaching concepts, teaching contents, teaching methods, examining ways and so on. Basic chemistry experiment course (BCEC) is one of the most important basic courses for the chemistry specialty and non-chemistry specialty and an important foundation for students to learn professional courses. For a long time, however, the design and contents of BCEC have continued using the mode of chemistry specialty experiment under the influence of the undergraduate course teaching system and teaching mode. This resulted in low learning activities and not ideal teaching effects for the students with non-chemistry specialty in learning BCEC. Therefore, it is imperative for us to reform the BCEC. In this paper, we analyzed the present situations of BCEC learning and teaching for the non-chemistry specialties, based on which, the measures and practice of the reform of BCEC in teaching concepts, teaching contents, teaching mode, teaching means and experiment evaluation were conducted and discussed.

2 ANALYSIS OF PRESENT SITUATIONS

Nanchang University College of Science and Technology was established in 2001. At present there are seven non-chemistry specialties at the college: pharmaceutical engineering, biological engineering, biological technology, polymer materials and engineering, environmental engineering, water supply and drainage, applied chemistry. Both the university chemistry experiment course and organic chemistry course open to the students of specialties concerned above. The students' total basic chemistry, experimental hours are 430 hours and their total annul man-hours of basic chemistry experiment are about 22000 hours. At this stage, there still are some problems during BCEC teaching in our college.

2.1 *Present students' situations*

For a long time, China's education system is based on and given priority to exam-oriented education, the experiment abilities of students are not satisfied. The students of independent college are admitted with an admission score about 100 points lower than the admit fractional line of ordinary college, so their basic knowledge is poorer than those of the ordinary college and key universities, especially their experiment abilities. According to our investigations, most of the non-chemistry specialty students used to do chemistry experiments in their studies in high schools, but their experimental levels and abilities are different due to different teaching levels of their local schools. A small part of the students never did chemistry experiments in high schools (mainly rural high schools), so they are strangers to the BCEC. These students really worry about BCEC due to their having not got a specification of experiment operation training before. On the other hand, some of students lack the correct understanding of the importance of the BCEC. They do not know the significance of learning and mastering basic chemistry, experimental skills for their future career. So they are often passively learning BCEC. For example, they do the experiment only according to what the textbook says and what the teachers write on the blackboard without carefully understanding the experimental principles and operation modes. This kind of learning method of depending on the gourd and ladle makes them not be able to really master experimental skills. During experiments, some of students' lack of patience due to longer working time and do not to do the experiments according to regulations due to saving time, even tamper with their experimental data and copy the others' experimental results.

2.2 *Present teaching situations*

At the moment, the experimental teaching mode of BCEC in our college is basically a replica of the mother school due to its particularities of experimental conditions limitation and sharing the highest quality resources of the mother school (Nanchang University). The traditional teaching mode of "infusion and imitate" is still used. The teachers only play an infusion and exemplary role and the students are in a passive position without a good interaction between teachers and students. They mechanically finish the experiments without having an active thinking of why to do so. The experimental contents are older and narrower and the experimental types are mainly validation experiments. The teachers give the students experimental schemes and the students do the experiments under the same operating conditions, which seriously limits the students to play their initiatives and creativity. The teaching means are simple and the teaching quality is not higher. The students are only asked to finish the experiments and submit the fixed format of their reports, which affects the development of the students' independent thinking and personalities. With the traditional teaching method and mode, a part of students can not still understand the principles and answer the thinking questions as well as analyze and solve the problems encountered during the experiments. In the light of the problems existed in our college, several effective measures are proposed and practiced as follows.

3 REFORM MEASURES OF BCEC

3.1 *Cultivating the right attitude of the students' learning BCEC*

In the light of the ideological problems existed in the students, we paid more attention to do the ideological works of the students The teachers shoulder the responsibilities of "teaching and educating people". In order to help the students to overcome their fuzzy understandings of ignoring or worrying about BCEC for cultivating their right attitude of learning BCEC, we give them a professional education to make them realize the important position of BCEC in their specialty and the significance of learning BCEC for their future carrier, to encourage them to set up the correct learning motivation, to let them know that they should not only learn a solid theoretical chemistry knowledge, but also master correct basic experimental operation skills for benefiting to their future work. According to the different levels of the students, we teach them the experimental scheme with a suitable speed in the class and pay more attentions to instructing the students who are not familiar with operations, especially those who used not to do the basic chemistry experiments in high school. In order to help some of students to ease their concerns of learning BCEC, the teachers use their spare time to contact with these students and listen to their ideas on learning BCEC and patiently do their ideological work for increasing their understanding of the importance of FCEC. When the students ask the teachers some questions during the experiments, the teachers always take the trouble to explain to the students and teach the students by their hand, especially to the students who are not familiar with basic operations or have difficulties in carrying out experiments. With the patient education and instruction of the teachers, the students increased their understanding of the importance of FCEC, set up their confidence of learning BCEC and inspired their activities to go into the study of the BCEC.

3.2 *Practicing open experiment teaching mode of the BCEC*

In the light of the traditional teaching mode of "infusion and imitate", the open experiment teaching mode (OETM) is being carried out in our college. The goal of independent college is to cultivate creative talents, however, the traditional teaching mode constrains the thoughts and hands of the students, which is not conductive to cultivate the student's creativity. The practicing of OETM is one of the important measures to realize the goal [Hao Z, F, 2003 and Cue H S, 2007]. The OETM gets rid of the traditional teaching mode of simply doing a verification experiment and realizes the shift from the "teachers' teaching as the center" to the " students' autonomous learning as the center". During the last semester, the open experiment teaching of "organic chemistry experiment" course was carried out in our college. The process includes (i) The teachers set eight experiment subjects for the students to select. (ii) The students select the subjects and independently design the experimental schemes through searching literature by themselves. (iii) After their finished experimental schemes are confirmed by the teachers, the students independently build the experimental devices, conduct the operations, treat the experimental data, calculate the target values, write the experimental reports and report to the teachers by PPT files. During the OETM, the students are allowed to discuss together, but not to copy each other. The types of open experiments can involve applying, comprehensive, researching and innovative experiments among which the comprehensive experiment is more suitable for the first-year of the college students. Through the open experiment teaching of organic chemistry experiment course, the students have got great improvements in subjective initiatives, innovation consciousnesses, literature searching abilities, the abilities of independently analyzing and solving practical problems and the basic chemical experiment operation skills. During this semester, the open experiment teaching of university chemistry experiment course is being conducted for the new students of 2014.

3.3 *Implementing the online teaching of the BCEC*

In the light of problem of traditional teaching means with boring and dull atmosphere, the implement of online teaching of BCEC is being realized in our college. The online teaching is conducted on the public network teaching platform established by the college. In the website of basic chemistry laboratory (BCL), the online teaching contents concerned with university chemistry experiment course and organic chemistry experiment course, are divided into five parts:(i) home page involving "personal profile of the head"; (ii) basic situations involving "project introduction", "personal profile of the head", "experimental team construction", "project declaration," and "project construction specification"; (iii) construction information; (iv) construction effectiveness and (v) resource recommendation. The students can click on the website of "Demonstration laboratory" and go into the web page of "Basic chemistry laboratory", where there are a lot of rich, vivid and lively teaching contents, most of which are prepared by our teachers. The students can benefit a lot from the network teaching platform, for examples, they can know the basic situations and development trend of the basic chemistry laboratory. They can download the multi-media courseware and "collected thinking questions and answers" to help them to well do the preview of each experimental project for improving their understanding of the experimental principles and thinking questions of each project. They can download the video materials of "basic chemical experimental operation skills" to repeatedly learn. The video material consists of about 23 lively demonstrations of chemical basic operation techniques, which can help the students correctly grasp the specifications of basic operation skills. The students can easily select the experimental project, submit their self-designed experimental schemes and give their experimental reports with PPT format online during their doing open experiments. The teachers can review and confirm the students' experimental schemes online. The network teaching method strengthened the interaction between teachers and students, made the experiment teaching vivid and lively and extended the students' time and space of studies, which greatly motivated the learning interests and enthusiasms of the students.

3.4 *Establishing scientific and objective experiment assessment mechanism*

In the light of the problem of imperfect basic chemical examination method, a scientific and objective basic chemical experiment assessment mechanism has been established and are in progress in our college. The assessment of the basic chemical experiment is the inspection and evaluation of the teachers' teaching quality and students' learning quality. The established assessment mechanism consists of four parts: (i) Inspection at ordinary times, which mainly inspects the students' understanding of the purpose, principle and thinking questions of each experiment project, beginning ability, ability of analyzing and solving practical problems, ability of treating experimental data and results, quality of written experiment report and experimental habits including safety, environmental protection, discipline, team work and public property consciousnesses [Liu K, 2010]. The grade of this part is 60 marks of total marks (100 marks) involving preview (10 marks), experimental operation (25 marks), experimental attitude (10 marks) and writing experimental report (15 marks). (ii) Theoretical examination, the purpose of which is to check whether the students have firmly grasp the theories about the principles, purposes, thinking questions of each experimental project to guide their experiments, The grade of this part is 10 marks of total marks. . (iii) Open experiment evaluation, which mainly check the students' abilities of literature review, scheme design, standard operation, experimental results, and experimental report as well as open reply with PPT file. The grade of this part is 20 marks of the total involving scheme design (4 marks), specification operation (4marks), data treatment (4 marks), experimental results (4 marks) and experimental report (4 marks). (iv) Basic operations assessment, which is carried out at the end of the course and mainly exam whether the students are able to do the standardized operations or not, such as acid-base titration, washing glass apparatus, vacuum filter, pipetting, evaporation, distillation, and so on. The realization of the assessment mechanism stimulated the students' activities and interests of learning BCEC, improved the students' experimental abilities of standard operation, increased the student's abilities of independently analyzing and solving practical problems and cultivated the students' innovation consciousness and abilities.

ACKNOWLEDGEMENT

This subject comes from the "Project on Demonstration Laboratory Construction" financially supported by Nanchang University College of Science and Technology.

RERERENCES

Chunxiang Zhang, Qingyun Zheng, Xiangyang Zhang, et al. 2014. Research on the Teaching Reform of Basic Chemical Experiment for non-Chemistry Specialty in Independent College. *Guangzhou Chemical Industry*, 42(6): 159–160.

Cui Hongshan, Xiong Ya. 2007.Exploration and research of evaluation system of basic chemistry experiment teaching. *Journal of Anhui University of Science and Technology* (social science edition), 995): 85–88.

Hao Zhengfang. 2003.Discussing the importance of preparing lessons before experiment to improve the quality of experiment teaching . *Journal of Shijiazhuang Vocational Technology Institute*, 5(4): 59–60.

Liu Kui, Wang Jianmin and Sun Hua.2010. "Exploration and research on establishing a comprehensive evaluation system of the experimental results". *GAO XIAO SHIYANSHI GONGZUO YANJIU*, 104 (2): 32–33,77.

Shuhong Zhai, Tao Chen, Chunxia Xue. 2013.Thinking about the teaching reform of basic chemical experiment in independent college. *Guangzhou Chemical Industry*, 41(2): 213–214.

Management, Information and Educational Engineering – Liu, Sung & Yao (Eds)
© 2015 Taylor & Francis Group, London, ISBN: 978-1-138-02728-2

Cultivation of students' comprehensive quality during teaching of fundamental chemistry experiment course in independent college

Zheng Ping Chen & Bo Quan Jiang
Department of Biology and Chemistry, Nanchang University College of Science and Technology, Nanchang, P.R.China

ABSTRACT: To cultivate the students' good comprehensive qualities during teaching of fundamental chemistry experiment course is an important task to realize the total training goal of independent college. This paper presents several measures to achieve the target which involves cultivating the students' creative thinking and innovation ability, cultivating the students' style of theory with practice, cultivating the students' good basic chemical experiment skills and cultivating the students' good experimental habits including setting up the consciousnesses of safety, environmental protection, discipline, public property and teamwork. The results showed that the implement of practicing the measures above made the students get great progress in creative thinking, combination of theory with practice, fundamental chemical experiment skills and experimental habits.

KEYWORDS: Fundamental chemistry experiments, comprehensive quality, cultivation, independent college, reform.

1 INTRODUCTION

Fundamental chemistry experiment course (FCEC) is one of the most important fundamental experiment courses in independent college. Nanchang University College of Science and Technology was established in 2001. Two FCECs of "university chemistry experiment" and "organic chemistry experiment "are opened to seven non-chemistry specialties: pharmaceutical engineering, biological technology, biological engineering, polymer materials and engineering, environmental engineering, water supply and drainage, applied chemistry in our college. FCEC is a compulsory basic course of chemistry and chemical specialties. Learning FCEC can make students master the skills of basic chemical experiment, the preparation and purification of inorganic substances, the preparation and separation of organic substances, the properties characterization of the compound, the measurement of chemical reaction velocity constant, et al. In recent years, with the rapid development of teaching and scientific research of independent college in China [Chunxiang Zhang,2014 and Yaping Yang,2012], the reform of FCEC has being smoothly carried out in our college. We pay more attention to the studies on cultivation of students' comprehensive qualities during teaching of FCEC. The comprehensive qualities mainly include creative thinking and innovation ability, style of theory with practice, basic experimental operation skills and experiment habits.

2 CULTIVATING STUDENTS' CREATIVE THINKING AND INNOVATION ABILITY

Innovation is the soul of a nation's progress and the power for the prosperity of a country. University is the cradle of cultivating innovative talents and cultivating the students' innovation ability is a major issue for each higher education workers to face. The characteristics of creative thinking mainly embodied in the following three aspects. The first one is novelty:innovative thinking is a kind of special way of thinking to obtain new ideas, new inventions and new breakthrough. The second one is uniqueness: the uniqueness of the innovative thinking is that it can unique ideas and vision to reflect outreach and pioneering in certain range. The third one is diversity: the diversity means the thinking is made from different angles to obtain variety of ideas and solutions for expanding choice and finally finding optimal answer. Cultivating the students' creative thinking and innovation ability is the goal of independent college. In order to cultivate creative talents, some of important measures have been taken in our college, such as setting up the advanced education idea, establishing a good experimental teaching team, building open teaching mode of basic chemistry experiment course [Boquain Jiang, Tng Liu, 2014 and Boquan Jiang Zhengping Chen, 2014], establishing scientific comprehensive evaluation system [Chen Hongyan, 2012 and Liu Kui,2010] and establishing perfect operational mechanism and management mode, during which, the realization of open teaching mode (OTM) of FCEC is a most important way to achieve this target. The OTM requires the students to independently search literature, design experimental schemes, build experimental devices, conduct operations, treat experimental data, calculate target values, write experimental reports and report to the teachers by PPT files.During the scheme design and doing experiments, when some of students have ideas different from what the textbooks and teachers said, the teachers always carefully listen to them and agree them to do the design and experiments by their own ways if their ideas are theoretically reasonable and practically feasible. The OTM realized the shift from the " teachers' teaching as the center" to the " students' autonomous learning as the center", increased the students' self-consciousnesses and initiatives of learning FCEC, improved the students' abilities of independently analyzing and solving practical problems and cultivated the students' creative consciousness and innovation ability.

3 CULTIVATING STUDENTS' STYLE OF THEORY WITH PRACTICE

The people's practice is different from the instinct activities of animals. Practice needs the guide of knowledge or theory. The practice without theoretical instruction is a blind practice. With the rapid development of modern science and technology, the roles of guidance, forecasting and promotion of theory to practice becomes more and more important. Scientific theory comes from practice and also accepts checking by practice and , in turn, guides practice. We are sure, on the one hand, that scientific experiment plays an important fundamental role on developing scientific theory, on the other hand, we can not ignore the important guiding role of scientific theory on scientific experiments because designing, conducting, analyzing, summarizing and explaining a scientific experiment can not leave the guidance of scientific theory. Scientific theory provides the principle of scientific experiment and guides scientific experiment. It is very clear from the description above, understanding and following the independent and dialectical relationship between theory and experiment is very important to cultivate a good study style of combining theory with experiment in the teaching of FCEC. In the college, both the theoretical and experimental courses of FCE are opened to the students at the same semester. The theoretical and experimental courses of university chemistry are separately opened to the students at the first semester and the chemistry theoretical and experimental courses of organic chemistry are opened to the students at the second semester, respectively. In order to ask the students to use the theory to guide their experiments, we require them to seriously learn theory and give them key interpretation of the contents corresponding to the experiments. Before experiments, we ask the student to make good previews for theoretically understanding the principle, purpose, thinking questions and operation steps of each experiment project. During experiments, we require the students to use theory to guide each experiment, for example, the use of the theoretical knowledge of quantitative in university chemistry experiments course, such as " calibration of hydrochloric acid concentration", " preparation and calibration of EDTA standard solution" and the other experiments related to quantitative analysis and the use of the theoretical knowledge of classifications of organic compounds, organic reactions and organic reaction mechanism in the organic chemistry experiments, such as "synthesis of diphenyl carbinol", " preparation of ethyl acetate system" and "synthesis of amino acid". For improving the students' understanding of the principles, purpose and thinking questions of each project, the teaches compiled the "collected questions and answers", which is uploaded on the network teaching platform of the college for the students to download to study before experiments.

4 CULTIVATING STUDENTS' GOOD BASIC CHEMICAL EXPERIMENTAL SKILLS

To learn FCEC well has an important role for the students to study subsequent experiment courses, such as biochemistry experiment, biotechnology experiment, polymer chemistry experiment and professional experiment courses. Therefore, skills training is the foundation, standard operation is the key and ability training is the target in FCEC teaching.FCEC teaching must focus on cultivating the students' rigorous study and precise scientific style and the students should firmly establish the concept of "quality is the first" and the concept should run throughout the whole experimental process. It should be clearly recognized that any error produced in each operation step in experiment will influence the final experimental results or product quality, for example, the non standard operation may result in an incorrect judgment of titration end point which makes a wrong concentration of the calibrated standard solution during the acid-base titration step, the non-standard operation may cause product loss during the vacuum or suction filtration, the non-standard operation may not result in expected experiment phenomena during the properties experiments, which causes a wrong judge and leads to a failed experiment. Thus, we ask the students to complete each step with diligence and attention according to the standard operating procedures. The teachers are required to strictly check the experiment scene and do not allow the students to arbitrarily temper with and make up their experimental data. During doing experiments, when the teachers find some of students who do not do the standard operations, they immediately go to the students to explain the operational principles, demonstrate the operations and correct the students' wrong operations until the students grasp the standard operation skills. For a few of students who made failed experiments, the teachers patiently help them to analyze why they failed in experiments and arrange suitable time for them to redo the experiments. When finding some of students who do not do their works with rapt attention, the teachers timely point out and ask them to focus on the experiments.We have understood deeply that the students are able to skillfully master experimental skills as long as they set up a correct learning attitude, form a good style of study and adhere to the principle of seeking truth from facts. The practice proved that the strict training in chemical experiments cultivated the students' rigorous style of study and meticulous spirit, which greatly improved the students' chemistry basic operation skills.

5 CULTIVATING STUDENTS'EXPERIMENTAL HABITS

During the teaching of FCEC, we not only pay more attention to cultivating the students' creative ability, style of theory with practice and good basic chemical experimental skills, but also focus on cultivating the students' good experimental habits. The experimental habits mainly include the consciousness of safety, environment protection, discipline, public property and teamwork.

5.1 *Safety consciousness*

It is important to cultivate the students' good safety consciousness for ensuring the experiments be smoothly conducted. In the first class of the FCEC, we give the students a lecture on safety and teach them to understand the importance of safety and to know the reasons of experimental accidents and how to prevent accidents.We require the students to strictly obey the rules and regulations related to safety. The safety consciousness is taken as one of the experiment examination contents.

5.2 *Environmental protection consciousness*

Environmental protection consciousness refers to the consciousnesses and sensitives of the people to treat the entire environment and related problems. During the experiments, we require the students to set up strong environmental protection awareness and ask them not to throw waste dump and waste liquid at random, to keep the laboratory clean and hygiene during experiments and to arrange the students on duty to clean the laboratory after experiments. The students' environmental production consciousness is taken as one of the experimental examination contents.

5.3 *Discipline consciousness*

To cultivate the students' good discipline consciousness is an important factor to keep the experiments be smoothly carried out. We strongly require the students to strictly obey the laboratory rules and regulations and they are not allowed to be late and leave early, play mobile phones, smoke, wear vests and slippers and have any behaviors of disciplinary violations in laboratory. The students' discipline consciousnesses is taken as one of the experimental examination contents.

5.4 *Public property consciousness*

To take good care of public property is the moral standard of social life, it is the base to safeguard social public life. Protecting the experimental instruments and equipment is the important guarantee to make the experiments be smoothly conducted. We require the students to understand the importance of cherishing public property and consciously foster the students' good quality of care for public property. During the experiments, the students are asked to do their experiments with carefulness and to operate each step according to the standard operation.The regulation on "Laboratory equipment provisions for compensation" was made for the students to obey.

5.5 *Teamwork consciousness*

Teamwork provides a chance to cooperate with classmates, it will make a friendly and enjoyable work environment, which is an important factor to make the experiments smoothly conducted. On the one hand, we ask the students to independently finish their experiments, on the other hand, we also require them to carry forward the spirit of teamwork. During the experiments, the students are asked to help each other. The students discuss together .when they meet some of difficult problems. The boys help the girls when a heavy work needs to be done. Some of students initiative to correct the wrong operation done by the others. Every body set up strong safety and environmental protection consciousness, which ensured the experiments of all class were smoothly carried out.

6 CONCLUSIONS

Four measures of cultivating the students' comprehensive qualities were proposed and practiced during the teaching of FCEC. The implementation of these measures cultivated the students' creative thinking and innovation ability, style of theory with practice, standard basic experimental operation skills and good experimental habits. However, the teaching reform is a long process, we will continue to adhere to the reform of FCEC teaching and to make deeper studies and explorations for constantly improving the comprehensive qualities of the students in our college.

ACKNOWLEDGEMENT

This subject comes from "the Project on Demonstration Laboratory Construction" financially supported by Nanchang University College of Science and Technology.

REFERENCES

Boquan Jiang, Ting Liu, Min Xu, et al. Establishment of Comprehensive Evaluation System of Fundamental Chemistry Experiment Course. 2014. *ICEMCT2014* Xiamen, China),447–450.

Boquan Jiang, Zhengping Chen, Lijuan Wan, et al. Reform of Basic Chemistry Experiment Course for Cultivating Creative Talents. 2014. *ICEMCT2014,* Xiamen, (China, 443–446.

Chen Hongyan ,Yang Jinzhao. 2012. An exploration in Open Teaching of the Basic Chemistry Experiment. *The Chinese modern education equipment*, (7):66–68.

Chunxiang Zhang, Qingyun Zheng, Xiangyang Zhang, et al. 2014. Research on the Teaching Reform of Basic Chemical Experimental for non-Chemistry Specialty in Independent College. *Guangzhou Chemical Industry*, 42(6): 159–160.

Liu Kui, Wang Jianmin and Sun Hua. 2010. "Exploration and research on establishing a comprehensive evaluation system of the experimental results". *GAO XIAO SHIYANSHI GONGZUO YANJIU*, 104 (2): 32–33,77.

YapingYang, ChuneLu. 2012. Reform and exploration of Basic Chemical Experiments Teaching in Independent College. *Science and TechnologyInformation*,(7):285–286.

Management, Information and Educational Engineering – Liu, Sung & Yao (Eds)

Exploration of physical education teaching reform from the perspective of multimedia

Ning Qi
Hebei University, Baoding, P. R. China

Le Wang
Hebei Finance University, Baoding, P. R. China

ABSTRACT: With the increasing deepening of the reform of the educational system, the physical education also constantly breaks through the previous teaching form, and fully implements the spirit of teaching reform. The traditional teaching content, teaching mode, and teaching idea has far from satisfied the social demand for high-quality talent, and in order to develop comprehensive and new types of talents, it needs to start from the common problems existing in the physical education class, and use the information-based teaching means and methods to improve the quality of teaching, improve students' physical fitness and sports skills, and make PE course play a greater role.

KEY WORDS: Physical education; Multimedia technology; Reform exploration.

1 INTRODUCTION

Physical education is an important part of the educational work in China. In recent years, due to the increasing concern about the health of people, the education department has paid high attention to the school activities of sports education, which has started from the student's physic fitness and provided much more comprehensive talent for the society. This paper mainly explores the current sports teaching mode, analyses the application value and effect of the multimedia technology in the teaching process, searches for ways to improve the teaching quality, and promotes the career of physical education to develop towards the direction of modernization.

2 THE APPLICATION EFFECT OF THE MULTIMEDIA TECHNOLOGY IN PHYSICAL EDUCATION TEACHING

2.1 *The rationality of multimedia technology being applied to physical education teaching*

First of all, the textbook content about sports is generally more machinery, so the students' attention to the knowledge of the textbook is generally low, which has brought a huge challenge to the teaching work. of sports teachers. Multimedia technology can simplify the knowledge on the textbook to make the professional boring language become intuitive and easy to understand. In the section of teaching martial arts, for example, the descriptions on palm, boxing, claw, hook, elbow in the textbook are relatively inane, so that students are difficult to grasp the essential of these actions. Producing them into animation effects of multimedia to make the students feel and imitate the operating essentials and action effect of characters in video, can make students better understand the theoretical knowledge. Secondly, it improved the teachers' teaching quality and work efficiency. Multimedia not only has improved students' learning ability of the sports knowledge, but also has become an important auxiliary tool for teacher's teaching. In the traditional teaching, physical education teachers mainly depend on the teaching experience to teach students the knowledge, skills, by which the content of teaching updates slowly, and the students' learning enthusiasm is not high. After the introduction of multimedia technology, the teacher can collect the outstanding coursework in the career of physical education online which has broadened the students' horizons, enriched the sports teaching content, and greatly enhanced the interest in the physical learning.

2.2 *The problems of multimedia technology existing in the application of the physical education teaching*

Although multimedia technology has gotten the favor of most students and PE teachers, there are still some problems existing in the application process which mainly show in the following aspects. First, the operation proficiency of the physical education teachers for the multimedia remains to be further improved. Physical education teachers generally tend to do outdoor sports; but do not pay much attention to the theoretical knowledge. Influenced by the traditional teaching concept, most PE teachers stick to the traditional demonstration teaching mode, but have not enough ability to accept the advanced technology, and even appear the phenomenon of exclusion, all of which reduce the teaching effect. Second, they did not understand the importance of the multimedia teaching methods. In the past, for quite a long time, the teacher had been the main body of teaching. The teaching content was set by the teacher, according to the arrangement of the outline, and students were in the position to accept passively. Some teachers think that the multimedia teaching makes the students master a lot of information so that distract their attention to learn sports skills and then influence the learning effect. Third, the application of the multimedia technology is too formal. Except for showing the textbook knowledge, vividly, multimedia can also increase the classroom atmosphere, which is the important way to attract students' attention. However, in practice, most teachers did not make full use of the computer network resources. They only turned the original content of the lesson plan into coursework, and copied the original teaching mode mechanically, which lost the meaning of the reform, so that students' learning demands were frustrated, learning motivation declined, and the teaching effect was unsatisfactory.

3 THE WAYS TO IMPROVE THE EFFECT OF PHYSICAL EDUCATION TEACHING REFORM

3.1 *Improving the teaching level of PE teachers*

The learning ability of students is often influenced by the teachers' teaching level, so improving the teachers' ability to rule the multimedia technology is an important way to accomplish the teaching goals of reform. The advanced teaching means are the basis to do the education work well, and the physical education teachers, should rapidly change their teaching idea and deeply understand the benefits of multimedia teaching. Next, it should correctly grasp the students' learning psychology, ask for opinions from the students, and promote the multimedia technology. Both students' curiosity about the multimedia technology and the desire to understand actively provided favorable conditions for the sports knowledge learning. Teachers should strengthen the ability to rule the students' psychology, seize the students' interest, use the advanced teaching means, and guide the students to discover problems independently. Finally, it should strengthen the teachers' ability to relearn. On the basis of deepening the existing knowledge, it should increase to absorb the elements of modern sports, keep the theory viewpoint and attitude of keeping pace with the times, understand the demands of modern students, and reduce the distance with the students. Schools should form an atmosphere of teaching new mode, and stimulate and inspire teachers' learning desire of multimedia technology through the teaching appraisal activities among teachers in different disciplines. It also should train the teachers in the school regularly, adapt the way that teachers attend lectures each other to complement each other and draw lessons from excellent teachers' teaching experience to achieve the purpose of improving the sports teachers' teaching ability and professional quality.

3.2 *Combining theory with practice, and focusing on physical exercise*

Unlike other disciplines, PE course has a strong practicality, and only after students actually practice the knowledge they mastered, they can we truly master the knowledge. In order to achieve a better studio effect, teachers must apply theory to practice, and increase the students' understanding of the content about the textbook. Teachers should divide the physical education teaching work into two parts according to the strong practicality, namely, the theory teaching and actual practice. Theory teaching should give full play to the role of multimedia technology in order to improve student's ability to master the sports knowledge. The range involved in the multimedia teaching is wider, so when choosing the materials, teachers should combine with the actual condition and choose the teaching case with a moderate degree of difficulty to improve the reliability. For example, teachers show students the punches routines and implementation effect of the five-step boxing through the multimedia, analyze the essential of every action step by step to let students feel and experience the strength and standard of punches, then play the whole set of boxing teaching video to make students clearly understand the ultimate learning purpose, and finally, make students transform the moves in their mind into practice. Due to having a preliminary understanding of the movement technology, the learning speed of students accelerates obviously, which not only saves teachers' teaching energy, but also improves the teaching efficiency to a great extent. For some special sections, teachers should also transform the teaching material content flexibly and create favorable conditions for students' learning and practicing activities. For example, when teaching the whip technique, teachers use the jump rope, the commonly used equipment activities, rather than a whip to create conditions for students to practice after class at any time.

3.3 *Innovating the teaching mode and the cultivating students' ability to use knowledge*

Interactive teaching is an effective way to improve students' participating enthusiasm, and it helps check and improve students' ability to master. Compared with the traditional teaching, the modern teaching methods pay more attention to the communication and exchanges between teachers and students. Every student has different sports ability and different degree and level of interest in sports. Therefore, in the process of teaching, it should vary from person to person, that is, teachers can appoint the students with standard actions as his assistant, which, on the one hand, can guarantee the practice progress, and on the other hand, can also have plenty of time to coach students whose practice level is relatively backward. The ultimate goal of teaching reform and the introduction of multimedia teaching means is to improve students' ability to observe and to apply the knowledge they have already mastered. Student have strong curiosity about new things, so teachers should take the psychological characteristics of students and adopt innovative teaching methods to attract their attention. Multimedia teaching show to students a more widely content and more nuanced interpretation of the knowledge points. In a sense, there are common characteristics for sports skills. For example, in the process of practicing a set of boxing, some movements and actions such as some step type, the shape of the hand, the direction of jumping and running can be used in the practice activities in other chapters. When learning, students should study deeply into the content and interpretation skill presented in the multimedia classroom, and in the process of practical training, they should pay attention to observe the teachers' demonstration actions, seeing the route from the dynamic movement, seeing the cooperation from the overall observation, seeing the posture from the static movement and summarizing the practice law to achieve the goal of use the knowledge flexibly.

4 CONCLUSION

Under the background of information, the PE teaching is no longer limited to the mechanized preaching, teaching and solving doubts, but more is to let students experience the joy of learning. At the same time of creating a relaxed learning environment for students, teachers achieve their goal of education.

REFERENCES

[1] Wang Yangfan: A Study on the Application of Multimedia Technology in the Colleges and Universities Physical Education. Sports World Vol. 02 (2014), pp. 134–135.

[2] Chen Aiping: Discussion about the Application of Multimedia in the Primary School Physical Education. Contemporary Sports Technology Vol. 33 (2012), pp. 53–54.

[3] Lv Weihua: Analysis about the Application of Multimedia Technology in Sports Teaching in China. New Technologies and Products Vol. 22 (2011), p. 244.

Management, Information and Educational Engineering – Liu, Sung & Yao (Eds)
 ISBN: 978-1-138-02728-2

Study of the application of network resources in English reading teaching

Hai Ying Jin
Teachers' College, Beihua University, China

ABSTRACT: With the population and development of information network technology, network resources are widely used in English reading teaching. Abundant resources on the Internet providc extensive learning resources for college students. Network resources are playing an important role in English reading teaching. This paper analyzes the application of network resources in English reading teaching.

KEYWORDS: Network resources; English Teaching; Reading teaching.

Since the beginning of the new century, English teaching has become the major course and reading teaching is the leading part of pro teaching. Reading is the integrated process of understanding a text and gaining information. Improving the ability of English reading is the main approach to master the new knowledge. Reading teaching has become the leading part in college English teaching. Bringing in network resources, teaching mode in English Teaching can effectively improve students' abilities of listening, speaking, reading and writing to improve students' English level.

1 THE STATUS QUO OF COLLEGE STUDENTS' ENGLISH READING TEACHING

The main English reading teaching methods in our colleges and universities are reading and writing. College English reading materials and teaching methods are relatively old-fashioned, which is not conducive to improve students' English reading level. College students' English courses are not much, their language foundation is weak, the teachers' explanations of English reading are not too thorough and students' lack of out-class reading training, which results in students' quite narrow reading surface and is difficult to relate English reading skills with language knowledge and results in low levels of the students' English reading. Based on this situation, college students' English reading, teaching must improve teaching mode and give up the traditional way of teaching. Use modernizing teaching methods to create new teaching ways and methods to promote the improvement of the students' reading level. In recent years, the development of network information resources has a great role in promoting English teaching and effectively improve students' English reading level.

2 INTRODUCE NETWORK RESOURCE ADVANTAGES IN ENGLISH READING TEACHING

2.1 *Abundant reading resources*

Network resources have no limitation of time and space and a variety of features, which contains resource information almost all over the world. From the formal point of view, it includes newspapers, magazines, news reports, pictures, charts and original books, etc. From the content point of view, it includes the knowledge of culture, history, geography, biology and sports and entertainment from the world. In English reading teaching process, you can use these resources to stimulate students' interest in reading to improve their reading level.

2.2 *The sharing of reading resources*

Internet IT can make educational resources shared on the Internet platform. English reading teaching can be shared and exchanged in the world by Internet and multimedia courseware, for example, some large sites provide bilingual materials and translation for students' searching. Using network resources for teaching can not only enrich the teaching content, but also enrich teaching methods, which makes teaching materials no longer limit to the traditional mode and can use the Internet, multimedia courseware and other forms of network resources to access a variety of reading information.

2.3 *The timeliness of reading materials*

Traditional English reading materials tend to be rigid, outdated, which is difficult to arouse students' interest in reading. While network resources content are continuously updated, which is a dynamic information system and information resources have aging characteristics, so reading materials on the Internet are the information closely related to today's facts so that students can find some strong timeliness reading material to learn, expand their knowledge, while increasing social knowledge, broaden cultural awareness, which can effectively cultivate students' cultural communicative competence.

2.4 *Relatively strong, independent learning*

Learning English reads through the information network technology can effectively implement student-centered self-learning mode. Targeted network courseware designation can meet the learning needs of all levels of learners of English reading. Extensive network resources can provide students the freedom to learn and students can choose their own reading resources according to their levels and conditions to improve students' reading motivation and interests in reading.

3 THE SPECIFIC APPLICATIONS OF NETWORK RESOURCES IN ENGLISH READING TEACHING

3.1 *Interactive teaching of materials and network resources*

English reading, teaching content is widely and students can easily produce boring, sensible in learning, thus reducing their interests in learning. While the timeliness characteristics of network resources can meet students' thirst for knowledge, network resource content is rich and has times flavor. Teachers should base on textbooks content when having lessons, on which basis, using network resource information expand their reading knowledge and effectively relate English textbooks with network resources to enrich students' knowledge and improve students' reading interest, for example, when students learn the New Edit English Language New *Unit one Personality*, teachers can allow students search for *the relationship between the color and your personality* and describe *personality* they consult and understand in detail to introduce the lesson learning. At the end of the course, teachers can also add some extra reading classes for students searching and learning under class to expand their horizons and develop students' interest in reading.

3.2 *Actively develop network reading*

Students' using network resources to improve their English reading level also needs active network read in the classroom. First, teachers can allow students to recommend their favorite e-books to the whole class students and say the reason it is recommended to adhere to the students' good habit of reading in English. In addition, teacher in the classroom enhance the guidance of the text reading and recommend some suitable level, high-quality reading materials to students, such as outstanding journal-Reader's Digest, excellent English website Times and so on. The recommendation should base on students' interests and their knowledge level to comply with individualized features to maximize the students' English reading level.

3.3 *Create English reading scenarios and interactive learning mode*

Learning English reading teacher in the classroom alone explain the content is not enough, a sense of language learning, language depends mainly on reasonable scenarios set up to improve students' English reading level is an effective way. In college English reading, teaching, teachers should make full use of coursework and extensive network of resources, design all kinds of contextual dynamic interactive activities. In the network environment such interactions mainly refers to the process of dynamic and static exchange student with a computer, students and students, between students and teachers, such as through the mutual interaction between them, so that students learn more knowledge of English. For example, discuss and exchange English national customs, cultural history, geography, and other topics about online establish English in the classroom through network resource and input them in an objective, true and natural language, so that students can grasp English charm in a real or virtual reality learning environment and cultivate interest in learning the language.

4 CONCLUSION

Network information resource like a boundless sea, where students can find their own way of learning methods. In the network environment, students' English reading, teaching must adhere to make the textbook as a foothold and network resources as a major expansion form to improve and deepen English reading to improve students' English reading level.

REFERENCES

[1] Chang Shixuan. *Make Full Use of Network Resources to Improve College English Reading Teaching* [J] Intellect. 2012 (23): 126–128.

[2] Tan Yan. *The Application of Network Resources in English Reading Teaching* [J] Examination Week 2010 (24): 120–121.

[3] Li Gaoyang. *Discussion and Analysis of Application of English Network Resources in College English Teaching* [J]. Era Education 2013 (5): 89–90.

[4] Shi Hongtao. [J] Study of network resources based college English reading teaching model. Science and Educational Wenhui 2013 (29): 87–88.

Management, Information and Educational Engineering – Liu, Sung & Yao (Eds)
© 2015 Taylor & Francis Group, London, ISBN: 978-1-138-02728-2

The application of multimedia technology in college English teaching

Nan Yang
WeiFang University of Science and Technology, Shou Guang, China

ABSTRACT: Multimedia technology plays an important role in college English teaching. It has changed the disadvantages of traditional teaching, but there are advantages and disadvantages. Obvious advantages are: it can give full play to students' highest body function, improve the effect of learning and ability cultivation, and solve the contradiction in the class. At the same time, it emphasizes the use of multimedia and despise in teaching, and the phenomenon such as the excessive use of multimedia and being turned into a demonstration class of courseware and the disadvantages of multimedia English teaching. The application of multimedia in college English teaching, makes the form of teaching and learning mode a diversification, improves the learning efficiency, and promotes the college students' English thinking ability.

KEYWORDS: Multimedia technology; College English teaching; Application; The pros and cons.

1 INTRODUCTION

With the development of the society, modern teaching technology facilities perfect gradually, applying multimedia teaching has been loved by many teachers, everybody wants to the means of information technology to improve the teaching of the deficiencies. The use of multimedia cannot only stimulate students' interest in learning, the more that they can truly become the main body of learning, change passive learning into active learning. With the deepening of quality education, college English teaching has shifted from teacher-centered to student-centered, teaching means and on teachers' teaching primarily to big to use multimedia technology, which makes classroom teaching more vivid and specific image. The use of multimedia to create good English communication environment, improve the students' interest in learning English, expand students' thinking space, greatly improving the efficiency of classroom teaching.

2 THE MEANING OF THE MULTIMEDIA TECHNOLOGY IN COLLEGE ENGLISH TEACHING

The application of multimedia technology in education teaching is the most promising and exciting development field. The progress of the society, the rapid increase of science and technology, knowledge and the expansion of population, be badly in need of training, can adapt to the times of personnel. Use of multimedia technology has high integration, good interactivity, large information capacity and feedback in time, will be a variety of information at the same time or alternating role of the learners' senses, fundamentally changed the shortcomings of traditional teaching, to make learning more fun, nature, and human nature. The significance of the application of multimedia technology in education teaching both deep and far.

1. Multimedia teaching of college English teaching content changed from abstract to intuitive, convenient for observation and understanding, facilitates learning and mastering the teaching material.
2. Change number of short, hard, and improve the teaching speed, saving, reduced the labor of the teacher.
3. With the advance of science and technology, multimedia technology combined with network technology, combined with simulation technology, combined with artificial technology and other high-tech application in the teaching of college English and all disciplines that will bring about a new revolution of human education career.

3 ADVANTAGES AND DISADVANTAGES OF MULTIMEDIA TECHNOLOGY IN COLLEGE ENGLISH TEACHING

3.1 *The advantages of multimedia technology in college English teaching*

3.1.1 *The application of multimedia technology can give full play to student's main body role*

Advocates of the subjectivity education, is the reality of the socialist modernization construction need, also are the trend of education reform in the world.

Subjectivity schooling process, it is the teacher guided students, independent learning and independent activity process. According to cognitive learning theory point of view, and must give full play to the initiative and enthusiasm of the students, students can get effective cognitive.

3.1.2 *Application of multimedia technology teaching composite to improve the effect of learning and training ability*

If let the student hear and see, through discussion, exchange, use your own words, keep knowledge will be greatly superior to the traditional teaching effect. Application of multimedia courseware teaching, can effectively stimulate students' interest in learning, make students produce strong desire to learn, to form a learning motivation, active participation in the teaching process, giving much information to the classroom, the students easy to accept, in a happy atmosphere, interactive discussion mastered the emphases and difficulties of teaching, the teaching effect is quite obvious.

3.1.3 *The application of multimedia technology can be effectively solve the contradiction in the class*

Computer multimedia technology, through the text, graphics, images, audio, animation and interactive network, can make the teaching process is illustrated, lively, more broad knowledge. The students go to the dynamic environment, the teaching content easier to understand and master, can greatly accelerate the learning process, improve the learning efficiency. More importantly, because the speeding up of the teaching development, in virtually increased the learning time for students. In the physics classroom teaching, it is hard to throw a lot of content that can solve a problem, such as satellite, the speed of the universe and the planets such as the movement of the sun, linear motion and force analysis, etc., if use the traditional teaching methods, even speak a lesson, students can also be difficult to understand, but will only deepen students fear of physics learning. This knowledge can use multimedia technology to let everybody see an intuitive situation of simulation, which not only can be content to be lively, stimulate students' interest in learning, can also increase the students' knowledge.

3.2 *The disadvantages of multimedia technology in college English teaching*

3.2.1 *Emphasizes the use of multimedia and content in teaching*

Some teachers at first tried to use multimedia auxiliary teaching of the computer, then his life, but only from, who has lost his formation of teaching style for many years, but don't know the computer multimedia course, there are other media incomparable superiority, but the other media and teaching means many of the features is that it cannot completely replace, such as physical, stick-figure function can not be ignored in the classroom. Therefore, teachers should not act blindly, but should be in accordance with the need to choose the appropriate teaching media and means. Reasonable use of multimedia and traditional media and means, play to their strengths. According to the actual students, teaching content, practices, properly chooses the media forms.

3.2.2 *Excessive use of multimedia and being turned into a demonstration class of courseware*

Reasonable use of computer multimedia auxiliary teaching can bring on the efficiency of English classroom teaching "geometry" leap. It is proved by the practice of the countess. However, deliberately pursue modernization, excessive use of multimedia, English class in a demonstration class of multi-media coursework, teachers became the projectionist, will bring the consequences of too much of a good thing.

4 THE APPLICATION OF MULTIMEDIA TECHNOLOGY IN COLLEGE ENGLISH TEACHING

The application of multimedia, it inspired the students' enthusiasm in learning English, at the same time makes it easier to college students' understanding of English. Still can effectively enhance the teaching effect and improve teaching efficiency, optimize the teaching process, ensure the realization of course objectives. The application of multimedia coursework in English teaching, bring a qualitative leap for English teaching, it is through the use of computer to text, images, sound, animation, information processing, such as sound, like, figure, and teaching system, to see, hear, touch, want to visualize the teaching a variety of ways. Broke the traditional teaching model, accords with the cognitive law of students, and promote the development of students' thinking in multidirectional.

4.1 *Multimedia teaching to promote college students' English thinking ability*

Thinking and language are inseparable, people's thinking is carried out through language, modern teaching technology to develop the students' ability of thinking in English has a great role in promoting, for intuitive in college English teaching has created a very good learning environment and conditions. Due to use all kinds of modern technology to create, to learn English can be ruled out as far as possible in the classroom in Chinese interpretation of the interference, which is beneficial to train students to think in English sound study habits.

4.2 *Multimedia teaching to college English teaching form of diversification*

The best English teaching the student can see only the teacher, teaching AIDS, a blackboard and a few simple implementations of the "a blackboard and a piece of chalk, a mouth and listen to" teacher infusion gives priority to the traditional teaching methods. Teachers' teaching on just to read, listen to the tape, such as dictation, recitation, check a series of links, make learning tool language in communication, people become really boring. Now, we can use multimedia to the modern teaching means, make the teaching of diversification, get rid of the same.

4.3 *The diversity of multimedia teaching, let the students' English learning*

Want to ask what's the hardest part of learning English? Is remembering words, this is the first big difficulty to learn English. Use of multimedia, can make the difficulty somewhat easier. As we know, recite to eyes, mouth, ears, hands, such as joint training, will reach the appropriate effect to recite. Only don't write back, back not only, only back not to see is to reach good effect, and make students more such problem of multimedia solved this problem for us, and the sound, image, animation, etc., make the word senses fully mobilize students, memory effect will be better. For example: in a physical memory word, you can use some brute image, let the students hear the pronunciation of the new words, both from the cashier and saw the physical, memory, faster, and higher interest.

4.4 *Speed up the pace of multimedia teaching to college english teaching, ensure English learning efficiency is also improved*

The traditional English teaching theory, tend to regard the teaching process as a simple linear one-way process of teachers, students and teachers are the main body of the whole teaching process, is the only source of information, while the student is the object that is a passive information receiver, teachers teaching the basic form is: explanation, practice, questions and feedback, according to the evaluation results is repeat and repeat or the end of the teaching practice. With multimedia, the students in the same computer man-machine communication, can put forward to illuminating, arouse the students to think about desire, constantly thinking, to explore, fast fast rendering, practice, feedback, etc., in this way, speed up the pace of teaching, improve the efficiency.

4.5 *Multimedia teaching to modernize the college English teaching*

In the habitual teaching mode, teaching in teachers' teaching, using chalk and blackboard teaching way. Modern multimedia teaching technology replaced the static, rigid teaching AIDS, and delicate audio-visual with excellent pictures and texts, injected fresh blood into classroom teaching, especially for the student's image thinking plays a unique role of guidance, stimulation.

5 CONCLUSION

Type in a word, English is neither understand course, need to understand; Nor the knowledge courses, to remember, English is a technical course, if you want to learn English well, the key is to practice, practice, and is inseparable from the scene of language practice. So the best use of multimedia technology in college English teaching, will greatly enrich the classroom teaching, as the college students' English learning situation, stimulate the students' interest in learning English, has set up a bridge for the college students' understanding of western culture. The characteristic of teaching of multimedia teaching, with its distinctively rich teaching contents, vivid teaching situation, promote the informatization of education technology, and gradually break "a blackboard, a piece of chalk and a mouth all listen to" infusion of traditional teaching methods of teachers, build up the new teaching mode. So the judicious use of multimedia teaching in college English classroom teaching can promote students' understanding of knowledge, increase classroom capacity, optimize the classroom teaching efficiency, to create an ecological classroom, is an effective auxiliary means of teaching. Let's continue to work hard on the use of multimedia technology in college English classroom teaching, teaching play to their potential, comprehensively improve the level of college English teaching, so that the multimedia technology in college English teaching continuously to make an ecological classroom.

REFERENCES

[1] Junhua Yi. Introduction to education technology. Beijing: higher education press. Beijing, 2009.
[2] Changshun Chen. Multimedia CAI courseware for tutorial. Nanjing: Managing university press. 2006.
[3] Tiequan Cai. Modern education technology curriculum. Beijing: Science press, Beijing. 2005.
[4] Zhengdong Zhang. Foreign language pedagogy. Shanghai: American foreign language education press. 2007.
[5] Ying Zhang. College English teaching. Changchun: the northeast normal university press. 2006.

Management, Information and Educational Engineering – Liu, Sung & Yao (Eds)
© 2015 Taylor & Francis Group, London, ISBN: 978-1-138-02728-2

The timer design basketball competitions

Ji Xiao Sun
Weifang University of Science and Technology, ShanDong ShouGuang, China

ABSTRACT: This design is a simple application of the pulse digital circuit, the design of the basketball competition 12 minutes and 24 seconds and the timer. The timer function is ready, can be directly reset, started, paused, is continuous and has an alarm function and application of the seven segment digital tubes to display the time. This paper presents the main function and application of the timer, studies the working principle of the timer, points out a module for design of the circuit, and finally, the designed circuit has carried on the overall performance of the test.

KEYWORDS: The counter; The timer. The decoding display circuit; 24 seconds.

1 INTRODUCTION

With the advent of the era of information, electronic technology is playing an increasingly important role in social life, the use of mold knowledge electric and electric design of electronic products has become an indispensable part of social life, especially in all kinds of sports, timer test performance athletes become an important tool. In many areas, the timer is widely used, such as in sports, timing alarm, the game's countdown timer, traffic lights, traffic lights, pedestrian lights, also can be used as a variety of pills, tablets, capsules in specified time remind drug use, and so on, there is the visible timer in the present society is how important.

2 THE MAIN FUNCTION AND APPLICATION OF THE TIMER

In the basketball match, the player's ball stipulated time can't more than 24 seconds, or foul. This course design of "basketball competition and 24 seconds timer", can be used in the basketball match, for players all time limit of 24 seconds. Once a player the ball for more than 24 seconds, it automatically alarms to determine the players foul.

24 seconds timer has the following functions: display the countdown function; Reset, start, pause, continuous function; 24 seconds timer for decreasing timing, timing interval to 1 second; Diminishing the timer timing to zero, the immortality of the digital display lamp, and a photoelectric alarm signals, etc. The timer has start, pause and continuous functions, you can easily implement breakpoints timing function, decreasing when the timer to zero, have a photoelectric alarm signals, also has extensive application value in social life.

3 THE WORKING PRINCIPLE OF BASKETBALL COMPETITION TIMER

3.1 *The working principle of description*

24 second timer solution reference block diagrams are shown in figure 1. It includes a second pulse generator, counter, decoding the display circuit, alarm circuit and control circuit module, etc. Which counter and control circuit is the main module of the system. Counter complete the 24 second timer function, and the control circuit to complete counter reset directly, start counting, pause/continuous counting, decoding display circuit without the lamp, timing, time to call the police.

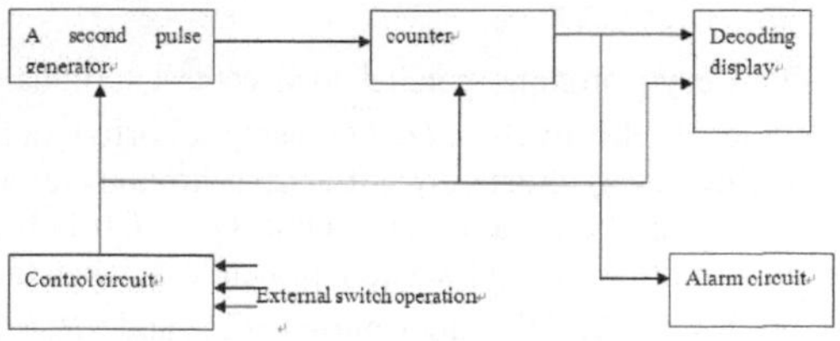

Figure 1. 24 seconds timer system design block diagram.

4 CIRCUIT DESIGN

4.1 *Basketball competitions of the 24 s timer circuit as shown in figure and 74 ls192 irfpa foot figure*

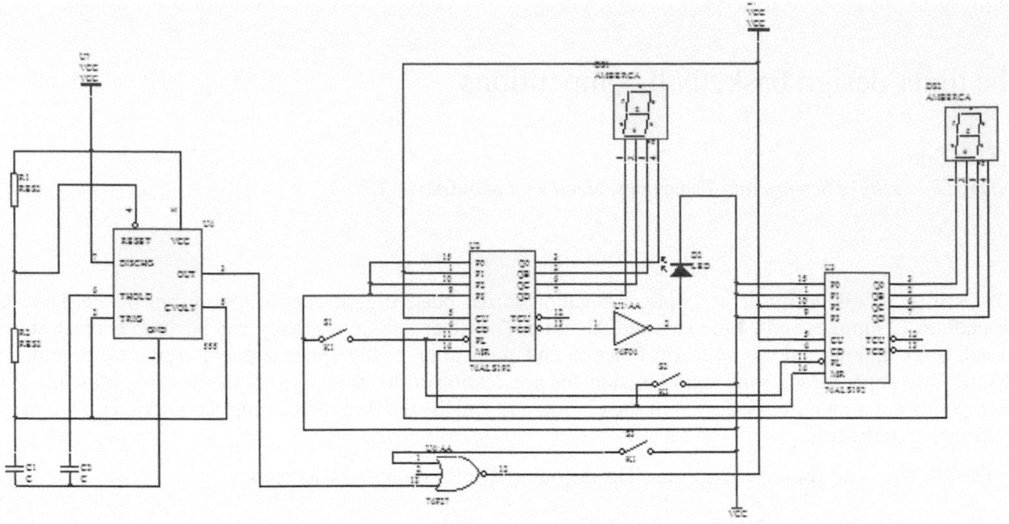

Figure 2. Basketball competition the 24 s timer circuit diagram.

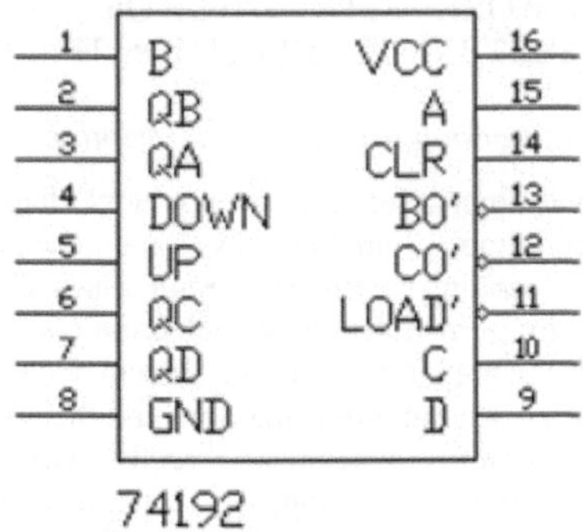

Figure 3. 74 ls192 irfpa foot figure.

$\overline{LD}$ is asynchronous parallel load control terminals (low level) effectively, $\overline{CO}$, $\overline{BO}$, carry, a borrow outputs (low level) effectively, CR is asynchronous reset, D3 – D0 is the parallel data input, Q3 – Q0 is the output terminal. 74192 working principle is: when the $\overline{LD} = 1$, CR = 0, if the clock pulse to CP_U, and $CP_D = 1$ are completed in the counter on the basis of the preset number counting function, when the count to 9 $\overline{CO}$ end sends out carry jump pulse; If the clock pulse to the CP_D end, and $CP_U = 1$, the counter count reduction functions performed on the basis of the preset number, when the count reduction to 0, $\overline{BO}$ end sends out a borrow jump pulse. Composed of 24 74 ls192 hexadecimal decreasing counter design principle is: only when low BO_1 end issued a borrower pulse, high counter to count reduction. When in full zero, high and low counter to 0, and $\overline{LD_2} = 0$ load end, counter parallel load, under the effect of CP_D the input clock pulse, the counter again into the next cycle count.

4.2 *The clock module*

To provide 74 ls192 counter with a temporal pulse signal, reduces the count, this design USES the multiple harmonic oscillation circuit, composed of 555, 555 job characteristics and its calculation formula, the output cycle of the pulse cycle as follows: T = 0.7 (R1 + r2) 2 C. As a result, we can calculate the various parameters through the calculate and determine the R1 take 15 k ohm, R2 take 68 k ohm, capacitance C for 10 uF, C1 0.1 uF,. So we got the relatively unchanging pulse, and the output cycle for 1 second.

4.3 *Auxiliary sequential control module*

Control circuit to complete the following four functions:

1 Directly open the "reset" switch, the counter out the lamp.
2 Closed when the "start" switch, the counter should complete load function, display shows 24 seconds; disconnect "start" switch, the counter to start the countdown.
3 When the "pause/continuous" switch in the "pause" position, the control circuit blocks the clock pulse signal CP, counter stop counting and displays the original number unchanged, "continuous" pause/switch is in "continuous" position, the counter to count.

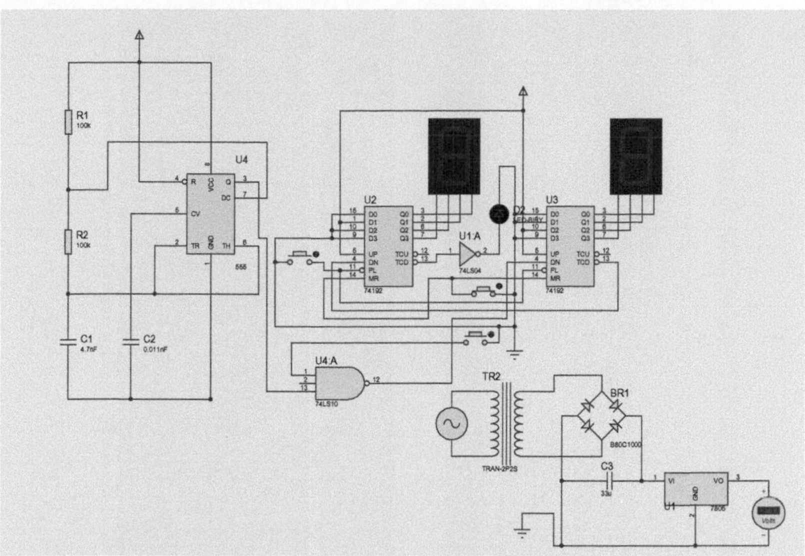

Figure 4. Circuit simulation diagram.

4 When the counter countdown to zero, the control circuit shall be published a warning signal, make counter the zero state unchanged, alarm circuit working at the same time.

When the count to zero, the two counter output a borrow more for low (0), so the design will be high a borrow $\overline{BO_2}$ feedback to the diode negative polarity, + 5 v power supply at this time thlessk resistor to make ledssignaloelectric alarm signals, alarm functions, and when the countdown, high output $\overline{BO_2}$ for (1), diode doesn't call the police.

5 THE WHOLE CIRCUIT PERFORMANCE TEST

1 First realizes basic load than 24 seconds regressive functions. The realization of the function of load is not hard. Will ~ the LOAD low level can be reached. As shown in figure 4. The simulation circuit.
2 Count to zero after stop counting and report to the police.

6 CONCLUSION

This design mainly through the modular thought, and gradually realizes the functional requirement of the design to achieve the required. Clock module provides a frequency of 1 hz to count the pulse signal, so as to realize the counter counting interval to 1 second; Counting, decoding display module is to achieve a subtraction counting functions; Alarm module is to achieve when the count reduction to the zero signal photoelectric alarm; Control module is mainly in order to realize the start of the timer, reset and pause/continuous function directly, including the direct reset, by the control switch control decoder blanking end, thus can realize display decoder out the lamp; Through pause/switch so as to realize continuous breakpoint timer function. At this point, the successful completion of the design.

REFERENCES

[1] ShiBai Tong, Cheng yin Hua g. Analog electronic technology foundation. [M] Beijing: Higher education press, 2006.
[2] Fusheng Dai. Electronic circuit designs and based on practice. [M] Beijing: National defence industry press, 2002.
[3] Boxue Tan. Principle and application of integrated circuits. [M] Beijing: Electronic industry press, 2003.
[4] Manqing Hua. Experiment and course design of electronic technology. [M] Beijing: mechanical industry publishing house, 2005.
[5] Shizhu Yan. Digital electronic technology base. [M] Beijing: Higher education press, 2008.

Management, Information and Educational Engineering – Liu, Sung & Yao (Eds)
© 2015 Taylor & Francis Group, London, ISBN: 978-1-138-02728-2

Analysis of the structure of university ideological and political online education platform

Yang Zhou
Hebei Software Institute, China

ABSTRACT: University ideological and political online education platform is the main object of education, and is an indispensable part of college ideological and political education. With the changing environment in our country, there are big shortages in the current college ideological and political online education platform. This paper first analyzes the current situation of university ideological and political online education platform, and then studies the improvement measures, hoping to provide some help for the relevant personnel.

KEYWORDS: Ideological and political education; online education platform; improvement measures.

1 INTRODUCTION

With the development of computer network technology, university education has a high dependence on the network, online education platform has played an important role in the college ideological and political education, how to build a more effective online ideological and political education platform under the new situation is the focus topic of this study, this paper mainly analyzes the structure of university online ideological and political education platform.

2 THE CONSTRUCTION OF UNIVERSITY IDEOLOGICAL AND POLITICAL ONLINE EDUCATION PLATFORM

University ideological and political online education platform is in line with the requirements of environmental changes, with the rapid development of Internet, the network culture status has been further improved, especially among young people, it is necessary to establish a unified educational platform to strengthen education. Construction of university ideological and political online education platform also complies with the requirements of digital technology, with the rapid development of cloud computing, Internet and mobile technology, the digital technology has become the development direction of information technology, education reform also needs to follow the direction of development.

Construction of university ideological and political online education platform is the optimal allocation of resources to achieve the ideological and political education requirements, open university resources can bring positive influence of social values and so on. In the ideological and political education work, theory and practice are important to the process of education, under the new situation of various cultural trends widely disseminated, have a great impact in education, the Internet itself is not self-cleaning function, the construction of university ideological and political education network platform is very necessary for implementing ideological and political education for college students.

3 PROBLEMS IN THE CONSTRUCTION OF UNIVERSITY IDEOLOGICAL AND POLITICAL ONLINE EDUCATION PLATFORM

The main body of university ideological and political education network platform is a network platform, the current development of university ideological and political online education platform is extremely uneven, there are many problems.

3.1 *Inadequate network infrastructure*

Most of the construction development of university ideological and political education network platform is relatively late, a lot of network equipment are relatively backward, affecting the stability of transport deformation of the network.

Due to local causes of uneven economic development of universities, resulting in time and scale of development of university ideological and political education network platform there are also very different, generally in the construction of the university's network infrastructure set up, the science and engineering construction investment is much higher also in agriculture, forestry, etc., ideological and political education network platform construction investment undergraduate colleges is higher than vocational colleges. In network security platform construction, initial construction for safety without much consideration, resulting in the presence of a lot of insecurity in network platform construction, many students were once hacked and infected with Trojan, etc..

3.2 *Inadequate management manners*

Currently the vast majority of university ideological and political education network platform management use rules and regulations, management personnel arrangements to start perfusion work. This management model is not suitable for the Internet era. With the network uses the crowd gradually increased, the quality of the management team limited staff and management in the form of very simple, can not effectively manage student online behavior. Management not in place also led to a lack of an incomplete school learning materials, or invalid information.

3.3 *Education website problem*

After years of development, all colleges and universities have established a university ideological and political education network platform, although the construction site flourish, but the column and the content of the site so inadequate, there is no ideological and political education sector design innovation, students generally are not interested. Many universities lack of ideological and political education network platform to discuss hot topics in the area of it and thinking, there is no educational function of social practice, although there are a lot of college ideological and political education network platform to establish a discussion board, in practical applications, there are no professional teachers guiding or educating, it's easy for students to form a misconception or view.

Site construction analysis, not college ideological and political education network platform sub-pages setting is not obvious, students have a difficulty quick inquiry, the impact of the network platform to play a role. The slowdown in the forum sets up web pages, content and themes, though they are very positive, but too scarce, there is no appeal, students rarely visit. The BBS campus building is a relatively hot construction sector, students generally have a high interest in participating, but there is a lack of supervision, failing to play the role of ideological and political education.

4 IMPROVEMENTS

For the problems in the above university ideological and political education network platform construction, this paper advises to improve in the following aspects.

4.4 *To strengthen infrastructure construction*

In university ideological and political education network platform, it needs to speed network infrastructure construction, renovate the campus network, accelerate technological change, it also needs to increase broadband investment, improve network speed, enhanced wireless network investment, to achieve docking of the campus network. Ideological and Political Education at the University of sinks in the network platform, hardware building needs attention, constantly upgrading servers to ensure reliability and scalability. For security problems, it needs to set up an emergency team of network security, do good network security work, monitor students' Internet behavior.

4.5 *To create a culture atmosphere beneficial to university ideological and political education*

Actively promote and highlight the campus network of ideological and political control in the freshmen, to stimulate the enthusiasm of students to participate. Schools need to further integrate education resources, expand the influence of network ideological and political education, increase the intensity of coverage, use certain technical means to regulate students' online behavior.

In the construction of university ideological and political education network platform, it needs to attract students' attention with exciting contents, can watch the exhibition network caucus activities columns to attract students' attention, also needs to expand the regulatory arrangements for full-time teachers, to intensify network propaganda and interactive efforts of other websites, improving the students' interest in participating.

University ideological and political education network platform construction needs to focus on and use a variety of popular web platforms, such as QQ group, flying letters groups and others, these sites have high popularity, members can participate in a wide range and active discussions. During the discussion in the QQ group, etc., are acquaintances of the world, so there is a strong provocative in some hot issues and common problems, need to pay attention to the guide, can not cause the student's reverse psychology. Construction of university ideological and political education network platform also needs to pay attention to innovation, for example, you can take full advantage of the micro - channel platform and micro-blogging platform to inspire students' enthusiasm to participate.

5 CONCLUSION

In summary, this paper first analyzes the current situation of college ideological and political online education platform, and then studies improvements. China's traditional university ideological and political education platform play an important role, the development of the Internet makes the role of traditional education platform effective, currently the construction of ideological and political online education platform also requires an innovative platform model, to strengthen the construction management.

REFERENCES

[1] Ruan Qi, Zhou Wei, Chen Shujun, etc.. *Investigation and Countermeasures of Dissemination Results of Ideological and Political Websites—Taking Southwest Jiaotong University "Frontier Net" for Example* [J]. Journal of Southwest Jiaotong University (Social Science Edition), 2012. 13 (02): 91–94.

[2] Wang Yan. *Cultural Norms Campus Network Platform for Innovative Ideas and Political Philosophy* [J]. *Industry and Technology Forum, 2014,7 (04): 122–123.*

[3] Huang He. *Ideological and Political Education of College Students after 1990s in Social Networking Community Platform Background* [J]. Heritage, 2011,32 (08): 44–45.

Management, Information and Educational Engineering – Liu, Sung & Yao (Eds)
© 2015 Taylor & Francis Group, London, ISBN: 978-1-138-02728-2

Analysis of colleges' sports cultural patterns

Hua Qian Lu
Academy Of Armored Force Engineering, Beijing, P. R. China

Hong Jie Zhang
Ordnance Engineering College, Shijiazhuang, P. R. China

ABSTRACT: With the country vigorously supporting vocational education, vocational colleges develop rapidly and their scale and impact continuously rise. At the same time, colleges' sports culture patterns have higher requirements. This paper analyzes the status of sports culture patterns from the connotation of sports cultural construction and carries out a new model of our college's sports culture.

KEYWORDS: Sports culture construction; status quo; mode.

1 INTRODUCTION

With the deepening reformation, quality education has become the main theme of physical education. Sport is an important part of our education and also the key element of quality education. But in the development of China's sports culture, there are many problems waiting for us to explore new ways to solve.

2 UNIVERSITIES SPORTS CULTURE CONNOTATION

In 1974, the renowned international sports terminology commission inaugural chairman, Dr. Ni Gua • Lai Kese, led the compilation of the Sports Vocabulary, in which the explanation of Physical Sports Culture is an important part of the generalized culture and the law and scope that integrate a variety of physical exercises that can be used to improve human biology and spirit capacity. The frequent concept of Sports Culture emerged in the mid-80s of the 20th century, which has become a unique cultural phenomenon in social life and social culture that covers the entire socio-cultural sport.

Construction of campus sports culture needs consideration of the cultural background to convert the strong physique destination of the traditional sports concept into an important means of improving human life quality. Mental health education is an important way to carry out students' all-round development and also an important component of our country moral education in colleges and universities. With the development of China's campus sports culture and shining campus sports culture will produce widespread positive impact on college students' mental health.

3 THE STATUS QUO OF COLLEGE SPORTS CULTURE

3.1 *Material culture construction*

Vocational colleges' sports material culture is an important safeguard for the development of colleges' sports culture from campus culture category, which mainly covers the sports venues, sports facilities and sports sculpture. Colleges' sports stadium is an important place for teachers and students exercise, recreation and organization activities and also an important basis for sports culture construction. With the vocational colleges' development, vocational education has transited into mass education and with the growing number of students, the original stadiums and infrastructure, the students teaching and training have been unable to meet.

3.2 *Cultural institution building*

The system is a public conduct code and achieve authority establishment in public compliance. Colleges' sports institutional culture mainly belongs to the middle part, specifically refers to the college sports systems, charters and organizational system. Campus sports activities the specification participation of teachers and students should reflect the values and spirit of school sports[1]. Cultural system is primarily a constraint file. Throughout the construction of cultural systems in China since the founding, the government continued to strengthen the construction of laws, regulations. Since the reform and opening up, China has announced a significant school sports laws and regulations.

3.3 *Spiritual culture construction*

With the further development of the education reformation, the status of vocational education further highlight and school sports also appears the transition from the previous simple pursuit of physical fitness to development and technology transfer and to physical education throughout students' lifetime under the guidance of new health concept. After seven vocational colleges' student survey, about 84 percent of students have scientific understanding of healthy concept. 76 percent of students do not have smoking habit and 68 percent of students do not have drinking habit. Drinking, which shows that students' understanding of health is not just the body having no disease[2]. At the same time, with social economic growth, psychological problems have become increasingly prominent. Students' health, especially mental health has caused widespread concern of the whole society. Most of the institutions have varying degrees of mental teachers, under the guidance of various forms of activities, through sports knowledge lectures, the students popularize mental health knowledge education.

4 COLLEGE SPORTS AND CULTURAL PATTERNS

A campus culture is important blood for universities to survive and develop and is the essence of universities, also the key factors constituting schools strength and competitiveness.

Construction of campus culture, is an important basis for the work of the institutions and forward-looking. College campus sports culture heritage campus culture important function, on the other hand, is innovation of college campus culture and the development model of campus sports culture is significant.

4.1 *Colleges sports cultural characteristics embodied through sports work*

Sports work is mainly instituted in daily physical activity training and curriculum development and student extracurricular school sports, and so on. Sports work exhibition process should reflect the vocational school characteristics as well as the school tradition. Sports programs can not only spread vocational sports cultural center, but also own the responsibility to promote school's physical culture. In curriculum designing process, make the curriculum become an important carrier of campus culture[3] in conjunction with the school's educational thinking and positioning institutions to become the school's display window, meanwhile, to achieve coordination and guidance of formal and scientific development through the sports culture carrier core to become the important contents of campus sports culture.

4.2 *Looking sports culture-building results from changes in the school*

The running time of many higher vocational colleges is short, but they have a long cultural predecessor. So many school's sports culture is also very bright and has relatively sharp features, which are important basis for vocational schools physical culture and has a vital role in sports and cultural development process. The formation of vocational physical sports culture must borrow school's history to produce their own brand characteristics, which requires us to dig deeper and comprehensively clear up.

4.3 *Using the school-enterprise cooperation and innovation and enrich campus sports culture*

School-enterprise cooperation is an important development direction of the vocational colleges and relates to vocational college's development and survival. Through school-enterprise cooperation, continuously explore the development of school characteristics and innovate educational model and scientifically select cooperation partners. Through school-enterprise cooperation mode, tap cooperation enterprise own content and characteristics and establish comprehensive relations and carry out schools and businesses work difficult break. Overall, the school-enterprise cooperation should strengthen cooperation and exchange with corporate to make students master the basic quality of enterprise requirements and go into corporate culture.

4.4 *Make industry culture as a supplement to expand campus sports culture space*

Industry culture is valued and behavior gradually formed in the long run. An industry at the time of formation will become a common philosophy of staff and stark industry characteristics. In college students' professional capacity training, we must focus on creating a culture industry[4], in the field development process, present a variety of campus sports culture development possible to deepen school-enterprise cooperation form and promote win-win situation of vocational colleges and industry.

5 CONCLUSION

Vocational colleges' sports culture construction has important implications for vocational education continued healthy development. We should make sports work content as a priority to construct unique cultural characteristics campus sports culture. Make changes in the school's history as the main line, through school-enterprise cooperation partners to constantly enrich the campus physical culture development results. Consider the status of the industry culture as a supplement to expand the campus sports culture space. Explain campus sports culture new connotations to build a new campus sports culture model of colleges' features.

REFERENCES

[1] Qiu Shuoping, Xu Jiuping. *On Students Physical Sports Study Human Culture Tendency* [J]. Sports and Science. 2010 (9) 31–33.

[2] Wei Qiuzhen. Reflections of Campus Sports Culture [J]. Hubei Sports Science. 2009 (4) 31–32.

[3] Chen Anhuai. *Physical Education Curriculum and Textbook Reformation by the Guidance of Quality Education* [J]. Shanghai Physical Education Institute 2011.1.

[4] Sun Zhaoming. On Colleges and Vocational Education Development Orientation [J]. Education Forum 2013. (7).

Management, Information and Educational Engineering – Liu, Sung & Yao (Eds)
© 2015 Taylor & Francis Group, London, ISBN: 978-1-138-02728-2

Application of visual simulation technology in business English teaching

Ting Yu
JiangXi College of Foreign Studies, JiangXi Nanchang, China

ABSTRACT: The visual simulation technology gives an immerse visual effect and attracts people to the creation of situations by creating a visual real virtual reality environment, making people emphatetic. Using the visual simulation technology in business English teaching can effectively stimulate the enthusiasm of students, improve student's motivation to learn, enhance the content of classroom interaction, and improve a student's learning efficiency. Meanwhile, the visual simulation technology creates an environment for students to better understand and use foreign trade English, it changes the traditional teaching model to improve the quality of English teaching in foreign trade, and enriches the practice of foreign trade English teaching to help improve the student's Business English comprehensive learning ability.

KEYWORDS: Visual simulation technology; Business English; teaching mode.

1 INTRODUCTION

Due to the rapid development of China's science and technology, more and more advanced equipment and technology are used in the field of education, enriching the teacher's teaching methods and improving the student's enthusiasm for learning and learning efficiency, improving the overall quality of education. Visual simulation technology is a typical example of China's foreign trade English teaching. Through the creation of relevant teaching environment, visual simulation technology makes students immerse in the created learning environments to learn expertise knowledge, improves the emotional and rational awareness of foreign trade English expertise, effectively promotes students to improve learning efficiency, enhances the learning proactive, while in the created environment, students are able to use their own ideas into the environment in mind, helping students cultivate divergent thinking. The important feature of visual simulation technology is reflected in the creation of transfer between immerse effect and feelings. From the visual and auditory sensory stimulation for students, improves student's understanding degree of trade English teaching, improving learning enthusiasm and promoting the improvement of teaching quality.

2 A THEORETICAL STUDY OF VISUAL SIMULATION TECHNOLOGY IN FOREIGN TRADE ENGLISH TEACHING

2.1 *Improve student's cognitive ability in foreign trade English*

In the process of learning foreign trade English, students can't just obtain knowledge taught by teachers in the classroom, which will not only allow students to feel boring about the curriculum so that students lose enthusiasm for learning, reduce learning efficiency, but also that the teacher impart knowledge depending on the teacher's own understanding of knowledge, can not let the students feel immediately the overall framework of knowledge. So visual simulation technology in Business English teaching makes up the deficiency of traditional teaching methods, attracts students to create situations through the creation of virtual situations, allows students to complete the construction of their own knowledge framework, and improves the awareness of trade application knowledge. If the acquisition of knowledge is out of using context, the knowledge learned by students is only the understanding of the textbooks, then the understanding of knowledge is isolated, one-sided. Business English teaching takes language learning as the important teaching target. Language teaching is inseparable from the words and dialogue learning, words and dialogue come from situations of life. Therefore, the visual simulation technology creates the language exchange situation, so that students learn English naturally. Through the creation of foreign language learning environment, visual simulation technology makes students to express their feelings in the environment, exchange foreign language learning in the environment actively, achieving the purpose of situation teaching.

2.2 *Interactive learning improves student's learning efficiency*

According to the relevant survey data, the process of students acquiring knowledge is that students acquire and discover knowledge independently, rather than passive learning in the classroom. Students are able to acquire knowledge only if the brain is involved in the interaction of knowledge. Through the creation of relevant context, visual simulation technology makes students to participate in the Business English learning, accomplish learning tasks through interactive learning thinking. And trade English language learning is mainly applied to the actual communicative activities, students conduct foreign trade through the creation of language in the context of the exchange of learning, improve their practical ability to use trade language. Through the professional language teaching situation, visual simulation technology makes students to conduct interactive learning in the context, master foreign language system, understand the latest developments of the knowledge of foreign languages and foreign trade language more deeply. In the context of students through interactive learning, increasing their exposure to trade foreign language, enriched trade foreign language learning content. By visual simulation technology trade language materials transformed into a virtual sound and image information converted from unity to acquire knowledge multifaceted way for three-dimensional way transmission of knowledge, students in English learning situations with visual and auditory comprehensive experience in foreign trade English knowledge, to acquire knowledge of the characteristics of foreign trade English, improve motivation to learn, improving the overall quality of teaching.

3 THE PRACTICAL APPLICATION OF VISUAL SIMULATION TECHNOLOGY IN BUSINESS ENGLISH TEACHING

3.1 *Visual simulation technology can be applied to individual items of English teaching*

Students learning foreign trade English are mainly about the practice of learning life related closely to the project of life. Create Business English learning environment through visual simulation technology, carry out a variety of experiential learning programs that can take advantage of learning resources. Students learn and practice continued in the context of the creation, learning in practice and practicing in learning. As a new mode of foreign trade English teaching, visual simulation technology in the foreign trade English teaching can improve the motivation of students to learn the knowledge, promote students to find access to knowledge actively, cultivate the student's innovative ability of foreign trade language learning, strengthen student's understanding of knowledge, improve their ability of trade language practical application, thereby improving the quality of teaching as a whole.

3.2 *The application of visual simulation technology deepens the reform of foreign trade English teaching*

Business English courses as an important part of business English professional courses, its teaching content should focus on the communicative and practice ability of foreign trade language, students should understand the relevant processes and professional knowledge of international trade, master some relevant regulations of international business, understanding deeply social culture and commercial culture of different English-speaking countries. Traditional teaching methods are only scripted, students only understand a little about the book knowledge of trade English, making it difficult to respond to different trade issues in the future work in business. Visual simulation technology in Business English teaching, by the guiding role of the teacher, students participate in the actual teaching activities, stimulating the learning enthusiasm of students, exercising student's practical ability. The application of visual simulation technology effectively reforms the traditional education methods and improves the quality of teaching.

4 CONCLUSION

Visual simulation technology completes teaching work through the creation of virtual situations, allows students to participate in teaching activities, enhances student's enthusiasm for learning, enhances student's awareness, effectively improves the efficiency of learning a foreign trade language, helping to cultivate comprehensive foreign trade English talents.

REFERENCES

[1] Zhu Lei,Zhang Jianqing. *Visual Simulation Technology Project in Business English Learning Application* [J]. China Educational Technology, 2011, 12: 110–113.

[2] Xing Wei, Zhu Lei. *Application of Visual Simulation Technology in Business English Teaching* [A]. IEEE Wuhan Branch, Wuhan University, Chongqing University of Posts and Telecommunications, Lanzhou University, University of Electronic Science and Technology, Shandong University, Central South University for Nationalities, Fuzhou University Institute of Engineering Information on Internet technology and Applications. *2012 International Conference Proceedings* [C]. IEEE Wuhan Branch, Wuhan University, Chongqing University of Posts and Telecommunications, Lanzhou University, University of electronic Science and Technology, Shandong University, Central South University for Nationalities, Fuzhou University, project information Institute: 2012: 5.

Management, Information and Educational Engineering – Liu, Sung & Yao (Eds)
© 2015 Taylor & Francis Group, London, ISBN: 978-1-138-02728-2

Development and design of art multimedia teaching system platform

Chen Jiang & Qiong Liu
Wuhan College, Zhongnan University of Economics and Law, China

ABSTRACT: In recent years, with the development of science and technology, multimedia teaching has been widely used in various subjects, especially the art teaching, through active use of multimedia teaching, the content of the art teaching is more visually displayed to students, to some extent has enhanced and improved the expressiveness of classroom teaching, promoting the updating of art teaching, new technologies, new theories and new concepts. Based on this, this article starts from the relevance of the new era in art teaching and multimedia teaching, explores actively the development and design of the multimedia teaching system in new era art teaching, providing a useful reference for the colleagues.

KEYWORDS: art teaching; multimedia teaching platform; design; development.

The 21st century is the era of information technology, computer technology has gradually entered the lives of ordinary people, and is playing an increasingly important role in people's learning, work and daily life, the relationship between computer-aided teaching and art teaching is becoming increasingly close, the introduction of the multimedia teaching in the new era art teaching increases the amount of teaching information within the unit time, comprehensively shows the application of graphics, sound, music, video, etc. in art teaching, vividly presents abstract concepts such as knowledge of the spatial, helps to increase students' interest in learning art, has a positive role to improve teaching effectiveness and quality of art, so the implementation of the media teaching system platform development and design has a very important practical significance to the new era aesthetic education.

1 ASSOCIATION BETWEEN ART TEACHING AND MULTIMEDIA TEACHING

Unlike other disciplines, art teaching in the new era is a failing education. Only in certain aesthetic scenario participating teachers and students can get the aesthetic emotion, which fully shows the importance of creation of scenarios in art education, the development of the new era, the high-quality art class should be the blend of emotion with circumstance, and should be really the present of truth and beauty, as well as combination of sound and painting, through the use of multimedia technology, promote students to think actively, positive play the main role of students in art teaching, making students the subject acceptable builder of information processing knowledge, use video, sound, figure means in multimedia technology to create situations, mobilize the enthusiasm and initiative of students to the maximize, promote the development of modern art teaching [1].

Basic Education Curriculum Reform Program (Trial) clearly states [2] that, promote the integration of information technology and curriculum, expand the teaching content and methods and provide a variety of teaching and learning environment for student learning and development is the focused task of teaching in the new era, clearly put forward the requirements of integration of information technology and art teaching, specific to the art curriculum, which is to promote mutual integration and contact of art teaching and classroom courses and IT, create a pleasant learning environment for students, improving the learning efficiency, so multimedia teaching has a positive role in promoting art teaching in the new era.

2 THE DEVELOPMENT AND DESIGN OF ART MULTIMEDIA TEACHING SYSTEM PLATFORM

2.1 *Selection phase of courseware*

Good courseware choice is the basis and the necessary conditions for the entire multimedia teaching system platform design and development, which is of great significance for the design and development of subsequent art teaching system platform. First, the subject selection must meet the needs of teaching, its purpose to select is to meet the needs of teaching of the majority of teachers, in specific content selection, should select more abstract difficult and important element, making selected content not only can be implemented using conventional teaching methods, but also can be described using multimedia technology, thus can better conduct the design of follow-up courseware; secondly, choose topics according to students' cognitive characteristics, the students at different learning stages have different cognitive characteristics, for example, child care curriculum design and development is different from the courseware for high school students, the difference between these aspects present in the courseware knowledge structure and expression form, etc.; finally, select subject according to the present objective conditions, objective conditions here mainly refer to the professional production technical level, financial support as well as users level, these objective conditions will constrain the choice of topics, make the choice of topics gradually reduced, more practical and more specific.

2.2 *Courseware script preparation stage*

The design and development contents in this part include script writing scripts making, the former is the basis of the latter, in particular design and development process, to describe the teaching content and presentation of each part according to the order of the teaching process is the text script, it is a manuscript form, usually written by the instructor, the implementation of the preparation must clear the role and significance of the courseware, carefully write the important teaching link, effectively integrate the prepared animation, text and sound, pictures and other data, and note the place where to show the effects and animations, and the production script different form this is constituted of three different parts, namely the main module analysis, courseware system structure and the production of script card, if specific classify the contents, it can be divided into the cover, interface design, material organizations, the arrangements of modules and technology application,etc.. Altogether, the form of the two branches of the same courseware scripts are basically same, both are descriptions of detailed information which users will see on the multimedia system interface, such as the interaction between computer and the courseware user, as well as teaching information presented by multimedia, etc., achieving the description and expression of such information and content.

2.3 *The design and development phase of courseware*

In the implementation of the development and design of art multimedia teaching system platform, the main task of this phase is to design the interface style, divide functional modules, and design programs and source code, and will select production software, etc., it mainly consists of the leader portion, the body portion and the end parts, then subdivide further, the body part can be divided into the teaching modules, imported modules, continuous test module and presentation module; interface style design needs to combine teaching content and formal beauty law, form of the visual aspects of the landscaping and standardization, containing the interface color scheme. Interface visual elements design and interface layout design require designers to use suitable color to solve the interface color matching, form a whole style of interface for the audiences who accept the teaching, layout design is the interfacial points. Line, surface and black, white, gray form design, form coordination between elements, making the audience in the process of human-computer interaction get the sense of beauty and order, improve the production of the system platform.

In the design and development of multimedia teaching system platform, the main required softwares are Flash, Director, Front-Page, Author-ware and PowerPoint and Geometer's Sketchpad, etc., among which Author-ware is a flowchart-based visual development tools, through a combination of sound, pictures and animation and text, use the process line to show the program flow, coupled with interactive icons to conduct interactive control, etc., but as a tool dedicated to produce presentation multimedia slides and slide, PowerPoint can be produced by the unit presentations, integrate together to form a complete courseware, conduct multimedia presentations teaching, which can be used to easily produce Word-Art, insert images and audio and others, enrich teaching content of presentation. During the development phase of courseware, in order to effectively reduce the development cycle and then improve the development quality, people should effectively enhance source code readability and reusability, designers should comply with courseware development norms, through effective communication and cooperation, improve the multimedia teaching system platform design and development, laying a solid foundation for China to better implement art teaching [3].

3 CONCLUSION

In summary, in the new era the art multimedia teaching system platform design and development is a complicated systematic project, requires to use a variety of the the newest science and technology currently,integrate the various needs of art teaching, perfect the use of multimedia teaching in art teaching in the new era by selecting courseware and writing curriculum scripts, provide strong scientific and technical support for the active promotion of art teaching, promoting the rapid development of China's education career.

REFERENCES

[1] Yao Yubo. *Cloud Computing Environment Design and Implementation of Multimedia Teaching Platform* [D]. Dalian Maritime University, 2013.

[2] Wu Weiying, *Network remote teaching system platform development and design of J2EE* [J]. Computer-based information (Theory), 2010,08: 52–53.

[3] Luo Wen,, Zhang Guobing. *Multimedia teaching system research and design cloud platform* [J]. Television technology, 2013,22: 47–50.

Management, Information and Educational Engineering – Liu, Sung & Yao (Eds)
© 2015 Taylor & Francis Group, London, ISBN: 978-1-138-02728-2

How mobile intervention education can revolutionize wellness market and patient self-efficacy

Dyna Y.P. Chao
Da'an District, Taipei City, Taiwan, ROC
Xinyi District, Taipei City, Taiwan, ROC

Tom M.Y. Lin
Da'an District, Taipei City, Taiwan, ROC

Ya Fan Yeh
Xinyi District, Taipei City, Taiwan, ROC

ABSTRACT: Diabetes has become a serious health and economic problem in Asia. The most recent survey has revealed 11.6% of Diabetes patients in China who cost about USD 25 billion a year. Furthermore, the population will increase to 143 million in 2035. Current healthcare management services, lack an interactive lifestyle and literacy assessment for designing a high-quality and positive-compliance care plan to the fresh diabetes patient. The approximate width of the desired health promotion method is relevant to accommodate self-management items. Hence, the different lifestyle and literacy preference is also the key elements of inference patient self-efficacy outcome. Moreover, lifestyle and patient self-care behavior in chronic disease is highly homologous in culture, geography and dietary habit.

This research aim is to analyze compliance preference of the new diagnostic patient, and discover specific requirements by using mobile health applications for identifying appropriate personalized services through pre-designed Diabetes Mellitus Interactive Management Framework (DMIMF). The DMIMF framework is based on theoretical model of behavior change (TTM) to build up the fundamental questionnaire. It includes a knowledge-based platform for digging out patient life-intervention requirements. Once the specific health improvement items are exposed, the professional group can provide effective intervention. Upon the DMIMF platform, the mobile-based inquiries, pre-assessment system is to collect patient Diabetes knowledge literature answer and compliance preference. Then, the readiness evaluation system is to discover a patient maturity level for education suite identification. Finally, the mobile physician dashboard system is the engagement tool to interact with patient for designing a personalized care plan.

The pilot was hiring 62 new diagnostic diabetes patients in the research group for analyzing their compliance preference by using mobile-based interactive application. The results indicate that different patients have preference for multiple and different self-care items. Moreover, the dietary care items are the top category in 117 times of follow-up visit. A patient who has blood pressure and BMI issues also indicates they have high motivation to change. In addition, mobile-based ICT technology is easy to accept in lower education, and older person; female behavior change of compliance rate is higher than male; young group patient has 75% improved compliance rate in dietary intervention. As the saying in the "Art of War" of Attack by Stratagem, "if you know the enemy and know yourself, you need not fear the result of a hundred battles", the Pilot result in Taiwan is strength to target the expanding market in the Asia market, especially in China. The result leads us pave the way for quickly adopting the personalized service combination in the new wellness market.

KEYWORDS: Compliance and Healthcare, Mobile Healthcare, Diabetes Education, Self-Efficacy, Patient Behavior, Dietary Habit, Care Plan.

1 INTRODUCTION

Chronic disease is not only the health problems but become the major economic burden in Asia. As the diabetes mellitus (DM), the most recent survey has revealed that 11.6% of adult in China have diabetes and it stands for about 114 million people [9]. The population of DM is forecasted to increase from 8.3% in 2013 to 10.1% in 2035 that accounts for 592 million people worldwide according to the International

Diabetes Federation [6]. In addition, epidemic studies have shown that the age of diabetes onset has decreased and it is not only prevalent in developed and developing countries, but in low and middle income countries [1, 10]. The huge population of diabetes patient has created a great financial burden on the medical system, family and society. For the health expenditure on diabetes, the overall disease spending is estimated to cause at least 548 billion United States dollar (USD) in 2013 that account for 10.8% of total health expenditure worldwide. It's expected to excess 627 billion USD by 2035 [6]. The key approaches for improving DM self-management to relieve the load have become the market demanded. An effective management of health care may not only prevent diabetes complications, reduce the overall disease cost and family's burden but improve patients' quality of life also.

Even the advance in applying information technology, most of the current hospital services are still mainly focus on the clinical data and status of hospital systems. This approach works well for managing acute diseases, but not effectively applied to chronic disease for lifestyle change is more critical for managing it. A high-quality and effective chronic disease care plan will need to include information from lots of hospital stand-alone systems, such as EHR (electronic health record), clinical management outcome, patient's BOI (body of information) measurement records, patient's knowledge and psychosocial assessment, and others. The current pain point of diabetes physician and health educator are lacking an integrated information system that could provide aggregate personalized information with both clinical and lifestyle to the particular patient. In this study, we designed a mobile-based interactive management system, Diabetes Mellitus Interactive Management Framework (DMIMF). It offers an integrated snapshot for patient-physician engagement and interacted wellness education through iPad and mobile-based application.

2 LITERATURE REVIEW AND HYPOTHESES

As the chronic disease is majorly associated with the patient's behavior in addition to the genetic and environmental factors, the effectiveness of the disease management depends not only on the patient's clinical status, but mostly on their lifestyle, disease knowledge, health literacy, belief and feeling. A previous study has shown that understanding of the patient's disease, literacy level, personal characteristic and readiness for action are required for health educator or care manager to assign adequate actions for patient to take [4]. Thus, it is important for a care provider to create a service environment that could facilitate the adoption of patient's behavior change according to the individual's condition and needs. As information technology has become a common way of supporting disease management, the attempt for providing evidence-based education tailored to individual patient has been conducted [8]. In this study, it mainly focuses on finding patient life-style and compliance preference, and utilizes the knowledge simulation platform to map diabetes maturity level for further patient-physician engagement entry point.

Hypothesis (1): Patient has own preference to manage their self-care item for improving the overall compliance rate and clinical outcome.

Considering the individual patient has different lifestyles, knowledge and feeling about the disease, diverse stage on action readiness and risk condition, this research hypothesized the first aspect of patients have their own preference on improving compliance of care item.

Hypothesis (2): DMIMF interactive system is positively impacting the patient compliance trend.

Personalized intervention and health education system will have a positive effect on the patient's compliance for chronic disease self-management. Attempts of finding the relation between compliance and health outcome have been studied and shown to be complex [5, 7]. Thus, we designed a DMIMF interactive system to observe the insight factor as second hypothesis.

Hypothesis (3): The specific attribute of patient characteristics and wellness status will influence compliance improvement.

The study on online services have shown that demographic profile like age, gender, education, income, and more affect the acceptance of the information technology applications [3]. The researchers hypothesized the particular patient has different progression in the disease pathway so that the adherence model will depend on the characteristics and status.

3 INFORMATION FRAMEWORK

In order to integrate the information collection described above and provide a research pilot environment for testing patient learning behavior, the aim of this framework is building up a service platform of diabetes education. Then, it can help physicians and health educator to understand patient specific needs and improvement area. DMIMF is not only aggregated clinical information from clinical databasc, it also provides a questionnaire generation mechanism and knowledge bank for collecting patient insight on DM knowledge literature, belief and emotion, healthy diet, glucose self-monitoring, and the influence of disease on productivity are also collected [Figure 1].

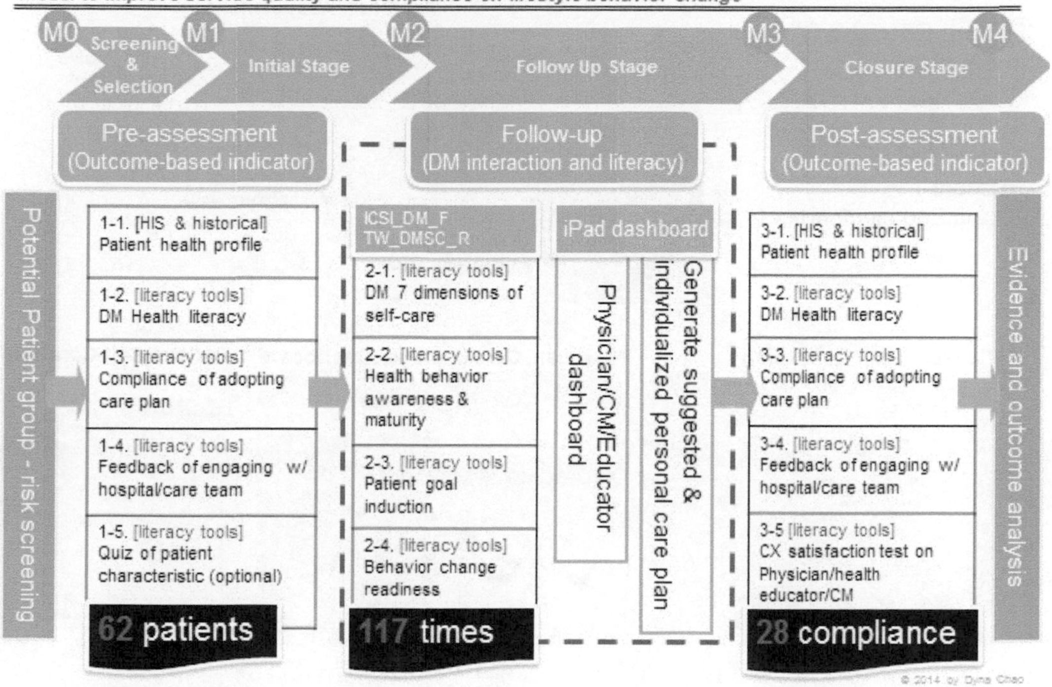

Figure 1. DMIMF (Diabetes Mellitus Interactive Management Framework).

4 INTERACTIVE PATIENT ASSESSMENT SYSTEM (MOBILE DEVICE, IPAD)

Collect patient's current health status, lifestyle and compliance preference. The assessment system includes pre-assessment and post-assessment application for research study. Patient feedback is stored in cloud-based DMIMF platform for understanding factors contributing to sustainable behavior change.

5 DM READINESS EVALUATION SYSTEM (MOBILE DEVICE, IPAD)

According to the American Association of Diabetes Educators [2], and evaluate patient mental readiness on DM disease, there are seven self-care behaviors are included as the diabetes disease important management criteria. These seven self-care behaviors are healthy eating, being active, monitoring, taking medication, problem solving, reducing risks and healthy coping. After the readiness evaluation, the score and level are transmitted into the physician dashboard for the next stage.

6 INTERACTIVE PHYSICIAN DASHBOARD SYSTEM (MOBILE DEVICE, IPAD)

The clinical status, results of patient assessment and readiness evaluation systems are presented on the physician dashboard for interacting with patient and assigning appropriate personalized care plan. Based on evaluation results and patient lifestyle preference, the system provides the suggestion of education program for each patient. Physician and/or health educator use the photo, one page disease brief, wellness article and self-exam tools to engage with the patient. It illustrates the correlation with disease progression and self-efficacy more friendly and easily.

7 RESEARCH DESIGN

The primary purpose of this study is to understand the patient insight might contribute to the adherence of diabetes self-management. Hence, a pilot includes the service flow [Figure 2] conducted in Taiwan for testing the effectiveness of the DMIMF and the mobile-based interactive system.

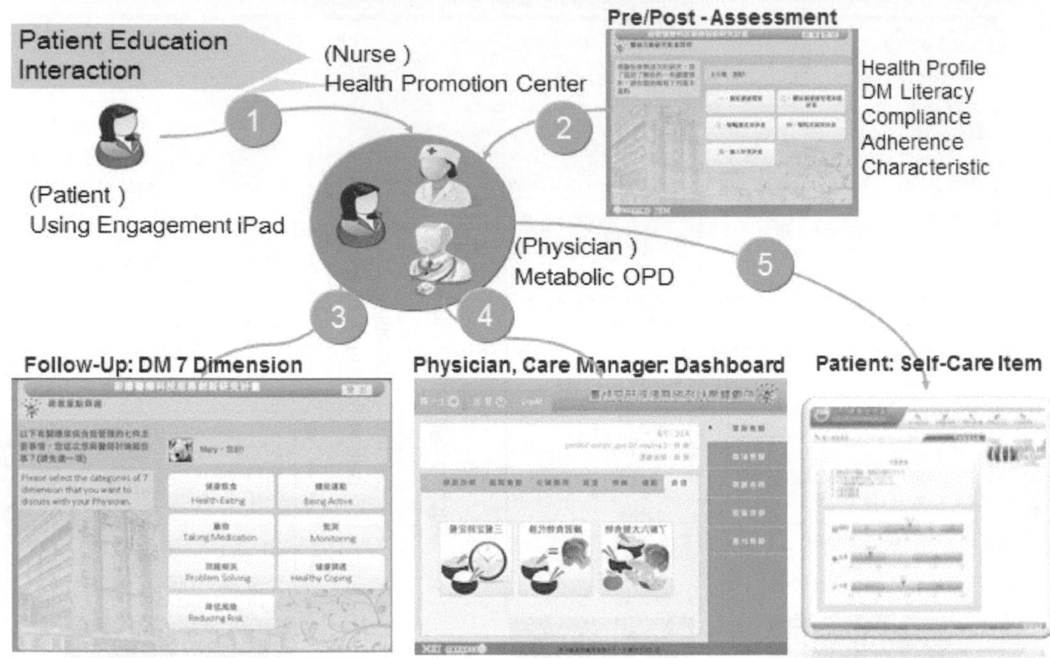

Figure 2. Patient engagement service flow.

Patient with newly diagnosed diabetes at 3 months were invited and recruited into the pilot group due to they need more disease self-management knowledge. All informed consents obtained and interviewed via professional team, including a physician, health educator, and service consultant. All recruited patients were provided with regular clinical advice and medication. The pilot performed over a period of 6 to 8 months during 2013. To better integrate with the wellness service flow, the interactive mobile applications were provided to the subject through a mobile tablet while they were waiting for doctor visits or health education. The mobile tablet, like iPad and a smart phone, are also served as a supporting tool for physician and health educator to access patient's integrated information. The patient compliance data were collected through the interactive mobile interface of the DMIMF as the patient entered in the pilot start phase for pre-assessment, and the closure phase of 6 to 8 months later for post-assessment. The patient self-evaluation scores are a taxonomic reference as one indicator, including dietary, exercise, medicine taking, blood glucose monitoring, blood pressure monitoring and health coping. Patient demographic data that highly relevant to health promotion factors are imported by hospital provide and insert into DMIMF for further use on interactive mobile systems. Participants were intervened with diabetes education on the 7 self-care behaviors. Physician and health educator conduct education, according to patient's interest and specific needs to enhance their ability on conducting self-management. The interactive mobile physician dashboard system improved patient's willingness of learning continuously. The educated knowledge was recorded in the physician dashboard system. Data analysis was conducted by IBM SPSS Statistics version 19.0. The methods of statistic in analytic use paired t test for pre-and post- assessment, and to test or chi square test for group comparison.

8 RESULTS AND DISCUSSION

There 62 participants join this study and fulfill 117 times mobile DMIMF intervention. After 6 months intervention, there are 28 patients complete post-assessment. Patients are free to choose which categories of DM self-management they are interested in. DMIMF includes 7 categories, healthy diet, being active, take medicine, monitoring, problem solving,

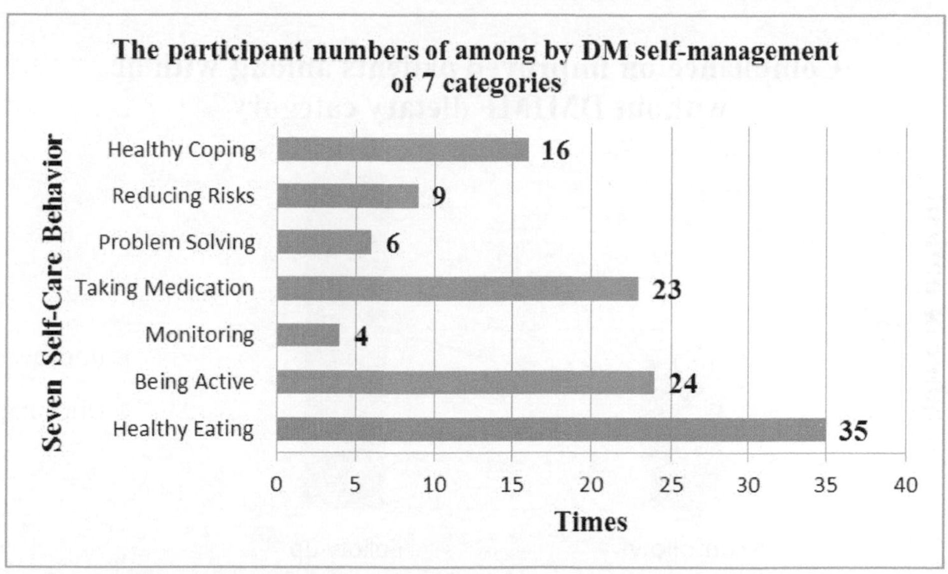

Figure 3. The participant numbers of among by DM self-management of 7 categories.

health coping and risk reducing. Among 117 of intervention experience, the top category is dietary which contribute 35 numbers [Figure 3].

The demographic characteristics of these subjects who have completed post-assessment are shown in the [Table 1]. The average age is 63.71 and the range is from 37 to 88 years old. There are 60% participants who complete post-assessment is over 60 years old. It means the age is not the barrier in mobile device usage. Besides, Education level is not the barrier to affect wellness, learning on a mobile device (like iPad). There are 79.5% participants are under or at the high school education level. To analyze the effect of the mobile DMIMF intervention, the compliance frequency between the pre-assessment and post-assessment are compared. The average compliance frequencies of dietary, exercise, blood sugar monitoring, blood pressure monitoring, and health, coping are increased after the intervention [Table 2], and the change of dietary compliance has reached statistical significance.

In general, there are 42.8% participants improve dietary compliance. Based on the compliance change, 28 subjects are then divided to improved-group (the compliance frequency increased after the pilot) and unchanged-group (the compliance frequency unchanged or decreased) for comparison. Need to be noticed, who experience dietary category of DMIMF has more percentage (47.4% vs. 33.3%) in improved-group [Figure 4]. That means participants pay more interest on dietary would trigger a healthy dietary behavior change.

Table 1. Demographic characteristics of subjects (N=28).

Characteristic	Subjects (%)
Age in years	
25≦ years <40	1 (3.6%)
40≦ years <50	3 (10.7%)
50≦ years <60	6 (21.4%)
60≦ years <70	9 (32.1%)
≧70	9 (32.1%)
Gender	
Male	17 (60.7%)
Female	11 (39.3%)
Education	
No education	1 (3.6%)
Elementary school	6 (21.4%)
Junior high school	4 (14.3%)
High school	11 (39.3%)
College	5 (17.9%)
Graduate	1 (3.6%)
Occupation (N=120)	
Unemployed	2 (7.1%)
Public Servant	1 (3.6%)
Office Worker	5 (17.9%)
Businessman	4 (14.3%)
Retired	15 (53.6%)
Others	1 (3.6%)

Table 2. Compliance frequency comparison between pre- and post- assessment.

Compliance	Average (pre)	Average (post)	P-value
Dietary	3.5 ± 1.0	3.8 ± 1.0	0.043
Exercise	3.0 ± 1.5	3.3 ± 1.4	0.130
Medicine taking	4.6 ± 0.7	4.4 ± 1.0	0.424
Blood glucose monitoring	3.2 ± 1.3	3.4 ± 1.3	0.326
Blood pressure monitoring	3.0 ± 1.4	3.2 ± 1.3	0.556
Health coping	3.8 ± 0.9	4.1 ± 0.9	0.062

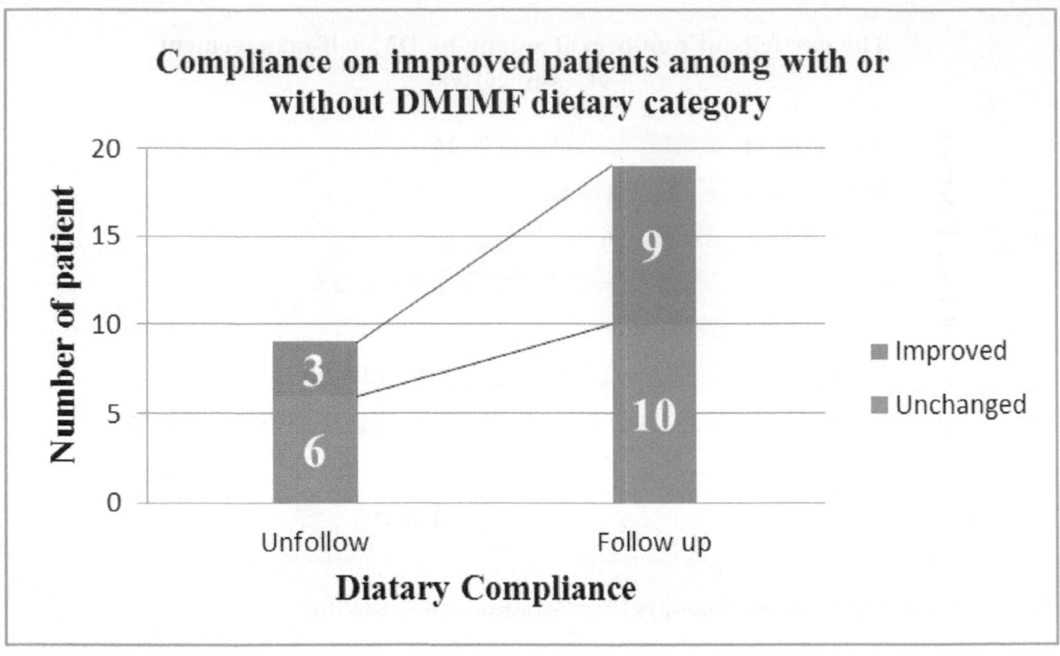

Figure 4. Compliance on improving patients among with or without a DMIMF dietary category.

Furthermore, characteristics would affect or trigger dietary behavior change are investigated. In the results show, patients have higher systolic blood pressure, higher diastolic blood pressure, higher BMI, and younger age is intended to change dietary compliance [Table 3]. It means patients have worse health indicators (blood pressure and BMI) are likely to change which implies DMIMF system is more helpful for type 2 DM patients with overweight or hypertension. Simultaneously, dietary compliance improved-group has significantly lower pre-assessment, dietary compliance and younger age. It is understandable because those patients who stand on the lower level of wellness recognition in pre-assessment stage so that they can reach the progression at the end of the post - assessment stage. Hence, the younger age patient is easier to motivate on changing their positive life habits in coming disease pathway.

Table 3. The difference between improved and unchanged dietary compliance in Pre-Assessment dietary compliance, blood pressure, BMI and age.

	Unchanged (Avg ± SD)	Improved (Avg ± SD)	t-test P-value
Pre-Assessment dietary compliance	3.9 ± 0.7	2.8 ± 1.1	0.005*
Systolic blood pressure	130.4 ± 17.3	136.7 ± 32.2	0.516
Diastolic blood pressure	74.4 ± 12.4	78.7 ± 18.4	0.466
BMI	24.8 ± 2.8	26.1 ± 3.1	0.272
Age	65.7 ± 11.1	61.1 ± 14.0	0.341

It is worth mention, the majority gender in improved-group is female (58.3%) and it is only 25% in unchanged-group [Table 4]. This result implied that an interactive tool, like DMIMF education system, could stimulate female to do some actions. Other health behaviors include smoking status, drinking status have no effect on dietary compliance change. The development of mobile devices with an increasingly influential presence in our lives, but only 28.6% participants have smart phone. The most likely reason is age, that because the average age is 63.71 years old. On the other hand, either do they have smart phone, they all join and finish this study. It means bring these resources closer to the elderly they can still enjoy the advantages.

Table 4. The trend effect of gender, drinking status, and smart phone usage among dietary compliance improved and unchanged groups.

	Unchanged N (%)	Improved N (%)	Pearson Chi-Square P Value
Gender			0.074
Male	12 (75.0%)	(41.7%)	
Female	4 (25.0%)	(58.3%)	
Drinking status			0.796
Non-drink	7 (43.8%)	4 (33.3%)	
Occasional	3 (18.8%)	2 (16.7 %)	
Daily drinker	6 (37.5%)	6 (50.0%)	
Smart phone usage			0.629
no	12 (75.0%)	8 (66.7%)	
yes	4 (25.0%)	4 (33.3%)	

The result indicates that the compliance rate increases after the intervention of DMIMF interactive education service in general. DMIMF may conducts some critical factors that might mostly affect patient's self-awareness by receiving personalized education and care plan through our system so that they can enhance compliance on disease self-management. On the increasing trend of DM population, there are 42.8% participants improve dietary compliance in this study. It can be expected that DMIMF could help more people when it deployed in widely market. As a consequence, the improvement of compliance by DMIMF could potentially help better control on health insurance cost and also establish stable customer relation with the hospital on customer retention. Moreover, patients with worse health indicators and female are preference in behavior change, especially when interactive mobile tool conduct. It also shows that mobile technology has limited barrier in healthcare service among elderly or lower technology usage people. In conclusion, personalized interactive education approach improved health behaviors compliance. Base on this study, the DMIMF system will be revised and redeployed in a near further, especially in Asia market, to get more consolidate evidence in the effect on health management.

REFERENCES

[1] Abdullah, N., Attia, J., Oldmeadow, C., Scott, R.J. & Holliday, E.G. (2014). The Architecture of Risk for Type 2 Diabetes: Understanding Asia in the Context of Global Findings. International Journal of Endocrinology, 2014, Article ID 593982, 21 pages, doi:10.1155/2014/593982.

[2] American Association of Diabetes Educators. (2008). AADE7 Self-Care Behaviors. Diabetes Education, 34, 445–449.

[3] Dholakia, R.R. & Uusitalo, O. (2002). Switching to Electronic Stores: Consumer Characteristics and the Perception of Shopping Benefits. International Journal of Retail and Distribution Management, 30, 549–469.

[4] Fransen, M.P., Beune, E.J., Baim-Lance, A.M., Bruessing, R.C. & Essink-Bot M.L. (2014). Diabetes Self-management Support for Patients with Low Health Literacy: Perceptions of Patients and Providers. Journal of Diabetes, doi: 10.1111/1753–0407.12191.

[5] Hays, R.D., Kravitz, R.L., Mazel, R.M., Sherbourne, C.D., DiMatteo, M.R. & Rogers, W.H. et al. (1994). The impact of patient adherence on health outcomes for patients with chronic disease in the Medical Outcomes Study. Journal of Behavior Medicine, 17, 347–60.

[6] International Diabetes Federation. (2013). IDF Diabetes Atlas, 6th edition. Brussels, Belgium: International Diabetes Federation.

[7] Kelders, S.M., Kok, R.N., Ossebaard, H.C. & Van Gemert-Pijnen, J.E.W.C. (2012). Persuasive System Design Does Matter: A Systematic Review of Adherence to Web-Based Interventions. Journal of Medical Internet Research, 14, e152. doi: 10.2196/jmir.2104.

[8] Ko, G.T., So, W.Y., Tong, P.C., Le Coguiec, F., Kerr, D. & Lyubomirsky, G. et al. (2010). From design to implementation - The Joint Asia Diabetes Evaluation (JADE) program: A descriptive report of an electronic web-based diabetes management program. BMC Medical Informatics & Decision Making, 10, 26. doi: 10.1186/1472–6947–10–26.

[9] Xu, Y., Wang. L., He, J., Bi, Y., Li, M. & Wang, T., et al. (2013). Prevalence and control of diabetes in Chinese adults. The Journal of the American Medical Association, 301, 948–959.

[10] Yeung, R.O., Zhang, Y., Luk, A., Yang, W., Sobrepena, L. & Yoon, K.H., et al. (2014). Metabolic profiles and treatment gaps in young-onset type 2 diabetes in Asia (the JADE programme): a cross-sectional study of a prospective cohort. Lancet Diabetes Endocrinology, doi: 10.1016/S2213–8587(14)70137–8.

Management, Information and Educational Engineering – Liu, Sung & Yao (Eds)
© 2015 Taylor & Francis Group, London, ISBN: 978-1-138-02728-2

Issues and countermeasures of college English multimedia teaching

Hong Bin Li
Ji Lin Business and Technology College, Chang Chun, China

ABSTRACT: With the continuous development of computer technology, the application of multimedia information technology teaching methods is more and wider. In particular, English teaching teachers better solved many drawbacks of traditional teaching, borrowing multimedia assisted teaching methods. However, the multimedia as the main auxiliary teaching mode still has some issues needing us to analyze aiming at a specific teaching situation, thus we come up with countermeasures.

KEYWORDS: College English; Multimedia; problems; countermeasure.

1 INTRODUCTION

In the constant advancement of higher education reformation work, multimedia technology is widely used. College English teaching through the application of multimedia communication technology, brought new ideas and teaching methods for students to enhance the teaching achievements. This supplementary teaching method has become the main method of teaching aids. Multimedia classroom teaching brought in a good income for teachers and students, at the same time, many factors and impacts on the implementation brought a lot of problems with the specific teaching job, which influenced the final outcome. Therefore, we should fundamentally analyze the reasons to solve specific issues in English multimedia teaching.

2 PROBLEMS IN ENGLISH MULTIMEDIA CLASSROOM

2.1 *Teacher's teaching philosophy and methods*

Because now there are many college English teaching classrooms using large classroom, which results in teacher's classroom activities more difficult to carry out, and indoctrination teaching makes student no interest in learning. The implementation of the method of multimedia teaching aids to some extent a good solution to this problem. However, through the teaching outcomes analysis, found that students of this teaching method is not very much. The main reason is due to the teachers in the actual teaching work still did not forget the "independent" role in the traditional teacher-centered teaching philosophy, it is difficult to mobilize the enthusiasm of students. Is teaching the course, some of the teachers because there is no modern methods of teaching theory, can not design a better student-centered teaching practice, teaching process is completely controlled by the teacher, monotonously through multimedia courseware to explain, not develop students' independent thinking and learning abilities. Lack of teacher-student interaction in the classroom makes teaching multimedia and more information becomes dull. Some teachers in order to complete the simple task of teaching students while ignoring the degree of acceptance, leading students overwhelmed with the large-capacity information. The classroom teachers monologue is not the place, is to help students gain more knowledge, problem-solving areas. Therefore, we should pay attention to the student in the actual teaching, so as to make organic union with assist teaching methods of multimedia.

2.2 *Information quality and operation of multimedia courseware*

The main work content of the modern multimedia features, is to present courseware materials to the students, the exchange in education is relatively weak. Some teachers in teaching, and students do not have good communication, multimedia courseware directly presented to the students complete, this approach loses a fundamental principle of individualized. Some teachers lower multimedia information literacy; teachers use multimedia information in teaching, their need for specific teaching information corresponding analysis; through screening, identification and other information on the quality of testing. Because, the rapid development of computer technology, so the use of multimedia teaching teachers targeted training work to be done; Also, because most of the English teachers are computer's "outsider", therefore, in the teaching process, the failure of the multimedia not only delays the progress of teaching, but also is likely to influence students' interest in learning, make teaching achievement decreased.

2.3 *Fail to understand the students' cognition*

Because of the different levels of the overall English proficiency of college students, it leads to cognitive differences to some extent. Coupled with the general form of the University of Large Classes in English, so that students change this difference is even more pronounced. Teachers in the actual teaching, and students do not have good communication, there will always lead to the teaching of all phenomena is busy. And some students can be attracted only to be attracted in the form of multimedia teaching, failing to do real knowledge absorption.

3 COUNTERMEASURES OF ENGLISH MULTIMEDIA CLASSROOM TEACHING

3.1 *Strengthen teaching methods and change teaching philosophy*

Teachers in the teaching activities should keep learning new teaching methodologies, strengthen knowledge of the multimedia teaching. While also should recognize the student-centered teaching orientation in teaching. To abandon the traditional teaching methods indoctrination, the teacher changes from the previous dominant mode is now coaching mode. Teachers in teaching, but also to communicate with students to pay more, change the previous mechanical classroom atmosphere. Mobilize the students' interest in learning from the perspective of the development of students' subjective learning behavior. According to the student's own actual situation, grasp the content of classroom teaching, in the right amount of scope makes students able to skillfully use the knowledge.

3.2 *Make the correct choice of coursework in accordance with students*

This question determines the success or failure of multimedia teaching activities. Teachers should select coursework appropriately according to the specific teaching content, neither too much nor too little. Combined with the students received an average level, choose some positive, creative strong courseware. In teaching courseware, can be discussed, collaborative approach to the game to help students learn communication. First, to establish a correct information teaching philosophy; to foster students' knowledge of cultural competence and the ability to choose other purpose line information resources. Through the rational use of information resources, create a teacher, multimedia and students trinity as the main teaching mode. Second, teachers need to constantly upgrade multimedia technology to master the basic operations technology. Regular organization of training, learning new skills, such as: makes written text, sound, graphics, and animation files, file conversion to understand the different formats, and different software environments the hardware applications, while also know failure treatment. After learning the different causes and treatment methods, people apply it to the actual teaching, prevent the equipment failure in teaching and the delay the teaching progress.

3.3 *Pay attention to students' personal cognition*

In the usual teaching activities, teachers should enhance communication with students, understand the students' specific cognitive knowledge in the sincere communication, adopt specific teaching methods in line with different cognitive conditions of students. This effectively reduces due to different circumstances which lead to students' cognitive learning activities perceptibly phenomenon incidence. Fully embodies the concept of individualized teaching. Teacher job title reserved for after-school way through multimedia learning content to help students master the key. Meanwhile, in the courseware teaching, teachers should not just focus on the form teaching of courseware, but conduct in-depth analysis, and in better communication process with students, strengthen training value of courseware for students' cognitive abilities.

4 CONCLUSION

Although the application of multimedia information technology in English teaching has good influence on it, the existing problems cannot be ignored. Through specific teaching practical activities, people specifically analyze the existing problems, adopt corresponding countermeasures,

REFERENCES

[1] Li Hewei. Students' Satisfaction Research of College Multimedia Classroom Teaching [D]. Henan University, 2013.

[2] Wang Yanbing. Existing Problems and Countermeasures of Current College English Teaching [J]. Education, 2013, (05), 40–41.

[3] Zhao Li. Interactive Model Exploration of College English Classroom Multimedia Teaching [J]. Reading and writing (Education Journal), 2013, (09), 13–14.

[4] Sun Xianhong. Information Technology and Curriculum Integration in College English Teacher Computer Self-efficacy Studies [D]. Shanghai International Studies University, 2013.

Management, Information and Educational Engineering – Liu, Sung & Yao (Eds)
© 2015 Taylor & Francis Group, London, ISBN: 978-1-138-02728-2

Multimedia college English listening and speaking class under meta-cognitive strategies theory

Yan Li & Yang Song
Jilin Communications Polytechnic, China

ABSTRACT: In terms of college English listening and speaking class, multimedia teaching is both an opportunity and a challenge. In the past people thought as long as the advanced network and multimedia technology is applied in listening and speaking teaching, it could improve the quality of teaching and improve the efficiency. However, it is not. New teaching model teaching effectiveness depends largely on the level of meta-cognitivity of learners. And now the majority of college students lack meta-cognitive awareness in English listening and speaking, learning, which resulted in the old teaching effect in the new teaching model. This article will introduce meta-cognitive theory and study its practical application in English listening and speaking teaching.

KEYWORDS: meta-cognitive strategies; meta-cognitive theory; multimedia university; college English listening and speaking class.

1 INTRODUCTION

Meta-cognition is the cognitive of cognitive, it refers to learners gradually achieve self-learning process through the development of learning programs on their own, monitor their own learning, self-assessment and other methods, is an ability to reflect the learner's own cognition. Metacognitive theory is one of the important elements of second language acquisition, meta-cognitive strategies are successful language scholars' self-management and self-learning behavior. The essence of meta-cognition is people's self-awareness and self-regulation of the cognitive activities.

2 METACOGNITIVE STRATEGIES THEORETICAL OVERVIEW

2.1 *Language learning strategies*

The definition of language learning strategies is summarized as "to obtain, store, regain and use language information learners conduct any operation, steps, plans and practices of behavior". Some people think that learning strategy is to learn to help themselves understand this person, learning or memory of a special thought and behavior information. In general, learning strategies is both a conscious mental language learners in language learning among control process, it is a positive ability to develop creative thinking process; is a very important method to determine the effect of language learning, but also a visual behavior of learner' own learning.

2.2 *Meta-cognitive theory*

Meta-cognitive is initially proposed by Flavel in the 1970s. Today 0'Malley & Chamot and other cognitive psychologists divide learning strategies into a meta-cognitive strategies, cognitive strategies and social affective strategies based on the process of studying cognitive psychology. Meta-cognitive strategies are successful self-management behaviors language scholars. Meta-cognition Chinese translation mainly anti-trial cognition, cognitive reflection, super cognitive, meta-cognitive four. Meta-cognitive knowledge and ability is the individual on their own cognitive processes regulating these processes. Meta-cognitive strategy is a typical learning strategy, referring to the students' effective monitoring and control for their own entire learning process.

Effective use of meta cognitive strategies can improve students' interest in learning, to enhance their self-confidence, improve learning attitude, learning to help them overcome anxiety, thereby improving students' listening comprehension.

Three elements of cognitive theory are: foreign language teaching should be student-centered; way of learning a foreign language should be combined and feedback each other of S (stimulation) and R (reaction); ensure regularity and creativity. Meta-cognitive theory is generally believed that meta-cognition includes meta-cognitive knowledge and meta-cognitive regulation. Meta-cognitive knowledge refers to knowledge associated with cognitive activity of individuals obtained. Meta-cognitive regulation refers to the individual continuous control monitoring conducted its own cognitive processes, regulating activities. It includes three capabilities: planning, monitoring and evaluation.

2.3 *Meta-cognitive theory and college English listening and speaking teaching*

The traditional teaching model, students over-rely on teachers, few students have meta-cognitive knowledge and ability of real self-learning, so this capability is needed in the process of slowly teaching training. Facts show that no meta-cognitive ability learners in fact, there is no learning direction people may not reflect on their own learning process, learning outcomes of people. Formation of meta-cognitive abilities they need to repeat strategy to strengthen the use of meta-cognitive knowledge, meta-cognitive awareness gradually formed in the subsequent study, and makes this learning process eventually becomes the student's own initiative conscious behavior. According to the learning mission objectives, achieve meta-cognitive strategies can be divided into three steps: planning, self-monitoring, self-assessment. On the English listening comprehension and oral expression, learning-related strategy can effectively improve heard heard that the level of meta-cognitive strategy which is very important. Only on the basis of the students have some cognitive strategies, the only better develop their cognitive abilities. College English teachers should recognize their mission: how to encourage students to develop a learning plan, teach them how to choose good learning strategies and self-monitoring and self-evaluation, to become independent learners.

3 META-COGNITIVE STRATEGIES TO INSPIRE COLLEGE ENGLISH LISTENING AND SPEAKING CLASS

3.1 *Role of teachers in meta-cognitive strategies theory teaching*

As a teacher, the first is to be a needs analyst, researchers social development's requirements of learners' English language proficiency, and then accordingly makes advanced, realistic teaching objectives; students do a good helper, students need to study, understand his door the starting point for analysis of their needs. Second, we must guide the students as soon as possible to find the best way to achieve learning objectives to guide its Metro good study habits and master effective learning strategies, develop their cognitive abilities; be managers of students' learning, which requires teachers should conduct learner-centered teaching, and gradually develop the students into the body learning.

3.2 *Strengthen targeted training to improve students' listening skills*

To help students acquire the necessary knowledge of voice, such as linking, strength, weak reading, assimilation sound, etc. Make students identify the tone from the beginning, the language units gradually increased to a sentence or chapter. Cultivate the ability to use listening strategies to understand does not mean that each word or sentence must all listen to understand. Sometimes do not understand individual words, students can speculate according to subject of listening material, background, context, etc.

In the teaching process, teachers should train the students' ability to choose primary information, and guide students to grasp the overall effect of the article. To appropriately combine the intensive listening with extensive listening, and have emphases.

Looking for meaningful teaching resources and rhythmic cycle with the use of multimedia, helping the students for their learning and memory.

3.3 *Purposefully cultivate meta-cognitive ability of students*

Network multimedia intervention makes students enjoy adequate learning resources, in this era of information explosion, teachers should guide students to reasonably receive a lot of useful information and learn to identify and recognize, abandon bad information. Which meta-cognitive knowledge and theories of learning is particularly important. In order to allow students to learn slowly become a leader, capable of autonomous learning, teachers must strengthen introduction and application of methods to explain meta-cognitive strategies. Professor He Ziran once said: "Foreign languages are learned, and methods and experience of foreign language vary, others' experience can be learned, but it cannot be applied mechanically, it is important to summarize learning method which is fittable for yourself and is effective." So no matter it is the teacher or the students themselves, they should focus on the training strategies and personalized training of learners' learning strategies.

4 CONCLUSION

In short, college English multimedia teaching of listening and speaking are carried out in student-centered teaching mode. Multimedia teaching creates a real listening language environment for students, also provides students with an independent study space. But at the same time it requires more for students to have a strong self-learning ability. Only giving full play to the guiding role of teachers, purposefully cultivate meta-cognitive ability of students, can help students to improve their level of English listening and speaking continuously, and promote students' comprehensive ability of listening and speaking, and help students improve their communication ability of using language knowledge and skills, so that in the process of comprehensive ability exercising students gradually grasp language learning rule, lay a solid foundation for the future effective use of learned language.

REFERENCES

[1] Wu Fei, Luo Shengjie, Teng Yuanjiang. Multimedia College English Listening And Speaking Class Study Under Meta-cognitive Strategies Theory [J]. Science Tribune, 2011,03 (32): 15–16.

[2] Liu Chunyan. Meta-cognition in College English Listening And Speaking Teaching Applied Research [J]. Changsha Railway Institute: Social Science Edition, 2011,16 (04): 99–100.

[3] Zhai Suqin. Meta-cognitive Strategies to Improve Students' English Hearing And Speaking Ability in The Application [J]. English Square (HEAD), 2013,09 (11): 45–46.

Management, Information and Educational Engineering – Liu, Sung & Yao (Eds)
© 2015 Taylor & Francis Group, London, ISBN: 978-1-138-02728-2

Multimedia network courseware development and application in preschool teacher training

Zhu Jing Lai
QiongTai Teachers College, China

ABSTRACT: Kindergarten is an enlightenment stage of life. Kindergarten teachers have a great influence on children. Therefore kindergarten teacher training, as part of improving the quality of preschool teachers, plays an important role in kindergarten teachers faculty. So, when we focus on early childhood and preschool teachers, at the same time we must focus on pre-kindergarten teacher's job and post-job training. This article, according to the training of kindergarten teachers currently, has discussed about the status of multimedia network courseware development and application in early childhood teacher training.

KEYWORDS: preschool teachers; training; status quo; multimedia network courseware; application.

Current society is the highly developed network era, as an integral part of people's lives and learning, the network is changing every aspect of people's lives. For training teachers for early childhood, with the gradual deepening of the kindergarten curriculum reform, teacher training content and methods must advance with the times, the development and application of multimedia network courseware has important implications for early childhood teacher training. And explore the multimedia network courseware development and application in early childhood teacher training in the former, we first have some understanding about the training status quo of kindergarten teachers.

1 PRESCHOOL TEACHERS TRAINING SITUATION

1.1 *Lack of systematic training, in the form of oversimplification*

Although with many people focus on early childhood education, many preschool and kindergarten teachers start to focus on early childhood teacher training, but there is still a lack of systematic training, child care nursery teacher training many blind obedience phenomena exists, rarely to develop an effective system for training programs just for training and training. On the other hand, the current early childhood teacher training and more training for teachers rely on explaining, in the form of over a single, boring, difficult to mobilize the enthusiasm of the training of teachers to learn, a lot of repetitive content, but led training inefficiencies, most teachers still at the learning stage imitation, no real understanding of new educational concepts and methods.

1.2 *Lack of targeted training, ignored the real purpose of the training*

Many kindergarten teacher training has always been a higher education administrative department to comply with the instructions of the arrangement, although a variety of training, but the lack of real training targeted to improve teaching standards. Many kindergarten teacher training is still stuck in the face stage, training is often the ability to re-light the knowledge, skills, light heavy weight morality, ignoring the real purpose of training for kindergarten teachers training needs and objectives rather vague understanding.

2 MULTIMEDIA NETWORK COURSEWARE PRACTICE IN EARLY CHILDHOOD TEACHER TRAINING

Preschool teacher training network environment, according to the difference in the course modules content, time and requirements, so that the preparation of textbooks becomes modular, multiple curriculum modules set. Teachers can participate in training based on their own needs, to be selected for their learning modules, the number of trainers module learning more to improve the overall level of its own will be higher. Such as public compulsory courses may include professional ethics module preschool teachers, education policy and law modules, and professional courses can be divided into kindergarten in addition to the basic theory, theory of reform, early childhood education research, but also can separate the teaching and research outcomes assessment, modern educational technology and other modules, training and teaching through these modules will be more targeted, training of kindergarten teachers can be more selective learning destination, not because of the monotony and repetition boring and lose patience and interest in learning. However, in the course of multimedia courseware modules in the design process, in addition to including a detailed course content, but also to deal with the courseware interface is reasonable or not, whether or not such as easy navigation to focus on design requirements, the training of teachers that children learn to provide greater convenience.

2.1 *Training system*

Through the network offers a range of early childhood teacher training management evaluation and other functions, such as providing teachers working arrangement, Tutor and other functions, in general, is to help establish a training system for training kindergarten teachers. An excellent multimedia network courseware, in addition to teaching the core part of the video and voice teaching, the learning process should also take full account control, Tutor, online exercises are essential training systems. Early childhood teacher training system inside the various components should be rendered non-linear relationship between the elements, namely mutual contact, interaction, and therefore, in addition to electronic Outsider multimedia network courseware to consider a combination of teaching content, but also in the training of teachers, childcare related roles and relationships of teachers, teaching content and teaching media as much as possible between the consideration.

2.2 *BBS module*

Early childhood teacher training in multimedia network courseware BBS architecture should be adopted, in accordance with the content of the training for different groups were established, training of kindergarten teachers clients only need to use a common Web browsers can directly enter the BBS, to find their own groups, participate in training early childhood teachers need to install courseware or other specialized software on the client can communicate with each other and answering questions between each other, help each other to form a harmonious beneficial training atmosphere. On the other hand, the training of teachers can always log management background courseware, which can easily be achieved on-line management in the background, such as updating the directory structure, upload pictures, updated with the latest information.

2.3 *Exam module*

A complete set of the training process, in addition to including teaching, but also should include examination and evaluation link. For early childhood teacher training system, the detection of the training effect in many ways, the examination is undoubtedly one of the most essential ways. Pass the exam, students can not only be ready for their own learning and to fully understand, at the same time more convenient trainers training progress and effectiveness of training participants have sufficient grasp. In general, the test module typically consists of three parts, one online exam, two online operations, three are independent practice. Pass the exam subsystem, training teachers for large or small, close or remote various examinations organizations, such as business exam teachers, learning tests, etc., in order to grasp the quality of early childhood teacher training, but also to better guarantee successful early childhood teacher training.

2.4 *Evaluation module*

Overall Evaluation Module relatively early childhood teacher training system is concerned, in fact, along with the construction process of the whole system, with important guidance and oversight role. By enabling the trainees to assess the effect of training are more fully understood, but can be targeted to adjust training courseware content and modules based on the actual results of the assessment, facilitate timely optimization of training methods. Secondly, the evaluation module also has a more important aspects, namely trainees after the training performance evaluation. The main component of the evaluation system for early childhood teachers to learn archives and institutions of learning archives, teachers of young children to learn through training, examination and other aspects of data collection and statistical analysis, able to provide a more intuitive leadership and organizational learning preschool teacher training and examination conditions. But also conducive to the training side of the kindergarten teachers grasp the knowledge points weaknesses were analyzed to assess and analyze the quality of the overall level of preschool teachers, and then make targeted improvements to the training mechanism, promote the upgrading of training effectiveness.

In carrying out early childhood teacher training, teachers should fully recognize the validity of the multimedia network courseware in early childhood teacher training, according to the specific circumstances of early childhood teachers, targeted to design a good courseware modules, teaching resources to better achieve optimization and information sharing, enhance early childhood teacher training effection.

ACKNOWLEDGEMENT

Hainan philosophy and social science planning project study on Hainan Province Preschool Teacher Training Vocational based on teachers' professional development (No.: HNSK (QN) 13–46).

REFERENCES

[1] Jing Han.Discussion about training for early childhood teachers' professional development[D] Hebei Normal University, 2010,12: 11–12.

[2] Aiguo Li.Exploration about network environment to support primary and secondary teacher training model[J] academy, 2014,04 (05): 46–47.

[3] Tao Wang.Reflections on the network courseware development process and its management[J] Chinese new technologies and products, 2009,11 (22): 89–90.

[4] Jinghui Zhang,Zhen Liu.Self-learning network courseware design and development [J] Chinese education technology and equipment, 2011,3 (03): 33–35.

Management, Information and Educational Engineering – Liu, Sung & Yao (Eds)
© 2015 Taylor & Francis Group, London, ISBN: 978-1-138-02728-2

Analysis on multimedia works and intellectual property protection issues

Nan He
Pingdingshan Industry Polytechnic College, China

ABSTRACT: In recent years, with the continuous development of multimedia technology, with a set of text, graphics, images, audio and digital features, multimedia technology has been widely used in learning and working and also in other various fields, there have been media works. Given that the current multimedia works are not able to enjoy the appropriate copyright protection, the interests of the creators of multimedia works are damaged. In this context, the research of multimedia works and intellectual property protection has an important practical significance.

KEYWORDS: multimedia works; intellectual property rights; protection.

In the period of rapid development of digital technology, multimedia technology has also been developed. Multimedia works, relying on multimedia technology to produce works, are produced, transmitted and preserved in the form of computers. While bringing cultural consumption the convenience, it brings interests damage to creators. This is mainly because multimedia works rely on networks to spread, it is difficult for the creators to control the flow of the whereabouts of works, it is difficult to own the copyright of the works. Therefore, it is very important to handle the current multimedia works and intellectual property protection.

1 THE CHARACTERISTICS OF MULTIMEDIA WORKS

While IT is developing continuously, digital technology as the representative of IT has been developing rapidly. As a form of information technology, multimedia converse the text, images and audio, value of using a computer, then process, transmit and store them, then to produce the final multimedia works. Compared with traditional single media works, multimedia works owned the larger information density, and contain a higher human wisdom. Our country's scholar Shi Yun believe that multimedia works are similar to a computer database, both are complex arranging the software arbitrarily, process animation, text, audio, images and other information, and then produce multimedia works.

Multimedia works have the following characteristics:

1.1 *Interaction and legal status*

Because multimedia environment works exists in a binary state, which can be added or deleted, combined and shifted, etc., has certain interactivity, which is the main function different from traditional works. It is because of the increasing acts of multimedia authoring environment, the use of multimedia works in the creative process but did not give a clear division of work, making it difficult to effectively protect the work. In addition, as long as multimedia works do not take measures to address copyright issues, people need to use the regulations in the existing laws to regulate it.

1.2 *Cooperative and the powers vested*

All copyrights are owned by the authors, in view that the multimedia work are of joint authorship, its rights should belong to all participants to share. Currently scholars have suggested that the work should be in accordance with the idea of editing to give the second creation of multimedia works. But it has also been proposed that multimedia works and construction works have the right to a standard of identity work, obtain the corresponding author's instructor qualifications according to the program and its implementation makers. From the current perspective, multimedia production is conceived and designed to work on the works, and creator's concept changes correspondingly, leading to uneven distribution of works right benefits.

1.3 *Integration and creative license*

Currently, the multimedia type of media involved in a large number of major media material, and creative works are created by those inspired people, most of the work have more advanced ideas. At present, imperfect copyright public system has seriously affected the media producer responsibility, if a large amount of material, the producers tend to make multimedia processing capabilities decline, and most difficult to use creative material producers.

1.4 *Diversity and reasonable application by using the method*

Not be able to use the traditional works of the work content, structure and form of expression be changed, such as: answer can enjoy intact, but also understand listening to music, reading novels. But the way the application showing a variety of multimedia works, making its property protection issues complicated. So far, scholars have proposed the establishment of the compensation system, giving the freedom to use for private purposes. So as to be acceptance of the application of some form of compensation, and to give a substantial return on their money, time and equipment.

2 MULTIMEDIA WORKS AND INTELLECTUAL PROPERTY ISSUES

Intellectual property issues of multimedia works are mainly reflected in the following two aspects: intellectual property issues involved in the production, intellectual property issues involved in the practical application.

2.1 *Intellectual property rights in production*

In the network environment, collect the existing works as a creative material, and use and index them, then the emergence of a large number of creative works, particularly multimedia works more. Currently, multimedia works will be external information and for its use as a material, in addition, will itself involve re-creation of works given to others. If the use of external information, by way of adaptation or reference, if given the copyright for external information material, is bound to overlapping rights. Because multimedia works may own the copyright, in part copyright may have lost their own, which requires its own structure should be fully considered in the library building, and reflect the different licenses and permissions for each element. If the copyright, trademark and patent law have equal relevance multimedia development, multimedia development should be avoided due to property disputes caused by the occurrence of the event. Furthermore, it should establish a system of collective management of multimedia works, through the establishment of such a management system can effectively solve the problem of protection of intellectual property, and has been applied in practice, and achieved some success. Finally, some scholars have proposed to establish an information system acquisition cost, the production of tender system works. A fee system in which access to information refers to the information superhighway network users may have access rights, and the obligation to pay the cost of giving. Such as: Switzerland has been neutral in the relevant statutory copyright license, which will be involved in copy shops and libraries, its users are managed by associations. The work produced by the tender refers to obtain the corresponding private interests under the relevant works contracts arranged by, and transfer their creations to the government, motivate creative again.

2.2 *Intellectual property rights in the practical application*

Multimedia works in practical application is divided into the protection of digital libraries, commercial business and home of a class of multimedia works. The protection of digital libraries, mainly provides computer software, television, films and other lending services to the public. Multimedia works offered on the market can be purchased from legitimate publications, and get prior permission of the rights otherwise infringement. The business operations are involved in the protection of intellectual property should get permission of the copyright holders. If able to use in a television program, the program should focus on holding my DVD recording to sell when the copyright. If unreasonably deal with the problem, it is difficult to enter into commercial channels. As for home multimedia works for a class, there are certain restrictions through existing copyright law, which was mainly due to a variety of traditional multimedia collection works in one, but in practice did not make relevant legal basis, its emendation should give full consideration to the related basis, and increase the corresponding network information media.

3 CONCLUSION

Given the current continuous development of multimedia technology, multimedia work increases, and problems involved in IPR protection have become an important issue to be addressed urgently. This paper describes the characteristics of multimedia works, then analyzes the problem involved in intellectual property protection, aiming at ensuring the interests of creators of multimedia works. In the next period of time, more ways should be looked for to solve the problem of intellectual property rights of multimedia works, optimize and improve the methods, and promote the healthy development of the multimedia industry.

REFERENCES

[1] Si Jie, Zhang Hua. *Multimedia works and intellectual property protection* [J]. Science Weekly Revision A, 2013, (2): 4.

[2] Li Ting. *Multimedia works to protect intellectual property rights* [J]. Heilongjiang Science and Technology Information, 2011, (25): 239–239.

[3] Wang Licheng, Zhou Yutao, Liu Hongkun. *A comparative study of digital works and intellectual property protection of digital works* [J]. Journal of Intelligence, 2005,24 (12): 91–92,96.

Management, Information and Educational Engineering – Liu, Sung & Yao (Eds)
© 2015 Taylor & Francis Group, London, ISBN: 978-1-138-02728-2

Preliminary application of multimedia technology in teaching football

Zhi Qiang Cai
Institute of Physical Education, Langfang Teachers University, China

ABSTRACT: Football teams playing against each other is not only a sport, but also very intense and full of fighting spirit. With the continuous development of football, football is loved by more and more people, while the difficulty of football action increases. Teaching in football, soccer technology is an important part of teaching, including teaching methods, the use of means for students to master techniques of speed and sound quality of teaching has a direct impact. In this paper, starting from the advantages of multimedia teaching, multimedia teaching in teaching football skills were discussed.

KEYWORDS: football skills teaching; multimedia; advantage; application; Study.

In the 21st century, information technology began to flourish and quickly became one of the main features of the current society, with its powerful multimedia technology advantages widely used in various fields, soccer teaching technology education is no exception. In football skills teaching, teaching technical movements, often through explanation, demonstration and organizations, how to develop lesson plans, determine the tasks to specific teaching methods and means to make a choice on the football technical education is extremely important. Multimedia teaching methods used in teaching football skills, greatly revolutionized football teaching modes, means and methods, and how to better integrate multimedia technology used in football teaching physical education teacher who is the current problems of common concern. And before exploring this question, we first have to understand the advantages of football skills to deal with the basic features and multimedia teaching.

1 THE BASIC CHARACTERISTICS OF FOOTBALL TECHNOLOGY AND THE ADVANTAGES OF MULTIMEDIA TEACHING

Football is a confrontation with field projects, mainly in skill-oriented. In the football players' athletic ability, and football skills can be described as the most important factor. From the conceptual point of view, technology is the generic term for specialized soccer football game action in order to achieve a certain purpose and methods carried out, but also constitutes a football offensive and defensive action system.

Football skills teaching process is complicated and will take some time, has its own particularity and regularity. In the traditional teaching football skills training, teachers will demonstrate in the action areas subject to many restrictions, tend to have greater randomness; partly from the student perspective, students in viewing angle and timing there is a big limitation. Many times because of fast action and high technology integrated football difficult, students often difficult to clearly demonstrate the action of the teacher's careful observation, and these will affect teaching football skills to improve teaching effectiveness and student skill levels. With multimedia technology in teaching soccer, teachers can not only get rid of dictation and simple way to explain the limitations of technology, and can effectively mobilize students sensory learning, improve student interest in learning has an important role. [1] of multimedia applications in the classroom football skills to optimize the teaching structure, the ability to carry out the negative factors that exist in football training integrated application, so that students have a more specific action intuitive understanding, while promoting the concept of the formation of the students complete the work there conducive players master the technical essentials in a short period of time, has great significance for the soccer breakthrough technology teaching.

2 THE CORRECT APPLICATION OF MULTIMEDIA TECHNOLOGY IN TEACHING FOOTBALL

Although multimedia has obvious advantages, but must correctly apply to fully play its role, teaching football skills in multimedia applications should note the following points.

2.1 *Good multimedia courseware*

Multimedia application technology in teaching soccer, the most critical and most important task is to do multimedia courseware. In football skills teaching, application of multimedia technology is not a simple, you need teachers in the classroom organization put in more effort, especially to deal with football skills courseware "integrity, ornamental, practical" in the production process to be attention. [2] Football technology teachers teaching football techniques before making multimedia courseware, the first response to gather a lot of relevant material, such as a live video game, professional football skills training tutorials and other information can be used in multimedia courseware, teaching basic needs of the teacher can set text, pictures, video, animation, such as one of football skills coursework. On the other hand, the theoretical and practical courseware courseware is teaching football skills essential multimedia courseware than two parts, in addition to teachers courseware for football concepts, theories and content rules football competition, football technical and tactical analysis, detailed analysis in addition, also in the relevant parts of the human body courseware technical action essentials and other content at any time interspersed. About combat courseware, teachers can organize multimedia courseware football skills through technical training for football or soccer game video cut, so that students of different football skills essentials to more realistic and intuitive viewing.

2.2 *Focused on the process of applying and stressed the difficulties*

Traditional teaching football skills have always been dependent on the teacher's demonstration or theory to explain the conduct, or less if the teachers on the demonstration, the students are usually difficult to action essentials accurate, in-depth understanding and knowledge; And when the teacher lectures and demonstrations over for a long time, will reduce the student's football skill training time. In addition, the drawbacks of traditional football skill lie in football technology demonstration, different actions and different difficulty require teachers to demonstrate multiple angles and positions change, and requires a lot of language about the students to a clear understanding of football skills. Multimedia teaching should be based on these shortcomings of traditional football skills teaching for teaching focus on football while emphasizing outstanding teaching difficult. That is in the multimedia teaching, teachers can increase the amount of information output to the students and the students 'all-round sensory stimulation to promote students' technical action structure and deepen the impression of the intrinsic link between action understanding and mastery, so that students can foster formation and mastered football skills. In addition, teachers have to deal with the more difficult of the football skills mastered part of comprehensive talks, the technical action by repeatedly multimedia playback and multi-angle switching, deepen students' memories of different football skills, so that students of different football skills features and actions to fully grasp the essentials.

2.3 *Clever use of multimedia video capabilities*

In football skills teaching, students are often unrealistic expectations of problems exist in the learning process, in their eyes,it is very simple for others to make different football action , but in their own practice preached many defects, many students are on their own football action. "out of shape, "the problem is not a profound understanding. In teaching teachers clever use of multimedia video capabilities, through cameras and other technical training situation for students to shoot the ball and records, and put it in a multimedia courseware, multimedia player by means of training engineering students in their play. [3] Through an intuitive video, it is more clearly recognized that students will have their own football action problems. For example, playing in front instep teaching, by the position of the support foot, the ball is high and so the problem is a common problem of students, teachers can use these techniques in the typical problems of football filmed, so that students of their own norms and action no clear perception. In addition, teachers can also exhibit high school students for the video and personality traits common to do a special in-depth explanation, promote student memories of football skills and mastery of action to strengthen again. When the use of multimedia assisted teaching football skills, teachers also can repeatedly play, slow, freeze technical action audio-visual materials, make a variety of short-term technical action to get rid of space constraints, the longer the time to start, and this technology analysis to explain interludes in which the action of the heavy and difficult to carry out outstanding students to action structure, location details, which are more deeply aware, until the students master the techniques of real football action.

In football skills teaching, multimedia technology is a very important teaching tools and methods, not only can greatly arouse the enthusiasm of students, while improving the effectiveness of teaching football skills are important, as a combination of theory and practice of teaching football skills opens up a new road, has far-reaching research and promotional value.

REFERENCES

[1] Ningjun Hou,Tongen Yang.complementary study with multimedia teaching and traditional teaching methods in physical education teaching [J] Sport science& technology, 2013,03 (20): 17–19.

[2] Hongwei Wang.The basic problem of teaching football skills to explore scientific training[J] intelligence, 2012,06 (15): 56–57.

[3] Jiawei Zuo.multimedia technology in teaching college football empirical research [J]. Jiujiang Vocational and Technical College, 2013,03 (15): 33–34.

Management, Information and Educational Engineering – Liu, Sung & Yao (Eds)
© 2015 Taylor & Francis Group, London, ISBN: 978-1-138-02728-2

Research of modern teaching long-distance system based on the streaming media technology

Xue Feng Liu
Public Experimental Center, Xuchang University, Xuchang, Henan, China

ABSTRACT: The streaming media have become the most distinct media in the Forth Era Media due to its numerous virtues. Indeed, it is a kind of "The Fifth Era Media". But its application in The Modern Distance Multimedia Instruction is not so optimistic. At the same time though, the modern distance multimedia instruction is promising. It is indeed of a new kind of technology to break away from the how-do-you-do. The thesis introduces the related knowledge of concepts of the modern long-distance education and the streaming media technology, it demands the basic framework of the design in modern long-distance education, and it also predicts the long-distance educational trend based on the streaming media technology.

KEYWORDS: streaming media; modern long-distance education; system.

1 INTRODUCTION

The modern long-distance education teaching is the perfect combination of modern information technology and education technology, which is open, flexible, a new type of teaching mode in a learning society. However, the problems related resulting distance education resource construction is also not allow to ignore. [1] On the one hand, multimedia online teaching resources seriously scarce; On the other hand, many multimedia teaching resources can't surf the Internet. One of the main factors is limited by the limited network bandwidth. The application of streaming media technology, which is faced with limited bandwidth, implementing distance education teaching video, audio, animation, multimedia courseware, and network course transfer is the best solution. This paper introduces the related concepts of modern distance education and relevant knowledge of streaming media technology, design and implement the modern distance teaching system based on streaming media technology. The practice has proved that the system is not only beneficial to make full use of teaching resources, but also conducive to the future construction of open education.

With the continuous development of modern communication technology and network technology Modern long-distance education is a new kind of education mode, which refers to the students and teachers, students and education institutions use a variety of media between the main means for remote education system, education and communication connection form of education. Streaming Media technology is to point to in the Internet use continuous time-based Media Streaming technology, and its key lies in the network data transmission and the client in parallel. Using streaming media, the client does not need to wait for the entire file will be able to play the download is complete, namely in the form of transmission and broadcast, such both neither takes local storage space, and greatly shortens the client waiting time.

2 STREAMING MEDIA

Streaming media refers to adopting the way of streaming broadcast media format on the Internet. It is an especially specific information format encoded by a media information sources, and generated by the streaming data is a unit with relatively independent data block transmission on the Internet. Streaming is the continuous sound and image information after dealing with the compression on the web server, and before the playing client does not download the entire multimedia files, only a partial content stored in the system will start to buffer, when actual network connection speed is less than the broadcast the speed of using information, broadcast program can remove this section of the contents of the buffer, avoiding disruption of the play, and to ensure the quality of the playing. This made up of data and related control information format, data format known as flow according to the format of data called packets, a multimedia file after special encoding compression forming constituted the file all the packets of data flow, which is called the streaming media. Streaming media have changed the traditional Internet, which can only show the static text and images of defects, can provide real interactive video class, and distributed processing, processing of large-scale concurrent on demand request. [2] It can be adapted to large-scale on demand environment. Streaming media have changed the traditional Internet can only show the static text and images of defects, can provide real interactive video class, and to launch a large-scale concurrent on demand request processing, able to adapt to mass on demand environment. The model is very similar with the current radio and television. Streaming media using a special data compression and transmission technology can make voice and video file is small, usually only 3% ~ 5% of WAV, AVI files, which is very suitable for publishing on the Internet a long sound and video clips.

3 PRINCIPLE OF STREAMING MEDIA TECHNOLOGY

- Digital compression: Common multimedia files do not support streaming. Due to the limitation of network bandwidth, it is needed to use special compression coding tools for audio compression coding, lower the quality of some audio and video in order to reduce file size, and generate the text format of streaming media transmission in the network to make smooth streaming media transmission.
- Caching technology: It is essential to the streaming transmission. For the server side, the part of the memory storage space as a cache is used to store a service path of each cycle service flow from the hard disk data according to the need, at a certain rate and service to the client in order to transfer. With cache, it can take the data back to the phase separation, output services to ensure smooth output bandwidth and client stream playback continuity.
- Access technology: streaming media server must at the same time for multiple clients or multiple flow retrieve data, which is more complex likely to multiple streams at the same time access to different parts of the same file copy. So reasonable flow media data access and management technology need to carefully consider when constructing a VOD system in order to meet more flow of real-time playback, storage system must be carefully considered support maximum flow, the size of the buffer and the number of disk access strategy and the organization of the file, and implements the data access optimization. The purpose of optimizing data access technology is based on improving the access of the whole system response speed and the number of access flow to consider.
- Streaming media transmission: streaming transmission process in general is this: After the user select a first-class media service, a WEB browser and streaming media server using HTTP/TCP exchange information in order to transmit the need of real-time data retrieved from the original information. Then on the client WEB browser starts the Audio/Video Player program (that is, the client's streaming media Player), using HTTP from streaming media server to retrieve relevant parameters on the Audio/Video Player program initialization, and these parameters may include directory information, real-time transmission of data coding type, etc. Audio/Video Player program and streaming media server run the information between real-time streaming protocols (RTSP) in exchange for Audio/Video transmission control information as needed. Implement streaming typically require a dedicated server and player. Foreign related vendors consider the streaming media market prospect is good, launch a fierce competition, and successively develop out of the streaming media products. At present, the influential three companies abroad and its development of streaming media products are: Real Networks Windows Media of Real System, the Microsoft Company and Apple's QuickTime (Table 1).

Table 1. Streaming media product comparison table.

Company name	Making product	Server products	Client products	Usage agreements
Real Networks	Real Producer	Real Server	Real Server	UDP, RTSP, RTP, RTCP, TCP
Microsoft	Windows	Windows Media	Windows Media Player	MMSU, MMST, MSBD, HTTP
Apple	Quick Time Pro	Quick Time Steaming Server	QuickTime 4/5	RTP, RTSP, SDP, FTP, HTTP

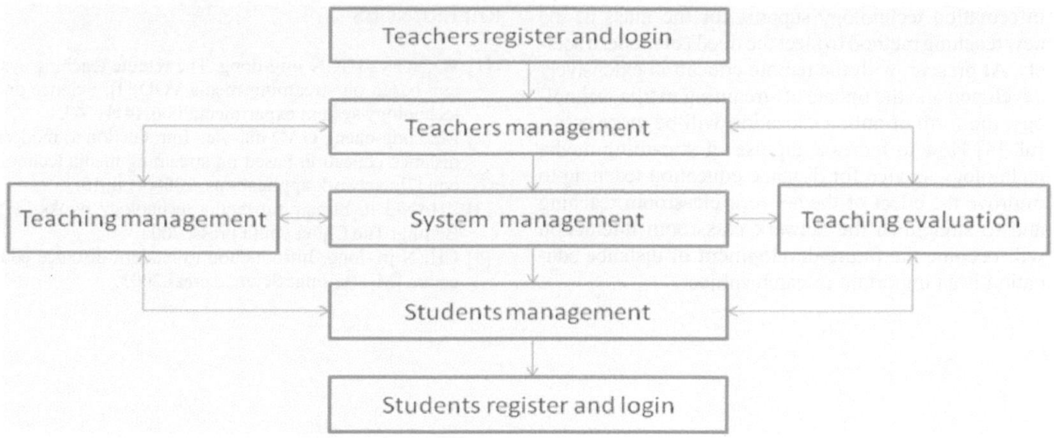

Figure 1. The software structure of modern distance education system.

4 THE MODERN LONG-DISTANCE TEACHING SYSTEM BASED ON STREAMING MEDIA TECHNOLOGY

Hardware system includes four basic parts: information to generate the streaming media system, server system, network system, the client system.

- Streaming media information generated system: Streaming media information is divided into real-time streaming media information and non-real-time streaming media information. Real-time streaming media information process is the teacher's class made streaming audio/video files, realizes the remote teaching live online.
- Server system: It is mainly composed of Web server and the streaming media server. The Web server provides Web site management and services. Streaming media server is used for streaming media storage and release. Balancing cost and effect, the remote teaching system adopts Windows Media server components, including: Windows Media Services 9.0, Windows Media Encoder9 Series Release Candidate, Windows Media Encoder 7.1, Windows Media Player 9.0 Build z903.
- Network system: It USES switched Ethernet, TCP/IP protocol, routing equipment support multicast.
- Client system: the client can through local LAN or the Internet or Intranet access streaming media information. The client in addition to install a browser, must also be installed streaming media player. At present, such as Internet explorer and Netscape web browser also supports streaming plugins.

Modern distance education platform system is mainly to complete the management of students, teachers and teaching, mainly divides into the remote teaching module, the remote teaching management module and the remote teaching evaluation module (Figure 1).

The function of the remote teaching module is divided into student management subsystem and management subsystem teachers. [3] Student management subsystem includes identity registration and certification, student status management students, students' course selection management, test management module. Teacher's management subsystem is mainly for the instructor to teaching management system, including: identity registration and certification, courseware uploading, teaching information release, the management of the teaching task. The function of the remote teaching management module is to complete the student registration management, grade management, students' course selection, teacher's courses arranging management and related professional setting and teaching and management, educational administration information management. Students through the system give the feedback to teachers, curriculum evaluation subsystem may at any time, courseware and opinions and Suggestions in the system, and the system gives these Suggestions feedbacks to the teacher or teaching staff in order to improve the teaching quality and effect.

5 CONCLUSION

Based on streaming media technology the modern distance education is not only changed the traditional Internet only show the drawback of the static text and images, reached the one-way courseware on demand, provided a truly interactive teaching and learning of the two sides of video class, change the traditional teaching mode, and established under the modern information technology support for the mass of the new teaching method to meet the needs of modern society. At present, with the remote education extensively developed and the update of streaming media technology, the form of online education will be more colorful. [4] How to increase the use of streaming media technology service for distance education teaching to improve the effect of the network classroom teaching and to strengthen the network classroom interaction will become the future development of distance education is an important research subject.

REFERENCES

[1] WANG yao SUN jing-dong. The remote teaching system based on streaming media VOD [J]. Science and technology system experiment.2006, (4)41–43.

[2] LI zhong-cheng GAO hui-yan. Introduction to modern distance education based on streaming media technology [J]. Network applications. 2006(4):8–10.

[3] ZHANG li. Streaming media technology books [M]. Beijing: The China youth press. 2002.

[4] CHEN jin-long. Introduction to modern distance education [M]. Beijing: Science press.2003.

Management, Information and Educational Engineering – Liu, Sung & Yao (Eds)
© 2015 Taylor & Francis Group, London, ISBN: 978-1-138-02728-2

Study on communication style and self-identity construction of vocational college students in English communities

Dan Yan
Environmental Management College of China, Hebei, Qinhuangdao, China

ABSTRACT: This study explores the role of an English community for vocational students to construct self-identity and the significance of self-instruction in foreign language learning from the perspective of social constructivism. The change of communication style and self-identity of vocational students in English communities are also studied in detail.

KEYWORDS: self-identity; social constructivism; communicative style; English communities.

1 INTRODUCTION

English communities play an important role in college students' English learning process. Vocational college students' communicative style and self-identity can change through participation in these communities. In this study, the signification of English communities in students' self-identity construction and its meaning to foreign language study is explored.

2 RESEARCH THEORY

"Self-identity" is a psychological concept. People build self-identities to understand individuals' physical and psychological characteristics, potential, personality, interests and social demands, and to find the integration point of personal needs and social needs. In that way, their own identity in the social environment can be built. Research on the relationship of language learning and self-identity construction has attracted the attention of many scholars in the field of language.

Lambert's "the Social Psychological Model" (1963a, 1963b, 1967, 1974)" is related to changes in bilingual development and self-identity theory "(Gardner, 1985: 132). This mode advocates L2 learning will affect learner's self-identity transformation, resulting in additional or subtractive bilingualism. Schumann's "acculturation model" (1975, 1978a, 1978b, 1986) is concerned that the process of second language acquisition will moisten the natural environment of learners, and it emphasizes that the social and emotional factors play a major causal role in second language acquisition (Schumann, 1978b). Schumann (1978a) believes enculturation is part of second language acquisition, but the extent of acculturation determines the level of the second language learning. Moreover, The Social Context Model (Clement, 1980) emphasized the importance of the cultural environment in second language acquisition. In a social environment in which one language plays a dominant role, L2 learners will also be affected by factors of two emotions: the will of moistening into the language and the fear of loss of cultural identity after enculturation. The Social Intergroup Model, also known as speech adaptation theory (Giles & Byrne, 1982) mainly focus on L2 learners in small language group members, and it believes that the degree of recognition of learners and L2 community members will determine the quality of their academic study. At present research on language learning and self identity is mostly limited to the second language context while studies in foreign language situations are comparatively less.

In China, Gao Yihong's "Productive Bilingualism" theory (1994, 1996, 2001, 2002) emphasized the positive role of mother language identity in L2 learning and the positive interaction and integration of different cultural identity to achieve value-added results so that learners' overall potential can be more fully exploited. Productive learning model discusses the possibility and importance of self-identity construction through learning a foreign language. The role of language (culture) study of people's growth is also emphasized. Most psychosocial studies on Chinese students' English-learning use quantitative research methods, and the research focus is the influence of "motive" and "learner factors" on language proficiency (Gui Shichun, 1986; Hao Mei, Hao Ruoping, 2001; Wen Qiufang Wang Haixiao, 1996; Wen Qiufang, 2001; Wu Yian, Liu Runqing, 1993; Xu Yulong, 1998). Research on motivation preset a static self-identity. Researches on cultural attachment

(Liu, 1997; Wang Xiansheng, 1997; Zhong Lili, 2000) notice the change of the learners themselves, but that is still confined to recognition of self-identity and cultural identity. Although few studies focus on the relationship between language learning and overall personality, but the study objective is the best foreign language learners, rather than ordinary college students (Gao, 2001) while studies of vocational students are even less.

3 RESEARCH METHODS

This research combines theories of social linguistics with practical situations of English communities and discusses what kind of self-identity members of English communities can build and how they construct it thus the meaning of such kind of construction to L2 learners' life can be found.

In order to verify whether self-identity changes with the internal and external situation of English communities and whether the communication style can be counted as a dimension of self-identity and what other factors can affect self-identity, the researchers decided to do a questionnaire, case survey and analysis based on several specific issues:

1 Will English learners' self-identity change due to attending English communities?
2 Will their communication style change due to attending English communities?
3 Are their self-identity and communication styles related?

Members of English Corner, the major English community in the Environmental Management, College of China are selected as research objectives of the survey. The contents of the questionnaire include the following five aspects: personal information, participation in community activities in English, impression of English communities, communicative style, the impact of English communities of self-identity and language emotional changes. Data of the questionnaires were analyzed by software SPSS (12.0). Stratified sampling method is also used in this research. Five members of English Corner are chosen as individuals for open interviews, according to their character, major, experience of English community (such as times of participation) and so on. A bottom-up analysis is also designed to focus on changes of communication style and construction of self-identity.

4 RESULTS AND DISCUSSION

This study attempts to find understanding what kind of self-identity foreign language learners of English Corner have, how they construct self-identity, and the significance of this construction to the daily life of foreign language learners. Meanwhile, the researchers would like to explore factors which are closely linked to self-identity, like communicative style and so on. Finally, the researchers reach the following conclusions:

The self-identity of learners in English Corner is diverse and mobile

In daily life, most language learners agree that the traditional communicative cultural norms have the function of discipline and supervision; while in English Corner foreign language learners sometimes go beyond the traditional Chinese cultural norms of communication and through conversion communicative style, drill and practice other self-identified. Some people practice the expected self-identity; some people merge the old self-identity with the new one; some people separate different forms of self-identity, while some people strengthen the existing self-identity. On the one hand, the research findings challenge the nature theory of self-identity and shows that self-identity and communication style are not affected by the experience. On the other hand, the result refutes the monism and stationary theory self-identity because the self-identity which English Corner participants develop is diverse, mobile, unstable, and even fractured.

Language learners of English Corner construct their self-identities in interaction with others

Traditional Chinese cultural norms, communicative rules of English Corner, the west (especially American) cultural norms and other factors influence language learners of English Corner, which are interpreted, understood, selected by the learners and transformed into important foundation of self-identity construction. Meanwhile, learners of English corner understand their needs and desires through self-reflection and self-communication, and adjust their behavior through obedience, utilization, negotiation and resistance of various forces, but also carry out ongoing construction of self-identity.

The significance English Corner has for language learners in their self-identity construction process

In everyday life, the social mainstream cultural norms, as the most powerful other factor, have a strong disciplinary function of individuals and become the most important basis for self-identity construction. The greater the risk of self-selection of other factors affected, the smaller the possibility of self-identity transformation. The feature that English Corner can hide the identity its members provide a situation of inclusion of a variety of positions. Foreign language learning in English Corner provides learners other factors as the cultural principles of the United States beyond the mainstream Chinese traditional cultural norms. Individuals can freely contact and select a different impact and practice new possibilities of self-identification thus open up space of dialogue and mutual exchange between the self and otherness. Meanwhile, the individuals face the desires and needs of self and can become persons of self-reflection, and on this basis, make autonomous choices. From this perspective, English Corner offers language Learners the possibility of self-identity transformation.

5 SUMMARY

This study shows that learners' changes of communicative style are positively correlated with their self-identity change. Use of a communication mode reflects the acceptance of the communicative body to corresponding self-identity while a shift in communication style, while communicative body's self-awareness will change with the shift in his communicative style.

ACKNOWLEDGEMENTS

This work was financially supported by the foundation of Hebei Educational Committee "Campus Language and Culture Construction under the Background of Language Variation" (Project Number: SQ141168) and Foundation of EMCC (Environmental Management, College of China) "Study on Communication Style and Self-identity Construction of Vocational College Students in English Communities" (Project Number: 2014025).

REFERENCES

[1] Gardner, R. C. & Lambert, W. E. Attitudes and Motivation in Second Language Learning [M]. Rowley, Mass: Newbury House, 1972.

[2] Hansen, J. G. & Liu, J. Social identity and language: Theoretical and methodological issues [J]. TESOL Quarterly, 1997, (31).

[3] Norton, B. Identity and Language Learning: Gender, ethnicity and educational change [M]. Harlow, England: Pearson Education, 2000.

[4] Peirce, B. N. Social identity, investment, and language learning [J]. TESOL Quarterly, 1995, (29).

[5] Price, S. Comments on Bonny Norton Peirce's Social identity, investment, and language learning–A reader reacts [J]. TESOL Quarterly, 1996, (30).

[6] Yim, Y. K. Identity and language learning: Gender, ethnicity, and educational change (Book review) [J]. TESOL Quarterly, 2001, (35).

Management, Information and Educational Engineering – Liu, Sung & Yao (Eds)
© 2015 Taylor & Francis Group, London, ISBN: 978-1-138-02728-2

The application of communicative approach in English teaching based on the IELTS speaking test

Juan Ma, Zuo Wei Huang & Xun Yu He
School of foreign language, Hunan University of Technology, Zhuzhou, China

ABSTRACT: In order to summarize the positive backwash effects from the speaking test of IELTS on the oral English teaching in universities, it analyzing the form, content and the standard of evaluation of the test and the questionnaire on the current situation of the IELTS candidates opinions towards the oral English teaching in China in detail, so as to find effective and specific strategies, like designing various kinds of teaching activities and making use of modern teaching equipment and establishing a supervision system of extracurricular speaking exercise for the improvement of the college oral English teaching. It is a productive test with a high level of reliability and validity that can test a candidate's language ability directly.

KEYWORDS: IELTS approach, speaking test, communicative, speech recognition, oral English teaching.

1 INTRODUCTION

With the development of the economic globalization, China needs to develop students who have the competence both with professional knowledge and English communicative ability. people are paying more attention to the use of a language instead of just learning its grammar. And communicative competence is of more value nowadays. Language is actually a means of communication. There is no doubt that foreign language learning is to cultivate students communicative competence of using a foreign language. However, how to apply communicative approach fully to English teaching practices in China still remains to be solved. And oral English teaching is a very important part of English teaching. The improvement of the teaching effects and the cultivation of students communicative competence are keys to oral English teaching. Communicative language teaching has therefore in recent years become a fashionable term to cover a variety of developments in syllabus design and, to a lesser extent, in the methodology of teaching foreign languages.

With the development of intercultural communication, more and more people intend to study or live abroad. Thus, IELTS, TOEFL, GRE and other international tests have become more and more popular in China. IELTS is a cosmopolitan language testing system and an essential condition to immigrate to or study in Commonwealth Nations or America. The immigration offices of British Commonwealth of Nations regard IELTS as the only standard to evaluate the English ability of immigrates.

2 METHOD

2.1 *Structure of IELTS speaking test*

Generally, there are three parts of the test. First part is introduction. After greeting each other, the examiner will ask some basic questions like the candidates life and background information.

The second part is individual speech. The examiner will give the candidate a topic card, a pen and a blank paper. The candidate will give 1 minute to prepare the speech in accordance with the questions listed on the topic card. After finish it, the examiner may ask one or two follow-up questions.

The third section consists of a discussion between the examiner and the candidate, generally on questions relating to the theme that they have already spoken about in the second part. This last section is more abstract, and is usually considered the most difficult. Final assessments would be made on a range of scales such as

Grammar, Pronunciation, Vocabulary, Communication Strategies and Task Achievement. The content of IELTS is both pragmatic and academic which consists of many occasions of study and life in reality. Through creating lifelike experiences the examiners and candidates can communicate in a more relaxed and natural way so as to complete the accurate evaluation of the candidates English proficiency and abilities.

2.2 *The design of questionnaire*

In order to make full use of the communicative approach in oral English class, the author designed a questionnaire. The questionnaire concerns about the understanding of the attitudes of the IELTS candidates towards the college oral English teaching on whether the oral English courses have helped cultivate their communicative competence. The content of the questionnaire includes the basic information of the participants, the attitudes towards the English teaching in college and in training agency and some comparisons.

The sample consisted of 100 IELTS candidates from 3 universities which are internationally recognized, prestigious academic institutions. The three universities are Fudan University, SHISU standing for Shanghai International Studies University and ECNU standing for East China Normal University. The researcher actually went to these universities to choose the students to help complete the questionnaires. Since the questionnaire is about the IELTS candidates attitude towards university oral English teaching, the students helping complete the questionnaire have all taken IELTS. As a result, the researcher handed out 100 questionnaires, and collected 92 with 91 valid. All of the participants took the IELTS. The participants are as follows(Table 1):

The number of the students who helped with the questionnaire is 92. There are 17 students from Fudan University, 48 from SHISH and 26 from ECNU.

Table 1. University.

		Frequency	Percent	Valid Percent	Cumulative Percent
Valid	Fudan	17	18.7	18.7	18.7
	SHISH	48	52.7	52.7	71.4
	ECNU	26	28.6	28.6	100.0
	Total	91	100.0	100.0	

For those candidates the researcher has surveyed, 52 of them are girls and 39 of them are boys.

Why do they take IELTS?From what shows above we can see that 53.8% of the candidates chose the IELTS for the further study and only 6.6% of them chose to immigrate. (Table2) There are other reasons why people take IELTS, such as working abroad and widening their insight. Have they attended IELTS training classes?

Table 2. Purpose.

		Frequency	Percent	Valid Percent	Cumulative Percent
Valid	Immigration	6	6.6	6.6	6.6
	Further Study	49	53.8	53.8	60.4
	Work	24	26.4	26.4	86.8
	Self-improvement	12	13.2	13.2	100.0
	Total	91	100.0	100.0	

From the above, we know that there are 58.2% of the candidates participated in the questionnaire have taken the IELTS training classes for the test. The percentage is a bit higher than that of people who chose to prepare the IELTS on their own.

The questionnaire is composed of 14 items. Every item is evaluated by 5-point Likert scale. Absolutely satisfied equals 5 points, probably satisfied equals 4 points, unable to judge equals 3 points, probably not equals 2 points and absolutely not equals 1 point. The 14 items are divided into 3 aspects which are about university education, training agency education and their comparison. The data collected from the questionnaires was mainly analyzed through the software of IBM.SPSS.Statistics.v19.0.

3 GENERAL RESULTS

According to the analysis of the first 2 parts, (by doing averaging of each question of a single person), we can conclude that the satisfaction level toward the teaching of both university education and training agency education of most students is on the level of relatively high with the percentage of 51.65 and 49.45 respectively.

These candidates are all quite satisfied with university education with the proportion of Fudan University slightly higher than the other two universities. In the meantime, the satisfaction level of Fudan University is the lowest towards training agency education. (The number below is the mean number)

The three universities are famous universities in Shanghai which possess their own English teaching characteristics and advantages. Therefore, the data I collected shows a moderate comparison between university and training agency education. When the researcher divided the candidates into different categories based on their purpose of taking IELTS(Fig.1), it is easier to

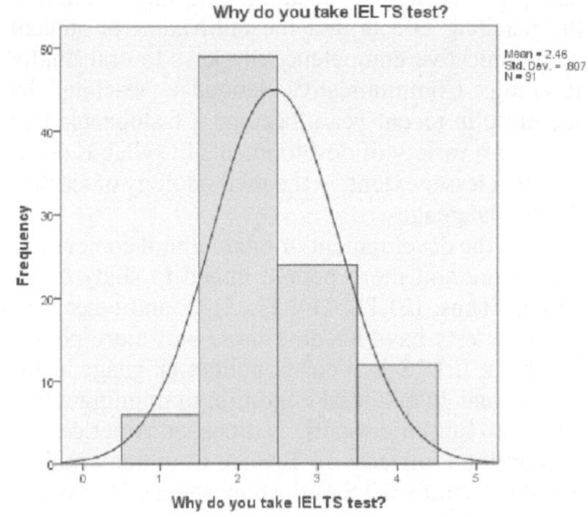

Figure 1. The chart of purpose.

Table 3. Satisfaction level.

level1level 2 level 3 level 4

X<=2 2<=X<=3 3<=X<=4 X>=4

Aspect	Level	Number	Percentage
Satisfaction	Lower	3	3.29
level towards	Low	22	24.18
university	High	47	51.65
education	Higher	19	20.88
Satisfaction	Lower	13	14.29
level towards	Low	18	19.78
training agency	High	45	49.45
education	Higher	15	16.48

notice that the differences in the satisfactory level become more and more pronounced. The number of candidates who want to immigrate is only 6.(Table 3) However, the 6 candidates have the highest rate of satisfaction of 3.63 towards university education. While for the candidates who want to seek for a better job by taking IELTS, their rate of satisfaction towards university education is the lowest with only 2.95points. The detailed information is as followed Table 4).

It is not hard to find out that most students aim to study abroad by taking IELTS and they usually think highly of both educational patterns. As for the reasons, I believe that students who eager to study abroad give higher priority to their study. Therefore, no matter it is university education or training agency education, they both can provide the students with plenty of knowledge and many learning skills. However, for the students who want to seek for a better job by achieving a satisfactory band in IELTS are actually not quite confident in their university study. They try to prove themselves through IELTS. As a result, they are not very pleased with their university education.

4 CONCLUSION

The basic theory of communicative teaching method has been one of the main trends affecting English

Table 4. Mean number of satisfaction level.

Category	Number	University	Training Agency
Fudan	17	3.25	2.95
SHISU	48	3.19	3.14
ECNU	26	3.21	3.11

Table 5. Satisfactory level based on purpose.

Category	Number	University	Training Agency
Immigration	6	3.63	2.98
Study abroad	49	3.27	3.35
Work	24	2.95	3.05
Self-improvement	12	2.97	3.37

teaching philosophy and methods in China after over three decades' research and practice of the experts and scholars since it was introduced to China in the late 1970s. It seems that the form of the IELTS speaking test is fixed, but the choices for topics are quite flexible. The communicative competence is the key to achieving a satisfactory band. IELTS speaking test has a wide range of topics which are close to real life. It also has a scientific grading system which is called the Analytical Scoring System. The examiner grades the candidates based on their performance in each section respectively. Because of all the features and advantages, IELTS then becomes a baton for some language tests and oral English teaching. The author designed the questionnaire mainly for the attitudes of the IELTS candidates towards the college oral English courses on whether the courses are able to help cultivate their communicative abilities. General results have been summarized from the basic information and presented.

ACKNOWLEDGMENT

This work is supported by the Hunan provincial philosophy&Socialist science Project(13WLH24), "Salingerian" heroes under the contest of Zen Buddhism.

REFERENCES

[1] Sysoyev, P. V. Individual's Cultural Identity in the Context of Dialogue of Cultures[M]. Tambov, Russia: The Tambov State University Press, 2001.

[2] Tajfel, H. Differentiation Between Social Groups: Studies in the Social Psychology of Intergroup Relations[M]. London: Academic Press, 1998.

[3] Tajfel, H. Human Groups and Social Categories: Studies in Social Psychology[M]. Cambridge: Cambridge University Press, 2008.

[4] Tarrant, M., North, A.C., Edridge, M.D. Social Identity in Adolescence[J]. Journal of Adolescence, 2001, (24): 597–609.

[5] Ting-Toomey,S. Communicative Resourcefulness: An Identity Negotiation Theory[A]. R.L.Wiseman & J.Koester (eds.). Intercultural Communication Competence[C]. Newbury Park, CA: Sage, 1993. 72–111.

[6] Ting-Toomey, S., Yee-Jung, K., Shapiro, R., Garcia,W., Wright, T.& Oetzel, J. Ethnic/Cultural Identity Salience and Conflict Styles in Four U.S. Ethnic Groups[J]. International Journal of Intercultural Relations, 2000, (24): 47–82.

Management, Information and Educational Engineering – Liu, Sung & Yao (Eds)
© 2015 Taylor & Francis Group, London, ISBN: 978-1-138-02728-2

The applied analysis of the multimodal PPT in the college English audio-visual course

Shu Hong Ling
Bohai University, China

ABSTRACT: With the development of the society, the multimodal interaction has become the main means and way for modern humans to communicate with each other. From the perspective of the present college English teaching situation, the traditional listening teaching mode has not met the needs of the development of the times, and cannot meet the demand of the students' listening and speaking. The multimodal PPT teaching mode is in favor of triggering the multiple senses of teachers and students, and applying it to the university English audio-visual teaching can effectively improve the students' learning interest, promote the communication and interaction among students and between teachers and students, so as to improve the efficiency of language learning and comprehensive application. This paper mainly overviews the multimodal theory and the advantages of multimodal PPT, and on this basis, analyzes the application practice of multimodal PPT in college English audio-visual courses.

KEYWORDS: College English; Audio-visual course; Multimodal PPT.

1 INTRODUCTION

The *College English Curriculum Teaching Requirements*, which was issued by China's Ministry of Education in 2007, further clarified the teaching target, and put forward to apply the multimedia technology, multimodal languages and multimodal discourse analysis theory and multivariate reading and writing teaching methods in English teaching. The necessity of applying the pedagogy of multimodal language comprehensive skills into college English teaching can be seen. In recent years, the information technology in China has obtained a rapid development, and along with the development of economic globalization and the penetration of cultural diversity, information transfer has gradually begun to present the multimodal development.

PPT is a kind of relatively simple and widely used multimedia resource in the classroom education and it is a tool throughout the entire classroom.

2 OVERVIEW OF THE MULTIMODAL THEORY AND THE ADVANTAGES OF MULTIMODAL PPT

Modal mainly refers to a way in which humans interact with the external environment through senses such as vision, hearing and so on, which can be divided into three kinds: single modal, double modal and multimodal. The single modal mainly refers to the interaction with a single sense, the double modal refers to the interaction with two senses, and the multimodal is the interaction with three or more senses. Since the 1990s, the multimodal discourse analysis theory has already appeared and emerged, which, on the basis of the theory of systemic functional linguistics and with some symbolic resources such as images, music, language, gestures etc., believed that other symbol systems, except for language, were also the source of the meaning and could be used in other communicating modals, except for language. Mr. Kress and Van leeuwen pointed out that the multimodal discourse analysis theory claimed that the discourse analysis should not be confined to the language itself, but it should also notice the social symbol with other forms, such as images, sound, movement, color, etc. [1]. After that, more and more domestic and foreign scholars have begun to focus on multimodal discourse theory and empirical studies. Multimodal discourse takes the multimodal discourse analysis theory as the instruction, and refers to the phenomenon of using a variety of feelings, such as hearing, vision, touch, etc., and fully uses many kinds of means and symbolic language such as language, image, sound, action and so on in the language communication. With the development of science and technology, PPT has been widely used in classroom education multimedia resources, and its advantages are mainly shown as follows: First, the operation is flexible. Teachers can change the size, color and position of the picture, text, symbol and so on, according to their own demands and preferences in the process of making PPT files, and when they need a dynamic video or when the additional data is too large, they can adopt the hyperlinks. At the same time, the teacher can also set the questions and answers shown according to the procedure, which can be applied to ask students questions in the class and to exercise the student's thinking ability. Second, the manufacture is simple. The production of PPT files only need to form fixed template according to the instruction, and modify on this basis in order to better display the related information of the teaching content. At the same time it can also insert some elements such as sound, audio, video etc., to greatly promote the teaching activities [2].

3 THE APPLIED ANALYSIS OF THE MULTIMODAL PPT IN COLLEGE ENGLISH AUDIO-VISUAL COURSES

3.1 *Preparation before class*

The production of PPT needs to be done before class. Before the multimodal teaching, teachers should fully collect the data and materials from many kinds of ways such as networks and databases, use voice, video, and pictures to produce the cultural phenomenon involved in the teaching content into courseware and show students in an all-round way at the beginning of the class. In addition, it also seriously analyzes the difficulties, important points and the test points of the teaching material and is able to point out the important points, refine thc difficulties and emphasize the test points, which asks for higher requirements regarding the teacher's preparation before class.

3.2 *The blackboard writing cannot lack in PPT teaching*

PPT is just a kind of multimedia software. If in classroom teaching teachers rely too much on PPT, but ignore the blackboard writing and textbooks, it will lead to students who cannot master the important points to a great extent, and cannot listen to the teacher's interpretation, being in a passive state in the process of accepting the new information and lacking subjectivity and interactivity, which will lead to the students' learning interest's decline for a long time. Therefore, in the process of teaching, teachers should pay much attention to the importance of blackboard writing, and avoid taking the PPT as a blackboard writing and show directly. The teacher should clearly distinguish between primary and secondary: show the more rich content in PPT, and reasonably apply the configuration of the picture and sound effects, which help teachers improve students' attention to the learning content. Besides, teachers should also notice the interaction in presenting the slides and the blackboard writing, and avoid switching back and forth among pages, which will have a harmful impact on the overall effect of the knowledge structure [3].

3.3 *The concrete application practice of the multimodal PPT*

First of all, before the start of the class, teachers should show the student video files and related pictures, or play some audio materials to students, according to the relevant contents of the course. These videos, images, and audio materials might show part of the contents related to the courses as well, which can cause students to guess and discuss it, and then make them understand the learning task by thinking. Secondly, in the process of collecting the multimodal materials, it should pay attention to its diversity and interesting parts, which, on the one hand, can help students expand horizons and increase their knowledge; and on the other hand, it also can effectively improve the students' learning interest and promote the communication and interaction among students. While teachers can understand the students' degree of mastering the cultural knowledge through the discussion among students, and on this basis, scientifically and reasonably adjust the course content and task designing. In this way the students' participation in the classroom learning has been increased [4]. At the same time, students can still do the situational imitation on the basis of the new cultural knowledge, and teachers can also exercise students by means of a debate, which can further consolidate the students' knowledge system. Finally, what should be stressed is that teachers should adhere to the principle of scientific and reasonable design in the multimodal design of the audio-visual teaching courses and promote the improvement of students' listening and speaking skills through various video resources. In the process, it should also pay attention to the difficulty, length, tone of voice, pronunciation etc. of the video [5].

4 CONCLUSION

Overall, the application of the multimodal teaching method in the teaching of college English audio-visual courses can make full use of various multimedia teaching resources to stimulate the students' various senses, which can not only make students master the taught knowledge and initiatively and easily finish the teaching task, but can also effectively stimulate the students' learning interest, improve the students' learning enthusiasm, and thus improve the teaching effect, which fully embodies the good prospect of the application of multimodal PPT in college English audio-visual courses.

REFERENCES

[1] Li Ping: Study on the Multimodal Teaching Pattern in the College English Audio-visual Courses. Trade Unions' Tribune (Journal of Shandong Institute of Trade Unions' Administration Cadres), Vol. 2 (2013), pp. 131–132.

[2] Zeng Qingmin: Study on the Effectiveness of Multimodal Audio-visual Teaching Pattern on the Development of Listening and Speaking Skills. Journal of PLA University of Foreign Languages, Vol. 6 (2011), pp. 72–76.

[3] Dong Mingjing: The application of Multimodal PPT in College English Audio-visual Courses. Theory Research Vol. 18 (2013) pp. 272–273.

[4] Liu Bin and Sun Xiaoli: Exploration of the Multimodal in College English Audio-visual Teaching. Course Education Research, Vol. 23 (2012), p. 12.

[5] Feng Xiuying: Possibility of Creating Multimodal Integrated Skills Pedagogy. Journal of Bohai University: Philosophy and Social Science Edition Vol. 4 (2012) pp. 114–117.

Management, Information and Educational Engineering – Liu, Sung & Yao (Eds)
© 2015 Taylor & Francis Group, London, ISBN: 978-1-138-02728-2

The discourse remodeling of ideological and political education in the new media age

N. Sha
Bohai University, China

ABSTRACT: With the continuous social development and progress of science and technology, the school education enters a new teaching environment. Chinese traditional ideological and political education courses have been replaced by new media technologies, and the new teaching methods have penetrated into every learning aspect. Traditional school ideological and political education courses largely can not meet the needs of modern society, with problems arising in the carrying out of the course curriculum, such as low efficiency teaching and lagging education which are the key issues. To improve the status of ideological and political course, the primary necessity is to value the student's right to speak, on the basis of which to achieve inter-subjectivity from subjectivity. This paper focuses on reshaping the right to speak of the ideological and political education in the new media age.

KEYWORDS: New Media Age; Ideological and Political Education; Remodeling.

The new way of media education is mainly a media form by using the modern technology as a supportive point of the information delivery. Though new media technology has been widely applied to every aspect of daily life, the implementation of ideological and political courses is not very mature thus the situation of unconvincing discourse appears. Therefore, it is necessary to improve the effectiveness of the right to speak which requires the use of new media on a certain scale to ensure the accessibility to discourse resources for the students' ideological and political education courses, on the basis of which to make reasonable discourse changes and improve the discourse effectiveness. The ideological and political education is a major course that universities as well as the government attach great importance to. And with the arrival of the digital media age, which has brought to the ideological and political education new vitality as well as new problems, there is great need of innovating current understanding of the situation and a sound grasp of the discourse, power of ideological and political courses, both of which constitute a most effective way to improve the quality of ideological and political courses.

1 THE LACK OF DISCOURSE POWER IN IDEOLOGICAL AND POLITICAL EDUCATION IN THE NEW MEDIA ENVIRONMENT

1.1 *The weakness in the means of traditional ideological and political education*

Traditional ideological and political education classes featured a very fixed delivery time and a limited class hour. Then ideological and political courses under the new media environment would be limited as they enable students to access information as well as some services through the network at any time, during which students can share different perspectives with each other. The implementation of the new media education exposes unreservedly drawbacks of the traditional ideological and political education. From aspect of the class time, traditional ideological and political curriculum could hardly allow students to access and share information at will on the whole courses, which produced some adverse effects.

1.2 *The limitation of the traditional lecturing way of education*

The most distinctive characteristic of the new media way of the ideological and political education is its personalization. Many new media ideological and political education provide students with various forms to get more information, such as blog, SNS and micro-blogging, etc., which greatly improve the students' individualized expressions and provide them with good channels and platforms. Compared education under the new media environment with the traditional education, the latter ideological and political education emphasizes the completeness of the knowledge system and the rationality and scientificity of the structure, as well as the content preciseness and seriousness while imparting knowledge, which creates a self-centered lecturing style and branches off with the personality development of the students. The didactic educational methods have lagged behind the due education under the new media environment.

1.3 *The lack of information establishment*

Under the new media environment appears contending phenomenon. That is, instead of directly flowing to the information absorber, the information and some oriented opinions in the network communication first need go through an intermediate links, which commonly controlled by those called "opinions leaders". By virtue of their rich experiences and knowledge accumulated, people who establish the information could attract information gatherers to follow them and thus making himself in the center position. Those information-establishing people could not only be able to strongly spread the information, but also change the direction of information flow, thus control the direction of public opinion on a large scale. The momentum and the number of participants of some opinions the network are very large, so that on a large scale, there is a lack of information establishment, which turns to be the responsibility of professional information builders.

2 THE RESHAPING OF THE IDEOLOGICAL AND POLITICAL EDUCATION DISCOURSE IN THE NEW MEDIA ENVIRONMENT

2.1 *Stimulating the education motivation and improving the number of discourses*

In this regards, it is necessary to develop a reasonable teaching team, a team composed of master teachers with new media technologies and sound understanding of students' thinking and behavior, which is an important safeguard to vigorously carry out the ideological and political education under the new media environment. The establishing of a strong faculty group largely requires a reasonable mechanism to motivate and organize teachers to engage themselves in ideological and political education. The major institutions make a good implementation of the ideological and political education plan in the new media environment and achieve good results. Regarding the scientific evaluation mechanism for education, it is necessary to evaluate teachers' job performance and set up some incentive awards or other ways to build a positive ideological and political education atmosphere. It will be beneficial to guide teachers to carry out ideological and political education by making most of the new media.

2.2 *Improving the professional accomplishment of the teaching team*

3 CONCLUSION

The future teaching mode is to abandon the traditional teaching method and develop students' individual self-study. Micro-class at the MOOC age will help students find learning problems themselves, and then by implementing strong oriented micro-courses to answer these questions. But to find the real problem is not easy. Under the ideological and political education curriculum reform, the core of education is an mode after the teaching subject exchange, while the core teaching guide still is the teacher. During the teacher training, training content should also be based on the actual content of the current education. Teachers should always pay attention to the problems in students' education to create an easy atmosphere for the students so that they can better develop their specialties through the new curriculum reform.

ACKNOWLEDGEMENT

The periodical research achievement of "Social science planning of Liaoning Province–the study of cyber language value" (Project Nr: L13DYY028).

REFERENCES

[1] Xiao Qingsheng, Ren Jiawei, Liu Chang. *The Construction of Scientific Discourse in College Ideological and Political Education under the New Media Context* [J]. Ideological and Theoretical Education, 2014, 04:83–86.

[2] Chen Ning, Zhou Xiang. *The Remolding of College Ideological and Political Education Discourse under the New Media Context* [J]. Beijing Education (Morality Education), 2011, 10:10–12.

[3] Ren Yan. *A Primary Discussion on the Ideological and Political Education Discourse under the Digital Age* [J]. Social Sciences Journal of Colleges of Shanxi, 2014, 05:63–65.

[4] Qin Xiulian. *A Research on the Effectiveness of Ideological and Political Education Discourse under the Network Context* [D]. Lanzhou University, 2013, 03:01–02.

Management, Information and Educational Engineering – Liu, Sung & Yao (Eds)
© 2015 Taylor & Francis Group, London, ISBN: 978-1-138-02728-2

The optimization study of Chinese language teaching based on multimedia technology

Xue Jun Li
Pingdingshan Industrial College of Technology, Zhanhe River District, Pingdingshan, Henan, China

ABSTRACT: With the rapid development of high and new technology, the computer auxiliates teaching in the classroom, to deepen the teaching reform, training students' ability, providing the new heaven and earth. Using the computer multimedia technology, optimizes the text presented way, improves the students' reading ability, reasonably play imagination, another poem image, implement the rapid positive transfer of knowledge.

KEYWORDS: Multimedia; Optimization; Classical poetry; teaching.

1 INTRODUCTION

At present, the computer aided teaching and promote the popularity, especially in Chinese classical poetry teaching in junior middle school, there are many teachers made a valuable attempt and exploration, there are many illustrated, well-made courseware, but in the concrete teaching practice, there are also many shallow and inefficient, for example, the excessive pursuit of vivid images presented, ignored the key of Chinese teaching, improve students' ability of language expression, asked the students to drawing according to the passage, language lesson into the art class. I think, the computer multimedia is a tool, our starting point should be how making good use of the tool service for Chinese teaching, on the back of the computer multimedia technology, should be a concept of modern Chinese teaching, such as cultivating the imagination of the students' ability to innovate, develop the students' personality and so on, not for the purpose of using tools and make teaching become a mere formality.

2 USING THE MULTIMEDIA TECHNOLOGY, CREATING LEARNING SITUATION, STIMULATE STUDENTS' INTEREST IN LEARNING

Classical poetry is very exquisite artistic conception and lasting appeal, therefore, teachers should set focus on the classical poetry teaching situation, to arouse the students' interest, guides the student to feel distinctly the poet's emotion. Multimedia can set text, voice, images, graphics, on the same interface, has affectionately, audio-visual blend, crisscross of the movement, the characteristics of strong appeal, broadening the horizons of the students, provides a vivid and specific image thinking material. Let's the student in learning ancient convincingly, vivid, intuitive understanding of poetry, to experience feelings, have the effect of addition, bring love. In the design of multimedia teaching software, choose several interconnected, before and after the echo of the lens, using multimedia, make the vision, close shot, panoramic appears alternately, provides students with rich and colorful images. Class, I simply by keystroke simple operations, you can see from the screen image associated with the text content: cloud hanging over the purple mountain, waterfall pouring from the top of the mountain, water transpiration; Instant, the waterfall falling rapidly along the steep and high wall, FeiZhuJianYu, momentum magnificent. And the underwater acoustic rumbled shaking each student's soul, stimulate their visual and auditory. In this case, the situation of poetry full show in front of the students, let them from the view of the overall preliminary experience illustrated in the figure and verse, arouse interest in learning poems. The teaching effect is generally illustrations can never achieve. If there are any students need to see, as long as with the mouse key, can enjoy yourself at any time to the required image. Students can be located in the audio-visual infection, read poetry, every taste, feels the scenery poem, beautiful language, causing the emotional resonance.

3 USING THE MULTIMEDIA TECHNOLOGY, REPRESENT THE POEM SITUATION, HELP STUDENTS TO UNDERSTAND THE POEM

Poetry is to provide the poet's feelings, and emotions must through image constitute a kind of artistic conception, and then with the help of language expression. The situation in classical poetry, if only rely on the teacher's description of the language is not sufficient. Only through the pictures reappear, become visible. "Looking at Luster Waterfall", written by a poet looks at the magnificent panorama of waterfalls seen when, praised the country's native land, the beautiful scenery of imagination and reality is naturally blended together. When teaching, I let the students watch the video, a waterfall at the top of the mountain torrents, water splash, water transpiration, sunlight, display fan receive the purple. The steep cliff; Long flowing a waterfall, splash water, luster waterfall panorama will appear in front of students. Thus the student to "purple smoke, hanging and flying" has the emotional understanding of the meaning of the word. So that the students deep investigation to understand the poem, though short, but the poet writes waterfall momentum to perfection! Is the poem painting, painting in poetry. This stimulates the student to love poems, love the motherland magnificent rivers.

Classical poetry is characterized by "the poem painting, painting in poetry, poetry and texts." A poem is usually a landscape painting, a picture of a pastoral scenery figure. From refining, which is mainly composed of image thinking and primary school students it is difficult to assess the true poetic words. A dynamic demonstration of multimedia courseware, can represent the situation, catch up with the defects.

4 USE OF MULTIMEDIA TECHNOLOGY, THE ESTABLISHMENT SITUATION, ENRICH THE STUDENTS' IMAGINATION

Classical poetry is colorful in writing, rich smell of flowers, is a poet when he is with tongue watering a common passion, those lovers of artistic conception far-reaching, are thus suggesting the poet's heart of emotion. Ye shelter once emphasizes classical poetry teaching is temperament, extended to imagine. In teaching classical poetry, therefore, can make use of the multimedia establishment situation, inspire the student to the wings of the imagination, the leap of linking up emotions, to omit the artistic conception of linking up, in this way can we truly understand the poet's feelings, into creating an artistic conception of the poet. Such as teach "mountain" the poem, to guide students to read first, chanting "far handgun oblique stony path, in white clouds deep YouRenGu" these two words, and then let the students according to the life of the accumulation of knowledge, use language to describe the autumn in the mountains, winding stone path, white clouds, the top of the mountain of the image of the dusty cottage; Then guide reading "parking sit love maple's woods night, leaves be red in February flower", the teachers are using courseware shows a red maple's woods, would create a picture of bright colors, levels in ancient landscape paintings, to inspire the student to draw about poetry: achylia showing Lin and stop watch is filled with an aura of a robust, poetry written in both pure and fresh and lively. In the process of layer of in-depth, the student to grasp the artistic conception of poetry, imagination and appreciate the beauty of the poems.

5 USE OF MULTIMEDIA TECHNOLOGY, OPTIMIZE THE READING TRAINING, AESTHETIC EDUCATION

Classical poetry due to the particularity of its artistic form, also has the language, music, painting, the beauty of emotional factors. These aesthetic factors in mining, multimedia technology have a powerful and incomparable advantages. Ancient rhyme match, the rhythm is bright, reads catchy, sweet, present a cadence beauty of music. Of classical poetry to read charm, however, is not an easy task. Because poetry is the language of passion and imagination, this often is "emotional" the cryptic clues. So in helping the students read a poem, on the basis of experience, we are about to read through the rhythm of the poem, the rhythm of poetry, to cultivate the students' language sense, emotional edification to their reading.

6 USE OF MULTIMEDIA TECHNOLOGY, IT'S REAL, WITH PASSION

The selected poems in the textbooks are mostly lyrics, some emotion in the scene, some scene, some's mind. Teachers can use multimedia in the teaching, with vivid images and profound artistic conception to mobilize the students' feelings. Guides the student to experience the feelings of the poet, struck a chord with ideology. Such as "choice" is a poem in the elementary school lower grade teaching materials, students today wealthy, most of them have no farm life experience, have a meal at ordinary times very serious waste phenomenon. So let students experience the farmers working hard and education they will cherish the work achievement, is this poem teaching difficulty. When teaching teachers using multimedia to show students a video like this: the farmer must sow, the sun sweating weeding, fertilizing... Last picture frames on the "how grain to be noon for the day, began sweating grain soils", don't need too much explanation, students are deeply realize rice hard-won and poetic sympathy for farmers and inspire their compassion and love for workers. Images and then play a set of life: many children play in a wheat field, and trampled the wheat seeding; A friend took left-over white steamed bread on the ground... Teachers are to guide students to team discussion: the scene. What do you want to say to video about children? Two corresponding, students naturally experience "behold dishes, each all pain" of the profound connotation.

7 BY ADOPTING THE TECHNOLOGY OF MULTIMEDIA, THE AUXILIARY CHANTING, TASTE FOR FUN

Classical poetry also has a sharp, words, grace, rhythm, harmony, rhythm and the characteristics of easy to remember, easy to read, especially suitable for pupils to read. When teaching classical poetry, the teacher should arrange ample time, utilizing a variety of forms such as reading aloud, background reading, let the students in reading, chanting scent, comprehension and expression. In the reading guide, the participation of multi-media means such as pictures, sounds, music to the student accurately, grasps the tone of the poem is very helpful.

8 USE OF MULTIMEDIA TECHNOLOGY, ENRICH THE INFORMATION, KNOWLEDGE INDEPENDENTLY

The large capacity of the computer network is a big advantage, not limited by time and space, it can provide a wide range of learning content and variety of presentation, can use hyperlinks to easily jump and consult, students can according to you need content and the way to obtain information, processing information, applying information. Therefore, in the teaching, I take full advantage of the characteristics of computer network greater participation, let the student through participation, man-machine participation behavior, participation, thinking and other ways to achieve their learning goals, inspire the enthusiasm of learners' active participation, give full play to their main body role, make the class full of vitality.

9 USING THE MODERN EDUCATION TECHNOLOGY, INCREASE THE DENSITY OF TEACHING, IMPROVE TEACHING EFFICIENCY

The teacher before using the multimedia technology in teaching, according to the lesson plan design programmed teaching software. Although the need investment a lot of time and energy, the elaborate design and preparation before class, however, can greatly improve the efficiency of classroom teaching. In class, the teacher as long as the click of a mouse, can in a short period of time to make the students see the clear picture, vivid video and the sound of music... Avoid teachers hands-on AIDS when the controls and attend, winning the teaching time, improve the teaching efficiency.

10 CONCLUSION

Auxiliary teaching using the audiovisual education, media, can effectively turn the abstract into concrete, become boring to fun, turn static to dynamic, create good atmosphere for the development of students' thinking, to better development, the potential of students' personality get full development, to produce learning drive play a great role in promoting, optimization of the classical poetry teaching, let the classical poetry teaching to glow the new vitality.

REFERENCES

[1] Xiangxian Xie. Language education [M]. Zhejiang: education publishing house, 2001123.

[2] Hua Zhang. Curriculum and teaching theory [M]. Shanghai: The educational publishing house, 2000,88–96.

[3] Ziran Chen. Talk about the application of multimedia courseware in Chinese language teaching [J]. Journal of Chinese construction, 2005, (8): 32–34.

[4] YunGe. Use a lot of media this "double-edged sword" [J]. Chinese Journal, 2005, (5): 21.

[5] Jianjun Xu. Chinese teaching with multimedia technology application [J]. Journal of secondary vocational education, 2003 (5).

[6] Angong Hu. Multimedia teaching is a double-edged sword [J]. Journal of Modern Chinese, 2004 (7).

Management, Information and Educational Engineering – Liu, Sung & Yao (Eds)
 ISBN: 978-1-138-02728-2

Value of network language in linguistics perspective

Li Wei Qu
Bohai University, China

ABSTRACT: Language development has very close ties with society. On the one hand, the social development will affect the development of language; on the other hand, the development and changes of language also reflect the social development and changes. With the information development of society and the improvement of people's living standards, the number of netizen increases sharply; network language has also been widely popular. Network language with varieties of forms not only develops the network literature, enriching people's lives and communications, but also makes the development of network culture more open and diversified. This paper mainly analyzes the generation background of network language and its types and characteristics, and from the linguistic point of view, explores and studies the value of the network language.

KEYWORDS: Linguistics; network language; background; definition; value.

1 INTRODUCTION

In recent years, along with the popularity of the network, as a special form of verbal language, the novel, vivid, concise and distinctive era featuring network language is applied more and more widely, its form are also more and more, and being more welcome and useful by more people. Development of language is done in the social environment and is influenced by it, at the same time, the development and changes of language is also a side reflection of social changes. In broad terms, the production, development and demise of Internet language is a linguistic phenomenon, but it is also a social phenomenon, from its development process, we can not only see the social and cultural changes and social life, but also can see the ideology and the value orientation change of social groups.

2 GENERATION BACKGROUND OF NETWORK LANGUAGE

Language is generated in the process of development of human society, and is always in dynamic development and change. Now there is a network of language has aroused concern and attention, there is a network of language is accompanied by the emergence and development of this new media network generated. Compared with the traditional form of language, network language is mainly used for virtual network platform and communication activities within the community, as well as a wide range of applications with the popularity of micro-blog, micro-channel network, it began a meteoric rise, has become universal features. Network with people's lives, so the network is not confined to the language network, but began to have an impact on people's daily lives unconsciously. For example, the first popular "Dandan body" and "Taobao body", and the later "TVB body" and "Smurfs body", and the modern popular "In those years body" and "Hold body", which are typical network language, with the characteristics of stylish, fresh and fast and certain social values. The main reason for the network to produce the language can be analyzed from the following aspects: First, the rapid development of the Internet to promote a wide variety of platforms application of various dating platforms, such as micro-blog, letters, etc., so that communication between people increasingly more frequently, increasing the people's access to and use of network language opportunities; Second, the people for the big stuff in the community have a stronger spirit of entertainment and social responsibility, especially young internet users, more like a network platform for their own express the views and opinions of freedom; Third, users relatively unassuming personality, unconventional behavior, emotional expression is also more open and active, thinking more active, and therefore the development of a simple, unique, satire, ridicule, joking, education, entertainment and various other forms of network language [1]. These all have laid a solid foundation for generation of diversified and open and inclusive language network.

3 TYPES AND CHARACTERISTICS OF NETWORK LANGUAGE

Network language can be divided into two types, one is the terminology which are related to computer networking or internet activities, such as online, dropped, home, links, bandwidth, downloads, flow and e-mail, and the other one is netizens media information symbols with BBS, chat rooms, online chat tools such as passing, and other ways to use when chatting online-Internet chat language, this paper explores the second type. Internet chat users generally require typing faster and simple language, easy humor, which formed some special symbols. For example, graphics table knowing: : D laughing, :) smiling, : -! Ridiculed, : - < forced smile, : - 7 furious, 8-> eyes wide open, and so on; digital knowing: 555 oo, oo, Kazakhstan 9- drinking, 100 or 10 - perfect, 56- bored, 9494 - that is, 7878- go go, etc; homophonic knowing: ou - I, Banzhu - moderator, Shuijiao - sleeping, Luzhu - the landlord, etc.; spelling and phonetic abbreviations: LP- Laopo (wife), LG-Laogong (husband), MM- Meimei (sister) (crush) and BB-Bye Bye (bye), etc; intentional overlap: look - look, the stuff - stuff, pretty - splendidly; unique to borrow, said: Frog - poor appearance of male worms, dinosaurs - poor appearance female worms, etc. [2]. Overall, the network language has the characteristics of fuzziness, wrongness, deformation and temporariness.

4 VALUE

4.1 *Communication value*

There are many similar network language "MM", "Yemeir (email)", "Konglong (dinosaur)" and other words with obvious simplification or joke, but with the standardized management of network and improvement of users' quality, main network languages begin to be generated from the hot news events and news phenomenon, which to some extent also accelerated the dissemination and discussion of the social phenomenon of news events, has a richer and more profound meaning of sociology and journalism. For example, a network language that appears relatively early, "Fuwocheng (push)", the word is derived from a 14-year-old girl drowned in the incident in Weng'an, mainly for teenage girls jumping into a river to hold a companion to indifference triggered a lively discussion of Internet users, " Fuwocheng (push)" began to be widely used in the network language, represented cited without issue, do not care, ignore, indifferent attitude of indifference and even unmoved, etc. [3].

In addition, the network language often can also cause people to focus on a certain type of social phenomenon. For example, in 2008, "Shanzhai" mobile phone is widely used, "Shanzhai (cottage)" is characterized by quick imitation, small workshops started and simply copy, it has become a household name as a word, and has aroused people's attention and discussion. Initially on the "cottage" hot mainly around the cottage phone rationality of the existence of the debate, followed by digital cameras, notebooks and other areas have begun to appear copycat phenomenon and cottage industries, thereby enabling the alleged plagiarism or reflect grassroots wisdom become a "cottage" of public opinion focus [4]. Thus, some network language as a summary and concentration of the news events, just a word can express a very deep meaning, also greatly enhance the event propagation speed.

4.2 *Social emotional expression*

Network language not only has a wealth of meaning, expressing concern about an event, while also is able to express their emotions through cold humor, pun, satire, etc., and thus has a broader sociology meaning and journalistic significance. The representative of Internet language with cold humor characteristic is "Dajiangyou (soy sauce)", the expression is the attitude of scorn that a matter does not relate to oneself; representation of a pun feature is "Duomaomao (hide and seek)", an expression of intense questioning and anger; representation of ironically feature is "Luoyoujia (naked oil)", the expression is trying to hide something [5]. Thus, network language not only to some extent enrich people's living space, but also played a role vent negative emotions. In addition, the network language plays a very important role in promoting the formation of public opinion force composed of media network and traditional media.

5 CONCLUSION

Internet language has timeliness, usefulness and diversity, is a reflection and portrayal of people's lives and thoughts. With the popularity of the use of the network, the use of network language is more and more frequent, and therefore draws more and more concern and attention from all aspects of society. By analyzing the value of a network language from the linguistic perspective, we believe that network language is an unconventional, unorthodox expression popular in a wide range, can greatly improve the spatial of Chinese freedom of expression, has important value in the dissemination of news events and social emotional expression of Internet users.

ACKNOWLEDGEMENT

The periodical research achievement of "Social science planning of Liaoning Province—the study of cyber language value" (Project Nr: L13DYY028).

REFERENCES

[1] Ning Shanshan. Chinese Network Buzzwords Research in Subculture Horizon [D]. Jishou University, 2013.

[2] Qi Yangyang. *Language Vocabulary Meme Network Analysis - Also on the Network Language Foreign Language Vocabulary Teaching* [D]. Liaoning Normal University, 2012.

[3] Duan Yuanyuan. *"Network Body" Evaluation Theory Perspective Language Phenomena Analysis* [J]. Southwest Agricultural University (Social Science Edition), 2013,11 (2): 74–77.

[4] Jiang Yuqing. *Memes Chinese Network Buzzwords Theory under Study* [D]. Qufu Normal University, 2012.

[5] Qi Xiangyi. Network Language: Public Discourse Practice and Discourse Game [D]. Guangxi University, 2013.

Management, Information and Educational Engineering – Liu, Sung & Yao (Eds)
 ISBN: 978-1-138-02728-2

The content construction and use form of resources applicable to the fragmentation of learning

Yi Fei Li
Bohai University, China

ABSTRACT: With the continuous development of digital technology, the use of mobile terminals to achieve fragmented learning becomes an effective aid in the form of extra-curricular learning. Unlimited space of time, in a relaxed atmosphere diversification, the fragmentation of the time utilized to enhance students' interest in learning efficiency and learning. This article from concept to start learning English fragmentation, fragmentation for the form and content of the English-learning made specific construction method, but also on the content of proposed application resources precautions.

KEYWORDS: Fragmentation of learning; resources; construction; use.

1 INTRODUCTION

The use of a mobile terminal as a fragmented learning, learning platform, you can a little bit of "fragmentation" utilized in student learning, leisure time, to achieve fragmentation, miniaturization, contextualized and personalized learning, which kind of learning in the form of interest is strong, strong interaction, it is worth to try and promote.

2 FRAGMENTATION LEARNING CONCEPTS

Fragmentation of learning is a new way of learning, mainly based learning resources and micro-terminal fragments of learning in a short period of time, focusing on small learning units. Fragmentation of learning resources have independence and short features, may be an explanatory picture, it can be a text message, the length of a short instructional video or a piece of news. Fragmentation of learning on learning with miniaturized carrier, the content reflects independence, learning when the time presents fragmentation characteristics, place of study are more random. Currently, with the continuous development of digital technology, the fragmentation of the application has been learning English education researchers great concern. Fragmented learning provides students with a new way to learn, but also for the professional development of teachers had a huge impact.

3 LEARNING RESOURCES, CONSTRUCTION DEBRIS

3.1 *Learning resources in the form of construction*

Content on the accumulation of resources, mainly in the form of "micro-blog blog + + micro letter", interact in fragmented learning.

1. Blog. Blog can support video, audio, images and hyperlinks content, can stimulate students in a variety of senses

Functional response, the formation of student learning styles under the "multi-tasking generation" of digital technology conditions. Teachers and students interact together to accomplish the task of learning to improve learner motivation in learning, play learning potential.

2. Microblogging. Fragmentation of learning in the form of microblogging, improved learning fast, rich content learning. And microblogging has the length and format of the content is not limited, to facilitate the consolidation and resource summary teachers. At lower spread costs, students can take advantage of. Fragmented time learning [1]. Microblogging use in English teaching, has been a favorite student population. Full play to the advantage of the rich resources of the network, changing spatial and temporal separation characteristics. Avoid the traditional teaching, problem solving often by students' mental, location, time factors limit. To

facilitate the efficient and effective form, it allows students and student, student and teacher exchanges on teaching issues increased.

3. WeChat. In addition, the micro-channel is also an effective form of learning is fragmented. On the official micro-letters, teachers can play a micro curriculum resource, students in the mobile terminal, using fragmented way of learning for learning's knowledge should be consolidated, learning requirements and learning problems posed by the teacher feedback. In this fragmented learning process, mainly between teachers and students to communicate through micro-channel group response.

3.2 *The content of the resource accumulation*

When carried out on the fragmented learning content design, the first step is to learn to fully consider the time, fragmented time, the random characteristics of space debris of the content and form of English, should be a small piece of information, and based on a certain theme, in difficult degree is relatively simple and easy. The second step, taking into account independent content, to facilitate students into learning state. Finally, the content in the organization and presentation of time, be filled with fun and attractive, consistent in order to meet students in learning English when studying personality characteristics, making the theme and content designed to attract and keep the attention of students in the learning time.

1. Finishing accumulated word fragmented learning. When accumulated, to consider different English-magnetic, and memory skills. English Vocabulary Builder, generally have the original root, then add the suffix, prefix components. For example, "normal" means that the usual, the prefix ab becomes "abnormal", which means "unusual." Again, "child", a child; "childish" means "childish." Compound nouns constitute "mooncake, bookshop" and so on. There are adjective compound words good-looking, hard-working ", etc. [2].

2. Sentence fragments accumulated learning. Finishing sentences when micro resources, taking into account accumulated after use - to read and write English. With particular emphasis on some of the English essay writing time, the accumulation of classic sentence, easy to use to go after the use of the. For example, the beginning of the essay: Most people are of the opinion that But in spite of..., I personally believe that... Now people in growing numbers are beginning to realize that... the end of the English essay: From what has been discussed above / Taking into account all these features / Judging from all evidence offered, we may safely draw the conclusion that... respectively formed different learning individual learning resources, the accumulation of finishing to students.

In addition, teachers need to learn the lesson of learning objectives, priorities and difficulties and the corresponding cultural background, using different forms of multimedia presentation in English learning platform fragmentation. The main contents of the book in conjunction with test points, to form an independent micro resources into the key elements related construction.

4 FRAGMENTATION OF LEARNING RESOURCES

4.1 *Teachers play a guiding role in the fragmentation of learning*

Use the form of microblogging, the aim is to achieve true equality exchanges. This requires teachers to play an active role in guiding the fragmentation of learning, the first student from the teacher began to abate equal dialogue characteristic of their identity, in order to ease the psychological burden of students have the right to speak [3]. But also pay attention to timely supervision and restraint, regulate the behavior of students in interactive learning platform to minimize avoid off-topic, or formality phenomenon.

4.2 *Student autonomy and personal use*

According to their level of learning and cognitive level, several independently repeated listening to look at these data, generate and their learning levels and learning styles of learning habits to achieve personalized learning. Students can take the form of questions and answers and debate on this platform, teacher feedback Bachelor preview and review the effect of making the teacher in the classroom can be effective for the important and difficult to explain. Students also actively submit teacher quiz content on this platform. For example, in reading and writing in English speaking practice, students can use the form of the video, will be submitted to the outcome of spoken language training to teachers to facilitate teachers to make targeted guidance, but also to facilitate effective communication between students.

5 CONCLUSION

Construction and application of the fragmentation of content resources, not only helps to effective communication between teachers and students, more conducive to individual learning applied to study the formation of personality habits. Teachers should use this new form of the same time, enrich their professional knowledge, research skills, teaching skills, in the digital context, to play a supporting role in English literacy training.

ACKNOWLEDGEMENTS

The periodical research achievement of "The general project of humanities and social sciences of education department in Liaoning Province—The Study and Practice on "Micro Learning" Teaching Model of Reading and Writing Course of College English in Digital Environment".

REFERENCES

[1] Yuanyuan Sun, Xiaoming Jiang, Shengquan Yu. Middle School English Teaching based handheld devices [J]. Modern Educational Technology, 2011,19 (3): 46–49.

[2] Yunfei Li, Minjuan Wang, Jiajun Wang, Weika Xiei, Ruimin Shen, Jason Wu. Mobile learning systems and related learning mode [J]. Open Education Research, 2012, (1): 152–158.

[3] Gaoxin Li. Microblogging application in language teaching [J]. Exam Week, 2010, (53).

Management, Information and Educational Engineering – Liu, Sung & Yao (Eds)
© 2015 Taylor & Francis Group, London, ISBN: 978-1-138-02728-2

Empirical study of user's behavior based on Folksonomy —taking Douban website as example

Hui Xiang Xiong & Chen Ling Wang
School of Information Management, Central China Normal University, Wuhan, China

Si Yuan Guo
Economic and Management School, Wuhan University, Wuhan, China

ABSTRACT: Taking Douban website as example, the paper analyzes the distribution rule of tags and the time change regulation of tags' usage. On this basis, the paper deeply analyzes the discovery of users group based on tags by constructing a co-occurrence matrix and by applying techniques such as the cluster analysis and the multi-dimensional scaling analysis and it also explores users' interests in order to recommend personalized information to them.

KEYWORDS: Folksonomy, tag, discovery of users group, personalized information service.

With the wide popularization of Web2.0, the Folksonomy technique becomes mature and users are getting used to describe information and express feelings through tags. A tag is a word selected by users on the basis of their own understanding of resources in order to describe a property, a characteristic, a function and so forth of resources. It can reflect both visible and invisible characteristics of resources. A tag can be characters actually existing in resources and can also be users' abstract understanding of resources [1]. Users use tags not only to list and describe resources, but also to express their subjective feelings about resources, such as their personal preferences or individual tasks [2]. Time and frequency of users' utilization of tags, to a certain extent, also reflect users' interests and the transfer of their interests. For users with same cultural or social backgrounds have a consistent understanding of specific articles and thanks to the wide use of functions such as the tag recommendation and the display of high-frequency tags in the system as well as effects of mutual influences among users, tags show a clustering trend which lays a foundation for the discovery and analysis of the user group.

However, information obtaining only from the searching portal of tags can no longer satisfy users' demands for information due to tags' unlimited usage and the rapid expansion of tag information. Through empirical analysis of user-defined tags in Douban Reading, the paper researches users' tagging behavior and co-occurrence situations and discovers users' interests in order to recommend personalized information to users.

1 TAG DATA COLLECTION

Douban Website is a typical Web2.0 site in China, which is consisted of Douban Reading, Douban Movie, Douban Community, Douban Music and other sections. In Douban Reading, users can retrieve books in three way: first by external characteristics of books, such as titles, authors, ISBN numbers, etc.; Second by the tag retrieval entry; third by browsing by categories, which are the results of the clustering of popular tags.

This study takes Douban Reading as an example to analyze empirically users' behaviors based on tags by statistical data, chart analysis, statistical analysis and so on.

In order to conduct the study, during the day time from January 26 to February 15 in 2011, 40 users were selected randomly from Douban Reading dataset, as shown in Table 1 and tag list data of each user were obtained by web browsing respectively from the "reading ", " read ", " want to read". Then, visit each user through resource lists. For example, if you visit http://www.douban.com/people/35072507, the page will show a list of resources tagged by the user with ID:35072507 . You can get the user's tag lists by entering Douban Reading and retrieve all resources marked by a certain tag through the tag retrieval entrance. For example, input "growth" into the tag retrieval entrance and the list of 1920 book resources sharing this tag is shown. Remove 40 repeated tags to obtain a sample data set containing 988 tags and put tags in descending order according to their frequency. Top 105 tags are shown in Table 2.

Table 1. User data collection [3].

User	Tags		Tags		Tags
Light blue dish	8	Nine tail black cat	191	Night rewelding	38
FrancesBenningt	5	Gentle and lovely dog	112	Liu Yi	4
Little Xi	50	Sumatra	100	Naturalness	55
Black and White Tannins	38	Plato's Eternity	10	Ben.S	21
Mo Jing	58	Scarecrow in city	16	Drawing Little Doudou	6
Heartwashing	39	DreamMaker	26	New comers' diary	23
Green train	2	gaolianhua	7	MAXWELL	15
Shu Nian	4	Douban sheep	55	Tossing tossing	3
Natalie	85	Maverick cat	20	Begin to dote on	9
Quiet good merle	10	Boy	41	Cotton socks Pipi	24
Hehe Li Hehe	54	Silly and stupid	7	Light	19
Red sweet soup	69	The night of love	14	Pullin	4
Cloud painting	134	Long	2		
Deadline	27	Fish	64		

Table 2. Sorting table of tags frequency.

NO.	Tag List	Frequency	NO.	Tag List	Frequency	NO.	Tag List	Frequency
1	novel	63	36	Naoko Takagi	10	71	Moon in that year	6
2	foreign iterature	43	37	Tai yan	10	72	Rumiko Takahashi	6
3	love	40	38	Wen Rui'an	10	73	economics	6
4	Britain	32	39	psychology	10	74	Luo luo	6
5	Taiwan	31	40	British literature	10	75	Ming dynasty's stories	6
6	Yi Shu	31	41	job market	10	76	biography	6
7	picture book	30	42	Mitsuru Adachi	9	77	software	6
8	Agatha Christie	29	43	Germany	9	78	poetry	6
9	comic	27	44	movie	9	79	Yan Geling	6
10	Han Han	26	45	read	9	80	Yu Hua	6
11	Japan	24	46	female	9	81	Zhang Yueran	6
12	growth	23	47	martial arts	9	82	politics	6
13	essay	23	48	suspense	9	83	China	6
14	inference	23	49	philosophy	9	84	super nice	5
15	Anne Baby	22	50	Anne	8	85	Czech	5
16	leisure	22	51	Cai Zhiheng	8	86	Management& inspiration	5
17	history	20	52	France	8	87	economy	5
18	prose	20	53	management	8	88	Leung Mantao	5
19	Professional book	20	54	Jimmy	8	89	Lung Yingtai	5
20	Joint Publishing	18	55	Lilian Lee	8	90	Qi Jinnian	5
21	Chinese literature	18	56	Milan Kundera	8	91	software engineering	5
22	biography	16	57	Nicholas Sparks	8	92	San Mao	5
23	Amy Cheung	16	58	Sun Xiaolin	8	93	commerce	5
24	architecture	15	59	fairy tale	8	94	Commercial war	5
25	America	15	60	think	8	95	SDX Joint Publishing company	5
26	Japanese iterature	15	61	Xiao Duan	8	96	Fyodor Dostoyevsky	5
27	boy	14	62	detective	8	97	Tartar in literary world	5
28	Wang Xiaobo	14	63	originality & life	7	98	literary review	5
29	literature	14	64	Guo Jingming	7	99	Hong Kong	5
30	Time-travel	13	65	classics	7	100	Little picture book	5
31	youth	13	66	Li Xinpin	7	101	romance	5
32	Eileen Chang	13	67	childhood	7	102	art	5
33	library	12	68	Wang Er	7	103	study economic well	4
34	Haruki urakami	10	69	use	7	104	Booker Prize	4
35	Feiwosicun	10	70	programming	6	105	Higashino Keigo	4

2 DISTRIBUTION LAWS OF TAGS

In order to describe the relationship between "sequence number" and "frequency" and to ex-plore more scientifically the distribution laws of sampling tag data, the author takes "sequence number " and "frequency " in descending order in Table 3 as the independent variable x and the dependent variable y to do the regression analysis. The author uses the SPSS statistical analysis software to firstly describe the data in the form of a scatter diagram and the result are shown in Fig. 1. Then it can be observed that by connecting scattered spots with a smooth curve, the image presents features of a power function. Thus, the author assumes that the functional relationship between x and y can be described as $y = ax^{-b}$. It can be obtained by regression curve estimation of SPSS that the value of "a" is about 113.7 and the value of "b" is about 0.732. Therefore, the expression is $y = 113.7x^{-0.732}$. Thereinto, R-square is 0.91, which means that this function is highly reliable, as shown in table 3.

Table 3. Model collection and estimation values of parameters.

equation	model collection					estimation value of parameters	
	R Square	F	df1	df2	Sig.	constant	b1
power	.910	9952.331	1	986	.000	113.652	-.732

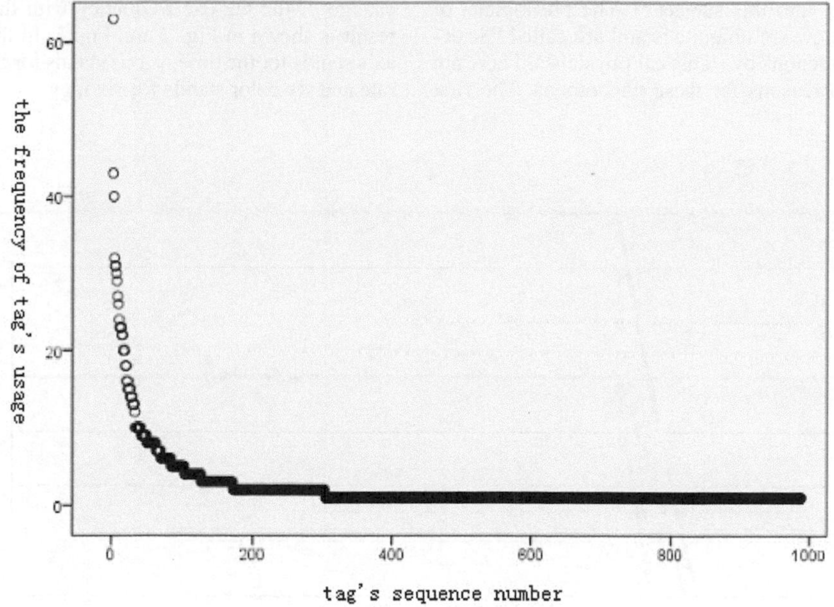

Figure 1. The scatter diagram of the frequency of sample tag's usage.

We can observe from Fig. 1 that tags among the top results are used more frequently, while they count for only a small part of the total number. Among the 988 tags owned by 40 sampling users, only 7 tags, making up 0.71% of the total amount, are used more than 30 times. Tags with larger sequence numbers are used less frequently and this general trend becomes stable after a short transition. 102 tags, making up 10.32% of the total amount, are used more than five times. At the ending part of the figure, the line tends to be parallel to the x-axis, showing a "long tail". And among all the tags, 685 of them are used only once, which counts for 69.33% of total tags.

For tags used over three times, there are several obvious characteristics, which could be concluded into three main categorical aspects:Frist,Types of the resource, like "novel", "picture book", "comic", "essays", "fairy tales", etc;Second,Authors of the resource, like "Yi shu", " Agatha Christie", "Han Han", "Zhang Xiaoxian", "Wang Xiaobo", "Zhang Ailing", etc;Third,Subjects of the resource, like "Youth", "History", "Architecture", "Psychology", "Female", "Business", "Marketing" , etc.

Although tags used only once or twice do include three kinds of tags mentioned above, these low-frequency tags are used less and the majority of them are more special and personal, like "*Sanlian @ Beijing*", "Culture Study", "Love", "Ancient Fashion", etc. Besides, compared to high-frequency tags, low-frequency tags contain more sentences, like "Granny Lao she, I miss you", "high school seminars that we do not know", "Do you like Brahms Williams", etc, for the reasons that it is less possible for sentences to be exactly the same and that sentences express rich emotions and are strongly personalized.

The statistics above tell that users' choices follow the Power Law. Researchers such as Thomas Vander Wal and Adam Mathes have also studied thoughtfully into this subject [4]. The phenomena of the Power Low are ubiquitous and are called "Scale-free Phenomenon" by statistical physicists. There are mainly two reasons for these phenomena. The first one is called "Preferential Attachment": people tend to use popular tags and this results in the Mathew Effect. The second reason is the growth property of tags: there are piles of tags with personality, which can only used by minors or even a single person. They are seldom used compared to the hot tag choices and the number of these tags is continuously growing because of its uniqueness [5]. Those two reasons also contribute a lot to the formation of the Long Tail in the figure.

3 THE TIME CHANGE REGULATION OF USERS' USE OF TAGS

In order to look into the change of the time pattern of tags' usage, a random sample of three users is selected from Table 1, namely "Hehe Li Hehe", "Nine tail black cat" and "Quiet good Merle". Then, six tags which are used most frequently are sorted in time order. According to the dynamic condition of the use of tags, an Excel spreadsheet is used to analyze the change of the tag-use frequency with the time. The result is shown in Fig. 2 and Fig. 3. In these plots, x axis stands for the time, y axis stands for the tag usage rate and six color stands for six tags.

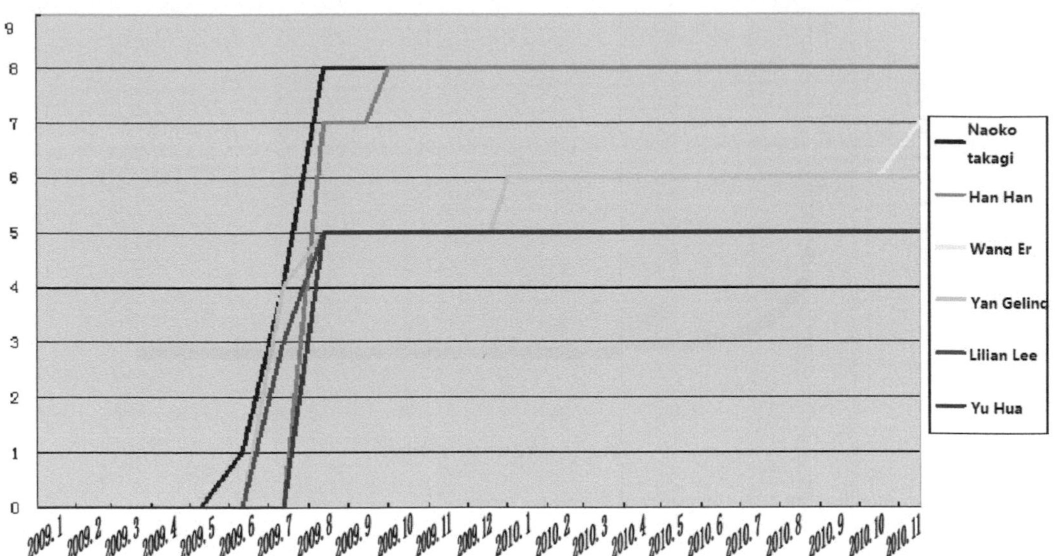

Figure 2. The tag's growth of "Hehe Li Hehe".

For lines representing growth, slopes stand for speeds of the growth. In a certain time period, steeper the lines faster the tag usage grows. Lines paralleled to x-axis stand for low or even no usage.

According to the usage of these three users, several common characteristics can be concluded as follows: First, there is a rapid growing time for each tag usage, which refers to the steep segment of the line; Second, there is a long tail after the steepest segment, counting for a large portion; Third, all tags have life cycles, which refers to a changing focus of users just like the change from "Professional book" to "Anne Baby". For a specific tag, the usage of it often clusters into a certain period of time with less usage outside this time period. Some tags even show only once.

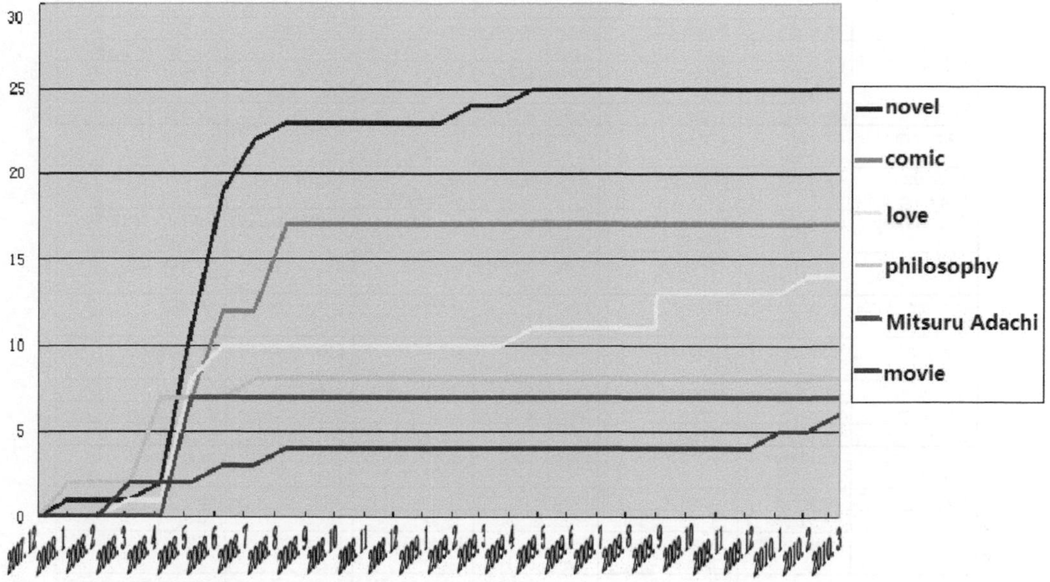

Figure 3. The tag's growth of "Nine tail black cat".

It can be observed from Fig. 2 that the trends of usages of these six tags share similar characteristics. Usages cluster into a time period from 2009/6 to 2009/9, indicating that "Hehe Li Hehe", during this time period, was interested in six authors' works represented by these six tags. Also, the frequency of usages of these six tags is similar, showing a rather equal concentration on these six authors. In this plot, six lines share the same trend, which to some extend shows a co-appearance of tags indicating the relevance of book resources. From Fig. 3, we can see that "Nine tail black cat" is interested in the resources represented by these six tags in a concentrated time period but with different focus and obvious preferences. For instance, he focuses more on novels and comics rather than movies and works of Mitsuru Adachi. However, there is a rather huge difference between Fig. 4 and Fig. 2, Fig. 3. The difference shows a scattered usage time of these six tags. Also, they are spending more time using "professional" tagged books compared to other equally focused tags.

4 DISCOVERY OF USERS GROUP BASED ON TAGS CO-OCCURRENCE

To a certain extent, users applying same or likely tags share common interests of reading and knowledge. The analysis of tags co-occurrence is a process in which users in a certain field are clustered into a community with related interests by contemporary pluralistic statistic methods [6]. If two (or more) resources share one (or more) identical tags, there must be a potential correlation between the resources. Co-occurrence strength, the number of tags that co-occur, is used to show how closely those tags correlate to each other. By indicating the existence and strength of the internal correlation among users, the analysis of tags co-occurrence mainly serves to help users find others who share same interests and to build user group with certain interests.

4.1 *Construction of co-occurrence matrix*

By random sampling, 20 users in Douban Reading are selected as research samples and all reading tags they used are counted to be arranged according to the frequency. As illustrated in table 4, 6 tags that are used most frequently by each user are recorded. Take the quantity of the same tags between two of the users as co-occurrence strength to construct a co-occurrence matrix for users' tags. As shown in table 5, this matrix is a symmetric matrix and the numbers on the diagonal are all 6.

Table 4. Users' tag data set.

User	Tag					
Light blue dish	growth(18)	Wang Xiaobo(5)	Love(3)	classics(3)	Jia Pinwa(2)	vagrancy(2)
Nine tail black cat	novel(25)	comic(17)	Love(13)	philosophy(8)	Mitsuru Adachi(7)	movie(6)
Fish	love(13)	Feiwosicun(10)	Anne Baby (6)	novel(6)	Tai Wan(4)	Lin Da(4)
Night rewelding	novel(10)	martial arts (7)	Love(6)	Naoko akagi(6)	Japan(5)	Foreign literature(4)
Little Xi	foreign literature(6)	China(3)	Maugham(2)	Fyodor Dostoyevsky(2)	Agatha Christie(2)	inference(2)
Mo mo	Britain(53)	China(29)	inference(28)	detective(25)	love(24)	growth(18)
Plain yogurt is better	Chinese literature(2)	novel(1)	inference(1)	Yu Hua(1)	Britain(1)	suspense(1)
Black and White Tannins	history(8)	politics(3)	Bo Yang(2)	Milan Kundera(2)	biography(1)	psychology(1)
Douban sheep	Ming dynasty's stories(7)	moon in that year(6)	Bloody Career(2)	China(5)	Yu Shicun(1)	The Love of the Hawthorn Tree(1)
orca	novel(18)	history(7)	Ming dynasty's stories (5)	China(5)	psychology(2)	programming(1)
Drawing Little Doudou	architecture	prose(4)	essay(4)	poetry(2)	picture book(2)	novel(1)
gentle and lovely dog	management(7)	job market(7)	suspense(6)	software(6)	programming(5)	commercial war(5)
Ming ming	job market(7)	America(7)	finance(5)	Han Han(3)	novel(2)	politics(2)
Roshihaku	novel(18)	psychology(11)	management(9)	finance(3)	business(2)	programming(1)
Ben.S	novel(5)	youth(3)	love(3)	vocabulary(2)	He Yuanwai(2)	literature(2)
New comers' diary	architecture(4)	construct(4)	my major(3)	architectural history(2)	materials and structures(2)	construction theory(1)
yeky	novel(5)	finance(4)	business(3)	Lin Da(3)	history(3)	essay(3)
Ce	psychology(24)	philosophy(10)	management(7)	novel(5)	business(4)	history(4)
moki	inference(18)	Agatha Christie(14)	foreign literature(11)	history(8)	Britain(6)	China(5)
Ray	novel(5)	user experience(5)	programming(4)	economics(3)	Internet(2)	financial anagement(1)

Table 5. Co-occurrence matrix for tags.

	Light blue dish	Little Xi	Black and White Tannins	gentle and lovely dog	Drawing Little Doudou	Nine tail black cat	Mo mo	Douban Sheep	Ming ming	New comers' diary	Fish	Plain yogurt is better	orca	Rashi Hoku	Night rewelding	moki	Ray	Ben.S	yeky	Ce
Light blue dish	6	0	0	0	0	1	2	0	0	0	1	0	0	0	1	0	0	1	0	0
Little Xi	0	6	0	0	0	0	2	0	0	0	0	1	1	0	1	4	0	0	0	0
Black and White Tannins	0	0	6	0	0	0	0	1	1	0	0	0	2	1	0	1	0	0	1	2
gentle and lovely dog	0	0	0	6	0	0	0	0	1	0	0	1	1	2	0	0	1	0	0	1
Drawing Little Doudou	0	0	0	0	6	1	0	0	1	1	1	1	1	1	1	0	1	1	2	1
Nine tail black cat	1	0	0	0	1	6	1	0	1	0	2	1	1	1	2	0	1	2	1	2
Mo mo	2	2	0	0	0	1	6	0	0	0	1	2	1	0	1	3	0	1	0	0
Douban Sheep	0	1	1	0	0	0	0	6	0	0	0	0	2	0	0	1	0	0	1	1
Ming ming	0	1	1	1	1	1	0	0	6	0	1	1	1	2	1	0	1	1	2	1
New comers' diary	0	0	0	0	1	0	0	0	0	6	0	0	0	0	0	0	0	0	0	0
Fish	1	0	0	0	1	2	1	0	1	0	6	1	1	1	2	0	1	2	2	1
Plain yogurt is better	0	0	0	1	1	1	2	0	1	0	1	6	1	1	1	2	1	1	1	1
orca	0	1	2	1	1	1	1	2	1	0	1	1	6	3	1	2	2	1	2	3
Rashi Hoku	0	0	1	2	1	1	0	0	2	0	1	1	3	6	1	0	1	1	3	4
Night rewelding	1	1	0	0	1	2	1	0	1	0	2	1	1	1	6	1	2	2	1	1
moki	0	4	1	0	0	0	3	1	0	0	0	2	2	0	1	6	0	0	1	1
Ray	1	0	0	1	1	1	0	0	1	0	1	1	2	1	2	0	6	1	1	1
Ben.S	0	0	0	0	1	2	1	0	1	0	2	1	1	1	2	0	1	6	1	1
yeky	0	0	1	0	2	1	0	1	2	0	2	1	2	3	1	1	1	1	6	3
Ce	0	0	2	1	1	2	0	1	1	0	1	1	3	4	1	1	1	1	3	6

4.2 *The analysis of co-occurrence*

The author uses contemporary pluralistic statistic methods, such as the cluster analysis and the multidimensional scaling analysis, to analyze the co-occurrence matrix of tags to discover the characteristics of user group in Douban Reading.

4.2.1 *Cluster analysis*

First of all, cluster the tags' co-occurrence matrix by CLUSTER in SPSS. CLUSTER is a pluralistic statistic method which analyzes tag data quantitatively according to their characteristics. The basic idea of CLUSTER is that individuals in the same class enjoy greater similarity while individuals in various classes have greater difference. Then, according to multiple observed objects, respectively cluster all samples or variables into different classes by various clustering algorithms. A class clustered indicates a group with close relationship—a group in which users share similar or same interests. The size of the class reflects directly the intensity of the group and the degree of concentration to the object of the group. Using cluster analysis towards data in table 5 and the result is shown in Fig. 5. The classification of the group is demonstrated in table 6.

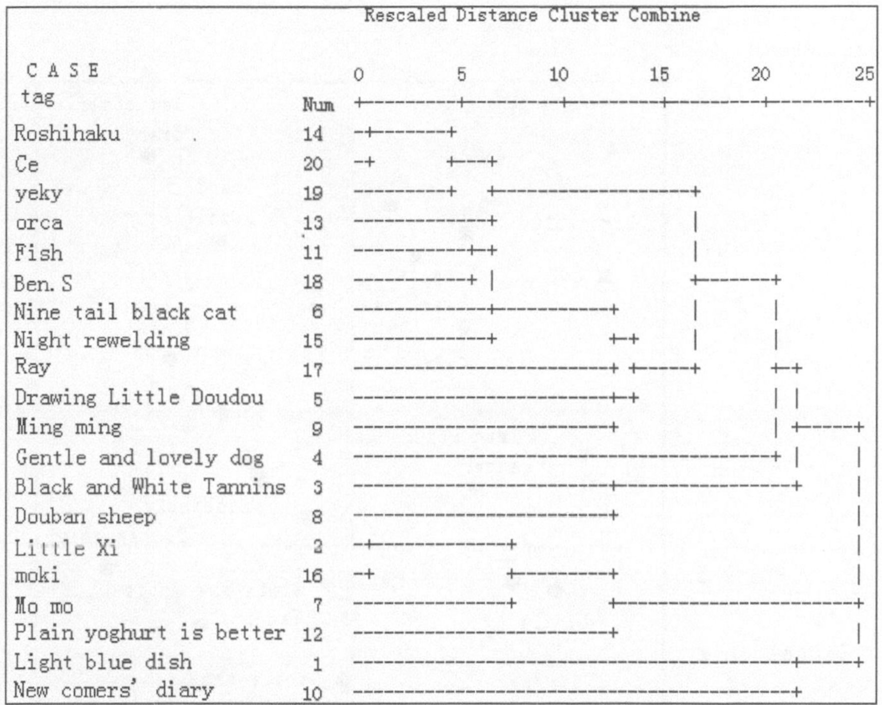

Figure 5. Result of the tags' cluster.

Table 6. Users group classification.

Group NO.	User	User interest group
1	Roshihaka, Ce, yeky, orca	novel, psychology, finance, economics, business, history
2	Fish, Ben.S, Nine tail black cat, Night rewelding, Ray, Drawing little Doudou	novel, love, youth, picture book
3	Ming ming, Gentle and lovely dog	Management, career, finance
4	Black and White Tannins, Douban Sheep	history, politics
5	Little Xi, moki, Mo mo, Plain yogurt is better	Chinese literature, Britian literature, detective, inference

It can be observed that Group 1, Group 2 and Group 5 have larger numbers of users, which indicates that sample users' attention towards books is big and concentrated and users are especially interested in novel, psychology, youth, picture book, economics and inference. To a certain degree, this observation also shows that users' attention to book resources has become a general trend. Nevertheless, Group 3 and Group 4 have fewer users, which indicates that sample users are less attracted by books on management, finance, history or politics. This result is also applied to general users in Douban Reading.

4.2.2 *Multidimensional scaling analysis*

Make a multidimensional scaling analysis using PROXSCAL in SPSS. The basic idea of the multidimensional scaling analysis is to transform high dimensional spaces into low dimensional spaces by a certain non-linear transformation and the transformed spatial pattern can approximately maintain the relation in original pattern. In the figure of the multidimensional scaling analysis, each point's position shows the similarity between objects and points with high similarity gather together to form a community. The closer a point stays to the center, the more objects it has relation with and the more important its position is in this area. Otherwise, the point is independent [7]. The outcome of the multidimensional scaling analysis of Table 5 is shown in Fig. 6.

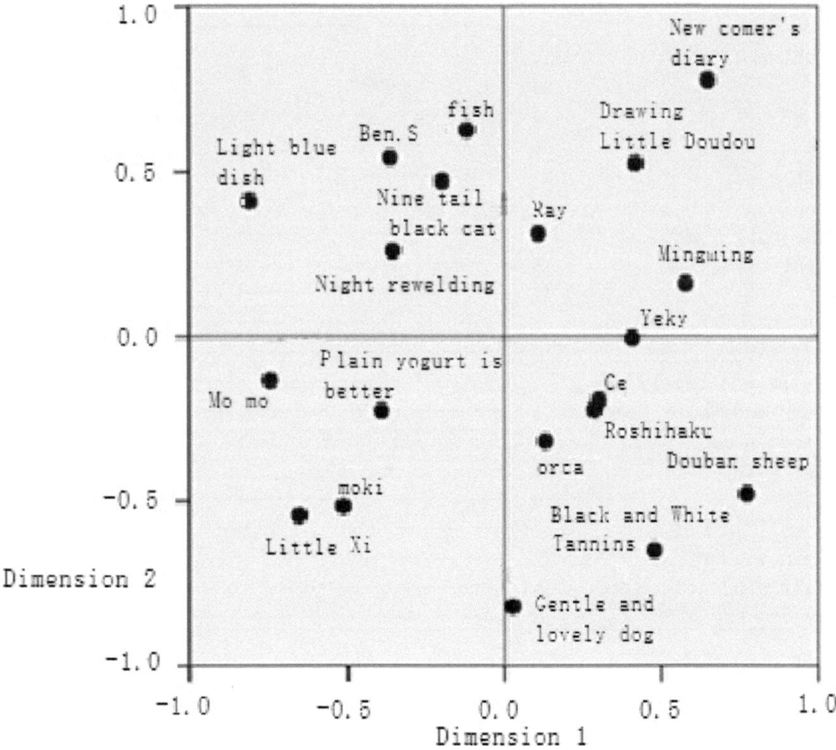

Figure 6. The result of the multidimensional scaling analysis.

It can be seen in Fig. 6 several groups with more obvious correlation, such as the group "Fish", "Ben.S", "Nine tail black cat", " Night rewelding ", "Ray"; the group "Momo", "plain yoghurt is better", "moki", "Little Xi"; and group "Ce", "Roshihaku", "Orca", "yeky". This observation is consistent with the result of the cluster analysis mentioned above. In general, however, users of Douban Reading have dispersed interests and correlations among groups arc faint. The underlying reason can be figured out by analyzing the sample data. First of all, books that users focus on belong to various fields, from novel, literature, love, inference to computer, Internet, economics, architecture, history and communication.

Tags have to have great coverage. Secondly, the co-occurrence strength we use is the quantity of same tags between every two users. However, the frequency that users make use of tags varies widely. The co-occurrence strength's failure to take each tag's weight into full consideration renders users' interests less obvious and dispersed.

5 CONCLUSION

With the development and popularity of Web 2.0 technologies, the rapid expansion of information on the Internet and the increasing cost for users to get valuable information, traditional keyword matching modes no longer fit the current development of the Web. Under such circumstance, users' need for personalized information becomes more pressing. Currently, Douban Reading mainly relies on the Internet registration system, log browsing and personalized information management system to track and record information actions of specific users in order to get their browsing behaviors, preferences, interest orientations and also changes of their interests and demands. Douban Reading is in the position to push information directionally and recommend personalized information. Additionally, Douban Reading also makes recommendation based on relevant users. For example, a user can find others with similar tastes in books through " the reading updates that I like". By pressing "follow his reading", a user gets to know other users' book reviews, books they want to read, books they are reading and books they read. Meanwhile, users' records will be tracked automatically by Douban. Douban relates users according to likely interests and recommends information according to users' records in order to achieve collaborative filtering[8].

However, based on researches mentioned above on tag's distribution, time change regulations of the use of tags and the discovery of users group, Douban Reading can adopt the recommendation based on analysis of tags so as to better discover users' preference according to users' personalized tags. There are two main methods: the one, by analyzing tags created by individual users, which requires not only the analysis on tags which users use frequently to find main interests, but also the discovery of tags which are meaningful only to individuals to cater to users' deep and unique needs. Also, analyze the time change regulation of tags to know dynamically the variance of users' interests and to recommend the most valuable information at the most proper time; the other, by analyzing simultaneously several users' several high-frequency tags. Construct a co-occurrence matrix and use the cluster analysis and the multidimensional scaling analysis to find user groups with similar interests in order to recommend more valuable information.

ACKNOWLEDGEMENT

This paper is supported by China national social science fund (#:12BTQ038).

REFERENCES

[1] J.Long. A Study on Tag-based Folksonomy on the Internet Peking University(2007).

[2] I. Cantador, I. Konstasb, J. M. Jose. Categorising social tags to improve folksonomy-based recommendations. Web Semantics: Science, Services and Agents on the World Wide Web.(2011).

[3] Information on http://book.douban.com.

[4] X.H.Li, W.Song and P.Yao. Research on Folksonomy Mechanism—taking CiteUlike as the example. Library Theory and Practice(2010).

[5] P.Wang. Network analysis of e-Learning co-occurrence based on folksonomy.China Educational Technology(2008).

[6] C.P.Hu, J.M.Hu and S.L.Deng. Analysis of Network Users Group Interaction and Research for Service Development Based on the Web 2.0. Journal of Library Science In China. (2009).

[7] D. M. Witten, R. Tibshirani. Supervised multidimensional scaling for visualization, classification, and bipartite ranking. Computational Statistics and Data Analysis. (2011).

[8] X.Fan. Research on online bibliographic recommendation system through the case study of Douban.com and National Library of China.Researches in Library Science(2008).

Management, Information and Educational Engineering – Liu, Sung & Yao (Eds)
© 2015 Taylor & Francis Group, London, ISBN: 978-1-138-02728-2

Ethical thinking on cognitive enhancement

Jian Feng Hu
Institute of Information Technology, Jiangxi University of Technology, Nanchang, China

ABSTRACT: The prospects of cognitive enhancement using neuroscience in healthy individuals have attracted considerable attention. Discussion about cognitive enhancement may at times obscure the ethical issues that are relevant today. This paper aims to contribute an adequate and ethically sound societal response to actual current developments.

KEYWORDS: Cognitive enhancement, Neuroethics, Smart Drugs, Neurotransmitter.

1 INTRODUCTION

Alzheimer's disease and Parkinson's disease, Attention Deficit Hyperactivity Disorder (ADHD), need effective drug treatment. Since some drugs can strengthen the brain, normal people who use drugs would be a superman and Wiseman? In fact, many drugs have been used to effect out of its mark, such as certain drugs that used to treat neuropsychological or other diseases may strengthen the normal cognitive and emotional functions. Due to the unclear mechanism for these long-term enhanced effects, these effects may be harmful, the restriction of the use seems very necessary. Even individuals with the capacity are allowed to take drugs to enhance cognitive, but they must be aware of the risk of long-term use of these drugs, and responsible for these consequences. To some extent, these problems are similar to some familiar situation with illicit psychoactive drugs. Morphine is a great help to the resulting from burns and other body disease, pain, but it is a mind-altering drug, which may lead to some social and spiritual issues [1]. Just because these painkillers will be abused. Why do we stop these related research? Why do we have to resist the use of drugs to make cognitive skills we change?

As many neural techniques, such as brain computer interface (BCI) technology, and the polygraph, may cause neuroethical problems to human [2], cognitive enhancement will also bring Neuroethics and a series of social problems. In this paper, the neuroethical problems caused by the cognitive enhancement were discussed.

2 THE BACKGROUND AND CURRENT SITUATION

Attention, perception, learning, memory, language, planning and decision-making, all can be regarded as a cognitive activity. Drugs leading to cognitive enhancement is an act of neuronal information processing process behind of cognitive activity.

Cognitive function is based on synaptic connections between neurons and neurons "synaptic" and information transmission. Neurons and information transmission between neurons are not immutable and frozen, but has great plasticity. While neuronal activity increases, the transmission of information between neurons and neurons capacity enhancement corresponding. Through the change of neuronal activity may alter cognitive ability. For example, glutamate is a neurotransmitter, it is to open the switch related to memory, and is conducive to the formation of memory. Ampakine, a class of chemical compounds, could enhance the activity of glutamate receptors, so as to enhance memory. The drug is developed with ampakine can be a cure for Alzheimer's disease, also can make people more alert, but not addictive. Another compound, D-cycloserine, with a combination of certain glutamate receptors, can selectively enhance forget, thereby suppressing some of conditioned reflex reaction, such as anxiety, addiction and xenophobia, help to eliminate unpleasant memories. Modafinil can stay awake, and no neuroticism, nervous, mental collapse and other side effects produced by caffeine or amphetamine, as long as you can take one tablet, energetically working continuously for 40 hours and not sleepy. Ritalin (Methylphenidate) can improve the ADHD children's learning performance, may also have a similar effect on normal children. A survey shows [3], the proportion of non medical use of cognitive enhancing drugs is as high as 20%, the Ritalin is the most commonly used cognitive enhancing drugs (about 62%), some cognitive enhancing drugs even can free online ordering. Therefore, cognitive enhancement, will become a kind of social phenomenon gradually.

3 COGNITIVE ENHANCEMENT: TAKING FREE

To allow or prohibit individuals taking cognitive enhancing drugs, there still exists a dispute. Taking cognitive enhancing drugs is an inevitable thing, it is impossible trying to completely restrict its application to realize. For "enhanced or not enhanced?" There are two questions worth discussing:

First, whether can take? If you want to participate in the interview or presentation, it is no doubt that your performance must be better in this day. On the contrary, if one day you just spirit depression, drug enhancing the memory and reaction force is very attractive. In fact, everyone loves wisdom. From this perspective, taking these drugs should be the rights and interests of the parties. As long as there are no strong side effects, it will not flush as prohibited drugs. But how to judge the side effects?

Second, whether to take? Once some drugs ("smart drugs") is believed to improve test scores or efficiency, many parents or employer will arrange or force their children or employee to take these drugs, in order to pass the exam or improve performance. Originally, taking "smart drugs" is a personal power, but not the obligation, but facing the parents or the employer they may not make much sense to tell. In this way, "bright future" will make people rush into danger, will take the "smart drugs" in knowledge or in violation of the wishes of the case.

4 COGNITIVE ENHANCEMENT: DRUG ADDICTION

Whether these drugs would not like drug addiction, but also is worthy of attention. These drugs can make people wise, but the effect was transient, and not take one time for all life "smart" (unless gene-modified drug); when the efficacy after the probabilistic intelligent degree is to return to the past, even worse than before. This kind of drug addiction has two kinds of situations:

Firstly, reasons of the drug itself. Some drugs will be exciting, lead to addiction. Because of the excitement is not directly related to "smart", that is to say, unlike stimulants that rely on excited to promote exercise potential of the athletes, excite is not required for the cognitive enhancing drugs. It can be solved by the improvement of the drug development process.

Secondly, reason of drug effect. Once the drug caused "smart", when the resistance is passed, who must again be taken in order to be "smart", then it will produce addiction. After drug withdrawal become no longer smart compared medication become clever seems more difficult to accept. "Smart" is also addictive, this addiction is the one of the most important problems of "smart drugs" in the future.

5 COGNITIVE ENHANCEMENT: INEQUALITY

Francis Fukuyama [5] said: "drugs are used to save people, not make the perfect person to become a god." Continuous use of refreshing drugs can improve the normal performance criteria, giving them an unfair advantage, changing intelligence gap between human taking drugs and human not taking drugs, even will make the negative effect for characters.

In fact, in the normal population, there are always some unusual people, they have unbelievable memory, or the ability of fast language and learning music, or with a special ability. In their minds, there may be something to make them have the cognitive ability of superman. Since these genius has some better than most people's brain structure, if the use of certain drugs can also achieve the same effect, why would we find taking drugs enhance cognitive wrong?

In order to improve test scores, do good presentations, will refreshing prescription drug taking behavior are same for athletes who injecting hormone in order to break the world record? Many people think that drugs used to improve intelligence are equal to cheat. If someone tries to get ahead by hard work, that even if the drug is no problem. But if someone takes a "smart drugs", pay little effort more than others has achieved great success, which is unfair.

Another inequality is from the price of this class of drug. If the price of these drugs is cheap, and everyone can enjoy, it will not have a problem; but if these drugs are expensive, so that only a few people can afford, it will form a vicious spiral, rich people taking drugs become more intelligent, more alert, more active, can earn more money; while the poor cannot afford to buy drugs, their intelligence is relatively backward, earn less.

6 COGNITIVE ENHANCEMENT: THE WAY OF LIFE

Now there are people begin to take the legal performance enhancers, such as musicians take drugs to relieve stage fright, also some people taking drugs to overcome the drowsiness, dozen spirit, and improve the working speed. If this kind of medicine can make the capable people as possible output efficiency higher, let everyone a better life, let people play the greatest potential, and does not constitute a direct damage to others, so why not?

Evolution of brain boosting drugs is similar to growth of plastic surgery operations. From the plastic surgery history, "neural plastic" development trajectory can be predicted, beginning to blame, which considered the vanity and a violation of natural performance. However, it is a mainstream up to now, which is self-modification and improve performance. Cognitive enhancement drugs are also so, as long as all humans taking these drugs, the human intelligence is promoted, then there are no inequality, it evolved into a way of life, become a part of earth culture.

7 CONCLUSION

At just 100 years ago, no proper regulation to control drugs-taking in the Olympic Games. At present, there may be a lot of people have been through the use of "smart drugs" received a lot of vested interests, and continue to get.

"Smart drugs" is the long anticipated, may therefore also opened a Pandora's Box. Smart mouse is a good example, mice changing to clever but pay physiological costs very "pain" — is very sensitive to the duration of chronic pain long. In the future, people will take "smart drugs", just like today cosmetic operation like a vitamin. It is quite common, but before there are many ethical and social problems to be solved.

REFERENCES

[1] Farah M J, Illes J, Cook-Deegan R, et al. Neurocognitive enhancement: what can we do and what should we do?[J] Nature Reviews Neuroscience, 2004, 5:421–425.

[2] Hu JF, Mao CL. Neuroethics—a perfect intersections of humanities and neuroscience[J]. Journal of Jiangxi University of Techonology, 2008, 2:15–17.

[3] Maher B. Poll results: look who's doping[J]. Nature, 2008, 452:674–675.

[4] Schermer M, Bolt I, De Jongh R, et al. The future of psychopharmachological enhancements: expections and policies[J]. Neuroethics, 11 February 2009 Published online.

[5] Fukuyama F. Our posthuman future: Consequences of the biotechnology revolution[M]. 2002, London: Profile books.

[6] Racine E, Forlini C. Cognitive enhancement, lifestyle choice or misuse of prescription drugs? Ethics blind spots in current debates[J]. Neuroethics, 4 September 2008 Published online.

Management, Information and Educational Engineering – Liu, Sung & Yao (Eds)
© 2015 Taylor & Francis Group, London, ISBN: 978-1-138-02728-2

A study on elevator Braille graduation system of junior high school in Chang-Hua county

Liang Tseng, Chun Mai Huang & Chih Yu Hsia
Feng Chia University, Taiwan R.O.C.

ABSTRACT: This article investigates the present situation, with regard to the lift Braille systems of 21 elevators of junior high schools in Chang-Hua county, with the research aim to probe into the different situations of the lift Braille system among them. Meanwhile, non-barrier laws and regulations standard are taken into consideration, to analyze the project about the facility size and function keys of elevators, to carry out present on-site investigation, and conduct comprehensive discussions. Then it is incorporated into a questionnaire and Analytic Hierarchy Process (AHP) to establish the standard and grade of barrier-free elevators in Chang-Hua county .The results of this research accessible elevator are: 1.To understand the Braille elevator signage systems of junior high schools in Chang-Hua county. 2.To conduct an analysis of the proportion of elevator sizes for the non-barrier laws and regulations standard. 3. To make comparisons of the main keyboard function, key configuration, partition, and floor. 4.To establish the ranking system and grading pattern (A、A+、AA、AAA) of elevators of junior high schools in Chang-Hua county.

KEYWORDS: Visual Impairment, Braille Signage System, Grading of Elevators.

1 INTRODUCTION

This article will survey a total of 21 elevator accessibility Braille signage system in the Changhua County campus to understand the various junior high school in the elevator Braille system.

1-1 Motivation and purpose: The motivation follows below: (1) Understanding the visually impaired in the Braille system of barrier-free campus environment. (2) To understand the current situation and accessibility of elevators in Chang-Hua junior high school. (3) To explore the elevator accessibility of Braille signage system problems and countermeasures in the campus. To analyze the current language system, definition, location, configuration and floor buttons. Barrier-free comparisons of four levels of design, construction, user and legislation as follows:

1 Design aspect: according to Taiwan elevator accessibility Braille signage system standard stipulates the elevator accessibility Braille signage system using phonetic symbols.
2 User aspect: Visually impaired uses the index finger pulp moving quickly from left to right to tactile literacy. Therefore the position of the elevator braille piece is located in discriminated manner, namely in the (A) right flank (B) left side (C) above side (D) underneath (E) pressed key interior underneath. Therefore, this article will depend on the user behavior and the investigation case compares, and will paste the piece position according to the Ministry of Interior standard braille proportion of in outside the 'pressed key left side.
3 Construction aspect: The general old elevator did not set the Braille patch. However, after the implementation of the elevator Braille system standard, manufacturers are required to fix it. The construction workers do not have knowledge of the elevator Braille system. Consequently, Braille stickers review content, as applied, location, etc. and the accuracy of proportion.
4 Law aspect: (A) The basic scale analysis: based on the "barrier-free building regulations" to sort out 15 elevator equipment shall comply with standard, and then with the survey were compared and analyzed. (B) Analysis of the main keyboard configuration: Taiwan's regulations do not regulate the relative size of the main keyboard, type, configuration, etc. Therefore, in this study, it will measure and compare the main keyboard configuration.

1-2. Sphere and object: (1) Scope of the study: the study area is Chang-Hua junior high school barrier-free elevator. (2) Object of study: there are 21 barrier-free elevators.

1-3. Methods and processes:

1 The investigation of Site: the survey includes [up] and [down] buttons, [open] and [close] buttons,

[floor] buttons, [alarm] buttons, elevator scale and keyboard positions.

2 Questionnaire: the questionnaire is taken visually impaired sound to collect the data. Also, there are four main content which is a user, Cognitive, Construction and Legislation, and 20 Secondary content. There are two types of questionnaire for experts and visually impaired individuals. There are 14 copies of the experts' questionnaire and 36 copies of the visually impaired' questionnaire.

3 Analytical Hierarchy Process: It is used in uncertainty and with the majority of assessment data. It helps designing the accessibility of building that concern about the use of the visually impaired.

2 BRAILLE SIGNAGE SYSTEM AND ACCESSIBILITY ELEVATOR LEGISLATION

Accessibility Elevator legislation: Braille labeling specifications (building accessibility design specifications): According to the design specifications of buildings accessibility lifts Chapter IV Section Braille labeling specified in Table 1 are (as shown in Table 1), compiled [up, down], [open, shut], [floor], [warning], which [up, down] Braille up (ㄕㄤˋ), down (ㄕㄤˋ) key phonetic Braille; [open, shut] Braille open (ㄎㄞ), shut (ㄍㄨㄢ) key phonetic Braille; [floor] digital floor Braille; [warning] alerts bell (ㄌㄧㄥˊ) phonetic Braille; refuge floors (main), telephone (tel), stop (stop) English Braille.

Table 1. Braille Signage System legislation table.

alphabet	Braille Signage	alphabet	Braille Signage	alphabet	Braille Signage	alphabet	Braille Signage
B1		1		up		★	
B2		2		down			
B3		3		open			
B4		4		close			

3 ANALYSIS OF SURVEY RESULTS

3-1 The levels of Investigation: there are 21 accessibility elevators in Chang-Hua County junior high school to analyze four levels as follows: The aspect of the legislation: (1). The basic scale analysis: based on the "barrier-free building regulations" to sort out 15 elevator equipment shall comply with the standard, then using the survey to compare and analyze. (2). Analysis of the main keyboard configuration: Taiwan's regulations do not regulate the relative size of the main keyboard, type, configuration, etc.

3-2 Analysis of the survey results of Braille system:

1 Analysis of the Design aspect: Analyzes various pressed keys braille content in order to conform to the laws and regulations of braille system proportion: On conforms to the standard ㄕㄤˋ to account for 76.2%, Under conforms to the standard ㄒㄧㄚˋ to account for 76.2%, Opens conforms to the standard ㄎㄞ to account for 76.2%, Closes conforms to the standard ㄍㄨㄢ to account for 61.9% (shown as Table 2).

2 Elevator host keyboard disposition pattern: The police shows key Alarm by representative A, floor key Floor by representative F, opens key Open-Close by representative O, time delays key Hold by representative h. According to the pressed key district, from top to bottom, this may be divided into namely AhFO, AFhO, AFOh, AFO, AOF, and FO six types respectively. In which the AhFO disposition pattern about 42.86%.

4 RESULTS OF THE QUESTIONNAIRE

4-1 Questionnaire objective: Questionnaire is designed for two groups of audience. One is targeted for experts including researchers, designers, architects and committee which comprises 14 copies of the questionnaire. The others are in Taichung visually impaired students, which makes up 36 copies of the questionnaire.

4-2 Questionnaire content: The construction surface construction discussions includes, respectively "the use", "the cognition", "constructs", "the laws and regulations"; Under four big construction surfaces, the again segmentation are 20 factors, separately does not discuss the correlation elevator factor one by one according to the isomorphism surface criterion holding importance. Each construction surface criterion content.

4-3 Analysis of the questionnaire: Questionnaire using Analytic Hierarchy Process (AHP) calculates the weight of each factor. (Shown as table 3).

1 The highest weight of user (0.328) is the operation keys that can be easily found (0.077).
2 The highest weight of awareness (0.258) is the meaning of Braille that understands easily (0.067).
3 The highest weight of construction (0.216) is Braille patch which should be prevented using the wrong device (0.057).
4 The highest weight of Legislation (0.198) is lift door switch time ≧ 5 sec(0.033).

Table 2. Design stratification plane - various pressed keys conform to the proportion table.

Braille system	Content	Analysis	Braille system	Content	Analysis	Braille system	Content	Analysis
[up]	Number of proportion	16	[down]	Number of proportion	16	[open]	Number of proportion	16
	Number of samples	21		Number of samples	21		Number of samples	21
	Statistical Analysis	76.2%		Statistical Analysis	76.2%		Statistical Analysis	76.2%
[close]	Number of proportion	13	[B1]	Number of proportion	3	[tel]	Number of proportion	9
	Number of samples	21		Number of samples	7		Number of samples	21
	Statistical Analysis	61.9%		Statistical Analysis	42.9%		Statistical Analysis	42.9%

Table 3. Questionnaire four big construction surface and 20 factor weight table.

Dimensions	factor	Weights	Dimensions	factor	Weights
User 0.328	1. Both sides of the entrance set the haptic device	0.070	**Construction** 0.216	1. Patch can prevent errors in the manufacture on the construction	0.062
	2. Braille patch set on the left	0.066		2. Braille patch have durability	0.055
	3. Each patch of Braille operation keys to partition	0.057		3. Must have enough space to the left of the button	0.057
	4. There should be no use of the elevator control or locked	0.058		4. Location and size of the patch should be unified	0.051
	5. Operation keys can be easily find	0.077			
Awareness 0.258	1. Braille patch should prevent wrong device	0.062	**Legislation** 0.198	1. Set up automatic opening device	0.021
	2. Braille patch design one piece	0.066		2. Entrance signage settings	0.029
	3. Operating panel should be applied Braille patch of space	0.067		3. Different materials to guide	0.029
	4. Braille patch should be normalized	0.062		4. Wheelchair turning space > 150cm	0.029
				5. Elevator door size should ≧ 90cm	0.028
				6. Lifts should ≧ 135cm depth	0.029
				7. Lift door switch time ≧ 5 sec	0.033

Table 4. Chang-Hua county countries elevator hierarchical table.

school	PI	Grade	Rank	school	PI	Grade		school	PI	Grade	Rank
zhangde	1.4891	AAA	6	Xihu	1.2188	A+	13	Siansi	1.4445	AA	10
Zhangxi	1.0124	A	19	Chen-	1.4733	AA	8	Lukang	1.1507	A	17
Zhangtai	1.0302	A	18	Dacun	1.4752	AA	7	Fenyuan	1.4613	AA	9
Jingcheng	1.5270	AAA	5	wanli	1.1528	A+	16	Yuanlin	1.5972	AAA	3
Xinyi	1.3184	AA	11	Pusin	1.5880	AAA	4	Tatung	1.5973	AAA	2
Homei	0.1614	A	21	Erlin	1.2252	A+	12	Tianjhong	1.1979	A+	14
Ho-exp	1.6559	AAA	1	Beidou	1.1946	A+	15	Wenxing	0.2267	A	20

5 CONCLUSIONS

1 In the levels of design: The highest percentage of [up] is 76.2% which is similar to [dn].
2 In the levels of user: standard 47.6% of the [up], [dn] is kept close to the left side of the button.
3 In the levels of construction:[B1] and [tel] are responsible for the rest, with 28.57% and 19.05%.
4 Accessible Elevator Braille plate recognition. [shown as Figure 1.]
5 AHP model analysis: The weight of user is 0.328, followed by awareness is 0.258. In addition, the weight of construction and regulation are 0.216 and 0.328.
6 Elevators' hierarchy is divided into four grades A 0~1.1507 -score, A+ 1.1507~1.2252 score, AA 1.2252~1.4752 score, and AAA 1.4752~2 score. [shown as Table 4.]
7 Demonstrates setting Braille clip site-position of Universial Design. [shown as Figure 2.]
8 In the main keyboard mode, the percentage of [AhFO] is 42.86%, followed by [AFhO] with 9.52%, and [AFOh] as 28.57%.

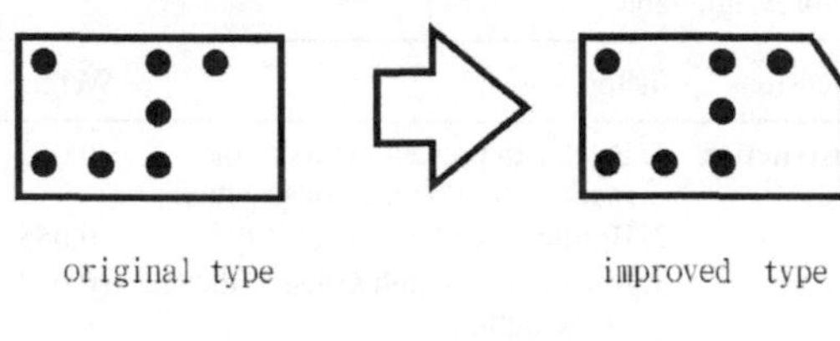

Figure 1. Braille patch site-position of avoid error.

Design aspect		Use aspect		Construction aspect	
Local county Braille	Universal-Design Language-system Braille	Local county Braille	Universal-Design Site-position Braille	Local county Braille	Universal-Design Produce-system Braille

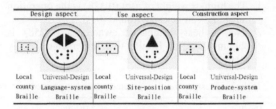

Figure 2. Braille clip site-position of U.D.

REFERENCES

[1] Tang Chen-Chen , Tseng Liang ,Hsia Chih- Yu, (2013). A Study of Elevator Braille Signage System in Ho Chi Minh City, Vietnam., Science Direct (Procedia-Social and Behavioral Sciences) ,85p139-p151.

[2] Liang Tseng, Chen-Chen Tang, Chuan-Jen Sun (2013) .A Study on the Braille Elevator Signage System in Public Buildings: The QFD Perspective., Original Research Article Science Direct (Procedia-Social and Behavioral Sciences) ,85p152–163.

[3] Su Mao-bin, "Research the public works non-barrier elevator braille system application - take *Hong Kong and Taiwan as the example,*" Feng-Chia University master the paper, in2010 /06.

[4] Wang Ming forgave, Tseng Liang, Chen-Chen Tang , Chuan-Jen Sun , "*Research the Taiwan populace transportation system non-barrier elevator Braille system - take Taiwan railroad Shan Xian and the submarine cable as an example,* "Republic of China constructs learns 24 session of 2nd building research results publication to meet the collection, in 2012/09.

[5] Tseng Liang, Chen-Chen Tang , " *Research the public works non-barrier facility elevator Braille system –take Aomen south the area and Taiwan as the example,* " the modern age builds for 2013/02.

Management, Information and Educational Engineering – Liu, Sung & Yao (Eds)
© 2015 Taylor & Francis Group, London, ISBN: 978-1-138-02728-2

Hypertext poetry creation under multimedia horizon

Guo Ying Li
Pingdingshan Industrial College Of Technology, HeNan, China

ABSTRACT: As for poetry spread, hypertext poetry is one of its changes, impacting on the form of poetry. Literature is bearing media changes, and the nature and expression of poetry changes too. With the emergence of new media, poetry is affected and certain changes happen, and provide a broad creation and development space. This article elaborates on hypertext poetry, and explore hypertext poetry creation under multimedia horizon.

KEYWORDS: Hypertext; poetry; multimedia; creation.

1 INTRODUCTION

With the popularity of computer network, people's lifestyles change, reading habits and aesthetic standards change. In this situation, the development layout of poetry gradually changes, the poetry transition from paper-based media to the media, creative way of poetry also transmits from paper-writing to electronic media writing. Thus, the network poetry emerges as a new form of poetry.

2 A HYPERTEXT POETRY OVERVIEW

Hypertext poetry is a kind of network poetry, in a broad sense, the network poetry is defined from the perspective of the media, referring to the poetry spread by network, mainly including networking and Pro screen poetry creation of text poetry [1]. The former refers to post the already written poems directly on the network, the nature of poetry does not change. The latter refers to that the author create directly on a computer and computer network, poetry texts have openness. And in the narrow sense, network poetry refers to a production method, through the computer multimedia technology, the author conducts creation of the digital text, which integrates text, images, sound and other elements, and hypertext poetry is a narrow network poetry.

Hypertext poetry includes the following forms, first, multimedia poetry texts, use multimedia technology to integrate text text, painting text, music text, animation text, etc., to form an organic whole, configure the matching pictures and animations to the mood of each poem, enhancing poetry spatially. Second, hyperlink poetry text, there are link points linking to media and texts in a text, mark the symbols and special words in the hypertext in some way, readers click on these symbols and enter different texts and read the relevant contents. In poetry creation, use hypertext technology, integrate different texts, break space limitations, perform the rich meaning of poetry, expand poetry space. Third, interactive poetry text, use computer technology, such as flash software, dynamic text and pictures, increase the dynamic nature of poetry text and increase interaction between readers and poetry text, and provide readers opportunities to participate in creation.

3 HYPERTEXT POETRY UNDER MULTIMEDIA HORIZON

3.1 *Mood, emotional specific expression*

In our traditional poetry, the poets pursue the painting and music beauty in poetry creation, in the paper media, paint picture by the text, express the poets emotions, convey thoughts and feelings to the readers. Meanwhile, in the creation, by oblique and rhymes, poems are so catchy and has a certain rhythm to achieve sound effects. However, due to limitations of the text and emotional complexity, the poet is sometimes difficult to make the creation of poetry to the desired effect, the reader's understanding has some limitations. In the present multimedia sight, the hypertext poetry text diversification provides conditions to convert poetry abstract mood to concrete poetry. Using multimedia technology in poetry creation, the poet configures suitable picture for poetry according to the specific emotions and ideas to be conveyed, of the poetry, add background music, thus enhancing the richness of poetry, provide readers with a visible and audible picture, help readers understand the poetry emotion. Make the poetry creation "poem paintings" through multimedia technology, and convert virtual mood to specific real mood. Meanwhile, according to poetry mood, configure the appropriate background music and sound effects to heighten the atmosphere and enhance the emotions. As in hypertext poetry *the Moon Dream*, configure to month centered pictures and music, create a dreamy and blurred mood for the reader to mobilize reader's emotions [2]. Meanwhile, in the multimedia environment, the creation of hypertext poetry should not only pursue the reasonable use of pictures and music, but should pay more attention to the correct expression of the body and spirit of poetry, strengthen the connotation creation of the poetry, should treat the the nature of poetry as the fundamental, use other elements on this basis to relieve against each other.

3.2 *Increase creative space and strengthen the creation quality*

Compared to traditional poetry material carriers, the network breaks the space constraints, has the ductility of time and openness of space. The limited material of traditional poetry shackled the presentation and dissemination of information. In traditional poetry creation, the poet creates by the carrier of pen, paper and ink, spread poetry in the form of manuscripts and prints, poetry is constrained by the space and time. In hypertext poetry creation, due to network threshold is low, and impact the status of official culture publications to some extent, the use of multimedia technology deconstructs the discourse hegemony, breaking the limitations of traditional publish poetry. In the network platform, the creators can post their own creation on the network anytime and anywhere, and create on a network platform, reduce the work reviewed and increase the creative space. Meanwhile, in the network platform, due to fastness and interactive of the network, the exchanges between creators increase, poets can conduct creative exchange anytime and release poetry dynamic timely, promoting poetry creation. In addition, in hypertext poetry, the readers can feedback to the author at any time, and communicate with the author, so the poet can better understand the shortages of the creation and enhance creative quality.

3.3 *Increasing the level of poetry, and enhance the reader's understanding*

In hypertext poetry, the use of hyperlink technology, the reader can enter a different link text and web page through a mouse click, reading the words in comments, content supplementary and background introductions. In hypertext poetry creation, the author adds a link point through setting color and underline to a word or text, thereby increasing the level and openness of poetry, add the poetry content to help readers understand. In this process, the poet should pay attention to the reasonable set of links point, avoid the loss of the artistic beauty of poetry due to excessive comments. Meanwhile, the quality of the linked content should also be enhanced, according to the reader's psychology, after clicking the mouse, the required novelty content can be obtained by readers, and the psychological expectations of readers can be met. In addition, in the hypertext poetry creation, the use of multimedia technology adjusts the existing poetry ideas of the readers to some extent degree, and brings readers troubles. In the animation, music and hyperlink settings, although increasing the level of poetry and enriching poetry content, the text jumps and dynamic can make readers produce anxiety easily and affect reading [3]. In the creation, all technology should service for poetic quality, good writing and reader's reading, should focus on the quality of the text, and then according to reading habits set jump time and link location reasonably. Meanwhile, for some poetry notes and text descriptions, people cannot simply copy and paste the data from the network, they should focus on secondary creation, and enhance the quality of content.

4 CONCLUSION

Hypertext poetry creation uses text, pictures, animations and music elements rationally, enhances poetry layering and richness, and breaks the traditional time and space constraints for poetry, provides convenient conditions for poets to create and publish poetry. Meanwhile, in the conditions of increasing creative space, the poet should focus on building the spirit of poetry, and strengthen the quality of poetry, uniform poetry quality and technology to ensure the long-term development of poetry.

REFERENCES

[1] Zhang Xiaowan, Li Quanlin. Hypertext, multimedia poetry experiments over the network - On Mao John PPS format Pictures Poetry aesthetic features [J]. Anhui University of Technology (Social Science Edition), 2010,12 (4): 67–71.

[2] Zhang Yu. Liberty song - the current development status of the network poem [J] Sichuan College of Education, 2011,27 (4): 57–59, 63.

[3] Liu Ping. Poetic art "Breakout" and the dilemma-On the creation of hypertext poetry moment [J] ginseng flowers, 2014, (1): 155–156.

Management, Information and Educational Engineering – Liu, Sung & Yao (Eds)
© 2015 Taylor & Francis Group, London, ISBN: 978-1-138-02728-2

Neuroathesetics issues on decision-making in different architecture

Jian Feng Hu
Institute of Information Technology, Jiangxi University of Technology, Nanchang, China

ABSTRACT: People spend most of the time indoors, and the physical characteristics of the building in which we live and work affects our mental state, and the different mental states will influence our decision making. Different physical characteristics of the building have an influence on our mental state, such as height, and decoration has a certain impact on perception, preference and neurasthenic.

KEYWORDS: Neurasthenics, Decision-making, Architecture.

1 INTRODUCTION

Neuroscience can build a bridge connecting the architecture and psychology. Neuroscience method can reveal the underlying mechanisms that explain how different physical characteristics of building lead to different results of behavior. The physical characteristics of building maybe affect memory, visual perception, the sense of space, and of course the aesthetic view. The integration of neuroscience and aesthetic lead to a new discipline, neuroaesthetics. Neuroaesthetics applies to architecture in several aspects, including preferences, space design, improve health and decision-making. This is an exploratory work, at present very little research on this aspect is found. Because of the limited space, this paper will be only for discussion, not for experimental analysis.

Decision making is a fundamental part of human life, which is also a basic survival ability. Decision exists everywhere, we have to make a lot of decisions every day, for example, what to eat, what to wear, what to do, what to play. When the traditional economics facing the decision-making of human behavior, other disciplines, such as psychology, cognitive neuroscience and Neuroeconomics tried to explore decision-making behavior of human being. Why do people sometimes will make the right decision, and sometimes they do not mind? How to deal with risk and loss? How to realize the benefit maximization?

2 THE CALCULATION PROCESS OF DECISION-MAKING

Dopamine is an important neurotransmitter and neuromodulator in the brain, which plays a decisive role when facing the decision, money, addiction and reward and so on. Dopamine is also the meal currency for weighing the pros and cons and assessing the value in the brain. Reinforcement learning based on dopamine loop is an important prerequisite for animals learning to make the right decision. Dopaminergic neurons calculate the relationship between reward and prediction according to the deviation between the actual reward and reward prediction. Gambling will raise dopamine levels, which reveals that uncertainty may be cause to induce dopamine release. This uncertainty induced increase in dopamine levels, may contribute to an increased risk for defensive behavior, and the prediction error after phase reaction may be enhanced and this reward related.

3 THE NEURAL BASIS OF GAIN AND LOSS

Kahneman and Tversky found, people's feeling for the same amount of gain and the loss is not the same, more pain of loss. In the brain, how to measure the gain and loss of decision? Studies show that it is not the same between gain process and loss process in the brain. When participating in risky bets, the dorsal medial system is related to loss process. When subjects make a decision leading to loss of income, the medial dorsal system plays a greater role in the computing process. Studies in the animal experiment show that, the current reward value may be predicted according to the established model by the previous reward experience. When monkey takes specific actions in order to obtain a juice reward, tegmental neurons in the substantia nigra and ventral midbrain are activated. fMRI experiments have revealed that a neural region related to monetary reward and economic reward overlaps with the neural region related to primitive demand (such as food), and expected reward may also be associated with basal ganglia. When making decisions, "reward circuit" is more sensitive to lose money on the brain. When people may take into account the losing, activated region will be closed. When losing bets increases, the regions related to fear and anxiety, such as the amygdala and insula, are not active.

4 DECISION-MAKING FOR RISK

Risk aversion is associated with the amygdala, the amygdala is associated with emotions. In animal experiments, continuous stimulating synaptic connections between amygdala and cortex, startle response will not produce. When the function of neural connectivity is restored, startle response will produce. The amygdala does not "forget" startle response of the past, and the cortex is trying to suppress the "memory". That is to say, there exist dispute of "sensibility" and "reason" between the amygdala and cortex. Risk decision and risk judgment are different. In the risk decision, the decision maker must select an option, which will force the decision maker in the brain to produce interactions between perceptual and rational. Studies found that normal subjects become more conservative after one choice, and patients will make a more favorable decision, ultimately win more money. Other studies have also found that even low levels of negative emotions may also lead to self control, which makes the best income decrease.

5 COMPARISON OF CONTOUR AND OPENNESS OF ARCHITECTURE

The architecture has different characteristics, such as shape, material, quality, color, size, style and so on. All these characteristics have an impact on the psychological state, and then decision-making. For simplicity, we begin from the indoor style. The line is undoubtedly the most tension, its complex visual image can create different styles of architectural space, through visual language influence on people's psychological state. The line has an extremely rich, expressive force and produce psychological hint. The straight line and the curve are also different, the vertical lines have a rigid feeling and cold feeling, the horizontal line with a sense of stability, slash with strong sense and sense of direction, curve with melodic and catchy.

Figure 1 shows different architecture with different contour (rectilinear and curvilinear) and openness (open and enclose). Previous studies reveal that viewing curvilinear spaces would activate neural networks associated with motor imagery or execution. People were more likely to judge curvilinear than rectilinear spaces as beautiful, which was associated exclusively with an increase in anterior cingulate cortex (ACC) activity. Activation in the amygdala could be maximal in relation to maximally curvilinear and maximally rectilinear spaces. Contour had no effect on decisions.

Figure 1. Example of comparison of contour (curvilinear vs. rectilinear spaces) and openness (open vs. closed).

Different contour in the context of landscapes might have an impact on decisions-making.

Similarly, color and size will also produce a different aesthetic feeling. These results indicate that, although the indoor contour will bring beauty, but has no actual effect on decision-making. Some people would like looking a specific place for negotiation, where it may make the guest very comfortable and very happy, but does not change his/her decision.

6 CONCLUSION

The architecture actually also bears the function of exchange between man and nature. The architecture is not only a product of science and technology, but also perhaps psychology and neuroscience. So neuroathesetics can be applied into the core architecture. Taking into account the people stay more and more time indoors, more and more decision-making also occurred in the indoor, the design of the architecture should also consider the neuroathesetics, in order to better decision making, but also to better work and life.

ACKNOWLEDGMENTS

This work was supported by Natural Science Foundation of Jiangxi Province [No 20142BAB207008] and project of Science and Technology Department of Jiangxi Province [No 2013BBE50051].

REFERENCES

[1] Schoenbaum G, Roesch MR, Stalnaker TA. Orbitofrontal cortex, decision-making, and drug addiction [J]. Trends Cogn Sci, 2006, 29:116–124.

[2] Sanfey AG, Loewenstein G, McClure SM, Cohen JD. Neuroeconomics: cross-currents research on decision-making [J]. Trends Cogn Sci, 2006(10): 108–116.

[3] Daeyeol L. Neural basis of quasi-rational decision-making [J]. Current Opinion Neurobiology, 2006, 16:1–8.

[4] Hu JF, Mao CL. Neural basis of decision-making [J]. Science and Technology Management Research, 2007, 11:264–266.

[5] Vartaniana O, Navarrete G, Chatterjeed A, Fiche LB, Lederf H, Modroñog C, Nadalf M, Rostruph N, and Skovi M. Impact of contour on aesthetic judgments and approach-avoidance decisions in architecture [J]. PNAS, 2013, 110:10446–10453.

[6] Leder H, Tinio PPL, Bar M. Emotional valence modulates the preference for curved objects [J]. Perception, 2011, 40:649–655.

[7] Vessel EA, Starr GG, Rubin N. The brain on art: Intense aesthetic experience activates the default mode network [J]. Front Hum Neurosci, 2012, 6:66.

[8] Weber R. Introduction to the special issue: Aesthetics and design? [J] Empir Stud Arts, 2012, 30:3–6.

[9] Vartanian O, Kaufman JC. Psychological and neural responses to art embody viewer and artwork histories[J]. Behav Brain Sci, 2013, 36:161–162.

[10] Di Dio C, Macaluso E, Rizzolatti G The golden beauty: Brain response to classical and renaissance sculptures [J]. PLoS ONE, 2007, 2:e1201.

Management, Information and Educational Engineering – Liu, Sung & Yao (Eds)

Analysis of educational innovation and schools' classroom teaching reform

Li Man Zhao, Yan Xu & Hao Zhang
Jiangxi Science & Technology Normal University, Nanchang, Jiangxi, P.R. China

ABSTRACT: Educational innovation is an important activity that our educational field develops, which directs at realizing goals specified by plans via innovative activities. Schools' classroom teaching reform should develop teaching activities energetically based on educational innovation, cognize important significance of educational innovation and classroom teaching reform profoundly, put specific measures of classroom teaching reform into practice feasibly and construct more perfect and more advanced modern classroom.

KEYWORDS: Principal Component Analysis, BP Neural Network, Fuzzy Neural Network Educational innovation, Classroom teaching.

1 INTRODUCTION

Progress and development of the era put forward higher requirements for quality of talent demands. As schools are educational institutions cultivating talent, the situation that they keep pace with development of the era and reform classroom teaching constantly becomes important content in the process of schools' development. Classroom teaching reform should be implemented based on continuous innovation, i.e., it should combine with educational innovative ideas practically and construct modern teaching classroom.

2 SCIENTIFICITY OF EDUCATIONAL INNOVATION

Educational innovation is an activity that uses a series of educational innovative methods to achieve specified goals when there are some educational objectives. Educational innovation is an important educational approach in schools' development. Via innovation of ideas, thoughts, forms and methods, schools' educational innovative goals can be realized.

Educational innovation has rich content, involving all aspects like schools' educational system, educational approaches and educational ideas. At the same time, educational and teaching activities are carried out by getting rid of old opinions, updating teaching methods and combining with modern methods in the process in which educational innovation is implemented. The essence of educational innovation is that schools get rid of the stale and bring forth the fresh in introspection and absorb new energy when they summarize experience, and innovative progress in all fields improve quality of classroom teaching, drive teaching quality to be improved effectively and exert students' potential sufficiently. In short, educational innovation is featured by advancement and scientificity to a large extent.

3 IMPORTANCE OF CLASSROOM TEACHING REFORM

To break current situations of traditional classroom teaching, change existing modes of classroom teaching, know teaching content and objectives more clearly and perfect structure of classroom teaching, schools' classroom teaching reform has become a development trend.

3.1 *Importance of classroom teaching*

Classroom teaching is a key platform where teachers and students communicate with one another. By classroom teaching, progress in students' learning can be promoted. Meanwhile, comprehensive impacts on all aspects of students are significant. Classroom teaching acts as an important composition link of schools' educational system. Moral education and courses can be carried out only under the condition that we ensure classroom teaching is implemented successfully. [1]

3.2 *Substantiality of classroom teaching reform*

To promote scientific perfection of classroom teaching, classroom teaching reform is a practice with substantial significance. In practice of classroom teaching, the purpose of using promotion of students' comprehensive development as the first goal and perfecting classroom structure is to optimize students' overall behaviors and their learning environment and improve their comprehensive strength in all aspects including learning, innovation and operation. [2] Classroom teaching reform plays an important consolidating role in schools' school-running features. By carrying out innovation for characteristic teaching methods and ideas, schools' teaching characteristics can be shown further. Classroom teaching reform lays particular emphasis on improving substantial thoughts about students' learning. Via innovative methods, schools can educate students to learn other cultures besides basic knowledge study and cultivate comprehensive development-based talent with balanced ability in all aspects.

4 KEY POINTS OF IMPLEMENTATION OF CLASSROOM TEACHING REFORM

Implementation of classroom teaching reform is an important measure of educational innovation. In the process in which innovative activities about classroom education and teaching are carried out, it is essential to aster key points of teaching reform and know innovative angles and direction of classroom teaching reform clearly.

Substantial content of classroom teaching should accord with demands of modern talent development, the idea that theories serve practice is used as a basic learning thought and students' humanistic quality is cultivated simultaneously. [3]

Encourage students to perform several learning methods jointly and carry out more communication activities in classroom, form interaction between teachers and students and cultivate students' innovative thinking and innovative ideas.

Another key point of classroom teaching reform is that schools should carry out developmental strategies for students in teaching, work hard to cultivate students' cooperative consciousness in classroom, encourage students' spirit of active participation and division of labor and train developmental talent that can adapt to social development.

In the process of developing classroom teaching reform, it is necessary to pay attention to fundamental changes in reform work in implementation of innovative emphases, as shown in the following figure:

Classroom teaching reform should work hard to realize fundamental changes in three aspects: (1) transferring attention from effectiveness of results to rationality of process and paying attention to how to realize students' different development; (2) analyzing teachers' teaching level and then mastering basic law of subject teaching.

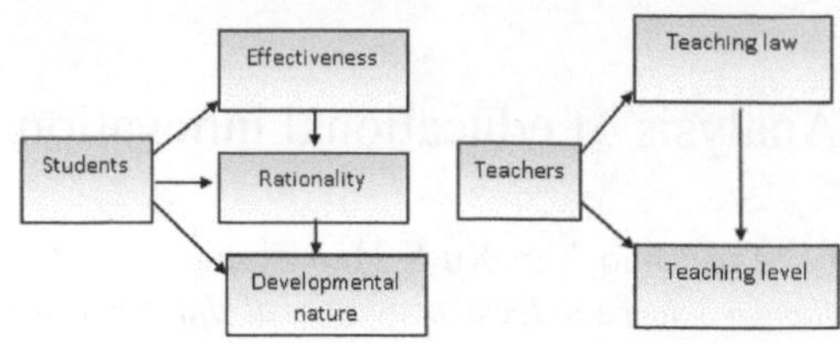

Figure 1. A mode chart about fundamental changes in classroom teaching reform.

5 WAYS TO DRIVE CLASSROOM TEACHING REFORM TO BE CARRIED OUT EFFECTIVELY

In the process in which worked related to classroom teaching reform is carried out and based on ideas of innovative education and the principle that students with comprehensive development are cultivated, it is essential to perform effective driving methods for classroom reform education and understand that core of classroom teaching reform is the first factor that should considered when classroom teaching reformed is implemented. In modern educational teaching classroom, schools should realize the purpose that modern classroom structure is perfected by training students' practical ability, cultivating their cooperative spirit and establishing their developmental thinking. When classroom teaching reform is carried out, it is essential to know core of innovative education clearly, i.e., training practical ability, cultivating cooperative spirit and establishing developmental thinking. Meanwhile, schools should carry out innovation for classroom teaching methods and enhance improvement in classroom teaching quality further.

5.1 *Training practical ability*

Learning about theoretical knowledge in classroom teaching aims at serving practical life better, so schools should strengthen training related to students' practical ability constantly in innovative activities of classroom reform. Take learning in physics classroom for example. Since knowledge related to Electricity has high abstraction, teachers should encourage students to do experiments in the process of teaching and master knowledge that they learn in theoretical process, which can improve students' practical ability and make students know improvement in their comprehensive ability simultaneously. [4]

5.2 *Cultivating cooperative spirit*

In traditional classroom teaching, cultivation about students' ability to cooperate and communicate is lacked. For modern society, shortage of cooperative spirit cannot adapt to development of the era. In classroom teaching reform, it is essential to encourage students to cooperate and communicate with one another and cultivate their cooperative spirit. [5] For instance, in order to learn profound emotion of a text in Chinese teaching (for example, the text Thunderstorm) better, we need analyze several characters' emotions. Thus, teachers may advocate that students can form groups voluntarily in classroom to play role performance and ask students to finish the task independently by independent image building, cooperation and communication. In the process of cooperation and communication, they can establish unity and enterprising spirit like cooperative consciousness, mutual help consciousness and sense of responsibility, which has important effect on students' healthy development.

5.3 *Establishing developmental thinking*

Improving students' developmental thinking is usually ignored in classroom teaching reform. There are students with different characters in classroom. In order to exert potential of all students more comprehensively, teachers should guide students to develop their own features, envisage different development and realize their free and comprehensive study in classroom teaching. For example, in history classroom, teachers will ask questions, for instance, 'expressing your own opinions', after students have learned historical events. At this moment, teachers should encourage students to express their opinions boldly. Teachers can give explanations to students with pertinence only when students express their real emotions boldly, which has significant impacts on students when they consider other questions and benefits them to adapt to the era with rapid development to a larger extent. [6]

5.4 *Enhancing innovation of classroom teaching methods*

Under educational innovation, schools' classroom teaching should pay attention to innovation of teaching methods. As information age develops, practice the development process 'informatization equipment — integration with courses — digital teaching and learning' and establish informatization teaching modes in classroom teaching, as shown in the following figure.

One the one hand, utilization of innovative classroom teaching methods can help students to concentrate their attention in classroom, arouse their learning interest and improve efficiency of classroom learning. On the other hand, it plays an important role in perfecting structure of classroom teaching.

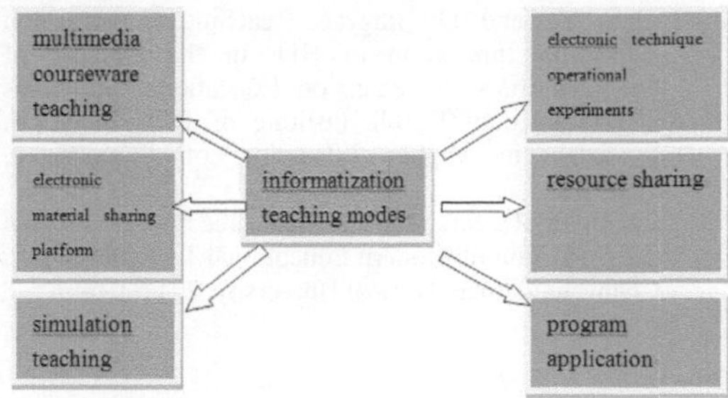

Figure 2. Digital teaching and learning' and establish informatization teaching modes.

6 CONCLUSION

According to overall discussion and situation analysis in this thesis, it is found that schools' development of educational innovation and implementation of classroom teaching reform are of great importance for cultivation of comprehensive and excellent future talent. In the process in which classroom teaching reform is developed, it is essential to know key points of teaching reform. At the same time, schools should train students' practical ability, cultivate students' cooperative spirit, establish students' developmental thinking and enhance innovation of classroom teaching methods in the process in which classroom teaching is implemented. The final purpose is to finish schools' classroom teaching reform, perfect structure of classroom teaching, promote improvement in students' comprehensive ability, build efficient and perfect modern classroom and provide more excellent talent for the society by educational innovation.

REFERENCES

[1] Zhang Miaomiao. Paying Attention to Students' Development and Focusing on Schools' Educational Innovation— Review about the Seminar 'Innovation of Classroom Teaching Reform and Students' Subjectivity Development' [J]. Journal of the Chinese Society of Education, 2010(06).

[2] Cue Yonggang and Fan Jianrong. Colleges' Classroom Teaching Reform and Innovative Talent Training [J]. Journal of Shanxi Economic Management Institute, 2011(12).

[3] Yu Lihong. Deepening Classroom Teaching Reform and Carrying out Innovation for Talent Training Modes —— Review about National Seminar on Innovation of Classroom Teaching Modes and Work Conference about Journal of the Chinese Society of Education in 2010 [J]. Journal of the Chinese Society of Education, 2010(11).

[4] Zhou Yu and Qi Jingyao. Teaching Reform and Education Innovation of HIT in the Context of Globalization— Research on Educational Strategies and Practice of Harbin Institute of Technology [J]. Research in Higher Education of Engineering, 2013(05).

[5] Liu Hean. Research on Classroom Teaching Reform in the Condition of Modern Educational Technology [D]. Changsha: Hunan Normal University, 2011(10).

[6] Yi Qizhi. Carrying our Innovation for Classroom Teaching and Letting Students Study Happily – Trial on Colleges' Classroom Teaching Reform [J]. Journal of Guangxi Normal University (Philosophy and Social Science Edition), 2010(10).

Management, Information and Educational Engineering – Liu, Sung & Yao (Eds)
© 2015 Taylor & Francis Group, London, ISBN: 978-1-138-02728-2

Research and thinking on applied talent training modes

Li Man Zhao, Yan Xu & Hao Zhang
Jiangxi Science & Technology Normal University, Nanchang, Jiangxi, P.R. China

ABSTRACT: In recent years, the educational world of China has improved its attention to applied education. School-running mode of schools is decided by applied talent training modes. Different types of training modes can be applied to the same type of talent, while their unique framework is needed for a specific mode. This article studies connotation and content of applied talent training modes and elaborates ways to improve applied talent training modes.

KEYWORDS: BP Neural Network, Applied talent, Training mode, Higher education.

1 INTRODUCTION

The educational world and the whole society begin to pay attention to talent training quality and problems existing in the process of talent training. Improving talent training quality constantly is a final goal of higher education and teaching system reform. According to related specifications of the country, we improve talent training mode in order that talent training schemes and training process can adapt to goals of applied talent training and satisfy social demands better. At the same time, this is a core idea about improvement in talent training quality. Providing talent, science and technology as well as social service for other places is a function of colleges of higher education, whose mainly school-running objective is to improve splendid talent, drive local economy to develop rapidly, be able to have high-tech applied talent, ensure development and promotion of local economic construction and usage rate of high and new technology, and improve enterprises' technological content and market share of products. Thus, strategic shift of talent training objectives is the content that colleges of higher education must realize. The strategic shift from academic education to improvement in capabilities can pay more attention to improving students' learning ability, entrepreneurial ability and employability in teaching process and educational concept.

2 CONNOTATION OF APPLIED TALENT TRAINING MODES

In social development, modern talent can be divided into four types including theory-based talent, engineering talent, skill-based talent and technology-based talent if we distinguish talent from different perspectives, for instance, we classify talent according to purposes of productions or work activities.

Ability is the center of the applied talent training mode which direct at cultivating applied professional talent [1]. In this mode, people require both talented people's ability to adapt to posts and communication ability among posts, i.e., talent not only needs both comprehensive professional ability but also has some comprehensive ability, employability, entrepreneurial ability, reproducing skills and creative techniques. On the basis of some scientific and reasonable theories, requirements for technology are higher than the ones for common skills. In another word, it need talent has some compound and comprehensive features and holds breakthroughs in the aspects of experience and technology. Before the 19th century, theoretical technology had only represented common technology and skills and the experience that has been accumulated for a long time was its main basis. After the 19th century, the most obvious feature of modern technology was theorization. Since the 20th century, technical function and nature has been clear gradually. Many technologies that only existed in production process in the past are widely applied to marketing management and service first. This method can broaden scope and methods of effect effectively. Thus, both non-material technology and material technology belong to modern technology.

Re-examination on technical talent and skilled talent plays an important role in higher education. Since motor skills were a major component of skilled talent in the past, secondary vocational education institutions cultivated skilled talent by themselves [2]. With modern scientific and technological level improves at present, motor skills are not a major component of most skilled talent's labor composition. Increasingly increased governance elements make skilled talent and supply of skilled talent more and more similar. Thus, with respect to selection and cultivation of teaching materials, higher vocational education should use applied talent and advanced technical talent as a training direction. This kind of talent is not white-collar workers or blue-collar workers but belongs to applied type, which is also called silver collar. The following figure shows application procedures of applied talent.

3 ANALYSIS OF BASIC APPLIED TALENT TRAINING MODES

Under some educational thoughts and methods, talent training uses training objective as a goal and applies some methods in the training process to make the systematic knowledge learned by students can be mastered more stably. Besides, it can improve corresponding ability, structural frame of quality and efficiency of operation. Currently, China proposes the following training modes for applied talent training according to different perspectives [3].

The first one is liberal education training model, which is carried out for professional education problems. Related scholars deem that this talent training model is liberal education for professional knowledge education, makes preparation for life-long education and lays a solid foundation according to basic training, scope of knowledge, degree of application ability and level of quality and innovative spirit etc. [4]

The second talent training mode is a personalized and multi-layer mode that uses ability cultivation as a key point. In accordance with related materials, it is shown that the talent training mode constitutes talent cultivation system, centers on ability training and is multi-layer. Personalized training mode is intensively reflected and goals of its theoretical course system are to perfect basic knowledge constantly under construction and perfection of basic theories in order to ensure that basic skills and professional skills of practical course system can be improved and to improve comprehensive ability of quality expansion system and broaden extension of the major.

The third talent training mode integrates knowledge, quality and ability effectively. Some scholars conclude two aspects of basic content for it. Firstly, it is a knowledge training model that uses wide scope of knowledge and splendid comprehensive ability, chooses a large scope, integrates courses, shortens teaching and combines primary and secondary thoughts. The other one is the mode that aims at expanding comprehensive quality including streakiness, gentility and excellent service ability.

4 CONTENT OF REFORM OF APPLIED TALENT TRAINING MODES

Currently, reform of applied talent training modes is mainly divided into three layers.

Reform of applied talent training modes can be mainly considered in three layers. Firstly, the first layer of reform of talent training modes means the whole school carries out the reform and is a practice that schools adapt to the society actively [5]. When economy, educational concepts, science and technology and educational system of the country have major reform, some colleges need ensure successful implementation of the whole school's reform of talent training modes to hold an invincible position in the reform. Its reference standard is social demands and uses overall structure of schools' majors for priority selection. For each major, including training objectives of added majors and re-oriented majors, it is essential to design new training specifications, formulate new talent training plans and choose new training methods.

With respect to the second layer of talent training reform, form of its reform is professional. It trains professional talent according to overall performance of the society and then hands talent to the society for inspection. Usually, schools' talent training quality cannot be accepted by the society and is thought that they cannot satisfy social demands or unify development trend of higher education. Thus, it is necessary to adopt professional reform of talent training modes and satisfy social demands sufficiently, for instance, using type of the major's talent, and mastered knowledge as reference. For these professional training goals, adjust reform methods of talent training modes to ensure professional training plans and training methods can adapt to social requirements better by adjusted methods and promote progress of the era and development of higher education under training specification and objectives.

Specifically, the third layer of talent training mode is reform of professional talent training methods and schemes, which also reflects social adaptability of this talent training process. When a result of professional talent training can evaluate originally oriented training objectives, quality of this talent training is not accepted by schools and cannot ensure schools' orientation and professional training goals. Thus, professional reform of talent training schemes and methods are needed. It uses schools' oriented and decided goals for major training and training specifications as reference standards, adjusts training schemes and methods of the major and makes it coordinates with schools' orientation and requirements of major training goals and specifications, i.e. coordinating with training goals and specifications better.

The three layers of reform of talent training modes are not constant. In the process in which higher education is reformed continuously, it is essential to carry out innovation for educational thoughts constantly in order to adapt to continuous improvement and optimization of talent training.

5 CONCLUSION

Thus, it is shown that diversity of higher education is decided by its educational diversity, and this kind of diversified higher education also needs diversified talent training modes to adapt to it. Under some situations, talent training modes are relatively stable are also are changed sometimes. In accordance with speed of economic and technological development as well as market demands for talent, dynamic changes in training modes can be decided according to their diversity. On the premise that basic rules of higher education are accorded with, talent training modes are optimized with social development and new content is injected constantly to satisfy social demands.

REFERENCES

[1] Yang Xinglin. Local Colleges' Applied Talent Training Modes Should Focus on Four Changes [J]. Journal of Yangzhou University (Higher Education Study Edition), 2011(2):135–136.

[2] Xi Chengxiao. On Innovation of Chinese Local Colleges' Applied Talent Training Modes According to Reform of Talent Training Modes in Developed Countries [J]. Journal of Ankang Teachers College, 2013(2):327–328.

[3] Sun Xiaojuan and Zhao Hongmei. Construction and Implementation of Applied Talent Training Modes of the Major Public Service Administration [J]. Heilongjiang Researches on Higher Education, 2011(5): 182–183.

[4] Hu Weizhong and Shi Ying. Applied Talent Training Modes of Australia and Their Enlightenment [J]. Open Education Research, 2011(4):186–187.

[5] Zhang Rixin, Liang Yuqing, Wang Lingjiang, Lai Li and Xiao Jun. Research on Applied Talent Training Modes of Undergraduate Courses and Their Practice [J]. Journal of Chengdu University (Natural Science Edition), 2011(5):156–157.

[6] Yang Chunchun and Liu Junping. Comparative Study on Applied Talent Training Modes of Undergraduate Education in China and Foreign Countries [J]. Journal of Nanjing Institute of Technology (Social Science Edition), 2011(5):156–157.

[7] Zhang Miaomiao. Paying Attention to Students' Development and Focusing on Schools' Educational Innovation— Review about the Seminar 'Innovation of Classroom Teaching Reform and Students' Subjectivity Development' [J]. Journal of the Chinese Society of Education, 2010(06).

[8] Cue Yonggang and Fan Jianrong. Colleges' Classroom Teaching Reform and Innovative Talent Training [J]. Journal of Shanxi Economic Management Institute, 2011(12).

[9] Yu Lihong. Deepening Classroom Teaching Reform and Carrying out Innovation for Talent Training Modes —— Review about National Seminar on Innovation of Classroom Teaching Modes and Work Conference about Journal of the Chinese Society of Education in 2010 [J]. Journal of the Chinese Society of Education, 2010(11).

[10] Zhou Yu and Qi Jingyao. Teaching Reform and Education Innovation of HIT in the Context of Globalization— Research on Educational Strategies and Practice of Harbin Institute of Technology [J]. Research in Higher Education of Engineering, 2013(05).

[11] Liu Hean. Research on Classroom Teaching Reform in the Condition of Modern Educational Technology [D]. Changsha: Hunan Normal University, 2011(10).

[12] Yi Qizhi. Carrying our Innovation for Classroom Teaching and Letting Students Study Happily — Trial on Colleges' Classroom Teaching Reform [J]. Journal of Guangxi Normal University (Philosophy and Social Science Edition), 2010(10).

Management, Information and Educational Engineering – Liu, Sung & Yao (Eds)
© 2015 Taylor & Francis Group, London, ISBN: 978-1-138-02728-2

Analysis on problems in strengthening the class teaching reform

Li Man Zhao, Yan Xu & Hao Zhang
Jiangxi Science, Technology Normal University, Nanchang, Jiangxi, P.R. China

ABSTRACT: The class teaching in China has been constantly transferring toward quality-oriented education. It is necessary to make targeted solutions to specific problems, so as to completely eliminate the drawbacks of traditional education. The concrete solutions include reinforcing the theoretical level of the teaching reform, determining the basic objectives of the reform, formulating the reform scheme, according to the problems, and completely strengthening the improving speed of the education system in China.

KEYWORDS: Principal Component Analysis, BP Neural Network, Applied talent, class teaching, deepen the reform, teaching problems.

1 INTRODUCTION

The social and economic development place increasing demands for education, and improving the quality of education is the necessity for social development, as well as the demands of developing school. Talent cultivation is the most important and most basic function of the teaching process in modern universities, thus how to strengthen the teaching quality is the major problem the education system faces. The practice proves that the class teaching directly influences students' learning quality, therefore, it is of great superiority to solve the drawbacks of the current class teaching, and formulate reform scheme, as well as improve the class teaching capacity in cultivating talents.

2 REFORM OF BASIC THEORY

Reform of basic theory provides guidance and objectives for teaching reform, and provides a solid theoretical support in promoting the process of reform. Class teaching process requires a clear starting point and actual action, emphasize scientific class teaching reform, and highlight the truth of reform. Only a profound theoretical basis can support the teaching reform. Class teaching is a very complex human phenomenon, including a variety of irregular areas. Solving problems need more subjects to participate, absorb the purified results of a variety of subjects and develop the basic subject education system. Teaching theory is a combination of diversity and integration. In recent years, the theoretical basis of relying solely on philosophy has been changed into open the field, and accepts the research findings of philosophy, psychology, sociology and other subjects enrich the theoretical resources of class teaching. In order to enhance the communication between subjects, research category should be subject to the characteristics of class teaching.

2.1 *Comprehensive development policy*

Marx's comprehensive development theory said that it is necessary to strengthen the personal quality of humanness, comprehensively discuss the inherent realization mechanism, independently develop and proof theoretical goals, according to free development, so as to provide basic values and developing direction for class teaching reform.

2.2 *Teaching cognition*

Strengthen students' exposure to experiences and cognitive characteristics by means of improving teaching cognition, combine internal activities with external activities, transform teaching development model, and provide students with methods to solve problems and ways to develop. All teaching activities should be based on theory, so as to present the nature of the subject through class teaching, and intensify talents training and knowledge education. Both the students and teachers are the teaching subjects, therefore, strengthening cooperation between teachers and students, promoting the progress of class teaching reform as well as intensifying the rational thinking capacity of teaching are of great significance.

3 PURPOSE OF TEACHING REFORM

Teaching reform mainly aims to enhance students' practical ability and comprehensive quality. With the acceleration of the social development, the traditional exam-oriented education has been unable to meet the market demands. Thus education and teaching reform are necessary to improve students' ability to develop in the future.

3.1 *Improve the applicability of knowledge*

Teachers should inspire and guide the students to cooperatively discuss the problems they met in learning, and formulate discussion goals, according to the subjects, introducing knowledge into daily life. Students should actively find problems, express and ask questions and solve problems by cooperating with groups or teams. In addition, students should flexibly apply the learned knowledge and get to know to apply the learned knowledge.

3.2 *Cultivate students' self-learning ability*

The best way of cultivating self-learning ability is to improve interest in learning and possess self-learning ability. Class teaching is important in improving students' creative ability, thus teachers should enhance students' interest in the subject, guide students to master new knowledge, strengthen the cultivation of learning methods as well as constantly improve their independence.

3.3 *Strengthen the depth of the class teaching*

Teachers should enhance the breadth of knowledge by means of situational teaching, and introduce new knowledge points by interacting with students. In addition, teachers should integrate various teaching resources, create learning environment, according to the specific subject, and improve the speed of solving problems. Moreover, they should introduce some difficulties in knowledge and discuss through practical activities.

3.4 *Enhance the class vitality*

Teachers should enhance the class vitality, so that students can exert their personality to the utmost extent, and learn knowledge of the class activities. This fully reflects the principle of innovation, in which students take the initiative in finding problems, solving problems by cooperating with groups and actively enhance the vitality and vigor of class teaching.

4 CLASS TEACHING REFORM SCHEME

As an important part of talent cultivation, class teaching reform is closely linked with most working linkage in school and also involves a lot of teaching elements. We should not just pay attention to the surface of the class teaching, instead, we should judge from the overall situation of the talent cultivation to find the reform purpose, explore teaching resources, and gradually deepen the penetration level of class teaching reform, combine reform methods with the ultimate goal, so as to achieve integration of motivation, effects and goals. Based on the theoretical teaching, we can divide class teaching into six interlinked factors: Teaching objectives, teaching methods, teaching evaluation, teaching contents, teaching framework, and teaching management. Therefore, we should take the teaching objectives and teaching methods as the breakthrough direction, so as to perfect and strengthen the teaching framework, teaching evaluation, teaching contents, and teaching management ability, focus on reform priorities, as well as coordinately promote the teaching reform.

4.1 *Determine the teaching objectives*

The primary objective is to change the outlook on talents and quality, cultivate all-around developed talents by teaching practice, establish a view of cultivating integrated talents, and enhance students' knowledge, practical ability, and comprehensive quality. In addition to improving students' professional knowledge, it is important to cultivate students' personality, so as to develop their creativity and comprehensive ability. It should also pay attention to the cultivation of personality factors of emotion, will, etc., so as to strengthen the cultivation of compound talents.

4.2 *Improve teaching diversity*

On the condition of enduring teaching activity, it should formulate specific subject objective, according to the specificity of the teaching objects, in view to strengthening heuristic teaching method, introduce teaching methods of interaction, case, exploration, etc. in class teaching, and enhance teachers' class teaching competence, as well as improves teaching diversity. Teachers should develop students' ability to draw inferences by inspiration method, so as to strengthen their rational cognitive ability, and highlight their creativity and exploring spirit, as well as pay attention to the exchange and experience between sensibilities. In addition, it should improve the emotional exchange, and strengthen teacher-student communication through the belt between teachers and students, and discuss problems by means of asking and answering, changing the teaching into a conversation between teachers and students, which is necessary to enhance students' ability of logical thinking and knowledge application. Determine the key teaching points by setting, teaching cases, change the inflexible knowledge imparting way of the traditional education, and provide students with self-learning methods and ability, strengthen the understanding of key points and principles, enhance their divergent thinking and guide them to learn by self-created. Inspire students' spirit by exploring teaching methods, present students' with the problems of the knowledge hierarchy, development direction, and different opinions, and guide them to solve problems by themselves, improve their interest in subjects, grasp the international academic perspectives and dynamic, and help students to cultivate habits of self-learning, self- thinking, self-questioning and self-solving. Class teaching can be divided into five steps, as shown in Fig.1

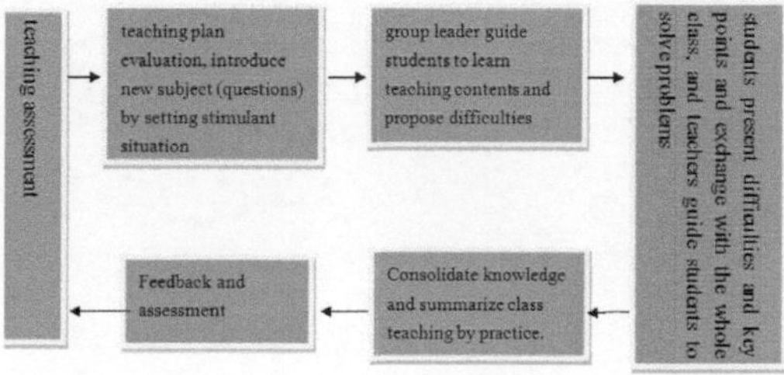

Figure 1. Five steps teaching methods.

4.3 *Mobilize the initiative of teachers*

Teachers are the key factors in improving the teaching quality, thus it is necessary to improve teachers' innovative spirit and initiative to cultivate innovative talents. In-depth class teaching reform requires motivating teachers' sense of responsibility and creativity, cultivating teachers' active creativity, and improving the teaching atmosphere on campus, so that all teachers can participate in class teaching reform. It is necessary to provide regular academic training for teachers, exert teachers' unique charm, target to improve young teachers' teaching ability, exert the top teachers' exemplary role, so as to create an excellent teaching team, complement teachers' capabilities and play good team effects.

4.4 *Comprehensive survey system*

Complete teaching system requires formulating detailed assessment system, achieving teaching objectives by supervision from 4 parties, comprehensively evaluating students, teachers, school and schedule of survey, formulating an annual report for educational quality, so as to improve the teachers' teaching initiative. As shown in Fig.2

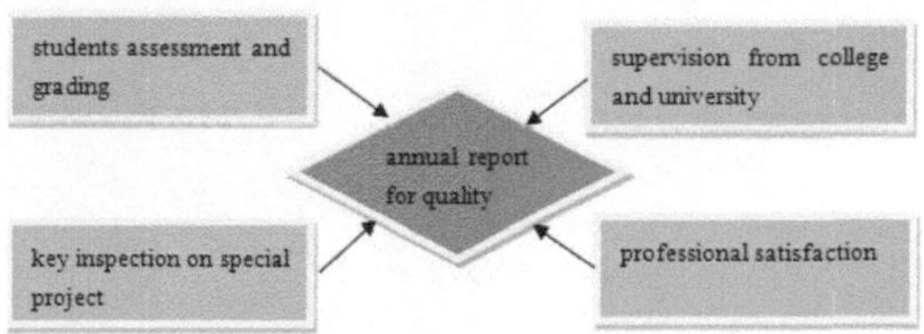

Figure 2. Comprehensive survey system.

5 CONCLUSION

Class teaching reform is the key to improve teaching quality, and is a consensus on the educational world. However, improving teaching quality is a long-term reconstruction project, cannot be accomplished in the short-term. We should strengthen the importance of class teaching reform in education, and be down to earth and take advantage of all favorable factors, so as to effectively improve the depth of teaching reform, and improve education and teaching system in China.

REFERENCES

[1] American, John Rawls, Political Liberalism [M]. Translated by Wan Junren, Yilin Press, Nanjing, 2010:141.

[2] Gao Youhua, Wang Yinfen, Research on Contemporary Class Teaching Reform and Development in Universities in America [J]. China Electric Power Education.

[3] Gao Youhua, Wang Yinfen, Research on Contemporary Class Teaching Reform in Universities in Germany [J]. Meitan Higher Education, 2010(5).

[4] Zhen Hong. Comparative Study on Class Teaching in Universities in America and China [J]. Higher Education Development and Evaluation, 2010(02).

[5] Pei Dina. Modern Teaching Theory 1 [M]. People's Education Press, 2011:175–288.

Section 3: Engineering management, production management, business and economics

Management, Information and Educational Engineering – Liu, Sung & Yao (Eds)
© 2015 Taylor & Francis Group, London, ISBN: 978-1-138-02728-2

Core technologies identification based on a citation-network model: A case of laser technology system

Han Lin You, Meng Jun Li, Jiang Jiang, Ji Li Luo & Jian Guo Xu
College of Information System and Management, National University of Defense Technology, Changsha, P. R. China

Fang Zhou Chen
College of Humanities and Social Sciences, National University of Defense Technology, Changsha, P. R. China

ABSTRACT: With the importance and quantity of technologies increasing, the core technology identification has become an important part of the technology Research and Development (R&D). However, the rapid increase of the size of the laser technology system brings a huge challenge. To solve this problem, a method using the importance of nodes analysis and Multi-Criterion Decision Making (MCDM) based on a citation-network model is proposed. As a case study, the analysis results of a laser technology system, which consists of 452 interdependent technologies, are displayed and discussed.

KEYWORDS: Laser technology; Citation-network model; Core technologies identification.

1 INTRODUCTION

Technological innovation is one of the most important driving forces in the development process of human being. In the 21th century, different kinds of modern technologies are used to improve people's life and the influence of technologies on the society and economy is increasing rapidly. As a result, the theories and methods of technology research and development (R&D) have become a focus research point of the management science and engineering.

As a common sense, a more advanced technology system with a bigger size tends to create more profits. However, due to the constraints of different kinds of resources, such as cost, schedule and knowledge, it's impossible to invest in every technology project. The key problem is how to select the technologies that are the most valuable ones to the whole system.

To improve the efficiency of the technology system, it's important to identify the technologies called core technologies, which have significant effects on the other ones. Core technology identification, which is based on technology relationships, is helpful to detect the key research field and capture the technology development chance.

2 LITERATURE REVIEW

With the R&D becoming a focus research field, lots of technology management approaches are proposed. Patent documents, which contain much technological and commercial information, are used to research the technological innovation and development (Chang et al. 2010).

A lot of methods are proposed to describe technology relationships by using patent analysis. Tseng (2007) proposed a method of patent co-word relationship analysis by using automatic keyword-abstraction based on patent text-mining. The cross impact analysis (CIA) is used to evaluate relationships of technologies based on quantitative analysis (Thorleuchter et al. 2010). An approach to structure weighted technology networks is proposed by using the frequency and context of keywords abstracted from patent texts (Chang et al. 2010). To research the innovation, knowledge-flow, a method to describe technology relationships by using patent citation is proposed (Lee et al. 2009).

Based on the patent-citation analysis, which is one of the most common relationship-analysis approaches, core technologies identification methods are proposed. Different kinds of patent-citation relationships are considered to select the most important technologies by using network analysis (Wartburg et al. 2005). ANP-based citation network analysis method is used to identify core technologies (Lee et al. 2009). Closeness centrality and betweeness centrality are used to analyze different kinds of the importance of nodes in the citation network (Kim et al. 2013).

Knowledge-flow is one of the most important properties of technologies. Citation data of patent texts contain a great deal of technology knowledge-flow information. Therefore, the patent-citation analysis is an appropriate approach to research the importance of technologies in the aspect of the knowledge-flow and the existing researches have proposed different kinds of effective methods. Although the kinds of centrality indices based on the network analysis are used to identify core technologies, it's difficult to access an integrated approach to evaluate which technologies are the most influential to the other ones. In this paper, an approach combining network analysis and multi-criterion decision making (MCDM) is proposed to handle the problem.

3 METHODS

In this section, a method combining network analysis and multi-criterion decision making (MCDM) is proposed to identify core technologies. Firstly, the citation data is obtained, with which the citation-network model is built, based on patent-text analysis. Secondly, 4 centrality indices, such as degree centrality, eigenvector centrality, betweenne--ss centrality and closeness centrality, are introduced and discussed. At last, an integrated assessment approach using the Technique for Order Preference by Similarity to Ideal Solution (TOPSIS) is proposed.

3.1 *Data collection*

The patent data is collected from the U.S. Patent and Trademark Office (USPTO) database. The related patent-texts are obtained by searching a given keyword and the needed data, such as patient ID, patent name, granted date and citation data, is abstracted by scanning the structured text using a text-mining method.

3.2 *Network analysis*

With the patent-data obtained by text-mining, a citation network model, which abstracts the patents as nodes and the citation relationships as edges, is built. Network analysis approaches are capable to handle technology system research problems. Therefore, assessment approaches of the node importance are used to select core technologies.

Centrality indices are widely used to assess the node importance in different kinds of networks. The advantages of this approach are not only easy to calculate and understand, but also comprehensively covering the factors of the node importance assessment. In the section, 4 centrality indices, such as degree centrality, eigenvector centrality, betweenness centrality and closeness centrality, are introduced and discussed to analyze citation-network model.

The degree centrality D_i means the neighbor count of the ith node and represents its direct influence to other nodes (Freeman. 1979). Degree centrality is an index to measure the direct influence of nodes to their neighbors. The nodes which have more neighbors, namely the high degree centrality value, tend to be core technologies. The equation representation of D_i calculating is shown as:

$$D_i = \sum_{\substack{j=1 \\ j \neq i}}^{j=n} A_{ij} \tag{1}$$

Where n = nodes amount; A_{ij} = adjacency matrix value.

The eigenvector centrality E_i is another index to measure influence of nodes based on their neighbors. However, the eigenvector neighbors are considered different in their importance rather than all equal, which is different from the degree centrality. As a result, a large number of poor neighbors are not enough to make sure high eigenvector centrality value. Besides, both the importance and the amount of neighbors are factors of the index value (Bonacich. 1972). The equation representation of E_i calculating is shown as:

$$E_i = \lambda^{-1} \sum_{j=1}^{n} A_{ij} e_j \tag{2}$$

Where λ = the maximal eigenvalue of the adjacency matrix; $e_j = j^{th}$ element value of the corresponding eigenvector.

Degree centrality is a classical and widely used index to analyze network focusing on neighbors. However, considering both amount and importance of neighbors are necessary to assess the node importance, the eigenvector centrality has better performance.

The betweenness centrality B_i means the counts of all the shortest paths of the network crossing the ith node (Freeman. 1979). This index is focused on the position of nodes in the network and widely used to select the key nodes in the aspect of the network diffusion. The more crossing shortest paths, one node has, the larger influence on the network model it possesses, namely the more important it is. The calculating equation representation of B_i is shown as:

$$B_i = \sum_{s \neq t \neq i} n_{st}^{i} \Big/ g_{st} \tag{3}$$

Where g_{st} = the count of those shortest paths connecting $Node_i$ to $Node_j$; n_{st}^{i} = the count of those shortest paths which cross $Node_i$.

Another index focusing on the position of nodes is the closeness centrality C_i (Sabidussi. 1966). The main factor of this index value is the average distance to other nodes rather than the number of crossing shortest paths. The shorter average distance is, the closer to network center the node is, namely the more influence it has. The equation representation of C_i calculating is shown as:

$$C_i = n-1 \Big/ \sum_{j=1}^{n} d_{ij} \tag{4}$$

Where d_{ij} = the distance from $Node_i$ to $Node_j$.

Betweenness centrality and closeness centrality respectively describe two different attributes of the position of the nodes. The former focus is on the shortest path while the latter one is on the average distance. As a result, it is difficult to judge which one is better. Furthermore, there is even less evidence to select one best index among eigenvector centrality, betweenness centrality and closeness centrality. The methods of multi-criterion decision making are needed to integrate those different kinds of index values.

3.3 *TOPSIS*

It is difficult to assess the fitness of the three centrality indices and select the best one. What is more, since the aspects of the three assessment indices focusing on are different, there is no effective information to compare the relative importance and calculate the weight vector, which is a key input datum in MCDM.

The Technique for Order Preference by Similarity to Ideal Solution (TOPSIS), one of MCDM methods, is selected to handle the problem (Georgiadis et al. 2013). Since the weight vector is an additional datum in TOPSIS, it is acceptable to consider all the weights are equal. It is easy to identify the positive ideal node-state and the negative ideal node-state with three types of indices. Considering the nodes as alternatives, the integrating importance assessment value of each node is calculated as the performances of alternatives based on the distances to the two ideal node-states. The algorithm is shown as follows:

$$\begin{aligned} &Define: Index_i^+ = \overset{n}{\underset{k=1}{Max}}(Index_{ik}), \\ &\qquad Index_i^- = \overset{n}{\underset{k=1}{Min}}(Index_{ik}) \\ &s_k^+ = \sqrt{\sum_{i=1}(Index_{ik} - Index_i^+)^2}, \\ &s_k^- = \sqrt{\sum_{i=1}(Index_{ik} - Index_i^-)^2} \\ &c_k = \frac{s_k^-}{s_k^+ + s_k^-} \end{aligned} \tag{5}$$

Where $Index_i^+$ = the best performance of the ith index; $Index_i^-$ = the worst one; $Index_{ik}$ = the ith index performance of the kth node; s_k^+ = distance to the positive ideal node-state; s_k^- = distance to the negative ideal node-state; c_k = the integrated assessment value.

To identify the core technologies with the highest integrated assessment value, the parameter ρ, which represents the proportion of the technology selection, is introduced. Based on the integrated assessment value c_i, the technologies with most influence to other ones are identified. The count of selected core technologies is calculated as follows:

$$CTnum = Min(X), X \geq \rho n \; and \; X \in N^+ \tag{6}$$

4 CASE STUDY

In this section, a case study of the laser technology system is displayed to illustrate the proposed methods. The system contains 452 laser technologies, of which data are abstracted from the patent-texts of the USPTO. The grant dates of the collected patents cover from 1971 to 2013. With the collected data, a citation-network model is built. The core technologies are identified and discussed using different kinds of assessment indices based on the network analysis result.

4.1 *Citation-network model*

Because the patent numbers are too long to be displayed in the network graph, the patents are sorted by download times and labeled with serial numbers from 1 to 452. The three earliest and three latest laser technologies is displayed in Table 1.

Table 1. Part of laser technologies description data.

No.	Patent ID	Patent Name	Granted date
250	3568087	Optically pumped semiconductor laser	1971
151	3569660	Laser cutting apparatus	1971
407	3576965	work pieces, particularly watch jewels by means of laser pulse	1971
138	7324867	Controller for a laser using predictive models of materials processing	2008
52	8242408	Masking device for laser machining system and method	2012
83	8324529	converged laser beam and laser machining method	2012

With the data of patent-texts, technologies are abstracted as nodes and citation relationships are abstracted as edges. The citation-network model of the laser technology system is drawn by using UCINET 6.0 (Borgatti et al. 1999) in Figure 1.

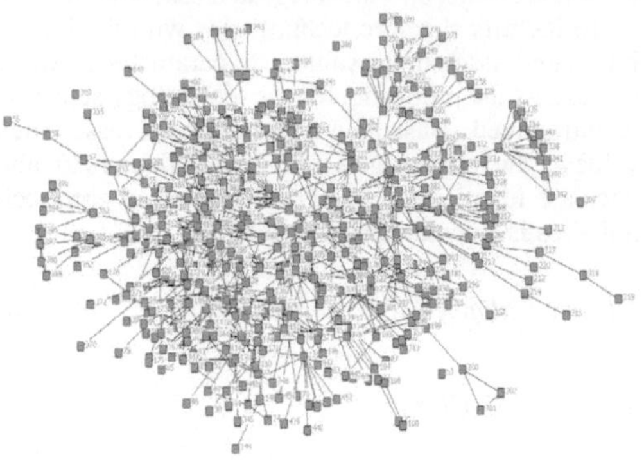

Figure 1. Citation-network model of the laser technology system.

4.2 *Core technologies identification*

Based on the citation network, the four centrality index values are calculated by using the Equation 1–4. With the index value data, three integrated assessment methods of the node's importance are proposed.

Firstly, the degree centrality, betweenness centrality and closeness centrality are integrated with TOPSIS (DBC). Secondly, the eigenvector centrality takes the place of the degree centrality to assess the node importance (EBC). At last, all the four types of index values are used to evaluate the influence of laser technologies (DEBC).

When the parameter $\rho = 3\%$, the analysis results of four centrality indices using the Equation 5–6 are displayed in Table 2.

Table 2. Analysis results of four centrality indices.

Degree		Betweenness		Eigenvector		Closeness	
ID	Value	ID	Value	ID	Value	ID	Value
137	19	123	16098.055	314	0.339	137	0.271
5	15	137	15198.771	123	0.263	123	0.263
225	14	223	12627.733	281	0.212	306	0.258
421	14	306	9266.0330	261	0.207	113	0.251
314	13	113	8257.7140	260	0.205	119	0.249
123	12	314	7804.5630	291	0.204	314	0.249
131	12	381	7743.3420	293	0.203	129	0.246
329	12	310	7095.4560	302	0.199	139	0.243
381	12	326	6886.8960	294	0.198	310	0.243
52	11	245	6331.6600	326	0.195	223	0.241
113	11	377	6153.2910	319	0.194	291	0.241
263	11	1	6088.7290	310	0.178	1	0.240
283	11	263	6028.6280	320	0.169	381	0.238
306	11	225	5583.8860	306	0.168	120	0.236

As the data in Table 2 shows, the analysis results of the core technology identification using four types of centrality indices are quite different and there is no reliable evidence to select the best one. Therefore, the analysis results using DBC, EBC and DEBC are compared. The performance curves are shown in Figure 2 and the analysis results are displayed in Table 3.

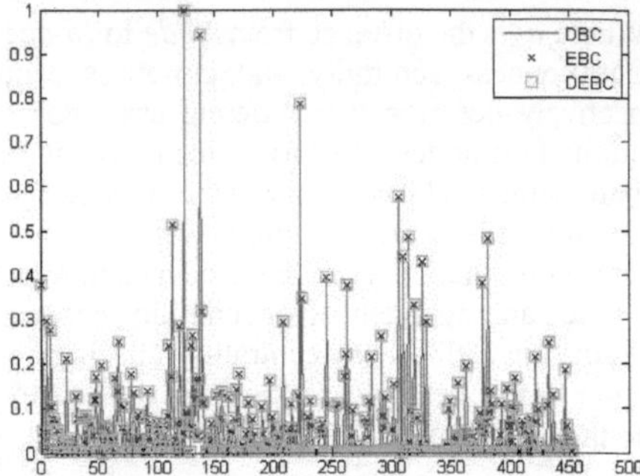

Figure 2. Performance curves of three integrated index values.

Table 3. Analysis results based on three integrated indices.

DBC		EBC		DEBC	
ID	Value	ID	Value	ID	Value
123	0.999565354	123	0.999995253	123	0.999565328
137	0.944137143	137	0.944137101	137	0.944137142
223	0.784425348	223	0.784426006	223	0.784425347
306	0.575599541	306	0.575599537	306	0.575599541
113	0.512963552	113	0.512963460	113	0.512963552
314	0.484814316	314	0.484814035	314	0.484814316
381	0.481011241	381	0.481011029	381	0.481011241
310	0.440764922	310	0.440764801	310	0.440764922
326	0.427809418	326	0.427809198	326	0.427809418
245	0.393318362	245	0.393318323	245	0.393318362
377	0.382238220	377	0.382238165	377	0.382238220
1	0.378227840	1	0.378227618	1	0.378227840
263	0.374494502	263	0.374494186	263	0.374494502
225	0.346867808	225	0.346867122	225	0.346867808

As the data in Figure 2 and Table 3 show, the performance curves are almost coincident and the analysis results of the core technology identification are totally coincident.

5 CONCLUSIONS

To identify core technologies of a technology system, an approach combining network analysis and multi-criterion decision making (MCDM) is proposed and a case of the laser technology system, including 452 technologies is studied to illustrate and validate the proposed approach in this paper. It is proved that the approach is effective to overcome the drawbacks of the conflict of different centrality indices and generate a coincident assessment result with the analysis data.

Assessment approaches of the node importance are considered effective to identify core technologies based on network models. In future researches, the assessment approaches and network models are key factors to handle the problem better.

ACKNOWLEDGEMENTS

This research is supported by the National Science Foundation of China under contract No. 71331008 and No. 71201168.

REFERENCES

[1] Chang, P. L., Wu, C. C., Leu, H. J. 2010. Using patent analyses to monitor the technological trends in an emerging field of technology: a case of carbon nanotube field emission display. Scientometrics 82, 5–19.

[2] Tseng, Y. H., Lin, C. J., Lin, Y. I. 2007. Text mining techniques for patent analysis, Information Processing and Management 43, 1216–1247.

[3] Thorleuchter, D., Poel, D. Van den, Prinzie, A. 2010. A compared R&D-based and patent-based cross impact analysis for identifying relationships between technologies. Technological Forecasting & Social change 77, 1037–1050.

[4] Lee, H., Kim, C., Cho H., Park Y. 2009. An ANP-based technology network for identification of core technologies: A case of telecommunication technologies. Expert Systems with Applications 36, 894–908.

[5] Wartburg, I., Teichert, T., & Rost, K. 2005. Inventive progress measured by multi-stage patent citation analysis. Research Policy 34, 1591–1607.

[6] Kim, E., Cho, Y., Kim, W. 2013. Dynamic patterns of technological convergence in printed electronics technologies: patent citation network. Scientometrics published online.

[7] Freeman, L. C. 1979. Centrality in social networks: Conceptual Clarification. Social Networks 1, 215–239.

[8] Bonacich, P. 1972. Factoring and weighting approaches to status scores and clique identification. Journal of Mathematical Sociology 2, 113–120.

[9] Sabidussi, G. 1966. The centrality index of a graph. Psychometrika, 31(4):581–603.

[10] Georgiadis, D. R., Mazzuchi, T. A., Sarkani, S. 2013. Using Multi Criteria Decision Making in Analysis of Alternatives for Selection of Enabling Technology. Systems Engineering, published online.

[11] Borgatti, S. P., Everett, M. G., Freeman, L. C. 1999. UCINET 6. 0 Version 1.00. Harvard: Analytic Technologies Publishers.

Management, Information and Educational Engineering – Liu, Sung & Yao (Eds)
© 2015 Taylor & Francis Group, London, ISBN: 978-1-138-02728-2

Innovation and scientific breakthroughs in artificial intelligence methods

Xi Zou
Wuchang University of Technology, China

ABSTRACT: In recent years, with the development of innovative computer technology and scientific methods, artificial intelligence technology has made a significant breakthrough. First, find a common corc mcchanism intelligence generated in a given condition "information - knowledge - Intelligent Conversion", thereby establishing a mechanism for simulation methods of artificial intelligence. Second, find the ecological structure of knowledge is instinctive knowledge in support of the "empirical knowledge - normative knowledge - common knowledge conversion" and thus to develop a vision of artificial intelligence research. Third, combine the common core mechanism generated by intelligence with and ecology structure of knowledge, found originally developed independently of artificial intelligence, harmony exception mechanism simulation under different conditions of knowledge are "structural simulation, functional simulation, behavioral simulation" methods, these three methods form a unified theory of artificial intelligence methods and research. This paper introduces the concept of artificial intelligence and a major breakthrough, discusses the importance of science and the development of innovative methods of artificial intelligence.

KEYWORDS: Artificial intelligence; breakthrough; scientific method; innovation.

1 INTRODUCTION

As society advances, the application of artificial intelligence is more and more widespread, through a long period of development and innovation, artificial intelligence technology has been perfected and makes a breakthrough. Guidance of the scientific theory of artificial intelligence technology is the key to progress, innovation of scientific methods makes independent, decentralized traditional research methods the harmony and unity, changes the original single-oriented research. Innovative scientific method is to break through the original scientific theory, to create a new technical approach, it has changed people's inherent thinking patterns.

2 THE CONCEPT OF ARTIFICIAL INTELLIGENCE

The word "Artificial Intelligence" was originally proposed on the Dartmouth Institute in 1956. Definition of artificial intelligence can be divided into two parts, namely "artificial" and "intelligent", where "artificial" means humanly manufactured or the people themselves have no high degree of intelligence to the point where you can create artificial intelligence; "smart" involves managing issues, such as consciousness, self, unconscious thinking and so on. It applies to computer science as well as biology, psychology, logic, philosophy and other disciplines, combines information technology, control technology and automation technology together closely. The main goal of artificial intelligence is to achieve the machine simulation of human action and smart thinking, learning, perception, etc., is to give the machine a study of intelligent activities.

3 THE BREAKTHROUGH OF ARTIFICIAL INTELLIGENCE TECHNOLOGY

3.1 *The discovery of the ecological structure of knowledge*

Like most biological, it has the innate instinct of knowledge when at the birth, masters new knowledge through the later continuous learning, and constantly perfects knowledge in practice, this is the ecological structure of knowledge. Realize the "empirical knowledge - normative knowledge - common knowledge" conversion on the basis of instinctive knowledge, people can see that the information theory is only the core of artificial intelligence theory. It can be seen that artificial intelligence technology mainly relies on the simulation of human intelligence, so the study of combinatorial methods of ecological structure and artificial intelligence is an important work to promote the development of artificial intelligence.

3.2 *Three mechanisms simulation methods*

Three mechanisms methods in traditional research mainly are functional simulation modeling, behavioral modeling, structural modeling, these three methods in the traditional research process are independent from each other, develop from each other, due to the "divide and conquer" methodology guidance, resulting in intrinsic link between these three has not been found, and thus the formation of the three pillars. This "divide and conquer" methodology makes the three simulation methods develop continuously in the competition, but it can not be combined with each other, unable to form a joint force.

3.3 *Find "intelligent generation common core mechanism" and "mechanism simulation methods of artificial intelligence"*

Artificial technical skill makes the machine able to simulate automatic information after determining questions of knowledge and preset target, to extract new knowledge to produce an intelligent strategy of human intelligence activities of technology. It can be defined as, extracts the necessary information from the unknown problem, prior knowledge and preset target, and then sums up the new knowledge to complement the existing information, and then under the guidance of the target, uses the obtained information and knowledge to generate smart strategy to solve problems, and implements the strategy to achieve the goal of solving the problem.

3.4 *The discovery of "mechanism simulation method is a unified simulation method of artificial intelligence"*

Organically combine the above mentioned three simulated mechanism with ecosystem structure of knowledge to form four new specific work patterns. 1, an artificial neural network model, information - empirical knowledge - empirical smart strategies conversion, using empirical knowledge to generate empirical smart strategy; 2, physical symbol systems, information - normative knowledge - normative smart strategy conversion, using the scale knowledge to generate normative type smart strategies; 3, perception action systems, information - common knowledge - common intelligent strategy conversion, using common knowledge to generate common smart strategy; 4, information - instinct knowledge - instinct intelligent strategies conversion. These four modes are exceptions and application under different knowledge conditions of structural simulation, functional simulation, behavior simulation.

4 THE SCIENTIFIC METHOD INNOVATION

4.1 *Innovation should be based on the scientific method*

It is called innovation because of no precedent to refer to, and therefore the result of innovation is uncertain. To try to ensure the advancement and practical feasibility of innovative approaches, people must adhere to the guidance of scientific theory, research on the basis of practical experience, eliminate uncertainty in the creation of new knowledge and new activities, concentrically express the practical application of deterministic regularity knowledge. Only using a scientific theory to guide innovation activities can provide the right direction for innovation activities in the innovation process detours.

4.2 *The innovative approach is an extension of the connotation of the scientific method*

Innovative methods can be understood from two perspectives, first, referring to the feasibility of various methods to promote technological innovation, innovative approaches that can be seen as a way; Second, referring to the scientific innovation and development of the known scientific methods, innovative approaches that can be seen as an activity. Innovative scientific methods must rely on innovative methods, only scientific innovation can provide a strong impetus for the development of artificial intelligence.

5 CONCLUSION

In summary, although the application of artificial intelligence technology is more and more widely, technical level has also gained a huge breakthrough, but many innovative methods of artificial intelligence don't combine together well, a serious impediment to the rapid development of artificial intelligence. In order to better serve humanity and speed up the development process of artificial intelligence, people are required to carry out innovative scientific methods. Therefore, people can say that innovative scientific method is the basic motivation to promote artificial intelligence technology, also it is the basic motivation to promote the social development.

REFERENCES

[1] Zhong Yixin. *Innovative Breakthrough In Artificial Intelligence And Scientific Method* [J]. Pattern recognition and artificial intelligence, 2012,03: 456–461.
[2] Yang Yang. *Innovative Methods of Scientific Method* [J]. Innovation and Technology, 2012,09: 28–29.
[3] Liu Jianjun. *Application of Artificial Intelligence* [J]. Modern industrial economy and information technology, 2013,14: 74–75.
[4] Liu Yanhua. *Innovative Scientific Methods to Enhance The Capability of Independent Innovation* [J]. Invention and Innovation (Comprehensive Edition), 2007,08: 4–6.

Management, Information and Educational Engineering – Liu, Sung & Yao (Eds)

Study on the application of artificial intelligent technology in intelligent building

Sui Xin Tang
Wuchang University of Technology, China

ABSTRACT: With the rapid development of computer science and technology, artificial intelligence technology emerged. As an emerging discipline, artificial intelligence mainly designs a more intelligent machine through simulating human intelligence and thinking process, instead of humans to conduct more efficient work. This paper made a detailed presentation on artificial intelligence technology and further discuss the application of artificial intelligent technology in intelligent building through analysis of intelligent building status quo, which provides a certain reference value for intelligent building future development.

KEYWORDS: Intelligent building; Artificial intelligence; Application; Development.

1 INTRODUCTION

In recent years, the rapid development of science and technology has been providing more and more support for the development of the construction industry. In order to improve their competitiveness in the industry field, construction units continue to explore more intelligent application technology to improve production efficiency and promote the generation of intelligent buildings. Intelligent buildings include systems, such as automation equipment, communications equipment and office automation equipment and form intelligent integration in the continuous development process. The juche idea of intelligent building is to make full use of advanced science and technology to improve the degree of human-computer integration at work and using intelligent machines instead of human conduct efficient work. The development of intelligent building is the reflection of constructing personnel making full use of the integrated system in practical work to better satisfy the building service needs.

2 THE INTELLIGENT BUILDING DEVELOPMENT STATUS QUO

In recent years, China's intelligent building has achieved rapid development and also has a certain status in the international community. However, there still exist some problems in the intelligent building development process, which has been plaguing the architect. Currently, there have been building automation systems in many buildings, which is automatically controlled by the machine through logical judgments. But the system during operation cannot logically think, judge and self-study and external environment has a great impact on it. In addition, the building automation system maintenance procedures are complicated and not easy for staff to overhaul it and its automation degree is not high enough.

In the large environment of science and technology and market environment, rapid development, to meet the requirements, intelligent building systems have gradually emergent many discrete systems. The disadvantage is that the application these discrete systems do not realize collaboration and unity in the intelligent building and cannot reach integrated control for construction purposes. At the time of operating and managing the systems, you need to master the training of different discrete systems personnel, which not only improve the training difficulty, but also increases the cost of business-to-employee training. Establish a unified system and integrate the current each discrete system, integration, to a large extent, can improve reliability of intelligent systems and promote the development of intelligent buildings.

3 STUDY ON THE APPLICATION OF ARTIFICIAL INTELLIGENCE TECHNOLOGY

3.1 *The application of expert systems technology in intelligent building*

Expert systems are the most practical significance results in artificial intelligence fields, which are much favored by construction companies since the production and are quickly put into production and made a profit as a business product. The expert system is based on a variety of expert knowledge controlling a variety of objects and controlling law expert and construct and run the system, which is an artificial intelligence computer program system, with the equivalent of a specialized field of knowledge and experience level of expertise and solve problems in this field. Through knowledge and experience within the system, reason, judge and solve complex problems. Expert systems break the traditional situation relying solely on mathematical models for system design and integrate knowledge model on the basis and effectively integrate knowledge information processing technology and controlling technology.

3.2 *The applications of artificial neural networks in intelligent building*

Since the artificial neural network is used in intelligent building, it has achieved good results in building systems modeling, learning control and program optimization. Its applications continue to expand. Artificial neural networks can achieve effective management on a modern building to ensure thousands of devices installed in building run safe and reliably. This is because the artificial neural networks have good learning and adaptive capacity. You can control, supervision and unsupervised to construction in two ways. A new neural network models using dynamic approach to reduce the model complexity and computer resources and hardware requirements, which is particularly suitable for small-scale intelligent buildings. However, the current computer artificial neural network model still exist issues such as real-time technic, while the computer astonishing speed will continue to improve the artificial network neural system in intelligent building. In the future, rational application of artificial neural network will make the building truly intelligent, and lower system cost costs will help the early realization of universal construction of intelligent building in the city.

3.3 *Intelligent decision-making system applications in intelligent building*

With the rapid development of computer science and technology and network information technology, the intelligent constructions continue to achieve data automatic control including collection information analysis, processing and storage. Introducing intelligent decision-making system in intelligent building helps to fully enhance the functionalization degree of intelligent building. Intelligent decision supporting system is a new information management technology, which is based on management science, operations research, cybernetics behavioral science and technically supported by advanced computer technology and network IT and help senior managers to solve problems, provide the necessary information and materials for them to help policy makers all-round understand structural basic materials and structural information on the data for of the building, and then formulate a more reasonable options to maximize construction enterprises economic benefits. The rapid development of the construction industry put forward higher requirements to the management work in building construction process and intelligent decision-making system came into being in this context. You can also say that building intelligent systems is the product making intelligent buildings develop from the automatic control to information management. Through intelligent decision supporting system, using a unified modular hardware and software architecture to simplify the complexity of management staff maintenance and management to realize the intelligent building science and efficient management and ensure construction activities safe and orderly conduct. Intelligent decision support system has very important application values in intelligent building, which can effectively improve the monitoring and management system throughout the building. Thus, the expansion of intelligent systems content and meaning has important practical significance.

4 CONCLUSION

In recent years, with the continuous development of intelligent buildings, artificial intelligence technology is constantly evolving. Expert systems, artificial neural network systems and decision supporting systems, etc., fully reflect the application value of artificial intelligence systems in intelligent buildings and vividly demonstrate its advantages. The application of artificial intelligence technology in intelligent buildings greatly reduces the application cost of intelligent building system and achieve optimal control and energy saving to make intelligent building develop towards a green and sustainable direction.

REFERENCES

[1] Wang Yongzhong. *Study on the Application of Artificial Intelligence Techniques in Intelligent Buildings* [J]. Science and Technology Information, 2009, 03: 343 + 342.

[2] Ma Jiehua. *Study on the Application of Artificial Intelligence New Techniques in Intelligent Buildings* [J]. Technology Innovation and Application, 2014,09: 228.

[3] Wu Xuanzhong, Ni Ziwei. *Study on the Application of Artificial Intelligence New Techniques in Intelligent Buildings* [J]. Fujian Construction Science & Technology, 2005,02: 45–46.

[4] Wang Jinxuan. *Study on the Application of Artificial Intelligence Techniques in Intelligent Buildings* [D]. Huaqiao University, 2004.

[5] Ai Hui, Xie Kangning, Xie Baizhi. *Discussion on Artificial Intelligence Techniques* [J] China Medical Education Technology, 2004, 02: 78–80.

Management, Information and Educational Engineering – Liu, Sung & Yao (Eds)

On the influence of Chinese paintings on ceramic paintings

Yang Gao
Jingdezhen Ceramic Institute, China

ABSTRACT: By exploring the development history of Chinese painting, ceramic painting always permeated with the Chinese painting style; they are closely linked with each other. Ceramic arts were given a rich emotional and cultural factor, which is subject to Chinese painting meaning and inspiration. The same or similar techniques and styles with Chinese painting techniques and genres can be find in ceramic paintings. In the artists' view, the essences of both are the same. They have a distinct national identity and regional characteristics, which are the best embodiment of aesthetic value and historical value of the art painting. This article elaborates the impact of Chinese painting on a ceramic painting, at the same time analyzing the common characteristics of Chinese painting and ceramics painting.

KEYWORDS: Chinese painting; ceramic painting; common characteristics; influence.

1 INTRODUCTION

Chinese painting is our historic heritage and subsequently developed ceramic painting is also more noticeable. Both of them mutually penetrate and impact each other. Chinese painting has a very important influence on later developed ceramic painting, which is also the focus of this article. Ceramic artists infused its unique features at the time of fully absorbing Chinese rich painting art form and style to create the perfect ceramic art and add more beauty to the ceramic art.

2 THE COMMON FEATURES OF CHINESE PAINTINGS AND CERAMIC ART PAINTINGS

First, the new era pottery is the most ancient painting arts and is the first integration achieved in the development process of ceramics and painting. Chinese paintings and ceramic paintings come from life and fusion develop in life, both jointly follow the aesthetic thinking and have many common features. Artistic creation should be able to achieve a high degree of integration and unity of mind and scenery. For example, Nostalgia makes the natural scenery as a creative theme, which describes the mountains, streams and farmhouses, etc., to show the kind scene of peaceful atmosphere and reveal a peaceful mind of writer and causing viewer's art sympathy and is a beautiful baptism.

Secondly, the Chinese painting occupies a certain arts position in the history of art and culture. The art of Chinese painting has always been applied thoroughly, quiet and elegant paintings mainstream were created, woks like The Journey of Mountains Line, Snow Creek Map embodied the writer's spirits and ideas in the works. Ceramic arts expressed the similar texture with Chinese paintings. Meanwhile, ceramic paintings have civil secular entertainment fun, such as Apocalypse Blue Characteristics Bowl, the scene in the painting leads to the endless reverie, the whole scene in the paintings is lifelike.

Again, Chinese painting is natural verve by ink language and can demonstrate writers' artistic aesthetics and inks are the reflection of writers' spirit. Influenced by traditional ideas, Chinese painting pay attention to ancient, elegant and natural ethereal taste in pen and ink, which is the sublimation artistic experience of things after all. Ceramic arts in the technique use of materials and color are almost the same. Watershed after the Tang Dynasty and porcelain in Ming and Qing Dynasties can make the viewers feel a strong taste of Chinese ink paintings. We can say that ceramic art is the new change under Chinese art painting continues. The fully integrated of ink and ceramic painting art Chinese painting, you can reach a higher realm of art, but also reflects a profound influence on Chinese painting ceramic paintings.

3 THE INFLUENCE OF CHINESE PAINTINGS ON CERAMIC PAINTINGS

3.1 *The impact of Chinese paintings on a ceramic painting new color and pastels*

Ceramic painting generally draws moss by wire and draws rocks with dyed-oriented. Pearl Mountain and Eight Friends fully demonstrated important influence of Chinese paintings to the new color and pastels of ceramic paintings.

Taking pastel flowers and birds as examples, its pastel art technology level is quite high, which uniquely combines Chinese paintings and ceramics pastels and was used in the creation of the painting. Representative works of ceramic art, Kingfisher Lotus, is a good example. Kingfisher on the rocks overlooks fish and lotus in the water, highlights a dynamic Mito, fully demonstrated interesting plants by curve to give unlimited reverie space to people.

3.2 *The influence of Chinese paintings on ancient ceramic painting color line drawing*

In ancient ceramic painting color line drawing, adding a line drawing technique of Chinese paintings. The application of crafts arts, decorative skills makes ceramic painting style ancient color line drawing more unique. Ancient color line drawing in ceramic painting often revealing a sense of simple elegance. Since the mid-20th century, ceramic painting ancient color line drawing made rapid development. Modern ceramic paintings, ancient color line drawing rhythmically apply composition, density patchwork and the materials and techniques have been improved and perfected to inject fresh ceramic art blood and the era has distinctive features and creative style. Ceramic painting ancient color line drawing inherited Chinese painting composition characteristics and colors matching, and infuse into the modern approach on this basis, and organically combine pastels and ancient color, its decorative stylc docs have some charm.

3.3 *The impacts of Chinese paintings on ceramic blue and white*

In ceramic paintings, blue and white inks have more similarities with Chinese painting tools. When artists draw for a blue and white ceramic decoration, they often subject to varying degrees of the influence of Chinese paintings by drawing features of outstanding Chinese painting works and line drawing characteristics. Ceramic paintings need the intervention of era scholars and aesthetic will, which has a very important impact on the quality of ceramic artists, making the blue and white in ceramic painting more similar art forms. Whether in Chinese paintings or ceramic paintings, the art forms blue paintings exhibiting are the production of the artist emotion and spiritual life. Blue and white landscape is a ceramic painting art form and also adds to the charm of ceramic art.

4 CONCLUSION

Exploring the art history during five thousand years, ceramic arts are enduring. Chinese paintings have very important impacts on the new color and pastels, ancient blue and white color line drawing of ceramic arts. Chinese paintings are the mainstream forms of ceramic art styles and there are both differences and many commonalities between them. They are the essence of history, culture and quint, which carry the Chinese long history and the traditional Chinese paintings and ceramic paintings, art forms are infused into a lot of elements with the times and national characteristics by modern artists, making these two arts can adapt fast-paced development of modern society. Chinese paintings and ceramic paintings have solid painting style and cultural heritage and infuse into the real art feeling of creators and infuse into the Chinese traditional deepening culture in art works, create art works to truly show reveal artistic charm of Chinese paintings and ceramic paintings. Ceramic artists should fully learn the essence of the success Chinese painting works, including its structure forms and composition style, and infuse the advantage into their own artistic creation. Ceramic artist is always close to nature and makes the harmonious natural environment as the creative team and continue to inspire their creative inspiration and make artwork develop towards multiple directions and also has the works own unique and continuously contribute powerfully to the progress and development of Chinese paintings.

REFERENCES

[1] Hu Min. *On the Performance of Chinese Paintings Techniques Suiting Ceramics Paintings*[J] Jingdezhen Ceramics, 2010,02: 17.

[2] Jiang Qiang, Wang Yong. *On the Integration of Chinese Paintings and Ceramics Painting Arts Charm* [J] Jingdezhen Comprehensive College, 2008,04: 103–104.

[3] Yan Huan, Tan Tao. *On the Influence of Chinese Paintings on Ceramic Paintings* [J] Big stage, 2011,10: 98–99.

[4] Ye Changxi. *On the Influence of Chinese paintings on ceramic paintings* [J] Jingdezhen Ceramics, 2012,03: 132–133.

[5] Liu Xuan, Yu Baochun. *Study on the Influence of Chinese Painting Line Drawing on Ceramic Paintings*[A]. *China Ceramic Industry Association. Ninth National Ceramic Art Design Innovation Competitions and the First Chinese Ceramic Art Conference Proceedings* [C]. Chinese Ceramics Industry Association: 2010: 6.

Management, Information and Educational Engineering – Liu, Sung & Yao (Eds)
© 2015 Taylor & Francis Group, London, ISBN: 978-1-138-02728-2

Did capital control and financial depth affect the demand of foreign reserves?

Dian Lei Lu & Zhi Xin Liu
School of Economics and Management, Beihang University, Beijing, China

ABSTRACT: We combined foreign reserves, capital control and financial depth with other economic variables in the OLS model and discussed the factors that affect the demand of China's foreign reserves. The results showed that 1) a considerable number of foreign reserves are passively accumulated, for instance, stronger capital controls and deeper financial system will lower the demand of foreign reserves; 2) the ratio of Hot Money/GDP is more significant than the FDI/GDP as a proxy of foreign reserves' protective demand.

KEYWORDS: Foreign Reserves, Capital Controls, Financial Depth.

1 INTRODUCTION

At the end of 2013, China held a huge stock of international reserves in the amount of US$ 3821.3 billion foreign exchange reserves, which is far more enough to prevent financial risks. Most research about China's foreign exchange reserves are basically admitting that international capital could flow in and out of China with few restrictions. However, the capital controls are not under consideration, which probably cause miscalculation on the optimal quantity and the opportunity cost of foreign exchange reserves. Therefore, this paper will show the significance theoretically and practically by integrating capital controls and other variables to re-estimate the foreign exchange reverses demand model.

2 RESERVE DEMAND FUNCTION

Most researches divided the demand of foreign exchange reserves into three levels: transactional requirement (R_1), precautionary requirement (R_2) and speculative requirement (R_3). However, apart from these requirements, the effect of capital controls and hot money should also be under consideration. Therefore, we establish this model as follows:

$$RES = R_1 + R_2 + R_3 + \sum P \quad (1)$$

Firstly, to cover the transaction requirement, we use the proxy of average propensity to import (API). The more the average propensity to import is, the bigger the effect suffered from outer shocks, therefore the more reserve holdings will be needed. As the proxy of economic opening, the coefficient of API should be positive.

Secondly, with the prosperity of Chinese economy, numerous international capitals enter into China by all kinds of investment; some of these are speculative capitals, also as known as Hot Money, which could cause turmoil by sudden flew. Thus the sufficient reserve holdings should be kept for precautionary demand. Some studies calculated Hot Money by Balance of Payment (BOP), that Hot Money equals to the newly added foreign exchange reserves minus FDI and Trade Balance [1], yet this measurement not only ignored the effect caused by the change of exchange rate and returns of foreign exchange reserves, but also presumed that no Hot Money in FDI. This paper will use a new approach which can avoid those problems above. We estimate that Hot Money should equal to the Newly Added Funds outstanding for foreign exchange minus the actual utilization of foreign capital and trade balance. Hot Money/GDP should have a positive relation with foreign exchange reserve holding for precautionary reasons. We also use FDI/GDP as an alternative variant. Because the larger amount of FDI returns comes back to the mother countries will reduce foreign exchange reserves.

Thirdly, many recent studies have blamed the short-term external debt (STED) as one of the reasons for financial crises (see [2] [3]). More importantly, the STED is considered to be an important indicator to measure the foreign exchange reserves by the People's Bank of India (PBC). At the end of 2013, the accumulative STED reached to more than $640 billion, as many as 7% of GDP or 70% of total external debt. A large of reserves will be used to pay back. Therefore, we include the relative size of STED (ratio of STED/GDP) in the reserve demand function.

Fourthly, most papers on reserves demand are presumed the free floating system (see [4] [5] [6] [7]). However, China has restrictions about international capitals; therefore some proper variables should be adopted for describing this kind of effect, such as the differential between international and domestic interest rate (ΔR) and the expected volatility of RMB exchange rate (ΔR). Wang & He (2007) [8] proved that there are long-term relations among short-term capital inflow and, EXVOL. Bai & Wang (2008) [9] also use the differential of interest rate to discuss the efficiency of capital controls under the framework of Interest Rate Parity Theory. The more effective the capital controls, the less influential the financial turmoil caused by capital inflow and outflow, which will decrease the demand of reserve holding. Therefore, the coefficient of and EXVOL should be negative.

Fifthly, we also use a new variable to keep the financial depth into consideration. When comes to financial depth and development, M2/GDP is commonly used by literatures, such as Mckinnon, 1989 [10]; Lin, 2003 [11], etc. But simply using that proxy cannot reflect the degree of China's financial depth correctly, as Edwards (1996) [12] pointed out that "under an immature market with restrictions on borrowing and lending money, there is always a negative trend between them (financial depth and M2/GDP)". Thus we will use another variable to represent the financial depth of China, Funds outstanding for Foreign exchange (FOFX)/GDP. On the one hand, over this period, 20% to 60% of monetary supply was contributed by FOFX, which shows some passiveness on monetary policy and imperfection of financial development; on the other hand, the variation of this ratio also reflected whether the central bank has enough financial instruments and assets to neutralize the excess FOFX. Compared with M2/GDP, FOFX/GDP will be more appropriate as a proxy of China's financial depth. The bigger the ratio, the more superficial the financial depth, and will need more reserves. Thus, the coefficient should be positive.

Sixthly, to satisfy the speculative demand of foreign exchange reserves, there will be an opportunity cost (OC) for this session of reserve holding. Therefore, we can use OC to measure the speculative demand of reserves, namely the differentials between 3-month Treasury Bill Rate and China's 3-month Redemption Rate. And a negative relation between OC and reserve holding demand are expected.

Then we establish the model as follows,

$$
\begin{aligned}
RES = {} & \gamma_0 + \gamma_1 OC_{t-1} + \gamma_2 HM_{t-1} + \gamma_3 EXVOL_t \\
& + \gamma_4 FOFX_{t-1} + \gamma_5 \Delta R_{t-2} + \gamma_6 STED_{t-1} \\
& + \gamma_7 API_{t-2} + \gamma_8 FDI_t + \varepsilon_t
\end{aligned} \tag{2}
$$

which RES means the ratio between the newly added reserves and GDP, OC is the differentials between 3-month Treasury Bill Rate and China's 3-month Redemption Rate, HM is Hot Money/GDP, EXVOL is the expected volatility of RMB exchange rate, FOFX is the ratio of FOFX and GDP, is the differential between benchmark interest rate of China' monetary market and US$ LIBOR, STED is STED/GDP, API is the amount of import/GDP, FDI is FDI/GDP. We will use the logarithm value on, EXVOL, STED and FDI.

3 EMPIRICAL RESULTS

The data used in this study are quarterly observations and the sample period spans from 1997: 01 to 2013: 04 base on the data's availability. All data are openly published by the People's Bank of China, Ministry of Commerce of PRC, General Administration of Customs, National Bureau of Statistics, State Administration of Foreign Exchange and Federal Reserve Bank of ST. Louis, USA.

We consider two traditional unit root tests, which are augmented Dickey-Fuller test (ADF) and Phillips-Perron test (PP), to confirm the stationary of variables. The result of unit root tests is presented in Table 1. All variables included in the model are found to be integrated of order one, i.e. I(1).

Given that variables are non-stationary, it is not justified to estimate the reserve demand function using OLSs because the results might be spurious. Therefore, we apply the cointegration test developed by Johansen (1988) [13] to investigate the presence of long-run relation among variables. The results of Johansen cointegration (Table 2) show that the null hypothesis of no cointegration is strongly rejected in favor of one cointegrating relation with plausible (normalized) coefficients. This implies that all variables share a common stochastic trend and do move together in the long-run.

Based on the estimated normalized cointegrating vector, the long-run relationship between foreign exchange reserves and its determinants is expressed below Table 3.

The results of regression showed that: 1) despite the conventional model or the new model, with introduction of the factors of capital controls and financial depth, the R-squared, Adjusted R-squared and F-statistic are significantly improved, which means the explanation of the model is better; 2) the coefficient of HM/GDP is more significant than that of FDI/GDP and the new model is superior to the conventional one, which mean that HM/GDP is more proper variable as an indicator for the productive demand of foreign reserves, 3) with the consideration of capital control and financial depth, the coefficients of demand in transaction, protection and speculation are, though still significantly, all witnessed huge drops by 60%, 70% and 40%, separately. They illustrated that huge amount of foreign reserves are not only accumulated actively, but passively, because of the immature of financial system, especially the foreign exchange management. 4) and we found that the strict capital controls also cause passively accumulation of foreign reserves.

Table 1. Results of unit root test.

At level	ADF	PP	At 1st difference	ADF	PP
RES	−0.4*	−4.47	dRES	−11.7***	−10.4***
OC	−2.45	−2.52	dOC	−2.84***	−2.85***
HM	−0.25	−1.29	dHM	−5.63***	−5.24***
EXVOL	−3.3	−2.54	dEXVOL	−4.32***	−4.33***
FOFX	0.42	−0.67	dFOFX	−4.56***	−3.93***
ΔR	−1.63	−1.62	dΔR	−3.47***	−3.39***
STED	−1.91	−1.91	dSTED	−4.08***	−4.11***
API	−3.6*	−3.6*	dAPI	−7.08***	−6.36***
FDI	−1.31	−2.88	dFDI	−4.70***	−2.82***

* (1) Optimal lags for ADF is determined based on AIC and for PP test it is Newey–West bandwidth selection using Bartlett kernel. (2) Probability values for ADF and PP test is as per MacKinnon one-sided p-values.
* Indicates significant at 10% critical level.
** Indicates significant at 5% critical level.
*** Indicates significant at 1% critical level.

Table 2. Results of Johansen cointegration test: reserve demand function.

Trace			Max-eigenv.		
Rank	Eigenv.	Trace-St.	Rank	Eigenv.	Trace-St.
None	0.896	182.8**	None	0.896	87.7**
1	0.7386	105.4**	1	0.7386	65.5**
2	0.5474	62.7	2	0.5474	49.1
3	0.4243	51.2	3	0.4243	37.3
4	0.3751	40.6	4	0.3751	23.5
5	0.2815	37.1	5	0.2815	19.4
6	0.2209	29.9	6	0.2209	10.1

* (1) VAR specification: optimal lag length selected using AIC, (2) deterministic trend assumptions of the cointegration test: intercept and trend in cointegrating relationship and no trend in VAR.* Indicates significant at 10% critical value.** Indicates significant at 5% critical value.

So the model of demand of foreign reserves is displayed as follows:

$$\begin{aligned} RES = 0.06 - 0.72OC_t + 0.05HM_t \quad & (3.298^{***})\ (-9.50^{***})\ (3.757^{***}) \\ -0.006EXVOL_t + 0.75FOFX_t \quad & (-6.664^{***})\ (9.656^{***}) \\ -0.008\Delta\ R_t + 0.01STED_t \quad & (-5.725^{***})\ (2.119^{**})\ (3.136^{***}) \\ +0.6API_t + \varepsilon_t \quad & \end{aligned} \tag{3}$$

The results of the reserve demand function show that hot money, funds outstanding for foreign exchange, short-term external debt and average propensity to import are statistically significant and have positive impacts on the demand of foreign exchange reserves; yet the measure of the opportunity cost, exchange rate volatility and differentials between international and domestic interest rate is found to be negative and significant, which all agrees with a priori reasoning. All the estimated parameters, except foreign direct investment, are significant at the 5% level and signs of all the coefficients are consistent with the theoretical explanations. Further, our results also show that foreign direct investment is not an important determinant of reserve demand in China.

Table 3. Results of Johansen cointegration test: reserve demand function.

	Conventional model	Conventional model with cc, fd	New model	New model with cc,fd
HM			0.177*** (0.039) [4.575]	0.046*** (0.012) [3.757]
FDI/GDP	0.007 (0.046) [0.146]	0.012 (0.017) [0.7]		
OC	−0.864** (0.288) [−2.997]	−0.734*** (0.157) [−4.672]	−0.513* (0.244) [−2.102]	−0.724*** (0.076) [−9.503]
STED	0.044*** (0.012) [3.669]	0.02** (0.007) [2.844]	0.027* (0.014) [1.913]	0.009* (0.004) [2.119]
API	1.142 (0.739) [1.546]	0.02 (0.396) [0.151]	1.422** (0.502) [2.831]	0.595** (0.19) [3.136]
FOFX		0.075** (0.026) [2.848]		0.747*** (0.077) [9.658]
EXVOL		0.007** (0.002) [3.207]		−0.006*** (0.001) [−6.664]
ΔR		0.01*** (0.002) [5.608]		−0.008*** (0.001) [−5.725]
R^2	0.787	0.977	0.864	0.99
Adjusted R^2	0.654	0.95	0.804	0.978
Prob	0.014	0	0.003	0

Note: cc and fd represent the factors of capital controls and financial depth separately; The figures in () are standard deviation and those in [] are t-statistics; *Indicates significant at 10% critical level. **Indicates significant at 5% critical level. ***Indicates significant at 1% critical level.

4 CONCLUSIONS

Firstly, this paper considered the issue of conventional measurement and 'internal drain' and used the yearly data from 2002 to 2013 to calculate China's adequate and excess foreign reserves, found that China held a lot of excess reserves and undertake considerable opportunity cost.

Secondly, we took capital controls and financial depth into consideration and renew the conventional model of foreign reserves demand, which caused the passive accumulation on foreign reserves demand.

Thirdly, compared to FDI/GDP, in the new model we noticed that HM/GDP is more significant and lead to a better outcome. In order to avoid the volatility caused by short-term speculative capital, HM/GDP is a more specific index to the protective demand of foreign reserves.

REFERENCES

[1] Lu, J. & Luo, W.Q. (2010) The Measurement and Analyze on Monthly Hot Money. Statistics and Decision, Vol. 19, 85–89. (in Chinese).

[2] Furman, J. & Stiglitz, J. (1998) Economic crises: Evidence and insights from east, Asia. Brookings Papers on Economic Activity, 2, 1–114. http://dx.doi.org/10.2307/2534693.

[3] Radelet, S. & Sachs, J. (1998) The East Asian financial crisis: Diagnosis, remedies, prospects. Brookings Papers on Economic Activity, 1, 1–74. http://dx.doi.org/10.2307/2534670.

[4] Bird, G. & Rajan, R. (2003) Too much of a good thing? The adequacy of international reserves in the aftermath of crises. The World Economy, 26, 873–891. http://dx.doi.org/10.1111/1467-9701.00552.

[5] De Beaufort Wijnholds, J.A.H. & Kapteyn, A. (2001) Reserve Adequacy in Emerging Market Economies, Working Paper No. 01/43, IMF.

[6] Liu, L.Y. (2008) Did Hot Money Promote the Stock Market and Real Estate Market? Journal of Financial Research, Vol. 10, 48–70. (in Chinese).

[7] Sheng, L.G. & Zhao, H.Y. (2007) Yields and Currency Composition of Foreign Reserves and Hot Money in China. China Economic Quarterly, Vol. 6 (4), 1255–1276. (in Chinese).

[8] Wang, S.H. & He, F. (2007) China's Short-term International Capital Flows: Status, Pathways and Factors. The Journal of World Economy, Vol. 7, 12–19. (in Chinese).

[9] Bai, X.Y. & Wang, P.J. (2008) The Effectiveness of Capital Control and Reform of Exchange Rate Regimes in China. Journal of Quantitative & Technical Economics, Vol. 9, 65–75. (in Chinese).

[10] McKinnon, R.I. (1989) Finance and Economic Development. Oxford Review of Economic Policy, Vol. 5, No. 1. Li, S.K. (2006) Foreign Exchange Reserves VS External Debt. Economic Herald, Vol. 10, 79–81. (in Chinese).

[11] Lin, Y.F., Zhang, Q. & Liu, M.X. (2003) Financial Structure and Economic Growth. The Journal of World Economy, Vol. 1, 3–21. (in Chinese).

[12] Edwards, S. (1996) Exchange Rates and the Political Economy of Macroeconomic Discipline. American Economic Review, vol. 86(2), 159–63.

[13] Johansen, S. (1988). Statistical analysis of cointegration vectors. Journal of Economic Dynamics and Control, 12, 231–254. http://dx.doi.org/10.1016/0165-1889(88)90041-3.

Management, Information and Educational Engineering – Liu, Sung & Yao (Eds)
 ISBN: 978-1-138-02728-2

Study on relationship between debt financing and performance for non-state holding listed companies

Jian Ru Zhang & Wei Xu

School of Management, Xi'an University of Architecture and Technology, Xi'an, China

ABSTRACT: This paper takes A-share non-state holding listed companies from the Shanghai and Shenzhen Stock Exchanges for the period from 2010 to 2012 as the research object and establishes leverage contribution rate index system as for the standard to measure the effect of liability utilization. According to the effect of liabilities utilization, the samples are divided into three categories that are excellent, secondary and poor respectively. Meanwhile, the relationship between total liabilities, operating liabilities, financing liabilities and performance is studied. The results show that the relationship between debt financing and corporate performance, to some extent, depends on reasonable utilization degree of enterprise to debt capital. As for the excellent enterprise, they are the positive correlation between total liabilities, operating liabilities, financing liabilities and performance; As for the secondary enterprise, however, between total liabilities, operating liabilities and performance are positive correlation. Financial liabilities are negatively related to the performance. At last, as for the poor enterprise, between total liabilities, financial liabilities and performance are negative correlation. Operating liabilities are positively related to the performance.

KEYWORDS: Debt financing; Performance; Leverage contribution rate; Empirical analysis.

1 INTRODUCTION

As the pioneer of accepting market test, the financing and the management decision-making behavior of the listed company are the key problems in the enterprise development. In our country, the listed companies obviously prefer to share financing. Therefore, the scholars also focus on study the relationship of share financing and corporate performance, and have the little research on debt financing. However, along with the deepening of the market economic system in china, the debt financing is playing more important role in corporate finance. Thus, based on above backgrounds, this paper investigate the relationship of debt financing and corporate performance to provide the practical guidance for the corporate debt financing.

Many scholars, at home or abroad, have investigated the relationship between debt financing and corporate performance from the perspectives of debt financing maturity structure and the proportion of debt financing. There are the following three aspects about the relationship between debt financing and corporate performance. Firstly, the debt financing and corporate performance have a significant positive correlation. From the perspective of the proportion of debt financing, Gilson (1989) concluded from his analysis that the proportion of debt financing within the reasonable scope can improve corporate performance. Zhang Ying and Zhang jianying (2012) got the conclusion through multiple regression analysis that although low the proportion of the debt financing, debt financing has a positive effect on corporate performance in the Shenzhen listed company. Secondly, the debt financing and corporate performance have a negative correlation. Yu dongzhi (2003) measured the financing structure with the asset-liability ration as independent variable and corporate performance with main business profitability and return ratio of total assets as dependent variable, which showed a negative correlation between them. Zhang rongyan (2012) investigated the impact of debt financing to corporate performance from the perspectives of debt maturity and debt category based on two cities, Shanghai and Shenzhen 100 listed companies. And, the results showed that different debt maturity structure and debt category have a negative effect on corporate performance. Lastly, a few scholars have got the conclusion different from the above two views. Viewing Tobin q as dependent variable and asset-liability ratio, current-liability ratio and long-term debt ratio as explained variables, Liu lei and Xue Jingjing (2013) use multiple regression analysis to show that capital structure and long-term debt ration have a negative effect on the corporate performance, but the current - liability ratio and corporate performance have a significant positive correlation. Huang lianqin and qu yaohui divided debt into financial liabilities and operating liabilities according to the different properties of debt and examined the difference of two financial leverages to corporate value-added ability and growth, which showed that the positive effect of operating liabilities leverage is more obvious.

To sum up, there are many different conclusions about the relationship of debt financing and corporate performance. Thus, those conclusions can't provide specific guidance for corporate financial decision. The main reasons are following two sides. One is the unreasonable assumptions. The debt utilization consistency of capital structure same samples is assumed and the impact of some important factors that are sample scale, growth stage, profitability to effective utilization of funds also is ignored. Meanwhile, most studies measured the financing structure of enterprises using a single variable-asset-liability ratio not considering the differences of the effect of financing liabilities and operating liabilities to corporate performance. Another point is soft constraints of debt financing to the investment behavior of state-owned enterprises due to corporate liabilities mostly coming from the four major state-owned commercial banks. Therefore, this paper selects non-state holding listed companies as samples and establishes leverage contribution rate index system as for the standard to measure the effect of liability utilization. According to the effect of liabilities utilization, the samples are divided into three categories that are excellent, secondary and poor respectively. Meanwhile, this paper researches the relationship of total liabilities, operating liabilities, financing liabilities and corporate performance to show the differences between debt financing and corporate performance of three different enterprises.

2 RESEARCH DESIGN

This paper takes A-share non-state holding listed companies from the Shanghai and Shenzhen Stock Exchanges for the period from 2010 to 2012 as samples (http://www.cninfo.com.cn/) and uses SPSS17.0 software to empirically analyze.

2.1 *Fundamental hypothesis*

There are two basic hypotheses in the empirical research: the relationship between debt financing and corporate performance depends on effective utilization of the corporate debt capital in a certain extent; operating liabilities have a positive effect on corporate performance for different enterprises, but financing liabilities and corporate performance have a positive correlation for excellent enterprises and they have a negative correlation for secondary and poor enterprises.

2.2 *Variable definitions*

Dependent variable: *ROE*(Return on Equity)- measure the corporate financial performance; Independent variable: *TFL*(Total liabilities), *FFL*(Financing liabilities), *OFL* (Operating liabilities), *LCR*(Leverage contribution rate), *EBIT*(Earnings before interest and taxes), *IRAT*(Interest rate of after taxes), *FL*(Financial leverage), *OL*(Operating leverage), *NFL*(Net financial liabilities), *NA*(Net asset), *O*(Operating liabilities), *NOA*(Net operating asset), *FLCR*(Financial leverage contribution rate), *NOPM*(Net operating asset profit margin), *OLCR* (Operating leverage contribution rate), *Growth*(Operating profit growth), *Size*(Natural logarithm of total assets with end-of-year book).

$$LCR = (EBIT\text{-}IRAT) \times FL$$

Where *LCR* is used to measure contribution rate of total liabilities to stockholder's equity. If *LCR* is greater than zero, total liabilities and corporate performance have a positive correlation. But, *LCR* is less than zero, the correlation between them is opposite.

$$FL = NFL/\,NA$$
$$OL = O/\,NOA$$
$$FLCR = (NOPM\text{-}\ IRAT) \times FL$$

Where *FLCR* is used to measure contribution rate of net financial liabilities to stockholder's equity. If *FLCR* is greater than zero, financing liabilities and corporate performance have a positive correlation. But, *FLCR* is less than zero, the correlation between them is opposite.

$$OLCR = NOPM \times OL$$

Where *OLCR* is used to measure contribution rate of operating liabilities to stockholder's equity. Therefore, operating liabilities always have a positive correlation with corporate performance because of *OLCR* greater than zero.

Based on this above conclusions. The samples were divided into three categories as follow:

when *LCR* and *FLCR* are greater than zero, the enterprises are successful;

when *LCR* are greater than zero and *FLCR* are less than zero, the enterprises are risk;

when *LCR* and *FLCR* are less than zero, the enterprises are failed.

2.3 *Model building*

This paper establishes three models and uses multiple regression analysis to study the relationship of total liabilities, operating liabilities, financing liabilities and corporate performance for three different enterprises mainly considering these significant factors-enterprise size, profitability and growth.

$$\text{Model 1: } ROE = a_0 + a_1TFL + a_2ROA + a_3GROWTH + a_4SIZE + \varepsilon_0$$

$$\text{Model 2: } ROE = a_0 + a_1FFL + a_2ROA + a_3GROWTH + a_4SIZE + \varepsilon_0$$

$$\text{Model 3: } ROE = a_0 + a_1OFL + a_2ROA + a_3GROWTH + a_4SIZE + \varepsilon_0$$

Where a_0 is constant term, a_i is the coefficient of regression equation, $i = (1,2,3,4)$, ε_0 is residual term.

3 EMPIRICAL RESULTS

3.1 *Descriptive statistics*

The samples have been composed of 712 different non-state holding listed companies which can be divided into three categories based on the effect of liabilities utilization. And, there are 278 borrowing successful companies, 248 borrowing risk companies and 186 borrowing failed companies. Meanwhile, all variables can be statistically analyzed for three kinds of enterprises and the statistical results are shown in Table1.

Table 1. The statistical results of all variables.

Variable name	Successful	Risk	Failed
ROE	0.1003	0.0558	−0.0210
TFL	0.4832	0.5260	0.5638
FFL	0.2396	0.1802	0.2990
OFL	0.2436	0.3457	0.2648
ROA	0.0734	0.0450	0.0372
GROWTH	0.6725	0.5329	−0.3351
SIZE	21.8883	21.9196	21.5206

From Table1, we can concluded three important points as follow. Firstly, the *TFLs* of three kinds of enterprises are 48.32%, 52.60% and 58.20% respectively and the standard deviations of them are about 0.15, which shows that non-state holding listed companies have a higher the scale of indebtedness and have a larger fluctuation range of liability ratio. Meanwhile, we can see that they are insignificant difference. Secondly, there are large discrepancy between financing liabilities(*FFL*) and operating liabilities(*OFL*) for three kinds of enterprises. As for the borrowing successful companies, the financing liabilities are close to the operating liabilities, which are 23.96% and 24.36% respectively. As for the borrowing risk companies, however, the ratio of the operating liabilities to total liabilities is about 66%. And, as for the borrowing failed companies, the ratio of the financing liabilities to total liabilities is up to 53.03% , which is highest in three kinds of enterprises. The analytical results of the above datas are much consistent with our hypotheses. Due to the impact of operating liabilities, the *ROEs* of the borrowing risk and failed companies have a large difference, which are 5.58% and -2.10% respectively. The results show that operating liabilities have a positive effect on corporate performance. Compared with the borrowing successful companies, the operating liabilities of the borrowing risk companies is higher but their *ROEs* is lower, which indicates that the financing liabilities of the borrowing risk and failed companies have a negative effect on corporate performance. Lastly, the variational trends of the independent variables(*ROA*, *GROWTH*) are same as the dependent variable(*ROE*), which shows that those control variables have a positive effect on the corporate performance. However, three kinds of enterprises' *SIZEs* are insignificant difference, which indicates that the enterprise size have no effect on the corporate performance.

3.2 *Regression analysis*

The regression results of the major variables based on three models are shown in Table2, Table3 and Table4 respectively.

Table 2. The regression results of model 1(TFL).

Adjust	R^2	*F*	Constant term	*B*	*P*
Successful	0.386	1.275	−49.947	0.527(1.325)	0.000
Risk	0.949	71.248	−9.372	0.221(3.145)	0.000
Failed	0.215	0.629	−8.789	−0.436(−1.148)	0.024

Table 3. The regression results of model 2(FFL).

Adjust	R^2	*F*	Constant term	*B*	*P*
Successful	0.451	1.310	−84.644	0.437(1.212)	0.000
Risk	0.932	37.969	−19.233	−0.060(−0.704)	0.000
Failed	0.211	1.934	−0.758	−0.614(−2.439)	0.000

Table 4. The regression results of model 3(OFL).

Adjust	R^2	*F*	Constant term	*B*	*P*
Successful	0.360	0.718	−43.356	0.040(0.119)	0.002
Risk	0.962	95.392	−14.991	0.218(4.087)	0.000
Failed	0.270	0.437	−11.030	0.273(0.752)	0.000

Note: Bracketed value indicates variable estimated parameter's *t* value.

Table 2 shows the result of regression analysis for model 1. As for the borrowing successful enterprises, the correlation coefficient of the total liabilities(*TFL*) and corporate performance is 0.527. And, as for the borrowing risk enterprises, the correlation coefficient is 0.221. The results show that the total liabilities (*TFL*) and corporate performance have a positive correlation in the two kinds of enterprises and the total liabilities(*TFL*) have a more positive effect on corporate performance in the borrowing successful enterprises. However, as for the borrowing failed enterprises, the correlation coefficient of the total liabilities(*TFL*) and corporate performance is –0.436, which indicates that the total liabilities and corporate performance have a negative correlation.

Table 3 shows the result of regression analysis for model 2. as for the borrowing successful enterprises, the correlation coefficient of the financing liabilities(*FFL*) and corporate performance is 0.437, which indicates that the financing liabilities and corporate performance have a positive correlation. However, as for the borrowing risk and failed enterprises, the correlation coefficients of the financing liabilities and corporate performance are –0.060 and –0.614 respectively, which shows that the financing liabilities (*FFL*) and corporate performance have a negative correlation in the two kinds of enterprises and the financing liabilities (*FFL*) have a more positive effect on corporate performance in the borrowing failed enterprises.

Table 4 shows the result of regression analysis for model 3. As for the borrowing successful, risk and failed enterprises, the correlation coefficients of the operating liabilities and corporate performance are 0.040, 0.218 and 0.273 respectively, which shows that the operating liabilities (*FFL*) and corporate performance have a positive correlation in all enterprises and the operating liabilities (*FFL*) have a more positive effect on corporate performance in the borrowing failed enterprises.

Based on the above discussion. As for the borrowing successful enterprises, we can see that total liabilities, financing liabilities and operating liabilities have a positive correlation with corporate performance. From Table 1, it is found that the scale of indebtedness for the borrowing successful enterprises is lower than other two kinds of enterprises. Thus, the performance of the borrowing successful enterprises can be improved by increasing indebtedness. As for the borrowing risk enterprises, the total liabilities and corporate performance have a positive correlation, and the financing liabilities have a negative effect on the corporate performance. This is because the non-state holding listed companies and the four major state-owned commercial banks are nonhomogeneous. However, as for the borrowing failed enterprises, the total liabilities and the financing liabilities have a negative correlation with corporate performance. Meanwhile, the ratio of the financing liabilities to the total liabilities is higher and the profit rate of capital is lower than the interest rate. Therefore, the financing liabilities have a more negative effect on the corporate performance. The operating liabilities have a positive effect on the corporate performance for three kinds of enterprises.

4 CONCLUSIONS

This paper takes A-share non-state holding listed companies from the Shanghai and Shenzhen Stock Exchanges for the period from 2010 to 2012 as the research object and researches the relationship between total liabilities, operating liabilities, financing liabilities and corporate performance. The following conclusions may be drawn:

1 The enterprises can measure the effect of their liability utilization through leverage contribution rate index system established by this paper and make sure reasonable debt financing to improve corporate performance;
2 The operating liabilities and corporate performance have a positive correlation for three kinds of enterprises and have a most significant positive correlation in the borrowing failed enterprises. Therefore, this kind of enterprises should try to use the operating liabilities to improve corporate conditions which are different from the conclusion got by Wang Qiong and Tang Zhen (2012) about state-owned listed companies;
3 The financing liabilities only have a positive effect on corporate performance in the borrowing successful enterprises which indicates that this kind of enterprises should give priority to increase bank loans to improve corporate performance.

REFERENCES

[1] Gilson, Stuart. Management turn over and Financial Distress[J]. Journal of Financial Economics, 1989, 25 (2) : 241–262.

[2] Zhang Ying, Zhang Jian ying. The empirical study on the debt financing and corporate performance based on the data analysis of listed companies in Shandong province[J]. Green Finance and Accounting, 2012(12): 33–35.

[3] Yu Dong zhi. Capital structure, governance of creditor's right and performance: an empirical analysis[J]. China Industrial Economy, 2003(1): 87–94.

[4] Liu Lei, Xue Jing jing. The influence of capital structure to corporate performance based on the empirical study of A-share state-owned listed companies[J]. Communication of Finance and Accounting, 2013(9): 62–64.

[5] Huang Lian qin, Qu Yao hui. Study on the difference between operating liabilities leverage and financing liabilities leverage[J]. Accounting Research, 2010(9): 59–66.

[6] He Ping. Study on the relationship between debt financing and corporate performance of state-owned listed companies based on Granger Causality test[J]. Communication of Finance and Accounting, 2009 (6): 71–74.

Management, Information and Educational Engineering – Liu, Sung & Yao (Eds)
© 2015 Taylor & Francis Group, London, ISBN: 978-1-138-02728-2

Obstacles in listening comprehension and its corresponding measures

Hong Mei Xing
Public Foreign Language Educational Institute, Beihua University, Jilin, Jilin, China

Lei Sun
Computer Institute, Changchun University, Changchun, Jilin, China

Yan Zeng
Beijing Webrate Technology Co., Ltd., Beijing, China

ABSTRACT: Listening is an important part of foreign language learning. According to the statistics from foreign language teaching experts W.M. Rivers and M.S. Temperly, in the social practice, 45% of language use is conducted through listening. As a result, both in daily life and language skills training which includes listening, speaking, reading and writing and translation, listening holds an extremely important position. Modern foreign language teaching emphasizes on listening practice, which is in order to meet communication needs of learners, and also reflects the regularity of learning a language - first of all, language learning depends on listening. The importance of listening is not only reflected in teaching, but also in the tests.

KEYWORDS: listening comprehension; obstacles; corresponding measures.

1 THE MEANING OF LISTENING COMPREHENSION

Listening comprehension is a complex process of understanding language. It is a complex psychological process that emphasizes the listener (language recipient) understands the language through auditory and it is also the interaction process between language and mental activity. British linguist Mary. Underwood divided complete listening process into three stages: the first stage is the stage in which the sounds go into a sensory store which is called input; the processing of the information from the short-term memory is the second stage, whose short term is processing; the third stage is transferring the information to the long term memory for later use) which is the output or response.

Listening can be regarded the process of explaining the content continuously in a certain language environment. "Hearing" differs from "understanding". The listener functions as a recorder that requires the listener not only remember "recording" but also can repeat what he has heard. In this process, the biggest problem for the foreign language learners is the three stages can't be completed, so as to achieve the effect of"listening" "understanding" and "response".

2 THE OBSTACLES IN THE PROCESS OF LISTENING COMPREHENSION

There are many factors that hinder the complete process of listening comprehension, which are summed up to three, namely: psychological barrier language barrier and cultural barrier.

2.1 *Psychological barrier*

First of all, the listener is used to translate the received target language signal into native language word by word, and to understand them in the native language. The essence of this process is decoding the input foreign language and recoding the understood information in the native language. The process of language decoding and encoding greatly reduces the speed of cognitive understanding. Second, in the process of listening comprehension, the listener holds unrealistic expectations on themselves. There often appear situations as listening obstacles, fatigue, mental tension and poor memory. Among them, time validity of short-term memory holds the strongest effect on listening comprehension level. Sometimes the listener has understood what he heard, but can't answer questions correctly because he has forgotten them. Third, it involves attention. Listening to an unfamiliar language's pronunciation, intonation, vocabulary, sentence, etc., easily leads to the listener's fatigue. Besides, the listener highly concentrates his attention, which is easy to make the brain into the state of extreme nervousness. At last, the listener listens passively. Most language learners regard listening comprehension as a process of receiving information passively. What it needs is just to sit listening, without the brain thinking actively. In fact, "listening" is active labor of mind. It requires the listener to apply his language and non-language knowledge into what he hears, in order to understand the intentions of the speaker. Thus, "listening" is an active thinking process. The learner's idea of passive thinking obviously becomes a major obstacle of listening comprehension.

2.2 *Language barrier*

Language barriers mainly include pronunciation, vocabulary, grammar, etc. Students lack the necessary basic knowledge of phonetics. Besides the difficulties in phonemes distinguish, they also have trouble in identifying language skimming (elisions) in the middle of the stream, liaison, assimilation, condensation, synthesis, stress, rhythm, intonation and other forms of language reading. In addition, the influence of the new words on "listen"is much greater than it is on reading. For some listeners, as long as an unfamiliar word comes into their ears, they become so nervous as the attention stabilizes on the word. At the same time, the following content come one after another. As a result, some words that they should have known will not be responded in their mind. The dictation can't be completed. In terms of grammar, unfamiliar and complex sentence structures of language material are different from the logic of native language expression, which would have interfered on auditory cognitive, causing the understanding obstacles.

2.3 *Cultural barrier*

In the process of listening comprehension, the listener's role of actively participating is very important. The listener's social and cultural background knowledge and the width of knowledge and pragmatic knowledge is as important as the listener's knowledge of the language. In the process of listening comprehension, the listener naturally contacts cultural information about the target language. If the listener's lack of understanding of the differences between two cultures, there must be effected on the deep understanding of the target language. Due to the knowledge width of the listener, if the field knowledge is not familiar, listening comprehension will be impeded. For example: some female listeners lack interest in the knowledge of politics, economy, sports, etc. and don't know much about them. Once this knowledge comes out in the listening material, they will be nervous, thus affecting the listening comprehension process.

3 STRATEGIES OF IMPROVING LISTENING COMPREHENSION OF FOREIGN LANGUAGES

3.1 *Build foreign language environment*

The principle of acquisition called by linguists applies to the improvement of foreign listening ability. Let yourself in a target language atmosphere constantly, as time passes, the listener will find you can understand more and more. Tens of thousands of Africans besides speaking their own tribal languages also use a trade language or a colony language. In the process of learning these languages, these people have never had a formal language education, but they easily master the language of other ethnic groups.

3.2 *Pay attention to listening training methods and techniques of foreign language*

First, grasp the main points. The listener should understand the main information of listening material. Some listeners try hard to grasp every word and every sentence, thus allocating their attention averagely. The result is "Grasp all, grasp nothing." To get the key words and language clues is the key to listening comprehension, since these words and phrases or sentences summarize the main idea of dialogues and passages.

Second, reasoning and speculating. This means on the basis of the context or additional information on listening material, such as intonation, tone and other language elements, the listener can make reasonable reasoning, speculating and judgment on the parts that can't be understood.

Third, forecast. According to question-stems, charts, options and the listener's experience, he can forecast what he will hear. This helps to understand the listening material quickly.

Finally, use the "ears" to listen. Some learners practice listening skills with the help of using text materials, trying to capture the relevant information and content. Reliance on outside of audio data information will reduce the listener's attention, with the passive coping mind.

3.3 *Reading is an important means to enhance the listening skills*

Reading can not only correct the incorrect pronunciation, maintaining a good foreign language pronunciation habit, but also can enlarge the vocabulary, expand knowledge width, which is of great benefits to improving your listening skills.

(4) Recite and write more
Reading and reciting can greatly promote foreign language learners to master the pronunciation, intonation, the syntax and vocabulary of the target language. Those are strengthened through writing.

In short, the improvement of foreign language listening skills is based on the improvement of comprehensive foreign language skills. Only after learners have a good ability of listening and fast reading, can they quickly write down what they hear. And comprehensive foreign language ability is not built in a day. This requires learners to practice and accumulate progressively.

REFERENCES

[1] Brian Seaton. A Handbook of English Language Teaching [D]. London: Terms and Practice. The Macmillan Press 1982 In English.

[2] Jeremy Harmer. The Practice of English Language Teaching [D]. London: Longman Press, 1983. In English.

[3] Goodman K.S A Psycholinguistic Guessing Game [J] journal of the Reading Specialists, 1967. In English.

Management, Information and Educational Engineering – Liu, Sung & Yao (Eds)
© 2015 Taylor & Francis Group, London, ISBN: 978-1-138-02728-2

Relationship between the ownership concentration and corporate performance of companies listed on gem

Zhen Hu & Jian Wei Zhang
China

ABSTRACT: The study found that the proportion of the largest shareholder and corporate performance is no significant correlation, and by controlling the growth rate of investment and asset-liability ratio, come as the company increased investment growth will weaken ownership concentration on corporate performance effects and excluding the impact of company size, the company's asset-liability ratio the greater concentration of ownership of the company, the stronger the correlation between the performance of these two conclusions.

KEYWORDS: Ownership concentration corporate performance investment behavior.

1 ISSUE RAISED

Since 1932 in the book "Modern companies and private property," the relationship between ownership structure and business performance became governance research firm in a long-lasting research topic. And since October 30, 2009, the first batch of 28 GEM companies focused on the Shenzhen Stock Exchange since, SMEs, emerging high-tech industries, especially those with independent innovation, quite the growth of high-tech enterprises has broader financing channels, while venture capital has also been a more convenient and efficient exit channel. GEM high price-earnings ratio may attract a large number of high-yield investors, while the ownership structure of the GEM listed companies is gradually changing. For the changes that occur within the company's ownership structure, its performance for the company, whether the beneficial effects are owners and operators of common concern. While ownership concentration changes will also affect the company's investment behavior, then under certain conditions on the investment behavior of companies, corporate performance will produce what kind of change has also been a problem business owners concerned.

2 RESEARCH HYPOTHESIS

2.1 *Theoretical assumptions*

This paper studies the impact of ownership structure on corporate performance, and the company's investment behavior as an intermediary variables to study how the investment behavior under certain ownership concentration affect the company's performance. The following assumptions:

H1: Performance of listed companies and the degree of the largest shareholder has significant correlation.
H2: As the company's increased investment growth will weaken the impact of ownership concentration on corporate performance.
H3: exclude the impact of company size, the company's asset-liability ratio, the stronger ownership concentration on corporate performance correlation.

2.2 *Variables selection*

Table 1.

Variable nature	Variable name	Variable definitions	Symbol
Explained variable	ROE	Net profit / Net assets	ROE
Explanatory variables	The ratio of the largest shareholder	The number of shares the largest shareholder / company total shares	X1
	The proportion of the top five shareholders	The sum of the top five shareholders / company total shares	X2
	The proportion of the top ten shareholders	The sum of the top ten shareholders / company total shares	X3

(*continued*)

Table 1. (*continued*)

Variable nature	Variable name	Variable definitions	Symbol
Control variables	Asset-liability ratio	Total Liabilities / Total Assets	Dar
	Fixed asset investment growth	Increase the number of fixed assets / total fixed assets last year	Y

Explanatory variables: paper selected ROE as a dependent variable that is a measure of corporate performance. ROE is a comprehensive indicator of relatively strong, and it can make shareholders' earnings at a glance. Financial analysis, ROE is applied DuPont financial analysis system, it is the financial analysis of a very important core indicators, no trade restrictions, you can use a very wide range.

Explanatory variables: According to previous literature, they use a proportion of the largest shareholder to measure the concentration of ownership.

Wherein the ratio of the largest shareholder X1:

X1 > 50%, the largest shareholder belongs absolute control;

30% < X1 < 50%, the largest shareholder is a relatively Holdings;

X1 < 30%, the equity structure are dispersed ownership structure.

In general, the largest shareholder in the relative control bit, supervision and encouragement operators have enthusiasm, can lead the company's business decisions. The ratio of the top five shareholders and X2, show the extent of checks and balances between the major shareholders; the proportion of the top ten shareholders X3, a measure of the distribution of the company's equity.

Control variables: asset-liability ratio Dar measure the level of company debt, fixed asset investment growth to measure the level of growth the company's fixed asset investment.

2.3 *Modeling*

Based on the assumption H1, can establish the following linear model:

$$ROE = \alpha_0 + \alpha_1 X_1 + \alpha_2 X_2 + \alpha_3 X_3 + \xi$$

Based on the assumption H2, can establish the following linear model:

$$ROE = \beta_0 + \beta_1 X_1 + \beta_2 X_2 + \beta_3 X_3 + \beta_4 Y + \xi_1$$

Based on the assumption H3, can establish the following linear model:

$$ROE = \gamma_0 + \gamma_1 X_1 + \gamma_2 X_2 + \gamma_3 X_3 + \gamma_4 Dar + \xi_2$$

3 EMPIRICAL ANALYSIS

This paper selects ended 2011-2013 Shenzhen GEM listed companies as samples, excluding net assets was negative or incomplete disclosure of the company, were screened out qualified companies 523. The data from the huge influx of information networks and GTA database, using Excel and SPSS19.0 for data processing.

3.1 *Regression analysis*

After the sample data descriptive statistics and correlation test, the correlation regression analysis, a weighted average net assets Net margin for the dependent variable, the ratio of the largest shareholder, the proportion of the top five shareholders and former the ratio of the top ten shareholders and for the explanatory variables, and consider asset-liability ratio, the investment growth rate of the control variables, using the least squares method and through parametric test, combined with the results of the regression model results are shown in Tables 2–4.

Looking at the overall test results from the model adjusted R2 of 0.868, indicating a good model fit, Sig multiple linear regression model is 0, indicating that the model is statistically significant; former five shareholders from the regression equation coefficients of view proportion, proportion of the top five shareholders, this year compared with last year percentage increase or decrease of assets and asset-liability ratio of Sig. values were 0.000,0.001,0.040, indicating coefficients of these three variables is very significant, the largest shareholder proportion Sig. > 0.05, say less obvious with sexual performance. Therefore, the negative hypothesis H1.

In the negative hypothesis H1, based on the adjusted models 2 and 3, the framework follows the model:

Model 2-1 $ROE = \beta_0 + +\beta_1 X_2 + \beta_2 X_3 + \beta_3 Y + \xi_1$

Model 3-1 $ROE = \gamma_0 + +\gamma_1 X_2 + \gamma_2 X_3 + \gamma_3 Dar + \xi_2$

2-1 and 3-1 according to the model, and the control variable Y are set in the range Dar made regression results shown in Tables 5 and 6. As can be seen from Table 5, when Y ≤ 0.5, the proportion of the ratio of the top five shareholders and top ten shareholders Sig. Value of 0, while in Y > 0.5 when, Sig. Values increased significantly, says with obvious significant weakened, supporting H2. Similarly, in Dar ≤ 0.5, the proportion of the top five shareholders and shareholding ratio of the top ten shareholders Sig. Values much larger than 0.5, the table was significantly decreased, while in Dar > 0.5 when, Sig. Values less than 0.5, say clearly the sex had a strong support H3.

4 EMPIRICAL RESULTS AND ANALYSIS

In this paper, 2011-2013 Shenzhen GEM listed companies' financial reports and data analysis to extract indicators carried out and verified by three assumptions. First, ownership concentration variables descriptive statistics analysis, the ownership is highly concentrated in China's current top five shareholders, and there are some companies exist due to the dominance system, proven correlation verification, denial of the largest shareholder holding proportion of shares and corporate performance were significantly correlated, but the sum of the top five shareholders, the sum of the top ten shareholders and corporate performance have a significant correlation, and then performed regression analysis, which was significantly correlated. Under the premise of a significant correlation, change the control variables, the investment growth rate and asset-liability ratio as a control variable to control, to observe the impact of ownership concentration on corporate performance, obtained with the increase of investment growth will weaken the impact of ownership concentration on corporate performance and exclude the impact of company size, the company's asset-liability ratio the greater concentration of ownership of the company, the stronger the correlation between the performance of these two conclusions, in order to verify the hypothesis H2 and H3.

5 RECOMMENDATIONS FOR THE OWNERSHIP STRUCTURE OF LISTED COMPANIES ON GEM

Through the above empirical research, raised the issue of equity for companies listed on the GEM of the concentration of some of the comments and suggestions in order to facilitate the company's continued effective development, specific measures are as follows:

First, under certain conditions, the company's investment behavior will affect the company's performance, therefore, can be a modest increase in investment to balance the impact of ownership concentration on corporate performance brings, both for the enlarged company, but also can restrict ownership is too concentrated on the business aspects of the company to bring negative effects to achieve the double-edged sword effect.

Third, the company's asset-liability ratio should be controlled within a certain range, asset-liability ratio is too large will not only bring short-term debt risk, increasing the burden on companies operating funds, but also to make too many changes in company performance depends on the ownership structure is not conducive to the company's operations.

REFERENCES

[1] Almeida, HV and Wolfenzon, DATheory of Pyramidal Ownership and Family Business Groups. Journal of Finance, 2006 (6).

[2] Faccio, M, and Lang, LH.P., 2002, "The Ultimate Ownership of Western European Corporations", Jurnal of Financial Economics, 65: 365–295.

[3] Jiang Lingyun, 2010, "Family ownership concentration of listed companies impact on business performance", Northeast Forestry University Outstanding Graduate Thesis.

[4] Yang fragrance, Hu Xiangli, Yu Lin affect ownership properties under different ownership structure of investment behavior - Empirical Evidence from Chinese Listed Companies in China Soft Science, 2010 (7): 142–150.

[5] Chen prosperity, Xu Wei An Empirical Study of large shareholders and corporate investment efficiency characteristic relationship Friends of Accounting, 2011 (1): 99–104.

Table 2.

Coefficient[a]

Model		Non-standardized coefficient		Standardized coefficient		
		B	Standard error	Trial version	t	Sig.
1	(Constants)	−1.236E-15	.081		.000	1.000
	Zscore(X_1)	.155	.118	.155	1.312	.193
	Zscore(X_2)	−1.334	.373	−1.334	−3.579	.001
	Zscore(X_3)	1.536	.337	1.536	4.555	.000
	Zscore(Y)	.366	.102	.366	3.579	.001
	Zscore(Dar)	−.077	.099	−.077	−.776	.040

a. The dependent variable: Zscore (weighted average ROE)

Table 3.

Coefficient[a,b]

Model		Y ≤ 0.5					Y > 0.5				
		Non-standardized coefficient		Standardized coefficient			Non-standardized coefficient		Standardized coefficient		
		B	Standard error	Trial version	t	Sig.	B	Standard error	Trial version	t	Sig.
1	(Constants)	−.158	.096		−1.654	.103	.344	.250		1.375	.085
	Zscore(X_2)	−1.671	.346	−1.618	−4.830	.000	-.292	.603	−.366	−.484	.334
	Zscore(X_3)	1.933	.347	1.870	5.571	.000	.614	.606	.775	1.013	.124
	Zscore(Dar)	−.110	.111	−.098	−.991	.025	.246	.193	.256	1.277	.117

a. The dependent variable: Zscore (weighted average ROE)

Table 4.

Coefficient[a,b]

Model		Dar ≤ 0.5					Dar > 0.5				
		Non-standardized coefficient		Standardized coefficient			Non-standardized coefficient		Standardized coefficient		
		B	Standard error	Trial version	t	Sig.	B	Standard error	Trial version	t	Sig.
1	(Constants)	−.001	.076		−.011	.991	−.221	.231		−.956	.350
	Zscore(X_2)	−.335	.291	−.454	−1.152	.053	−2.123	.711	−1.523	−2.986	.007
	Zscore(X_3)	.391	.302	.498	1.294	.009	2.616	.656	2.054	3.987	.001
	Zscore(y)	.316	.094	.393	3.342	.001	.361	.179	.267	2.019	.056

a. The dependent variable: Zscore (weighted average ROE)

Management, Information and Educational Engineering – Liu, Sung & Yao (Eds)
 ISBN: 978-1-138-02728-2

Problems and corresponding measures of real estate enterprises' capital structure

Tong Wu
Real Estate Economy, Langfang Polytechnic Institute, China

ABSTRACT: Whether the capital structure of an enterprise is perfect or not directly affects the financial risks that the enterprise faces as well as its long-term development. This paper firstly has a brief introduction to the capital structure of an enterprise including its definition and influencing factors, and then carries on the discussion to problems and specific optimization measures of China's real estate enterprises' capital structure which will be of great help for its improvement.

KEYWORDS: Real estate enterprises; Capital structure; Problems and measures.

According to modern financial theories, capital structure is a crucial issue for an enterprise to make financing decisions. With the development of the real estate enterprise, more attention has been paid to the selection and adjustment of the capital structure, but in the course of its development, it will inevitably encounter problems. The key to solve the problem is to probe into it. In recent years, Chinese government has issued a series of regulatory policies implemented step by step in order to achieve affordable housing prices, which has intensified the urge to improve the real estate enterprises' capital structure.

1 AN OVERVIEW OF REAL ESTATE ENTERPRISES' CAPITAL STRUCTURE

1.1 *Definition of an enterprise's capital structure*

An enterprise's capital structure refers to the relationship between composition and proportion of capital resources. An enterprise's capital is generally composed of long-term debt capital and equity capital. Therefore, capital structure refers to the proportion accounted for by the long-term debt capital and equity capital in the enterprise's total capital. The capital structure of an enterprise directly or indirectly affects the enterprise's capital cost, market value and the efficiency of management. The ultimate goal for an enterprise's investment decision is to optimize the allocation of limited capital and create value as much as possible for the enterprise. The real estate enterprise belongs to a typical capital intensive enterprise. Compared with other types of enterprises, it has a relative longer investment cycle, a slower capital turnover and is easily influenced by relevant national policies. Hence it will bring challenges for the real estate enterprise's long-term and stable development.

1.2 *Factors affecting real estate enterprises' capital structure*

The capital structure of a real estate enterprise is closely related to business development. Listed companies in Shanghai and Shenzhen from 2003 to 2011 have once been chosen as research objects to explore factors that impact on real estate enterprises' capital structure. The findings are as follows: First, there is a positive correlation between an enterprise's scale, its income tax and its capital structure. Second, there is a negative correlation between an enterprise's profitability, non-debt tax shield, collateral value of assets and its capital structure. Third, an enterprise's growth, short-term debt repayment ability and its ownership structure have little impact on the capital structure. Obviously, there are numerous factors affecting an enterprise's capital structure that need to be optimized to promote the development of the enterprise.

2 PROBLEMS EXISTING IN CHINA'S REAL ESTATE ENTERPRISES' CAPITAL STRUCTURE

2.1 *Unreasonable financing structures*

According to capital structure theories and actual situations of business development, an enterprise's capital should focus on internal financing, but for the majority of our real estate enterprises, the financing structure is mainly external financing. This situation is caused by the characteristics of the real estate enterprise. First, the investment cycle of a real estate enterprise is relatively long. It requires a very long period from buying land to housing construction to final sales which need to invest a large amount of funds to ensure normal operation; Secondly, the real estate enterprise's initial capital chain is great, and in a long investment cycle it has no cash inflows as its supplement, the enterprise's own capital accumulation capability hasn't been improved throughout the investment cycle, so only relying on the enterprise's own capital does not meet the needs of real estate projects. Therefore, an enterprise has to rely on external financing as its capital supplement. The inflow of foreign capital indeed alleviates a real estate enterprise's development, so it can use the principle of financial leverage to obtain greater benefits by using the supplement capital. However, compared with the internal financing, external financing is less stable. In the enterprise's capital structure external financing's proportion is too high, which increases the enterprise's financial risk and easily causes the break of its capital chain. Besides, in recent years, in order to curb the excessive rises of housing prices, China has adopted a series of macro-control measures, including strictly limit foreign financing channels of real estate enterprises, which makes the external financing environment increasingly serious and forces a number of real estate enterprises to rely on sales capital which intensifies the unreasonability of the financing structure.

2.2 *A high asset-liability ratio*

The above analysis shows that the financing structure of real estate enterprises in China is unreasonable which lacks of internal financing. In order to obtain sufficient capital, the majority of real estate enterprises use bank loans, which leads to the higher rate of asset-liability ratio. According to statistics, the real estate enterprises' asset-liability ratio is generally higher than other industries, which is very unfavorable for asset intensive enterprises. It is extremely easy to cause financial crisis. Any problems in the capital chain are likely to get the enterprises in trouble, even impact on the entire real estate enterprises' development. At the same time, it may even affect the upstream enterprises such as cement, reinforcing steel and other manufacturing industries, or the banks serving as a creditor of the real estate enterprises. With the improvement of national macro-control policies, the real estate enterprises' financing channels are becoming narrower and financial situation continues to be worse which makes the entire real estate enterprises face a critical period of development.

2.3 *Mainly relying on short-term debt*

It is the most appropriate to have half the proportion of short-term debt in the total debt of an enterprise. However, at present China's real estate enterprises' short-term debt ratio is higher; this is because short-term debt's financing cost is relatively low. The real estate enterprise expand short-term debt with the lowest cost, but at the same time increase the total debt needed to be repaid in a short term and also increase the enterprises' pressure of repaying debt. Mainly relying on short-term debt increases the enterprise's financial risks and operational risks, which causes the enterprise to face challenges for a long-term and sound business.

2.4 *A small proportion of bond financing in debt*

The majority of the real estate enterprises in China have a relatively small proportion of bond financing. Two main reasons are as follows: First, bond financing is difficult for the real estate enterprises. In order to reduce financial risks, our country has made very strict rules for the qualification of bond issuance and has controlled it by checking which causes the difficulties of financing through the issuance of bonds. On the other hand, until now the bond market in China has not obtained great development. It is still difficult for financing even the enterprises issue bonds through layers of check. Second, it is relatively easy for the real estate enterprises to loan money from banks. China has issued a series of preferential policies which reduces the difficulties loans for enterprises. Therefore, bank loans have accounted for the largest proportion of the real estate enterprises' total debts. The bank loan not only includes that has directly loaned from the bank by the real estate enterprises, but also includes the pre-sales payments paid by consumers for house purchase. The proportion of debt financing in the aggregate liability is minimal, less than one percent. The unreasonable debt structure also reflects the unreasonable capital structure.

3 MEASURES FOR THE OPTIMIZATION OF REAL ESTATE ENTERPRISES' CAPITAL STRUCTURE

3.1 *Optimization of capital structure and expansion of diversified financing channels*

In recent years, China has successively introduced some policies to regulate the property market. The improvement of those regulatory policies has a huge impact on the development of China's real estate market. Real estate enterprises have a large fund demand, but insufficient channels of internal financing and restrictions on external financing channels have made the problem of capital shortage more prominent. Only relying on bank loans as a major way of financing can't adjust to the national tight money policy. Under such a grim situation for financing, real estate enterprises must innovate financing channels and reduce financing costs. The diversified approaches of financing ensure the normal link of a capital chain. The more common approach includes the expansion of real estate funds and the increase of the bond financing ratio. Real estate enterprises can develop by means of raising funds lying idle in society which reduces risks for the bank and also promotes the optimum distributions of those funds.

3.2 *Innovation of financing mode and adjustment of financing structure*

The current financing mode of China's real estate enterprises, the proportion of debt financing far exceeds equity financing. China's listed real estate companies account for less than one percent of the real estate enterprises' total number which fully illustrates that equity financing is not the main way of the real estate enterprises' financing. In view of this situation, China's real estate enterprises need to innovate financing mode and establish a new one which places debt financing first and equity financing second so as to inject new vitality into China's real estate enterprises. The above mentioned ways to adjust financing structure can maintain the main position of banks in the real estate enterprises' financing and can also make the stock market play an important role. They can optimize the corporate governance structure and financing structure. By optimizing the distribution of resources, they can expand the investors' benefits.

3.3 *Control of expanding scale and speed*

China's real estate enterprises mostly adopt a rapid expansion mode. Debt ratio is high in the operation process. The majority of real estate enterprises invest large amounts of money to purchase land. But the construction of the project cannot keep up because of a relatively tight capital chain, which will impact on the sales. If the capital cannot be withdrawn as soon as possible, it will not only cause a lot of land lying idle, but also exacerbate the problem of the capital chain. Therefore, real estate enterprises must enhance their awareness of risks and prepare beforehand the financial budget according to the average level of the industry, which determines the capital indicators of purchasing land and project construction. The indicators can reflect how much money will be invested every year for the purchase of land or for the construction of projects. They are also used to reflect whether the enterprises' capital chain tenses up due to the idle land or the withdrawal of sales capital indicating normal operation. So they provide the basis for the enterprises' further decision-making.

3.4 *Optimization of companies' equity structure*

The improvement of the enterprise's capital structure lies in its own status of development and development plan for a period of time in the future. Equity simply means that the shareholders' ownership of the enterprise. A scientific equity structure is the foundation for the enterprise's existence, as well as for its long-term development. In China most of the real estate enterprises' property rights structure is not clear enough, so it's necessary to perfect it. First of all, property relations should be identified to absorb outstanding technical talent who become shareholders with their skills and to enhance the enterprise's comprehensive competitiveness. Secondly, when an enterprise develops to a certain period and encounters bottlenecks, it can separate the ownership and the power of operation to reform the organization structure for the development of the enterprises. Lastly, the enterprise's governance structure should be innovated to realize the goal of shared governance by the board of directors, managers, shareholders and other stakeholders.

3.5 *Adjustment of enterprise development strategies and establishment of a cooperative mechanism*

As everyone knows, high risks also mean high returns. The real estate business is a typical representative of high risks and high returns. In order to minimize the risk of their own, the real estate enterprises can cooperate in the same industry and establish a perfect cooperative mechanism to resist the market risk together. In practice, the specific operational method is that different enterprises jointly invest in a project which is developed by way of equity cooperation. The cooperation is not only limited to spread risks, it can also be used to cooperate with the local real estate enterprise when opening markets in new cities. They can use each of their advantages to solve problems in the development of a project. In addition to the ways of project cooperation, they can also choose capital cooperation. Cooperation has clearly become an important choice for many enterprises when facing difficulties.

4 CONCLUSION

To sum up, with the implementation of a series of macro-control policies, some problems have become more prominent in the real estate industry, such as unreasonable financing structures in China's real estate enterprises' capital structure, a high assets-liability ratio, mainly relying on short-term debt, a small proportion of bond financing, etc. Some measures need to be taken in order to improve the development of the enterprises, such as optimization of the capital structure, expansion of diversified financing channels, innovation of the financing mode, adjustment of the financing structure, control of the expanding scale and speed, adjustment of enterprise development strategies, establishment of a cooperative mechanism, and optimization of the companies' equity structure, etc.

REFERENCES

[1] Zhixue Li. A Study on China's Real Estate Enterprises' Capital Structure and Governance Effect[J], China Management Informationization, 2010, (05): 67–68.

[2] Hongling Ge. Empirical Analysis of Features and Influencing Factors of China's Real Estate Enterprise Capital Structure – Case Study of Listed Corporation [J]. Yellow Sea Academic Forum, 2010, (01): 67–68.

[3] Min Guan. Empirical Study of Influencing Factors of Real Estate Enterprises' Capital Structure –Case Study of Listed Corporation in Hong Kong [J]. Special Administration Region Economy, 2011, (01): 9–10.

Management, Information and Educational Engineering – Liu, Sung & Yao (Eds)
© 2015 Taylor & Francis Group, London, ISBN: 978-1-138-02728-2

Effects of the application of human care theory on the care of cancer patients

Ya Li Sun & Ling Gao
Nursing College of Beihua University, Jilin city, China

Tian Rong Yu
Affiliated Hopital of Beihua University, Jilin city, China

ABSTACT: Objective: To investigate the effect of the application of human care theory on the life quality of cancer patients in household life after they receive a community nursing care by community nurses. Method: The quality life of 93 patients with cancer living in the six communities in the city was assessed and the assessment was based on a life quality scale (SF-36). The community nurses care for these patients by using the theory of human care and the life quality of the patients was observed after the care. Results: The results showed that there was a significant difference in the score of the eight dimensions in the quality of life before and after the patients' care (<0.01 or <0.05), suggesting that the theory of human care can be used for the care of patients with cancer and the application of the theory can effectively improve the life quality of the patients, strengthen their problem-solving abilities and make them have a positive attitude to face their lives.

KEYWORDS: Community nurses; Theory of human care; Patients with cancer; Quality of life.

1 INTRODUCTION

With the progress of the times and the rapid development of medical science, cancer has become a serious threat to human health and a tremendous psychological and psychological burden due to it has been put on the patients and their families to cause the significantly decreased the quality of life of the patients with cancer [1, 2]. Because of the advance in medical technology, the survival period of patients with cancer has been significantly prolonged. At the present time, the quality of life has been concerned by whole society more and more. In this study, the change in the quality of life of patients with cancer was observed and studied after the implementation of humane care was given to the patients with cancer [3].

2 MATERIAL AND METHODS

2.1 *General information*

93 Cases of patients with malignant tumor were selected and they were diagnosed before March 2011, were selected. After the combined treatment in a hospital, they live their home lives, they can take care of themselves, their vital signs were stable, there was no communication barrier in language for them, all of them had received an education, at least in a junior high school, and most of them were voluntary to participate in the community health educational activities. The average age was (57 ± 7.2) years. There were 59 male and 47 female patients, including 15 cases of pancreatic cancer, 18 cases of breast cancer, 16 cases of cervical cancer, 6 cases of rectal cancer, 22 cases of gastric cancer, 11 cases of lung cancer and 5 cases of esophageal cancer. The selected requirement for the community nurses who would implement the human care include that they had to have at least more than 10 years of clinical experience, and they should have gained some understanding in the theoretical connotation of human care, the knowledge of human care, the art and skill for communication with patients, establishing a concept of humane care and the status of human care. 12 nurses who met, the selected requirement were appointed to carry out the humane care for the patients in their own communities.

2.2 *Methods*

The 93 patients selected were assessed on the existing quality of life. Based on the quality life scale (SF-36) [4], the assessment was conducted according to 8 dimensions and 35 items. The eight dimensions included the physiological function (10 items), physical functions (4 items), general health (5 items), physical pain (2 items), life vitality (4 items), social contact function (2 items), emotional functions (3 items) and mental health (5 items). The score of the physiological function was 10~30 points, the score of physical pain was 2~11 points, and all the other items were 1~5 points. The score for each item was positively proportional to the quality of life.

The happiness indexes of 93 Patients selected were assessed and assessment was conducted based on Memorial University of Newfoundland scale of Happiness (MUNSH) [5]. MUNSH contains 24 items. Among the 24items, 10 items are considered to reflect the positive and negative emotions, 5 items the positive emotion (PA), 5 items the negative emotion (NA), 14 items positive and negative experiences, 7 items the positive experience (PE), and 7 items the negative experience (NE). The total happiness index: PA–NA + PE–NE. The determination of score was based on the following rule: that the answer to each item was "yes" was recorded 2 points, that the answer to each item was "I don't know" was recorded 1 point, and that the answer to each item was "No" was recorded 0 point; that the answer to item 19 was "the residence where he or she is living now" was recorded points and "any other residence" was recorded 0 point; that the answer to item 23 was "satisfactory" was recorded 2 points and "unsatisfying" was recorded 0 point. The range of the total average point was from –24 to+24points. In order to be easy to calculate the score, 24, as the constant, was added, and the score range was from 0 to 48; the higher the score the higher the happiness index.

The community nurses implemented their human care by applying the 10 factors in the human care described in the theory of human care set up by Dr. Watson J. The patients were made three appointments a week, the time for each appointment was not less than 60 minutes, the care was implemented in a small course or separately aiming at the particular case for different patients and the appointment time could not be missed. Through collective psychological counseling, interactive discussions, so that patients understand their illness, the correct treatment to their own circumstances, emotional self-regulation, the surrounding environment to adapt skills in collective environments enable patients to understand the original is not the only bear the same pain yourself, there are other people, enabling them to care about each other and promote each other, to share the joy, confidence to overcome the disease, the ultimate goal is to enhance the patient's overall quality of life and well-being index.

During the care, the nature of human care was particularly emphasized, science and humanities knowledge was integrated and used to communicate with the patients, and the human care was completed based on 10 factors proposed in the theory of human caring. 10 factors for human included forming a value system of the human altruism; inculcating a trust and hope; developing a sensitivity of themselves and others; building up a relationship in helping, trusting and caring; encouraging and accepting the expression of positive emotions and negative emotions from the clients ; applying scientific methods to solve problems in the decision-making system; providing a psychological, social and spiritual environment to support, protect and correct them; helping meet individual needs; allowing the presence of existentialism, phenomenology and spiritual power. The community nurses should follow the nature of human care to give the patients with cancer a psychological support and cultural care, answer any questions for them or explain the related successful stories and so on.

After the community human care was implemented over 3 months, the life quality of the 93 patients was assessed one more time still based on the life quality scale (SF–36) And well-being Scale (MUNSH) were evaluated.

Statistical method. SPSS10.0 soft ware was used to analyze the survey data in this study. Least significant difference method using LSD and SNK analysis of variance test procedures between the data before and after pairwise comparison of survey data processing and analysis to $P<0.05$ was considered significant differences.

3 RESULTS

Community nurses patient quality of life and happiness index has changed before and after the humane care, including quality of life, physical function, vitality, social functioning, role emotional, mental health after making five dimensions of human care theory $P<0.05$, significant differences, as shown in Table 1. Cancer patients after human care happiness index theory of positive emotions, positive sexual experience increased average $P<0.05$ significant difference, negative emotions, negative experience somewhat reduces the average $P<0.05$ significant difference, happiness comparison of the total average score of $P<0.05$ were significantly different before and after care theory, as shown in Table 2. Through the application of quality of life scale (SF–36) and the Scale of Happiness (MUNSH) quality of life for cancer patients and happiness index evaluation, has been fully proved theory of human caring for cancer patients can improve patient quality of life and the happiness index of patients with cancer.

The comparison before the human care and after the human care for the patients is shown in Tables 1 and 2.

Table 1. Comparison in SF-36 score before the human care and after the human care for the patients (points) x±s.

Dimensions	Average sore at the first assess	Average sore after the human care	*P* values
Physiological function	13.21±6.12	16.91±4.62	<0.05
Physiological functions	6.06±2.22	7.81±1.82	<0.01
General health	10.03±3.57	15.52±4.89	<0.05
Physical pain	6.25±3.09	8.94±7.99	<0.05
Life vitality	8.71±4.28	14.29±5.97	<0.05
Social contact function	6.42±5.83	8.97±2.72	<0.01
Emotional functions	4.14±3.36	5.76±1.79	<0.01
Spiritual health	8.63±4.18	13.11±2.79	<0.01

Table 2. Comparison in happiness between before and after the implementation of the human care theory(points) x±s.

Items	Average sore at the first assess	Average sore after the human care	*P* values
Positive emotion	2.75±2.9	4.82±2.5	<0.05
Negative emotion	7.24±2.3	4.31±3.7	<0.05
Positive experience	4.03±2.7	6.82±4.89	<0.05
Negative experience	11.25±2.7	9.25±2.6	<0.05
Total happiness score	12.59±2.8	22.08±3.7	<0.05

4 DISCUSSION

It is well known that the daily life of cancer patients is limited, their life skills are decreased significantly, their social activities are reduced, they always lack of interpersonal communication and emotional support, often show a negative attitude to the present life, and moreover, the disease can cause the decrease in their body resistances to directly affect the life quality of them. Community nurses applied the human care theory to give a respect to the personality of the patients and look after them with a personalized care, which can give the patients an individualized treatment. Psychological counseling and therapy is medical workers and patients with benign impact on the psychological state of the patient's pathology of clear guidance, to promote physical and mental health of patients with an effective therapy [5.6]. Community nursing intervention models can alleviate the symptoms of cancer patients, improve functional status and overall quality of life of patients [7]. Another study showed that stress directly affects the individual's level of depression and affect the quality of life of individuals, society and the functions of the individual can reduce the degree of depression, improve the quality of life of individuals [8]. Other studies show that the overall health of cancer patients by society, family and other factors that support more cancer patients perceive better the overall health of the more improved [9]. The emphasis on patient-centered and the care service which advocates to serve for the clients with love and patience, and in an all-around way to improve the life quality of patients with cancer significantly. In this study, the results showed that there were significant differences in the factors which can be used to assess the life quality of patients with cancer, such as the physical function, social function, emotional function and mental health after the theory of human care was applied for the care of the patients with cancer ($p<0.01$), and there were also significant differences in physiological function, general health, physical pain and life vitality ($p<0.05$), indicating the effect of human care on the cancer patients in a home life. Therefore, the theory and the method of human care should be applied universally, and all nurses should understand the meaning of human care theory and the knowledge of human care in their future work. It is believed that the application of the human care theory will be a new development in nursing.

REFERENCES

Frick E PanzerM.Depression and quality, of life of cancer patients andergoing radiation therapy. a corss-sectional study in community hospital out patient centre. Eur J Cancer, Vol. 16(2007), p. 130~136.

GolantM,A im an T,M artin C,et al M anaging Cancer side effects to improve quality of life a cancer psychoeducation program.Cancer Nursing, Vol. 26(2003), p. 36~44.

Watson J. Nursing:Human science and human care. New York: National League for Nursing, Vol. 1(1985), p. 1–3.

Li Lu, Wang Hongmei, Shen Yi. Development and psychometric tests of a Chinese version of the SF–36 Health Survey Scales .Chinese Journal of Preventive Medicine, Vol. 36(2002), p. 109~113.

Yang Shaoping. Study effects of psychological intervention on quality of life of cancer patients Sichuan Medicine, Vol. 31(2010), p. 135 – 137.

WANG Jian-fang, Zhou Jianhong, Ma Xiuqiang community nursing intervention model for cancer patients. Journal of Nursing, Vol.27(2010), p.881 – 884.

Wang Minlan psychological nursing intervention in patients with breast cancer resection application. Chinese Pharmaceutical Guide, Vol.14(2012), p. 496 – 498.

Yost KJ, Hahn EA, Zaslav sky AM, et al.Predictors of health related quality of life in patients with cocrectal cancer. Heath Qual life Outcomes , Vol. 6(2008), p. 66–75.

Zhou Jianhong, He Baolun, Huang Lixin, etc. community quality of life of cancer patients and social support correlated. Chinese Chronic Disease Prevention and Control, Vol. 17(2009),17 p. 305–307.

Management, Information and Educational Engineering – Liu, Sung & Yao (Eds)
© 2015 Taylor & Francis Group, London, ISBN: 978-1-138-02728-2

A review of research on Huashan Rock Paintings and Huashan tourism

Xin Hua Ma
Department of Economics and Management, Guangxi Normal University for Nationalities, Chongzuo, Guangxi, China

ABSTRACT: This paper provides a systematical overview of the progress of research on Huashan Rock Paintings (HRP) and the associated tourism. Critical future research needs are also identified. Current research can be described in the following perspectives: 1) origin of HRP including historical age, author, reason, and content of HRP (i.e., when, who, why, and what); 2) value of HRP, its tourism development and conservation; 3) technologies for protecting HRP;4) application of HRP for World Heritage listing. Future research should focus on balancing the HRP tourism development, World Heritage listing, and conservation.

KEYWORDS: Huashan; Rock painting; Progress; Prospect.

1 INTRODUCTION OF HUASHAN ROCK PAINTINGS AND HUASHAN SCENIC AREA

Huashan literally means "flower mountain". Several mountains in China are called Huashan, and they are located in Zibo, Shandong; Kaihua, Zhejiang; Dawu, Hubei; Anqing, Anhui; and Chongzuo, Guangxi. Among these "flower mountains", some were named after their flower-like shapes, others were named because of its prominence of flowers and blossoms. Uniquely, Huashan in Chongzuo, Guangxi was named because of the presence of large amounts of rock paintings on many cliffs that are located on both sides of the Zuojiang River. In ethnic Zhuang local dialect, there is a word "pyalaiz", which is pronounced Ba Lai and means that there are many decorative patterns on the mountains. These decorated mountains are paraphrased Huashan in mandarin Chinese. Many rock paintings are discovered in the Zuojiang River valley in Pingxiang, Longzhou, Ningming, Chongzuo, and Fusui of Guangxi. Because the rock paintings on Huashan in Ningming County have the largest dimensions, the largest varieties of portraits, and the richest contents, the cliff frescoes on the Zuojiang River basin are collectively referred to as the Huashan Frescos, also known as Huashan Rock Paintings (HRP).

Huashan Rock Paintings were listed as a Protected National Key Cultural Relics by the State Council in 1988, and in that year Huashan Scenic Area became one of National Scenic Areas. In order to obtain more domestic and international financial and technical support, to protect HRP better, and to promote the recognition of the HRP, authorities have started to apply for World Cultural Heritages for HRP from the year 2003.Three levels of the government, including Guangxi Zhuang Autonomous Region, Chongzuo City, and Ningming County, have spent a lot of human and material resources in preparation for successful declaration HRP as one of World Cultural Heritages in 2016. Especially the application effort for the "Inscription" of HRP as one of World Cultural Heritages has started since 2013. Huashan Rock Paintings have drawn extensive attention from the society. Therefore, it is necessary to compile recent research progress on HRP and Huashan tourism in academia, and identify future research directions.

2 RESEARCH PROGRESS ON HRP AND HUASHAN TOURISM

Although records of HRP can be spotted in ancient literature, such as *Continued Natural History* authored by Li Shi in the Song Dynasty, *A Collection of Anecdotes* compiled by Zhang Mu in the Ming Dynasty, *Ningming State Archives* compiled in the ninth year of the reign of Emperor Guangxu. But it was after the establishment of the People's Republic of China when a study on HRP was started. It peaked in 1950–60s, but it entered a trough during the Cultural Revolution period. Since the late 1970s, especially since the 1990's, the research on HRP entered a new age. At present, research on HRP and Huashan tourism are mainly as the following: when the rock paintings were painted, who painted them, why they were painted, what content was painted, what the value of HRP is, what causes destruction of HRP, and which technology can be used to protect HRP, how to develop Huashn Scenic Area and protect HRP, and application of listing of HRP as one of World Cultural Heritages.

2.1 *When were HRP painted*

Liang Renbao thinks that HRP were painted in ancient time or medieval time[1]; Shi zhongjian[4], Tan Shengmin, Qin Cailuan et al [5], think that HRP were painted as early as in the Spring and Autumn period an as late as in the Western Han Dynasty and Eastern Han Dynasty; Song Zhaolin [6] believes that HRPs were painted from the Warring States period to the Eastern Han Dynasty; Zhang Shiquan [7] believes that HRP were painted as early as in the Eastern Han Dynasty, as late as in the Southern Dynasties or in the Sui Dynasty; Lan Baiyong [8] thinks HRP were painted from the late Western Han Dynasty; Yang Qun[9] thinks that HRP were painted from the late Warring States period to the early Western Han Dynasty; Huang Zengqing [10], Liu Jie [11], Chen Hanliu [12] et al think that HRP were painted in the Tang Dynasty; even some other scholars think HRP were painted in the Taiping Heavenly Kingdom or the period of Sino-French war [1]. However, researchers and scholars have not reached a consensus on exactly when these HRP were painted.

2.2 *Who painted the HRP*

Huang Huikun [13] believes that it were the ancestors of the Miao Nationality and Yao Nationality who had painted the HRP, because dog is Totem of the Miao Nationality and Yao Nationality, and there are pictures of dog on these HRP; Mo Duoqing [14] believes that HRP are the works of West of the original Barbarians on the time of they moved to Zuojiang river; Qiu Zhonglun[15], Chen Mingfang [16], Long Bin[17], Yang Qun[9], think that it were the ancestors of Zhuang Nationality who was also named Luo Yue or Luo tribe that painted HRP. Among these opinions, a lot of experts have come to an agreement that HRP were painted by ancestors of Zhuang Nationality.

2.3 *Why were HRP painted and what contents were painted on HRP*

Liang Renbao thinks that the ancestors of Zhuang Nationality painted HRP to commemorate a war victory scene [1];Guang Min[2] thinks that the ancestors of Zhuang Nationality painted HRP to worship the God of water, it is on the basis that the ancestors of the Zhuang Nationality mainly lived on hunting and fishing, frequently contacted with water, and the water threatened their lives, but they could not overcome the water god, thus they have a fear of water, and thought that water has the spirit, and worshiped the water; Jiang Yongxing[18] thinks that HRP depicted real scenes of the ancestors of Zhuang Nationality who were worshiping their ancestors, and the Holy Land was located in Huashan. Liang Tingwang [19] thinks that HRP depicts scenes of offering sacrifices to frog god, which is based on the fact that there are images of frogs and drums on the rock paintings, and in the mythology of Zhuang frog is the patron saint of Zhuang Nationality, and the bronze drum is a Talisman with which frog god can reach heaven, the cliff is the habitat of the soul of the frog god, and also is the road from the frog god lead to heaven, so ancestors painted the patron saint on the cliffs. Lan Duomin[20] believes that Zuojiang River rock paintings are graphs of the frog god map; Meng Fei [21] thinks that the rock paintings depicted the life scenes of the ancestors of Zhuang Nationality who were hunting, because the four legged animal in rock paintings is a wild beast, and similar to murals in Cangyuan which are located in the Tropic of cancer, murals in Cangyuan reflect hunting scenes that the ancestors of the Wa Nationality were hunting. Xie Shouqiu [22] believes that the rock paintings on the Zuojiang are religious altarpieces with which the ancient Luo people called back the spirits of the dead because of wreck up and Requiem, where there are ancient rock paintings there are the sacrificial site in the Zuojiang River Basin. Many scholars think that the contents of the HRP are related to some type of worship in ancient times.

2.4 *Value of HRP*

Guo Hong et al[23] analyzed artistic characters and artistic value of the rock paintings in Zuojiang River; Jiang Riqing,Yu Zhaowen[24] analyzed its long history culture, distinctive and rich connotation of Zhuang culture, superb art of painting, finally pointed out that HRP are treasures of the Zhuang culture and a gourmet of the world rock painting and attentions should be paid to their protection and rational development; Shen Fumin [25] analyzed the characteristics of ecology and aesthet on HRP of primitive art; Li Ping [26] did an interpretation on national culture connation of HRP; Song Xiaoyu[27] analyzed the sports value contained in rock paintings; Chen Jianqiang discussed[28] on how to excavate the artistic value of HRP as much as possible.

2.5 *Causes to damages of HRP and HRP protection technologies*

Huang Huaiwu, Xie Riwan, and Zhang Bingfeng [29] point out that HRP have been damaged seriously due to natural forces and environmental changes, and summarize the methods of inspection of HRP; Guo Hong, Han Rufen, Zhao Jing, Huang Huaiwu, Xie Riwan, Lan Tiyong[30] analyze the source and quality of water, and study on the role of the water in the process of weathering damages of the rock paintings, and put forward some measures about controlling water and methods of waterproof for HRPs; Guo Hong, Han Rufen, Zhao Jing, Huang Huaiwu, Xie Riwan, Lan Tiyong [31] also analyze the status of petroglyphs physical weathering and weathering mechanism, discuss the conservation method and requirements of repair material, and on the basis of environment characteristics which are required for saving the rock paintings, research the best time to the govern physical weathering; Wang Zhiliang [32] analyzes the reason of HRP pigment falling off and fading; According to the conditions of hydrogeology and geology in Huashan area, Xu Lijun, Fang Yun et al[33] analyze the types of damages of seeping water and hazards and its formation mechanism, and put forward some countermeasures which includes preventing leakage, establishment of surface drainage system, and building the vertical wall shielding eave; According to the problems present in Huashan tourism areas and the regulations on the protection of tourism resources requirements, Huang Jianqing, Hu Hengsheng et al [34] proposed to establish Huashan tourism resource management information system. Ma ChunXiao[35] analyzes the railway project of Nanning to Pingxiang going through Huashan Scenic Area, and put forward that there are much work to be done, such as environmental impact assessment, environmental protection measures, and minimizing the influence of the project on the scenic spot. Huang Yuqin[36] discusses on using 3D laser scanning technology and GIS technology to gather data and analyze the data.

2.6 *Tourism development in Huashan and its conservation*

Huang Jianqing, Wei Qianhong [37][38][39] analyze tourism resources and tourism market of Huashan tourist area, and study on development and innovation of tourism products in Huashan tourist areas from tourism product brand strategy to tourism products innovation. Sun Yanzhong, Yao Lei[40] explain the necessity and feasibility of in-depth development of cultural tourism in Huashan scenic area, and they propose that: in order to avoid the "fast food" type of shallow layer development and its destructive effect on Huashan culture, Huashan culture connotation should be discovered further, the Huashan Cultural tourism brand should be built around the core of HRP, gradient development should be implemented to protect tourism resources, and in the end a win-win will be achieved between Huashan Cultural Heritage and the sustainable development of Huashan scenic areas; Liao Yang and Meng Li[41] study from the angle of anthropology, and put forward that we should pay attention to the real carrier of the culture of the nationality when developing the folk culture resources of Zhuang in Huashan, should ensure that we correctly display on nationality culture and keep its authenticity; Chen Hongling [42] analyses the present situation of Huashan tourism development and reasons of the destruction of the HRP, and puts forward specific counter-measures on how to develop tourism product in Huashan tourist areas; Pan Qi [43] analyses the status and the role of HRP ,and proposed the Huashan location of cultural brand and create Huashan culture brand; Based on the experience management process model and analyzing the Huashan Scenic Resources and tourist experiences demand, Zhao Jie [44] positions tourists experience, and advises building World Rock Painting Park and designing the experience environment and experience products and experience service in the World Rock Painting Park, in the end, he discusses how to implement the tourism experience and evaluation of the effect and how to improve it.

2.7 *Application of Huashan Rock Paintings for world heritage listing*

Yang Bingzhong [45] thinks that it is necessary to apply for World Heritage for HRP. Listing HRP as World Heritages facilitates the upgrade from ornamental value to research value and to wealth value, and it is of a strategic significance for promoting the cultural development of Guangxi. Chen Xuepu [2] describes important acceptance rules for World Heritage, and analyzes the obstacles which will impact on the application of HRP for World Heritage, and he pointed out that HRP has to accept the fierce competition and go through strict examination and approval procedures because of limited quota, although the conditions of HRP are consistent with the first, third, and sixth requirement for inscription. Finally, Chen Xuepu puts forward, two stratcgics for applying for World Heritage: 1) broaden HRP to Huashan culture and increase the inscription component, 2) applying for rock paintings in Guangxi, Yunnan, Sichuan, Guizhou, Xinjiang and other provinces instead of applying for Huashan alone. According to images of frogmen and bronze drum in HRP, He Ying [46] proposes the concept of the Huashan Cultural Circle with a broad view of including some Southeast Asian countries. It is worthwhile to explore the field of application for cultural landscape, and the application for intangible cultural heritage is a successful direction for World Heritage listing. Xie Yongxin [47] discusses ways to make people pay more attention to the application of HRP for World Heritage and measures to excavate the historical and cultural value. Xie Yongxin proposed to correctly handle the development and conservation issues. Lei Mengfa [48] rethinks the protection and development of HRP from the perspective of the application for World Heritage listing. Guan Xie [49] studies archives of World Heritage, and analyzes files need to be collected for the application of HRP for World Heritage, and also analyzes how to gather the files.

3 CRITICAL REVIEWS AND PROSPECTS

In conclusion, most experts have been mainly focusing on studying the value of HRP from the perspective of anthropology, ethnology, literature, and aesthetics; on studying physical and chemical techniques and methods for protecting HRP; on studying developing tourist resources, raising recognition, attracting tourists, creating economic benefits for the locality from tourism, marketing and other aspects. However, only a few experts have studied the problems of tourism development in the Huashan Scenic Area and protecting HRP from the perspectives of World Heritage. Even though some scholars have done some research on it such as the following: Xie Yongxin [47] analyzes on how to make people pay more attention to the application of HRP for World Heritage listing and how to excavate the historical and cultural value, and then proposes to correctly handle the development and protection issues. Lei Mengfa [48] rethinks the "protection and development of HRP from the perspective of the application for World Heritage listing. However, these two studies have just scratched the surface. A deeper analysis is needed and a systematic strategy is yet to be defined.

As early as in the 1990s, governments at all levels in Guangxi started to work on the application of HRP for World Heritage. Although Ningming HRP was included in 2007 in the *Chinese Tentative List of Application for World Heritage* [47] and application for World Heritage approval in 2016 is currently ongoing, there are still a lot of work to do. Even if HRP is listed as piece of World Heritage, HRP must accept strict supervision and inspection from the World Heritage Committee every three years. The world heritage included in the *World Heritage List* must be strictly ensured that the authenticity and integrity of the world heritage are protected. If not, the world heritage will receive a "yellow card" or "red card" warning from the World Heritage Committee, even be removed from the World Heritage List. There are some precedents in China that those who have joined the *World Heritage List* have received the "yellow card" warning, and even worse abroad that items been removed from the *World Heritage List*. In order to avoid recommitting the same error, HRP are still in the process of applying for world heritage, which has not been world heritage, but we must balance the issues of protection and development with the standards of the world heritage and requirements from the World Heritage Convention and the Guide to Action for The World Heritage. Therefore, there are two areas of research on HRP in future: On one hand, we should continue to excavate the cultural connotation and value of HRP, and should continue to make efforts for protection technology and control measures of HRP. On the other hand, we should follow the conventions of the protection of world heritage authenticity and integrity which are asked by the Protection of the World the Cultural and Natural Heritage Convention. Future research topics may include: how to protect the authenticity and integrity of Huashan culture which centers on HRP; how to balance the issues of the protection and conservation of Huashan culture and the development of Huashan scenic area.

4 ACKNOWLEDGMENTS

This work is supported by the Humanities and social sciences research project of Guangxi Zhuang Autonomous Region Education Department (Grant No.SK13LX487) and the young and middle-aged backbone teacher scientific research startup project of Guangxi Normal University for nationalities (Grant No.2012RCGG003).

REFERENCES

[1] Lv Wenjie. Reviews on research of legend about Huashan rock paintings [J].Journal of Nanning Teachers College, 2009 (2).

[2] Chen Xuepu. On Huashan culture in view of World Heritage and intangible cultural heritage [J].Journal of Guangxi Normal University for Nationalities, 2010,27.

[3] Guangmin. On the ancient cliff frescoes of Zhuang nationality [J].Central College for nationalities, 1978(4).

[4] Shi Zhong Jian. On the relationship between Guangxi rock paintings and Fujian stone inscriptions [J]. Academic Forum, 1978(1).

[5] Tan Shengmin,Qin Cailuan. Investigation on Zuo Jiang cliff paintings [J].Ethno-National Studies, 1985(5).

[6] Song Zhaolin. Investigation on Guangxi cliff frescoes [J].Cultural Relics World, 1986(2).

[7] Zhang Shiquan. Discussion on some issues related to Guangxi hole burial [J].Ethno-National Studies, 1982(4).

[8] Lan RiYong. Discrimination the earliest time of painting Zuojiang cliff painting [J].Southern Heritage, 1997(1).

[9] Yang Qun.New Investigation on Guangxi Zuojiang cliff painting [J].Guangxi ethnic studies,1986(3).

[10] Huang Zengqing. on the ancient cliff frescoes in the west of Guangxi Zhuang Autonomous Region and the time of painting these cliff frescoes[N].Guangxi daily,1957-3-9.

[11] Liu Jie. Development of XiOu family and the murals spreaded in Ningming,Chongzuo and long Jin [N]. Guangxi daily, 1957-7-26.

[12] Chen Hanliu.Guangxi Ningming Huashan Frescos–the symbol of Zhuang ancient language [N].Guangxi daily, 1961-9-18.

[13] Huang Huikun.Investigation on Huashan cliff paintings from ethnology, also discussion on nature's and clan of Huashan cliff painting in Guangxi[J].Journal of Yunnan Nationalities University, 1985(1).

[14] Mo Junqing. Discussion on the subject of Zuojiang cliff paintings [J].Ethno-National Studies, 1986(6).

[15] Qiu Zhonglun. On the clan of Zuojiang cliff painting [J].Academic Forum,1982(3).

[16] Chen Mingfang.On the clan of Guangxi Huashan Frescos[J].Guangxi Ethnic Studies, 1988(4).

[17] Long Hua Bin. Huashan mural research notes [J].Journal of Guangxi University for Nationalities, 1986(3).

[18] Jiang Yongxing.Zhuang ancestors worship—exploration on the theme of Huashan Frescos [J].Journal of Guangxi University for Nationalities, 1985(2).

[19] Liang Tingwang. Huashan cliff paintings–The holy land of sacrifices for Frog God [J].Journal of South-Central University For Nationalities, 1986 (8).

[20] Lan Duomin. Zuojiang cliff painting should be the frog God figure [J]. Journal of the Central Institute for nationalities, 1986(3).

[21] Meng Fei. Exploration on the topic of Huashan mural [J].Journal of Guangxi University for Nationalities (Philosophy and Social Science), 1990(3).

[22] Zhang Ying. Experts believe: Guangxi Zuo Jiang cliff painting is likely painted for the sacrifice to the dead in the shipwreck. http://news.hexun.com/2011-09-14/133370719.html?from=rss.

[23] Guo Hong, Huang Huaiwu, Xie riwan, Lan Riyong.On features and values of Guangxi Zuojiang rock painting art [J].Southeast Culture, 2004(2).

[24] Jiang Riqing, Yu Zhaowen. Analysis on the connotation of the historical and national culture included in Huashan rock paintings in Guangxi [J].New Western, 2011(35–36).

[25] Shen Fumin. Discuss on ecological aesthetic characters of primal arts–an Example of Huashan Paintings [J]. Guizhou Ethnic Studies, 2008(3).

[26] Li Ping. Interpretation on zuojiang Huashan mural national cultural [J].Journal of Huazhong Normal University (Humanities and Social Sciences), 2011(3).

[27] Song Xiaoyu. Analysis on the sports value contained in Huashan rock painting [J].Science and Technology Vision, 2014 (1).

[28] Chen Jianqiang. On the development of Guangxi Huashan mural art [J].Technology and Arts, 2014(2).

[29] Zhang Bingfeng. Survey on Huashan rock paintings disease [J].Chinese Cultural Relics Research, 2009(9).

[30] Guo Hong, Han Rufen, Zhao Jing, Huang Huaiwu, Xie Riwan,Lan Riyong. The water in Huashan rock paintings of weathering diseases the role and countermeasures [J]. Science of Conservation and Archaeology, 2007,19(2).

[31] Guo Hong, Han Rupan,Huang Huaiwu,Xie Riwan,Lan Riyong.Study on physical weathering mechanism and intervention of Huashan rock painting in Guangxi province [J].Research on cultural heritage Science and Technology, 2004(00).

[32] Wang Zhiliang. Analysis on the cause of the pigment of Guangxi Huashan rock painting fall off and fade[J].Science of Conservation and Archaeology, 2011,23(2).

[33] Xu Lijun, FangYu, Wang Jinhua, Peng Pengcheng et al. Research on the water permeation disease and environmental cure of Huashan rock art [J]. Safety and Environmental Engineering, 2006(6).

[34] Huang Jianqing, Hu Hengsheng, Liang Haihua.On tourism information system Huashan tourist area in Guangxi [J].Resources and Living Environment, 2007(18).

[35] Ma Chunxiao. Landscape sensitivity based highway location choice: Huashan resort example [J].Planners, 2011(12).

[36] Huang Yuqin. 3D laser scanning technology used in the protection of Huashan rock paintings [J].Chinese Heritage Scientific Research, 2012(1).

[37] Huang Jianqing, Hu Hengsheng, Wei Qianhong.On development of tourism culture in Huashan[J].Journal of Maoming University, 2007,17(4).

[38] Huang Jianqing, Wei Qianhong. On development and innovation of tourism products of Huashan in Guangxi[J]. Resources Environment and Development, 2007(4).

[39] Huang Jianqing,Hu Hengsheng,Wei Qianhong. Research on the development and utilization of tourist resources in tourism area of Huashan in Guangxi [J].Journal of Hengyang Normal University, 2006,27(6).

[40] Sun Yanzhong, Yao Lei. Thought on deep exploitation of cultural tourism of Ningming Huashan in Guangxi [J].Anhui Agricultural Sciences, 2011, 39(30).

[41] Liao Yang, Meng Li. The ethnic cultural tourism resources development in the perspective of anthropology–taking Zhuang Huashan culture as an Example [J].Social Scientists, 2009(7).

[42] Chen Hongling. Research on Ningming Huashan tourism product development oriented [J].Market Modernization, 2007 (515).

[43] Pan Qi. Create Huashan culture brand [J].Academic Forum, 2004(1).

[44] Zhao Jie .The experience management model applied in tourism—taking Huashan scenic spots in Ningming as an example [J].China Business & Trade, 2013(14).

[45] Yang Bingzhong. The application of Huashan rock paintings and Guangxi cultural development [J].Journal of Guangxi Normal University for Nationalities, 2010, 27(1).

[46] Yang Bingzhong. Descission on the application of Huashan rock paintings for World Heritage [C].Guangxi: Guangxi People's Publishing House, 2010:22–32.

[47] Xie Yongxin. Issues arising from the application of Huashan cliff paintings for World Heritage [J].Journal of Guangxi Normal University for Nationalities, 2010,27(2).

[48] Lei Mengfa. Rethinks the protection and development of Huashan rock paintings from the perspective of the application for World Heritage [J].Coastal Enterprises and Science & Technology, 2010(11).

[49] Guan Xie. Research on archives collection for the application of Huashan rock paintings in Ningming for the World Heritage [J].Problem Exploration, 2010(6).

Management, Information and Educational Engineering – Liu, Sung & Yao (Eds)
© 2015 Taylor & Francis Group, London, ISBN: 978-1-138-02728-2

An analysis for the new high-tech photoelectric equipment under the free-replacement & pro-rate warranty strategy

X. Zou, Y.X. Jia, X.Y. Li, X. Liu & J. Zhou
Department of Equipment Command and Management, Mechanical Engineering College, Shijiazhuang, China

ABSTRACT: The characteristics of the new high-tech photoelectric equipment are diverse, integrated, and complex in the commercial area. A mode of the simplest free-replacement warranty for the new high-technical photoelectric equipment cannot satisfy the desire of the customers. In order to improve the quality of the warranty strategy for the new high-tech photoelectric equipment, a new mode for the warranty was improved by the theory of free-replacement & pro-rate warranty. Then, the cost-availability model of one-dimensional warranty combining the free-replacement & pro-rate warranty strategy was proposed in the application of an analysis on a laser distance range finder. At last, the model was proved to be effective and practical by the case study.

KEYWORDS: Photoelectric equipment; Warranty; Strategy.

1 INTRODUCTION

As the development of photoelectric technology, a larger number of the high-tech photoelectric equipment was put into the market and paid more attention by customers. The function of the photoelectric equipment was much more integrated and complicated. In order to meet the requirement of customers in support, it's necessary to bring the manufacturers' technical service level and skill to keep the equipment in a steady, available condition. So, customers often had signed a contract with contractors in program warranty. Within the warranty, customers could perceive the potential failure mode in the photoelectric equipment and carry out the preventable work to decrease the possibility of the failure by the contractor. The work which could make the best use of the equipment age and reduce the cost of the maintenance task, had practical implications.

Furthermore, it's necessary to apply the free-replacement & pro-rate warranty strategy to carry on the preventable work in the warranty period of the high-tech photoelectric equipment for the sake of improving the availability and intact rate. The paper presented a viewpoint about a one-dimensional warranty combining the free-replacement & pro-rate warranty strategy based on the theme of discipline in the high-tech photoelectric equipment warranty. The strategy proved by the example of an analysis in a laser distance range finder for surveying engineering.

2 THE CONCEPT OF FREE-REPLACEMENT & PRO-RATE WARRANTY AND ITS IMPLICATIONS

2.1 *Warranty and pro-rated warranty*

Warranty is that the contractors make commitments to customers with any specific service about ensuring the product operating normally in a specified period. The warranty contract presented that the compensation or offset must be paid for the customers if the failure had occurred or would occur in a specific period or in the prescribed conditions.

Prorate warranty is that customers will pay for the cost of the failed components or preventive work in the certain proportions and the manufacturers will detect the condition of the product and replace the failed components in time. The strategy often used for the warranty of consumables and the cheap material, such as replacing the tires in the automobile industry. In the strategy, the contractors would share the risks of the warranty cost. The main work carried out by the manufacturers.

2.2 *The characteristics and problems of the high-tech photoelectric equipment warranty*

The high-tech photoelectric equipment had more complex system structure, more integrated function and was used by the wide customer groups. There were some features and problems as follows: ①Possibilities of the related failure were large. Compared with the traditional single system, it was hard to predict the failure occurrence. In case of a certain type of a laser distance range finder, the equipment couldn't lose in the normal state. The surface phenomenon of the failure was in the photoelectric technology. But the actual reason for the failure was the caused by the steel foundation wearing. It led to deviate from the correct position and losing axis were shifting out of the installing condition. The failure is the mechanic failure mode. ②The equipment distributed more randomly. The cost of the warranty was much higher than the cost of the traditional warranty. If the manufacturers and customers only pay attention to the corrective warranty, it couldn't satisfy the need of support and maintenance. Meanwhile, if the program hadn't carried out the preventive warranty, it would waste a large scale of the human and material resources, and would have a deep influence on the customers' usage.

2.3 *The applicable implications of the free-replacement & pro-rate warranty*

Now, the existing strategy for the high-tech photoelectric equipment warranty was the corrective warranty in free-replacement model. In order to prevent potential serious failure consequences and remove the failures, it's necessary to introduce the improved free-replacement & pro-rate warranty strategy on the basis of the pure free-replacement warranty. This strategy could take the initiative to introduce preventive warranty and reduce the risks of the failure. It also could be satisfied with the customers' needs for the availability and support. Finally, the implementation of the strategy could increase the efficiency of the equipment system and effectiveness in logistic procurement cost.

3 A MODEL FOR FREE-REPLACEMENT & PRO-RATE WARRANTY STRATEGY AND CONCLUSIONS

3.1 *The assumption of the model*

The free-replacement & pro-rate warranty strategy was that contractors paid the corrective maintenance costs for failure and shared the preventive warranty costs with customers in proportion. The contractors carried out the corrective maintenance and preventive maintenance at the same time in the warranty period. A model was proposed in the free-replacement & pro-rate warranty strategy to estimate the rationality of the strategy.

It assumed that a laser distance range finder was in the free-replacement & pro-rate warranty strategy. In warranty period W, N was the numbers of the preventive warranty which carried out in the moment $T_i \ (i = 1, 2, \cdots N)$. The rate of repair $M(M_{\min} \le m \le M_{\max})$ depended on the contractors. $\theta(m)(0 \le \theta(m) \le 1)$ was the function of the repair effectiveness. The function was monotonically increasing. The formula was shown as follows:

$$\lambda(t^+) = \lambda(t^- - \theta(m) \times t^-) \tag{1}$$

The earlier system failure rate of the high-tech photoelectric equipment meets the rule of earlier changing in the "tub curve":

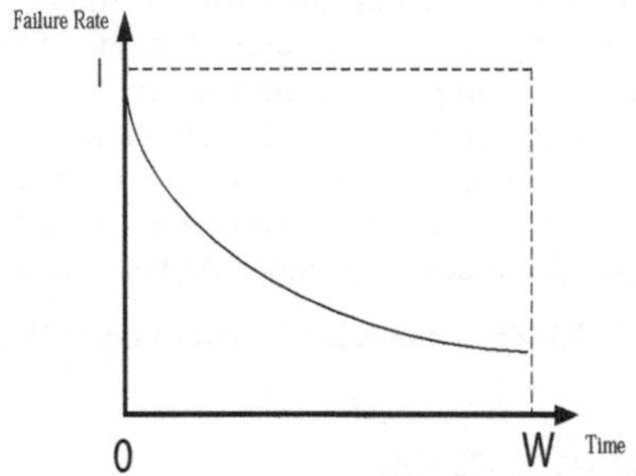

Figure 1. The change of the earlier failure rate during the warranty period.

It assumed that the proportional factor of the pro-rate warranty was $\alpha = 1 - t^2/W^2$. The contractors' cost in the preventive warranty was $C_1 = C_p(1 - t^2/W^2)$. The figure of the cost was showed as follows:

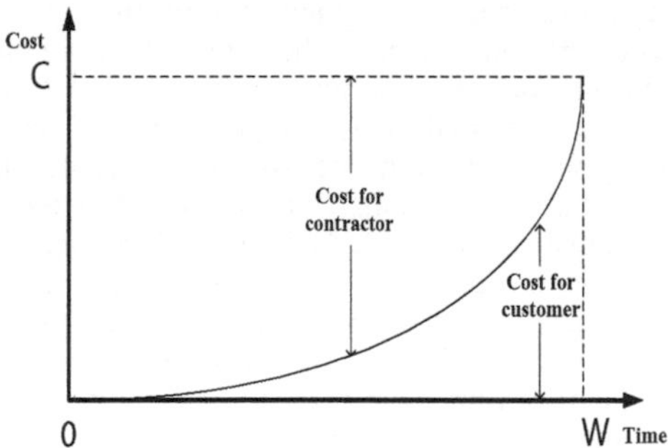

Figure 2. The strategy for the pro-rate warranty during the warranty period.

3.2 *The foundation of the cost model*

For the laser distance range finder system, the contractors had a preventive warranty strategy based on age replacement model. The component would be tested and replaced for a certain interval. T Each preventive warranty time was T_p. The cost of the certain interval T was the cost of the corrective warranty C_f and the cost of the preventive C_p.The all warranty cost for contractors during the warranty period was C:

$$C=\sum_{j=1}^{n} C_p\{1-\frac{[jT+(j-1)T_p]^2}{W^2}\}+\sum_{i=1}^{n}\int_{(i-1)(T+T_p)}^{iT+(i-1)T_p}\lambda_i(t)\mathrm{d}tC_f+\int_{n(T+T_p)}^{W}\lambda_{(n+1)}(t)\mathrm{d}tC_f \quad (2)$$

In the formula, n was the numbers of imperfect warranty during the warranty period, $n=\left[W/(T+T_p)\right].[*]$ was presented rounding numbers. $\lambda_i(t)$ was the failure rate in the imperfect preventive warranty interval i_{th}:

$$\lambda_i(t)=\lambda(t-(i-1)\times\theta(m)\times T) \quad (3)$$

3.3 *The foundation of the availability model*

The expected availability A during the warranty period could describe as follows:

$$A=\frac{W-[nT_p+\sum_{i=1}^{n}\int_{(i-1)(T+T_p)}^{iT+(i-1)T_p}\lambda_i(t)\mathrm{d}T_f+\int_{n(T+T_p)}^{W}\lambda_{(n+1)}(t)\mathrm{d}T_f]}{W} \quad (4)$$

3.4 *The foundation of the cost-availability model*

The best cost-availability E_c during the warranty period could describe as follows:

$$E_c=\frac{C}{AW}$$

Calculating and comparing with formula (2), (3) and (4), the best cost-availability per unit could be gotten. The decision makers should choose the best strategy for the equipment.

4 AN ANALYSIS FOR THE LASER DISTANCE RANGE FINDER SYSTEM WITH THE FREE-REPLACEMENT & PRO-RATE WARRANTY STRATEGY

In the warranty strategy of a laser distance range finder system, for example, the models were set up. The initial strategy was single free replacement. The cost, availability, best cost-availability are shown as following:

$T=W=1080\,\text{day}\,(3\text{ years})$, $C=36150$ Yuan, $A=0.8830$, $Ec=37.9074$ Yuan / Days

The mean time of system down for corrective maintenance was much longer than ever before. It had much more influence in the usage of the customers. So, the minimum warranty was carried out if the system failure occurred between check intervals. When the degree of imperfect warranty was $\theta(m)=0.9$, the cost of the warranty, availability, cost-availability showed as following figures:

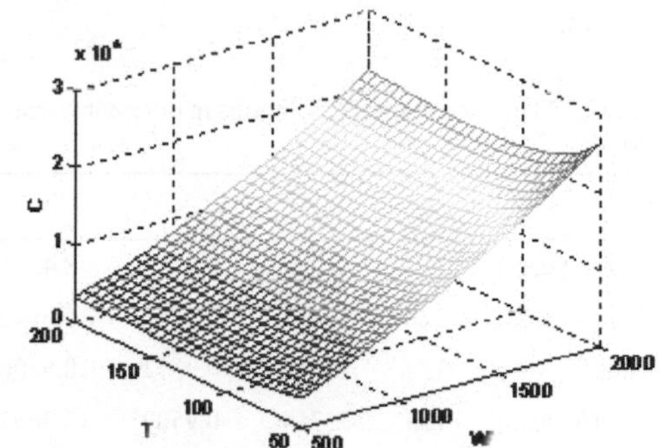

Figure 3. The model for the cost of the warranty.

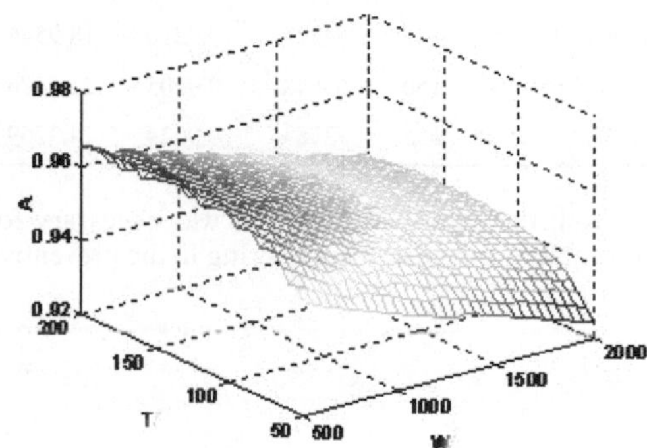

Figure 4. The model for availability.

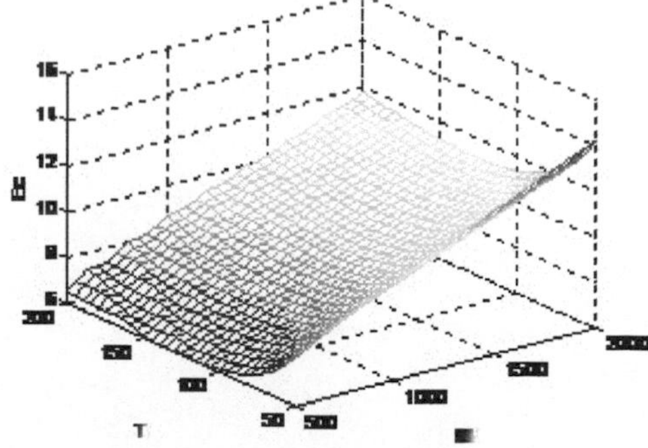

Figure 5. The function of cost-availability per unit time.

As shown in the figure, if the number of warranty period *W* was the constant value, the cost of a warranty and the cost-availability per unit time would decrease slowly and then increased. The availability would increase slowly in the beginning and decrease finally in the figure. To optimize the warranty cost, there were different statistics about imperfect preventive warranty interval *T*, availability A, cost-availability *Ec* when the equipment had been set for the different warranty period. The specific results were shown as follows:

Table 1. The corresponding solutions in different warranty periods.

W	*T*	*C*	*A*	*V*
1080 (3 years)	130	8405	0.9766	8.4521
1440 (4 years)	128	13784	0.9689	9.6142
1800 (5 years)	132	19118	0.9522	10.9606
2160 (6 years)	133	23714	0.9509	12.3692
2520 (7 years)	134	30266	0.9399	13.8267
2880 (8 years)	137	37537	0.9321	15.4055
3240 (9 years)	139	45522	0.9260	17.1685
3600 (10 years)	145	54229	0.9210	18.9546
3960 (11 years)	150	63648	0.9105	21.5476
4320 (12 years)	155	73783	0.9024	24.3269

From the data in the table, it was necessary for contractors and customers to bring in the preventive warranty to decrease the cost and improve the availability of the components in the equipment. For example, the warranty cost decreased obviously and availability increased by 10%. These proved that the improved free-replacement & pro-rate warranty strategy could be satisfied with need of the customers on the basis of the initial free-replacement warranty strategy. The purposes of the models were achieved.

REFERENCES

[1] W.R. Blischke and D.N.P. Murthy. 1992, Product warranty management-I: A taxonomy for warranty policies[J]. European Journal of Operational Research, 62: 127–148.

[2] W.R. Blischke and D. N. P. Murthy 1994, Warranty cost analysis[M]. Marcel Dekker, 246–312.

[3] W. R. Blischke and D.N. P. Murthy 2006, Warranty Management and Product Manufacture[M]. London, 103–110.

[4] X.Y Li, Y.X Jia. 2014, Study on the Warranty Mode for the New-tech Equipment under the Condition of Military-Civil Integration[J]. Journal of Mechanical Engineering College, Vol. 26(3).

[5] M. Shafiee, S. Chukova. 2013, Maintenance models in warranty: A literature review[J]. European Journal of Operational Research, http://dx.doi.org/10.1016/j.ejor.2013.01.017.

[6] L.C Wang. 2010, Research on Warranty Period of High and Advanced Technology Equipment Based on Preventive Warranty Policy[D]. The Mechanical Engineering College.

[7] J. Xie, 2005, Warranty cost and Length under Warranty Policies[D]. The University of Tianjing.

Management, Information and Educational Engineering – Liu, Sung & Yao (Eds)
© 2015 Taylor & Francis Group, London, ISBN: 978-1-138-02728-2

Effect of Schisandrae chinensis lignin on blood glucose of diabetic rats

Xiao Tong Shao, Guang Yu Xu, Guang Xin Yuan, Hong Yu Li, Pei Ge Du & Li Ping An
College of Pharmacy, Beihua University, Jilin, Jilin, P. R. China

Fu Xiang Ding & Mei Zhen Fan
Affiliated Hospital, Beihua University, Jilin, Jilin, P. R. China

ABSTRACT: Purpose: The aim of the present study was to investigate the hypoglycemic effects of the Schisandrae chinensis lignin on diabetes mellitus rats by streptozotocin induction. Methods: Type 1 diabetes mellitus rats models were set up by intraperitoneally injection of streptozotocin; The Fasting Blood Glucose (FBG), insulin levels (FINS) along with MDA content, SOD and CAT activities in serum were measured after the rats were treated with the Schisandrae chinensis lignin for 6 weeks. Pathological morphology effects of Schisandrae chinensis lignin on rat pancreas were observed by HE staining. Results: The results showed that Schisandrae chinensis lignin significantly decreased fasting blood glucose and MDA level, but increased FINS level as well as SOD and CAT activities. Schisandrae chinensis lignin could relieve pancreatic pathology change, islet volumes were becoming larger, islet B cell populations were increased. Conclusion: Schisandrae chinensis lignin had hypoglycemic effects on diabetic rats, strengthened the antioxidant ability of rats.

KEYWORDS: Schisandrae chinensis lignin, Type 1 diabetes mellitus rats, Blood glucose, Insulin.

1 INTRODUCTION

The dried ripe fruits of Schisandra chinensis (Turcz.) Baill., are officially listed in the Chinese Pharmacopoeia as Wuweizi in Chinese and mainly used as a tonic and sedative. Schisandra contains lignans, polysaccharides, volatile oils, fatty acids, vitamins and amino acids etc. [1]. Schisandra lignans is one of the most important ingredients with pharmacological activities [2]. Lignin was the main composition of Schisandra chinensis oil, had hypoglycemic effects on type 1 diabetes mellitus rats induced by alloxan, but the mechanism is unclear[3]. Therefore, it could provide theoretical basis for the development of hypoglycemic drugs by understanding the hypoglycemic effect and mechanism of Schisandrae chinensis lignin.

2 METERIALS AND METHODS

2.1 *Reagents*

Schisandrae chinensis lignin was made by our lab. Streptozotocin was purchased from Sigma Aldrich (USA).Kit to measure blood glucose was obtained from Beijing BHKT Clinical Reagent Co., Ltd (Beijing, China).Iodine [125I] insulin radioimmunoassay kit was purchased from Tianjing Nine Tripods Medical & Bioengineering Co., Ltd (Tianjing, China).Kits to measure SOD, MDA, and CAT were provided from the Nanjing Jiancheng Chemical Factory (Nanjing, China).

2.2 *Animals*

Six-week-old male Wistar rats were housed in an environmentally controlled breeding room (temperature: 20±2°C, humidity: 60±5%, 12-h light/dark cycle). All rats were provided with free access to tap water. All procedures were approved by the Ethics Committee for the Use of Experimental Animals of Jilin University.

2.3 *Animal model and experimental groups*

Rats were injected intraperitoneally (i.p.) with streptozotocin (75 mg/kg), 2 week later, blood samples were collected by tail incision for fasting blood glucose measurements by glucose oxidase peroxidase. Rats with a fasting blood glucose of ≥ 7.8 mmol/L was considered diabetic. The control and diabetic rats were then randomly divided into 4 groups: 1) control group (CON, rats treated with matched saline), 2) control Schisandrae chinensis lignin-treated group (CON+SCL, control rats treated with Schisandrae chinensislignin 0.5 mg/kg), 3) diabetic model group (DM, diabetic rats treated with saline in a matched volume), 4) diabetic Schisandrae chinensis lignin-treated group (DM+SCL, diabetic rats treated with Schisandrae chinensis lignin 0.5 mg/kg). Schisandrae chinensis lignin was administered via oral gavage daily for 6 weeks.

2.4 *Measurement of glucose and insluin parametes*

After a 12-h to 16-h fast, rats were anesthetized with 20% urethane (100 mg/kg). Blood samples were obtained from abdominal aorta, centrifuged (3,500×g, 10 min, 4 °C), and the supernatant was used for measurement of glucose and insulin parameters. Blood glucose was estimated by a commercially available glucose kit based on the glucose oxidase method. Insulin was measured by radioimmunoassay method, respectively.

2.5 *Determination of superoxide dismutase (SOD) and catalase (CAT) activities, and lipid peroxidation levels (MDA)*

MDA was determined by TBA method, SOD was detected by the method of oxidase, and CAT was measured by ammonium molybdate method. Specific procedures were operated as kit instructions.

2.6 *HE staining*

Pancreata were removed, fixed in 10% formalin solution. After fixation,the samples were embedded in paraffin and cut into sections 4 μm thick which were then stained with haematoxylin and eosin by routine methods. Pancreata histopathology was examined by light microscopy.

2.7 *Statistical analysis*

All data were expressed as mean±S.E.M. "n" denotes the sample size in each group. Statistical analysis was performed using one-way analysis of variance (ANOVA) with post hoc test for multiple comparisons. SPSS software (version 13.0 for Windows) was used for statistical analysis. $P < 0.05$ was considered statistically significant.

3 RESULTS

3.1 *Serum biochemical parameters*

Mean values of the serum biochemical parameters from normal and diabetic rats were summarized in Table 1. Diabetic rats showed higher FBG and FINS levels compared with those of control. Treatment with Schisandrae chinensis lignin significantly decreased FBG and FINS levels compared with untreated diabetic group. Schisandrae chinensis lignin did not affect these parameters in control rats.

Table 1. The levels of fasting blood glucose and serume insulin of rats in various groups (n=8, $\bar{X} \pm S$).

Group	*FBG* (C_B/*mmol•L*$^{-1}$)	*Insulin* (λ_B/*mIU•L*$^{-1}$)
CON	4.71 ± 0.52	16.32 ± 1.52
CON+SCL	5.08 ± 0.53	17.61 ± 1.57
DM	19.06 ± 1.57*	10.57 ± 1.48*
DM+SCL	11.32 ± 0.35△	14.82 ± 1.71△

*$P < 0.05$ *vs* CON group; △$P < 0.05$ *vs* DM group.

3.2 *Effect of schisandrae chinensis lignin on MDA, SOD and CAT levels in serum*

Schisandrae chinensis lignin treatment significantly lowered MDA content and upregulated SOD and CAT levels (P < 0.05) in serum compared with the untreated diabetic group, Schisandrae chinensis lignin did not affect these parameters in control rats, as shown in Table 2.

Table 2. The serum MDA level and SOD, CAT activity of rats in various groups (n=8, $\bar{X} \pm S$).

Group	*MDA* (C_B/*μmol•L*$^{-1}$)	*SOD* (λ_B/*U•mL*$^{-1}$)	*CAT* (λ_B/*U•mL*$^{-1}$)
CON	6.47 ± 0.57	270.30 ± 13.39	6.79 ± 0.56
CON+SCO	7.01 ± 0.64	260.42 ± 14.54	5.82 ± 0.66
DM	15.22 ± 1.91*	163.41 ± 14.42*	1.41 ± 0.78*
DM+SCO	9.18 ± 0.92△	202.52 ± 15.15△	3.36 ± 0.54*△

*$P < 0.05$ *vs* CON group; △$P < 0.05$ *vs* DM group.

3.3 *Pancreata pathology morphology analyses*

Under a light microscope, the structure of rat islets in the control group and the control rats treated with Schisandrae chinensis lignin group was integrated, the size of them was larger and the boundary was clear; Islets of rats in the diabetic group presented various sizes, most of them were narrow and their boundaries were not clear. Compared with the model group, the volume of islets was enlarged, the boundary was clear, and the numder of pancreatic β-cell was increased in the Schisandra chinensis lignin treated group (Figure 1).

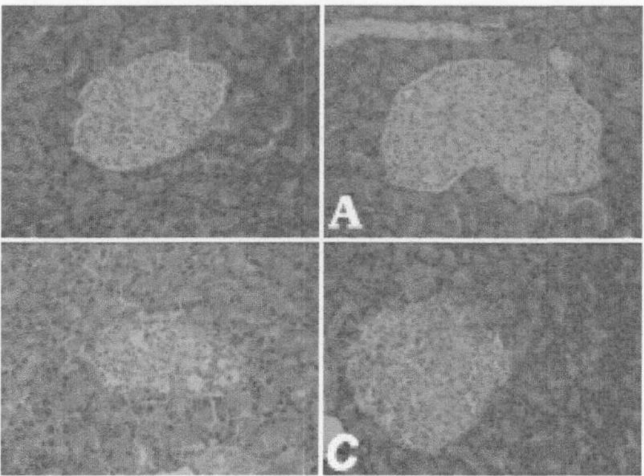

Figure 1. Pancreata pathology morphology analyses A as the structure of rat islets in the control group; B as the structure of rat islets in the control rats treated with Schisandrae chinensis lignin group; C as the structure of rat islets in the model group; D as the structure of rat islets in Schisandra chinensis lignin treated group. All the pancreata histopathology chang was examined by light microscopy(×200).

4 DISCUSSION

Development of traditional Chinese medicine with hypoglycemic effect has great significance for the treatment of diabetes. It is clear that lignin is the main hepatoprotective ingredient of Schisandra, but hypoglycemic effect of lignin is rarely reported [4]. Diabetes is a complicated disease caused by genetic and environmental factors. Diabetes is characterized by pancreatic β-cells dysfunction [5]. In the decompensated β-Cells, fasting blood glucose (FBG) was increased, further aggravating the pancreatic β-cell dysfunction. Uncontrolled pancreatic β-cell dysfunction leads to β-cell failure and cessation of insulin secretion results in diabetes [6,7]. In this study, serum FBG and FINS content in diabetic model group were significantly decreased, indicating severely impaired function of pancreatic β-cells. The impaired β-cells function might be closely associated with genetic and environmental factors, and was probably induced by oxidative stress. The declined antioxidant mechanisms elevated the oxidative stress and generated reactive oxygen species (ROS)[8]. SOD and CAT are one of the major antioxidant in the body, and play an important role in cleaning ROS. MDA is a lipid peroxidating by-product, and a representive of damage extent by ROS, especially the extent of lipid oxidation [9]. The results revealed that the MDA level was significantly increased but the SOD and CAT activity was decreased in serum of the diabetic rats, suggesting that increased ROS was due to enhanced oxidative stress. The diminished activity of expression associated withantioxidant enzymes such as SOD and GSH-Px in pancreatic β-Cells, increases the damage by oxidative stress. High levels of ROS may lead to large-scale non-specific oxidative damage including disruption of membrane integrity in pancreatic β-Cells leading to apoptosis[10]. Schisandra chinensis lignin could make the blood glucose of rats returned to normal, the serum insulin levels increased, the MDA content decreased, and the SOD and CAT activity enhanced. Pancreata pathology morphology analyses showed that the number of pancreatic B cells were increased, and necrosis had decreased. Therefore, the hypoglycemic mechanism of schisandra chinensis lignin may be reduce the degree of oxidative damage, increase the number of pancreatic B cells,restore islet cells function and promote insulin secretion. The underlying mechanism needs further study.

ACKNOWLEDGEMENTS

This work was funded by project "2013–199" supported by the Education Department of Jilin Province, project "201262503" supported by Scitech Department of Jilin City and project "20122082" the Health Department of Jilin Province.Authors to whom correspondence should be addressed; Peige Du: Tel:+86-432-64608278; Fax: +86-432-64608281; E-mail:dupeige 2001@126.com;
LipingAn: Tel: +86-432-64608278; Fax:+86-432-64608281; E-mail:dupeige2001@126.com.

REFERENCES

Panossian, A., Wikman, G.: Pharmacology of Schisandra chinensis Bail.: an overview of Russian research and uses in medicine. Journal of ethnopharmacology 118(2), 183–212 (2008).

Won, D.Y., Kim da, Yang, H.J., Park, S.: The lignan-rich fractions of Fructus Schisandrae improve insulin sensitivity via the PPAR-gamma pathways in in vitro and in vivo studies. Journal of ethnopharmacology135(2),455–462(2011).

AN, L.p., WANG, Y.p., LIU, X.m.: Effect of Fructus Schisandrae oil on blood glucose of type 2 diabetic rats induced by Streptozotocin. Chinese Traditional and Herbal Drugs 43(3), 52–556 (2012).

Yu HY, Chen ZY, Sun B,et al.Lignans from the Fruit of Schisandra glaucescens with Antioxidant and Neuroprotective Properties.J Nat Prod,2014,77(6): 1311–20.

Pu Y, Lee S, Samuels DC, et al. The effect of unhealthy β-cells on insulin secretion in pancreatic islets. BMC Med Genomics, 2013, 6 Suppl 3:S6.

Kim, W.H., Lee, J.W., Suh, Y.H., Hong, S.H., Choi, J.S., Lim, J.H., Song, J.H., Gao, B., Jung, M.H.: Exposure to chronic high glucose induces beta-cell apoptosis through decreased interaction of glucokinase with mitochondria: downregulation of glucokinase in pancreatic beta-cells. Diabetes 54(9), 2602–2611 (2005).

Bensellam, M., Laybutt, D.R., Jonas, J.C.: The molecular mechanisms of pancreatic beta-cell glucotoxicity: recent findings and future research directions. Molecular and cellular endocrinology 364(1–2), 1–27 (2012).

U.K. prospective diabetes study 16. Overview of 6 years' therapy of type II diabetes: a progressive disease. U.K. Prospective Diabetes Study Group. Diabetes 44(11), 1249–1258 (1995).

Weir, G.C., Laybutt, D.R., Kaneto, H., Bonner-Weir, S., Sharma, A.: Beta-cell adaptation and decompensation during the progression of diabetes. Diabetes 50 Suppl 1, S154–159 (2001).

Sheng Li Xue Bao. Progress in the role of oxidative stress in the pathogenesis of type 2 diabetes, 2013, 25;65(6):664–73.

Management, Information and Educational Engineering – Liu, Sung & Yao (Eds)
© 2015 Taylor & Francis Group, London, ISBN: 978-1-138-02728-2

The application of PCR-DGGE to study on the dominant bacteria of chilled mutton during storage time

Y.B. Zhou, W.S. Xu & Q.J. Ai
Beijing University of Agriculture, Beijing, China

D. Q. Zhang
Institute of Agro-products Processing Science And Technology CAAS, Beijing, China

ABSTRACT: This study was designed to explore the main bacterial flora of chilled mutton. Traditional microbial culture and polymerase chain reaction-denaturing gradient gel electrophoresis (PCR-DGGE) were used to examine the quality guarantee period of chilled mutton and its dominant bacteria during 4°C storage condition.The results showed that the total number of colonies increased with the extension of storage time at 4°C pallet storage conditions, and after 7 days, it reached 1.4×10^6CFU/g and exceeded the specified value. PCR-DGGE demonstrated that the dominant spoilage bacteria in chilled mutton were mainly *Pseudomonas*, *Brochothrix sp*, *Lactobacillus sp*, *Psychrobacter* and *Bacillus sp*. Among them, *Pseudomonas* and *Brochothrix* were the main advantage of chilled mutton spoilage bacteria.

KEYWORDS: Chilled mutton, Total number of colonies, Storage time, PCR-DGGE, Main flora.

1 INTRODUCTION

With the rapid development of economy, meat production in China increased rapidly. According to the National Statistics released data showed, in 2013, China's total output of meat reached 85,350,000 tons, up 1.7% over the previous year. The mutton, because of its nutrient rich, the output reached 4,080,000 tons, a growth of 2.0% [1-2]. However, the meat was readily contaminated for microorganisms because of its high water content and abundance of essential nutrients, the environment and artificial factors also led to quality loss in process of livestock slaughter and processing.

Chilled meat was meant that the slaughtered carcass was continued cooling treatment, not only the center temperature dropped to 0-4 °C within 24h, but also keep the temperature in subsequent processing, transportation and marketing process [3]. China's meat market mainly included temperature meat, chilled meat, frozen meat and meat products. Based on the various advantages of chilled meat, the state pointed out that in 2015 to make chilled meat ratio increased to 30% [4]. So, it was an inevitable trend in the development of chilled meat industry in China.

Most of traditional microbial detection was used to study the bacterial diversity and the main flora of chilled meat. But this method was time-consuming, and the majority of microorganisms could not be cultured by conventional methods, only 0.1-3% of the bacteria could be cultivated. So, this traditional method should be as an aid method to combine with other advanced methods so that it reflects the true information on microbial community structure objectively and comprehensively.

The PCR-DGGE technique, as a culture-independent method, was applicable to monitoring bacterial population dynamics. It was put forward by Fischer and Lerman in 1979 [5]. In 1993, Muyzer applied this technique for the first time to the study of microbial ecology [6]. As one of the methods of molecular biology research on microbial community structure, DGGE technique did not take training methods, but the direct extraction of total DNA from samples, to avoid the loss of those difficult to culture or not cultured microorganisms, and the detection rate of DGGE was faster, compared to the traditional culture methods, could reaction samples of microbial composition more quickly and directly [7]. In this study, PCR-DGGE was used to investigate the changes in the composition of the bacterial population of tray-packaged mutton during chill storage. Application of DGGE technology in the food microbiology research have been reported. As Jiang Yun [8] and Li Miaoyun [9] studied in the dominant spoilage bacteria from chilled pork by using PCR-DGGE technique. Cen Lujia [10], Jia Wenting [11] was also used to study the changes of the microorganism of meat in this method.

This paper made the total number of colonies as reference indexes to judge the storage time of chilled mutton in the condition of 4 °C.According to the samples of total bacterial DNA and used PCR-DGGE methods to study on the main flora of chilled mutton, so as to understand the microorganisms which cause the chilled mutton spoilage.

2 MATERIALS AND METHODS

2.1 *Sampling*

The fresh mutton thigh meat, were removed from mutton carcasses at 24h postmortem in a meat plant, was divided into several small pieces (approximately 10g). All of the sampling were tray-packaged and stored at 4 °C. Each group was analysed at 0,1,3d of storage.

2.2 *DNA extraction*

First of all, for each sample time,10 grams of meat was diluted in 90ml saline peptone water, pat 1min, 50ml of dilution was centrifuged for 5min at 4000×g, the supernatant was transferred to a 50 ml sterile centrifuge tube and a further centrifugation was carried out at 4000×g for 5min, 1ml supernatant was centrifuged for 3min at 12000×g, the sediment was used for further analysis.Then, the total bacterial DNA was extracted by TIANamp Bacteria DNA Kit (TIANGEH, China) according to the manufacturer's instructions, and suspended in 100ml of TE. DNA solution was estimated by 1.0% agarose gel electrophoresis.

2.3 *PCR reaction*

Primers 338f(5'-CGCCCGCCGCGCGCGGCGGG CGGGGCGGGGGCACGGGGGGACTCC TACGGGAGGCAGCAG-3') that contained GC clamp and 518r(5'-ATT ACC GCG GCT GCT GG-3') were used to amplify the V3 regions of the bacterial 16S rRNA gene. PCR reaction system (50μl)included 25μl of Taq polymerase, 1 μl of primers GC338f and 1μl 518r, 2μl of DNA dilution 21μl of UV-sterile water. The PCR program was used: 95°C for 5 min,and 35 cycles of 95°C for 30s, 56°C for 30s, and 72°C for 45s, and 72°C for 8 min final extension. The result was analyzed by 1.0% agarose gel electrophoresis. PCR products were stored at -20°C.

2.4 *DGGE analysis*

The PCR amplicons were analyzed with DGGE that using the D-Code system (Bio-Rad).Briefly, an 8% polyacrylamide gel (acrylamide: bis-acrylamide= 37.5:1) containing a denaturing gradient of 30–80% urea-formamide was electrophoresed at 200V for 7 h at 61°C.After electrophoresis, gels were stained for 30 min with nucleic acid dye(Andy Safe,America) and digitized with a UV transilluminator2000 (Bio-Rad) and using the Bio-Rad Quantity One software for image acquisition to save photos.

2.5 *DGGE band extraction and sequencing*

Bands were excised from the DGGE gel and incubated overnight in 40μl of ddH2O to allow the DNA to redissolve. 2μl of DNA was used for re-amplification with primers GC338f and 518r for a second round of DGGE. Then, purified bands were again excised and DNA was eluted as described above.Until the pure single bands were obtained,the purification steps were went off. Primers 338f and 518r without GC clamp were used to re-amplify the DNA. PCR products were quantified by electrophoresis on an agarose gel (1.0%) and store at 4°C. The results of sequencing was worked in Sangon Biotech Company. These sequences were identified with the Advanced BLAST similarity Search option in the GenBank DNA database.

3 RESULT

3.1 *Colony counting*

Table 1. Effects of different storage time on the total number of chilled fresh mutton surface colony.

Time(d)	0	1	3	5	7	9
colony counting CFU/g	1.7×10^3	3.1×10^3	1.6×10^4	1.3×10^5	1.4×10^6	3.9×10^7
Lg(CFU/g)	3.23	3.49	4.20	5.11	6.15	7.59

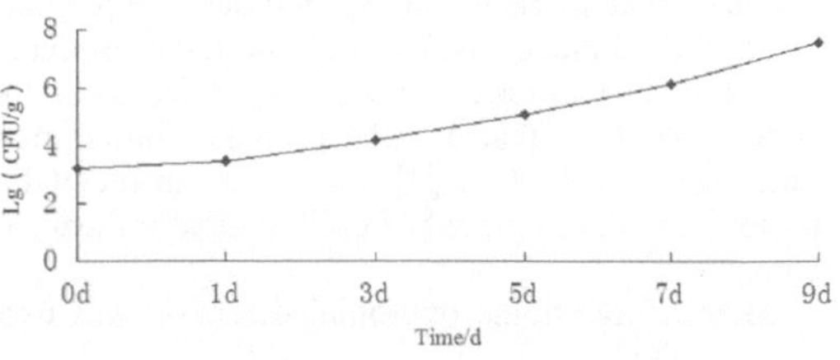

Figure 1. The change in the total number of chilled fresh meat surface colony under different storage time.

Table 1 and Figure 1 show that when the samples store for 7 days, the colony had reached 1.4×10^6 CFU/g. The general evaluation standard regulated that fresh meat was below 10^4 CFU/g, sub meat was 10^4 CFU/g~10^6 CFU/g, Metamorphic meat was 10^6 CFU/g. The increase of the total number of colonies was not obvious in the initial storage period, after three days, it begins to increase rapidly, when up to 7 days, the sample colony had reached 1.4×10^6 CFU/g, according to the evaluation criteria, the chilled mutton had been spoiled.

With the passage of time, the meat quality was declining, the color gradually deep and accompanied by the smell of meat, the water holding capacity also decreased, surface structure could not be restored after pressing. So according to the total number of colonies and sensory judgment shows that the chilled mutton could be preserved for 6 days under the condition of 4°C.

3.2 *DNA extraction*

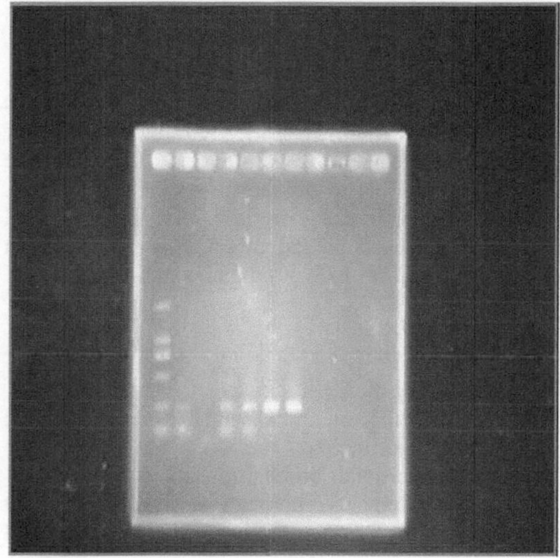

Figure 2. The electrophoresis map of amplification of bacterial 16S rDNA V3 zone.

It was determined that the chilled mutton in 4 °C storage conditions can be preserved about 6 days, according to the total number of colonies method. Therefore, extracting the bacterial total DNA as template directly which stored for 0, 1, 3,5,7,9d, choose the primer GC338f and 518r to amplify the sample,then the products was detected by 1% agarose gel electrophoresis, the results shown in Figure 2. Compared with the marker, we can see that its molecular weight was about 180bp, amplified bands and all the samples are lighter, so the next PCR amplification conditions DGGE experiment could be continued.

3.3 *PCR-DGGE result*

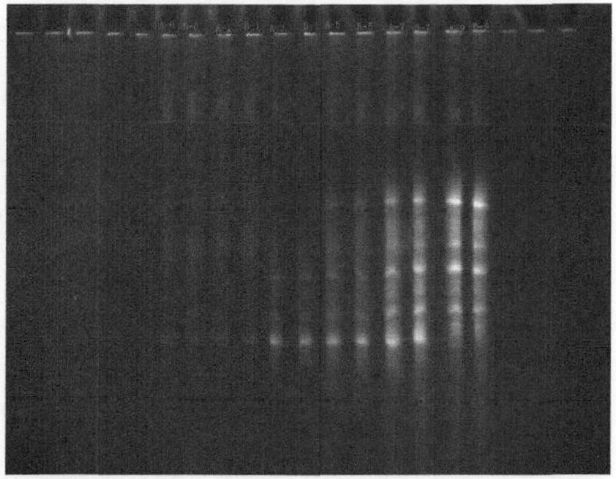

Figure 3. DGGE patterns of different storage period of bacterial DNA.lane A0, B0,stored for 0 days;lane A1, B1, stored for 1 days;lane A3, B3,stored for 3 days;lane A5, B5,stored for 5 days;lane A7, B7,stored for 7 days;lane A9, B9,stored for 9 days.

Figure 3 shows the complexity and variability of bacterial flora in chilled mutton.As we know,the DGGE spectrum, the same electrophoretic bands of different position represent different microbial species [12], the microbial diversity and abundant could be found by the DGGE method. DGGE patterns of bacterial DNA from the direct extraction of chilled mutton as shown in Figure 3, each sample was repeated 2 times. During the storage initial period,we could see some very dark band, it may be due to the slow reproduction rate of bacteria in low temperature storage conditions, the amount of bacteria was not obvious. But with the extension of storage time, strip increase obviously and become brighter. It can be speculated that, with the increase of storage time, some bacteria hold sway and then lead to the meat spoilage.

Selecting the bright band to recover, each sample was repeated DGGE purification by gel slices until the formation of a single strip, every single band was testing by 1% agarose gel electrophoresis, the results shown in Figure.4.The PCR amplified products was sent to the SANGON biological Company to acquire the sequence result, and then continue the retrieval and homology analysis in the Gene Bank database.

Using the single band as the template which was recovered from DGGE gel, through PCR amplification, part of them can be amplified successfully, and then used in the following experiments, But some of them did not amplify. Maybe in rubber cutting process, some sample loss, or DNA content was too low to cause part of the sample did not amplify successfully.

Table 2. Comparison of partial 16S rDNA sequencing fragments of dominating microbe similarity from DGGE bands.

Band No.	Closest relative(s)	Identity(%)	Accession No.
1	*Brochothrix sp.*	98%	gb|KC618438.1|
2	*Pseudomonas sp.*	99%	emb|HF546529.1|
3	*Rhodococcus sp.*	99%	gb|KF447662.1|
4	*Lactobacillus sp.*	99%	gb|AF157036.1| AF157036
5	*Bacillus sp.*	98%	gb|KF669527.1|
6	*Psychrobacter sp.*	96%	gb|GQ169112.1|

The PCR amplified sequences and the known sequencing results were compared with the Genbank database in order to understand their similarity. The sequencing results as shown in Table 2, during storage at 4 °C, The bacteria in chilled mutton mainly were *Pseudomonas* sp, *Brochothrix* sp, *Lactobacillus* sp, *Bacillus* sp, *Rhodococcus* sp and *Psychrobacter* sp. Among them, *Pseudomonas* sp and *Brochothrix* sp where the dominant bacteria from chilled mutton surface.

Pseudomonas was a common kind of aerobic spoilage bacteria in meat, it could grow quickly at refrigeration temperatures. When the temperature was the only or the main factor limiting conditions, the growth rate of *Pseudomonas* was faster than other bacterial 30% [13]. It was demonstrated to be one of the dominant spoilage flora in chilled mutton independent of packaging methods. This agrees with the study of Dainty and Mackey [14] who reported that Pseudomonas sp. accounted for up to 90% of the spoilage flora at refrigerated storage.

And *Brochothrix thermosphacta* was discovered to play an important role in meat and meat products. It can grow rapidly in meat food [15]. Through the sequencing results, *Brochothrix thermosphacta* had always existed during the storage time. This result was in accordance with earlier studies.

4 DISCUSSION

The experiment applied to traditional culture methods to observe the changes of the total number of colonies of chilled meat during storage at 4 °C conditions. The result showed that the total number of colonies in the early storage of change was not obvious, it may be due to low temperature conditions, and the fresh-keeping film oxygen effect makes the reproduction of microorganisms was not so fast. After 3 days, bacterial grew rapidly, and when storage 7days, the total number of colonies had exceeded the prescribed standard and the meat was rotten metamorphism.

In order to analyze the dominant bacteria of chilled mutton surface composition, the total DNA would be extracted directly from the chilled mutton surface, then the PCR amplification products were analyzed by DGGE electrophoresis, electrophoresis separation can be observed clearly, a different bright band will be cut, recycling and send test sequence. The results reflected the advantage bacteria in chilled mutton. They were mainly *Pseudomonas* and *Brochothrix thermosphacta.* At the same time, the application of PCR-DGGE technology also detected *lactobacillus*, *thermophilic bacteriastrain*, *Rhodococcus sp* and *Bacillus.*

In the course of the experiment, the experimental method is feasible, but the process operation should reduce the unnecessary error. Firstly, DGGE gel conditions need to master, such as in the experiment the comb part of the glue is not always coagulation. Secondly, the conditions of running gel time required to observe again. Thirdly, in the gel extraction process, because of some DNA content is not high, resulting in a band is not obvious, there will be loss of sample inevitably.Finally, the concentration of DNA in the recovered is too little, it will result in the problem of unable to obtain the sequencing results. Therefore, during the experimental process should control the environmental effect of temperature on gel process and try to avoid the loss of sample.

5 CONCLUSIONS

Through the experiment, PCR-DGGE and traditional microbiological methods was useful to analyze the changes of microbial composition and to provide real-time information about the state of chilled mutton, it would provide a rapid analysis method for microbial content in the food and reflect the kinds of microorganism samples directly. The predominant bacterium were *Pseudomonas sp.*, *Brochothrix sp*, *Lactobacillus sp.*, *Psychrobacter, Rhodococcus sp* and *Bacillus sp* under aerobic conditions.

ACKNOWLEDGEMENTS

This research was funded by the public sector (Agriculture) special funds for scientific research projects (201303083). The authors thank the professor of the microbiology Lab, College of Food Science and Engineering, Beijing University of Agriculture, for their technical assistance.

REFERENCES

[1] Zheng, C.L.2003.The Nutritional Value of Mutton and the Influencing Factors to Mutton's Quality[J].Meat Research,Vol.p.47.

[2] Information on http://www.stats.gov.cn/.

[3] Yao,D.Yu,C.Q.2007.Research Progress on Preservative Method of Chilled Meat,Academic Periodical of Farm Products Processing[J],Vol. 6 p.9.

[4] Deng, F.J.2010.The Research Report of Meat Industry Development Strategy During the Twelfth Five[J]. Meat Research,Vol. 8,p.3.

[5] Fischer S.G.,Lerman L.S.1983.DNA fragments differing by single base-pair substitutions are separated in denaturing gradient gels:correspondence with melting theory[J].Proceedings of the National Academy of Sciences.Vol. 80, p.1579.

[6] Muyzer G.1999.DGGE/TGGE a method for identifying genes from natural ecosystems[J].Current Opinion in Microbiology.Vol.2,p.317.

[7] Muyzer G,SMALLA K.1998.Application of denaturing gradient gel electrophoresis (DGGE) and temperature gradient gel electrophoresis (TGGE)in microbial ecology[J]. Antonie Van Leeuwenhoek International Journal of General and Molecular Microbiology. Vol.73, p.127.

[8] Jiang,Y.Gao,F.Xu,X.L.2011.Changes in the Composition of the Bacterial Flora on Tray-Packaged Pork during Chilled Storage Analyzed by PCR-DGGE and Real-Time PCR[J].Journal of Food Science.Vol. 76,p.27.

[9] Li, M.Y. Zhou,G.H.Xu, X.L.2006.Changes of bacterial diversity and main flora in chilled pork during storage using PCR-DGGE[J].Food Microbiology. Vol.23,p.607.

[10] Cen,L.J.TANG,S.H,HAO,X.Q.2012.PCR-DGGE Analysis of Dominant Bacteria in YakMeat during Chilled Storage[J].Meat Research,Vol.26, p.36.

[11] Jia,W.T.Jiang,C.H.Li,K.X.2013.Study on the dynamic changes of microorganisms in 2 different processing of slaughter lamb during storage by PCRDGGE[J]. Science and Technology of Food Industry.Vol. 34,p.73.

[12] Ercolini G.2004.PCR-DGGE fingerprinting:novel strategies for detection of microbes in food review[J].Journal of Microbiological Methods.Vol. 56,p.297.

[13] Nychas G.J,Skandamis P.N, Tassou C.C,Koutsoumanis K.P.2008.Meat spoilage during distribution[J]. Meat Science.Vol. 78, p.77.

[14] Dainty,R.H.Mackey,B.M.1992.The relationship between the phenotypic properties of bacteria from chill-stored meat and spoilage processes[J].Journal of Applied Bacteriology,Vol. 73,p.103.

[15] Gill,C.O.Newton,K.G.1977.The development of aerobic spoilage on meat stored at chill temperatures[J]. Journal of Applied Bacteriology.Vol. 43, p.189.

Management, Information and Educational Engineering – Liu, Sung & Yao (Eds)
 ISBN: 978-1-138-02728-2

Protective mode for Sichuan traditional bamboo weaving craft

Y.J. Luo & W.Y. Zhang
Fashion Institute of Sichuan Normal University, Chengdu, China

ABSTRACT: Sichuan traditional bamboo weaving craft is one important part of our traditional non-material heritage, thus it is of great importance to inherit and develop the bamboo weaving craft. The Sichuan bamboo weaving craft has a rather long history, unique features and marvelous market potential, and its deficiency appearing in the process of its development should be guided and corrected with scientific methods. This paper studies three aspects, namely the development situation of Sichuan traditional bamboo weaving, its status analysis and protective mode, so as to provide theoretical basis and enlightenment for the development of traditional bamboo weaving craft.

KEYWORDS: Bamboo weaving; Inheritance; Protect.

1 INTRODUCTION

Bamboo weaving craft is one important ancient handcraft in our country, and its inheritance and development is of vital value for the carrying forward of Chinese culture. Nowadays, new technology springs out continuously. Handcrafts like bamboo weaving gradually lose their original position. Sichuan is an important region for the rise and development of bamboo weaving. Therefore the core of this paper researches the historical development of bamboo, using modern technology to better protect traditional bamboo craft, inspiriting new vitality.

2 DEVELOPMENT STATUS OF SICHUAN TRADITIONAL BAMBOO WEAVING CRAFT

2.1 *Qingshen bamboo weaving*

Qingshen bamboo weaving has a really long development history, and plays a critical status in the bamboo history of our country. Qingshen county is located in the southwest of our country, and is next to Leshan, Jiajiang county, Renshou county and Jingyan county. The geographic position is unique and its climate belongs to the subtropical monsoon climate. This climate brings warm weather, moist air, abundant rain and provides natural conditions for the growth of bamboos.

The bamboo with the largest yield is "Sinocalamus affinis", and a wide variety of different types, including more than ten different species. The most representative bamboos like "Sinocalamus affinis", "Baifen bamboo" and "Qingmian Sinocalamus have few bamboo joints and long tubes, one of the longest could be more than one meter. Besides, the fiber of these bamboos is rather tough, and it can be split into very thin and fine bamboo wire. Doubtless, these bamboos can be used as the best material to make rather fine bamboo handiwork of top grade.

As early as 2000 years ago, the development of bamboo craft in Qingshen county has started. At the beginning, people use bamboo weaving to management of water conservancy. Later, bamboos were used to weave various common household life appliances.

In ancient times, Qingshen bamboo weaving has reached prime time for several times. Fine HE "mandarin fan" was listed into tributes for the royal family, and now it is restored in the national palace museum of Shenyang city. And modern Qingshen bamboo weaving has also some fame in the country. The chance was in the early 30s, when the development of sericulture produced a great demand for bamboo weaving products, which was therefore a great motivation for the bamboo weaving industry. At that time, Qingshen bamboo weaving has realized mass production, and its bamboo weaving products were sold at home and abroad. Thus, bamboo weaving industry developed greatly, and it became a famous bamboo weaving brand in the country.

With the continuous development of technology, Qingshen bamboo weaving craft is no longer restricted to the making of articles of daily use, but gradually becomes a kind of bamboo weaving culture possessed only by Sichuan people. Qingshen bamboo weaving craft is endowed with new historical meaning. Since Qingshen bamboo weaving has been impacted by external culture and industrialization, and mass production also poses new challenges for handicraft, its production efficiency and aesthetic taste have to change with the influence of modern culture. Under the circumstances, many local handicrafts-men begin to reflect on how to unite bamboo weaving craft with modern art. They are determined to change the current situation of traditional bamboo weaving, and hope bamboo weaving craft would get a rocketing development, combining practicability with artistic aesthetics, so that more excellent Qingshen bamboo weaving handicrafts would be designed and created. Various new weaving ways are created, and bring new opportunities for Qingshen bamboo weaving art. The bamboo weaving, handicrafts which were originally used in farm work become some real artistic works.

Local bamboo weaving artists started factories of bamboo weaving crafts. New productions and new designs of bamboo weaving crafts spring out with surprising creativity. Recently, bamboo weaving industry constantly expands, many local bamboo weaving enterprises have their products exported to foreign countries, such as Japan, Korea, Russia, and western countries. Thus doubtless the industry has become one leading group in the local agricultural economy.

2.2 *"LIU bamboo weaving craft"of Qu county*

"LIU bamboo weaving craft"of Qu county has a history of almost a thousand years long, thus occupies a rather crucial position in Sichuan bamboo weaving culture. The prosperous development of "LIU bamboo weaving craft" of Qu county depends on the natural resources richly endowed by nature. Qu county has abundant rain, mild climate, and prosperous sinocalamus affinis. This kind of bamboo has many good features for bamboo weaving, therefore almost every family grows this kind of bamboo and do bamboo weaving works which promotes the development of the local economy. Thus gradually "LIU bamboo weaving craft"of the Qu county of Sichuan features has come into beings.

In 2003, Ministry of Culture officially named Qu county of Sichuan "hometown of Chinese bamboo weaving art", which has a great significance to the bamboo weaving culture of Qu county. With the renovation and innovation of many years, now the county has many various bamboo weaving products. Well-known common bamboo weaving life items are involved, but more importantly, create artistic products with modern meanings like fans, baskets, basins, hangings, flower receptacles, colored drawings can also be made from bamboo weaving works. "LIU bamboo weaving craft"of Qu county does not only develop in the aspect of its variety, but also improves its weaving methods, adding ways like wearing, screwing, inserting, locking, covering, winding and so on. Thus the bamboo weaving technique is diversified. In 1972, the fourth generation of "LIU bamboo weaving" Liu Jiafeng created Jacquard weaving method which is very representative.

Nowadays, the main products of "LIU bamboo weaving" are: bamboo weaving screen, double grained bamboo weaving, jacquard porcelain body bamboo weaving, bamboo weaving artistic works and so on. Among the jacquard porcelain body bamboo weaving becomes more artistic with the renovation. Graphic weaving and stereoscopic weaving are better combined, and new and novel topics are created like dragon and phoenix hoping for prosperity, dragon plum bottle, national treasure, family happiness and so on

"LIU bamboo weaving craft"of Qu county basically only passes down to male but not female. And later on it evolved into that the craft can only be passed down from teachers to apprentices. So the essence of bamboo culture has been passed down from generation to generation. Because only few people could get to know the core technique, it is not conducive to spread widely. From 1980, classes for training the technique of bamboo weaving has made it possible for more people from the society to have a thorough to learning.

3 PROBLEMS EXISTED IN THE PROTECTION OF SICHUAN TRADITIONAL BAMBOO WEAVING CRAFTS

3.1 *The market of traditional handicrafts is gradually disappearing with the improving of economy and life*

Bamboo weaving industry develops amazingly in the country, Chinese bamboo weaving industry is launching into a period of accelerating development. Enterprises transform the art, combining modern technology with bamboo weaving techniques. However, more and more traditional technique is disappearing from the market, and the products from industrious production remain rough-wrought. There are advantages and disadvantages in mass production, on one hand, bamboo weaving crafts mass production could realize the maximum production and minimum cost; on the other hand, current mass production is not good for the inheritance and development of traditional craft. The reason is that there is no much technology involved, and rough-wrought products are the main products. Thus, from aesthetics, the comprehensive utilization rate of quality, no much needs to be highlighted.

3.2 *Traditional bamboo weaving craft is facing the risk of no successors*

Now people who grasp this traditional craft are mainly old, lower educational handicrafts-men. Most local young generations prefer to make a living in more developed cities rather than learn this traditional craft. The production of traditional craft is based on the minimum unit of one family. One generation passes the central knowledge to another generation. With the development of modern society, this way of inheritance can no more satisfy the need of the society. Thus the market is narrowing gradually, and bamboo weaving craft is losing its important status in daily life. Therefore, the current situation of traditional bamboo weaving craft is worrisome, and poses much more serious challenge to the inheritance of the craft.

3.3 *The products do not satisfy the fashion need of the customers.*

Traditional bamboo weaving mainly produces daily items, like pack basket, dustpan, basket, fan, table, chair and so on. But in the society, these kinds of items have no longer been used by many customers. Most products have really low expressive force, creativity, additional value and sense of design. Within the aesthetic standard of modern consumers, no much competition is exhibited by these artistic works.

4 THE PROTECTIVE MODE OF SICHUAN TRADITIONAL BAMBOO WEAVING TECHNIQUE

4.1 *Establish study institutes*

The inheritance way of Sichuan traditional bamboo weaving industry is simple, basically from teachers to apprentices. To improve the method of inheritance, many study institutes should be established, and the techniques usually passed down in an oral way could be written down into theories. As the media of study, propagation, promotion, study institutes make it possible for more people to understand the culture, and deepen the exploration and promotion of bamboo weaving techniques. With activities like exchange meetings, exhibitions and various media, people could know more about the culture.

4.2 *Design creativity*

The design of traditional bamboo weaving is single, and weaving method is conventional. The most common methods are herringbone and cross. Now there are many new methods created for bamboo weaving methods. Yet, the efforts should be continually made to increase its additional value and artistic aesthetics.

4.3 *Develop bamboo weaving craft with new techniques and methods*

Bamboo weaving with full color has been a hard problem for the craft. Products of this craft have been two colors, black and white for centuries. It is even harder to make the perfect color transition. But now the problem has been solved by bamboo weaving artists in Qingshen county, with computer-aided design program bamboo weaving ways to use new tools to improve and innovate.

4.4 *Broaden propaganda channels*

Using modern information technology tools to promote brand and make more people know the culture. Online marketing network could be set up, and personalized customization could be increased. New type of electronic commerce and micro-letter sales could be applied to expand markets. The craft should have precise orientation, and should be combined with fashionable daily items, home furnishings, high-end gifts and so on. With the guidance of modern aesthetic culture, the uniqueness and culture sense of traditional handicrafts should be stressed, and the fame and market of Sichuan bamboo weaving products should be improved. In the meantime, overseas market should be expanded, and as a non-material cultural heritage, it should become well-renowned in the world.

4.5 *Seek the development of tourism*

Sichuan is one large tourist province in west China, with famous scenic spots like Mount. Emei, Jiuzhaigou Valley, Aba Ganzi autonomous regions and the Bamboo Sea in southern Sichuan province. These places are famous for their unique regional landscapes. Sichuan bamboo weaving craft should not only improve its weaving methods, but also is supposed to combine with tourism. The development of tourism will promote development of Sichuan bamboo weaving industry, Get huge economic returns in the simultaneous development of cultural heritage, and achieve a win-win for cultural protection and economic development.

ACKNOWLEDGMENT

The corresponding author of this paper is Wanyu Zhang. This paper is supported by the 10th Student's technology innovation project of Sichuan Normal University.

REFERENCES

[1] Jiang M H, Li K N. The inheritance and development of Qingshen bamboo art. FEITIAN, 2011,(16).

[2] Fei Y M. The Research of Qingshen'S Bamboo—weaving Art [D]. Soochow University, 2008.

[3] Qing S D, Du Z H. Protection and Development of "Liu's Bamboo Utensils" in Quxian County. Journal of Sichuan University of Arts and Science.2013, 23(2).

[4] Chen X H, Chen X L. Heritance of Bamboo Culture and Modern Package Design. PACKAGING ENGINEERING. Vol28.NO.8 2007(08).

Management, Information and Educational Engineering – Liu, Sung & Yao (Eds)
© 2015 Taylor & Francis Group, London, ISBN: 978-1-138-02728-2

Research on the system structure and cultivation of ecological personality

Jing Wei
School of Management, Hefei University of Technology, Hefei, China

Rong Wei
School of Marxism, Hefei University of Technology, Hefei, China

ABSTRACT: Ecological personality is the psychological quality of an individual in understanding, experiencing, and processing of the relationship between man and nature, which is featured by uniqueness and consistency in terms of thoughts, feelings and behavior patterns. This paper uses systems research methods to resolve the structural elements of the ecological personality, and then summarizes the complex characteristics of ecological personality. Based on theoretical analysis of social subjects' ecological personality, as well as practice requirements of social subjects' wisdom, emotion and behavioral intentions in coordinating relationships between human and nature, the research concludes that sublimating Cognitive-Affective-Volitional processes consisting of promoting different mental states from cognition to identification, from awe to enjoyment, and from internalization to externalization, are the basic ideas of optimizing subjects' ecological personality.

KEYWORDS: Ecological Personality; System Structure; complex characteristics; Sublimation of Cognitive-Affective-Volitional.

1 INTRODUCTION

Development of human civilization has experienced several stages: a passive attachment to nature, transforming and conquering nature, and living in harmony with nature. The improvement of science and technology has been accompanied by excessive exploitation of natural resources, destruction of the natural ecology and other negative issues. Protecting the home planet, constructing the ecological civilization are the huge responsibility shared by all human beings, and the subjects undertaking the task should possess Ecological Personality. Ecological Personality is a comprehensive psychological state shown in the process of individual understanding, experiencing and dealing with the relationship between man and nature. From the perspective of psychology, personality is psychological qualities which make people remain consistent in different situations and in different periods, which is a continuously changing process. Under the modeling of individual's inner needs and cognition, as well as external pressure of the social environment, personality shows a series of stable and unified thoughts, feelings and behavior patterns distinct from others. Social individuals' overall personality tends to be a synthesis of a plurality of individual characteristics. Ecological Personality is one of the whole pictures to show the personality dimension. Ecological Personality is an organic system which contains a certain ecological structure and function and acts as an intermediary of subject's existence. It has a significant predictability to the subject's level of ecological awareness and practice effect.

2 SYSTEM CONSTITUTION OF ECOLOGICAL PERSONALITY

Analyzing from the viewpoint of System Science, human psychological phenomenon entails both structure and function. American psychologist Izzard's studies suggest that personality has six relatively independent and interacting subsystems, which are: homeostasis, internal driving force, emotion, perception, cognition and action systems (Stallman. 2006).

This viewpoint integrated the individual mental processes and personality psychology into the personality structure research, highlighting that the individual personality differentiation is built on the basis of commonality psychological process, and embodying modern research trends of taking both factor analysis and overall comprehensive study into consideration. Inspired by ideas like Izzard's, using subject's internal psychological activities and its external behavior concord development as research clues can logically decompose an ecological personality system with elements of physiology, cognition, motivation, emotion, and behavior, etc. The elements themselves and the process of working are affected by external conditions. (See Figure 1)

Physiological ingredients are organic basis and material carriers of subject's ecological personality development, because certain cortical and subcortical neural structures control and guide the individual ability and reaction in perceiving nature. Through comprehensive research, Professor Richard Depi from Cornell University found that: there are three neurobiological systems associated with personality, which are behavior contribute system, control system and select and distinguish system. They play different roles in driving goal-oriented behavior, controlling information flow in the brain, as well as making decisions, coordinating relationships and stopping activities, etc. (L.W, Peng. 2011). Subjects' different psychological and behavioral profiles formed by the interaction of these three neurobiological systems correspond to a large number of theoretical results of personality traits types, which have a strong and convincing explanation on neurobiological effects that produce personality differences. The biological activity of the ecological personality system inputs material energy for the function of a system, while protecting the system's dynamic development.

Cognition, motivation and emotion are internal psychological components of the development of the ecological personality system, intermediating the generation of individual ecological behavior. The cognitive function of ecological personality is embodied in "Seeking the truth", that is, the subject has the wisdom to maintain ecological balance, to master the scientific method of rational exploitation of the nature. Motivations are those implicit or explicit tendencies that could drive us to pursue a series of specific goals (Robert. E. Franken. 2005). The motivation function of ecological personality is embodied in "Choosing the good", namely the subject extends the horizon of moral solicitude from mankind to the whole nature, optimizes the moral quality of the subject, and enhances the life realm of self-improvement. The emotion function of ecological personality is embodied in "appreciating the beautiful", that is the subject experiences natural beauty in a positive and optimistic state of mind, hence enhances their own aesthetic consciousness and ability, and reaches a perfect fusion of subjective and objective aesthetic.

Social subjects express their responsibility of caring for the environment with all the different behavior styles under the influence of their internal psychological mechanisms of "Seeking the truth", "Choosing the good" and "appreciating the beautiful". Subject's practice behaviors are the results and the externalization of the internal elements of their ecological personality system, and also the ultimate symbol to check whether the individual has internalized the concept of harmonious coexistence between human and nature into the value orientation to guide their own behavior.

In short, the internal components of subjects' ecological personality system structure interact with each other. Wherein, components of emotion and cognition co-occur and reinforce each other. Emotion can launch, interfere, organize or destroy the development process and behavior of ecological cognition. While assessment of things by the ecological cognitive components can start, transfer or change emotional reactions and experiences. Social individuals'dependence on nature is too strong, or their motivation to change nature is too strong or too weak will all lead to a negative effect on people's cognitive effects, while moderate motivation will lead to the best. Healthy and optimistic emotion derived from a closeness to nature itself is a positive force, which is able to drive the subject to bear the responsibility of ecological civilization construction and practicing the ecological behavior. It can not only enhance the dynamic behavior of the subject, but also activate some "blind spot" in their cognition. Internal components and external representations of the Ecological Personality system together constitute unique personality of individuals.

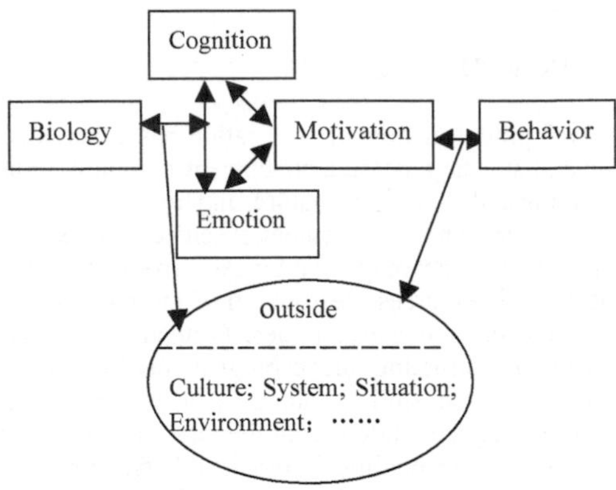

Figure 1. Ecological personality system structural diagram.

3 SYSTEM CHARACTERISTICS OF ECOLOGICAL PERSONALITY

As an organic system with dynamic development, ecological personality shows multidirectional system characteristics.

Firstly, integration and uniqueness. Ecological Personality characteristics are neither results of a simple sum up of the various components or elements nor patchwork, but the results of the integrated development of the biological and social, subjectivity and objectivity, the internal and external. Psychologist

Carl Jung, particularly emphasized the unity and integrity of personality, who believed that individual developed the diversity, coherence and harmony on the basis of complete personality, and if the components of the system are segmented, or in conflict with each other, it is easy to form a distorted split personality as a result (Xue. Zheng. 2007). The coordinated functioning process of Ecological Personality systems reflects more integration features of the system, while the externalized psychological behavior patterns of ecological personality system's operating mechanism reflect more unique characteristics of the Ecological Personality System, demonstrating the unique style of different individuals.

Secondly, interaction and complexity. Interactivity and complexity characteristics of the subject's Ecological Personality System are mainly reflected in the dynamic relationship between the structure and the constituent elements of the system. Ecological personality is shaped by the interaction of nature and nurture. Components inside the Ecological Personality System can never operate independently, for there is a strong coupling action among the function of each component, so that subject's Ecological Personality System exhibits a higher level of flexibility and coordination. Interactive features of the Ecological Personality System reflect interwoven interaction of each component inside the system as well as the interaction between the two levels inside and outside the system. The connotation of subject's Ecological Personality components is complex and diverse, and the combination of the components is interwoven, thereby forming a multi-level system transmission path, showing multi-dimensional ecological psychology and ecological action of the subjects.

Thirdly, evolutionarity and stability. People were not born to be a "whiteboard", and the tendency of genetic temperament which has a biological attribute restricts the direction of individual's personality development to some extent, making individual exhibit common personality characteristics in different situations rather than a moment-to-moment changes, reflecting the stability of personality system development. The ecological Personality system uses self-regulatory function to maintain or restore the structure and function of the system, so as to ensure the stable development of the system. Evolutionarity is the essential attribute of the Ecological Personality System, which is in constant evolving and changing process under the role of many elements. Although each person is a non-renewable gene pattern vector, resulting in complex and diverse characteristics of people, due to the fact that human psychology is always changing and developing, and even adults' personality can change dramatically under internal and external influences, personality mutations may occur under extreme external environment intervention. Therefore, the Ecological Personality System has integrated features of both evolutionary development and relatively steady performance.

Moreover, regularity and nonlinearness. The functioning of the Ecological Personality System has its regularity: development of the system is in harmony with the external environment. Namely, individuals' cognitive, emotional and behavioral patterns match the social and cultural environment, the institutions, and the economic status. Internal components of eco-system are interdependent but the development of the components may not be balanced, each component separately exhibiting certain advantages and disadvantages. Although system components cannot replace each other, their functions can compensate for each other to protect co-evolution of the various components in the Ecological Personality System and to achieve dynamic operation of the system.

Nonlinear characteristics of the development of the subject's Ecological Personality System, on the one hand, reflects the non-independent relationship of the interaction between the various components and elements of the system. That is to say, the interaction between various components is not proportional or linear, therefore, no symmetrical action-reaction relationship is expected. On the other hand, nonlinearness is reflected in the non-linear relationship between the subject's Ecological Personality System and intermediate variables which affect overall system development. To be more specific, a series of internal and external factors of the system, such as biological, environmental, cultural, institutional factors fail to predict proportional change in the performance of the subject's personality. Demonstration of the power of an individual element, or the non-coherent development of certain elements is the display of the nonlinear feature.

4 CULTIVATION OF ECOLOGICAL PERSONALITY

Significance of cultivating Ecological Personality is that it can lead to stable performance and optimal quality of ecological personality. Social subjects' wisdom, emotion and behavioral intentions of coordinating relationships between human and nature, and the sublimating Cognitive-Affective-Volitional process consisting of promoting different mental states from cognition to identification, from awe to enjoyment, and from internalization to externalization, are the basic ideas of optimizing subjects' Ecological Personality.

From cognition to identification is the conscious approach to cultivate ecological personality. A profound understanding of the relationship between human and nature is the prerequisite for the formation of the ecological personality. Enhancing social subjects' awareness of knowledge on the natural ecosystems as well as the relationship between human and nature through school education, social media and other ways is a valid cultivation method. It is also important to make the subjects comprehend the serious consequences of human-nature conflicts, hence to improve their awareness of the importance of ecological protection, out of strong internal driving force. There from subjects' environmental awareness and consciousness can be improved, and they will reflect on the sustainable development of human society, thus forming the values of respecting nature and complying with nature (L.W, Peng. 2011).

From awe to enjoyment is the emotional way to cultivate ecological personality. Eco-emotional experience reflects the relationship between the natural world and human needs, which is built on the basis of people's understanding of the nature. Nature supports human existence. Using both positive and negative facts in the history of human development can facilitate the cultivation people's sense of awe for nature, so as to curb people's undesirable behaviors of destroying the environment. Only in awe will there be moderated, and human desire can be accordingly regulated. Sense of awe is a low-level emotional experience of the harmonious relationship between human and nature. The refreshing feeling people experience in the process of enjoying the nature is the highest level of emotional experience which embodies the harmonious relationship between human and nature. Delight of enjoying the nature enables people to absorb spiritual resources from the natural experience, to alleviate the psychological conflict and stress. Experiencing the nature is the best way to cultivate the sense of enjoyment and delight.

From internalization to externalization is the behavioral way to cultivate ecological personality. Internalization is the process in which the social subjects transfer the experiences of learning and experiencing the harmony between human and nature of their internal psychology, then form a new psychological structure. Through education and practical activities as well as environmental influence, the social subjects identify with, pick out and accept the views and norms of harmonious coexistence between man and nature, and incorporate them into their own ideology and morality construction, thereby transform them into their own individual consciousness and ideological faith. As long as people consciously act on their internalized ideological concepts and value criteria, and form positive behavioral patterns, can the true purpose and effect of cultivating the ecological personality be achieved.

ACKNOWLEDGEMENTS

Thanks to China National Philosophy and Social Sciences Youth Fund (11CSH041) for the financial support; thanks also go to my partner, professor Rong Wei, without her effort, this thesis could not be accomplished. She is also the corresponding author of the paper. E-mail: tian_an2001@163.com.

REFERENCES

L.W, Peng. 2011. Ways to model Ecological Personality. *Journal of Jishou University (Social Sciences Edition)*.

Stallman. 2006. *Emotional Psychology*. Beijing: Light Industry Press of China.

Robert. E. Franken. 2005. *Human Motivation*. Xian: Shanxi Normal University Press.

Xue. Zheng. 2007. *Personality Psychology*. Guangzhou: Jinan University Press.

Management, Information and Educational Engineering – Liu, Sung & Yao (Eds)
© 2015 Taylor & Francis Group, London, ISBN: 978-1-138-02728-2

The analysis to invalid handling of social insurance agreement

Xian Bin Wang
College of Engineering, Chengdu University of Technology, Leshan City, Sichuan Province, China

ABSTRACT: Social insurance is a part of the social security system. Therefore, it is a mandatory legal obligation for both enterprises and employees to pay social insurance premiums. A social insurance agreement with the purpose of evading social security obligations between an enterprise and an employee is invalid.

The paper points out that the proper invalid agreement handling methods by the judicial authority benefits the enforcement of the social insurance system, by comparative analysis between law-abiding cost and law-breaking cost for enterprises.

KEYWORDS: Social insurance agreement, Law-abiding cost, Law-breaking cost.

1 INTRODUCTION

In March, 2011, Party B (surnamed Qiu) signed a labor contract with party A(a limited company), article 6 of which stipulates that "Party A pays Party B social insurance premiums in cash along with other payments as wages, owing to Party B's mobility." A pay sheet provided by Party A shows that wages of Party B consist of basic wage, five social insurance premiums (namely, old-age pension, unemployment insurance, health insurance, industrial injury insurance and maternity insurance) performance wage and overtime wage. In February 2013, Qiu submitted an arbitration application to the labor dispute arbitration committee, asserting that his company didn't pay for his social insurance premiums and demanding termination of the labor contract and compensation from the company. The labor dispute arbitration committee didn't support Qiu's compensation claim. So Qiu sued to the court.

The court said, according to The Article 72 of Labor Law, which says that "employing unit and laborers must participate in social insurance and pay social insurance premiums in accordance with the law", paying social insurance premiums is a legal obligation for both employing unit and laborers. In this case, the company shall not take the full liability as the defendant's failing in paying social insurance premiums for the complaint wasn't caused by its unilateral mistake. Therefore, the court didn't support Qiu's compensation claim, either.

The agreements on social insurance prescribed in the contract between the complaint (Qiu) and the defendants(the limited company) in this case are equivalent with the social insurance agreement referred by this paper. And the definition and handling ways of the judicial authority in terms of the force of the agreement are the research problems of this paper.

2 THE NATURE OF THE SOCIAL INSURANCE AGREEMENT

A social insurance agreement is an agreement of an enterprise and an employee on whether or not the social insurance premiums are paid. Such an agreement may appear in many different forms, for example, in the form of a term of a labor contract, or of a letter of commitment for laborers, etc. To cut down operating costs, enterprises sometimes reach agreements with laborers on not paying for their social insurance. The social insurance agreement in this paper refers to an agreement within which both parties agree on not paying for social insurance. Such kind of agreement runs counter to relevant legal provisions. Therefore, it is not difficult to understand the above mentioned statement of the judgment. The definition of the force of the agreement is closely linked with the nature of social insurance. And the handling way of it has a tremendous impact on the implementation of the social insurance system.

2.1 *The state mandatory feature and social security function of social insurance*

Since its born in Germany in 1880s, the social insurance system has been recognized throughout the world and become a part of social security system of various countries. At present, social insurance is recognized by all countries as a mandatory social system implemented by the state. The liability subject of social insurance is the state or the society. The insurance aims at providing material aids for laborers in some inevitable cases such as oldness, illness, injury and disability, unemployment, death and so on. When any of the above risks occurs, the laborer should obtain material compensation in accordance with the law. The social insurance system is not only a state mandatory one, but also a system with social security functions. Therefore, the system shall not be excluded by any private agreement.

China's labor law stipulates that "employing unit and laborers must participate in social insurance." Therefore, social insurance is a mandatory system implemented by the state in accordance with the law. The social functions of the social insurance system include: providing compensation for the loss of income, stabilizing social order, promoting economic growth and regulating fair distribution. Thus, such a system performs as a tool maintaining social security, and plays an important role in promoting harmonious development of both the society and the state.

2.2 *Something that should be clarified as invalid-agreements for the purpose of evading social insurance payment obligations*

A labor contract is an agreement by which a labor relationship between an enterprise and an employee is established and the rights and obligations of both parties are specified. As something with legal effects, contents of such agreements should define the rights and obligations of both parties according to legally required liberty clauses or default clauses. During the fulfillment of a labor contract, a laborer is attached to a company and deprived of his/her independence to a large extent, due to the nature of subordination of the contract. Therefore, to protect the interests of laborers, laws, including the labor law and labor contract law have specified mandatory terms. Agreements based on the exclusion of such mandatory terms prescribed by relevant parties have no legal effect.

The payment of social insurance is a legal obligation for both enterprises and laborers. Enterprises must go through the formalities of social insurance for employees to meet the requirements with the demands of perfecting and implementing the state's social security system. This is about safeguarding not only the individual interests of employees, but also the overall interests of the state and the society. Therefore, some agreements on social insurance infringe the interests of the whole society and the state. It is both a right and responsibility for the Ministry of Labor and Social Security to deal with the collection and payment of social insurance premiums from enterprises and laborers. Since the relationship between the collections and payments of social insurance premiums doesn't belong to the private law relation, neither enterprises nor employees have the right to make any agreement with the exclusion of it. According to the "mandatory provisions on activities contrary to law and administrative regulations" stipulated by Paragraph 3, Article 3 of the labor law, "agreements with social insurance between a laborer and an enterprise should be recognized as invalid. However, in the above mentioned case involving Qiu, the court didn't express any opinion on the validity of the agreement between Qiu and his company. But in the case involving the labor dispute between Li Yanli and Beijing Xiyuan Hotel, the Beijing First Intermediate People's Court pointed out that it is a mandatory legal obligation for an employer to pay social insurance premiums for its employees. The obligation shall never be changed out of the will of the enterprise or laborer. Therefore, the court demanded that the agreement was invalid.

2.3 *Handling methods after an agreement is claimed as invalid*

In terms of handling methods after an agreement is claimed as invalid, in the judicial practice, there is a view to agreeing on the handling method by the court in the case involved Qiu. According to the view, the laborer's signing the agreement was a mistake. So it was ruled that the company doesn't pay Qiu any compensation. On the other hand, the pay sheet shows that Qiu's company has paid him his social insurance premiums in cash. So the court didn't follow Article 1 of the "Judicial Interpretations of Applicable Laws in Handing Labor Disputes (3)" by the Supreme People's Court to rule that the company pay compensation to Qiu. However, there exists another view agreeing on the handling method of the Beijing First Intermediate People's Court in the case involving the labor disputes between Li Yanli and Beijing Xiyuan Hotel. According to such a view, the company should compensate for the laborer's losses. In addition, a third view claims that the agreement between both parties is invalid and that the company should be ruled to pay compensation and make amends to the laborer in accordance with Article 38 and Article 46 of the Labor Law.

3 ECONOMIC ANALYSIS OF HANDLING METHODS AFTER AN AGREEMENT IS CLAIMED AS INVALID

As a profit organization, a company must operate with "cost-profits" as its budget goal, in the hope of making the most profits at the least cost. Cost calculation is also involved in the decision by a company on whether to arrange for their worker's social insurance. If the cost of arranging social insurance is less than that for taking the illegal way of not arranging for that, companies will be willing to do that, otherwise, they would rather choose the latter.

3.1 *Cost statistics for companies' payment of social insurance premium*

Chinese enterprises must pay for their employees' social insurance, including old-age pension, unemployment insurance, health insurance, industrial injury insurance and maternity insurance, so as to guarantee that employees get material support in accordance with the law in case they suffer certain risks. According to the current law and regulations of China, the social insurance premiums paid by enterprises for their employees should cover:

A. According to social insurance law and decision on establishing a unified basic old-age insurance system for enterprise employees of the State Council, enterprises should take 20% of staff total wages to pay for their basic old-age insurance.
B. According to decision on establishing a basic medical care system for employees in urban areas of the State Council, enterprises should take 6% of staff total wages to pay for their medical insurance.
C. According to regulations on unemployment insurance, enterprises should take 2% of staff total wages to pay for their unemployment insurance.
D. According to regulations on worker's compensation insurance and Notice on the issue of industrial injury insurance premium rate, enterprises should take 0.5-2% of staff total wages to pay for their industrial injury insurance (this paper adopted 0.5% as a statistical standard).
E. According to trial procedures for maternity insurance for enterprise employees, enterprises should take 1% of staff total wages to pay for their maternity insurance.

In all, enterprises should take 29.5% of staff total wages to pay for their social insurance. And this amount of money is referred to as a law-abiding cost. In this paper, "A" and "a" are used to denote law-abiding cost and sum of wage units. The law-abiding cost of an enterprise can be calculated based on the following equation: $A=a\times 29.5\%$.

3.2 *Risks for enterprises from the evasion of social insurance formalities*

Enterprises, which didn't arrange social insurance for their employees should take some disadvantageous risks in case social insurance matters occur or an employee/employees claim rights. And their law-breaking cost is from such risks. According to China's current laws and regulations, the risks enterprises should take from the evasion of social insurance formalities include:

A. Enterprises pay economic compensation to laborers in accordance with Article 38 and Article 46 of the labor law. "The economic compensation shall be paid to a laborer by the rate of one month's salary for each full year the laborer worked. Any period of above six months but less than one year shall be deemed as one year. The economic compensations that are paid to a laborer for any period of less than six months shall be one-half of the monthly salary. Here the term monthly wage refers to the laborer's average monthly wage for the 12 months prior to the termination or ending of the labor contract. "Therefore, enterprises which didn't arrange social insurance for their employees should pay economic compensation equaling the product of the laborer's average monthly wage for the 12 months prior to the termination or ending of the labor contract and his/her years of working.
B. Enterprises pay compensation when social insurance matters occur. Compensation cost of industrial injuries is higher than any that for any other matters. At present, some enterprises engaged in high-risk operations, arrange an independent industrial injury insurance for their employees, for the purpose of avoiding possible risks. Therefore, they are excluded from the enterprises discussed in terms of compensation cost for industrial injuries in this paper. Compensation cost of old-age insurance is the second highest. When a laborer's claim for damages, the judicial authority will rule that the enterprise pay damages according to the should-be cumulative amount of the laborer's individual account, which is relatively small. Unemployment insurance is the third highest. If a laborer is unable to receive unemployment insurance compensation due to the enterprise's failing to arrange for his/her unemployment insurance, he/she may get compensation of up to two years, which is calculated according to local minimum-wage standards. In addition, labor and social security departments may provide job training and job opportunities as a replacement of unemployment insurance compensation.

Both medical insurance and maternity insurance are only effective during the existence of the labor relations. Thus, no making up is available after the termination or ending of the labor contract. Therefore, enterprises don't have to take a disadvantageous risk on these two insurances in juridical practice. They are excluded from those risk factors analyzed in this paper, too.

Based on above two points, the law-breaking cost for an enterprise, which is equaling the product of the laborer's average monthly wage for the 12 months prior to the termination or ending of the labor contract and his/her years of working, can be calculated with the help of the following equation: $B=b\times n\times p$. Here "B" represents the overall cost of the risk an enterprise takes for not arranging social insurance for its employees; "b" represents the laborer's average monthly wage for the 12 months prior to the termination or ending of the labor contraction represents the total years of working for the laborer. It should be noted, though, that not all employees would apply for arbitration or institutes legal proceedings to have the enterprise take its responsibility. Therefore, here opportunity cost "p" should also be taken into consideration.

3.3 *Economic analysis of enterprises' violation of the social security system*

According to the principle of "cost-profits", an enterprise would voluntarily abide by the law when its law-abiding cost is lower than its law-breaking cost. And when it goes the opposite way, e.g. the law-abiding cost is higher than its law-breaking cost, an enterprise puts the law aside. And in case the law-abiding cost is equivalent to its law-breaking cost, based on the principle of "the less trouble the better", the enterprise may fail to abide the law. Therefore, the performance of the social insurance system (M) is determined by the ratio of A and B. Its calculation equation is as follow:

$$M = \frac{B}{A} = \frac{b \times n \times p}{a \times 29.5}$$

Based on the above statement, it can be known that if an enterprise would try to circumvent the law when M≥1 and be voluntary to abide by the law when M<1. And several conclusions can be obtained by analyzing the equation:

A. Increases with the enhancement of the living standards of the society. Therefore, the higher the living standards, the higher the law-abiding cost in terms of arranging social insurance for employees for an enterprise. But several points should be taken into account: first, in a given period (the period for both arranging the insurance according to the state's statistics and a stable wage period is 1 year), the law-abiding cost for an enterprise is fixed and stable. Besides, the wages of employees get increased when the law-abiding cost is enhanced. This results in the enhancement of the law-breaking cost.
B. The value for B denoting the law-breaking cost will increase with the working years of employees' and with the enhancement of their wages. Since newly recruited employees gain relatively low wages and don't have so many working years, the enterprise's law-breaking cost is relatively low for them. From this perspective of view, it is not difficult to understand why companies are reluctant to arrange social insurance for their employees.
C. The higher, the wages of employees, the more the law-breaking cost of an enterprise. As in most enterprises the wage standards for management personnel are relatively high, enterprises would arrange social insurance for them.
D. Because not all employees who didn't have their enterprises arrange for their social insurance would claim economic damages, the value of p varies between 0 and 1. Whether an employee would claim economic damages depends on his/her legal knowledge. That is to say, for employees with strong legal consciousness who would rather claim compensation, p=1. And in such case the enterprise has to pay compensation. While for employees with weak legal consciousness, p=0. Therefore, the law-breaking cost for their enterprises is 0, too. The strong legal consciousness of management personnel is one of the factors enabling their enterprises to arrange for their social insurance. The ordinary employees' doing of giving up claiming their rights from the low wage level and weak legal consciousness of ordinary of them turn the law-breaking cost of their enterprises for them to 0. This is why enterprises would give up the arrangements of social insurance for their ordinary employees. From this point of view, the increase of employees' legal consciousness is helping in spreading the social insurance system.

In combination of the above mentioned case involving Qiu, it can be known that enterprises do not have to pay compensation to employees with whom they reached an agreement on social insurance and to whom they pay the social insurance premiums in cash. In this sense, it is the judicial authority that reduces the value of M to 1. Because of this, enterprises will choose not to arrange for their employees' social insurance. Under such circumstances, the social insurance system is just an empty talk. What laborers lose is possible benefits from the insurance, both for the present and for the future. This is very unfair to them.

4 CONCLUSION

Agreements on not arranging for their employees' social insurance between enterprises and their employees for the purpose of evading social insurance obligations should be ruled as invalid. Some enterprises promise to pay their employees' social insurance premiums in cash, resulting in the fact that the wage structure of employees is split and the number of their social insurance items is increased. However, employees will not gain more benefit in the process. So such agreements are false ones. According the author of this paper, when handling cases involving enterprises' not arranging for their employees' social insurance, the court should first declare that relevant agreements are invalid and then require enterprises to make up compensation for their employees. At the same time, employees should enhance their consciousness frights safeguarding and prosecute their enterprises when necessary. In this way, the implementation of social insurance system will be promoted, because the law-breaking costs for enterprises are increasing.

REFERENCES

[1] Kang Shaoda, Chen Jinxiang, Fu Chunlei, December-2011. *Study of the Perfection of Migrant workers' Social Insurance System in Hebei Province [J] in Macroeconomic Management*: 57–58.

[2] Li Zhiming, Peng Zhaiwen, July-2012. *Redefinition of the Concept of Social Insurance [J] in Academic Study*: 6–50.

[3] Tang Qinghui, 2007. *The Economic Analysis of Labor Contract Law [D]*. Jilin University

[4] Xulin, 2005. *Social Security Studies [M]*. Tsinghua University Press, Beijing Jiaotong University Publishing House: 102.

[5] Yang Junwen, 2007. *Comparative Research on Social Insurance System between USA, German and Japan [D]*. Jilin University.

Management, Information and Educational Engineering – Liu, Sung & Yao (Eds)
© 2015 Taylor & Francis Group, London, ISBN: 978-1-138-02728-2

An empirical study on the components of management quality of industrial enterprises—a case of industrial enterprises in Fujian province

Hai Ling Duan, Zu Ping Zhu & Chi Dou
College of Economics and Management, Fuzhou University, Fuzhou, Fujian, China

ABSTRACT: Taking industrial enterprises in Fujian province as an example, the paper aims to provide useful references to improve the management quality of industrial enterprises and gain management bonuses. By using the method of Gray Relational Analysis, we research the relational degree between the components of management quality and management, and make a detailed analysis of the contribution made to the management quality of the factors. It concludes that the contribution differs from component to component and the order of dimension is as follows: relevant work experience, average tenure, education level, team size, CEO dominance, tenure heterogeneity and political relations. According to the results, some suggestions are proposed: firstly, to enrich the management team resources, with emphasis on employing experienced and highly educated talents and expanding management team sizes; secondly, to improve the management team structure, especially to extend the average tenure of management team members; thirdly, to build good political relations.

KEYWORDS: Industrial Enterprise; Management Quality; Grey Relational Analysis.

1 INTRODUCTION

With the advance of global economic integration, enterprise's management environment increasingly manifests the ultra-competitive environment characteristic of complexity, discontinuity and uncertainty. At the same time, as a social system's complexity, enterprise continues to improve which leads to the various contradictions of nature and society, spirit and material, control and innovation. The highly complicated and changed external environment and the intertwined contradictions ask higher requirements for Enterprise Management. In addition, with the rapid growth to lower-middle growth of Chinese economy, management dividend becomes the main driving force of Chinese economy development instead of demographic dividend and market dividend. This means that the quality and bonuses of management have become the important issues of current research.

Although some scholars have studied the quality of management, most researches still concentrate on the evaluation model [1-3], influence factors [4-5], the effect of management quality on enterprise performance [6-8]. There are little researches about the constituent elements of the quality of management. Understanding the influence of each constituent of management quality is, not only an essential part of the constituent elements of management quality, but the basis of improving the quality of management and reaping the management bonuses. Therefore, examining the degree of association between each constituent of management quality and the quality of management is a key issue in current research under the hyper-competitive environment. In view of this, the article, taking industrial enterprises in Fujian province as an example, using the method of GRA, aims to find the contribution degree and the important degree of each component of management quality to manage quality.

2 INDEX SELECTION AND MEASUREMENT AND DATA SOURCE

2.1 *Index selection and measurement*

Management quality refers to the impact and effect of management activity which management personnel had done for achieving the business goal, it is a kind of evaluation of the management process. Chemmanur and Paeglis (2005) first put forward the conception of management quality, and divided the conception into a management team resource, management team structure, external reputation of management personal [9]. Then, based on it, some scholars deem that top management team quality are constituted by human quality, structure, external reputation and economic incentive [10]. Besides, the elements of management quality contain team resource, team stability and team-oriented innovation [11]. All above studies believe that team resource and team structure are the most important parts of managing quality. In the view, the article will also use the precious elements—team resource and team structure as our dimensions. Considering the viewpoint is the same with Switzer and Bourdon [12], our will be on their basis and combine with the special circumstance of China, selecting the secondary indicators of management team resources from management team size, education level, related working experience and political relationship, and selecting the secondary indicators of management team structure from average tenure, tenure heterogeneity and CEO dominance.

Measurement of each variables is as follows: ① management quality (x_0), take weighted average rate of return on equity(ROE) as its mapping variable; ② management team size (x_1), meaning the total number of members of management team; ③ The education level of management team member(x_2), meaning the ratio of the total number who holds an MBA degree or above in the management team; ④ relevant experience of management team number(x_3), meaning the ratio of management team numbers with manager position or above in other companies before they entered the sample enterprises; ⑤ political relationship of management team members(x_4), meaning the ratio of management team number's deputies or CPPCC member in the management team size; ⑥ The average tenure of management team member(x_5), meaning the ratio of total tenure of management team member in management team size; ⑦ the heterogeneity of tenure(x_6), meaning standard deviation coefficient of management team member's tenure, which is equal to the standard deviation of the management team member's tenure divided the tenure's mean value; ⑧ CEO dominance(x_7), meaning the ratio of CEO's salary and bonuses in other members' average salary and bonuses.

2.2 *Data source*

Firstly, taking mainly operating revenue of all listed companies (23) which are the top industrial enterprises (2012) from the Statistical Yearbook in Fujian Province in 2013 as the original sample; secondly, rejecting 3 companies that ROE is negative and 5 companies that information disclosure are incomplete, then selecting 15 listed companies with complete information disclosure; thirdly, checking the news section of Flush stock quotation and the information news section of 2013 annual report from the periodical information of CNINF and searching CSMAR and WIND databases to improve the related information; finally, obtaining and determining the relevant data (table 1).

3 GRA OF MANAGEMENT QUALITY OF INDUSTRIAL ENTERPRISES

Gray System Theory (GST) is a cross-disciplinary subject pioneered by professor Ju Long DENG in 1982, and it is a theory to deal with small samples, poor information and the problems of uncertainty. The basic idea of gray relational analysis is according to the similarity degree of sequence curve geometry to determine the close relationship among sequence. The closer the curve, the stronger the correlation, and vice versa [13]. The analytic steps are as follows:

The first step: To determine the reference and comparative sequence. Taking the mapping variable of management quality—weighted average of return rate on net assets as a reference sequence $X_0 = \{X_0(k)|k = 1,2,\ldots,15)$, taking management team size (x_1), education level(x_2), relevant experience(x_3), political relationship(x_4), the average tenure(x_5),the heterogeneity of tenure(x_6), CEO dominance(x_7) as comparative sequence $X_i = \{X_i(k)| k = 1,2,\ldots,15\}$ $(i = 1,2,\ldots,7)$.

Table 1. Management quality and its elements in industrial enterprises.

Sample	1	2	3	4	5	6	7	8	9	10	11	12	13	14	15
X_0	7.65	12.28	11.42	26.1	33.34	16.98	12.37	4.96	12.1	25.6	2.87	12.8	1.26	14.3	4.38
X_1	22	25	25	17	15	24	23	17	18	14	20	16	23	15	18
X_2	0.45	0.2	0.44	0.24	0.53	0.33	0.52	0.12	0.5	0.43	0.2	0.5	0.3	0.27	0.28
X_3	0.55	0.64	0.92	0.71	0.8	0.63	0.87	0.53	0.89	0.71	0.75	0.63	0.78	0.4	0.56
X_4	0.23	0.04	0	0.24	0	0	0.17	0	0.11	0	0	0.13	0	0	0
X_5	5.82	6.36	5.24	6.94	5.4	7.92	4.13	5.29	5.61	3	6.15	4.25	3.57	7.33	3.22
X_6	0.55	0.46	0.52	0.63	0.23	0.37	0.43	0.21	0.18	0	0.47	0.34	0.42	0.73	0.69
X_7	4.77	3.5	3.58	2.92	0.98	3.26	2.17	2.13	1.33	2.04	2.52	2.93	1.7	2.3	2.15

The second step: To make the index of reference and comparative sequence being dimensionless. The different indexes of sequence or the number of space are in large different size in non-time series, but people can set a number artificially under the same index to make the magnitude equal[14]. So the article will use normalization to quantify the criterions.

The third step: To solve difference sequences. Put $\Delta i(k)=|x_{0'}(k)-x_{i'}(k)|$, including $\Delta_i=(\Delta_i(1),\Delta_i(2),\cdots\Delta_i(k))$, i = 1,2,…7,k = 1,2,…,15.

The forth step: To solve the maximum and minimum difference between two levels. Put $M=\max_i\max_k\Delta_i(K)$, m = $\min_i\min_k\Delta_i(k)$, $\Delta_i=(\Delta_i(1),\Delta_i(2),\cdots\Delta_i(k))$, i = 1,2,…,7, k = 1,2,…,15. Solve out the results, M = 4.3529, m = 0.

The fifth step: To calculate correlation coefficient. According to follow formula1(below)

$$\eta_i(k)=\frac{\min_i\min_k|x_0(k)-x_i(k)|+\varsigma\max_i\max_k|x_0(k)-x_i(k)|}{|x_0(k)-x_i(k)|+\max_i\max_k|x_0(k)-x_i(k)|}$$

(i = 1,2,⋯,7) (1)

to solve out the correlation coefficient of the reference and comparative sequence. For convenient calculation, we use $\varsigma = 0.5$ as resolution number, and accurate the calculation results to 4 decimal places.

The sixth step: To solve correlation degree. According to the formula 2(below)

$$\gamma_i(k)=\frac{1}{15}\sum_1^{15}\eta_i(k) \quad (i = 1,2,\cdots,7) \qquad (2)$$

to solve out correlation degree of the reference and comparative sequence.

The whole calculation process is completed by Gray System Theory Modeling Software (GTMS3.0). Due to the limited space, we just list the final results. As show in table 2.

The seventh step: To sequence correlation degree in a descending order. Table 2 tells us: $\gamma_3>\gamma_5>\gamma_2>\gamma_1>\gamma_7>\gamma_6>\gamma_4$. Comparing the value of correlation degree can get that the related experience and average tenure are the largest contribution to the constituent elements of management quality, and then education level and team size are less important, and last is the political relationship.

Table 2. The correlation degree between management quality and its element.

indexes	Team size (γ_1)	Education level (γ_2)	Related experience (γ_3)	Political relationship (γ_4)	Average tenure (γ_5)	Tenure heterogeneity (γ_6)	CEO dominance (γ_7)
correlation degree	0.7232	0.737	0.755	0.6547	0.7381	0.6967	0.7022

4 CONCLUSION AND SUGGESTION

In conclusion, management quality is made up of many elements, it is a multidimensional conception. On the basis of GRA results, contribution differs from component to component and the order of dimension is as follows: related experience, average tenure, education level, team size, CEO dominance, heterogeneity of tenure, political relationship. To sum up, the extent of the contribution to the management team resources is greater than the management team structure. The recommendations are as follows:

1 To rich management team resource. By hiring highly educated and abundant relevant work experience, talents and expanding the size of the management team to enrich management team resources, we can improve management quality and get management dividend.
2 To improve the management team structure. By extending the average tenure of management team number, and weakening CEO dominance, and increasing the heterogeneity of tenure to improve the management team structure, we can improve management quality and get management dividend.
3 To establish a good political relation. Although the contribution of political relationship of management team members of the management quality is the lowest, establishing a good political relation with the government is still a factor that can't be ignored under Chinese national conditions.

ACKNOWLEDGEMENTS

We thank the 2014 International Conference Editor for exemplary editorial guidance throughout the review process. The paper funding comes from the National Natural Science Foundation Project, which name is Enterprise Management Quality:

Property, Measurement, Optimization and Evolution (71171054). We are also grateful for the guidance by professor Zu Ping ZHU. The authors of the paper include Hai Ling DUAN, Zu Ping ZHU and Chi DOU.
Hai Ling DUAN is a doctoral candidate in College of Economics and Management at Fuzhou University. Her interests are enterprise management theory and management quality. Zu Ping ZHU (PhD) is the president of Fuzhou University Zhicheng College, and the professor in College of Economics and Management at Fuzhou University. His interests include enterprise management theory, management quality and quality management. Chi DOU is a master candidate in College of Economics and Management at Fuzhou University. Her interests are enterprise management theory and management quality. If you have any further questions, please contact duanhailing20@163.com.

REFERENCES

[1] Zhu, Z. P. 2007. Research on quality evaluation model for enterprise management. Journal of Fuzhou University: Philosophy and Social Sciences 21(3): 26–32.
[2] Yahagi, S. 1992. After product quality in Japan: management quality. National Productivity Review11(4):501–515.
[3] Liao, S. W. 2006. Study on evaluation of management quality. Guangzhou: South China University of Technology.
[4] Bloom, N. & Van Reenen, J. 2007. Measuring and explaining management practices across firms and countries. The Quarterly Journal of Economics 122(4): 1351–1408.
[5] Van Reenen, J. 2011. Does competition raise productivity through improving management quality? International Journal of Industrial Organization29(3): 306–316.
[6] Bettman, J. R. & Weitz, B. A. 1983. Attributions in the board room: Causal reasoning in corporate annual reports. Administrative Science Quarterly 28: 165–183.
[7] Schweiger, H. & Friebel, G. 2013. Management quality, ownership, firm performance and market pressure in Russia. Open Economies Review 24(4): 763–788.
[8] Rahaman, M. M. & Zaman, A. A. 2013. Management quality and the cost of debt: Does management matter to lenders? Journal of Banking & Finance 37(3): 854–874.
[9] Chemmanur, T. J. & Paeglis, I. 2005. Management quality, certification, and initial public offerings. Journal of Financial Economics 76(2): 331–368.
[10] Zhou, B. Y. 2013. Research management quality factors impact on business performance. Tianjin: Tianjin University of Finance & Economics.
[11] Shao, C. J. 2010. Company executive team quality and IPO underpricing—Based on the SME board and GEM empirical data. Shanghai: Fudan university.
[12] Switzer, L. N. & Bourdon, J. F. 2010. Management quality and operating performance: Evidence for Canadian IPOs. International Journal of Business 16(2): 1–34.
[13] Deng, J. L. 2005. The Primary methods of Grey System Theory. version 2. Wuhan: Huazhong University of Science and Technology Press: 2–8, 74–86.
[14] Fu, Y. 1992. Grey System Theory and Its Applications. Beijing: Science and Technology Literature Press: 189.

Management, Information and Educational Engineering – Liu, Sung & Yao (Eds)
© 2015 Taylor & Francis Group, London, ISBN: 978-1-138-02728-2

The limitations status and countermeasures analysis of animation creation in Hebei province

Ting Zhao, Feng Xia Qi & Yan Mei Zhang
College of Art and Design, Hebei Normal University, China

Lei Liang
Hebei Institute of Communication, China

Gang Li
College of Art, Hebei University of Economics and Business, China

Dong Li
College of Journalism and Communications , Hebei Normal University, China

ABSTRACT: Since the introduction of animation industry support policies, the development of the cartoon industry in Hebei province has experienced for eight years. So far, there are still many problems in technology and creativity in Hebei animation works. This paper starts with the creative technology and creative resources of Hebei animation, and analyzes the animation status and development strategy research in our province. It is believed that the Hebei animation should suit the local conditions, to maintain cooperation with large animation enterprises at home and abroad to processing the relationship, for the accumulation of capital means the growth of existing animation enterprises, learning advanced technology and experience at home and abroad at the same time. Strengthening the cooperation between colleges and enterprises, pay attention to the cultivation of talents for classification, suitable and willing to learn digital 3D and interactive technology personnel specialized training for the quick mature, so as to adapt to the society and the needs of the enterprise. In addition, Hebei province animation creation should be from our own cultural tourism resources, not only can produce original animation works, but also can make an important contribution to protect and promot the tourism resources in this province.

Since 2006, Hebei Province issued the implementation opinions on promoting the animation industry to accelerate its development, our province has experienced a rapid development period of eight years. In the past eight years in our province, it has established three animation industry bases in Baoding, Shijiazhuang and Handan, as well as the emergence of many animation enterprises and a number of award-winning works, but now the overall level of the animation industry in Hebei province is not satisfactory. In the research field of the animation industry, many scholars regarded the animation as a business industry to study, while ignoring the creative nature of animation. Our province animation creation is affected by many unfavorable factors, although the technical progress has been made, especially the three dimensional animation technologies has obtained a certain level, but it still has a big gap compared with the international digital animation technology. And the lack of original animation works in our province, especially in the animation creation theme, the lack of cultural self mining depth, especially Hebei rich tourism and cultural resources. Therefore, this paper will start from the perspective of creative technology and creative resources, analysis of the current situation and the development strategy of animation in our province.

1 THE LIMITATIONS AND COUNTERMEASURES RESEARCH OF ANIMATION CREATION TECHNOLOGY IN HEBEI PROVINCE

In the field of animation, the proportion of film special effects and the application of digital 3D animation technology in the product display area is larger and larger. In recent years, Pixar produces several 3D commercial animation films, such as the "cars" series, "flying Pixar" have won the audience's approval, also gain considerable box office. In the field of film and television special effects taking Weta digital in New Zealand for example, the production of the "Lord of the rings" series, "the Hobbit" series films, and worked with Spielberg shooting a 3D version of the "Tintin". Among them the large number of digital special effects scenes and characters are based on digital 3D theory and technology. But the product display areas, 3D panoramic display and 3D digital roaming technology develop rapidly. A virtual product form is displayed in the digital interactive three-dimensional panoramic device, which can make the audience have more comprehensive demonstration experience than from the photo and video product. Today, three-dimensional panoramic display technology is mostly used in product launches and Museum restoration show hidden treasures. 3D digital roaming technology, including 3D modeling and computer interactive technology, realizes the migration, freedom in virtual model under way. The technology has been matured, and more applications have been made in Beijing, Shanghai and other developed areas of real estate projects and some theme park in exhibition display.

The development of the animation industry in our province is in the crack of Beijing and Tianjin. Although experienced eight years of development, Shijiazhuang, Baoding has established animation industry bases, and has formed a production, learning and research, interrelated industry pattern, in the province to form a broad impact, but with the Beijing first-tier city animation development there is a huge gap, the gap formed by many factors, we think that the reason but the most important is the promotion and the application development of 3D digital animation technology is insufficient. They combine through the introduction of advanced equipment, software and technical personnel training way has walked in the most front-end of digital 3D technology development and application. Relatively, there is only Hebei Academy of Fine Arts in our province introducing a motion capture system from Edison animation in 2013 and carrying out the teaching work. In addition, Shijiazhuang Information Engineering College is taken as the base of the animation industry Incubation Park in the introduction of the 3D digital animation engine; specifically for 3D digital game production research work is teaching and practice. Aiming at the digital 3D animation technology progress of Hebei has not yet commenced from the whole, the speed is slightly slow, we should study more from advanced areas. For example, our province animation enterprises can cooperate with Beijing Disheng Animation Technology Co., Ltd, the latter has been organized with the International Animation Association (ASIFA) to establish contact, and has set up for animation processing cooperation and international animation enterprise online platform, make the international animation orders and processing work to achieve a direct contact with the production of the platform on the Internet, makes the animation processing more convenient, also can make the 3D animation processing in our province level of making rapid promotion.

In many years of teaching practice and communication, it is found that, in most universities in Hebei Province, the training goals of animation professional also remain in the cultivation of talents with comprehensive technical ability, and two problems caused by this training goal of talent training, namely the graduates of the professional and technical level is low and lack of innovative capacity. In "Hebei Animation Education Forum" held in April 25, 2014, the organizers and participants of institutions also present graduates from school that have the difficulty in technology to achieve the work needs and creative planning capacity is insufficient.

Hebei province animation professional colleges design two-dimensional, three-dimensional equal in cultivating mode, the overall training mode bias of two-dimensional animation talent. This training mode emphasizes too much on talent and technology comprehensively and lacking the division for the personality technology, and also far away from the social practice demand. At the beginning of the personnel training, colleges and universities are based on 2D animation teaching as the basis, but lay the foundation shall then personality characteristics and social practice for students of the need for classification training. Excessive pursuit of comprehensive technology for cultivating talents, will cause in the process of learning to care for this and lose that talent.

The two-dimensional paperless animation is operated based on drawing combined with two-dimensional hand-painted plate and appropriate software, so hand-painted performance technique is paying more attention in creation. And the 3D animation software is very different, the most popular 3ds MAX or Maya and virtual reality software Virtools, Quest3d software commands have tens of thousands of, make a 3D model and movement is more complex and requires a skilled operation. The cultivation of students who majored in Animation 3D software general college teaching is from the beginning of the third year, that is to say, a student in the university after two years, only to face the arduous task of learning and practice, but also master the complex three-dimensional software operation, time is very limited. While the graduates entering society, need to focus its work on a post, these positions to a comprehensive technology request is not high, they can continue to learn the relevant animation technology in the long-term work in the future.

Therefore, the personnel training should be carried out in accordance with the classification of student interest and social needs, so that students can concentrate more, teaching can be more flexible, and practice more closely. For the demand of large three-dimensional software, can arrange students to early contact and learning, and appropriately increase the number of teaching. In addition, schools can also be considered and social training institutions of cooperation in running schools, not only can improve the students' operation skills quickly, but can also raise the need of talent for enterprise directional, so as to solve the employment problem of graduates. In the "Hebei Animation Education Forum", Dyson animation for cooperation in running schools and colleges conducted about, its purpose is to enhance the students' practical ability, can be the competent related work after graduation as soon as possible. In the previous training, both animation pipeline collaboration skills, also have 3D software technology operation training, and has obtained the good effect.

2 THE CREATIVE RESOURCE LIMITATIONS AND DEVELOPMENT COUNTERMEASURES RESEARCH OF ANIMATION CREATION TECHNOLOGY IN HEBEI PROVINCE

The reasons for lacking the creativity are the following aspects:

1 The creation is less, the processing is multi. The animation productions in our province have been in the downstream link for a long time, which just do generation processing for large companies. This is related to the development of international animation, America, Japan, South Korea animation companies consider from the angle of making precision and cost, tending to China and India animation companies to cooperate. Large foreign companies are responsible for the creative, original painting creation, such as part of the province, while similar animation company is responsible for the animation processing work. Does the animation production cost limit do the original reason, our province animation company dimensions are lesser, can undertake multiple expenses broadcasting, publicity and promotion of research and development of peripheral products etc.. This is the objective factors restricting our province animation development, is our province animation enterprise long-term in machining stage. We need to find their own way, make our province animation has its own characteristics, to embark on the road of development.

2 Subject of creation are limited. Our province animation development is in the initial stage. There are 22 enterprises specializing in the animation production, and 10 enterprises have been identified by the Ministry of culture. Although most of them are processed, but there are some great companies trying to produce original animation. *"Douding happiness diary"* produced by Hebei Maya film company has been broadcast in the CCTV children's channel; "Douding happiness diary" produced by Hebei Maya film company has been broadcast in the CCTV children's channel; "Zhao Yun and click the box" produced by Shijiazhuang deepness animation technology company is awarded in the eighteenth session of the Shanghai TV Festival "Asian animation venture capital will". Baoding Zhongke head Digital Technology Co Ltd in 2009 produced 3D animation "the king of milu deer", won the thirteenth session of the China Film Huabiao Award "excellent Animation Award" the first European and Belgian international stereoscopic film festival "best long form stereoscopic film festival" Golden Crystal award". These films have been reflected the animation production level in our province, but also reflects the common problems in our province animation creation — subject limitation, reason lies in the lack of creativity and play as well, our province animation creation has experienced for 8 years, only three influential works, because of lack of creative themes play less limited technical advantage and improved. Our province tourism culture resources are abundant, but the three film, only positive definite Zhao Yun and our province tourism culture related. Douding story is somewhat similar to the "big head son and small head dad", "deer" is a product of learning from the 1994 Disney animated "Lion King". Although these two cartoons produced influence all over the country, but has not become our province animation business cards, nor will improve our province animation influence.

3 In personnel training, there is a of lack of creative talents. Long term since, domestic animation has been effected by the Japan and US animation, from the story to the image design of learning in the Japanese beauty, not like the *"Confucius"* with a kind of Chinese brand works. Colleges tend to focus on Technology in animation talents training, ignoring the cultural. Causes the student to China traditional culture, local culture and even don't know. Therefore, when writing the narrow field of vision, the lack of domestic or regional works of cultural connotation, and copying, drawing the Japan US animation obvious traces. Now hit the *"Qin moon", "Qin and Han heroes",* although the plot is taken from the Chinese ancient, but the design of characters or with the Japanese

animation and network game shadow; *"Polly"* cop car with Japanese animation *"irongut train man"* of the shadow, which figures are Europe and the United States of. Visible, the domestic animation is mostly from Japan US animation, the lack of cultural self mining depth.

Relatively, *"Elk king"* built by Baoding Zhongke head and CTV interactive media cooperation is the more influential animation works. It combines the China animal protection of endemic elk with traditional mythology books *"ShanHaiChing"* about *"Yunmeng water"* scene together the story is the destruction of environment and protection against as external cues, aesthetic feelings will love as clues, to the animation as the media will Chinese of traditional culture and modern civilization organic blend. Compared with the *"pleasant goat and grey wolf", "naughty little horse jumping"* and other children's animation, *"Elk king"* will dig out of Chinese classical culture, has become in recent years has the cultural connotation and realistic significance of excellent animation works.

How to explore our province animation creative resources better? *"Zhao Yun and click box" produced by* Shijiazhuang Deepcg Animation in 2011 is a good representative. First of all, the works selected the famous general in Three Kingdoms - Hebei historical and cultural celebrities Zhao Yun as a hero, Zhao Yun hometown - Hebei historical and cultural city of Zhengding, important tourist cultural heritage Dafosi and animation organic fusion. Through the design of new animated characters, to make the image of Zhao Yun lively, and animation and cultural tourism resources combined, realize the Hebei cartoon brand operation and cultural tourism resources in the promotion of the win-win effect. From a technical point of view, *"Zhao Yun and click box"* characters and scenes are to achieve the full effect of 3D, make the characters look more lively, more realistic three-dimensional scene. This work is not the depth of the animation will be the first 3D animation technology and tourism culture resources with the attempt. In early 2008, deepcg animation had made 3D animation "arch bridge", the CCTV movie channel broadcast has caused wide concern. Therefore, in 2011 the creation of *"Zhao Yun and click box"* in the eighteenth session of the Shanghai TV Festival "Asian animation venture capital will" become the Asian range ten eventually winning one of the works. This also proves that the limitations of Hebei animation to seek survival and development to break their own creative resources in the crevice, finding suitable road and method of their own.

According to the actual situation of our province animation, it is thought that it should suit one's measures to local conditions, to maintain cooperation with large animation enterprises at home and abroad to process the relationship, for the accumulation of capital means the growth of existing animation enterprises, learning the advanced experiences at home and abroad at the same time. Strengthening the cooperation between colleges and enterprises, pay attention to the cultivation of talents for classification, and the willingness to learn for three-dimensional and interactive technology personnel specialized training, to quickly mature, so as to adapt to the society and the need of the enterprise. In addition, Hebei province animation to from our own cultural tourism resources, not only can produce original animation works, can make an important contribution to protection and promotion of tourism resources in this province.

The historical and cultural tourism resources are very abundant in our province. Take the ancient capital of Handan to Yanzhao culture as an example, it is beyond count of the idiom, the related historical legend is a huge number of. Idioms and historical stories with a widely recognized advantages of resources, many idioms are originated from the monopoly of resources in our province, as long as we find the appropriate animation enterprise skills to exploit this resource, we can discover the treasure of this culture. Taking Shandong Province as an example, in 2010 launched the "cartoon" Confucius, a counter old Confucius old image, but the performance of young Confucius wisdom, all this point more in line with the aesthetic appeal of children, is the local traditional culture into the animation example. Three dimensional animation "in 2008 launched the Zhaozhou Bridge" based on "the Luban built the Zhaozhou Bridge" legend, young Luban as the breakthrough point, about the Zhaozhou Bridge story.

Overall, the development of our province animation should start from the realistic conditions of their own, to keep learning, the latest animation technology as a means, to find suitable for their own development road. At the same time, our province tourism culture is rich in resources, and needs to be extended through the appropriate channels. Combining the two, to the three dimensional animation technology as the support, to the tourism, cultural resources as the animation creation and adaptation of the background, so that our province animation works have distinctive regional characteristics, and to show my province's tourism culture resources, help to promote the tourism culture in Hebei.

ACKNOWLEDGMENT

2014 Hebei Province Social Science Fund Project *Digital 3D animation technology and Hebei culture, tourism development of mutual benefit* Project No.:HB14YS013

Group Members: Zhao Ting (lecturer of College of art and design of Hebei Normal University), Qi Fengxia (lecturer of College of art and design of Hebei Normal University), Zhang Yanmei (lecturer of College of art and design of Hebei Normal University), Liang Lei (lecturer of Hebei Institute of Communication), Li Gang (lecturer of College of Art of Hebei University of Economics and Business), Li Dong (Postgraduate student of College of Journalism and Communications of Hebei Normal University)

REFERENCES

[1] Liu Yi, Zhang Yan. China new period of cartoon industry and cartoon marketing[M]. Beijing:China Drama Press, 2005(12).

[2] Lu Bin, Zheng Yuming, Niu Xingzhen. Report on the development of Chinese ajimation industry(2011)[M] Beijing:Social Sciences Academic Press,2011.

[3] Lu Bin, Zheng Yuming, Niu Xingzhen. animation Blue Book, the report of Chinese animation industry development.(2011)[R]. Social Sciences Academic Press,2011.

[4] Ning Kun, Wang Yang. Research on the development of animation industry in Hebei[J]. Hebei Academic Journal,2012,(4).

Management, Information and Educational Engineering – Liu, Sung & Yao (Eds)

Study on endowment insurance for new generation migrant workers—based on the survey in Lanzhou city

Xiao Hui Wu & Jian Min Han
College of Humanities, Gansu Agricultural University, Lanzhou, Gansu province, China

ABSTRACT: The endowment insurance problem of the new generation migrant workers has attracted more and more attention in the process of China urbanization. On the basis of the endowment insurance investigation from the new generation migrant workers, this paper analyzed the insuring situation of endowment insurance of these migrant workers, and at the same time, through the related data model, the Logistic regression effect of variables that Impact the migrant workers to participate the endowment insurance were studied. The result showed that the properties of the company, the stability of salary payment, the choice of acknowledging and importance, awareness of policy has significant influence on their choice of participating the endowment insurance. This paper also puts forward some suggestions for improving the endowment insurance of the new generation migrant workers.

KEYWORDS: New Generation Migrant Worker; endowment insurance; investigation; insurance behavior.

1 INTRODUCTION

With the rapid development of urban economy, the rural labor force which is a large scale and the instability was pushed to cities gradually. Urban migrant workers present a phenomenon that the new take places of the old, that to say the older generation of urban migrant workers returning gradually to hometown and subsequent young urban migrant workers fill the vacancy. Compared with the old ones, the new generation migrant workers present different social characteristics in many aspects such as outlooks on employment and the planning of their lives[1-3]. Concerning the demand of economic development, the new generation migrant workers have more knowledge than the older ones, their degrees and some practical skills are more acceptable than previous generations. What more, new generation migrant workers have stronger learning ability and they can easily adapt to city life[4, 5]. So, under these circumstances, new generation migrant workers play the role of the main force of urban economic and social development[6, 7].

For the new generation migrant workers, they own their ideals despite they are followers of their elder generation. When entering the city, new generation migrant workers can clearly feel that there is a huge gap between their former life in villages and present urban life in respect of physical, spiritual and cultural level[8]. What's more, the message explosion makes them tired of rural life and look forward to city life[9, 10]. They also intensely hope for Integrated into the city life, are part of the city, get equal treatment and respect and create better conditions for next generation[11-13]. But under the condition that too many migrant workers flock into cities, serious social problems would be bursting out if they cannot enjoy equal rights such as social security and pension security with citizens[14, 15].

Three aspects are used to analyze the current situation of new generation migrant workers' endowment insurance. From a personal point of view, although the new generation migrant workers have a higher cultural level compared with the previous generation, but gaps between new generation migrant workers and urban employment do exist. In recent years, even they began to have some understanding of the endowment insurance, but they overlook the importance of endowment insurance because they believe that they are young adults now. From enterprise perspective: most companies think that this part of the fees which account little part of companies' earnings would increase the burden of the enterprise itself. From an institutional perspective, the current endowment insurance system is to planning at a higher level and stipulates that endowment insurance can be get after your payment years add up to fifteen years which is harder for new generation migrant workers[16]. This is now facing the dilemma of the lower number of Insurance population and higher number of surrender. Based on these situations, it is particularly important to explore the problems of endowment insurance for new generation migrant workers.

2 INVESTIGATION AND ANALYSIS OF NEW GENERATION MIGRANT WORKERS' BASIC SITUATION IN LANZHOU

In this article, the required data are obtained from the questionnaire survey which is distributed in Lanzhou city, Gansu province in 2014. questionnaire was distributed to agricultural registered permanent residence, Non agricultural registered permanent residence and migrant workers. New generation migrant workers of the 80's generation in Lanzhou were mainly investigated. 350 questionnaires were distributed in this survey, and 336 recycled. After eliminating unqualified questionnaires, 330 effective questionnaires were got, percent of pass is 98.2%. During this investigation, there were 135 people (40.9%) confirmed that they attend the endowment insurance, 172 people (52.1%) did not attend endowment insurance, the other 23 (7.0%), not sure whether they attend endowment insurance. The following part investigates and analyze new generation migrant workers' basic situation in Lanzhou.

2.1 *Investigation of basic situation*

In all new generation migrant workers that have been inquired which are born in 1980 there are 168 males which account for 50.9% and 162 women which account for 49.1%. 244 of them have gotten married, 84 unmarried and 2 divorced. The detailed information about the degree and income is specified in table 1.

Table 1. Degree and income condition of new generation migrant workers.

Item	Class	Number	Percentage (%)
education	primary school	11	3.3
	middle school	144	43.6
	senior high school	95	28.8
	primary school	51	15.5
	Bachelor Degree	29	8.8
monthly income	≤1500	7	2.1
	1501-2000	120	36.4
	2001-3000	136	41.2
	3001-4000	45	13.6
	≥4000	22	6.7

2.2 *Work unit*

The nature of the work unit and sign of labor contract has great influences on new generation migrant workers' joining of endowment insurance. According to the statistics, concerning the nature of enterprise, 49 of them work in state-owned enterprise, 264 of them work in private enterprise, and the rest work in units of other properties. 29.7% of the new generation migrant workers want to join the insurance while their work unit makes that impossible. 22.3% of new generation migrant workers fail to sign a contract with their work unit.

2.3 *Policy understanding*

The cognitive situation survey reflects that 34.5% of the new generation migrant workers think that endowment insurance is particularly important, 60.1% think is important, only 5.4% of respondents think endowment insurance is not important. With policy understanding, however, only 4.8% of respondents said they are very familiar with the endowment insurance policy, 53.2% of the respondents have a rough understanding of endowment insurance policy, and 31.2% of respondents have a general understanding of endowment insurance policy, while 10.8% of respondents are blind to endowment insurance policy. By contrast, we find that although more than 90% of the respondents thought the endowment insurance is important, most of the new generation migrant workers are not entirely clear to endowment insurance policy.

2.4 *Settle-in-city tendency*

From table 2, the new generation migrant workers have a great willingness to Settle in the city, the proportions of settling down in the city, not clear and must return to native place are 52.4%, 31.8% and 15.8%, respectively, this suggests that the new generation migrant workers are yearning for city life, and also strongly want to integrate into the city, and become part of the city. For years of working, at the same time, only 56.7% of the new generation migrant workers are more possibly to work more than 15 years (be sure or in all probability to work more than 15 years). in the context that endowment insurance should be paid more than 15years, according to the survey, new generation migrant workers who work more than 15 years account low proportion of samples that acquired.

Table 2. Condition of new generation migrant workers' settle-in-city tendency and years of working.

item	class	number	Portion (%)
settle-in-city tendency	settle in city	173	52.4
	Not clear	105	31.8
	go hometown	52	15.8
Years of working	work more than 15 years definitely	58	17.6
	More possible to work more than 15years	129	39.1
	Less possible to work more than 15years	105	31.8
	Definitely not work more than 15years	38	11.5

3 ELEMENT OF NEW GENERATION MIGRANT WORKERS' JOINING ENDOWMENT INSURANCE BEHAVIOR

In this paper, Spss16.0 and Logistic regression analysis methods are used to explore the element of new generation migrant workers behavior of joining endowment insurance. Five dimensions, personal situation, working unit, stability, cognitive level and settle-in-city tendency are chosen, 13 variables included. Interpretation of the dependent variables and variables are as follows

3.1 *Dependent Variable*

Whether joining the endowment insurance is chosen as dependent variable and is separated into two categories, the first category: join the endowment insurance which is assigned 0, the second category: did not attend endowment insurance which is assigned 1.

3.2 *Explaining variable*

In this paper, personal basic situation, work units, stability, cognition level and settle-in-city tendency are chosen as explaining variables, a total of 13 variables included. The variable assignment results are shown in table 3. The following five types of explanatory variables are presented respectively.

3.2.1 *Personal basic situation*

New generation migrant workers' personal basic situation consists of gender, education, marital status and income. The four variables have obvious effect to the mode of decision-making, the outlook on life, values, the direction of life, and the ability to pay, thus affect new generation migrant workers' attitude toward endowment insurance.

3.2.2 *Work unit*

New generation migrant workers' work units are divided into unitary nature and whether to sign labor contract. The nature of the unit and whether to sign a labor contract are external factors that determine whether new generation migrant workers would attend endowment insurance.

3.2.3 *The stability*

Stability is mainly referred to frequency of work changing, the frequency of city changing and wage stability. Frequency of work changing and frequency of city changing to mainly reflect the stability of work which can influence the continuation of new generation migrant workers' endowment insurance; Wage stability decides whether new generation migrant workers can pay pension every month timely.

3.2.4 *The cognitive level*

New generation migrant workers' cognition level can be divided into policy understanding and awareness of the importance of endowment insurance.

3.2.5 *Settle-in-city tendency*

New generation migrant workers' tendency to stay in the city tends to be measured by two variables: the willingness to settle in the city and years of working. The willingness to settle in the city decide the migrant workers' attitude toward the premium of endowment insurance, years of working can decide whether migrant workers can pay enough for continuous 15 years.

3.3 *The empirical analysis model*

Table 3. Independent variable assignment list.

Independent variable name	variable name	Variable type	Variable assignment
Basic situation	sex (X1)	dummy variable	female=0; mail=1
	education (X2)	Ordinal variables	Primary school=1, middle school=2, senior school=3, college=4, university=5
	Marital status (X3)	dummy variable	unmarried=0, married=1
	Monthly income (X4)	Ordinal variables	≤1500 =1, 1501-2000 =2, 2001-3000 =3, 3001-4000 =4, ≥4000 = 5
Work units	unit nature (X5)	Ordinal variables	State owned=1, private=2, individual=3, foreign-owned=4, joint venture =5, other=6
	labor contact (X6)	dummy variable	unsigned=0, signed=1
stability	frequency of work changing (X7)	Ordinal variables	≤1month=1, 1-3months=2, 4-6months=3, 7-12 months =4, 1-2year=5, ≥3years=6
	frequency of city changing (X8)	Ordinal variables	≤1month=1, 1-3 months =2, 4-6 months =3, 7-12 months =4, 1-2year=5, 3years=6
	stability of wage payment (X9)	Ordinal variables	Monthly paid=1, quarterly paid=2, year paid =3, not clear=4

(*continued*)

Table 3. Independent variable assignment list. (*continued*)

Independent variable name	variable name	Variable type	Variable assignment
Cognitive level	policy knowledge (X10)	Ordinal variables	Absolute understanding=1, rough understanding =2, basic understanding=3, hardly understand=4
	importance awareness (X11)	Ordinal variables	especially important=1, important =2, not important =3
settle-in-city tendency	willingness to stay in city (X12)	Ordinal variables	Settle in city=1, not clear=2, back to hometown=3
	years of working (X13)	Ordinal variables	≥15years=1, more likely to work over 15years =2, less likely to work over 15years =3, ≤15years =4

3.4 *Analysis of influence factors of new generation migrant workers' joining endowment insurance behavior*

Table 4. Logistic regression of new generation migrant workers' joining endowment insurance behavior.

Independent Variable name	Variable name	regression coefficient (B)	significance (Sig.)
Basic situation	sex (X1)	−0.326	0.252
	Education (X2)	0.502	0.214
	marital status (X3)	−0.268	0.273
	Monthly income (X4)	1.322	0.080
Work unit	Nature of unit (X5)	−1.852	0.014
	labor contract assignment (X6)	0.287	0.153
stability	Frequency of work changing (X7)	0.030	0.452
	Frequency of city changing (X8)	0.080	0.731
	Stability of wage paying (X9)	0.34	0.021

Table 4. Logistic regression of new generation migrant workers' joining endowment insurance behavior. (*continued*)

Independent Variable name	Variable name	regression coefficient (B)	significance (Sig.)
Cognitive level	Policy understanding (X10)	0.275	0.016
	Cognitive of importance (X11)	0.734	0.001
settle-in-city	Willingness of stay-in-city (X12)	−0.423	0.061
	Years of working (X13)	0.628	0.128

In This article, not attend endowment insurance is assignment 0, attended endowment insurance assigned. Logistic regression method analysis is used to study how 13 variables, as listed in the table above, impact new generation migrant workers' joining of endowment insurance. The selected variable standard is set 0.05, exclusion criteria is set 0.10. Using significant (Sig) to select variables that have significant impact in the model. Combined with the actual analysis, the regression results are as follows:

3.4.1 The personal basic situation of new generation migrant workers does not have a significant effect on endowment insurance joining. A significance level of variables is as follows: gender: 0.252, education: 0.214, marital status: 0.273, monthly income: 0.080. all values are bigger than 0.05. Though monthly income, compared with gender, education, and marital status, are low, which means has an impact on endowment joining, all these factors cannot get through of Significance test. So personal basic situation has little effect on endowment insurance joining

3.4.2 The nature of work unit which significance level is 0.014, has significant influence on endowment insurance joining while labor contract signature has no significant effect on endowment insurance joining. Informal work unit damages new generation migrant workers social security rights. Informal employments which cannot supply endowment insurance make the study of the nature of the work unit necessary.

3.4.3 The stability, in addition to wage payment stability, has no significant effect on new generation migrant workers' endowment insurance joining. Because endowment pension should be paid by month in 15 continuous years, many new gencration migrant workers confront with problem of difficulty in pension payment for their wages are paid by year or project.

3.4.4 According to the investigation and analysis of cognitive level, the importance, awareness has great influence on new generation migrant workers' endowment insurance joining. The importance of trust level detection of a significant degree of important awareness is 0.001, far less than 0.01, while the significance level of policy understanding is 0.015, which varies between 0.01 and 0.05.

3.4.5 Analysis of settling-in-city tendency shows that settle-in-city willingness and years of working have no significant effect on new generation migrant workers' endowment insurance joining.

To sum up, the nature of work unit, wage payment stability, policy understanding and important awareness have significant impact on new generation migrant workers' endowment insurance joining in Lanzhou city while other impacts are not significant.

4 SUGGESTIONS ON OPTIMIZING NEW GENERATION MIGRANT WORKERS' ENDOWMENT INSURANCE

4.1 *Applying principles of flexibility and availability*

4.1.1 *Reduce expends base reasonably*

Comparing with migrant workers' income, current endowment insurance premium base is excessive which makes the new generation migrant workers don't have the ability to bear. This situation results in new generation migrant workers' lack of motivation to join endowment insurance. New generation migrant workers' income is finite, and their basic life will be influenced after the burden of the pension. To eliminate the dilemma to new generation migrant workers, premium base should be reduced in order that more new generation migrant workers can afford.

4.1.2 *Set up flexible premium methods*

The current premium method of endowment insurance for town workers is from work unit which is hard for migrant workers whose work are not stable. Migrant workers who work on the construction site, for example, their wages are generally paid according to the project length which may pay by a year or more. In this case, migrant workers, in a very long period, have no money to pay insurance cost; it is also the limitations of the present system. Therefore, flexible premium methods will be suitable for migrant workers, which will improve the migrant workers' pension problem.

4.1.3 *Adjust timing of contribution payment reasonably*

As is known to all, the current system regulates that pension can be get after paying continuous 15years. Endowment insurance is a long process; it only can be get after paying the time of contribution payment that prescribed. According to the endowment insurance law: male workers that over 60, female workers over 55 and accumulative total pay fully 15 years can get the pension. From the survey of migrant workers' work cycle, migrant workers' average work time in the same place is only five years or so, and construction industry workers are less. In this case, in order to absorb more new generation migrant workers, time of contribution payment should be adjusted.

4.2 *Strengthen the construction of relevant laws and regulations*

The trend of phenomenon that young city migrant workers without endowment insurance are normalized. Although laws stipulate that work units should pay insurance for the workers, the implement is poor for ambiguous provision and invalid provision. Now companies are not paying endowment insurance for new generation migrant workers, whose legal awareness is inefficient to protect themselves Thus, it is imperative to introduce laws to protect the new generation migrant workers' rights which are also appealing by new generation migrant workers.

4.3 *Improve the importance, awareness of new generation migrant workers and firms to endowment insurance*

Many enterprises, as new generation migrant workers mentioned above, exist the same problem of low awareness of endowment insurance. Enterprises should avert faulty notion of traditional family endowment and land endowment patterns. Private enterprises should change their faulty ideas which regard new generation migrant workers as cost burned. Notion should be built that new generation migrant workers own more specialized knowledge and skills which will bring more profits through prevent the outflow of skilled workers and saving capital of training new staff by paying endowment insurance for new generation migrant workers in the long run. For new generation migrant workers, payment of endowment insurance gives them a sense of belonging which make them contribute more for the enterprise.

4.4 *Improve transferring of new generation migrant workers' endowment insurance*

The current regional endowment insurance system conflict with migrant workers instability of employment. Every social worker hopes to be own endowment insurance, and willing to pay the relevant expenses. However, for the new generation migrant workers, it is difficult to avoid frequently changing jobs. This will cause endowment insurance suspend after work transformation. Waste is caused when work changed for the limit of regional planning which caused the problem of low joining rate. So the solve of transferring problem will guarantee migrant workers joining endowment insurance.

4.4.1 *Establishment of id and social endowment data network*

Current planning level of endowment insurance for urban working group remain at city in most China's cities, only a small part of the region achieves provincial planning. To fundamentally solve the problems of flow of the labor and the endowment insurance mechanism, transfer of endowment insurance should be solved first. Migrant workers' transfer of work unit leads to the unsustainable of endowment insurance and lose the function of lifetime guarantee after they returning home. So, to improve the system of endowment insurance for migrant workers, it is critical to solve the problem of transfer. The rationality and validity of transfer of Payment should be valued. Labor and social security departments can use resident identity CARDS to establish a database of migrant workers social endowment, so that the migrant workers endowment insurance account can be operated and transferred nationwide.

4.4.2 *Establishing effective transfer of endowment insurance for urban and rural migrant workers*

From the perspective of new generation migrant workers, they may be settled in cities after many years of work in the city, the government should transform new generation migrant workers' identity into urban workers, remove the previous migrant status slowly. So the government should reasonably transfer new generation migrant workers' endowment insurance for rural to endowment insurance for urban. The government should also take part of the responsibility, bearing the transfer cost of new generation migrant workers in the process of combination of endowment insurance for urban working group and rural residents, and remove the barriers that The transferred set to them, for the achievement of mutual communication, connection and transformation between endowment insurance for urban and rural, in order to achieve the transition of endowment insurance for rural to urban with the speeding up of urbanization and industrialization process gradually. There are many places have imposed the endowment insurance for rural residents now, residents of these areas should transfer their endowment insurance to endowment insurance for urban working workers as a whole after they get into city. But the capital of pooling account are too small comparing with pooling account of endowment insurance for urban working workers . Based on the situation that parts of our country has achieved the provincial planning, the local government should appropriate part of local fiscal budget to pooling account as repaying of industry to agriculture. If local fiscal budget has no money to fill the gap, the central government should take the final out responsibility.

5 CONCLUSION

New generation migrant workers are special group which formed in the Coordination of Urban and Rural Development and will make great contributions for the building of the Well-off Society in All-round Way. This paper studied the status of migrant workers endowment insurance and then analysis influence factors to the joining behavior of new generation migrant workers' endowment insurance. finally Suggestions to improvement the endowment insurance for new generation migrant workers is put forward. The results of the study show that: work units, stability, and cognitive level have bigger influence on new generation migrant workers joining of endowment insurance, while personal basic situation and settle-in-city tendency are small. In this article, suggestions are put forward from aspects of principle of flexibility and validity, legislation building, improvement of new generation migrant workers and work units awareness, transfer mechanism improvement of endowment insurance for new generation migrant workers. With the coordinate efforts of the government and residents, the problems in endowment insurance for new generation migrant workers will eventually be resolved.

ACKNOWLEDGEMENT

The National Social Science Fund. Project: "study on rural social security system of Northwest minority areas", 2008, (08XMZ009).

REFERENCES

[1] Liyun Huang. New generation migrant workers value in urbanization [M]. Beijing: Social sciences academic press, 2012.

[2] Hua Wang, Jinyu Zhang. Endowment insurance joining willingness and impact factors of new generation migrant workers –based on investigation in Nantong and Shijiazhuang city [J]. Northwest Population Journal, 2013, 04:95–99.

[3] Zhiying Liu, Chao Liu. Study of generational differences of migrant workers and endowment insurance analysis [J]. Journal of wuhan university (philosophy and social sciences, 2011, 06:81–88.

[4] Lei Zhu, Fang Li. Generation differences of migrant workers endowment insurance joining—based on the investigation of Nanjing city [J]. rural economics journal, 2012, 05:86–90.

[5] Jinqiu Dong, Shuang Liu. Migrant worker: Social support and urban integration [J]. Journal of south China agricultural university (social science edition,2014,02:41–48.

[6] Xionghui Leng, Na Yi. Study on new generation migrant works expense behavior—based on the investigation of Jiangxi province [J]. Journal of jiangxi agricultural university (social science edition,2012,03:62–66.

[7] Mengyi Wang, Zhaoyu Yao. Study on consumption behavior and impact factor of new generation migrant worker—based on questionnaires of Nanjing city [J]. Journal of hunan agricultural university (social science edition,2014,01:43–48.

[8] Xiaoli Liu, Jing Zheng. New generation migrant workers identities and impact factors study [J]. Journal of south China agricultural university (social science edition,2013,01:45–50.

[9] Jianli Zhang, Xueming Li, Li Zhang. Study of new generation migrant worker urbanization process and spatial diversity [J]. China Population Resources and Environment, 2011, 03:82–88.

[10] Fei Zhang. Analysis of new generation migrant workers urbanization and elements [J]. Population Studies journal, 2011,06:100–109.

[11] Jianrong Liu. The system guarantee of the new generation of migrant workers [J]. Study Forum journal, 2008,07:67–70.

[12] Zhuo Tang. Some problems of new generation urbanization of migrant workers in China [J]. Journal of jiangxi agricultural university (social science edition,2010,02:16–21.

[13] Lei Liu, Honggen Zhu, Lanyuan Kang. Migrant workers stay will influence factors analysis, based on the data from Shanghai, guangzhou, shenzhen 724 survey [J]. Journal of hunan agricultural university (social science edition,2014,02:41–46.

[14] Chunhua Yang. Study on problems of new generation migrant workers [J]. Issues in Agricultural Economy, 2010,04:80-84+112.

[15] Lixia Xia, Jun Gao. Social security of the process of new generation migrant workers urbanization [J]. urban development research, 2009,07:119–124.

[16] Lanyuan Kang, Honggen Zhu. element of Work chosen of migrant workers under the circumstance of"labor shortage"—based on the view of generation difference [J]. Journal of jiangxi agricultural university (social science edition,2013,04:479–485.

Management, Information and Educational Engineering – Liu, Sung & Yao (Eds)
© 2015 Taylor & Francis Group, London, ISBN: 978-1-138-02728-2

Conditions and strategies research on construction of international tourism destination in Sichuan, China

Rong Jia
Department of Tourism Management, Chengdu University of Information Technology, China
Yinxing Hotel Management College, Chengdu, Sichuan, China

ABSTRACT: This paper firstly explains the connotation of international tourism destination, through analysis of international tourist destination construction conditions of Sichuan province and comparative analysis of development conditions with 3 neighboring provinces in Western China, pointed out the conditions and shortages of international tourist destination construction in Sichuan, proposed development strategies of international tourist destination construction in Sichuan.

KEYWORDS: Sichuan; International tourism destination; Development strategy.

1 THE CONNOTATION OF INTERNATIONAL TOURISM DESTINATION

What is the international tourism destination? This will be the first to speak from the definition of tourism destination. In recent ten years, people who clearly proposed the concept of tourist destination is only British scholar Dimitrios Buhalis, he believes that the tourism destination is a unique entity determining the perception of geographical areas by tourists, and can provide the political and legal protection for tourism marketing and planning; Leiper think destinations can be interpreted as a traveler to stay for some time, and experience the rich local attraction places; Cooper believes that destination is focused to meet the needs and services of tourists. Although all three destination concepts expression somewhat different, it is not difficult to see that the tourism destination is a geographical area which area has unique overall tourism image, can attract tourists to come, can satisfy basic living needs and personalized travel needs of tourists , has a sound service system of regional management and coordination mechanism. What kind of tourism destination can be called an international tourism destination? The world famous tourist destination has a wide geographical distribution and different image, to propose an accurate and accepted definition is difficult. There are two main categories of domestic literatures on international tourism destination, one kind is research literature on the international tourist island of Hainan, the other is research literature on construction of international tourism city. Since Sichuan and Hainan are provincial-level regions, the more we made reference to the understanding and interpretation of "international" words is about Hainan relevant literatures. The Secretary of China National Tourism Bureau Shao Qiwei thinks: international tourism island means the high degree internationalization, high environmental quality, high service standards, comprehensive supporting facilities. Hainan provincial Party Secretary Wei Liucheng pointed out: internationalization is to comprehensively improve the tourism infrastructure and social infrastructure, management, service level, to provide international standards of service. Governor of Hainan Luo Baoming also pointed out that: construction of the international tourism island is to let the international tourists enjoy the first-class services, at the same time, allowing more domestic tourists enjoy the international standard of tourism service without going abroad. All those three government officials talked about internationalization of internal destinations from the perspective of tourism supply, namely through internal factors of tourism destination such as tourism products, tourism service facilities, tourism infrastructure, tourism destination management. This paper argues that, achievement of international tourism destination must be considered from two aspects of tourism supply and tourism demand, continuing to explore the international tourist market is also an important strategic step in the internationalization of tourist destinations.

2 ANALYSIS CONDITIONS OF BUILDING AN INTERNATIONAL TOURIST DESTINATION IN SICHUAN PROVINCE

2.1 *The features of tourism resources*

2.1.1 *The original ecological outlook highlights*

First of all, as far as natural tourist resources in Sichuan, " Ancient" (ancient Shu and culture of the

Three Kingdoms), "Secret" (religious culture of Mount Emei, Le, Qingcheng), "Simplicity" (Qiang, Tibetan, Yi and other ethnic minority customs) and "wild" (original ecology of mountains and rivers) together constitute Sichuan's irreplaceable tourism development foundation.

Secondly, from resource- market- products angle, Sichuan compared with neighboring provinces, its historical culture resources less than Shaanxi, its religious culture resources less than Tibet, its Ethnic customs less than Yunnan, only natural tourism resources is rich and unique. In eastern Sichuan is famous Sichuan Basin, terrain is low, altitude is generally between the 300 to 400 meters, low mountains and hills interspersed, products is rich; The west is full of plateau and mountains. This kind of terrain the landscape creates different types of natural environment and unique scenery, What's more, its biological diversity, perfect ecological environment make Sichuan becoming one of the most abundant tourism resources in China. The success of Sichuan panda habitat application of World Heritage demonstrates the unique charm of the original ecological tourism once again. The ancients said that "the world landscape view in Shu", so the original ecological outlook highlights Sichuan tourism resources.

2.1.2 *"Quick rhythm" Chengdu urban style under the concept of modern tourism resource*

Chengdu, capital of Sichuan province's tourism revenue was ranked first, Chengdu is also the first station of Sichuan for inbound tourists. As the most important tourist hub in Sichuan, Chengdu's development speed is amazing. On 2010 October, well-known financial magazine " Forbes" in America release a research report about the fastest city over the next ten years. With strong development path and rapid rise of the global high-end industry, Chengdu entered the list of "the fastest city" and reached the top. In April 9,2012, American "Fortune" magazine and Chengdu Municipal People's Government jointly announced that the twelfth Fortune Global Forum will be held in Chengdu, this is a Fortune Global Forum held in Shanghai, Hongkong, Beijing, the first choice in China's central and western hinterland city. On 2013 December, the latest issue of "the first finance and economics (blog) weekly" classified 400 National Cities, 15 cities such as Chengdu, Hangzhou, Nanjing is becoming new First-tier city, Chengdu ranks the first. In January 3, 2014, the first press conference by Chengdu Government Press Office in 2014. Chengdu's press spokesman Chen Fu introducts that, 2013 Fortune Global Forum and the twelfth World Chinese Entrepreneurs Convention held successfully, greatly improving the visibility, reputation and international influence of 'city of wealth and success' in Chengdu, more than half of the world's top 500 enterprises settled in Chengdu. Multinational enterprises settling will promote development of business and MICE tourism, enhance image and international reputation of Chengdu, thereby promoting the overall development of tourism in Sichuan province.

2.2 *Regional comparisons on international tourism destination condition of Sichuan*

2.2.1 *Preliminary conditions for construction of international tourism destination*

In recent years, tourism economy grows rapidly in Sichuan. In 2011 the total tourism income is 244.9 billion yuan, in 2012 the total tourism revenue increased to 328 billion yuan, in 2013 was 387.7 billion yuan, a great increase, and ranked the seventh in china and the first in western region.Compared with 2008,the tourism economy increases 280 billion yuan in 5 years. First half of 2014, in the harsh economic environment, Sichuan's tourism industry is still steady advance, under the background of inbound tourism downturn, the number of inbound tourism in Sichuan achieved the growth of 15.8%.

As the only tourism standardization model Province, Sichuan sets standards to deal with tourists rush in china firstly. In 2013 "eleven" golden week, Jiuzhaigou were stranded because of the large number of visitors at the peak. In order to avoid similar incidents happening again, Sichuan study and formulate the " tourist attraction standards at tourist peak" which became China's first standard at peak period of visitor management in tourism industry, filled the 5A-class tourist scenic service quality standard blank at tourist peak in our country, made service function and quality in Sichuan reaching international level at tourist peak.

In 21st century science and technology change rapidly, the construction of international tourism destination must place tourist information, service network in the equally important position with tourist traffic. Sichuan is currently building "2+1" intelligent tourism demonstration zone, Chengdu, Mianyang, Leshan will be bulit intelligent tourism city, Jiuzhaigou, Dujiangyan, Huanglong and the orther seven scenic spots will be built intelligent tourism scenics. Enhancing tourism information application level is also one of the attractive factors to contruct international tourism destination in Sichuan.

2.2.2 *Comparative difference exists between the western provinces in inbound tourist market*

The first step to construct an international tourism destination for Sichuan is getting absolute advantages in competition with China's western provinces.

Unfortunately, although rich tourism resources in Sichuan, serious imbalance have existed between domestic tourism market and inbound tourism market in tourism development. In 2013, Sihuan's number of domestic tourists ranked fourth in China, but inbound tourists' number was fifteenth. Even in the western China, inbound tourists' number is lower than Yunnan, Guangxi, Shaanxi, As shown in table 1. From 2008 to 2012, reception of international tourists grow rapidly, improved from about 700000 passengers in 2008 to 2273400 passengers in 2012, but compared with other three provinces there are gaps, especially compared to Yunnan Province, the number of international tourists received only half of Yunnan in 2012. But overall, the international tourism market development potential of Sichuan is huge, growth rate ranked first in these four provinces. Due to the gaps of inbound tourists' number compared with other provinces, foreign exchange income also failed to go beyond. At present, the inbound tourism level of Yunnan ranked first in western provinces, from 2008 to 2012 tourism, foreign exchange income of Yunnan (as Table 2 shown) has maintained a growth rate of more than 12%, especially in 2011 and 2012, growth is more than 20%, with an important contribution to the economic growth of Yunnan Province, Yunnan focuses on the development of the tourism industry. Comparing Shaanxi with Guangxi, inbound tourism reception number in Guangxi is more than Shaanxi, but tourism, foreign exchange earnings are less than Shaanxi, this phenomenon shows that Guangxi province should promote per capita consumption and stayed days of inbound tourists, also gives great inspiration to expand inbound tourism market in Sichuan province.

In addition, Shaanxi Xi'an and Guangxi Guilin are the China's first introducing best tourism city to the world which were investigated by the World Tourism Organization in 2003. Yunnan Lijiang is also well-known in recent years, liked by many overseas visitors, while there is a gap in core tourism city 's notability in Sichuan province and these three provinces.

Table 1. 2008-2012 reception of inbound tourists comparison in Sichuan and Shaanxi, Guangxi, Yunnan (million).

Province	2008	2009	2010	2011	2012
Yunnan	250.22	284.49	329.15	395.38	457.84
Guangxi	201.02	209.85	250.24	302.79	350.27
Shanxi	125.73	145.08	212.17	270.41	335.24
Sichuan	69.95	84.99	104.93	163.97	227.34

According to the National Tourism Bureau official website statistics http://www.cnta.gov.cn

Table 2. 2008-2012 international tourism income comparison in Sichuan and Shanxi, Guangxi, Yunnan (million).

Province	2008	2009	2010	2011	2012
Yunnan	10.08	11.72	13.24	16.09	19.47
Shanxi	6.60	7.71	10.16	12.95	15.97
Guangxi	6.02	6.43	8.06	10.52	12.79
Sichuan	1.54	2.89	3.54	5.94	7.98

3 STRATEGIC CHOICE OF INTERNATIONAL TOURISM DESTINATION CONSTRUCTION IN SICHUAN

According to the research on the understanding of "international" word, it reflects in two aspects: tourism supply internationalization and tourism market internationalization. Based on these two points, strategies for building international tourist destination in Sichuan should act from two pieces.

3.3 *Tourism supply internationalization strategy*

Tourism supply internationalization is mainly embodied in tourist products, Regional tourism cooperation, tourism infrastructure and other aspects of internationalization.

3.3.1 *Brand strategy of tourism products*

Sichuan should strive to create several international well-known tourist route brands and tourism festival brands. Tourist route products can be combined with original ecology, image, creating Panda Home exploration tour, looking for a Shangri-La trip, creating a different world heritage tour; Tourism Festival should focus on building China Chengdu International Festival of intangible cultural heritage, China Chengdu International Food Festival, Liangshan Yi International Torch Festival, Dujiangyan Water Festival and so on.

3.3.2 *Strategy of regional joint development*

The State Council document No.31 promote integration of regional tourism as the key, Sichuan should scizc this opportunity, actively seeking tourism cooperation with neighboring provinces, participating in regional tourism cooperation. For example, build 317,318 world tourism destination with Tibet; achieve free tourism cooperation, share regional tourism resources, tourist information with Chongqing, Shaanxi, Yunnan and other neighboring provinces (city).

3.3.3 *Strategy of tourism infrastructure consolidation*

Tourism infrastructure consolidation is the pilot work of tourism developers. In order to achieve an international level of Sichuan's tourism, comfort and convenience traffic network, tourism information construction and perfect tourism public service facilities are indispensable. On one hand, Sichuan should consider the construction of external transportation system, especially the open and incremental of international non-stop flights; optimization of internal tourism traffic system, perfection of tourists distributing and transferring system adapted to main traffic facility's capacity, to adapt to the visitors center and transit system, forming safe, convenient and comfortable stereoscopic traffic network. On the other hand, while strengthening the basic capacity of building tourism information, establish multi-level and open tourism information application system, for example, construct global-oriented, comprehensive and accurate, timely updated, intelligent and convenient public travel information service system with strong credibility.

3.4 *Strategy of tourism market internationalization*

Tourism market internationalization is mainly reflected in the choice of target market positioning and internationalization of tourism marketing and promotion.

3.4.1 *STP strategy on inbound tourism market*

Domestic tourism market and inbound tourism market in Sichuan developed unbalanced, in order to achieve the goal of international tourism destination construction, inbound tourism must be paid great attention to and be put into more prominent position. Sichuan must focus on international tourism market segmentation research, according to the tourism demands of segmented markets to develop a strong, targeted tourism product, to carry out targeted marketing activities. Sichuan must deepen Taiwan market, consolidate Hong Kong and Macau market, focus on developing Japan and South Korea markets, the ASEAN market, which is closer to China, take the American and European markets into account.

3.4.2 *Strategy of overall image*

During the marketing of tourist destination, a unified and clear overall tourism image is very important. Sichuan must constantly deepen the "World Sichuan, Panda Hometown " image in the marketing process, highlight the original ecological landscape, the original ecological town, the original ecological nation and the original ecological country through integration of excellent natural and cultural tourism resources of Sichuan, and continuously improve well-knowness and attractiveness of Sichuan original ecological tourism brand at home and abroad. At the same time, strengthening the publicity of Sichuan tourism image, actively playing the role of expo, sports events, festivals, the International Summit Forum, striving to set up offices in Japan, Korea and other key foreign markets, accelerating the establishment of Sichuan tourism website with main tourist source-language version, innovating Sichuan tourism brand advertising form at mainstream media home and abroad, to highlight long-term and sustainable publicity effect.

ACKNOWLEDGMENTS

This research was supported by two projects. Social sciences key research base of Sichuan Education Department—Sichuan Tourism Development Research Center funded project (LYC14-19); Chengdu University of Information Technology, Yinxing Hotel Management College research funded project (YXK2014-09).

REFERENCES

China city new classification list: Chengdu among the "new line" first financial weekly list of [N]. City, 2013-12 -19.

Liu Youli. "The world's fastest-growing city", "the researchers Forbes" columnist Joel Kotkin first Lai Rong [N]. Chengdu Business Daily, 2011-04 -19.

Sichuan Province People's government. Sichuan Province "1025" tourism development planning of [Z], 2011-12-05.

Wang Yue, cold grandeur. The world's top 500 enterprises settled in Chengdu Tianfu Morning Post super half of [N]., 2014-01-03.

Xu Navy. The international tourism island construction standard of inbound tourism perspective and evaluation system research — take Hainan Island as the example of Nanjing Normal University doctoral thesis based on [D], 2011,3.

Zhang Li Ming. Department of tourism destination variable mode research [J]. Journal of Southwest Jiao Tong University system and space to play, 2005,6 (1): 78–83.

Zhong Hangming, Yu Xuecai. Foreign tourist destination [J]. research review of tourism science, 2005,19 (3): 1–8.

Management, Information and Educational Engineering – Liu, Sung & Yao (Eds)
 ISBN: 978-1-138-02728-2

The business model of system optimization for The Fifth Party Logistics (5PL)

Shu Feng Wang
Guangdong Baiyun University, Guangzhou, P.R.C.

ABSTRACT: The 5PL refers to "system optimization" logistics service providers, which have some logistics assets (light asset type) and they use the system optimization theory, electronic commerce and information network technology to conduct the overall coordination and logistics operation in the supply chain. Providing the service for customers such as supply chain integration, process optimization and resource collaboration, is the core idea of the fifth party logistics theory system. Research has shown that, system integration and construction of the new business model, can realize logistics system optimization, supply chain management integration, logistics solution implementation, integrated logistics resources collaboration. The market system for 5PL, between the service elements of supply side, service products of the demand side, has the characteristics of volatility, oriented, interactive and fusion.

KEYWORDS: The fifth party logistics (5PL); system optimization; business model.

1 INTRODUCTION

This paper focuses on the discussion of the fifth party logistics business model. The research is based on two premises: the author has engaged in the research on the logistics industry practice for more than 30 years; the author presided over the subject of "international container transportation business area management"(COSCO) and "logistics enterprise management system"(Grandbuy Group) subject to a successful case. The theory of long-term thinking as well as successful practical case, constitute the research foundation of the fifth party logistics series theoretical results.

According to the latest research results, the author has put forward different views on 5PL connotation, looking forward to peer discussion. One is 5PL as a part of logistics assets (asset light) service providers, participating in the actual operation of supply chain logistics services, is the upgraded version for 3PL service; the second is based on two kinds of viewpoints of the information service of logistics and virtual logistics, further propose the research theory of 5PL is "system integration logistics" business model.

2 THE CONNOTATION OF 5PL

2.1 *The major views of foreign*

Morgan Stanley Group (2001), [1] consider 5PL is the logistics supply chain information network based on electronic commerce. It covers all parties in the supply chain which emphasize information ownership; With the development of first party logistics to 5PL, the logistics, asset logistics service provider owned reduced continuously, the ability of information control was strengthened. Gunasekaran and Ngai [2] believe that 5PL is the electronic commerce, logistics network based on global operation. Vasiliauskas and Jakubauskasy. [3] Abraham (2002) thinks, [4] 5PL is the virtual logistics. 5PL is a bridge between the traditional 3PL and the model of 4PL, which prompted the existing technology and infrastructure of first party logistics, to drive the cost transfer from supply chain to the virtual enterprise organization.

2.2 *The major views of domestic*

Mingke He (2004) think, [5]5PL is the logistics service providers of information services. 5PL provide logistics information service, providing supply chain logistics information service regional range, multi industry, many enterprises.

Zuqi Feng (2004) thinks, [6]5PL is the logistics information service provider. It provides e-commerce technology to support the entire supply chain in the actual operation and can be combined with the interface with the executive members as well as providing collaborative service for enterprises in the supply chain.

Xiongfei Lu (2008) thinks, [7] 5PL is the logistics service providers to provide comprehensive operation of electronic commerce solutions. Improve the efficiency of the supply chain in the actual operation by using information technology, and can effectively all member enterprises in the supply chain collaboration service. Construction of a service oriented architecture (SOA) electronic commerce information service platform based on 5PL.

Xingzhong Wang (2009) thinks, [8] 5PL refers to itself does not have the logistics operation, physical products, but it manages the whole logistics network, virtual logistics service providers.

Shufeng Wang (2014) thinks, [9] 5PL refers to a part of the logistics assets (asset light), provides a plurality of integrated supply chain management service for the customer, with system integration, process optimization, resource coordination function attributes, is a system of integrated logistics service provider.

3 FIFTH PARTY LOGISTICS SYSTEM OPTIMIZATION

3.1 *The core thought of 5PL theoretical system*

The latest research shows, system of 5PL market, supply side factors between services, demand side service product, with volatility, oriented, interactive, the fusion characteristic. 5PL, providing a plurality of integrated supply chain management service for the customer, system integration, process optimization, coordination of resources, is the theoretical system of important core. By optimizing the system, and building a new business model, can realize the optimization of logistics system, supply chain integration, logistics solutions implementation, overall logistics resource synergy.

The development process of 5PL, constitutes the different functional logistics service provider. 5PL evolution process with various Service Providers Association. see table 1.

Table 1. 5PL evolution and various service providers association matrix.

Evolution Process \ Service Providers	5PL Carrier service provider	5PL forwarder service provider	5PL transport service provider	5PL warehouse service provider	5PL comprehensive service provider	5PL technology service provider	5PL intermediary agent service
Information Network L.	√	√	√	√	√	√	√
Virtual Logistics						√	√
System Integration L.	√	√	√	√	√	√	
Process Optimization L.	√	√	√	√	√		
Collaborative Logistics	√	√	√	√	√	√	√
Supply Chain Integration L.	√	√	√	√	√	√	√

The service content of 5PL, with e-commerce, network and information technology, the overall coordination of the whole supply chain, and provide new supply chain logistics solutions.5PL system connotation and extension of function structure model as shown in figure 1.

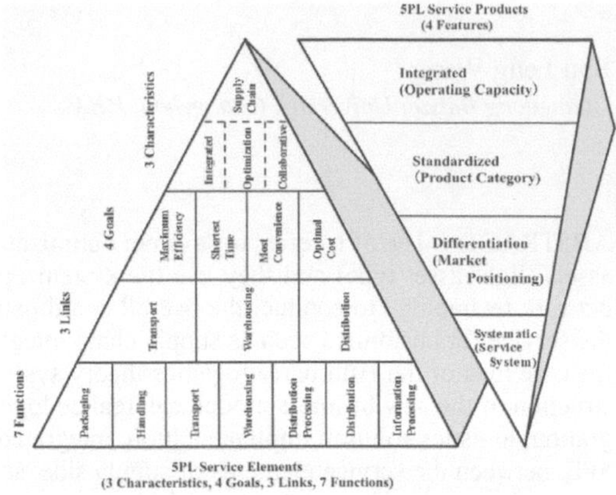

Figure 1. The theory of 5PL function structure diagram.

5PL service element. Mainly consists of three characteristics of integrated, optimization and collaborative; Four goals of the highest efficiency, the shortest time, the most convenient, the optimal cost; Three links of transportation, warehousing and distribution; Seven functions of packaging, handling, transport, storage, distribution processing, distribution and information processing. as shown in table 2.

Table 2. The 5PL service elements list.

Category	Service Elements						
3 Characteristics	System integration	Process optimization	Resource collaboration				
4 Goals	The maximum efficiency	The shortest time	The most convenient	The optimal cost			
3 Links	Transportation	Warehousing	Distribution				
7 Functions	Packaging	Handling	Transport	Storage	Distribution processing	Distribution	Information processing

5PL services products (4 characters). Service products have four characteristics, can be integrated, standardized, systematic, differentiation. Integrated (operation), 5PL is based on IT technology for each link of the supply chain customer portfolio, the actual operation platform system into the client, collect items in order to achieve the real-time dynamic information, tracking, monitoring, evaluation, feedback operation information, to satisfy the demand of customer. See table 3.

Table 3. The 5PL service products list.

Features	Service products
Integrated (Operating Capacity)	5PL is based on IT technology for each link of the supply chain customer portfolio, the actual operation platform system into the client, collect items in order to achieve the real-time dynamic information, tracking, monitoring, evaluation, feedback operation information, to meet customer service requirements
Standardized (Product Categories)	5PL through benchmarking management, systematic connection, can promote the standardization of logistics
Differentiation (Market Positioning)	5PL through the logistics system planning technology, through a combination of qualitative and quantitative analysis method, to find the accurate market positioning
Systematic (Service System)	5PL through the top design, construction between a user can seek a variety of combined service system, a multi interface, multi user, cross regional, no time limit of the platform of logistics service

3.2 *5PL system integration advantage*

The realization of the optimization of supply chain integration. To participate in the actual operation, the standardization of logistics information system integration provider, integrated logistics information system to a system of a system. As long as any one of the links in the supply chain, can be installed in the logistics information system, seamless integration with the upstream and downstream of its own, any information in this system is open, transparent.

4 THE 5PL BUSINESS MODEL

In the development of economic globalization, industry market environment, vigorously develop modern service industry, has become a new mode of economic development. Relevant data shows, the modern service industry market is huge, modern logistics demand is increasing, especially for the systems with integrated business model of fifth party logistics service market demand more urgent, become the future trend of the development of the new logistics service market. By expanding the service function of the logistics, for enterprises provide effective supply chain management solutions, for Multinational Corporation to provide "one-stop logistics services solutions", can be implemented by "logistics service provider" to "supply chain service provider" change, is the logistics enterprises an important means to gain sustainable competitiveness. [10]

4.1 *Fifth party logistics operation mode*

5PL operation mode is mainly embodied objects dynamic and logistics resources in the supply chain of information integrated. 5PL promotes the development of modern logistics industry, that has to do a good job of informatization and, information system plays an important role in the supply chain.

[Case 1] American dry cargo storage company (D.S.C) has more than 200 customers a day and, received a lot of orders, information systems need good. To this end, the company will be many table compiling computer programs, a large amount of information can be quickly input and transmission.

[Case 2] A European distribution company, set up an effective customer information feedback (ECR) and just in time (JIT) system. Do we produce what the customers want, rather than to produce something customers to buy. Turnover warehouse goods per year of about 20, if the use of customer feedback and the most effective means, can be increased to 24 times! So, can greatly increase the throughput of the warehouse. Through the JIT system, which can be quickly obtained from a retail store sales feedback information.

4.2 *Fifth party logistics system structure*

In 5PL system, logistics enterprises will play a greater role, which can support the whole supply chain and can integrate all the members of the cooperative service for the enterprises in the supply chain. Adding the basic data layer on the basis of the fourth party logistics system, and the database update and exchange information timely; in the data analysis layer, the third party logistics will get the customer and supplier information, then conduct an analysis and sorting, so as to obtain the best integration scheme. 5PL system structure is shown in figure 2.

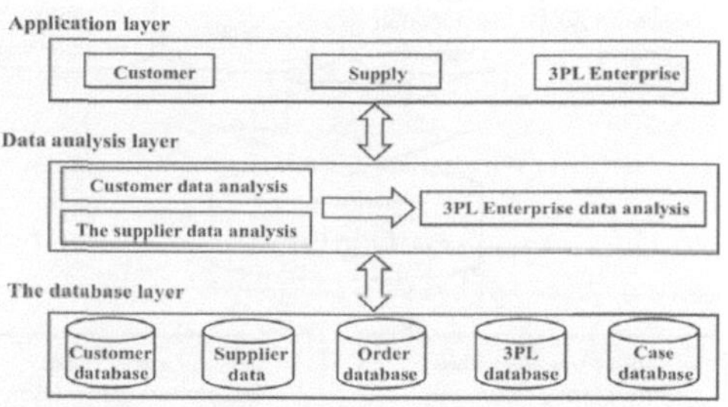

Figure 2. The 5PL system structure diagram.

5PL based on IT technology, focuses on the whole supply chain system. Fifth party logistics is a kind of standardization of logistics information system providers, if any one link in the supply chain, can be installed downstream of the seamless connection between the logistics information system and their. Any information in this system is open and transparent, it is a layer with a level of logistic system. Future development of the traditional business model, is based on 5PL operation mode innovation.

4.3 *Application of the middleware of GIMIS positioning system in 5PL*

Logistics management object is always moving in space, will result in both MIS and GIS fusion and

GIMIS system. The GIMIS system will map engine, engine and workflow engine combination, can accurately identify the logistics demand and provider of spatial information, and on this basis to provide fifth party logistics service.

Through the mobile phone number recognition logistics personnel information, determine the items of information through a mobile phone to scan the bar code information, determine the location information from the GPS module in the mobile phone, realize the logistics of goods location, ßselect the map can be determined the region's articles, and can sell the goods and related logistics personnel and vehicles, the whole process flow monitoring logistics. ViewSonic Q6 mobile phone set intercom, positioning and barcode scanning in a body, can be used as experimental prototypes. Can these features to object linking and embedding user control (OCX) approach to the third party call.

Fifth party logistics business process and integration with GIMIS: review of 5PL business processes, development of the fifth party logistics system and workflow engine based on MIS, and the GIMIS middleware integration, to build 5PL Based Middleware, workflow engine, positioning engine, map engine function modules provide a third party call to object linking and the embedded user control mode. See figure 3.

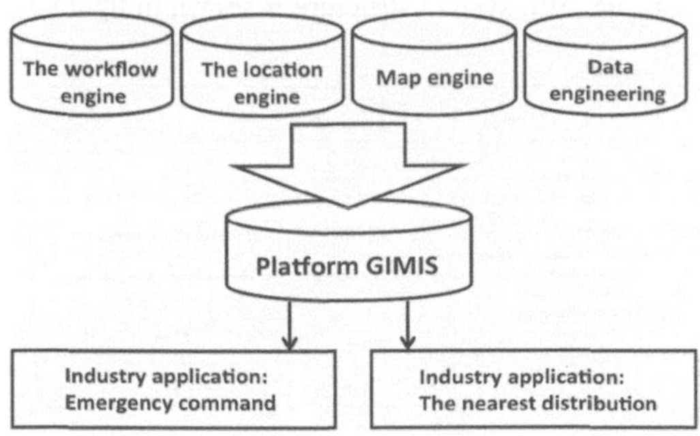

Figure 3. 5PL business process and the integration of GIMIS diagram.

The integration of GIMIS middleware and group communication module based on 5PL. Development of GIMIS based telephone dispatching system of 5PL. The system is embedded workflow engine for 5PL, the integration of personnel, vehicles, goods identification, scheduling, tracking and other functions, and has a variety of communication channels, including telephone, intercom, SMS, data etc..

5 CONCLUSION

Through the system optimization of fifth party logistics, construction of the new business model. The service of SCM integration, process optimization and resources coordination for the customer, is the core idea of 5PL theory system.5PL refers to have some logistics assets (asset light), using the system optimization theory, E-commerce and information network technology etc., As well as the logistics service provider of System optimization of the overall coordination and logistics operation in the plurality of supply chain.

5PL development into a new pattern, study in the new stage. Further research should be based on the three basic theory of system engineering, SCM, system optimization, Using the system theory and method etc., to make further progress in the aspects of concept, mode of operation, information platform construction.

ACKNOWLEDGEMENT

Fund Project: Guangdong Province philosophy and social science "Twelfth Five Year plan" planning project (GD12XGL02).

Author introduction: Shufeng Wang (1963-), male, From Wendeng city of Shandong Province, P.R.C., Associate professor of Management School of Guangdong Baiyun University , research direction of logistics engineering, logistics management, supply chain management, system optimization theory etc..

REFERENCES

[1] Morgan Stanley Group. China logistics Spot the Early Bind[R]. Hong Kong October 5,2001.

[2] Gunasekaran A, Ngai E. The successful management of a small logistics company[J]. International Journal of Physical Distribution & Logistics Management,2003, Vol.33, No.9,825–842.

[3] Vasiliauskas, Jakubauskasy. Principle and Benefits of Third Party Logistics Approach When Managing Logistics Supply Chain [J]. Transport, 2007, Vol.22, No. 2, 68–72.

[4] Abraham A.5PL: The concept of a Virtual Organization with Zero Party Logistics[R]. Council of Logistics Management Eastern Michigan Roundtable, March 5,2002.

[5] Mingke He. The logistics system theory [M]. Beijing: Higher Education Press, 2004:290–295.

[6] Zuqi Feng. Fifth party logistics emerge as the times require [EB/OL]. http://www.ycwb.com/gb/content/2004-06/08/content_704540.htm.2004.

[7] Xiongfei Lu. Advances in technology and application of electronic commerce logistics information platform on [C]. computer based on 5PL, 2008 (volume). Hefei: University of Science & Technology China press, 2008:285–289.

[8] Xingzhong Wang. Study on logistics and procurement [J]. connotation and research development of 5PL,2009(36):10–12.

[9] Shufeng Wang. "Fifth party logistics" theory is applied in emergency logistics [J]. China business and market , 2014, 28 (2): 41–45.

[10] Qingbao Zeng. The analysist of logistics operation in 2013 and prospect in 2014[EB/OL]. http://www.chinawuliu.com.cn/lhhkx/201403/07/284816.shtml.

Management, Information and Educational Engineering – Liu, Sung & Yao (Eds)
© 2015 Taylor & Francis Group, London, ISBN: 978-1-138-02728-2

Legal governance of food and drug crimes

Guo Feng Ding & Chong Zhao
Law School , Kunming University of Science and Technology, China

Xue Feng Zhai
Faculty of Foreign Language & Culture, Kunming University of Science and Technology, China

ABSTRACT: In recent years, with the increase of social development and criminal crackdown, the food and drug crime has gradually emerged, and has been increasing. In order to effectively control the food and drug crime, we should deeply analyze its risk factors, current situation and its causes, which can help us seek a reasonable path to legal governance.

KEYWORDS: Food and Drug Crime; Risk Factors; Legal Governance.

1 FACTOR OF FOOD AND DRUG CRIME

Risk factors of crime are an important part of the risk factors prevention paradigm. 20th century late, There were pioneers just like Hawkins David and Richard Catalan o who introduced the risk factor prevention paradigm into criminology from the view of medical and public health. The basic idea of this paradigm is to identify and limit the risk factors as well as to identify and enhance protective factors. This is mainly because the risk factors indicate increasing the likelihood of post-implementation of crime, however, protective factors presages reducing the possibility of post-implementation of crime. The risk factor of the food and drug crime refers to, which is closely related to the occurrence of food and medicine crime, and a number of factors which have the role of foreshadow for food and drug crimes.

In terms of risk factors to the occurrence of food and drug crimes, Yu Qiwu held that "The reason affecting fluctuations of product quality can be summarized as man, raw materials, equipment, methods, and environments in these five factors". And Felson believed that any crime is necessarily the result of the combined effect of three factors, which is a potential motive for the crime perpetrators, missing a potential target and effective supervision. It can be seen that risk factors related the food and drug crime are mainly concentrated in the main, the entire process of production and sales operations and external environments. Using an analysis of McKinsey logic tree can be achieved on specifics of the food and drug crime of risk factors. This method is hierarchically listing sub-problems of the food and drug related crime to all matters relating to the implementation of the food and drug crime associated exhaustively. Risk factors of food and drug crime mainly include: owning a large illegal interests, unscrupulous dealers, various sales channels, the low-income consumer groups, the illicit origin of raw materials, fake, anti-fake, anti-counterfeit awareness of consumer knowledge and weak government supervision, counterfeiting technologies and the means of covering up criminal acts.

This paper used analysis of Spearman to analyze 71 food and drug crimes judgments coming from the "China Court Net", which is to explore the risk factors and the dangers degree of the food and medicine.

Table. Correlated coefficient between risk factors of the food and drug crime and the degree of harmfulness of crime

Note: (1) the total sample is equal to 71. The number of risk factors refers to the number of risk factors present in each case, the degree of harm is thinking of a measure as the amount that case involved, the numbers involved behind are the amount each case involved (more than million yuan).

(2)When ** confidence level (twin probe) is 0.01, the correlation is significant.

Variables								
The number of risk factors								
5 4 8 7 5 5 5 4 5 4 6 7 6 5 5 6 7 7 6 4 5								
6 5 7 7 5 6 6 6 4 3 4 7 3 2 2 3 3 2 6 5 4								
6 6 5 6 6 6 5 7 5 3 7 5 7 8 7 5 4 4 4 6 6								
4 5 4 5 6 6 5 5								
Harmfulness Level								
185	50	300	1000	1000	260	346	93	36
12	25	3000	5000	7000	200	300	600	1000
400	300	300	200	3000	695	9900	61	300
35000	164	5	9	0.7	50	44	8	0.9
4	0.8	7.5	200	44	160	640	22	238
1500	1580	20000	100	24000	148	4	54	12
15000	10700	500	969	30	32	42	100	800
34	300	300	340	400	780	340	300	
Sig. (Bilateral/twin probe)								0.000
Standard error								0.072
95% Range of Confidence 0.806 (Upper limits) 0.528 (lower limits)								
The correlative coefficient								0.686* *

From the above chart, an analysis of 71 total samples showed that value of Sig. (bilateral)0.000 is less than 0.01, from upper limit 0.806 of 95% range of confidence to lower limit 0.528 of the range of confidence. Therefore, the risk factors of the food and drug crime and the degree of harmfulness of the food and drug crime is a strong positive correlation (0.686), which means that the number of risk factors in food and drug crime is , the more severe the degree of its harmfulness is, the fewer number of risk factors in food and drug crime is, the lower the degree of its harmfulness is.

2 SITUATION AND ITS CAUSES ANALYSIS OF FOOD AND DRUG CRIME

2.1 *Situation of food and drug crime*

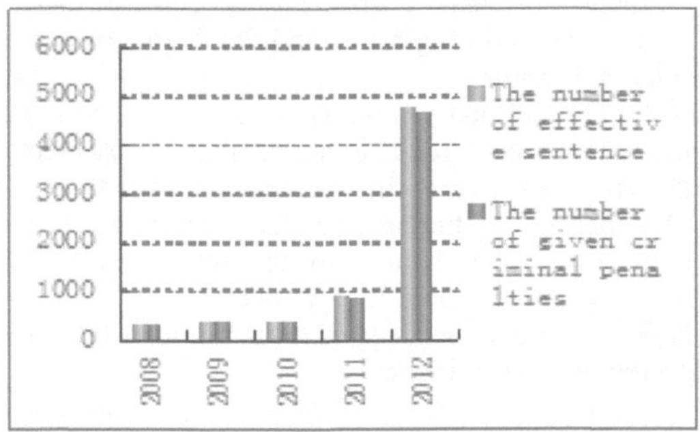

In recent years, the food and drug crime has been increasing exponentially. From 2008 to 2012, the national court accepted the food and drug safety crimes 6449, of which concluded 6180, 6789 people sentenced criminals. Five-year number of cases increased 1.11 times average annual, of which in 2012 increased by 5.39 times in comparison with 2011. Among them, the courts accepted 1953 food cases that included producing, selling toxic and hazardous and not-meeting with safety standards, of which 1789 case concluded, and had 2399 people sentenced criminals. The cases increased of 94.11% average annual. In the accepted cases, the cases of production and sales of toxic and hazardous food were 1518, of which were nearly accounting for 77.73% of all food crimes. The cases in 2012 showed a significant growth trend, the admissibility of the case throughout the year increased 2.21 times in comparison with 2011, the cases the offenders sentenced increased 2.32 times. All levels of court accepted 4496 cases in production, sales of counterfeit, substandard drugs, of which 4391 cases were concluded, 4386 criminals were sentenced. The number of the cases increased 1.19 times in average annual, the cases grew 8.13 times in 2012 more than in 2011.

2.2 *Cause analysis of the food and drug crimes*

First, a greater profitability is induced factors of food and drug motive for the crime. In judicial practice, the subject of food and drug offenses accounted for the overwhelming majority are not high education level of farmers, the special circumstances of the transition period also make their living space be squeezed, imbalance between a strong culture in pursuit of money and a weak culture of the means of compliance regime, plus social structural tensions and food and drug crime can bring greater interests in reality, these factors together will inevitably lead the farmers(vulnerable groups) to choose implementing the food and drug crime. "The occurrence of any criminal behavior is based on their criminal psychological activity of the crime subject as the foundation, offender's requirements is its source and foundation. When the needs combine with the specific satisfied objects, tools and instruments, namely, which is motivated to promote individual to implement the behavior of meeting their needs."

Second, unscrupulous dealers are the bridges of "intersection" with food and drug crime and the emergence of potential victims. In general, the purchase goods sources of legal dealers are legitimate, of course, because we do not deny that individual dealers will choose to purchase illegal channels for the sake of the interests. Food and drug offenders are often difficult to carry out the conversion between use value and value through legitimate dealers. Therefore, unscrupulous dealers will be their best choice, illegal dealers are mostly small retail traders, supermarkets or other self-employed, who are located in the suburban area markets or the fringe of the city, and these people are vulnerable groups. This has resulted in a phenomenon that the groups of implementing the food and drug crime will use vulnerable groups as unscrupulously dominated dealers to sell goods and medicine, and use vulnerable group as consumers to achieve the transformation of use value and value. This is ironically a "self-help of vulnerable groups".

Third, the low-income consumer groups are basic carriers that food and drug offenders will achieve a conversion of use value and value, while the fraud technology is a prerequisite for the food and drug offenders achieving their purposes. The consumption ability of low income groups is weak while it is compared with the higher income groups, and the food and drug offender's purpose is to get as much as profitable benefits, they inevitably turn their own criminal targets into higher income earners. If they want to successfully realize the transformation of use value and value in the higher income earners, they must resort to fraud technology, counterfeiting technology of agricultural product mainly relates to a method and tool which can change the properties of the physical and chemical material. The fraud technology of food and drug crime has its own characteristics, it is not difficult to obtain, especially for counterfeiting technology of agricultural products, such as sodium nitrite, phosphate, carmine, red yeast rice powder and beef flavor of the old mother pork processing into beef.

Fourth, recognizing, preventing and striking counterfeit awareness of consumers and government supervision are weak, which is an important reason for the persistence of food and drug crime. Not only the low-income group exits the problem that their recognizing, preventing and striking counterfeit awareness is weak, but the same phenomenon exists in higher-income earners. As the victims of food and drug crime, they are carrying the intent of offenders to obtain a benefit; only when they are recognized, preventing and striking counterfeit awareness is weak, the offender's purpose of crime can be achieved. Therefore, recognizing, preventing and striking counterfeit consciousness of consumers is an internal constitutive element of food and drug crime, its importance determines its position in the control of food and drug crime. The weak of government supervision is one of the reasons rampant in food and drug crime. Government should crack down on crime, and "earn their efforts to solve the incentive"; only when recognizing, preventing and striking counterfeit consciousness of consumers is improved and unscrupulous dealers are cracking down, the rate of food and drug crime can be minimized.

3 IMPROVE CONTROLLING LEGAL MEASURES OF FOOD AND DRUG CRIMES

3.1 *Establish and improve the system of food and drug law*

We should modify the "Food Safety Law", "Drug Administration Law" and other systems and regulations, establish a sound legal system, to set more stringent liability provisions for offenders of food and medicine, so as to effectively prevent food and drug crime. In addition, we should intensify the legal crackdown, increase the cost of breaking the law, in order to deter potential offenders.

3.2 *Improve regulatory measures of food and drug*

Unscrupulous dealers play an important role in the food and drug crime, law enforcement agencies should update the regulatory philosophy and improve regulatory efficiency. Regulators should resolve the shortcomings of the previous segmented regulation, which had ever become a weakness in the regulation of unscrupulous dealers, and establish the overall concept, concept of cooperation, and the concept of legal rule for food and drug regulation, lead to form the supervisory mode of food and medicine safety, which regulators is led, the relevant departments are coordinated with, and the rule of law is protected, and maximize the effectiveness of supervision.

3.3 *Protect the vulnerable groups*

To govern food and Drug crime fundamentally, the government needs to pay attention to vulnerable groups, and safeguard its legitimate rights and interests. First of all, the income of vulnerable group is low, they can only choose inexpensive items, which often makes them become victims of food and medicine crime. Therefore, we should care for vulnerable groups, and protect their basic living, improve their medical condition, make institutionalization for the protection of their rights and interests. Secondly, the vulnerable group lacks basic knowledge of self-protection, which often makes them be deceived, so it is necessary to strengthen the knowledge of publicity, popularity and improve their life skills.

3.4 *Stop actions of counterfeiting technology*

Fraud, mixed technology plays an important role in the food and drug crime, therefore, we should increase efforts to crack down on counterfeiting technology. Criminal law combats the crime of food and drug, which not only needs us to pay attention to producers, operators of food and medicine, but also the group who provided the manufacturing process of toxic and hazardous food and drugs or counterfeiting technology, needs to be included in the object of legal regulation, and give it severe sanctions.

4 CONCLUSIONS

In recent years, the rate of food and drug crime is growing, and there are deep-seated reasons. Risk factors in food and drug crime have a lot, which is positively correlated with the degree of harm, that is, the more the number of risk factors is, the more the degree of its harmfulness is severe. Improving a sound legal system and strengthening supervision and sanctions, which can prevent and reduce offenses of food and drug.

REFERENCES

[1] David P. Farrington. Explaining And Preventing Crime: The Globalization of Knowledge–The American Society Of Criminology 1999 Presidental Address[J]. Criminology, 2000, 38(1):1–24.

[2] Jin Gaofeng. Analysis of Chinese Crime Situation and Criminal Policy in 2012 [J]. Journal of Chinese People's Public Security University(Social Science Edition), 2013,(2):1–10.

[3] YU Qiwu. Quality Management[M]. Beijing: the Capital University of Economics and Business Press, 2012:281.

Management, Information and Educational Engineering – Liu, Sung & Yao (Eds)
© 2015 Taylor & Francis Group, London, ISBN: 978-1-138-02728-2

On the path to improvement in the environment judicial protection system

Guo Feng Ding & Chong Zhao
Law School, Kunming University of Science and Technology, China

Xue Feng Zhai
Faculty of Foreign Language & Culture, Kunming University of Science and Technology, China

ABSTRACT: Harmony of human society and the natural resources needs judicial activities to maintain the status of environmental justice, but conservation in China is not optimistic: environmental cases are too few, and setting up environment judicial bodies is imperfect, which issues serious constraints to improve their level of judicial protection. In order to effectively control the environment, we should accelerate the pace of reforms, enhance the credibility of the judicial protection of the environment, strengthen the environmental efforts of criminal justice protection, improve the organization of environmental justice, and set up a sound system of the environment judicial protection.

KEYWORDS: Environment; Judicial Protection; Public Interest Litigation.

1 INTRODUCTION TO THE CONNOTATION OF ENVIRONMENTAL JUDICIAL PROTECTION

Environmental justice protection refers to the action that the judiciary who use the substantive law and procedural law to deal with various types of environmental disputes, to protect citizens' environmental rights. Its common way is environmental litigation, there is a small amount of environmental disputes involving judicial mediation, settlement of judicial enforcement. Judicial protection of the environment is different from traditional civil, administrative and criminal judicial activities, which has its own characteristics: first, litigant of the role of litigation is diversified, which litigant status is different in "the official sues the people", "people sue government officials" and "people sue people". Second, the range of plaintiff qualification is extensive, which is including the public, NGO administrative authorities and People's procuratorate. Third, the subject matter of the litigation is special, which is a new, cross-type subject matter of litigation, including environmental claims that has been infringed or violated by the proposed hazard.

2 THE LACK OF CHINA'S ENVIRONMENTAL JUDICIAL PROTECTION AND ITS REASONS ANALYSIS

2.1 *Few cases in environmental justice*

China's environmental situation is grim, citizens are serious discontent about environmental problems,

Table 1. Environmental petition and court of first instance concluded various environmental statistics cases (2006–2009).

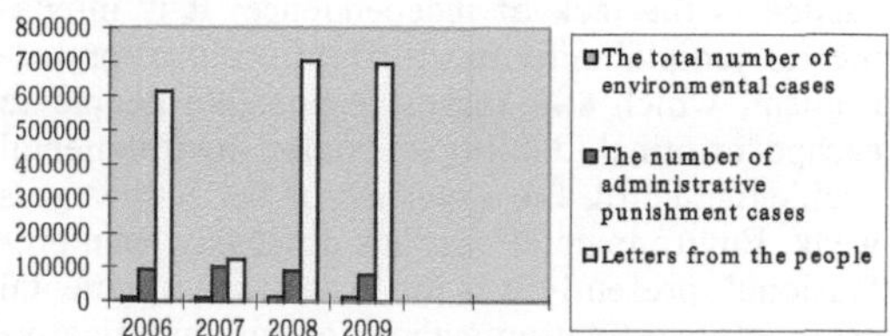

according to statistical data, there are approximately 700,000 environmental petition cases every year. However, environmental litigation cases are significantly low, even if the number of the environment cases is in 2009, its number was only 15,197, which is less than the number of 1/50 environmental petition cases. In environmental tort disputes, citizens are more likely to resolve this problem through petitions, very little to resolve in legal channels. The main reasons are included: Firstly, the credibility of the judiciary is lacking. The loss of credibility of the judiciary is just as cutting off the last straw of the civil rights of relief, strangles the survival way of citizens. Secondly, the cost of judicial relief is very high. In Environmental justice proceedings, the huge litigation costs and expensive appraisal fee for Citizens are unbearable. Finally, the environmental legal system is not perfect. Current environmental judicial protection lacks specialized substantive and procedural system.

2.2 *Inadequacy of environmental criminal law protection.*

Table 2. The number of environmental criminal cases in China.

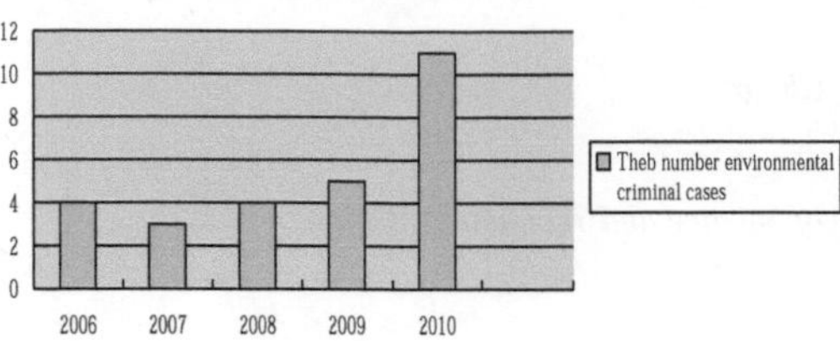

China's environmental criminal cases are scarce, almost hovering in the single digits. According to related research, in the period of 1998-2007, there are 477 cases in major environmental pollution accidents, of which only 37 cases are considered crimes. The main reasons are including: First, the convergence of environmental justice and environmental law enforcement isn't smooth, benign coordination mechanism is not established, the prosecution cannot implement legal supervision, public security bureau is difficult to find the environmental source cases and to investigate them. Second, Justice is the lack of independence. It is in-enable for judicial activities to fight on local protectionism, which still makes responsible people to escape criminal liability in major environmental accidents. Third, the expertise of the judiciary is weak. Environmental case has obviously been professional, presently it is the lack of environmental criminal investigation authorities and investigative ability, and is difficult to effectively file and investigate, the environmental professional judiciary is more rare.

2.3 *Imperfect settings of environmental justice agencies*

Since to the end of March 2013, there has been 134 various environmental trials organizations. China's trials agencies of environmental protection have taken shape, but it is still facing difficulties of inadequate source cases. According to reports, a widespread environmental courts suffered the embarrassment that "making bricks without straw". For example, the Kunming Intermediate People's Court has been established environmental court, has received a total of 20 environmental cases so far, which are including three administrative cases, seven civil cases, ten criminal cases. There are many reasons for rare source cases: first, the prosecution body is narrow, the main scope of the case is not clear. Environmental

Table 3. The number of specialized judicial bodies in China's environmental protection.

Level / Name	The Higher People's Court	The Intermediate People's Court	The Basic People's Court
The Trial Court	2	17	51
The Collegian Panel	0	4	47
Circuit Court	0	1	9
Dispatched Tribunal	0	0	3
Total	134		

protection courts suffered upsets because the range of the plaintiff was mainly too narrow. Second, the trial organization of environmental protection is lack of professionalism. In environmental cases have complex, professional features, which makes its proceedings needing environmental science and technology support, Expertise of judge's requirements is high, but the environmental expertise of judges in China is uneven, the existing expertise of judges is difficult to effectively deal with complex environmental cases. Third, the scope of case jurisdiction under the protection of environmental organizations is not clear. Our specialized judicial institutions of environmental protection set up inside various people's courts, cases jurisdiction mainly based on administrative division, this jurisdiction mode is unscientific, which could easily lead to the vacuum of jurisdiction or cross-zone of powers. Fourth, the supporting system is imperfect. Convergence of environmental courts and environmental agencies has not yet formed systems, many places established environmental court in isolation, which make its role greatly reduced. In addition, the personnel and the proceedings in the specialized environmental judicial organization have not formed a system, trial activities are not normative.

3 IMPROVE THE SPECIFIC PATH OF THE ENVIROMENTAL JUDICIAL PROTECTION IN CHINA

3.1 *Enhance the credibility of environmental justice*

Lack of credibility of the judiciary makes its rules of protection of the environment jurisdiction that cannot be highlighted, seriously erodes system of environmental justice, and results in failure of the judicial system. To enhance the credibility of environmental justice needs severely punishing judicial corruption, guarantees judicial fairness, impartiality; we should strengthen information disclosure, ensure judicial transparency; we should strengthen judicial independence, and reduce the external intervention to justice. With the advent of setting environmental courts and filing environmental public interest litigation, judicial protection of the environment will encounter new development opportunities. Environmental issues are related to public safety and property, life and health. In the process of the establishment and improvement of environmental courts, environmental public interest litigation, we should strive to be fair, impartial, transparent and find a sense of public trust in the judiciary.

3.2 *Enhance the efforts of environmental criminal justice protection*

3.2.1 *Strengthen investigation of environmental cases*

First, it should enhance the investigative skills. Environmental cases, professional make investigations are faced with the need to enhance the professional and technical issues, and timeliness of investigations have characteristics, environmental cases evidence of the need for timely filed, save. Evidence of timely detection and preservation requires the use of certain environmental science and technology, the investigating authorities in their efforts to enhance staff expertise, the need to strengthen cooperation with scientific research institutions, for the detection of environmental cases for technical force. Second, we should strengthen law enforcement and inspection, the source environment to expand the case of criminal cases. The investigating authorities wait for green environmental crimes cases transferred executive practices seem passive, negative and not conducive to environmental protection work. The investigating authorities should be proactive law enforcement, explore the establishment of environmental crime reward systems, and actively expand the source case, the fight against environmental crime.

3.2.2 *Increase the efforts of sanctions*

With the high-tech development, environmental criminal justice cannot effectively deal with the increasingly serious environmental problems. in addition, punishment, responsibility for environmental crime is more light, which made the crime cannot be effectively curbed. In view of this, with the development of society, environmental crimes should be improving the legal system, expanded the offense, increased the responsibility of punishment. In order to prevent such crimes, we should deter such crimes with severe criminal sanctions.

3.3 *Set scientific, environmental Judiciary*

3.3.1 *To expand the prosecuting body*

Article 55 of the Civil Procedure Law does not provide the environmental litigation right for community groups which are irregularities. Social groups, especially environmental NGO has strong strength, well-organized, strong technical force, who files the environmental public interest litigation in favor of environmental protection. In addition, Plaintiff qualification of the administrative department of environmental protection and prosecutors should not be ignored, because of their limited capacity, their human and material resources is difficult to meet environmental requirements, who should bear the complementary role of environmental public interest litigation. When citizens and social groups are unwilling or unable to sue, they should use the view of protecting the social welfare, positively file the environmental public interest litigation. The entrance of citizens and social groups will broaden the main scope of the environmental public interest litigation, the law should be limited to the abusive complaint behavior of citizen, while it is certain to citizen's environmental litigation right in certain environments.

3.3.2 *Enhance the professionalism of judicial officers*

To increase the quality of environmental public interest litigation cases, the court need improving the professionalism of the environmental court judge, which is mainly from internal and external aspects to enhance. Internally, we not only should positively select the staff own background of environmental science knowledge as a judge, but also enhance their professional training; externally, we should strengthen cooperation with environmental research institutions, and seek their technical support. In addition, improvement to professional quality of environmental judicial officers should fully learn the professional experience and practice abroad.

3.3.3 *Seat arrangement of the environmental court*

The development of China's environmental protection professional trial organization is the lack of direct legal basis. "Court Organization Law" has made clear that the railway court, forest court, maritime court is specialized, but the environmental specialized trial organization for environmental protection belongs to a new thing, its legal status has not been explicitly defined. The development of professional trial organization for environmental protection is an inevitable trend of judicial construction, we should speed up legislation to deal with it.

4 CONCLUSIONS

The environmental judicial protection is an important way to solve environmental problems, China's system of environmental judicial protection has begun to take shape. To further enhance the efficiency of the environmental judicial protection, we should effectively get rid of threshold to the environmental judicial protection, intensify the crackdown on environmental crime, improve the settings of environment Judiciary.

REFERENCES

[1] Zhao Jun. Modification of "Environmental Law" and Determination of the Environmental Public Interest Litigation's Plaintiff[J]. Environmental Economy, 2012(11):17.

[2] Xie Wei. The Role of the Judiciary in Environmental Governance: Germany's Consideration[J]. Hebei Law, 2013(2):84.

[3] Deng Yifeng. Research In Environmental Litigation System[M]. China Legal Publishing House, 2008:126.

Management, Information and Educational Engineering – Liu, Sung & Yao (Eds)
© 2015 Taylor & Francis Group, London, ISBN: 978-1-138-02728-2

Ultimate ownership structure and stock price crash risk: Evidence from China

Zhi Jian Zeng & Yi Jun Zhang
College of Business Administration, Hunan University, Changsha, Hunan, China

ABSTRACT: Using a sample of A-share listed firms in China for the period 2004~2013, this paper empirically investigates the effect of the ultimate ownership structure on stock price crash risk from the perspective of the second corporate agency cost theory. We provide strong evidence that the separation of control power and cash-flow right is positively associated with stock price crash risk. And there is a significant negative correlation between cash-flow right and stock price crash risk. We also find that stock price crash risk is related with the nature of the ultimate controlling shareholder. Compared with non-state-owned listed companies, the state-owned companies may face lower stock price crash risk.

KEYWORDS: Ultimate ownership structure; Stock price risk; Agency problem.

1 INTRODUCTION

In recent years, stock price crash risk has attracted increasing attention from a broad spectrum of parties, including academics, practitioners, and legislators. For instance, Jin and Myers (2006) provide empirical evidence from 40 countries showing a positive correlation between stock market opaqueness and market-wide stock price crash risk. The seminal paper has motivated several follow-up studies that examine the association between stock price crash risk and earnings management, large controlling shareholders' tunneling behavior, and corporate tax avoidance. However, a huge body of research documents that ownership structure affects managerial incentives and therefore exacerbates/mitigates agency problems between controlling and minority investors, which affects firms' information environment and stock returns. This paper brings together these two strands of literature by addressing the important but hitherto underexplored question of whether ultimate ownership structure matters in explaining stock price crash risk. In particular, it focuses on three important corporate governance characteristics in an environment where ownership is concentrated, namely, the ultimate cash flow rights of controlling shareholders, the separation of voting and cash flow rights and the nature of ultimate owner.

2 HYPOTHESES DEVELOPMENT

Due to the extensive use of control-enhancing mechanisms, large shareholders are endowed with enhanced control compared to their interests in the firm, which may give rise to agency conflicts with minority investors. To hide any egregious opportunistic behavior, the ultimate owner may opt for poor disclosure policies by either reducing information disclosure to outsiders or publishing unintelligible, untimely, or irrelevant information. So we claim that the excess control undermines the corporate informational environment. A huge body of researches documents that the probability of stock price crashes is negatively associated with the quality of corporate information environment. Drawing on the above discussion, we posit the following hypothesis.

H1. Stock price crash risk increases with the excess control of the ultimate controller.

Conversely, as the ownership stake of controlling shareholders increases, their interests in the firm become more aligned with those of minority shareholders, which curtails the potential extraction of private benefits. Consequently, the incentive to adopt a poor disclosure policy diminishes, resulting in increased firm transparency. Based on the previous discuss, we present our testable hypothesis as follows.

H2. Stock price crash risk decreases with the cash flow rights of the ultimate controller.

Due to China's socialist environment, the property of enterprise must be considered when we study on the ultimate ownership structure. As state-owned enterprises and non-stated-owned enterprises have different political and economic background, the agency problem and corporate information environment must be different. Specifically, Chinese SOEs receive much support from the government, which will reduce the motivation of earnings management. What's more, as they carry great social responsibility and political cost, the information disclosure transparency is relatively high. In summary, we present the third hypothesis.

H3. Compared with the non-state-owned enterprises, the SOEs face a lower stock price crash risk.

3 EMPIRICAL METHODOLOGY

3.1 *Sample and data source*

Our study uses a sample of A-share listed firms in China, all the data were collected from the China Stock Market and Accounting Research (CSMAR) database. We compute the divergence of control rights and cash flow rights by using the same method as La Porta et al. (1999). We exclude (1) financial services firms, (2) firms whose ultimate shareholders cannot be identified, (3) firms with fewer than 30 weeks of stock return data, and (4) firm–year observations with insufficient financial data to calculate control variables. Following common practice, we winsorize all the variables (except for the dummy variables) at both the 1st and 99th percentiles to mitigate possible data errors and influential extreme observations. Our final (unbalanced) sample consists of 10709 firm–year observations for 1576 unique firms.

3.2 *Variables*

3.2.1 *Dependent variables*

This study employs two measures of crash risk (*CR*), which are constructed following previous studies in the crash risk literature. We first estimate firm-specific weekly returns, denoted *W*, as the natural log of one plus the residual return from the expanded market model regression for each firm and year:

$$R_{i,t} = \alpha_i + \beta_1 R_{m,t-2} + \beta_2 R_{m,t-1} + \beta_3 R_{m,t} + \beta_4 R_{m,t+1} + \beta_5 R_{m,t+2} + \varepsilon_{i,t} \quad (1)$$

Where $R_{i,t}$ is the return on stock i in week t and $R_{m,t}$ is the value-weighted A-share market return on week t. The firm-specific weekly returns for firm i in week t are measured by $W_{i,t}$ = Ln(1+$\varepsilon_{i,t}$), where $\varepsilon_{i,t}$ is the residual in Eq. (1).

Our first measure of crash risk is the negative coefficient of skewness, *NCSKEW*. Specifically, *NCSKEW* for a given firm in a fiscal year is calculated by taking the negative of the third moment o firm-specific weekly returns for each sample year and dividing it by the standard deviation of firm-specific weekly returns raised to the third power. Specifically, for each firm i in year t, we compute *NCSKEW* as:

$$NCSKEW_{i,t} = -\left(\frac{n(n-1)^{3/2} \sum W_{i,t}^3}{(n-1)(n-2)(\sum W_{i,t}^2)^{3/2}} \right) \quad (2)$$

Where n is the number of weekly returns during year t. A higher value for *NCSKEW* corresponds to a stock being more "crash prone" and vice versa.

The second measure of crash risk is down-to-up volatility, DUVOL, which is computed as follows. For any stock i in year t, we separate all the weeks with firm-specific weekly returns below the annual mean(down weeks) from those with firm-specific weekly returns above the period mean (up weeks) and compute separately the standard deviation for each of these subsamples. We then take the log of the ratio of the standard deviation of the down weeks to the standard deviation of the up weeks. Thus we have

$$DUVOL_{i,t} = \log \left(\frac{(n_u - 1) \sum_{DOWN} W_{i,t}^2}{(n_d - 1) \sum_{UP} W_{i,t}^2} \right) \quad (3)$$

Where n_u and n_d are the number of up and down weeks, respectively. A higher value of *DUVOL* indicates greater crash risk and vice versa.

3.2.2 *Independent variables*

The procedure of identifying ultimate ownership and control patterns follows the approach outlined by La Porta et al. (1999). In each layer of the control chain, we identify the direct owner of the firm, the direct owner of this direct owner, and so on, until we reach the ultimate owner, that is, a shareholder who maintains at least 10% of a firm's voting rights without being controlled by anyone else. We compute the ultimate cash flow rights of the largest controlling shareholder (*UCF*) as the sum of the products of direct cash flow rights along the different ownership chains and its ultimate control rights as the sum of the weakest links across all these chains. Excess control (*EXC*) is defined as the difference between the ultimate control (*UCO*) and cash flow rights of the largest controlling shareholder, scaled by ultimate control rights (*UCO-UCF*)/*UCO*. State-owned enterprises (*SOE*) is a dummy variable that equals one when the ultimate owner is state-owned statue.

3.2.3 *Control variables*

Following previous studies in the crash risk literature, we include a set of control variables deemed to be potential predictors of crash risk. Stock return (*RET*) is the mean of firm-specific weekly returns over the fiscal year, times 100. Stock return volatility (*SIG*) is the standard deviation of firm-specific weekly returns over the fiscal year, times 100. Leverage (*LEV*) is the ratio of long-term debt over book assets. Firm size (*SIZE*) is the natural log of the firm's market equity. Return on assets (*ROA*) income before extraordinary items divided by lagged total assets. Market-to-book (*MB*) is the ratio of market equity over book equity and is included to control for the growth status of the firm. Audit opinion (*OP*) is a dummy variable that equals one when the audit opinion type is the standard and unqualified one. To explain the potential serial correlation of *NCSKEW* or *DUVOL* for the sample firms, the lag value of crash risk is controlled. We also include firm and year dummies to control for firm and year fixed effects.

3.3 *Empirical models*

To investigate how ultimate ownership structure is associated with firm-specific future stock price crash risk, we estimate the following regressions that link our measures of crash risk in year *t* to our proxies for excess control, ultimate cash flow rights and the state of ultimate controller in year *t*-1 and to a set of control variables in year *t*-1:

$$CR_{i,t} = \alpha + \beta_1 EXC_{i,t-1} + \beta_2 CR_{i,t-1} + \gamma ControlVariables_{i,t-1} + \varepsilon_{i,t} \quad (4)$$

$$CR_{i,t} = \alpha + \beta_1 UCF_{i,t-1} + \beta_2 CR_{i,t-1} + \gamma ControlVariables_{i,t-1} + \varepsilon_{i,t} \quad (5)$$

$$CR_{i,t} = \alpha + \beta_1 SOE_{i,t-1} + \beta_2 CR_{i,t-1} + \gamma ControlVariables_{i,t-1} + \varepsilon_{i,t} \quad (6)$$

Where the dependent variable, *CR*, is proxied by *NCSKEW* or *DUVOL*; our primary independent variables are *EXC*, *UCF* and *SOE* as discussed above; we also include a series of control variables (discussed in Section 3.2). If H1 is valid, β_1 in Eq. (4) will be positive and significant; if H2 is valid, β_1 in Eq. (5) will be negative and significant; If H3 is valid, β_1 in Eq. (6) will be negative and significant.

4 EMPIRICAL RESULTS

We examine the effect of ultimate ownership structure variables on the future stock price crash risk by running the regression models discussed above. The use of the *NCSKEW* and *DUVOL* measures offers a robust test of our hypotheses. Table 1 and Table 2 present regression results of using *NCSKEW* and *DUVOL* as stock price crash risk measure respectively.

4.1 *Regression results of using NCSKEW as the crash risk measure*

Table 1. Regression analysis on the effect of ultimate ownership structure on crash risk proxied by *NCSKEW*.

	CR_t		
	(1)	(2)	(3)
EXC_{t-1}	0.043*		
	(0.067)		
UCF_{t-1}		−0.145***	
		(0.001)	
SOE_{t-1}			−0.027**
			(0.037)
RET_{t-1}	1.598***	1.571***	1.593***
	(0.000)	(0.000)	(0.000)
SIG_{t-1}	0.096***	0.094***	0.096***
	(0.000)	(0.000)	(0.000)
LEV_{t-1}	0.107***	0.104	0.109***
	(0.005)	(0.006)	(0.004)
$SIZE_{t-1}$	−0.016**	−0.014**	−0.014**
	(0.011)	(0.025)	(0.027)
ROA_{t-1}	0.380***	0.363***	0.360***
	(0.001)	(0.002)	(0.003)
MB_{t-1}	0.050***	0.049***	0.049***
	(0.000)	(0.000)	(0.000)
OP_{t-1}	−0.134**	−0.129**	−0.132**
	(0.012)	(0.016)	(0.014)
CR_{t-1}	0.039***	0.039***	0.039***
	(0.000)	(0.000)	(0.000)
Cons	−0.336**	−0.371**	−0.371**
	(0.032)	(0.018)	(0.043)
Industry	Yes	Yes	Yes
Year	Yes	Yes	Yes
N	10709	10709	10709
Adj- R^2	0.084	0.084	0.084
F	26.70***	27.66***	27.41***

The t-statistics reported in parentheses. Here*, **, and ***indicate statistical significance at the 10%, 5%, and 1% levels, respectively.

4.2 *Regression results of using DUVOL as the crash risk measure*

The first column of Table 1 and Table 2 present the impact of excess control on crash risk. The coefficient of *EXC* in Table 1 is 0.043, which is significant at 0.1 level. The coefficient of *EXC* in Table 2 is 0.055, which is significant at 0.01 level. The results show that the future stock price crash risk is positively correlated with excess control in China. The findings support H1. The second column of Table 1 and Table 2 report the impact of cash flow rights on crash risk. The coefficient of *UCF* in Table 1 is −0.145, which is significant at 0.01 level. The coefficient of *UCF* in

Table 2. Regression analysis on the effect of ultimate ownership structure on crash risk proxied by *DUVOL*.

	CR^t		
	(1)	(2)	(3)
$EXC_{t\text{-}1}$	0.055***		
	(0.001)		
$UCF_{t\text{-}1}$		−0.137***	
		(0.000)	
$SOE_{t\text{-}1}$			−0.025***
			(0.008)
$RET_{t\text{-}1}$	1.100***	1.076***	1.095***
	(0.000)	(0.000)	(0.003)
$SIG_{t\text{-}1}$	0.063***	0.061***	0.062***
	(0.000)	(0.000)	(0.000)
$LEV_{t\text{-}1}$	0.059**	0.058**	0.062**
	(0.034)	(0.040)	(0.026)
$SIZE_{t\text{-}1}$	−0.019***	−0.018***	−0.018***
	(0.000)	(0.000)	(0.000)
$ROA_{t\text{-}1}$	0.165*	0.151*	0.148*
	(0.059)	(0.084)	(0.092)
$MB_{t\text{-}1}$	0.031***	0.030***	0.030***
	(0.000)	(0.000)	(0.000)
$OP_{t\text{-}1}$	−0.010**	−0.096**	−0.098**
	(0.011)	(0.015)	(0.013)
$CR_{t\text{-}1}$	0.029***	0.028***	0.029***
	(0.003)	(0.004)	(0.003)
Cons	−0.004	−0.022	0.023
	(0.974)	(0.846)	(0.838)
Industry	Yes	Yes	Yes
Year	Yes	Yes	Yes
N	10709	10709	10709
Adj- R^2	0.080	0.081	0.080
F	26.10***	26.38***	26.00***

The t-statistics reported in parentheses. Here*, **, and ***indicate statistical significance at the 10%, 5%, and 1% levels, respectively.

Table 2 is −0.137, which is significant at 0.01 level. The results show that the future stock price crash risk is negatively correlated with the ultimate cash flow rights in China. The findings lend strong support to H2. The third column of Table 1 and Table 2 present the impact of statue of ultimate controller on crash risk. The coefficient of *SOE* in Table 1 is −0.027, which is significant at 0.05 level. The coefficient of *SOE* in Table 2 is −0.025, which is significant at 0.01 level. The results show that SOEs face a lower future stock price crash risk in China. H3 is confirmed.

For the control variables, lag value of crash risk (CR_{t-1}), standard deviation of stock mean returns (RET_{t-1}), stock returns (SIG_{t-1}), Leverage (LEV_{t-1}), Return on assets (ROA) and market-to-book ratio (MB_{t-1}) are positive and significant. That is, past crash risk, past return, and past total risk on stock return, past leverage, past return on assets and past market-to-book ratio are all positively related to crash risk. We find that the firm size ($SIZE_{t-1}$) variable is negative and significant. That is, a larger Chinese firm, on average, tends to have a lower crash risk or vice versa. We also find that audit opinion (OP_{t-1}) is negative and significant, that is, a company has a higher information transparency, on average, tends to have a lower crash risk or vice versa. We contend that these variables capture some firm characteristics. Thus, they may be useful to control for firm fixed effect in the regression models.

5 CONCLUSION

This study investigates the effect of ultimate ownership structure on the future stock price crash risk in a sample of 1576 China firms from 2004 to 2013. It provides strong evidence that the separation of control and cash flow rights is positively associated with the future crash risk. Additionally, the study shows that stock prices are less likely to crash when ultimate controller own a large fraction of cash flow rights. Another important finding is that compared with non-state-owned listed companies, the state-owned companies may face lower stock price crash risk.

ACKNOWLEDGMENT

The research is supported by the National Natural Science Foundation of China under Grant No.71340014, the Social Science Foundation of Hunan Province of China under Grant No.09YBA037, the Foundation for Innovative Research Groups of the Natural Science Foundation of Hunan Province of China under Grant No.09JJ7002, the Program for Chang Jiang Scholars and Innovative Research Team of Ministry of Education of China under Grant No. IRT0916.

REFERENCES

Boubaker S, Mansali H, Rjiba H. 2014. Large controlling shareholders and stock price synchronicity. *Journal of Banking & Finance* 40: 80–96.

Chen S, Wang K, Li X. 2012. Product market competition, ultimate controlling structure and related party transactions. *China Journal of Accounting Research* 5(4): 293–306.

Haggard K S, Martin X, Pereira R. 2008. Does voluntary disclosure improve stock price informativeness? *Financial Management* 37(4): 747–768.

Hutton A P, Marcus A J, Tehranian H. 2009. Opaque financial reports, R2, and crash risk. *Journal of Financial Economics* 94(1): 67–86.

Jin L, Myers S C. 2006. R2 around the world: New theory and new tests. *Journal of Financial Economics* 79(2): 257–292.

La Porta R, Lopez-de-Silanes F, Shleifer A. 1999. Corporate ownership around the world. *The Journal of Finance* 54(2): 471–517.

Lee C F, Kuo N T. 2014. Effects of ultimate ownership structure and corporate tax on capital structures: Evidence from Taiwan. *International Review of Economics & Finance* 29: 409–425.

Xu N H, Li X R, Yuan Q B, et al. 2014. Excess perks and stock price crash risk: Evidence from China. *Journal of Corporate Finance* 25: 419–434.

Kim J B, Li Y, Zhang L. 2011. Corporate tax avoidance and stock price crash risk: Firm-level analysis. *Journal of Financial Economics* 100(3): 639–662.

Shen Y, Jiang D, Chen D. 2014. Large Shareholder Tunneling and Risk of Stock Price Crash: Evidence from China. *Frontiers of Business Research in China* 8(2): 154–181.

Management, Information and Educational Engineering – Liu, Sung & Yao (Eds)
© 2015 Taylor & Francis Group, London, ISBN: 978-1-138-02728-2

The ways of synergistically developing e-commerce in Beijing-Tianjin-Hebei region

Zheng Rong Chen
Langfang Polytechnic Institute, China

ABSTRACT: Beijing-Tianjin-Hebei region, as the most developed economic zone in northern China, plays an important leading role. In the synergistic development setting, in-depth cooperation of e-commerce in Beijing-Tianjin-Hebei region, can promote system optimization and green integration of relevant industries in this area, which will lead consumers to form a low-carbon green consumption concept and lifestyle. As to the synergistic development of e-commerce, Beijing-Tianjin-Hebei region has good fundamental conditions, but there are also some problems. Therefore, this article focuses on how to effectively develop e-commerce and puts forward some constructive suggestions.

KEYWORDS: Beijing-Tianjin-Hebei region; Synergistic development; e-commerce.

Beijing-Tianjin-Hebei region is the most developed economic zone and the largest industrial concentrated district in northern China. The region is also the material distribution hub of the northeast region and the hinterland of the Central Plains. In addition, it is the main access to the sea in Inner Mongolia, Shanxi, Shaanxi. The surrounding region makes full use of this favorable condition to vigorously promote and output its local products. On the other hand, it also introduces products needed by way of these hubs.

To promote the synergistic development of Beijing-Tianjin-Hebei region is an inevitable choice to solve outstanding contradictions and problems existing in the three regions and is also an intrinsic requirement for optimizing the layout of national regional development and social productive forces to build a new economic growth pole and form a new mode of economic development. To synergistically develop e-commerce in Beijing-Tianjin-Hebei region is one of the strategic policies of economic development.

1 CURRENT SITUATION OF SYNERGISTICALLY DEVELOPING E-COMMERCE IN BEIJING-TIANJIN-HEBEI REGION

1.1 *Basic conditions of synergistically developing e-commerce*

Beijing, Tianjin and Hebei have a better basic condition to synergistically develop e-commerce. In 2011, a horizontal comparison was made between the three regions and other provinces. The results were as follows: the Internet users' popularizing rate in Beijing reached 70% ranked first in the country and the Internet users' growth rate was high; the Internet users' popularizing rate in Tianjin was 55.6% ranked sixth and the Internet users' growth rate was also high, but compared with the other 16 provinces, it was relatively slower; while in Hebei province, the Internet users' popularizing rate was 36.1%, accounting for a third of the entire province. Though it was low, the Internet users' growth rate in the country was very high. To analyze from the growth of Internet users, Beijing had the best foundation to develop e-commerce, followed by Tianjin. Although the number of Internet users in Hebei Province was relatively small, it had great potential because of its big development space and fast speed of development.

In 2013, the volume of trade on e-commerce in Beijing amounted to 797.5 billion yuan ranked first in the country. In Tianjin, its volume of trade on e-commerce in 2013 was over 300 billion yuan, which accounted for 3% of the country. But in the overall rankings of the domestic development level of e-commerce, it was not included in the previous ten. On the first half of 2013, the volume of trade on e-commerce in Hebei Province achieved 400 billion yuan and the purchase amount of online shopping was 39 billion yuan.

1.2 *Domains of the synergistic development of e-commerce*

The synergistic cooperation in Beijing-Tianjin-Hebei region has made a lot of new progress in the field of e-commerce. Here mainly introduce the following 3 aspects:

i. Synergistic development of tourism e-commerce

Beijing is the capital of China; Tianjin is an important port city in China; Hebei is surrounded by Beijing and Tianjin. So there is a natural, close connection in the three regions in terms of economy, society and culture. In tourism resources, the three has always been the tourist generating regions and destinations. Some of Beijing's enterprises value the tourism resources and the potential of the tourism market in Beijing-Tianjin-Hebei region. They invest and construct tourism projects in Tianjin and Hebei. For example, in March 2009, "Beijing-Tianjin-Hebei one-card in travel" was launched in Tianjin for the first time which was a travel pass launched jointly by Beijing, Tianjin and Hebei. In 2009, visitors visited 58 scenic spots by using this card. The launch of the card was a landmark of the three regions' tourism cooperation which created an obvious stimulating effect on the trans regional tourism and played a role in the expansion of domestic demand in the tourism industry and integrated the tourist attractions' resources of the three region in a market-oriented way to promote the prosperity of the tourism industrial chain.

ii. Synergistic construction of the logistics transportation network

The accelerated development of e-commerce in Beijing, Tianjin and Hebei rely largely on the service ability of the transportation and logistics system. At present, the governments of the three have closer cooperation and jointly promote the construction of various transport capacity in the region. The region contains 35 expressways and is connected by 280 national and provincial trunk roads which basically form a three-hour of urban traffic network covering Beijing, Tianjin and 11 municipal-level cities in Hebei province. Beijing, Tianjin and Hebei open more than 900 road passenger transport routes. Coastal ports' carrying capacity as designed in Tianjin and Hebei is 705 tons, accounting for 16% of the whole nation. Traffic integration of Beijing-Tianjin-Hebei region has shown its initial shape. The public transportation line in Beijing has extended to Zhuozhou, Langfang and Sanhe in Hebei province. Municipal Highway Bureau in Tianjin actively strengthens the communication and coordination with Hebei Province. They jointly make a plan for constructing 45 interface highways. Moreover, the transportation departments of the three have established a joint coordination mechanism in terms of the expressway network toll, toll by weight and the control of overload on highways.

iii. Good conditions of application of e-commerce in large enterprises

Data shows that about 30% of large enterprises in Beijing-Tianjin-Hebei region have established e-commerce system and have a higher level of e-commerce. Online sales of goods and services account for about 31% of the total sales; online purchase of goods and services account for about 16% of the total amount. The development direction of e-commerce in some large enterprises has now being changed from its basic application like the release of online information, purchasing and sales to a full range of business like online design, manufacturing and planning management in the companies up and down the supply chain, which makes the synergistic development between enterprises significantly improved. At present, large enterprises like Sinopec, CNPC (China National Petroleum Corporation), Sinochem and COFCO (China Oil&Foodstuffs Corporation) have carried out different forms of e-commerce. The development of e-commerce is directly driven by information industries which will have an obvious effect on the efficiency and quality of basic industries, manufacturing industries and traditional service industries.

2 EXISTING PROBLEMS IN THE SYNERGISTIC DEVELOPMENT OF E-COMMERCE IN BEIJING-TIANJIN-HEBEI REGION

2.1 *The existence of multi stakeholders between regions*

Due to different administrative relationships and traditional economic system, there is "local protectionism" in Beijing-Tianjin-Hebei region which segments the market and resources and hinders the natural flow of production factors and the maximum optimal allocation and also ignores the role of comparative advantage to form a kind of vicious competition in the region.

The local government, as a subject of interest, in order to gain its interests, strives for preferential policies which not only makes it play a game with the central government but also has an intense competition with other local governments. The excessive competition between local governments in the region will have a negative impact on the regional economic development of the whole society which can even make macro economy out of control.

In addition, there is a lack of a long-term management mechanism for administrative officials who have a short term of office. In order to make achievements in a short term, each government has a very strong performance desire and pays more attention on the recent development and prefers some projects with little investment but quick result. Encouraged by tax revenue and economic growth, the government is particularly active in developing some high taxes, high profit projects. It becomes quite difficult for the government to think about long-term regional benefits from the perspective of its local development which is an important reason for the failure of cooperation among the local government.

2.2 *The overall level of applying e-commerce is not high*

The application and service of e-commerce for most enterprises in Beijing, Tianjin and Hebei mainly rely on the release of online information, information communication and off-line trading. The breadth and depth of the application and service of e-commerce are not enough. Many e-commerce enterprises still remain in a primary form like online stores and portals. E-commerce services are mostly related to the purchasing and sales of the enterprise chain. The small service coverage cannot extend the core business processes and customer relationship management to the Internet to meet the customers' requirement for the products and services. The unscientific website positioning cannot shift the "big and comprehensive" mode to a professional subdivided business portal. There is a low degree of interaction and synergy in the intra industry and inter industry. The existence of these problems causes a big gap which can't play a role of e-commerce and can't promote the transformation of economic development patterns. Although some websites of e-commerce in Beijing-Tianjin-Hebei region have made some good attempts and profitable explorations, the overall level of application is not high. For those e-commerce networks that have been built, they don't actually enter the practical stage in terms of management, technical standards, information content, communication speed and resource sharing.

2.3 *The lack of e-commerce talents*

The core competitiveness of an enterprise is a composition of forces which impacts on the survival and development of the enterprise. The cluster of the human capital is a key factor. At present, the process of enterprise informatization has become more and more rapid. E-commerce has become an important means of enterprise management. However, due to the lag of talent training, talent shortage has become the most basic and urgent problem of developing e-commerce in China.

The survey of the demand of e-commerce talents in enterprises shows that the majority of small and medium-sized enterprises which have applied e-commerce have few interdisciplinary talents who can understand marketing and operation of e-commerce. Talent determines the future of the network economy. There is an urgent need of interdisciplinary talents who not only understand business management but have network knowledge with the development of e-commerce. The majority of small and medium-sized enterprises pays less attention and invests less money in the introduction of talents because of funding, system and some other reasons which becomes a bottleneck restricting the development of e-commerce.

3 THE WAYS OF SYNERGISTICALLY DEVELOPING E-COMMERCE IN BEIJING-TIANJIN-HEBEI REGION

3.1 *Strengthening the government's overall planning and formulating security measures*

The government makes policy guidance and support clear and improves the consciousness of developing e-commerce in the three areas to create a good environment for the great-leap-forward development of e-commerce. To improve basic facilities, promote e-commerce application and cultivate emerging industrial formats will bring many advantages of gathering future e-commerce resources.

Strengthening the government's overall planning, policy guidance, market norms and coordinated promotion can fully optimize the development environment of e-commerce. The effective and rapid development of e-commerce can be promoted by way of fully arousing the enthusiasm of the enterprises and society to develop e-commerce and making the enterprises play a major role in setting up platforms, innovating patterns and deepening the application and also by allocating market resources.

A statistical indicator system of synergistically developing e-commerce needs to be formulated. Relevant standards and norms should be stipulated explicitly. A statistical system of e-commerce should be established gradually. E-commerce statistics will be brought into the economic statistical system. E-commerce enterprises' directory should be improved and those above the quota should be brought into the scope of statistics. To improve the mechanism of releasing information and regularly release reports on the development of e-commerce in the three areas can provide a basis for e-commerce macro management and government decision-making.

3.2 *Innovating services and upgrading the industrial chain*

Relying on high-end industries, top talents and capital platforms in Beijing, its surrounding areas like Langfang, Baoding, Tangshan, Chengde and Zhangjiakou build big data and industry belts of e-commerce. The transformation and upgrading of conventional industries in Hebei can make use of the means of combining e-commerce and modern finance. The industries can also integrate the three main industries and rely on industrial clusters to create a new industrial pattern. Meanwhile, they can take advantage of the position advantage, port advantage, industry advantage and policy advantage to deepen the application of e-commerce in the field of staple commodities, advantageous manufacturing industries and foreign trade. They can mainly focus on developing B2B (business to business) transaction, and positively develop B2C (business to consumer), C2C (consumer to consumer), O2O (online to offline), mobile e-commerce and other new patterns of e-commerce. They need to actively attract leading enterprises of e-commerce and famous enterprises to be settled in Tianjin where they can construct a high-level demonstration base of e-commerce.

3.3 *Cultivating e-commerce talents*

Hebei province attaches great importance to the cultivation of e-commerce talents in the Twelfth Five-year Development Plan which points out that government guidance needs to be strengthened; the interaction of schools, social training institutions and enterprises should be promoted; school education, continuing education and on-the-job training need to be carried out in various channels; institutions of higher learning, higher and secondary vocational colleges are encouraged and supported to improve course education of e-commerce and to actively exchange and cooperate with enterprises which will quicken the cultivation of interdisciplinary, skilled and practical talents of e-commerce.

The government should be involved in the vocational training system of quality assurance and development, course offered, examination rules and core materials' certification and promoting innovation, etc. According to different positions, it should formulate unified acceptance standards and financial subsidies standards. It should also guide the vocational training to be "quality-oriented".

Vocational colleges should closely cooperate with e-commerce enterprises and allow them to participate in the decision of teaching and examination content and let the enterprise's senior director of human resources and career development mentors involve in the students' vocational counseling, employment guidance, interview skills and other aspects of training. At the same time, the enterprise should provide more opportunities of internship for students in order to cultivate more skilled talents.

Author: Chen Zhengrong(1968.5-), associate professor, Langfang Polytechnic Institute, Department of Business Management, research field: e-commerce theory and practice.

REFERENCES

[1] Cao Baogang. Study on the synergistic development of Beijing-Tianjin-Hebei [M]. Hebei University press, 2009.

[2] Lv Hongliang. Rise of e-commerce and innovation of the disabled employment patterns[J]. Modernization of shopping malls, 2010 (17).

[3] Deng Xinjie, Zhu Xiaorong. Analysis of e-commerce of Yiwu Business Technology Institute[J]. Practice and exploration, 2010 (6).

[4] Tang Canqing, Dong Zhiqiang, Li Yongjie. Study on research status of employment relations and future prospect[J]. Foreign economy and management, 2011 (9).

[5] Feng Yanhui. The role of e-commerce in foreign trade of China. Knowledge economy.2012.

Management, Information and Educational Engineering – Liu, Sung & Yao (Eds)
© 2015 Taylor & Francis Group, London, ISBN: 978-1-138-02728-2

On the mental health diathesis of netizens

Rong Wei & Mei Ling Jin
HeFei University of Technology, China

ABSTRACT: Netizens' Mental Health Diathesis is some inherent, relatively stable psychological quality by their rational use of the network and avoid the hazards. This paper studies using structural analysis, research methods to tease out the expressive dimension of the Mental Health Diathesis of Netizens, at the same time summarize the composite features of the Mental Health Diathesis of Netizens. On the basis of the above theoretical studies, combined with operating rules of the network society and its psychological demands of netizens, using practice research method, producing results that expanding the scope of teaching, increasing thematic training, building team education, strengthening the screening and tracking of mental health diathesis, expanding dissemination channels of health care knowledge, methods and techniques of guiding self-psychological adjustment, etc. are important ways to cultivate Netizens' Mental Health Diathesis.

KEYWORDS: Netizens; Psychological Quality; Mental Health; Diathesis.

1 INTRODUCTION

With the expansion of the breadth and depth of netizens' activities in cyberspace, the dual impacts of the net, positive and negative, which they have met has become even more significant. During the course of social networking, online business, network of political and other activities, netizens are, on the one hand, enjoying a carefree brought by weaker constraint of networks, on the other hand, also bearing atmosphere pressure created by network public opinion, which raised a lot of demands of netizens 'mental health datasets. Netizens' Mental Health Diathesis refers to some inherent, relatively stable psychological quality that the individual can reasonably use the Internet and avoid the hazards under the combined effect of genetic and environmental, these psychological qualities influence or determine psychology and behavior of individual netizens. Netizens' Mental Health Diathesis is an "alloy" of congenital and acquired with a strong plasticity. Netizens, interacted with other individuals or groups in a virtual space, by choosing, changing, adapting, and continuously improving their coordinated cognitive ability and mental adaptive capacity with other individuals or groups in the network virtual space, which make the netizens to comply with the network laws and regulations, have a good network of ethics, to maintain a relatively stable mood, to use the Internet to promote their skills and academic levels, to be proactive in the face of stress in the network and so on, show good mental health quality.

2 PERFORMANCE DIMENSIONS OF NETIZENS' MENTAL HEALTH DIATHESIS

Netizens' Mental Health Diathesis is the psychological phenomenon of network individuals in virtual space, and this psychological phenomenon includes the quality about cyber self which point to the internal, as well as networks personality qualities, motivation for Internet use, cognitive abilities on network information and network individuals' coping style in virtual space, etc. which point to the external. I referenced Shen Deli's structural division of Mental Health Diathesis[1], and established the performance dimensions of netizens' mental health diathesis as network adaptability, network interpersonal quality, network personal qualities, network power systems, cyber self, network cognitive abilities and network coping style.

1 Network Adaptability
   Network Adaptability dimensions mainly include: Network Emotional Adaptation, reactive state of psychology and behavior caused by emotional changes; Networks Interpersonal Adaptation, reactive state of psychology and behavior raised by relationships changes in the network; Network

Learning Adaptation, reactive state of psychology and behavior caused by using the Internet for learning activities; Network Environment Adaptation, reactive state of psychology and behavior caused by the conversion of network and the reality environment.

2 Network Interpersonal Quality
Network Interpersonal Quality dimensions mainly include: Interpersonal Skills in the Network, refering to network using individual's expression ability, enthusiasm, etc.; Interpersonal Regulation in the Network, referring to the netizens using communication skills to achieve interpersonal ability when interacting with other individuals, groups; Interpersonal Perception in the Network, refering to the netizens' understanding of relationships between both sides of the network exchange object and its impact.

3 Network Personality Quality
Netizens' Personality Qualities we concerned about in the study are those that can reflect mental health-related personality qualities in the network. Reference to definition of personality quality theory latitude given by domestic scholars on the basis of the research on Cartel's sixteen personality factors[2], consolidated related theories, we believe that quality of the network personality dimensions mainly include: Netizens' Inside and Outside Tendencies as well as their acceptance in cyberspace, etc.; Netizens' Will Factors, namely self-restraint capacity of netizens, the ability to govern behavior and to overcome difficulties; Optimistic-Pessimistic, ntizens' emotional tone in cyberspace; Netizens' Independent Factor, thinking and behavior independence of netizens of not affected by complex rhetoric in cyberspace.

4 Network Drive System
Network drive System dimensions mainly include: Target Drive, the online behavior of netizens is purposeful or planned; Growth Learning Motivation, the online motivation of netizens is guided by the purpose of certain related growth needs; Entertainment Motivation, the online motivation of netizens is guided by the purpose of entertainment needs.

5 Cyber Self
Cyber Self dimensions mainly include: Individuals' Self-Cognition and Self-Evaluation in the network. Network Self-Cognition includes three aspects: General Self-Perception, netizens' overall subjective perception of themselves in the network; Physical Self, netizens' subjective perception of body image of themselves on the network; Emotional Self, netizens' subjective perception of emotions, the mental activity of themselves in the network. Network Self-Evaluation includes two aspects: General Self-Efficacy, netizens' general perceived ability and belief of themselves in the network; Self-Esteem, netizens' degree of self-acceptance and self-evaluation of the network.

6 Cyber Self
Cyber Self dimensions mainly include: Individuals' Self-Cognition and Self-Evaluation in the network. Network Self-Cognition includes three aspects: General Self-Perception, netizens' overall subjective perception of themselves in the network; Physical Self, t netizens' subjective perception of body image of themselves on the network; Emotional Self, netizens' subjective perception of emotions, the mental activity of themselves in the network. Network Self-Evaluation includes two aspects: General Self-Efficacy, netizens' general perceived ability and belief of themselves in the network; Self-Esteem, netizens' degree of self-acceptance and self-evaluation of the network.

7 Network Coping Style
Network Coping Style dimensions mainly refers to coping styles netizens adopted when facing conflict stressors in the network, Psychologist Janis has proposed five kinds of coping styles which has certain representativeness, including Continuing of Conflict-Free, netizens do not react to stress stimuli; Changing of Conflict-Free, netizens accept all reasonable proposals in the face of stress stimulator and change behavior without conditions; Resistant Escape, netizens react by taking rationalization, delaying, denying, blaming others, and other psychological and behavioral to avoid pressure when confronting with stressors; Excessive Vigilance, impulsive reactions and decisions made by netizens when facing with stress stimuli; Vigilant, after systematic understanding of the information, netizens take reasonable and flexible way to solve the problem when confront with stressors, which is the most mature coping style.

3 CHARACTERISTICS OF NETIZENS' MENTAL HEALTH DIATHESIS

1 Biological and Social
Netizens' Mental Health Diathesis is one of the many psychological phenomena of individuals', from the study of genetics, evolution, nervous system physiological mechanism formed by physiological function and psychological quality, etc., Netizens' Mental Health Diathesis can not get rid

of the biological basis, although biological can not directly determine netizens' mental activity and behavioral outcomes, it can affect psychological development trends and development threshold limit value of netizens. The formation and development of Netizens' Mental Health Diathesis are on the basis of natural attributes, also deeply branded with the mark of the role of the network society, and individuals has already not a purely organism since entering the social environment after birth, netizens' activities in the network society make their consciousness, mental, spiritual and behavioral with a certain traces of network social impact. netizens constantly adapt and change individual's psychological structure through acquisition, migration, conversion and other forms, and reflect the biological and social unity of their mental health diathesis.

2 Implicit and Explicit
Netizens' Mental Health Diathesis is a series of psychological quality after the formation of the three interactions: on the basis of congenital physical qualities, consolidated individuals' acquired knowledge and stimulation of the external environment. Human psychological quality puts man's natural attributes as conditions for development, constrained by individual nervous system, characteristics of brain as well as sense organs and organ movement characteristics, with some implicit, as subordinate concept of psychological quality, mental health diathesis also has implicit. Meanwhile, the effect of netizens' mental health diathesis has to be reflected through netizens' explicit emotional, behavioral, cognitive, etc. in network virtual space, therefore netizens' mental health diathesis exists in implicit attribute and expresses in explicit way.

3 Stability and Derivative
Once formed, Netizens' Mental Health Diathesis would present in the organic body in a certain structure and format, with relative stability. Netizens' Mental Health Diathesis is inherently stable psychological quality formed after multiple conversions of the network information through the "internalization - externalization - internalization" by individual citizens, they usually will not occur significant changes due to the appearance or disappearance of any single event at a particular time. But netizens' mental health diathesis is also not rigid solid, with changes in the network environment, the process of netizens interaction with other network individuals and groups, has certain feedback regulation force to their psychology health diathesis.

Netizens' psychological and behavior often derived from psychological quality that contribute to adapt to the network environment better from its biological properties, thereby enhance the subject's behavior initiative.

4 CULTIVATING THINKING OF NETIZENS' MENTAL HEALTH DIATHESIS

The basic goal of cultivate netizens' mental health diathesis is to guide netizens to actively adapt and maintain psychological balance, to promote active development, and to stimulate creativity.

Mental health diathesis cultivating is not only for the individual who has already appeared obvious network psychological disorders, it also needs to face broader health netizen groups, to promote the improvement of netizens' psychological function, to optimize netizens' psychological development conditions, therefore, mental health diathesis cultivating should not only take the scope of cultivation on "face" into account, but also should pay attention to the characteristics of nurturing on "point".

1 Extensive set- up of netizens' mental health programs
Taking schools, communities, enterprises, NGO, etc. as the carrier, setting up network mental health diathesis development programs in online or offline organizations, encouraging netizens to participate in the study relevant courses, especially for student groups, it is necessary to integrate network mental health diathesis development programs into education and teaching activities in school. Mental health quality education classes which face to netizens should pay more attention to the learner's participation and practice, to meet the development needs of different learners, in addition to teaching and popularization of relevant psychological knowledge, it should be flexible in applying a variety of educational tools in teaching activities, teaching netizens in accordance with their different ages and backgrounds to realize sufficient interaction and psychological exchanges between teachers and students, to help netizens master principles and skills of interacting with others and handling problems in cyberspace, to maintain a healthy state of mind, so as to achieve the purpose of the course of education through knowledge sharing, exchange of experiences, cognitive remediation and other methods.

2 Organizing Featured Training enhancing the quality of mental health
Featured Training is specialized training carried out in the process of netizens' network mental health development which focuses on some specific issues. Featured training takes promoting healthy and balanced development of the quality

of the overall structure of netizens' mental health diathesis as a fundamental goal; and puts the development of certain advantage elements in neitizens' mental health diathesis as specific goals. Featured training can be divided into systematic training and specialized training, systematic training aims at systematic cultivation on netizens' overall mental health diathesis, specialized training points to training for key issues.

Featured training of cultivating netizens' mental health diathesis should be in a democratic, open and relaxed atmosphere, to create question context which is valuable, comply with the development level of various objects' network mental health diathesis, and cause participants to think deeply, guiding trainees master scientific methods, procedures, and policies on how to think, act, experience and reflect. Featured training emphases on targeted and interactivity, to reach emotional resonance and excitation among learners, between instructors and learners, in addition, educators should guide students take an instant reflection, induction, summarize on training proceeds during training, in order to consolidate the results of the training.

3 Strengthening team building of Mental Health Diathesis Education

Constructing an education team with a higher network psychological theory literacy and stronger practical ability is the basic condition for cultivating netizens' mental health diathesis. National and provincal departments of mental health centers, Family Planning Commission and other relevant departments of various provinces and cities should consider expanding and building education teams of network mental health diathesis as one of the important work.

Through policy guidance and support to improve the professional standards of network mental health education team, optimizing education teacher structure, using special, concurrent post, recruited and other forms, to protect the stability of educational personnel in the main field of education such as schools, communities, etc.; to encourage staff with qualifications on mental health diathesis counseling and treatment to expand diversified forms of service and education forms, to meet the growing needs of the community.

4 Regularly Survey on Netizens' Mental Health Diathesis

Human psychology is always in constant change and development process, understanding netizens' mental health level contributes to take targeted cultivation measures. Carrying out regular mental health diathesis surveys and evaluation to netizens on community-based, establishing netizens' mental health diathesis files, distinguishing different levels of mental health diathesis according to the survey results, particularly concerning groups with lower level of mental health diathesis, and keeping track maintenance and feedback, meanwhile targeted screening individuals and groups whose individual dimensions of mental health diathesis is more prominent, imposing appropriate interventions or incentives measures, will provid the basis for formulating feasible training program of netizens' mental health diathesis.

5 Expanding distribution channels on Mental Health Diathesis Knowledge

Creating multiple channels of information dissemination to popularize health knowledge of mental health in the real world as well as network platform.

Regularly or irregularly organizing seminars about network mental health diathesis knowledge, choosing excellent video lectures as a learning resource recommended in the network; conducting online discussion contrapose constantly emerging issues in the network, to clarify viewpoints in the debate, to raise the cognitive level of netizens; carrying out social contest on mental health related knowledge, to intensify propaganda; collecting and rewarding propaganda works with the theme of network mental health diathesis; taking World Mental Health Day and other anniversaries as an opportunity to prepare a wealth of activities, in order to enhance people's awareness of an active interest in health care knowledge of mental health diathesis. Relying on the majority media to create a correct understanding of mental health issues, the courage to face and solve the problem of social atmosphere.

6 Improving and unimpeding psychological counseling channels, guiding the netizens to master the methods and techniques of self-regulation

Using a variety of ways for netizens' psychological counseling and guidance, including individual counseling and group guidance in real-world environment, conducting online tutoring by creating psychological counseling site, at the same time through letters, phone calls, SMS, QQ, MSN and other forms as well, target to provide netizens with timely and effective guidance and services on mental health.

Cultivating netizens' mental health diathesis, external measures play a role of booster while self-regulation plays a role of source of power. Therefore, there is a need to apply psychological counseling as well as featured training and other ways to guide netizens acquiring the necessary self-regulation methods and techniques, including relaxation techniques, aerobic exercise regulation, target transfer strategies, psychological catharsis, break the inertia of thinking, optimistic attribution, positive self recognition and so on. Effective self-regulation can reduce the stimulus intensity subject accepted and its associated occurrence of cognitive, to replace threatening or passive feeling with non-threatening peaceful feeling, form a healthy state of mind [3]. Therefore, using these regulating methods can help ease the psychological pressure of netizens, make them examine their own more objectively, enhance independent consciousness and self-discipline, take a dialectical view of contradictory, form a flexible way of thinking, conduct rigorous logical reasoning, improve netizens' ability of independent thinking and behavior.

ACKNOWLEDGEMENTS

Thanks to financial support of the 2014 National University Outstanding Young Political Theory Teachers merit aid projects and Major teaching reform project of Anhui Province (2013zdjy016).

REFERENCES

[1] Shen Deli, Ma Huixia. 2004. On Mental Health. Psychological and Behavioral Research.

[2] Zhang Dajun, Chen Xu.2009. Chinese college students' mental health quality survey. Beijing: Beijing Normal University Press.

[3] Written by Brian Luke Seaward forward, Translated by Xu Yan, etc.2008. Stress Management Strategies. Beijing: China Light Industry Press.

Management, Information and Educational Engineering – Liu, Sung & Yao (Eds)
© 2015 Taylor & Francis Group, London, ISBN: 978-1-138-02728-2

Analysis of consumer behavior in e-commerce environment

Zhi Sheng Dong
International College, Qujing Normal University, Qujing Yunnan, P. R. China

Gu Sheng Zhu
Editorial Department of JQJNU, Qujing Normal University, Qujing Yunnan, P. R. China

ABSTRACT: As a new way of shopping, online shopping will become one of the major ways of people's daily shopping in the future society, because the internet will become highly developed in the future. People choose to do online shopping, mainly because of the abundant resources and cheap prices of online goods. Online shopping consumers tend to form these consumer behaviors, namely, relying on online shopping, focusing on the shopping experience, relying on evaluation and enjoying evaluating etc.

KEYWORDS: E-commerce; Consumer; Consumer behavior; On-line shopping.

1 INTRODUCTION

In recent years, with the increasing number of Internet users and mobile phone Netizen, the explosive development of online shopping in China caused by the improvement of online payment security, online payment convenience and the rapid development of the logistics industry, China's e-commerce has developed rapidly. According to the 34th China Internet Development Report, the number of internet users in China reached 632 million, and the number of mobile phone Netizen reached 527 million up to June 2014 [1]. Online shopping in China has gained tremendous progress due to the large base of internet users and the development of mobile interconnection technology. According to the Ali Institute, the sales volume of Taobao and Tianmao reached 35.019 billion Yuan on 11th November 2013, which is also called "Double-Eleven Festival", and the sales volume of Taobao and Tianmao reached 1.4 trillion Yuan in 2013. According to the report of the McKinsey Institute, it can be predicted that China's online shopping market size will reach 4.2 trillion Yuan by 2020. It can be foreseen that online shopping will be bound to become an important way of people's daily life consumption in the future. Under the e-commerce environment, consumer purchasing behavior has changed in some aspects because of the different characteristics of online shopping and the traditional way of shopping. Analyzing consumer behavior under e-commerce environment has both theoretical significance and practical guiding significance.

2 MOTIVATION ANALYSIS ON CONSUMER ONLINE SHOPPING IN E-COMMERCE ENVIRONMENT

2.1 *To meet niche requirement*

With the advent of Abundance Economic Era, consumers are increasingly demonstrating a wide range of values. More and more people are taking pride in the value orientation of the non-mainstream, which makes the popular products no longer have the same high status in the minds of consumers as it did in the past. Moreover, as the improvement of people's living standards, more and more consumers are pursing personal, customized niche products which can reflect their unique personalities. Different consumers have different niche demands and different demands for the product. The massive amount of online products ensures that different consumers can find what they want on the internet. In fact, the reason why online shopping has been able to achieve rapid development is that consumers are becoming more and more dependent on online shopping because they can easily find the products they need on the internet.

2.2 *Convenient, efficient and save time*

With the development of the society, people's work pressure and life pressure are increasing in more and more cities, and people's pace of life is becoming faster and faster. People's time is becoming more and more valuable. Meanwhile, traffic congestion has become more and more serious in some large and medium-sized cities, which brings a lot of inconvenience when people go out. It takes people longer and longer time to go out to buy daily necessities. But for most urban residents, especially young people, they need more time to work, study, entertain and rest. So spending a lot of time going out to buy daily necessities has become a burden for them. We can say that the physical cost, time, cost and spiritual cost have exceeded the monetary cost spent when people go out to buy daily necessities in some cities. Online shopping can be a good solution to this problem. At present the delivery staff will provide home delivery service, so consumers can get products on their doorstep. What's more, online payment has become faster and securer. People can finish all the procedures from selecting goods, placing an order to complete the payment within a few minutes if they buy daily necessities, which saves a lot of time for people.

2.3 *Abundant resources of online goods*

Abundant resource of online goods is an important reason why consumers are willing to do online shopping. With the development of Internet, network has covered every corner of the world. In the near future, with the emergence and development of some global network platforms, such as Tianmao International, Amazon and others, online shopping products will come from all over the world. Moreover, with the development of technology and the widespread application of cold-chain logistics in the logistics industry, the last barrier of sales and transportation of whole categories of online products will be broken down. People can buy fresh products on the Internet, which will greatly enrich online product resources. People can buy whatever they can buy in real life on the internet. Moreover, suppliers of online products come from all over the world, and many suppliers sell local specialty. It is difficult for most consumers to buy these products if there is no network. Actually, online shopping provides more product categories for consumers to choose.

2.4 *Cheap price of online goods*

For the same product, the price of it on the internet is much lower or lower than the price of it in real life. In real life, products will come into the hands of consumers after going through several different levels of distributors and retailers, which increases the transferring costs, loading and unloading costs, warehouse storage costs at different levels of distributors and retail store rent costs, etc. And distributors and retailers at all levels want to earn profit, which makes the costs of products in real life much higher than the price of online products. Furthermore, the price of some products in some remote areas is much higher than the price of it in other economically developed coastal areas because of the underdeveloped economy, the relatively isolated information and the relatively underdeveloped business in these remote areas. Online shopping allows people in these remote areas to have the same price level with consumers in other places of the country. Although shipping cost will be a little more expensive, the price of the same products on the internet will be much lower than the price of it in real life. In order to attract consumers' attention and shape good brand image of high quality and reasonable price, e-commerce enterprises will introduce one or several so-called "popular series", the price of which will be rather low and enterprises have little profit on these products. The reason why enterprises take this action is that they want to attract consumers' attention. If consumers happen to need this kind of products, they will feel that online products are very cheap. Moreover, with the improvement of online shopping, the quality of online products will be gradually guaranteed.

2.5 *Convenient price comparison in online shopping*

One of the reasons why many consumers are keen on online shopping is that it is easy to make price comparison in online shopping, which makes consumers feel that they have transparent consumption. When buying products in real life, consumers will be confused about the price because it is difficult for them to make price comparison. Consumers will subjectively think that the middleman have earned a lot of profit, especially when buying some less frequently used products. In real life, when selecting a product in a store, people have to go to other places if they want to make price comparison. With the development of the internet, information searching cost has become significantly lower. For the consumer, the information searching cost approximates to zero, which can be neglected. For example, when a consumer finds an appliance on Taobao, they can immediately and easily make price comparison on Jingdong, Amazon and other websites if they are not sure whether the price is reasonable or not, and the switching cost for price comparison approximates to zero. What consumers need to pay is just a little time costs, which is far less than the price comparison costs in real life.

2.6 *The long tail theory is applicable at online shopping market*

The concept of Long Tail was first put forward by Chris Anderson, the chief editor of American Wired Magazine, in an article titled The Long Tail in October 2004. It was put forward to describe the business and economic model of some websites such as Amazon and Netflix [2]. In the Long Tail theory, it is believed that as long as the storage and distribution channel are large enough, the market share of the products with slack demand or poor sales can be equal to or even larger than the market share of some best sellers whose amount is not very large. Consumers on the internet are hundreds of millions of Chinese consumers, and with the development of the internet, consumers on the internet will be billions of consumers from all over the world. Therefore, the large base of consumers ensures the sales of niche products, and e-commerce enterprises should ensure that their products are attractive to consumers with niche needs.

3 CHARACTERISTICS OF CONSUMER ONLINE SHOPPING BEHAVIOR IN E-COMMERCE ENVIRONMENT

3.1 *Easy to become dependent on online shopping*

In recent years, it has been frequently reported that consumers have addicted to online shopping, and there is a teasing expression, widespread in the society, which goes like this, "Online shopping makes three generations poor and Taobao ruins one's life" [3]. There are many real cases about online shopping. For example, a lady in Suzhou has been complained for her online shopping addiction, and she cut off her finger by herself. A lady in Zhejiang was crazy about online shopping, and her husband was so angry that he wanted to divorce. A man in Heilongjiang fainted after continuous online shopping for 18 hours [4]. This actually reflects the current status of many online shopping consumers. They have addicted to online shopping. When they first get to know online shopping, they just buy what they need, or they just get around in shopping websites to see whether there is something they need when they are bored. Later, when they are addicted to online shopping, they often want to search for shopping websites over a certain period of time. If they cannot do online shopping, they will feel uncomfortable and be anxious. Certainly, most consumers will not be addicted to online shopping, but what cannot be denied is that consumers have become dependent upon online shopping, which will become tools for e-commerce enterprises to build brand loyalty.

3.2 *Shopping at any time and at any place*

People cannot go shopping at any time in real life because there is a time limit, and usually large supermarkets close at nine p.m., while people can buy what they need on the internet at any time and at any place. If people just want to buy some daily necessities on the internet, they can finish the purchasing process by themselves, and there is no need to consult. With the rapid development of smart phones, the improved performance of mobile phone hardware and the rapid development of relevant software applications, mobile phone online shopping has developed rapidly in the mobile terminal business market. Mobile phone online shopping users reached 144 million by the end of 2012 [5]. Meanwhile, with the development of society, people's lives have become increasingly diverse and the pace of life has gradually become faster. People have less and less long and whole period of time and their free time has gradually been fragmented. For example, they will have fragmented time when they are encountering traffic jam, when they come to work in advance, when they have a break during the meeting, when they are waiting for the latest customers and when they are queuing up at the bank [6]. It is good for those people because on the one hand, they can make full use of the fragmented time, on the other hand they can finish the daily shopping task. What's more, those people can do online shopping at any time and at any place if they like. According to statistics, some orders were placed after eleven p.m. in the evening, and some consumers even placed orders at two or three o'clock in the morning.

3.3 *Focus on shopping experience*

Many consumers mainly want to buy niche products on the internet and they want to meet their spiritual needs. They place great emphasis on shopping experience and will switch to another store as long as they feel a little uncomfortable with the online shopping process, because the cost of switching from one store to another approximates to zero. In fact, many consumers take online shopping not only as a way to buy necessities, but also as a pleasure and a new lifestyle [7]. We can say that when consumers are doing online shopping, they focus more on shopping experience than on some major indicators in traditional shopping, such as product quality and price/performance ratio etc.

3.4 *Relying on evaluation, and enjoying evaluating*

Online shopping platforms, represented by Taobao, generally use customer evaluation mechanism, which makes the consumers form the habit of referring to the former consumers' comment before they buy products on the internet. From the perspective of psychology, it is consistent with consumer psychology. When consumers want to buy a product, they do not want to be "the first person to eat crab", which means the first one to buy the product. They want others to have a try first and they want others' comment. If the comment on the product is favorable, they will choose to buy the product and vice versa. They will refuse to buy the product if the comment on the product is bad. From a practical perspective, the evaluation mechanism is conducive to the sale of products. Consumers can find problems about the product after they use it and can remind other consumers noticing these problems, which is helpful for improving product satisfaction. Moreover, the vast majority of consumers prefer making comments after shopping on the internet. Ninety point two percent of consumers make comments on the goods in the comment area on the original shopping website and nearly ten percent of consumers make comments on the goods in the website community of the original shopping website, still some other consumers make comments on the goods in other websites or in their blogs [8], which in fact meets a lot of consumers' demand to express their views.

3.5 *Consumer satisfaction is related to logistics*

With the rapid development of e-commerce, logistics industry has developed rapidly. However, the logistic industry still cannot meet consumers' demand. Under normal circumstances, consumer satisfaction is related to the relevant attributes of the product, while it is different from online shopping. In online shopping, consumer satisfaction has a certain relationship with logistics, and the level of logistics affects consumers' shopping experience. Even if consumers are quite satisfied with the product they bought online, they will be unsatisfied with this experience of online shopping if the logistics service is not good and vice versa. Even if consumers are not very satisfied with the product, they will still give favorable comment if logistics service is very good. Thus, it is important for e-commerce enterprises to choose logistics enterprises with high level of service as its cooperative partner.

3.6 *Online consumers with richer shopping experience consult less*

E-commerce enterprises engaged in online sales generally contain the following departments, namely, network retail department, network distribution department, art designing department, logistics and storage department, order processing department and customer service center and other departments. Among these departments, customer service center is a very important department, which needs more staff. When consumers are choosing products on the internet, they will have some questions about products and it will increase the probability for consumers to buy the product if their questions can be answered timely, but the more customer service staff hired, the higher the cost will be. Generally speaking, online consumers with a richer shopping experience consult less because generally they have strong purchasing power and they can make objective judgments about the quality and price/performance ratio of the product according to the introduction of the product and former consumers' comments on the product. E-commerce enterprises should try to pursue this kind of consumer, and they should provide a detailed introduction of different aspects of the product for the consumer and think about what consumers might care about from the consumers' point of view, which can not only save the cost of hiring customer service staff but also attract consumers with strong purchasing power.

4 SUMMARY

Consumer behavior will inevitably have some new features under e-commerce environment. Internet will be one of the major sales methods in the future society, so enterprises should attach great importance to this sales channel. They should spend great energy to grasp the characteristics of consumers, consumer behavior and consumers' motivation of online shopping. In order to enhance the competitiveness of enterprises, they should also make their marketing strategy accord with consumer behaviors.

ACKNOWLEDGEMENTS

This work is supported by Economics Teaching Team of Yunnan province and Dominant characteristic of Yunnan Province key disciplines of Applied Economics and is also supported by Study on the pricing strategies and methods of E-commerce enterprises based on B2C network platform (2014C124Y). Thanks for Limei Yuan, a teacher in Qujing Normal University, who helped me collect some materials and correct the grammatical mistakes of this paper.

REFERENCES

[1] China Internet Network Information Center. Statistical Report on Internet Development in China (2014.07) [EB / OL].

[2] Qingjie Wu, Tao Hong & Jun Ma. A review of the long tail theory [J]. Journal of Zhoukou Normal University Vol. (01) 2010, pp. 124–125.

[3] Ye Jian & Cheng Yuanzhou. People will Be Addicted to Online Shopping, People Should Be Rational When Doing Online Consumption [N]. China Business Times, 2012-01-22 (003).

[4] Liu Min. Shopping Online No Matter It Is Useful or Useless, Is Online Shopping Obsessive-compulsive Disorder an "Epidemic"? [N]. Chongqing Commercial News, 2012-11-28 (A06).

[5] China Internet Network Information Center. China's Online Shopping Market Research Report 2013 [EB/OL].

[6] Zhong Lin. Mobile Communications of Scientific Journals in a Time-fragmentation Environment [J]. Publishing Journal Vol. (01) 2014, pp. 89–92.

[7] He Heping, Zhou Zhimin & Liu Yanni. Review on Research Fronts of Online Shopping Experience [J]. Foreign Economics and Management Vol. (11) 2011, pp. 42–51.

[8] Wen Xiaoqing. Study on Consumer Online Shopping Behavior [J]. Lanzhou Academic Journal Vol. (06) 2009, pp. 91–93.

Management, Information and Educational Engineering – Liu, Sung & Yao (Eds)
© 2015 Taylor & Francis Group, London, ISBN: 978-1-138-02728-2

A new distance minimization model for portfolio selection with fuzzy returns

Yan Ying Zhao
School of Mathematics and Information Science, Anshan Normal University, Anshan, China

ABSTRACT: This paper researches portfolio selection problem in fuzzy environment. We use a more accurate distance measure to reflect the divergence of fuzzy investment return from a prior one. Firstly, a new mathematical model is proposed by expressing divergence as distance, investment return as expected value and risk as variance, respectively. Secondly, the crisp form of the proposed model is also provided when investment return is the triangular fuzzy variable. Finally, a numerical example is given to illustrate the effectiveness of the proposed model.

KEYWORDS: Portfolio selection; distance; credibility theory; fuzzy variable.

1 INTRODUCTION

One of the hottest points in applied finance is portfolio selection which is to select a combination of securities that can best meet the investor's goal. The first mathematical model was proposed by Markowitz [1], in which expected return and variance were used to describe investment return and risk, respectively. However, the mean-variance model has limited generality since variance considers high returns as equally undesirable as low returns. Thus, Markowitz proposed semivariance as an improved measure of risk.

Generally speaking, the extension of Markowitz model is defined by minimizing the risk and maximizing the investment return. However, Kapur [2] introduced an entropy maximization model and a cross-entropy minimization model. The objective of the cross-entropy model is to minimize the divergence of the random return from a prior one.

In the above literatures, security returns are considered as random variables. Since the security market is complex, in many cases, security returns are hard to be well reflected by historical data. Thus, many researches argued that we should find another theory to solve the portfolio selection problem in this situation. With the introduction of fuzzy set theory and credibility theory, many scholars began to employ them to describe and study fuzzy portfolio selection problems. Numerous models containing fuzzy variables are proposed. For example, Bilbao-Terol et al. [3] extended the mean-variance model to fuzzy environment. Li [4] extended the Kapur cross-entropy minimization model to fuzzy environment. These models in [4] were solved by using a hybrid intelligent algorithm which is designed by integrating numerical integration, fuzzy simulation and genetic algorithm.

Distance between fuzzy variables is an important concept in fuzzy theory. Many scholars gave different definitions of distance between fuzzy variables, such as Hamming distance, Euclidean distance and Minkowski distance, etc. In this paper, we use the distance measure of [5] to reflect the divergence of fuzzy investment return from a prior one. Comparing to the other distance measures, the distance measure of [5] is more accurate.

In this paper, our motivation is that the divergence of fuzzy investment return from a prior one is measured by using the distance of [5]. Based on this idea, we establish a new distance minimization model by defining investment return as expected value and risk as variance, respectively.

The remainder of the paper is organized as follows: some preliminary concepts of credibility theory are briefly recalled in Section 2. The concept of distance between fuzzy variables is introduced in Section 3. In Section 4, we will propose a new model by minimizing the distance between fuzzy variables. In Section 5, the crisp form of the new model will be presented. Section 6 gives a numerical example to illustrate availability of the proposed model. Finally, a brief summary is given in Section 7.

2 NECESSARY KNOWLEDGE ABOUT CREDIBILITY THEORY

Copy After Zadeh [6] initiated the concept of fuzzy set via membership function in 1965, he further indicated possibility theory [7]. Many scholars, such as

Dubois and Prade [8], made their great contribution to its development. In 2002, B. Liu and Y.-K. Liu [9] defined a self-dual credibility measure to describe a fuzzy event. In order to develop an axiom system similar to the theory of probability, Liu founded the credibility theory in [9], which is a branch of mathematics for studying fuzzy phenomena.

Definition 1 [9] Let be a fuzzy variable in membership function. The credibility measure is defined as

$$Cr\{\xi \in B\} = \frac{1}{2}\left(\sup_{x\in B}\mu(x) + 1 - \sup_{x\in B^C}\mu(x)\right), \tag{1}$$

for any set B of real numbers. It is easy to see that credibility is self-dual.

In order to make a more general definition of expected value of a fuzzy variable, Liu and Liu [10] defined the expected value of ξ.

Definition 2 [10] Let ξ be a fuzzy variable with membership function μ. The expected value of a fuzzy variable is defined as

$$E[\xi] = \int_0^{+\infty} Cr\{\xi \ge r\}dr - \int_{-\infty}^0 Cr\{\xi \le r\}dr, \tag{2}$$

provided that at least one of the two integrals is finite. If the fuzzy variable ξ has a finite expected value, then its variance is defined as

$$V[\xi] = E\left[(\xi - E[\xi])^2\right]$$

3 DISTANCE MEASURE BETWEEN FUZZY VARIABLES

Definition 3 [5] Suppose that ξ and η are fuzzy variables with membership functions $\mu(x)$ and $v(x)$, and the λ- cut of ξ and η are $u_\lambda = \left[\mu_L^{-1}(\lambda), \mu_R^{-1}(\lambda)\right]$ and $v_\lambda = \left[v_L^{-1}(\lambda), v_R^{-1}(\lambda)\right]$ for all $\lambda \in [0,1]$, respectively. Then the distance measure between ξ and η can be defined by

$$\begin{aligned} D^2[\xi,\eta] &= \int_0^1 D^2(\mu_\lambda, v_\lambda)d\lambda \\ &= \int_0^1\left\{\left[\left(\frac{\mu_L^{-1}(\lambda)+\mu_R^{-1}(\lambda)}{2}\right) - \left(\frac{v_L^{-1}(\lambda)+v_R^{-1}(\lambda)}{2}\right)\right]^2 + \right. \\ &\left. \frac{1}{3}\left[\left(\frac{\mu_R^{-1}(\lambda)-\mu_L^{-1}(\lambda)}{2}\right)^2 + \left(\frac{v_R^{-1}(\lambda)-v_L^{-1}(\lambda)}{2}\right)^2\right]\right\}d\lambda \end{aligned} \tag{3}$$

If we use the distance to measure the degree of divergence, then we can obtian a new distance minimization model.

4 DISTANCE MINIMIZATION MODELS

Let x_i be the investment proportions in securities i, ξ_i the fuzzy returns of the i th securities, $i = 1, 2\cdots, n$, respectively. In this paper, we use the distance, expected value, and variance to measure the degree of divergence, return and risk, respectively. Then we have the following model

$$\begin{aligned} \text{minimize} \quad & D[x_1\xi_1 + x_2\xi_2 + \cdots + x_n\xi_n, \eta] \\ \text{subject to:} \quad & E[x_1\xi_1 + x_2\xi_2 + \cdots + x_n\xi_n] \ge \alpha \\ & V[x_1\xi_1 + x_2\xi_2 + \cdots + x_n\xi_n] \le \beta \\ & x_1 + x_2 + \cdots + x_n = 1 \\ & x_i \ge 0,\ i = 1, 2\cdots, n. \end{aligned} \tag{4}$$

The first constraint ensures the expected return is no less than some given value α, and the second one assures that risk does not exceed any given level β the investor can bear. The last two constraints imply that all capitals will be invested to n securities.

5 CRISP FORM

In this section, we propose the crisp equivalent of the optimization model. In order to simplify models, the objective function $D(x_1\xi_1 + x_2\xi_2 + \cdots + x_n\xi_n, \eta)$ of model (4) is replaced by $D^2(x_1\xi_1 + x_2\xi_2 + \cdots + x_n\xi_n, \eta)$ in crisp form.

Theorem 1. Assume that each security return is the triangular fuzzy variable denoted by $\xi_i = (a_i, b_i, c_i)$ $(i = 1, 2, \cdots, n)$. Let the prior fuzzy investment return $\eta = (a', b', c')$ be a triangular fuzzy variable, then the model (4) can be transformed into the following crisp form:

$$\begin{aligned} \min\ & \frac{1}{3}\left[\left(\sum_{i=1}^n x_i(2b_i - a_i - c_i) - (2b' - a' - c')\right)\right. \\ & \left.\left(\sum_{i=1}^n x_i(a_i + b_i + c_i) - (a' + b' + c')\right)\right] \\ & + \frac{1}{18}\left[\left(\sum_{i=1}^n x_i(a_i - c_i)\right)^2 - (a' - c')^2\right] \\ & + \frac{1}{2}\left[\sum_{i=1}^n x_i(a_i + c_i) + (a' + c')\right]^2 \end{aligned} \tag{5}$$

$$\text{s.t.} \quad \sum_{i=1}^{n} x_i b_i \geq \alpha$$

$$\left(\sum_{i=1}^{n} x_i c_i - \sum_{i=1}^{n} x_i a_i\right)^2 \leq 24\beta$$

$$x_1 + x_2 + \cdots + x_n = 1$$

$$x_i \geq 0, \; i = 1, 2 \cdots, n$$

6 NUMERICAL EXAMPLE

In this section, a numerical example is given to illustrate the availability of the new model.

Table 1. The asymmetrical fuzzy returns of 10 securities.

Security i	Fuzzy return ξ_i
1	(-0.4, 2.7, 3.4)
2	(-0.1, 1.9, 2.6)
3	(-0.2, 3.0, 4.0)
4	(-0.5, 2.0, 2.9
5	(-0.6, 2.2, 3.3)
6	(-0.1, 2.5, 3.6)
7	(-0.3, 2.4, 3.5)
8	(-0.1, 3.3, 4.5)
9	(-0.7, 1.1, 2.7)
10	(-0.2, 2.1, 3.8)

Example 1. Assume that each security return is the triangular fuzzy variable denoted by $\xi_i = (a_i, b_i, c_i)$ $(i = 1, 2, \cdots\cdots 10)$, where the parameters a_i, b_i and c_i are determined based on the estimated values of financial experts. The data set is given in Table 1. Suppose that the minimum expected return the investor can accept is 2.15 and the bearable maximum risk is 0.65. In addition, the prior fuzzy investment return is $\eta = (-0.2, 2.3, 4.0)$. From the model (5), we can obtain a simple optimization model, and employ fmincon in Matlab 7.1 to solve this model. The numerical results are given in Table 2.

In order to obtain the minimize distance of the investment return from the prior return η when the portfolio satisfies the return and risk constraints, the investor should assign his or her capital according to Table 2. The corresponding objective value is 0.162. In addition, the investment return is $\xi = (-0.19, 2.57, 4.02)$. Based on the computational result, we can see that the investment return ξ and η are similar. This indicates that our model is effective. However, solving our model is easier than solving the cross-entropy model.

Table 2. Investment proportion of 10 securities (%).

Security i	Allocation	Security i	Allocation
1	0	6	0
2	0	7	0
3	3.6	8	35.5
4	3.2	9	3.2
5	1.5	10	52.9

7 CONCLUSIONS

In the paper, a concept of distance between fuzzy variables was introduced for measuring the divergence of fuzzy investment returns from a prior one. By defining the risk as variance, a distance minimization model was proposed. In addition, the crisp equivalent of the optimization model has also been provided. Finally, the result of the numerical example illustrated the availability of the new model.

REFERENCES

[1] Markowitz, H. 1952. Portfolio selection. Journal of Finance 7:77–91.

[2] Kapur, J. & Kesavan, H. 1992. Entropy Optimization Principles with Applications. Academic Press, New York, 1992.

[3] Bilbao-Terol, A., Perez-Gladish, B., Arenas-Parra, M. & Rodriguez-Uria, M.V. 2006. Fuzzy compromise programming for portfolio selection. Applied Mathematics and Computation 173:251–264.

[4] Li, X. Qin, Z.F. & Ka, S. 2010. Mean-variance-skewness model for portfolio selection with fuzzy returns. European Journal Operational Research 202: 239–247.

[5] Liem Tran & Lucien Duckstein. 2002. Comparison of fuzzy numbers using a fuzzy distance measure. Fuzzy Sets Systems 130:331–341.

[6] Zadeh L A. Fuzzy sets. Information and Control, 8 (1965) 338–353.

[7] Zadeh L A. 1978. Fuzzy sets as a basis for a theory of possibility. Fuzzy Sets and Systems1:3–28.

[8] Dubois D. & Prade H. 1988. Possibility Theory: An Approach to Computerized Processing of Uncertainty. New York: Plenum.

[9] Liu, B. 2007. Uncertainty Theory, 2nd ed., Springer-Verlag, Berlin.

[10] Liu, B. & Liu, Y.K. 2001. Expected value of fuzzy variable and fuzzy expected value models. IEEE Transactions on Fuzzy Systems 10:445–450.

Management, Information and Educational Engineering – Liu, Sung & Yao (Eds)
© 2015 Taylor & Francis Group, London, ISBN: 978-1-138-02728-2

Practice and research of project-driven teaching mode based on tutorial system in computer science and technology specialty

S.Y. Cheng & X.M. Zhou
College of Math and Computer, Jiangxi Science & Technology Normal University, Nanchang, China

ABSTRACT: In order to improve the practical ability and comprehensive ability of students, we propose a project-driven teaching mode based on the tutorial system in computer science and technology specialty. The junior is the study object of the new teaching mode. In the new teaching mode, the tutor is the core, the student is subject and project research is main line. This paper introduces the research background, the content framework, main features and implementation in detail. The practice shows that the project-driven teaching based on tutorial system can greatly improve students' ability of solving problems, raises their overall qualities in practice and enhance the employment rate.

KEYWORDS: Computer specialty; Tutorial system; Project-driven mode teaching; Talents training.

The train object of computer science and technology specialty is to cultivate talents in computer fields, which have the ability to research and develop the computer system after years of study and training. In recent years, with the development of information technology, the need of computer professionals is increasing. [1][2] How to train high quality and skilled talents is the primary issue to professional teachers.

1 RESEARCH BACKGROUND

Recently we are dedicated to training compound talents in computer specialty, who can research, design and develop a computer system. So we had many experiments such as co-sponsoring, practicing in Software Company, however for various reasons, it failed. After exploration and thought, in view of the structure of teaching staff, curriculum system and students' knowledge structure, we proposed project-driven teaching mode based on tutorial system. The theory foundation of the new teaching mode is scene cognition, social constructivism theory and Pragmatism Education. We have carried out the education reform project since 2011. The professional teachers change into professional instructors. Theory combined with practice skill is emphasized. In the educational reform, project development is the main line; practice base is the software company. Individualized teaching strives to meet the need of various students. Tech talk, technical communication and project support are launched. The students learn the theory and skill of project development from tutors.

2 CONNOTATIONS

How to understand the project-driven teaching mode based on tutorial system? Is it an effective method of heightening the students' skills of practice? We can find the answer from the educational theory. Constructivist learning theory considers that with the help of others, students can obtain knowledge by means of construction method instead of imparting from teachers, which presented by the famous psychologist J. Piaget [3]. Project-driven teaching mode based on constructivism changes traditional teaching. Teacher centered changes into student centered; textbook centered changes into project centered; course centered changes into practical experience centered. [4] So adopting project teaching in computer teaching can easily bring teachers and students' ability into full play. It also improves the students' levels of ability and practical experience of the implementation of the project. And the upstanding team work spirit develops among students. All of these can produce positive and profound influence. From the view of application process of project-driven teaching, it conducts tutor applying comprehensive cases or tasks with practical value to teach. The teacher divides a project into several modes. According to the working process, every mode refines to serial tasks. The task is relevant to the appropriate knowledge point. Consequently, students can catch the application and connotation of knowledge by project-driven teaching. Moreover, this teaching method expands learning and application of profound technology. Thanks to the project-driven teaching, students have the ability to put knowledge into action. The vocational ability and comprehensive ability of students' reaches new heights.

3 CONTENT FRAMEWORKS

Project-driven teaching mode based on tutorial system is an important part of the training plan. It is launched in the third year of undergraduate education. The reform process takes one year. The whole education process is in the laboratory. The tutors guide the student to study professional courses. [5] They use, flexible individualized teaching mode so that students improve the software developing ability, project management skills and teamwork ability rapidly.

3.1 *Teacher-student relationship*

It goes without saying that the relationship between teachers and students is very important. In the tutorial system, the relationship is more important and particular than traditional teacher-student relationship. We cultivate good relations quickly by the following ways. We know the students' study, life condition and physiological state by means of looking up students' personnel file. Meanwhile, the documentation of university student personal information is made. The content of personal information documentation includes basic status chart and well designed questionnaire. We can obtain the students' extensive information, family background and learning experience by the personal information documentation. In order to keep track of the students, we have built up the bridge of communication such as QQ group, public mailbox, blogging and tweeting.

3.2 *Cultivating student based on tutorial system*

1 Requirement of tutors

The tutors must have the ability to do scientific research independently in some respects. They must organize and guide students to develop projects. And they promote teaching all around the project.

2 Tutors' allocation

Every tutor has a software studio and an assistant who comes from software enterprise.

3 How to carry out the teaching work?

A tutor is in charge of 6 students or so. Every student is appointed to different project team and then study relevant professional knowledge about the project. During the developing process, the tutors help student to finish the task. Especially tutors teach students according to their aptitude and help each student to make development plan in the light of personality difference.

3.3 *Project-driven teaching*

Thanks to good relation between students and tutor, the project-driven teaching mode based on tutorial system is effective in improving students' professional and comprehensive quality. We discuss the target, characteristics and implement method.

1 Target

During the teaching, students must consolidate the professional theory and grasp professional practice skill. The definite learning route must be made; conscious learning habit must be formed; the spirit of team work also must be developed. By means of project driven teachings, the students not only acquire knowledge but also lubricate relationship between teachers and students.

2 Characteristics

In the teaching reform, the projects come from the tutor's actual project or optional research project. The tutors explicate project in detail, so students can be kept informed of all process, and then come to know the whole mechanics of the project develops little by little. In order to easy to communicate, the number of learning team is about 6. Projects division is connected with course study. The students will take more time to theory, learning who will take part in the graduate entrance exam, while the students who will take up an occupation will take more time to practice skill learning.

3 Curriculum

Several points of Curriculum for attention:

Compulsory courses must be reduced greatly and optional course must be decreased. The content, venue, date and time of optional course are arranged flexibly. [6] This can guarantee a widely choices to students and tutors. Every optional course is to help the development of the project; its implementation is in tutor's studio.

4 Detailed implement method

Phase 1 Basic Stage

The target of phase 1 is to ensure students acquiring the theory and practical knowledge needed for project development.

In order to guarantee the efficiency of study, we plan a reasonable academic progress. Project team discusses about the studio once a week. The students report study progress and existed question by PPT or demo. Then around the question, group discussion is carried out. The tutors are responsible for helping queries.

Phase 2 Practice stage

According to principle of work division, students participate in project genuinely in this phase. Different students have different personality characteristics, different ability and performance, so they carry on different specific task of the project.

5 Individualized teaching method

Many methods can be adopted to realize individualized teaching. There are formal class teaching, technical seminars, surveys, experimental and individual coaching. Firstly tutors impart basic knowledge to students and help them block out a plan of study. Then tutors guide and excite students to study independently. Thus the ability of learning, solving and teamwork can be trained and strengthened. Tutors must work out a teaching play and detailed schedule about teaching method mentioned above at the beginning of each semester. Of course tutors can make an adjustment based on fact.

3.4 *Cloverleaf assessment methods*

The assessment methods include stage assessment and final assessment in project-driven teaching. Different appraising officers are adopted in the different state assessment.

1 The tutor of project checks the quality and schedule of project task in contrast with the original plan. Then a score will be made to assess how well the student performs. The assessment result will be saved in the student's archival repository.
2 The final assessment will be made by 2 proficient tutors at least. Students in a team make a technique, defense about the finished project. The panel of defense assesses development ability, documentation processing ability and coordination and organizing abilities comprehended on the basis of students' defense and achievements. The panel of defense will give a score, then save the score and all data of the project into archival repository. Finally the project tutor gives the comprehensive score with a certain percentage of stage assessment and final assessment.

4 MAIN CHARACTERISTIC

4.1 *Enterprization of teaching environment*

Teaching construction in project-driven teaching is software laboratory and training center. In this place, students take part in practical exercises. All of these serve for project development. Students learn knowledge with a clearly defined goal. Specific task and fixed staff can improve learning efficiency. Students must follow the process and standard of software development to develop programmes.

The laboratory not only has a superior environment for study and research, and still has modern popular development tools to make developing project which comes from practical application. These projects, which are developed by students, are mainly classical curriculum design, tutors research project. It is requested that the finished projects must have a certain value in practical application. The operation mechanisms of enterprization can shift from passive learning of traditional education to active learning, shift from study aimlessly to study with clear aim, shift from study prosaically to study with interest. The mechanisms strive for market-oriented accomplishment. Our students can stand the test and succeed because their innovative spirit and practice ability are improved greatly. When students graduate from university, they can work in a market without difficulties. [7]

4.2 *Function of tutors*

The function of tutors is vital to project-driven teaching. The relationship between teacher–student is harmonious. Tutors not only impart knowledge to students, but also manage the development of the project. They help students to realize themselves. Students consult with tutors about academic and employment planning. During the teaching process, tutors instruct students by various ways, such as face-to-face teaching, network and telephone. Individualized teaching can be achieved because tutors teach student knowledge with a highly individualized approach.

This teaching mode improves tutors' professional skill simultaneously. Teaching benefits teacher as well as students. The relationship between teachers and students is a conversational one, which is equal and democratic.

4.3 *Curriculum teaching*

The arrangement for curriculum teaching services for project development and implement. Project development plays a leading role in curriculum system. Theory and practice skill imparted by tutors are all for developing projects. Meanwhile examinations are all around the application in project development.

5 ACHIEVEMENTS

From September 2011 to October 2013, project-driven teaching mode based on tutorial system has already implemented twice among Class 2013 and class 2014. The student is subject, project research is the main line and the tutor is the core of the teaching. Autonomic learning, self learning and mutual learning are in harmony and unification. Students' enthusiasm for study is rising to unprecedented heights. And good results have achieved in practice.

1 Learning target is clearer. Due to the goal - orientation, students explore and practice voluntarily; the degree of professionalization is improved.
2 Students' learning is from passive learning to active learning. Practice skill is emphasized. Scope is given to the students' initiative and creativeness.
3 Students are proficient in various skills and theory about project development. Students have the qualifications for a programmer. Employment rate of 2013 was in front in Jiangxi province.
4 The students won great glory in national and province's contest. From 2011 to 2014, students won a first prize and a second prize in the National Mathematical Modeling Contest. More than 10 Students have won first prize in the province's contest. A great number of students got good marks in technological innovation & vocational skills contest in the province. There are lots of other prizes which I won't go over one by one.
5 Greater demands are being placed on the tutors. The tutors must qualify themselves, not only for management and leadership, but also for Engineering and teaching. So they can organize and control classes well, enhance the enthusiasm and initiative of students.

6 EXPERIENCES

1 The strong support of the college: owe to plenty of funds and policy given by the college, the project-driven teaching reform progresses smoothly. Curriculum design also supports the reform greatly.
2 Adequate preparation: Careful preparatory work for project-driven teaching is made. Abundant information and educational reformation practice such as curriculum reform, examination reform, are made to prove the feasibility of this reform. The preparations ensure the success of reform. Before pilot reform, educational rules and workflow are all revised after many discussions. Mobilization meetings have been held many times.
3 Tutors' devotion: tutors have put lots of time and effort to keep the project-driven teaching reform going. Individualized teaching, project development management and guide are all long and hard work.
4 Innovative management system: innovative management system is the guarantee of our implementation of education reform. A management group is set up. The charge man of the group is the dean of the Computer faculty. The group is responsible for instructional supervision, project inspection and assessment.

7 CONCLUSIONS

The purpose of project implementation, teaching management and evaluation system is all for creative education. Scientific project system and project content create favorable conditions for project-driven teaching. The new teaching mode improves students' practice ability, vocational ability and comprehensive ability. The successful implementation of project-driven teaching mode based on the tutorial system satisfies the need of qualified computer talents.

The practice has shown that the new teaching mode aroused the teachers' and students' initiative. The students' practical skill and comprehensive ability are strengthened obviously, employment rate is also enhanced. Moreover the reform trains a large number of double - professional teachers.

At present, the project-driven teaching mode based on tutorial system is still on the stages of development. The supporting mechanism needs to polish gradually. Curriculum system, platform building and industry-academy- research cooperation are all worth exhaustive researching.

ACKNOWLEDGEMENTS

Project is supported by the Education Research Foundation for 12th Five Years Plan in Jiangxi province (No. 4YB071).

REFERENCES

[1] M.H. Huang, G. Lei, B. Guo. Reseach on Project-driven Teaching Mode. Modern Education Management. 2007(6), pp: 29–32.
[2] M.J. Si. An Exploration and Practice of Software Engineering Applied Talents Training . Chinese Science and Technology Information, 2011(12),176–182.
[3] H.M. Shen. The Practice and Study of the Program "Specialty Tutorial System" in Computer Specialty. Computer knowledge and technology, vol.7 no.12, 5683-5684, 2011.
[4] Y.G. Curriculum Guidelines for Undergraduate Degree Programs in Software Engineering. Chinese Vocational and Technical Education, 2012(14).
[5] M.X. Zeng, Q.P. Zhou, X.B. Wang. Construction of a "Project" Teaching System for Software Engineering Majors. Research and Exploration in Laboratory, vol .32, no.5, 2013.
[6] J. Huang, Q. Liu, Z. Gao. A New Model Research and Practice on Training Talents of Software Engineering Based on University Industry Cooperation. Computer Engineer and Science, 2011(1):70–73.
[7] Q. Jing, S.D. Dong, X. Liu. Employment Oriented Software Engineering Teaching Quality Assessment System. Journal of Chong Qing Three Gorges University, 2011(3), 141–143.

Management, Information and Educational Engineering – Liu, Sung & Yao (Eds)
© 2015 Taylor & Francis Group, London, ISBN: 978-1-138-02728-2

Work position effects on physiological aspects of machining workers

I Gede Oka Pujihadi & I Ketut Widana
Mechanical Engineering, Bali State Polytechnic, Bali, Indonesia

I Nyoman Budiarsa
Mechanical Engineering, University of Udayana, Bali, Indonesia

ABSTRACT: In this research a procedure to evaluate the physiological-based machining workers working position has been developed using a cross-sectional as observational study design. The methods applied are observation, interview and measuring. The subjects of the research are practicing students amounting to 21 students with average ages of 19.5 ± 0.67. Body mass index on the average of 21.33 ± 2.13, considered normal. Referring to the analysis of the statistical test of the Wilcoxon Signed Ranks Test, the difference of effect of work position is significant, namely $p<0.05$ towards musculoskeletal disorders (MSDs) before and after working. The quantity of the average complaint after working is 44.52 ± 9.28. The musculoskeletal complaint is felt 100% of skeletal muscles with details as follows: (a) 76% is waist ache; (b) 71% left tarsus ache; (c) 67% back pain; (d) 62% felt stiff in the upper neck, lower neck and left calf ache. The result of the Wilcoxon Signed Rank Test shows that there are significant differentiation effects of a standing work position, namely $p<0.05$ towards fatigue generally before and after working. The degree of fatigue effects is that 100% of the practicing students feels tired. Based on the questionnaire, 30 items of general fatigue are grouped into 3 (three), namely: (a) Question group 1 – 10 shows the attenuation of activity of 77%; (b) questions 11-20 show the attenuation of motivation of 86% and; (c) questions 21-30 show the description of general physical fatigue of 53 %. The degree of working pulse is on the average of 110.00 ± 10.44 ppm (pulse per minute) which can be categorized into the medium workload. By means of Paired t test, the result is $p<0.005$. The concentration, consisting of the speed, correctness and Constance will also be decreased each 15.23%, 11.20% and 16.33%. It shows that there is a significant difference of a standing work position effect towards the musculoskeletal disorders, fatigue, workload and concentration when having a rest and working with the practicing students. The efforts of working station repair, short term-rest and supplying drinking water are able to decrease musculoskeletal disorders, fatigue, work load and increase concentration as well as increase the work productivity.

KEYWORDS: ergonomic; position; musculoskeletal; fatigue; workload; concentration.

1 INTRODUCTION

The working practices involve turning, cutting, scraping, welding, grinding training and so on. Almost all working practices are conducted in standing position. With eight hour session a day, it can be predicted that there will be a lot of disorders, especially the subjective ones such as the musculoskeletal disorders, general fatigue and concentration as well as the workload. In order to be able to compete, hence, the industry has to be able to give the best service to the customers, have a comfortable working atmosphere, interesting and friendly performance of the students, fast service, and the products fulfill the customers' expectation. Consequently, the efficiency and productivity of work must be accelerated optimally in order to reach the above goal. The improvement of the work productivity can be reached by pressing all kinds of input into the minimum level and increasing output into the maximum one [1]. The input, especially related to resources, has to be employed in an optimal fashion. In order to reach such condition, the students must be facilitated with comfortable, safe, and efficient work facilities. The work facilities comprise of, work station, work environment and work organization that is in accordance with the capability, skill and limitation of students in the hope that the productivity can be reached at the highest level [2]. It is a study of the psychology of the machine workers need to be developed, the evaluation of the influence of the working position before and after work that could potentially result in musculoskeletal disorders, general fatigue, workload and workers' concentration has been explored, it is becoming a major base for providing information about the parameters important in formulating motion studies on the psychology of the machine working in a range of complex conditions and broader data.

2 ANALYSIS AND RESULTS

2.1 *Subject condition*

The descriptive analysis results of average, stretches of time, standard deviation of the subject characteristics that involve age, height, weight, body mass index and work experiences is presented in table 1 below.

Table 1. Subject Characteristic.

No	Variable	N	Average	SD	Stretches of Time
1	Age (year)	21	19.48	0.68	18.00 – 21.00
2	Height (Cm)	21	157.48	3.98	150.00 – 166.00
3	Weight (kg)	21	56.62	3.47	49.00 – 67.00
4	Body Mass index	21	22.88	1.98	19.88 – 29.77

The average age of subjects is 19.48 ± 0.68 years old, which means within productive ages. Body mass index (BMI) is a comparison of weight (kg) and height quadrate (m). The average body mass index of subjects is 22.88 ± 1.98 kg/m2, which shows a normal body mass. According to reference [3], body mass index of the Indonesian is considered to be normal if it reaches an average value of 18.5 – 25 kg/m^2, therefore a body mass index of the subjects is considered to be normal as it is within the value range.

2.2 *Work position*

The practical activity begins at 08.00 up until 15.00 WITA every day for 5 day work time. The work process comprises cutting, forming and finishing. The amount of students or students observed are 21 students who are all male, aged 18-21 years old, being on the third semester. The standing work position is frequently performed by the students at the cutter station. They rarely perform the work with sitting position as they consider it can slower the finishing process of working. They do not realize that such condition can have an effect on the musculoskeletal disorders, fatigue, concentration and workload. According to [4], the standing position is an alert position physically and mentally, therefore the work activity performed is faster, stronger and more careful. Basically, standing is more tiring than sitting and the energy spent when standing is more 10-15% compared to sitting. To minimize the effect of musculoskeletal disorders, fatigue, concentration and workload, consequently the work must be designed in such a way that it doesn't reach forth, bend down, or performing unusual positions of the head.

2.3 *Musculoskeletal disorders*

To find out the musculoskeletal disorders (MSDs) of the students at the cutter station, one of the ways is by filling questionnaire of Nordic Body Map before and after working with the Likert scale scored from 1 to 4. From the tabular data, the musculoskeletal disorders are analyzed descriptively and by normality test supported by the application program of SPSS 15.00 for Windows. The result of data tabulation of musculoskeletal disorders before and after working with statistical analysis can be seen in table 2 below.

Table 2. Results of Descriptive Analysis of The Musculoskeletal Disorders (Msds).

No	Variable	N	Average	SD	Normality test K-S test
1	Musculoskeletal disorders before working	21	28.67	1.06	p = 0.002
2	Musculoskeletal disorders after working	21	44.62	9.47	p = 0.515
3	Difference before and after working	21	15.95	9.59	p = 0.000

The Table 2 above shows that data of musculoskeletal disorders before working is not distributed normally p = 0.002 ($p<0.05$). As there is one of data is not distributed normally, therefore Non Parametric test is applied namely the Wilcoxon Signed Test. The result is, there is a significant difference standing work position effect towards musculoskeletal disorders before and after working with the students with p = 0. 000 ($p<0.05$). The average amount of effect of standing work position towards musculoskeletal disorders is 44.52 ± 9.28. Musculoskeletal disorders felt according to the percentage per item of disorders, with the details (a) 76% is waist ache; (b) 71% left tarsus ache; (c) 67% back pain; (d) 62% felt stiff on upper neck, lower neck and left calf ache. Such condition results from the standing work position of the students that is performed continuously and repeatedly. The complaint of skeletal muscles generally occurs as the muscle contracts exceedingly due to the excess of workload and long duration of loading

[5]. The muscle disorders may not occur if the muscle contraction ranging from 15-20% of the maximum muscle power. If the contraction of the muscle is over 20%, so the blood circulation to the muscle will reduce according to the contraction level that is influenced by the capacity of energy needed. The oxygen supply to the muscle decreases, the carbohydrate metabolism process is blocked and as a result the accumulation of lactate acid occurs which results in muscle aches [6].

2.4 *Fatigue*

To obtain data of fatigue, the questionnaire is used which contains 30 items of general fatigue before and after working. The results of the questionnaire apply the Likert scale with scores from 1 to 4. The result of tabular data and general statistical fatigue test before and after working on the students is obtained with the descriptive analysis and normality test. For more details, the analysis results of the general fatigue before and after working are clearly defined on table 3

Table 3. Results of Descriptive Analysis of Fatigue.

No	Variable	n	Average	SB	Normality test K-S test
1	General fatigue before working	21	30.00	0.00	
2	General fatigue after Working	21	53.90	6.71	P = 0.17
3	Difference between before and after working	21	23.90	6.71	P = 0.17

Seen from Table 3, it is ascertainable that one of the data of general fatigue before working is not distributed normally as p is zero, therefore the general fatigue data is tested non-parametrically with the Wilcoxon Signed Rank Test. The data analysis is revealed that there is a significant difference of standing position effect towards the general fatigue before and after working with the students, in which p = 0. 000 ($p<0.05$). Based on the questionnaire of 30 general fatigue items, it can be grouped into 3 (three) namely (a) group of questions 1-10 showing the attenuation of activity of 77%, (b) group of questions 11-20 showing the attenuation of motivation of 86% and (c) group of questions 21-30 showing the general physical fatigue description of 53%.

The fatigue results from the body condition that accepts excessive workloads, continuously, repeatedly and also the standing position as well as the uncomfortable working environment. The fatigue will be recovered, if a short-term rest is applied to the temporary fatigue. The permanent fatigue will be recovered if a one day sleeping rest is taken [5].

2.5 *Workload*

The quantity of the workload of the students can be discovered by calculating the pulse when having a rest and working with the ten-pulse method. The results of the calculation of the pulse when resting, and when working, then are analyzed with statistical tests. The data are analyzed descriptively and then continued with normality tests. If the data is distributed normally, the Paired T test is applied and if the data is not distributed normally then the Wilcoxon Signed Ranks Tests is applied. For more details, table 4, shows the results.

Table 4. Descriptive Analysis Results and Normality Test of Pulse When Resting and Working.

No	Variable	N	Average	SD	Normality Test K-S Test
1	Pulse when resting	21	72.27	8.15	p = 0.108
2	Pulse when working	21	110.78	17.80	p = 0.145
3	Working pulse	21	38.51	18.84	p = 0.504

2.6 *Concentration*

The result of research indicates that all of the indicators are decreased. Speed average work at period of I (resting) is 8.24 ± 1.05 and at period of II (after working) is 9.72 ± 1.56, or happened by the speed decrease work equal to 15.23%. Meaning analysis with the test t-paired indicate that the value $p < 0.05$ owning meaning that speed, average work at second period differ to have a meaning. Correctness average hereafter conduct the work at period of I is 19.34 ± 6.68 and at period of II is 21.78 ± 5.54 or there is decrease equal to 11.20%. Meaning analysis with the test t-paired indicate that the value $P < 0.05$. The matter of this means that correctness average at second period differ to have a meaning. Average Constance hereafter conduct the work at period of I is 4.51 ± 1.48 and at period of II is 5.39 ± 1.70, or decrease 16.33%.

3 CONCLUSION

Based on the research and discussion above, conclusions were obtained as follows

1 Based on Wilcoxon signed rank test, it shows that there is a difference of effect of standing work position significantly towards the musculoskeletal disorders before and after working with the students with $p = 0.02$ ($p<0.05$). The musculoskeletal disorders are suffered according to the percentage per item of complaint of ache with the details (a) 76% is waist ache; (b) 71% left tarsus ache; (c) 67% back pain; (d) 62% felt stiff on upper neck, lower neck and left calf ache.

2 Based on the analysis of Wilcoxon signed rank test, it is ascertainable that the difference of standing work position effect is significant towards general fatigue before and after working with the students with $p = 0.002$ ($p<0.05$). Based on the questionnaire of 30 items of general fatigue can be grouped into 3 (three) namely : (a) group of question 1-10 showing activity attenuation of 77%; (b) group of questions 11-20 showing a motivation attenuation of 86% and; (c) group of questions 21-30 showing general physical fatigue description of 53%.

3 Based on *paired t test*, it is ascertained that there is a difference of pulse beat while having a rest and working with the students with $p = 0.000$ ($p<0.05$). The degree of the effect of the *standing* work position towards the workload on the students is on the average of 110 ± 10.44 ppm and can be categorized into a medium workload. Hard and soft of the workload can be accepted by the students depending on the length they perform the activity of work which is adjusted to their capability. The workload can be influenced by the continuous, repeating works and the *standing* position while working, as well as the working environment that is hot.

4 Concentration will decrease really if job attitude do not be natural in a condition. Concentration consisted of the speed, correctness and Constance will experience of the degradation of each 15.23%, 11.20% and 16.33%.

REFERENCES

[1] Chavalitsakulchai, P., Shahnavaz, H. 1991. Musculoskeletal Discomfort and Feeling of fatigue among Female Profesional Workers : The Need for Ergonomics Consideration. Journal of Human Ergology. 20 : 257–264.

[2] Manuaba, A. 2006. A Total Approach In Ergonomics Is A Must To Attain Humane, Competitive And Sustainable Work System And Products. In : Adiatmika and Putra, D.W. editors. Proceeding Ergo Future 2006 : International Symposium On Past, Present And Future Ergonomics, Occupational Safety and Health. 28 - 30th August. Denpasar : Department of Physiology Udayana University – School of Medicine. p. 1–6.

[3] Caple, D. 2006. Ergonomic – Future Directions. In : Adiatmika and Putra, D.W. editors. Proceeding Ergo Future 2006 : International Symposium On Past, Present And Future Ergonomics, Occupational Safety and Health. 28 - 30th August. Denpasar:Department of Physiology Udayana University – School of Medicine. p. 7–11.

[4] Sutalaksana, I. Z., dalam Tarwaka, dkk. (2004). Ergonomi Untuk Keselamatan, Kesehatan Kerja dan Produktivitas. UNIBA PRESS, Surakarta-Indonesia. 2000.

[5] Suma'mur, Higiene Perusahaan dan Kesehatan Kerja. PT.Toko Gunung Agung, Jakarta, 1995.

[6] Grandjean, E., Fitting the Task to the Man. : A Textbook of Occupational Ergonomics. 4 th Edition. Taylor & Francis Ltd.,London, 1993.

Management, Information and Educational Engineering – Liu, Sung & Yao (Eds)
 ISBN: 978-1-138-02728-2

The determinant of profitability—empirical evidence in productive services enterprise of China

Y.Y. Chen & Z.K. Bao
School of Economics, Dalian University of Technology, P.R. China

ABSTRACT: We use the enterprise census database investigated by the national bureau of statistics and employ recently developed the Shapley value decomposition framework to disassemble the specific factors' contribution to profitability difference and combine it to calculate the contribution of industry and enterprise effect from the whole and segment sample. The empirical results reveal both industry effects, which mainly caused by market structure and market demand expansion, and enterprise effect chiefly influenced by ownership types, unproductive expenditure and financing channel, play a significant role in shaping profitability difference. In the whole sample, industry effect contributes 42.71% and 51.12% respectively in return on sale and Per Capita Create Profits model. In addition, the finding clearly indicates that the industry effect increases in low marketization regions or high administrative barriers industries, enterprise effect vice versa.

KEYWORDS: Productive service enterprise; Shapley value decomposition; Industry effect.

1 INTRODUCTION

In the process of promoting strategic adjustment of the economic structure and speeding up the economic transformation, the Chinese government has made a series of productive service industry supporting policies. Therefore, the productive service industries, such as business service, research and development service, information service, have great opportunities for further development, which are attracted continuously investment by the industrial capital and financial capital. According to the national bureau of statistics, the fixed asset investment of the productive service reached 3.91trillion RMB in the year of 2010, which was about 4 times than the level of 2003 and the average annual growth rate was about 46.61% recently decade. As a consequence, what is the micro productive service enterprise profitability under the background of supporting policy and whether it shows obvious differences among the industries? If so, how is formation mechanism? To answer that question, not only can inspect the validity of past a series of supporting policies, but also can influence policy trend of productive service industry in the future.

In this study, we employ the methodology of Shapley value decomposition (Shorrocks, 1999) and combine with wan (2004) decomposition method. Recently, the Shapley decomposition has been widely used in the labor economics, particularly in disassembling of income inequality. The advantages of the Shapley decomposition method doesn't only disassemble the contribution of industrial effect and enterprise effect, but also can explain the specific influence factors such as the scale of enterprise, government correlation degree provided in the model. The method is appropriate for any profitability difference index and any profitability decision equation.

2 SAMPLE AND METHODOLOGY

2.1 *Sample*

We employ a set of unique survey data the year of 2008 which investigated by the national bureau of statistics covering 30 provinces, municipalities, and autonomous regions of china mainland. The investigation data set accounts for 67.26 percent of all amount 12 industries in China and includes more than 60 indexes, such as enterprise established time, business income, operating profit and the quantity of workers, and so on, which provides enough support with our study.

2.2 *Methodology*

Based on the shortage of statistical method employed by previous literatures, this paper uses the Shapley value framework raised by Shorrocks (1999), and combines with Wan (2004) decomposition method to make up for the shortcomings of the statistical methods. The advantage of the Shapley value decomposition method not only can explain the specific factors' influencing enterprise profit rate, such as firm size, government relationship and market structure, but also can merge the contribution of enterprise effects and industry effects from the differences in enterprise profit rate while previous researchers couldn't resolve that problem. Meanwhile, Shapley value decomposition method is suitable for any profit rate deciding equation and indicators of measuring profit rates differences, and could handle the residual error's contribution to enterprise profit rate differences (Wan, 2004). Currently, the Shapley value decomposition method has been widely used in the field of labor economics, particularly in decomposition of income inequality (Wan and Zhang, 2006; Wan at all, 2007; Zhao and Lu, 2009).

Based on Shapley value decomposition of regression, we disassemble the determinant of profitability differences in the following two steps: First, by building the formation model of enterprise profit rate, we estimate the coefficients and significance of every explanatory variable, then exclude non-significant variables to identify the profit rate deterministic model; second, we ultimately obtain the contribution of each explanatory variable to enterprise profit rates differences by using calculation indicators of profitability difference in profit rate deterministic model. This paper adopts profit rate formation model as follow:

$$Profit_{ij} = \alpha_1 + enterprise(x_{ij})\beta + industry(y_j)\gamma + \mu_i \quad (1)$$

Where the indices i and j denote enterprise and industry respectively; profit is the profit rate of enterprise, which use Return on Sales(ROS)and Per Capita Create Profits(PCCP)respectively; enterprise and industry is enterprise effects and industry effects, the specific indicators including are as follows:

Enterprise Effects

1. The size of the enterprise (size), calculated by the number of enterprise employees at the end of the year. 2. Human resources (resource), we measure by the ratio of employees with a university degree or above to total enterprise employees. 3. The level of enterprise diversification (diversity), we measure by the logarithmic value of the number of enterprise activity units. 4. Financing channels (financing), since we can't accurately measure the financing channels for enterprises, the paper employs enterprise financing costs (cost) and asset-liability ratio (debtasset) as a proxy variable of the financing channels, and use the value interest payments accounted for total equity capital to measure enterprise financing cost; 5. The established year of enterprise (age), we measure by the logarithmic value of the established year of enterprise. 6. Type of ownership (owner), This paper introduces three dummy variables to reflect that whether the firm is a foreign-invested enterprises (foreign), Hong Kong, Macao and Taiwan-invested enterprises (hk) or state-owned enterprises (state); 7. Unproductive expenditures (unproductive), measured by the logarithm value of operating management costs in this paper.

Industry Effects

1. Market structure (structure), we use industry concentration to reflect the market structure, taking about the feature that most producer services enterprises are in small size, we use CR50 specifically, the proportion of the size of the top 50 enterprises to the size of the industry; 2. The level of market demand expansion (demand), we measure by the growth rate of industry value-added; 3. Industry intellect intensive level (intellect), referring to Yuan and Chen (2011), we measure by the proportion of industry intermediate goods inputs coming from high-tech manufacturing industry

The core idea of the Shapley decomposition method is as follows: first, get the mean of independent variable X in the profitability deterministic model, then substitute the average of variable X for actual variable X in the profitability deterministic model to calculate the profitability data, recorded as Y1, whose difference coefficient denotes by f (Y1). Second, the profitability difference contributed by variable X is [f(Y0)-f(Y1)]/f(Y0), where f(Y0) indicates difference coefficient calculated by actual data. If the formula [f(Y0)-f(Y1)]/f(Y0) >0, it shows that variable X has a positive contribution and expands profit margin difference, vice versa.

Prior to decompose the profitability difference, it is necessary to ascertain regression equation. We have concluded some variables haven't displayed the significant correlation in profitability determinant through the table 5, 6 and 7. In view of this, the decomposition equation deletes that variables and choose significant variable only, so, the decomposition equation is as follows:

$$Profit_{ij} = \alpha_1 + enterprise(x^*_{ij})\beta_i + industry(y^*_{ij})\gamma_i + \mu_i \quad (2)$$

Where x* and y* represent significant factors in enterprise and industry respectively.

3 RESULTS

Table 1 reports the Shapley decomposition result of the profitability deterministic model. From table1, the proportion of explanation in return on sales model is above45%, regardless of all sample or subdividing samples, and that in per capita creates profit model is between 20% and 30%. Since the return on sales and per capita create profit denotes two aspects of profitability in the productive service enterprise, we can believe the design of the profitability deterministic model is reasonable.

Table 1. Shapley Decomposition Results Based Profitability Deterministic model.

	Subdividing Sample							
All Sample			High level		Low level		Eastern	
	ROS	PCCP	ROS	PCCP	ROS	PCCP	ROS	PCCP
	(1)	(2)	(3)	(4)	(5)	(6)	(7)	(8)
Resolution	52.6	22.8	48.1	24.1	57.0	21.0	49.1	22.9
Diversity	-6.1	1.3	—	—	−15.8	2.0	−2.1	1.4
Owner	24.7	10.1	15.1	11.0	32.1	10.0	37.1	17.6
Age	-3.4	-4.9	-8.8	-5.6	0.1	3.5	1.3	—
Size	8.9	12.3	8.9	14.5	8.8	11.1	10.5	8.4
Financing	13.3	10.7	6.0	—	16.1	6.1	12.1	18.7
Unproduct	13.8	12.5	3.2	23.1	13.6	6.8	7.1	18.5
Resource	5.0	7.6	4.8	-6.0	6.1	12.5	3.8	9.4
Subtotal	57.2	49.8	32.4	37.0	61.1	52.2	69.9	74.3
CR50	17.9	21.1	31.1	30.5	10.7	19.7	8.2	10.9
Intellect	3.3	12.9	2.0	14.0	8.2	11.1	2.1	4.7
Demand	21.3	17.0	34.2	18.2	19.8	16.8	19.7	10.0
Subtotal	42.7	51.1	67.4	62.8	38.8	47.7	30.0	25.7
Total	100	100	100	100	100	100	100	100

On the one hand, the industry effect of return on sales and per capita create profit in the productive service enterprise contributes 42.71% and 51.12%, respectively. The results provide strong support for the view that industry related factors have important influences on profitability. The conclusion intensively contrasts with the weak industry effect of previous scholars (Mcgahan and Porter 1997, 2002; Rumelt, 1991). In productive service enterprise of China, the contribution of industry effect can be roughly shown in three parts: the first path is market demand, in the process of market transition, the judgment of future development opportunities is easily recognized according to the track of highly market-oriented countries (Lin, 2007). Thus, once a certain industry appears policy induced opportunities or phased development condition, the new entrants will perform a great opportunism behavior to enter into the industry excessive just as "wave phenomenon". That excessive entrance results in strong homogeneity of enterprise competitive advantage and don't have heterogeneous resources, so, the source of profitability difference is embodied in the industry level largely. We can verify that path by the contribution of variable demand in Shapley decomposition results, which displays 21.38% and 17.07% respectively. The second path is market structure. Because of institutional obstacles, policy and other reasons, the degree of entry regulation between productive service industries presents different obviously. For example, market access threshold in the geological exploration industry, post and telecommunication, finance, insurance is too high to accept potential investors and industries were monopolized by minority incumbents. As such, the enterprise with them presents higher profitability than the ones which degree of entry regulation isn't serious. In other words, the difference of market entry regulation leads to the difference of enterprise profitability. We can verify that path by the contribution of variable CR50 in Shapley decomposition results, which displays 17.99% and 21.15% respectively. The third path is the knowledge intensive degree of industry, which generates an industry incentive effect, demonstration effect and correlation effect to promote spread of knowledge and technology spillover among enterprises intra-industry. As a result, the high degree of knowledge intensive brings about the overall competitiveness of the industry to a leading level. Unfortunately, at present, contribution of knowledge intensive degree to profitability difference is extremely limited, only 3.38% and 12.90% respectively in the results.

On the other hand, enterprise effect of return on sales and per capita create profit in china's productive service enterprise contributes 57.29% and 49.88%, respectively. It shows that enterprise effect is also relative important of profitability difference determinant. Specifically, the contribution of the ownership types is larger than any other enterprise factor, 24.73% and 10.12% respectively. Since the local producer services can't provide high technology service for multinational manufacturers to cause foreign productive service enterprise connect with them easily, and local manufacturers are lack of marketization and profession under which the demand for productive services is at a low level and cause local manufacturers tend to choose local services. Accordingly, it forms parallel and independent supply and demand cycle between domestic productive services and local services and is consistent with yang et al. (2011), which lead to the main differences of enterprise profit.

Unproductive expenditure is in second place of enterprise effect in sharping profitability difference, 13.81% and 12.54% respectively. Cai et al. (2005) shows that the restaurant and entertainment expenditure in domestic firms not only purpose to establish the relationship with customers or suppliers, but also has been used to pay the government official protection fee or bribe fee and so on, which particularly prominent in producer services due to enterprise products are characterized by the form of a service or a scheme in service, then directly impress the formation of profitability differences. Financing channels expenditure is in third place of enterprise effect in sharping profitability difference, 13.35% and10. 78% respectively.

Then, we can find, from the contribution of variance of enterprise effect, the accumulation of non-operative play the significant role in shaping profitability differences of China's productive service enterprise, such as social network, relationship, special channel, unproductive expenditure and so on. However, the function such as operation, diversity and age, enterprise scale and resource are so limited. So, it partly explains the cultivation of the core competitiveness is obviously insufficient in China's productive service enterprise.

From the point of regression results of different group samples, the industry effect in the region of higher marketization or industry of low barriers to entry is relatively small, the contribution of the former is 30.09% and 25.70%, respectively, and the latter value is 38.84% and 47.75%, both which is about almost 33% lower than the result of the overall sample. On the contrary, the enterprise factor contribution is relatively large, for example, the enterprise factor of per capital profit index in the eastern region is reached 74.30%. It shows that it presents the characteristic of the intensity industry effect or inferior enterprise effect in reign of higher marketization or industry of low barriers to entry, which is almost consistent with previous research conclusion (Mcgahan and Porter, 1999,2002,2005; Rumelt, 1991).

In general, under the environment of highly marketization or low administrative barriers, the conclusion is almost consistent with the enterprise resource school, which means enterprise profitability differences are more from the enterprise level and the less or almost no industry factor. In addition, the paper extends the content of theory of enterprise resource view, namely, under the environment of highly marketable reigns or low administrative barriers industries, it presents the intensity industry effect which is caused by market demand and entry regulation.

4 CONCLUSION

Promoted by a series of government support measures, Chinese productive service enterprises have higher profitability compared with industrial enterprises over the same period. Nevertheless, productive service enterprise' profit margin difference exists strong industrial effect that is caused by market structure and market demand, particularly in the regions of a lower degree of marketization or industries with higher administrative entry barrier, which is partly demonstrating the interference of administrative monopoly factors on enterprise efficiency. Furthermore, enterprise effect, which is the major factor contributing to productive service enterprise's profit margin, becomes especially prominent in the industries with administrative barriers to entry and in the areas with higher degree of marketization, the differences of ownership factors, non-productive expenditure and financing channels are the major factors that cause enterprise effect. Human capital, the duration the firm established, enterprise scale and other factors indicate limited contribution to the profit margin differences, which displays that Chinese productive service enterprises have obvious weaknesses in the cultivation of core competitiveness.

ACKNOWLEDGEMENTS

The paper is supported by the National Natural Science Foundation of China (71373033), the Social Science Fund Project of Liaoning Province of China (L12DJY045).

REFERENCES

Gu .N. Effects and channels between producer services and profits of industries-empirical research based on cities' panel data and SFA mode[J].China Industrial Economics,2010,(5):48–58.

Gao .JM, Li XH. Theoretical and empirical study on the interactive mechanism between producer services and manufacturing[J].China Industrial Economics, 2011, (6):151–160.

Chen .JJ, Chen .JJ. The research on the co-location between producer services and manufacturing —the empirical analyses based on the 69 cities and regions in Zhejiang province[J].China Industrial Economics,2010, (5):141–150.

Chen .DZ. Development level, structure, and impact of producer services in china :an international comparison-based on input output approach[J].economic research journal, 2008, (1):76–88.

Chen .JJ, Chen .GL, Huang .H. The study of producer service industry cluster and its influencing factors under the new economic geography-empirical evidence from 222cities in china.[J].Management World, 2009, (4):37–49.

Shorrocks. A. Decomposition procedures for distribution analysis: a unified framework based on the Shapley value[M]:University of Essex,1999,254–285.

Schmalensee. R. Do markets differ much?[J].American Economic Review,1985,75(6):71–112.

Rumelt. R. How much does industry matter?[J].Strategic Management Journal,1991,12(3):19–43.

Mcgahan.AM, Porter.ME. The persistence of shocks to profitability[J]. Review of Economics And Statistics, 1999, 81(1) :321–335.

Mcgahan.AM, Porter.ME. What do we know about variance in accounting profitability[J]. Management Science, 2002, 48(7) :657–681.

Misangyi VF, Elms H, Greckhamer T. A new perspective on a fundamental debate: a multilevel approach to industry, corporate and business unit effect[J].Strategic Management Journal, 2006, 12 (3):115–142.

Management, Information and Educational Engineering – Liu, Sung & Yao (Eds)
 ISBN: 978-1-138-02728-2

The measure and countermeasures to the bubble of the real estate in Shanghai

K.S. Xiao, C.M. Xia & R.C. Yang

Key Laboratory of Organic Chemistry in Jiangxi province, Jiangxi Science and Technology Normal University, Nanchang, P.R. China

ABSTRACT: As a representative of first-tier cities in China, the real estate market shows signs of a bubble in Shanghai, and there are a lot of speculative buyers at present. So, it is very important to measure the bubble of the real estate in Shanghai. In this paper, we should use the index of the real estate bubble to measure the bubble of real estate price in Shanghai, and then to judge the possibility of the existence of the real estate bubble in Shanghai, and aim to provide theoretical and the practical policy basis for the government to regulate and control real estate bubble.

KEYWORDS: The bubble of the real estate; Measure; Countermeasures; Policy proposal.

1 INTRODUCTION

1.1 *Research background*

In recent years, with the growing prosperity of the real estate market, the bubble is one of the most heated topics among people besides prices. As a representative of first-tier cities in China, with the implementation of the Chinese housing reform and the acceleration of housing commercialization, the real estate industry develops fast in Shanghai, and has become an important pillar industry which promotes economic growth in the city. With industrial development, the Shanghai housing prices also show excessive growth. It is very important to measure the speculative bubble in the real estate market in Shanghai.

1.2 *Current situation of Shanghai's real estate*

Official data shows there were 13,089.47 Million square meters of residential property under construction From January to August 2014 in Shanghai, year on year growth of 3.7 per cent. New floor areas under construction in the real estate sector were 1,479.13 million square meters, year on year growth of -10.5 per cent. The real estate of completion was 1,211.24 million square meters, year on year growth of -10.0 per cent. There were 7,681.29 million square meters of residential property under construction, year on year growth of 0.2 per cent, the new floor area were 5,256.80 million square meters, the area of completion were 834.48 Million square meters, year on year growth of -1.8 per cent. Real estate fixed asset investment were 1,875.03 billion Yuan, year on year growth of 7.0 per cent. Investment in residential real estate were 1,046.82 billion Yuan, year on year growth were 9.0 percent. The constructions of real estate investment were 1,048.39 billion Yuan, year on year growth of 8.8 per cent. In the first quarter, the new supply of new commodity residential of Shanghai was turning up from down. In March the new supply of new commodity residential of Shanghai nearly twice than before, there were 41 projects obtain presale permit ion, involving 9801 sets of commodity housing, 80% of the project push plate volume were more than 100 sets in Shanghai. Data shows that the total pushing plate increases 97% than the previous month in March. In the first 3 months of consecutive decline, new supply became growth. The volume of new commodity residential fell into about 30% than the same period of last year. Residential market turnover rose after the first drop. New commodity residential transactions were 5409 sets; the dealer area was 652 thousand square meters in January. The volume falls down than last month, new commodity residential transactions were 3,502 sets, and the dealer area was 421 thousand square meters in February. The deal and the area were the lowest since the February of 2012. Market turnover increased significantly in March, new commodity residential transactions were 8296 sets, the deal area were 939 thousands square meters, turnover were 24,300 million Yuan, transaction size and turnover respectively increase 1.23 times and 1.24 times than last month. The situations of volume declining for 5 consecutive months were changed. New commodity housing transaction size and volume of transactions were 2,012 thousand square meters and 51,290 million Yuan in the first quarter, Compared with the same period of last year decreased by 34.8% and 29.3%. Housing sales prices continue to rise in Shanghai.

In the first quarter, new commodity residential house prices had risen by 1.3%. Since last July, the new commodity housing quarter-on-quarter growth continued to fall, until this February, increasing price was down to 0.4%. In March, new commodity residential prices showed a steady growth trend, increased 0.4%. Look from the year-on-year, new commodity housing price had declined for three consecutive months, 1-3 month rose respectively 20.9%, 18.7% and15.5%, it dropped month by month. The old apartment prices continued to rise, ring up rise and fall. In the 1-3 month, the month-on-month price of second-hand housing were up 0.1%, 0.6% and 0.2%; the year-on-year prices rose month by month, rose respectively 13.2%, 12.1% and 9.5%.

2 MODEL SELECTION

In recent years, the frequent occurrence because of the real estate market bubble caused great disasters to social stability and economical development. More and more people focused on the real estate market and the real estate market bubble. Particularly in the real estate bubble theory and measure method, there was a gratifying progress. Now, the popular and applicable measure methods of real estate bubble were a statistical test method, index evaluation method and the theory of price. In this paper, we should use the index evaluation method to measure the real estate bubble of Shanghai.

In the world, there are lots of methods to measure real estate bubble indicators. Such as the ratio of the real estate development investment and the whole society fixed assets investment, real estate, the ratio of the completion of area and the total construction area, the ratio of the real estate housing price and income, the ratio of growth rate of real estate investment and gross domestic product (GDP) etc. In this paper, we use the ratio of growth rate of real estate investment and gross domestic product (GDP) to measure the real estate bubble of Shanghai.

3 MODEL EATABLISHMENT AND ANALYSIS

3.1 *Foam standard*

We can conclude whether it is too much investment of real estate during to expect too high the price in future with a comparison of the growth rate of real estate investment and GDP growth rate. The growth rate of real estate investment and GDP growth rate should be matched in the normal economical society. The difference in the growth rate of real estate investment and rate of GDP is not great in the course of health economical development. The real estate market will be more prosperous and investment growth rate is relatively high, the ratio can reach 15%-30%, in the early stage of economic development and rapid development stage. After the rapid development, in the period of steady development, the speed of growth and economic development will be unified, generally maintained at around 5%-15%. Therefore, the real estate investment growth rate and GDP growth rate should be relatively high in the recovery and prosperity stage, can reach 2-4, the ratio is too high then it is more bubble of the real estate market. It is to say, the growth rate of real estate investment is far greater than the growth rate of GDP, It means too much social housing investment, hot investment will bring a high price of growth of real estate and leading to the emergence of the real estate bubble

Table 1 shows, by the international financial crises in 2008 and 2009, the real estate investment growth was very slow in Shanghai, much lower than the growth rate of GDP growth. But it turned back and grew rapidly in the first half of 2010, even 3 times as a developing country; the real estate industry had a very important effect on the development of regional economy. It shows that it was the overheated real estate investment and had bubble of the real estate market since 2010. According to the views of the professor Lang Xianping of Hong Kong Chinese University on Guangdong TV financial Lang eyes show, the standard of the ratio of growth rate of real estate investment and gross domestic product (GDP) (P) index to evaluate the real estate bubble was:

Table 1. Judgment standard of real estate bubbles.

P value	P≤1	2≤ P<1	2<P
Foam state	no	mild	serious

3.2 *The four operation stage of the Shanghai City real estate market in 2005-2013*

From table 2 and figure 1 we could see during the 2005 to 2013, the real estate investment growth rate was through the following four periods comparing with the growth rate of GDP.

The first period was from 2005 to 2009. Relative to the growth rate of GDP, the growth rate of real estate investment grew slowly. In 2005, house market developed steadily, and the real estate investment of Shanghai growth closed to GDP growth. In 2006 to 2009, the growth rate of real estate investment was significantly slower than the speed of economic growth. The period, housing market was in the benign development.

The second period was 2009 to 2010. Real estate investment was booming. Real estate investment growth rate during this period was significantly faster than the speed of economic development. Especially in 2010, the rate of real estate investment was rising more than two times than the rate of economic growth in the third period, it was from 2010 to 2011. In January 10, 2010, the State Council promulgated the country eleven, which can tame the housing market. And in April 17th, it released a more stringent" new country of ten ", while Shanghai was promulgated the"the most ruthless "restriction order all over the country. In a sense, these contained the real estate investment rising too fast, and Shanghai also stops the excessive growth of footsteps. The growth rate of real estate investment has been greatly reduced compared with the speed of economic growth.

The fourth period was from 2011 to 2013, the Shanghai real estate investment appears historic rebound. To 2013, the real estate investment was rising more than three times than the economic growth.

3.3 *The operating characteristics of the real estate market of Shanghai City*

In table 2, it depicts the price bubble trajectory and trend of the real estate market in 2005-2013year in Shanghai. From the foam running track, the real estate market of Shanghai has two features.

First, the real estate market has experienced a bubble phase, from 2005 to 2009 in Shanghai; the remaining time was the foam operation period. In the 2006 to 2009, the ratio of growth rate of real estate investment and GDP was less than one, indicating that it was a period of benign development of the real estate. But after 2010, due to various reasons, the real estate market developed rapidly. The real estate bubble appeared slowly bulking up. In 2010, the foam index was 2.23; it was a serious bubble period. In 2011, due to the national macro-control, Shanghai's real estate bubble index dropped to 0.88, return to a normal level. But in the period 2012-2013, the real estate bubble index rebounded. In 2012, the real estate bubble index was 1.5. It was the mild foam, but in 2013, the real estate bubble index reached to the highest level in history, the Shanghai real estate market was in a serious bubble operation stage.

Second, there was a relationship between the real estate market bubble and the national macro-control efforts. After a period of operation of In the real estate market, the bubble began to rebound. In 2008 the state promulgated the "country of ten", January 10, 2010, the State Council promulgated the "country of eleven" and the April 14, 2010 to 17, the State Council promulgated to curb some city commercial housing prices rise too fast, "four new country" and "ten new country", the Shanghai real estate bubble began to shrink. By the year of 2011, it had been in no bubble operation stage of real estate. But along with the national regulation's relaxation, began a strong rebound in 2012 and 2013, the Shanghai real estate was in a serious bubble operation stage.

Table 2. The rate of real estate investment growth rate and gross domestic product (GDP) growth rate from 2005 to 2013.

Year	P	A	B
2005	0.9622	0.14	0.1455
2007	0.3561	0.051	0.1432
2008	0.2915	0.053	0.1818
2009	0.375	0.026	0.069
2010	2.2285	0.314	0.1409
2011	0.8799	0.104	0.1182
2012	1.537	0.079	0.0514
2013	3.054	0.215	0.0704

Data source: the people's Republic of China Statistics Bureau website

P: the ratio of growth rate of real estate investment and gross domestic product (GDP)

A: real estate investment growth

B: gross domestic product (GDP) growth

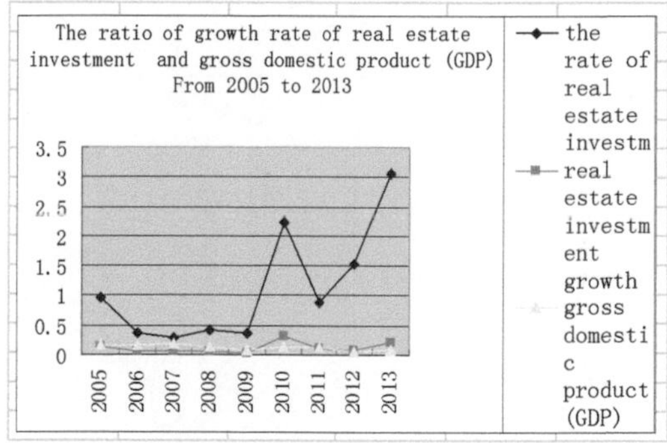

Figure 1. The ratio of growth rate of real estate investment and gross domestic product (GDP).

4 POLICY RECOMMENDATIONS

Because the long-term real estate bubble will affect the healthy development of the real estate market, the bursting of the housing bubble will cause great damage to the national economy. In China, it is once appeared a large degree of the bubble. After 2012, the first and the second tier city had the bubble expasion trend therefore, this paper holds that government need to take reasonable measures to rein in the expansion of the housing bubble, including: strengthen macro-control and management; reform the system of land transfer, land resources to strengthen management; to promote the protection of housing construction; the adjustment of the real estate tax system; improve the real estate market information disclosure system; dynamics strengthen supervision of real estate finance; and the development of the capital market and broaden the investment channels and so on.

ACKNOWLEDGMENT

The corresponding author of this paper is Yang ruchun. This paper is supported by Education Department of Jiangxi Province (GJJ12583) the Training Program of Jiangxi Youth Scientists.

Management, Information and Educational Engineering – Liu, Sung & Yao (Eds)
© 2015 Taylor & Francis Group, London, ISBN: 978-1-138-02728-2

Research on MICE enterprise performance evaluation system based on the balanced scorecard

Bin Wang
School of Management, Lanzhou University, Lanzhou, China

ABSTRACT: The MICE industry is a new engine for economic development and plays an increasingly significant role in national and regional growth. In order to evaluate the performance of MICE enterprises in China, this paper constructs a three-layer performance evaluation system based on the balanced scorecard. Then, through an Analytic Hierarchy Process (AHP), the paper calculates the weights of related evaluation factors. Finally, some suggestions for MICE industry's development strategies and managerial optimization are offered with the aim to further enhance their performance management.

KEYWORDS: MICE; performance evaluation system; balanced scorecard.

1 INTRODUCTION

The MICE (Meeting, Incentive, Conference, and Exhibition) industry is now playing an increasingly significant in the national and regional economic development, and a rich literature is forming. Wang et al. (2011), for instance, proposed a four-dimension approach to analyze the economic effect of the MICE industry. Those are MICE consumption, employment, industrial promotion and contribution to the perfection of city functions. They also pointed out that the MICE industry had remarkable impacts on the entrepreneurial activities.

At the same time, as performance appraisal is the "barometer" and "security" of corporate operations, it is essential that a set of standardized performance evaluation system be constructed to monitor and improve its management, and to ensure a healthy and smooth of domestic MICE industry. However, taking into account of its present development status, China's MICE industry is still at the preliminary stage. What is more, few researchers are focused on the construction of performance evaluation system of the MICE industry, so it is of theoretical and practical value to build such a performance evaluation system.

In view of these, this paper constructs a three-layer performance evaluation system based on the Balanced Scorecard (BSC) with a special attention to the MICE industry.

2 LITERATURE REVIEW

Kaplan & Norton (1992) first proposed the concept of the balanced scorecard in "The Balanced Scorecard: Measures That Drive Performance", where they defined the four dimensions of the balanced scorecard, namely, the finance, customer, internal business processes, learning and innovation. Later, Kaplan & Norton (1996) in a paper entitled "Using the Balanced Scorecard as a Strategic Management System", they added four transformation steps of vision and strategy for BSC, which include transforming vision, communication and contact, business plans, feedback and learning. Furthermore, to demonstrate BSC's usage in strategic management, Kaplan & Norton (2000), in their paper "Having Trouble with Your Strategy? Then Map It", proposed the concept of strategic map. Later, the theory of the balanced scorecard was further improved and applied in a substantial number of academic research papers and publications.

In addition, as the MICE enterprises are usually joint-ventures with state capitalism or government contribution in China, they need not only to assume economic responsibilities, but also to fulfill the social responsibilities of creating social welfare assets. Compared with the classic performance appraisal methods, the BSC is more suitable to evaluate MICE's performance for its convenience, objectivity and representativeness. Thus, the BSC is applied as the basic approach of this paper.

3 PERFORMANCE EVALUATION SYSTEM OF MICE ENTERPRISE BASED ON THE BALANCED SCORECARD

3.1 *High frequency indexes analysis*

Through the software of ROST, some related literatures are analyzed (Du 2010; Luo 2010; Pan & Wu 2014; Qiao et al. 2007; Shan & Shen 2010; Shang et al. 2005; Sun & Peng 2009; Xing 2010; Ying et al. 2007; Zhang et al. 2009; Yue & Jin 2011). The result, according to the similarity and high frequency criteria, is shown in the table below (Table 1):

Table 1. High frequency indexes analysis.

Goal (A)	First level index (B)	Second level index (C)
Enterprise Performance	Financial (B1)	ROE(C11)
		ROI(C12)
		Cost and Expense Ratio (C13)
		Revenue(C14)
	Customer (B2)	Customer Satisfaction Rate (C21)
		Customer Retention Rate (C22)
		Customer Acquisition Rate (C23)
		Customer Profitability Rate (C24)
		Market Share (C25)
	Internal Business Process (B3)	Product /Service Quality (C31)
		R&D (C32)
		Internal Management (C33)
	Learning and Innovation (B4)	Employee Satisfaction (C41)
		Employee Training (C42)
		Employee Ability (C43)
		Information System (C44)

Source: author reorganizes according to ROST software analysis results.

3.2 *Reconstruction of indexes*

According to the specific circumstances of the MICE industry, the new indicators are re-screened and added by an expert team (including 6 MICE company executives and 2 industry experts) on the basis of the high frequency indexes (Table 2).

Firstly, the financial dimension aims to reflect the concept of "increase revenue, reduce costs, and improve profitability". Accordingly, the second level indexes are divided into three categories: the revenue, the cost and the profit. As the measure of the MICE industry in regard is similar to other services industries, the third level indexes are constructed through the high frequency indexes.

Secondly, the customer dimension adopts the view of "big customer", and thus this dimension includes not only external customers, but the internal customers and social groups as well.

Thirdly, the internal business processes is orientated to evaluate the financial health and customer value creation. The dimension is divided into 3

Table 2. The performance evaluation system of the MICE enterprise based on the BSC.

Goal (A)	First level index (B)	Second level index(C)	Third level index (D)
MICE Enterprise Performance	Financial (B1)	Revenue (C11)	Prime Operating Revenue (D111)
		Cost (C12)	Cost and Expense Ratio (D121)
		Profit(C13)	ROE (D131)
			ROI (D132)
	Customer (B2)	External Customers (C21)	Customer Satisfaction Rate (D211)
			Customer Retention Rate (D212)
			Customer Acquisition Rate (D213)
			Customer Profitability Rate (D214)
		Internal customers (C22)	Employee Satisfaction (D221)
		Social Responsibility (C23)	Charitable Activities (D231)
			Environmental Protection (D232)
	Internal Business Process (B3)	Platform Construction (C31)	Information Construction (D311)
			MICE Platform(D312)
		Employee Ability (C32)	Professional Staff Ratio (D321)
		Internal Management (C33)	Team Work (D331)
			Service Quality (D332)
	Learning and Innovation (B4)	Employee Construction (C41)	Employee Expostulation Adoption Rate (D411)
			Employee Training Rate (D412)
			Individual Scorecard Construction (D413)
		Information System (C42)	Information Capacity Development (D421)
		External Cooperation (C43)	Industry Alliance Construction (D431)
			Industry Benchmarking Learning (D432)

aspects: foundation platform construction, employee ability, internal management efficiency.

Finally, learning and innovation focuses on the improvement of the company's overall quality through constant learning and innovation. And it is not only to improve staff quality and information ability, but also to equip them with a globalized view. Therefore, the second level indexes includes employee construction, information system and external cooperation.

3.3 *Importance rank and consistence check*

As the analytic hierarchy process (AHP) is relatively effective in weighting (Chang & Jiang 2007), this paper applies the process in its performance evaluation system construction. Its steps is shown below.

3.3.1 *Construction of judgment matrix*

The judgment matrix of the first level indexes is established by the experts team.

$$P = \begin{bmatrix} 1 & 2 & 2 & 2 \\ 1/2 & 1 & 1 & 2 \\ 1/2 & 1 & 1 & 3 \\ 1/2 & 1/2 & 1/3 & 1 \end{bmatrix} \tag{1}$$

3.3.2 *Weight calculation*

Firstly, this paper calculates the product of each row in the judgment matrix P, where M_i is obtained (i stands for the row, i = 1, 2, 3, 4).

$M_1 = 1 * 2 * 2 * 2 = 8$, $M_2 = 1/2 * 1 * 1 * 2 = 1$, $M_3 = 1/2 * 1 * 1 * 3 = 1.5$, and $M_4 = 1/2 * 1/2 * 1/3 * 1 = 1/12$.

Secondly, this paper opens the fourth power of M_i, where

$$W_i = \sqrt[n]{M_i} \tag{2}$$

$$w_i = W_i / \sum_{i=1}^{n} W_i \tag{3}$$

And w_i (i stands for the row, i= 1, 2, 3, 4) is the weights of the first level indexes.

W_1=1.68, W_2=1, W_3=1.11, W_4=0.54;

w_1=0.39, w_2=0.23, w_3=0.26, w_4=0.12.

Finally, it can be obtained,

$$\lambda_{max} = \frac{1}{\text{n}} \sum_{i=1}^{n} \frac{(Pw)_i}{w_i} \tag{4}$$

So, λ_{max}=(4.13+4.04+4.04+4.33)/4=4.14.

3.3.3 *Consistence check*

This paper checks the validity of the weights by the general consistency index (CI), the random consistency ratio (CR) and the average random consistency index (RI), where,

$$CI = (\lambda_{max} - n) / (n - 1) \tag{5}$$

$$CR = CI / RI \tag{6}$$

Then CI and CR are obtained, where CI=(4.14−4)/(4−1)=0.14/3=0.05, CR=0.05/0.9=0.06.

Table 3. Value of RI.

N	1	2	3	4	5	6	7	8	9
RI	0	0	0.58	0.9	1.12	1.24	1.32	1.41	1.45

The result according to the above steps is shown Table 4, and other matrices have the same algorithm (Table 5).

Table 4.

A	B1	B2	B3	B4	weight
B1	1	2	2	2	0.39
B2	1/2	1	1	2	0.23
B3	1/2	1	1	3	0.26
B4	1/2	1/2	1/3	1	0.12

λmax= 4.14 , CI= 0.05 , CR= 0.06

Table 5.

Judgment matrix	CI	CR	λ_{max}	w_i
A*	0.05	0.06	4.14	0.39/0.23/0.26/0.12
B1*	0.03	0.05	3.05	0.2/0.49/0.31
B2	0.03	0.05	3.05	0.49/0.31/0.2
B3	0.03	0.05	3.05	0.33/0.41/0.26
B4	0	---	3	0.4/0.4/0.2
C13	0	---	2	0.5/0.5
C21	0.05	0..06	4.16	0.34/0.24/0.17/0.24
C23	0	---	2	0.5/0.5
C31	0	---	2	0.67/0.33
C33	0	---	2	0.5/0.5
C41	0.03	0.05	3.05	0.49/0.31/0.2
C43	0	---	2	0.33/0.67

*A stands for the judgment matrix A, w_{B1}=0.39, w_{B2}=0.23, w_{B3}=0.26, w_{B4}=0.12 (As Table 4).

*B1 stands for the judgment matrix B1, w_{c11}=0.2, w_{c12}=0.49, w_{c13}=0.31.

Generally, when CR < 0.1 or λ_{max} = n, CI = 0, the judgment matrix is considered to good. In this paper, all of the judgment matrices have a satisfactory consistency.

Thus, the comprehensive weights of all indexes are calculated according to the weights of the evaluation factors (Table 6):

Table 6*. Comprehensive weight of each index.

A	B	Comprehensive Weight	C	Comprehensive Weight	D	Comprehensive Weight
MICE Enterprise Performance	B1	0.39	C11	0.08	D111	0.08
			C12	0.19	D121	0.19
			C13	0.12	D131	0.06
					D132	0.06
	B2	0.23	C21	0.11	D211	0.04
					D212	0.03
					D213	0.02
					D214	0.03
			C22	0.07	D221	0.07
			C23	0.05	D231	0.03
					D232	0.03
	B3	0.26	C31	0.09	D311	0.06
					D312	0.03
			C32	0.11	D321	0.11
			C33	0.07	D331	0.03
					D332	0.03
	B4	0.12	C41	0.05	D411	0.02
					D412	0.01
					D413	0.01
			C42	0.05	D421	0.05
			C43	0.02	D431	0.01
					D432	0.01

*The comprehensive weights are rounded with minor deviations.

3.4 *Result analysis*

3.4.1 *The analysis of first level indexes*

From the table 6, it can be observed that the comprehensive weight of the financial index is 0.39. So, it shows that the finance plays an important part in the performance, where it meets the ultimate goal of enterprise operations. In addition, the comprehensive weights of the customer and the internal process are similar. Thus, it reflects the tenet of the customer-orientation and the improvement of the internal management.

However, the weight of the learning and innovation is relatively weak, so the enterprises should adjust their practical operations to build a learning and innovative organization in order to guarantee their competitive advantage.

3.4.2 *The analysis of second level indexes*

The enterprises should focus on the key indicators and control the key links. Among them, the weights of the cost, the revenue, the external customers and the employee's ability are higher than 0.1. It shows that, firstly, these enterprises should pay attention to their cost and revenue to ensure an effective budget management. Secondly, they should emphasize the external clients in order to create more market value.

And to meet the demand of external customers, these enterprises should place importance on their employees' ability. This will also be helpful to improve employee satisfaction to make their business operations more efficient and effective.

3.4.3 *The analysis of third level indexes*

The weights of the cost and expense ratio and the professional staff ratio are higher than 0.1. This reflects the fact of a poor cost-control in the industry and the lack of the professional talents. Therefore, the MICE enterprises should construct a set of scientific cost-control systems. Besides, they should put more efforts to cultivate professional talents.

4 CONCLUSION

First of all, this paper analyzes the high frequency indexes of the performance system of various related literatures. Secondly, the paper excludes some inappropriate indicators and constructs the appropriate indicators by an expert group. Finally, this paper uses the method of AHP to calculate the weights of all indexes, and the relevant indicators and weights are adjusted through the consistency check. The performance evaluation system will help the MICE enterprises to evaluate their development. And the MICE enterprises should pay more attention to their performance system and adjust the indexes according to own situation.

However, this paper still has some limitations. Firstly, the expert team should be of a more diversified nature. Secondly, the indexes of this performance evaluation system are not all-inclusive, and the paper can add more suitable indicators to make the system more operational. In future, this paper will adopt other methods to make up for these deficiencies and defects, so as to make the research be of a more application value.

REFERENCES

Chang, J. E., & Jiang, T. L. 2007. Research on the Weight of Coefficient through Analytic Hierarchy Process. *Journal of Wuhan University of Technology (Information & Management Engineering), 29*(1), 153–156.

Du, J. H. 2010. Study on the Application of the Dynamic Balanced Scorecard Based on System Dynamics in Enterprises in Service Sector. *Seeking Truth, 37*(4), 59–63.

Kaplan, R. S., & Norton, D. P. 1992. The Balanced Scorecard: Measures That Drive Performance. *Harvard Business Review,* (1), 71–79.

Kaplan, R. S., & Norton, D. P. 1996. Using the Balanced Scorecard as a Strategic Management System. *Harvard Business Review*, *74*(1), 75–85.

Kaplan, R. S., & Norton, D. P. 2000. Having Trouble with Your Strategy? Then Map It. *Harvard Business Review, 78*(5), 167–176.

Luo, J. Q. 2010. Enterprise Strategic Performance Evaluation and Model Building. *Journal of Xidian University (Social Sciences Edition), 20*(6), 26–30.

Pan, L. S., & Wu, Z. Y. 2014. Research on Performance Evaluation of State-owned Science and Technology Enterprises Based on Balanced Scorecard. *Journal of Hefei University of Technology (Social Sciences),28*(2), 28–31.

Qiao, J., Qi, X. L., & Chu, J. S. 2007. Study on Company's Performance Measurement Based on Balanced Scorecard—With Special Reference to A Telecom Corporation Jiangsu. *China Industrial Economy,* (2), 110–118.

Shan, G. Q., & Shen, F. 2010. Empirical Study on the Performance Evaluation System of Enterprises Based on Balanced Score Card. *Science and Technology Management Research, 30*(16), 243–246.

Shang, R. B., Tang, Z. H., & Wen, G. B. 2005. Utilizing the Balanced Scorecard for R&D Performance Evaluation. *Science of Science and Management of S.&T., 26*(4), 15–18.

Sun, J., & Peng, H. X. 2009. Study on BPO Performance Evaluation Based on the Balanced Scorecard. *Journal of Intelligence, 28*(z1), 139–141.

Wang, X. W., Zhang, Y. L., & Wang, J. N. 2011. A Study on the Mechanism of MICE Economic Effects:An Entrepreneurial Activities As Pathway View. *Tourism Science, 25*(4), 49–57.

Xing, D. M. 2010. Reconstruction and Application of Enterprise Performance Evaluation System Based on Balanced Scorecard. *Journal of Changchun University(social science edition), 20*(7), 16–20.

Ying, K. F., Gong, J., & Xue, H. X. 2007. Empirical Study on Virtual Enterprise Performance Evaluation Based on the Balanced Scorecard. *Science of Science and Management of S.& T., 28*(9), 197–200.

Yue, J. Y., & Jin, S. Y. 2011. Construction of Enterprise Performance Evaluation System Based on the Balanced Scorecard. *Heilongjiang Foreign Economic Relations & Trade,* (9), 112–115.

Zhang, H., Li, Y., & Cao, J. A. 2009. Research on the Multiple Levels System of Balanced Scorecard. *Science & Technology Progress and Policy, 26*(8), 73–76.

Management, Information and Educational Engineering – Liu, Sung & Yao (Eds)
© 2015 Taylor & Francis Group, London, ISBN: 978-1-138-02728-2

A comparative analysis of the change in Dongba culture inheritance modes

Hou Liang Kang & Yu Ting Yang
Tourism and Culture College, Yunnan University, Lijing, Yunnan, China

Fan Wang
Yunnan University, Kunming, Yunnan, China

ABSTRACT: With the changing times, the traditional inheritance of the Dongba culture has gradually disappeared. In investigating the basic conditions of Gongba in Lijiang at present, we found some problems: First, the regional distribution of Dongba suffered a sharp decline; Second, the number of Dongba who have a systematic mastery of Dongba culture is smaller; Third, inheritors' age gap leads to fewer and fewer young Dongba. So the Dongba culture inheritance by school training mode becomes necessary. By comparing and analyzing the traditional inheritance and school training mode, we find that only by improving the training of Dongba culture school, establishing preserved areas, enhancing the current Dongba's skills and techniques, and fully mobilizing Dongba's enthusiasm to create a cultural inheritance atmosphere, can Dongba culture be protected better.

KEYWORDS: Dongba inheritance, inheritance mode, inheritance change, Dongba culture.

1 INTRODUCTION

"Dongba" is a folk name for the priests of traditional Naxi Dongba religion who are responsible for inheriting Dongba culture, and a religious equivalent from themselves is "benbo" ($py^{33}mby^{31}$), which in English means "teacher" or "instructor" or "the wise"[1]. Dongba religion is a national religion all Naxi people are supposed to believe in, and it is thought of as a post-original religious form by scholars. By absorbing the essence of diverse religious forms including Tibetan Bon Religion, Tibetan Buddhism, Chinese Buddhism and Taoism, Dongba religion has formed a special culture system without losing its own characteristics[2-5].Dongba, responsible for inheriting Dongba religious culture, have maintained Dongba culture continuously inherited and publicized by dictation and written books. In 2003, Dongba ancient books written in Dongba character were listed as one of items in the World Heritage Memory, Dongba therefore feel more responsible in finishing their task. Today, the traditional way of inheriting Dongba religion has been drastically changed and is even in danger of disappearing because of the rapid economic development of the society. So in order to protect and better develop Dongba culture, the traditional inheritance has to be replaced by the school training mode. We hope our comparative analysis of the above two inheritance modes' advantages and disadvantages could benefit a better inheriting and developing of Dongba culture.

2 BASIC INHERITANCE CONDITIONS

By analyzing the number of Dongba (Dongba Society applicants, title-awarded Dongba by the Society, and attendees of Dongba religious rituals), their age structure, and their living areas, the basic inheritance situations can be clearly and better understood.

2.1 *The educational backgrounds of Dongba society applicants*

In 2010, Dongba Society was established and took in some villagers and town people with a basic mastery of Dongba culture knowledge to offer a better training to the existed Dongba and to better protect Dongba culture. According to the information offered by Dongba Society: by the year of 2011, the total number of its members is 90, among whom the Dongba receiving primary education are 52 with 58.9% of the total; the Dongba receiving Junior education are 32 with 35.5% of the total; and the Dongba receiving senior education and college education are respectively 3 with the same 3% [6]. Additionally, members are mainly from rural areas, and the average age of Dongba skilled at ten or over-ten sacrificial rituals is over 45 years old; the Dongba proficient at writing and reading Dongba characters are only 20 people, each of whom is over 50; while there are only 4 Dongbas mastering about three simple sacrificial ceremonies, reading and writing some simple characters and words. So it's clear to see the problem of age gap in Dongba culture inheritors and the young Dongba skillfully mastering Dongba characters and sacrificial rituals are extremely few, which endangers the continuous inheritance of Dongba culture.

2.2 *Dongba ritual participants' age structure and geographical distribution*

For the Dongba in Lijiang, "Dongba Religious ritual" is a big annual occasion held in Lijiang scenic spot of Jade Water Villa on the fifth day of every lunar year, when all Dongba will get together to offer their ancestor— Dongbashenluo, to recite scriptures and to perform their special Dongba dancing. Since the first recovered Dongba religious ritual, the one held in 2013 was the 13th time. Dongba religious ritual not only offers a noble chance for old Dongba to show their superb artistry and for young Dongba to consolidate what they have learnt, but offers a valuable platform for the Dongba from different areas to share with and learn from each other.

An interview with Dongba participating in the ritual of 2011 shows that all religious ritual participants are male. Their general age distribution are as follows shown in Figure 1[7]: Dongba above 80 are 3; Dongba between 60 and 79 are 7; those between 40 and 59 are 13, those between 20 and 39 are 12; and there are 8 young Dongbas under the age of 19 who are from the Dongba school in the Jade Water Villa. In the respect of geographical distribution, most Dongba come from Daju village, Xin village, Lashi village and Tai'an village, and their culture therefore gets a better development. Besides, with more and more Dongba religious rituals being held, the number of Dongba participating in rituals is increasing year by year, accompanied by the decreasing number of old Dongba and the increasing number of young Dongba, still young Dongba are relatively a few.

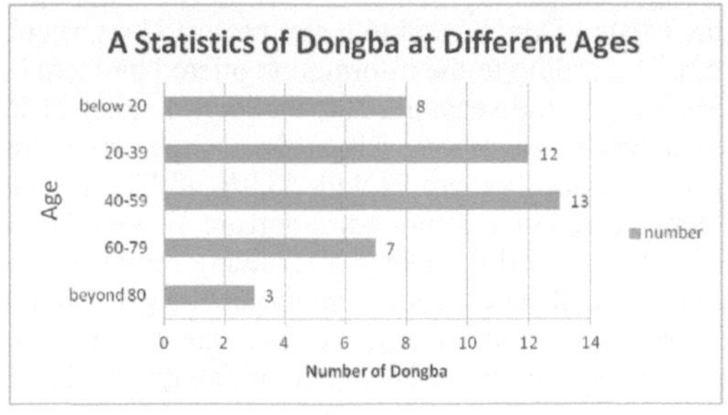

Figure 1. Age structure of 2011 ritual Dongba participants.

2.3 *A survey of title-awarded Dongba*

In 2012, authorized by the People's government of Yulong Naxi Autonomous County, Lijiang Dongba Culture Inheritance Society and the Inheritance Base of Dongba Culture in Jade Water Villa carried out Dongba degree assessment. By having a comprehensive assessment of Dongba's time length of learning skills, their mastered skill varieties, their proficiency of holding Dongba sacrificicial ceremonies, and their contributions to culture protection, 123 Dongba were awarded with different titles[8] as shown in Figure 2: 6 Dongba were awarded with the title of Senior Mage; 30 Dongba, Mage; 40 Dongba, inheritors; 47 Dongba, learners. Of the 123 awarded Dongba, 29 come from Labo village, Ninglang county; 13 come from Tacheng village, Yulong county; 6, Daju village; 5, Ludian village; 5, Jinshan village; and 4, Taian village. The above survey indicates that Dongba culture was preserved and developed better in the above 6 villages, and Dongba activities were usually held in Jade Water Villa, Labo village, Tacheng village and the scenic spots of Lijiang city. Meanwhile, Lijiang's prosperous tourism indirectly moves the holding of some relevant Dongba activities.

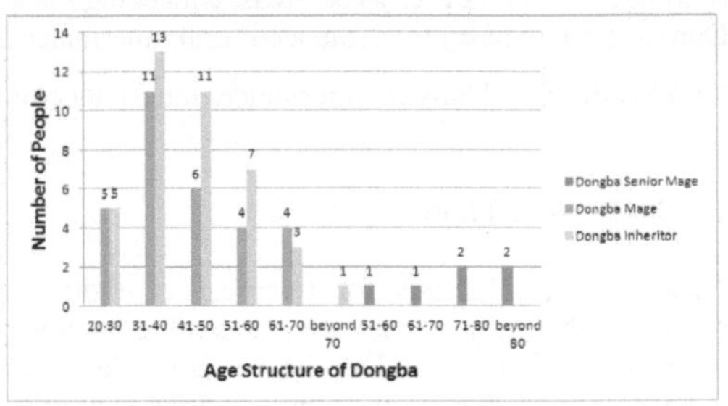

Figure 2. Histogram statistics of awarded Dongba's age structure.

3 THE CHANGE OF DONGBA CULTURE INHERITANCE MODE

3.1 *The traditional mode of Dongba culture inheritance*

For thousands of years, Dongba in Lijiang have been living in villages, as usual, for every village, there is one senior Dongba whose major work is farming like other ordinary villagers and only holds some rituals on traditional days including climatic and seasonal days, customs, weddings and funerals, etc. There are two traditional inheritance modes: hereditary and master-apprentice system, where one's own ancestors and fathers are usually their "teachers". Teachers live with their students, doing farming together with "students" at daytime and teaching them scripture reciting and writing, and ritual holding by words and mental inspiration. Master-apprentice inheritance system has no time or place limitation. Whenever holding rituals, "students" are indispensible assistants of "teachers". Therefore, Dongba educated by traditional inheritance mode usually begin studying earlier at very young age and learn for a longer time of over ten years on average, so these Dongba can master solid and full fundamentals. However, they couldn't hold sacrificial rituals until teachers become old or die. In turn, the next generation of young Dongba is cultivated in the same way as shown in Figure 3. For the traditional inheritance mode, students could inherit all their teachers have taught and, better inherit and develop the essence of Dongba culture, while; the students inspired by life-devoted teaching are as few as one or two.

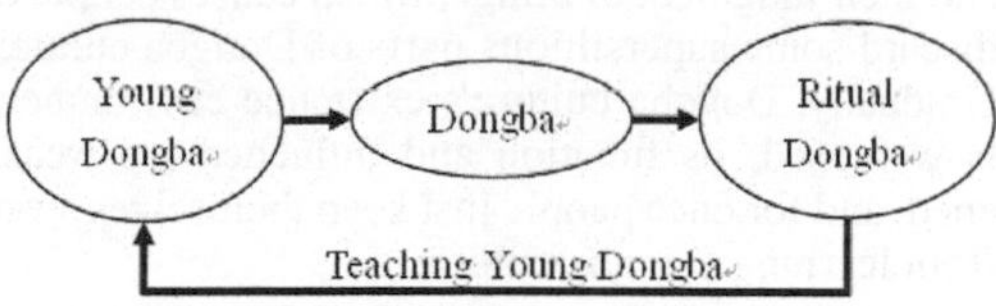

Figure 3. Traditional inheritance mode.

3.2 *Dongba school training mode*

Since the 1980s, the two traditional inheritance modes of hereditary and master-apprentice system have gradually disappeared, which marked an end to the traditional inheritance of Dongba culture. The first Dongba school opened in 2003 signaled the change into Dongba school training. Until now, there have been two Dongba schools in Lijiang established to teach Dongba culture: one is in Xinzhu, Ludian village, the other is in Jade Water Villa. In the two schools, young men with a basic knowledge of Dongba culture are trained to consolidate their mastery of Dongba culture in a short term. By the beginning of 2011, there are 43 graduated Dongba. In terms of Dongba school's advantage, the first is its obvious target: Dongba school aims to train learners to master professional knowledge and skills indispensible for Dongba culture; the second advantage is its wide range of trained targets, in other words, Dongba school has a relatively low entrance requirement; the third is its high training efficiency, because learners could learn more about Dongba culture in a shorter time. Compared with its advantages, Dongba school's disadvantages are also very obvious: due to short training time, learners' learning cannot be supervised well, moreover, given the truth that learning Dongba culture is a systematical long-term process, learners can only master some simple sacrifice holdings and some primary knowledge like sculpture and writing Dongba characters, which restricts Dongba culture's inheritance and full development.

In order to function well, school applies five-year boarding enclosed education, where two Dongba are responsible for teaching the current 8 students whose major learning subjects include scripture, Chinese, art, sculpture, P.E., labor, etc. Students are supposed to have seven classes at daytime from every Monday to Saturday with 45 minutes for each class, and have one-hour self-taught class in the evening. Students in Dongba school can be given a free accommodation and a monthly subsidy of 200 yuan, what's more, every graduate could find a relevant job in the nearby scenic spots according to the arrangement of school. The above whole process can be shown in the following Figure 4. All students are from poor families in the rural areas, but they have had certain knowledge of Dongba culture, so their willing learning motivations and actions help to better the learning result. In November, 2012, granted officially by Culture, Broadcasting, Television Press and Publication Bureau of Lijiang City, Dongba Inheritance school in Jade Water Villa was registered in Civil Affairs Bureau of Lijiang City in the name of Lijiang Dongba Culture School, becoming a Donba culture education organization with an independent corporate qualification, which marks an official beginning of school-mode inheritance.

Figure 4. Employment situations of school graduates.

3.3 *Advantages and disadvantages of two inheritance modes*

Compared analysis show that learners trained by the traditional mode need a longer time to acquire more Dongba culture to be qualified, despite their rich knowledge and strong application ability, the number of students each teacher teaches is very small; in contrast, learners educated in the school are bigger in number for each term, and they are taught at more fixed time and in a more standard way, which guarantees the relatively high qualification of graduated Dongba, though they are less capable of practice. In addition, since every village has the resource for Dongba culture inheritors, traditional mode can train more young Dongba, which really benefits culture inheriting and developing like table 1 indicates. While, because of only one independent-recruited Dongba school in Lijiang, school training mode may increase the number of young Dongba, but its total number is still less. Despite the above disadvantage of school training mode, school education, as one of the ways of culture inheriting and developing, is bound to be an effective way to cultivate young generations of Dongba in the present period.

4 SURVEY OF THE REASONS OF THIS CHANGE

4.1 *Chang of Dongba's living environment*

In the past, Donba all lived in villages, mainly doing farming while doing something indispensible for a village like holding religious rituals, selecting auspicious time by divination and offering treatment by prescription, for which Donba win great respect and love from villagers and a high social status in turn. Today, for the facilitated transportation, popularity of scientific knowledge and introduction of high-efficient economic crops, villagers make more money and get more chances to communicate with the outside, which help to reduce their reliance on Dongba; besides, the skills and arts Dongba have mastered cannot help them make a fortune, for which many young men are not willing to learn Dongba knowledge, what's worse, some young Dongba choose to quit school to go out for money. This largely accounts for the interruption of Dongba culture inheritance. Another impact is from the reform and opening-up, since then, Lijiang city has been enlarged and its economy has got rapid development, which moves a large number of free workforce in the village to enter the city, leaving only the old, the weak, women and their young children who cannot finish some tedious sacrificial rituals like the Shenluowu ritual that lasts seven days and seven nights. This directly stops the better inheriting and developing of Dongba culture. Rapid development of the society greatly improves people's mind and their judgment of things, which causes people to discard some superstitious parts of Dongba culture. Gradually, Dongba culture's existence environment is worsened, its function and influence are weakened, and for once people just keep themselves away from learning Dongba culture.

Table 1. A comparison of two inheritance modes.

contents mode	Teaching Spot	Teaching mode	Class time	School age	persons	School years	Learning contents	Employment	Advantages	Disadvantages
Hereditary	yard	Independent training	During working	At early age	1-2	Above 10 years	Scripture Rituals; Dongba dancing	farming and working as a priest	Solid basic knowledge; More knowledge about rituals	Small number of trained Dongba
School training	Dongba school	Class teaching	Every week	11-15	Around 8	Five years	Scripture Rituals; Dongba dancing; Chinese	Doing Related work	More Dongba Learners; High-rate talent	Less chances to practice

4.2 *Change of school children's learning enviroment*

In the past, it was very hard for the school children in distant country with backward primary education to receive a systematic education and get rich knowledge. So many parents desired to send their children to learn rich culture and knowledge from Dongba with a high social position, which for most families was a great honor, and in particular, which offers adequate sources of learners to inherit and develop Dongba culture. Now, with the implement of birth control policy and nine-year compulsory education, every couple has only one child and parents hope their children to study in a town or city for the best education, even though it is far away from their living villages. Heavier study burden and less free time keep school children away from learning about Dongba culture, consequently, traditional inheritance of Dongba culture has to be interrupted.

4.3 *The impact from the tourism and culture industry*

In recent years, the tourism and culture industry in Lijiang has developed prosperously, especially, Dongba sacrificial activities, Dongba sculpture and Dongba character writing have got market share, which triggered a spectacular Dongba culture craze[9]. Unprecedentedly large need for Dongba performers in the scenic spots and theme performance venues moves Village Dongba to step out into these commercial spots to perform Dongba dancing, write and sell Dongba calligraphy and painting. Dongba's knowledge and skills decide their different social roles, generally, there are there kinds: first is Dongba for tourism, who have a higher Chinese level and are more capable of adapting to the outside society, but master less sacrificial rituals; second is aged senior Dongba holding sacrificial rituals, who know a little or nothing about Chinese and are less capable of adapting to the outside, but could hold more rituals with their solid knowledge and rich experience; the third kind of Dongba can do whatever the former two kinds do, they can both adapt to tourism market and go back to hometown to hold rituals whenever it's needed. In number, that of the first is larger, that of the second and third is relatively smaller. Our interview shows that more and more Dongba are going out for business with an increasing salary of 3000 yuan for one person each month at the price of less or little time to go back home for sacrificial activities. To cater to the commercial need, some parts of the traditional Dongba rituals or sacrificial activities have to be cut, which fragments the original complete state of Dongba culture, accelerates Dongba's departing from their hometown, and then causes the disappearing of well-ordered inheritance in many villages, although tourism development offers a good chance for Dongba to show themselves to the world and a correspondingly good change of their original life.

5 REFLECTIONS UPON THE CHANGE OF DONGBA CULTURE INHERITANCE

There are many reasons resulting in the end of traditional inheritance mode, whereas the development of times is the rooted one. Dongba culture in Lijiang didn't get a full, systematic or continuous inheritance and protection before the reform and opening-up. Specific measures include establishing Dongba Research Center, Dongba Museum, Inheritance Society and Inheritance Base, opening Dongba Culture School and having degree assessment as well as promulgating "Dongba Culture Protection Regulations of Lijiang Naxi Autonomous County, Yunnan" and "Naxi Dongba Culture Protection Regulations of Yunnan Province". Dongba school opened with the new times is another way to continue the traditional inheritance of Dongba culture. Though we need time to wait to judge whether the current school scale and teaching way are good enough for the inheritance and developing of Dongba culture, Dongba culture itself can't wait. Otherwise, to inherit and develop Dongba culture may be impossible and "endangered Dongba culture" may become "died culture" right after the death of Dongba[10].

To have a successful transformation to the traditional inheritance, we need to take some actions as follows: 1) absorb the advantages of both traditional inheritance and school training to adjust and improve curriculum, training mode and teaching method; 2) to offer tourism Dongba short-term training courses on a regular basis to improve their relevant culture knowledge and strengthen sacrifice Dongba' s communication skills; 3) to set up test bases in the villages with well-maintained Dongba culture for inheritance and propaganda, which could help people have a deeper understanding of Dongba culture and alarms people to inherit Dongba culture by protecting; 4) it is not only Dongba school that should inherit and propagandize Dongba culture, but the whole society that should join together to finish the job. Only when creating an active atmosphere through joint efforts of the whole society, can Dongba culture be well inherited, developed and protected.

REFERENCES

Shihua Mu (2001). *Dongba Religion and Naxi Culture*. Central University for Nationalities. Beijing (2001).

Limin He (1991) *On the Nature of Dongba Religion – A Record of Dongba Culture Theory*. Yunnan People. Yunnan. (1991).

J.F.RockNakhi & N–g. (1952) *Culture and Related ceremonies*. Rome. (1952).

Fuquan Yang. (1998) *Multi-culture and Naxi Society*. Yunnan People. Yunnan. (1998).

Rao Wang (1993). Minorities and Taoism–Reading Notes of History. *Central College for Nationalities*. Beijing (1993).

Dongba Society (2011). *Membership Application List of Dongba Society*, http://dongba.lijiang.com/ (2011).

Houliang Kang & Dejing Li & Qinghai He (2013). *An Empirical Survey of Dongba Dancing of Naxi People in Lijiang*. Scientific P.E. Journal. 12(2013) pp. 22–24.

Dongba Society (2012). *Degree Award list of Dongba and Daba*. http://dongba.lijiang.com (2012).

Diya Hu (2012) *Endangered Culture and Education – An Educational Analysis of the Change in Dongba Culture Inheritance*. National Education Research. 5(2012). pp. 58–62.

Aiguo Chen (2012). *An Interpretation and reflection upon "Culture Disappearance Discourse"*. Culture and Art Research. 1(2012). pp. 9–14.

Management, Information and Educational Engineering – Liu, Sung & Yao (Eds)
© 2015 Taylor & Francis Group, London, ISBN: 978-1-138-02728-2

The necessities and approaches of cultivating the entrepreneurial culture of China's new undergraduate colleges and universities

Qing Song Li
Biological and Chemical Department, Chongqing University of Education, Chongqing, China

ABSTRACT: The birth and development of China's New Undergraduate Colleges and Universities are journeys of founding undertakings. Cultivating entrepreneurial culture is conducive to resolving the survival problems, achieving transformational development, building new university spirit, and cultivating applied talents of New Undergraduate Colleges and Universities. So China's New Undergraduate Colleges and Universities must be aware of this, integrate entrepreneurial culture into their development process.

KEYWORDS: New Undergraduate Colleges and Universities; Entrepreneurial Culture; Necessity; Approach.

1 INTRODUCTION

Worldwide, every important historical period of higher education development, it will appear new-type colleges and universities which are very different from the old ones. Chinese New Undergraduate Colleges and Universities, which is also known as Founded Application-oriented Undergraduate Colleges and Universities, are the new products of popularization and internalization of Chinese higher education in recent decade.

Scholar Gu Yongan thought the new features of New Undergraduate Colleges and Universities embodies in the following 3 aspects. Firstly, the new historical missions, especially the function of serving the society. Secondly, new quality standards, which is varied because of the diversification of needs of modem society. Thirdly, new educational model, which calls for cooperating extensively in society, especially the mode of production-study-research. It is the strategic choice for enhancing the new features to cultivating entrepreneurial culture.

Every organization is endowed with its expected value and responsibilities on the day of its birth. The birth and development of China's New Undergraduate Colleges and Universities are hard journeys of founding undertakings. In order to satisfy the society exception, New Undergraduate Colleges and Universities must seize the opportunity based on reality, pioneer and innovate to achieve breakthroughs.

2 MEANINGS OF ENTREPRENEURIAL CULTURE

Starting an enterprise means to start a career. In fact, it is a dynamic process of discovering the value and creating wealth. (Wen Huron, & Zhang Bin. 2011) In broad sense, starting an enterprise is the autonomy of maintaining the achievements of predecessors. (Zhou Hang. 2008)

In a broad sense culture refers to all material and spiritual achievements by human being. The entrepreneurial culture of New Undergraduate Colleges and Universities means starting an education career and creating material and spiritual wealth. The core connotation of it is the spirit and ability of self-dependence. (Gu Yongan. 2012)

3 THE NECESSITIES OF CULTIVATING ENTREPRENEURIAL CULTURE

3.1 *The requirement of solving survival problems and achieving transformational development*

Comparing with old universities, new ones are on a sticky wicket, Old universities are advancing rapidly on the fast lane, while the new ones are stumbling at the starting line. One of the severe problems which troubles New Undergraduate Colleges and Universities is funded shortage and sluggish financial channel.

As the social transformation, the running environment changes greatly. The government reduced the input to higher education relatively, which force colleges and universities themselves to obtain money by market means. Most New Undergraduate Colleges and Universities are lack of financing channels, and it is hard for them to introduce funds for development by themselves. (Zhang Sheping. 2013)

As Rowling Tingle, American education administration specialist, said school expenditure is just like the backbone of educational activities. Expenditure is the most basic material foundation. Burton Clark also thought universities cost lot, and good universities cost more. It costs much funds in all aspects. For example, they perfect the infrastructure such as purchasing instruments and equipment, introduce and keep talents largely, and provide a favorable environment of work, study and living for teachers and students, etc. So financial matter is a common problem New Undergraduate Colleges and Universities facing. This life-and-death circumstance rouses their entrepreneurial consciousness, and impel them to cultivate rely on their own efforts.

Besides, New Undergraduate Colleges and Universities have no opportunities, no enough time, and are not necessary to follow the old routine which old universities took. Transformational development is the inevitable choice for them. Transformational development of New Undergraduate Colleges and Universities includes two significant aspects. First, in the field of talents cultivation, from college level to the undergraduate level. Second, in the field of the orientation and type, from traditional patterns (research-oriented, or academic) to the new pattern (application-oriented). And these two transformations depend on the core change of university functions, which highlights the function of serving the society. The transformational development appeal entrepreneurial culture.

3.2 *The requirement of building new university spirit*

University spirit is group consciousness, conceptual work, and psychology, which is sedimentary deposits for a long period of time, and it concentrates to reflect the value pursuit of teachers and students, which is the spiritual pillar of behaving, working, and pursuing their studies. (Dong Tajian. 2009) On one side, traditional university concentrates on eternal value such as freedom, democracy, humanity, critique, and innovation and so on. On the other side, it regards university spirit as spiritualization. While after further expanding of university functions, we should increase new spirit and ideas, which is called the spirit of service or participation.

New university spirit is the essential and nucleus distinguishing features. On one hand, the new university spirit of New Undergraduate Colleges and Universities is the contemporary expression of traditional university spirit. On the other hand, it has its new features. It is the development impetus and important assignment for New Undergraduate Colleges and Universities to cultivate new university spirit energetically, foster the tradition of universities, and endow university spirit with era characteristics. (Kong Fanmin. 2006)

The core of entrepreneurial culture is the spirit and ability of self-dependence. Entrepreneurial culture has embodied of the new university spirit. Favorable entrepreneurial culture will bring motivation and power of development for New Undergraduate Colleges and Universities. So under the guidance of entrepreneurial culture, New Undergraduate Colleges and Universities must have the courage to put forward new ideas, and try new schemes, build qualified universities, which is the spiritual home, scientific cradle, cultural base, and think tank, in order to meet the challenges of intending characteristic development.

3.3 *The requirement of cultivating applied talents*

Some people think universities only care about how money turns into knowledge. In actual fact, in modern society, it is not. Universities care about not only how money turns into knowledge, but also how knowledge transforms into money.

Cultivating talents is the core work of New Undergraduate Colleges and Universities. New Undergraduate Colleges and Universities provide not only professional knowledge, ability cultivation, and skill training for students to seek a livelihood, but also devote to the guidance of attitude, emotion and values, lead students possess morals such as honesty, integrity, responsibility. Entrepreneurial culture is good for the knowledge, ability and morals spreading, and also is benefiting to cultivate career orientation of students, and encourage their innovative spirit and creation enthusiasm.

What is more, the new demand of cultivating talents for New Undergraduate Colleges and Universities is socially adaptable entrepreneurial education. Entrepreneurial education aims at cultivating an entrepreneurial spirit and consciousness, developing entrepreneurial characters and abilities. The development of entrepreneurial culture can carry forward entrepreneurial education. Firstly, New Undergraduate Colleges and Universities promote cultivating an entrepreneurial culture to the height of educational principle, and make graduates strive for progress with determination, have the courage to create, be fearless of difficulties, and be adept in grasp opportunities. Second, New Undergraduate Colleges and Universities develop the entrepreneurial activities as supportive mechanisms, and lead graduates regard starting a career as an important direction of personal development.

4 METHODS OF CULTIVATING ENTREPRENEURIAL CULTURE

4.1 *With the core of cultivating distinctive entrepreneurial culture, establish and improve the long-term mechanism*

Distinctive university culture is the essence of distinguishing one university from another. The entrepreneurial culture of New Undergraduates Colleges and Universities should be various, which lies in surveying the present situation and latent capacity of the schools soberly and objectively based on the school situation. On the basis of positioning the schools accurately, clearing and defining their own development objectives, and choosing a reasonable developmental pattern, leaders and controllers establish an entrepreneurial culture fitting for the schools, based on which cultivate distinctive entrepreneurial culture.

The process of cultivating an entrepreneurial culture reflects patterns of treating people and doing things of teachers and students, which can't be completed in short time. It takes historical deposition and is the achievements of cultivating for long range. So New Undergraduate Colleges and Universities must insist on cultivating the distinct culture based on position themselves rationally, establish and improve the long-term mechanism to provide a system of cultivating culture.

4.2 *Emphasize on enhancing humanistic education*

Humanism is the essential requirement of university culture, in the same way it is also the requirement of entrepreneurial culture. The key to promoting qualities of New Undergraduate Colleges and Universities all-around the ground, and shaping healthy personality of students and advancing their mental shackles is enhancing humanistic education. Humanistic education means not only having the specialized courses of humanistic education, but also protruding the human significance in specialized courses. First of all, New Undergraduate Colleges and Universities should enhance the humanistic attainment of teachers, and put forth effort on developing teams of professors with personal charm, rigorous scholarship, and profound scientific attainments. Secondly, enhancing humanistic subject building, and strengthening cultural quality education. New Undergraduate Colleges and Universities should explore effective methods from in-class to outside class and from school to society, making the most of scholastic and regional cultural resources, bring the humanistic education into the teaching program. Thirdly, structuring syncretic courses of humanity and science.

New Undergraduate Colleges and Universities should also increase the contents of an entrepreneurial class in employment guidance courses, and perfect educational system of entrepreneurial knowledge and theory. At the same time they should lay stress on the entrepreneurial technical abilities and extended instruction, tamp entrepreneurial abilities, emphasize cultivating them managerial knowledge. Leaded by social practice, they promote social practice normalized and scientific, organize and hold all kinds of entrepreneurial plan competitions, and encourage teachers and students start a new career and become self-employed. They should strive to develop educational services, and make education industry to manage. In theory and practice, they should pay attention to cultivating entrepreneurial consciousness, the shape entrepreneurial spirit which is also called business man's spirit. Entrepreneurial spirit, briefly speaking, is the summation of a sense of opportunity and innovation, and rational spirit of adventure. Moreover, they should attach importance to moral education, put forth effort to set up entrepreneurial character, which pays attention to not only economic benefits, but also forging a self-personality and self-character. (Zhang Xuesong. 2008)

4.3 *Perfect material culture*

Entrepreneurial Culture includes material culture, which is material insurance of university culture, and external reflection of university spirit. It requires New Undergraduate Colleges and Universities think highly of the work of campus design, to think over architectural layout, building decoration, campus afforestation, and designs of sights, to strengthen the construction of high-tech business incubators and practice bases, in order to embody an entrepreneurial culture always and everywhere, and protruding harmonious unification of university spirit and material.

4.4 *Make the best of regional humanistic and economic resource*

As an integral component of regional culture, New Undergraduate Colleges and Universities are usually affected by regional culture, and be provided with regional feature. (Zhuang Yan. 2012) New Undergraduate Colleges and Universities should make the most of regional cultural human and economic resources, form their own characteristic, and develop it. It can not only fulfill their missions, explore local preeminent cultural resources adequately to remedy schools themselves, but also can quicken pace of cultural construction, at the same time it can react on local cultural construction development and renewal.

4.5 *Launch studies of entrepreneurial culture development*

Modern university cultural spirits innovate based on cultural heritage and enrich. For New Undergraduate Colleges and Universities, it is the requirement of university regional culture development, accelerating entrepreneurial culture exchange and innovation, promoting entrepreneurial culture construction, cultivating applied talents keeping abreast of the times, and improving capabilities of running schools. By the means of lectures, meetings, visits, they can study domestic and foreign advanced universities precious experience about cultivating entrepreneurial culture. In the process, firstly, they should seize the opportunity by emancipating the mind, grasp the important trend of popularization, openness and internationalization of higher education development, step onto the new development road by prolongation and innovation. Secondly, facing changes of social environment strives for characteristic, scientific, and spanned development under the new development idea. Thirdly, they should refuse arguments and prattle, do seriously and conscientiously, explore down to earth, and promote school pioneering development.

5 CONCLUSION

Transformational development of New Undergraduate Colleges and Universities is a process of profound qualitative change, which requires material insurance, systematic guarantee, and cultural support. The importance of the first two aspects can be seen obviously, and embodies in practical work well, while the cultural level change usually falls through, so cultural support should catch more attention. In view of the reality of transforming and reform, New Undergraduate Colleges and Universities should pay attention to cultivate entrepreneurial culture based on inheriting intrinsic university culture. Entrepreneurial culture should become significant form and important dependence, and spiritual culture and collective value of transforming the development of New Undergraduate Colleges and Universities. It provides fruitful soil and prosperous precedents that higher education reform deeply, connection between universities and society enhances, and New Undergraduate Colleges and Universities stress on application-oriented characteristics. New Undergraduate Colleges and Universities should integrate entrepreneurial culture into their development, accomplish the mission of their transform development through persistent pioneering behavior.

REFERENCES

Dong Tajian. 2009. *Outline of New Undergraduate Colleges and Universities Cultural Construction*: 18. Chengdu: University of Electronic Science and Technology of China Press.

Gu Yongan. 2012. *Discussion of Transformational Development of New Undergraduate Colleges and Universities*: 159. Beijing: China Social Science Press.

Kong Fanmin. 2006. The Road of Building Application-oriented University: 155. Beijing: Beijing University Press.

Wen Hourun. & Zhang Bin. 2011. *Practical Course for College and University Students Hunting Jobs: College and Univerisity Students Vocational Development and Employment Guidance*: 164. Beijing: Higher Education Press.

Zhou Hang. 2008. *College and University Students Employment and Entrepreneurship*:167. Chongqing: Southwest Normal University Press.

Zhang Sheping. 2013. *Structure and Choices of New Undergraduate Colleges and Universities Characteristics*: 33. Beijing: Science Press.

Zhang Xuesong. 2008. *Study of College and University Students Employment: Colloquium of Chongqing College and University Students Employment Work*: 209. Chongqing: Chongqing Education Commission.

Zhuang Yan. 2012. *Study of New Undergraduate Colleges and Universities Core-competitiveness in Economic less-Developed Areas*: 225. Haerbin: Heilongjiang University Press.

Management, Information and Educational Engineering – Liu, Sung & Yao (Eds)
 ISBN: 978-1-138-02728-2

The quality evaluation of logistics park based on extenics

Jia Hui Feng & Mao Sheng Yang
College of Management, Xi'an University of Architecture and Technology, Xi'an, Shaanxi, China

ABSTRACT: The paper, combined with the basic characteristics of logistics park, using the extenics, established a set of logistics park quality evaluation index system and extension model with the review department of the government project as the main assessment. By calculating the relative weights of the indexes and the correlation degree of the evaluation level in order to sort the quality of the regional logistics park, reflecting the quality level of the logistics park to be evaluated in the selected sample, in order to help the government department or the relevant enterprise investment and management in the logistics park to better adjustment and improvement.

KEYWORDS: logistics park; extenics; quality evaluation system; evaluation model.

1 INTRODUCTION

With the function of modern logistics industry in the domestic development increasing, as a new form of industry, logistics park is vigorously developed. As of 2012 September, there are 754 logistics park has been built or in the construction. In many areas, with logistics park at the base, hope to drive the development of the logistics industry in the entire region. More and more logistics park from the planning and construction gradually into the actual operation, the operation effect and development level is concerned with more and more people. In the construction process of the logistics park, there has formed a relatively mature theory, such as the logistics park planning theory [1], the feasibility evaluation theory [2], the business performance evaluation theory [3], the logistics park function evaluation theory [4] etc.. But it's rarely considered that the logistics park from the investment and construction to the actual operation brings the impact to social and environment, the whole process of the logistics park's quality. This led directly to the relevant government departments or enterprises can't invest as well as the operation and management of the logistics park for better adjustments and improvements, and will not help the future development of the logistics park. Therefore, with the construction of logistics park operation, scientific and comprehensive analysis and evaluation of the logistics park quality has become the urgent need of the logistics industry. On the basis of existing research, this paper combines the basic characteristics of the logistics park, consulting the experts, leaders and related personnel, determine a set of logistics park quality evaluation system and evaluation standard, with the government project review department as the evaluation subject, and use the extension evaluation theory to evaluate the quality of logistics park.

2 THE CONSTRUCTION OF LOGISTICS PARK QUALITY EVALUATION SYSTEM

2.1 *The construction of logistics park quality evaluation index system*

In constructing logistics park quality evaluation system, besides considering the basic characteristics of logistics park, but also reflect the government's macro guidance, and the information collection of the quality evaluation is difficult and complex. Considering these factors, combined with the construction principles of the evaluation system of systematic, scientific, operational, the combination of qualitative and quantitative[5], referencing the evaluation system of key industry project quality of Shaanxi province[6] and surveying the experts, leaders and related personnel, The main analysis was made from the five aspects of resources cost, emissions, tax contributions, investment intensity and integrated drive, and a logistics park quality evaluation index system has been set up as shown in Table 1.

2.2 *The determination of the index weights*

Because of the Indexes' influence of the logistics park quality evaluation index system in the total target of evaluation is not consistent, the weight was determined reasonable or not will have a decisive influence on the results of the logistic park quality evaluation. So this paper will use the combination of qualitative and quantitative analysis to determine the weight of each index. In this paper, through questionnaire, invite relevant experts to compare the Indexes to give judgment matrix between each evaluation index, combined with AHP, by calculating and trade-offs, and ultimately get the weight values of each index relative to the overall goal as shown in Table 1.

Table 1. Logistics park quality evaluation index system and index weight.

Primary Indexes	Weight	Secondary Indexes	Weight
Resources cost A1	0.24	The amount of land per unit value C1	0.09
		The consumption of energy resources per unit value C2	0.08
		The water consumption per unit value C3	0.07
Emissions A2	0.16	Water, gas, slag, acoustic emission concentration C4	0.04
		CO_2 emissions C5	0.03
		Nitrogen oxide emissions C6	0.03
		COD emissions C7	0.03
		Ammonia emissions C8	0.03
Tax contributions A3	0.28	Local tax allocation proportion C9	0.08
		The enterprise tax allocation proportion C10	0.07
		The total business income per unit area C11	0.07
		Per capita tax proportion C12	0.06
Investment intensity A4	0.14	The investment per unit area C13	0.075
		Proportion of fixed assets to total assets C14	0.065
Integrated drive A5	0.18	Job placement C15	0.07
		Technology innovation level C16	0.06
		Project output capacity C17	0.05

3 THE CONSTRUCTION OF LOGISTICS PARK QUALITY EVALUATION EXTENSION MODEL

3.1 *The determination of matter-element for evaluation*

According to matter-element theory [7], to evaluate the quality of logistics parks should analyze every factor involved firstly. Based on the primitives' definition in extenics and logistics park quality evaluation index system, the corresponding data of the factors effect on the logistics park quality represented by matter-element is as follows

$$R=(N,c,V)=\begin{bmatrix} N, & c_1, & v_1 \\ & c_2, & v_2 \\ & \vdots & \vdots \\ & c_i, & v_i \end{bmatrix} \tag{1}$$

Where, N is the quality of the logistics park to be evaluated; V is the measured data of each Index c corresponding value of the logistics park to be evaluated.

3.2 *The determination of the classical field*

If the evaluation effect of each Index divided into j levels quantitatively, the classical field matter-element matrix is as follows.

$$R_{oj}=(N_{oj},c,V)=\begin{bmatrix} N_{oj}, & c_1, & V_{oj1} \\ & c_2, & V_{oj2} \\ & \vdots & \vdots \\ & c_i, & V_{oji} \end{bmatrix}=\begin{bmatrix} N_{oj}, & c_1, & \langle a_{oj1},b_{oj1}\rangle \\ & c_2, & \langle a_{oj2},b_{oj2}\rangle \\ & \vdots & \vdots \\ & c_i, & \langle a_{oji},b_{oji}\rangle \end{bmatrix} \tag{2}$$

Wherein, N_{oj} is the grades of ordinal j, c_i is the evaluation index of ordinal $c_i(i=1, 2, \ldots, n)$, $V_{oji}=\langle a_{oji},b_{oji}\rangle$ is the classical field of each evaluation index corresponding to the N_{oj}.

3.3 *The determination of segment field*

The segment field R_p of the logistics park quality evaluation is as follows.

$$R_p=(N_p,c,V_{pi})=\begin{bmatrix} N_p, & c_1, & v_{p1} \\ & c_2, & v_{p2} \\ & \vdots & \vdots \\ & c_i, & v_{pi} \end{bmatrix}=\begin{bmatrix} N_p, & c_1, & \langle a_{p1},b_{p1}\rangle \\ & c_2, & \langle a_{p2},b_{p2}\rangle \\ & \vdots & \vdots \\ & c_i, & \langle a_{pi},b_{pi}\rangle \end{bmatrix} \tag{3}$$

Here, $c_i(i=1, 2, \ldots, n)$ represents the factor affecting the quality of logistics park, the evaluation index. V_{pi} is the range of the evaluation index c, the segment field.

3.4 *The clculation of correlation functions*

The correlation degree indicates the degree of relevance between two objects. The correlation between the object to be evaluated, and which in level j is higher, the object to level j is closer. Therefore, it can be used to decide the quality level of the object to be evaluated. The correlation degree between the matter-element model to be evaluated and classical field expressed in elementary correlation function is as follows:

$$k_j(v_i)=\begin{cases} -\dfrac{\rho(v_i,v_{0ji})}{|v_{0ji}|}, & v_i \in v_{oji} \\ \dfrac{\rho(v_i,v_{0ji})}{\rho(v_i,v_{oi})-\rho(v_i,v_{0ji})}, & v_i \notin v_{0ji} \end{cases} \tag{4}$$

$k_j(v_i)$ is the correlation of the eigenvalue v_i of ordinal c_i to the *j-th* evaluation level.

Here, $\rho(v_i,v_{oji})=\left|v_i-\dfrac{a_{oj}+b_{oj}}{2}\right|-\dfrac{b_{oj}-a_{oj}}{2}$, which indicates the distance betweeen the point v_i and the classical field $v_{oj}=[a_{oj},b_{oj}]$. $\rho(v_i,v_{pji})=\left|v_i-\dfrac{a_{pj}+b_{pj}}{2}\right|-\dfrac{b_{pj}-a_{pj}}{2}$, which indicates the distance between the point v_i and the segment field $v_{pj}=[a_{pj},b_{pj}]$.

$|v_{0ji}|=|a_{0ji}-b_{0ji}|$ is the mold of section $v_{0ji}=\langle a_{0ji},b_{0ji}\rangle$.

3.5 *The determination of comprehensive evaluation level*

If the weight of the evaluation index c_i is ω_i, then the correlation degree of object N to be evaluated to evaluation level j is as shown below.

$$K_j(N)=\sum_{i=1}^{n}\omega_i k_j(v_i) \tag{5}$$

A larger value of $K_j(N)$ indicates a higher correlation degree of the object N to be evaluated to the evaluation level j. The category of object N to be evaluated belongs to is the evaluation level $maxK_j(N)$ corresponding to.

For the evaluation objects at the same level, bias level eigenvalue j^* is usually used to distinguish their differences. According to the value of j^*, the merits of the evaluation objects can be sorted. The larger j^* is, the better the quality of the logistics park will be.

$$j^*=\frac{\sum_{j=1}^{m} j\overline{K_j}(N)}{\sum_{j=1}^{m}\overline{K_j}(N)} \tag{6}$$

$$\overline{K_j}(N)=\frac{K_j(N)-\min K_j(N)}{\max K_j(N)-\min K_j(N)} \tag{7}$$

Where, m is the total number of species of evaluation level. $\overline{K_j}(N)$ is the standardized result of the multi-index comprehensive correlation $K_j(O)$.

4 THE EMPIRICAL ANALYSIS

Based on the logistics park quality evaluation system, select A, B, C, 3 logistics park as the research object, carry on the extension evaluation respectively, and in accordance with the quality in order.

4.1 *The division of evaluation standard*

Through the related data statistics and standardizing of the three logistics park about evaluation index, inviting relevant experts to score with centesimal system, based on the average score of each index, make evaluation standard in accordance with the rules of Table 2.

Table 2. Evaluation standard rules.

Level	Scores Range
Excellent	[100. $\bar{x}+(100-\bar{x})\times 50\%$]
Good	($\bar{x}+(100-\bar{x})\times 50\%$, $\bar{x}\times 90\%$]
Middle	($\bar{x}\times 90\%$, $\bar{x}\times 60\%$]
Poor	[$\bar{x}\times 60\%$, 0]

Then the scores range of the index of each level is as shown in Table 3.

Table 3. The classification of logistics park evaluation indexes' level.

Index	Level			
	Excellent	Good	Middle	Poor
C1	[100,87.22]	(87.22,67.00]	(67.00,44.66]	(44.66,0]
C2	[100,66.88]	(66.88,30.38]	(30.38,20.26]	(20.26,0]
C3	[100,72.83]	(72.83,41.09]	(41.09,27.39]	(27.39,0]
C4	[100,88.23]	(88.23,68.82]	(68.82,45.88]	(45.88,0]
C5	[100,87.95]	(87.95,68.31]	(68.31,45.54]	(45.54,0]
C6	[100,81.59]	(81.59,56.87]	(56.87,37.91]	(37.91,0]
C7	[100,81.30]	(81.30,56.34]	(56.34,37.56]	(37.56,0]
C8	[100,88.72]	(88.72,69.69]	(69.69,46.46]	(46.46,0]
C9	[100,89.29]	(89.29,70.73]	(70.73,47.15]	(47.15,0]
C10	[100,81.41]	(81.41,56.54]	(56.54,37.69]	(37.69,0]
C11	[100,94.46]	(94.46,80.02]	(80.02,53.35]	(53.35,0]
C12	[100,89.97]	(89.97,71.95]	(71.95,47.96]	(47.96,0]
C13	[100,93.60]	(93.60,78.47]	(78.47,52.32]	(52.32,0]
C14	[100,92.75]	(92.75,76.95]	(76.95,51.30]	(51.30,0]
C15	[100,84.48]	(84.48,62.06]	(62.06,41.37]	(41.37,0]
C16	[100,85.76]	(85.76,64.38]	(64.38,42.92]	(42.92,0]
C17	[100,86.90]	(86.90,66.43]	(66.43,44.28]	(44.28,0]

4.2 *The determination of the extension evaluation matter-element*

According to Table 3 and Formula (2) and (3), the classical field matter-element and segment field matter-element of each level can be determined as follows.

The classical field of each level:

$$R_0 = \begin{bmatrix} & N_{01} & N_{02} & N_{03} & N_{04} \\ c_1 & [100.87.22] & (87.22,67.00] & (67.00,44.66] & (44.66,0] \\ c_2 & [100,66.88] & (66.88,30.38] & (30.38,20.26] & (20.26,0] \\ c_3 & [100.73.83] & (72.83,41.09] & (41.09,27.39] & (27.39,0] \\ c_4 & [100,88.23] & (88.23,68.82] & (68.82,45.88] & (45.88,0] \\ c_5 & [100,87.95] & (87.95,68.31] & (68.31,45.54] & (45.54,0] \\ c_6 & [100,81.59] & (81.59,56.87] & (56.87,37.91] & (37.91,0] \\ c_7 & [100,81.30] & (81.30,56.34) & (56.34,37.56] & (37.56,0] \\ c_8 & [100,88.72] & (88.72,69.69] & (69.69,46.46] & (46.46,0] \\ c_9 & [100,89.29] & (89.29,70.73] & (70.73,47.15] & (47.15,0] \\ c_{10} & [100,81.41] & (81.41,56.54] & (56.54,37.69] & (37.69,0] \\ c_{11} & [100,94.46] & (94.46,80.02] & (80.02,53.35] & (53.35,0] \\ c_{12} & [100,89,97] & (89.97,71.95] & (71.95,47.96] & (47.96,0] \\ c_{13} & [100,93.60] & (93.60,78.47] & (78.47,52.32] & (52,32,0] \\ c_{14} & [100,92.75] & (92.75,76.95] & (76.95,51.30] & (51.30,0] \\ c_{15} & [100,84.48] & (84.48,62.06] & (62.06,41.37] & (41.37,0] \\ c_{16} & [100.85.76] & (85.76,64.38] & (64.38.42.92] & (41.92,0] \\ c_{17} & [100,86.90] & (86.09,66.43] & (66.43,44.28] & (44.28,0] \end{bmatrix}$$

The segment field:

$$R_p = \begin{bmatrix} N_p & c_1 & [0,100] \\ & c_2 & [0,100] \\ & c_3 & [0,100] \\ & c_4 & [0,100] \\ & c_5 & [0,100] \\ & c_6 & [0,100] \\ & c_7 & [0,100] \\ & c_8 & [0,100] \\ & c_9 & [0,100] \\ & c_{10} & [0,100] \\ & c_{11} & [0,100] \\ & c_{12} & [0,100] \\ & c_{13} & [0,100] \\ & c_{14} & [0,100] \\ & c_{15} & [0,100] \\ & c_{16} & [0,100] \\ & c_{17} & [0,100] \end{bmatrix}$$

After the values of the indexes of the three logistics parks to be evaluated are standardized, according to the formula (1), the matter-elements of A, B, C three logistics park to be evaluated are as follows.

$$R_1 = \begin{bmatrix} N_1 & c_1 & 96.15 \\ & c_2 & 98.51 \\ & c_3 & 22.07 \\ & c_4 & 82.26 \\ & c_5 & 91.56 \\ & c_6 & 95.68 \\ & c_7 & 55.12 \\ & c_8 & 81.36 \\ & c_9 & 89.62 \\ & c_{10} & 53.85 \\ & c_{11} & 91.00 \\ & c_{12} & 92.99 \\ & c_{13} & 99.02 \\ & c_{14} & 83.60 \\ & c_{15} & 82.86 \\ & c_{16} & 59.17 \\ & c_{17} & 99.38 \end{bmatrix}, \quad R_2 = \begin{bmatrix} N_2 & c_1 & 89.29 \\ & c_2 & 1.040 \\ & c_3 & 18.56 \\ & c_4 & 55.89 \\ & c_5 & 40.32 \\ & c_6 & 24.76 \\ & c_7 & 35.59 \\ & c_8 & 54.29 \\ & c_9 & 96.15 \\ & c_{10} & 96.15 \\ & c_{11} & 96.92 \\ & c_{12} & 98.60 \\ & c_{13} & 95.26 \\ & c_{14} & 97.90 \\ & c_{15} & 47.62 \\ & c_{16} & 92.08 \\ & c_{17} & 63.86 \end{bmatrix}, \quad R_3 = \begin{bmatrix} N_3 & c_1 & 37.88 \\ & c_2 & 1.73 \\ & c_3 & 96.34 \\ & c_4 & 90.25 \\ & c_5 & 95.82 \\ & c_6 & 69.12 \\ & c_7 & 97.09 \\ & c_8 & 96.64 \\ & c_9 & 50.00 \\ & c_{10} & 38.46 \\ & c_{11} & 78.82 \\ & c_{12} & 48.32 \\ & c_{13} & 67.30 \\ & c_{14} & 75.00 \\ & c_{15} & 72.38 \\ & c_{16} & 63.33 \\ & c_{17} & 58.18 \end{bmatrix}$$

4.3 *The calculation of correlation degree between each index and each evaluation level*

Logistics park A as an example, according to the formula (4) to calculate the correlation between its evaluation indexes and evaluation level as shown in Table 4.

Table 4. The correlation degree between evaluation indexes and evaluation level of logistics park A.

	Level			
Index	Excellent	Good	Middle	Poor
C1	0.30	-0.70	-0.88	-0.93
C2	0.04	-0.96	-0.98	-0.98
C3	-0.39	-0.20	-0.06	0.19
C4	-0.23	0.26	-0.46	-0.69
C5	0.30	-0.30	-0.73	-0.85
C6	0.23	-0.77	-0.90	-0.93
C7	-0.37	-0.03	0.06	-0.28
C8	-0.28	0.39	-0.39	-0.65
C9	0.03	-0.03	-0.65	-0.80
C10	-0.37	-0.06	0.14	-0.26
C11	-0.28	0.24	-0.55	-0.81
C12	0.30	-0.30	-0.75	-0.87
C13	0.15	-0.85	-0.95	-0.98
C14	-0.36	0.42	-0.29	-0.66
C15	-0.05	0.04	-0.56	-0.71
C16	-0.39	-0.11	0.24	-0.28
C17	0.05	-0.95	-0.98	-0.99

According to the formula (5), the comprehensive correlation degree of logistics park A to each level is (-0.07, -0.26, -0.53, -0.68). Similarly, the comprehensive correlation degree of the logistics park B and C to each level are shown in Table 5.

Table 5. The results of logistic park quality evaluation.

Logistics Park	Excellent	Good	Middle	Poor	Attribution Level
A	-0.07	-0.26	-0.53	-0.68	Excellent
B	-0.01	-0.39	-0.44	-0.51	Excellent
C	-0.28	-0.23	-0.15	-0.35	Middle

It is shown that the level of quality logistics parks A and B are in the same evaluation level, are excellent. In this case, according to the formula (6), the bias level eigenvalue j^* of them can be calculated, and the result of their quality sort is as follows.

$j^*_A=1.61, j^*_B=1.38.$

Then, the sequence of three logistics park quality from excellent to poor is: A> B> C.

5 CONCLUSION

The application of extenics in the research of logistics park quality evaluation provides the government a feasible evaluation method to assess the logistics park. But this paper is confined to the primary extension evaluation. In the future study, the multi-level extension comprehensive evaluation method is hoped to be used in the research to make a better decision.

Secondly, in the extension priority degree evaluation method, the value of the comprehensive correlation, that is the priority degree, can be used to select the optimal scheme, can also find the sub-optimal scheme. It can provide decision makers with more reference information than other method.

ACKNOWLEDGEMENT

University philosophy and social science characteristic discipline construction special fund projects of Shaanxi province (E08003).

REFERENCES

[1] ZHOU Rong-xia (2008)," An Integrated Appraisal Study of the Programming & Construction Scheme Based on Logistics Park", Wuhan University of Science and Technology, pp. 25–38.

[2] MA Kun(2010),"Research on Logistics Park Feasibility Evaluation", Southwest Jiaotong University, pp. 12–53.

[3] FENG Xiao-jing(2012)," Research of Logistics Park's Performance Evaluation on PCA-DEA Model", Chang'an University, pp. 22–37.

[4] XIANG Bin(2010)," Research on the Evaluation of Logistics Park's Competitiveness", Beijing Jiaotong University, pp. 10–45.

[5] ZHANG Jing-rong(2010),"Study on the Post-evaluation of Social Economic Benefits of Logistics Projects Based on Fuzzy AHP", Logistics Technology, pp. 62–69.

[6] Shaanxi Provincial Development and Reform Commission(2013)," the Documents Compilation of Promoting the Key Projects Construction in Shaanxi Province", pp. 32–33.

[7] YANG Chun-yan & Cai Wen(2000)," Study on Extension Engineering", Engineering Science, vol. 12, No.12, pp. 90–96.

Management, Information and Educational Engineering – Liu, Sung & Yao (Eds)
© 2015 Taylor & Francis Group, London, ISBN: 978-1-138-02728-2

Empirical study on the influence of city development level on the price of commercial housing

Jian Bin Wang & Gui Ying Deng
School of Management, Xi'an University of Architecture and Technology, Xi'an, China

ABSTRACT: This paper, using the relevant statistical data from 2003 to 2012 in Zhengzhou, founded the index system of city development, to study the influence of city development on the price of commercial housing. The empirical study result shows that urban population status, commodity housing output level and residents' consumption level are the direct factors influencing commercial housing prices, and the influence degree increases, in which housing output has a negative effect on housing price; social and economic development level, infrastructure, education and medical treatment level and commodity residential investment level are the indirect factors. Through the analysis of the overall effect it is known that residents' living consumption capacity and urban population condition are demand leading factors, residential investment and other factors are supply leading factors.

KEYWORDS: urban development level; commercial housing prices; structural equation modeling.

1 INTRODUCTION

Housing is the important carrier of urban development. Commercial housing price is the core problem of the residential market, the full understanding of the impact of commodity housing price factors and mechanism, and summarizes its regularity, to enhance the government macro regulation and control prices, stable housing prices, promote the healthy development of the urban social economy, has a positive practical significance and good guidance function. By constructing the urban development level index system, this paper studies the urban development the influence of various factors on the commodity residential house prices and the mechanism of action, and provide a theoretical basis and effective method for the regulation of commercial housing prices in China.

2 ESTABLISH THE INDEX SYSTEM AND MODEL OF URBAN DEVELOPMENT

2.1 *The index system of urban development level*

Review of relevant literature from the two aspects of index system for city development and the relationship between commercial housing price and the city development level, determine the index system of the city development level.

1 The relationship between city development and commercial housing price. The domestic scholars, Liu Yachen analyses the coordination index of the Shenyang City real estate industry and the national economy[1]. Building subsystems of economy, society, population, resources, environment, the commodity housing, Yang Xiaodong analyses each subsystem and the coordinated development of commodity housing price mechanism, confirm the effect on housing prices have a positive impact[2]. Foreign scholars Berger and Gabriel found that the development of the urban economy, the improvement of labor productivity and the quality of life of the residents have certain influence on prices for the urban development and commodity residential market research in Russia, America[3-4]. Y F Ho, H L Wang forecast and explain the variation of housing price from the aspects of city economic development[5].
2 Establish an index system of city development. Li Song constructed the index system of urban development of urbanization of population, urbanization and economic to the research harmonious relationship between land use and city development[6]. Yang Chenglin and Zhang Liao analyze the city economic development level from the four aspects of level of economic development, resource supply and ecological environment [7]. Du Shufang measure MengDong urban development dynamic mechanism by constructing the evaluation index system of nine aspects of social development, economic development, infrastructure, people's living, etc.[8]

Based on previous studies, this paper puts forward seven measurements of the city development level index of the social economic development level, residents living consumption ability, the condition of urban population, commodities, commodity residential housing development investment output level, infrastructure, education and health level, and taking Zhengzhou city as the research object to analyze the effect of the city development level of influence on the price of commercial housing.

2.2 *Establish model*

Structural equation model was analyzed based on the covariance matrix of variables, on the basis of a variety of traditional statistical methods, mainly using the linear statistical modeling techniques, by using confirmatory factor analysis, path analysis, regression analysis and other statistical methods, the analysis can measure relationships between variables and implied, and verify the hypothesis whether to set up technology.

1 Structure model. Structure model illustrates the effect relationship between the latent variables and exogenous latent variables. The equation for the:

$$\eta = B\eta + \Gamma\zeta + \zeta \tag{1}$$

Type: B expresses the coefficient matrix of endogenous relationships between latent variables, Γ expresses the coefficient matrix between exogenous latent variables and endogenous latent variables, ξ expresses the residual vector.

2 Measurement model. It reacts between the actual measurement model observed and latent variables, describing the relationship between latent variables of ξ, η and the actual measured index of X, Y.

$$Y = \Lambda\eta + \varepsilon \tag{2}$$

$$X = \Lambda\xi + \delta \tag{3}$$

3 MODEL APPLICATION

3.1 *Data sources*

This paper uses the statistical data of Zhengzhou city in 2003-2012 years, mainly from 10 years statistical data of the "statistical yearbook of Henan province" and "Zhengzhou national economic and social development statistical bulletin", the connotation of the variables are consistent with the defined by statistical yearbook or bulletin. On the basis of the analysis of particularity and variables of the city development and formation of commercial housing price, for qualitative analysis on the relationship between city development factors and commercial housing price, its index system includes 8 categories of 25 specific variables.

3.2 *Empirical research*

3.2.1 *Research hypothesis*

Considering that the structural equation model was statistical method which used to test and verify the feasibility and suitability of building model in the hypothesis conditions[12]. According to the index system to analyze the relationship between urban development and housing prices were constructed in this study, clearly put forward the following research hypotheses.

H1: social and economic development has a positive influence on education and health level; H2: education and medical treatment level has a positive influence on residential development investment; H3: social and economic development has a positive influence on infrastructure; H4: infrastructure has a positive impact on the housing investment; H5: residential development investment has a positive influence on residential output level; H6: residential output level has a negative effect on housing price level; H7: urban population situation has a positive influence on commodity housing price; H8: urban population situation has a positive influence on social and economic development; H9: residents consumption ability has a positive influence on commodity residential house price; H10: residents consumption ability has a positive influence on residential development investment.

3.2.2 *Model operation*

Entry and descriptive analysis and processing of the original data by using SPSS statistical software, the correlation matrix is obtained from observed variables, the data is imported into AMOSS17.0 software, operations using the maximum likelihood method to obtain the result of running the program.

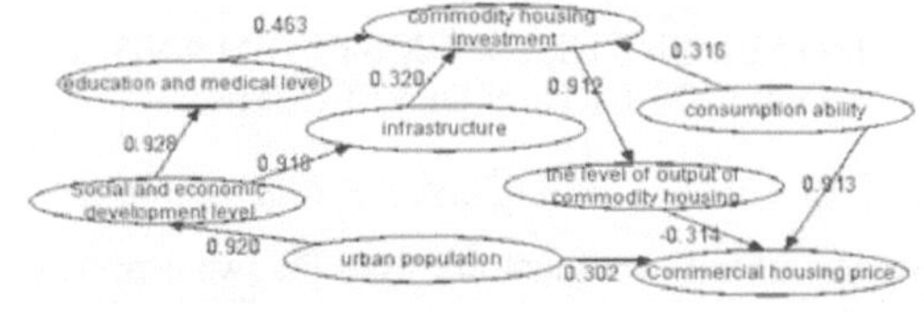

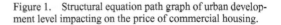

Figure 1. Structural equation path graph of urban development level impacting on the price of commercial housing.

By computing the model whether load coefficients of latent variables and the observed indexes and endogenous and exogenous latent variables between latent variables statistically significant results of path coefficient of P values are less than 0.05, shows that all the parameters in the model estimation is of high significance level, so you can accept the null hypothesis.

3.3 *Model evaluation*

The use of reliability coefficient and validity of the model of measurement and evaluate the reliability and validity of latent variables.

1 The reliability test (Reliability). It's mainly used for testing the reliability of data collection, reliability test, including internal consistency reliability test and synthetic reliability test. The former uses Cronbach alpha coefficient Measurement, judging standard for coefficient is close to 1, the higher reliability, when greater than or equal to 0.7 belong to high reliability; The latter uses Composite Reliability (CR) as a measure. The composite reliability computation formula is as follows:

Cronbach's alpha coefficient formula:

$$\alpha = \left(P/(P-1)\right)\left[1-\sum_{i=1}^{p} Var(y_i)/Var\left[\sum_{i=1}^{p} y_i\right]\right] \quad (4)$$

Composite reliability calculation formula:

$$CR = \left(\sum \lambda_i\right)^2 / \left(\left(\sum \lambda_i\right)^2 + \sum \Theta_{ii}\right) \quad (5)$$

Type: p is the number of the observation variable, y_i is observed variables; λ_i shows the load coefficient of variables on the corresponding latent variables, Θ_{ii} is for measurement error.

2 The validity of the test (AVE). The average variance extracted is used to reflect its convergent degree of a latent variable estimated a set of observation variables.

The calculating formula of AVE:

$$AVE = \sum \lambda_i^2 / \left(\left(\sum \lambda_i^2\right) + \left(\sum 1 - \lambda_i^2\right)\right) \quad (6)$$

Type: λ_i is the load coefficient of significant variation in the corresponding latent variables. Evaluation of reliability and validity of the model is shown in table 1.

This paper uses SPSS20.0 statistical software test the reliability and validity of the measurement model and the latent variables , reliability and discriminant validity of the results as shown in table 1.

4 CONCLUSION

By solving the model structural equation, empirical analysis of the influence of city development of the price of commercial housing reaches the following conclusions:

(1) Direct factors of commercial housing price

Various factors influencing the level of urban development on the price of commercial housing, the residents living consumption ability, situation of the urban population and housing output level are directly influencing factor, their path coefficient respectively were 0.913, 0.302, -0.314. Residents living consumption ability had the largest effect on housing prices, and there is positive correlation with housing price and also explains that high price of the urban residential is influenced by demand; the influence of the output level for the commodity residential is second, and there is a negative correlation between them, the improvement of residential output has the trend of bringing down the commodity residential house price when the leading demand of housing market is real demand; urban population situation has the minimum influence on the price of commercial housing, urban population increase plays an promoting role in housing prices, which has a positive effect.

(2) Indirect factors of commercial housing price

Social and economic development level, infrastructure, education and medical level and commodity housing development investment effects are indirect factors, their impact on domestic prices and degree are not identical. ① Firstly through the infrastructure and education and medical factors, social and economic development level indirectly affect the price of commodity housing, their influence degree were 0.928, 0.918, and have significant role; ② the urban infrastructure, education and healthcare are first acting on the commercial housing development investment, and they have an indirect influence on housing price again through the influence of commodity housing output level. They determines the residential investment environment and provides the external effect for ; ③ residents living consumption ability has a significant influence on commodity residential house price, and has certain positive effect on the development of commercial residential investment; ④ the development of commercial residential investment has a significant positive effect on the level of output of commodity housing (influence coefficient is 0.912), and indirectly effects on residential commodity prices through the impact on the level of output of commodity residential house.

(3) Overall effect of commercial housing price

By calculating the direct influence and the indirect influence to determine the total effect that factors of urban development level are influence on the price of commercial housing. The total effect of calculated values shown in table 2.

By table 2 analysis: residents living consumption ability, the condition of urban population have positive total effect for the commodity residential house price,the existence of a positive effect; Other factors have negative total effect on the commodity residential house price, and there is negative influence.

According to the positive and negative values of the total impact of commodity housing price effect coefficient, the urban development level of each factor is divided into two kinds: one kind is the positive effect, namely the commercial housing price changes with the influence factors change and changes in the same direction; one kind is the negative effect, namely the commercial housing price changes with the factors affecting the change while the reverse change. From the perspective of the mechanism of supply and demand on the price, two factors of residents living consumption ability and the urban population are defined as the dominant factor demand, 5 factors of residential development investment etc are defined as the leading type of supply factors.

Table 1. Results table of the Reliability and validity.

	Social and economic development level	The condition of urban population	Commercial housing price level	Residents living consumption ability	Commodity housing development investment	The level of output of commodity housing	Infrastructure	Education and medical level
Cronbach 'α'	0.960	0.983	0.951	0.962	0.968	0.965	0.979	0.891
CR	0.971	0.981	0.955	0.973	0.979	0.975	0.974	0.947
AVE	0.893	0.968	0.949	0.895	0.941	0.907	0.876	0.901

Table 2. The total effect of urban development factors on the price of commercial housing.

Social and economic development level	The condition of urban population	Residents living consumption ability	Commodity housing development investment	The level of output of commodity housing	Infrastructure	Education and medical level
–0.229	0.163	0.865	–0.278	–0.321	–0.146	–0.082

REFERENCES

[1] Liu Yachen, Wu Zhenhu, Wang Huan, Coordination degree between the real estate industry and the national economy in Shenyang City based on principal component analysis[J]. *Building economy,*(4),pp.84–88,2011.

[2] YANG Xiaodong, WUYongxiang,Commercialized residential Buildings Price Model from the Perspective of Urban Coordinated Development in Shanghai[J]. *China soft science,* (01),pp.160–170,2013.

[3] Ho Y F,Wang H L,Liu C C.Dynamics Model of Housing Market Surveillance System for Taichung City[C]. *Proceeding of the 28th International Conference of the System Dynamics Society.Seoul,Korea: Korean System Dynamics Society*,pp. 971–978,June2010.

[4] [4]Gabriel S A, Mattey J P, Wascher W L. Compensating differentials and evolution in the quality-of-life among US states[J]. *Regional Science and Urban Economics*,33(5), pp.619–649, June 2003.

[5] Berger M C, Blomquist G C, Sabirianova Peter K. Compensating differentials in emerging labor and housing markets: estimates of quality of life in Russian cities[J]. *Journal of Urban Economics*,63 (1),pp25–55,2008.

[6] LI Song,ZHANG Xiaolei,LI Shoushan,DU Hongru, Spatial and temporal differentiation of the systemic harmonious degree between land use and urban development in Xinjiang[J]. *Journal of arid land resources and environment,*(03),pp.23–30,2014.

[7] ZHANG Liao,YANG Chenglin, Analysis on Urban Agglomeration Sustainable Development Level and Influencing Factors: Evidence from China's "Top Ten" Urban Agglomeration[J].*Statistics and Information Forum,*(01),pp. 87–93,2014.

[8] Shufang,Empirical research on the Dynamic Mechanism for Eastern City Development of Inner Mongolia[J]. *Research on financial and economic theory,* (01), pp.61–68,2013.

Management, Information and Educational Engineering – Liu, Sung & Yao (Eds)
© 2015 Taylor & Francis Group, London, ISBN: 978-1-138-02728-2

Research on the spatial effects of regional economic development to rural labor transfer in China—based on spatial econometrics perspective

Chun Yang
Institute of Agricultural Economics and Development, Chinese Academy of Agricultural Sciences, P.R. China

ABSTRACT: This paper, based on spatial econometrics perspective, using 2258 county statistics of China in 2010, on the view of the national county and the eastern regional, central regional and western regional, analyses the spatial effects of regional economic development to rural labor transfer. The results show that regional economic development has a positive effect on rural labor transfer. From a national perspective of the county, if county regional economic development increased by 1%, there will be 0.7737% contribution to rural labor transfer, and adjacent counties of rural labor transfer have space interactive effects. From the eastern, central and western counties of view, the impact of regional economic development to rural labor transfer in eastern regions was significantly higher than the central and western regions. The role of space interaction effect in the western region was significantly higher than the eastern and central regions. Then put forward the following recommendations, realize the coordinated development of regional economy, help to promote the transfer of rural labor. It is useful to formulate regional policy to promote rural labor transfer. It is also necessary to build the perfect rural labor employment and social protection mechanisms.

KEYWORDS: Regional Economic Development; Rural Labor Transfer; Spatial Effects.

1 RESEARCH BACKGROUND

Since the reform and opening up, China's regional economic development has shown significant differences that further form the three large "eastern, central and western" economic zones. The areas with high level of economic development are mainly concentrated in the southeast coastal regions, while those with a low level of economic development are mainly concentrated in the western regions, and the central regions are between the two. With regional economic development, rural labor force transfer also shows the higher transfer degree in the eastern coastal areas than the central and eastern regions. The areas with a high degree of rural labor transfer are mainly concentrated in southeast coastal areas and a few of the central regions. The areas with a low degree of rural labor transfer are mainly concentrated in the central and eastern regions. Since 2004, the annual Top Document of the government always puts emphasis on the Three Rural Issues. It is put forward that based on "farmers' income growth" as a starting point, the final solving of Three Rural Issues and the socialist new rural construction cannot be realized without the smooth transfer of rural surplus labor (Cheng, 2007). At present, the domestic research on rural labor force transfer is mainly focused on the role and impact of rural labor force transfer on economic growth, regional differences, rural economy and rural development (Zhang, 2013; Jia, 2012; Zhou, 2006; Yang, 2011), as well as the economic effect of rural labor transfer and other influencing factors (Guo, 2007; Liu, 2011). However, domestic research is insufficient on the role and influence of economic development on rural labor transfer, but only on the empirical analysis of panel data in 17 cities of Shandong Province (Zhang, 2011) and the national economic growth at the provincial level, industrial structure and rural labor transfer (Cheng, 2007). At the same time, the related research is all based on the traditional OLS regression analysis, but little on county level and with little considering the spatial effect. At present, the interregional correlated influence of economic and social development is more and more significant. The county economy is considered to be the basic regional economy unit in the national economy. It is more representative and typical to study the impacts of regional economic development on rural labor force transfer considering the spatial effect based on a more micro-county level. Therefore, combined with the shortage of the existing research, based on spatial econometric analysis, the national statistics in 2258 counties in 2010 are used to analyze the degree of influence of regional economic development on rural labor transfer from the perspectives of national county level and the eastern, central and western county level. Research results play an important role in accurately grasping the degrees of influence of regional economic development in national level and the eastern, central and western county level on rural labor transfer. In the meantime, it also can guide the direction of county economic coordinated development with the vital significance of achieving the scientific, reasonable and effective national rural labor force transfer.

2 VARIABLE SELECTION AND DATA SOURCES

In order to understand how the degree of influence of regional economic development is on rural labor transfer, and what are the circumstance in China's eastern, central and western regions, research variable indicators must first be selected. In this study, the quantity of rural labor transfer is selected as the explained variable, and the level of regional economic development is taken as the explained variable for the research on spatial correlation.

The national county data in 2010 are used as research data, in which the quantity of rural labor force transfer (denoted as rulabtrf) refers to the total rural workforce in the rural population minus the quantity of the labor force engaged in animal husbandry fishery; county economy is expressed by the county level of GDP (denoted as reggdp). In data, the quantity data of county-level rural workforce and the labor engaged in animal husbandry fishery are sourced from China Social and Economic Statistical Yearbook 2011 (the data of 2711 counties are collected, of which the data of 2695 county are effective). The county-level regional gross product data are sourced from China Regional Economic Statistics 2011 (the data of 2258 counties are collected, of which the data of 2258 counties are effective. According to the analysis demand of data matching analysis, the data of 2258 counties are used for empirical analysis).

The eastern, central and western counties are divided in accordance with the following standards. The eastern regions include Beijing, Tianjin, Hebei, Liaoning, Shanghai, Jiangsu, Zhejiang, Fujian, Shandong, Guangdong and Hainan; the central regions include Inner Mongolia, Shanxi, Jilin, Heilongjiang, Anhui, Jiangxi, Henan, Hubei and Hunan; the western regions include Sichuan, Chongqing, Guizhou, Yunnan, Tibet, Shaanxi, Gansu, Qinghai, Ningxia, Xinjiang and Guangxi. It is needed to explain that due to different division standards of the eastern, central and western regions, in order to meet research requirements, Inner Mongolia was divided into a central region, which is beneficial to combine the spatial econometric model for an empirical analysis of spatial effect.

3 EMPIRICAL ANALYSIS

3.1 *Spatial statistical test of rural labor transfer*

3.1.1 *Global autocorrelation analysis of rural labor transfer*

Firstly, GRODA software is used to test the spatial autocorrelation of the quantity of rural labor transfer in China's 2258 counties in 2010. The Moran I index of rural labor transfer is 0.2494, and the normal statistic Z values of the Moran I index are greater than the critical values of the normal distribution function in the 0.01 level (1.96). It is suggested that China's county-level rural labor transfer has gathering characteristics in the spatial distribution with a significant positive autocorrelation, i.e., the counties with higher levels of rural labor force transfer relatively tend to be close to each other, while those with lower level of rural labor force transfer tend to be adjacent.

In the eastern, central and western counties, the spatial autocorrelation test results of the quantity of rural labor transfer show that the Moran I indices of rural labor transfer are 0.2211, 0.4708 and 0.4080 respectively, and the normal statistics Z values of Moran I index are all greater than the critical value of the normal distribution function in the 0.01 level (1.96). It is shown that the rural labor transfer of China's eastern, central and western counties has gathering characteristics in the spatial distribution with significant positive autocorrelation.

3.1.2 *Local autocorrelation analysis of rural labor transfer*

Further combining with the local spatial autocorrelation analysis, the local distribution characteristics of county rural labor transfer in 2010 are shown. The regions with High - High gathering are distributed in the counties of Shandong, Jiangsu, Henan, Anhui, Hubei, Hunan, Chongqing, Guizhou in the Yangtze river coast and coastal Shanghai, county Zhejiang, Fujian, Guangdong and Guangxi, which show gathering forms with positive correlation, indicating that the regions with high labor transfer are surrounded by the same kind of regions. The regions with Low - Low gathering are distributed in the most counties of western Gansu, Ningxia, Xinjiang, Qinghai, Tibet, Sichuan and partial counties of central Inner Mongolia, Heilongjiang, Jilin and Liaoning, which show gathering forms with positive correlation, indicating that the regions with low labor transfer are surrounded by the same kind of regions.

3.1.3 *Local autocorrelation analysis of rural labor transfer*

Further combining with the local spatial autocorrelation analysis, the local distribution characteristics of county rural labor transfer in 2010 are shown. The regions with High - High gathering are distributed in the counties of Shandong, Jiangsu, Henan, Anhui, Hubei, Hunan, Chongqing, Guizhou in the Yangtze river coast and coastal Shanghai, county Zhejiang, Fujian, Guangdong and Guangxi, which show gathering forms with positive correlation, indicating that the regions with high labor transfer are surrounded by the same kind of regions. The regions with Low - Low gathering are distributed in the most counties of western Gansu, Ningxia, Xinjiang, Qinghai, Tibet, Sichuan and partial counties of central Inner Mongolia, Heilongjiang, Jilin and Liaoning, which show gathering forms with positive correlation, indicating that the regions with low labor transfer are surrounded by the same kind of regions.

3.2 *Spatial effect analysis of regional economic development on rural labor transfer*

The above analysis shows that the inter-county rural labor transfer has obvious spatial correlation (spatial gathering phenomenon). Based on county level, the spatial effect analysis of regional economic development on rural labor transfer shall be in need of the spatial econometric model that considers spatial effect for estimate.

3.2.1 *The national county-level analysis*

1 Modeling

The spatial lag (error) model of county-level analysis has the following general forms:

1) Spatial lag model

$$\ln \text{rulabtrf} = \rho W\text{rulabtrf} + \beta_0 + \beta_1 \ln reggdp + \varepsilon$$

Where, rulabtrf is explained variable matrix; reggdp is the explanatory variable matrix; ρ is the spatial regression coefficient of the explained variable vector; W is the spatial weighting matrix; ε is the random error term vector, β_0 is the constant term and β_1 is the coefficient.

2) Spatial error model

$$\ln \text{rulabtrf} = \beta_0 + \beta_1 \ln reggdp + \varepsilon \qquad \varepsilon = \lambda W\varepsilon + \mu$$

Where, rulabtrf is explained variable matrix; reggdp is explanatory variable matrix; ε is random error term vector; λ is the spatial error coefficient of explained variable vector; W is the spatial weighting matrix; μ is the random error vector with normal distribution; β_0 is the constant term and β_1 is the coefficient.

2 Model estimate result

Estimation is performed using GEODA0.9.5 software. Combined with LM estimated results, it is determined to choose the spatial lag model or spatial error model. Model estimation results are shown as follows. The GDP coefficient of regional economic development is 0.7737 with significant test results, showing that when other conditions remain unchanged, every 1% increase of county regional economic development will contribute 0.7737% to rural labor transfer. It is also indicated that county regional economic development has an obvious spatial correlation to rural labor transfer. The correlation coefficient of spatial error in the model (λ) is 0.5679, also passing the significance test of level 0.01. It is proven that the regional rural labor transfer has a strong spatial interactivity, besides the spatial correlation with county regional economy. With other factors besides regional economic development of the adjacent county (such as the development of urbanization, industrial structure adjustment and the per capita area change of cultivated land), the error term is used to produce a certain effect, i.e., the county-level rural labor transfer has mutual influences in space, performed as spatial interaction of inter-county rural labor transfer .

3.2.2 *The eastern, central and western county-level analysis*

The modeling in the eastern, central and western regions was the same as above. GEODA0.9.5 software was also used for estimation, and model selection was also combined with LM estimation results. Model estimation results are shown as follows:

As viewed from the comparative analysis of eastern, central and western regions, the eastern-county regional economic development has the biggest influence on rural labor transfer, higher than the central and western regions, which is consistent with the eastern economic development level that is higher than the central and western regions, leading to the economic development greatly driving rural labor transfer. Spatial interaction effect in the western regions is more significant than eastern and central regions, which is basically consistent with traditional agricultural production in western regions and the large mutual influence between peasant household economic behaviors. The spatial interaction effect between the adjacent counties is more apparent (see Table 1).

4 CONCLUSIONS AND POLICY RECOMMENDATIONS

This study results show that the regional economic development has a positive role in promoting rural labor transfer. From the perspective of the county,

every 1% increase in county-level region economic development will contribute 0.7737% to rural labor transfer. The rural labor transfer in adjacent counties has a spatial interactive effect. As viewed from the eastern, central and western counties, the eastern-county regional economic development has a significantly higher degree of influence on rural labor transfer than the central and western regions, while the spatial interaction effect of western regions is obviously higher than those of eastern and central regions.
Combined with the research results, the recommendations are proposed as follows:

4.3 *To realize the harmonious development of regional economy and help to promote rural labor transfer*

Therefore, as viewed from the county-level economic development, there should be a push to further strengthen the coordinated development of China's county-level economy in the future, so as to realize the comprehensive advancement of rural labor force transfer.

4.4 *To develop regional policies and effectively implement the rural labor transfer*

In the future, during the policy formulation of rural labor transfer promoted at county level, there should be a push to fully combine the characteristics of regional gathering pattern (High - High gathering and Low - Low gathering) to develop regional policies, so as to implement the policy effect.

4.5 *To further develop rural economy and urban industry, construction industry, and the tertiary industry so as to realize rural labor force transfer in western and central regions*

Furthermore, there should be a push to develop the county town economy, especially the urban industry, construction industry and the tertiary industry, sufficiently create jobs, and realize the rural labor transfer.

4.6 *To full play the existing regional economic development advantage and realize the effective transfer of rural labor force in the eastern regions*

In the future, there should be a push to further integrate the eastern economic advantage, formulate scientific and reasonable rural labor transfer plans, and gradually realize the transfer and absorption of labor in the region and surrounding areas.

4.7 *To establish and improve the transfer and employment of rural labor as well as the social security mechanism*

Therefore, in the process of rural labor force transfer, there should be a push to build and improve the transfer and employment of rural labor as well as the social security mechanism, including the household registration system, employment training, city service system and social insurance covering rural migrant workers, etc.

Table 1. The Model estimation results.

Estimated area	National County in China	Eastern County in China	Central County in China	West County in China
Estimation model	SEM	SLM	SEM	SEM
Constant term	0.5027 (0.0000)	0.7916 (0.0000)	0.9924 (0.0000)	0.9298 (0.0000)
Regional economic development	0.7737 (0.0000)	0.8050 (0.0000)	0.7503 (0.0000)	0.7306 (0.0000)
Interactive effects of space	0.5679 (0.0000)	0.2310 (0.0000)	0.4927 (0.0000)	0.6307 (0.0000)
R-squared	0.8862	0.9181	0.8461	0.7607
LogL	−6798.92	−1783.14	−1892.03	−2171.19
AIC	13601.81	3572.28	3788.05	4346.38
SC	13614.11	3586.94	3797.82	4356.19
B-P test	1493.42 (0.0000)	694.45 (0.0000)	758.57 (0.0000)	541.98 (0.0000)

REFERENCES

[1] Zhang Yan, Song Shanmei. The research on the influence of rural labor transfer on regional economic development. Theory And Contemporary, 2013 (1) : 42–43.

[2] Jia Wei. An analysis on the influence of rural labor transfer on economic growth and regional disparities. China Journal of Population Science, 2012 (6) : 55–65.

[3] Zhou Pinghua. The influence of rural labor transfer on rural economic development in China . Special Zone Economy, 2006 (12) : 138–139.

[4] Yang Xiangfei. A research on western rural labor transfer and rural development questions, Ph.D. Thesis, Lanzhou University, 2011.

[5] Guo Shengrong. A research on economic effect and influence factors of rural labor transfer in Jiangsu, Ph.D. Thesis, Nanjing University of Aeronautics and Astronautics, 2007.

[6] Liu Ke. A research on the influential factors of rural labor transfer in Ya'an, Master's Degree Thesis, Sichuan Agricultural University, 2011.

[7] Cheng Mingwang, Shi Qinghua. Economic growth, industrial structure and rural labor transfer - based on an empirical analysis of the data from 1978 to 2004 in China. Economists, 2007.5:49–54.

[8] Zhang Zhixin, Li Ya, Ren Xin. Rural labor transfer with differences in regional economic development - based on an empirical analysis of Shandong 17-city panel data in 1999-2008. Dongyue Tribune, 2011.10:129–133.

Management, Information and Educational Engineering – Liu, Sung & Yao (Eds)
© 2015 Taylor & Francis Group, London, ISBN: 978-1-138-02728-2

MBA comprehensive quality evaluation based on the Grey theory

Li Xia Zhang
Management College, Tianjin Normal University, Tianjin China

Hui Liu
Social and Science office, Tianjin Normal University, Tianjin China

ABSTRACT: The comprehensive quality analysis of MBA is an important guiding index for the comprehensive development of graduate student. Based on the summary of relevant researches on MBA comprehensive quality, this paper analyzed serials of evaluation factors, tested comprehensive quality with 20 MBA students by applying Grey Correlation Analysis at random. This paper analyzed those factors in terms of factor ranking, student classification. This will be conductive to MBA management.

KEYWORDS: MBA; Comprehensive Quality; the Grey Theory.

1 LITERATURE REVIEW

Quality is explained in Ci Hai, " the original characteristics and foundations in some aspects, and the growth of accomplishment in practice training of people or things". Quality includes, by definition, congenital genetic factors and learning practice for the cultivation of the day after tomorrow. In psychology, quality includes congenital physiological characteristics and the growth of people's cultivation in practice at the same time, such as political quality, cultural quality, etc, the quality in psychology refers to the accomplishment that can be obtained through practice and study the day after tomorrow. Comprehensive quality refers to one's inborn,inherited gifted elements, such as the function characteristics of the nervous system, sense organs, sport organs and other genetic predispositions, and all individual qualities which is formed in the environmental influence and educational impact the day after tomorrow.

There is no students (comprehensive) quality evaluation in abroad, instead, student evaluation is mainly a survey of students learning effect, etc. since the 1980s, China has brought in the research and practice of work of education measurement and evaluation, the early education evaluation work was developed in Shanghai education bureau. In 1983, based on the theory of education objective taxonomy. Shanghai education bureau established and implied the first junior high school plane geometry standardized assessment scheme in China, which opened the domestic teaching target evaluation at home. Since then, the research perspective by scholars extended from the students' academic attainment evaluation gradually to the morality and physique evaluation.

From the results of consulting CNKI literatures, the studies is divided into two classes, postgraduates comprehensive quality evaluation index system and postgraduates comprehensive quality evaluation. For example, Zhang Yunhua et al (2006) from scientific research and professional theory and other four aspects to conduct a comprehensive evaluation for postgraduates, by quantificating nine indicators to build the postgraduates' comprehensive quality evaluation index system; Zhang Qing-yi&He Ying(2007) constructed postgraduates comprehensive quality evaluation system multivariate model . In the aspects of postgraduates' comprehensive quality evaluation, He Rui et al(2005)& Jiang De-long et al(2011) used fuzzy mathematics method, discussing the fuzzy evaluation model of postgraduates' comprehensive quality, constructing the fuzzy postgraduates' comprehensive quality evaluation system to evaluate the students' comprehensive quality fussily. Currently, relevant researches on professional postgraduate degree are relatively rare, some scholars evaluate the professional postgraduates' comprehensive quality in education, military, clinical medicine, ports and training disciplines.

In short, the study of current scholars about postgraduates' comprehensive quality reflects in evaluation system, methods and models, from the perspective of research methods, most scholars use analytic hierarchy process(AHP), fuzzy comprehensive evaluation method and statistical analysis method to evaluate the postgraduates' comprehensive quality. As it can be seen from the literature review, there are scholars having studied the postgraduates' comprehensive quality evaluation, which mostly from the perspective of the whole postgraduates, while relevant studies of professional postgraduate degree are confined to a few professional fields. Literatures have yet to see that scholars have comprehensive quality evaluation of such kind of professional postgraduate degree like master of business administration (MBA).

2 EVALUATION INDEX SYSTEM OF MBA

Regarding to the quality standards of MBA, Watson, S.R.(1993) says that MBA should have intelligence, professional skills, interpersonal skills, knowledge of the organization and the ability to analyze the management. Mintzberg, H.(2004), shows from the opposite that MBA can't be a "young, poor man, even has a little experience, yet only through analysis and technological form " to manage, so as to avoid the wrong consequences. Gu Yongcai(1997) thinks that MBA is a multidimensional, composite and the three-dimensional structures of talents, and he puts forward a kind of "plane" MBA talents' standards, which uses various parts of the aircraft structure on behalf of the MBA essential ability quality.

These quality standards include: (1) the common knowledge of the culture in social life; (2) industrial and commercial management expertise and other related disciplines of professional knowledge; (3) professional frontier knowledge; (4) political ideology, moral character and methodological knowledge and some modern management ideas ; (5) foreign language level; (6) the operation and usage of computer and other advanced science and technology tools; (7)practical knowledge; (8) aesthetic knowledge; (9)non-intelligence factors;(10)social using environment.

This paper puts forward a MBA comprehensive quality index model, on the basis of "plane" type model by Gu Yongcai and references to the research achievements of other scholars, as it is shown in figure 1

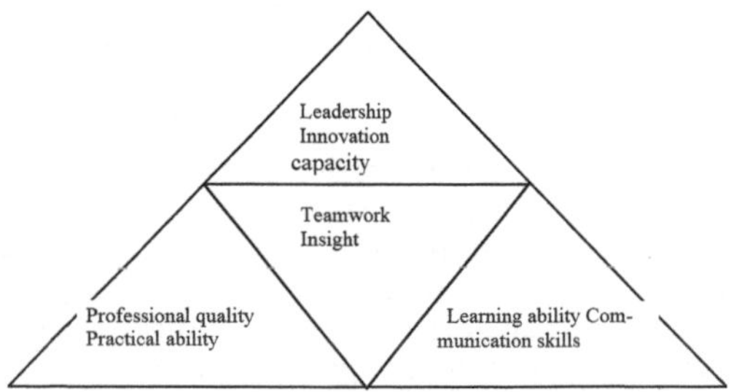

Figure 1. MBA comprehensive quality index model.

This pyramid model is divided into 4 levels, 8 indexes, the following two triangles show MBA students' basic quality requirements, in the middle of the triangle shows MBA students' core quality, the top of the pyramid triangle shows MBA students high quality on the basis of the basic quality and the core quality.

MBA professional quality refers to the degree of master of business administration basic of professional disciplines and professional theoretical knowledge. Practical ability is a ability to use professional skills solve practical problems quickly. Learning ability is defined according to the problems, learning knowledge skills and needs quickly, so as to improve their comprehensive quality and ability continuously. Communication skills refer to in-depth communication, which focus on efficiency and results of efficient collaboration through communication . Teamwork refers to MBA students in teams who are of mutual trust, mutual support, responsibility and mission to promote to reach team goals. Marketing insight refers to the sensitivity of the MBA students that react to market changes . Leadership means to encourage guide and inspire others to practice positively and initiatively. Innovation refers to the MBA students apply knowledge and theories which have been learned in the marketing practices, to continue to provide economic value, social value, ecological value, new ideas, new theories and new methods.

3 GREY THEORY CORRELATION ANALYSIS

Grey correlation analysis methods are quantitative research methods which compare interrelation of the factors that interact, influence and restrict mutually included in the system. The basic idea of Grey correlation analysis methods is to determine the connection of the comparative sequence curve and the reference sequence curve is close or not according to their similarities. Associated sequence reflects the close order of the comparative sequence and the reference sequence . Its essence is to analyze and calculate the similar or dissimilar levels of associated sequences, the more consistent target sequences express evolving situation, the greater correlation it has; Conversely, the smaller correlation it has. Therefore, we can use grey correlation to analyze the pros and cons of the evaluation objects.

The specific steps of grey correlation analysis are as follows:

3.1 *Identify analysis sequence*

On the basis of qualitative analysis to the research questions determining the comparative sequence (evaluation object) $X_1=\{X_i(k) \quad i=1,2,\ldots,N; k=1,2,\ldots,L\}$ and reference sequence (evaluation criteria) $X_0=\{X_0(k)|k=1,2,\ldots,L\}$.

3.2 *Non-dimensionalization of variable sequence*

General raw data series have different dimensions or magnitudes, in order to ensure reliability of the results, we should carry out non-dimensional process. Commonly used methods are the average valued, the initial valued and normalized, etc.

3.3 *Calculate the difference sequence, the maximum difference and the minimum difference*

Using the formulas to calculate the difference sequence.

$$\Delta_i(k) = | X_0(k) - X_i(k) | \quad (1)$$

$$\Delta_{max} = \max | X_0(k) - X_i(k) | \quad (2)$$

$$\Lambda_{min} = \min | X_0(k) - X_i(k) | \quad (3)$$

3.4 *Calculate the grey correlation coefficient*

The correlation coefficient between X_i (k) and X_0 (k) is:

$$\delta_{0i}(k) = (\Delta\min + \rho\Delta\max)/(\Delta_i + \rho\Delta_i) \quad (4)$$

Among them, $\rho \in (0,1)$ $\rho \in (0,1)$ is the resolution factor, in general grey correlation applications, we usually take $\rho \in (0,1)$ $\rho = 0.5$ to simplify the calculation.

3.5 *Calculate the grey correlation*

Too many correlation coefficients will make information more fragmented and not easy to compare. We can use averaging method to deal with it, so that the information which the correlation coefficients reflect epitomizes. Define the correlation of comparative factor X_i and the reference factor X_0:

$$\gamma_{0i} = (1/L)\sum_{K=1}^{L} \delta_{0i}(K) \quad (5)$$

3.6 *Sorting analysis*

Correlation epitomizes the correlation coefficients. Therefore, correlation describes the differences between the comparative sequence curve and the reference sequence curve in geometry, in order to describe the closeness of the relationship between the sequences. Put N correlation in descending order to get a associated sequence, we can judge the order of close degree of X_i to X_0.

4 MBA QUALITY EVALUATION USING GREY CORRELATION

In the research process of this article, to obtain quantitative data conveniently, we refine each specific indicator of the model of MBA comprehensive quality indexes into the second level indicators, using a MBA management center of a Normal University School management college as the research object which this article relied on, due to limited space, we selected 20 masters of business administration (represented by the letters A-T respectively) randomly, marked for the 20 objects of study by professors and scholars related in the MBA management center, and evaluated the evaluation objects and indexes, as well as the sequence analysis of matrix according to the research contents.

Because the original data sequences have different-order magnitudes, in order to guarantee the

Table 1. Correlation coefficient and correlation degree.

Indicator	Professional a	Practical b	…	Leadership g	Innovation h	longitudinal correlation
1	0.55	0.32	…	0.49	0.57	0.48
2	0.44	0.39	…	0.49	0.48	0.43
3	0.73	0.50	…	0.49	0.50	0.46
4	0.33	0.27	…	0.58	0.44	0.38
5	0.70	0.50	…	0.43	0.55	0.45
6	0.37	0.38	…	0.49	0.38	0.37
7	0.55	0.38	…	0.53	0.48	0.44
…	…	…	…	…	…	…
19	0.55	0.30	…	0.59	0.50	0.44
20	0.44	0.33	…	0.56	0.44	0.40
horizontal correlation	0.89	0.78	…	0.91	0.88	

*1-20 represent 20 MBA students

*a-h represent 8 indexes of MBA comprehensive quality. a represent Professional quality, b represent Practical ability, c represent Learning ability, d represent Communication skills, e represent Teamwork, f represent Insight, g represent Leadership, h represent Innovation capacity

reliability of the analysis results, we should carry out non-dimensional process, then, we can do the correlation analysis.

4.1 *Horizontal analysis*

The horizontal analysis is the analysis of the ranking of association among each MBA comprehensive quality index. In specific analysis, we use each MBA students' maximum value of every item of comprehensive quality evaluation index as the reference sequence, every evaluation index as the comparative sequence, according to step (1) (2) (3) (4), calculate the correlation coefficient of each comparative sequence relative to the reference sequence. According to the formula (5),we can get the ranking of the importance of MBA's comprehensive quality index ordered in the degree of association finally.

According to the correlation formula we can calculate: $\gamma_{01} = 0.89$; $\gamma_{02} = 0.78$; $\gamma_{03} = 0.83$; $\gamma_e = 0.85$; $\gamma_f = 0.80$; $\gamma_g = 0.91$; $\gamma_h = 0.88$; Therefore, the ranking of the importance of MBA comprehensive quality indicators results is: $\gamma_g \succ \gamma_a \succ \gamma_h \succ \gamma_e \succ \gamma_c \succ \gamma_f \succ \gamma_d \succ \gamma_b$

4.2 *Longitudinal analysis*

The longitudinal analysis is based on the ideal MBA students who are very good at every indicator, and we use the students comprehensive quality index value as the reference sequence. In the specific application, we use the maximum of value of each student as the reference sequence and each comprehensive quality index value of 20 MBA students as the comparative sequence, according to step (1),(2),(3),(4), we calculate the correlation coefficient of each comparative sequence relative to the reference sequence. Finally, using formula (5) we can get order the final grey correlation in the degree of association, and classify the grades of 20 students roughly.

What in table 1 are the correlation coefficient and correlation degree calculated by the longitudinal analysis and horizontal analysis. According to the results of calculation,we can classify 20 students into 3 classes using correlation degree $\gamma_1 \leq 0.4$, $0.4 < \gamma_2 \leq 0.5$, $\gamma_3 \leq 0.5$ as threshold value:

The first students' class of correlation degree ranking ($\gamma_1 \leq 0.4$) : 20, 12, 4, 6, 14

The second students' class of correlation degree ranking ($0.4 < \gamma_2 \leq 0.5$) : 1, 3, 5, 15, 19, 7, 2, 13, 16

The third students' class of correlation degree ranking($\gamma_3 > 0.5$.) : 9, 8, 18, 17, 10, 11

5 RESULTS ANALYSIS

According to the horizontal correlation analysis results, the ranking of the importance of MBA comprehensive quality indicators shows that leadership, professional ability, innovation capacity are the main factors affecting MBA comprehensive quality of the 20 students in this case; teamwork, learning ability, insight are the important factors as well. Therefore, we should strengthen the training and exercise of these aspects by a series of ways in the process of MBA training.

The results of the longitudinal correlation analysis inspire managers from another perspective. Through the longitudinal correlation ranking, students can be classified, in this paper, the comprehensive quality of the first class is lower overall, and the third class is higher. We can be more refined and targeted in the process of MBA training. For example, in this case we can get such a conclusion directly, we should strengthen the training of the first students' class in order to further improve their comprehensive qualities.

6 CONCLUSION AND DISCUSSION

This paper uses the grey correlation analysis to evaluate and compare the comprehensive quality of the master of business administration (MBA), the empirical results show that (1) the grey correlation analysis method can rank all the evaluation indicators that influencing the MBA comprehensive quality. According this,manager should adjust it's management pattern in practice. (2)We can get the ranking of all of the MBA students and classify them, aiming to cultivate and manage them on this basis. At the same time, the grey correlation analysis method of MBA comprehensive quality has the advantages that operate simply, highly efficient, so we can compute large samples. In practical application, we can establish the second-level index to obtain more scientific and accurate data combined with the actual situation.

ACKNOWLEDGEMENTS

This research was supported by (1) the humanity and social science project of Tianjin under Grant TJGL10-881 (2) science of education project of Tianjin under grant HE4046 (3) Tianjin normal university project under grant 52WZ1104.

REFERENCES

[1] Wang Zhi-xiang; Xu Lan-ping. Comprehensive quality training of graduate students thinking [J]. Journal of Education Exploration, 2009, (4) : 65–66.

[2] Yang Hang-chang. Research on Master of Business Administration(MBA) Education's Market Adaptability [D]. NING BO University, 2006.

[3] Zhang Yun-hua;Mao Dong-sheng. Study of Measuring system of Graduates'Colligrate Diathesis in High School [J]. Journal of NING BO University (Education Edition), 2009, 31 (5) : 64–67.
[4] Zhang Qing-yi, He Ying. Graduate students comprehensive quality evaluation system of multivariate model construction [J]. China's adult education, 2007, 21:61–62.
[5] He Rui, He Bo. Fuzzy evaluation system of masters' comprehensive quality [J]. Journal of Liaoning Technical University (Social Science Edition), 2005, 7 (5) : 554–556.
[6] Jiang De-long, Yin Shu-ping, Shi Li, Luo Song-tao. Application of fuzzy synthetic decision on comprehensive quality evaluation of postgraduate [J]. Computer Engineering and Design, 2011, 32 (9) : 3208–3212.
[7] Watson, S.R.The place for universities in management education [J].Journal of General Management, 1993, 19(2):14–42.
[8] Mintzberg, H. Managemers, Not MBAs: A Hard Look At the soft practice of managing and management development [M]. San Francisco: Berrett-Koehler publishers, inc., 2004:7–11.
[9] Gu Yong-cai. MBA ladder engineering [M]. Beijing: China City Press, 1997–27.
[10] Deng Julong. Grey theory [M]. Wuhan: Hua Zhong University of Science &Technology Press, 2002.

Management, Information and Educational Engineering – Liu, Sung & Yao (Eds)
© 2015 Taylor & Francis Group, London, ISBN: 978-1-138-02728-2

The exploratory research on the management mechanism of operating university assets

Wei Min Wang & Liang Hui Jiang
Department of Technology Economy and Management, Xi'an University of Architecture and Technology, Xi'an, China

ABSTRACT: This article briefly describes the generation and the previous main reforms of the university's operating assets, and analyzes the current situation and main issues faced with the management of the university's operating assets. Combined with the guiding ideology of SASAC on the management of the state-owned assets and the market-oriented reform of the state-owned enterprises, it makes some exploratory research on resolving the main operating problems and makes exploratory research on the management mechanism.

KEYWORDS: Operating assets, Management mechanism, Ownership structure.

1 GENERATION AND PREVIOUS REFORMS

1.1 *Generation of the university operating assets*

The university operating assets are state-owned assets of universities which are used to invest, lend and guarantee under the premise of meeting the need of teaching and research for colleges. It mainly contains logistics group and asset management company assets. State-owned Assets Supervision and Administration Committee (SASAC) issued "the management implementation approach of transferring non-operating assets of the public institutions with operating assets" in 1995. As public institutions, universities have achieved operating assets represented by university-run enterprises in that process. Besides, colleges have achieved logistics assets in the process of logistics socialization.

1.2 *Precious main reforms of the university operating asset management*

According to the reform regulation of university logistics socialization implemented in 1999, colleges should separate logistics service from school and implement logistics socialization. The reform of university logistics socialization has been improving the logistics service quality and increased the value of university logistics assets. But there are also some problems existed. For example, the asset ownership is difficult to determine and the operating model is hard to specify.

With regard to problems existed in the university-run companies, such as unclear ownership and operational risks, Ministry of Education required colleges to transfer university-run companies to university-owned companies. At the same time, the Ministry of Education asked university-owned companies to establish a modern enterprise system which has clear ownership, scientific management, etc.

2 CURRENT SITUATION AND MAIN PROBLEMS

2.1 *Lower capital gains rate*

According to the statistics (2010) form The Science and Technology Development Center, the total asset of the country's top ten university asset management company has reached to 165.4 billion yuan, involving school-affiliated industrial assets, equity ownership interest in holding equity interests and ownership right interests. However, the return of operating university asset is relatively low. The total income of asset operating companies above was 126.6 billion yuan, and the total profit was 6.67 billion yuan, and the profit margin was 5.26%. It was lower than that of the state-owned enterprises in the same period, which was 6.97%, and lower than that of the private enterprise in the same period.

2.2 *Excursive management system*

According to "the Ministry of Education guidance on positively develop and manage the university technology industry", universities should put all operating assets under the university asset operation companies. The companies are responsible for the operation of the university operating assets, and university is no longer set up other asset management institutions. But, the operation work still be of bull management and unclear responsibility. Based on the survey on the current situation of the provincial university asset management institutions in a northwest province, some universities still put their dressing rooms under the management of the estate management department, and others put these operating assets under the management of the asset management department. Overall, many universities still don't put all their operating assets under the management of the university asset operation companies. The intangible asset management is still a bull. There are no institutions to be responsible for the intangible asset management. Some universities put their intangible assets under the management of the financial departments, and some universities put those under the management of the technology department. Based the survey of 20 vocational colleges, the number of the universities taking independent asset management institutions model is 4, accounting for 20%. However, the number of the universities taking long lead management institutions model is 9, accounting for 45%, and the number of the universities taking non-independent asset management institutions model is 7, accounting for 35%.

2.3 *Unclear property rights*

Unclear property rights are one of the basic problems of operating assets management colleges have been facing. On the one hand, when universities transferred non-operating assets to operating assets, universities didn't perform relevant procedures according to the law, leading to unclear ownership. On the other hand, operating assets management agencies hadn't enough attention to the property assessment and property income was not clear in the asset management process involving external rental, lending and foreign investment, etc.

2.4 *The lack of professional teams*

The lack of the operating university asset management professionals is an important factor which restricts management. The university operating assets have special properties. On the one hand, it requires keeping increasing the value of the state-owned asset. On the other hand, as a public institution, colleges must take the interests of the relevant subjects into account. From the point of schools, due to the relative lack of business sense, colleges didn't give full attention to asset operation and be lack of motivation to the introduction of professionals. To a certain extent, colleges, even views asset management and asset management company as agencies which are used as dispersing staff and easing the pressure on employment.

2.5 *Single ownership structure*

The establish of university asset management company has achieved the purpose effectively setting up a firewall between the school-owned companies and the universities. But, as the scale of the university asset management companies has been increasing, state-owned sole asset management companies generated a series of problems due to the single ownership structure, such as lower asset utilization, the lack of inner competition and effective reward system. In terms of relative social capital, single ownership structure leads to asset management companies uncompetitive in the competitive market environment, facing great risk. In addition, with respect to the tasks universities undertake, promoting national industrial restructure and accelerating scientific and technological achievement, it isn't enough by only relying on limited university asset.

2.6 *The lack of intangible asset management*

Compared to the management of the fixed asset, the management of the intangible assets is the lack of laws and regulations' support. The Ministry of Education and each province has published "university fixed asset management approach" in 2005, but there are not intangible asset management rules promulgated. There are not an appropriate institution which has enough rights to manage intangible assets. Besides, due to the incomplete data of property transactions, property assessments and property income can't be specifically identified during the transactions of intangible assets.

3 OPTIMIZATIONAL INQUIRY OF THE MANAGEMENT MECHANISM

Recently, the State-owned Assets Supervision and Administration Committee (SASAC) published "suggestions on the market-oriented reforms of the state-owned assets management and state-owned companies". The document pointed that we should form or restructure the state-owned capital investment companies, and these companies would be responsible for the asset operation directly. It indicated that we should increase the proportion of market-oriented corporate hiring managers and promote the market-oriented reform and shareholding reform. Represented by university-owned companies or the asset management companies, the university operating asset is operating assets. It has great importance taking measures to implement limited market-oriented reforms.

3.1 *To rationalize the management system*

To rationalize the management system is the basis of improving the operational asset management. Universities should establish institutional settings, including asset management committee, the asset management division and asset operation company. As investors of operating university assets, asset management committee enjoys investor's rights and fulfills the investor's obligations. As the permanent agencies of the asset management committee, the asset management department is responsible for managing state-owned assets. It should be divided into operating asset management agency and non-operating asset management agency. The operating asset management agency is responsible for the operating asset management, and non-operating asset management agency is responsible for non-operating asset management. Asset operation companies carry out specific operation and management of university operating assets.

3.2 *Establish and improve the performance evaluation system*

Performance evaluation focuses on work completed evaluation in accordance with standards, and implements incentive measures or improves works on these basis. Establishing and improving the performance evaluation has great significance in improving the management of operating university assets. We should take the special nature of operating university assets when we establish the performance evaluation of the operating university asset. Firstly, the characteristics of the operational asset require its preservation and appreciation. Secondly, as public institutions, the management of the operating asset must take the interests of the schools and related subjects in consideration.

Balanced Scorecard theory can be used to establish the evaluation system of the operating university asset management. The financial norm includes profitability, asset quality conditions, debt risk, operation and growth. The stakeholder norm includes student practice, school faculty practice, social contributions. The internal process norm includes mechanisms, decision-making, risk control. Learning and growth norm includes development and innovation, human resources, incentive system.

3.3 *Introduce social capital and improve capital structure*

The university asset operation companies are responsible for asset operation. With regard to single ownership structure, it is reasonable to improve the capital structure and management mechanism of the asset operation companies by introducing social capital. The introduction of social capital can expand the size of the asset management companies, and ease the fund gap of asset operation to a certain extent. Besides, it is more important that they seek for capital gains of social capital can increase the yields of the university operating assets. But, it is primarily for the university asset management committee or asset management companies keeping their effective control of existing or newly established business when we introduce social capital.

In practice, state-owned companies have effectively resolved some malpractices such as lower capital gains and work inefficiently through joint-stock reforms. At the same time, some university-hold listed companies, for example, Founder Technology, Fudan Fuhua, have provided effective reference for university asset operation companies.

3.4 *Hire market-oriented talent and improve the management team*

It is the common feature of university operating assets and corporate assets what increase their value.

To a large extent, the management of the university operating assets is similar to that of corporate capital. An excellence and effective management team is the premise for the success of an organization or enterprise. To enhance the management of the university operating assets requires raising the quality of the management team.

We can increase the quality of the management team by increasing the proportion of market-oriented talents and introduce the work competition mechanism in thc practice of the university operating asset management. At the same time, we can establish the appropriate post system, equity incentives and other measures to improve the overall quality of the management team.

3.5 *Pay the same attention to the intangible assets and fixed assets*

The scientific research and technology patents are of great importance for an organization or country. These intangible assets will bring huge economic benefits in the era of knowledge economy. Universities have rich research resources, making universities to get higher economic returns with a low cost. The university operating asset management should pay enough attention to both fixed assets and intangible assets. Colleges should establish high effective intangible asset management institutions and develop specific management rules. At the same time, the government should summarize and count the property transaction statistics of various universities. Combined with the social enterprise transaction data, the government can establish the property right transaction database which can provide effective reference for the determination of the property assessment and property benefits.

4 CONCLUSIONS

Since the State-owned Assets Supervision and Administration Commission issued "the implementation approach of transferring non-operating assets of the public institutions with operating assets" in 1995, the university operating asset management has been achieving remarkable success, through the restructure of the university-run companies and logistics socialization. However, there are still many problems in the university asset operation, and taking market-oriented reforms and rationalizing the management system is an effective approach.

REFERENCES

[1] Jie Zhibo. The current problems and solutions in the management of the university operating assets. Economic Management 2008(5):117.

[2] Li Naipeng, Zhang Kuiping. The innovative exploration on the management system and performance evaluation methods of the university operating assets. Chifeng college Journal2012(9):149–151.

[3] Fan Qingwen, Liu Li, etc. The exploration on supervision and management of the university operating assets. Laboratory Technology and Management 2007(11):149–151.

[4] Gao Wei, Li Jingjing. The design of the university balanced scorecard performance evaluation index. Economist 2013(9):14–16.

[5] Wei Liping. The thought of the university asset operating companies' corporate governance structure. Contemporary Economic 2009(11):118–119.

[6] Zhu Wenhua. The analysis of the state-owned university operating asset management. Financial Management 2008(10):46.

[7] Li Qiyong. The reform exploration on the management system of the state-owned assets. University Logistics Research 2010(2):23–24.

[8] Wang Qi, Wen Xinghuo, Chen Hui. To improve the management system and mechanism of the state-owned university asset management. Laboratory Technology and Management 2008(10):179–183.

[9] Liu Gan. The property analysis of the restructure of the state-owned university asset management system. Shaanxi Normal school Journal 2005(1):85–89.

[10] [10] Huang Li, Zha Zhiling. The research of the university intellectual property operation. 2003(10): 98–102.

Management, Information and Educational Engineering – Liu, Sung & Yao (Eds)
© 2015 Taylor & Francis Group, London, ISBN: 978-1-138-02728-2

Research on the coordinated development of Chinese industrial structure and employment structure

Shu Ru Liu & Dong Xu Zhang
School of Management, Xi'an University of Architecture and Technology, Xi'an, China

ABSTRACT: This paper is based on combing the industrial structure and employment structure of current research, using coordination coefficient overall to analyze the coordinated development of Chinese industrial structure and employment structure, comparable with international standards. The results showed that: Since the reform and opening up, the degree of coordination of Chinese industrial structure and employment structure has been greatly improved, but the employment structure obviously has lagged behind the development of the industrial structure. The paper puts forward the countermeasure suggestions to promote the labor transfer and expanding employment to improve the level of employment structure, and promote the coordinated development of Chinese industrial structure and employment structure.

KEYWORDS: industrial structure; employment structure; coordination coefficient.

1 INTRODUCTION

The composition of the industrial structure refers to the industry and the connection between the industries and proportion relations. The employment structure refers to the amount of labor and the proportion occupied by the various departments of the national economy and their interrelationships. A country's economic development is not only characterized by the growth of the economy, at the same time, inevitably accompanied by the evolution of the industrial structure. As a carrier of the employment structure, industrial structure, to a certain extent determines the level and evolution direction of employment structure; at the same time, the reasonable employment structure also promotes the adjustment and development of industrial structure [1].

Since the reform and opening, China from the traditional division of labor, capital deepening and economies of scale of traditional economic mode into the pursuit of innovation driven economy. The new economy mode is bound to lead the industry structure change. According to the Petty Clark theorem, in the process of evolution of industrial structure, the relative proportion of the first industry of the national economy income and labor force gradually declined; the relative proportion of the second industry of national economic income and labor rise, with the further development of the economy and the relative proportion of the third industry in the national income and labor also began to rise [2]. Eventually, the distribution of national income and labor force transferred from the first to the second industry and tertiary industry. Domestic scholars who have many years of research showed that: in the process of economic development, the evolution of industry structure and employment structure has a strong correlation. The development of national economy, not only values the GDP value-added, but pays more attention to the optimization and upgrading of industrial structure. Therefore, the research on industrial structure and employment structure is the coordinated development is related to the healthy development of national economy.

2 METHODS REVIEW

For the coordinated development of industry structure and employment structure research, most of domestic scholars are given priority to with the empirical, using different methods of measurement for a regional industrial structure and the relationship between employment and quantitative study, using the method basically has the following kinds:

Structure deviation analysis. The method mainly reflects the proportion of the added value of industry and the degree of the corresponding differences in the proportion of labor is the first choice when researching the coordinated development of analytical tools.

Structure of employment elasticity coefficient analysis. The method mainly reflects the economic growth on the capacity to absorb labor force.

Comparative labor productivity. This method is mainly to objectively reflect the labor productivity of a department. [3].

In addition to the above three methods, the domestic scholars also introduced some new methods of scientific research. Dahong Chen from the perspective of the relationship between industry structure and employment structure puts forward the research structure of the two correlation regression analysis method [4]. Dong-dong Liu, according to the characteristics of the industrial structure of multi-factor variable used the method for gray correlation analysis of the connection degree of industry structure and employment structure [5]. Yicai Zhang, Rongguo Yan used the entropy principle to construct transformation coefficient of industry structure and employment structure [6].

3 MODEL BUILDING AND EMPIRICAL ANALYSIS

3.1 *Model building*

In order to study the regional industrial structure similar problems, the United Nations industrial development organization international industrial research center structure similarity coefficient was proposed [7]. It can reflect the degree of similarity or difference between the regional industrial structures in general.

The formula is as follows:

$$S_{ij} = \sum_{k=1}^{n} X_{ik} X_{jk} \Big/ \sqrt{\sum_{k=1}^{n} X_{ik}^2 \sum_{k=1}^{n} X_{jk}^2}$$

Where: X_{ik}, X_{jk} shows the region k sector i and j industrial structure proportion, S_{ij} shows similarity coefficient i and j region industrial structure. $0 \le S_{ij} \le 1$, Coefficient greater, indicating that the more similar the two regional industrial structure. On the contrary, it means that two regional industrial structures do not converge.

Industrial structure similarity coefficient can not only from the static to the dynamic analysis, the development trend of regional industrial structure similarity between can also from a single structure to promote a variety of different structure comparison coefficient [8]. Gravity structure with three industrial structures and employment structure for data coordination coefficient calculation, analyze the degree of coordination of Chinese industrial structure and employment structure.

The formula is as follows:

$$H_{se} = \sum_{i=1}^{n} S_i E_i \Big/ \sqrt{\sum_{i=1}^{n} S_i^2 \sum_{i=1}^{n} E_i^2}$$

Where: S_i and E_i shows the industrial structure and the closure of the I industry proportion, H_{se} shows the coordination coefficient of industrial structure and employment structure. $0 \le H_{se} \le 1$, The larger coefficient indicates a higher degree of coordination between the two. On the contrary, it indicates that the lower the level of coordination between the two.

3.2 *Empirical analysis*

According to the coordination coefficient formula of the design, calculation of 1978–2012 years of Chinese industrial structure and employment structure coordination coefficient in Table 1 and figure 1.

Table 1. 1978–2012 years of Chinese industrial structure and employment structure coordination coefficient table.

Years	Coordination Coefficient	Years	Coordination coefficient
1978	0.6979	1996	0.7903
1979	0.7424	1997	0.7832
1980	0.7353	1998	0.7805
1981	0.7663	1999	0.7661
1982	0.7877	2000	0.7505
1983	0.7941	2001	0.7447
1984	0.8120	2002	0.7365
1985	0.7873	2003	0.734
1986	0.7857	2004	0.7659
1987	0.7905	2005	0.7767
1988	0.7837	2006	0.7936
1989	0.7694	2007	0.8159
1990	0.7953	2008	0.8286
1991	0.767	2009	0.8448
1992	0.7428	2010	0.8576
1993	0.7345	2011	0.8765
1994	0.7575	2012	0.8916
1995	0.7772		

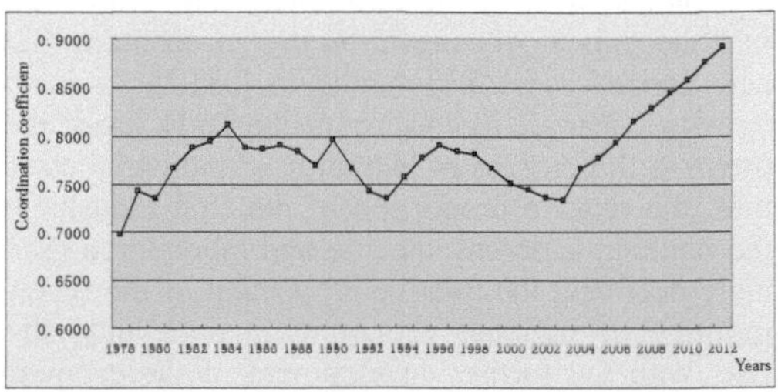

Figure 1. Chinese industrial structure and employment structure coordination coefficient trends.

It can be seen from table 1, since the reform and opening up, the coordination degree of industrial structure and employment structure in China has been greatly improved, coordination coefficient changed from 0.6979 in 1978 to 0.8916 in 2012, the coordination coefficient of growth rate as high as 27.75%.

See from Figure 1. the coordination coefficient of industry structure and employment structure in China wavy development trend. Since reform and opening up in the middle of 1980 s, the coordination coefficient has been in a state of sustained growth. It has much to do with the national implementation of the household contract responsibility system, the new system of agriculture greatly enhanced the enthusiasm of farmers, improve work efficiency, large labor has been released. Meanwhile, beginning in the early 1980s to promote the role of light industry development policy, making the secondary industry in a short time provide a lot of labor jobs, enough to absorb the labor force, so that employment levels have been substantially improved. In the late 80's, the coordination coefficient in the slow decline state, is due to the policy release labor productivity is a one-off. To rectify the development strategy of heavy industry, light and heavy industry make a balance of basic development momentum. Since 1992, the coordination coefficient again continue to rise, is because of the increase, as per capita income level resident gradually increased the demand for consumer goods, pull the household electrical appliances as the core of mechanical and electrical industry is developing rapidly. Since 1997, the coordination coefficient has shown a trend of decline again, is due to the outbreak of the financial crisis swept through Asia, have significant impact on the Chinese foreign trade market, resulting in a large number of workers unemployed, seriously affect the coordinated development of industry structure and employment structure. Since 2003, the coordination coefficient keeps rising trend, this is because for the first time in 2003, the central government to create jobs as economic macro-control goal, then, the party's 16 big reports specifically employment is vital to people's livelihood, the party's 18 big reports emphasized promote higher quality employment. This suggests that the healthy development of national economy, not only need the reasonable industrial structure, also need to have that meet the needs of employment structure.

In order to study our country industry structure and employment structure coordination development level, this paper selects some countries are compared and analyzed. According to the World Bank to the national standard of classification, the selected country is divided into low-income countries and lower middle income countries, middle-income countries, developed countries, and calculate the coordination coefficient of each country, the concrete calculation results are shown in table 2.

Table 2. China and part of the country's employment structure and industry structure comparison table in 2010.

			Industrial Structure (%)		
sort	State	GDP per capita	Primary Industry	Secondary industry	Tertiary Industry
Low-income countries	Cambodia	760	36	23.3	40.7
Lower-middle-income countries	India	1340	16.2	28.4	55.4
	Philippines	2060	12.3	32.6	55.1
	Egypt	2440	10.1	29	60.9
Upper-middle-income countries	China	4260	9.5	44.6	45.9
	Malaysia	7760	10.6	44.4	45
	Mexico	8930	4.1	34.8	61.1
Developed countries	USA	47240	1.2	20	78.8
	Japan	42130	1.2	27.4	71.5
	Britain	38560	0.7	21.7	77.6

(Continued)

Table 2. (*Continued*)

sort	State	Employment structure (%) Primary Industry	Secondary industry	Tertiary Industry	Coordination coefficient
Low-income countries	Cambodia	54.2	16.2	29.6	0.9361
Lower-middle-income countries	India	51.1	22.4	26.5	0.7379
	Philippines	33.2	15	51.8	0.9087
	Egypt	28.2	25.3	46.3	0.9407
Upper-middle-income countries	China	36.7	28.7	34.6	0.8567
	Malaysia	13.3	27.6	59.2	0.9433
	Mexico	13.1	25.5	60.6	0.9835
Developed countries	USA	1.6	16.7	81.2	0.9989
	Japan	3.7	25.3	69.7	0.9993
	Britain	1.2	19.1	78.9	0.9994

From the level of international industrial structure, China's industrial structure belongs to the category of the middle-income countries, its industrial structure evolution trend basically accord with the general rule of world industrialization accelerating stage structure change. It can be seen from table 2, compared with developed countries, the proportion of primary industry output value in our country is larger, the first industrial production value accounting for 9.5% of the whole national economy (developed in 2% or less), but it occupies 36.7% of the workforce. This is because in agricultural production in China, due to the low level of mechanization, low labor productivity, have invested a lot of labor. Our proportion of secondary industry output was significantly higher than other countries in the world, even higher than the United States, Japan and Britain and other developed countries, in line with the level of development of Chinese industrial structure is in the middle stage of industrialization. However, the second industry output value in the national economy accounted for the proportion of 44.6% only absorbed 28.7% of the workforce; this indicates that the status of the secondary industry and labor absorption capacity does not match. Since the reform and opening up, while the third industry has been in a steady growth, but compared with other countries, Chinese output of the third industry is not only lower than middle-income countries Mexico is below lower middle income countries, Egypt, India and Philippines and so on. This shows that the third industry cannot adapt to the first and the second industry development, coordination degree of industrial structure and employment structure in China to be raised. From the view of coordination coefficient, coordination coefficient of developed countries is basically the same, infinitely close to 1, this shows that the industrial structure and employment structure in developed country is very harmonious. While the coordination coefficient of medium, low-income countries have great differences. Although China is the scope of the middle-income countries, the coordination coefficient is lower than that of low income countries Kampuchea and lower middle-wincome countries Philippines and Egypt. The structure of industry of our country is similar to those of Malaysia, China's per capita GDP (2010) for $4260, the per capita GDP of Malaysia reached $7760, was 1.8 times of that in China so much. Meanwhile, the coordination coefficient of China is 0.8567 far less than 0.9433 in Malaysia. It showed that the trend of the evolution of employment structure of our country lags behind the evolution of the industrial structure, employment structure and hysteretic hindered the development of economy, industry structure and employment structure visible coordinated development plays an important role in promoting economic development.

4 CONCLUSIONS AND RECOMMENDATIONS

Since the reform and opening up, the coordinated degree of industrial structure and employment structure has been significantly improved in China. However, compared with the world level, the evolution of the employment structure has lagged industrial structure in China. In particular, the first industry in GDP contribution and its absorption is out of proportion to the amount of labor, and agriculture is relatively low labor productivity to a great extent, led to the evolution of employment structure lags behind the industrial structure [9]. Secondly, secondary and tertiary industry absorbing ability and status of labor is not matching. Solve the coordinated development of economy in China; the most important thing is to solve the lag of employment structure. Therefore, the paper puts forward some countermeasures to promote the labor transfer and expanding employment.

Promote the transfer of labor force; it is first necessary for rural infrastructure investment. Reasons for the low productivity of agricultural labor in the final analysis is due to unequal rural and urban infrastructure in education, health, employment conditions, social security and so on, leading to the presence of the majority of farmers and citizens in the information, culture and quality, income, and many other aspects of the larger gap. Secondly, to increase investment in human resources in rural areas, improve the overall quality of education of farmers, establish and improve labor market information sharing platform to facilitate the promotion of the transfer of the labor force.

Expanding employment, the key is to improve the capacity to absorb secondary and tertiary industry's workforce. At the present stage in China, the second industry in the whole economic growth is still mainly supported role. For this reason, while vigorously developing technology-intensive, capital-intensive industry, still need proper development-intensive industries to absorb a lot of labor to promote the stable development of the employment structure. From the international development trend, the tertiary industry accounted for in a dominant position in the national economy. Although in tertiary industry, there has been an upward trend in China, there is still much room for development. To this end, vigorously develop the tertiary industry, we must break the administrative monopoly service areas, so that more capital inflows, diversification of market competition, thus providing a large number of jobs, thereby increasing the level of employment and the promotion of industry structure and employment structure and coordinated development.

5 NOTES

The special fund project of key discipline construction of Shaanxi Province"Management Science and Engineering"(E08001), Project of the special funds for key research base construction of philosophy and Social Science in Shaanxi higher education institutions (DA08046), Featured discipline of philosophy and Social Science of Shaanxi higher education institutions "Shanxi Province Construction Economic and Management"(E08003), Real Estate Economic Technology and Management of Shaanxi Province (E08005),Ministry of education of humanities and social science research project planning funds" Under the restriction of resources and environment evaluation of the rationalization of industrial structure and adjust the policy research in China"(13YJA790070),The soft science project in shaanxi province" Coordinated development of industry structure and resource utilization in Shanxi Province"(2014KRM14), National soft science project" Industry transfer perspective "the guanzhong - tianshui economic zone" equipment manufacturing industry cluster research collaborative growth path"(2013GXS4D152).

REFERENCES

[1] Fengqing Li, 2009," Chinese industrial structure and employment structure coordination measure research", Technology Management Research,No.11,pp112–114.

[2] Gongpu Yang, Dawei Xia, 2005," Modern Industrial Economics", Shanghai University of Finance and Economics Press,pp180–181.

[3] Haoxin Bian, 2011," The relationship between Chinese industrial structure and employment structure analysis", Western Economic Management Forum ,No.2,pp95–97.

[4] Dahong Chen, 2007," Correlation between Chinese industrial structure and employment structure", Industry and Technology Forum,No.3,pp26–28.

[5] Dongdong Liu, 2013," Development of non-balanced industrial structure and employment structure in Hebei Province", Oriental Culture,No.1,pp265–266.

[6] Yicai Zhang, Rongguo Yan, 2007," An Empirical Analysis of the industrial structure, the degree of industrialization and regional disparities", Xi'an University of Architecture and Technology Journal: Social Sciences, No.1,pp25–29.

[7] Su Dongshui,2000, Industrial Economics, Higher Education Press, pp301–302.

[8] Wensen Wang, 2007," Industrial structure similarity coefficient in the application of statistical analysis", Statistical science,No.10,pp49–50.

[9] Yunyong Wu, 2008," Again concerning employment structure and industrial structure upgrade in China are not synchronized", Liaodong college journal,No.6,pp51–54.

Management, Information and Educational Engineering – Liu, Sung & Yao (Eds)
© 2015 Taylor & Francis Group, London, ISBN: 978-1-138-02728-2

Social background study of the social policy in the British welfare state infancy

Hao Hua Zhao
College of Humanities, Harbin University, China

ABSTRACT: As the birthplace of western social policy, British experience in implementing social policies are drew by a lot of countries. This text sorts from the background of the social policy appearance in British welfare state infancy, analyzes the contents of the social policy in British welfare state infancy.

KEYWORDS: Welfare International; infancy; social policy.

Emergence of social policy is inseparable from the profound social background, must begin in-depth study of the social background of the country, in order to understand why the state implements such social policies in certain period. All kinds of social policy in the British welfare state infancy do not come out of thin air, which has a more profound social background support. Because the social policies in British welfare state infancy cover many aspects of social policy, it has helped the British post-war difficulties. Due to the effective implementation of welfare state policies, the British economy can recover in a short period of time prior to the station level, will alleviate poverty in the UK, which is important for narrowing the gap between the rich and the poor. This paper takes social policy in the British welfare state infancy as the basis for the study, analyzes the social background of the countries at that time, summarizes relevant content about the social policy of British welfare state infancy.

1 A BRIEF ACCOUNT OF THE EMERGENCE BACKGROUND OF THE SOCIAL POLICY IN BRITISH WELFARE STATE

British social policy legislation starts from *Poor Law* in 1601, the law implements the poor hospital model, uses this form to create a poor shelter to allow the able-bodied poor to labor. Because of extremely poor living conditions in Shelters, most people cannot accept this relief mode. During the 18th century, due to the growing number of the poor, the implemented poor shelter management model is more unable to meet the needs of development. In 1782, the British began to implement the *Gilbert Law*, adopted outside the hospital for poor relief mode for relief, but the institutional shelved because of the cost issue. In 1834, England adopted the *New Poor Law* by the Parliament, expecting to solve the problem of poor relief. The set the Act no longer took moral character as a benefit criterion, reformed the concept of "poor". But the *New Poor Law* still retains a very strong authoritarian parents ingredients, the starting point of the law is to better maintain social order. During the implementation period *Poor Law*, poverty was seen as the result of laziness, the so-called Poor was with punishment purpose. The principles of that period are " people who are willing to work to get work, people who do not want the work to be punished, people who cannot work to get the bread." Under the guidance of this principle, the government put together a joint relief with forced labor. The series ways had not resolved the British poor at the time, but also completely intensified class contradictions. The nature of the New, Old, Poor Law is completely different from that of charity medieval implemented by religious institutions in the Middle Ages, they are the first of its kind legislation based on social assistance personnel, and clear the responsibility of society to the poor through legislation, has a certain reference effect for the future implementation of social policy.

2 THE CONTENTS OF THE SOCIAL POLICY IN BRITISH WELFARE STATE INFANCY

2.1 *A comprehensive social security legislation*

British welfare state era social policy is to rely on social legislation to be implemented. After the war the British strongly require social reforms, Labor government follows the trend of the times, uses the Beveridge's report as a basis for the implementation of the relevant legislation. This is different legislation to 1946 implementation of the "National Insurance Act" and the "National Security Law" and 1948's "National Relief Act" is the most important. In the "National Health Care Act" clearly states that "outside the project to remove the fee provisions of this Act, all medical services free to all." The implementation of the Act, can guarantee everyone can fully enjoy medical insurance, amounting to achieve health equality modules in large part. "National Insurance Act," the insurance expanded into universal social insurance, while the unemployment insurance extended to various industries. After policyholders facing unemployment, cost of living can receive payment from the insurance summary, ensure subsistence needs of their families, to ensure that all British nationals under what circumstances will not fall below the national minimum standard of living. 1948 adopted the "National Relief Act" as a supplement to "National Insurance Act", the actual implementation of the Act is that for some reason can not pay the insurance amount due to different citizens also have the right to enjoy the national relief. These three laws in the UK, from the surface, it has created a modern social security system, insurance has covered every citizen, insurance programs also rose from the cradle to the grave state of development.

2.2 *The implementation of the hospital nationalization*

In the aspect of health care, British Attlee Government implements hospital nationalization, to ensure that more people can enjoy the opportunity of free treatment. Minister of Health Beavan circle was responsible for the work. He managed the hospital's institutional innovations envisaged under the Ministry of Health set up District Hospital Authority hospitals under the Hospital Authority set up management committees have been set for all institutions University School of Medicine at the center of each. National health care spending, mostly from tax countries. For the salaries of doctors, is greater than the use of the basic salary plus subsidy policy, subsidy cost is calculated based on the number of patient visits a doctor, played a large part in this mode and then the incentive effect. In addition, the implementation of the hospital when nationalization, nor restrict the opening of a private doctor, but must accept the supervision of a doctor opening committee. Less than two years, the British government puts the country's hospitals nationalized. The implementation of the system ensures that the Master lived in the UK, does not need to obtain insurance eligibility to accept free or low-cost, comprehensive medical services, ensuring that everyone can enjoy health care.

2.3 *The reform of the housing sector*

In the housing sector, the British Labor government obtains limited success, the person in charge of housing reform is greater than all, he implemented reforms in housing and no medical care so successful. Due to the destruction of the war for a long period, the country house with 1939 reduced by 70 million units, the situation Station anterior chamber shortage unknown. No one could have predicted, the number of the first three years after the war into the marriage will rise 11 percent over the previous three stations, newborn birth rate also increased 33%. People did not think that lifestyle changes will lead to a substantial increase in housing demand. All these factors have led the Labor government did not achieve 5 million housing units to be solved in the short term promise. However, when the government also has some achievements in 1945–1951, a total of 157,000 built simple houses built 806,000 formal housing, repair back to back houses 330,000, these achievements largely alleviated the war after the housing shortage problem, unfortunately, this did not solve the housing problems of Britain was fundamental. At this time, the British government introduced the Housing Act in 1946 and 1949, effectively abolished the relevant regulations in which local government builds only for the working class. This period also implemented a rent Management Act, which has great significance to safeguard the interests of private houses lessee.

3 CONCLUSION

Throughout the development process of the social policy in the British welfare state, the social policy implemented by the country in the infancy protects the need of most middle and lower class people to a certain extent, ensures that the problem of poverty in postwar Britain eased, plays an important role in maintaining social stability.

ACKNOWLEDGEMENT

Harbin University youth (MSc) scientific research fund project: " the social background of the bud social policy of the British welfare state". Project number: HUYF2013–021.

REFERENCES

[1] Zhao Haohua. *On The Social Policy of The British Welfare State Infancy* [J]. Knowledge economy, 2012, (23): 68–68.

[2] Wei Dake. *British Aging Society Can Influence Policy Development And Institutional Pension Service Transformation* [J]. City building, 2014, (5): 34–36.

[3] Robert Pinker [English], Liu Jitong. *"Citizenship" And The Theoretical Basis of The "Welfare State": TH Marshall Welfare Thought Summary* ① [J]. Social Welfare (Theory), 2013, (1): 8–16.

[4] Geng Wei. *Influence of Webers British Welfare State* [J]. Intellect, 2012, (12): 121.

Management, Information and Educational Engineering – Liu, Sung & Yao (Eds)
© 2015 Taylor & Francis Group, London, ISBN: 978-1-138-02728-2

Problems and countermeasures of tourism management major hotel practice management

Zhi Ping Bai
Hunan Vocational College for Nationalities, China

ABSTRACT: The hotel internship is a very important part in training tourism management majors, it helps students to summarize work experience, has great significance in the convergence of the completion of study and work, however, there still exist some concepts and institutional problems in tourism management major hotel practice management, only to solve these problems can better play the role of internships. This paper will start from the facing problems, propose solutions countermeasures from aspects of institutions, hotels and students.

KEYWORDS: Hotel internship; management; management strategies.

1 INTRODUCTION

The rapid development of modern tourism makes the increasing demand for tourism management graduates, cultivating tourism management talents, featuring good moral character and professional quality is the training objectives of the institutions, and the practice can enhance the application ability of theoretical knowledge in practice, obtain a wealth of experience to help students better go to work. The hotel also attracts a number of highly educated labor force reserves, attracts potential employees reserve personnel. However, all kinds of factors lead to the emergence of practice management problems in the actual course of the internship, influence the training effect, and how to solve these problems is particularly important.

2 PROBLEMS IN HOTEL PRACTICE MANAGEMENT OF TOURISM MANAGEMENT MAJOR

2.1 *The intern's own psychological perception problem*

Currently the hotel management students have not experienced heavy labor before hotel internships, they are not used to a strong command, cannot give themselves correct position after hotel internship, cannot conduct hard working, rebel against superior orders, increase impulsiveness especially in an unhappy work environment. Intern had a better vision for the future holding into the hotel internships, internships only to find work tedious, work intensity, and not on the use of complex knowledge, sometimes facing censure guests, enthusiasm will be gradually eliminated, even loss of interest in work. Interns can not establish a correct concept of the internship, cannot devote themselves to work, cannot really accumulate experience, cannot reach the goal of the internships.

2.2 *Management system problems of internship company for interns*

For internships, the establishment of rational and orderly management system relates to whether it will establish a stable and healthy internship environment, has great significance to the internships of interns. However, individual internships from their own interests, focusing on talent needs to consider their own, ignoring the real needs of the intern, and even the intern as a lower cost alternative, as lower wages receive cheap labor. Let interns do not have the technical content of any common physical work, would not achieve the role of the professional internship experience has helped accumulated. Because the employer's lack of management systems, management confusion, leading to a series of complex issues, interns placed in inconsistent with their professional sectors, internship copes official, late and leave early, cut corners in work hours and reduce the quality of service, lack of job enthusiasm.

2.3 *Guide and assessment problems of institutions*

Institutions not long intern internship management planning, such as the practice base is not careful, there is no long-term stable cooperation, no coordination with the hotel to develop training programs, between the interests of all hotels, schools, students less than a unified conflict, resulting in confusion hotel management. Institutions did not do a follow-up guidance for interns, not combined with the performance of students in internships to ask questions, solve problems, students will not improve the discussion. In addition, the institutions nor the intern various posts in the internship process, various aspects of the lack of monitoring and mentoring, internships, internships for students basically subject to the hotel, there is a problem it cannot be resolved in a timely manner. There is no uniform, standardized assessment after the end of the internship, so that does not reflect the true state of student internships, unable to form a feedback on teaching, not doing benign interaction between the institutions and hotels.

3 HOTEL INTERNSHIP MANAGEMENT COUNTERMEASURES OF TOURISM MANAGEMENT MAJOR

3.1 *Guide interns to establish a correct psychological concept*

Before practice does mobilization work, introduce in detail the specific circumstances of internships, working environment, internship precautions, so interns have a general understanding of the practice, prepare mentally to avoid the depressed mood due to the big gap to psychological expectations. Also, show internships importance for the future development of the students' work so that students attach importance to internship opportunities, a correct orientation to practice, to work after not pleasure-seeking, to have clear learning objectives, change their attitude and adapt to the hotel as soon as possible to work with enthusiasm into work in the face of difficulties, accept the challenge, hard to learn to work experience as the main purpose, straighten internship mentality. Interns also consciously change their ideas, good role reversal, to overcome the "unrealistic expectations" psychological consciousness from the grassroots level, do not over pursuit salary, accumulate experience, prepare for the follow-up work.

3.2 *Internship company to strengthen the management of internship work*

Develop a scientific interns management system is particularly important in the students' internship process, develop specific regulations, according to the student's majors and positions. According to the needs of students arrange jobs for them, designated qualified instructors guide students. Schools, hotels, student identification tripartite agreement binding develop long-term planning, the task should not arbitrarily change internships, internships stable as possible, to provide opportunities for qualified interns come into contact with the hotel management really practical power sector. The hotel can establish intern file management system, the intern's performance is recorded in the personal file, improve management systems.

3.3 *Institutions to strengthen the guidance and perfect appraisal system*

For institutions, they should do the whole track to guide the work of the interns, select high level and promising enterprises, sign long-term and stabile agreement, avoid disputes. Appropriate adjustments to the property assessment system to do the school assessment system corresponding convergence Colleges arrangement with the hotel should try to meet the syllabus institutions. Timely guidance to the student internship process problems, given the right guidance, but also on the intern after the internship overall performance reviews, review and exchange practical experience, the establishment of internships feedback system, standardized assessment system. Institutions jointly held a seminar with the hotel, providing a full range of evaluation for interns to combine the usual assessment, theoretical knowledge examination, assessment and other end of the internship, and improve multi-party assessment mechanism.

4 CONCLUSION

In summary, the hotel internships have great significance for tourism management majors, people must pay attention to practice and cherish the opportunity to practice, but the current management system still exists some problems to be solved, which requires colleges and universities, hotels and students to strengthen coordination and cooperation, change the concept of practice continuously, strengthen the management of practice, improve practice performance appraisal system, identify problems timely and deal with problems, improve their professional quality, delivery of qualified students to achieve the purpose of the internship.

REFERENCES

[1] Li Weili. Tourism Management Major Hotel Management Interns Exploration [J]. *Agriculture and Technology*, 2010,02: 115–119.

[2] Qin Bingzhen. Vocational Tourism Management Internship Major Hotel Management Problems and Countermeasures [J]. *Shandong Youth Administrative Cadres College*, 2010,03: 146–149.

[3] Yu Tao. Hotel Tourism Management Professional Practice Problems and Countermeasures [J]. *Henan*

Vocational Technical Teachers College (Vocational Education Edition), 2009,05: 129–130.

[4] Hua Li. Analysis of Hotel Internships Common Problems and Countermeasures of Tourism Management Students [J]. *public business*, 2009,16: 157–158.

[5] Liao Yanfang. Probe of Hotel Internships Problems and Countermeasures of Tourism Management Students [J]. *Consumer Guide*, 2008,18: 125–228.

Management, Information and Educational Engineering – Liu, Sung & Yao (Eds)
© 2015 Taylor & Francis Group, London, ISBN: 978-1-138-02728-2

Analysis of Eco-tourism development status and trend

Hui Ying Li & Xiao Lin Pei
Shijiazhuang Institute of Railway Technology, China

Ya Hui Liu
Shijiazhuang Engineering School, China

ABSTRACT: Eco-tourism has become the world's tourism hot spot, developing rapidly in the world, when it has entered the stage of maturity and stability, it sets off a fashion style even in the tourism industry. However, with its continuous development, a variety of issues also appears. Combined with China's actual situation, this paper understands the overall development status and growth trend of Eco-tourism, puts forward its own proposals in order to promote Eco-tourism to develop towards a more scientific direction.

KEYWORDS: Eco-tourism; development; status; trend.

1 INTRODUCTION

China's tourism industry is booming in recent years, and has become a green sunrise industry. It brings together ecological civilization, expands domestic demand, stimulates economic development and narrows the gap between rich and poor, etc., it can be said to be a very promising strategic industry. At present, the development of tourism demand gradually tends to feature and original ecological leisure trips and special tours.

2 ECO-TOURISM RELATED CONCEPTS

In 1983, the concept "Eco-tourism" was first proposed by Mexico expert Sherberrose, who was Special Adviser to the World Conservation Union. Later, until the early 20th century, with the proposal of the concept of sustainable development, Eco-tourism has been researched and practiced widely. All along, there are divergent views on the concept, but the goal of ecotourism gets a uniform recognition that maintaining natural resources and biodiversity, getting close and protecting nature to achieve sustainable development of tourism; secondly, it is to highlight ecological education for tourists, make travel managers pay more attention to environmental protection.

3 CHINA'S ECO-TOURISM DEVELOPMENT STATUS

China is the world's most richest country in tourism resource, Eco-tourism, mainly develops in nature reserves, ecological parks and other local places, especially the construction in Forest Park area. In 1982, Zhangjiajie National Forest Park was established, marking China has its first national forest park, layed a good foundation for the development of Eco-tourism. After that, Forest Park speed greatly improved, to 1997, our country already has 926 nature reserves, of which there were many mountains and wetlands. Establish a number of national parks relying on state-owned forest farms, service systems are basically formed, and have become the new tourist destination, promoting the development of Eco-efficiency and social benefits.

Recently, the development of China's tourism industry is focused on the construction of forest parks, to carry out special ecological tourism, making Forest Park become an important base for Eco-tourism construction. In addition, in order to adapt to the development of Eco-tourism, our many provinces have proposed the construction of the province's goal of Eco-tourism, Eco-tourism country started our demonstration area construction projects, trying a combination of language market through government guidance referrals, build a number of Eco-tourism boutique items. Although China has made some achievements in ecotourism development in so far, many problems are exposed in the development process, let's take a look at the problems of Eco-tourism.

4 THE PROBLEMS OF ECO-TOURISM

4.1 *Lack of proper awareness of Eco-tourism*

The concept of Eco-tourism is comparatively generalized in China, many people understand it one-sided and inaccurate, and are lack of proper awareness of Eco-tourism, including related knowledge restrictions, poor environmental awareness, etc.. Eco-tourism should be the premise of ecological protection, environmental education and literacy as the core, at the same time close to nature, appreciation of nature, to enhance awareness of protecting the natural environment. In our Eco-tourism in the region, there are still free to trample the grass, climbing pick flowers, and other uncivilized behavior has seriously affected the normal growth of Eco-tourism area plant updates. In addition, some tourists do not hesitate to consume ecological resources to reap short-term benefits at the expense of causing environmental damage and waste of resources. There are some guides who are lack of knowledge of ecological protection, failing to effectively transfer ecological protection awareness to tourists. These have become resistance to the development of ecological tourism.

4.2 *Overdevelopment and lack of planning*

There should be close and detailed planning before the development of tourism resources and tourism, but many local governments have neglected this point, make quick success in development, lack systematic planning and develop in blind and extensive approach, resulting in a lot of unnecessary waste of resources. Our protected areas are divided into three core areas, buffer zones and experiment, of which only the experimental area can be tourism activities. And according to surveys, some nature reserves even carry out tourism activities in the core area, seriously violate the relevant provisions, cause great stress to the environment of the tourist protection area.

4.3 *Serious guest overload problem*

International Eco-tourism area should strictly limit the number of tourists, but this did not make scientific planning in our country, there is no limit to the number of tourist visitors travel restrictions lead to too much too close, causing a variety of environmental, economic contradictions. Such as Jiuzhaigou, Zhangjiajie Nature Reserve, are overcrowded during holidays, Jiuzhaigou area this year is because an excessive number of visitors and generate disputes. Excessive number of visitors to protected areas is bound to cause excessive trampling, and it will produce more garbage, erosion original ecological protection zones, not only seriously affected the quality of tourism, while protecting valuable ecological areas is also a serious blow, and Eco-tourism seriously contrary to the original intention.

5 ANALYSIS OF ECO-TOURISM DEVELOPMENT TREND

Our Eco-tourism has just started, in conjunction with China's national conditions, at the same time learn from foreign experience, exploring the Eco-tourism road with Chinese characteristics.

5.1 *Improve the legal construction, focus on ecological protection*

Construction and improvement of laws and regulations is essential for sustainable development of Eco-tourism, Eco-tourism and to do so to develop rules to follow. Many foreign-developed architecture, Eco-tourism have developed a relatively strict rules, and set up a special body to oversee the implementation. For example, in 1916 the United States was founded on the Yellowstone National Park Act, the development of relevant laws, the park management into the legal track. This initiative is undoubtedly the most powerful protection of Yellowstone National Park, the practice in many countries to follow suit after the development of Eco-tourism protection laws related to the protection of national Eco-tourism zone. And yet our laws and regulations in this regard. We should proceed from the long-term interests, the development of relevant laws, the Eco-tourism activities in the legal system, and promote the healthy development of Eco-tourism.

5.2 *Popularize ecological knowledge, change people's attitudes*

Development and management of Eco-tourism is inseparable from the tourism development manager, the first of their ideological education is particularly important to strengthen publicity and education work, so that they recognize the essential meaning of Eco-tourism, to establish a correct concept of Eco-tourism, while taking measures prevent sabotage tourism environment behavior. Secondly, should the ecological education into national education programs and publicity through various forms of ecological knowledge, universal awareness of the ecological importance of environmental protection, improve environmental awareness of tourists. The third guide training should strengthen knowledge about Eco-tourism, Eco-tourism allowed to learn the knowledge, which will tour the process of passing it to tourists, and enhance tourists' environmental awareness.

5.3 *Strengthen control and coordination of ecotourism, environmental capacity*

As already introduced to the problem of overloading on tourists, ecotourism, environmental capacity issues cannot be ignored. We can fully understand and detailed investigation of environmental capacity of each Eco-tourism zone, and saturation through relevant media to tell the tourists that have occurred to change the tourism strategy choice other tourist places; Secondly, in the peak travel period can be taken to improve the tickets, etc. to reduce the cost of tourists; another can open up new Eco-tourism area to reduce the pressure of the original ecological zones; Furthermore ecological zones can partition management, such as an Eco-tourism zone is divided into distance zones, moderately open areas, normally open area and so on. These measures can be well controlled environmental capacity, and thus serve the purpose of protecting the ecological area tourism environment.

6 CONCLUSION

China's tourism industry has been developing rapidly in recent years, although has made some achievements, the Eco-tourism industry is still in its infancy, people are still lack of scientific knowledge for Eco-tourism. Now, with the development of Eco-tourism, some problems are increasingly apparent. We should address these issues, base on China's national conditions, learn from foreign experience, try to find solutions to problems, step out of Eco-tourism misunderstanding. As China's economic development, the development of Eco-tourism with Chinese characteristics will become an inevitable trend.

REFERENCES

[1] Chen Zhongxiao, Peng Jian. *Connotation Analysis of Ecotourism* [J]. Guilin Tourism College, 2001 (01).

[2] Chen Zhikui. *Research of Ecotourism Resource Utilization Strategies* [D]. Chinese Geology University (Beijing), 2012.

[3] Zhang Guangsheng, Wang Yan. *Nature Reserve Ecological Education and Its Implementation* [J]. Ecological economy. 2002 (12).

Management, Information and Educational Engineering – Liu, Sung & Yao (Eds)
© 2015 Taylor & Francis Group, London, ISBN: 978-1-138-02728-2

An analysis on the necessity of business and management education for design students

Yan Li, Xin Luan & Hua Xin Chen
China

ABSTRACT: Nowadays, business and management abilities have become the basic skills required by professional designers, business and management knowledge has also become an important part of design class student education. In this paper, the necessity for design class students to carry on the business and management education is discussed, combined with not only the present situation of the domestic and foreign design class teaching, but also the need of today's enterprise.

KEYWORDS: Design; business and management education; person cultivation.

1 INTRODUCTION

Nowadays, design is sharing an increasingly close relationship with business activities, which requires outstanding business managers to be capable of design thinking and designers to have business and management ideas. Consequently, it is particularly important for design students to develop business and management skills. Nonetheless, there has been a tendency of emphasizing on design practices and ignoring related theories in China's design education field. The economic, social and cultural elements in design courses are not being paid enough attention to, while the cultivation of students' managerial ability, operational ability, leadership and necessary knowledge of laws are not properly valued and popularized. Designers nurtured under current education pattern work as highly-skilled product-development professionals, rather than managers of development projects. Therefore, it is of great importance to educate design students with business and management knowledge.

2 PRESENT SITUATION ANALYSIS OF DESIGN CLASS STUDENTS AT HOME AND ABROAD ON BUSINESS AND MANAGEMENT EDUCATION

2.1 *Business and management education development situation of foreign design class students*

Some UK, USA, Japan and other foreign universities have realized the importance of business and management education in early times, and have introduced a teaching plan into the design business and management courses. Britain was the first country to carry design business and management education in the world. First, it opened design courses in the MBA program to introduce the management and design class for design students; then opened management courses in design majors, so that students can understand the design category management. Nowadays designing business and management education in the UK is relatively mature, many British universities opened design management courses. Higher design colleges and universities in the United States have learned education method from the management of the professional, then established the design of business and management education system from a multidisciplinary perspective. This kind of brand-new education idea of multidisciplinary also inspired many countries to rethink of design education. The design business and management education in the United States have the following three characteristics; starting from the management professionals bidirectional education for experienced operators and managers and professional designers; setting a large number of management courses, paying attention to cultivate the professional designer's ability to lead and grasp of the business and legal language. Japan's design business and management education started in the design colleges and universities, referencing the method of design colleges in U.S, achieving the professional designers and managers of the bidirectional education model. Japan did not take a lot to set the management mode of teaching subjects, but take how to use design in business management as the core, set up multi-disciplinary subjects to complement management knowledge. This also explains the design of higher education in Japan is the disciplines as the origin, explore new areas of design and related disciplines connected so as to nurture new design talents.

2.2 *The development of Chinese design class students of business and management education*

Professional design classes setting up in China, which has more than 1000 institution of higher learning. The number is huge. But China's current design education mostly born out of traditional art education or engineering education (such as industrial design). There has been a long history of design education in our country which has paid more attention to the design practicing skills than design theories, especially the economic characteristics of design have been ignored for a long time, domestic and international economic environment has undergone great changes, the design industry is also closely moving along , designers encounter many problems in the development , meet a lot of confusion. But recognize that the relevant sectors: good design is not just a design, but also a business and management job. It's an inevitable trend to combine design with enterprise management, Enterprises (especially design-oriented companies) come to realize that only they combine the design with planning, technology production and circulation together, they can constitute a unified development system in order to promote continuous innovation, therefore , the demand for design and business management personnel are increasingly urgently.

3 THE IMPORTANCE OF BUSINESS AND MANAGEMENT EDUCATION

3.1 *Enterprises need designers with business and management skills*

As a result of design's increasingly close ties to business, design targets are no longer confined to product design or its relevant fields. A wide range of business activities have been involved in design targets, including production, sales, exterior design, organizational management, market development and product development, i.e. the enterprise is integrally planned as a design target. To ensure a coordinated design program, the designer is required to be acquainted with knowledge concerning business operations, project procedure, risk control, and team management, beyond essential design skills. Therefore, business and management abilities have been an important tool in the management of an enterprise.

During its global expansion, Starbucks has developed a suite of integrated design strategies to keep design themes consistent in brand value, and to convey the consistency in service experience through designs. For instance, Starbucks's "Global Creative Processway" (a design procedure) (Figure 1) does a good job in illustrating the significant role that business and management knowledge plays in a design procedure. As shown in the picture, "Global Creative Processway" refers to a suite of creative design procedure, incorporating the role of designers, writers, clients and managers. In this way, Starbucks' products fit in the user's intention, technology and the environment and serve to promote the corporate image, instead of being simple products. As According to IBM's general manager, a good design means a successful enterprise. Moreover, a high level of business and management offers guarantee for the success of an enterprise.

At present, most designers in Chinese enterprises tend to focus on design, lacking enough consideration for management factors, which will surely lead to parochialism and limitations. During the design process, neglect of either design or management will cause problems for production and influence future development and design of the product. Therefore, enterprises need designers who possess business and management competence. In addition, unlike designers, enterprise leaders who only concentrate on market efficiency and economic benefits, often adopt conservative, obsolete techniques and materials to reduce costs, which goes against the autistics and innovation of designs. Some of them even consider it unnecessary to make efforts on "appearance". Enterprises lack vitality and creativeness due to the unawareness of the significance of design. In contrast, Apple's unwavering success can be largely attributed to Jobs' personal talents. He goes beyond an enterprise manager to a great designer, bringing management and design into perfect integration. Therefore, in the operation of enterprises, design and management are inseparable from each other[1].

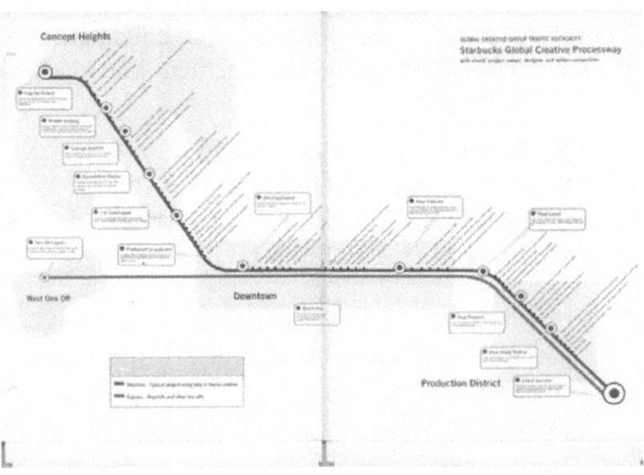

Figure 1. Starbucks's "global creative processway".

3.2 *It is necessary for universities to train design students in business and management skills*

Nowadays, the market demands for all-rounder designers who are capable of market researches, layout, planning and design, besides their design conception and performance. Therefore, it is necessary for universities to add business and management elements into design education.

Collect and organize educational materials and build a database of cases. Apart from traditional instructions, discussions and case studies, foreign universities also emphasize on student-oriented discussions in the course of design business and management. Students give presentations of research reports on design management, after which more students are involved in the discussion. Most cases are based on American design works. As those cases employed in the study happened in the past, it is difficult for students who have not experienced them firsthand to get an in-depth understanding. Hence multiple teaching methods are utilized, including role play, cross project and design auditing. Chinese students are used to receive knowledge passively, rather than to take the initiative to acquire new knowledge. Universities should take the opportunity of students' coursework every year to investigate renowned enterprises at home and abroad in terms of design strategies, corporate image management, operation procedure of design projects, risk management of design innovation and management mode of brand operation. Based on those investigations, a report should be published and made into archives, serving as learning materials for future classes. Meanwhile, students' ability of expression and communication will also be improved through the process of publishing and communication.

Reinforce university-enterprise cooperation. Universities may get connected with renowned enterprises and design companies such as Haier, Lenovo, and New Plan, to organize a series of activities such as site visits and lectures given by corporate professionals, through which students broaden their horizons and get acquainted with the present situation of design business and management in domestic enterprises.

Arrange different course contents for different levels of students. According to the focus of public elective courses as well as the features of undergraduate and graduate students of design major, course contents should be customized to meet the needs of different levels of students, in order to benefit students in a real sense. Up to now, a number of Chinese universities have tried or planned to offer the course of design and business management. Central Academy of Fine Art, Zhejiang University, Shanghai Jiaotong University, Dalian Polytechnic University and Shandong University of Art & Design have enrolled postgraduates of design management. Tsinghua University and Shanghai Jiaotong University organize conferences on design business and management every year to discuss new concepts, patterns and approaches in the design field, with participants of entrepreneur, educators, scholars, managerial personnel, researchers and design teachers at home and abroad. Such activities have contributed to academic exchanges as well as corporate propaganda.

4 CONCLUSION

Design class students' business and management ability is closely connected with the enterprise. To educate design class students in business and management should be combined with China's education system, grasp the teaching purpose and direction, perfect its own structure system, strengthen them the principle of combining theory with practice, and based on the market, focus on the development of business and management industry design, cultivating qualified design business and management talents for the society.

REFERENCE

[1] WENKE KANG,XIN CUI Discuss the Importance of Design Management to the Enterprise[J] Journal of Northwestern Polytechnical University, 2001, 21(1).

Management, Information and Educational Engineering – Liu, Sung & Yao (Eds)
 ISBN: 978-1-138-02728-2

The impact of common promotion versus novel promotion on consumer response

Hui Zeng
School of Economics and Management, Southwest JiaoTong University, Chengdu, Sichuan Province, China

ABSTRACT: Based on the promotion framing theory, we examine the effect of a common promotion (e.g., fifty percent off) versus a novel promotion (e.g., one Yuan worth two Yuan) on consumer response in China. The empirical results showed that the former promotion results in higher perceived value, perceived quality, purchase intention and brand loyalty than the latter one. Product involvement moderates the effect of promotion framing on consumer response by showing that it is significant at low product involvement, but the effect is eliminated at high product involvement.

KEYWORDS: Promotion Framing; Consumer Response; Product Involvement.

1 INTRODUCTION

According to framing effects, the same information with different presentation could make consumers produce different understanding (Kühberger, 1995). Based on that, Kim and Kramer (2006) studied the effect of the novel (pay 80%) versus common (get 20% off) discount presentation to consumers in U.S, demonstrating that the novel promotion resulted in higher perceived saving and higher purchase intention than the common promotion. Meanwhile, Kramer and Kim (2007) also compared fluent promotion (gain-frame coupons) and novel promotion (loss-frame coupons) in U.S, differently, showing that the fluent promotion was better than novel promotion on deal perceptions. According to those literatures, many scholars focus on the effect of promotion framing on consumers' perceptions and purchase intention, while the conclusions are not coincident. Therefore, it is worth further to explore relevant studies. "Fifty percent off" promotion is very commonly used in China, and "one Yuan worth two Yuan" promotion is novel in China. However, little research studied the effect of them on consumers' responses in China. From the literature review, we also find that many studies pay attention to the consumer responses of perceived value and purchase intention (e.g. Kim and Kramer, 2006; Kramer and Kim, 2007), little research focuses on consumer perceived quality and brand loyalty in a sales promotion context. Therefore, this study will examine the impact of the common promotion (50% off) versus novel promotion (one Yuan worth two Yuan) on consumers' responses, including perceived quality, perceived value and purchase intention and brand loyalty in China, to fill these research gaps with considering the boundary effect of product involvement.

2 HYPOTHESES

2.1 *The effect of promotion framing on consumer responses*

According to previous research, there are two possibilities of consumer responses to sales promotion framing (e.g. Kim and Kramer, 2006; Kramer and Kim, 2007). On the one hand, consumers prefer to novel promotion because of the depth of information processing for the novel stimulus (e.g., Aakerand Williams, 1998). On the other hand, consumers like common promotion because of processing fluency which makes the message easier to process (e.g., Lee and Aaker, 2004).

In this paper, we propose that if the perceived processing fluency drives the effects of promotion framing on consumer response, "fifty percent off" promotion is better than "one Yuan worth two Yuan". Meanwhile, if the perceived novel drives the effect, "one Yuan worth two Yuan" promotion will be better than "fifty percent off". Thus, we draw the hypotheses:

H1: Perceived fluency generates better consumer responses of "fifty percent off" promotion than "one Yuan worth two Yuan".

H2: Perceived novelty generates better consumer responses of "one Yuan worth two Yuan" promotion than "fifty percent off".

2.2 *The moderating effect of product involvement*

Product involvement is defined as the continuous potential value owned by a product or service (Vaughn, 1986). Different levels of product involvement can affect the consumer's level of information processing (Cacioppo and Petty, 1979). When consumers face high involvement product, they will positively seek more information of the product. But when consumers face low involvement product, they are more passive to receive the product information (Vaughn, 1986). Therefore, different levels of product involvement will lead to different effects of promotional frame on consumers. Thus, we propose the following hypothesis:

H3: Product involvement moderates the effect of promotion framing on consumer response.

3 METHODOLOGY

3.1 *Constructs and measurement*

The study uses a 2 (promotional frame: fifty percent off vs. one Yuan worth two Yuan) ×2 (product involvement: high vs. low) between-subjects design, forming four different questionnaires. Clothing represents a high involvement product, while washing product represents low involvement product. The independent variables, perceived value, perceived quality, purchase intention and brand loyalty, are adopted by Minghua Jiang (2003). The manipulated variables of processing fluency and novelty are modified from Lee and Aaker (2004) and Holbrook (1981) respectively. The measurement of product involvement is proposed by Vaughn (1986). All items are measured on seven-point Likert-type scales ranging from 1 to 7 ("1" represent strongly disagree, "7" represent strongly agree). Demographic variables were presented in the last section of the questionnaire. Collection of data took place at three universities in the city of Chengdu, China. The total respondents were 250, but after discarding of non-useful and incomplete questionnaires, the valid respondents were 201.

3.2 *Manipulation check*

The manipulation check results show that the mean score of novelty (M_1=3. 359) in "one Yuan worth two Yuan" promotion is significantly higher than that in "50% off" promotion (M_2=2. 904), meanwhile, the mean score of processing fluency (M_1=4. 572) in "one Yuan worth two Yuan" promotion is significantly lower than that in "50% off" promotion (M_2=5. 021). The product involvement mean score of clothing (M_1=4.703) is significant higher than the score of washing product (M_2=3.100). Above all, our manipulation checks are successful.

3.3 *Tests of hypotheses*

We use SPSS16.0 to analyze our collected data. Firstly, we do the MANOVA analysis to examine the impact of promotional frame on perceived value, perceived quality, purchase intentions and brand loyalty. The results are shown in Table 1. indicating that the promotional frame has significant impact on the consumer perceived value (F=4.458, P<0.05), perceived quality (F=4.626, P<0.05), purchase intentions (F=4.692, P<0.05) and brand loyalty (F=4.466, P<0.05).

Table 1. The results of MANOVA analysis: Promotional frame on consumer response.

Source	Dependent Variable	Type III Sum of Squares	df	Mean Square	F	Sig.
Promotion frame	Perceived value	3.758	1	3.758	4.458	.036
	Perceived quality	3.771	1	3.771	4.626	.033
	Purchase intention	5.028	1	5.028	4.692	.031
	Brand loyalty	3.345	1	3.345	4.466	.036

From Figure 1. we can see that the mean scores of perceived value, perceived quality, purchase intentions and brand loyalty in 50% off promotion are all significantly higher than those in "one Yuan worth two Yuan" promotion.

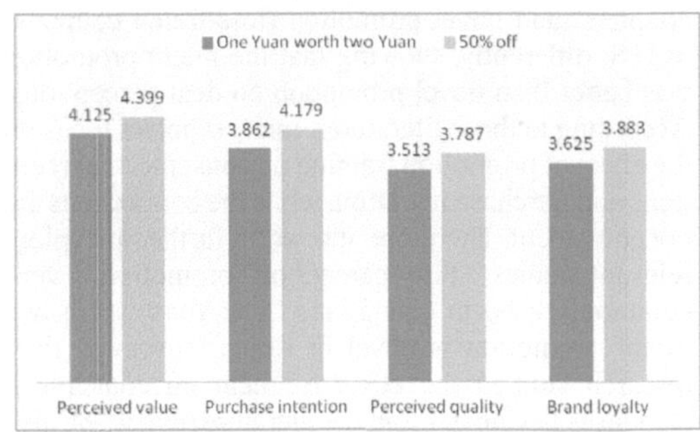

Figure 1. The mean scores of consumer response in two promotions.

Then, we do the MANOVA analysis of processing fluency to test H1. The results revealed that processing fluency has an impact on perceived value (F=4, 370, P<0.05), perceived quality (F=2. 465, P<0.05), purchase intentions (F=2. 544, P<0.05) and brand loyalty (F=2. 279, P<0.05) in Table 2. Thus, H1 is supported.

Table 2. The results of MANOVA analysis: Processing fluency.

Source	Dependent Variable	Type III Sum of Squares	df	Mean Square	F	Sig.
Processing fluency	Perceived value	34.773	11	3.161	4.370	.000
	Perceived quality	20.831	11	1.894	2.465	.007
	Purchase intention	28.153	11	2.559	2.544	.005
	Brand loyalty	17.844	11	1.622	2.279	.012

To test H2, we also do the MANOVA analysis. The results showed that novel promotion does not have the significant impact on perceived value (F=0.983, P>0.05), perceived quality (F=0.910, P>0.05), purchase intentions (F=0.638, P>0.05) and brand loyalty (F=0.543, P>0.05) in Table 3. Thus, H2 is not supported.

Table 3. The results of MANOVA analysis: Novelty.

Source	Dependent Variable	Type III Sum of Squares	df	Mean Square	F	Sig.
Novelty	Perceived value	12.655	15	.844	.983	.475
	Perceived quality	11.409	15	.761	.910	.554
	Purchase intention	10.737	15	.716	.638	.841
	Brand loyalty	6.427	15	.428	.543	.913

In order to test H3, we do the MNOVA analysis. The results showed that the moderating effects are supported in perceived value (F=4.708, P<0.05), perceived quality (F=6.995, P<0.05) and purchase intention (F=4.117, P<0.05), but failed in brand loyalty (F=0.023, P>0.05). Thus, H3 partly supported. From our analysis results, we can get the conclusions that framing effects will be obtained at low product involvement, but the effect is eliminated at high product involvement. Note that the effect of promotion framing on brand loyalty is not moderated by product involvement.

4 GENERAL DISCUSSION

We explored the effects of the common promotion versus novel promotion on consumer response by investigating 201 university students in China. Our study mainly referred to the study of Kim and Kramer (2006) and Kramer and Kim (2007), but supplemented and extended their study findings in China.

According to our data analysis, we draw conclusions that the common promotion presentation (e.g. 50% off) results in higher perceived value, perceived quality, purchase intention and brand loyalty than the novel promotion presentation (e.g. One Yuan worth two Yuan). Product involvement moderates the effect of promotion framing on consumer responses of perceived value, perceived quality and purchase intention: it is eliminated at high product involvement, and the effect is significant at low product involvement. The findings of our paper contribute to the promotion frame study about the effects of promotional framing on consumers' responses, and those findings also have some practical implications to companies. Companies should pay attention to the processing fluency character when design the type of sales promotions. When companies face those two sales promotions of 50% off and one Yuan worth two Yuan, the former one is suggested. Companies also should consider their product character.

There are some limitations existing in our study. First, we just compare two sales promotions which stand for processing fluency and novel promotion in China. We should extend this study to other countries to verify our study findings. Second, we do not analysis relationships between perceived value, perceived quality, purchase intention and brand loyalty. Thirdly, we just do an experimental study with subjects who are university students, therefore we should do the field study and survey real consumers in some shopping malls.

REFERENCES

[1] Aaker, J.L. and Williams, P. 1998. Empathy versus pride: the influence of emotional appeals across cultures, Journal of Consumer Research, 25 (2): 241–261.

[2] Cacioppo, J. T., and Petty, R. E. 1979. Effects of message repetition and position on cognitive response, recall, and persuasion. Journal of personality and Social Psychology, 37(1), 97.

[3] Holbrook, M.B. 1981. Integrating compositional and decompositional analyses to represent the intervening role of perceptions in evaluative judgments, Journal of Marketing Research, 18(4): 13–26.

[4] JIANG M H ., DONG W M. 2003. The Empirical Study of the Impact of Discount Amount of Price Promotion on Brand Equity. Journal of Peking University (Philosophy and Social Sciences), 40(5): 48–56. In Chinese.

[5] Kim, H. M, Kramer, T. 2006. "Pay 80%" versus "get 20% off": The effect of novel discount presentation on consumers' deal perceptions. Marketing Letters, 17: 311–321.

[6] Kramer, T., Kim, H. M. 2007. Processing fluency versus novelty effects in deal perceptions [J]. Journal of Product & Brand Management, 16(2): 142–147.

[7] Kühberger, A. 1995. The framing of decisions: A new look at old problems. Organizational Behavior and Human Decision Processes, 62(2): 230–240.
[8] Lee, A.Y. and Aaker, J.L. 2004. "Bringing the frame into focus: the influence of regulatory fit on processing fluency and persuasion", Journal of Personality and Social Psychology, 86(2): 205–218.
[9] Vaughn, R. 1986. How advertising works: A planning model revisited. Journal of Advertising Research, 26(1): 57–63.

Management, Information and Educational Engineering – Liu, Sung & Yao (Eds)
© 2015 Taylor & Francis Group, London, ISBN: 978-1-138-02728-2

Analysis and countermeasure on employment mentality of contemporary undergraduates

K. Wang & Y.M. Cheng
College of Electrical & Information Engineering, Beihua University, Jilin City, Jilin Province, China

ABSTRACT: During two sessions in 2014, the proposals which involve employment and social security reached 200 cases, among which the employment problem for undergraduates was again the social focus. In 2014, the amount of graduating students in China broke through 7,000,000, and university students' employment has already become a major problem concerning national and social stability, as well as impacted vital interests of thousands of households. Among the graduates, the employment mentality of all forms influences their selection. This paper mainly analyzed the job hunting mentality from undergraduates, hoping to help them know psychological misunderstandings of their own and plan their career in a right way.

KEYWORDS: undergraduate, employment mentality.

1 EMPLOYMENT MENTALITY OF CONTEMPORARY UNDERGRADUATES

In 2014, the amount of graduating students in China broke through 7000,000, which was called "the most difficult year for employment". With the intense competitive status in an undergraduate employment market day by day, the undergraduates' employment mentality has changed a lot. Taking control of their own correctly and walking out psychological misunderstanding is the first step for undergraduates to march toward success in getting a job. Through the internet investigation over the employment mentality to 30,000 undergraduates in our province, the following psychological problems to undergraduates during job hunting are prominent.

1.1 *Psychological misunderstanding caused by increasing employment pressure, mental conflict increase*

Facing the intense competition, the undergraduates are easy to generate psychological misunderstandings. They compare mutually, lack their own thoughts, or divorce from reality, cannot adapt to society. Their emotions are expressed as anxiety, loss, fear, etc. Some of them are in prominent mental conflict, which express as possessing great ideal, but cannot face reality correctly, focusing on realizing value of life while lacking of mental preparation for hard work, the yearning for fair competition however expecting themselves become privileged, wishing to able to work independently but desiring others to help.

1.2 *Lacking occupational knowledge cause vague, unstable employment object and direction*

At present, due to the colleges and universities do personal training according to their own mode, which makes serious disjoint between major, course setting and market requirement, the undergraduates' major does not fulfill the needs of the job, deficient in occupational knowledge and skills. The Contemporary Undergraduates are generally lacking of "employability", which leads to "the undergraduates don't know how to hunt job". This circumstance is expressed as seeking quick success and instant benefits, lacking of lofty idealism, fickle, changing loyalty frequently, uncertain occupational object for a long time.

1.3 *The deviation in self-assessment by the generation after 90s leads inadequate employment expection*

Due to the particularities in terms of age structure, cultural quality and group consciousness among undergraduates, unique mental structure and personality are formed. They always overvalue themselves, and hope to seek out the place of self-realization, play a major role in social stage as soon as possible. While the reality is always making them greatly disappointed. Some of them even lack of confidence, they blame everyone and everything but themselves, and are not good at competing.

1.4 *Lacking of employment skills leads insufficient to employment ability*

Many undergraduates lack of necessary common sense in occupation and skills in employment, they feel shy during self-recommendation, and are short of the ability to adapt and self-confidence. The insufficient skills in preparation of documentation, examination and mentality lead to self-promotion that cannot reach to expected objective, which makes them incapable of leaving a favorable first impression to the employer.

2 SEVERAL SUGGESTIONS FOR IMPROVING MENTALITY OF JOB HUNTING FROM CONTEMPORARY UNDERGRADUATES

2.1 *Changing the traditional employment idea is the priority for ideological educating undergraduates in employment*

Chinese higher education has already transformed to popular education from elite education, while the undergraduates turn to the independent choosing profession from the guarantee job assignment in the past, this is the necessity during historical development. Under the influence by traditional views of "those who work with their brains will rule", the undergraduates would not prefer "labor" work. So "going to the grass roots, going to the places needed by motherland" is always turning out an empty slogan. In addition, parents influence their children a lot; they would rather their children be "neet", than let their children have a rough time, this phenomenon makes many positions without employees. This should be solved by guiding undergraduates to combine social responsibility with individual demand, classifying recognition of becoming a useful person into the track of overall development demand in society, and strengthening the guidance in terms of professional ideal education, professional moral education, unemployment policy education and value orientation in choosing a job. That will be enabling the undergraduates to set up world outlook, view of life and employment which are abreast of the times. Then not only they could have a general situation of collectivism, but also the mind of contribution to country, society and individuals; they will not only see national rules about unemployment policy, but also development tendency in society; so they will not only would pursue self value realization, but also strive for contributions for the society. In Tao, ying, 2012. Education guides the undergraduates to handle well relation among country, individual and collective during choosing a job, which should be the emphasis in ideological education during occupational guidance. This is not merely the demand of national and social development; this is also a demand and foundation for undergraduate to correctly cognize, complies with society and stick to good mind.

2.2 *Strengthen occupational guidance, complete career planning of personalized*

The college students are youth pioneers with ideals, ambitions, innovative spirit and aggression. So the employment guidance should not be all in the same key, it needs the tutor to do career planning guidance of personalized to students. In that way the students could set up career consciousness of personalized, correct thinking way and prepare actively combined with their characteristics, do well in job hunting and employment preparation from thought, idea and action, be confident when facing social competition and challenge. One successful career planning will not only offer undergraduates to know their personality characteristics, interests and specialties, but also help them to have a clear view over social reality and development tendency, all of which could assist them to reexamine themselves, and reasonable position themselves combined with social practice. Wang, Zhen (2011) pointed out that it is meaningful and popular on how to make personalized guidance over career planning combined with students' self-development and personal characteristics. The 18–21 years old undergraduates are in a period of drastic psychological changes, they need to face a series of issues like studying, making friends, job selection and employment, so it will not be beneficial to them unless these issues were being solved as well. While the career planning of personalized could assist them to overcome psychological weakness, cognize themselves bitterly, establish the stable self-concept. According to the comparison and analysis to one's advantages, disadvantages and practical situations, one could determine his career objective which fits for himself, choose the development route of the professional career of his own, formulate educational training plan suitable for himself and promote employment competitiveness, thus to realize professional ideal and life goal.

2.3 *Listing and numbering taking an active part in social practice, adapting transform of social roles ASAP*

For this period, the right way is not to regard oneself as an undergraduate, but set strict demands on themselves, taking the standard of occupational cognition, devote one's own work, open-minded study by the aim of boosting working competency, fostering unique opinion which used for challenging work, annealing oneself and enhancing oneself. During the time at school, the undergraduates should seek opportunity to attend social practice, enrich the experience. In this way after they are on the job; they could be a social man with an independent qualification which is accepted by society, they could also adapt, transform of professional roles, be taking charge of a department alone and assume certain responsibility ASAP.

2.4 *Strengthening vocational skills, quality-oriented education, promoting psychological diathesis on employment*

After professional learning, the undergraduates should be able to master basic theories and methods in his subject and major in detail and systematic. By way of these theories and methods to guide practice is professional skill, which includes manipulative ability, analyzing problem ability, solving problem ability, learning ability, innovation ability, etc. Feng, Yifu (2011) stated that the professional skills are key elements for employers to choose undergraduates. However, some of them have not mastered professional skills solidly, they even lack of the most basic common sense in their major. Therefore, they will be suffering from difficulty and frustration during employment. The undergraduates should start with enhancing employability skills, reinforcing occupational qualities, education and cultivating good psychological quality over employment. The excellent psychological quality is not cultivated overnight. We should not focus the emphasis on higher education step; we should start employment, education and psychological education from middle school, even primary school to train their excellent quality.

2.5 *Actively launching consultation work on employment psychology, cultivating healthy employment psychology*

Referring to the psychological consultation, it aims to offer a service for undergraduates when they are meeting psychological confusion in employment. It is not only an important content in employment guidance, but also a key part which composed psychological counseling work in colleges. To develop the psychological consultation work on employment and strengthen psychological guidance will conduce to: cultivate undergraduates' healthy psychology, enhance mental health status and maintain a favorable mentality over employment; overcome psychological disorder, exclude psychological crisis; get out of trouble, walk out mental misunderstanding during the job selection; objectively cognize difficulties they faced, build confidence. The final objective is to help undergraduates obtain a job successfully. Wang, Haiyan (2011) and Cui, Jiaping (2007) thought that the colleges should guide the undergraduates to: know and evaluate them correctly, grasp the expectation value in choosing a job in a right way, give full play to their advantages, adopt good points and avoid shortcomings, choose an occupation which suitable for themselves to develop talent and aspirations. The colleges should also guide the undergraduates to actively attend the competition, do well in mental preparation, adjust mentality and goal, realize own value by "independent choosing profession". In addition, the colleges should help the undergraduates to: enhance self-adjustment capacity and psychological enduring capacity, learn how to adjust personal emotion, to treat frustration in a right way and keep optimistic. All of the above will exclude all kinds of unhealthy mentality, avoid psychological conflict, keep a positive mindset and win in employment competition.

In a word, only by joint efforts of society, colleges and universities, students could effectively help undergraduates adjust employment mentality, walk out employment psychology and get a job successfully.

ACKNOWLEDGEMENT

Thanks are due to corresponding author Yanming Cheng for assistance with the surveys and valuable discussion.

REFERENCES

Cui, Jiaping, 2007. Analysis on Psychological Disorder of Employment among College Graduates and Countermeasure Research . *Leading Edge9*:107–109.

Feng, Yifu, 2011. Try Discussion on Main Experience and Practice of Enhancing Undergraduates Employability. *Business Condition.*6:46–47.

Tao, ying, 2012.Thoughts on Education and Guidance over Employment View to College Students. *Guide of Scitech Magazine*15. 68–69.

Wang, Haiyan, 2011. Brief Discussion on Psychological Problems and Countermeasures of Employment among Undergraduates . *Managerialist1*:105–106.

Wang, zhen, 2011.Personalized Guidance on Career Planning during Talent Cultivation. *Vocational & Technical Education Forum*29:55–56.

Management, Information and Educational Engineering – Liu, Sung & Yao (Eds)
© 2015 Taylor & Francis Group, London, ISBN: 978-1-138-02728-2

The application of industrial and engineering technology in business management overall optimization

Xiang Jun Ji & Li Hong Ma
School of Economics and Management, Hebei University of Science and Technology, Shijiazhuang Hebei, China

ABSTRACT: In the process of rapid economic development, competitiveness among Chinese enterprises is also increasing. It becomes a global issue that our current businesses face, that is, how to achieve business management overall effective optimization and thereby expand market competitiveness. A relatively effective and straightforward solution to this problem is to increasingly promote the development and application of industrial engineering. Using scientific and effective industrial engineering technology rationally analyses and studies enterprises daily production and management activities to find an appropriate solution corresponding to the problem, enabling enterprises to achieve overall optimization of management on the situation as soon as possible. This paper focuses on the corresponding analysis on traditional industrial engineering technology and departure from the production, actually considering the existing problems to find the corresponding measures.

KEYWORDS: industrial engineering technology; business management; production management; overall optimization.

1 INTRODUCTION

Industrial Engineering (IE) originated from the 20th century in America, and was defined based on relevant research theories of predecessors such as Jill Brace and Taylor. For specific industrial and engineering technology mainly are industrial product design and manufacturing processes, including: specialized division of labor, the functions of the organization, as well as work measurement. Due to the rapid development of the current market economy, people's thought level improve further correspondingly at that time, so people constantly enrich the interpretation of the meaning of industrial engineering. From its born till now, the industrial engineering mainly experienced four stages of development: scientific management period, industrial engineering period, management operations research period, as well as the systematic industrial engineering period. For industrial engineering in different periods, its own problems handling capacity and efficiency are not the same, at the same time, among the various regions of importance for industrial engineering degree there are obvious differences, the situation is not optimistic with respect to our current development of industrial engineering.

2 THE APPLICATION OF INDUSTRIAL ENGINEERING TECHNOLOGY

For the current industrial engineering (IE) which is mainly through the production line, as well as corporate layout and production environments to optimize, making the production personnel, equipment, materials, venues and other system produced a series of combinations to get the appropriate state improvement. Promote internal information flow, and finally comprehensively upgrade the production efficiency. To IE technology, it should be applied to the industrial production of each step currently, this technology is a production management measure that is scientific and systematic, through the actual production work and its combination with IE technologies, each other, enable the elements of enterprises production have gained very good application, in reasonable and effective configuration, speed up the overall optimization goals.

3 IE TECHNOLOGY HELPS COMPANIES ACHIEVE THE OPTIMIZATION DESIGN AND IMPLEMENTATION

The current overall management optimization objective in internal enterprise is mainly based on the systematic investigation and analysis, first of all, conduct the production and deposit transport analysis research by IE technology, then, optimize the design correspondingly. Take a seamless steel company in our country as an example, combine IE optimization to analyze the specific transport and production.

3.1 *Deposit analysis in order to achieve classification optimization of the business product*

Annual deposit yields a seamless steel products shipped in tens of tons, which makes the pressure of its major warehouse manager for the product can be reasonable classification, as well as orderly and efficient completion of work placement at the same time, which directly issues related to the overall credibility of the enterprise, as well as get the comprehensive competitiveness of enterprises. For the reality of development, due to the development of enterprises is accelerating, leading to confusion Warehouse uncommon to daily production and operation of enterprises poses a serious problem, for in this case, we need to use IE technology companies more scientific the analysis, change management status of daily business through scientific and rational design provides a good solution for the enterprise. For previous product storage place, after which the production is complete, direct into the storehouse, there is no appropriate classification of the product number, is a chronological manner in accordance with product placement, this way it is easy to put so that their transit time is longer, and that regular scheduling order chaotic phenomenon occurs during transport, resulting in some product placement location not up to standard. This is also appropriate to make some products are shipped relatively low efficiency, the phenomenon is likely to cause extrusion products, so that the flow rate of the product is reduced, not only damaged the company's own business reputation, but also reduces the economic efficiency of enterprises. Therefore, to solve this problem is imminent, by the relevant principles of IE technologies, application design for the design and improvement of appropriate programs, specific methods include: Delete some unnecessary rules, merging some ways, orderly production and save transportation arrangement, simplify management measures. From four aspects of the appropriate scientific enterprise storage products, the specific program improvements are as follows: for different products, design appropriate placement area, using a sufficient number of classification bent, place the product area is divided into several or several areas of the same size and use English letters or Arabic numerals to distinguish. According to this classification method, different products will make the appropriate place, and then paste the appropriate identification plate on each product's packaging, which can help employees effectively handling the transfer and reduce errors when transferring village products save transport phenomenon, which saves time, but also improve work efficiency, so as to effectively solve the product because they do not store brought by rank corresponding loss, holistic optimization of the company's existence transportation management, also contributed to the production management optimization implementation.

3.2 *Production analysis in order to achieve production line optimization of the enterprise*

For production analysis of enterprises, whose main business is to make the appropriate mobile product line and distance analysis. Through this analysis, to find problems in production, to find solutions to specific problems corresponding optimization measures, so that the production efficiency has been a corresponding increase. Originally spend for a seamless steel roller bearing production is within the A-1 zone, but due to its distance from a given unit reducing far, resulting in a formal production, easily lead time is too long, and the human and material resources than big. So, for this phenomenon, a specific set of IE analysis techniques, to improve the efficiency of the production of products from the starting line of each of their production have made the corresponding sound research, while the lines corresponding to the respective calculated invalid transfer coefficient, whereby the optimal production lines obtained, wherein, for the production of an invalid formula the coefficient moves the moving distance of the line is equal to the total effective product minus moving distance, and then divided by the effective movement of the respective distances have the production line of the corresponding shift factor is invalid. Moreover, its specific values invalid transfer coefficient, is reduced as much as possible, preferably less than 1 for the production of the best route in accordance with the design criteria is to ensure the physical production personnel cost, low production efficiency. Under ensure that all factors of production location to place reasonable conditions, to try to choose the best route.

4 COMBINE THE IE TECHNOLOGY WITH IT TECHNOLOGY TO IMPLEMENT NEW IE TECHNOLOGIES

By combining IT technology and IE technology, to create a new IE technology, for this new IE technology, it starts from IE technology, through the appropriate control and design tools to plan their production. In the specific business management, IE technology-based management ideas and talent capability, coupled with

IT technology and information in systematic auxiliary benefits to help companies achieve innovation management to continuously improve production efficiency while reduce costs, increase management level. IE technology in the construction of such an early stage, and the need to strengthen the overall awareness of the importance of long-term corporate planning. For its design, the needs of enterprises to improve their overall IE personnel from the point of view of production efficiency, combined with the actual situation and make the appropriate practice, while in which to constantly sum up, in order to achieve enterprise information management and systematic pick up the pace. Due to the current status quo of enterprise development, making the new technology by the industrial enterprises IE extensive attention. Whether it is product development, production or sale of products or product through IT technology and a high degree of information systemic circulation, with IE technology, related product design from the characteristics of the market demand and technological development point of view, the product of the production process more systematic. The IE technology and IT technology combined with each other, and achieve overall optimization of production management transformation, to help industrial enterprises achieve a high-quality management procedures.

5 CONCLUSION

In short, the promotion of industrial engineering technology to achieve overall optimization of enterprise management, first of all, conduct comprehensive analysis of materials, personnel, equipment, and space of the internal business production, combine with the actual production situation through IE technology, design a reasonable optimization measures of the corresponding problem, improve enterprise management, make enterprise production management fully optimized.

REFERENCES

[1] Qi Ershi, Liu Hongwei. Localization Studies of Industrial Engineering And Application of Analysis [J]. Journal of Management, 2010,11: 1717–1724.

[2] Ma Hanwu. Facilities Planning and Logistics System Design [M]. Beijing: Higher Education Press, 2010: 28–29.

[3] Lan Jianyi. Industrial Engineering Curriculum System Overall Optimization Studies [J]. Education Forum, 2012, (34), 173–175.

[4] Liu Runquan. On The Application And Development of Industrial Engineering In Enterprise Management [J]. Glamorous China, 2010, (17), 25.

Management, Information and Educational Engineering – Liu, Sung & Yao (Eds)
© 2015 Taylor & Francis Group, London, ISBN: 978-1-138-02728-2

Celebrity endorser scandal and companies' reaction

Yuan Zhang, Shan Li & Wei Li
Business School, Sichuan University, Chengdu, PR China

ABSTRACT: Over the years, more celebrity endorsers are widely involved in publicizing negative events. Because of the associative link between celebrities and endorsed brands, the negative repercussion can transfer to the endorsed enterprises and brands, which cannot be underestimated. Also companies' respondent will affect the consumer's perception of the brand. This paper mainly summarizes the domestic and foreign related literatures about the spokesperson's undesirable events, and then discusses future opportunities in this field.

KEYWORDS: celebrity endorser, scandal, brand asset, companies' reaction.

1 INTRODUCTION

Recently, celebrity endorser scandals are exposed frequently, such as Edison Chen's Pornographic event, Jaycee Chan's drug abuse event, Zhang Wen's derailment event and so on. Meanwhile, consumers' online reviews are hot and fierce. When the celebrity endorser scandal breaks out, most endorsed companies have to find effective methods to deal with the tarnished spokesperson, working hard to reduce the deleterious effect. The responses from the endorsed company may be different based on various people, or in a different situation.

Taking Edison Chen's Pornographic event, for example, it would be found out that both Pepsi and Levis terminated the contract with Edison Chan immediately, and Disney shut down the advertisement of Gillian Chung. However, the endorsement situation of Cecilia Cheung seemed smoother. Not only the previous advertisements of Dayun motorcycle were still working, but more post-incident endorsement offers become available, such as Jieeryin and Southern Decoration.

In academic fields, the consumer and social psychology has formed a basis of knowledge about celebrity endorser scandals. Among them, numerous researches aim at advertising effectiveness of endorsers. Since of the frequent exposure of spokespersons' negative events, more scholars have begun to study on the effect of celebrity endorser scandals, consumers' perception and companies' reaction and so on. In this paper, we examine the main findings of the present literatures in celebrity endorser scandals, and put forward the limitation and opportunities for further research.

2 CELEBRITY ENDORSER SCANDAL'S CONCEPT AND CLASSIFICATION

2.1 *Concept of celebrity endorser scandal*

In the field of marketing and psychology, similar words on negative news would be "undesirable event", "bad thing", "negative information", "negative news", "scandal" Etc. Haskings (1981) pointed that negative news referred to undesirable things that happened to someone. Shruti (2009) noted that similar to product harm crisis, once the scandal breaks out, it would cause complaint, angry and condemn. As a result the support rating of the celebrity endorser and related parties would become lower.

2.2 *Classification of celebrity endorser scandal*

The first classification is based on the star professional attribute. Bailey (2007) see the difference of negative events from the type of celebrity: entertainment stars, sports stars, entrepreneurs stars, experts and scholars stars, politicians, stars. Among them, the athlete, celebrity type is mostly studied.

Secondly, it is according to the degree of blameworthiness. From the perspective of attribution, there is a tendency to judge the blameworthiness of involved parties (Louie et al. 2001). Jacobson (2001), Obermiller (2002) divided negative events into high responsibility events (high blame event, high culpability incidents) and low responsibility of the event (low blame event, low culpability incidents). Notably a high degree of responsibility (blameworthiness) is referred to scandals which are controllable and personal negligence.

3 EFFECT OF CELEBRITY ENDORSER SCANDAL

3.1 *Meaning transfer mode*

McCracken (1989) proposed the "meaning Migration Model" (Meaning Transfer Mode), depicting that the endowed cultural and meaningful significance of celebrity endorsers would be transferred to products, and then to consumers via endorsement. According to this model and empirical researches, it could be found that, celebrity endorser scandals will significantly reduce the public's trustworthiness and preference towards the endorser, and then low the advertising effectiveness. Thereby it would damage the public's confidence in endorsing goods, and then affects the sales and earnings. Till (1998) and White (2009) also proved that the bad influence can be spread to the brand image through the associative link. In addition, Shangwei, Lishan and Liwei (2010) concluded that the actual negative impact could be adjusted by the resilience of spokesperson's established image and the brand strength.

3.2 *Scandals may have a positive effect*

In general, scandals about the endorser would harm the celebrity and the brand. However, based on some actual cases, Louie, Kulik and Jacobson (2001) and Money et al. (2006) found the impact path of negative events is not static. Sometimes the scandals even enhance the image of their spokesmen and endorsed firm value, such as Nancy Kerrigan incident. Besides the slight negativity, it is also found out that a high matchup between the nature of the brand and the nature of the scandal may produce goodwill. For instance, an arrest record could gain more respect in hip-pop during the young man (Kamins, M. A. 1990).

3.3 *Celebrity blameworthiness for scandals*

Louie et al (2001) pointed that blame is a crucial dimension of perception. It suggested that, in the situation of very high blame events (such as alcoholism), the majority of consumers think that the spokesperson should be condemned, and therefore feel angry and disgust toward the endorser and then harm the firm value. On the contrary, for the low blame events (like illness, etc.), consumers prefer to show more compassion, sympathy and affection. Thus, with proper media exposure and visibility, positive perception and attitude will be aroused towards the spokesperson and the brand.

3.4 *Effect for other parties in the same industry*

From the perspective of spillover effects and brand association, scholars discovered that, in the same industry, the brand similar to the scandal brand will be negatively affected; in contrast the dissimilar or unrelated brand will be positively impacted (Bush, 2004).

4 COMPANIES' RESPONSE FOR CELEBRITY ENDORSER SCANDAL

4.1 *Judgment factors*

Mostly, companies would make the (dis)associative decision depending on the nature of the scandals. For instance, the celebrity involved in a murder or rape would be immediately dismissed, on the contrary, the celebrity is a victim of an illness would be retained. But, in a real situation the boundary of scandal the nature is not so obvious and clear. Thus, oftentimes companies need to take consumer perceptions of blameworthiness into consideration.

4.2 *Strategy selection*

Louie and Obermiller (2002) suggest that if it is a potential company-celebrity relation, consumers may not pay attention to the company's reaction, but focus on the situation facing the star. Despite of the negative incident, companies would enhance their brand image by selecting a low blame endorser, since there would be some championing and compensation about the person.

In an established company-celebrity relation, the focus is primary on the company's response, and whether the judgment is based on the blameworthiness of the endorser. Most studies reveal that companies should retain low responsible or low blameworthiness spokesperson, but abandon the high responsible or high blameworthiness spokesperson.

5 LIMITATION AND FUTURE EFFORTS

5.1 *Fewer research on different types of celebrity, attributes of products*

The majority of existing studies is aimed at sports star, rarely on other types of celebrity spokesperson, such as political figures, actors or other figureheads. Besides Edwards (2009) discussed the spokesperson gender have different effects on negative events, another spokesperson characteristic differences (such as celebrity type) are not studied. Additionally, different product attributes (such as FMCG and durables) research are even more rarely.

5.2 *Very few studies involved of the consumer network comments*

The vast majority of existing studies is based on the experimental method. More and more scholars began to focus on the customer online reviews, but only Shruti (2009) and Petty (2011) used semantic analysis, with the Internet blog to discuss the impact of celebrity endorser scandal. Thus, it would helpful to take advantage of consumer online reviews to dig the negative impact of the event.

5.3 *Limited research on strategies of endorsement companies*

When star negative events occurred, Louie and Obermiller (2002) thought that coping strategies for companies could be defined two kinds: Keep spokesperson or abandon spokesperson.

Table 1. Typical celebrity endorser scandals.

			Enterprise strategy	
Celebrity	Scandals	Endorsement	Retain	Abandon
O.J.Simpson	Murder	Hertz	√	
Sharon Stone	Slander China	Dior		√
Edison CHEN	Pornography	Levis		√
Gillian ZHONG	Pornography	Disneyland		√
Cecilia ZHANG	Pornography	Jieeryin	√	
Liu Xiang	Out of Race	Nike	√	
Annie YI	Derailment	Eastern Shop		√
Phelps	Drug abuse	Mazda Motor	√	
Angela ZHANG	Drug abuse	Meters bonwe		√
Zhang Ziyi	Donation fraud	Maybelline		√
Tiger Woods	Affair	Nike		√
Han HAN	Ghostwriter	Eslite	√	
Natalie ZHAO	Family violence	Healthy drinks	√	
Yang SUN	Drive without license	Hyundai		√
Daimo LEE	Drug abuse	Dongfeng Nissan		√
Zhendong KE	Drug abuse	Disneyland		√
Haibo HUANG	Prostitute	Chery Automobile		√

From the Table 1. we could see that the response of the company towards the tarnished celebrity may be inconsistent with the blameworthiness rule exactly. Such as Liu Xiang, 2008 "out of the race", Nike released a new ad which takes Liu Xiang's drop-out as the theme within 12 hours. On the contrary, VISA chose to replace Liu Xiang's ad next day.

Many real cases have involved doubtful and complex culpability. For instance, a celebrity's derailment or domestic violence has been alleged but not proven. Or if the endorser candidate holds in very high esteem, consumers may not downgrade him by an accident, which would also affect the company's reaction. Perhaps for a national-renowned celebrity, the endorsed company would take totally opposite strategy in different district. Situation likes Edison Chen's Pornographic incident, different parties may be different treated by endorsement companies. Such as Therefore, in addition to responsibility attribution, more factors affecting the company's response should be explored.

ACKNOWLEDGMENTS

This research was supported by China Postdoctoral Science Foundation (2013M542285); SRFDP (20100181120031); the Fundamental Research Funds for the Central Universities (skqy201224); and Department of Humanities and Social Sciences Education in general project (09XJC630007) .

Shan Li: Corresponding author. E-mail addresses: lishan@scu.edu.cn

REFERENCES

Ainsworth Anthony Bailey. 2007. Public Information and Consumer Skepticism Effects on Celebrity Endorsements: Studies among Young Consumers. *Journal of Marketing Communications June*: 85–107.

Brian.Till et al. 1998. Endorsers in Advertising: The Case of Negative Celebrity Information. *Journal of Advertising*(spring):67–82.

Bush, A. J., Martin et Al. 2004. Sports celebrity influence on the behavioral intentions of Generation Y. *Journal of Advertising Research*(44): 108–117.

Gupta. S. 2009. How do consumers judge celebrities' irrespons ible behavior: An attribution theory perspective. *Journal of Applied Business and Economics* 10 (3):1–14.

Haskins,J.B. 1981 The trouble with bad news., *Journal of Newspaper Research(3)*:3–16.

Kamins, M. A. 1990. An investigation into the "match-up" hypothesis in celebrity advertising. *Journal of Advertising*(19): 4–13.

McCracken, G.. 1989. Who is the celebrity endorser? Cultural foundations of the endorsement process, *Journal of Consumer Research* 16(3) : 21–30.

Petty, R. D et al .2009. The use of dead celebrities in advertising and marketing: Balancing interests in the right of publicity. *Journal of Advertising*(38): 37–49.
R.Bruce Money et al.2006. Celebrity Endorsements in Japan and the United States: Is Negative Information All That Harmfui? *Journal of advertising research.*
Steven .M.Edwards et al. 2009. Does Gender Impact the Perception of Negative Information Related to Celebrity Endorsers. *Journal of Promotion Management*15:22–35.
Shanwei et al. 2010. The impact of celebrity endorer scandals. *Financial research* 324(11):116–124.
Therese A. Louie et al. 2001. When Bad Things Happen to the Endorsers of Good Products. *Marketing Letters* :13–23.
Therese A. Louie et al. 2002. Consumer Response to a Firm's Endorser (Dis)Association Decisions. *Journal of Advertising(4)*: 41–52.
White, D.W. et al. 2009. The effects of negative information transference in the celebrity endorsement relationship. *International Journal of Retail & Distribution Management*37 (4): 35–42.

Management, Information and Educational Engineering – Liu, Sung & Yao (Eds)
© 2015 Taylor & Francis Group, London, ISBN: 978-1-138-02728-2

The analysis of standardized administration in enterprise financial accounting

Jin Zheng & Ling Duo Su
Langfang Pyrotechnic Institute, China

ABSTRACT: The competition among enterprises becomes increasingly fierce with the fast development of the domestic market economy in recent years. The enterprises are devoted to improving the efficiency of operation and management and to strengthening the financial management system so as to increase the competitiveness of the enterprise in the market.

KEYWORDS: Enterprise; Financial accounting; Standardization.

As the development and improvement of market economy, the enterprises in China confront with opportunities and challenges. Enterprise managers should standardize the management regulation and quicken the modern construction of the enterprises in order to gain superiority in the fierce market competition. Financial management, as the core of an enterprise, is not only the focus for it, but also the key element to realize the modern construction of the enterprise. Therefore, enterprises ought to establish scientific and reasonable financial management system and improve the efficiency both in enterprise management and in economy through financial accounting.

1 BRIEF INTRODUCTION OF ENTERPRISE FINANCIAL ACCOUNTING

Financial accounting originally refers to the accounting for the economic activities the accounting entity has already happened or completed, i.e. The general name of bookkeeping, calculation and accounts submitting. Modern financial accounting adds the financial accounting before the event and within the event to it. Financial accounting can give feedback immediately on the operation of the plan or budget and make instant adjustment to the schedule so as to guarantee it to work for the object expected. The enterprise managers are able to discover the problems existed in the management process on time and conduct research to solve them in order to increase the competitiveness of the enterprise and to make sure its stable development.

2 DRAWBACKS EXISTED IN TENERPRISE FINANCIAL ACCOUNTING

2.1 *Imperfect financial accounting system in enterprises*

Enterprise financial accounting plays an important role in the overall management and operation as the significant basis of helping managers find out and solve problems. However, there still lacks of the complete accounting system and accounting procedure in actual management among most of enterprises, which leads to the disorder of internal management and supervisory system and consequently the problems frequently occurring. The problems are reflected in the following aspects: firstly, a unclear entity in financial accounting. Financial officers often confuse personal assets and enterprise assets, which results in misappropriating assets. Secondly, low quality of the accounting officers. The financial accountants are quite important in the enterprise and they must be equipped with professional knowledge and skills as well as experienced in dealing with accounting affairs. However, in fact, because some Small and Medium Enterprises lack of complete management regulation and staff institution and many managers directly employ their relative, the staff undertaking such an important position are usually lack of professional knowledge and skills as well as some experience. They can't satisfy the requirement enterprises set of professional accountants, and they are incapable of the financial accounting in enterprises. It will seriously affect the accounting management or even result in great loss.

2.2 *Non-standard account establishment*

Account establishment is necessary to the self management for the enterprises, therefore, it should take the enterprise's nature, scale and volume of business into consideration. However, the books of most enterprises are usually prepared to respond to the regulation that financial institutions must establish an account to use invoice so that many accounts are substituted by invoices and are confused in this process. Some enterprises, even establish different accounts to evade paying taxes and make fake accounting information. Some don't observe the general principle in financial accounting, drawing expenses in advance do not comply with amortization and the income do not match the expenses. The incomplete management system makes it possible that some enterprise leaders have the right to manipulate the accounts, which exerts serious influence to the quality of accounting information.

2.3 *Weak internal control over financial accounting*

The internal system in enterprises consists of audit system, accounting check system and the examination and approval system in revenue and expenditure, etc. In fact, the phenomenon of disorder in labor division and violation of laws and regulations within the enterprises happen frequently, which seriously affect the normal conduction of financial accounting and leads to severe economic loss.

2.4 *Mistakes of basis information in accounting*

Financial accounting involves and deals with the data and information related to enterprise economic activities. As a result, the authenticity and reliability of the information is the prerequisite to the successful financial accounting. As the enterprises have been expanded greatly, the economic activities are becoming more and more complex, which subtly increases the difficulty of accounting work. The accounting officers need to collect more and more complicate data and take records and make an analysis. The technical personnel have to devote too much energy and patience during this process. Actually, due to some objective and subjective reasons, some errors, still can't be avoided in data processing. The truth of information in financial accounting can not be guaranteed.

3 STANDARDIZED MANAGERIAL MEASURES IN ENTERPRISE FINANCIAL ACCOUNTING

3.1 *Strengthening education and training of financial officers*

The professional skill, ethics and working experience of financial officers can directly decide whether the accounting work in an enterprise operates successfully or not. Enterprises should take a comprehensive consideration of the above elements when arrange these people and set reasonable number according to the nature, scale and volume of business of the enterprise. Meanwhile, enterprises should also stick to the people-based principal, caring for them in work and life, giving them reasonable salary, benefits and bonuses, positively coordinating their work with other departments. As to the professional skill, enterprises should require them to study hard to improve their professional capability and to get familiar with the basic knowledge and skill in financial accounting. Enterprises can organize some professional training for these staff on regular and conduct examination for those who participate the training. Those who get high marks can be awarded while those who fail the examination should be eliminated on time. In this way, the accounting officers are encouraged to improve their working ability, maintaining the quality of accounting. At the same time, enterprise managers ought to attach importance to the strengthening trustworthiness of accounting officers. they lack of correct guidance in professional ethics so that many of them are not honest, making fake accounts for the pursuit of personal advantages because there are no relatively complete ethic system in accounting.Enterprises should strengthen the education of honesty for the accounting officers continually and organize them to learn the related laws and regulations regularly, hoping that they could standardize their professional ethics by the obligation and responsibility in Accounting Law and improve their own professional ethics.

3.2 *Establishing complete enterprises accounting managerial system*

Complete management system is the basic foundation of enterprise self-management and effective operation. It is necessary to standardize financial accounting and establish complete and reasonable managerial system. The primary task is to build relatively complete external supervision system. Therefore, the government should perfect the laws and regulations related to enterprise accounting management and monitor the financial accounting work and enterprise activities by the legal system. The government ought to standardize the regulation of charge, make the task delegation known to the public and provide overall guidance in the development of bookkeeping in intermediate agents. The phenomenon of some enterprises violating laws and regulations during the process of financial accounting and operation should be punished by the government immediately accordingly to give a warning to them. Meanwhile, the standardization of financial accounting can be realized by strengthening social supervision theory. As a result, capital verification from the third party and accounting audit could be made. The enterprises should also build and improve the managerial system domestically. The enterprises could build internal accounting management system, a system of personal responsibility, financial procedural system, system of internal check, financial analysis system and so on. They can strengthen each link in accounting works to formalize and systematize the internal accounting and guarantee the increase of work capability.

3.3 *Tightening the enforcement of accounting system*

Complete accounting management system lays a solid foundation for the standardization of enterprise accounting. the system could play a role to the most extent so long as it is carried out practically. Therefore, enterprises must deal with every problem in financial accounting strictly according to standardized procedures, making sure that each link can be effective and correct. Only in this way, can the enterprises realize the standardization in financial accounting and make right and scientific decisions based on accurate and reliable information when deciding on important policies.

3.4 *Strengthening the internal control over financial accounting*

The internal control of enterprises consists of audit system, cost accounting system and financial revenue and expenditure approval system, etc., which could effectively improve the problems existed in financial accounting and increase the quality and efficiency of the accounting management. Policies made by managers within the enterprise play an irreplaceable and important role in enterprise operation and management. Therefore, managers should attach more importance to the internal control over financial accounting and promote the accuracy of decisions through improving the quality of accounting information.

4 STRENGTHENING THE PROFESSIONAL ETHICS OF ACCOUNTING OFFICERS

As the direct executor of financial accounting, the professional integrity and ethics of accounting officers have direct impact on the work. The enterprises must take some measures to strengthen their ethics in order to guarantee the normal management and standard operation.

4.1 *Establishing complete accounting professional ethics system*

The professional ethics of accounting officers need the constraint of complete ethics system as enterprises need a complete system to standardize management and operation. Western countries are more experienced in this aspect. China, basing on the western experience, could relate the reality and establish complete and reasonable accounting professional ethics which the ethics and trustworthiness construction can follow. In addition, the definition of the ethics should be concrete and detailed so that the accounting officers are able to actually find out feasible guidance when meet some problems, improving the operability of standardization in accounting professional ethics.

4.2 *Strengthening credibility education*

The accounting management industry in China is still developing, so the professional skill is simply emphasized while ethics ignored in the selection of accounting officers. In the long term, it will lead to low-quality accounting information and stagnant development. Therefore, China must strengthen the credibility education in daily work and make credit and ethics to be an important standard in the audition. The credibility education ought to be placed on the key position in educating accounting officers in order to promote the professional quality. Meanwhile, accounting officers should be strict to themselves according to the professional ethics and abide by the laws and regulations, trying to keep a good ethic attitude and order in the accounting industry.

4.3 *Perfecting the internal and external supervision system*

Credibility education needs to be strengthened and consolidated internally and externally. On one hand, certified accountants office, firstly standardize the behavior of the certified accounting personnel from regulation to basically prevent some actions of breaking law and regulation, e.g. fake accounting. On the other hand, external supervision institutions should increase the enforcement and positively enhance the awareness of the importance and necessity in credibility construction and create a better outer atmosphere. Moreover, the nation also should strictly crack down the actions of violating law and regulation as to maintain deterrence to other accounting personnel and to make them be aware of the danger of these illegal behaviors.

5 CONCLUSION

Financial accounting plays an important role in enterprise management. Enterprises are able to take full control over the variety of economic activities and expenses and expenditure only with accurate and complete accounting information, exchanging them into substantive value and profit through correct and scientific strategies. Standardizing financial accounting system is a long-term and significant work. Only in this way, can the enterprises make scientific and right decision for the future direction and make reasonable strategies so as to realize the sustainable development of the enterprises.

REFERENCES

[1] Lianhong Zhao. A brief analysis of the problems and solutions in enterprise financial accounting [J]. Business Finance &Accounting. 2012(05):64–66.

[2] Lizhen Cao. A view on the management in enterprise financial accounting under new situations [J]. Journal of Chinese collective economy.2010(25):39–42.

[3] Wenhua Wang. Standardized managerial strategies in enterprise financial accounting [J]. Modern economic management. 2011(08):105–107.

Management, Information and Educational Engineering – Liu, Sung & Yao (Eds)
 ISBN: 978-1-138-02728-2

Tourism development in Zuojiang River area from the perspective of world heritage

Xin Hua Ma
Department of Economics and Management, Guangxi Normal University for Nationalities, Chongzuo, Guangxi, China

ABSTRACT: Huashan Rock Paintings (HRPs) on cliffs on both sides of Zuojiang River are being nominated for inscription on the World Heritage List. Therefore, the tourism development of Zuojiang River should follow the rules for World Heritage sites. Attention should be paid to the protection and inheritance of HRPs and the associated ancient civilization which was created by the people of Luoyue who were the ancestors of the Zhuang nationality. This paper expounds the concept and classification of World Heritage sites; the qualifications and evaluation standards for inscription on World Heritage sites; the protection principles, systems of supervision, and inspection of World Heritage sites. The author also analyzes the significance and qualification of HRPs in Zuojiang River area being nominated for inscription on the World Heritage List. Finally, the paper puts forward some suggestions for tourism development of Zuojiang River from the perspective of World Heritage.

KEYWORDS: World Heritage; Perspective; Tourism Development; Zuojiang River.

1 INTRODUCTION

Due to economic development, wars, natural disasters, and many other various reasons, more and more human heritage is facing the threat of destruction and disappearance. However, some heritage host nations still ignore the protection of heritage because of the lack of funding and technology for the protection. In order to protect the precious heritage and wealth of human beings, promote the international cooperation in such aspects as protection funds and technology, *The World Heritage Convention* has been passed by the United Nations Educational Scientific and Cultural Organization (UNESCO) in Paris in 1972. Besides, UNESCO also has established the World Heritage Committee specifically to be responsible for the implementation of this convention. Many national and regional governments have signed on as state parties of *the World Heritage Convention*, and actively nominate items for inscription on the World Heritage List. By the 38th World Heritage Committee meeting, which was ended on June 25, 2014, 1007 items have been listed on the *World Heritage List*. Since China became one of the charter nations in 1985, the work of nominating for World Heritage was brought to the attention of the governments at all levels, and then an upsurge of nominating for World Heritage appeared. China currently has 47 World Heritage items, ranking the second place in the world (after Italy which has 50).

Representing the ancient Luoyue culture of Zhuang nationality, HRPs has already been applied for the World Heritage since the beginning of 2003, and were listed on the *China World Cultural Heritage Tentative List* in 2007 and 2012. Six government entities that belong to three levels of government offices, including Guangxi Zhuang Autonomous Region, Chongzuo City, and four counties/districts such as Ningming, Longzhou, Fusui, and Jiangzhou, are all working together and preparing actively for nominating HRPs for 2016 World Cultural Heritage listing. Especially from the beginning of 2013, comprehensive efforts have initiated for the HRPs nomination. These efforts have drawn a great deal of attention in every community of the society. In China, while all levels of government departments and tourism scenic areas are working hard towards the declaration of the World Heritage sites, they also use these activities to create economic benefits by attracting tourists. On the other hand, people like to visit heritage tourism resources. These activities have already become a trend. Whereas, in the past, many problems appeared when World Heritage sites in China were in the process of developing World Heritage resources. Huashan Rock Paintings are located sporadically on the cliffs of Zuojiang River, so the tourism development of Zuojiang should be based on the perspective of World Heritage.

2 INTRODUCTION ON WORLD HERITAGE

2.1 *Concept and classification of world heritage*

World Heritage refers to the world heritage items confirmed by UNESCO and the World Heritage Commission. These heritage items are rare, and currently an irreplaceable wealth of the human world are cultural relics and the natural landscapes recognized by all mankind that these items have outstanding significance and universal value. World Heritage sites/ items include world natural heritage, world cultural heritage, mixed cultural and natural world heritage, world cultural landscape heritage, world memory heritage, world oral and intangible heritage of humanity [2].

2.2 *Rules of world heritage*

2.2.1 *Strict and meticulous evaluation standards*

World cultural heritage and world natural heritage are defined clearly in the first and second articles of the *World Culture and Nature Heritage Convention* [3], The 24th article in the *Guide to Action for Implementing the World Heritage Convention* (referred to as the *Guide to Action*) specifies the evaluation criteria of the world's cultural and natural heritage. There is a total of ten assessment criteria for world heritage site: 6 evaluation criteria for world heritage site and 4 criteria for assessment of the world natural heritage. Nominees must have outstanding universal value of the property, manifested in at least one of the ten criteria. Otherwise, it will no longer be eligible to be nominated as a World Heritage site, let alone become a World Heritage site. World Heritage criteria are strict and meticulous.

2.2.2 *Principles of world heritage protection*

For World Heritage items, authenticity and integrity are the basic premises for nominating an item to be listed on the World Heritage List. Protecting the authenticity and integrity is a basic principle for guiding the development and exploiting of the World Heritage items.

2.2.2.1 The authenticity principle

"Authenticity" was first proposed in The Venice Charter. Since then, the Guide to Action pointed out that "a cultural heritage to be listed in The World Cultural Heritage List should meet at least one criteria about outstanding universal value given in The World Heritage Convention and with authenticity", each established item should "meet the judgments of authenticity in design, materials, techniques or surroundings, personality and inscape, etc. " The Nara Document pointed out further that: "Aspects of the sources linked to authenticity judgments may include form and design, materials and substance, use and function, traditions and techniques, location and setting, and spirit and feeling, and other internal and external factors. The use of these sources permits elaboration of the specific artistic, historic, social, and scientific dimensions of the cultural heritage being examined." Authenticity in this document obviously applies to cultural heritage at that time, while it also applies to natural heritage, whose authenticity shall be protected as well. For example, those natural heritages aimed at protecting special ecosystems must not randomly have alien species introduced or existing species extirpated.

2.2.2.2 The integrity principle

Integrity is a measure of the wholeness and intactness of the heritage and its attributes. It is stated in The *Guide to Action* that World Heritage items should be examined for the conditions of integrity in terms of assessing the extent to which the following properties are met: a) all elements necessary to express its outstanding universal value; b) adequate size to ensure the complete representation of the features and processes which convey the heritage item's significance; c) adverse effects of development or negligence. It also provides the specification for four criteria that are related to natural heritage integrity. The first one is integrity of physical space, for example there should be a peripheral buffer zone to protect a World Heritage site, moreover the boundary between heritage core region and buffer region shall be definitive. The second one is integrity of environment, which means historical or cultural (culture concept) or natural ecological environment. In conclusion, integrity refers not only to natural heritage but also to cultural heritage.

2.2.3 *The supervision mechanism of the world heritage*

The World Heritage List is not a lifelong system. A heritage item successfully joining *The World Heritage List* does not mean that it can "sleep without any worries". We all know that the World Heritage not only has strict selection conditions and evaluation standards, but also more continued thorough supervision and inspection system which includes reactive detection and systematic detection. *The World Heritage Convention* and *Guide to Action* all require the charter nations to have a regular comprehensive professional assessment for the status of World Heritage in their own countries and submit a detailed report [6] to the World Heritage Committee (this is the systematic detection). The World Heritage Committee organizes experts to inspect and evaluate the World Heritage item, and then takes corresponding measures according to the situation of the assessment. If the situation is too bad and difficult to change, the international community will supervise and help the host nation. If the World Heritage is threatened by serious special dangers, it will be listed in *the List of World Heritage in Danger.* At last, if the heritage is indeed unable to stop the heritage value loss and deterioration of the environment, it will be removed from *World Heritage List* and *the List of World Heritage in Danger* (this is reactive detection) [7]

3 ON TOURISM RESOURCES IN ZUOJIANG RIVER HUASHAN AREA AND APPLICATION FOR WORLD HERITAGE OF HRPS

3.1 *General situations of tourism resources in Zuojiang River Huashan area*

Zuojiang Huashan is located in South Asian tropical regions and rich in tourism resources: 1. The great variety of natural tourism resources. Zuojiang and its branch river Mingjiang have peculiar karst landscape and beautiful scenery. Many river reaches have peaks skyrocketing, cliffs confronting, and mist floating. The scenery is like poem and painting, it is not the Lijiang River but like the Lijiang River, it is not small Three Gorges but exceeds small Three Gorges. It is called the Thyme Landscape Gallery. 2. Strong Zhuang folk customs tourism resources. The Zhuangs' population accounts for more than 90% in the Zuojiang River basin. It is one of the famous Zhuang nationality concentrated residences. There are many Zhuang villages built along the river. The people in the village are simple and sincere. Zhuang custom is strong, of which the most famous is Singing Fair Day. 3. The valuable HRPs culture tourism resources. In Zuojiang and Mingjiang, a large number of ancient rock paintings are on rock cliffs on both sides of the rivers. Because the size of Ningming rock paintings is the biggest, portrait is maximum and the content is the richest, rock paintings on Zuojiang River and its branches are collectively called as HRP or Huashan cliff painting. The contents of these HRPs are mostly portraits, there are also bronze drum, bronze bell, horses, dogs, other animals, and some weapons. Mysterious ancient rock paintings and unique Zhuang nationality ethnic styles bring out the best to each other, add radiance and beauty to each other. According to historical records, HRPs were found by people in the Southern Song Dynasty period. Mysteries have existed for a long time on who the artists were, when, how and through what way they were painted, and what contents were painted. For decades after the founding of P.R. China, experts and scholars in China and abroad have come to explore and research the area to solve the riddle of HRPs. At the same time the HRPs also attracted a large number of tourists to visit.

3.2 *Significance of nominating HRPs for world heritage*

The HRPs being nominated for the world cultural heritage have been finalized after several adjustments. Currently, a total of 38 of the most typical cliff paintings are included in the scope of nomination. These HRPs are bound together in one package in the name of "Zuojiang HRPs Cultural Landscape" for the nomination. These HRPs are located in four counties (or district) including Ningming, Longzhou, Jiangzhou and Fusui. The significance of nominating HRPs for World Heritage is summarized below. Firstly, the nominated world heritage site is to obtain international certification, which has the brand of high authority and credibility. Therefore, HRPs can provide visibility of Zuojiang River basin's rock paintings after the nomination. It can not only bring the development of the whole Huashan Scenic Area, but also radiate the surrounding areas and enhance the awareness of people to protect HRPs and its implication of the LuoYue Culture. Secondly, once the nomination of HRPs is successful, it will fill the Chinese world heritage "family" a blank. That is, the World Heritage list has included 27 rock-art heritage sites until 2013 and even though China has many ancient rock paintings heritage sites, but none of them are listed on the World Heritage List. At last and most importantly, if it's successfully nominated for the World Heritage, it will have more opportunities to protect the cultural heritage of HRPs. For more than 2000 years, HRPs have been suffered from the weathering (erosions caused by rain, sun and wind), part of the rocks appeared cracking, cracks, water seepage, shedding and weathering damages. Huashan rock paintings are precious wealth of ancient LuoYue people who were the ancestors of Zhuang nationality, they are supposed to be protected. Through nominating for World Heritage and inscription on the *World Heritage list*, we can get financial, technical and other support from the international community.

3.3 *Qualifications of HRPs for world heritage*

Huashan rock paintings are one of the world's largest ancient paintings. It has the characteristics of imposing manner. Rock paintings, mainly created by adopting the method of projection monochrome flat painting. It reflects the unique ancient Luoyue artistic skills and characteristics which provide the real basis for the culture and history study of dying in the history of the ancient Luoyue (9). It conforms to the first and the third of the ten requirements that are included in *The World Heritage Convention* for nominating HRPs for World Heritage. Huashan rock paintings were drawn on the cliffs of Zuojiang River and its branches, such as Mingjiang, Pingerjiang and Black River. They have a unique style of painting. But painting on the rock cliffs is so difficult. It reflects the ancient superb artistic creativity. Therefore, it conforms to the requirements of the first criterion of "classic work that embodies human creativity". Huashan rock paintings display the scenes of sacrifice, worship and routine life of ancestors that lived more than two thousand years ago. They depict countless objects which were drawn vividly, such as the mage of figures, animals and plants and articles of drums and so on. They showed a picture of two thousand years ago that ancestors of the Zhuang Nationality were in daily life and prayed that God blessed the good weather for crops and sacrificial activities. Huashan rock art also presents the reality of what has been lost in the customs and habits. These can meet the requirements of article three of the standard "present the relevant existing or have disappeared cultural traditions and unique civilization evidence".

4 SUGGESTIONS FOR ZUOJIANG RIVER TOURISM DEVELOPMENT

Although Huashan rock art has not yet become a World Heritage site, the tourism development should be prepared for the future. People should actively learn from the advanced experience of the tourism development and heritage protection of the majority of the World Heritage sites. At the same time, we should learn the painful lessons of tourism development and heritage protection in a few World Heritage sites. We must not step on the "yellow" or "red" line related to World Heritage rules. We should do everything according to the rules of the World Heritage. Therefore, the author of this paper puts forward the following suggestions on tourism development in Zuojiang River area.

4.1 *Strengthening advocacy and education on the knowledge of world heritage*

According to the world heritage protection experiences in developed countries, they focus on carrying out publicity and education about heritage knowledge to the public. The World Heritage is a system's engineering, the declaration of HRPs cultural heritage is really a good situation to carry out publicity and education awareness of heritage for the government department, the Huashan Scenic Area and the regional distribution of the surrounding villages' residents of HRPs, so we can make people understand what is the relevant knowledge of World Heritage, qualification, supervision and inspection system of World Heritage. In addition, through the publicity, we can also make people know more about significance of declaring world cultural heritage, how to protect the HRPs, and the relevant laws and regulations for protection of HRPs. Popularizing the knowledge of World Heritage is conducive to forming a good atmosphere for the protection of the heritage of the HRPs, to better promote the nomination of HRPs for world cultural heritage, and further promote Zuojiang River tourism development.

4.2 *Fine interpretation of the rules of world heritage, scientific planning*

Zuojiang tourism planning is not only planning for development but also planning for protection. Huashan rock paintings are a unique artistic achievement. In the process of its development, we can't destroy its artistic and cultural value. According to the requirement of the World Heritage Convention authenticity and integrity should be strictly protected. Experts should be retained to carry out scientific planning: On one hand, re-examining the previously proposed and approved *outline of Guangxi Huashan Scenic Area overall planning (2012–2030)*; On the other hand: re-evaluating the following draft proposals: *Huashan Tourism development plan* (launched in 2014) and "Ningming HRPs core scenic zone control detailed planning" "Guangxi Zuojiang Basin tourism concept planning" (launched in 2013) "Guangxi Zuojiang Basin tourism master plan" (launched in 2013) and so on. For the experts in planning these projects, we must let them follow the concept of World Heritage Protection, also we should check strictly when reviewing these plans, and we must assess whether these plans are scientifically sound and are advantageous to the protection of the future of the World Heritage. Many problems in other World Heritage sites are due to poor planning. For example, in 1980s and 90s, the Tower of Gong and Drum outside the gates of Zhangjiajie national Forest Park in Wulingyuan center view (also in the Jinbian river's upstream), a large number of tourist accommodation and catering reception facilities were constructed, resulting in this area being seriously "urbanization, artificiality, commercialization". Also because there were no sewage treatment facilities, the water of Jinbian river was seriously polluted by these reception facilities.

4.3 *Tourism development construction should strictly follow the laws and regulations related to heritage protection*

Under the background of market economy, the tourism products created by tourism scenic spots need to meet the market demands. However, HRPs are cultural heritage, we must abide by relevant laws and regulations if the area of HRPs is developed for tourism products. At present, several laws apply for the protection of HRPs, such as the *Protection of World Cultural Heritage Management Measures*` and the *Guangxi Zhuang Autonomous Region Rock Paintings of Zuojiang River Protection Measures.* Any tourism products violating relevant laws and destroying the authenticity and integrity of cultural heritage should never be developed despite the fact that they can meet the needs of tourists' preference. For example, "Approaching Huashan" large-scale tourism project, also known as the "Impression Liu Sanjie" companion, was proposed with the intention to bring live the ancient Luo Yue civilization of the ancestors of the Zhuang nationality. The project proposed six unique elements, including "Heading for Huashan",

"Luo Yue Rhyme", "Tu Zhai Style", "Rock Story", "Stage in Moonlight", and "Ceremony on the Cliff"[10]. The project was creative, but it required some construction on the side of the Zuojiang River basin for the proposed night shows and activities, it bounded to cause some damages to the whole environment of HRPs' cultural heritage because of setting stage especially the stage lighting. Besides, it acts against the provision of "should not use a strong light directly irradiating Huashan rocking paintings" in *the Guangxi Zhuang Autonomous Region Rock Paintings of Zuojiang River Protection Measures.* Finally the "Approaching Huashan" large-scale tourism project had to be terminated. Accordingly, we should carefully consider the development of tourism products of Huashan and never forget the lessons of this project.

When a planning is approved, documented by relevant government departments, it is the legal ground for the exploitation and construction of the planned area, and it can regulate and restrict all behaviors for the exploitation and construction of the planned area. When developed, it should be managed strictly, and be carried out strictly following the plan in the construction of scenic spots to avoid many mistakes other heritage sites in China has made such as *artificialization*, *commercialization*, and *urbanization*. Huangshan Mountain, Mountain Tai, Wulingyuan, and Jiuzhaigou are among those that have made this kind of mistakes. These three "-izations" phenomena in World Heritage sites demonstrate the construction of scenic spot did not obey the policy. The Huashan area of Zuojiang should take care of the development of reception facilities about eating, living, and traveling entertainment. Besides, protect the local natural tourism resources, cultural tourism resources, especially the civilization of Zhuang ancestors Luoyue such as the HRPs.

4.4 *Market appropriately and control volume of tourists*

The main purpose of HRPs as a cultural heritage is not only to protect it, but also the civilization of Zhuang ancient Luoyue reflected by the HRPs. Contrary to increase the tourists' numbers through advertising, the managers of heritage site should identify specific target consumers, control the volume of visitors, and raise the standards of heritage protection at the same time. The planning and implementing of marketing activities affect the tourist traffic of heritage sites. The HRP scenic spot's marketing activities should consider the environmental capacity and traffic if they don't want to be destroyed by an influx of tourists. We should adopt conservative tourism marketing means to stabilize and equalize the tourist volume to the scenic spots when it is in the period of the tourist peak season. Measures should be in place to avoid the situation that the tourism is overheated and surpasses its tolerating capacity. Moreover, we should create some other marketing means to attract tourists when it is in the off-season. Let Huashan heritage tourism resources have a balanced and rational use whenever in the peak or trough season.

5 CONCLUSION

World Heritage is the most precious common property of mankind. The sites/items included on the World Heritage List definitely are elite heritage items because of the complex and strict eligibility and evaluation criteria. Whether it represents the human cultural heritage with splendid civilization that has endured many ages, or the fabulous natural heritage carved by the hands of the creator, they all need everyone's appreciation and protection. China is a large country, and its culture has a long history, has created countless brilliant cultural heritages. Huashan rock paintings located in the southern border of China and in Zuojiang River basin are being nominated for World Cultural Heritage. In order to avoid repeating mistakes in other World Heritage sites, we should strictly follow the rules of the World Heritage in the process of tourism development and construction to facilitate the successful nomination for inscription on the World Heritage List, effective protection of HRP cultural heritage, long-lasting radiance of World Heritage, and sustainable tourism development in the Zuojiang River area.

ACKNOWLEDGEMENTS

This work is supported by the Humanities and Social Sciences research project of Guangxi Zhuang Autonomous Region Department of Education (Grant No. SK13LX487) and the Young and Middle-Aged Key Faculty Scientific Research Startup project of Guangxi Normal University for Nationalities (Grant No.2012RCGG003)

REFERENCES

[1] Li Xianhua.On the legal protection of World Heritage [EB/OL].http://old.chinacourt.org/public/detail.php?id=150422,05–02–16.

[2] Jiang Jinghong. World heritage local special legislation condition [J]. Journal of Sichuan Academy of Police Officer, 2007, 19 (4):92–95.

[3] Wang Xuxiao Inquiry into the related problems of contemporary aesthetic value—aesthetics analysis on

world heritage education—Also on the necessity of the world heritage education become aesthetic education curriculum [J]. Journal of Hebei University(Philosophy and Social Science), 2012, 37(1): 1–5.

[4] Zhang Chengyu, Xie Ninggao.The Principles of Authenticity and Integrity and the Conservation of the World Heritage[J].Journal of Peking University (Philosophy and Social Sciences), 2003(3): 62–68.

[5] Zhang Chengyu.Authenticity and Integrity:Questioning and Rethinking[J]. Southeast Culture, 2012, (1): 27–34.

[6] Li Huaming. Research on countermeasures of "World Natural Heritage" sustainable use of legal protection in China [D].Beijing: the Central Institute for nationalities, 2006.

[7] Ma Xinhua,Li Wei.Analysis of the principles of tourism product development in the world heritage sites[J]. Tourism Research,2010, 2(3): 57–60.

[8] Huang Xuean, Huang Jun. Huashan rock paintings applications for world heritage: the final sprint [EB/OL]. http://www.gx.xinhuanet.com/2013-10/25/c_117866277.htm.

[9] Chen Xuepu.On Huashan culture in view of World Heritage and Intangible Cultural Heritage [J]. Journal of Guangxi Normal University for Nationalities, 2010(1):5–9.

[10] Hu Lingling, Gan Xiaojun. "Go to Hua Shan" Refresh "Huashan culture pattern of night" to reproduce the Zhuang folk [EB/OL]. http://www.gxnews.com.cn/staticpages/20051015/newgx434fff37-475760.shtml,2005-10-15.

Management, Information and Educational Engineering – Liu, Sung & Yao (Eds)
© 2015 Taylor & Francis Group, London, ISBN: 978-1-138-02728-2

A literature review on warranty and maintenance

X.Y. Li & Y.X. Jia
Mechanical Engineering College, Shijiazhuang, China

Y.B. Zhang
The Air Force Xi'an Flight Academy, Xi'an, China

ABSTRACT: One way of providing the assurance that the products will perform satisfactorily over their useful lives to customers is through a product warranty. In warranty servicing, corrective and preventive maintenance actions have a significant impact on the total costs and exert a great influence on the product performance. As warranty and maintenance is highly important in the context of products, they receive greater attention of researchers from different disciplines and the literature on the relationship between them are vast. So in this paper, a review of the literature which links warranty and maintenance was carried out, and then a framework to define new topics for future research was developed.

KEYWORDS: Warranty; Corrective Maintenance; Preventive maintenance; Literature review.

1 INTRODUCTION

In recent years, since the product warranty can be a powerful incentive in selling a product and a satisfaction warranty provides an opportunity to woo the customers, it plays an increasingly important role in industrial transactions. So reducing the warranty servicing cost and improving the product performance has become a great issue for warranty service provider. One of the possible ways to achieve the above goals is by making optimal decision on the maintenance strategies in the warranty period. Maintenance is usually classified into corrective maintenance (CM) and preventive maintenance (PM). Maintenance is significant in the warranty context since it has a major impact on expected warranty servicing costs and the product performance. So in this paper, we focus our attention on the link between warranty and maintenance and review the limited articles dealing with this topic, which are mainly published between 2002 and 2013.

2 WARRANTIES AND MAINTENANCE

2.1 *Product warranty*

Warranty assures the buyer that the product will perform its intended function under normal conditions of use for a specified period of time. In this paper, we consider warranties for both new and second-hand products. And new products can be divided into the following four categories: consumer durables, consumer nondurables, industrial and commercial products, and specialized defense related products. In the context of warranty, there are four different perspectives: consumer perspective, manufacture perspective, public policy perspective and third-party warranty servicing provider perspective. Warranty plays different roles in these perspectives, but is common in the characteristic of being protectoral. According to different types of product sold and the purpose of satisfying a different perspectives' need, there are many different types of warranties show as Tab 1. Whichever warranty policy chosen to guarantee the product, it will have an impact on the product cost and reliability. The consumers prefer to the products demonstrate high reliability. However, it will increase the product additional cost. So warranty service provider must achieve a proper trade-off between the additional cost and product reliability.

2.2 *Maintenance*

The critical issues for the consumer are product performance. One of the ways of satisfying the consumer's need is adopting proper maintenance policies. As indicated earlier, maintenance policies can be defined as actions to PM and CM. CM actions can be divided into improved repair, perfect repair, imperfect repair, minimal repair, worse repair and worst repair. And PM actions can also be divided into clock-based maintenance, age-based maintenance, condition-based maintenance, opportunity-based maintenance and design-out maintenance.

Table 1. Classification of warranties.

Standards of classification	Types of warranties
Warranting which part of the product	Component warranty; system warranty
The number of the product	Single item warranty; fleet warranty(cumulative warranty)
Warranty policy dimension	One-dimension warranty; two dimension warranty
Renewing mechanism	Renewing warranty; non-renewing warranty
Maintenance cost	Free replacement warranty; pro-rata warranty;rebate warranty
Precondition for maintenance	Complete warranty; conditional warranty
Time span of warranty period	Base warranty(BW) ; extended warranty(EW)
Complexity of the warranty policies	Simple warranty; combination warranty
The maintainability of the product	Repairable item warranty; un-repairable item warranty

3 REVIEW OF LITERATURES

The literature involving warranty and maintenance can be organized into warranty servicing with only CM actions, warranty servicing involving both CM and PM action and maintenance outsourcing during the warranty period. In the literatures, the relation between warranty and maintenance is always described by mathematic models. These models are proposed depending on some of these factors. In the following sections, we will review these models by the categories organized by year.

3.1 *Warranty servicing involving only CM*

In the period between 2002 and 2004, Chen presented a maintenance/replacement policy under two-dimensional warranty. Iskandar studied two different repair-replace strategies for items sold with a two-dimensional free replacement warranty. Bai presented discounted warranty cost model for repairable series systems assuming the impact of repair actions on components' failure time is minimal. Huang proposed a Bayesian decision model for determining the optimal warranty policy for repairable products. Chukova present a framework to analyzing the lifetime of a unit which undergoes multiple repairs. In the 2005 and 2006, Iskandar developed a strategy characterized by four parameters. Bai studied a repair-limit risk-free policy and provided the first and second moments of the warranty cost per unit sold through censored quasi-renewal processed. Sheu split the warranty period into two intervals in which only minimal repairs can be undertaken, separated by a middle interval in which no more than one replacement is allowed. Wu presented a decision model for manufacturing to determine the optimal price and warranty length to maximize profits. Jiang proved Jack's model was correct and showed a more general model where the repair cost is random. Chen investigated an imperfect production system with allowable shortages for products sold with free minimal repair warranty. Chukova consider a rectangular warranty region and divide it into three disjoint sub-regions. In the period between 2007 and 2009, Manna proposed a model that applied to automobile part or component. Yun looked at two new warranty servicing strategies. Manna considered the problem of calculating warranty cost with a rectangular two-dimensional policy with minimal repair. Jack proposed a strategy can be applied to the single components, the modules and the whole product. From 2010 until recently, Pan developed a continuous parameter Markov chain model for a warranty action determination policy for a machine. Park proposed warranty cost models on the quasi-renewal processes, and introduced altered quasi-renewal and mixed quasi-renewal processes to obtain the expected value of warranty cost. Rao worked on the decision to replace or repair depending on a variety of factors. Vahdani developed a renewing free replacement warranty policy for a multi-state deteriorating repairable product with N working states and N failure states. Banerjee analyzed the cost of a new two-dimensional warranty servicing strategy that probabilistically exercises a choice between a replacement and a minimal repair to rectify the first failure in the middle interval. Su proposed two types of extended warranty policies from the manufacturer's perspective.

3.2 *Warranty servicing involving both CM and PM*

In the period between 2002 and 2006, Wang developed a cost model using a Markov chain to jointly determine the production cycle, process inspection intervals, and maintenance level. Kim set up a model to determine when preventive maintenance carried out at discrete time instants by assuming the cost of preventive maintenance is borne by the buyer. Yeh analyzed the effects of a renewing free-replacement warranty on the optimal age replacement policy, and developed a mathematical model to derive the optimal preventive maintenance warranty policy for repairable products to jointly determine the optimal lot size and product inspection policy when products are sold with free minimal repair warranty. Bai presents full-service warranty for repairable multi-component systems under which a perfect maintenance action will be performed to reduce the chance of future system failure. Pascual established a cost optimization model to determine optimal levels of preventive maintenance. In the period between 2007 and 2010, Giri studied the problem of inspection scheduling in an imperfect production process. Huang considered a periodic preventive maintenance program is performed during the warranty term to slow down the product deterioration employing a non-homogeneous Poisson process. Chien analyzed the optimal age-replacement for a product. From 2011 until recently, Wu studied a general periodic preventive maintenance policy for a single buyer. Tsai established a model to find the optimal PM schedule which minimizes the expected total cost over the operation time interval within a given finite operation time period. Chien studied on the effects of a free-repair warranty on a periodic replacement policy with a discrete time process. Sana studied on an imperfect production system with allowable shortages due to regular preventive maintenance. Bouguerra developed an extended warranty cost model considering six different maintenance policies.

3.3 *Maintenance during the post warranty*

From 2009 until recently, Jung studied the optimal periodic preventive maintenance policies following the expiration of the warranty from the user's perspective. Jung developed a replacement model following the expiration of the warranty that optimizes a two-attribute value function. Kim studied the optimal periodic preventive maintenance policies of a second-hand item following the expiration of the warranty. Gopinath and Anisur defined lifetime, developed lifetime warranty policies and models for predicting failures and estimating costs for lifetime warranty policies. Park presented new warranty cost models subject to warranty and post-warranty period, minimal repairs, and various types of warranty policies including free repair/replacement warranty and pro-rata warranty for k-out-of-n systems.

3.4 *Maintenance outsourcing in warranty period*

There is a tendency that the manufacture outsources the product warranty to a third-party. The main reasons that motivate companies to outsource the warranty are reducing costs, improving service quality and concentrate on their core competencies. Maintenance outsourcing is one of the main contents in the warranty outsourcing. From 2002 until recently, Hakan considered maintenance was outsourced to independent contractors. A performance-based incentive contract was offered to each contractor. He also studied learning effects on maintenance outsourcing, considering a situation in which a manufacturer offered a short-term outsourcing contract to an external contractor. Behnam presented two novel models for the maintenance outsourcing problem under uncertainty to obtain maximization of the profit and reliability and get minimization of the outsourcing cost. Antonio examined the large body of existing research on outsourcing, and assessed the status of research on outsourcing the maintenance of medical devices.

4 CONCLUSION

This paper has overview maintenance and product warranty at first. Then eighty-one articles were subsequently selected for their relevance to maintenance and classified them into four main categories. According the review results, the scope for future research can be carried out in the following areas: warranty servicing models which include both PM and CM for items covered by two-dimensional, research on the relationship among product warranty, maintenance and product performance, maintenance models with non-zero maintenance time, warranty servicing for multi-components which have multi failure mode, and studying the problems involved when warranty servicing is outsourced.

REFERENCES

Bai&Pham(2004), Discounted warranty cost of minimal repair series systems, IEEE TRANSACTIONS ON RELIABILITY VOL.53 NO.1,.37–42.

Bai&Pham(2005), Repair-limit risk-free warranty policies with imperfect Repair, IEEE TRANSCTIONS ON SYSTEM, MAN, AND CYBERNETICS-PART A: SYSTEM AND HUMANS, VOL.35, NO.6, 765–772.

Bai&Pham(2006), Cost analysis on renewable full-service warranties for multi-component systems," European Journal of Operational Research 168 492–508.

Banerjee&Bhattacharjee(2012), Warranty servicing with a brown-proschan repair option, Asia-Paci.c Journal of Operational Research Vol. 29, No. 3, 1240023(1–13).

Bouguerra et al(2012),"A decision model for adopting an extended warranty under different maintenance policies," International Journal Production Economics 135 840–849.

Chattopadhyay&Rahman(2008), Development of lifetime warranty policies and models for estimating costs, Reliability Engineering and System Safety 93 522–529.

Chen et al(2006), Optimal production run length for products sold with warranty in an imperfect production system with allowable shortages, Mathematical and Computer Modelling 44 319–331.

Chen&Popova(2002), Maintenance policies with two dimensional warranty, Reliability Engineering and System Safety 77 61–69.

Chien(2008), A general age-replacement model with minimal repair under renewing free-replacement warranty, European Journal of Operational Research 186 1046–1058.

Chien(2010), Optimal age for preventive replacement under a combined fully renewable free replacement with a pro-rata warranty, International journal Production Economics 124 198–205.
Chien(2010), The effect of a pro-rata rebate warranty on the age replacement policy with salvage value consideration, IEEE TRANSACTIONS ON RELIABILITY, VOL. 59, NO. 2, 383–391.
Chien(2012), The effects of a free-repair warranty on the discrete-time periodic replacement policy, International Journal Production Economics 135 832–839.
Chukova et al(2004), Warranty analysis: An approach to modeling imperfect repairs, International Journal of Production Economics 89 57–68.
Chukova&Johnston(2006), Two-dimensional warranty repair strategy based on minimal and complete repairs, Mathematical and Computer Modelling 44 1133–1143.
Giri&Dohi(2007), Inspection scheduling for imperfect production processes under free repair warranty contract, European Journal of Operational Research 183 238–252.
Cruz& Rincon(2012), Medical device maintenance outsourcing: Have operation management research and management theories forgotten the medical engineering community? A mapping review, European Journal of Operational Research 221186–197.
Huang&Fang(2008), A cost sharing warranty policy for products with deterioration, IEEE TRANSACTIONS ON ENGINEERING MANAGEMENT, VOL. 55, NO. 4, 671–627.
Huang&Zhuo(2004), Estimation of future breakdowns to determine optimal warranty policies for products with deterioration, Reliability Engineering and System Safety 84 163–168.
Iskandar&Murthy(2003), Repair-replace strategies for two-dimensional warranty policy, " Mathematical and Computer Modelling 38 1223–1241.
Iskandar et al(2005), A new repair-repalce strategy for items sold with a two-dimensional warranty, Computers & Operations Research 32 669–682.
Jack et al(2000), Optimal repair-replacement strategies for a warranted product, International Journal Product Economics 67 95–100.
Jack et al(2009), “A repair-replace strategy based on usage rate for items sold with a two-dimensional warranty,” Reliability Engineering and System Safety 94 611–617.
Jiang et al(2006), On a conjecture of optimal repair-replacement strategies for warranted products,” Mathematical and Computer Modelling 44 963–972.
Jung&Park(2003), Optimal maintenance policies during the post-warranty period, Reliability Engineering and System Safety 82 173–185.
Jung et al(2008), Optimization of cost and downtime for replacement model following the expiration of warranty, Reliability Engineering and System Safety 93 995–1003.
Kim et al(2004), Warranty and discrete preventive maintenance, Reliability Engineering and System Safety 84 301–309.
Kim et al(2011),Optimal Maintenance Policies During the Post-Warranty Period for Second-Hand Item, IEEE.
Manna et al(2007),A use-rate based failure model for two-dimensional warranty, Computers & Industrial Engineering 52 229–240.
Manna et al(2008), A note on calculating cost of two-dimensional warranty policy, Computers & Industrial Engineering 54 1071–1077.
Pan&Thomas(2010), Repair and Replacement Decisions for Warranted Products Under Markov Deterioration, IEEE TRANSACTIONS ON RELIABILITY, VOL. 59, NO. 2, , 368–373.
Park&Pham(2010), Warranty cost analyses using quasi-renewal processes for multicomponent systems, IEEE TRANSACTIONS ON SYSTEMS, MAN, AND CYBERNETICS-PART A: SYSTEMS AND HUMANS, VOL. 40, NO. 6, 1329–1340.
Park&Pham(2010), Altered quasi-renewal concepts for modeling renewable warranty costs with imperfect repairs, Mathematical and Computer Modelling 52 1435–1450.
Park&Pham(2012), Warranty cost analysis for k-out-of-n systems with 2-D warranty, IEEE TRANSACTIONS ON SYSTEMS, MAN, AND CYBERNETICS-PART A: SYSTEMS AND HUMANS, VOL. 42, NO. 4, 947–956.
Pascual&Ortega(2006), Optimal replacement and overhaul decisions with imperfect maintenance and warranty contracts, Reliability Engineering and System Safety 91 241–248.
Rao(2011),A decision support model for warranty servicing of repairable items, Computers & Operations Research 38 112–130.
Sana(2012), Preventive maintenance and optimal buffer inventory for products sold with warranty in an imperfect production system, International Journal of Production Research 50(23), 1 6763–6774.
Sheu(2005), Warranty strategy acconts for bathtub failure rate and random minimal repair cost, Computers and Mathematics with Applications 491233–1242.
Su&Shen(2012), Analysis of extended warranty policies with different repair options, Engineering Failure Analysis 25 49–62.
Tarakci et al(2006), Maintenance outsourcing of a multi-process manufacturing system with multiple contractors,"IIE Transactions 38, 67–78.
Tarakci et al(2009), Learning effects on maintenance outsourcing, European Journal of Operational Research 192 138–150.
Tsai et al(2011), Optimal maintenance time for imperfect maintenance actions on repairable product ,"Computers & Industrial Engineering 60 744–749.
Vahdani et al(2011), Warranty servicing for discretely degrading items with non-zero repair time under renewing warranty, Computers & Industrial Engineering doi:10.1016/j.cie.2011.08.012.
Vahdani et al(2012), Two fuzzy possibilistic bi-objective zero-one programming models for outsourcing the equipment maintenance problem, Engineering Optimization Vol. 44, No. 7, 801–820.
Wang&Sheu(2003), Determining the optimal production–maintenance policy with inspection errors: using a Markov chain, Computers & Operations Research 30 1–17.

Wu et al(2006), Determination of price and warranty length for a normal lifetime distributed product, International Journal Production Economics 102 95–107.

Wu et al(2011), On a general periodic preventive maintenance policy incorporating warranty contracts and system ageing losses, International Journal Production Economics 129 102–110.

Yeh et al(2005),“Optimal age-replacement policy for nonrepairable products under renewing free-replacement warranty,” IEEE TRANSACTIONS ON RELIABILITY, VOL.54 NO. 1, 92–97.

Yeh&Chen(2005), Optimal preventive maintenance warranty policies for repairable products with age-dependent maintenance cost, International Journal of Reliability, Quality and Safety Engineering Vol. 12, No. 2 111–125.

Yeh&Chen(2006), Optimal lot size and inspection policy for products sold with warranty, European Journal of Operational Research 174 766–776.

Yun et al(2008), Warranty servicing with imperfect repair,” International Journal Production Economics 111 159–169.

Management, Information and Educational Engineering – Liu, Sung & Yao (Eds)
© 2015 Taylor & Francis Group, London, ISBN: 978-1-138-02728-2

The utilization of story inspiration at the introduction stage of management courses

Xiu Lai Gu
Bengbu Automobile NCO Academy, Bengbu, China

ABSTRACT: Story inspiration is that teachers select a flexibly appropriate management story, according to teaching purpose and contents, and inspire students to think and analyze the implicit knowledge of management science to realize inferring other things from one fact and implementing divergent thinking. As a kind of important teaching method, Story Inspiration has important significance at the introduction stage of management courses.

KEYWORDS: Story illumination, management teaching, utilization.

1 THE PURPOSE OF STORY INSPIRATION UTILIZATION AT THE INTRODUCTION STAGE OF MANAGEMENT COURSES

1.1 *Attracting students' attention rapidly by stories*

Story inspiration methodology means teachers give stories at the beginning of classes. With the teachers' vivid description of stories' contents and plots, students' disturbed attention have been moved to teachers quickly and have been attracted to the plots, so that students can attach their attention to understanding the contents and plots, which lay a good foundation for the real management teaching.

1.2 *Developing students' thought and intelligence by story inspiration*

With Story Inspiration Method, teachers describe the plots and content, and invite students to think and found implicit knowledge concerning Management Science, so enlighten students' thought and expand intelligence, and exert the subjective initiative of the students to learn and dig their inner wisdom.

1.3 *Making students understand the contents of the course by story inspiration and grasp the given contents from a strategically advantageous position*

That buries the foreshadowing and does strong bedding for the further study of corresponding knowledge of Management Science.

2 THE WAY OF THE UTILIZATION OF STORY INSPIRATION AT THE INTRODUCTION STAGE OF MANAGEMENT COURSES

Because Story Inspiration has been used at the introduction stage of management courses with purpose and pertinence, it needs strong method. Without certain skills, its teaching quality will be not as decent as expected. Combining with teaching practice, the writer thinks that teachers must know the following stages when using Story Inspiration. First of all, according to teaching purpose and contents and the characteristics of the teaching object, teachers should carefully select appropriate stories of management. This stage is very important. If the chosen stories can not accurately reflect teaching contents, the effect of teaching will be discounted. The stories mentioned in the essay include philosophical fable stories and the stories of history, life, drama, humor, and literature, etc. Secondly, teachers ask some questions about the stories, that is, put one or two problems for thinking. Thirdly, teachers, according to the given story, actively encourage students to think, analyze and discuss the given problems, and put forward relevant management knowledge contained in the story. Finally, teachers encourage the students to speak their mind freely about the problems they had considered. After the students' speech, teachers should summarize, synthesize and evaluate the views of students, and then introduce the formal teaching content of the management course (Especially for the stories which cannot be understood easily).

3 THE ANALYSIS OF THE OBTAINED TEACHING EFFECT OF STORY INSPIRATION ON THE INTRODUCTION STAGE OF MANAGEMENT COURSES

Before the beginning of the formal course, teachers must describe the stories for students, and ask questions to make students think, analyze and discuss questions, which have an essential effect to inspirit students' learning interest and cultivate their ability of independent thinking and analysis. With several period teaching practices, it is proved that story inspiration at the introduction stage of management courses will cause a good teaching effect and is welcome by students. Specifically, it has the following aspects.

3.1 *Inspiriting students' learning activity and interest*

It is well known that the stories themselves have strong attraction, and especially the carefully selected management stories themselves are more attractive. For example, when explaining the section, "the concept of management", teachers can choose the fable story, Kangaroos and the Cage, to lead the concept that we understand normally what is management in order to inspirit students to further explore and know the real meaning of management. When understanding the concepts of management, effective combination of the organizational resource, teachers can explain it with "the Operational Plan of Lion", which stimulate students to learn how to give full play to their advantages and strengths of each person, and realize the effective combination of organization resources so as to maximize the organization function. Moreover, at the chapter, Policy Decision, the fable story, Act in a way that Defeats the Purpose, can be selected, which can illustrate the importance of policy decision and excite students' interest in the management process, policy decision. By using the philosophical fable stories, teachers can grasp students' curiosity and further stimulate their learning interest. So, students can change their learning from hardship to happiness, and change their passive learning to active exploration.

3.2 *Impressing deeply the knowledge students have studied*

Nowadays, in the process of communication among people, as for the evaluation of a person, the first impression often is particularly important. For the understanding of management knowledge, its introduction, just like the first impression in the process of communication among people, is very significant. In the chapter, Organization Environment, the introduction can be selected as The Story about the Wolf –child. In the chapter, the Art of Communication in Leadership, the story, The Scholar to Buy Firewood, can be introduced. At the Summary of Leader, Three Parrots can be used. Such combination of stories and knowledge can lead students to have a one-to-one impression. In future, when the story about the wolf –child is mentioned, they will remember the corresponding knowledge of organizational environments. As for the story of Three Parrots, they remind of the corresponding knowledge of the Summary of Leader. For students, the introduction process is a process of thinking, analyzing, discussing and discovering management knowledge. Such knowledge obtained from students' brain activities will be firmer, deeper and clearer.

3.3 *The stories of course introduction are accompanied by the process of asking questions, thinking and discussion*

Which provide opportunity for students to exercise and improve their language expression ability, and the ability to think and analyze problems. For instance, when the contents are mentioned, how to correctly use the excitation method to award what is necessary and award what is needed, the story, The rich man and the beggar, can be introduced, which can explain the importance of using correctly the method of simulation. Teachers can say,... then ask that as for a manager or leader, how they can fully mobilize the enthusiasm of the work with award and give full play to the functions of reward. Students themselves can draw their conclusion by their own thinking, analysis and discussion. When awarding, the manager or leader must award what is necessary and improve the effect of awarding in accordance with different objects. Teachers can comprehensively evaluate the students' views and put forward the importance of inspiration or awarding, which is to award what is necessary. In this way, students have a chance to think, analyze and discuss, which can greatly improve teaching effectiveness and effect, so that students can improve their ability of thinking and analyzing.

3.4 *By the management science teaching with story introduction*

Students can easily understand the following knowledge, reduce the distance among the management phenomenon or knowledge in theory and the real life, and further improve the post ability. For example, in the section of Object Management, at the beginning of the class, teachers should introduce the fable story, Heel, to show that details and the weak link cannot be ignored in object management, which pull the student back from the process of theoretic study to real post need. From the study of the fable story, students can deeply understand the importance of object management and the problems need to pay attention to the object management in a future post. So students can understand visually and concretely the concepts of object management and shorten the distance between theory and practice and lay a solid foundation for the future post.

Management, Information and Educational Engineering – Liu, Sung & Yao (Eds)
 ISBN: 978-1-138-02728-2

How to break down ghost towns predicament in China

Xiu Hua Tian & Qiu Ying Ge
Business Institute, Anhui University of Finance and Economics, Bengbu, Anhui, China

ABSTRACT: As the urbanization develops very fast, an increasing number of new cities, which are constructed according to new rules and high standards, is emerging in China. Some of them are known as ghost towns, which used to describe the cities being apparently built on a whim with high vacancy rates and no lights in the evenings. The recession caused ghost towns in the U.S., while the reasons for China are rapid urbanization, vague positioning of the city function and inadequate industry development. The solutions are to take the new-type urbanization as the strategic focus for economic restructuring, to develop characteristic industries and market and to improve the quality of public service with further market-oriented reforms in China.

KEYWORDS: Ghost towns; Real estate; Urbanization; Vacant properties.

1 INTRODUCTION

The emergence of cities is a sign that human society has moved toward more maturity and civilization, and also is an advanced form of human social life. Since China's reform and opening up, many Chinese cities have begun building new areas and new city districts to relieve population explosion, land shortage, traffic congestion and environmental pollution in the older sections of the cities. According to the China Statistical Yearbook of 2013, China's urban built-up area, that is the area of an urbanized region being smaller than the area of administrative division, increased from 12856 square kilometers in 1990 to 45566 square kilometers in 2012, which equals a 354% of the increase. It's almost a sidelight about an expanding China with an active building market that there are 330063 national first-grade registered architects working in China.

Table 1. China's urban built-up area.

Year	1990	1995	2000	2010	2011	2012
ABD	12856	19264	22439	40058	43603	45566

Source: The China Statistical Yearbook of 2013
Note: ABD means Area of Built Districts (sq.km)

Beijing City, as the capital of China, the changes of real estate development can be a microcosm of the nationwide development of real estate. The development investment of the Beijing City real estate is 568.4 billion yuan from 1991 to 1995, but the one is 10863.2 billion yuan from 2006 to 2010, a 19-fold increase. Land acquisition cost is 161.5 billion yuan from 1996 to 2000, while the one is 3642.0 billion yuan from 2006 to 2010, about a 23-fold increase.

Table 2. Beijing City real estate development investments.

Period of time	1991–1995	1996–2000	2001–2005	2006–2010
REDI	568.4	1979.5	5974.0	10863.2
LAC		161.5	993.6	3642.0

Source: The Beijing Statistical Yearbook of 2013
Note: REDI means Real estate development investments (billion yuan); LAC means Land acquisition costs (billion yuan);

However, rapid urbanization has late effects, and a flood of ghost towns with empty houses is the most prominent representative. The term Ghost Town originally expresses the cities with mythology and the supernatural, but now with the new semantic migration phenomenon in the process of urbanization, the word, especially means urban areas with high vacancy rates and low occupancy rates in real estate in the process of urbanization (Nie and Liu, 2013). Because most of the apartments appear empty and the wide streets are almost deserted, Kangbashi District, in Ordos City, China's Inner Mongolia, earns the tag of "ghost town". So the ghost town actually is an empty city, lack of people and industries. About every big city has built a whole new area which the planning area and population are even higher in older neighborhoods, boosted by rapid urbanization in China.

Not only Ordos, there are ghost towns with empty houses like Kangbashi in other places of China, arousing reflection and reconsidering around the political circle and society as a whole. Taking CNKI (China National Knowledge Infrastructure Database) as the retrieval source, we can find 23 articles in 2009, 34 articles in 2010, 61 articles in 2011, 75 articles in 2012 and 171 articles in 2013 by searching for the words "ghost town" across the reference topic. Thus, it can be seen that the whole society has paid high attention to the ghost towns of empty houses.

2 CHINA'S GHOST TOWNS

When one now thinks of China's ghost towns, Ordos City in Inner Mongolia usually come to mind. Kangbashi district includes many large office towers, administrative centers, government buildings, museums, theaters, schools' playing fields and exhibition centers, located near Dongsheng District that is Ordos's important center of economy, finance, transportation, information and culture and has about 500, 000 inhabitants which is about a third of Ordos population. The original planning of the population in the New District Kangbashi is 1,000,000 inhabitants, but the current population is only 80, 000. Ordos has made brilliant achievements as a Chinese version of Dubai while nowadays is enduring its slump: coal resource depletion, investors staying away and an empty city formed.

Another typical ghost town is Chenggong New Area, in China's southern city of Kunming, Yunan Province. According to the reports on xinhuanet.com (on December 6, 2013) and Xinhua Daily (on February 28, 2014) in China, Yunan Province has decided the construction of the modern new Kunming as early as 2003, and the planning for Chenggong is to become Kunming's political, cultural and financial center. The planning control area of Chenggong New Area is 153 square kilometers when completed in 2020 with a population of 1,000,000. Chenggong New Area has built 17 neighborhoods which only two of them achieved occupancy levels of more than 90% and yet is building 15 neighborhoods with a total area of 3 million square meters. Because of similarly deserted roads, high-rises and government offices, Chenggong has raised the alarms of ghost towns.

As many as a dozen of other Chinese cities just like Ordos with sprawling ghost town annexes have been reported, but all have certain common characteristics, i.e., vast expense, luxury buildings, wide and quiet streets, and high housing vacancy rate. Certainly these ghost towns are primarily in three or four line city which cannot absorb a large employing population because of lacking resources and lagging economics, thus these cities have not sufficient strength to continue to support the real estate markets.

3 LITERATURE REVIEW

Chen (2013) points out that ghost towns have constantly increased in number and been widely criticized in China. Nie and Liu (2013) indicate that the ghost town is broadly concerned not only because of its universality, but also the severe waste of land resources, which results in huge economic losses and serious damage to hinder positive urban developments. Sha (2014) thinks that China's property (real estate) boom has made substantial contributions to economic growth and considerably improved the housing conditions of urban residents. But there has been too much speculation, which has caused property prices to deviate from their true value. A direct consequence of high speculative housing purchases is soaring house prices, which, in turn, have made housing unaffordable to some social groups. In recent years, there have been two increasingly common phenomena: one is 'ant tribes' who cannot afford the high prices and must share a tiny room with their peers; and the other is dangerously high vacancy rates in 'ghost cities' with no residents. Legény and Špaček (2014) find that China currently is moving towards a recession in the building market and the ghost cities are coming into being. They also question how strong the architectural market demand in these countries really is, because the proportion of poor people is large. The housing price changes diffuse from the city center to the suburbs and those changes in the city center housing bubble cause price movements in the suburbs (Teng et al., 2013). Because the housing market acts as a reward system for rent seeking activities, thereby influencing investment incentives, the housing bubble size in the suburbs is larger than that in the city center, verifying the prediction in theory that activities of rent seekers contribute to the bubble contagion. Teng et al. (2013) explain further that the current vacancy rate in the suburbs is higher than that in the city center, suggesting that bubble contagion leads to overdevelopment, which then results in the appearance of so-called ghost towns. Ghost towns are to be sorted into disaster "ghost towns", declining type "ghost towns" and planning type "ghost towns".

This phenomenon of China ghost towns is not unique, and for various reasons, many countries have been the problems of urban decay and residents migrating (Yang, 2013). In America, the economic vitality and employment of some traditional industrial cities remains depressed. For instance, Detroit was in bankruptcy as its automotive sector lost competitiveness. As another example, a mining community could become a ghost town when the mine's minerals are exhausted. (Hall, 2013) But Yang (2013) thinks that government-led has a more significant effect on the formation of ghost towns than financial ruin. As the "Great Recession" of 2007 to 2009 has taken a great toll on housing markets in most cities and metropolitan areas in all parts of the country, Follain (2010) defines the concept Declining City which means that the people have left, but the houses, apartment buildings, offices and storefronts remain, an extreme form being a ghost town, that is a town that lost its reason for being.

In order to put an end to ghost town, city managers and planners can have the rational plan, focus on the city's cultural and ecological harmony, implant emerging industry, and transfer urban function, etc. (Nie and Liu, 2013). With trilateral efforts between the central government, the local governments and the real estate agencies, the real estate industry in China will be developed healthily, and the citizens who really need houses will realize their dreams (Chen, 2013).

4 THE PROBLEMS WITH GHOST TOWNS

4.1 *Economic problems*

Local real estate and economic development are unlikely to be sustained because of inadequate industry support. During the process of new town planning and construction, local governments fail to take introducing industries into account, so jobs are scarce in a new town. A large number of houses left vacant are an inevitable outcome of industries hollow. So given enough time, this could be dangerous for the continuity of local governments and real estate industry. Real estate functions mismatches, which mean the real estate investment, not as the real estate expense, could run so far out of control. Then the real estate bubble gets larger and larger, and a collapse seems increasingly to be true. Moreover, the construction of new towns was accomplished by means of bank credit, so a collapse of the real estate bubble will easily bring about regional financial risks.

4.2 *Government problems*

Local government is the operator of constructing cities, so no matter where the ghost town appears, local government has played a dominant role. According to a report on the debt of 36 local governments from China's National Audit Office on June 10, 2013, the debt ratios of 9 China province level capital cities are over 100% for large-scale urban development at the end of 2012. Local government can rake in selling land to developers, and then by drafting development plans for the land the government can hike its value several times over. Many local governments are no longer relying so much on tax revenues, and are instead depending much more on selling land to generate revenue. However, local governments maybe make seemingly remarkable achievements by using land sales income to speed up the urbanization process. But even if the local GDP growth remains strong for the time being, local people, maybe doubt the capacity of local government and are disgusted by its greed because of high unemployment and lots of unsold homes.

4.3 *Social problems*

Encroaching urbanization should be a natural result of the economic development, and the urbanization is for the sake of people's happiness, further economic development and the improvement of people's living standard. But ghost towns of empty houses can cause or increase social unrest and distrust. First, because homes were grossly overpriced, fueled by binge borrowing, the broad masses of working-class and low-income households with rigid demand could not afford to purchase a house or be a house slave who has to work hard for the mortgage loans. Most migrant workers are living in temporary housing assigned by their employers, or in rental units. Second, a small number of very rich people and speculators are hoarding a large number of properties to fuel price rises, being bound to produce serious real estate bubble without regulation. Third, ironically, some peasants whose land has been expropriated for commercial development live in high-rise buildings as compensation. But they, most of whom are elderly, will have a hard time adapting new ways of living because of their lack of skills to work with the new environment. The real fear, however, is that for the past two decades or more, urbanization in some parts of China cannot significantly improve most people's quality of life, especially people's quality of spiritual life, contributing to this divided society and being unable to carry out everything local government promises.

5 GHOST TOWNS FORMATION MECHANISM

There are many thrusts to promote the formation of Ghost towns, including the governments, developers, investors, financial institutions and consumers, as shown in Figure 1.

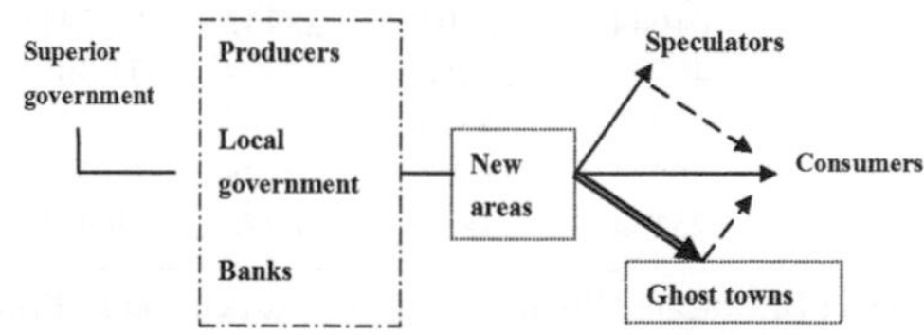

Figure 1. Many thrusts to promote the formation of Ghost towns.

Residential buyers for non-investment purpose who are on the low end of the real estate supply chain may buy homes within their capability. If the end-consumers do not respond to the enthusiasms from a beneficial community of local governments and developers, then he a new town will only turn to ghost one because of market mechanisms. Therefore, no consumer demand is the final decision factor for the formation of Ghost towns, but the governments, producers, speculators and financial institutions play a role in people's decision to buy a house.

5.1 *Governments*

The governments are necessary to be divided into local government and its superior government. As far as the superior government (for example, the central government) is concerned, there are at least three key reasons resulting in the formation of ghost towns. First, local officials are judged largely on their ability to increase economic growth, that is, officials' performance is based on the GDP. China is at the moment of the social transformation which refers to transferring the system, transforming economic growth pattern and adjusting economic structure. But a check-up system of officials' career achievement, which has an obsession with GDP figures, investment attraction and infrastructure and fixed asset investment, is little changed in the past a very long time, making local governments more aggressive to look for ways to work on big projects, big investments and repeated construction at a low level. Ghost towns are probably the most extreme manifestation of these activities. Second, the disequilibrium of government's routine power and financial power causes local governments trying to look for other sources of income, and local officials quickly realized that income from land can be a major source of fiscal income in China.

Table 3. Government expenditure and revenue by region total (100 million yuan).

Year	PBE	PBR	DER	RDR
2012	107188	61078	46110	75.49%
2011	92734	52547	40187	76.48%
2010	73884	40613	33271	81.92%
2009	61044	32603	28442	87.24%
2008	49248	28650	20599	71.90%
2007	38339	23573	14767	62.64%
2006	30431	18303	12128	66.26%
2005	25154	14884	10270	69.00%

Note: PBE means Public budgetary expenditure; PBR means Public budgetary revenue; DER means Difference between the expenditure and the revenue; RDR means Ratio of the difference to the revenue

As shown in Table 3. public budgetary expenditure by China's 31 provinces, municipalities and autonomous regions is much higher than public budgetary revenue, and the ratio of the difference between the expenditure and the revenue to the revenue is more than 60%, hitting records with 87.24% in 2009. Commercialization of assigned land brings local governments a high amount of extra-budgetary revenues. So local governments have been especially keen to buy low (from relocated households) and sell high (to property developers) to serve three purposes: bolstering the city's image, raising GDP growth and getting considerable sums of land-transferring fees. Third, regulatory policy of real estate markets is still not perfect. Local governments blindly sell land, approve real estate projects and don't care vacancy rate. The Less effective performance of real estate information system may cause corruption and rent seeking behavior.

Moreover, at the beginning of the planning development zone, policymakers might ignore an over-all city planning and the scientific instruction. In fact, many new towns with luxury appearance are not practical because the surrounding environment, traffic, property management and so on. A growth mode of heavily reliant on the housing market is not sustainable due to weak internal demand and insufficient job opportunity in the new urban district without the supports of other businesses and services. Some questions about why broad boulevards are unimpeded by traffic, office buildings stand vacant, and pedestrians are in short supply at all in the Kangbashi New Area have brought attention to whether Ordos City should plan and build it at all. The following Tables 4 shows the comparison of Ordos and Shanghai.

Table 4. The comparison of Ordos and Shanghai.

City	Years	RP	ABD	RRR
Ordos	2007	154.8	99.7	1.55
	2012	200.4	250.2	0.80
Shanghai	2007	2063.6	885.7	2.33
	2012	2380.4	998.0	2.39

Note: RP means Resident population at year-end (ten thousand men); ABD means Area of Built Districts (sq.km); RRR means The ratio of Resident population at year-end to Area of built districts

Shanghai is a municipality directly under the central government of China, and a prosperous international metropolis. Shanghai is located in the Yangtze River estuary, adjacent to the East China Sea in the east, lying across the sea from the Japanese island of Kyushu, bordering on Hangzhou Bay in the south, neighboring on Jiangsu and Zhejiang provinces in the west. Shanghai together with Jiangsu and Zhejiang provinces form China's largest economic zone - Yangtze River Delta Economic Circle, and Shanghai is the core of the Circle. Ordos' urban sprawl is more rapid that the area of built districts increased from 99.7 square kilometers in 2007 to 250.2 square kilometers in 2012, which equals a 151% of increase, while the ratio of the resident population at year-end of the area of built districts declined to 0.80 from 1.55 within 5 years. Shanghai's urbanization increased by 13%, while the ratio also increased by 0.06 from 2.33 to 2.39 within 5 years. For that matter, the problems of population-density and urban function overload simply do not exist in Ordos City. So, planning and building Kangbashi New Area lacks of scientific proof.

5.2 *Producers*

Following the rule of maximization of profits, real-estate companies build houses and sell houses for the cycle business activity. The developers only think whether the sale of the realty products would be successfully and effectively or not, but whether the houses for the purpose of reside or invest have taken things lightly. Real-estate companies might ignore the business ecosystem indulging in strong short-run demand. For instance, at present, the park, hospitals, schools, employment, and traffic have become important factors to consider for buyers. But the real estate companies couldn't go with this special demand in time, resulting in serious homogenization on house design, item theme and so on. Because local governments which rely more on the land revenue provide preferential policies and financing platform for the developers, real-estate companies are excited to build a newly developed area of the city on a large scale with blind disorder and lack of planning. Under the circumstances of the diversified demand tendency and changeability of real estate markets in China, once the risks from those factors take into effect, some developers may be in great loss.

5.3 *Real estate speculators*

There are two types demand of consumption and investment when one purchases house on. The investment and speculative demand intensified China's real estate market supply and demand imbalance in recent years. On the one hand, investors couldn't find a reliable alternative investment because of the Renminbi external revaluation and internal depreciation in a down economy; on the other hand real estate is being a tools people loving because of its preserving and increasing value, high rate of paying back over the years. The speculators and investors don't place much value in the houses in itself (quality and living conditions), but switch operation as a means of making profit. The vicious speculation of property market speculators, such as paying loan by loan, also further pushes up the housing price by using flaws of the finance system. Namely, the building boom is driven by frenzied investors, not the "solid" demand for residential housing. So, Teng et al. (2013) think that producers are agents who contribute to the aggregate output of a region, whereas rent seekers make no contribution.

5.4 *Banks*

There is a close relationship between the prosperity of the real estate market and bank credit in China, such as heavily relying on bank loans. The imbalance of supply and demand results in monopoly profits, which induces banks to extend excessive credit to real estate market to make a greater profit margin from both developers and buyers. The real estate trade is a typical intensive trade of fund and without huge capital input the builders and developers can't move a step. The over-expansion of bank loans to the real estate sector plays a very important role in the forming process of ghost towns in China.

6 HOW TO RESOLVE THE PREDICAMENT

China released the National New-type Urbanization Plan (2014~2020) in March, 2014 to guide the healthy development of China's urbanization. In order to promote the healthy development of urbanization in China, the two problems must be solved by adhering to the path of urbanization with Chinese characteristics: turn exiting ghost towns into a thriving center for businesses and residents and avert more ones. The ways of cleaning up the existing ghost towns are (1) to accelerate the development of supporting industries as a magnet for employment around it and then absorb those empty homes by the workers; (2) to consummate it peripheral public facilities to bring the convenience to community residents' life and employment; (3) to impose taxes on the vacant residential property over a specified period of time in order to prompt the developer to reconsider the relationship between house-building and consumption and help take some speculative froth of real estate; and (4) to wipe out ghost towns as an assessment system, which would be applied to local officials (Feng, 2013).

The ways to avoid increasing the size of ghost towns are as follows.

6.1 *Interventions at national level*

The examining system of local officials' achievements should be improved to avoid an only-GDP-oriented mistake towards more sustainable development model. The new development model shifts observation points away from infrastructure and fixed asset investment towards the number of Blue Sky Days, water quality, traffic congestion situation, public security, health service, etc. The Central Government encourages local governments to accelerate the upgrading of industrial transformation, thus to drive sound and rapid development of economy. The rules with respect to land exploitation must be strictly implemented to avoid land abuses. Land profit is an important cause of the property industry treated as an important grasp on economic growth for many local governments. Local governments' financial resources certainly are not too dependent on land sale income by balancing the duty and financial power. The nationwide real estate information management system should be set up and run to make personal housing information more transparency, to restrict speculative investments effectively and to prevent corruption in the real estate sector. The Chinese housing security system should put even more emphasis on low-income housing in a way that not only ensure the basic living conditions of low-income people, but also offer options on commodity houses for wealthy and middle-class families. The housing finance market should be enhanced regulations to abstain demand of speculation. And the approval of building new urban district should be suspended or be more stringent to ensure the reliability and rationality of the urbanization plans.

6.2 *Interventions at local level*

Local governments could not continue to see real estate as the most important growth sectors and should change the unbalanced pattern of economic development by making full use of local advantage resources of nature, culture, history and geography to facilitate a multi-pronged economy. The expanding gap between the rich and poor should be reduced in order to increase the purchasing power of the middle-low-income class. Urbanization is the product of social and economic development and also is the only way to gradually realize the modernization in China, but urbanization does not inevitably lead to an economic boom. So, the less-developed cities with a very low population density which have devoted a serious amount of resources to built new urban districts will no doubt have a recipe for disaster.

6.3 *Decisions at corporate level*

The builders and property firms should make an accurate assessment of the housing market to for locating correctness. It is an undeniable fact that China's property market has still a very large room for development with the new-type urbanization as long as the corporate avoid homogeneity. There is every indication that at present commercial real estate lagging behind the residential sector and municipal facilities being really behind urban expansion have turned some new towns into ghost towns.

7 CONCLUSIONS

The proximate causes of ghost towns, taking shape may have been a serious oversupply of commercial and residential property, but the ultimate reason is the product of the dislocation between politics and business. Local governments and developers have formed a close-knit community of interests which tends to create havoc in the market. In turn, ghost towns are the punishment by the marketplace. In addition, the financial sector has quickly become a booster of ghost towns because real estate enterprises have limited sources of funds mainly from bank loan and property presales. So, deserted ghost towns are against the general rule of urbanization because building the new urban districts became a game of the authority and capital without considering the actual demand in the region. And deserted ghost towns are inconsistent with the path of economic progress when rigid consumption demand is pushed out of the housing market by speculative demand. Local governments should take new-type urbanization as the strategic focus for economic restructuring and the fundamental approach to the removal of urban-rural dual structure; develop characteristic industries and market; and improve the quality of public service with further market-oriented reforms.

ACKNOWLEDGEMENT

Financially supported by the provincial academic research projects in Anhui, China (Grant No.: 2008sk214).

REFERENCES

Chen Y H., 2013, 25(12):31–35, On Ghost Towns-Uncovering the causes of ghost towns and restraining the property bubbles [J]. Social Sciences Journal of Universities in Shanxi.

Feng H N., 2013–08–08, Accelerating the process of new-type urbanization behind wiping out ghost towns [N]. Xinhua Daily Telegraph.

Follain J R., A study of real estate markets in declining cities [J]. Research Institute for Housing America (Special Report), Retrieved from http://www. rockinst. org/pdf/cities_and_neighborhoods/2011–01–06-Real_Estate_Declining_Cities. pdf, 2010.

Hall J., 2014, 16(1):20–25Africa's internal diaspora: Africa-wide-continental overview [J]. Africa Conflict Monthly Monitor, 2013: 4–7.

Legény J, Špaček R. Trapped by crisis: the plight of architects in Europe [J], Global Journal of Engineering Education.

Nie X Y, Liu X J., 2013, 29(4):112–117, Types of "Ghost Towns" in the process of urbanization and countermeasures [J], Journal of Nantong University (Social Science Edition).

Sha K., 2014, 42(3): 391–392 Destructive construction and constructive conflicts [J], Building Research & Information.

Teng H J, Chang C O, Yu C M, et al., January 5, 2013, Rent Seekers vs. Producers in Cities: Contagious Housing Bubbles Force Housing Price Diffusion on Urban Overdevelopment [J]. Available at SSRN: http://ssrn.com/abstract=2196798 or http://dx.doi.org/10.2139/ssrn.2196798.

Yang T., 2014–03–04, The urbanization path-from ghost towns breaking out perspective [N], Guangzhou Daily.

Management, Information and Educational Engineering – Liu, Sung & Yao (Eds)
© 2015 Taylor & Francis Group, London, ISBN: 978-1-138-02728-2

Approach on the "last kilometer" problem with the commercial mode

Chun Hang Wang, Zhe An, Yu Xiang Liu & Li Feng Yao
Tangshan Vocational & Technical College, China

ABSTRACT: The biggest resistance of Chinese farmers' information life is they cannot get and use agricultural information in their homes; the government calls this problem the "last kilometer" problem. This approach analyzed the problem and found a commercial mode by analyzing a company, which used its own business channel to set up an information network. Compared with the other promotion mode, the commercial mode has many advantages such as sustainability, self learning, and self development, and it is an effective mode.

KEYWORDS: Agricultural information; Mode; "Last kilometer" problem.

The "last kilometer" problem refers to the phenomenon that the agricultural information can't transform to the village fluently, the farmers can't get the information product directly, the investment of the system can't produce the expected benefits and the farmers can't use the information to get more money. At present, many researches have been done to solve the problem, but almost every mode, use non commercial modes, which have many issues.

1 THE ISSUES OF NON COMMERCIAL INFORMATION EXTENSION MODEL

1.1 *The benefit between the promotion, organization and the information system does not agree*

It is a basic rule in economics that the benefit must meet the need of the investor. But in the most information promotion modes big benefit gaps really exist. In the view of government, the information promotion activity is a long time project and very difficult to quantify. So the government is more willing to invest in the hardware of the information system and not very care about the soft part. In the angle of non benefit organization, it's not easy to find a way to touch the farmers' group and the information service is not like emergency as others, for example, health.

1.2 *It's very difficult for the farmers to get an analysis of the correctness of the information*

Though the basic information equipment system and the agricultural information database have been established by the government department, the source of the information and utilization of the information have still been problems. One side, the education level of the farmer is not very high, on the other, it is difficult for them to use the equipment of information and the difficulty is they can't analyze the right or wrong of the information available from many information resources. The other side, the government mechanism makes the authoritative information update slow, so the farmer can't get the newest and fastest useful information.

1.3 *The share of information equipment in rural areas is not high*

Based on the survey of the CNNIC (China Internet Network Information Center), until April 2013, rural Internet users accounted for only 27% of the number of Internet users in the country[3]. Furthermore, the statistical data do not reflect the number of Internet users engaged in agricultural production. The author investigated in a large county of vegetable production in Hebei province and found that in better economic conditions village only 10% of family have a computer, the statistic data in the villages that not rich enough are even lower.

Based on the above three points, an efficient information promotion system must have three traits. First, it must have an effective information resource. The information in the system must be correct and new. Second the farmer can get the information easily, they did not need to operate the complex machine or only several very simple steps can they get the information they want. Third the farmers did not need to pay more money for the information. We found a company which has established a system like that.

2 THE RE MODEL

RE company is an agricultural chain enterprise, with more than 260 direct chain stores, covering more than 70% administrative villages of the county, sells agricultural material to the farmers. Beside this, the company built a promotion system of agricultural information by using its business channel, explored a new way of information promotion.

2.1 *Build the information strongholds with chain stores*

There are more than 260 commercial chain stores in the RE company, all the stores are located in the village. The company set an integrated machine and a printer in the every sore, all the machine connected with the website of the company, the farmer can use the machine to inquiry the information, if they have any question about how to use the machine they can ask the staffs in the store who have been trained by the headquarters. If the farmers have questions that can't be solved by the information in the network, they can connect to the experts through the visual system, ask questions and show the leaves or branches to the expert by microphone and camera.

2.2 *Build the support system in the company headquarters*

A department of information was established in the company's headquarters, receiving, analyzing all kinds of problems in information system which was proposed by farmers. The questions will be sent to authorities or the relevant experts, and the solved results will be fed back to the information system. The other missions of the department are responsible for system updates, upgrades, maintenance and works on the staff technical training. At the same time, the department also contracts with experts in the universities and research institutes, hire them to judge the accuracy of the information that published on the website, send the newest technology information to the company and have remote interaction with the farmers. Through these methods the company can guarantee the accuracy, progressiveness, authority and timeliness of information that be released.

2.3 *Newspapers in the store*

As an effective complement to the computer information network, the company sponsored a newspaper called *New Agriculture*, which focus on agricultural hot points and practical technology. The papers were sent to the chain stores with the logistics, free to farmers.

2.4 *A win-win result*

Through the above market operation model, both the farmers and the company obtain good incomes. As an example, in 2010, three technologies were promoted through the information network: Apple high photosynthetic efficiency of pruning, Set high-quality bag and Reflecting film. The orchard which used the three technologies got high quality fruit rate reached more than 80%, compared to not using higher more than 1 times, and greatly increased the income of farmers. At the same time the company's sales are promoted. In 2008, the sales increased more than 20%. With the combination of the main business with information network, the RE company gets a win-win result.

3 COMMERCIAL MODE

From the example of the RE company can be seen above, we found a commercial agriculture information promotion mode that contains following features:

3.1 *A perfect organization*

One reason for the fail of the Information promotion system is the unreasonable structure of the organization. In commercial mode, comprehensive agricultural information was flowing through the channel of information networks and traditional business chain combination, which formed a relatively reasonable structure form of organization. The organization headquarters is the core and resource; stores are strongholds and terminal information release systems. The whole organization is stable and perfect.

3.2 *Solved the problem of information equipment fault*

At the commercial mode the farmers do not need to buy the information equipment, they can inquiry information, learn the latest technology, print related information content in the branch store when they buy the agricultural material or any time they like. Equipment maintenance and update are taken by the company professionally; farmers do not need to worry about information equipment problems.

3.3 *Solved the problem of information analysis and recommendation*

The company cooperates with the county forestry bureau, bureau of science and technology and other government departments. It also cooperates with teaching and scientific research units and higher education schools, so they can invite experts in agriculture as information personnel, screening and analyze information to ensure the accuracy of information dissemination.

3.4 *Ensuring the stability access to information system*

The farmer seldom study agricultural information systematically, they usually want the direct answer when meet a problem. While if it's difficult to find suitable results, the farmers will not keep the enthusiasm of access to the information system. RE agricultural information network combined with agricultural sales, the farmers can access the network stably.

3.5 *Win-win and sustainable development*

The benefits of information system promotion are from the sale of business production. So the mode avoids the drawbacks of input inconsistent of the government-led mode and other promotion mode, make a sustainable development of agricultural information systems.

4 COMPARISON OF TYPICAL MODE CASES

According to the information promotion problem, many domestic provinces and cities have tried a variety of patterns, the following table chose three typical cases in the carrier, approach, advantages and disadvantages to compare with the commercial model (*Table 1*).

Table 1. The comparison of typical mode cases.

Area	Mode	Carrier	Method	Advantages and disadvantage
Gansu province	School	Pupils and teachers	The government let the teacher print the message and the pupils take the message home	Advantages: correct information, teachers and students have a strong sense of responsibility. Disadvantages: Do not understand the needs of farmers, the cost is higher, not conducive to the spread of interest distribution mechanism, if the focus of government changed, difficult to sustain.
Hunan province	Order agriculture	Agricultural Technology Extension Station staff	Free consultation, fees printing	Advantages: accurate, strong pertinence, staff enthusiasm Disadvantages: Increase farmer's costs, bring a series of problems such as financial management.
Gaocheng city Hebei province	The government	The government administration units	Different departments are involved in the dissemination of information, establish information dissemination Hall	Advantages: accurate, comprehensive, one-stop service. Disadvantages: The government guidance is obvious, once the work center of gravity changes, is difficult to maintain, must focus on the incentive problems of staff
Commercial mode	The combination of dual network of mutual benefit and win-win	The enterprise	The establishment of information integrated in each branch with the new agricultural newspaper supplement	Advantages: Accurate information for sustainable development, organizational security information system, to solve the three problems in fault Disadvantages: Information just to the enterprise

5 IMPROVEMENT AND SUGGESTIONS

5.1 *Speed up the promotion of commercial mode*

Through several years of construction, integrated information service network has been formed basically by the enterprises such as the RE company. But the coverage of the network is still narrow, the rural information resources are still scarce, farmers still expect for more information. Acceleration of the construction of rural information service network, acceleration of the promotion of new varieties of agricultural technology has to be done. The government departments should deal with the promotion and corporate support, through the project, investment, and other means to accelerate the key construction, strengthen the mode.

5.2 *The information tendency*

The main body of the commercial mode is enterprise, the information promoted by them is inevitable with the tendency for their own business products, and this is the inherent disadvantages of the commercial mode. It can be solved through the following two ways, one is the government departments strengthen the market supervision and guidance of information promotion efforts, has issued relevant policy documents, standardize on promotion; two is integrating the agricultural industry association and the enterprise to one information release platform to release the correct information.

5.3 *The challenge of the new technology*

With the rapid popularization of smart mobile phone and price decreases of the 3G network, mobile phone has become the main equipment of farmers, the utilization rate as high as 78.9%[3], face this new change, agriculture information promotion network should make corresponding adjustment, one is based on the integrated machine in chain stores ,gradually transferred the machine layout to the information collection, collation, publication; two is to comply with the development trend of mobile intelligent information terminals, develop application on mobile phone to meet the different requirements for the production of large producers and agricultural economic man and the normal farmers.

6 CONCLUSIONS

The "last kilometer" problem is the main problem facing the present agricultural informatization, compared with other extension mode of agricultural information promotion, commercial mode more close to the farmers, has more development space and the ability of sustainable development. The practice of RE company shows commercial model can effectively solve many problems of information promotion such as hardware issues, training issues, the utilizing of information and so on. It is an effective way to promote information. But the system's innate information tendency, the challenges of new technology is worth noting.

ACKNOWLEDGMENTS

This paper is the research result of Ministry of science and technology project " Pollution-free facilities vegetables production and demonstration (Section of the state agricultural [2013]514)".

I would like to extend my sincere gratitude to Ms. *Zhe An*, for her instructive advice and useful suggestions on my research. I am deeply grateful for her help in the completion of this thesis. I am also deeply indebted to all the other authors and Science and Technology Bureau of Laoting County who help me a lot on investigating and collecting material.

REFERENCES

[1] Wang Yanxia 'Last mile', the cause of the problem and the solution countermeasure [J]. Journal of Northeastern University: Social Science Edition, 2005, 8 (3) 180–182 pages.

[2] Lin Tao On the agricultural informatization construction "the first one hundred meters problem. Journal of Southwest Agricultural" [J], 2006, 19 (5), 969–973.

[3] Development statistics report of Internet (2013 July) China Internet Network Information Center.

[4] Liu Lili He Yuan solving the agricultural information landing puzzle computer world [N] 2005, 12, 05.

Management, Information and Educational Engineering – Liu, Sung & Yao (Eds)
 ISBN: 978-1-138-02728-2

Empirical analysis of traditional retail's transformation to E-commerce

Bo Zhang, Hai Jun Zhang & Jing Tian Zhang
School of Information, Beijing Wuzi University, Beijing, China

ABSTRACT: In the last five years, China's online shopping scale was in a rapid growth. In 2013, total retail sales online has accounted for 7.9% of the total retail sales of the whole society. E-commerce enterprises represented by Tmall, Jingdong Mall, Suning Tesco etc. formed a large impact on the traditional retail industry. They squeezed the traditional retail market and snatched a large number of medium and high customers, forcing traditional retailers to combine online and offline business models. The impacts of e-commerce are different for different retailers and retailers should get involved in e-commerce, according to their local conditions respectively to exploit their own advantages.

KEYWORDS: Empirical Analysis; Traditional Retail; E-Commerce.

1 E-COMMERCE IN RETAIL BUSINESS AND DEVELOPMENT OF TRADITIONAL RETAIL COMPANIES

In 2013, China maintained a momentum of continued rapid growth in electronic commerce. The trade volume of E-Commerce exceeded 10 trillion yuan and achieved an increase of 26.8% from a year earlier. China has become the world's largest online retail market. In 2013, the size of Internet users reached 302 million in China, and the annual volume of Internet retail exceeded 1.85 trillion yuan and had an increase of 41.2%, which accounted for 7.9% of the total retail sales of social consumer goods.

In the second quarter of 2014, the volume of the online retail market reached 651.172 billion yuan in China, which had an increase of 13.14% compared to the first quarter of 2014. According to the data released by the National Bureau of Statistics, it showed that in the 2nd quarter of 2014, total retail sales of social consumer goods in China have reached 6.2118 trillion yuan, and the online retail market accounted for 10.48% of the total retail sales of social consumer goods in China, while the figure of the first quarter is 9.28%, which means the penetration of online shopping reached a new high.

E-commerce as a strategic industry played an important role in changing the mode of economic growth, promoting industrial restructuring and upgrading, and accelerating the modernization of circulation. E-commerce has become the main ways to boost domestic demand, to increase consumption, and to promote employment.

Online shopping saved total circulation cost because it cuts down trading links and reduced frictional cost, which became one of the fastest-growing emerging retail in recent years. After more than 10 years of development, China's online shopping industry has made great achievements and formed a trend of substituting traditional retailing. Meanwhile, in the next few years, the online shopping industry will continue to grow at 20% per year, and will have a growing substitution effect to traditional retail. Expected in 2017, online shopping trade size will reach 4.45 trillion, accounting for 12.4% of total retail sales of social consumer goods [1], which is shown in Figure 1.

Figure 1. Online shopping trade size.

In 2013, China has 300 million online shopping users, which accounts for 48.9% of total Internet users in China while the figure of 2012 is 42.9%. In recent years, the proportion of online shoppers to the total Internet users has a rapid growth in China, and online shopping is becoming a common Internet surfing behavior. The increased proportion is the key reason to promote the rapid development of China's online market.

The next few years, with the growth of China's Internet users slowing, the increase of online shopping users will have a certain extent reduce, but the online shopping users will maintain steady growth as a whole [1], which is shown in Figure 2.

Figure 2. The number of online shopping users in china.

2 IMPACT OF E-COMMERCE ON TRADITIONAL RETAIL

E-Commerce and traditional retail are engaged in a zero-sum competition, where one side must fall for the other to rise, because they are in the same market, strive for the same consumer group, and split the same "cake". E-Commerce has formed a huge impact on the traditional retail market.

2.1 *The deal size of the traditional retail market is squeezed*

Online shopping transactions in 2013 accounted for 7.9% of the total retail sales of social consumer goods, which had an increase of 1.6% compared with the proportion in 2012. It is estimated that by 2017 the proportion of online shopping transactions to total retail sales of social consumer goods will reach 12.4%, and the deal size of the traditional retail market is squeezed further, and this trend is likely to continue.

2.2 *Medium and high customers shift from traditional retail market to the online market*

In 2013, China has 300 million online shopping users, 80% of which are mainly young, white-collar urban professionals. Meanwhile, 7% of them are high-end customers, which spent more than $10,000 one year, and accounting for 40% of the overall online consumption. The loss of these high-end customers and the loss of part of high-end merchandise sales made the traditional commercial and traditional retail fall into a passive position deeply.

2.3 *Impacts on different traditional retailers*

According to the data published by iResearch, it shows that in recent years, among of the goods most frequently purchased by China's online shopping users, clothing, shoes, hats, and bags ranked first, followed by digital products, books and audio-video products. Clothing, shoes and hats are major businesses of traditional department stores, and these businesses have a higher profit, the main appealing for users to purchase them online is the low price. Digital products, books, and audio-video products are standardized commodity which does not require user's personal experience and the online prices are low. These products are the main business of the traditional department store, so E-Commerce has the most impact on traditional department stores and household appliance stores.

Supermarkets mainly engage in daily necessities, such as food, household department, baby products and etc. These commodities are in great demand by customers and have a smaller profit, which makes online shopping of these commodities has not obvious advantages. Especially for large stores, they can meet the user's one-stop shopping demand, and E-Commerce has difficult to replicate this advantage in the short term, so E-Commerce has limited impact on supermarkets and hypermarkets.

The convenience store is set up for customer's convenience, which operates small commodities urgently needed by customers, so, E-Commerce has minimal impact on convenience store. Moreover, because the convenience store is located in the community, close to the people, it has an irreplaceable role in the development of E-Commerce.

3 TRANSITION OF TRADITIONAL RETAILING TO E-COMMERCE

Faced with the impact of E-Commerce, traditional retail shifts to E-Commerce is a general trend. Compared with online retailers, traditional retailers have their own advantages such as operational cost control, financial operation ability, brand value, marketing, technology, supply systems, distribution channels and etc. So, traditional retailers and online retailers complement each other to combine online and offline operation modes can make up the defects of the domestic credit system, distribution network and etc.

Department stores have the advantages of good reputation, which are older shoppers' paradises. But in recent years, under the impact of E-Commerce, department stores lost a lot of customers and became a "shopping with mother" place. Coupled with the rising cost of rent, manpower, logistics, department store is facing enormous difficulties. The department stores which focus on low-end customers should take some effective measures. On the one hand, they should actively adjust the commodity composition to attract young, white collar customers, and they should enhance the customer experience and improve service quality. On the other hand, they should actively develop E-Commerce, and combine online and offline services.

Large-sized chain supermarkets have high competitiveness and brand effect because of their many chain stores everywhere, low price, and great variety of goods. They have a guarantee for the best quality, after-sale service, and vendor reputation. They also have their own distribution center and a strong logistics system which ensures a low-cost, timely, and smooth distribution. Based on this, they can better solve distribution problems faced by online retailers and will attract a large number of online customers. Among of so many traditional retail stores, large-sized chain supermarkets have the best advantages to develop E-Commerce. Large supermarkets primarily services the surrounding community, they should take an active role in developing regional community E-Commerce, and provide "online shopping, home delivery" services (i.e. one-stop shopping services) based on low price to eliminate customer's worries and further retain customers.

Outstanding features of convenience stores are: (1) they are located in the community; (2) they are widely distributed; (3) they provide 24 hours and 7 days services. Convenience stores should actively cooperate with E-business to exploit these geographical advantages to become an E-business's distribution site, and they can also sell their goods online.

4 EMPIRICAL ANALYSIS

4.1 *Suning built integrated network platform*

Suning Tesco strengthens virtual networks as well as physical stores, and continually improves the online market share. In 2011, Suning Tesco with total sales income of 5.9 billion yuan and more than 10 million registered members ranked in the top three domestic E-Commerce companies. The first half of 2012, Suning's total sales reached 47.19 billion yuan with an increase of 6.69% from a year earlier, while the online revenues reached $5.28 billion yuan, jumping 105.53% from a year before.

4.2 *Shanghai Bailian group co. Ltd. will do omni-channel retailing*

Since the 4th quarter of 2011, retail industry represented by department stores faced a downward inflection point of earnings. Under the double pressure of shopping center's better consumption experience and E-Commerce's low commodity prices, faced with increasing cost of rent, manpower, and logistics in circulation, traditional department stores entered a low profit era. Although gross profit margins of department store can reach about 25% up to 30%, the net interest rate only maintains below 5%. Since 2011, Pacific Department Store has closed two stores in Beijing, and the Shanghai No.1 Shopping Center closed Huaihai Road store, which highlighted the quandary faced by retail business.

Facing the impact of internet shopping, Shanghai Bailian Group Co. Ltd. owned by Bailian Group improved the quality of goods and provided more merchandises that need to be tried on by customers, meanwhile, the quality of services were enhanced to lure the public to enjoy shopping. Moreover, the company actively tried to do E-Commerce and promoted the integration of online and offline businesses. The company's Medium-term Report of 2014 pointed out that the company will start the Omni-channel strategy to promote the integration of online and offline businesses, and the company should be defined as an Omni-channel retailer, which focusses on goods, covering department store, shopping center, and outlets, whose target customers are 24–35 year-old consumers which are consistent with the core consumers of offline business [2].

4.3 *Auchan hypermarket's mode—"order online, pickup in-store"*

Many merchants of the traditional online shopping mall must stick with goods by themselves, which induced no guaranteed quality and different price of a similar product, so, consumers sometimes spend the same money may not buy the same quality of goods. Auchan supermarket buys goods directly from the producers, which not only ensures quality at a lower price, but also provides perfect services, so, it is a good choice for consumers.

Shanghai Branch of Auchan supermarket mainly engaged in online retail of daily consumer goods, which provides delivery services in the urban areas of the city, different delivery points with different fees. The consumer can also pick up the goods in store by himself. Consumers can choose different payment modes, such as payment in store, cash on delivery, payment by Oney card issued by Auchan, and payment online.

Suzhou Branch of Auchan supermarket launched automatic pickup machines for the online shopping consumers, where the consumer can pick up the goods in store when he has placed an order online 2 hours ago, and the pickup process will take only 5 minutes. This model of "order online, pickup in store" provides a reference model for many physical retailers.

The offline process of this "order online, pickup in store" model is: step 1, consumers drive to pickup area and input online reservation phone number on automatic pickup machine, then it will enter the payment interface; step 2, consumer can pay the order by bank card, then the invoices will be printed out and the machine will display the number of parking space, if the consumer already paid order online, the machine will display the parking space number directly when the consumer entered his phone number; step 3, consumer drives the car to the parking space, then the goods will be transported to the parking space and loaded on the car by the distribution staffs. The total process only needs 5 minutes.

4.4 *Transformation of Haode & Kede chain convenience stores*

Haode & Kede Chain Convenience Company, owned by Agricultural-industrial-commercial Supermarket Group, has more than 2000 chain stores, covering 18 cities in Jiangsu, Zhejiang and Shanghai. The company is one of the largest chain convenience companies in China, whose stores open 24 hours a day and have an advanced management specification.

In recent years, as China's largest chain convenience company, Haode & Kede faced enormous pressure, such as customer churn incurred by E-businesses and high costs, whose slim profit margins are squeezed again and again. In order to survive and develop, Haode & Kede keeps up with market changes, and takes adjustments to continually seek new growth points.

Since last June, the service, i.e. "order online, pickup in store", was provided by Haode & Kede as a starting point for exploring online-to-offline business model. Besides Chblt.com company owned by Agricultural-industrial-commercial Supermarket Group, Haode & Kede enhanced its cooperation with other E-companies. In April of 2014, due to all the more than 2000 chain stores of Haode & Kede are straight camp stores, the company has a higher control and strong executive power, so the company was selected by Tmall as the physical receiving station in Jiangsu, Zhejiang and Shanghai. More than 2000 convenience stores owned by Haode & Kede, began officially to provide services for consumers of Tmall in the 18 cities in Jiangsu, Zhejiang and Shanghai. The cooperation between Tmall and Haode & Kede helps Tmall to solve the "last 1 km" issue. Meanwhile, it also exploits the advantages of straight camp stores to increase consumer traffic, to expand the proportion of life services business, and to speed up the stores' transformation from merchandise sellers to life services providers.

Because the convenience stores of Haode & Kede open 24 hours a day, they are widely distributed, and they have a good reputation, this mode is welcomed by online consumers. Since the launch of this service, the business increased daily, and the received goods from the stores in one day accounted for 10% of those in Tmall.

ACKNOWLEDGEMENTS

This paper is supported by the fund of Scientific Research Project of Beijing Educational Committee [No.0351405712, study on the regional E-Commerce system to serve business services in Beijing], the fund of Beijing Social Science [No.14JGC103], the Statistics Research Project of National Bureau [No.2013LY055], and youth fund of Beijing Wuzi University [No. 054130341600].

REFERENCES

iResearch, 2013.2. 2014 China annual monitoring report of online shopping industry.

Bailian Group Co. Ltd. 2014.07. 2014 Medium-term report.

Management, Information and Educational Engineering – Liu, Sung & Yao (Eds)
© 2015 Taylor & Francis Group, London, ISBN: 978-1-138-02728-2

Application of project management in the organizational change of technology-based small and micro enterprises

Lin Na Che &Yun Cheng
Liaoning Economic Management Cadre Institute,Shenyang , Liaoning, China

ABSTRACT: The number of technology-based small and micro enterprises is increasing rapidly in China, which has an impact on the development of the social economy. However, the technology-based small and micro enterprises fall behind in management system, human resource allocation and capital funding. The whole world competition is increasingly fierce at present, the organizational change of technology-based small and micro enterprises is imperative. Project management provides a new pattern for the organizational change of enterprises. The thesis analyzes the influencing factors of the organizational change of technology-based small and micro enterprises internally and externally. Moreover, it also analyzes the advantages of project management in the organizational change of technology-based small and micro enterprises comparing with traditional management pattern and how project management applies in the organizational change of technology-based small and micro enterprises. It presents the implementation of the project management is an effective mode which will be widely used in technology-based small and micro enterprises.

KEYWORDS: Organizational change; Project management; Technology-based small and micro enterprises; Enterprise group.

1 INTRODUCTION

Changing the current market environment and economic policies have changed the stable and orderly competition environment in the past and organizations are suffering challenges of many factors, such as products, technology, environment, policy and so on. Only organizations carry out change activities constantly according to the changes of internal and external environment, can they be able to survive and grow in a competitive environment.

According to the research, 80% jobs are offered by middle and small-sized enterprises which account for 99% of the number of enterprises and 50% state tax revenue are paid by middle and small-sized enterprises. The small and micro enterprises are the main component of middle and small-sized enterprises. The technology-based small and micro enterprises are increasing rapidly with the society's dependence on technology. Technology-based small and micro enterprises assume an important place in our scientific and technological achievements and have a significant impact on the social and economic development.

Larry H.P. Lang, the economist, proposed that small and micro enterprise is the general term of small enterprise, micro-enterprise, family-owned enterprise, individual industrial and commercial household. Technology-based small and micro enterprises are special enterprises with a mission of technology innovation. The technical content of their products is relatively high with core competitiveness and they can continue to launch new marketable products and expand the market [1].

In the rapidly changing and competitive environment, technology-based small and micro enterprises are facing many changes in the environment, technology and policy. In order to maintain the development, change can be one of its greatest features. However, the organizational change of technology-based small and micro enterprises is a complex and ever-changing process and how to ensure the rationality and effectiveness of change is the key.

2 THE INFLUENCING FACTORS OF ORGANIZATIONAL CHANGE IN TECHNOLOGY-BASED SMALL AND MICRO ENTERPRISES

Organizational change, by its nature, is the process that the organization change and innovate its various elements to meet internal and external environmental changes and realize the organizing goal [2]. Now no organization is under particularly stable environment and large and small enterprises are facing changes. Technology-based small and micro enterprises are no exception.

Technology-based small and micro enterprises mainly involve high and new technology industry such as electronic information, bio-pharmaceutical, environmental protection and energy saving, new energy development and so on. The biggest feature is the emphasis on research and development. They are a kind of newly rising enterprises integrating research and development, production and market [3]. Its organizational change is a behavior affected by the internal and external environment. It is the inevitable choice for technology-based small and micro enterprises to seek long-term scientific and technological development. Finding and analyzing the influencing factors of organizational change internally and externally and analyzing what the model can facilitate organizational change are very important.

2.1 *The analysis of internal influencing factors of organizational change in technology-based small and micro enterprises*

2.1.1 *The impact of internal personnel allocation of technology-based small and micro enterprises on organizational change*

Technology-based small and micro enterprises need high-quality talents in technology, management and production, and have a higher requirement of knowledge. How to attract and retain high-quality talents in technology-based small and micro enterprises is a top priority. Organizational change will inevitably cause internal resistance, or even lead to greater loss of talents. Therefore, for technology-based small and micro enterprises, the reasonable personnel allocation will drive the change, improve the acceptance of organizational change culture, promote organizational change, reduce the internal staff fluctuation and reduce the loss of key personnel.

2.1.2 *Technology-based small and micro enterprises internal management system impose restrictions on organizational change*

Most technology-based small and micro enterprises are started by an individual or a small team lacking the normal internal management system. Most of them have internal core leaders and their internal management style is mainly affected by the personal style of the main leadership lack of formal and long-term development planning. When enterprises carry out organizational change, personal style and staff's emotional recognition play a leading role. There are no formal organizational change structures and mechanisms to control the smooth development of every period of organizational change to ensure the successful implementation of organizational change. And reasonable internal management system will improve the efficiency of organizational change and be conducive to carry out organizational change.

2.2 *The analysis of external influencing factors of organizational change in technology-based small and micro enterprises*

External influencing factors of organizational change include social environment, regulations and policies, human environment and so on. For the technology-based small and micro enterprises, the main external environments include the financing environment of the enterprise and policy support of government.

2.2.1 *The impact of financing environment on organizational change in technology-based small and micro enterprises*

Financing always restricts the development of small and medium-sized enterprises. For technology-based small and micro enterprises, pursuing research and development of the high technology need to invest a large amount of fund, because high-tech industry need a large amount of fund in every step of the research and development, production to marketing. In carrying out organizational change, enterprises shall concern that whether the change would bring too much turmoil to affect the original capital maintenance of the enterprises. If the capital chain ruptures, the enterprises will develop unstably even go bankrupt. The unstable financing environment will lead to organizational change of technology-based small and micro enterprises suddenly to interrupt or have difficulty to carry out. The enterprises will be unable to focus on the conduct of organizational change. However, the fund of entrepreneurs of technology-based small and micro enterprises comes from personal savings or private loans, and normal credit channels are deficient. The financial support of government for technology-based small and micro enterprises shall be strengthened.

2.2.2 *The impact of government support policies on organizational change in technology-based small and micro enterprises*

Our government should formulate reasonable policies and regulations to protect the legitimate rights and interests of technology-based small and micro enterprises. Our country promulgated "the Law of Promoting Small and Medium-sized Enterprises", but lack of implementation details which can use legislative protection for small enterprises of U.S. for references. Since 1953, the United States has promulgated "Small Business Act", "Equal Opportunities Act", "Small Business Investment Act", "Small Business Economic Policy Act" and other laws and regulations to protect the competitive and equal position of small enterprises from the legislative level [4].

China's current policy of supporting system of technology-based small and micro enterprises is not perfect. The government now has a lot of preferential tax policies for small enterprises, but too restrictive in terms of implementation. The financial support of technology-based small and micro enterprises should be strengthened, such as the establishment of a special fund for technology-based small and micro enterprises and the introduction of a series of policies to encourage entrepreneurs of technology-based small and micro enterprises.

Without government's support policy, the technology-based small and micro enterprises are difficult to seek long-term development, are restricted when formulating changing strategy are easily affected by the external policy environment when carrying out organizational change and even can't prepare resources for change, which may lead the standstill of to the management of technology-based small and micro enterprises.

3 THE ADVANTAGES OF PROJECT MANAGEMENT MODEL OVER TRADITIONAL MANAGEMENT MODELS IN ORGANIZATIONAL CHANGE OF TECHNOLOGY-BASED SMALL AND MICRO ENTERPRISES

In China, more and more enterprises begin to pay close attention to the application of project management in the enterprises. Project management has been widely used in many industries, such as software, IT industry, manufacturing, and so on, and also has been widely used in science and technology enterprises. The effective development of enterprises and the improvement of core competitiveness depend on enterprise change and innovation in organization and management.

Traditional management models are mostly functional organization and have a natural resistance to change. The project management itself has the characteristics of uniqueness, purpose, innovation and integration. These characteristics are consistent with organizational change, and organizational change is an activity with creativeness, integration and purpose. Therefore, we can regard organizational change as a project. We say that the goal of project management is to achieve the dynamic management of the whole process of the project and to achieve their goals, which are consistent with the goals of organizational change. Organizational change is also to implement management of change and to achieve goals. In order to complete organizational change project, we should overcome the organizational resistance, play the organizational power and take advantage of project management model. Project management plays an important role in achieving the successful goal of organizational change (as shown in figure 1).

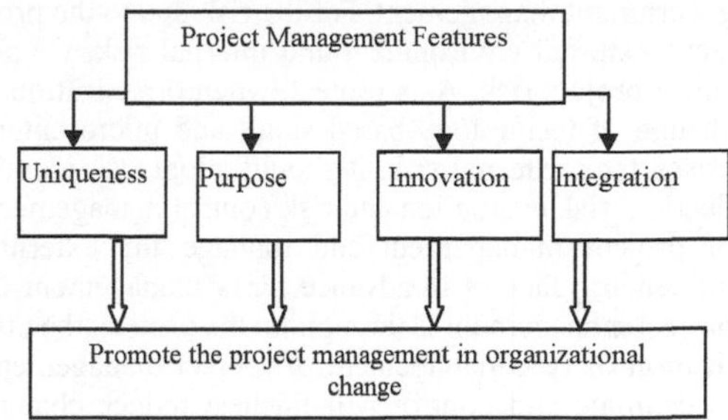

Figure 1. Schematic diagram of project management to promote organizational change

3.1 *Project management is a new form of organizational change in technology-based small and micro enterprises*

Project management provides a new form of development model for organizational change of technology-based small and micro enterprises as to the organizational change of internal factors such as staffing and internal management system. Enterprises pay more attention to the project management team in project management model than in the traditional management model where the staff is in a state of resistance. It is more conducive to staff to accept change and strengthen the staff's compatibility in technology-based small and micro enterprises if the head person of change centralizes personnel spread in an organization to manage and organize cultural transmission.

Moreover, only by constantly improving organizational management system of technology-based small and micro enterprise, can we increase enterprise's benefits, reduce costs and improve product quality and the core competitiveness. The project management model shows a significant advantage in organizational change of technology-based small and micro enterprises which can help enterprises achieve change goal through time, cost and objectives of management plans. Project management itself is the integrating process of management from the start, plan to implementation and control. It conforms with organizational change risk and the characteristics of dynamic management requirements and improves the efficiency of organizational change.

3.2 *Project management reduces the external risk of organizational change in technology-based small and micro enterprises*

Project management will help technology-based small and micro enterprises to better face external environment impact and fierce market competition on the side of the external financing environment and government management. For the risk due to the project's external environment and internal risk, we all call it project risk. As a project, when organizational change of technology-based small and micro enterprises face external risk, it should adopt risk identification, risk evaluation and risk control management of project management and manage the external influencing factors in advance. Risk management in project management is throughout the project, the utilization of risk management of project management to estimate and control will furthest reduce change resistance of technology-based small and micro enterprises from two aspects of the external financing environment and government policies and management regulations. Moreover, risk management of project management is throughout every process of project startup, plan, implementation, control and feedback, which can help technology-based small and micro enterprises to control every step of organizational change and to adopt corresponding measurements to minimize risk factors of organizational change.

4 APPLICATION ANALYSIS OF PROJECT MANAGEMENT ON THE TECHNOLOGY-BASED SMALL AND MICRO ENTERPRISES IN ORGANIZATIONAL CHANGE

4.1 *The necessity of technology-based small and micro enterprises organizational change*

In the current competitive market environment, technology-based small and micro enterprises organizational change is imperative, which is the only way to ensure the healthy development of enterprises. Our technology-based small and micro enterprises are in the development stage, growing rapidly in the number and size, but the growth is troublesome. Compared to mature and large enterprises, technology-based small and micro enterprises lie far behind in management system, staffing and capital financing. This form of organization of project management provides a new model for the organizational change, but at the time of the introduction of project management, technology-based small and micro enterprises should be innovative applications according to their actual situation. Any kind of management model will be difficult to function correctly, if out of the specific situation.

4.2 *Project management is a feasible and effective way to organizational change*

In the analysis of the influencing factors and project management model of the advantages of science and technology-based small and micro enterprises, it can be inferred the introduction of project management can be an effective and feasible way to organizational change. The practical application of technology-based small and micro enterprises can be used in organizational change. Technology-based small and micro enterprises focus on research and development, mainly involved in high-tech industries, with the decision facing the constantly changing in its own. Conduct organizational change in change. So technology-based small and micro enterprises, organizational change itself is constantly changing, constantly an adjusting process. And the project management itself is a management in the face of change. Seeing the organizational change in the technology-based small and micro enterprises as a project for project management, will greatly reduce the risks. To estimate of risk, and to carry out project time management, project management change, project risk management change in the organizational change, to ensure effectively to achieve organizational change project objectives.

4.3 *Application model of project management in organizational change*

We say that organizational change is based on the pursuit of performance targets. They try to change the objects, including the behavior of members of the organization, the culture, the structure, the strategy, and the relationship between organization and competitive environment [5]. The project management is carried out in technology-based small and micro enterprises in organizational change, it changes the behavior of members of science and technology-based small and micro enterprises, organizational structure, the relationship between the organization and the environment and so on. When applying science and technology project management in small and micro enterprises organizational change, it can change from technology-based small and micro enterprises organizational structure change, organizational management change, organizational strategic change, organizational culture change and other aspects.

Project management has several stages, including the definition of decision-making, program design, implementation, control and target completion, technology-based small and micro enterprises organizational change which can be divided into several stages of the same management, and conduct scientific management at each stage in order to achieve its organizational changes carried out at each stage correctly. For conducting project management in science and technology-based small and micro enterprises at the same time, time management, cost management, quality management and risk management can control the progress of change, the risk of change and the cost of change.

5 CONCLUSION

Today, the only unchanged is change. Organizational change has become a kind of necessary means of survival and development of technology-based small and micro enterprises. To find and develop a new way of organizational change is the trend of enterprise's organizational change. Project management embodies the great advantage in modern enterprise management. The business activities of enterprises or organizational change activities can be seen as the project which provides a new management model for organizational change. The implementation of the project management is an effective mode, which will be widely used in technology-based small and micro enterprises. However, how to combine the scientific management ideas and management tools into technology-based small and micro enterprises organizational change, how to improve the methods of project management is the key to success. Gradually to build a practical application mode of project management in organizational change of technology-based small and micro enterprises, and at the same time constantly summarize the correct application method.

REFERENCES

[1] Wang Junfeng . Wangyan. Research on development issues of Small and micro businesses in China. [J]. Business research,2012 (9).

[2] America] Thomas. Organization change and development [M]. Beijing: Tsinghua University press.

[3] Li Jianlin. Zhaoling. Study on the system of financial support in high-tech Small and micro enterprise of China . Contemporary economy ,2013(1):36–39.

[4] Liu Hesheng. Policy Research on the promotion of the development of the high-tech Small and micro enterprise in china.M Beijing,China.

[5] Denis. Rock, Yang Aihua, Wang Lizhen, Li Yingxia translated, "project management" (Ninth Edition), the electronic industry press.

[6] Li Min. Enterprise social capital and organizational innovation [J]. commercial research, 2005, (21):13 ~ 16.

[7] Wang Yifeng, Wu Yaping. Development of small micro enterprises based on life cycle theory ,Science and technology progress and policy2013, 30(2): 1 ~ 4.

[8] Kurt J P, Cohen D S. the heart of Revelation [M-I. Liu Xiangya, translation. Beijing: Mechanical Industry Press, 2003.

[9] Liu Luo, Chen Shuwen. [J]. Inspection of the loan customer manager's job performance structure model in Small and micro enterprise 2012, 30 (2):75 ~ 79.

[10] Baker, D. Strategic Change Management in Public Sector Organi—zations[M]. GB: Chandos Publishing, 2007.

[11] Helfat C E. Dynamic Capabilities: Understanding Strategic Change in Organizations[M]. Singa—pore: Blackwell Publishing, 2007.

Management, Information and Educational Engineering – Liu, Sung & Yao (Eds)
© 2015 Taylor & Francis Group, London, ISBN: 978-1-138-02728-2

Media reporting, company IPO and audit fees—empirical evidence from Chinese listed companies

Feng Niu & Ming Li
School of Economics and Management, Southwest Jiaotong University, Chendu, China

Ya Lu Li
School of Economics and Management, Henan Polytechnic University, Jiaozuo, China

ABSTRACT: Using the data from 2009–2012 of 89 companies listed on the small board of Shenzhen, the paper examines the relationship between media reporting and audit fees from both dimensions of media coverage and media supervision. The empirical results show that: media coverage can increase the audit fees significantly, the more the media coverage, the more the audit fees, and find no significant impact of the media supervision on audit fees, which is mainly due to the fewer negative reports in China, and the fact that the supervisory role the media plays is limited.

KEYWORDS: Media Coverage; Media Supervision; Company IPO; Audit Fees.

1 INTRODUCTION

With the rapid development of modern communications and network technologies, including new media and personal media (Gillmor, 2006),the power of the media's influence on the economic and social development is growing. As the fourth right, the media is considered to be an effective alternative important institutional arrangement of justice to realize the protection of investor(Dyck et al., 2008; Liu et al., 2014), received wide attention in academia. For example, in the Chinese capital market, Shengjingshanhe IPO event, a secondary listing incident of Lili electronic alleged , Suzhou permanent defeat GEM events, the occurrence of new land fraud incidents and other events listed, let people see the media play an increasingly important role in protecting investors. The general logic that media plays a role in investor protection is as follows: First, the media exposes the financial problems that may exist in companies. Then, due to the authority and influence of the media, causing regulatory authorities involved in the investigation. Next, according to the findings of regulatory authorities, the companies restructure irregularities or investors to sue companies. In the company's IPO process, the media reporting of the company will be attracted wide attention, and even lead to more regulatory scrutiny, thereby increase the auditor's audit risk. To reduce this risk within the acceptable range, auditors will increase the workload, thus affecting the auditor charges.

The existing research on media and IPO focused on the study of IPO underpricing (Cook et al, 2006;. Bhattacharya et al, 2009.), Few studies the influence of media reporting on auditor fees in the company's IPO process. In China, since the 2009 IPO restart, intermediary service revenue of major investment banks, accounting firms and financial public relations firm, etc., achieved tremendous growth, just like the same enthusiasm of financial market, and audit fees constitute an important part of IPO financing costs and intermediary service charges.Studies on the relationship between the media reporting and audit fees have important practical significance to reduce IPO financing costs and improve the efficiency of capital markets financing.

Based on this, using the data from 2009–2012 of 403 companies listed in the small board of Shenzhen, the paper examines the influence of media reporting on audit fees from the two dimensions of media coverage and media supervision, studies the relationship between media reporting and audit fees.

The remainder of this paper proceeds as follows. Section 2 The second part makes the literature review and proposes hypotheses. Section 3 describes the sample selection and research data sources, builds the model and gives the definition of variables. Section 4 makes empirical analysis of the impact of media coverage on the audit fees, and tests the proposed assumption. Section 5 concludes the study.

2 LITERATURE REVIEW AND HYPOTHESIS

Early theorists of the media focused on the relationship between asset price changes and media information in capital markets(Barber and Odean, 2008), since the 1990s, began to focus on the corporate governance role played by the media through the influence of public opinion(Thompson,2013). By examining the media's role in corporate governance, Dyck and Zingales (2008) found that the media can effectively reduce the private benefits of control, which made the corporate governance function of media to get more attention. Since then, scholars have studied the corporate governance role of the media from different perspectives, such as revealing the accounting scandals (Miller, 2006), prompting companies to correct the behavior against the interests of outside investors (Dyck et al, 2008),exposing the inefficiency of the Board (Joe et al.,2009). Using a unique sample, Chinese scholars, LI peigong and SHEN Yifeng(2010) verified that the media still has a corporate governance functions under special transition economy. KONG Dongmin et al (2013) found the media coverage reflects the significant governance functions in all levels of Chinese listed companies from the perspective of protecting the interests of minority shareholders.

Some academics have also studied the relationship among media (or media sentiment),stock price and the company's IPO under-pricing. Niederhoffer and Victor (1971) and Cutler et al.,(1989) earlier studied the market reaction to the news reports. Next, the scholars have explained how the media affect the stock price from perspectives of investor attention(Odean, 1999;. Barber et al, 2008), investor sentiment (Hong and Stein, 2007; Tetlock, 2007), the number of traders (Dyck and zingales, 2004)and so on. With the development of behavioral finance theory, research on the relationship between media coverage and IPO prices gradually become a hot academic research(Cook et al., 2006; Bhattacharya et al., 2009).Cook et al (2006) examined the relationship between media coverage and IPO, and found that media coverage pushed the IPO underpricing by affecting investor sentiment. With Taiwan's IPO samples, Jang (2007) verified the more the media coverage, the lower the company's IPO underpricing level. Bhattacharya et al (2009) found that media coverage does not explain the difference between the earnings of the Internet and non-Internet companies risk adjusted through through studying the IPO in the Internet bubble period.Liu et al (2014) also studied the impact of media coverage before the IPO on expectations of future earnings and long-term value of the company.

The above literature play an important role for our in understanding the Corporate governance functions and investor protection role of media thoroughly. Meanwhile, by combing the literature, we find, the reason why the media can play a role in investor protection and has the function of corporate governance to influence stock prices, the key lies in reporting and disclosure of media, which largely reduce information asymmetry between the company's management layer and external investors, thereby reducing the information risk of investors' transactions(Fang and Press, 2009).Using a total of 2.2 million news data of the Dow Jones news online for 29 years, Tetlock (2010) analyzed the mechanism of media to improve the enterprise information environment and eliminate the information asymmetry, and confirms it. And the reason why IPO company received extensive media attention, probably because the company itself is better quality, higher credibility, so investors will maintain high interest for them. Therefore, the company received widespread media attention has the better financial foundation, the risk faced by auditors in service is lower, so the audit fees are low. On the other hand, since the media reporting for IPO companies improved asymmetry condition between investors and companies, so the requirements for the quality of financial data would inevitably be increased, and the the litigation risk and audit risk faced by auditors increase correspondingly. To control the risk within an acceptable range, the auditor has to increase the company's IPO audit work, which pushed up the audit fees.

Thus, we propose the following two competing hypotheses:

Hypothesis 1a: The more the media reporting, the stronger the media coverage, the lower the audit fees

Hypothesis 1b: The more the media reporting, the stronger the media coverage, the higher the audit fees

Different to the general media coverage, negative media coverage includes some use of questioning, criticism, negative evaluation even neutral language to point out the problems of companies (Liang Hongyu, etc., 2012). these negative reports reflect the IPO's possible financial fraud, inflated assets, transfer of benefits and other issues. Obviously, negative media coverage has more significant economic consequences. Because of the media supervisory or negative public pressure caused by negative media coverage, media supervision will bring a significant impact on the company's IPO audit fees. Since media supervision will increase the difficulty of auditors' work and audit risk, thereby increasing working hours, or even raise fees, ultimately leading to increased audit fees.

Thus, we present hypothesis 2: The more negative media coverage, the stronger media supervision, the higher audit fees.

3 SAMPLE SELECTION, DATA SOURCES, AND VARIABLE DEFINITIONS

3.1 *Sample selection and data sources*

The paper selected small plates as samples issued in China Shenzhen Stock Exchange from June 2009 to December 2012, and received a total of 89 samples excluding missing data companies. Company media reports data was obtained by manual sorting based on CNKI of "full-text database of China's major newspapers". Other relevant financial data was obtained from the bulletin information GTA (CSMAR) database and the China Securities Regulatory Commission issued.

3.2 *Model building and variable definitions*

3.2.1 *Model building*

By following OLS regression equation, this article studied the relationship between media reports and audit fees to test the hypothesis 1 and 2.

$$\mathrm{CFR}_i = \beta_0 + \beta_1 Media\mathrm{X}_i + \beta_2 \mathrm{Assets}_i + \beta_3 \mathrm{Lev}_i + \beta_4 \mathrm{LDBL}_i + \beta_5 \mathrm{ROE}_i + \beta_6 \mathrm{Pro}_i + \beta_7 \mathrm{Rep}_i + \sum Ind_i + \varepsilon_i$$

MediaX= MediaT or MediaN.

According to previous studies, the control variables were selected as follows: Assets, Leverage(Lev), Current Ratio(LDBL), Rate of return on net assets(ROE), Scale fund-raising(Pro), Underwriter Reputation(Rep) and Industry dummies(Ind).

3.2.2 *Explained variable*

Audit fees, the variable measures the fee charged by CPA for their audit service to IPO companies, we use the proportion of fees charged by auditors with a total underwriting fees as the proxy variable, CFR expressed.

3.2.3 *Explanatory variables*

Media reporting is the Explanatory variable, This article measures it from the two dimensions of media coverage and media supervision, expressed as MediaT and MediaN respectively. Negative media reporting mainly refers to the media news reports of the companies using questioning, criticism, negative evaluation even neutral language to reveal the problems of the companies(Liang Hongyu, etc., 2012), theoretically speaking, the reporting is not conducive to the company's share price.

MediaT means the total amount of reports during the data declaring the application to market and the data of issuing stocks.

MediaN means the total amount of negative reports during the data declaring the application to market and the data of issuing stocks.

3.2.4 *Control variables*

Table 1 shows the definitions of control variables.

Table 1. Data Definition. This table describes the main control variables used in our analysis.

Variable	Description
Assets	Total assets
Lev	The previous year's total liabilities before listing / total assets
Ldbl	Current Ratio before IPO, Current Assets / Current Liabilities
Roe	ROE before IPO, Premarket Net profit / Net assets
Pro	The actual amount of fund-raising
Rep	Underwriter reputation, IPO underwriter's market share (in the period 2009–2012, the ratio of each underwriter's volume to the total amount, due to the small number of bits about underwriting, so below the median of its reputation underwriters takes 0)
Ind	Industry dummies, a total of 19 sectors, taking 18 dummy variables

4 EMPIRICAL ANALYSIS AND DISCUSSION

4.1 *Descriptive statistic*

Table 2 shows the descriptive statistics of the main study variables.

According to the results in Table 2. Audit fees are only small proportion of the total underwriting fees , just 7.4%. Within the statistical range, the mean number of total amount of media reporting and the negative media reporting were 28.7 and 3.3. Thus, China's media reporting is still relatively small, especially the negative report. Therefore, the supervisory role of media negative report is limited.

4.2 *Regression analysis: Media reporting and audit fees*

Table 3 reports the regression results of Media reporting and audit fees. According to the Table 3 (1) Column, the coefficient of MediaT was 0.0007 (T value of 3.19), and was significantly at 1% level, in control of the assets, leverage, current ratio, rate of return on net

Table 2. The descriptive statistics of main variables.

Variable	N	Mean	SD	P50	Min	Max
Cfb	89	0.074	0.055	0.066	0.014	0.431
MediaT	89	28.697	22.216	25	3	171
MediaN	89	3.326	4.572	2	0	39
Assets	89	88383.09	69354.44	23859.33	23859.06	50861.8
Lev	89	0.504	0.144	0.515	0.210	0.781
Ldbl	89	1.892	1.009	1.57	0.69	5.66
Roe	89	0.305	0.151	0.263	0.097	1.088
Pro	89	64665.94	33799.97	55000	3870	200000
Rep	89	0.033	0.026	0.026	0	0.084

assets, scale fund-raising, underwriter reputation and industry dummies. Empirical results show that media coverage in IPO process have a significant impact on audit fees, the more media coverage, the higher audit fees, the empirical results support the hypothesis proposed 1b. Media reporting reflected the media's coverage on the forms' IPO, which makes investors on the company's financial information requirements increase. Thus, the risk of auditors increases, and eventually pushes up audit fees.

As can be seen from Table 3(2) column, the coefficient of MediaN was -0.0001 (T value of -0.09). Empirical results indicate that media negative reporting has no significant effect on audit fees during the company's IPO process, and the hypothesis 2 has not been supported by empirical results. Theoretically, although media supervision will increase the public pressure and audit risks faced by auditors, thereby causing an increase in audit fees, however, empirical results did not confirm the significant impact of the media supervision on audit fees. The reason of such a result may be due to the small number of negative media reporting in China, which can be verified from the descriptive statistics in Table 2. The mean number of negative media reporting was only 3.3 times per firm.

4.3 *Notes robustness and sensitivity analysis*

In order to ensure the reliability of research findings, we take the following robustness tests: (1)Excluding the samples that negative media reporting equal to 0, carry out the multiple regression analysis on the relationship between media reporting and audit fees; (2) Expand the Media's statistical range form a month before the company's IPO application to issuing tme, recalculate MediaT and MediaN, and do the same multiple regression; (3)use the proportion of fees charged by auditors with the scale of fund-raising as explained variable, and make the same multiple regression. After making four robustness test, the basic conclusions of this article unchanged. Due to limited space, data robustness test results are not listed in the text, the reader, if desired, can contact the author on request.

Table 3. Media supervision and audit fees.

Varible	(1)	(2)
MediaT	0.0007***	
	(3.19)	
MediaN		-0.0001
		(-0.09)
Assets	0.0000***	0.0000***
	(2.81)	(3.05)
Lev	-0.0684	-0.0833
	(-1.40)	(-1.60)
Ldbl	-0.0059	-0.0078
	(-0.88)	(-1.10)
Roe	0.0063	0.0080
	(0.17)	(0.20)
Pro	-0.0000***	-0.0000***
	(-3.93)	(-4.12)
Rep	-0.0075	0.0013
	(-0.03)	(0.01)
Cons	0.1216***	0.1552***
	(3.43)	(4.20)
Ind	Control	
Adj. R2	0.2713	0.1799
N	89	89

Note: t-statistics in parentheses, "***", "*", "*" represent 1%, 5% and 10% significance level.

5 CONCLUSION

Using the data from 2009–2012 of 89 companies listed in the small board of Shenzhen, the paper examines the influence of media reporting on audit fees from the two dimensions of media coverage and media supervision.The study found that the media coverage can increase the audit fees significantly, but the media supervision has no significant impact on audit fees.This paper complements the theoretical study on media and IPO, and provides clear recommendations to relevant government departments to develop policies and regulations.

(This research was partially supported by the Social Science Planning Project in Henan Province, Item Number:2013BJY030).

REFERENCES

Cook D O, Kieschnick R, Van Ness R A. On the marketing of IPOs[J]. Journal of Financial Economics, 2006, 82(1): 35–61.

Dyck A, Volchkova N, Zingales L. The corporate governance role of the media: Evidence from Russia[J]. The Journal of Finance, 2008, 63(3): 1093–1135.

Fang L, Peress J. Media Coverage and the Cross-section of Stock Returns[J]. The Journal of Finance, 2009, 64(5): 2023–2052.

Gillmor D. We the media: Grassroots journalism by the People, for the People [M]. [S.l.]: [s.n.], 2006.

Liu L X, Sherman A E, Zhang Y. The Long-Run Role of the Media: Evidence from Initial Public Offerings[J]. Management Science, 2014.

Management, Information and Educational Engineering – Liu, Sung & Yao (Eds)
© 2015 Taylor & Francis Group, London, ISBN: 978-1-138-02728-2

Collection uncertainty and treatment center capacity choice

Yan Fen Mu
School of Economics and Management, Southwest Jiaotong University, Chendu ,China

ABSTRACT: This study analyzes the effects of collection volume uncertainty on recovery center capacity choices. It is found that collection volume uncertainty will not change the optimal capacity choice if the collection volume variation is low while the capacity cost is high; otherwise, the optimal recovery center capacity under collection volume uncertainty will be larger than is the case when deterministic mean collection volume is considered. The conclusion is robust with respect to the different market structures considered in this study. The moderating effects of commercial revenue, capital cost and recovery center operation cost on recovery center capacity choice are qualitatively the same in the cases of uncertain collection volume and deterministic collection volume.

KEYWORDS: WEEE; Collection; Treatment center; Capacity Choice.

1 INTRODUCTION

The exponential increase in electrical and electronic equipment (EEE) generation in economies is a growing concern. The average lifespan of computers in developed countries has dropped from 6 years in 1997 to just two in 2005 and mobile phones have a life-cycle of <2 years, according to Greenpeace (2010). If these trends continue, the generated Waste electrical and electronic equipment (WEEE) will reach higher volumes. So, the WEEE problem is present on regulatory agendas and take-back legislation mandates producer responsibility for collection and recycling of e-waste. Producers or regulators may have to set up infrastructures and design collection and recycling networks to comply with WEEE take-back laws, which have capacity to make decisions in order to manage WEEE to achieve desired policy targets (e.g., collection and recycling targets). Strategic and operational decisions with respect to investments in new treatment facilities are needed. Optimization models can assist in ensuring that these investment strategies are economically feasible.

If demand can be perfectly forecasted, treatment center can simply invest the capacity in advance so that the right amount of capacity is available on time. The long capacity lead time can cause efficiency loss if there is substantial demand uncertainty: since capacity mismatch caused by forecast error cannot be remedied in short term, with imperfect planning either capacity over-supply or shortage. Therefore, one must assume that the future may be different from what seems most likely at present. A number of articles have focused on uncertainties and imprecise data in relation to waste modeling and on the development of WEEE management models.

Stochastic programming is widely used when assuming perfect foresight. This is done, for example, in Cui et al. (2011), who minimize possible loss using the interval-based MinMax focusing on waste flows and capacity. Xu et al. (2010) use the interval-parameter stochastic robust optimizing waste flows and revenue from WTE. Tan et al. (2010) study superiority-inferiority-based inexact fuzzy stochastic programming approach for solid waste management under uncertainty. Li and Chen (2011) focus on the fuzzy-stochastic-interval linear programming for supporting municipal solid waste management.

Fuzzy programming is another much used optimization method that assumes perfect foresight. In Gomes et al. (2011) the focus is on waste management of WEEE, optimizing the location of both collection and sorting centers with offset in a case from Portugal. Guo and Huang (2009b, a) minimize costs in waste handling using inexact fuzzy-stochastic MIP (mixed integer programming). Srivastava and Nema (2011) minimize costs of waste flow and landfill through fuzzy parametric programming method.

Interval analysis is used, for example, Cui et al. (2011) minimize possible loss using the interval-based MinMax focusing on the waste flows and landfill. Srivastava and Nema (2011) have included a review of types of waste management optimization models such as interval programming and fuzzy linear programming. Finnveden et al. (2006) have illustrated the possibilities and limitations of a range of different waste management models.

There are many other models which we have not listed above such as MIP, dynamic optimization and so on. However, none of these studies has formally modeled demand uncertainty and its implications for treatment center capacity investment. Few had started to focus on environmental factors as is very important in most of today's models. This article focuses on the decision-making process and actual sustainability in terms of the integrate analysis of the economic, environmental and the social aspects.

This study aims to fill such a research gap by analyzing the effects of demand uncertainty on treatment center capacity choice. Compared to the previous articles, the contributions of this study are multifold: (1) unlike most previous models considering capacity and price choices simultaneously, in our study only capacity is chosen with uncertain demand which follows a continuous probability distribution. After the demand pattern is observed, the price is then chosen. (2) Our study benchmarks the behavior of a welfare-maximizing social planner versus that of profit-maximizing enterprise(s). (3) We explicitly model the situation of multi-treatment centers, where treatment center providing differentiated services may compete with each other or they may be under the control of the same authority. These improvements allow us to offer important management and policy insights to the WEEE industry in particular, and to infrastructure investments of treatment firms in general.

The paper is organized as follows: Section 2 presents the basic economic model analyzing a monopoly profit-maximizing treatment center. Section 3 benchmarks the case when a monopoly treatment center aims to maximize social welfare. Section 4 investigates the case in which two treatment centers provide (imperfectly) substitutable services. The last section summarizes and concludes the paper.

2 CAPACITY CHOICE OF A PROFIT-MAXIMIZING TREATMENT CENTER

To focus on the effect of collection volume uncertainty on treatment center capacity decision, we consider that a risk-neutral, monopoly treatment center needs to make capacity decision in advance so as to accommodate future treatment volume and that, when making capacity decision, it faces a linear (inverse) demand function:

$$(1) \qquad P = X - bq \qquad (b > 0).$$

In (1), P is the treatment center charge (fees for WEEE treatment), q is the dispose volume, and X is a random variable that captures the demand forecast error, because demand cannot be precisely estimated. Let $f(x)$ be the density function of X, and $F(x)$ be the corresponding distribution function. For modeling tractability, X is assumed to follow a uniform distribution in the interval $(\underline{x}, \bar{x})$ and so $f(x) = 1/(\bar{x} - \underline{x})$. Clearly, $(\bar{x} - \underline{x})$ measures the degree of collection volume uncertainty/variability, with a large difference between $\bar{x}$ and $\underline{x}$ indicating high collection volume variability/uncertainty. Further, the treatment center derives an (average) net profit h from each recycled WEEE. Examples can be observed for large appliances, where component reuse is not common, but recycling of steel and copper can generate net profits. while the associated environmental benefit is v. Both h and v are assumed to be positive and exogenously determined. In addition, it is assumed that the treatment center has constant marginal cost c for treating each WEEE.

The treatment center's decision process is modeled as follows:

Stage 1: the treatment center decides its treatment capacity K based on the distribution of collection volume shifter $f(x)$. With the unit cost of capital being r, total capacity cost is rK.

Stage 2: After capacity is installed, the treatment center observes the actual collection and treatment volume X and then sets its service charge P.

We are not modeling a cash flow in multiple periods, the discount rate is not considered in this study, because of which will bring no additional insights.

First study the case of a monopoly treatment center that maximizes profit. To solve the optimal capacity, we adopt with the backward induction, and we study the situation in the second stage, which is characterized by the following optimization problem,

$$(2) \qquad Max_P \Pi \mid K, X = (P + h - c)q - rK \ s.t. \ q \le K.$$

That is, the treatment center maximizes its profit Π given both the capacity K invested in the first stage and the realized collection volume shifter X. The constraint $q \le K$ reflects the treatment center's inability to schedule more WEEEs than the maximum.

Differentiating the corresponding Lagrangian function with respect to P, the first-order condition (FOC) requires $X - 2P - h + c + \lambda = 0$, where λ is the Lagrangian multiplier. If the treatment center capacity constraint is binding thus that $q = K$ and $P = X - bK$, the condition $\lambda \ge 0$ implies $X \ge 2bK - h + c$. The corresponding treatment center profit is

$$(3) \qquad \Pi_1 = (X - bK + h - c - r)K.$$

Otherwise if the capacity constraint is not binding thus that $q < K$, then Lagrangian multiplier $\lambda = 0$. Solving the FOC we have $P = (X - h + c)/2$ and

$q = (X + h - c)/2b$. In addition, the capacity constraint $q < K$ implies $X < 2bK - h + c$, with the corresponding treatment center profit specified as

$$(4) \qquad \Pi_2 = \frac{1}{4b}(X + h - c)^2 - rK.$$

In summary, conditional on the capacity choice in stage 1, when the realized collection volume is sufficiently large (in the sense that $X \geq 2bK - h + c$) the treatment center will charge a high price at which the capacity is just fully utilized. Otherwise, if the collection volume is not large enough (when $X < 2bK - h + c$), the profit-maximizing treatment center will set a treatment center charge P which induces partial treatment center utilization.

In stage 1 the treatment center maximizes its expected profit based on the probability distribution of collection volume shifter, which can be specified as

$$(5) \qquad Max_K \mathrm{E}[\Pi] = \mathrm{E}[(P^* + h - c)q^* - rK],$$

where P^* and q^* are, respectively, the optimal treatment center charge and treatment volume obtained from stage 1 as derived above. Defining $\hat{X} = 2bK - h + c$, the treatment center's expected profit can then be specified for the following three cases depending on the ranges of capacity K to be chosen.

Clearly, the treatment center will choose a capacity to achieve the maximum (expected) profit of the three cases analyzed above. Comparing the profits among those cases leads to the following result:

Proposition 1. *For a monopoly treatment center maximizing its expected profit,*

If $r \geq (\overline{x} - \underline{x})/2$, *the optimal treatment center capacity is* $K^{us} = \frac{1}{2b}(\frac{\overline{x} + \underline{x}}{2} + h - c - r)$, *which will always be fully utilized.*

Proposition 2. *For a monopoly treatment center maximizing its expected profit,*

If $r < (\overline{x} - \underline{x})/2$, *the optimal treatment center capacity will be increased from* K^{us} *to the level of* $K^{ul} = \frac{1}{2b}(\overline{x} + h - c - \sqrt{2r(\overline{x} - \underline{x})})$, *which may be partially or fully utilized depending on the actual collection pattern X observed after the capacity being invested.*

Note that in K^{us}, the superscript u stands for "uncertainty" and superscript s for "small" collection volume uncertainty, referring to the degree of variability $\overline{x} - \underline{x}$ being less than or equal to $2r$. Similarly, in K^{ul}, the superscript l for "large" collection volume uncertainty, referring to the degree of variability $\overline{x} - \underline{x}$ being greater than $2r$. Furthermore, it is shown

$$(6) \qquad K^{ul} - K^{us} = \frac{1}{2b}(\sqrt{(\overline{x} - \underline{x})/2} - \sqrt{r})^2 > 0.$$

Propositions show that greater uncertainty over future collection volume will increase the investment in capacity for a risk-neutral monopoly treatment center. Also note that for the case of $r \geq (\overline{x} - \underline{x})/2$, the optimal capacity is equally influenced by the upper and lower bounds of collection distribution.

3 SUMMARY AND DISCUSSION

There is substantial collection volume uncertainty in the WEEE industry. However, collection uncertainty has not been adequately considered in economic investigations on the choices of treatment center capacity and pricing. While a few studies on infrastructure investment provide valuable insights on capacity choice under collection uncertainty, the cases of profit-maximizing firm has not been analyzed. The different natures of capacity and price decisions have also not been recognized in most of previous studies. This investigation aims to fill such a gap in research by benchmarking capacity choices under collection volume uncertainty. In the present paper, treatment center revenue has explicitly been recognized, and treatment center price has been set conditionally on the observed collection pattern and prior capacity invested. We found that collection volume uncertainty will not change if collection variation is low while capacity cost is high; otherwise optimal treatment center capacity under collection uncertainty will be larger than the case when deterministic mean collection is considered. This result is robust with respect to the different market structures considered in this study. These analytical results suggested that it is important to consider the effects of collection uncertainty in treatment center planning process.

The above material should be with the editor before the deadline for submission. Any material received too late will not be published. Send the material by airmail or by courier well packed and in time. Be sure that all pages are included in the parcel.

REFERENCES

Cui, L., Chen, L.R., Li, Y.P., Huang, G.H., Li, W., Xie, Y.L., 2011. An interval-based regret-analysis method for identifying long-term municipal solid waste management policy under uncertainty. Journal of Environmental Management 92, 1484–1494.

Xu, Y., Huang, G.H., Qin, X.S., Cao, M.F., Sun, Y., 2010. An interval-parameter stochastic robust optimization model for supporting municipal solid waste management under uncertainty. Waste Management 30, 316–327.

Finnveden, G., Bjorklund, A., Ekvall, T., Moberg, A., 2006. Models for waste management: possibilities and limitations. ISWA 2006 "Waste Site Stories" ISWA.

Gomes, M.I., Barbosa-Povoa, A.P., Novais, A.Q., 2011. Modelling a recovery network for WEEE: a case study in Portugal. Waste Management 31, 1645–1660.

Guo, P., Huang, G.H., 2009a. Inexact fuzzy-stochastic mixed integer programming approach for long-term planning of waste management – Part B: case study. Journal of Environmental Management 91, 441–460.

Guo, P., Huang, G.H., 2009b. Inexact fuzzy-stochastic mixed-integer programming approach for long-term planning of waste management – Part A: methodology. Journal of Environmental Management 91, 461–470.

Li, P., Chen, B., 2011. FSILP: fuzzy-stochastic-interval linear programming for supporting municipal solid waste management. Journal of Environmental Management 92, 1198–1209.

APPENDIX

Case I. If $\hat{X} \leq \underline{x}$ or, equivalently, $K \leq (\underline{x}+h-c)/2b$, the expected treatment center profit is, by (3), specified as $\mathrm{E}[\Pi]=\int_{\underline{x}}^{\overline{x}} \Pi_1 f(x)dx$. Solving the corresponding first-order condition yields $K=\frac{1}{2b}(\frac{\overline{x}+\underline{x}}{2}+h-c-r)$. Note that this solution is optimal if and only if it is compatible to the specified range. Substituting the solution to expression $K \leq (\underline{x}+h-c)/2b$, we obtain $r \geq (\overline{x}-\underline{x})/2$. Otherwise if $r < (\overline{x}-\underline{x})/2$, it can be shown that $\partial \mathrm{E}[\Pi]/\partial K > 0$. Thus the optimal capacity is the corner solution $K=(\underline{x}+h-c)/2b$.

Case II. If $\underline{x} < \hat{X} \leq \overline{x}$ or equivalently $(\underline{x}+h-c)/2b < K \leq (\overline{x}+h-c)/2b$, the expected treatment center profit is, by (3) and (4), equal to $\mathrm{E}[\Pi]=\int_{\underline{x}}^{\hat{x}} \Pi_2 f(x)dx + \int_{\hat{x}}^{\overline{x}} \Pi_1 f(x)dx$. Solving the corresponding first-order condition yields $K=(\overline{x}+h-c \pm \sqrt{2r(\overline{x}-\underline{x})})/2b$. The solution $K=(\overline{x}+h-c+\sqrt{2r(\overline{x}-\underline{x})})/2b$ is not compatible to the range defined for this case. The other root $K=(\overline{x}+h-c-\sqrt{2r(\overline{x}-\underline{x})})/2b$ will be the optimal capacity if and only if $(\underline{x}+h-c)/2b < K \leq (\overline{x}+h-c)/2b$, which can be simplified to $r < (\overline{x}-\underline{x})/2$. Otherwise if $r \geq (\overline{x}-\underline{x})/2$, it can be shown that $\partial \mathrm{E}[\Pi]/\partial K \leq 0$ and thus the optimal capacity is $K=(\underline{x}+h-c)/2b$. However, $K=(\underline{x}+h-c)/2b$ is not within the range defined for Case II (it is within the range defined for Case I instead). Therefore, there is no optimal capacity when $r \geq (\overline{x}-\underline{x})/2$.

Case III. If $\hat{X} > \overline{x}$ or equivalently $K > (\overline{x}+h-c)/2b$, the expected treatment center profit is $\mathrm{E}[\Pi]=\int_{\underline{x}}^{\overline{x}} \Pi_2 f(x)dx$. The corresponding first-order condition is $\partial \mathrm{E}[\Pi]/\partial K = -r < 0$, implying that there is no optimal capacity in this case.

Management, Information and Educational Engineering – Liu, Sung & Yao (Eds)
© 2015 Taylor & Francis Group, London, ISBN: 978-1-138-02728-2

Political connections, enterprise growth and risk taking: Empirical evidence from private listed enterprises of China

Chi Xie & Da Gang Yin
College of Business Administration, Hunan University, Changsha, Hunan, China

ABSTRACT: Proceeding from the background of economic transition and institutional environment of China, this paper analyzes the relevance of political connections, enterprise growth and risk taking in theory. Meanwhile, this paper empirically examines the effect of political connections on the enterprise growth and risk taking, based on the experimental data from 2010 to 2013 of Chinese private listed enterprises through multiple regression analysis. The empirical results show that political connections have a significant negative influence on the enterprise growth, that is to say, political connections conducive to enterprise growth. And political connections have a significant positive influence on the risk taking of enterprises, suggesting that the enterprises with political connections take greater risks in the growth process than their non-connection peers.

KEYWORDS: Political Connections; Enterprise Growth; Risk Taking; Private Listed Enterprises.

1 INTRODUCTION

At present, China is in a special historical period of economic and social transformation, the government functions are gradually converting. But the government still plays the role of resource allocation, enterprises will often seek to establish a "political connections" with government in order to get more policy information and social resources.

The influence of political connections on private enterprise performance and growth caused the long-term concern of management science and economics scholars. However, scholars have not drawn the same conclusion about the relationship between political connections and the growth of enterprises. Some studies show that political connection is valuable, it could bring to private enterprise financing convenience, reduce the cost of financing, access to preferential tax and improve business performance and many other obvious benefits. Some scholars expounded the negative effects of political connection from the angle of social burden, that political connection enterprises to assume more social "obligation", its profitability and corporate value lower than non political connection enterprises. The inconsistency on this problem, prompted us to make further study on "political connection".

Meanwhile, enterprises survival and development in the complex and uncertain environment is always accompanied by risks. That classical economics theory pointed out that the behavior of entrepreneurs to take risks in the pursuit of profit opportunity is a basic factor to promote the long-term economic growth. The manager is the key factor that decides the development of the enterprise, whether the enterprise is willing to undertake the risk depends on the degree of managerial risk preference. So, will managers spend time and money to build political connections to assume greater risk in pursuit of excess profits? In view of this, this paper also analyzes the influence of political connection with risk taking of enterprises.

2 HYPOTHESIS DEVELOPMENT

2.1 *Political connections and enterprise growth*

The relationship between political connections and firm performance mainly has two competing perspectives: "relationship" view and "intervention" view.

In "relationship" view, scholars often draw a conclusion that political connection is valuable to the enterprises. With companies from 42 countries as samples for analysis, Faccio found political connection enterprises' market influence and the debt ratio were significantly higher than non political connection enterprises, and could obtain a higher tax preference. Mian and Khwaja stated that political connection enterprises can obtain more bank loans than non political connection enterprises. After studying different countries, Faccio advocated the value of enterprises increased significantly after the establishment of the political connection, and political connection enterprise is easy to get help from the government when it is in financial distress. These findings show that political connection is indeed a valuable resource.

In China, a large number of listing enterprise executives have a political background, and the research on political connection as the theme gradually on the rise in recent years. Wu and Rui supposed the income tax rates of enterprises with political connections is significantly lower than non-connection peers. Zhang and Huang proposed that enterprises with political connections can obtain more diversified resources, access to government regulation industry easier.

Based on the above analysis, this paper puts forward:

Hypothesis 1: Political connections of private enterprises can improve their performance, it is conducive to the growth of enterprises.

On the other hand, in " intervention " view. The government through political connections as a means of intervention on the enterprises, will increase the policy burden of the enterprises. Vickers and Yarrow, Shleifer and Vishny proposed that the intervention of government would interfere with managers' decision-making process and reduce the operating performance of the enterprises [12,13]. Faccio concluded that although political connection enterprises' market influence, the ratio of debt and tax incentives are higher than non political connection enterprises, enterprises through the "Rent-seeking" campaign to establish political connection cost time and money may be offset by the benefits of political connection. Claessens, Feijen and Leaven proposed that political connection enterprises need to inherit the policy burden and the existence of non efficiency investment caused of political connection enterprises' Tobin'Q lower than non political connection enterprises.

Based on the above analysis, this paper puts forward:

Hypothesis 2: Political connections of private enterprises will reduce their performance, it is not conducive to the growth of enterprises.

2.2 *Political connections and risk taking*

At present, the literature about the influence of political connection on the risk taking is still rare. Fisman and Faccio confirmed that the political connection enterprises can get more bank loans, tax incentives and higher market share, the ability to respond to changes in the external environment is also stronger. Zhang and Huang found that political connections could significantly reduce the risk of the stock market in the process of diversification through the research on Chinese stock market. Ma and Meng proposed that executives with a good political background may enhance the enterprise's ability to bear the risk.

Based on the above analysis, this paper puts forward:

Hypothesis 3: Political connections enterprises will take lower risks in the growth process than non-connection enterprises.

However, Faccio advocated that the managers in political connection enterprises do not necessarily have the corresponding ability of entrepreneurs. Despite the political connection enterprises can get more bank loans, tax incentives and government financial subsidies, managers are likely to obtain a greater amount of loans because of blind or excessive debt financing. The debt ratio is too high to lead enterprises to bear the financial risks, thereby affecting the overall risk of enterprises. At the same time, the government has the motivation and ability to intervene in the business activities of the enterprise because of the unique environment of China. Claessens, Feijen and Leaven proposed that political connection enterprises have more non efficiency investment because of the policy burden, then increase the enterprise business risk.

Based on the above analysis, this paper puts forward:

Hypothesis4: Political connections enterprises will take greater risks in the growth process than non-connection enterprises.

3 RESEARCH DESIGN

3.1 *The definition of political connection*

The present study has no consistent standard for how to measure political connection, different literatures use different methods according to its research purposes. Among them, the dummy variable method has become the main method to measure the political connection. In this article, I say a company is connected with a politician if one of the company's large shareholders or top officers is: (a) a member of parliament, (b) a deputy to the National People's Congress, (c) a member of the CPPCC.

3.2 *Sample selection and data sources*

The data of the Chinese private listing company during the period from 2010 to 2013 are selected as research samples. Firms in the following samples are removed: (1) financial, insurance firms and the likes, (2) ST (Special Treated) and PT (Particular Transfer) firms, (3) the state-owned firms through equity transfer and into private firms, (4) the firms whose accumulated years of the listed companies less than three years, (5) the firms whose information is incomplete. The data are obtained from the China stock market and CSMAR database, the background information of the chairman, general manager and the actual controller through our manual collection to get. Data processing software is EVIEWS 6.0.

From Table 1. we can see that nearly 50% companies of the samples have political connections, political connection is a common phenomenon in the private listing company.

Table 1. Political connection of private listing company.

Year	2010	2011	2012	2013	N
Effective sample	321	545	674	673	2213
Political-connected company	169	267	327	328	1091
Proportion (%)	52.65	48.99	48.52	48.74	49.30

3.3 *Model and Variables*

To test the relationship between the political connection with firm growth, the model is as follows:

$$Growth = \alpha_0 + \alpha_1 Pol + \alpha_2 ProR + \alpha_3 Lev + \alpha_4 Size + \sum_{i=1}^{4} \alpha_{1i} Year_i + \sum_{j=1}^{11} \alpha_{2j} Ind_j + \varepsilon \quad (1)$$

To test the relationship between the political connections with risk taking, the model is as follows:

$$Beta = \beta_0 + \beta_1 Pol + \beta_2 ROA + \beta_3 Tobin'Q + \beta_4 Lev + \beta_5 Size + \sum_{j=1}^{4} \beta_{1j} Year_j + \eta \quad (2)$$

Growth: The growth ability represents the company's sustainable development. In this article we use the annual revenue growth rate as the index of weighing firm's growth. The index value is greater, the better for the firm's growth.

Pol: "1" represents politically-connected firms, "0" represents non-politically-connected firms.

ProR: Higher profit rate means that the firm can obtain sufficient funds, helps to promote the firm's growth.

Level: It is measured by the asset-liability ratio, which is an important indicator for measuring the quality of the company. The company to maintain the moderate debt ratio is advantageous for the company to improve its performance.

Size: It measures the degree of the scale by the logarithm of listed companies' total assets. The bigger the company size is, the more adequate resources the company has.

Beta: annual Beta coefficient, it measures the degree of risk taking, the index value is greater, the greater the risk taken by the company.

ROA: it represents the return on total assets, it is measured by net profit divided by total assets.

Tobin'Q: it is measured by market value divided by replacement cost, the index value is greater, the lower the risk taken by the company.

4 EMPIRICAL RESULTS AND ANALYSIS

4.1 *Descriptive statistics*

In all 2213 samples, the average annual revenue growth rate is 58.14%, indicates that the better growth of private enterprises in China. Judging from the Beta coefficient, there are great differences between enterprises, but the average value close to 1. That Chinese private listed enterprises undertake the risk of the stock market is low. There are 49.40% of samples have political connections, political connection is a common phenomenon in the private listing company. In addition, the average profit rate is 11.19%, the average asset-liability ratio is 37.68%, the average value of enterprise scale is 21.2994, the average rate of return on total assets (ROA) is 5.46%, Tobin'Q is 2.0176.

4.2 *Regression results analysis*

Table 2. The regression results of political connection and firm growth.

Variables	coefficient	P-value
Constant	0.7024***	0.0000
Pol	-0.0641**	0.0195
Size	-0.0278	0.1447
Lev	0.0862**	0.0117
ProR	0.3819***	0.0000
Year	YES	YES
Ind	YES	YES
F-value	26.26***	0.0000
Adj. R2	0.1772	
N	2213	

Note: Numbers in bracket is t statistic; *p<0.1, **p<0.05, ***p<0.01; the same below.

Table 3. The regression results of political connection and risk taking.

Variables	coefficient	P-value
Constant	1.1374***	0.0000
Pol	0.0263**	0.0355
Size	-0.0059	0.4681
Lev	0.1328***	0.0008
ROA	-0.8866***	0.0000
Tobin'Q	-0.0018	0.7466
Year	YES	YES
Ind	YES	YES
F-value	16.7663***	0.0000
Adj. R^2	0.0500	
N	2213	

In Table 2. it can be seen that from the results of Eq. (1). In the model the dependent variable is the annual revenue growth rate, the independent variable is political connection. The political connection (Pol) regression coefficient is -0.0641, it is significantly correlated with the growth variable at 5% level respectively. That means political connection has a significant negative influence on the enterprise growth, which is consistent with the hypothesis 2.

In Table 3. it can be seen that from the results of Eq. (2). In the model the dependent variable is the Beta coefficient, the independent variable is political connection. The political connection (Pol) regression coefficient is 0.0263, it is significantly correlated with the risk taking variable at 5% level respectively. That means political connection has a significant positive influence on the risk taking of enterprise, which is consistent with the hypothesis 3.In order to test the reliability of results, we use the variance of weekly stock returns of the enterprises to replace the Beta, and the new regression's results still support the hypothesis 3.

5 CONCLUSION

In this research, we examine the effect of political connections on the enterprise growth and risk taking based on the experimental data from 2010 to 2013 of Chinese private listed enterprises through multiple regression analysis. The results show that political connections conducive to enterprise growth. And political relations have a significant positive influence on the risk taking of enterprise, suggesting that the enterprises with political relations take greater risks in the growth process than their non-connection peers.

ACKNOWLEDGMENTS

This work was financially supported by the National Natural Science Foundation of China (71373072), the Foundation for Innovative Research Groups of the National Natural Science Foundation of China (71221001) and the National Soft Science Research Project of China (2010GXS5B141).

REFERENCES

[1] Fisman R. Estimating the Value of Political Connections[J]. *American Economic Review*, 2001(91): 1095–1102.

[2] Goldman E, Rocholl J. Do Politically Connected Boards Affect Firm Value[R]. *Working Paper*, 2006.

[3] Boubakri N, Cosset J and Saffar W. Political Connections of Newly Privatized Firms[J]. *Journal of Corporate Finance*, 2008(14): 654–673.

[4] Faccio M. Politically-Connected Firms: Can They Squeeze the State?[R]. *Working Paper*, 2002.

[5] Claessens S, Feijen E, Leaven L. Political Connections and Preferential Access to Finance: The Role of Campaign Contributions[R]. *Working Paper*, 2006.

[6] Krueger A. The Political Economy of the Rent-seeking Society[J].*American Economic Review*, 1974, 3(64): 291–303.

[7] Chiu M M, Sung W J. Loans to Distressed Firms: Political Connections, Related Lending, Business Group Gffiliations, and Bank Governance[R]. *Working Paper*, 2004.

[8] Khwaja A, Mian A. Do Lenders Favor Politically Connected Firms? Rent Provision in an Emerging Financial Market[J]. *Quarterly Journal of Economics*, 2005, 120(4): 1371–1411.

[9] Faccio M. Politically-Connected Firms [J]. *American Economic Review*, 2006, 96(1): 369–386.

[10] Vickers J, Yarrow G. Economic Perspectives on Privatization[J]. *The Journal of Economic Perspectives*, 1991, 2(5): 111–132.

[11] Shleifer A, Vishny R. Politician and Firms[J]. *Quarterly Journal of Economics*, 1994, 109(4): 995–1025.

[12] Fan J P H, Wong T J and Zhang T. Politically-Connected CEOs, Corporate Governance, and Post-IPO Performance Of China's Newly Partially Privatized Firms[J]. *Journal of Financial Economics*, 2007, 84(2): 330–357.

[13] Niskanen J, Niskanen M. The determinants of firm growth in small and micro firms—Evidence on relationship lending effects[R].*SSRN Working Paper*,2005.

[14] Montgomery C A, H Singh. Diversification Strategy and Systematic Risk[J]. *Strategic Management Journal*, 1984, 5(2): 181–191.

[15] Barton L S. Diversification Strategy and Systematic Risk: Another Look[J]. *Academy of Management Journal*, 1988, 31(1): 166–175.

Management, Information and Educational Engineering – Liu, Sung & Yao (Eds)
© 2015 Taylor & Francis Group, London, ISBN: 978-1-138-02728-2

Corporate governance of state-owned commercial banks in China

Lin Pan
China West Normal University, Nanchong, China

ABSTRACT: Finishing the shareholding system reform, Chinese state-owned commercial banks have already established modern corporate governance systems with many defects. Based on the analysis of the current situation of state-owned commercial bank corporate governance, the paper puts forward suggestions to improve the level of corporate governance.

KEYWORDS: State-owned commercial banks; Corporate governance; Governance structure.

1 INTRODUCTION

Since China's accession to WTO, Chinese state-owned commercial bank reform has achieved periodical victory, in which the establishment of the corporate governance system plays an important role. Note that corporate governance of Chinese state-owned commercial banks still exists many defects, and the global financial crisis has caused different influence on Chinese commercial banks, the corporate governance institutional arrangement has yet to be further improved.

2 CHARACTERISTICS OF COMMERCIAL BANKS CORPORATE GOVERNANCE

2.1 *Capital structure*

Compared with other enterprises, the proportion of commercial banks' equity capital is significantly lower. For example, according to the Basel Agreement's standard of international banking capital supervision, the minimum requirements of their equity capital of commercial banks are up to 8%, which means that a commercial banks' debt ratio can be as high as 90% or more. Capital of commercial banks is mainly from the depositors and other creditors, which mainly relies on deposits. The characteristic of high liabilities determines that the operational risk of commercial banks is higher than other enterprises. The supervision and management of the creditors generally do not take the initiative to participate in the commercial banks, commercial banks while at the state of high liability, faced with greater liquidity risk, the lack of supervision and control makes the creditors governance mechanism of commercial banks fail. In addition, if there is the risk of bankruptcy, bank shareholders, according to their contribution only assume limited responsibility, most of the rest losses will be borne by depositors and other creditors, which makes the possibility of bank shareholders and executives using short-term behaviors, damaging the interests of the depositors and other creditors. Therefore, in the research of commercial banks' governance problems, interests of depositors and other creditors should be concerned.

2.2 *Information asymmetry*

Characteristics of information asymmetry in commercial banks are mainly manifested in two aspects. Firstly, information asymmetry exists among depositors, regulators and the managers of commercial banks. Generally, the theory of corporate governance analyzes information asymmetry among shareholders, directors and executives. Commercial banks' creditors, small shareholders are in the information inferiority, can't restrict large shareholders and executives effectively. Commercial banks provide special products with high risk, which cause more serious information asymmetry problems than tangible goods in common, so that depositors and regulators are in the information inferiority, the external supervision effect was weakened. Secondly, product markets have information asymmetry problems. The quality of credit and other monetary products is difficult to observe in the short term, so that commercial banks' operating performance is more difficult to assess, the depositors or shareholders are difficult to detect, which weaken the external effects of governance mechanism of commercial banks.

2.3 *Supervision*

The financial sector is a key sector of the economy, while commercial banks are the core of the financial system. Different with the general industries, risk of banking is contagious, which can affect the whole economic system. Once a banking crisis occurs, it will cause serious damage to the national economy. The Asian financial crisis and America subprime crisis are examples. Although the scope and intensity of regulation in banking change from different periods and different countries, due to the particularity and vulnerability of the banking industry, governments actively supervise commercial banks with the extremely strict supervision system. There are two main objectives for banking supervision, protect the interests of depositors, and to prevent systemic risk. The supervision department's behaviors substitute and weaken the banks' corporate governance mechanism to a certain extent.

3 OWNERSHIP STRUCTURE OF STATE-OWNED COMMERCIAL BANKS

Equity diversification characteristics of state-owned commercial banks have many positive effects on the management of banks. Finishing the shareholding system reform, as the bank's owners, shareholders have the right to vote for the board of directors and board of supervisors members, the right to know the bank's operating conditions, and the residual claim right. The problem of client's absence before reform has been solved to a certain extent. The general meeting of shareholders is the highest organ of company power, shareholders entrust corporation property of the board of directors, as the legal representative of the commercial bank, the board of directors entrust the daily management right to the managers, so that incentive and restraint mechanisms between the general governance subject form in general. With the equity diversification, non state-owned shareholders balance with large state-owned shareholders, to reduce irrational behaviors of state-owned shareholder. Clear property right is advantageous to keep the state-owned asset value, to make clear of risk undertakers, also changes the soft budget constraint condition to a certain extent. The government undertakes limited liability to its investment in state-owned commercial banks, which prompts the bank managers to improve the risk prevention consciousness and responsibility consciousness. In addition, the implementation of foreign strategic investors measures also brings many positive effects for the state-owned banks, such as the balance of the state-owned shareholders, bringing the company advanced Corporate governance mechanism, products and risk management techniques.

However, the shareholding system reform did not solve all the problems. After the reform, from the actual situation, the ownership structure of state-owned commercial banks still maintains state-owned shares a single big holder, not only the traditional agency problem is not resolved, the new agency problem emerges between large state-owned shareholders and small shareholders or creditors. The current academic circles on the government's ownership exists two kinds of different views, namely "development view" and "politicians view". The developmental view argues that government shareholders are conducive to the bank to control risk, and increase public confidence in the banking system, which is conducive to the development of the banking and financial system. Politicians' view argues that state-owned bank shareholders tend to pursue personal political goals, political objectives and multiple principal-agent problems will reduce operating efficiency of state-owned banks.

Whether government shareholders are positive or not, and how about the reasonable proportion of government shareholding, there has been controversy in the academic circle. At present, China is still in the transition period, the capital market environment and legal system are not mature. In order to focus on the allocation of resources, macroeconomic regulation and control, to maintain the stability of financial security and the promotion of economic growth, as the largest shareholder of state-owned banks, the government often intervenes the operation of banks. It should be recognized that, in China, the government as the largest bank shareholders need to be kept for a long time to complete the process of equity dilution in order to avoid capital market shock. Currently, more popular view is, moderate concentration of government ownership structure is the most reasonable choice. In the circumstance of poor legal protection of investors, the appropriate concentration of equity is helpful to improve the efficiency of corporate governance, which has been supported by a number of researches.

4 THE GOVERNANCE STRUCTURE OF STATE-OWNED COMMERCIAL BANKS

4.1 *Foreign strategic investors*

Foreign strategic investors and the state-owned commercial banks have carried out extensive cooperation in many fields, which play a positive role in corporate governance. The Construction Bank of China is the first state-owned commercial banks to introduce foreign strategic investors, for example, in 2005 June, the Construction Bank of China and Bank of America officially signed an agreement on foreign investment and cooperation. According to the agreement, Bank of America purchases shares of the Construction Bank of China into two stages, and provide strategic assistance in the field of corporate governance, financial management, risk management and human resource management. But, state-owned commercial banks are still in the primary stage of imitating strategic investors' governance system at present. For operational perspective, due to regulatory restrictions, the shareholding ratio of the foreign strategic investors is lower than 10%, which cannot form effective incentive to participate in the governance of the state-owned banks, the effect of governance by foreign strategic investors is being questioned.

4.2 *Board of directors*

At present, the composition of Chinese state-owned commercial bank's board members includes the executive directors, non-executive directors and independent directors. In fact, the duty of directors is fuzzy, the coordination between the board and the managers is not smooth. Independent directors as outside directors to participate in corporate governance are beneficial. However, the independent directors of state owned commercial banks are mostly scholars or industry celebrities, they are usually busy, cannot focus on the internal affairs of banks, even participate in the board of directors through the phone video mode. Foreign independent directors know little about Chinese conditions and cultures, which hinders the effective functioning of independent directors.

4.3 *Board of supervisors*

In Chinese state-owned commercial bank corporate governance structure, the board of supervisors plays a certain role. However, the board of supervisors does not have a clear function definition, part of the functions are the same with committees under the board of directors. It is difficult for board of supervisors to form effective restriction to outside directors, whose role is to weaken. The state-owned commercial banks generally introduce employee supervisors, the proportion accounted for about 30% of the board of supervisors, but the majority of employee supervisors often has middle-level positions in the bank, the representativeness has to be strengthened.

4.4 *Party committees*

According to China's special political system background and economic environment, the party committees play a very important role in Chinese state-owned banks' decisions, which is different from the western commercial banks. For example, the party committees can appoint or remove the managers, the party committee members, most are decision-making personnels also, and even the party committee replaces the board of directors to make decisions on important matters, while most secretaries of the party committees are also chairmen of the board. Such reality makes how to correctly treat the party committees at the bank decision making and the role of corporate governance a key problem of state-owned commercial bank governance structure.

5 SUGGESTIONS OF IMPROVING CORPORATE GOVERNANCE

5.1 *Adjustment of ownership structure*

Ownership structure determines the basis of state-owned commercial bank corporate governance. Government as the single-largest shareholder played an active and important role in the early stage of economic development, but such controlling shareholder causes low efficiency and excessive intervention. The reality of national conditions determines that the state-owned equity can not completely exit in the short term, but to give the government shareholding ratio fell to a suitable level. To achieve this goal, government should deepen the reform of non tradable shares, reduce the state-owned shares gradually, and finally realize the full circulation of stock; the introduction of institutional investors, institutional investors have relatively more professional investment and supervision experiences, have more incentives to participate in the bank governance than scattered small shareholders; increase the proportion of decision-making personnel shareholdings, using equity incentive to reduce agency cost, which makes the bank more diversified ownership structure.

5.2 *Improving internal governance mechanism*

In the construction of the board of directors, most directors of state-owned commercial banks represent state-owned shares with complex principal-agent relationship, so that the board of directors is unable to fully function. Therefore, the system of board should adjust with ownership structure adjustment, ensure the independence of the board, let inside directors and outside directors can effectively participate in bank governance, to improve the function of professional committees. In the construction of the board of supervisors, banks should strengthen the monitoring system of the board of supervisors as the center, to further clarify the duties and responsibilities of the supervisory board, improve the comprehensive quality and professional ability of supervisors, to ensure supervisors have the right to know. The Party committee has its political core position, the relationship between the party committee and decision makers should be clearly defined, to reform the personnel system, introducing market mechanism, make excellent professional managers get in state-owned commercial bank management. For foreign strategic investors, strict restrictions should be relaxed, so that they have more incentives to participate in corporate governance of state-owned commercial banks.

5.3 *Improving external governance environment*

Firstly, the external supervision should be strengthened. Regulators should actively guide the state-owned commercial banks' governance, evaluate the capital structure, non-performing assets, internal control systems, audit systems, risk management systems in time, protect the interests of depositors, maintain the stability of the financial system. Secondly, information disclosure should be strengthened. Market discipline is one of the three pillars of capital supervision established by new Basel agreement, which aims to let stakeholders (including bank shareholders, depositors, creditors) participate in bank governance. The main way of stakeholder governance base on the bank's information disclosure, so the banks should be required to disclose information timely and comprehensive. Thirdly, the construction of the legal system should be accelerated. Compared with the western developed countries, China's legal system is not perfect, especially in the protection of small investors, which leads to tunneling. As the core mechanism of banking external governance, the perfection of the legal system can greatly improve the external governance environment of the state-owned commercial banks, and create good conditions for the optimization of corporate governance.

ACKNOWLEDGMENTS

I would like to express my gratitude to all those who have helped me during the writing of this paper. I gratefully acknowledge the help of Professor Cao. I do appreciate his encouragement and professional instructions. This paper is supported by MOE (Ministry of Education in China) Project of Humanities and Social Sciences (No. 14XJC630004), Scientific Research Fund of Sichuan Province Education Department (No. 14SB0094), and Fundamental Research Funds of China West Normal University (No.13D037).

REFERENCES

[1] Gorton, G., Rosen, R., Corporate Control, Portfolio Choice, and the Decline of Banking, Journal of Finance, 1995, 50(5): 1377–1420.

[2] Marcus, Deregulation and Bank Financial Policy, Journal of Banking and Financ, 1984, 8(4): 557–565.

[3] Agusman, A., D. Gasbarro, Zumwalt, Bank Moral Hazard and the Disciplining Factors of Risk Taking: Evidence from Asian Banks during 1998–2003, FMA European Conference, Stockholm, 2006.

[4] Houston, J., C. James, CEO Compensation and Bank Risk: Is Compensation Structured in Banking to Promote Risk Taking? Journal of Monetary Economics, 1995, 36(2): 405–31.

[5] Shrieves, Dahl, The Relationship between Risk and Capital in Commercial Banks, Journal of Banking and Finance, 1992, 16(2): 439–457.

[6] Rime, Capital Requirements and Bank Behavior: Empirical Evidence for Switzerland, Journal of Banking and Finance, 2001, 25(4): 789–805.

[7] Christophe, Bank Capital and Credit Risk Taking in Emerging Market Economies, Journal of Banking Regulation, 2005, 6(2):128–145.

[8] Jacques, Nigro, Risk-Based Capital, Portfolio Risk, and Bank Capital: A Simultaneous Equation Approach, Journal of Economics and Business, 1997, 49(6) :533–547.

[9] Konishi, M., Yasuda, Y., Factors Affecting Bank Risk Taking: Evidence from Japan, Journal of Banking and Finance, 2004, 28(1): 215–232.

[10] Amihud, Y., Lev, B., Risk Reduction as a Managerial Motive for Conglomerate Mergers, Bell Journal of Economics, 1981, 12(2): 605–617.

Management, Information and Educational Engineering – Liu, Sung & Yao (Eds)
© 2015 Taylor & Francis Group, London, ISBN: 978-1-138-02728-2

Strategic research on the internationalization of Chinese enterprises: A WTO perspective

Huo Can Wang
Management School, Donghua University, Shanghai, China
Institute of International Business, SUIBE, Shanghai, China

ABSTRACT: This article aims to explore the strategic choices in the rising process of Chinese enterprise internationalization based on an analysis of the unique interaction relationship between the Chinese enterprise internationalization and the WTO multilateral trading system. The process of Chinese enterprise internationalization started with the revolution of the multilateral trading system and creation of the WTO in the last 20 years of the last century, and 13 years since China's accession to the WTO, its trade and FDI have rapidly risen, and the regime and concept relative to the Chinese enterprise internationalization have been through great changes. In recent years, particularly since the 2008 crisis, Chinese enterprises internationalization was facing the severe strategic environment and many challenges, therefore, China needs a more forward-looking "WTO strategy" in order to make the future process of the Chinese enterprise internationalization better and more stable, and the first principle of the strategy is multilateralism.

KEYWORDS: Internationalization; Chinese Enterprises; WTO; Strategic Research.

1 INTRODUCTION

In the last 20 years of the last century, the four processes of the development of Chinese enterprise internationalization, the revolution of the multilateral trading system and creation of the WTO, China's resumption to the GATT/accession to WTO negotiations, and the new round of economic globalization encountered occasionally and interact mutually, and as a result, a legendary chapter of rapid rise of Chinese enterprise internationalization has been composed in a strong resonance with the WTO multilateral trading system(MTS) in the first decade of this century.

During the 13 year period after China entered into the WTO, the internationalization of Chinese enterprises has achieved a remarkable rise, which is reflected in the rapid development of trade and investment, and the huge changes of trade and investment system. This phenomenon is very rare among other WTO members. However, only a few researchers had paid in fact attention to and study on the special interactive relationship between the Chinese enterprise internationalization and the WTO multilateral trading system, such as Zhu (2000), Quan (2001), Chang & Fu (2008), and Alon at all. (2009). This article will explore the special interaction, also referred to "the strong coupling relationship", between the Chinese enterprises internationalization and the WTO, and then will point out the WTO-direction, strategic choices of Chinese enterprises in the future process of their internationalization under the WTO ecological environment. It will focus on four aspects that are why does the unique interactive relationship between the Chinese enterprise internationalization and the WTO multilateral trading system emerge, how does it reflect, what is the strategic environment of Chinese enterprises internationalization, and what strategy of internationalization should Chinese enterprises choose in WTO perspective.

2 THE HISTORICAL COINCIDENCE BETWEEN CHINESE ENTERPRISE INTERNATIONALIZATION AND THE CREATION PROCESS OF THE WTO

In 1979, China unlocked the door of reform and opening up, Chinese enterprise internationalization is started; And in the same year, the 7th round of GATT—Tokyo round also had just concluded, and soon later, the western powers like the United States and the European Union, the leaders of the MTS, began brewing a new round of multilateral trade negotiation, for seeking to keep up with the new economic globalization tide based on modern science and technology such as electronic computer and information technology, and also to satisfy the global operation needs of their multinational companies. In 1986, the GATT Uruguay round that started up the multilateral trading system reform process officially launched, and China in the same year also put forward a formal application to resume the original contracting party status in GATT, started the GATT-status resume negotiations. After the WTO created in 1995, the GATT resumption negotiation of China had to transfer into WTO accession negotiation. In fact, since China established the objective of the reform toward socialist market economy in 1992, the Chinese enterprise internationalization started to enter a new stage of rapid growth, and entered into the rapid rise of the stage since China became the WTO formal member in 2001. Therefore, the internationalization of Chinese enterprises began to grow under the unique historical conditions, and was born with countless ties with the multilateral trading system. Since 2001, the strong resonance effects between Chinese enterprise internationalization and the WTO, which was a rarity in the history of the multilateral trading system have been seen. The resonance effects mainly reflected in the four dimensions of trade, investment, institution and ideal.

3 THE STRONG RESONANCE EFFECTS BETWEEN WTO AND CHINESE ENTERPRISE INTERNATIONALIZATION SINCE CHINA'S ACCESSION

3.1 *Rapid rise of trade and FDI*

China's accession to the WTO has greatly promoted progress pace of the Chinese enterprise internationalization. First of all, China's trade liberalization commitments for accession to the WTO did help its export and import growth. China soon became the world's largest exporter and the world's second largest importer at 8th and 9th year after accession to the WTO respectively. From 2002 to 2011, China's exports grow from 325.6 billion US$ to 1.9 trillion, increased of 583%, the average annual growth rate was up to 22.35%. At the same time, the world merchandise exports rose from 6.5 to 18.29 trillion US$, an increase of 281.56%, the average annual growth of 12.19%. China's export share of the world exports was from 4.3% in 2001 to 10.4% in 2011. During the same period, China's imports grow from 295.17 billion US$ to 1.74 trillion US$, increased of 590.67%, the average annual growth rate was up to 21.82%. At the same time, the world merchandise imports rose from 6.66 to 18.4 trillion, increased by 276.12%, the average annual growth was up to 11.95%. China's exports and import growth rates are nearly 2 times of the world.

For FDI dimension, the strong coupling relationship between the Chinese enterprises internationalization and the WTO is the most prominent in China's OFDI flow during the more than 10 years after China's accession to WTO.

In general, the degree of coupling between Chinese enterprise internationalization and the WTO is stronger than that of other major WTO members.

To measure the degree of strong coupling between Chinese enterprise internationalization and the WTO, we set up a strong coupling index (SCI), which based on the average annual growth rate of import and export trade or FDI during a period (such as 10 years after China's entry into WTO, 2002–2011), by which we can calculate the SCI of Chinese enterprise internationalization relative to the other WTO members. The calculation formula is:

$$SCI_{CN} = \frac{\text{average annual \% Change of China's trade or FDI}}{\text{average annual \% Change of Member's trade or FDI}} \times 100 \quad (1)$$

We measured the China's SCIs of export, import and OFDI in table 1 relative to the other main WTO members. It is shown that the China's SCIs are all stronger than that of other main members. For example, the China's SCI of export range from low 112.59 of Russian Federation to high 321.07 of France. It is also shown that China's SCI based on OFDI flow is stronger than that of other main members with range from the low 118.35 of the Russian Federation to high 2455.5 of France. It means of China's annual growth rate of OFDI flow is very higher than other main members.

Table 1. SCI for China's trade and OFDI compared to selected WTO members.

	Average annual growth rate, 2002–2011 [%]			China's SCI* [others=100]		
Members	Export	Import	OFDI	Export	Import	OFDI
China	21.64	21.82	45.73			
USA	8.8	7.32	12.73	245.91	298.09	359.31
Germany	10.18	11.01	11.91	212.57	198.18	383.91
Japan	7.86	10.9	14.31	275.32	200.18	319.49
France	6.74	9.08	1.86	321.07	240.31	2455.50
UK	6.71	7.08	8.19	322.68	308.19	558.24
Korea, Rep. of	14.63	14.74	28.55	147.92	148.03	160.15
Italy	8.34	9.49	18.81	259.47	229.93	243.16
Russian Federation	19.22	20.39	38.64	112.59	107.03	118.35
Canada	6.69	8.23	7.15	323.37	265.26	639.50
World	12.19	11.95	13.64	177.52	182.59	335.26
Developing economies	15.98	16.01	27.66	135.42	136.29	165.33
Developed economies	9.44	9.54	10.56	229.24	228.72	433.05

Source: WTO (2014) for export and import; UNCTADstat (2014) for OFDI

3.2 *Institutional change and idea renovation*

The strong coupling between Chinese enterprises internationalization and the WTO multilateral trading system also reflect in institutional dimension. First of all, in a very short period of time, the China central and local governments cleaned up, included in abolishing, amending and reformulating, tens of thousands of pieces of laws, regulations, policies and measures related to Chinese enterprise internationalization that were inconsistent with WTO rules and China's WTO accession commitments. Second, the WTO basic principles such as transparency and openness have promoted China's legislative and governmental affairs public. Again, the WTO has pushed China's transition to a market economy, although it cannot be simply attributed to the pressure from accession to the WTO, the accession undoubtedly promoted the transformation process to a market economy (Lu & Gao 2012). In the end, the WTO has influenced on the concept of internationalization, the global view and the strategic consciousness of Chinese enterprise.

4 NEW CHALLENGES FACING CHINESE ENTERPRISES INTERNATIONALIZATION UNDER THE WTO ECOLOGICAL ENVIRONMENT

As mentioned above, in 13 years of China's accession to WTO, Not only the international trade and FDI in China have experienced rapid rise, but also the regime and concept relative to the internationalization of Chinese enterprises have changed dramatically, all of these indicate that the WTO had great influence on Chinese enterprises internationalization. In the next decade, from the view of the new process or issues of the WTO and the new situations of economic globalization and international competition under the WTO system, the future process of Chinese enterprises internationalization will face many huge challenges, which in summary including ten aspects as follow:

4.1 *The Doha round impasse*

The Doha round has for 13 years, and are still deadlocked. This is currently the biggest dilemma facing the WTO. Although the Doha round of protracted gives China more time to exercise and learn to participate in the multilateral trade negotiations, it make the interests of the Chinese enterprise internationalization difficult to deliver as soon as possible, and how to push to finish the Doha round is also a challenge for China.

4.2 *Regional and multilateral investment rules*

In the WTO, TRIMs agreement is only a short and small agreement, but many multilateral investment rules hide in GATS and the Government Procurement Agreement, which are constantly advancing; At the regional and bilateral level, high standard investment rules such as in the TPP will have a huge impact on the future process of the Chinese enterprise internationalization. The BIT between China and the United States is in the midst of talks.

4.3 *Rules for SOE*

At present, the rigorous restrictive rules for state-owned enterprises (SOEs) are promoting at the bilateral and/or regional level, and it is likely to keep up with the pace at the multilateral level. As is known to all, China is still the country that the state-owned enterprises are still in the leading position in the internationalization process of firms. As a result, the trend will have a huge impact on the future process of Chinese enterprise internationalization.

4.4 *Competitive neutrality*

The spearhead of competitive neutrality directly points to the so-called unfair competition between state-owned enterprises and non-state-owned enterprises, so its objective is to reorganize the existing rules or system of international economy by the way of constraining the state-owned enterprise, and to guarantee fair competition between the different enterprises. In recent years, the United States and the European Union are promoting to develop the framework of competitive neutrality rules, trying to put the terms and conditions of limiting the state-owned enterprise competitive advantage into bilateral or multilateral trade and investment agreements. Therefore, as a state-owned enterprises dominant country, China will face enormous challenges.

4.5 *Anti-dumping and countervailing*

Anti-dumping and countervailing strength against China have significantly increased in 12 years since China's accession to the WTO (see in table 2), especially in the six years since the crisis in 2008, for example, the average number of the global anti-dumping measures into the final implementation in 2008–2013 fell 18.9% than that of 2002–2007, on the contrary, that of China increased by 18%.

Particularly notable is that countervailing against China in recent years is rising fast. It is shown in table 2, the average annual number of countervailing initiations and measures against China in 2008–2013 increased by 384.6% and 1533.3%, respectively than that of 2002–2007, which exceeded highly than that of global.

Table 2. Global intensity of anti-dumping and Countervailing against China in 2002–2013.

	Total 2002–13	Annual average 2002–07	Annual average 2008–13	% change 2008–13 to 2002–07
Global anti-dumping Initiations	2599	222.3	210.8	−5.2%
Of which: against China	725	56.7	64.2	13.2%
Global anti-dumping measures	1777	163.5	132.7	−18.9%
Of which: against China	534	40.8	48.2	18.0%
Global Countervailing Initiations	191	9.5	22.3	135.1%
Of which: against China	76	2.2	10.5	384.6%
Global Countervailing measures	108	6.2	11.8	91.9%
Of which: against China	52	0.5	8.2	1533.3%

4.6 *New rules of trade in service*

Due to the Doha round impasse, service trade liberalization had to suspend steps at the multilateral level, but negotiations of the agreement on trade in services (TiSA) are carrying out at plurilateral level, and China has also joined the negotiations. Due to China's trade in service is still relatively weak in the world, therefore, the process of service trade liberalization will affect many emerging service industries in China.

4.7 *New trends of TRIPS-plus*

The new trend of TRIPS-plus is obviously in recent years at Bilateral and regional levels, such as in many FTAs and in ACTA. This reflects the interest differentiation and even opposition related to TRIPS between developed and developing countries. China's enterprises will face more pressure from TRIPS-plus trends in the internationalization process.

4.8 *GPA*

According to the WTO commitments, China is negotiating to join the government procurement agreement (GPA). The government procurement market is very huge in the world and also in China, it involve products, services, investment and so on, so Chinese enterprises must be good ready for a strategic response to join GPA as early as possible.

4.9 *A new situation of regional trade liberalization*

The "new generation of free trade agreements", such as TPP (The Trans-Pacific Partnership) and TTIP (Transatlantic Trade and Investment Partnership) not only have challenged in the WTO, but also to China's participation in the multilateral and regional trade liberalization. Due to the complexity of China's relations with the surrounding neighbors, as well as the particularity of Chinese enterprise internationalization, China's participation in regional trade liberalization is limited by many factors. The future process of Chinese enterprise internationalization must effectively cope with this challenge.

5 STRATEGIC CHOICE OF CHINESE ENTERPRISE INTERNATIONALIZATION UNDER THE WTO ECOLOGY

In light of the special strong coupling relationship between China's enterprise internationalization and the WTO multilateral trade system, and considering the many challenges and strategic environment facing in the future process of Chinese enterprise internationalization, China needs a more forward-looking "WTO strategy" in order to make the future process of the Chinese enterprise internationalization go better and stable under the international competition ecological environment constructed by the WTO multilateral trading system. We argue that the first principle of the "WTO strategy" which should be selected in the rise process of Chinese enterprise internationalization is multilateralism. It means the WTO multilateral trading system is the priority of Chinese enterprises and Government in strategic decision-making for internationalization. The core target of the strategy is to maintain and promote the WTO multilateral trading system running more effectively. To realize this strategic target, two main paths of the strategic implementation should be elected that of learning for application and effective participation. This strategy can be implemented effectively by raising the participation ability of China's firms and government, improving and playing the function of industry associations, and strengthening effective cooperation between government and enterprises.

6 CONCLUSIONS

China's trade and OFDI achievements have shown in more than a decade since China's entry into WTO that there is a strong coupling relationship between

China's enterprise internationalization and the WTO multilateral trading system which is different from other WTO members. From a strategic consideration, therefore, in the future process of the internationalization of China's enterprises, China should firmly continue to hold high the multilateralism flags, maintain and promote the WTO multilateral trading system, so as to ensure the strong coupling effect between Chinese enterprise internationalization and the WTO multilateral trading system to play more fully.

REFERENCES

Alon,I. & Chang, J. et al. 2009. *China Rules: Globalization and Political Transformation*. Palgrave Macmillan.

Chang, L. & Fu, X.G.2008. WTO and internationalization of China's enterprises. *International Economic Operation*(8):36–40.

Lu, N. & Gao, H.J. 2012. China and the WTO: Retrospect and prospect of global perspective. *Journal of Tsinghua University* (6):5–17.

Quan, Y. 2001. Accession to the WTO and strategy of internationalization of China's enterprises. *Asia-Pacific Economic Review* (2):59–63.

UNCTAD. 2014. UNCTADstat, Information on http://unctadstat.unctad.org/wds/ReportFolders/reportFolders.aspx?sCS_ChosenLang=en, 2014–10–20.

WTO.2014. WTO statistics database. Available at: http://stat.wto.org/StatisticalProgram/WSDBStatProgramHome.aspx?Language=E. 2014–10–10.

Zhu, G.X. 2000. WTO and internationalization of China's enterprises. *Study of Technical Economy and Management*(5):8–10.

Management, Information and Educational Engineering – Liu, Sung & Yao (Eds)
© 2015 Taylor & Francis Group, London, ISBN: 978-1-138-02728-2

The cloud resource allocation algorithm based on double auction and artificial fish swarm

Shuo Xu, Hong Juan Liu, Guo Qi Liu & Yu Jia Zhang
Software College, Northeastern University, Shenyang, Liaoning, China

ABSTRACT: At present, since there are many resource providers and consumers in cloud resource allocation, the bidirectional auction model is adopted to allocate cloud resources. Furthermore, the allocation scheme is optimized by the artificial fish swarm algorithm. The carbon emissions are virtualized to be cloud resources. Based on CloudSim toolkit, some simulation experiments and performance analysis are executed for auction and allocation mechanism of cloud resource. The results demonstrate the method has some advantages in user satisfaction, turnover rate and user gains.

KEYWORDS: double auction; artificial fish swarm; cloud resource allocation; carbon emission permits.

1 INTRODUCTION

Carbon emissions trading is a market mechanism to reduce global greenhouse gas emissions and reduce global carbon dioxide emissions. It combines the scientific issues of climate change, the technical problem of reducing carbon emissions and the economic problem of sustainable development, solving the problem of science, technology and economy issue by using the market mechanism (Cramton et al. 2002). Cloud computing is a kind of new resources and payment mode based on grid computing, utility computing, virtualization technology SaaS Application. This paper creatively virtualizes carbon emissions trading as cloud server, making the carbon emissions trading have the advantage of easy using and cloud computing specialty. Through the analysis of the cloud resource provider and consumer behavior characterize of the buyer and seller standard, we determine the price update strategy, design the overall auction process, ultimately determine the transaction price. Moreover, the using of artificial fish swarm algorithm to optimize the allocation of resources. Simulation results show that this method has some advantages in user satisfaction, turning over and user gains.

2 THE CLOUD RESOURCE ALLOCATION METHOD

2.1 *Design of double auction mode*

The cloud resource allocation process includes 3 participants: the cloud resource provider CRP, the cloud resource consumers CRC and the organizers Broker.

In a double auction model, according to the demand for resources, cloud resource consumers CRC issued corresponding purposes. The cloud resource provider CRP participates in the auction by providing appropriate resources. Auction organizer Broker is responsible for the supervision and control of the entire auction process. When certain conditions are met, the auction began (Shi et al. 2010).

After the start of the auction, CRP and CRC began to auction following auction organizer according to the auction rules. In each round of the auction process, if the trade success, the transaction price will be stored in price matrix *P*. Otherwise, the CRC and CRP pay for them are updated. In this auction round, until it reaches an end of the auction. The CRP and CRC have reached an agreement on the price, can undertake according to clinch a deal the price auction organizer the allocation of resources. The auction of the overall flow chart is as shown in Figure 1.

The specific auction process includes:

Step1: Auction organizer will CRP ascending order according to the bid price from low to high, the CRC in descending order bid price from high to low;

Step2: Matching CRP queue sellers and CRC queues buyers bid price, if the price is not higher than the buyer, seller will clinch a deal the price is determined based on the two sides bid price, and will be stored in all price matrices, otherwise turn Step4;

Step3: Buyers and sellers who participate in the auction are free to choose whether to continue to participate in the next round of the auction, algorithm USES the random number decision;

Step4: Buyers and sellers did not attend trade, in accordance with the bid price update rules update bid price;

Step5: Check whether the conditions reach the auction ends, if does not reach the end condition turn Step1, otherwise the auction ended.

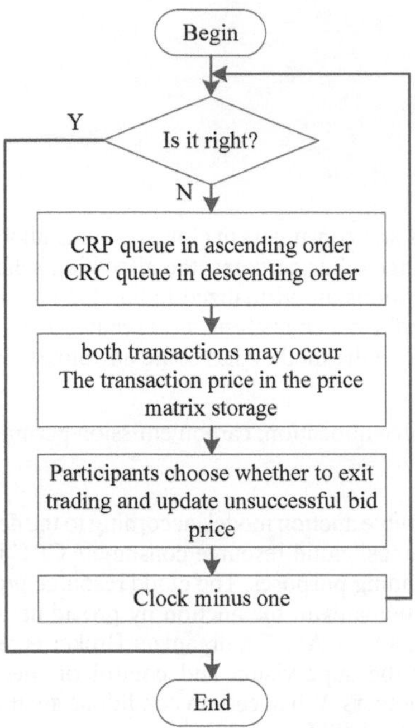

Figure 1. The flow chart for auction process.

In each round of the auction, when the seller price is not higher than buyers are, the CRP and the CRC will clinch a deal valence determined by the formula (2.1).

$$\text{Trade Price} = kBP_c + (1-k)BP_p \tag{2.1}$$

The value of the parameter k in the formula (2.1) can be calculated by the formula (2.2).

$$k = \frac{P_c}{P_c + P_p} \tag{2.2}$$

$P_c = CP_c - BP_c$ and $P_p = BP_p - CP_p$ are both the price between the bid price and the reserve price. The setting of parameters k makes the transaction price be closer to the side whose price is more reasonable. To some extent, it can reduce the participants a hostile bid.

Before the start of each round of the auction, auction organizer will check whether the end conditions for the auction reach or not. If the end condition is satisfied, we should immediately finish the current auction. Then, resource allocation is done according to the transaction price (Zeng et al. 2011). In the proposed auction mechanism, the end condition for auction has the following four:

1 The clock is 0, e.g. the maximum auction round number;
2 Number of sellers is 0;
3 Number of buyers is 0;
4 Both sellers and buyers who participate in the auction of bid price updates to the reserve price and the seller the minimum value is greater than the maximum buyers bid price.

In the above four conditions, as long as there is a condition met, the auction organizer will stop the auction.

2.2 *The design of the artificial fish algorithm*

This paper uses the artificial fish algorithm (Wang et al. 2008) to solve the optimization problem of cloud resource allocation. It is to solve a kind of scheme to maximize the objective function value, represented by formula (2.3).

$$\begin{cases} \max f(Q) \\ s.t. \\ Q_{ij} \in \left[l_{ij}, u_{ij}\right] \\ \forall i\left(\sum_{j=1}^{n} Q_{ij} \le TQ_p^i\right) \\ \forall j\left(\sum_{i=1}^{m} Q_{ij} \le TQ_c^j\right) \end{cases} \tag{2.3}$$

In the formula (2.3), l_{ij} and u_{ij} are the lower bound and upper bound of Q_{ij}, M represents the total number of sellers, n represents the total number of buyers, the objective function $f(Q)$can be expressed as

$$f(Q) = \alpha \times \frac{\sum_{i,j=1}^{I,J}\left(\frac{P_{ij}}{AP_p^i} + \frac{OP_c^j}{P_{ij}}\right)}{I \times J} + \beta \times \frac{\sum_{i,j=1}^{I,J}\left(\frac{Q_{ij}}{TQ_p^i} + \frac{Q_{ij}}{TQ_c^j}\right)}{I + J} + \gamma \times \frac{\sum_{i,j=1}^{I,J} Q_{ij} \times \left(CP_c^j - CP_p^i\right)}{I + J}$$

Finally obtained $f(Q)$ value allocation scheme for the best Q is the final solution, the organizers will allocate resource according to this scheme.

(1) The initialization parameter setting and material

In the artificial fish swarm algorithm, the parameters include: maximum number of iterations G_{max}, that is the number of algorithm cycles, the total amount of material M, the artificial fish's vision scope visual, the moving step of artificial fish step, congestion factor δ, Y as the value of the objective function. The distance between individual artificial fish is expressed as $d_{ij}=\|X_i\text{-}X_j\|$.

As for the physical initialization, using a random algorithm to generate M fish in the solution space, and the computation of their function value, the function value of maximum material retained in the best fish Q_{best}.

(2) Foraging behavior

Set up the current state of artificial fish X_i, within its field of view of a randomly selected state X_j, if $Y_i<Y_j$, then move a step further to the front direction. Conversely, we should randomly select X_j, and judge whether the condition satisfied or not. After repeated several times, if still not satisfied forward condition, the random movement step.

The distance between artificial fish a and fish b is calculated using Euclidean distance, as shown in equation (2.4).

$$\left\|Q^a - Q^b\right\| = \sqrt{\sum_{i,j=1}^{m,n} (Q_{ij}^a - Q_{ij}^b)^2} \tag{2.4}$$

(3) The swarming behavior

Suppose the status of artificial fish is X_i, explore the number of partners nf and the center position X_c in the current neighborhood ($d_{ij}<visual$). If $Y_c/nf>\delta Y_i$, it shows that partner center has more food and is not enough. Then the fish run a step forward towards the direction of the center position of partners. Otherwise, they will execute foraging behavior.

(4) The following behavior

Suppose the current status of artificial fish is X_i, and explore the partner X_j whose Y_j is the biggest in the current neighborhood (d_{ij} $<visual$). If $Y_j/nf>\delta Y_i$, it shows the status of X_j have food with higher concentrations and its surroundings not too crowded , it goes in X_j direction step; otherwise executing foraging behavior.

(5) Update bulletin board

The bulletin board is used to record the individual state of artificial fish. In the process of optimizing the artificial fish, every action is completed the inspection status and bulletin board their own state. If its state is better than the bulletin board, will notice board state rewritten as its state. This allows the bulletin board to record the history of optimal state.

The overall flow chart of the artificial fish swarm algorithm as shown in Figure 2.

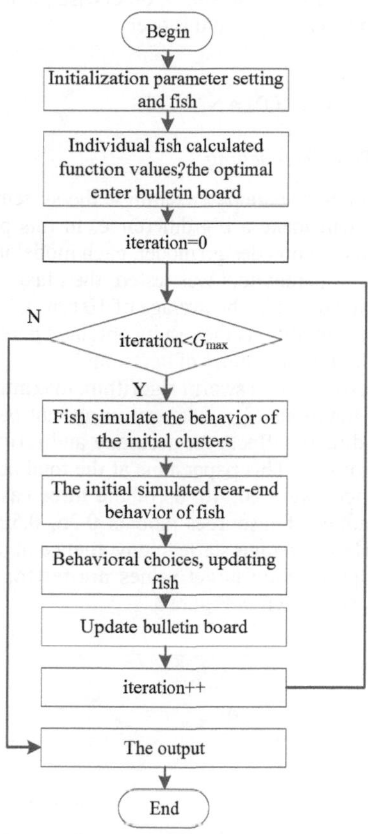

Figure 2. The flow chart of artificial fish swarm algorithm.

The specific process design algorithm is as follows:

Step 1: Initialize the parameters.

Step 2: The fish is initialized by chaos. The N fish is generated in feasible solutions space ($N>Np$). Choosing the better N_p individuals as the initial fish $X(0)$.

Step 3: Calculating the functional value $X(0)$ of each individual, choose the best into the notice board.

Step 4: If the current iteration is less than the maximum number of iterations, go to step9; otherwise go to step5.

Step 5: For each one artificial fish X_i behavior were simulated clusters and rear-end behavior, the corresponding X_{next}.

Step 6: Behavioral choices. Select the optimal X_{next} executor act as a final, new and after swimming groups $X(t)$.

Step 7: The best comparison bulletin board $X(t)$ in the new group, if better than a bulletin board, bulletin board is updated.

Step 8: When reached the maximum number of iterations is set, go to step 9; otherwise go to step 5.

Step 9: The algorithm is over.

3 SIMULATION ANALYSIS

3.1 *Parameter setting*

With different parameter settings, the system performance will make a big difference. In this paper, on the basis of good design model, each model in several important parameters was tested, then take the average. Each datum is the average of 10 runs taken under the same conditions the results obtained herein.

(1) Maximum number of iterations

In artificial fish swarm algorithm, maximum number of iterations G_{max} is a very important parameter, and it directly affects the accuracy and efficiency of the algorithm. This paper aims at the total number of substances were 20, 40, 60 of the three cases. Both the amount of resources ratio is 0.25, 0.5, 0.75, 1, 1.25. Here, average started convergence algebra and final optimization target values are tested. The test results are shown in Figure 3.

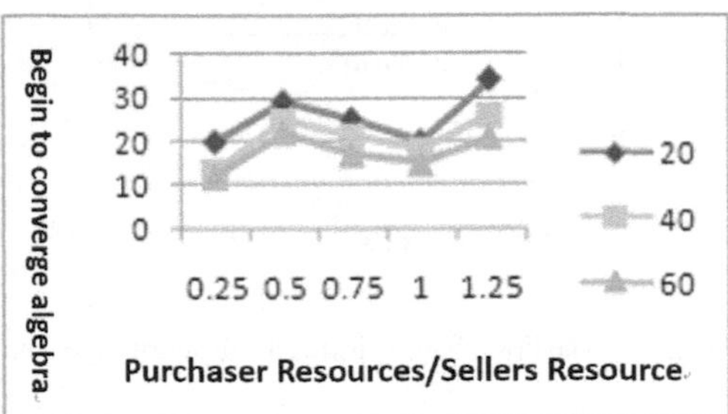

Figure 3. The variation of Started convergence algebra.

From Figure 3 you can see that the more material number increases, the fast convergence speed is. Moreover, convergence speed is faster when the amount of buyer's resources and the amount of the seller is equal or less. In general, it began to converge algebraic from 10 to 35 ranges. So in this paper, the maximum number of iterations G_{max} is 40, In the premise of ensuring the convergence results, try to reduce the amount of calculation.

(2) Select the number of substances

In addition to the maximum number of iterations, the number of M material size is also very important. At the same time, the more the number of substances, the higher the accuracy of the algorithm, and need the greater the amount of calculation, so choose an appropriate amount of material for improving the overall performance of the algorithm is very critical. In this paper, the total amount of material for each of these three cases 20, 40, 60, when buyers and sellers resources than is 0.25, 0.5, 0.75, 1, 1.25, finally optimize the size of the test target, the test results are shown in Figure 4.

As can be seen from Figure 4 the resource ratio of the buyer and the seller is 0.75. The number of the target material is the amount of substance 40 is very close to the target value 60. While its overall performance is relatively good, moderate amount of calculation, and therefore the total amount of material herein M value is 40.

The 40th cycles:

The current optimal trading volume:

0	2	0	0
0	1	0	0
3	12	0	0
0	5	0	0

Price satisfaction: 0.533893352863843

Turnover rate: 0.34598214285714285

Profit: 1.475E-4

Target value:0.22004262393024646

Figure 4. The variation of target value.

The values of some parameters of double auction and the artificial fish swarm algorithm are shown in table 1.

Table 1. The related parameters of double auction and artificial fish swarm algorithm.

Parameter name	Value
The maximum number of iterations	40
Number of fish	40
View of artificial fish	0.3
Artificial Fish step	0.05
Price satisfaction factor	0.25
Turnover rate coefficient	0.25
Profit factor	0.5
Auction round number	20

The price satisfaction factor *c_coe* in Table 1will also be required in the use of dividing 10^n. Thus, three evaluation index on the same order of magnitude, according to the actual value of the exponent in the case may be.

3.2 *The experimental results*

The matrix of bidirectional auction is shown in Figure 5.

Number of buyers: 4

Number of sellers: 4

Price:

0.00	115.50	0.00	0.00
0.00	116.93	0.00	0.00
115.38	115.99	0.00	0.00
0.00	118.77	0.00	0.00

Figure 5. The matrix of bidirectional auction.

The cloud resource allocation algorithm results using artificial fish swarm algorithm are shown in Figure 6.

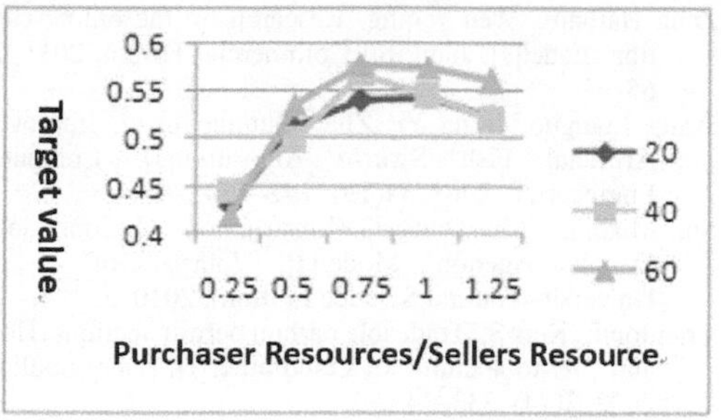

Figure 6. The results for resource allocation algorithm.

3.3 *Performance Analysis*

In order to test the performance of the artificial fish swarm algorithm in cloud resource allocation, the proposed resource allocation method is compared with the allocation algorithms in literature (Zhao et al. 2012). In this paper, for two allocation programs, three evaluation indicators and a target value are compared in different quantity buyers and sellers resources ratio.

(1) User price satisfaction

The comparative results in user satisfaction between this method and the algorithms in literature (Zhao et al. 2012) are shown in Figure 7. As shown from Figure 7, the performance in user satisfaction of this method is far superior to the algorithms in literature (Zhao et al. 2012), because each user in the algorithms of literature (Zhao et al. 2012) are possible for to all the resources of the transaction. Therefore, the number of users for final transaction is much less than those in the artificial fish swarm algorithm, leading to the final price satisfaction value is very small.

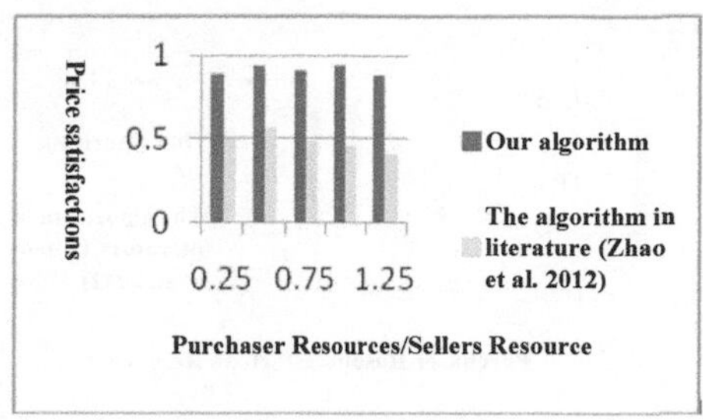

Figure 7. Comparison of price satisfaction between this method and literature (Zhao et al. 2012).

(2) Turnover rate

The turnover rate is defined as the ratio of the amount of successful trading resources and the total amount of resources. The comparative results in turnover rate between the two methods are shown in Figure 8. From Figure 8, there is a little difference between the two algorithms on the turnover rate. The algorithm in the literature (Zhao et al. 2012) is slightly better than the proposed method.

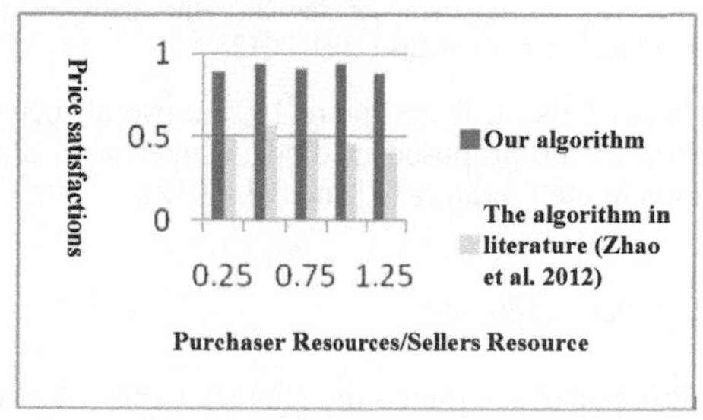

Figure 8. Comparison of turnover rate between this method and literature (Zhao et al. 2012).

(3) User benefits

The comparative results in user benefit between this method and the algorithms in literature (Zhao et al. 2012) are shown in Figure 9. We can see that when the amount of resources required for buyers is much smaller than the amount of resources needed by the seller, the user benefit of literature (Zhao et al. 2012) is greater than the proposed method. However, with the increase in buyer demand, the user gains the value of this method is gradually approaching or even exceeding the literature (Zhao et al. 2012) algorithms.

(4) Target value

The size of a single index value does not indicate whether the algorithm is good or bad. The final decision on the merits of the algorithm is the final target size. The comparison of target value between the two methods is shown in Figure 10.

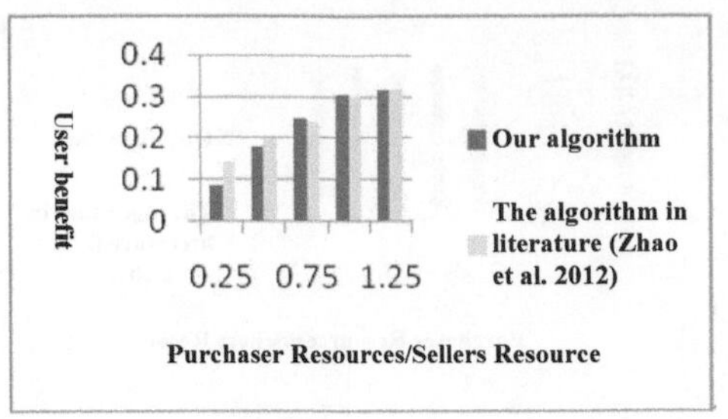

Figure 9. Comparison of user gains between this method and literature (Zhao et al. 2012).

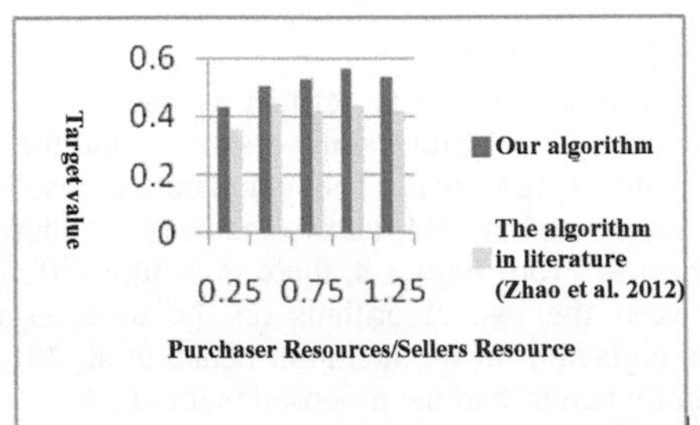

Figure 10. Comparison of target value between this method and literature (Zhao et al. 2012).

It can be seen from Figure 10, the overall performance of the proposed method is superior to algorithms in the literature (Zhao et al. 2012).

4 CONCLUSIONS

In this paper, combined the characteristics of cloud resources, we designed a double auction process for carbon emissions trading to ensure the fairness of buyers and sellers effectively. The artificial fish swarm algorithm is creatively applied on optimizing the allocation cloud resources. By transforming the merits of resource allocation scheme to sake the maximized value of the function, the optimal allocation scheme is thus determined. Compared with the algorithm in the literature (Zhao et al. 2012) in user price satisfaction, turnover rate and user benefits, the results show that the overall performance of the proposed method is superior to the algorithms of literature (Zhao et al. 2012).

ACKNOWLEDGMENT

The corresponding author of this research is Hongjuan Liu.

REFERENCES

Zeng Hailiang, Wen Yaqing. Research on the online auction model[J], Jiang Su Commercial Forum, 2011, 2: 65–67.

Wang Lianguo, Hong Yi, Zhao Fuqing, et al. Improved Artificial Fish Swarm Algorithm[J]. Computer Engineering, 2008, 34(19): 192–194.

Shi Mianjun, Liu Dongxi. Construction of Continuous Double Auction Model[J], Journal of Jishou University(Natural Science Edition), 2010, 2: 1–4.

Cramton P., Kerr S.. Tradeable carbon permit auctions: How and why to auction not grandfather[J]. Energy policy, 2002, 30(4): 333–345.

Zhao Wei, Peng Yong, Xie Feng, et al. Modeling and Simulation of Cloud Computing: A Review[A]. 2012 IEEE Asia Pacific Cloud Computing Congress[C]. IEEE, 2012, 20–24.

Management, Information and Educational Engineering – Liu, Sung & Yao (Eds)
© 2015 Taylor & Francis Group, London, ISBN: 978-1-138-02728-2

The influence of economic bubbles by the Wall Street Crash

N. Ye
Department of Finance, Xinyang Agricultural College, Xinyang, China

ABSTRACT: The international economic situation is now starting to get better after the financial crisis, but still has many unstable and uncertain factors with the deeper problems remaining unsolved. In order to understand the situation of the current complex global economy, we must focus on the Wall Street Crash, which is the biggest financial disaster of the 20th century, and which is the experience and lesson for nowadays economic growth. This text analyzes the irrational exuberance by the Wall Street Crash of the 1930s, and points out three important reasons of the Wall Street Crash to make bubbles occur frequently. Nowadays, with the development of science and technology, how to avoid the burst bubble economy is a question worth thinking about.

KEYWORDS: economic bubbles; Wall Street Crash; the Great Depression.

1 INTRODUCTION

"What about the present economic situations in the world. Is this a bubble or a recovery?" It is a question often asked after people have had the worst years of the great economic depression. Actually, it's a fact that people always worry about it even though nobody can make sure what will happen next. At all events, the economy has been one of the most important things for countries to pay attention.

Let's recall the history, there were some important events about economic bubble, for instance Tulip mania, South Sea Bubble, Japanese asset price bubble and so on. But now, we will focus on Wall Street Crash, which is the biggest financial disaster of the 20th century, and which is the experiences and lessons for nowadays economic growth.

2 BUBBLE AND BUBBLE ECONOMY

Why stock price was growing fast just in one day without any reasons? Why the great financial crash could be happening immediately after economic boom? If people want to know the answers clearly, they should understand what the bubble mean is.

A bubble, a situation where market prices are unsustainably high, it means a bubble involves a rapid and unfounded rise in prices. Another way to describe it is: trade in products or assets with inflated values.

"The bubble economy is an economic phenomenon of the virtual capital of excessive growth and the continued expansion of the relevant transaction increased from the physical capital growth and the industrial sector growth". The economic bubble is a kind of speculative finance, resulting in socioeconomic borrowed prosperity, and finally the bubble must burst, leading to social convulsion, and even economic collapse.

3 THE GREAT DEPRESSION BY WALL STREET CRASH

3.1 *The procedure of economic bubbles*

We know that there are too many economic bubbles can cause socioeconomic borrowed prosperity. And how bubbles occurred?

At first, bubble usually happen in the stage of a more relaxed policy for banks and rapid economic growth, then it appears socioeconomic borrowed prosperity after government takes measures of reduced interest and deregulation to stimulate investment and consumer demand.

Next, the text will make a lucid explanation by irrational exuberance by relating 1929 Wall Street Crash to a historical bubble.

At the end of First World War, people had to face the threat of an economic depression hangs over the world. For the American government, the first important thing at the present time was that it could change negative status into economic recovery. So the president of America, Hoover believed that they must focus on the special interest programs and expansion of governmental power, so it can increase the overall economy. For a long recession during 1920, under the current policy of easy credit, The Federal Reserve expanded credit by setting low market interest to many large banks, which led to the amount of money supply went up to attract more and more investors to credit by soaring stock.

After a bull market that had lasted most of the decade, the famous crisis began on Thursday 24 October 1929. The stock prices would fluctuate dramatically. All of investors began to know unexpected issue would be coming soon. On Black Thursday, investors couldn't wait for decreasing the price of share and stock, so they tried to get out of the market, because most of them had got all her money in stocks and shares. Some of Wall Street's top bankers had to face the difficult steady; they tried to put the large amount of money into the big companies to avoid the current depression of the stock market.

The fighting around the stock market continued for a week, the Black Monday would be coming soon with the Dow ending the day 13% lower. When people couldn't stand this kind of condition, what is worse, the Dow declined another 12% on Black Tuesday. Everyone became crazy, they didn't know what to do and how should do.

Unfortunately, Millions of Americans had put money in banks to get more money in boom years. When many people still fervently believed that they can be rich by keep ever-increasing stock, the stock market crashed as soon as possible.

John, K.G. (1997) reported that in Wall Street, many Wall Street citizens had a deep faith in the power of incantation. They immediately felt the danger after the stock market fell, because this situation had a strong impact on income and employment. It had to be prevented.

3.2 *The reasons of economic bubbles occuring*

In fact, bubbles could be everlasting, although some economists didn't think that bubbles occur. It's just like the potential danger, people can't see and touch it, but once there is a sudden drop in the stock market, the bubble should be burst. Under the modern economic condition, there are three important reasons of the Wall Street Crash to make bubbles occur frequently.

3.2.1 *Because of liberalization and globalization of financial market*

For Wall Street Crash, some investors' thought are based on a series of wrong assumptions, which put money into the stock market with full of bubbles. So it leaded more and more bubble around the stock market. However, investors to rely on the "greater fool" argument to believe incorrectly that they might get the money up when the bubble hadn't burst, but the greatest fools bought the high price of stocks from the greater fools; they couldn't wait for another greatest to buy from their hands, so both bubble and stock market burst.

3.2.2 *The formation and development of financial instrument derivatives*

Some economists consider that the economic bubble occurs in the long term because of its objective cause. Economic bubble is a general economic phenomenon. It occurred based on some fictitious economic factors, such as bonds, futures, financial securities and speculative financial transactions. The development of financial derivatives is an unprecedented opportunity for the financial industry. On the one side, the increasing extension, variety, complexity of the financial derivative instruments must post great challenges to audit ability and audit instruments. On the other side, Foster, J. B. (2008) supported, it can take advantage of a market economy to stimulate the economy and promote competition for achieving more profit. Such speculative mania must bring a rapid increase in the quantity of debt and a rapid decrease in its quality. However, financial instrument derivatives will be made more bubbles around the whole financial market; it changes the total demands of social development.

3.2.3 *Lack of restraint mechanism for bubbles*

If people wanted to control the development of bubbles, the first one thing was monitoring and controlling total activities for the growth of the economic bubble. The government cannot control the speculative financial transactions between industries, also it can easily agree with socioeconomic borrowed prosperity, even it might find out the costly mistake resulted in severe loss after the bubble burst. The bank still had no power to limit the trade. It's just an intermediation according to the customers' instructions. So if the useful monitored system can deal with payment activities, people can find out problems to avoid the financial crisis.

4 THE IMPACT OF WALL STREET CRASH

According to an analysis of the Wall Street Crash, the end of economic bubble is the crisis. From the history we know Wall Street Crash by the end of October led to the consequence of the Great Depression of the 1930s cover the whole world. An economic crisis engulfed the entire capitalist world. Some historians blamed that America central bankers can't protect the small banks' benefits by their responsibilities. Janszen, E. (2008) said some economic experts believed the "bubble cycle" and the only way to solve this problem is looking to another bubble. What's more, it emphasized the fact that the virtual economy cannot move too far away from the real economy.

1929 Wall Street Crash, as the biggest financial disaster of the 20th century, it's been 80 years since the Wall Street Crash happened. It still has a deep potential impact on modern society.

After the Great Depression of the 1930s, it reached the peak of US's economic exuberance during the period 1966–1985, which the average of the gross domestic investment as a percent of GDP is 23.08%. It showed that it provided an opportunity to stimulate economic growth by the Second World War. According to the impact of the economic cycle, lending and investment abilities tightened that continued to the present as the boom times fade. Especially the financial collapse of 2008, it caused a global recession. Compare with during 1966–1985, the average of 21.50% in 1991–2010. As a result asset price bubbles have burst and restrain economic growth for a long time (Fig. 1).

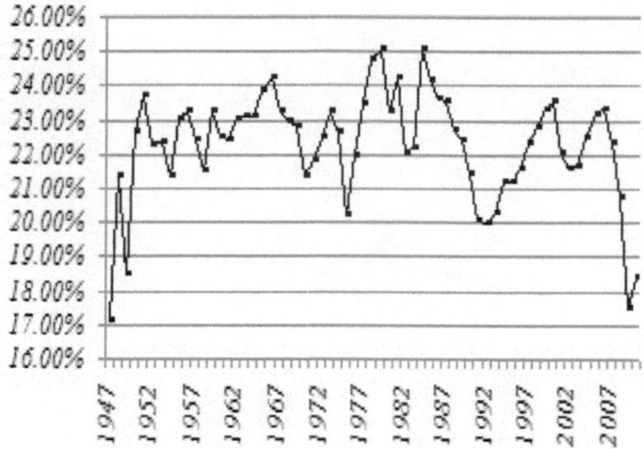

Figure 1. Gross domestic investment as a percent of GDP.*

* Financial Accounts of the United States - Z.1, Federal Reserve Statistical Release, Board of Governors of the Federal Reserve System.

5 HOW TO PREVENT THE BUBBLE ECONOMY

With the rapid social and economic development, lots of financial innovation in derivatives triggered speculative mania in real estate and stocks, which breeding the different kinds of bubbles, such as stock market bubble, housing bubble, credit bubble and so on. The immediate priority now is taking measures to prevent the bubble economy.

5.1 *Strict control of inflation of stock*

The rising stock prices encouraged more people to invest in the stock market, and then people hoped shares with high prices to bring more benefits. Speculation leaded to sky-high prices and created economic bubbles. On the one hand, some stock holders gained the short term bubble profits. On the other hand, lots of bubbles not only harmed to medium and small investors, but also harmed the national interest. Thus, to set the stock price ceiling should become the top priority to do.

5.2 *Reduce the real estate bubbles*

In 2002, financial analyst Pomboy, S. (2002) pointed out a theory as "The Great Bubble Transfer," which indicated speculative bubbles by real estate market could cover the losses for the bursting of the stock market bubbles. In fact, the excessive expansion of bank credit was a major reason for proper bubbles. In either case, all bubbles would explode and would be into the economic downturn. Especially real estate bubbles could pose a threat to the whole economic system and bring a lot of problems to hinder the operation of the market economy.

From government sectors, make sure to implement relative tax policy and property policy and raise the transaction cost of speculators. And it effectively reduced the speculative behaviors in the real estate market, thereby control real estate bubbles.

From commercial banks, should improve the sound financial system. It must monitor loan program approval and support optimizing loan approve process. The aim is to reduce speculators in the real estate market through control loan funds.

5.3 *Strengthen the economic infrastructure and enhance innovation*

Commercial banks should improve the sound financial system. It must monitor loan program approval and support optimizing loan approve process. The aim is to reduce speculators in the real estate market through control loan funds. Asset bubbles were caused by the unreasonable expansion of the virtual economy, and led to burst at the end. At that time, it resulted in large damages to the economy, so local government must strengthen the economic infrastructure to prevent bubbles for the unreasonable capital structure. Moreover, the government must improve the capacity of development of technology and innovation of technology, and then it can resist the impacts of all bubbles.

6 CONCLUSION

Eventually, we can answer these two foregoing questions. Why stock price was growing fast in one day without any reasons? Why the great financial crash could be happening immediately after economic boom?

Too much bubble was full of stock market led the stock price increased dramatically in one day. This kind of false prosperity attracted more and more investors to put an entire fortune into the stock market to gain their best interests. Unluckily, their wishes came to nothing. They never came to anything in the end. The economic boom was the false prosperity. When the bubble burst once, everything people owned has gone up the spout. At the end, the great financial crash could be happening.

1929 Wall Street Crash is a harrowing experience. Everyone needs to realize what had occurred and remember it. More importantly, we should try to find out more useful measures to avoid negative effects by bubble and maintain the steady economic development, that's the first important challenge to face in recent years.

REFERENCES

Foster, J.B. 2008. The financialization of Capital and the Crisis, *Monthly Review:* volume 59, issue 11(April)

Gohn, K.G. 1997. The great crash, 1929, *Pitman Publishing Corporation*: 89–90.

Janszen, E. 2008. The Next Bubble, *Harper's (February 2008)*: 39–45.

Pomboy, S. 2002. The Great Bubble Transfer, *MacroMavens*: 8–10.

Management, Information and Educational Engineering – Liu, Sung & Yao (Eds)
© 2015 Taylor & Francis Group, London, ISBN: 978-1-138-02728-2

Analysis of tourism consumer psychology under the influence of Chinese people's traditional culture

Xiao Li Ni & Bo Xue
Qinhuangdao Institute of Technology, Qinhuangdao, Hebei, China

ABSTRACT: Extensive and profound Chinese traditional culture has a long history. Correct understanding and scientific analysis of China's traditional culture under the influence of tourism consumption psychology and behavior characteristics, for tourism enterprises in the new product design, development, service, promotion and marketing strategy and has important practical significance. This paper analyzes the tourist consumption psychology from the traditional culture, and analyzes the behavior characteristics of tourism consumers.
KEYWORDS: Traditional cultural; Tourism consumer Psychology; Consumer behavior.

1 INTRODUCTION

Chinese traditional culture refers to the Han nationality as the main body, composed of multi-ethnic Chinese nation in the long history of the development process creation. It is based on thousands of years of small-scale peasant economy, the patriarchal clan system as background, the Confucian ethics as the core of traditional agricultural culture, kinship and cultivation culture. Extensive and profound Chinese traditional culture has a long history, Thrive in the context of the culture of the Chinese nation, its values, thinking mode, lifestyle and consumption idea has its uniqueness.

2 THE INFLUENCE OF TRADITIONAL CULTURE IN THE TOURISM CONSUMPTION PSYCHOLOGY

2.1 *Doctrine of traditional culture*

Confucian doctrine of the mean is a basic proposition, Dali scientist Zhu Xi said: moderation is the "not partial of that, and that's not easy In layman's terms, the doctrine of the essence of things is the development process has certain standards and norms, If you exceed or fail to meet the standards and norms are not conducive to the development of things, so no matter what should be taken impartially, to reconcile compromise attitude. Everything about "degree", opposition beyond the conventional oppose fundamental change, emphasizing the continuity and stability.." On the individual, the Chinese people will tend to see themselves as a member of a group, trying to comply with group norms, and strive to be consistent with the demeanor of everyone, avoiding prominent individuals, Otherwise it will get attention and discussion, which is what most Chinese people do not like, but also to avoid. This cultural awareness reaction in the tourism consumer behavior, is to emphasize the importance of maintaining a consistent and society, with the flow, catch the trend, on par with others, tourism convergence of consumer behavior; Not in favor of excessive consumption, focus on traditional, could not understand new things, do not pay attention to unconventional tourism consumption.

2.2 *Focus on the traditional culture of ethics*

The most basic function of traditional Chinese culture as the patriarchal system is to maintain and strengthen the basis of kinship, emphasizing the relationship between people and between groups, with emphasis on blood kinship and blood-based derived relationships. It attaches great importance to the core idea of the relationship between people's relationship with the West pay attention to the relationship between man and nature, emphasizing the concept of individuality and freedom in stark contrast, And a direct result of Chinese and Western cultures differ in many ways. The reaction in the tourism, consumer behavior, the Chinese people's tourism consumption of the natural family unit is carried out. In the aspect of tourism products and services information transmission and communication, oral communication than formal information, communication can let more Chinese tourists believe and accept.

2.3 *The traditional culture of good face*

Hobby face psychological is typical of Chinese cultural characteristics. Is the Chinese special pay attention to their image in others heart status, pay attention to "face". Chinese face saving is various, both pay attention to oneself give oneself of face, face to others, also pay attention to other people to leave their faces. Reaction in tourism consumption behavior, Chinese tourists are to the tourism consumption, "decent" value consistent with their status, consistent with others around the seeking of consumption and consumption. Meanwhile, travel consumer behavior will as a group member is different from the experience, I feel very face.

2.4 *The culture of valuing loyalty over money*

Heavy note of friendship and spiritual value, despise material interest, emphasizing the relationship between people and moral, is the other big characteristic of Chinese culture, is also one of the main differences between Chinese and western cultures. Confucius said: "Gentlemen loyalty, small man profits." The values of valuing loyalty over money makes the Chinese people are very pay attention to the emotional communication between people. Keen to exchange all kinds of gifts in the interpersonal communication, to strengthen the mutual relationship, this kind of cultural mentality caused many people read the friendship between people is very important. In the tourism consumption behavior of valuing loyalty over money is, tourism product emotional gifts and souvenirs.

The above four aspects concentrated and embodies the basic spirit of Chinese traditional culture and to the shaping and building of the Chinese nation, its impact on Chinese tourism consumption psychology and behavior are various.

3 TRAVEL CONSUMER PSYCHOLOGY AND BEHAVIOR CHARACTERISTICS

3.1 *Pay attention to human feelings and seek common tourism consumption motives*

The Chinese group feels strong, pay attention to the specification, pay special attention to interpersonal relationships, stressed the importance of good human relations. Reflected in the tourism consumption behavior is that most people in the society, the general concept of tourism consumption and tourism consumer behavior to standardize and restrain their consumption behavior. A person to buy and what kind of tourism products consumer often the first thing to consider other people's comments and evaluation, Even if your product is very popular with, if that it does not accord with the requirement of group norms, purchase and consumption of tourism products can make the person has a "stand out" and the feeling of incompatible with others, so they will consider to give up the purchase behavior, So its tourism consumption behavior has obvious social orientation and the characteristics of "people orientation". This is the background of western culture, the outstanding individual rights and value, emphasizing individuation, diversification, with "self orientation" as the characteristics of tourism consumption behavior is significantly different. Therefore, we are learning and using for reference western marketing model, marketing experience, should be particular case is particular analysis.

Of course, as the exchange of Chinese and western culture, the collision and fusion, along with the deepening of the reform and opening up in China in recent years, tourist consumption psychology and behavior of the Chinese people are changing so quietly, People in the consumption of seeking common backdrop, started to buy and use some embody and reflect the character of tourism, Especially in the young there is an obvious novelty and unique personalized consumption trend. To this, tourism marketers must give enough attention, On the basis of the traditional tourism product marketing strategy, pay attention to the new situation, new change, pay attention to the new trends of people's consumption at any time, so that in the new tourism product design, development and service at any time to meet the new demand of tourists.

3.2 *Tourism consumption concept of thrift*

For thousands of years, the Chinese people have been advocating the thrifty consumption idea, opposes any form of wasteful and deficit spending. Reflected in people's consumption behavior is to be more cautious in spending, shopping, rich planning, attaches great importance to the accumulation, not is costly, opposed to spending, Consumption idea has now if not tomorrow, scoff at such spending is lack of planning behavior, see it as no mind, no life. People are used to saving money to buy things, not used to borrow money or loans to buy things, argues that the cost of living will "carefully", "maintenance", to achieve "more moon", "more than" year after year. In general, they are more used to buy necessities, and enjoy the luxury or buy this kind of tourism products necessary living consumption tend not to row in the forefront of the family or personal consumption expenditure sequence table. Reflected in the middle-aged and old consumers, they always try to save money, on their children to school, get married, buying a house, pension, etc. In western developed capitalist countries, most people are comfortable with the comforts life, consumer spending is often beyond their income level, So for tourism expenditure proportion and much higher than the domestic consumers.

3.3 *Subtle national character and humility travel consumer behavior*

Due to the "warm, good, purge, thrifty," and "harmony" of the traditional Chinese cultural characteristics, makes the Chinese tend to be introverted, reserved, and solemn, speak civilization, polite, Don't like westerners make public and visible. In the tourism consumption activity, people often carefully worded, moderate and pragmatic attitude, If you think the other's words and deeds have wrong, most people who did not like Europe pointedly, but take the best possible care, the best way to tactfully tell each other, Avoid hurt each other's face, and let the other side is anathema. Occasionally encountered something meat, often also a bear it, or just learn relief similar to remind yourself, pay attention next time.

3.4 *The tourism consumption principles of family-based*

Due to the influence of traditional culture, the Chinese family values, family dependence, family sense of responsibility is stronger than westerners, the Chinese family is often a consumer unit. In China, personal travel consumer behavior often is closely related to the activities for the whole family, rather than just individual, isolated. Therefore, in the predominantly Chinese tourism consumption market, individual consumer behavior should not only consider their own needs, and to consider the consumption needs of the whole family. This situation makes the tourism enterprises and marketing personnel in such aspects as product development and marketing service should pay attention to tourism products to the satisfaction of all family members, try to make the product to suit the needs of all members in the home, everyone is satisfied, all like it.

3.5 *Pay attention to an intuitive judgment way of tourism product purchase decisions*

Chinese people in terms of thinking often rely on personal experience and intuition can be extrapolated to explain the field's understanding of things is often used is an intuitive extrapolation symbolic analogy, introspective experience, chaotic record, simple vague values and methodology. Therefore, when the Chinese to buy travel products often have to get a good overall impression, And then find the corresponding basis of its overall performance, see the impression is correct, very few or rational analysis of the tourism product of meticulous. That is to say, Chinese tourists when buying a tourism product often is used broadly and intuitive judgment method, use the fuzzy thinking and comprehensive thinking. Unlike western tourists used to put a detailed analysis and comparison, one by one, to analysis the stand or fall of each link of tourism products, Then comprehensive analysis of these links to draw a general impression. Insight into the eastern and western tourism consumers in the purchase decision in the differences of thinking mode, will help the tourism marketing personnel to carry out the marketing work of science.

4 SUMMARY

Tourism consumption activity essentially is a kind of cultural consumption activity, Cultural factors influence the formation tourism motivation and consumption behavior, and these factors in different levels of different degrees of influence on tourism consumption behavior, Cultural factors decide the tourism consumption idea and behavior standard of the individual. Chinese traditional culture creates and influence the tourist behavior and specific consumer behavior, Through social ethos, reference groups such as dominance and influence the development direction of tourism consumption demand. With the development of Chinese traditional culture, people's travel consumer behavior will show new trends. However, the correct cognition and scientific analysis of China's traditional culture under the influence of tourism consumption psychology and behavior characteristics, For tourism enterprises in tourism product development, service promotion and marketing strategy and so on has the important practical significance.

REFERENCES

FuJianHua;Chinese and foreign tourism consumption idea of contrast and marketing strategy analysis, Travel through (second half); (2012), 11.

Chen Ji; under new media the background of China's tourism marketing research, The central university for nationalities. In (2011).

Zhu Huying;; Theory of tourists irrational consumption behavior, Business research; (2007), 12.

Taodan Yan; Tourists Ethics Perspective of Harmonious Tourism,Guangdong Business School; (2011).

Wang Zhongyu;; Domestic tourism market segmentation research review, Modern agricultural science and technology; 04, (2008).

Management, Information and Educational Engineering – Liu, Sung & Yao (Eds)
© 2015 Taylor & Francis Group, London, ISBN: 978-1-138-02728-2

Association between urbanization and economic growth in Zhejiang province: 1978 to 2012

Hai Sheng Chen
The Party School, Zhejiang Provincial Committee, CPC, P.R.China

ABSTRACT: Based on per capita GDP in Zhejiang province from 1978 to 2012, the proportion of non-agricultural population, non-agricultural industry output value proportion and the industrial added value based on time series data, using econometric analysis tools Eviews6.0, urbanization and economic growth of Zhejiang province association has carried on the empirical research; econometric analysis results show that the three indexes in reflecting the connotation of urbanization China-Africa agricultural industry output value proportion is the most important factors that influence the economic growth of Zhejiang, is recommended to promote the industrialization and the urbanization process of population, should pay more attention to the effective gathering and structure of the non-agricultural industries to upgrade.

KEYWORDS: Urbanization; Economic growth; Non-agricultural industries.

1 A LITERATURE REVIEW

Domestic demand is an important motivation of economic growth, in the international economic situation increasingly complex circumstances, how to release the domestic demand is a problem of the current provinces.In the 12th five-year plan and the central economic work conference, a new generation of central collective leadership puts forward with the aid of the policy of stimulating domestic demand of urbanization. Before that, about the role of urbanization on economic growth, the scholars have discussed more. The article from the perspective of zhejiang domain further study the relationship between urbanization and economic growth, trying to show an economic profile of the eastern coastal provinces.

Although most studies tend to think that there is a long-term cointegration relation between economic growth and urbanization, but there are differences on the empirical results. Jia Yunyun sail (2012) and generation (2011), and other people think that urbanization can promote economic growth, chukhung to et al. (2011) calculated every 1% increase of urbanization in our country, can maintain economic growth of around 7%.Rocky (2006) found that the overall lags behind the economic growth of China's urbanization, but shi-bin zhang (2010) and wang chao (2012) that the government should not increase the rate of urbanization as the final goal, but more should pay attention to economic growth and urbanization of the endogenous relationship, this is consistent with some empirical analysis result, lagging Su Fa gold (2011) found in different period, the urbanization and the agricultural economic growth have individual causality, the influence degree change according to the length of time. Yang jie et al. (2012) and xiao-juan qin (2011) study of gansu province economic growth effect on urbanization more obvious conclusion, but Mr. Zhang, lian-qing (2012) years in shandong's investigation found that the province economic growth and urbanization level there is a long-term and stable relationship, but the urbanization effect significantly greater than the impact of economic growth.This conclusion, with Yang jie and xiao-juan qin people shows in different provinces, the relationship between economic growth and urbanization are different.At the same time, most of scholars before will examine the object location and a handful of provinces across the country, and the research about the relation between economic growth and urbanization of zhejiang is not enough, so the article may be able to fill the blank slightly.

2 INDEX SELECTION AND TEST METHOD

2.1 *Index selection*

For measurement of urbanization, the article on the basis of reference to existing research results, select the proportion of non-agricultural population (X1) and non-agricultural industry output value proportion (X2), respectively in the process of urbanization of zhejiang province population agglomeration and non-agricultural industries agglomeration situation; In view of the industry the important influence to the urbanization at the same time, choose the proportion of gross value added of industry of Zhejiang province (X3) as an endogenous variable of urbanization in the description.This paper select the per capita GDP (Y) in zhejiang province as described main indicators of economic growth.At the same time, in order to more in line with the general form of economic growth theory, this paper adopts the logarithmic form of per capita gross domestic product is LnY as investigation form.

2.2 *The test of the model*

The article establish error correction model (ECM) analysis:

$$\Delta Z_t = \sum_{i=1}^{p} A_i \Delta Z_{t-1} + \partial \beta' X_{t-1} + \theta D_t + \varepsilon_t (t = 1,2,3,\text{KT})$$

Among them,

$\Delta Z_t = (\ln Y_t, X_{1t}, X_{2t}, X_{3t})$, *Ai* is matrix; ∂ is the error correction term coefficient vector; β is the cointegration vector; Dt is the deterministic item; Θ is the coefficient vector.

3 THE EMPIRICAL ANALYSIS

3.1 *Unit root test*

Since 1978, Zhejiang economic growth per capita maintained a high speed.The logarithm of per capita gross domestic product showed a rising trend. According to the image can be judged lnY is a non-stationary time series.For the proportion of non-agricultural population X1, non-agricultural industry output value proportion X2 and X3 of the total value added of the industrial and other variables can also be similar conclusions.

Considering the article USES the time series data, but most of the time series data is not smooth, not smooth time series for regression analysis may lead to spurious regression results, so we first of all variables, stationarity test of using ADP (Augmented Dickey Fuller) test and PP (Fillips - Perron) test for unit root test, inspection form for (C, T, L), respectively, paragraphs drift and time trends and lag order, using measurement software Eviews6.0 is analyzed, the results such as table 1.From table 1 columns 2 and 3, according to all levels of vector are not rejected has a unit root hypothesis, so can think LnY per capita gross domestic product, the proportion of non-agricultural population X1, non-agricultural industry output value proportion X2 X3 are stationary series, of the total value added of the industrial time trend is obvious. From the fourth column of table 1 and 5 columns can be seen that the variable before the first order difference time series is not smooth, but after the first order difference under the condition of 1%, 5% and 1% of the critical value is a stationary time series, therefore, the variables are first order list the whole sequence, satisfy the I (1) process.For non-stationary variables, should use cointegration test analysis method.

Table 1. Unit root test results.

variable	ADP	PP	ADFd	PPd
LnY	−1.997768	−1.129241	−4.352451***	−3.211531**
X1	0.587344	0.951816	−2.914866*	−2.894855*
X2	−2.886112	−1.537230	−5.418954***	−8.934717***
X3	−2.459610	−1.753472	−6.626223***	−6.578661***

Note: (1) the unit root test equation is contained in paragraphs drift and trends, explained variable lags number according to AIC and SC minimum standards;(2) ADFd and PPd are first order difference from variable ADFd and PPd test values;(3) *, * * and * * *, respectively test values are less than the critical value of 10%, 5% and 1% confidence levels.

3.2 *Cointegration test*

Cointegration is for long-term equilibrium relationship of non-stationary economic variables statistical description.In the face of each variable to carry on the cointegration test, respectively.Per capita GDP and urbanization in zhejiang province from single factor to measure metrics, in the original hypothesis of cointegration vector rank under the condition of zero, in addition to non-agricultural industry output value proportion X2, the other two variables, the proportion of non-agricultural population X1 and industrial added value proportion X3 did not reject the null hypothesis, that is to say, only the non-agricultural industry output value proportion and zhejiang per capita gross domestic product there is a cointegration relationship.This shows that zhejiang province since the reform and opening, per capita GDP and non-agricultural industry output value proportion are developing very quickly. Despite the X2 and lnY have first-order non-stationary, but the specific linear combination between them et = lnY -10.6247 - X2 is smooth.And from the point of multivariate combination, in addition to non-agricultural Population X1 and the combination of the industrial added value proportion X3 support null hypothesis, other combinations, such as the proportion of non-agricultural population X1 and non-agricultural industry output value proportion X2, non-agricultural industry output value proportion X2 and X3 and X1, X2 and X3 of the total value added of the industrial combination reject the null hypothesis, that measure the single factor of urbanization and per capita GDP of causality is unlikely, and the interaction between various factors and per capita GDP shows good cointegration relationship, the cointegration relationship helps to explain the phenomenon of long-term growth of lnY.From lnY, X1, X2, and X3 cointegration relationship, non-agricultural industry output value proportion X2, and other combinations were reflected the strong cointegration relationship, this shows that in the process of urbanization of zhejiang province, the development of non-agricultural industries can not only by creating jobs to absorb into the city, to raise the proportion of non-agricultural population, but also accelerate the industrialization and urbanization process, increase the per capita output.Thus, while the diversity of the connotation of urbanization determines the growth of GDP per head is necessarily the result of many factors, but specifically X2 non-agricultural industry output value proportion is the most important.

3.3 *Granger causality test*

In 1978–2012 period, the proportion of non-agricultural population is not the granger reason of changes in per capita gross domestic product, per capita gross domestic product at the same time the proportion of non-agricultural population effect is not obvious, can be thought of in the whole sample interval, the change of non-agricultural population in Zhejiang province and the interaction of economic growth is very small, at least in statistics is not very clear, that in the process of urbanization of zhejiang province, accompanied by the Labour of the single direction is not harmonious state of city industry, absorbing labor force tend to be the supporting role of economic growth, most of the department and some high technological content, good economic benefit and capital-intensive enterprises demand for labor is rigid, lead to the relationship between the urban and rural population change and economic growth present certain dislocation;the industrial added value for the variable proportion X3, under 5% significance level, the results show that the same is not the granger cause of the change of the per capita GDP, and per capita GDP change on the influence of the industrial added value also is not very big, so can be judged in zhejiang's economic structure, the role of the industry is not the most obvious, and this is closely related to the situation of zhejiang province.In zhejiang province in 1978 and 2012 meter 34 years, industrial output value accounted for the proportion of the province's GDP rose from 37.96% to 44.25%, the average annual growth of only 0.18%, and the third industry is developing rapidly, data show that in 1978, only 2.311 billion yuan output value of tertiary industry in Zhejiang province, accounting for 18.68% of the entire province GDP, by 2012 the tertiary industry output value reached 1.568113 trillion yuan, 45.24%, and 34 years the proportion of steadily rising, the average annual growth of almost 0.78%, is the proportion of industrial output grew by an average ratio of 4 times the left and right sides.Therefore, in the reform and open policy for 30 years, stimulating rapid economic development in zhejiang province is the pull of the third industry, the important factors in industrial production and industrialization process of rather than itself driven; the proportion of non-agricultural industries output value X2, whether alone or combination of variables, with per capita GDP LnY granger causality test, the results are rejected the null hypothesis, namely the X2 in full sample interval and the relationship between economic growth significantly, in statistics can be found the non-agricultural industry output value proportion and LnY there is one-way granger causality.

4 CONCLUSION

Using cointegration test and granger causality tests the correlation between urbanization and economic growth of zhejiang province was analyzed, and corresponding results are obtained as follows:(a) from the cointegration test result shows that in the process of single factor test, only the non-agricultural industry output value proportion and Zhejiang per capita gross domestic product there is a cointegration relationship. And the other two measure the proportion of non-agricultural population urbanization and industrial added value is to accept the null hypothesis is not there is a cointegration relationship.Research from multiple factors, the proportion of non-agricultural population and non-agricultural industry output value proportion, non-agricultural industry output value proportion and the proportion of the added value of industrial and X1, X2 and X3 combination with economic growth there is a cointegration relationship, this shows that although the measure the single factor of urbanization and per capita GDP is the causation of probability is small, but the interaction between various factors to per capita gross domestic product showed good cointegration relationship. (2) from the granger causality test, the range of all samples, whether single or combination of variables, the proportion of non-agricultural industries output value in the granger causality test with per capita GDP, the results are rejected the null hypothesis, so on statistics can maintain non-agricultural industry output value proportion in zhejiang province is granger cause economic growth change, and vice versa. Suggesting that agriculture, though important, but the agricultural output value accounts for the province's share of GDP has declined, the agricultural technology promotion and the inflow of capital, talent, makes people become surplus labor originally engaged in agricultural production, through the corresponding training, into the industrial, construction, service industries, thus creating greater than the agricultural labor value, promote economic growth. So while the diversity of the connotation of urbanization determines the economic growth must be the result of joint action of many factors, but specifically X2 or non-agricultural industry output value proportion is one of the most important. In promoting industrialization and population urbanization process, the effective should pay more attention to non-agricultural industries gathered and structure upgrade, to adapt to changes in economic conditions, stable economic growth.

Management, Information and Educational Engineering – Liu, Sung & Yao (Eds)
© 2015 Taylor & Francis Group, London, ISBN: 978-1-138-02728-2

Survey of new generation employees' enterprise democratic participation: Compared to traditional staff

Yu Hua Xie, Guo Lan & Pei Pei Chen
College of Business Administration, Hunan University, Changsha, Hunan, China

ABSTRACT: In this paper, we choose 1117 employees in Guangdong, Hunan, Hubei, 3 provinces and 9 enterprises as the object of investigation. We analysed the new generation of employees in enterprise democratic participation (the expression of public opinion, to participate in supervision, participation in management, participation in decision making) status, and compared the new generation employees with traditional employees in the enterprise democratic participation differences. The study found that, the new generation employees' democratic participation level was significantly higher than that of the traditional staff, but generally remained in the low level of participation. Moreover, the democratic participation of employees intergenerational difference is significant.

KEYWORDS: Enterprise Democratic Participation; New Generation Employees; Traditional Staff.

1 INTRODUCTION

Democratic management of the enterprise is to safeguard the legitimate rights and interests of workers, construction of harmonious labor relations, to promote the sustained and healthy development of enterprises, strengthen the building of grassroots democracy and an important support force. At the beginning of 2012, the Joint Commission for Discipline Inspection, the CCCCP and issued "enterprise democratic management regulations", this is 26 years in China for the first time in the six departments jointly issued rules to form a comprehensive system of norms of democratic management in enterprises, and the non-public enterprises should also be clear practice democratic management.

In foreign countries, the research about the enterprise democracy can be traced back to the nineteenth Century birth of the industrial democracy thought. With the complexity and importance of the labor relations problem highlights the democratic theory of the enterprise, the growing prosperity, the research mainly focused on the participation of the performance, influencing factors and form. Participate in performance mainly through the management efficiency measure.

In China, state-owned enterprises restructuring, enterprise democratic management is part of the management system in state owned enterprises. The rapid development of restructuring state-owned enterprises and foreign investment, the private economy, the impact of democratic management in enterprises and the traditional system, and bring new labor relations, labor conflict has become an acute social problem.

Through the collation of the literature research enterprise democratic participation of existing, it is not difficult to see that: whether from management or political point of view, enterprise democratic participation are the objective requirements of the development of enterprises and people.The existing research and analysis of the importance and necessity of enterprise democratic participation mainly from the angle of theory, the research focuses on labor relations, trade union organizations and trade union workers is the work report, and from the point of view of the employee's own feature description of enterprise democratic participation present situation, to explore its implementation forms of research lack. The demand of the new generation of employees in the enterprise is realized, the lack of research in academic circles. This paper makes exploration.

2 RESEARCH DATA

This 9 enterprises, 1 private enterprises, 3 state-owned enterprises, 5 for Sino foreign joint ventures: the 1 China Japan joint venture enterprise, the 1 Sino Italian joint venture, the 1 China UK joint venture company, 1 Sino German joint venture, 1 Sino South Korea joint venture, joint ventures in China are basically state owned shares; 1 for the financial and insurance enterprises, 8 for the manufacturing industry enterprise.

In order to ensure the reliability of the data by using the method of questionnaire investigation, comprehension, outside the factory investigation, depth interview method. The questionnaire adopts Likert 5 subscales, measure the behavior and employee satisfaction of enterprise democratic participation of employees. Outside the factory survey was conducted research in business door and dormitory area random intercept workers, one by one to explain to investigators question, again by the respondents to fill out. In depth interview was aimed at some enterprise management, the human resources department or the trade union organization, partial employee interviews, involved in the operation of state enterprises from the deep understanding of democracy.

3 RESEARCH RESULTS AND ANALYSIS

During the period of planned economy, enterprise democracy includes not only the masses of workers in the production, coordination, management at different levels in the behavior pattern of activity, also contains the bear and realize the behavior of the crowd Democratic Workers of various organization forms. In the modern enterprise system, the connotation of democratic management is involved in the control of the enterprises, workers as a control to the enterprise of material production, but also for the performance of enterprise decision of supervision and participation in. Chinese industrial democracy and employee involvement is mutual learning and integration of traditional state-owned enterprise workers' democratic participation and Western employees in foreign enterprises to participate in the enterprise democratic participation, Chinese enterprise characteristics of 12 main forms.

3.1 *The expression of public opinion*

The expression of public opinion is the most basic level of democratic participation, expression, individual employee behavior based views include: reflect the problem difficult to complain to the boss, the Ministry of human resources consulting, a suggestion box or BBS employee opinion expression system. Investigation shows, the enterprise internal basically established the expression channels, but the system utilization rate is not high. "31.9% of respondents had no" or "little" reflected to the boss difficult problem, the 13.2% "never" or "little" to the human resources department advisory complaints, 26.3% "never" or "little" through the suggestion box or BBS to convey the views of. During the interview, the majority of staff said, the human resources department has a specialized staff to deal with all kinds of personnel issues, quick feedback, so the employees encounter problems or suggestions tend to seek the human resources department; if the problem is simple, easy processing, they are more willing to communicate with your direct supervisor; as for the comment box or BBS, they that the lack of a person responsible for, this form of too much formality, of little use.

Table 1. Independent samples T test results of the new generation employees with traditional employee opinion expression.

	Total	Mean NG	Tra	Test P	Significant >p
Opportunity of reflecting problem	2.94	3.07	2.72	0.000	YES
Department of human resources consulting	3.42	3.53	3.21	0.000	YES
Suggestion box or BBS	3.14	3.29	2.87	0.000	YES
Expression of public opinion	3.17	3.30	2.84	0.000	YES

3.2 *Participation in supervision*

To participate in supervision is higher level of participation form, refers to the staff through the expression of opinion to participate in supervision on enterprise management and the managers, is the combination of organizational behavior and individual behavior, including basic opinion polls, the democratic appraisal management, democratic life meeting or consultation meeting, the publicity of factory affairs etc. Investigation shows, in addition to the publicity of factory affairs (mean =2.95), the level of participation of opinion polls, democratic appraisal of managers and the democratic life of the grass-roots level are relatively low (mean = 2.44,2.15,2.05). Respectively 61.4%, 74.1%, 78.2% of the respondents "never" or "little" participated in the poll, grass-roots democratic appraisal and democratic life.

The new generation employees desire is higher, be good at concern through various channels of enterprise information, understand the management of the enterprise, making important decisions, so the new generation employees in the publicity of factory affairs on the involvement of high (mean =3.05). However, the new generation employees in the Democratic review (mean =2.00), democratic life meeting (mean =1.94) on participation is very low, in the "little" level. The independent samples T tests were found (Table 2): the new generation employees and employees participate in the supervision of traditional no difference in overall and grass-roots opinion polls, but the traditional democratic appraisal of managers and employees in the democratic life meeting participation is significantly more than the new generation employees, significantly less than the new generation employees in making public participation. Further analysis found that, after 70 (mean =2.54), after 80 (mean =2.41) employees to participate in supervision involvement were significantly higher than 90 employees (average =2.17).

Table 2. The new generation employees and traditional employees participate in the supervision of independent samples T test results.

	Total	Mean NG	Tra	Test P	Significant >p
Grassroots public opinion survey	2.44	2.45	2.43	0.743	NO
Democratic appraisal management	2.15	2.00	2.42	0.000	YES
Democratic life meeting	2.05	1.94	2.24	0.000	YES
Publicity of factory affairs	2.95	3.05	2.77	0.000	YES
Participate in supervision	2.39	2.35	2.44	0.077	NO

Table 3. SOE democratic Council, democratic life meeting independent samples T test results.

	Total	Mean NG	Tra	Test P	Significant >p
Democratic appraisal management	2.67	2.48	2.77	0.136	NO
Democratic life meeting	2.31	2.28	2.32	0.823	NO

3.3 *Participation in management*

Participate in the management of employee participation refers to the specific enterprise management activity, compared with the indirect expression of public opinion supervision and participate more directly, deeply, including autonomous work teams, rationalization proposals. Autonomous work teams originated in Toyota mode, in our country is more common in manufacturing enterprises. Investigation shows, autonomous work teams form the participation level is generally low, the mean is only 2.42, lower than the "general" level, only 26% of the respondents "often" or "always" by autonomous work teams to participate in the management. On the rationalization proposal, part of the large enterprises clearly defined several should raise employee each month, and rewards to take advice. Therefore, the staff in the reasonable suggestions on the highest participation levels (mean =3.37), only 13.4% of respondents said "never" or "little" rationalization proposals. If the rationalization proposal was adopted, 59.2% of the respondents think about their performance "almost no influence or effect of" small "".

The new generation employees a strong sense of participation, the team tends to a positive performance, ideas and suggestions of their own contribution. The new generation employees, there are 28.3% "often" or "always" in autonomous work teams, 40.8% "often" or "always" rationalization proposals. The independent sample T tests were found (Table 4): the new generation employees and traditional employees to participate in significant differences existed between the management; the new generation employees in autonomous work teams, reasonable suggestions on participation is significantly higher than that of the traditional staff. Further analysis found that, participation in management, 70 (mean = 2.85), after 80, after 90 (mean =3.03) (mean =2.92) employee involvement were significantly higher than those of 60 (mean =2.64) after 80 employees; employee involvement was significantly higher than that of 70 employees.

Table 4. The new generation employees with traditional employee participation in the management of independent samples T test results.

	Total	Mean NG	Tra	Test P	Significant >p
Autonomous work teams	2.42	2.50	2.27	0.004	YES
Rationalization proposals	3.37	3.46	3.25	0.000	YES
Participation in management	2.91	3.01	2.76	0.000	YES

3.4 *Participation in decision*

Participation in decision making is the highest form of participation, that employee participation in decision making and management of major activities, enterprises share the profits of enterprises, including the Congress of workers and trade unions, collective wage negotiation.

The new generation employees of the Congress of workers and Trade Unions (mean =1.77), the collective salary negotiation (mean =1.84) on the level of participation is low, 48.3% said "never" participated in the trade unions and workers congress, the 77.3% "not heard" or "just listen to the people around said" the collective wage negotiation, 6.9% signed salary collective agreement. Independent samples T test was found (Table 5): the new generation employees and traditional employees have significant differences in participation in decision making; the new generation employees participate in collective wage negotiations on a slightly higher than the traditional workers, and no difference in the Congress of workers and trade unions on the participation and the traditional staff. Further analysis found that, after 80, after 90 (mean =1.81) (mean =1.77) employee involvement was significantly higher than that of 60 employees involved in the decision aspects (mean =1.51).

Table 5. The new generation employees with traditional employee participation in decision making independent samples T test results.

	Mean			Test	Significant
	Total	NG	Tra	P	>p
Congress of workers and trade union activities	1.77	1.77	1.77	0.988	NO
collective salary negotiation	1.73	1.84	1.53	0.000	YES
Participate in decision	1.75	1.80	1.65	0.002	YES

4 CONCLUSION

This study from the public opinion expression, participation in management, participate in supervision, participation in four dimensions decision with discussion enterprise democratic participation status, and focus on the new generation employees compared with traditional employee in democratic participation differences. The study found that: first, enterprise democratic participation is not optimistic. Employee participation level from high to low is the expression of public opinion (mean =3.17), participation in management (mean =2.91), to participate in supervision (mean =2.39), participation in decision making (mean =1.75). The staff in the lowest degree of involvement in the form of suggestion box, to the human resources department consultation etc. in good condition, while in high degree of participation form participation is rare, especially the trade unions and workers' congress, the collective wage negotiation etc. Secondly, there are differences in the new generation employees and employees in the present situation of traditional enterprise democratic participation. In addition to participate in supervision, the public opinion expression, participation in management, participation in decision-making, the new generation of employee involvement level were significantly higher than the traditional staff. Moreover, in four dimensions, intergenerational differences apparent participation of employees.

Enterprise democratic participation channels should be from low to high perfection. The expression of public opinion supervision, participation effect on employee satisfaction greatly, enterprises should focus on the strengthening of the channel construction on these two aspects: optimizing the internal communication channels, improve the efficiency of the expression of public opinion; to expand participation in supervision, enhance its enforcement, public evaluation system and evaluation result. At the same time, enterprises should also pay attention to a higher degree of participation of channel construction and improvement, enhance employee satisfaction: to establish and improve the participation in decision-making channels, improve the system of workers' Congress, the trade union and collective wage negotiation system; widening participation management range, let more employees to participate in enterprise management.

REFERENCES

[1] Wagner J A. Participation's effects on performance and satisfaction: A reconsideration of research evidence [J]. Academy of management Review, 1994, 19 (2):312–330.

[2] Griffin R W. Consequences of quality circles in an industrial setting: A longitudinal assessment [J]. Academy of management Journal, 1988, 31 (2):338–358.

[3] Kochan T A. Worker Participation and American Unions. Threat or Opportunity? [M]. Kalamazoo: WE Upjohn Institute for Employment Research, 1984:185–197.

[4] Jirjahn U, Smith S C. What factors lead management to support or oppose employee participation—with and without works councils? Hypotheses and evidence from Germany [J]. Industrial Relations: A Journal of Economy and Society, 2006, 45 (4):650–680.

[5] Glew D J, O'Leary-Kelly A M, Griffin R W and Van Fleet D D. Participation in organizations: A preview of the issues and proposed framework for future analysis [J]. Journal of Management, 1995, 21 (3):395–421.

[6] Othma R, Arshad R, Hashim N A and Isa R M. Psychological contract violation and organizational citizenship behavior [J]. Gadjah Mada International Journal of Business, 2005, 3 (7):137–153.

[7] Shadur M A, Kienzle R and Rodwell J J. The Relationship between Organizational Climate and Employee Perceptions of Involvement—The Importance of Support [J].Group & Organization Management, 1999, 24 (4): 479–503.

[8] Lawler E E, High-Involvement Management. Participative Strategies for Improving Organizational Performance [M]. San Francisco: Jossey-Bass Inc., 1986: 113.

Management, Information and Educational Engineering – Liu, Sung & Yao (Eds)
© 2015 Taylor & Francis Group, London, ISBN: 978-1-138-02728-2

A research on one product from the special pedagogical approach of the game development specialty

Cheng He
Department of Computer Engineering, Zhongshan Polytechnic, China

ABSTRACT: The focus of this paper is on how to develop and teach several primary curriculums to students who major in the game development specialty on innovation processes for sustainable procedures of the complete Super Mario Game product, from product definition to sustainable manufacturing and business models throughout different courses. The pedagogical approach of the proposed curriculums comprises a set of modules for facilitating the students to quickly learn the game programming skills by accordingly implementing different components of a large-scale game respectively in nine curriculums. The well-designed game composites can effectively develop the multidisciplinary skills required for a successful design and development of sustainable game products. From the results, this pedagogy demonstrates its effectiveness and feasibility for students to acquire the ability to apply design and development principles in the construction of a game system of varying complexity.

1 INTRODUCTION

It has been demonstrated that teaching computer science concepts based on programming interactive graphical games motivates and engages students while accomplishing the desired student learning outcome [1, 2]. Good game programmers today are primarily problem solvers, who know how to acquire a mental model of a complex software environment and to solve technical problems in order to achieve a timely goal. Broad survey courses in video game design cover artistic, technical, as well as the sociological aspects of video games [3]. Students will learn about the history of video games, archetypal game styles, computer graphics and programming, user interface and interaction design, graphical design, spatial and object design, character animation, basic game physics, plot and character development, as well as psychological and sociological impact of games. Students will design and implement a large scale industrial video game in interdisciplinary teams of 3–4 students as part of a semester-long project.

2 COURSES DESIGNED FOR GAME DEVELOPMENT SPECILATY

The primary nine courses for game development specialty students are listed in Table I. The primary game product for almost all courses is the same Super Mario Game but with various modules dispersed into different courses with their own emphasizes. This game places its focus not only on the game design but also on its implementation so as to gaining a final game product. This game focuses on the technological aspects of game development by not only covering various algorithms ranging from graphics to artificial intelligence, networking and sound but also primarily exploring the technical aspects of game development. As such, the Super Mario game will involve many computer science topics, not just computer graphics. The whole development process will emphasize the tools and algorithmic techniques that are critical components of the game. A game due at the end of the study is a good, timely goal. Part of the students' task will be to understand and augment components of the tool chain as needed to make their game work.

3 GAME PRODUCTION SELECTION

The game production will combine to form a fully functional simulation of the arcade game Super Mario. Super Mario was one of the first games to use vector graphics [4]. The game was based on a simple premise: Super Mario World and his dinosaur companion, Yoshi, are looking for the dinosaur eggs Bowser has stolen and placed in seven castles. Many secrets exit aid Mario in finding his way to Bowser's castle, completing over 70 areas and finding about 90 exits. With multiple layers of 3D scrolling landscapes, find items including, a feather that gives Mario a cape allowing him to fly, and a flower so he can shoot fireballs.

No.	Course name	Main teaching contents
1	Game Planning	Design Super Mario story, game play, game level
2	The Basic Game Art	Design Super Mario game GUI, Character, Player, Turtle, map, etc.
3	Game Flash Design	Character Animation Design and Drawing
4	Object Oriented Programming	Design and Implement Game classes in the game
5	Windows Programming	Design the game program structure
6	OPENGL Technology	Implement the game displaying with OpenGL Graphical Engine
7	Data Structure and Algorithm for Game Programming	Implement Quad tree data structure and A-Star algorithm which use in Super Mario and useful in other programming
8	Game Project Design and Implementation	Design and implement game level and game evolution algorithm
9	Game Deployment and Testing Technology	Integrate all components to complete the game and make integration testing

The project consists of about nine modules while each module will guide students to add a new component to the game product Super Mario. This will include rendering individual elements of the game, for example, marios of various sizes, enemy turtles, enemy plants, fire bullets, and explosions in lit, 3D form, implementing a simulation engine to allow the game to progress over time, and to track the motion of enemies, shots, and turtles, implementing collision detection to allow the enemy turtles to be shot, and to detect collisions between game character, both player and enemies, and designing a basic user interface layer to allow the player to interact with the game, to keep score, and to maintain high scores [6].

4 ONE PRODUCT FOR ONE SPECIAL PEDAGOGICAL APPROACH PROCEDURES

The primary task of the "One product for One Specialty" pedagogical approach is to decompose the product into several modules which should disperse into different courses. The basic procedures are illustrated on figure 1 and Figure 2. The Super Mario has divided into nine separate but related modules which will be discussed in nine different kinds of courses including game art, programming language, game design and game testing.

For the first task in the product, students will learn how to write a synopsis of the game there are creating which focuses on the game design. As we know, students are supposed to learn to produce a side-scrolling

Figure 1. Super Mario Game GUI.

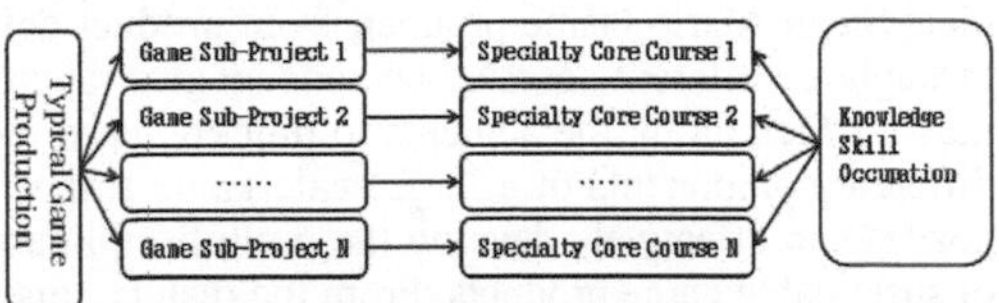

Figure 2. Game modules decomposition.

video game in the tradition of Super Mario Bros. as their main tutorial project during their study period. However, simply cloning an existing game is not as much fun as putting their own spin on it, and students certainly hope they will! For this problem, students need to write up their ideas for the game they will produce given the constraints of a 2D side scroller. The game design covers all of the following areas briefly, including background story, such as the overall goal of the game, the various actors, their motivation, some features of the environment which makes clear what the overall graphical theme will be but also what kinds of harmonious music or sound effects, the sequence of levels, the outcome of each level, the basic mechanics of the game, the player's performance, powerups, dexterity, cleverness, or firepower,, bonus lives or med packs.

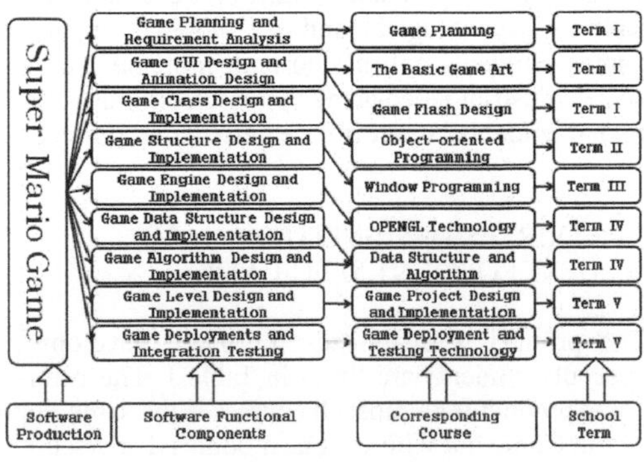

Figure 3. Game product components decomposition.

In the course Basic Game Art We talked about the various approaches to game graphical user interface necessary for the game in class including splash screen, a title screen, a menu screen, a playing screen and a high-score screen. Ideally, students have some graphical things related to the game going on, for example an artist's rendition of either a typical event in the game or some kind of intriguing encounter with a boss.

The course named Game Flash Design will concentrate on the sprite animation to add to the player character as well as the enemies. For example, certain enemies could explode if they either fall too far or are attacked by the player in the right way. Having the various characters react to what is happening to them makes for a much more believable game experience. What kinds of animation you will eventually need depends on the specifics of your game, of course, but you will certainly need a basic walk cycle. In this course, students can learn how to make animations for all the characters including Mario, turtle, plants and bullets etc.

The course named Object Oriented Programming will emphasize elemental OOP design — using encapsulation to divide a large scale industrial game product into many friendly little independently testable problems. This course will help students set up the primary classes which will be used in the later modules of the game Super Mario including the sprite class, game engine class, enemy class, bullet class, map class and sound effect class. The students are especially encouraged to write classes that implement Super Mario quickly rather than just writing classes that purely implement Super Mario. This course makes a nice exercise in OOP decomposition — dividing the rather large Super Mario problem into several non-trivial classes that cooperate to solve the whole thing.

In this Windows Programming video game programming course, the student will be guided to learn about fundamental graphic concepts such as double buffering, sprites, animation and timing, and masking. The lecture and tutorial of this course will help students implement the Super Mario game with the graphical interface using win32 API functions, such as timing, animation, sprites to design and implement the game. Also, students can learn generic programming, object oriented design paradigms useful in gaming such as entity management, scene management, state machines, message system and singletons. Further, other gaming technologies, including basic collision detection, entity movement and interaction, area triggers are also covered in the course.

The topics covered in the course OPENGL Technology include some topics related to game programming. The First part is a working knowledge of basic OpenGL, including textures, displaying sprites, animation, text, and rendering. The second part is a working knowledge of physics in the game's programming context, including basic movement and interaction. The third part is a working knowledge of OpenGL 3D rendering, including geometry, models, cameras, textures and lighting. The module related with the game Super Mario is the development of the game using the OpenGL game engine. The students will learn how to render individual elements of the game such as enemies of various sizes, the player's character design and graphical model, bullets from enemies and projectiles from the player, laser blasts, and explosions in lit, 3D form.

In the course of Data Structure and Algorithm we guide students to learn algorithm to implement collision detection capabilities of the tiles in the platform and another central conceptual algorithm of most side-scrolling platformers is that the player character has the ability to jump, whether from platform to platform or across obstacles and enemies. Jumping implies that there's a clear notion of up and down in the game, and therefore some sort of gravity that will pull things toward the bottom of the screen unless they are on solid ground. To make a jump look realistically we will have to apply some form of acceleration since you need to get a nice trajectory.

The primary teaching contents in the course of Game Project Design and Implementation are game level design. In almost all side-scrollers, levels are bigger than the screen, and you need to put together your very first level now. The students will be illustrated to a tile-based approach to level design and rendering with example code in this course. Also the technology to tackle the problem of smoothly scrolling the level from left to right in this game product will be discussed in the lecture while introducing some kind of camera abstraction to add smoother transitions without changing too many of the basics. Thirdly, skills for implementing a rendering system that supports various layers, for example a background layer, the actual game layer, a foreground layer, a heads-up-display layer, etc. will be presented to the students in this course.

The course Game Deployment and Testing Technology will guide the students to fix bugs and polish their game product. The primary teaching contents of this course are software testing technologies which involve the execution of a software component or system to evaluate one or more properties of interest. In general, these properties indicate the extent to which the component or system under test, including meeting the requirements that guided its design and development, responding correctly to all kinds of inputs, performing its functions within an acceptable time, being sufficiently usable, and running its intended environments, and achieving the general result its stakeholders desire.

5 CONCLUSION

Teaching courses with the contents related to the same game production with different components for teaching game development technology has presented its effectiveness [7]. Students illustrate sky-scraping interests while studying game development skills [8]. One product for one specialty pedagogical approach culminates students' interests and confidence about learning game development technologies by learning how to develop Super Mario Game and complete another assignment game completed by themselves.

REFERENCES

[1] E. Sweedyk, M. deLaet, M. C. Slattery, and J. Ku_ner. Computer games and cs education: why and how. In SIGCSE '05, pages 256{257, 2005. ACM Press.

[2] U. Wolz, T. Barnes, I. Parberry, and M. Wick. Digital gaming as a vehicle for learning. In SIGCSE '06, pages 394{395, 2006. ACM Press.

[3] Yang, C-S. and Yu, J. W. (2011). Using Incremental Worked Examples for Teaching Python and Game Programming. International Conference on (CSEIT 2011).

[4] Schaffer, S., Chen, X., Zhu, X., and Oakes, C. (2012). Self-efficacy for cross-disciplinary learning in project-based teams. Journal of Engineering Education, 101, 82–94.

[5] Vos, N., van der Meijden, H., & Denessen, E. (2011). Effects of constructing versus playing an educational game on student motivation and deep learning strategy use. Computers & Education, 56(1), 127–137.

[6] Yen-Lin C., Chuan-Yen C.Yo-Ping H.Shyan-Ming Y. (2012). "A Project-based Curriculum for Teaching C++ Object-Oriented Programming", 9th International Conference on Ubiquitous Intelligence and Computing and 9th International Conference on Autonomic and Trusted Computing, 667–672.

[7] Wang, X. (. O. P. H. I. E. (2010). Teaching programming skills through learner-centered technical reviews for novice programmers. Software Quality Professional, 13(1), 22–28.

[8] Berglund, A., Daniels, M., & Pears, A. (2006). Qualitative research projects in computing education research: An overview. Proceedings of the Eighth Australasian Computing Education Conference (ACE2006), Hobart, Tasmania, Australia, January 2006.

Management, Information and Educational Engineering – Liu, Sung & Yao (Eds)
© 2015 Taylor & Francis Group, London, ISBN: 978-1-138-02728-2

Study on the influence factors of U.S. federal government performance audit methods

Yuan Qing Mei, Wei Wei Dong, Li Zhang & Jie Yang
Department of Economics and Management, Chengxian College, Southeast University Nanjing, China
Department of Teaching, Chengxian College, Southeast University Nanjing, China

ABSTRACT: U.S. federal government performance audit methods have their own trajectory changes. Along these trajectory changes, we could analyze the promotion caused by changes in the method of U.S. Government Performance Audit. After the trajectory and promotion studies in the United States government performance audit methodology changes, we used the United States governmental financial reporting data from 2002 to 2012 related to empirical study, and concluded: Federal government financial position, performance of the U.S. federal government budget management, the U.S. Federal Government Financial basis of accounting reports and the U.S. Government Auditing Standards changes have some positive effects on the U.S. government performance audit approach. This helps our government performance audit work to bring enlightenment.

KEYWORDS: Government performance audit methodology, Change, governmental financial reporting, regression model.

1 INTRODUCTION

The United States is one of the most advanced country about government performance audit in the world. Performance audit has accounted for the Supreme Audit institutions more than 85% of the total work [1]. After analysis, motivation and trajectory changes of the government performance audit method, we analyzed empirically the effect of various factors affecting the implementation of the performance. We could draw the conclusion of the experience of the ongoing reform of government accounting.

2 U.S. FEDERAL GOVERNMENT PERFORMANCE AUDIT METHODS

American Accounting Association and U.S. Government Accountability Office state the concept of Performance Audit formally. The definition is selected in this paper. Performance audits are defined as audits that provide findings or conclusions based on an evaluation of sufficient, appropriate evidence against criteria. Performance audits provide objective analysis to assist management and those charged with governance and oversight in using the information to improve program performance and operations, reduce costs, facilitate decision making by parties with responsibility to oversee or initiate corrective action, and contribute to public accountability [2].

After analyzing the data from the site of the United States Government Accountability Office, we propose the four stages about changes in the U.S. government performance audit methodology.

2.1 *Prototype stage*

Audit methods at this stage is mainly to pay for government expenditure agencies based on vouchers, and then the Auditor General checks accounting documents and receipts to ensure the legality of revenue and expenditure. This audit reflects the concept of the audit in the late 19th century and early 20th century. In addition to some of the issues , it focused on the financial records of the postaudit. This phase is characterized by the emergence of consciousness and government performance audit ideology based audit methods of invoice checking.

2.2 *Growth stage*

This stage is in the development of the Auditor General to audit by the integrated into the economy, the effect of the audit period.

In order to adapt to this change and improve comprehensive audit, the Auditor General has taken a number of concrete measures to adapt to changes, such as start trying to develop a series of technical guides. In 1952, the first comprehensive audit manual "comprehensive audit guide" was published, including audit authority and responsibility, the audit objectives and policies, the implementation of a comprehensive audit, management and control of nature. It was replaced by Project Manual and General Policy manual in 1981. This phase is characterized by: an object into statement audit, internal control systems, taking into account the efficiency of resource use began.

2.3 *Stage of development*

June 1972, the Auditor General under the authority conferred by legislation, coined the government agencies, projects, activities and responsibilities of the Auditing Standards, which provided an audit of three parts: financial and compliance audits, the economy and efficiency audits, the effectiveness of the project audit. It had been modified in the year of 1981, 1988, 1994, 2003, 2007 and 2011. Eventually, it was renamed the American National Standards on Auditing. In the 1988 revision, performance audit terminology was first used explicitly. Audit types entirely are divided into two types of financial audits and performance audits. Performance audits of clear goals included economy and efficiency, project audit and compliance audit. Thus, the U.S. Government began moving into a complete performance audit results-oriented, project evaluation as a means of the high speed development period, the performance audit to become the dominant type of government auditing.

2.4 *Mature stage*

July 7, 2004, according to the Audit Department of Human Resources Reform Act, General Accounting Office was renamed the Government Accountability Office (GAO). After the name changed, specifically its responsibility, fair and credible core values, indicating the Government Accountability Office, help realize the constitutional responsibility of Congress to improve the performance of the federal government. Auditors were no longer government accounting officers and mere supervisors, but the important builders and active participants in the national administration, who should facilitate the work of the federal government to improve the performance of the Congress and the American people to fulfill the public accountability. The Government Accountability Office assesses at this stage was in the project development stage into the popularity of the performance audit.

3 HYPOTHESES

In order to study factors that the government performance audit method changes influenced, we analyzed the performance audit method change from the dynamic influence factors. Therefore, we could state the following four hypotheses.

After the economic crisis in the early 20th century, with the increasing of public spending and stateowned enterprise's development needs, the government takes up a huge fiscal capital, adding to the taxpayers' burden and causes the public to the attention of the public resources saving and efficiency. The social public sector of entrusted economic responsibility requires more and more high. It calls for the government's public expenditure performance audit. Therefore, GAO expands the scope of review in the United States from pure financial audit gradually to performance audit, which strengthen the supervision of public resources and gain the maximization of benefit. Tax revenues and nontax revenue cause the increase of government revenue, which lead to the financial position of the U.S. federal government has a positive effect on the performance audit methods. As a result, hypothesis 1 could be achieved.

H1. The financial condition of the federal government has a positive impact on the government performance audit method.

Performance budgeting is on the ideas of new public management and under the background of democratization of a way of the pursuit of the efficiency of the budget management in the United States.

In the 1990s, serious financial deficit of the federal government began to develop a performance budget, performance management, assessment that is associated with the budget. The federal government each have a set of administrative and legislative participation in budgeting and auditing system. The budget process includes four stages: (1)the budget preparation: requests for internal budget allocation and institutions and then be incorporated into the President's budget; (2)budget approval: congress and the President issued legal license tax and spending; (3)budget implementation: the organization and regulations within the limits of the execution budget in congress; (4)audit and evaluation: internal and external assessors and budget activities auditor to determine the financial and performance results. [3] [4](2012) In the process of the entire budget, the first three stages have a direct impact on the fourth stage of the audit and evaluation. Thus, we can extrapolated hypothesis 2.

H2. The federal government performance budgeting management effect of the implementation of government performance auditing method have a positive impact.

Government spending in the federal budget based on the basis of cash accounting, but income completely based on cash basis. Comprehensive government financial report use the accrual basis to measure various fees, the costs of these financial resources, government owned assets and debt. Federal government calculated the net operating costs for the government financial report, which equals the government revenue minus the government fees.To the extent that the federal government used the accrual basis accounting elements measured, the effect of the implementation of the federal government performance audit method also can produce certain effect. This effect is positive.

Namely the liabilities are measured on the accrual basis. The degree of uses on the accrual basis is higher, the federal government performance audit method for the implementation of the results is better. Hence, this can be extrapolated hypothesis 3.

H3. The federal government financial report using the extent of the accrual basis has a positive impact on the implementation of the performance audit method.

Since 1920s, U.S. federal government, through a series of the audit law, effectively promoted the development of the different period of government audit and the change of the government performance audit method. In addition, changes of government auditing standards, directly affects the change of the United States government performance audit method.

Each time the change of government accounting auditing standards will directly affect the financial and the specific operation of the performance audit method, the change of the specific operation details will influence the effect of the implementation of the performance audit method. Therefore, we propose the hypotheses 4.

H4. The change of the us government auditing standards for several years after the U.S. government performance audit method the effect of the implementation will produce certain positive impact.

4 DATA AND METHOD

The analysis is based on time series data which is derived from U.S. Government Accountability Office official website. The sample data comes from Financial statement of U.S. federal government financial report from 2002 to 2012. The U.S federal government consolidation financial report includes six financial statements. There are statements of net cost, statements of operations and changes in net position, reconciliations of net operating cost and unified budget deficit, statements of changes in cash balance from the unified budget and other activities, balance sheets and statements of social insurance and changes in social insurance amounts. The U.S. Government's fiscal year begins on October 1 and ends on September 30.

The method of OLES regression is finished through the software SPSS.10. The variables are described in Table 1. And the linear model is shown in Eq.1.

$$MC = \beta_0 + \beta_1 NP + \beta_2 BM + \beta_3 AC + \beta_4 GAS + \mu \quad (1)$$

Method Change. Using the net annual cost accounting for the proportion of net assets to measure the effect of the implementation of government performance audit about each year U.S. Government Accountability Office. It can be shown in Eq. 2

Method change = net cost of the government operation / beginning and ending net assets (2)

Net position. We get net assets by using the Federal Government's total assets minus total liabilities, which used to measure the financial condition of the government after a year of America federal government operation. The federal government American financial situation itself will affect the American government and asked the implementation methods of the government performance audit accountability office. It can be shown in the Eq.3.

Net position = total liabilities / net assets of the federal government (3)

Budget Management. We can get the differences in net cost USA federal financial reporting in the federal budget and the federal government to measure the variable, as shown in Eq. 4.

Budget Management = federal budget / net cost (4)

Accrual. American federal government financial report uses the accrual basis. Through the federal budget with a total net assets ratio, we could measure the degree that the measurement of accrual based as shown in Eq.5.

Accrual = federal budget / net assets (5)

Government Auditing Standards. America government auditing standards occurred two times larger changes in the last ten years, respectively in 2003 and 2007 in 2004 and 2008, the value of the observed variable GAS is 1, the other 0 observed variables. When U.S. government auditing standards change, the next year GAS equals 1; otherwise, GAS equals 0.

Table 1. Description of variables.

Variable	Description
MC	Method Change
NP	Net position
BM	Budget Management
AC	Accrual
GAS	Government Auditing Standards

5 REGRESSION RESULT

Table 2 shows the coefficients of the regression model and the test of significance.

As shown in the table, the value of the constant sig. is 0.374, which is larger than the value 0.05, the constant is little effective on the regression. This indicated that the effect of the federal government performance audit methods America influenced by external factors.

The hypothesis 1 is hardly proved right. As is shown in the table 4, the sig. value is 0.621, which is more than 0.1, indicating the independent variable could not be explained the dependent variable. Probably the measurement of net position feature could be improved in future studies.

The factor of budget management gives the opposite conclusion. The value of sig. is zero, which is less than 0.05, indicating a correlation between the method change and budget management. But the coefficient -1.092 shows the contrary to the proposition of hypothesis 2. Nevertheless, when budget management becomes more, government performance audit methods change will be harder.

The accrual impact on the establishment of standards effectively. As is shown in the table 4, the sig. value is zero, indicating a positive correlation between the accrual and government performance audit methods change. The hypothesis 3 is proved right.

Yet the hypothesis 4 is not proved right. As is shown in the table 4, the sig. value is 0.436, which is more than 0.1, indicating the independent variable could not be explained the dependent variable. Probably the measurement of Government Auditing Standard feature could be improved in future studies.

Table 2. Coefficients of the model.

model	Sig.	Coef.
constant	0.374	/
NP	0.621	−0.065
BM	0.000	−1.092
AC	0.000	1.707
GAS	0.436	0.056

6 CONCLUSION

Four dynamic factors can be concluded from the changes of US Government's performance auditing ways. On this basis, we construct a government performance audit method implementation model, including U.S. federal government's financial position, performance budget management, financial reporting basis and the government auditing standards. We make use of data from the financial report of the U.S. federal government (2002–2012) , and have had some conclusions.

The financial position of the federal government has a positive impact on the effects of performance audit methods. This conclusion haven't been strictly proved, but to some extent, the financial situation of the federal government can affect performance auditing directly or indirectly. The reason is that the U.S. government's financial situation affects the sources of funding of the U.S. Government Accountability Office to conduct performance auditing. U.S. Government Accountability Office applies monies needed for U.S. national budget authorities directly. The Congressional Budget Committee allocates monies directly to the U.S. Government Accountability Office. U.S. Government Accountability Office takes responsibility to use its budgetary without government control and intervention of Administration.

The U.S. federal government adopts accrual in its financial reports. The degree of using accrual has a positive effect on ways of performance auditing. Governmental financial reports under the cash basis not fully reflected the movement of government funds, and the information reflected are incomplete. But accrual can not only provide more comprehensive and accurate information concerning income and expenditures, but also offers information about assets and liabilities to make managers clearly understand changes in financial costs as well as the changes in time and space [5]. As a result, financial reports based on accrual may better reflect financial conditions of the U.S government, providing strong auditing evidence for the latter part of the performance audit, bringing better results for the implementation of the method of performance auditing.

7 CONCLUDING REMARKS

The impact of net position, budget management, accrual and government auditing standards on the U.S. government performance audit methods change is the main content of this paper. Yet there are other control variables, such as whether the country has inflation or not, the country's business development policies will also have some impact on the U.S. government performance audit methods change.

ACKNOWLEDGEMENTS

This work is supported by the department of Education in Jiangsu Province under grant 2012SJD790020. And the author would like to thank his collaborators from and support of Chengxian College Southeast University.

REFERENCES

[1] LI Lu, Approach Change of Governmental Performance Audit in USA and Its Enlightenment, Journal of Zhongnan University of Economics and Law,177(2009)51–59.

[2] Audit Research Institute, American government auditing standards (2003 Revised Edition),China financial and Economic Publishing House, Beijing, 2004.

[3] Ren Xiaohui, Performance budget reform process and Enlightenment of USA federal government, Shanghai University of Finance and Economics , School of public economics and management, The financial supervision, (2012) 19 period, pp27–32.

[4] Liu Jidong , USA federal government performance budgeting process and Enlightenment, Modernization of Management, Guangdong Provincial Department of Finance, (2004) 05 period, pp60–62.

[5] Xiang Wenwei, The state (government) of China US Comparative Study of auditing standards, friends of accounting,(2012) 16 period, pp4–7.

Management, Information and Educational Engineering – Liu, Sung & Yao (Eds)
© 2015 Taylor & Francis Group, London, ISBN: 978-1-138-02728-2

The exploration about framework of tourism security ethics standards of high-risk groups' self-help leisure activities in the wild

Liang Zhao
Research Center of Philosophy of Science & Technology, School of Marxism, Northeastern University, Shenyang, China

Cheng Cheng
Institute of Aesthetic and Critical Theory, Zhejiang University, Hangzhou, China

ABSTRACT: The concept of "high-risk groups' self-help leisure activities in wild", represents the current development which changes from the traditional tourism to new tourism format gradually. It focuses on security. Two nature relationships are closely around the activities. We put forward rational safe coexistence including guaranteed safe coexistence and responsible safe coexistence, and further extend the rational security concept including activities members, group members, nature, other groups or individuals. What's more, we combine the features of high-risk groups' self-help leisure activities in the wild, and fully refine the model about the corresponding safety coexistence. It will be of great significance in the future of high-risk groups' self-help leisure activities in the wild. By setting up tourism security ethics framework of exploration and basic establishment, it's expected to effectively avoid accidents.

KEYWORDS: high-risk groups' self-help leisure activities in the wild; safety problem; framework of tourism security ethics; exploration.

1 THE SO-CALLED HIGH-RISK GROUPS' SELF-HELP LEISURE ACTIVITIES IN WILD AND ITS MAIN CONTENT

If the prosperity of "new tourism format" represents a postmodern way of life and thought, the safety problem is a fundamental problem which lies in front of it. It has huge hidden dangers and developmental disorders. "High-risk groups' self-help leisure activities in the wild" is specially focused on the new concept of safety problems. It reflects various core issues which the "new tourism format" meet, but the traditional tourism doesn't face. This concept covers the following contents specifically:

Table 1. The specific connotation of "high-risk groups' self-help leisure activities in the wild".

Be divided in the organization form of groups	Registered travel club and non-registered travel club, profit travel club and non-profit travel club, raising funds travel club and non-raising funds travel club, etc
Be divided in the tools of transportation	Hiking, driving, cycling, on the water, in the air, etc
Be divided in the detailed content of leisure	Climbing, crossing, voyager, desert, camping, caves, forests, mountains, snow and ice, rafting, lakes, coastal and island, high altitude, spectacle, adventure, etc
Be divided in the type of danger	Environmental risk (season, geography, hydrology, meteorology, disasters, pollution, etc.), human risk (selection, preparation, organization rescue, etc.)
Be divided in the degree of danger	Immediate danger, potential danger, accident risk (injury, disease, etc.), relative risk (gender, age, physical condition, etc.)

After 20 years' development, "extreme tourism" has already included the land, sea and air. It constitutes a subset of the field high-risk groups' self-help leisure activities in the wild in the very great degree; and nowadays so-called "risky tourism project"[1] is popular which qualitatively differ from high-risk groups' self-help leisure activities in the wild. The conception of "high-risk groups' self-help leisure activities in the wild" is put forward, which both have clear pertinence and strong recapitulate. In essence, during high-risk groups' self-help leisure activities in the wild, when they are in danger, the self-help groups are both only subjects and their own unique stakeholders. This is the essence of this kind of group as unique social relationship. It also has difference between this and other leisure activities.

2 THE EXPLORATION OF THE FRAMEWORK OF TOURISM SECURITY ETHICS STANDARD OF HIGH-RISK GROUPS' SELF-HELP LEISURE ACTIVITIES IN THE WILD

2.1 *The inspiration of tourism ethics to high-risk groups' self-help leisure activities in the wild*

Safety is the lifeline of tourism, is the most important link of the tourism industry. Adventure is not equal to taking risks. High-risk existing facts should be treated carefully. In April 2013, *The tourism law of the People's Republic of China* is established and perfect the safeguard mechanism of the safety of tourists,[2, 3] but there is still no clear relevant laws and regulations about the safety problem of the high-risk groups' self-help leisure activities in the wild. Besides, there is no rational, strong and effective action in the process of high-risk groups' self-help leisure activities to save the security loopholes and hidden trouble of constraints. The only applicable problem is just about all involved tourism and leisure activities of safety regulation, safety risk warning, safety emergency management, safety prevention, etc. It is difficult to play a fundamental role.

Under the reality of absence of pertinent laws and regulations, the scholars bring up some ideas, trying to compensate it through a variety of software and hardware facilities, the mechanism and institution construction, system and network construction. One of the representative views as follows: Because of frequent for environmental accident, equipment market chaos limited conditions, lack of insurance mechanism, personnel quality, lack of emergency, we strengthen safety knowledge, specify standard equipment, strengthen the market order insurance mechanism construction, create construction of national rescue network, promote normal personnel organization;[4] At the same time, we must prepare weather forecast, medical assistance and facilities such as safety early warning and control;[5] we draw lessons from foreign advanced experience, the government strongly encourages and covers all kinds of hard and soft knowledge reserve, providing infrastructure construction, the traffic is convenient, specialized service building, professional organization support, aiming at market scale and profit to form a virtuous circle;[6] we explore establishing private outdoor professional rescue system, to relate government departments closely and complement each other after the accident, which can increase aid flexible mobility, reduce social cost and improve the effect of relief;[7] we have the full implementation of destination qualification admittance management system and information publishing platform in high-risk groups' self-help organizers of leisure activities;[8] and some scholars do research on accident rescue, monitoring, early warning and risk assessment, traffic evaluation, safety management system, fully combining computer network technology, mobile communication technology and other high and new science and technology, which has gradually become the trend of such research.

These external constructions are unable to replace of ethics construction, which can fundamentally prevent, avoid and during the high-risk groups' self-help leisure activities. "Ethics" in English comes from the Greek words "ethos", and "ethos" can translate as "custom", "moral", even "faith". Ethics concerns about human beings themselves, and the rules and orders that coordinate and deal with the relationship of society and social norms of behavior about man and nature, man and society. It discusses what is good, what is bad, what is right or wrong, what is good and evil, justice and crime, and moral responsibility and moral duty. Ethics and the ethics subject are inseparable. If morality is defined as all of the code of conduct which society or a culture recognizes, ethics is the moral life systematic thinking and research, the ethics subject constructs a guidance law standard system, and carries on a strict evaluation theoretically. Broadly speaking, all specifications, customs, systems, etiquettes, rules, laws, statements, are included in the category of ethics. Having a view of the high-risk groups' self-help leisure activities in the wild and setting up travel safety ethics standard explores pioneering work. It accords with the needs of the current situation. And the legal discussion above, this activity fully concentrated on the safety ethics between man and oneself, people and people, people and groups, man and nature and group and groups. Of course, the safe ethics here involves related concept of tourism ethics.

The earliest domestic definition about tourism ethics is defined as: tourism ethics is all the codes of ethics that people should follow in tourism activities, is the guide of good and evil about travel behavior. It is also the core software about the development of tourism industry, mainly dealing with people and nature, between people and cultural relics, companions, and a series of complex relationship of body and mind. These relationships eliminate tension and confrontation to achieve harmony and unity. Dealing with the relationship between human and nature, it mainly emphasizes we should respect for nature and love nature. Dealing with the relationship between human beings, it focuses on mutual respect, mutual relationship, mutual help. Dealing with the relationship between man and oneself focuses on rich material world and full spiritual world and their balanced development.[9] This definition is used today and recognized widely. Some scholars give powerful supplement on the definition of its connotation and theoretical framework: Tourism ethics is the new application subject of ethics. The subjects are tourists and tourism stakeholders. The basic question is the relationship between tourism interests and morality. The tourism ethics theory system includes ethical awareness of tourism ethics, tourism ethical relationship and ethical activity.[10] Some scholars pointed out that tourism ethics is a production during the development of the tourism industry, people are thinking about how to improve themselves and create a new civilization. It not only involves that people in the tourism activities should follow the code of ethics, showing travel standards of ethics and how to use the code of ethics to guide and regulate the behavior of individuals and groups, but we also should understand complex tourism activity under ethical perspective, and carries on ethics interpretation.[11] And some scholars regard benefit balance principle and the principle of respect for life as the basic principles of tourism ethics.[12]

In addition, the "responsible tourism" of tourism ethics is remarkable. The study of "responsible tourism" inquiry by the negative effect produced in the process of tourism activities. Overseas researches on the formation think ethical responsibility is the core, and it becomes an independent research system which is based on ecology, economics, sociology and anthropology theory relatively.[13] The core issue of the responsible tourism is to adopt a way which can maximize positive impact and minimize the negative impact of tourism development. It argues that subjects have a self-discipline of responsible behavior constraints, rather than expect the ethical behaviors of others to achieve the goal of sustainable development. Based on the development of responsible tourism, subjects in travel must have the correct ethics of tourism, whose behaviors will take more concern on the impact of tourism.[14] At the end, some scholars put forward that the tourists will change moral behaviors from the life world in the tourism world. Those who have high maturity, high moral qualities are not easy to change. Conversely, other tourists are vulnerable influenced by some factors such as tourism environment, fellow tourists, fellow tourists, tourist groups, tourism motivation, tourism characteristics such factors.[15]

Domestic tourism ethics and leisure ethics research is still in its preliminary stage of development, its basic idea can be used as the establishment the overall direction of high-risk groups' self-help leisure activities tourism safety ethics framework. And the concrete structure about the framework of tourism safety requires some strong enrichment related safety ethics.

2.2 *The exploration of the framework of tourism security ethics standard of high-risk groups' self-help leisure activities in the wild in tourism safety ethics' view*

Safety is to predict the activities of the inherent risks and potential danger in all areas, and it's a kind of state method, means and action to eliminate these dangerous, including safety analysis, safety evaluation, safety measures, accident analysis, etc. These security behaviors objectively contain the moral or immoral problems.[16] Safety of high-risk groups' self-help leisure activities belongs to the so-called non-traditional security. Non-traditional security is on the basis of traditional security, just further reveals the center value of "human security", and "the safety of the people" is a security when a person is staying away the threat of the violence and non-violence. It's also a free situation where we can avoid the threatening of the rights, safety, and even life.[17] Article 3 of *Universal Declaration of Human Rights* says everyone has the right to life, liberty and security, *International Covenant on Civil and Political Rights* also points out that everyone has the right to have personal freedom and safety. But when the personal freedom and personal safety even life safety conflict, or when the personal freedom and go against security threats even life-threatening freedom, what we can do? "The first law of human nature, is to maintain own survival. The humanity's first concern is our deserved own care"[18], and life should be a higher degree of freedom. The key problem is how to achieve this higher free.

According to Kant, as limited rational beings, human choose their own behavior freely. We prefer to the satisfaction of our own emotional desire and give up the moral subject. It's the irrational factors of leads to security issues.[19] He also argues that there is an unsocial sociality in humanity, making people want to handle it all blindly by yourself, and therefore the property will be met with resistance everywhere, as himself can understand it, he tends to be the resistance of others"[20] The security problem of the human is made. To achieve higher free in high-risk groups' self-help leisure activities, we must set up positive, spontaneous ethical consciousness based on the rational and take positive safety ethics action. In other words, more secure freedom only can get from rational. Only through rational, the special leisure desire we want during high-risk groups' self-help leisure activities can be satisfied properly.

In traditional concept of security, accidents shall be controlled by hard technology countermeasures, but practice proves that soft ethical countermeasures in many cases are more reliable, especially in the face of special high-risk groups' self-help leisure activities. Ethical soft countermeasures will control it effectively. The implementation path is set up rational basis and self-motivated consciousness and safety culture. Safety culture is the summation of attitude and spirit related to security which is present in the groups and individuals. It more focuses on structuring, understanding, standardizing the knowledge, technology and value concept system. Control and avoid safety accident caused by errors or mistakes is the central idea.[21] Safety ethical consciousness includes three key elements: First, the moral consciousness that respect for the love of life become the basic composition of subjects' concept. Second, the security subjects must form moral behaviors about formation of respect for life, love life. Third, the security subjects must have self-discipline, when they pursue self-development, and consciously abide by the safety code of ethics and guidance the security subjects' own behavior.[22] In the end, the security participation consciousness and responsibility consciousness is indispensable, and private groups are important platforms about risk decision-making and security resource allocation.[23]

Integrated with the related tourism ethics concept and safety ethics content, we can basically establish the framework of tourism safety ethics standard of high-risk groups' self-help leisure ethics framework is as follows. First of all, to all the persons who join in high-risk groups' self-help leisure activities (principal stakeholders):

Table 2. Participants code of high-risk groups' self-help leisure activities in the wild.

	Rational safe coexistence	
	Guaranteed safe coexistence	Responsible safe coexistence
With their own	Hardware equipped; Carry enough necessary items; Related knowledge in place; Unexpected preparedness in place; Pay attention to daily targeted quality training and the accumulation of experience; Ensure food safety.	Have a rational knowledge of specific situation, rational choice, would rather conservative never overestimate oneself circumstance; Gradual activities, not aggressive; Refrain from any threat to their own activities; To be able to save his life.
With the group	Timely mutual understanding; In a timely manner to establish mutual trust and effective communication; In time on any safety reminders, supervise and urge each other, share any safety information in a timely manner.	Pay attention to team work, obey the unified coordination and experience in obedience to authority, not unusual or act alone; Don't blindly follow when necessary, not superstitious, adhere to the independent thinking and objective discussion; Care about each other, mutual care, mutual tolerance, try my best to help the weak; Refrain from any threat to group activities; Distress calm, rational trying to rescue.
With the nature	Fully master all kinds of natural conditions, adjust measures to local conditions; Low carbon environmental protection.	Awe, not to conquer, not for lack of security risk; Adaptation and rapid response to natural conditions change.
With other group or individual	Actively sharing knowledge and equipment, and related information.	Actively provide relevant experience, and actively absorb relevant lessons.

Second, as for the organizer of a group of high-risk groups' self-help leisure activities:

Table 3. Organizers code of high-risk groups' self-help leisure activities in the wild.

	Rational safe coexistence	
	Guaranteed safe coexistenc	Responsible safe coexistence
With the group	Fully grasp the members; Fully grasp all the knowledge, information, experience and possible problems, and timely sharing; Ensure that the organization experience; Ensure the psychological quality good; Ensure that solve the problem of members, members make up deficiency; Fully predict the worst and timely sharing, preparing; Clear power, responsibility and obligation in advance; Urge members of the reasonable insurance; Ensure the food safety.	Design activities according to the members of the situation, if necessary, stick to stop; Not because any interests blindly compromise or going her own way; Decision science, decisive; Timely check members, security guidance in a timely manner, timely reminder safety guidelines; The correct process control activities; Sharing security information in time. Timely mediation members; Perceived risk in advance, avoiding risk in advance; Distress can organize members to the greatest degree of danger.

3 CONCLUSION

Starting from the security ethics and tourism ethics, we scan it at root from the sight of ethics. By setting up tourism security ethics framework of exploration and basic establishment, it's expected to effectively avoid accident. What's more, it can promote more targeted provision, important theoretical significance and practical significance in the present moment. The framework of tourism security ethics standard of high-risk groups' self-help leisure activities in the wild preliminarily establishes, closely around the essential relationships between organizers and members, members and members. We put forward rational safe coexistence including guaranteed safe coexistence and responsible safe coexistence, and further extend rational security concept including all activities members with themselves, group members, nature, other groups or individuals. And under the idea, we combine the features of high-risk groups' self-help leisure activities in the wild, fully refined the model about the corresponding safety coexistence. It will be of great significance in the future of high-risk groups' self-help leisure activities in the wild.

ACKNOWLEDGEMENTS

The research is supported by the Fundamental Research Funds for the Central Universities (N120314007) in China and Postdoctoral Sustentation Fund from Northeastern University.

REFERENCES

[1] Xie Chaowu, "High-risk tourism projects and safety management system research in China", Vol.118, 2011, pp.133–138.

[2] Yang Fubin, "ten system innovation of 'tourism law'". Journal of law, Vol.10, 2013, pp.19–28.

[3] Kong Lingxue, "Discuss about the Travel Law's all-round protection mechanism for tourist security", Journal of travel, Vol.28, No.8, 2013, pp.29–30.

[4] Li Wen, "My opinion about "Donkey Friends" self-help tourism development", Journal of Inner Mongolia Finance and Economics College (Comprehensive edition), Vol.8, No.2, 2010, pp.125–128.

[5] Liu Tianhu,Jin Hailong etc, "Mountaineering adventure tourism security guarantee system research", Productivity research, Vol.2,2010, pp.100–102.

[6] Zhu Xuan, "The experience and enlightenment about abroad backpack tourism development t", China tourism news, 16 March 2011, P.11.

[7] Li Yingzhou, Fang Liang, "Study of self-help travel security problem", Journal of social scientists, Vol.135, 2008, pp.89–92.

[8] Deng Hao, "Self-help travel's legal problems and countermeasures research", East China University of Political Science and Law, 2010.

[9] Li Jian, "Thinking about tourism ethics", The Guangming Daily (theory), 11 April 2000.

[10] Xia Zancai, "An introduction to tourism ethics concept and theoretical framework". Journal of travel, Vol.19, No.2, 2003, pp.30–34.

[11] Liu Haiou, "The outline of tourism ethics", Journal of Hunan Normal University, Vol.2, 2007, pp.19–22.

[12] Shi Qun, "The construction of tourism ethics principle", Tourism BBS, Vol.3, No.3, 2010, pp. 265–268.

[13] Zhang Fan, "Overseas study's 'responsible tourism' dimension review", Tourism BBS, Vol.3, No.5, 2010, pp. 589–594.

[14] Zhang Fan, "Government management of behavior based on responsible tourism", Journal of tourism science, Vol.26, No.2, 2012, pp.10–18.

[15] Lin Jing, "The moral problems research in travelling world", Dongbei University of Finance and Economics, 2012.

[16] Zhang Changyuan, WuZhuo, "A beginning study of safety ethics", Journal of industrial safety and

environmental protection, Vol. 30, No.12, 2004, pp.35 to 37.
[17] Zhang yan, "'the safety of the people' and 'environmental rights'", Wuhan University of Science and Technology (Social science edition), Vol.13, No.6, 2011, pp. 636–642.
[18] Rousseau,He Zhaowu, "Social contract theory", Beijing: The Commercial Press, 1994.
[19] Lin Guozhi, ZhanTing, "Kant's security ethics thought and its modern significance", Northern review, Vol.220, 2010, pp.130–133.
[20] Kant, Miao Litian, "The principle of moral metaphysics", Shanghai: Shanghai People's Publishing House, 1986.
[21] Feng Haoqing, "Safety ethics is the soul of the safety culture", Journal of Wuhan University of Technology (Social science edition), Vol.23, No.2, 2010, pp.150–155.
[22] Liu Xing, "Safety and moral qualities: missing and construction", Chinese journal of safety science, Vol.19, No.3, 2008, pp.88–94.
[23] Liu Xing, Cui Fang, "Several problems about safety ethics discussion", Journal of North China Institute of Science and Technology, Vol.4, No.3, 2007, pp.109–113.

Management, Information and Educational Engineering – Liu, Sung & Yao (Eds)
© 2015 Taylor & Francis Group, London, ISBN: 978-1-138-02728-2

A study of the disclosure of the corporate social responsibility

Fang Peng Wu
Wuhan Business University, China

ABSTRACT: The disclosure of the corporate social responsibility is indispensable, and it should be gradually standardized and internationalized. Every country should continue to improve laws, regulations and systems of the disclosure of the corporate social responsibility in order to better urge the enterprises to fulfill their social responsibilities and to protect natural environment for the survival of all human beings. The enterprise should regularly offer information on social responsibility to government departments, units, administrators, investors, consumers and the public, etc. The information should be open and transparent and open and provide services for their rights to know and decision-making.

KEYWORDS: Enterprises; Social responsibility; Disclosure.

The enterprise is not innate; it was born after the application to national administrative departments for industry and commerce according to legal procedures and being examined and approved by registration authorities. Since the birth of the enterprises, as a member of society, the responsibility for society, environment and stakeholders must be shouldered which should greatly outweigh the responsibility of being a man. For example, in Europe, the product is produced in a way of being responsible for the society and the environment. It is necessary for the enterprise to disclosure the corporate social responsibility. On the one hand, the disclosure of the corporate social responsibility needs to come under the supervision of relevant society, units and the public; on the other hand, it gives service to the enterprise. If the enterprises take their social responsibilities in the process of production or management, they can promote social development and can also maintain a healthy development of themselves at the same time.

1 THE CONNOTATION OF THE CORPORATE SOCIAL RESPONSIBILITY

Corporate social responsibility (CSR) refers to legal responsibilities taken by enterprises for their production and management, profit-making and shareholders and the responsibility undertaken for their employees, consumers, communities and the environment. The nature of the corporate social responsibility is the moral restriction of the enterprise on its own economic performance. CSR is not only the enterprise's tenets and business principles, but also a set of evaluation system used to restrain the enterprise's behavior of production and management. Corporate social responsibility goes beyond the traditional concept of making profit which used to be the only goal. CSR lays emphasis on human value in the course of production and management and pays attention to human health, safety and entitled rights; it goes beyond the range of only being responsible for shareholders and emphasizes social responsibilities for shareholders, employees, consumers, communities, customers, governments, etc. The most basic responsibility for an enterprise is its legal responsibility, including complying with the laws of the country, not violating business ethics, protecting the community and the environment and giving supports and donations to social programs for public good and so on.

2 DELIMITATION OF DISCLOSURE CONTENT OF CORPORATE SOCIAL RESPONSIBILITY

The information of an enterprise needed to be learned by national government departments, enterprises, employees, investors and the public is the disclosure content of the corporate social responsibility. In particular, the information that has received a universal concern by the public must be disclosed. At present, there are different delimitations of the disclosure content of the corporate social responsibility among different countries and scholars. As for the disclosure content of the corporate social responsibility, in UK it is specified to the category of environment, energy, consumer and community, charity and donation, etc.; American Institute of Certified Public Accountants has reduced it to four categories, namely human resources, natural resources and environment, product and service, community participation; the French government has issued a law; in accordance with the provisions when the enterprise has more than 750 people, it should draw up a social balance sheet which can reflect relevant information through currency including product quality, environmental protection, employees' job satisfaction degrees, contributions to the community, etc.

Many Chinese scholars have different views on the disclosure content of the corporate social responsibility. For example, Lu Daifu thinks that corporate social responsibility includes the responsibility for employees, creditors, consumers, protection and reasonable utilization of the environment and resources, economic development of the community, social welfare and social programs for public good. Wang Jiacan holds that corporate social responsibility is actually to undertake the responsibility for stakeholders who have a close relationship with the enterprise. The enterprise should strive to fulfill the stakeholders wish and meet their various requirements. Jin Sichang considers that the disclosure content of an enterprise should include financial contributions, the development and the use of human resources, contributions to the communities, contributions to improve the ecological environment, responsibilities for competitors, contribution to provide product quality and after-sales service and so on.

In summary, although the delimitations of the disclosure content of the corporate social responsibility vary among different countries and scholars, considering the similarities in their content and the requirements for providing information of an enterprise to the user, the author thinks that the main disclosure contents of the corporate social responsibility include:

I. Main financial information of an enterprise: It is the information concerned by government authorities, investors, creditors and business units. For example, investors focus their attention on the corporate earnings and dividend payouts; creditors pay close attention to the enterprise's capacity to repay loans; tax departments attach importance to the turnover and taxable amounts of an enterprise; banks and suppliers pay attention to an enterprise's balance of deposits.

II. The enterprise's information on human resources: Social workers hope to learn the enterprise's employment information, such as information on recruitment positions, number, wages, welfare, five social insurance and one housing fund, security, promotion, training and personal development.

III. Information about the consumption of resources and environment: Governments, community residents and the public pay close attention to the information about "the three wastes (waste gas; waste water; industrial residue)" of the enterprise, the noise, and the impact on natural environment due to the consumption of resources.

IV. Information of product quality and services: The administration for industry and commerce and consumers hope that the enterprises can have a lawful and honest operation, so that they can produce and sell products of good quality, low prices, perfect functions, high safety performance, first-rate after-sales services, etc.

V. Information of social programs for public good: The society and the public focus their attention on the enterprises' information about their donations to the charity, the disadvantaged groups, the disaster areas, the services of the surrounding community, etc.

3 THE DATA OF THE DISCLOSURE OF THE CORPORATE SOCIAL RESPONSIBILITY

In the above mentioned five items, the main financial information of an enterprise can be written based on the data in the accounting statement; the enterprise's information on human resources can be written according to the annual employment plan and the existing wage and welfare system; information of social programs for public good can be written on the basis of the corporate funds; the third and the fourth items are analyzed as follows:

I. The consumption of resources and environment

a. Waste gas treatment: In the process of production and management, enterprises will consume resources and emit waste gas. The country should compel the enterprise to treat waste gas which will be transformed into resources through compulsory legislation; government authorities should regularly supervise and inspect the enterprise and issue inspection data as the basis of the disclosure content. Gas report index includes the enterprises' monthly emissions of waste gas, capacity of monthly waste gas treatment, and value of the enterprises' waste gas treatment equipment.

b. Waste water treatment: In the process of production and management, enterprises will use water resources and discharge waste water which can be treated by two means including being treated by the enterprises themselves and by professional agencies charging for centralized processing. Government authorities need to regularly supervise and inspect the enterprise that have treated waste water all by themselves and provide inspection data as the basis of the disclosure content. Waste water report index includes the enterprises' monthly discharge of waste water, capacity of waste water treatment, and value of the enterprises' waste water treatment equipment.

c. Waste residue treatment: In the process of production and management, some enterprises will produce certain amount of waste residue. If the amount is large and it can be used as raw materials of certain products, the enterprises should utilize it comprehensively and process it into certain products; if there is a small amount of waste residue that consists of many different kinds, it can not be used in other new ways, and then it is for garbage disposal. The assessment index includes the enterprises' monthly output of waste residue, monthly handling capacity of waste residue.
d. Noise treatment: In the process of production and management, the enterprise needs to take those factors into consideration, for example, whether the machines produce noise, or the noise reached a certain decibel to have an impact on the surrounding residents. Government authorities should regularly check to the enterprises' noise and issue inspection data as the basis of the disclosure content.
e. Consumption of resources: When producing products, the enterprise has to consume raw materials, electric energy (or coal, gas) and water resources. The products of the enterprise are produced by consuming these resources. The consumption of these resources can be reflected by some indexes like monthly utilization rate of available raw materials, the consumption of water (coal) for monthly output, the consumption of electricity (gas) for monthly output.

II. Product quality and services

In order to supervise the enterprises' product quality, the government should establish a specialized network for them. Those having obtained business licenses must register the network which can make it convenient for the enterprises to release information about their products, accept the public's supervision and evaluation, and report counterfeit products, in order to make information communication easier between the enterprises and the public.

a. Product quality: Product quality is the life of an enterprise. It is also related to the consumers' health, safety, and even their lives. Therefore, the enterprise must produce qualified products in accordance with the standards of product quality and take responsibilities for consumers. Now the phenomenon of counterfeiting is pervasive in the society, the enterprises should strengthen their management such as inquiring the products' anti-fake batch number and the products' license for distribution. When consumers purchase the products, they can inquire the products' batch number on the Internet or telephone to find out whether they are genuine or not. If the products are fake, consumers can report instantly either through the telephone or the Internet. They can inform the enterprise of the purchasing time, location and stores to help it quickly crack down on counterfeit products and maintain its reputation and interests. Product quality can take the standard of grading as a reference: i. The products' instruction and function introduction (20 points); ii. Whether the product passes the inspection and its safety (20 points); iii. The manufacturer and its address (20 points); iv. Effective date and batch number inquiry (20 points); v. Informants' hotline telephone and website (20 points).
b. Product services: Product services include sales service, installation and debugging services, technical consultation services and maintenance services. Product services can take the standard of grading as a reference: i. Whether the product's packaging is good or providing home delivery service (20 points); ii. Whether the sales service is good and satisfactory (20 points); iii. Whether the product has technical consultation and warranty (20 points); iv. Whether the maintenance and parts fees are reasonable (20 points); v. Whether it provides with information feedback and complaints (20 points).

III. Community services

In the process of production and operation, enterprises can not have an impact on the surrounding communities' environment; on the other hand, they should provide services for the surrounding community by utilizing their own resources in order to have a harmonious development. When filling in the forms for reporting social responsibility, the enterprise should distribute questionnaires to the residents in the surrounding community at the request of the government authorities for supervision and should collect data to calculate average scores and to fill in the forms. Its standards of grading are as follows: i. Whether the enterprise provides the residents with products, raw materials and other related resources (20 points); ii. Whether the enterprise offers skills training, maintenance and other services to the residents (20 points); iii. Whether the enterprise lets out waste gas and waste water affecting the community (20 points);

iv. Whether there is waste residue and noise influencing the community (20 points); v. The environment inside and outside of the enterprise (20 points).

4 DISCLOSURE STATEMENTS OF THE CORPORATE SOCIAL RESPONSIBILITY

Governments, enterprises, public institutions and the public can learn an enterprise on the basis of the disclosure of the corporate social responsibility which is an important way to enhance the competitiveness of an enterprise and is helpful to improve the enterprise's reputation and establish a brand image and is an important resource for information-needers. The forms for reporting the corporate social responsibility may refer to the following indexes:

I. A main index of the management of an enterprise: a. general assets; b. turnover; c. deposits held in banks; d. accounts receivable; e. borrowed funds; f. account payable; g. net profits; h. earnings per share; i. dividend per share; j. minimum wages; k. maximum wages; l. per capita wages; m. per capita welfare.

II. The index of social programs for public good: a. donations to disaster areas; b. donations to the charity; c. donations to the disadvantaged groups.

III. The index of assets value for the disposal of "the three wastes": a. assets value for waste water treatment; b. payments for waste water treatment; c. assets value for waste residue treatment; d. payments for waste residue treatment; e. assets value for waste gas treatment.

IV. The index of the consumption of resources: a. utilization ratio of raw materials; b. the rate of output value of consuming water; c. the rate of output value of consuming coal; d. the rate of output value of consuming electricity; e. the rate of output value of consuming gas.

V. The index of treating "the three wastes": a. the annual discharge value of waste water (kg.); b. the annual handling capacity of waste water (kg.); c. the annual discharge value of waste gas (m^3); d. the annual handling capacity of waste gas (m^3); e. the annual output value of waste residue (kg.); f. the annual handling capacity of waste residue (kg.); g. the noise affecting residents (db).

VI. Grading by employees and the public: a. grading the staff's working environment; b. grading the corporate environmental; c. grading the product quality; d. grading the product services; e. grading the surrounding communities.

VII. The index of the consumption of resources: a. the consumption of main raw materials(kg.); b. the consumption of water resources (kg.); c. the consumption of coal resource (kg.); d. the consumption of electricity resources (degree); e. the consumption of gas resources (m^3).

VIII. The index of human resources: a. recruitment positions and number; b. whether pay five social insurance and one housing fund; c. new employees probationary period; d. whether to train the new and old employees.

5 INSTITUTIONALIZATION OF THE DISCLOSURE OF THE CORPORATE SOCIAL RESPONSIBILITY

Today our country is ruled by law. In order to make the enterprise fulfill social responsibility and disclose social responsibility in a practical and realistic way, it is necessary for the country to establish a legal system and set up government authorities for supervision. At the same time, the government should create certain conditions for the enterprises to fulfill their social responsibilities and should evaluate and examine regularly on the enterprise's index of fulfilling the social responsibility. The enterprise should disclose regularly the performance of the social responsibility in order to jointly protect the natural environment for all human beings and to make the country, the society, the enterprise and each individual person have a harmonious development. In the near future, the forms for reporting the disclosure of the corporate social responsibility will become an important part of the enterprise's accounting statements. The disclosure of the corporate social responsibility will gradually be internationalized.

REFERENCES

[1] Jin Sichang. Research on the information disclosure of social responsibility. The economist, 2007 (4).

[2] Liu Hongqi. The conception of the disclosure of accounting information of corporate social responsibility. Northern economy, 2010 (6).

[3] Deng Guojie. A discussion on accounting measurement and the disclosure problem. China Securities and futures, 2010 (12).

[4] Liu fragrance. A brief analysis of information disclosure of corporate social responsibility. Technology Review, 2011 (6).

Management, Information and Educational Engineering – Liu, Sung & Yao (Eds)
© 2015 Taylor & Francis Group, London, ISBN: 978-1-138-02728-2

Color extraction from typical Hakkas earth building and applied research relating to landscape

Hui Huang
Shenzhen Polytechnic, China

Ruo Fei Gao
Chiba University, Japan

ABSTRACT: This paper takes Dafudi Building, Yijing Building, and Chengqi Building of the typical Hakkas earth buildings as research objects. Based on field survey and literature survey, and incorporated with digital techniques, the authors have extracted background colors, integral colors, entranceway colors, and decoration colors which represent earth buildings. According to landscape application, they have made further screening so as to determine earthy yellow, black, gray, brown, etc. as common colors used in Hakkas landscapes.

KEYWORDS: Hakkas earth building, color, landscape.

1 BACKGROUND AND PURPOSE

Although Hakkas architecture goes by the name of 'Orient Ancient Rome'[1], Hakkas landscapes are short of a mature pattern and system. In recent years, there appear large-scale Hakkas theme parks and Hakkas residential districts. Meanwhile, traditional architectural form is urgently required to come up in a manner of satisfying present-day inhabitation and aesthetic orientation. Hence, corresponding landscapes seem even more in lack of design instruction and examples to learn from. On the other hand, the common colors, materials, and space constitution of traditional Hakkas earth buildings can serve as a good entry point for studying today's Hakkas landscapes.

Colors are important elements affecting human sensory organs, directly influencing different complicated receptions like joyfulness, peacefulness, excitement, blueness, easiness, and heaviness. Furthermore, the colors in landscapes have an effect on the viewer's state of mind and inherits historic and cultural reserve. This paper starts with color and nails down the common colors used in Hakkas landscapes, laying the foundation for more in-depth researches on their materials and space constitutions.

2 PAST RESEARCHES

Domestically, the study of Hakkas elements of culture and arts largely focuses on architectural element at the present time [2–5]. Taking "Hakkas Enclosing Houses" [6] as an example, the book unfurls the elegant demeanor of Hakkas residential buildings, evolution of Hakkas culture, relation between Hakkas culture and other cultures in different parts of China. Aside from architectural arts, the artistic forms of Hakkas like characters, pictorial arts, dramas, clothing, and statuary draw attention from scholars in the same way, who have organized Hakkas culture and arts into literature and left behind a large collection of papers [7–9]. However, to date there have been instructive documents and studies on what impact Hakkas architecture has on landscapes [10,11]. Given that, this paper makes an attempt to look after a language for instructing the design of Hakkas landscapes in modern times.

3 RESEARCH METHOD

The authors went to Fujian and Guangdong for twice, in March and June of 2014, respectively. We took pictures of existing earth buildings on site and conducted a questionnaire survey. Later on, from July to August of 2014, we straightened out the pictures and survey data while collecting and organizing related document literature. In the end they, by means of the pictures and the software Photoshop, carried out extraction of color frame by frame.

4 TYPICAL HAKKAS EARTH BUILDINGS

4.1 *Dafudi building (fig. 1)*

Dafudi Building is a generic term for any mansion constructed by one of Hakkas people who filled a post as senior official in ancient China. In general, a Dafudi Building has a pleasant environmental setting, most probably being located in a Hakkas village surrounded by beautiful hills and streams. Standard architectural composition of a Dafudi Building can be sketched out as follows: three main rooms arranged in a horizontal line, four rows of houses in parallel with central axis, nine halls, and 18 courtyards. Its plain layout is centered with quadrangle dwellings, with principal rooms facing south as well as eastern and western wing-rooms being axially symmetric. Beam columns, cornices, and window lattices of the building as well as wood joints are carved with ornamentation; where walls connect tiling of a hall have mural painting.

4.2 *Yijing building (fig. 1)*

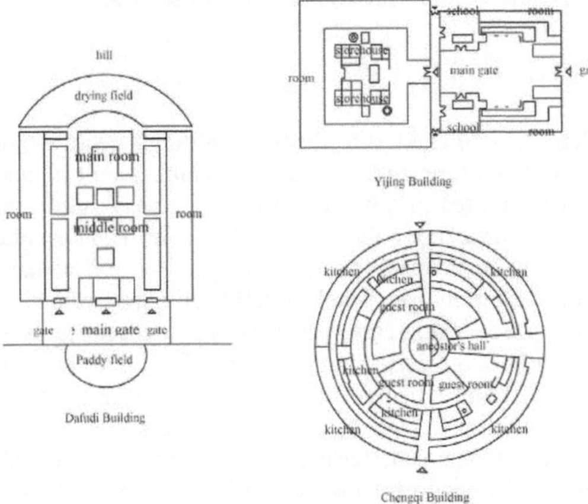

Figure 1. The plane of typical hakkas earth buildings.

Yijing Building is situated in Longyan City, Fujian Province, and it looks quadrate in a planar graph. The whole building has a standardized composition: the left and right ends of main building vertically connect a four-storey building for each and connect the four-storey front building which is in parallel with the main building, thus circling a giant square tower in the shape of "□". There is a group of in small shape of "□" inside the giant square architecture, together forming a unique plane composition in the shape of "回". Ancestral room is at the center. There are two schools on the left and right sides of front building; there is a level ground paved with stones at the center of each school; and there is a gate tower in front of each level ground. Behind main building are accessory constructions like garden, fish pond, pestle room, and oxtall.

4.3 *Chengqi building (fig. 1)*

Situated in Yongding County, Fujian province, Chengqi Building has a big size. It looks like an official headgear, impressing one with its decency and pomp. This building was created according to five element theory, the Eight Diagrams, and nine-grid pattern while incorporating the principles evolved by the Book of Changes. Its round main building is the oldest, biggest circular architecture in actual existence with most inhabitants in the world. Hakkas people esteem "Man is an integral part of nature" and geomantic omen and they thus selected site of residence between natural hills and streams, which was surrounded by farmland. Besides, for the purpose of defense, tall and big arbors were all removed around the architecture.

5 ARCHITECTURAL COLORS OF HAKKAS EARTH BUILDINGS (FIG. 2)

Background colors

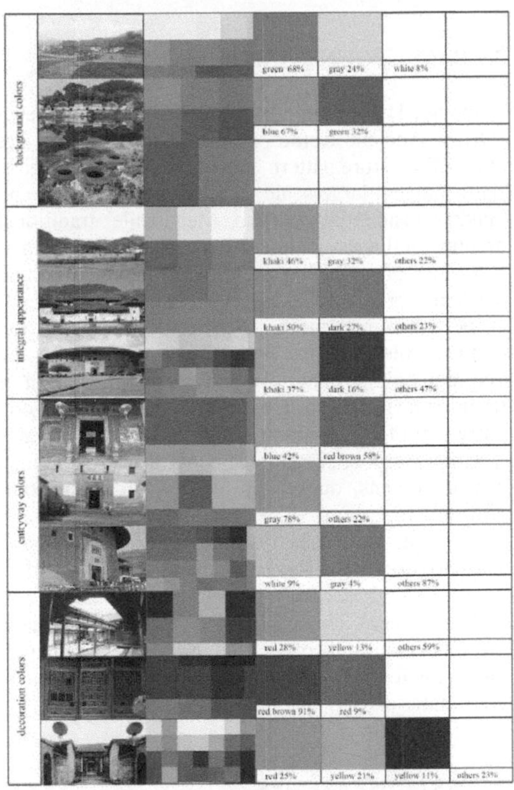

Figure 2. Architectural colors of hakkas earth buildings.

According to the principle of "Be surrounded by mountains and rivers to attain Chi energy"[12], most

Hakkas villages chose a traditional architecture setup of "Front wall back slope", implying waters in the front and hills in the rear. Hakkas earth buildings dissolve in natural colors, beautiful hills and streams, in gigantic size, setting various hues and opponent colors in an extra-large space and thus presenting a decent color tone.

5.1 *Integral appearance*

Like its shape, the color of architecture functions to show its existence in the first place and acts as one of the significant factors influencing its first impression. Traditional Hakkas architecture generally chooses the colors of the exterior wall as follows: lime-pointed wall footing made of ash black bank gravel, black dome and hood, and high yellow mud wall. Black and khaki look brilliant in a green environment and generate a strong light sense. These effects bring about coordination and comparison between architecture and the natural environment.

5.2 *Entryway colors*

Main gate of the Hakkas earth building is an important place to conduct Hakkas etiquette and custom. When seen from appearance, the Hakkas gate tower has artless eaves and are built with black tiles and blue bricks; wood carving is used for the beams and square columns beneath the eaves. The mud wall around entranceway is painted with whitewash, forming a color contrast with earthy yellow exterior wall. Wooden door is covered with reddish-brown algam; door head is carved with granite; red couplets in white wall strike the eye.

Hakkas earth building seldom uses strong colors on a large area. Instead, it principally gives priority to white, gray or primary colors of materials, and decorates key points with relatively lustrous colors like magenta, black, gold, and green so as to exaggerate variance in facades.

5.3 *Decoration colors*

On the aspect of decoration, Hakkas earth building usually applies glowing colors to break the mold of big constructure in homochromy. Main constructional materials inside the earth building are black brick and timber; the colors of timber components like beam, ceiling, column head, and bracket have such dominant hues as red, yellow, cyan, and green which symbolize a thriving family, peace and happiness.

From a perspective of cold and warm integral colors, internal cold-warm contrast is realized on column, wall, door, and window, namely widespread graphite as environmental color set against an appropriate area of gorgeous colors thereby creating an effect of strong contrast. In order to accomplish a good transition from shining bright colors to background colors, bright-dark rendering technique is in common use.

6 APPLICATION OF COLORS IN LANDSCAPES

6.1 *Imitations of decorative colors and background colors*

Hakkas people generally inhabit hilly areas and construct their buildings in a manner of leaning against hills and facing waters. However, such kind of natural background has great limitations in urban areas.

When designing any landscape, appropriate landscape structures are key to realistically embodying Hakkas culture. Yet there are limited applications of decorative colors and decorative structures in landscaping; besides, those applications are restricted by area and type of landscape project.

6.2 *Application of entranceway colors in landscapes*

At the entranceway of the Hakkas earth building, ash gray is in strong contrast with earthy yellow of wall, highlighting the identifying function of entranceway. White and gray as dominant hues convey tranquility and conciseness while forming a color contrast with earthy yellow exterior wall. In modern landscaping, especially in the design of entranceway landscape, application of the materials coloring black, white, earthy yellow, gray, etc. should be noticed so that the frugal basis of Hakkas culture can be materialized.

6.3 *Application of integral colors in landscapes*

Earthy yellow wall and black eave together constitute the dominant hues of Hakkas earth building which is thus enabled to spotlight artificial colors in natural background. Along with it, the spirit of Hakkas people in ancient times, standing alone in nature, is also illustrated to a certainty. Therefore, in landscaping which is oriented with Hakkas culture, the fundamental tone for the entire hard landscape should be earthy yellow and black; the former expounds how Hakkas ancestors were independent of nature and the latter describes how they coordinated with nature.

7 CONCLUSION

Based on the investigation and analysis of the colors of typical Hakkas earth building at three locations, four levels of colors have been extracted: blue mountains and green rivers as environmental setting, integral colors with earthy yellow and black as dominant hues, entranceway with emphasis on gray and white, and interior decoration embellished with the gorgeous colors like red, yellow, and cyan.

Application in landscapes should look after both sides of limited use (background colors and decoration colors) and widespread use (entranceway colors and integral colors of hard landscape), thereby laying the foundation for further correlated researches on the materials embodying Hakkas culture in landscapes.

REFERENCES

[1] Guo Zhikun, Zhang Zhixing: *Orient Old Castle: Hakkas Earth Building of Yongding, Fujian*. Shanghai People's Publishing House, 2008. P.10.

[2] Wan Younan: *Traditional Architecture and Culture in Southern Jiangxi*. Jiangxi People's Publishing House, 2013. P360.

[3] Wu Qingzhou: *Hakkas Architectural Culture in China (Volume I, II)*. Hupei Educational Publishing House, 2008. Volume I P270, Volume II P328.

[4] Liao Dong, Tang Qi: *Unscrambling the History and Architecture of Hakkas Earth Building*. Mount Huangshan Publishing House, Anhui, 2013.P150.

[5] Xu Hui: *Trip to Chinese Antique Buildings: Fujian Hakkas Earth Building*. Jiangsu Science and Technology Publishing, 2014.P208.

[6] Huang Chongyue: *Hakkas Enclosed Houses*. Publishing House of South China University of Technology, Guangzhou, 2006.P191.

[7] Xie Chongguang: *Review of Hakkas Culture*. China Social Sciences Publishing House, Beijing, 2008.P511.

[8] Tan Yuanxiang, Huang He: *Introduction to Aesthetics of Hakkas Culture*. Publishing House of South China University of Technology, Guangzhou, 2001.P308.

[9] Yang Honghai, Ye Xiaohua: *Hakkas Arts and Charms*. Publishing House of South China University of Technology, Guangzhou, 2006. P206.

[10] LIU Peilin: Study on the identification of Hakkas traditional village's landscape gens and analysis in the perspective of geography, Human geography, Jun.2009, P40–43.

[11] Zhang Xianzhong: The Harmonious Beauty of Human and Nature in Hakka Enclosed-house, World Heritage, Jun.2014, P92–99.

[12] YANG Zhouwei: On the Application of the Learning of the Change in the Construction of the Capital City of the Regimes of Nanzhao and Dali in the Tang and Song Dynasties, Studies of Zhouyi, 2011, P72–76.

Management, Information and Educational Engineering – Liu, Sung & Yao (Eds)
© 2015 Taylor & Francis Group, London, ISBN: 978-1-138-02728-2

Application research of biotechnology in animal husbandry

Jin Xiu Mu
Weifang University of Science and Technology, ShanDong WeiFang, China

ABSTRACT: The article discusses the biotechnology in livestock and poultry breed resources development, animal genetics and breeding, animal feed, animal disease prevention and diagnosis, the application of biotechnology for germplasm resources conservation and utilization of fine livestock, breed improvement, improve the feed nutrition value and fast and exact diagnosis and treatment of diseases provides a broad prospect. Through biotechnology can promote faster development of animal husbandry in China.

KEYWORDS: Biological technology; Animal husbandry; Feed development; Disease diagnosis.

1 INTRODUCTION

Biotechnology development is rapid, broad application fields, including medicine, food, agriculture and animal husbandry, fishery, chemical industry, energy, metallurgy, marine industry and environmental protection, etc. Animal husbandry is an important field of biotechnology. Genetic engineering as the core, including cell engineering, enzyme engineering and fermentation engineering of biotechnology with the progress of modern animal husbandry and veterinary have close relations, become an important part of the animal husbandry technology revolution. These technology research, development and application of excellent germ plasm resources conservation and utilization of livestock, breed improvement, improve the feed nutrition value and fast and exact diagnosis and treatment of diseases provide a broad prospect. The establishment of the modern biological technology, pioneered a new way of livestock production.

2 BIOTECHNOLOGY AND LIVESTOCK BREEDS

Most livestock or poultry genetic resource are in the agricultural production of herds and flocks to keep, that is currently in live animals such as pigs, cattle and poultry to save. In biotechnology, livestock and poultry variety resources, mainly has the following two ways: one is to use frozen embryos and reproductive cells technology, nowadays there are rats, rabbits, ten many kinds of animals such as cattle, sheep embryo freezing transplant success, some of which kinds of refrigeration technology has been programmed, and commercialization of the kit. The second is using molecular biological techniques to establish herds and flocks of the DNA gene library. The so-called gene library, is contained in the genome of a living in a group of different segments of DNA recombinant clone. Overall is the preparation of cloned livestock breeds gene library, save the genomic library is to save the livestock breeds. Through biotechnology traits, you can keep good varieties of livestock and poultry to protect endangered animals.

At present, many developed countries have built livestock frozen sperm and embryos, the research and application of cryogenic sperm cryopreservation livestock have fast progress in more than 50 years. Using biotechnology can be simplified for introducing a method of embryo transplantation and embryo-freezing technology. The combination of improved a variety of introduction can be simplified as the introduction of frozen embryos, not only easy to transport, quarantine procedures simple, low cost, and the offspring of introduction to the ecological environment adaptability and enhanced disease resistance. At present, cattle and sheep embryo transfer and freeze technology have become the international, regional genetic resources of good exchange cheap and easy way.

3 BIOTECHNOLOGY AND ANIMAL GENETIC BREEDING

Scientists use genetic engineering through a certain method of artificial restructuring of exogenous DNA into cells or embryonic cell receptor animals or the receptor genes in the genome of a DNA excision, so that the receptor animals for a change in the genetic information, to produce animals with exogenous DNA fragments, and this change can be passed on to offspring. Breeding using this technology can break the species barrier, breakthrough the limitation of genetic relationship, cultivate nature and conventional breeding is difficult to produce, added: there are special excellent characters of animal species, and effectively improve livestock productivity, to improve the performance and quality of livestock and poultry production. Pickett, etc in 1978, for example, transgenic technology was employed to get the eight bovine growth hormones, transgenic pig, this gets a big, daily gain fast feed conversion rate is high. Biotechnology can also change the conventional breeding long time needed for the need for more hybrid weakness, to speed up the breeding process, satisfy people's demand for livestock and poultry products. Fast breeding process, satisfies people's demand for livestock and poultry products. Reported in 1997, the British people with somatic cell nuclear transfer technology for the first time, the sheep somatic – mammary gland cells, the nuclear transfer to another nuclear oocyte cytoplasm, develops in the receptor sheep, and gave birth to the clone sheep, caused the sensation. The use of genetic engineering methods to certain bodily growth hormone gene transfer in bacteria, and then produced by the bacteria to breed a large number of useful hormone. These hormones in animal metabolism process improve the body protein synthesis and fat consumption, so as to speed up the growth and development, namely, without any increase in the consumption of feed cases, improve the yield and quality of livestock and poultry. In addition, utilizing biological technology to cultivate disease-resistant varieties, enhance its resistance to disease and internal and external parasites, thereby avoiding livestock diseases, to the economic consequences of the livestock production.

4 BIOTECHNOLOGY AND THE DEVELOPMENT AND UTILIZATION OF FEED RESOURCES

4.1 *Sweeteners*

Sweeteners can enhance chicks and piglet appetite, born chicks drink a definite concentration of sugar water can improve the survival rate of newborn chicks, and can improve the stress state of chicken feed intake, improving palatability. Now commercialization application of dipeptide sweeteners: aspartame (aspartame) is through the synthesis of a new type of sweetener of biotechnology.

4.2 *Enzyme preparation*

Enzyme preparation is a kind of high efficient biological active substances with enzyme properties. The enzyme was used as a feed additive and has decades of history, substrate feeding enzyme preparation can directly decompose, supply the body's nutrients; stimulate the secretion of endogenous digestive enzyme hydrolysis of plant cell walls of nutrient release in the cell; damage the soluble non-starch polysaccharides in feed, lower the viscosity of the intestinal contents, increase the digestion and absorption of nutrients; participate in animal endocrine regulation, promote anabolism. With the development of enzyme engineering technology, a variety of enzyme have been found that there are more than 5000 kinds. More than 20 were used as feed additive enzymes, mainly including protease, lipase, amylase and pectinase, saccharified enzyme, cellulose enzyme, phytase enzyme system. Past industrial enzymes produced by microbes in the nature of natural enzyme screening on the basis of the production, and now the enzyme production can use genetic engineering technology, make some small, hard to cultivate some microbial enzymes, by selecting the genetically modified (gm) to some lower host microbial growth requirement of production. Some enzymes in the body can be obtained by transgenic technology to the plant body. As the phytase production is the point of molecular biology techniques, production of the phytase gene was isolated, and then inserted after these gene amplification suitable expression vector, and a large number of phytase was produced. Phytase production by this method compared with the conventional method production strains or enthusiastic strain of production, production can improve the 50 ~ 100 times.

4.3 *A new type of feed protein*

Serious shortage has become a worldwide problem. Protein feed from microbial fermentation to produce single cell protein (SCP) is an important way to solve this problem. SCP nutrient-rich, elevated protein content, amino acid contained components and complete balance, and there are a variety of vitamins, high utilization rate of digestion, and the wide raw material sources, microbial breeding fast, low cost and high efficiency. It can be utilized to produce single cell protein of microorganisms such as bacteria, fungi, yeast and algae. A lot of raw materials produce single cell protein, such as wine, monotonous glutamate, starch, paper making, sugar, pharmaceutical and other industrial wastes. All kinds of plant straw, shells, sawdust and other agricultural and sideline products processing by-products, etc. Such as cell and yeast using methanol, ethanol, methane, and paraffin production SCP: many of using waste material into SCP, such as rice straw, bagasse, citric acid wastes, the stone of the fruit, syrup, animal dung and dirt, etc.; With a mixture of starch by-product as raw materials, to produce single cell protein by solid fermentation, using algae produce SCP (e.g., chlorella, cyanobacteria). With the production of single cell protein research in our country, the main products are feed yeast and spirulina protein. Shanghai yeast works through the specific biological technology can develop into rich trace elements of microorganisms, such as selenium yeast, yeast zinc, protein content was 62% ~ 79%, rich in carotene, algal protein, sodium alginate and class active substances such as insulin. Different types of feed protein production is developing rapidly in recent 10 years, and biotechnology applied in the feed industry is one of the most potential fields. Its development will provide the industrial and agricultural waste into high nutritional value of feed resources.

4.4 *Probiotics*

Probiotics is also called the beneficial bacteria agent, is the microbial bacteria and its corresponding substances directly feeding animals, participate in animal gastrointestinal tract microbiota ecological balance and maintain the normal function of the gastrointestinal tract, so as to achieve the purpose of animal health and production performance. At present, it has been confirmed that the production of forage microbial additive species mainly include: Lactic acid bacillus, streptococcus, bacillus, nitrobacterium, bacteria and yeast, etc.

The beneficial bacteria agent is widely used in animal husbandry, it can solve the problem of antibiotic residues as feed additives instead of antibiotics, and in animal to the problem of antibiotic resistance, has the inestimable function of the development of animal husbandry.

5 BIOTECHNOLOGY AND ANIMAL DISEASE PREVENTION AND DIAGNOSIS

Biotechnology has played more and more important role in animal disease diagnosis, prevention and treatment. In recent years, the human is no longer limited to detect proteins in body fluids to the diagnosis of disease, the change of the concentration of sugar and other material, but can be applied to molecular biological techniques, from molecular level detection and analyse the causes of some diseases, traces the disease development process, also can also to the infection of pathogenic microorganisms identification, classification, and screening effective drug treatment, etc. Modern molecular diagnostic technology mainly refers to the application of immunology and molecular biology methods to pathogenic substances for diagnostic testing. Such as: enzyme-linked immunosorbent assay and DNA diagnosis technology.

At present, utilized in the production of livestock and poultry vaccine still is given priority to with conventional vaccines, for the prevention of disease has played a positive role. But there are a lot of problems, because the conventional vaccine is based on a large number of cultivation of pathogenic microorganisms in production. Often due to causes such as the pollution immunity effect is not stable, sometimes even cause the failure of the immune, bring potential threat vaccinated healthy animals. With the development of biotechnology, now we have been able to describe the vaccine from the molecular level and its inducing mechanism of the body's immune response, and at the molecular level to design more accurate vaccine development plan; Through the analysis of the active ingredients, harmful ingredients of vaccine and unnecessary ingredients, improve the vaccine efficacy and safety. It can be achieved by recombinant DNA technology needed for the in vitro synthesis antigen protein molecules, and according to the have to be modified and restructuring. Application in recent years is given priority to with lymphocyte hybridoma technology and recombinant DNA technology of modern biological technology research and production of new vaccine, can overcome the defects of conventional vaccine, the vaccine to the pathogenic bacteria, yeast or animal cells to produce, so as to avoid the traditional way of the mass culture of pathogenic microorganism, overcomes the drawback of these vaccines exist a series of.

6 CONCLUSION

To sum up, the application of biotechnology in animal husbandry broad prospects, plays a more and more important role in the development of animal husbandry. So we need to further improve the level of biotechnology research, speed up the development of animal husbandry and biological technology research, make full use of the advantage of our country, using foreign advanced aspects of biotechnology development and promoting its application, with high and new biotechnology to promote the development of animal husbandry in China.

REFERENCES

[1] Xueping Li. The application of biotechnology in modern livestock production progress [J]. Journal of China animal husbandry and veterinary, 2007 (8) : 58–61.

[2] Haijun Xiao. Introduction to the application of biotechnology in animal husbandry and veterinary [J]. Journal of Inner Mongolia livestock science, 2003, 6:40–41.

[3] Xuyong Zhao. The application of modern biotechnology in animal husbandry, and prospects for development [J]. Journal of animal husbandry engineering college of zhengzhou, 2004, 24 (2): 97–99.

[4] Ping Tian. The application of biotechnology in animal nutrition and feed research [J]. Journal of ennui agricultural science, 2007, 35 (2): 374–375.

[5] Tao Chen. The application of biotechnology in animal husbandry and veterinary overview [J]. Journal of human animal husbandry and animal medicine, 2007, 28 (6): 10, 21.

Management, Information and Educational Engineering – Liu, Sung & Yao (Eds)

The kinematics analysis of elite tennis athletes double backhand topspin technique

Jun Guo & Ji He Zhou
Graduate Student Faculty, Chengdu Sport University, Sichuan, Chengdu, China

ABSTRACT: Backhand technique is as important as forehand technique in tennis basic technique. Backhand is divided into double backhand and single backhand. Double backhand has the advantage of good stability etc., thus most tennis players use the double backhand. By the end of April 14, 2014, there are seven players in the top 10 players whose backhand is a double backhand, three players use a single backhand. Double backhand is a main means of base line attack. Therefore, it is the inevitable requirement of tennis backhand technology to constantly improve the double backhand.

This paper is the discussion and analysis of the double backhand for the ATP champions tour Chengdu Open contestant called Marat Mikhailovich Safin who was the grand slam champion of the U.S. Open and Carlos Moya who was the grand slam champion of the French Open, and to find out some regular patterns and features, to demonstrate the principle of kinematics of tennis, aimed at providing some methods for tennis athletes' training and competition in the future, as well as make some statistics for the coaches to guide the tennis players in the backhand tactics.

KEYWORDS: tennis; backhand; kinematics.

1 THE RESEARCH OBJECT AND METHOD

1.1 *Object of study*

This study selects ATP Champions tour Chengdu open Spain athlete Marat Mikhailovich Safin and Swede athlete Carlos Moya as the research object, the basic conditions are shown in table.

1.2 *Research methods*

Two JVC9800 cameras and a 24-pointed framework which are fixed at a certain point are then used to record Marat Mikhailovich Safin and Carlos Moya double backhand in Chengdu International Tennis Exchange Center. The camera has a length of 1.25m, and their main axis angle is 65°. One of them is located about 15m behind the player to the right. Another is placed in front of him 20m or so to the right. 3D Single TEC analysis system has been adopted for the analysis of the video recorded. Meanwhile, the process will be studied one motion after another with the purpose of obtaining reliable statistics. Furthermore, in order to satisfy the need of this research, three measuring points are added which cover the top of the racket, the tennis and the projected angle of the shoulder and hip.

2 RESULTS AND DISCUSSION

Analyzed and discussed to Marat Mikhailovich Safin and Carlos Moya double backhand technology, in order to reveal the kinematic characteristics of the high-level athletes double backhand, at the same time analyzes its action principle.

2.1 *The division of action stage*

On the technical features of the tennis double backhand action, this study will divide double backhand into the following three phases (Figure1) .

Table 1. The object of study.

Name	Height	Weight	Grip	ATP Record	ATP Singles ranked	Highest record
Marat Mikhailovich Safin	193 cm	88 g	right hand	single chamption:15	1	doubles champion: 1
Carlos Moya	190 cm	84 kg	right hand	single chamption:20	1	doubles champion: 1

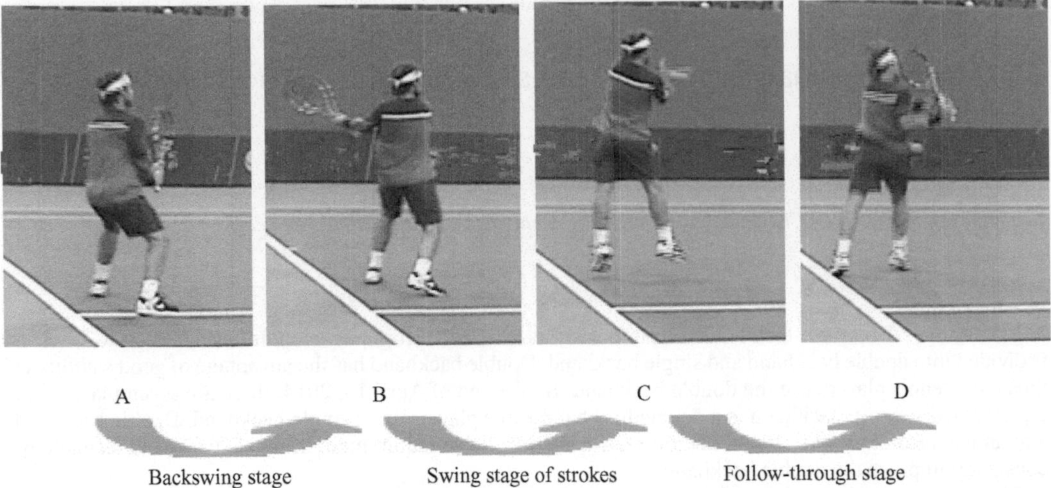

Figure 1. The division of double backhand technology stage.

Backswing stage (A–B): starting from the moment of feet stand stable to the moment of the top of the racket began to decline.
Swing stage of strokes (B–C): from the moment of the racket began to decline to the moment of hitting the ball.
Follow-through stage (C–D): from the moment of hitting the ball to moment of following-through to the right shoulder.

2.2 *Backswing stage*

The main purpose is based on the hip joint shaft to driven upper body to the left, increase the torsion of the body, for hitting the ball fully prepared.

At the end moment of back swing, the angle of the shoulder and elbow reflect the athlete's extension status of arm relative to the trunk. By Table 2, it is found that Safin's left and right shoulder angle is 35.6°and 74.7°at the end moment of back swing, Moya is relatively small. The modern tennis link theory requires the forearm distance trunk farther more, do the full extension. In the knee joint angle data, two athletes' left knee angle smaller than right knee, they concentrated their weight on the left leg before hitting the ball, left leg plays a major supporting role.

2.3 *Swing stage of strokes*

The purpose of this stage is based on the attitude after back swing, using of lower limbs and stretching in the trunk of rapid reverse, form the best muscle initial state.

Table 2. Characteristic parameters of back swing.

	Shoulder joint angle (°)		Elbow joint angle (°)		Knee joint angle (°)	
Name	left	right	left	right	left	right
Safin	35.6	74.7	145.6	151.4	133.2	139.7
Moya	28.5	69.7	152.7	157.2	135.3	141.3

Table 3. Characteristic parameters of swing stage of strokes.

	Maximum joint velocity (m/s)			
Name	left shoulder	left elbow	left wrist	Maximum head speed (m/s)
Safin	2.17	3.22	5.17	26.9
Moya	2.15	3.20	5.11	25.4

The maximum speed of Safin's left shoulder, left elbow and left wrist appeared in 0.1 seconds before hitting, the maximum speed of Safin's left shoulder is 2.17m/s, left elbow is 3.22m/s and left wrist is 5.17m/s (Table 3). Moya's left shoulder, left elbow and left wrist appeared in 0.2 seconds before hitting, the maximum speed of Moya's left shoulder is 2.15m/s, left elbow is 3.20m/s and left wrist is 5.11m/s. Through the analysis of the data, it is found that two elite athletes' linear velocity increase in turn, the main joint reach maximum before hitting, this compliant with whiplash movement rules.

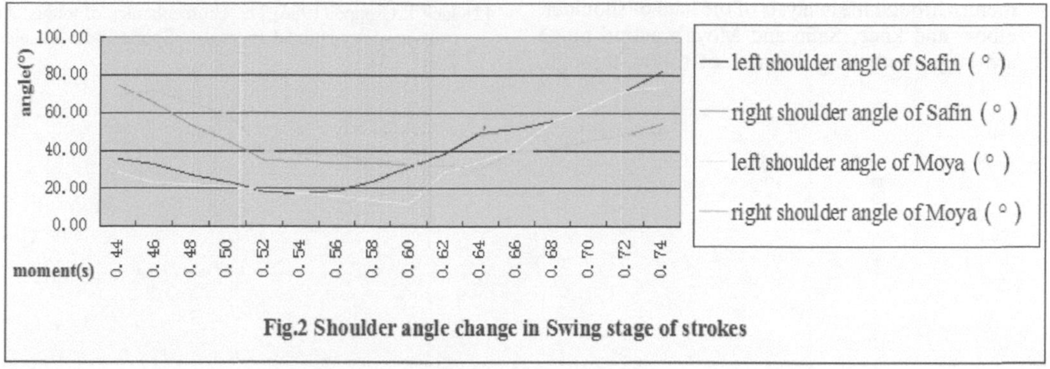

Fig.2 Shoulder angle change in Swing stage of strokes

Figure 2 shows that Safin and Moya take 0.28 s in this stage, Safin's left and right shoulder from 35.6° and 74.7° in the beginning to 82.4° and 54.1° in hitting. Moya's left and right shoulder from 28.5° and 69.7° in the beginning to 74.1° and 49.7° in hitting. The study found that two elite athletes' right shoulder angle are gradually decreases, it can make the rotating radius decreases, thus obtaining the larger angular velocity.

2.4 *Follow-through stage*

The main purpose of this stage ensures the ball to target direction in hitting,which determine the direction, angle and depth.

Table 4. Characteristic parameters of follow-through stage.

	Shoulder joint angle (°)		Elbow joint angle (°)		Knee joint angle (°)	
Name	left	right	left	right	left	right
Safin	82.4	54.1	72.4	31.1	48.1	39.4
Moya	74.1	49.7	84.3	24.6	42.6	30.5

Safin's left and right shoulder angle is 82.4° and 54.1° at the begin moment of follow-through stage, left and right elbow angle is 72.4° and 31.1°, left and right knee stretch range is 48.1° and 39.4° (Table 4). Moya's left and right shoulder angle is 74.1° and 49.7° at the begin moment of follow-through stage, left and right elbow angle is 84.3° and 24.6°, left and right knee stretch range is 42.6° and 30.5°. The study found that two elite athletes' arm have obvious forward movement, this technical is to increase the time of the racket effect the ball, thus better control the ball placement and depth.

3 CONCLUSION

Double backhand is atypical of the whip. Through the analysis and discussion of Safin and Moya think that the action of Safin and Moya's technology model are worth our using for reference, its action model main indicators are as follows:

1 In backswing stage, Safin and Moya's right shoulder angle was bigger at the end of this stage, it shows that the range of right elbow is bigger, this action can make a racket in a relatively high initial position, so the athlete can have great potential energy, to create conditions for hitting of the power.
2 In swing stage of strokes, Safin and Moya take 0.28 s in this stage, the maximum speed of Safin's left shoulder, left elbow and left wrist appeared in 0.1 seconds before hitting, in contract, Moya's left shoulder, left elbow and left wrist appeared in 0.2 seconds before hitting, two elite athletes' linear velocity increase in turn.

The study found that two elite athletes' right shoulder angle are gradually decreases, and it shows that the rotating radius decreases in hitting process, and the ability of body control of the ball is improved.

3 In follow-through stage, the study found that two elite athlete arms have obvious forward movement. Through the analysis of the data of shoulder, elbow and knee, Safin and Moya's action range are large, with long distance of follow-through, conform to the principle of modern tennis technology.

REFERENCES

[1] Bruce Elliott,Machar Reid,Miuel Crespo.ITF Biomechanics of Advanced Tennis[M].itfltd,2003.

[2] Jack L.Groppei(1986).The Biomechanics of tennis: An overview.International Journal of Sports Biomechanics. 2:141–155.

Management, Information and Educational Engineering – Liu, Sung & Yao (Eds)
© 2015 Taylor & Francis Group, London, ISBN: 978-1-138-02728-2

Analysis and design of eco-tourism risk management system

Yan Mei
Chengdu University of Technology, China

ABSTRACT: In recent years, with the development of society and the economy, the continuous progress of human society, the voice of ecological environment gets louder and louder, and promotes the development of global tourism. In this context, eco-tourism and sustainable tourism projects have been proposed and favored by travel enthusiasts, and are becoming an important part of tourism development. As eco-tourism project investment has a greater risk, enhancing risk management is important. Therefore, this article is a brief analysis of the eco-tourism risk management system designed to promote eco-tourism and sustainable development.

KEYWORDS: ecological tourism; risk management systems; analysis; design.

The target of eco-tourism is a natural area, based on the concept of sustainable development and ecology and tourism activities, enjoy the natural beauty, wildlife and related cultural characteristics. It can effectively change the effective circulation of ecosystems and to benefit local residents and tourism businesses. The current point of view, tourism development also has its own characteristics, and in the process of investing in the development of damage to the environment, if not handled properly, it will cause serious ecological risks, and ultimately affect the development of eco-tourism projects. Thus, in the development of eco-tourism project, we should avoid the risk, strengthen risk management, thereby reduce the risk and loss of ecological tourism project development, and promote eco-tourism and sustainable development.

1 ECO-TOURISM RISK

Eco-tourism risk refers to the entire ecosystem or threatened ecological landscape features outside, thereby reduces certain elements within the system or their own health, productivity and genetic structure. Eco-tourism refers to the risk analysis of eco-tourism by one or more stress factors to assess their possible line. The reason for the ecological risk analysis is to be the negative effect of a scientific evaluation of environmental damage caused, and thus provide a scientific basis for the eco-tourism system. From the current perspective, ecological risk is ecology or environmental chemistry and risk-based on science and other disciplines, the main research about ecology issues. From the current point of view, the risks can be divided into ecological risk caused by biological engineering, and by the invasion, etc. Some scholars and experts have analyzed the ecological risks, got receptor analysis, hazard assessment and risk characterization. The mean of the receptor assay is determined by both qualitative or quantitative methods to determine, analyze, the system will be analyzed in the key eco-tourism species, communities and populations, the selection of appropriate risk receptors, the risk of the entire eco-tourism final reaction system, implementation analysis, calculation and grasp of objectives. The hazard assessment is the use of eco-pathology of various toxicity tests to analyze indicators and determine the likelihood of risk. Risk analysis of eco-tourism, coupled with the complexity of the ecosystem is able to determine the receptor, the wind vector characterization and non-deterministic processing, and ultimately determine the ecological risks.

So far, the risk of major ecological approach entropy associated with the exposure – reaction. Relevant concentrations of indicators which entropy method is mainly used for eco-tourism system in a specific environment pollution levels if there is a certain sense, is also set up to protect a particular receptor, and thus there is a certain relationship with the environmental concentrations measured. However, due to their relatively low cost, and rely on indicators and their standard readily available, so the risk of eco-tourism in the analysis, its advantages over other evaluation indicators. Exposed - Reaction applicable assessment data in the system of eco-tourism, eco-tourism system suitable for the assessment of the number of risks.

2 B. ECO-TOURISM PROJECT RISK MANAGEMENT SYSTEM

Since the 1950s and 1960s, risk management has arisen. Risk management is an important part of project management implementation, identify risks in project construction projects, analyze risk and predict risk, and thus brings greater impact. Risk management of eco-tourism project, to be managed for multiple projects, tourism projects to assess the potential risks. Investment management of eco-tourism project, which is a new and effective multi-project management and compared with the traditional project management can play an important role in eco-tourism risk management.

Eco-tourism project risk management systems follow the "input - output" indicator design ideas, and then different indicators project on this basis, depending on the evaluation object set. The "input - output" design basis, mostly through the years of practice and put in, and then select Manage indicators in line with eco-tourism through this project. From the current perspective, eco-tourism project has more features, and is related to the interests of the different aspects of risk management indicators built, most of these need to coordinate all aspects. Since the goal of eco-tourism projects has been achieved, only eco-tourism has the function of protecting the environment and promoting local economic development, and it will be able to achieve the promotion of eco-tourism development requirements. In addition, eco-tourism is a new type of tourism products, the traditional tourism activities, and has less impact on the surrounding environment, eco-tourism, and to enhance the environmental awareness of tourists, it has a greater difference. Since the objectives of eco-tourism are shaping the behavior of visitors, and there is a certain sense of tourists and local residents in the Code of Good Practice and scientific environment, and be able to increase the growth of the knowledge of visitors and residents, and ultimately serve to protect the environment and the concept of morality. Therefore, this project should be based on eco-tourism purposes, thus enhancing the indicators of local residents and eco-tourism.

When evaluating the risk of eco-tourism, it includes the following elements: First, the management of eco-tourism project management body. Before eco-tourism activities, when their managers develop and construct ecological tourism resources, the establishment of appropriate accommodation facilities in the tourism process is equipped with appropriate business and services. Because of eco-tourism and traditional mass tourism are different, they not only pay attention to market-oriented operations, but also shoulder the burden of protecting nature. Therefore, you should include the depth and sustainability of tourism management. Secondly, it is rich in natural and cultural resources. Mostly based on local tourism activities than the special natural and cultural attractions to attract tourists, and no special items for the tourist destination, its tourist activities can not be expanded. Because renewable natural resources are difficult, therefore, should focus on its natural resources protection. Finally, eco-tourists. Since tourism is the main tourist generating and developing eco-tourism is mainly due to the special needs of eco-tourism. Ecotourists and traditional mass tourist essentially different. For eco-tourists, the emphasis was given in coordinating the whole process of travel for tourists and visitors to pay attention to self-management capabilities. The level has a direct impact, and the effect of management degree Theme Manager tourists eco-tourism experience tourists will also have an impact. Eco-tourism risk management system is shown in Figure 1. Factors should also fully recognize the risks of eco-tourism, and many factors are presumed not to be predictable and to find and research should be looking for links between risk events and other events, and ultimately find the risk-associated information. In addition, the list shows the establishment of risk into a substantial risk identification stage, and there is potential for objectivity and risk, and the establishment of various project-risk indicators.

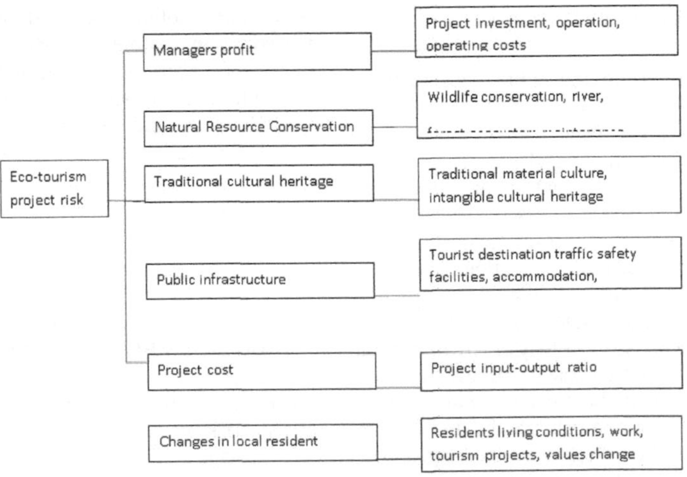

Figure 1. Risk management evaluation of eco-tourism map.

3 CONCLUSION

In conclusion, eco-tourism is not only a manifestation of social progress, but also economically important to the development of performance. The eco-tourism in the development process will inevitably lead to unexpected problems, which need to strengthen their risk management, promote the development of eco-tourism. This paper will be on a brief analysis of eco-tourism risk management, assessment methods and risk indicators of eco-tourism in order to cast a brick to attack jade.

ACKNOWLEDGEMENT

Annual research project for social sciences of Sichuan province in 2012 (SC12LY04).

REFERENCES

[1] Tao Li.Ecosystem management study tour of ecological risk[J] Anhui Agricultural Sciences, 2006,34 (24): 6652–6653.

[2] Yujun Zhang,Ling Shi,Yiqi Jia etc.. Significance and potential risks, such as ecotourism Nature Reserve[J] Central South University of Forestry Science and Technology (Social Science Edition), 2013,7 (1): 7–10.

[3] Tiancheng Shang.Ecosystem Management and Ecological Risk tourism analysis[J] Arid Land Resources and Environment, 2008,22 (5): 91–94.

Management, Information and Educational Engineering – Liu, Sung & Yao (Eds)
 ISBN: 978-1-138-02728-2

Analysis of local governments' response to public emergency under new media environment

Yi Qun Wang
Yinzhou District, Ningbo, Zhejiang, China

ABSTRACT: This paper explored the problems existing in local governments' response to public emergency and online public opinions in the era of microblog, and proposed diversified governance measures for local governments to enhance their abilities in terms of improving their ability of pacifying online public opinions, establishing an equal dialogue mechanism to ensure orderly political participation, making greater efforts in information disclosure, and building an online participation in government and political affairs system, etc.

KEYWORDS: Public Emergency; Online Public Opinions; Social Management.

Microblog, as a newly emerging open social platform, allows users to release and share information within 140 characters, and has changed the traditional communication method and people's way of social contacts. At the end of 2013, the number of Chinese netizens reached 618 million in which microblog users were 281 million and accounted for 45.5%. ① A survey shows that 96% of microblog users learn about and expressing their opinions about public social events and emergencies through microblog. ② It indicates that microblog has become an important information platform and public opinion front that integrates information, views and will of the people. At present, China is going through economic and social transformation, which is a key period. People's strong sense of political participation is intertwined with their confusion, anxiety and dissatisfaction with social contradictions, thus publicizing individual appeals and forming greater social influences. Microblog, with its huge internet users, spreads virally and rapidly develops into an important medium for people to participate in public emergencies. The Ningbo PX incident happened because some villagers in Wantang of Zhenhai wanted to be included in the compensation for land acquisition by Zhenhai SinoPec but it's rendered by microblog into public appeal for environmental protection and triggered a large-scale mass incident.

In the era of microblog, how should local governments efficiently and timely respond to public emergency? How to effectively steer online public opinions? How to build government credibility? All these have gradually become an important part of the government governing ability construction.

1 PROBLEMS EXISTING IN NINGBO LOCAL GOVERNMENT'S RESPONSE TO THE ONLINE PUBLIC OPINIONS ABOUT THE PX INCIDENT

1.1 *The concept of official oriented still exists in local governments and crisis awareness missing in social management*

Local governments ignore the strong influences of microblog. Currently, with regard to most of online public opinions caused by public emergencies, local governments or related departments start responding only after the public opinions were heated. In the PX incident, the local government noticed the public surging emotions right after the collective petition, but it failed to handle it first time. It's after two days later that it announced "Clarification about the Refining-Chemical Integration Project in Zhenhai", admitted that collective petition happened and promised to "follow the strictest emission standards". It's far slower than the "prime four hours" to deal with public opinions. Much as the explanation was made, the effects were barely satisfactory. Later on, "microblog users can't send pictures", "Zhenhai" and "PX" were listed as sensitive words in Ningbo, which all the more stimulated the public aversion because the government's "aggressiveness" and "arrogance" catalyzed the incident. So far the obsession of hierarch is still deep-rooted in Chinese people's mind and the official oriented thinking can only weaken the government's service functions in the face of online public opinions and make the public confused about the principal and subordinate status in public management.

1.2 *The top-down public policy execution mode becomes a hidden concern of emergencies*

The top-down public policy execution mode grants certain autonomous administration power to local governments and officers at the basic level, but usually it evolves into bureaucratism and makes the policy makers, decision makers and performers stand high above the masses, thus causing damages to people's benefits and rights. Quite a few local governments have been caught in the twirl of public opinions due to errors in the policies they make. In recent five years, several PX incidents have occurred. However, some local governments failed to learn lessons from them and make more prudent decisions. Worse, they even hid more things from the public and intended to conduct activities secretly. The internal orders of government departments and obedience were simply applied to the relation between governments and the public, which reduced the recipients of policies to a passive status. It not only led to the deviation of policies from public interest, but also went against the philosophy of service-oriented government and citizens first as well as infringing upon citizens' interest and laying hidden dangers for the occurrence of emergencies.

1.3 *Imperfect information disclosure mechanism damages the government credibility*

As citizens' political participation awareness is getting stronger, information disclosure has become an integral part of modern social and political civilization. Whether a government is transparent or not and whether information is open or not is an important standard for the public to measure a government's credibility. Establishing an information interaction channel with good communication can not only placate the public feelings and clarify facts but also build the government credibility. In the PX incident, Ningbo media maintained silent. The Ningbo official microblog "Ningbo Announcement" and Zhenhai official microblog "Zhenhai Government" seldom replied to hundreds of thousands of replies from the internet users. The public which was unaware of the truth had to resort to the internet for the incident progress so public opinions and rumors prevailed the microblog. Although the government clarified the fact through microblog, many people lost their calmness under the flow. At that time, it was pointless to talk about the truth. The public dissatisfaction with the local government not only damaged the government credibility, but also severely affected the government image.

2 SOLUTIONS FOR LOCAL GOVERNMENTS TO ENHANCE THEIR ABILITY TO RESPOND TO PUBLIC OPINIONS AND STRENGTHEN SOCIAL MANAGEMENT

2.1 *Expanding social management fields and realizing diversified governance*

Simulated online social management has turned into a significant part of social management innovation. Concerning the development trend of online public incidents in recent years, online public opinions are no longer purely online behaviors; instead, they are reflections of netizens' realistic intentions and behaviors on the internet and influences the real society. Netizens are the subjects of online public opinions which integrate both the simulated and realistic management. In the process of management, local governments should strengthen double management and construction of the simulated online society and the real society, build a diversified governance system and carry out consistent laws and regulations. It should expand social management fields, change the past traditional way of relying on administrative means and control only, and build benign interaction between self-management and restraint of society and the public and government administrative governance. Besides, it should combine government administrative governance and social services, materialize orderly and lively diversified governance, and build a socialist market economy, democratic politics, advanced culture and a social management system with Chinese characteristics that is in line with the requirements of harmonious society.

2.2 *Optimizing social management mechanism and enhancing ability to pacify public opinions*

Different from traditional media, microblog can promote public opinions to heat up soon, increase social unstable factors and disturb government work based on the WEB 2.0 interactive communication. It appears especially important for local governments to consider how to optimize social government, innovate online public opinion guide mechanism and enhance their ability to resolve online public opinions. Hence, the following measures shall be taken: proactively respond to online public opinions and regard the internet as an important channel of listening to the public and understanding the public will; open a political affairs microblog and build it into a platform for listening to the public opinions and dialoguing on an equal basis; pay great attention to and assess the influences of public opinions, actively reply to social opinions and realize "systemized communication" and "normalized interaction" between the government and netizens; set up and improve the news release mechanism, public opinion monitoring mechanism, full media communication mechanism, crisis coping mechanism and public opinion guide mechanism, etc., and explore the inherent rules; give scope to the role of the internet as the "opinion leader" and direct self-education of netizens.

2.3 *Making good use of the subjects of social management and establishing an equal dialogue mechanism to ensure orderly political participation*

With netizens' rising enthusiasm about participation in political affairs, the governments shouldn't ignore and become "laissez-faire", let alone control, suppress and "crack down" in a rude way with regard to the issue of netizens' participation. Rather, they should start with the concept of building a service-oriented government and work out a benign interaction means between government decision making and netizens' participation in political affairs. Regards to social management innovation, local governments should take the initiative to build the social atmosphere of political participation, put up the access to political participation, provide related laws and systems guarantee for online participation, expand netizens' paths of participating in social management, and intensify the organizational degree of netizens' political participation.

2.4 *Changing social management concept and making greater efforts in information disclosure*

Today, the internet has become the distributing center of ideologies and cultural information and the amplifier of social public opinions. Greater government information disclosure and transparency through the social influences of the emerging medium is the benchmark for government governing ability and the most immediate demonstration of a service-oriented government as well. Citizens' increasingly strong awareness of political participation is the inevitable outcome of the on-going development of the internet and information disclosure is a critical standard for them to measure a governments' credibility, which poses greater requirements for building "transparent governments" and government affairs disclosure. Local governments need to change their old social management concepts, respect citizens' right to know, participation right and right of supervision, and internalize it into their consciousness. The old-fashioned thought of "Don't wash one's dirty linen in public" should be changed. Only information disclosure can maintain the government image, shape their credibility in the public and help with government governing and lasting political stability of the society.

2.5 *Optimizing social management environment and building a new system of online participation in government and political affairs*

The appearance of online participation in government and political affairs has expanded citizens' means of political participation and enabled them to gain opportunities and access to direct dialogues with the governments. Local governments need to optimize social management environment, normalize, systemize and standardize the online political participation, build a democratic and equal dialogue platform between the government and the public, solve social contradictions from the source, construct a simulated online social management model and form benign interaction between the government and the public. Rather than regarding the netizens' will as "a great disaster", it should internalize it as favorable governing thinking and actions to obtain the public understanding and support. In addition, it should make the most use of the internet to guide orderly political participation of the public, let the internet become a strong force that drives social governance, find out a new path of online political participation to address social contradictions and problems, develop it into an effective carrier to resolve all kinds of social contradictions and problems in the era of microblog, maintain social harmony and stability, and build favorable social environment for economic development.

1 Data source: The 33rd "Statistical Report on Internet Development in China"
2 Ni Lin. "Study on the Spreading Characteristics and Influences of Microblog". Journal of Shanghai Business School, 2011, No. 2.

ACKNOWLEDGEMENTS

The corresponding author of this paper is Wang Yi Qun. This paper is supported by (Analysis of Zhejiang social online public opinion and public events under the New Media Environment) (Project number: 1320131027).

REFERENCES

[1] Cao Jinsong. Network in Politics and the Social Management Practice Innovation [J]. Social Sciences in Nanjing, 2011(4):97–103. (in Chinese).
[2] Zhu Sibei. Public Opinion Crisis in Emergencies and Exploration of the Response Mechanism [J]. Press Circles, 2011(2):47–49. (in Chinese).
[3] Zhu Sibei. Public Opinion Crisis in Emergencies and Exploration of the Response Mechanism [J]. Press Circles, 2011(2):47–49. (in Chinese).
[4] Yan Shushan. Analysis and Thinking of Local Governments' Ability to Respond to Public Opinions [J]. Guide of Sci-tech Magazine, 2010(14):160. (in Chinese).

Management, Information and Educational Engineering – Liu, Sung & Yao (Eds)
© 2015 Taylor & Francis Group, London, ISBN: 978-1-138-02728-2

Study on the association of different sources of dietary fiber and colorectal cancer

Dong Jie Wang
Langfang Health Vocational College, Langfang, Hebei, China

Yan Rong Liu
Langfang City People's Hospital, Langfang, Hebei, China

Li Bin Yang
Langfang Health Vocational College, Langfang, Hebei, China

ABSTRACT: This paper reviews various different sources of dietary fiber characteristics, sums up the physiological functions of dietary fiber and analyzes different effects of dietary fiber on colorectal and the forming relationship of dietary fiber separate groups to colorectal cancer. The results showed that dietary fiber has a protective role in colorectal cancer and closely related to intake and type.

KEYWORDS: Dietary fiber; Colorectal cancer; Association study.

1 INTRODUCTION

Dietary fibers are favorable at the forefront of food and nutrition science, which are mostly found in plant foods and generally divided into non-water-insoluble fiber and soluble fiber and cannot be directly decomposed by the body's digestive enzymes and attributed to non-nutritive carbohydrate compounds. After the decades of development, the relationship between dietary fiber and cancer has become a hot research in nutrition and disease prevention field.

2 CHARACTERISTICS OF A VARIETY OF DIFFERENT SOURCES OF DIETARY FIBER

Dietary fibers are food nutrients that are generally difficult to be digested, and adequate intake of fiber can also prevent cardiovascular disease, cancer, diabetes and other diseases. Fiber can clean the digestive wall and enhance digestion. At the same time, fiber foods can dilute and accelerate the removal of carcinogens and toxic substances to protect the fragile digestive tract and prevent colon cancer.

3 PHYSIOLOGICAL FUNCTION OF DIETARY FIBER

3.1 *Dietary fiber and constipation*

Dietary fiber can make intestinal peristalsis and promote substances mobile and is convenient to discharge waste. Using water absorption of dietary fiber wet tract and promote the formation of defecate to cure constipation. Niu Guangcai use 2.5, 5 and 10g / (kg.d) sand marc constipation dietary fiber feed mice. Tests showed that high-dose group and the control group have significant difference and meet statistical significance.[1] Liu Dan and Yuan Yaozong use wheat cellulose particles (7g / d) treat constipation patients. Tests showed that patients with constipation significantly improve and improve stool quality and defecation comfort. Gu Qing and Jiang Guohong do dietary fiber intervention experiment for constipation patients by use of placebo, the results show that dietary fiber does have a function to improve constipation situation.

3.2 *Adsorption of dietary fiber*

Dietary fiber has water absorption advantages and can dilute the concentration of substances in the intestine and further reduce the damage of harmful substances to rectum. Bile acids are substances produced by metabolism of human cells, which is one of the colorectal cancer risk factors. Li Haiyun through vitro simulation experiments illustrates that dietary fiber can adsorb bile acids and bile salts.[2] Nitrite ions, tertiary amines and secondary amines react chemically to produce a carcinogenic nitrosamines.

3.3 *Dietary fiber promotes the growth of probiotics in the gut*

Intake of dietary fiber can promote the growth of probiotics in the gut. It is well-known that probiotic is beneficial to make the body to maintain intestinal health and play a significant role in the prevention and treatment of intestinal diseases. [3] A large number of domestic and foreign scholars did verification on this incident. Scholars like Azuma use Balb / c do the intestinal experiment for mice. Results show that small doses of dietary fiber can promote normal growth and the reproduction of the intestinal flora, which maintained intestinal health balance.

4 RELATIONSHIP BETWEEN DIETARY FIBER AND COLORECTAL CANCER

4.1 *Impact of dietary fiber on colorectal*

In the research front, a large number of experiments test that not all of the dietary fiber can change the carcinogens role of intestinal bile acid. Dietary fiber has a protective effect on the rectum and closely related to the intake, types and sources of dietary fiber; non-water-soluble dietary fiber has more significant advantages than soluble dietary fiber in terms of protective effect. Vegetables dietary fiber is an important material to reduce the incidence of colorectal cancer, however, cereals, fruits decline in dietary fiber protection.

4.2 *Relationship of each separate group of dietary fiber and formation of colorectal cancer*

1 Cellulose. A large number of studies have found that cellulose can have a certain effect on cancer prevention.
2 Pectin. On the contrary of the effect of cellulose, pectin has the effect of promoting the formation of canker. But by appropriately modifying essence, pectin also has anti-cancer effect. Pienta and Inohara successfully made citrus pectin that has anti-cancer effects by modifying its essence. The method is that first put citrus pectin in a high pH and then process it in the low pH, which make pectin to have the role of preventing cancer. Mechanism of pectin is because of its galacturonic acid strand at a high pH and can obtain the non-branched side chains at low pH, thereby change the properties of pectin and have effect with the cells after going into the blood.
3 Hemicellulose. The component of this substance is relative complex and there are few studies of the individual components. Currently, hemicellulose has anti-cancer effect on research results.
4 Lignin and keratinocytes. A large amount of experiment found that lignin and keratinocytes can enhance the production of cancer.

5 CONCLUSION

Dietary fiber is an essential human nutrient, whose intake sources are diversification, inexpensive and have broad application prospects. Dietary fiber is the current hot study field in nutrition science and dietary fiber is researched by detail at home and abroad and makes a lot of research results.[4]However, the research of these scholars largely is based on the extraction and detection and only for some single properties and physiological functions of the body needs and lack of in-depth study of the nature of things and things modification. In summary, for human health, studies of dietary fiber need a deeper study to meet the needs of social development.

Note: This article is for 2014, Langfang Technical Support Programs Issue, Item Number 2014013052)

REFERENCES

[1] Niu Guangcai. *Animal Experiment Study on Sand Marc Dietary Fiber Sand Pomace Laxative Effect* [J] Food Science, 2011, 32 (12): 293–296.

[2] Li Haiyun, Wang Xiuli. *Study on Lychee Shell Water-insoluble Dietary Fiber Adsorption of NO2- and Sodium Cholate*[J]. Food Research and Development, 2006, 27 (8): 167–170.

[3] Jiang Ye, Liu Jun, Ren Hongyu. *Probiotics and Gastro-intestinal Diseases* [J] . World Journal of Gastroenterology, 2011,19 (17): 813–1818.

[4] Li Yihe. *Dietary Fiber Research Status and Development Trend* [J] Modern Agricultural Science and Technology, 2010 (6): 349–350.

Management, Information and Educational Engineering – Liu, Sung & Yao (Eds)
© 2015 Taylor & Francis Group, London, ISBN: 978-1-138-02728-2

Research on the national fitness and sports key technology

Chang Wu Sun, Jing Chai & Meng Meng Lei
Hebei Finance University, Hebei, Baoding, China

ABSTRACT: The government emphasized the development of sport. Our country's athletics level is also constantly upgrading. From giving full play to public service functions of sports perspective, two key problems focused on the current development of sports dispute. The first is fitness, the second is the issue of priority between competitive sports, in the view of the system, the necessary connection between exposing them to exist between the delicate relations. Sports will be an indispensable factor for national construction and development; countries needs to develop, society needs to progress, and the national health body is indestructibly backing, and sports can effectively inspire the national spirit and enhance national unity. But some differences exist between the national fitness and sports competition. In today's global economic integration, competitive sports will face serious challenges in the future. Therefore, the deepening sports system is the inevitable result of the development of competitive sports. Actively promoting the national fitness is to carry forward the glorious tradition of sports culture, but also fully embodies the sports Huimin, people's important content.

KEYWORDS: National fitness Sports Development Suggestions.

1 INTRODUCTION

Since reform and opening up, China's overall competitive strength enhanced obviously, and the levels also significantly increased. In consecutive four Olympic Games, china ranks the first from the fourth in the gold medal list. The results obtained show that our country implements the system of competitive sports, which has achieved the obvious effect in a short product of the planned economy period. With the in-depth development of competitive sports in the period of market economy, our country implements the system of competitive sports, which has exposed many unfavorable to our country sports enterprise development. In the implementation of the National System of reality contradictions, it is necessary to deeply study the impact of sports enterprise that our country implements the system of competitive sports and social sports development imbalances, coordinated development in our country sports enterprise road reflect contradictions increasingly obvious. Only to realize the perfect unity of the national fitness and competitive sports, to promote the great progress and development of sports enterprise in our country.

2 THE CURRENT SITUATION OF THE DEVELOPMENT OF NATIONAL FITNESS

In 2008, the success of the Beijing Olympic Games will be China's sports undertakings onto a new level. The country set up a "National Fitness Day", this is a new symbol of China's national fitness. In addition, on October 1, 2009, China has promulgated and implemented the "National Fitness Regulations". This initiative also highlights the confidence and determination of our government to the development of fitness. In addition, our country frequently raised a hot wave in the construction of a large stadium. The 11th national games held in 2009, the 16th Asian games held in 2010, in 2011 the 26th world university games, these games develop the momentum of the construction of sports venues. Thus, China's national fitness causes a benign development momentum.

3 THE DEVELOPMENT TREND OF COMPETITIVE SPORTS

From the founding of China, our country sports enterprise develops the soaring, up to now, our country take part in various sports competitions of the world, there are more than one sports formed gold monopoly situation. In 1984, Xu Haifeng got the win, refresh the history of the world sports meeting Chinese zero gold record. Games held in London in 2012, the Chinese team once again created the success that has won the gold medal 38, silver 27, bronze medals 23, the total number of medals ranked second in the world. These are sufficient to prove the Chinese athletic career has developed to the full swing of the situation.

4 THE CONTRADICTION AND REASONS BETWEEN SPORT AND NATIONAL FITNESS DEVELOPMENT

4.1 *The contradiction between the national fitness and athletic sports*

As sports enterprise growing, our country paid more and more attention to the development of sports. Compared with competitive sports, the status of national fitness is relatively weak, National fitness still didn't get the relevant state department, the urban and rural limited fitness venues and facilities and the increasingly popular fitness team contradiction increasingly sharp, high emotional fitness severely impaired people.

4.2 *The "uncertain" behind the national fitness and athletic sports contradictions*

What is the reason lead to sharp contradictions between them. Investigate its reason, mainly for the following aspects:

First of all, the cultivation of the professional sports team lack of sound mechanism, to some extent hindered the development of the national fitness sports. China's athletics is a grassroot of professional teams from all layers of the delete selection, and then through a long-term professional training, so that these sports gradually become professional sports "patent movement", while others can only serve as everyday life fitness auxiliary movement. The lack of enthusiasm for the athletes present, China has become a common problem in major sports school, so long to develop, is bound to affect the development of China's sports undertakings.

Secondly, the hysteresis seriously hindered the development of sports enterprise in our country. In order to improve competitive sports, nation blindly spends enormous human and material resources, which makes this even more fitness crowd dissatisfaction, but also greatly dampened the enthusiasm of fitness and enthusiasm. This also contributed to fitness and competitive sports contradictory breeding.

5 HOW TO USE SPORT TO ACTIVATE THE NATIONAL FITNESS

Since joining the WTO, Sports in China has attracted more and more people are concerned, the status of competitive sports has been improving continuously. But at the crucial moment of developing sports, to learn from each other, the only way of developing China's sports is further undertakings should go further.

5.3 *Perfecting the allocation of resources for the development of competitive sports*

In order to make valid activation of athletic fitness, the most effective way is to change the traditional single sports mode in a gradually diversified direction. Do not blindly chase the development of competitive sports. The development of social sports should also be given appropriate support and encouragement. Talent and professional skills training social sports college sports talents professional skills-based, effective promotion of social investment and the perfect combination of sports and competitive sports.

5.4 *Effectiveness to promote the unification of the competitive sports and the national fitness*

The success of the Beijing Olympic Games pointed out the direction for the development of sports enterprise in our country. In recent years, our country blindly pursues sports record and ignores the development of the national fitness, which makes the national fitness and sports development of serious imbalance. Our country to realize leap from sports to sports power, we must encourage and national fitness should get balanced development of competitive sports. Under the background of market economy, the development of undertakings of physical culture and sports should rely on market macroeconomic regulation and control, more sports, hand in hand with the market economy invisible visible hand, so as to realize the win-win of the market economy and the sports enterprise.

5.5 *Mining new business opportunities, the national fitness boost the development of the sports economy*

In full swing of the national fitness, the national fitness has unconsciously by sports fitness development as the main symbol of the lifestyle, the gym, all kinds of gym has gradually become the main content of people's life. Sports industry privatization has become the main trend of social development, private sports industry firmly grasp the pulse of the development of the national fitness market, actively take slightly wrong, catering to the development of the market. In this respect, the sports of private industry in the development of the sports economy play an important role. However, compared with developed countries, the development of the sports market of our country is far from enough, still need to dig deep, with diversified sports private industry to fill the blank of the existence of national fitness.

6 CONCLUSION

Sport and the national fitness are the important parts of our country sports enterprise. There are links but different. The national fitness is the basis of the implementation of sports in our country power, and competitive sports are our country sports power booster. Without the development of the national fitness, it is impossible to have the rise of competitive sports, the two complement each other, therefore, only from the perspective of the national fitness boost the development of undertakings of physical culture and sports, to achieve the leap from sports country to sports power in our country.

ACKNOWLEDGEMENTS

The authors of this paper are Sun Changwu, male, Associate Professor, Master's degree, Research: Physical Education, Hebei Finance University; Chai Jing,female, Lecturer Master's degree, Research: Physical Education, Hebei Finance University.

This topic is the subject of Hebei Province Science and Technology Department. Item Number:13455710.

REFERENCES

[1] W.Q.Yu and Y.W.Xu. Self-organization theory perspective of non-olympic sports development path [J]. Journal of Wuhan sport university, 2014, 48(1): 24–28.

[2] X.X.Huang and K.Jiang. Physical education in colleges and universities to promote bayu wushu culture inheritance thinking [J]. Journal of southwest normal university (natural science edition) , 2014.

[3] F.Y. Li, J.M. Xing. The perspective of ecological civilization in colleges and universities sports for the promotion of college students' health [J]. Journal of Wuhan sports university, 2014, 48(1): 83–86.

[4] J.Wang, P.L. Xia. Based on the Web of Science international sports policy research focus in the visualization analysis [J]. Journal of Shenyang sports university, 2013 (1): 32–36.

[5] W.W. Liu. The traditional martial arts into the necessity and the way to our school system research [J]. Journal of Beijing sport university, 2013 (1): 97–101.

[6] Z.Q. Zheng, Z.M. Liu, H.S. Liu. China power sports image build path and difficulties [J]. Journal of Wuhan sport university, 2013, 46(12): 29–33.

Management, Information and Educational Engineering – Liu, Sung & Yao (Eds)
© 2015 Taylor & Francis Group, London, ISBN: 978-1-138-02728-2

Part biological characteristics study on various diameter seeds in Chinese cabbage

Rong Zou & Man Lian Wang
Guangxi Institute of Botany, Guangxi, Zhuangzu Autonomous Region, China
Chinese Academy of Sciences, China

ABSTRACT: Vegetable seed coating method is one of the important measures to improve the quality of seed and commercialization. Meanwhile, the selection of a seed is an important prerequisite for improving the coating effect. This test mainly focuses on researching the proportion of different diameter class seeds, 1000-seed weigh, indoor germination rate. The aim of this paper is to provide theoretical evidence for the selection of seeds. This paper is divided into three parts. Part I introduces the background of writing this paper and the materials and methods needed during testing. The second part is the core content of this paper; focusing on analyzing the experimental results. The last part combs the conclusions reached in this paper.

KEYWORDS: diameter, Jincai 3, biological characteristics.

1 INTRODUCTION

The origin place of cabbage is located in the Eastern Mediterranean and China. After 1970s, the rapid expansion of the cultivation area occurred in North China. The cultivation area and consumption amount are the most among all kinds of vegetables. With China's increasing demand for cabbage, improving cabbage's output became a common concern. As we all know, to guarantee a high yield of Chinese cabbage selecting good seed is an essential factor, naturally, how to choose the good seed is the focus of our research.

2 THE TEST METHODS OF DIFFERENT DIAMETER CLASS PROPORTION

Weigh 10g seed with scales 1, divide seed into different diameter classes. A diameter class: ≥10 mm; B diameter class: 6< diameter class <10 mm; C diameter class: 3< diameter class <6 mm; D diameter class: diameter class <3.0 mm.

3 THE TEST METHODS OF 1000-SEED WEIGH OF DIFFERENT DIAMETER CLASS SEED

The A, B, C, D different diameter class seeds were put into paper bags, placed in dark storage at room temperature. Each time test, with coning and quartering method split out the needed seeds according to the requirements of diameter class. Take 8 seeds randomly, 100 capsules per serving, put into 8 paper bags, weigh with scales 2 respectively, measure the quality of different diameter class seed then converted into 1000-seeds quality, 4 samples used for emergence test.

4 THE TESTING RESULTS

Table 1. The proportion of different diameter class in Taiyuan Erqing.

Class	Seed diameter/ mm	2010		2011		2012		2013	
		quality/g	percent/%	quality/g	percent/%	quality/g	percent/%	quality/g	percent/%
A	≥10	1.1	13	1.8	18	0.9	9	0.9	10
B	≥5<10	7.6	74	6.5	65	6.3	63	7.1	70
C	≥3<5	1.2	10	1.5	15	2.4	24	1.9	18
D	<3	0.4	3	0.2	2	0.4	4	0.1	2

5 THE PROPORTION OF DIFFERENT DIAMETER CLASS

From Table 1, we can see that the proportion of a diameter class is 10%-17% in Taiyuan Erqing, the proportion of B diameter class is 61%-72%, and the proportion of C diameter class is 12%-27%, the proportion of D diameter class is 2%-6%. Though the particle size ratio of Taiyuan Erqing is different, the proportion of B diameter class was the highest in Taiyuan Erqing seed, the proportion of A, B, C diameter class is accounted for 94%-99%.

From Table 2, we can see that the proportion of A diameter class is 4%-12% in Jincai 3, the proportion of B diameter class is 44%-56%, the proportion of C diameter class is 5%-11%, the proportion of D diameter class is5%-11%. The proportion ratio change taint of Jincai 3 is different from Taiyuan Erqing, though the proportion of B diameter class is the highest in Taiyuan Erqing seed, and the ratio is relatively small compared with C diameter class seed, the proportion of A, B, C diameter class is accounted for 89%-95%.

6 1000-SEED WEIGH OF DIFFERENT DIAMETER SEED

From table 3 we can see that 1 000- seed weigh of A class is 3.646-4.162 g in Taiyuan Erqing, 1 000- seed weigh of B class is 2.917-3.331 g, 1 000- seed weigh of A class is 2.098 -2.440 g, 1 000- seed weigh of A class is 1.393-1.534 g.

From Table 4, we can see that 1000-seed weigh of A class is 3.674-3.911 g in Jincai 3, 1000-seed weigh of A class is 2.780-3.043 g, 1000- seed weigh of A class is 1.987 -2.146 g, 1000- seed weigh of A class is 1.987 -2.146 g. This showed that it is reduced with the decrease of seed diameter, The average 1 000-seed weight ratio of A, B, C classes was 2.7, 2.1-2.2 and 1.5 times than D class. In the same diameter class, 1000-seed weigh of Jincai 3 slightly smaller than that of Taiyuan Erqing.

Table 2. The proportion of different diameter class in Jincai 3.

Class	Seeds diameter/ mm	2010		2011		2012		2013	
		quality/g	percent/%	quality/g	percent/%	quality/g	percent/%	quality/g	percent/%
A	≥10	0.4	6	1.2	12	0.4	6	0.6	6
B	≥5<10	4.3	43	4.5	45	5.2	50	5.6	53
C	≥3<5	4.2	42	3.5	35	3.8	36	3.3	36
D	<3	1.2	9	0.8	8	0.6	8	0.5	5

Table 3. 1000-seed weigh of different diameter seed in Taiyuan Erqing.

Seed diameter class	2010	2011	2012	2013	Average
A	4.221	3.912	3.231	4.121	3.121
B	3.312	3.323	2.121	3.123	3.412
C	2.432	2.211	2.213	2.312	2.646
D	1.512	1.423	1.213	1.141	1.978

Table 4. 1000-seed weigh of different diameter seed in Jincai 3.

Seed diameter class	2010	2011	2012	2013	Average
A	3.841	3.711	3.674	3.911	3.712
B	3.042	2.782	2.901	3.029	2.932
C	2.143	1.911	2.033	2.139	2.073
D	1.511	1.334	1.350	1.412	1.407

Table 5. Indoor germination energy and seed germination percentage of different diameter class in Taiyuan Erqing.

Seed diameter class	2010		2011		2012		2013	
	Germination energy	Germination percentage	Germination energy	Germination percentage	Germination energy	Germination percentage	Germination energy	Germination percentage
A	84.8	86.3	83.0	83.0	85.8	85.8	84.5	88.8
B	88.0	89.3	94.8	94.8	93.3	93.3	90.0	93.8
C	86.3	88.6	81..5	81.5	88.8	88.8	83.3	88.8
D	72.8	74.3	71.3	71.3	88.5	88.5	77.1	83.3

Table 6. Indoor germination energy and seed germination percentage of different diameter class in Jincai 3.

Seed diameter class	2010		2011		2012		2013	
	Germination energy	Germination percentage	Germination energy	Germination percentage	Germination energy	Germination percentage	Germination energy	Germination percentage
A	87.6	89.2	80.3	81.9	85.2	87.2	84.1	89.5
B	85.3	86.1	90.2	93.2	89.1	92.4	87.2	92.2
C	85.2	86.4	88.1	90.2	83.2	86.4	81.3	85.1
D	79.5	80.7	70.6	73.1	75.5	78.3	72.4	76.2

7 INDOOR GERMINATION ENERGY AND SEED GERMINATION PERCENTAGE

From Table 5, we can see that the germination energy (88.0%-94.8%) and seed germination percentage (89.3%-93.8%) of B diameter class is the highest in Taiyuan Erqing from year to year. The germination energy (88.0%-94.8%) and seed germination percentage (89.3%-93.8%) of D diameter class is the lowest. The germination energy and seed germination percentage of the C diameter class and A diameter class are located between B and D.

From Table 6, we can see that the germination energy (85.0%-90.3%) and seed germination percentage (86.5%-93.8%) of B diameter class is the highest in Taiyuan Erqing. The germination energy (70.5%-79.1%) and seed germination percentage (73.0%-80.6%) of D diameter class is the lowest. The germination energy and seed germination percentage of C diameter class and A diameter class ar located between B and D.

8 CONCLUSIONS

The effects of climate and cultivation management lead to the proportion changes of different diameter. The proportion of B diameter class was the highest in Taiyuan Erqing seed; the proportion of A, B, C diameter class is accounted for 96%-99%. The proportion of B diameter class was the highest in Jincai 3, the proportion of A, B, C diameter class is accounted for 89%-95%.

The average 1000-seed weight ratio of A, B, C classes was 2.7, 2.1-2.2 and 1.5 times than D class. In the same diameter class, 1000-seed weigh of Jincai 3 slightly smaller than that of Taiyuan Erqing.

The germination energy (85.0%-90.3%) and seed germination percentage (86.5%-93.8%) of B diameter class is the highest in Taiyuan Erqing. The germination energy (70.5%-79.1%) and seed germination percentage (73.0%-80.6%) of D diameter class is the lowest. The germination energy and seed germination percentage of C diameter class and A diameter class are located between B and D.

ACKNOWLEDGEMENT

It is a project supported by the funded scientific project of Guangxi (11107010-2-3). The corresponding author is Jiang Yun-sheng.

REFERENCES

[1] Fan Shuangxi. Research status and application of vegetable seed processing[J]. Hunan agricultural science, 1994, 10 (4):34–36.

[2] Wu Zhixing. The Encyclopedia of vegetable seeds. Nanjing: Jiangsu science and technology publishing house, 1993:270–325.

[3] Agricultural Sciences Institute of vegetable of China. Vegetable cultivation in China.Beijign: Chinese Agricultural Science Bulletin, 1987:389–408.

[4] Zhang Wenfeng, Yue Yuqin. The process of vegetable seed before sowing[J]. North Garden, 1993(1), 37–39.

Management, Information and Educational Engineering – Liu, Sung & Yao (Eds)
 ISBN: 978-1-138-02728-2

Brainwave analysis of positive and negative emotions

Fu Chien Kao, Shin Ping R. Wang, Chih Hsun Huang, Chih Chia Chen & Yun Kai Lin
Department of Computer Science & Information Engineering, Da-Yeh University, Taiwan

ABSTRACT: Emotion is the generic term for various subjective cognitive experiences, and a psychologically and physiologically synthesized state generating under a variety of perceptions, thoughts, and behaviors. In general, emotion can be categorized into Joyful, Angry, Protected, Sad, Surprised, Fear, Satisfied and Unconcerned; eight types of positive-negative emotions. This paper from the perspective of cognitive neuroscience investigates the difference of human brainwave of positive and negative emotions (i.e. Joyful and Angry emotions). The experiment uses acoustic stimuli to stimulate the positive and negative emotions of the test subjects and uses Electroencephalogram (EEG) to extract test subjects' frontal lobe brainwave. The extracted brainwave is further transformed into a frequency domain signal where sub-band energy is calculated, characterized, and finally digitally encoded for analysis.

KEYWORDS: Brainwave, Cognitive Neuroscience, Emotion.

1 INTRODUCTION

Emotions can be classified into innate "basic emotions" and "complex emotions" which are acquired through learning. Basic emotions are innate and closely related to human survival instinct. In contrast, complex emotions have to be learned through human interaction and hence each individual owns a different number of complex emotions and has a different definition of them. Emotion has been described as an abrupt response to the internal or external important events and a person always takes the same response to the same event. Emotion lasts for a very short duration. It collaborates with the actions of language, physiological, behavioral and neural mechanism [1]. Human emotions are also derived from biological functionalities/survival instinct and strengthened through evolution. It provides simple solutions to frequent problems that early human had to confront, such as fear causes evasion [2-3].

Emotions are both a subjective experience and an objective physiological response. It has its own purpose and also a social expression. The five basic elements are cognitive assessment, physical reactions, feelings, tendencies, expression, and action [4-5]. Beside the subject who is experiencing emotional shift, bystanders can also learn subject's emotional shift through observation. However, are there any other ways to learn people's emotional shift beside observation and interaction? Is there an effective scientific approach to identify the inner emotional shift of people? This study identifies positive and negative psychological emotions using brainwave variation. It focuses on analyzing two types of positive and negative emotions: Joyful, Angry. Figure 1 depicts the relation between the above emotions in three dimensions. The figure expands the emotions into three-dimensional space using three orthogonal axes: the positive or negative, the strength, and the transformation of the emotion.

This research, based on the cognitive neuroscience, uses brainwave sensor to extract the brainwave signal of the test subjects while they are performing the induction of emotions. The extracted measurements are further analyzed, compiled statistics for its distribution over the brainwave characteristic frequency bands, and finally the characteristic frequency bands of emotional brainwaves are digitally encoded to come up with a metric for human emotion identification.

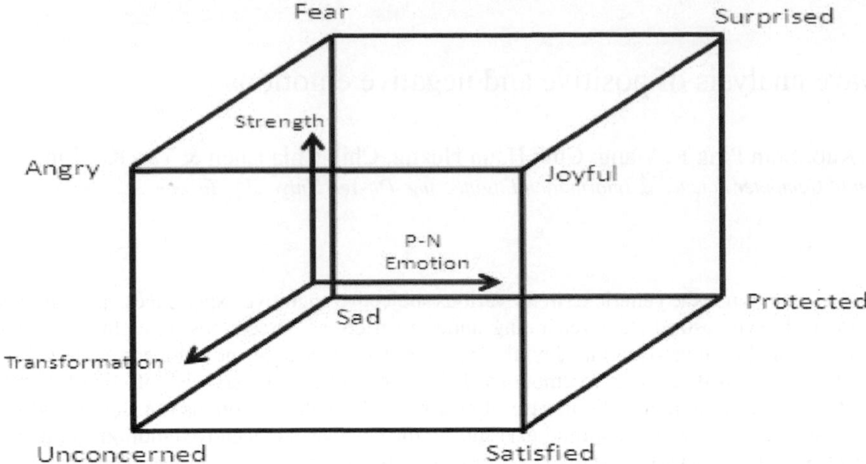

Figure 1. 3D Emotions graph.

2 DESIGN OF EEG SENSOR MODULE

The adopted relevant brainwave EEG functional block diagram and sensor module are shown as in Figs. 2 and 3 [6]. The brainwave sensor proposed in this research is not only small in size, convenient to carry and easy to operate, but also is low in price, and is applicable to being used in various industries in the future compared to the medical grade electroencephalograph.

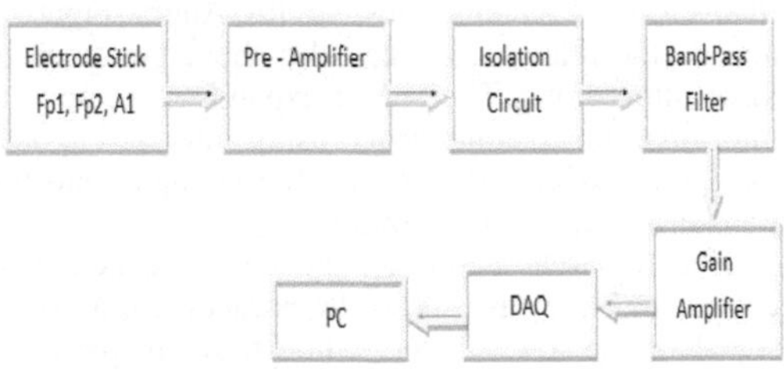

Figure 2. EEG functional block diagram.

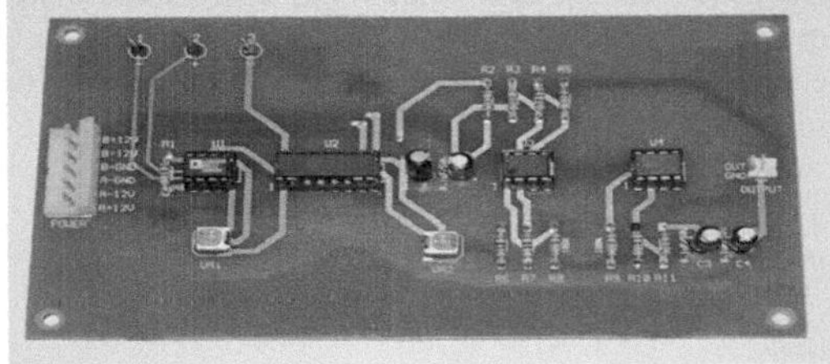

Figure 3. EEG sensor module.

3 SYSTEM IMPLEMENTATION AND STATISTIC ANALYSIS

This research follows the perspective of cognitive neuroscience, develops characteristic frequency bands of brainwave for identifying positive and negative emotions of the human brain. The research experiment accomplishes this by extracting test subjects' brainwave under different emotional response. The extracted measurements are then analyzed and compiled statistics for its distribution over the brainwave characteristic frequency bands with respect to different emotional brainwaves, and finally the digital codes of the emotional brainwaves are encoded. The relevant framework for characteristics analysis of the emotional brainwave frequency bands are described as follows:

1 After installation of a brainwave sensor, the electrode patches are attached to the participants and then the acquisition program is used to capture emotional brainwave signals.
2 After being converted by the ADC module of the sensor, the emotional brainwave signals are sent to PC and saved as Excel or .txt format through the USB port.
3 The emotional EEG analysis GUI provides brainwave analysis for the data in the format of Excel or .txt. The time-domain part of the GUI provides the strength change in time for the original emotional brainwave signals. The emotional brainwave signals are then processed by Fast Fourier Transform (FFT) formally. In this research, the percentages of amplitudes of the sectional brainwave frequency band are used to calculate the sectional brainwave energy.

4 Statistics and analysis of the corresponding characteristic frequency bands energy.
5 Difference comparison and digital encoding of brainwave characteristic frequency bands.

4 ANALYSIS OF EMOTIONAL BRAINWAVE CHARACTERISTIC BANDS

This experiment conducted in a coherent environment, applies a set of acoustic emotional stimuli to stimulate an emotional shift and measures brainwave variation with respect to different stimuli of the test subject. The acoustic stimuli are for the medical experiment purpose and have to be registered for downloading. It contains 2 types of stimuli of emotions: Joyful, Angry [7]. It takes three different sounds to stimulate each emotion shift for the required brainwave measurement. Before the experiment started, 15 test subjects are set to listen three different stimuli, then subjects pick a stimulus that best fit his/her current emotion and uses the stimuli to perform necessary test stimulation. In the experiment, a sound is played three times consecutively for each emotion. It takes approximately 100 seconds for each emotional test. Before the test, the subject takes a 10-second break to calm down. The test starts by playing a 20 seconds of acoustic stimuli and having the brainwave of the subject recorded at the same time, it then immediately followed by a 10-second break. The above sequel is repeated for three times and a total of 100 seconds for the complete test.

5 ANALYSIS METHOD OF EMOTIONAL BRAINWAVES

The relevant emotional brainwave energy calculation method is described as follows. The average of the total potential amplitude of different frequency bands for 15 participants is calculated so as to obtain the energy of the zone frequency band and the total energy using Eqs. (1) and (2). In the above equations, B is the zone frequency bands, f is the start frequency of each frequency band, n is the end frequency of each frequency band (the frequency sampling interval is 0.01Hz), and E is energy of each frequency band. E_T is the total energy of the four zone frequency bands from 0.2Hz to 25Hz. The energy percentage of α, β, θ and δ is respectively (E_B/E_T) %. The energy percentage of the subzone frequency E_Δ is namely the percentage of the energy in the individual subzone and the energy in the total frequency band, as shown in Eq. (3) [8].

$$E_B = \sum_{f}^{n} Power_f \tag{1}$$

$$E_T = \sum_{f=0.2}^{25} Power_f \tag{2}$$

$$E_\Delta(\%) = \frac{E_\Delta}{E_T} \tag{3}$$

Based on the energy distribution of main brainwave characteristic frequency band by analyzing the subject's emotions characteristics, it establishes the digital encoding of emotional brainwaves. The average energy percentage of the frequency band of each zone is calculated and the relevant characteristic frequency bands are found out according to the energy level when the participants were tested for their emotional response. Table 1 shows distinguishable brainwave characteristic band encoding of positive and negative emotions (Joyful and Angry). In addition, the measured negative emotion (Angry) has a greater total brainwave energy than the positive emotion (Joyful).

Table 1. Difference comparison of joyful and angry emotions at α brainwave.

P-N Emotion Freq. Bands		Joyful emotion			Angry emotion		
		Energy (E_Δ%)	Energy	Digital code	Energy (E_Δ%)	Energy	Digital code
	8~9	3.76	0.21		3.55	0.26	
α	9~10	9.38	0.53		9.27	0.69	
Alpha	10~11	11.39	0.64	α(01100)	11.63	0.86	α(01110)
(8~13)	11~12	3.62	0.20		4.13	0.31	
	12~13	3.31	0.19		3.27	0.24	
Total Bands Energy		100.0	5.60		100.00	7.40	

6 CONCLUSIONS

This research forms the perspective of cognitive neuroscience, extracts and computes the emotional brainwave energy using a brainwave sensor. The emotional brainwave energy data are further analysis and characterize for different emotions. The experiment uses medical acoustic stimuli to stimulate brainwave responses of different types of positive and negative emotions. The experiment shows that negative emotion has a greater energy compared to the positive emotion. The research shows the processed brainwave characteristic band digital encoding technique can effectively identify brainwave of positive and negative emotions (Joyful and Angry).

ACKNOWLEDGEMENT

This research is supported by the grant NSC 102-2511-S-212-001-MY2 from the National Science Council of Taiwan.

REFERENCES

[1] Fox, Elaine. Emotion Science: An Integration of Cognitive and Neuroscientific Approaches. New York: Palgrave MacMillan. 2008: 16–17. ISBN 0230005179.

[2] J. C. Gaulin, and D. H. McBurney: Evolutionary Psychology. Prentice Hall. 2003. ISBN 978-0-13-111529-3, Chapter 6, p 121–142.

[3] Ekman, Paul: An argument for basic emotions. Cognition& Emotion.1992, **6**: 169–200.

[4] K. R. Scherer: What are emotions? And how can they be measured?, Social Science Information.2005, 44: 693–727.

[5] M. F. Mascolo, K. W. Fischer, & J. Li, (2003). Dynamic development of component system of emotions: Pride, shame, and guilt in China and the United States. In R. J. Davidson, K. R. Scherer, & H. H. Goldsmith (Eds.), *Handbook of affective sciences* (pp. 375–408). New York: Oxford University Press.

[6] F. C. Kao, J. H. Jhong: Analysis of Brainwave Characteristic Frequency Bands under Different Physiological Statuses, *INFORMATION-AN INTERNATIONAL INTERDISCIPLINARY JOURNAL*, Vol.16, No. 9(B), pp.7249–7259, October, 2013.

[7] "The International Affective Digitized Sounds (2nd Edition; IADS-2): Affective Ratings of Sounds and Instruction Manual," NIMH Center for the Study of Emotion and Attention, Gainesville.

[8] F.C. Kao, Y.K. Lin and C.C. Hung: Brainwave Analysis during Learning, *ADVANCED SCIENCE LETTERS*, Vo.19, No.2, pp. 439–443, 2013.

Management, Information and Educational Engineering – Liu, Sung & Yao (Eds)
© 2015 Taylor & Francis Group, London, ISBN: 978-1-138-02728-2

The school comprehensive management system based on SQL design and construction

Guo Xian Jiang
WeiFang university of Science and Technology, Shou Guang, China

ABSTRACT: Design school comprehensive management system, analyzes its demand and significance first. System requirements, include functionality, confidentiality, rationality, efficiency, accuracy and scalability. The profile of the system is designed on the basis of overall structural design and function module design and then, under the premise of following basic principles, such as design the interface style of the system, the integrated management system based on Client/Server (Client/Server) mode, uses JSP + JavaBean + Servlet programming to complete.

KEYWORDS: Integrated management system; Functional; Confidentiality; Demand analysis.

1 INTRODUCTION

Current with the rapid development of computer technology and network technology, various schools and units in each work information demand are higher and higher, to a unit of management work, in the process of informatization reform emerged many new problems and challenges. Traditional management work, no matter from the efficiency, cost, accuracy, and other points of view has a lot to improve, how to make use of information devices and the Internet promotion, has become an urgent problem.

2 SYSTEM RESEARCH PURPOSE AND MEANING

The arrival of network age, the school management way of object, the environment, such as major changes have taken place, the old management mechanism cannot effectively solve various problems arising from the management under the network condition, the innovation of school management is imperative.

Set up to adapt to the network, a flexible of suitable integrated management system based on B/S structure of the school, turn past single-user single operation for multiple users to participate in the network system, can give full play to the function of the campus network, make between different departments, different campuses, Shared data more convenient, data integrity and consistency of the strengthen, increase between intercollegiate, school management center with the functions of communication, to further improve the school informationization level and efficiency of the staff.

Setting up a comprehensive objective and scientific school management system, to strengthen the school teaching and administrative management, promote the development of students and the school, the development of society has important significance. Integrated management system involves the school administration office, scientific bureau, the personnel department and other departments, is geared to the needs of teachers, students and the administration of the work; With the comprehensive management of the school as the core; In order to improve the school work quality and efficiency as the goal; Can realize schools all aspects of the integration of resources, make the application of information technology from discrete departmental applications on the comprehensive level of the whole school and even higher levels of application.

3 SYSTEM REQUIREMENTS ANALYSIS

School management system module decomposition, the use of JSP + JavaBean + Servlet programming to complete the design. The system structure is as follows:

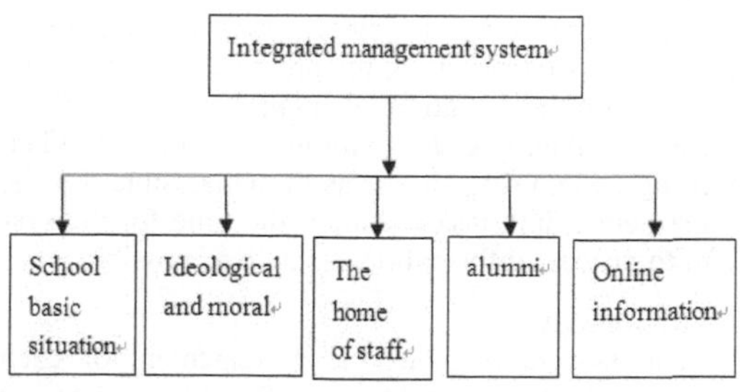

Figure 1. School management system structure.

3.1 *The functional requirements*

Integrated management system is mainly to complete the school comprehensive management and development, its main goal is for each department information, teaching resources, curriculum information software and hardware facilities, schools and related information for management and maintenance, lessen the pressure of school management, improve the efficiency of school management. Through the investigation into the business of school management, management can be achieved by the system management and the maintenance of information about the school. School teachers can use this system for each function department information, teaching resources, school of software and hardware facilities, such as information about the school department of entry and look at it. Students can utilize this system for school-related information and teaching materials to browse and view, and can with school administrators, teachers and classmates on the platform for communication. System's overall mission is to ensure the whole school information management systematization, standardization and automation.

3.2 *System requirements*

The integrated management system is based on Client/Server (Client/Server) mode, in order to adapt to different management personnel's work demand, system also must meet the below requirements:

1 Functional

The system should have the basic management system, able to complete user registration, user permissions distribution, functional management and announcement and a series of elementary functions. It must be convenient for the user management, and improve the operational efficiency and accuracy.

2 Confidentiality

Before a user login the system, it must be a relevant authentication, then, depending on the identity of the related user, it must show related identity to the user interface, to ensure the rigor of permissions.

3 Efficiency

When designing the system, the relevant modules in the first of all, on the premise of meeting the function on the design, implementation are simple as far as possible, in order to improve efficiency. When writing code, using simple as far as possible, for the most demanding tasks, shorten the time for the system to process information as much as possible.

4 Rationality

The system involves the management of each department, the management of multiple data, so the use of personnel is more, and the data are more complicated. Avoid duplication of effort and the data redundancy, and in order to efficiently implement the data sharing, the process of design system should understand related modules time order and the use of the constraint relationship between them, as much as possible to meet the user's habits.

5 Accuracy

To have good precision, accuracy reflects the management working process.

6 Scalability

The system should adapt to the network, support more users at the same time operating requirements, and has expanded its second and perfect on the function demand, truly achieve the Browser/Server (Browser/Server) mode.

4 SUMMARY SYSTEM DESIGN

4.1 *System overall structural design*

Integrated management system mainly includes five important subsystems, namely the school basic information system, the ideological and moral construction, home of faculty, alumni and online information system. According to the different system to design the corresponding function module.

Implementation of main functions, include the basic situation and the development goal of display, department of the information management module, the student community management module, etc. In order to guarantee the timeliness of data and increase the maintainability of the system, and the related data stored in database, management via the backend interface can be according to the need for such as add, modify, and delete operations, to complete the update of data.

4.2 *Ideological and moral construction function module*

Through the classification management, ideological and moral construction system is divided into units and departments and students. Each level has a corresponding management and the relevant administrative authority. Unit management includes the unit for the management of ideological and moral, political study, and so on and so forth, the corresponding functional modules are as follows: the announcement information module, learning module, the theory of the practice management module, etc., and the corresponding information displayed on the corresponding page. The department is mainly management department faculty members, there are activity project modules, study discussion modules, data management modules, management modules, etc. Student management is primarily aimed at students' party and league member activists. The corresponding functional modules are data management module, party school management module, information announcement module and honor list management module, etc.

4.3 *Alumni function module*

Alumni record is in a class or a collective schoolmate's home address, contact information, phone number, personality, language and so on basic information, convenient contact, in order to achieve mutual understanding among classmates, makes progress together. It mainly includes the announcement module, student information module, management of registration module, query module, the students' login module, message management module, etc.

4.4 *Online information management module*

From information about the courseware and the study made in the form of online video, everyone can rely on to set the user name and password after logging in, can learn online, online query, and download online learning activities. It should include the following modules: online learning, search, download, course management, assessment management, etc.

5 SYSTEM INTERFACE DESIGN STYLE

Interface style refers to the appearance of the system and is utilized to communicate with users, components and programs, and so on. Interface style set from appearance has been creative in order to achieve the purpose of attracting eyeballs, and based on the relevant principle of graphics and layout, so as to make the system design into a unique art. This system pursuit is concise, simple, more practical. This system follows the following basic principles:

1 The user guide principle

First has been clear about the users of an integrated management system for school administrators at different levels, stood in their point of view and stand up to consider design website, necessary to achieve the desired effect.

2 Layout of the control

According to psychologist George Aren Iller, studies have shown that people who once the amount of information at around 7 bits are advisable. This principle is commonly used in the website construction, general website the above section to choose between five to nine. This system layout according to this idea, gives users an easy application environment.

3 Visual balance

Text and images of reasonable collocation, can give a person a kind of visual balance, such ability according to people's reading habits.

4 The harmonious consistency

Numerous separate pages can be together as a whole. It is key to keep the consistency principle, a little. Consistent with the structural design, they can let your visitors to the site image that has a profound memory; Consistent navigation design can let browse quickly and effectively into the part of the need in the site; Consistent operation design, can let your visitors to quickly learn the various functions of the entire site operation. To undermine this principle, can mislead viewers, and lets the entire website show desultorily, give a bad impression.

6 CONCLUSION

From the point of view of the current social development, the proportion of modern information technology in all walks of life is also more and more big. The traditional management is no exception. We just study the school management system, no fuses in together, the whole campus resources such as digital library, student status management, educational administration and so on, although there is such shortcomings and defects, but anyway, for the use of the campus network, improving the management level is also obvious, realizing the school various aspects of the integration of resources, making the application of information technology application from discrete department into composite applications based on Internet, after all, comprehensive management work in the office automation, informationization, is a step forward.

REFERENCES

[1] Hui Jiang. Software requirements management a use case method [M], China electric power press, 2004.

[2] Jianyi Li. The principle and application of database - PowerBui older edition [M], Tsarina University press, 2006.

[3] JianXiaShou, Guohong Mao. Database principles and applications of case tutorials, mechanical industry publishing house, 2005.

[4] Jian Zou. Simple - development, management and application instance, SQL Server 2005 people's posts and telecommunications publishing house, 2008.

[5] Haifan Zhang.introduction to software engineering (fifth edition), Tsarina University press, 2008.

[6] DingXiao.model and method of software engineering, Beijing University of posts and telecommunications publishing house, 2008.

Management, Information and Educational Engineering – Liu, Sung & Yao (Eds)
© 2015 Taylor & Francis Group, London, ISBN: 978-1-138-02728-2

The design and construction of tourism information management system

Li Hua Wu
Weifang University of Science and Technology, Shouguang, China

ABSTRACT: The design and development of the travel information management system provide a platform for operations, The information can be managed and classified on the basis of travel information on specific aspects of the modules for the division, including the system administrator module, the travel information module, the hotel information modules, traffic information module, the module to solve common problems, Dalian own festival module and exit system module. And the function of these modules add, edit, delete, select are finished.

KEYWORDS: Visual Basic.NET, C/S Structure, The Travel information system.

1 INTRODUCTION

Tourism information management system manages resources of tourism information in the system. Along with the increase in tourist information on the type and quantity tourism information management difficulty is also increased. Along with the popularization of computer, people are willing to have a habit through the computer to access to information resources, people can in the shortest possible time receive information, suitable for their own travel plan is then developed, both economically and improve work efficiency, to achieve the twice the result with half the effort.

2 SYSTEM DEVELOPMENT TOOLS

2.1 *Introduction to visual basic.NET*

The continuous development of the Internet and the wide application of the future will be based on the network at the center of the world. In the face of the coming world, Microsoft company officially released in 2002 on the technology of the revolutionary significance of network computing platform - Microsoft.NET (referred to as.Net), announced the beginning of a new era of. In June 2000, the United States declared Microsoft.Net strategy. Visual Basic. NET is an important aspect of Microsoft's latest development suite Visual Studio.NET, simple, efficient and suitable for programming even beginner to learn. It is part of the Visual Studio.NET supporting a variety of programming languages, is the first launch based on Visual Studio.NET. The .Net Framework application development tools are not independent development tools, but with a variety of high-level languages be integrated into the Visual Studio.NET. It not only inherited the use of Visual Basic 6.0 simple, powerful, high efficiency rate, etc, and also add the function of "inheritance", using Visual Basic programming in the real "object-oriented programming tools, therefore it is one of the most outstanding application system development tools, and is currently the most popular. NET development tool.

2.2 *Introduction to access 2000*

Access 2000 was developed by the Microsoft company, under the Windows operating system, object oriented, the event-driven mechanism is a new type of relational database management system. Using it, users do not have to write any code, just through simple and intuitive visual operation, management tasks can be completed for most of the database. Access 2000 provides table generator, query builder, report designer and many other convenient visualization tool operation, as well as the database wizard, table wizard, query wizard, form wizard, the report wizard such guide, can be easily constructed magma computer company, the function of the software center perfect database management system. In addition, it also provides a Visual Basic database development management for applications (VBA) programming language, and is advantageous for the advanced users progress function of more perfect database management system.

3 DEMAND ANALYSIS AND OVERALL DESIGN

3.1 *Specific analysis of demand*

According to the overall functional requirements, the specific functional requirements are described as follows:

1 The function of the tourism information, bus information needs.

When the query to the attractions of the related content, depending on the tour bus, bus information of scenic spots interactive query at the bus information module, can also accord to the circuit via the attractions information query.

According to the attractions, information can be updated or changed in the bus, such as add, modify, and delete operations.

2 The hotel's functional requirements:

Hotel information, as an integral part of the tourism industry, in the system can do the corresponding query and management, listed in the system level of the hotel, and hotel information, and can query the nearest site information.

According to change update hotel information, to ensure the latest sex.

3 The function of information service requirements:

Because this system is for the durian tourism system, so for the convenience of information query, in this system, provides the corresponding modules, such as traffic information and the characteristics of durian holiday for flight information, long-distance passenger information and trains are made a detailed introduction, for the common problems and questions answer also solved in this function.

3.2 *Modular design of the system*

According to the analysis of system requirements, we can divide system, such as The system administrator module, tourism information module, the hotel information module, the scenic spot bus module, the other traffic management module, tourism service module and exit the system module (as shown in figure 1).

1 The system administrator module

The system administrator module is mainly to the system administrator information for maintenance, including:

Administrator information query: Browse the relevant information of the administrator.

Administrator information to be added: Adding a new administrator makes it a system administrator.

Password change: The user name as the primary key, immutable, only can change the password.

The administrator to delete: According to the information, can delete the user name of the users of the system.

The functional significance of the system administrator module in each module is an increase of traffic in the system, increasing the administrator can facilitate better management system and maintenance.

2 Tourism information management module

Tourism information management module includes: Add resort information query module, information module, modify attraction information, delete module and attractions. The specific functions are as follows:

The name of the attraction information query: According to the tourist information spots, type or scenic spots belong to the corresponding query. Also can query directly all attraction information. At the same time one can query the corresponding site bus information.

Add attractions: to input further attractions.

Change attraction information: Change the attractions to make changes in a timely manner.

Delete attraction information: to remove attractions.

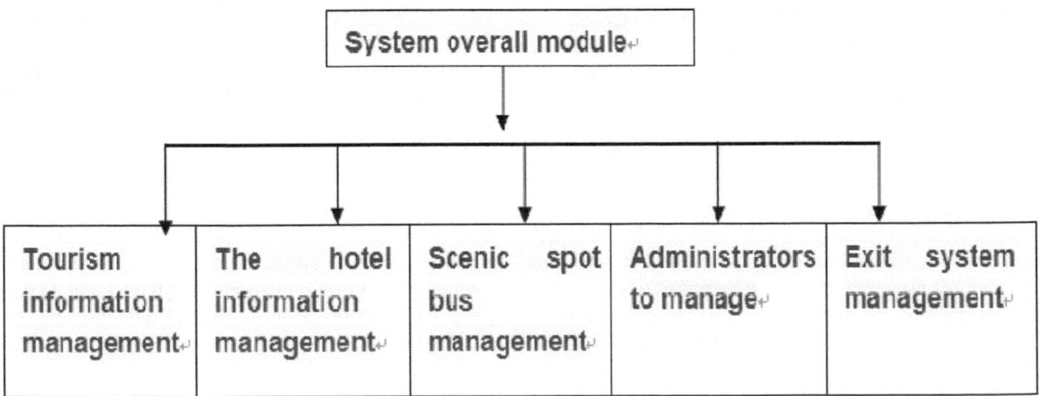

Figure 1. System module figure as a whole.

Tourism information management module of each sub module in the actual meaning is when the administrator in the management of tourism information, can quickly and easily add attractions information system, at the same time to maintain and management of information, and connected with the data in the database, it is not only convenient for visitors to browse, but also improves the accuracy of the information. Administrators can also query of attractions, scenic spots to ensure that information in the accuracy, timeliness, so as to assure the accuracy and completeness of the system.

3 The hotel information management module

Mainly to the hotel information management and maintenance.

Hotel information module includes: add hotel information query module, information module, information modify module and hotel information delete module, specific function as follows:

Hotel Information query: According to the name of the hotel, or hotel belongs to the level of the corresponding query. Also can query all hotel information directly, and can also query according to the hotel information to nearby scenic spots.

Hotel information to add, add additional hotel information.

Hotel information changes: For each change in the hotel-related information to make changes in a timely manner.

Hotel information delete: delete for hotel information.

Hotel information management module of each sub module in the actual meaning is when the administrator in the management of the hotel information, you can quickly and easily add the hotel information system, at the same time to maintenance and management of information, when this hotel name or resettlement, also can timely modify the relevant information system and improve the accuracy of the information. Administrators can also query directly to the hotel information, guarantee the accuracy of the hotel information, so as to ensure the accuracy and completeness of the system. Information to include and modify with dynamic background database connection.

4 Scenic spot bus information management module

Mainly to the scenic spot bus information management, including bus information management at the same time.

Resort bus information module includes: the bus information query module, resort bus information and module, bus information modify module and scenic spot bus information delete module. Specific functions are as follows:

Scenic spot bus information query: According to the bus information query through the scenic spots or bus itself.

Scenic spot bus information to add, can to enter the basic information of the bus into the database, can also accord to the attractions of bus information input and perfect.

Scenic spot bus information change: According to the attractions to increase or delete or change bus lines itself to modify the basic information of the bus accordingly.

Scenic spot bus information delete: According to the scenic spots to delete or change or cancel the bus information deleted from the database.

Scenic spot bus information module of each sub module in the actual meaning is when the administrator to administer, add the bus's own information in the corresponding table in the database. At the same time information in the table according to the change of attractions information make a conforming change, the dynamic connection between table and table is completed. At the same time also can accord to the change of oneself to the corresponding modify, and delete, when information changes, for scenic spot bus information also can produce a corresponding change.

5 Exit the system module

The system user can shut down the system according to your own need. Shut down the system when prompted dialog box, select the system will be safer after quit.

4 CONCLUSION

The design of the harvest. By the CLIENT/SERVER (CLIENT/SERVER, C/S) bookstore management information system on the basis of the framework design, I consolidate the knowledge learned in Visual Basic.NET, and can be more skillfully to VB.NET commonly used for tool design and development of some simple procedures. Moreover, database knowledge also has been consolidated and enhanced, which is found in this design. Because in the original system requirements analysis collected and prepared for the material shortage, cause in the process of the actual development of the program, there is many comprehensive consideration due to the lack of unnecessary problems, increased the application design and development time. In addition, since there is not any good use user-defined functions and processes, that some could have code still need to repeatedly write briefly, and increased the amount of writing code, but lower the reuse of code. It also increases the number of unnecessary code in the development process. In the aspect of database design, owing to the negligence of respect is designed in conceptual structure makes the connection between the database and table is not flexible, resulting in the emergence of redundant data in the database.

REFERENCES

[1] Jinqiang Wang. Visual Basic.NET Tutorial. Beijing: Tangent university press, 2004.
[2] Aihong Tong,Kai Liu. VB.NET application tutorial. Tsingtao University press, 2005–01.
[3] Like Zhang. Design and development of Visual Basic. NET instance. Mechanical engineering press, 2005–2.
[4] Sheng Wang. Visual Basic.NET database development. Beijing: Tangent university press, 2005.
[5] Craig Eddy. Timothy Buchanan Chinese Access 2000. Machinery industry press, 2003.
[6] Peizeng Gong, Zhiqiang Yang. Visual Basic.NET knows experiment and test. Beijing: Higher education press, 2005.

Management, Information and Educational Engineering – Liu, Sung & Yao (Eds)
 ISBN: 978-1-138-02728-2

Numerical simulation of the process of fruit tree growth research

Ming Zhen Ma
Weifang University of Science and Technology , Shouguang, China

ABSTRACT: Fruit quality, high quantitative research, the agriculture ecology and the establishment of orchard production, stable high yield population structure, not only have important significance, but also have important application value. This article is about the opposite condition and growth index on the basis of full investigation and test, using the modern mathematical methods and computer methods, and the process of the evolution of the fruit tree structure to build the fruit trees ecological response to the environmental factors, physiological processes to establish mathematical model method to carry on the research.

KEYWORDS: Fruit, growth dynamic, mathematical model, the growth rule.

1 INTRODUCTION

Using mathematical method, the physiological processes of plant growth and development of quantitatively describe the relationship between the physiological process and ecological factor interactions, is a botanist, plant ecophysiology home one of the common pursuit of goals. Since the 1980s, along with the computer technology, mathematical statistics and mathematical analysis methods of development and wide application, many with investigation and observation can't solve the problem of accurate interpretation. Based on mathematical biology, biophysics, biochemistry, theory foundation of the ecological physical model in fruit trees has been widely used in the development and production.

2 THE TREE STRUCTURE CONSTRUCTION AND THE MATHEMATICAL MODEL OF EVOLUTION

2.1 *Apple long tip growth dynamic model*

Mechanism of membrane in the Logistic growth model was built on the basis of a detailed discussion on the improvement, first set up different Joe anvil of Fuji apple tree potential long tip growth dynamic unified model:

$$y = \frac{\mathrm{arctg}(\beta t)}{b_2 + b_3 e^{-\alpha t}}$$

b1, b2, alpha, beta, parameters to be estimated, given a set of observation data, the methods of LS estimates are available, and t for the growth and development time, as revealed by both theoretical derivation and data fitting to describe apple new tip growth rule, the model is better than the current widely used Logistic growth model. The model in terms of potential control tree has wide application prospect.

2.2 *Mathematical model of dried apples week change rule*

When the load(x) and the girth of the trunk(y) always maintain the proper proportions, mathematical models available:

$$y = y_0 + at - b\,\mathrm{arctg}(ct) + \frac{b}{2c}\ln(1 + c^2 t^2)$$

$y(0) = y_0$ for transplanting seedling dry weeks, when a (> 0) as the young period the growth rate of dry fast growth stage dry weeks, b (> 0), c (> 0) for the decline rate parameter of $\frac{dy}{dt}$. The girth of the trunk with the time changing law is expressed as:

$$y = 1.4679 + 6.7609t - 5.2966t\,\mathrm{arctg}(0.06t) + 44.1387\ln(1 + 0.0036t^2)$$

Parsing model shows that sapling stage work weeks of year growth should be 6.0 cm; Orchard for 10 to 15 years, keep the length of the trunk girth growth in 1.6–2.5 cm each year, can make the orchard continuous output.

2.3 *The relationship model between the load and the trunk surrounded*

On the basis of the existing models and research results, a mathematical model is established, the model represents the link between (X) and (Y):

$$y = b_0 + b_1(1 - \frac{b_2}{b_1}x)x^2$$

B1 (> 0) reflects the linear relationship between y and x2, $1 - \frac{b_2}{b_1}x$ reflects the limitations in a linear factor model with the x^2y in a linear growth. Using type guide on fruit production, scientific planning to reduce DaXiaoNian phenomenon, improve the economic benefit. If considering the influence of the organic matter content (z) on the load, the change rule of load can be expressed as:

$$y = b_0 + b_1x^2 - b_2x^3 + b_3z$$

2.4 *To build the dynamic model of the leaf act*

According to the leaf form the dynamic law of the tent of meeting, set up spring shoots stop long ago, Fuji Apple Group leaves of different tree potential screen form the unity of the dynamic mathematical model of the:

$$y = b_0 + b_1\sqrt{t} + b_2 \ln t$$

Which b0, b1, b2 are be estimated parameters, given a set of observation data, can be obtained by the LS estimation. Solving domain analysis showed that the model of Joe anvil of Fuji Apple Tree potential is more than 70% of the leaf area formed before in early may, more than 80% of leaf area formed before the in mid-may. Fertility has formed by 80% before the leaf area in early May, nearly 90% of leaf area before mid-May formation; Quantitatively distinguish the various potential Fuji apple tree leaves the dynamic difference during the intervals.

3 FRUIT TREES ECOLOGICAL RESPONSE TO ENVIRONMENTAL FACTORS OF THE PHYSIOLOGICAL PROCESSES OF NUMERICAL SIMULATION

3.1 *Photosynthesis, stomatal conductance coupling model*

California north of black walnut as sample, will Farquhar, single leaf photosynthesis, physiological and biochemical model proposed by combined stomatal conductance model B - B, photosynthesis, stomatal conductance coupling model is established. In appropriate conditions, the water model simulation results with the field test results that have good consistency. In gas transmission theory, the coupling model is one of the sub models of the above – mesophyll intercellular CO_2 concentration (Ci) model, namely:

$$C_i = C_s - a_1(C_s - \Gamma)(1 + \frac{VPD}{VPD_0})$$

Among them, Cs, VPD, respectively, atmospheric CO_2 concentration and saturated vapor pressure difference, a1, VPD0 as characteristic parameter, Γ for CO_2 saturation point. In sewage, sludge is appropriate, the temperature is 25.0 °C, photosynthetic active radiation quantum flux density of 1000.0 mu, mol, m - 2-1 s conditions, when the CO2 concentration doubling (from 350 to 700 mu mol, mol, 1), leaf photosynthetic rate increased by 25.3% and 26.2%, respectively.

Atmospheric CO_2 concentration and temperature constant (350 mu respectively mol, mol, 1 and 25.0 °C), the photosynthesis of photosynthetic active radiation response meeting Michaelis - Menten response curve, the photosynthetic active radiation is small, black walnut photosynthetic rate is greater than the eastern part of northern California, with the increase of photosynthetic available radiation, the less the difference. Under the condition of typical sunny day, the diurnal variation of the photosynthetic rate in both two maxima (respectively at 10 and 16 or so), midday photosynthetic rate is low.

3.2 *Canopy photosynthesis model structure and groups*

Ross and Nilson leaf Angle distribution model is based on the canopy of the geometric structure numerical value, and is set up a canopy structure model with high resolution; According to the nature of the direct radiation and scatter radiation and its transmission characteristics in the canopy, respectively, established direct radiation and scatter radiation within the canopy transmission submodel. Canopy leaf by light of the situation can be divided into "flare area" and "shade" area, respectively, to calculate the rate of leaf photosynthesis, two area further refine and improve the canopy photosynthesis multilayer model theory. Under the condition of typical sunny day, northern California black walnut unit land area, total dry matter of synthetic (CH2O) is about 88.6g·m-2d-1, and to the east is about 63.4 g·m-2d-1, the former is about 28.5% more than the latter.

3.3 *Dry matter accumulation and distribution model*

Under the appropriate water condition, on the basis of conservation of mass and concentration gradient theory, a dry matter accumulation and allocation model is established. Assume that each organ assimilation substance concentration with the obtained is directly proportional to the amount of the photosynthate, exported the crown layer between the roots and the relation between the dry matter transmission flux (Kl) :

$$K_l = \frac{G_c(T)P_c W_s}{W_s + G_c(T)W_l}$$

Among them, G (T) at time T of photosynthate from the canopy to the roots of conductance, PC for daily total canopy photosynthesis, Wl, Ws respectively "production organs" and "consumption organs" total dry matter accumulation. Model estimation results compared with the field test results. dW_l/dW_s (the ratio of the canopy, roots, increment of dry matter) not only vary with growth period, also change with environment factors, in the early leaf screen built dW_l/dW_s is larger, the average is about 2.0, then the photosynthetic product is mainly used for growth and development of the blade, in the middle of the night to build, the average of the dW_l/dW_s is about 1.5, the period of dry matter distribution in the canopy, root is balanced, built in the late in the tent of leaf, dW_l/dW_s straight down, the dry matter distribution center is a major shift to underground part and leaf growth tends to stop.

4 CONCLUSION

The article through the establishment of mathematical model, analyses the physiological processes of the growth of fruit trees and research, makes the process of fruit tree growth have a certain class handling, people promote the growth of fruit trees, and increase the yield and fruit quality.

REFERENCES

[1] Shuhan Cheng, Huairui Shu, Qinping Wei.The mathematical model of Red Fuji increasing[J]. Mathematical statistics and management, 1999, 18 (3): 1–4.

[2] Xuerong Xu. Some curve model of saturated growth trend research [J]. Journal of agricultural system science and integrated research, 1997, 13 (1): 4, 9.

[3] Hongbing Deng,Qingkong Wang. Korean pine, research and application of changing leaves, pine high growth model [J]. Journal of learning forestry science and technology, 1997, (5) : 24–27.

[4] Peizhen Liu. Fruit tree trunk cross-sectional area of the simple calculation method [J]. Journal of fruit science, 1991, 8 (2): 127–128.

[5] Yongnian Luo. Adult apple orchard population structure index and the investigation of the load [J]. Journal of shadow fruit trees, 1982, (1): 12–17.

Author index

#0488 - 141116 - C0 - 246/174/45 - PB - 9781138028555